Springer Complexity

Springer Complexity is an interdisciplinary program publishing the best research and academic-level teaching on both fundamental and applied aspects of complex systems – cutting across all traditional disciplines of the natural and life sciences, engineering, economics, medicine, neuroscience, social and computer science.

Complex Systems are systems that comprise many interacting parts with the ability to generate a new quality of macroscopic collective behavior the manifestations of which are the spontaneous formation of distinctive temporal, spatial or functional structures. Models of such systems can be successfully mapped onto quite diverse "real-life" situations like the climate, the coherent emission of light from lasers, chemical reaction-diffusion systems, biological cellular networks, the dynamics of stock markets and of the internet, earthquake statistics and prediction, freeway traffic, the human brain, or the formation of opinions in social systems, to name just some of the popular applications.

Although their scope and methodologies overlap somewhat, one can distinguish the following main concepts and tools: self-organization, nonlinear dynamics, synergetics, turbulence, dynamical systems, catastrophes, instabilities, stochastic processes, chaos, graphs and networks, cellular automata, adaptive systems, genetic algorithms and computational intelligence.

The two major book publication platforms of the Springer Complexity program are the monograph series "Understanding Complex Systems" focusing on the various applications of complexity, and the "Springer Series in Synergetics", which is devoted to the quantitative theoretical and methodological foundations. In addition to the books in these two core series, the program also incorporates individual titles ranging from textbooks to major reference works.

Editorial and Programme Advisory Board

Péter Érdi
Center for Complex Systems Studies, Kalamazoo College, USA and Hungarian Academy of Sciences, Budapest, Hungary

Karl Friston
Institute of Cognitive Neuroscience, University College London, London, UK

Hermann Haken
Center of Synergetics, University of Stuttgart, Stuttgart, Germany

Janusz Kacprzyk
System Research, Polish Academy of Sciences, Warsaw, Poland

Scott Kelso
Center for Complex Systems and Brain Sciences, Florida Atlantic University, Boca Raton, USA

Jürgen Kurths
Nonlinear Dynamics Group, University of Potsdam, Potsdam, Germany

Linda Reichl
Center for Complex Quantum Systems, University of Texas, Austin, USA

Peter Schuster
Theoretical Chemistry and Structural Biology, University of Vienna, Vienna, Austria

Frank Schweitzer
System Design, ETH Zurich, Zurich, Switzerland

Didier Sornette
Entrepreneurial Risk, ETH Zurich, Zurich, Switzerland

Understanding Complex Systems

Founding Editor: J.A. Scott Kelso

Future scientific and technological developments in many fields will necessarily depend upon coming to grips with complex systems. Such systems are complex in both their composition – typically many different kinds of components interacting simultaneously and nonlinearly with each other and their environments on multiple levels – and in the rich diversity of behavior of which they are capable.

The Springer Series in Understanding Complex Systems series (UCS) promotes new strategies and paradigms for understanding and realizing applications of complex systems research in a wide variety of fields and endeavors. UCS is explicitly transdisciplinary. It has three main goals: First, to elaborate the concepts, methods and tools of complex systems at all levels of description and in all scientific fields, especially newly emerging areas within the life, social, behavioral, economic, neuro- and cognitive sciences (and derivatives thereof); second, to encourage novel applications of these ideas in various fields of engineering and computation such as robotics, nano-technology and informatics; third, to provide a single forum within which commonalities and differences in the workings of complex systems may be discerned, hence leading to deeper insight and understanding.

UCS will publish monographs, lecture notes and selected edited contributions aimed at communicating new findings to a large multidisciplinary audience.

Vladimir G. Ivancevic · Tijana T. Ivancevic

Complex Nonlinearity

Chaos, Phase Transitions, Topology Change
and Path Integrals

With 125 Figures and 1 Table

Dr. Vladimir G. Ivancevic
Human Systems Integration
Land Operations Division
Defence Science & Technology Organisation
PO Box 1500, 75 Labs
Edinburgh SA 5111
Australia
Vladimir.Ivancevic@dsto.defence.gov.au

Dr. Tijana T. Ivancevic
School of Electrical and Information
Engineering
University of South Australia
Mawson Lakes Boulevard
Mawson Lakes, S.A. 5095, Australia
Tijana.Ivancevic@unisa.edu.au

ISBN: 978-3-540-79356-4   e-ISBN: 978-3-540-79357-1

Understanding Complex Systems ISSN: 1860-0832

Library of Congress Control Number: 2008925879

© 2008 Springer-Verlag Berlin Heidelberg

This work is subject to copyright. All rights are reserved, whether the whole or part of the material is concerned, specifically the rights of translation, reprinting, reuse of illustrations, recitation, broadcasting, reproduction on microfilm or in any other way, and storage in data banks. Duplication of this publication or parts thereof is permitted only under the provisions of the German Copyright Law of September 9, 1965, in its current version, and permission for use must always be obtained from Springer. Violations are liable to prosecution under the German Copyright Law.

The use of general descriptive names, registered names, trademarks, etc. in this publication does not imply, even in the absence of a specific statement, that such names are exempt from the relevant protective laws and regulations and therefore free for general use.

Cover design: WMXDesign GmbH

Printed on acid-free paper

9 8 7 6 5 4 3 2 1

springer.com

Dedicated to Nick, Atma and Kali

Preface

Complex Nonlinearity: Chaos, Phase Transitions, Topology Change and Path Integrals is a graduate–level monographic textbook. This is a book about *prediction & control* of general *nonlinear dynamics* of high–dimensional *complex systems* of various physical and non-physical nature and their underpinning geometro–topological change.

The book starts with a textbook–like expose on *nonlinear and chaotic dynamics* (Chapter 1). After an introduction into attractors and chaos, a brief history of chaos theory is given. Then, temporal chaotic dynamics is developed, both in its continuous form (nonlinear differential equations) and in its discrete form (nonlinear iteration maps). Spatio–temporal chaotic dynamics of nonlinear partial differential equations follows with some physiological examples. The Chapter ends with modern techniques of chaos–control, both temporal and spatio–temporal.

The *dynamical edge of chaos* physically corresponds to the *phase transitions*. Therefore, Chapter 2 continues exposé on complex nonlinearity, from the point of view of phase transitions and the related field of *synergetics*. After the introduction and classification of equilibrium phase transitions, a brief on Landau's theory is given (providing a background for order–parameters and synergetics). The concept is subsequently generalized into non–equilibrium phase transitions, together with important examples of oscillatory, fractal and noise–induced transitions. This core Chapter of the book also introduces the concept of partition function, together with its general, path–integral description. After that the basic elements of Haken's synergetics are presented, and subsequently developed into synergetics of *attractor neural networks*.

While the natural stage for linear dynamics comprises of flat, Euclidean geometry (with the corresponding calculation tools from linear algebra and analysis), the natural stage for nonlinear dynamics is curved, *Riemannian geometry* (with the corresponding tools from tensor algebra and analysis). In both cases, the system's (kinetic) energy is defined by the metric form, either Euclidean or Riemannian. The extreme nonlinearity – chaos – corresponds to the *topology change* of this curved geometrical stage, usually called configuration manifold. Chapter 3 elaborates on geometry and topology change in relation with complex nonlinearity and chaos.

Chapter 4 develops general nonlinear dynamics, both continuous and discrete, deterministic and stochastic, in the unique form of *path integrals* and their *action–amplitude formalism*. This most natural framework for representing both phase transitions and topology change starts with *Feynman's sum over histories*, to be quickly generalized into the *sum over geometries and topologies*. This Chapter also gives a brief on general dynamics of fields and strings, as well as a path–integral based introduction on the chaos field theory. The Chapter concludes with a number of non–physical examples of complex nonlinear systems defined by path integrals.

The last Chapter puts all the previously developed techniques together and presents the *unified form of complex nonlinearity*. Here we have chaos, phase transitions, geometrical dynamics and topology change, all working together in the form of path integrals. The concluding section is devoted to discussion of hard vs. soft complexity, using the synergetic example of human bio-mechanics.

The objective of the present monograph is to provide a serious reader with a serious scientific tool that will enable them to actually *perform* a competitive research in modern complex nonlinearity. The monograph includes a comprehensive bibliography on the subject and a detailed index. For all mathematical questions, the reader is referred to our book *Applied Differential Geometry: A Modern Introduction*. World Scientific, Singapore, 2007.

Target readership for this monograph includes all researchers and students of complex nonlinear systems (in physics, mathematics, engineering, chemistry, biology, psychology, sociology, economics, medicine, etc.), working both in industry (i.e., clinics) and academia.

Adelaide, V. Ivancevic, Defence Science & Technology Organisation,
Feb. 2008 Australia, e-mail: Vladimir.Ivancevic@dsto.defence.gov.au

T. Ivancevic, School of Mathematics, The University of Adelaide,
e-mail: Tijana.Ivancevic@adelaide.edu.au

Acknowledgments

The authors wish to thank Land Operations Division, Defence Science & Technology Organisation, Australia, for the support in developing the *Human Biodynamics Engine* (HBE) and all the HBE–related text in this monograph.

We also express our sincere gratitude to the *Springer* book series *Understanding Complex Systems* and especially to its Editor, Dr. Thomas Ditzinger.

Contents

1 Basics of Nonlinear and Chaotic Dynamics 1
  1.1 Introduction to Chaos Theory 1
  1.2 Basics of Attractor and Chaotic Dynamics 16
  1.3 Brief History of Chaos Theory 25
    1.3.1 Poincaré's Qualitative Dynamics, Topology and Chaos . 26
    1.3.2 Smale's Topological Horseshoe and Chaos of
          Stretching and Folding 34
    1.3.3 Lorenz' Weather Prediction and Chaos 44
    1.3.4 Feigenbaum's Constant and Universality 47
    1.3.5 May's Population Modelling and Chaos 48
    1.3.6 Hénon's Special 2D Map and Its Strange Attractor 52
  1.4 More Chaotic and Attractor Systems 55
  1.5 Continuous Chaotic Dynamics 67
    1.5.1 Dynamics and Non–Equilibrium Statistical Mechanics .. 69
    1.5.2 Statistical Mechanics of Nonlinear Oscillator Chains ... 82
    1.5.3 Geometrical Modelling of Continuous Dynamics 84
    1.5.4 Lagrangian Chaos 86
  1.6 Standard Map and Hamiltonian Chaos 95
  1.7 Chaotic Dynamics of Binary Systems 101
    1.7.1 Examples of Dynamical Maps 103
    1.7.2 Correlation Dimension of an Attractor 107
  1.8 Spatio–Temporal Chaos and Turbulence in PDEs 108
    1.8.1 Turbulence 108
    1.8.2 Sine–Gordon Equation 113
    1.8.3 Complex Ginzburg–Landau Equation 114
    1.8.4 Kuramoto–Sivashinsky System 115
    1.8.5 Burgers Dynamical System 116
    1.8.6 2D Kuramoto–Sivashinsky Equation 118
  1.9 Basics of Chaos Control 124
    1.9.1 Feedback and Non–Feedback Algorithms for Chaos
          Control 124

XII   Contents

        1.9.2 Exploiting Critical Sensitivity 127
        1.9.3 Lyapunov Exponents and Kaplan–Yorke Dimension 129
        1.9.4 Kolmogorov–Sinai Entropy 131
        1.9.5 Chaos Control by Ott, Grebogi and Yorke (OGY) 132
        1.9.6 Floquet Stability Analysis and OGY Control 135
        1.9.7 Blind Chaos Control 139
        1.9.8 Jerk Functions of Simple Chaotic Flows 143
        1.9.9 Example: Chaos Control in Molecular Dynamics 146
    1.10 Spatio-Temporal Chaos Control 155
        1.10.1 Models of Spatio-Temporal Chaos in Excitable Media . . 158
        1.10.2 Global Chaos Control 160
        1.10.3 Non-Global Spatially Extended Control 163
        1.10.4 Local Chaos Control 165
        1.10.5 Spatio-Temporal Chaos–Control in the Heart 166

2 Phase Transitions and Synergetics 173
    2.1 Introduction to Phase Transitions 173
        2.1.1 Equilibrium Phase Transitions 173
        2.1.2 Classification of Phase Transitions 175
        2.1.3 Basic Properties of Phase Transitions 176
        2.1.4 Landau's Theory of Phase Transitions 179
        2.1.5 Example: Order Parameters in Magnetite Phase
              Transition 180
        2.1.6 Universal Mandelbrot Set as a Phase–Transition Model 183
        2.1.7 Oscillatory Phase Transition 187
        2.1.8 Partition Function and Its Path–Integral Description . . . 192
        2.1.9 Noise–Induced Non–Equilibrium Phase Transitions 199
        2.1.10 Noise–Driven Ferromagnetic Phase Transition 206
        2.1.11 Phase Transition in a Reaction–Diffusion System 218
        2.1.12 Phase Transition in Negotiation Dynamics 224
    2.2 Elements of Haken's Synergetics 229
        2.2.1 Phase Transitions and Synergetics 231
        2.2.2 Order Parameters in Human/Humanoid Biodynamics . . 233
        2.2.3 Example: Synergetic Control of Biodynamics 236
        2.2.4 Example: Chaotic Psychodynamics of Perception 237
        2.2.5 Kick Dynamics and Dissipation–Fluctuation Theorem . . 241
    2.3 Synergetics of Recurrent and Attractor Neural Networks 244
        2.3.1 Stochastic Dynamics of Neuronal Firing States 246
        2.3.2 Synaptic Symmetry and Lyapunov Functions 251
        2.3.3 Detailed Balance and Equilibrium Statistical Mechanics 253
        2.3.4 Simple Recurrent Networks with Binary Neurons 259
        2.3.5 Simple Recurrent Networks of Coupled Oscillators 267
        2.3.6 Attractor Neural Networks with Binary Neurons 275
        2.3.7 Attractor Neural Networks with Continuous Neurons . . 287
        2.3.8 Correlation– and Response–Functions 293

3 Geometry and Topology Change in Complex Systems 305
3.1 Riemannian Geometry of Smooth Manifolds 305
3.1.1 Riemannian Manifolds: an Intuitive Picture 305
3.1.2 Smooth Manifolds and Their (Co)Tangent Bundles 317
3.1.3 Local Riemannian Geometry 328
3.1.4 Global Riemannian Geometry 338
3.2 Riemannian Approach to Chaos 343
3.2.1 Geometrization of Newtonian Dynamics 345
3.2.2 Geometric Description of Dynamical Instability 347
3.2.3 Examples 361
3.3 Morse Topology of Smooth Manifolds 368
3.3.1 Intro to Euler Characteristic and Morse Topology 368
3.3.2 Sets and Topological Spaces 371
3.3.3 A Brief Intro to Morse Theory 380
3.3.4 Morse Theory and Energy Functionals 382
3.3.5 Morse Theory and Riemannian Geometry 384
3.3.6 Morse Topology in Human/Humanoid Biodynamics.... 388
3.3.7 Cobordism Topology on Smooth Manifolds 392
3.4 Topology Change in 3D 394
3.4.1 Attaching Handles 396
3.4.2 Oriented Cobordism and Surgery Theory 402
3.5 Topology Change in Quantum Gravity 405
3.5.1 A Top–Down Framework for Topology Change 405
3.5.2 Morse Metrics and Elementary Topology Changes 406
3.5.3 'Good' and 'Bad' Topology Change 408
3.5.4 Borde–Sorkin Conjecture 410
3.6 A Handle-Body Calculus for Topology Change 411
3.6.1 Handle-body Decompositions 414
3.6.2 Instantons in Quantum Gravity 418

4 Nonlinear Dynamics of Path Integrals 425
4.1 Sum over Histories 425
4.1.1 Intuition Behind a Path Integral 426
4.1.2 Basic Path–Integral Calculations 437
4.1.3 Brief History of Feynman's Path Integral 445
4.1.4 Path–Integral Quantization 452
4.1.5 Statistical Mechanics via Path Integrals 460
4.1.6 Path Integrals and Green's Functions 462
4.1.7 Monte Carlo Simulation of the Path Integral 468
4.2 Sum over Geometries and Topologies..................... 474
4.2.1 Simplicial Quantum Geometry 475
4.2.2 Discrete Gravitational Path Integrals 477
4.2.3 Regge Calculus 479
4.2.4 Lorentzian Path Integral 481
4.2.5 Non-Perturbative Quantum Gravity 486

	4.3	Dynamics of Fields and Strings............................511
		4.3.1 Topological Quantum Field Theory 511
		4.3.2 TQFT and Seiberg–Witten Theory 515
		4.3.3 Stringy Actions and Amplitudes.....................528
		4.3.4 Transition Amplitudes for Strings 532
		4.3.5 Weyl Invariance and Vertex Operator Formulation.....535
		4.3.6 More General Stringy Actions.......................535
		4.3.7 Transition Amplitude for a Single Point Particle.......536
		4.3.8 Witten's Open String Field Theory 537
		4.3.9 Topological Strings 554
		4.3.10 Geometrical Transitions 569
		4.3.11 Topological Strings and Black Hole Attractors 572
	4.4	Chaos Field Theory.......................................578
	4.5	Non–Physical Applications of Path Integrals 580
		4.5.1 Stochastic Optimal Control 580
		4.5.2 Nonlinear Dynamics of Option Pricing 584
		4.5.3 Dynamics of Complex Networks 594
		4.5.4 Path–Integral Dynamics of Neural Networks 596
		4.5.5 Cerebellum as a Neural Path–Integral................617
		4.5.6 Dissipative Quantum Brain Model 623
		4.5.7 Action–Amplitude Psychodynamics 637
		4.5.8 Joint Action Psychodynamics 651
		4.5.9 General Adaptation Psychodynamics 654
5	**Complex Nonlinearity: Combining It All Together** 657	
	5.1	Geometrical Dynamics, Hamiltonian Chaos, and Phase Transitions 657
	5.2	Topology and Phase Transitions 664
		5.2.1 Computation of the Euler Characteristic 666
		5.2.2 Topological Hypothesis..............................668
	5.3	A Theorem on Topological Origin of Phase Transitions 670
	5.4	Phase Transitions, Topology and the Spherical Model 673
	5.5	Topology Change and Causal Continuity 680
		5.5.1 Morse Theory and Surgery 682
		5.5.2 Causal Discontinuity................................687
		5.5.3 General 4D Topology Change 689
		5.5.4 A Black Hole Example 690
		5.5.5 Topology Change and Path Integrals.................692
	5.6	'Hard' vs. 'Soft' Complexity: A Bio-Mechanical Example 693
		5.6.1 Bio-Mechanical Complexity 694
		5.6.2 Dynamical Complexity in Bio–Mechanics 697
		5.6.3 Control Complexity in Bio–Mechanics................700
		5.6.4 Computational Complexity in Bio–Mechanics 705
		5.6.5 Simplicity, Predictability and 'Macro-Entanglement' ...706

         5.6.6 Reduction of Mechanical DOF and Associated
                Controllers 707
         5.6.7 Self–Assembly, Synchronization and Resolution........ 709

References 713

Index 831

1
Basics of Nonlinear and Chaotic Dynamics

In this introductory Chapter we develop the basis of nonlinear dynamics and chaos theory to be used in the subsequent Chapters. After a basic introduction into attractors and deterministic chaos, a brief history of chaos theory is given. Then, temporal chaotic dynamics is developed, both in its continuous form of nonlinear ordinary differential equations (ODEs) and in its discrete form (nonlinear iteration maps). Spatio–temporal chaotic dynamics of nonlinear partial differential equations (PDEs) follows with some physiological examples. The Chapter ends with modern techniques of chaos–control, both temporal and spatio–temporal.

1.1 Introduction to Chaos Theory

Recall that a popular scientific term *deterministic chaos* depicts an *irregular and unpredictable* time evolution of many (simple) deterministic dynamical systems, characterized by nonlinear coupling of its variables (see, e.g., [GOY87, YAS96, BG96, Str94]). Given an initial condition, the dynamic equation determines the dynamic process, i.e., every step in the evolution. However, the initial condition, when magnified, reveals a cluster of values within a certain error bound. For a regular dynamic system, processes issuing from the cluster are bundled together, and the bundle constitutes a predictable process with an error bound similar to that of the initial condition. In a chaotic dynamic system, processes issuing from the cluster diverge from each other exponentially, and after a while the error becomes so large that the dynamic equation losses its predictive power (see Figure 1.1).

For example, in a *pinball game*, any two trajectories that start out very close to each other separate exponentially with time, and in a finite (and in practice, a very small) number of bounces their separation $\delta x(t)$ attains the magnitude of L, the characteristic linear extent of the whole system. This property of sensitivity to initial conditions can be quantified as

1 Basics of Nonlinear and Chaotic Dynamics

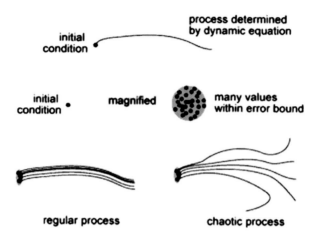

Fig. 1.1. Regular v.s. chaotic process.

$$|\delta x(t)| \approx e^{\lambda t}|\delta x(0)|,$$

where λ, the *mean rate of separation of trajectories* of the system, is called the *Lyapunov exponent*. For any finite accuracy $|\delta x(0)| = \delta x$ of the initial data, the dynamics is predictable only up to a finite *Lyapunov time*

$$T_{Lyap} \approx -\frac{1}{\lambda}\ln|\delta x/L|,$$

despite the deterministic and infallible simple laws that rule the pinball motion.

However, a positive Lyapunov exponent does not in itself lead to chaos (see [CAM05]). One could try to play 1– or 2–disk pinball game, but it would not be much of a game; trajectories would only separate, never to meet again. What is also needed is mixing, the coming together again and again of trajectories. While locally the nearby trajectories separate, the interesting dynamics is confined to a globally finite region of the phase–space and thus the separated trajectories are necessarily folded back and can re–approach each other arbitrarily closely, infinitely many times. For the case at hand there are 2^n topologically distinct n bounce trajectories that originate from a given disk. More generally, the number of distinct trajectories with n bounces can be quantified as

$$N(n) \approx e^{hn},$$

where the *topological entropy* h ($h = \ln 2$ in the case at hand) is the growth rate of the number of topologically distinct trajectories.

When a physicist says that a certain system "exhibits chaos", he means that the system obeys deterministic laws of evolution, but that the outcome is highly sensitive to small uncertainties in the specification of the initial state. The word "chaos" has in this context taken on a narrow technical meaning. If a deterministic system is locally unstable (positive Lyapunov exponent) and globally mixing (positive entropy), it is said to be *chaotic*.

While mathematically correct, the definition of chaos as "positive Lyapunov exponent + positive entropy" is useless in practice, as a measurement of these quantities is intrinsically asymptotic and beyond reach for systems observed in nature. More powerful is Poincaré's vision of chaos as the interplay of local instability (unstable periodic orbits) and global mixing (intertwining of their stable and unstable manifolds). In a chaotic system any open ball of initial conditions, no matter how small, will in finite time overlap with any other finite region and in this sense spread over the extent of the entire asymptotically accessible phase–space. Once this is grasped, the focus of theory shifts from attempting to predict individual trajectories (which is impossible) to a description of the geometry of the space of possible outcomes, and evaluation of averages over this space.

A definition of "turbulence" is even harder to come by. Intuitively, the word refers to *irregular behavior of an infinite–dimensional dynamical system described by deterministic equations of motion* – say, a bucket of boiling water – *described by the Navier–Stokes equations*. But in practice the word "turbulence" tends to refer to messy dynamics which we understand poorly. As soon as a phenomenon is understood better, it is reclaimed and renamed as: "a route to chaos", or "spatio–temporal chaos", etc. (see [CAM05]).

First Motivating Example: A Playground Swing

To gain the initial idea of the principles behind chaos and (strange) attractors, let us consider a common *playground swing*, which physically represents a *driven nonlinear pendulum*. Its dimensionless differential equation of motion (to be explained later in more detail) reads

$$\ddot{\theta} + \gamma \dot{\theta} + \sin\theta = F\cos(w_D t), \tag{1.1}$$

where θ is the *pendulum angle* in radians, overdot represents the time derivative (as always), γ is the *damping parameter*, F is the *amplitude of the driving force*, while w_D is the *frequency* of that force.

Now, it is common to use the small angle approximation for this equation with $\sin\theta \sim \theta \ll 1$. This *linearization* allows the equation to be analytically integrated, but at the same time physically means that the pendulum will either undergo regular motion or, without a driving term, eventually stop swinging altogether. For the type of motion that we desire to study, this approximation is invalid and hence the equation can no longer be solved analytically.

Instead, in nonlinear dynamics, we would rewrite the original second–order ODE of the pendulum motion (1.1), either as a 2D *autonomous system* (suitable for numerical integration),

$$\dot{w} = -\gamma w - \sin\theta + F\cos(w_D t), \qquad \dot{\theta} = w, \tag{1.2}$$

or, as a 3D autonomous system,

$$\dot{w} = -\gamma w - \sin\theta + F\cos\phi, \qquad \dot{\theta} = w, \qquad \dot{\phi} = w_D, \qquad (1.3)$$

where the new variable, ϕ, is called the *phase of the driver*.

The dynamical variables of the system, in this case the pendulum's angle θ and angular velocity w, are the coordinates defining the system's *phase–space*. In the 2D case (1.2), the variables can be plotted to display a *phase portrait* of the system's dynamical behavior. By varying the parameters of the equation for the nonlinear pendulum and then plotting the resulting phase portrait a wide range of behavior can be observed.

The point (0,0) of the pendulum's phase plane is called an *attractor*. Roughly, *an attractor is a 'magnetic set' in the system's phase–space to which all neighboring trajectories converge*.[1] That is, an attractor is the subset of the phase–space with the following properties (see, e.g., [Str94]):

1. it is an invariant set (i.e., any trajectory that starts in it – stays in it for all time);
2. it attracts all trajectories that start sufficiently close to it;
3. it is minimal (it cannot contain one or more smaller attractors).

Ordinary (or, regular) attractors are *stable fixed–points* (which can exist both in linear and nonlinear dynamics) and *stable limit cycles* (which can exist only in nonlinear dynamics). Finally, in chaotic dynamics the most important geometrical objects are *strange attractors*, also called *chaotic attractors* or *fractal attractors* (that is, attractors with non–integer dimension), which are *special attractors that exhibit sensitive dependence on initial conditions*. A strange attractor typically arises when the phase–flow undergoes *stretching*, *squeezing* and *folding*. Trajectories on a strange attractor remain confined to a bounded region of phase–space, yet they separate from their neighbors exponentially fast.

Now, by approximating the nonlinear equations with linear forms and assuming that their behavior will not differ substantially around critical points it can be shown that the damped pendulum with initial conditions θ_0 and w_0 will spiral in towards the point $(2n\pi)$ closest to it, where n is an integer. A useful analogy could be to think of a series of bowls (imagine, somehow, that there are no gaps between them) with the points of $\theta = 2n\pi$ at the centers. These are *equilibrium points*, where the potential energy is the lowest. The peaks of the walls separating these 'bowls' or, *basins of attraction* are at the points $((2n+1)\pi, 0)$, representing unstable critical points called *saddle points*. At these saddle points the pendulum would be pointing vertically upwards and the merest push will send it back down again.

Suppose we now use a non–zero driving force F. We start off with a value for g that's not too strong, the intention is just to overcome the energy loss due to damping. If we plot the phase diagram we see that the attractor is no

---

[1] Opposite of an *attractor* is a *repeller*.

longer a single point at (0,0) but now a closed, almost elliptical curve, called the *limit–cycle attractor*. A limit cycle is an *isolated closed trajectory*. If all neighboring trajectories approach the limit cycle, we say that it is *attracting* or *stable*. Otherwise, it is *repelling* or *unstable*. In our case, the pendulum is swinging back and forth tracing out the same path, undergoing regular motion.

If we increase the driving force a bit more, we now find that instead of just tracing out one loop, the pendulum must now swing through two loops until it reaches the same point again on the phase diagram. The pendulum's *period has doubled*. Increase a bit the driving force F and the period doubles once more: 4 loops appear. This doubling continues as F is increased until a point is reached where, in order to return to the same point, the pendulum must pass through an infinite number of swings. In other words, the motion of the pendulum ceases to be regular and becomes *chaotic*.

Using the so–called *Poincaré section* (taking 'snapshots' of the phase–space at time intervals equal to $t_n = \frac{2n\pi}{w+\phi}$, where n is an integer and ϕ is the above phase term), we find that regular motion with a single period means that only one point is plotted. A period doubling adds another point, and each extra period doubling adds two more points. When a chaotic state is reached then instead of points Poincaré section consists of long 'wavy' lines, composed themselves of bunches of lines. If we magnified one of these lines we would see that this was also composed of another bunch of lines. In fact we could continue the magnification indefinitely and we would still see much the same thing. Hence, the pendulum's attractor is now a fractal, just like the celebrated *Lorenz attractor* (see below).

The unique character of chaotic dynamics may be seen most clearly by imagining the system to be started twice, but from slightly different initial conditions (in case of human motion, these are initial joint angles and angular velocities). We can think of this small initial difference as resulting from measurement error. For non–chaotic systems, this uncertainty leads only to an error in prediction that *grows linearly* with time. For chaotic systems, on the other hand, the error *grows exponentially* in time, so that the state of the system is essentially unknown after very short time. This phenomenon, firstly recognized by H. Poincaré, the father of topology, in 1913, which occurs only when the governing equations are nonlinear, with nonlinearly coupled variables, is known as *sensitivity to initial conditions*. Another type of sensitivity of chaotic systems is *sensitivity to parameters*: a small variation of system parameters (e.g., mass, length and moment of inertia of human body segments) results in great change of system output (dynamics of human movement).

If prediction becomes impossible, it is evident that a chaotic system can resemble a stochastic system, say a Brownian motion. However, the source of the irregularity is quite different. For chaos, the irregularity is part of the intrinsic dynamics of the system, not random external influences (for example, random muscular contractions in human motion). Usually, though, chaotic systems are predictable in the short–term. This *short–term predictability* is useful in various domains ranging from weather forecasting to economic forecasting.

Recall that some aspects of chaos have been known for over a hundred years. Isaac Newton was said to get headaches thinking about the 3−body problem (Sun, Moon, and Earth). In 1887, King Oscar II of Sweden announced a prize for anyone who could solve the n−body problem and hence demonstrate stability of the solar system. The prize was awarded to Henri Poincaré, who showed that even the 3−body problem has no analytical solution [Pet93, BG79]. He went on to deduce many of the properties of chaotic systems including the sensitive dependence on initial conditions. With the successes of linear models in the sciences and the lack of powerful computers, the work of these early nonlinear dynamists went largely unnoticed and undeveloped for many decades. In 1963, Ed Lorenz from MIT published a seminal paper [Lor63, Spa82] in which he showed that chaos can occur in systems of autonomous (no explicit time dependence) ordinary differential equations (ODEs) with as few as three variables and two quadratic nonlinearities. For continuous flows, the *Poincaré–Bendixson theorem* [HS74] implies the necessity of three variables, and chaos requires at least one nonlinearity. More explicitly, the theorem states that the long–time limit of any 'smooth' two–dimensional flow is either a fixed–point or a periodic solution. With the growing availability of powerful computers, many other examples of chaos were subsequently discovered in algebraically simple ODEs. Yet the sufficient conditions for chaos in a system of ODEs remain unknown [SL00].

So, *necessary condition* for *existence of chaos* satisfies any autonomous continuous–time dynamical system (a vector–field) of dimension three or higher, with at least two nonlinearly coupled variables (e.g., a single human swivel joint like a shoulder or hip, determined by three joint angles and three angular momenta). In case of non–autonomous continuous–time systems, chaos can happen in dimension two, while in case of discrete–time systems – even in dimension one. Now, whether the behavior (a flow), of any such system will actually be chaotic or not depends upon the values of its parameters and/or initial conditions. Usually, for some values of involved parameters, the system behavior is oscillating in a stable regime, while for another values of the parameters the behavior becomes chaotic, showing a *bifurcation*, or a *phase transition* – from one regime/phase to a totally different one. If a change in the system's behavior at the bifurcation point is really sharp, we could probably be able to recognize one of the celebrated polynomial *catastrophes* of R. Thom (see [Tho75, Arn92]). A series of such bifurcations usually depicts a *route to chaos*.

Chaos theory has developed special mathematical procedures to *understand* irregularity and unpredictability of low–dimensional nonlinear systems, including Poincaré sections, bifurcation diagrams, power spectra, Lyapunov exponents, period doubling, fractal dimension, stretching and folding, special identification and estimation techniques, etc. (see e.g., [Arn78, Arn78, Arn88,

Arn93, YAS96, BG96]). Understanding these phenomena has enabled science to *control* the chaos (see, e.g., [OGY90, CD98]).

There are many practical reasons for *controlling* or *ordering chaos*. For example, in case of a distributed artificial intelligence system, which is usually characterized by a massive collection of decision–making agents, the fact that an agent's decision also depends on decisions made by other agents – leads to extreme complexity and nonlinearity of the overall system. More often than not, the information received by agents about the 'state' of the system may be 'tainted'. When the system contains imperfect information, its agents tend to make poor decisions concerning choosing an optimal problem–solving strategy or cooperating with other agents. This can result in certain chaotic behavior of the agents, thereby downgrading the performance of the entire system. Naturally, chaos should be reduced as much as possible, or totally suppressed, in these situations [CD98].

In contrast, recent research has shown that chaos may actually be useful under certain circumstances, and there is growing interest in utilizing the richness of chaos [Gle87, Mos96, DGY97]. Since a chaotic, or *strange attractor*[2] usually has embedded within it a dense set of unstable limit cycles, if any of these limit cycles can be stabilized, it may be desirable to stabilize one that characterizes certain maximal system performance [OGY90]. The key is, in a situation where a system is meant for multiple purposes, switching among different limit cycles may be sufficient for achieving these goals. If, on the other hand the attractor is not chaotic, then changing the original system configuration may be necessary to accommodate different purposes. Thus, when designing a system intended for multiple uses, purposely building chaotic dynamics into the system may allow for the desired flexibilities [OGY90].

Within the context of *brain dynamics*, there are suggestions that 'the controlled chaos of the brain is more than an accidental by–product of the brain complexity, including its myriad connections' and that 'it may be the chief property that makes the brain different from an artificial–intelligence machine [FS92]. The so-called *anti–control of chaos* has been proposed for solving the problem of driving the system trajectories of a human brain model away from the stable direction and, hence, away from the stable equilibrium (in the case of a saddle type equilibrium), thereby preventing the periodic behavior of neuronal population bursting. Namely, in a spontaneously bursting neuronal network in vitro, chaos can be demonstrated by the presence of unstable fixed–point behavior. Chaos control techniques can increase the periodicity of such neuronal population bursting behavior. Periodic pacing is also effective in entraining such systems, although in a qualitatively different fashion. Using a strategy of anti–control such systems can be made less periodic. These techniques may be applicable to *in vivo* epileptic foci [SJD94].

---

[2] *Strange attractor* is an attracting set that has zero measure in the embedding phase–space and has fractal dimension. Trajectories within a strange attractor appear to skip around randomly.

Within the context of *heart dynamics*, traditionally in physiology, healthy dynamics has been regarded as regular and predictable, whereas disease, such as fatal arrythmias, aging, and drug toxicity, are commonly assumed to produce disorder and even chaos [Gol99, AGI98, IAG99, KFP91]. However, in the last two decades, laboratory studies produced evidence to show that:

1. The complex variability of healthy dynamics in a variety of physiological systems has features reminiscent of deterministic chaos; and
2. A wide class of disease processes (including drug toxicities and aging) may actually decrease, yet not completely eliminate, the amount of chaos or complexity in physiological systems (decomplexification).

These postulates have implications both for basic mechanisms in physiology as well as for clinical monitoring, including the problem of anticipating sudden cardiac death. In contrast to the prevalent belief of clinicians that healthy heart beats are regular, recent research on the inter–beat interval variations in healthy individuals shows that a normal heart rate apparently fluctuates in a highly erratic fashion. This turns out to be consistent with deterministic chaos [Gol99, AGI98, IAG99, KFP91].

Similar to the brain (and heart) dynamics, human biodynamics represents a highly nonlinear dynamics with several hundreds of degrees of freedom, many of which are naturally and nonlinearly coupled (see [II05, II06a, II06b]). Its hierarchical control system, neural motor controller, necessarily has to cope with the high–dimensional chaos.

Nevertheless, whether the purpose is to reduce 'bad' chaos or to induce 'good' ones, researchers felt strongly the necessity for chaos control [CD98].

Basic Terms of Nonlinear Dynamics

Recall that nonlinear dynamics is a language to talk about dynamical systems. Here, brief definitions are given for the basic terms of this language. All these terms will be illustrated at the pendulum example (see Introduction).

- *Dynamical system:* A part of the world which can be seen as a self–contained entity with some temporal behavior. In nonlinear dynamics, speaking about a dynamical system usually means to speak about an abstract mathematical system which is a model for such an entity. Mathematically, a dynamical system is defined by its *state* and by its *dynamics*. A pendulum is an example for a dynamical system.
- *State of a system:* A number or a vector (i.e., a list of numbers) defining the state of the dynamical system uniquely. For the free (un–driven) pendulum, the state is uniquely defined by the angle θ and the angular velocity $\dot{\theta} = d\theta/dt$. In the case of driving, the driving phase ϕ is also needed because the pendulum becomes a non–autonomous system. In spatially extended systems, the state is often a *field* (a scalar–field or a vector–field).

Mathematically spoken, fields are functions with space coordinates as independent variables. The velocity vector–field of a fluid is a well–known example.
- *Phase space:* All possible states of the system. Each point in the phase–space corresponds to a unique state (see Figure 1.2). In the case of the free pendulum, the phase–space has 2D whereas for driven pendulum it has 3D. The dimension of the phase–space is infinite in cases where the system state is defined by a field.
- *Dynamics, or equation of motion:* The causal relation between the present state and the next state in the future. It is a deterministic rule which tells us what happens in the next time step. In the case of a continuous time, the time step is infinitesimally small. Thus, the equation of motion is an ordinary differential equation (ODE) (or a system of ODEs):

$$\dot{x} = f(x),$$

where x is the state and t is the time variable (overdot is the time derivative – as always). An example is the equation of motion of an un–driven and un–damped pendulum. In the case of a discrete time, the time steps are nonzero and the dynamics is a map:

$$x_{n+1} = f(x_n),$$

with the discrete time n. Note, that the corresponding physical time points t_n do not necessarily occur equidistantly. Only the order has to be the same. That is,

$$n < m \quad \Longrightarrow \quad t_n < t_m.$$

The dynamics is *linear* if the causal relation between the present state and the next state is linear. Otherwise it is *nonlinear*. If we have the case in which the next state is not uniquely defined by the present one, this is generally an indication that the *phase–space is not complete*. Thus, there are important variables determining the state which had been forgotten. This is a crucial point while modelling a real–life systems. Beside this, there are two important classes of systems where the phase–space is incomplete: the *non–autonomuous and stochastic systems*. A non–autonomuous system has an equation of motion which depends explicitly on time. Thus, the dynamical rule governing the next state not only depends on the present state but also at the time it applies. A driven pendulum is a classical example of a *non–autonomuous system*. Fortunately, there is an easy way to make the phase–space complete: we simply include the time into the definition of the state. Mathematically, this is done by introducing a new state variable: t. Its dynamics reads

$$\dot{t} = 1, \quad \text{or} \quad t_{n+1} = t_n,$$

depending on whether time is continuous or discrete. For the periodically driven pendula, it is also natural to take the driving phase as the new state variable. Its equation of motion reads

$$\dot{\theta} = 2\pi w,$$

where w is the driving frequency (so that the angular driving frequency is $2\pi w$). On the other hand, in a *stochastic system*, the number and the nature of the variables necessary to complete the phase–space is usually unknown. Therefore, the next state can not be deduced from the present one. The deterministic rule is replaced by a stochastic one. Instead of the next state, it gives only the probabilities of all points in the phase–space to be the next state.

- *Orbit or trajectory:* A solution of the equation of motion. In the case of continuous time, it is a curve in phase–space parametrized by the time variable. For a discrete system it is an ordered set of points in the phase–space.
- *Phase Flow:* The mapping (or, map) of the whole phase–space of a continuous dynamical system onto itself for a given time step t. If t is an infinitesimal time step dt, the flow is just given by the right–hand side of the equation of motion (i.e., f). In general, the flow for a finite time step is not known analytically because this would be equivalent to have a solution of the equation of motion. For example, Figure 1.2 shows the *phase–flow* of a *damped pendulum* in the $(\theta, \dot{\theta})$–phase–plane.

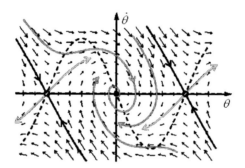

Fig. 1.2. Phase–portrait of a damped pendulum: Arrows denote the phase–flow, dashed line is a null–cline, filled dot is a stable fixed–point, open dot is an unstable fixed–point, dark gray curves are trajectories starting from sample initial points, dark lines with arrows are stable directions (manifolds), light lines with arrows are unstable directions (manifolds), the area between the stable manifolds is basin of attraction.

Phase Plane: Nonlinear Dynamics without Chaos

The general form of a 2D vector–field on the phase plane (similar to one in Figure 1.2) is given by

$$\dot{x}_1 = f_1(x_1, x_2), \qquad \dot{x}_2 = f_2(x_1, x_2),$$

where f_i ($i = 1, 2$) are given function. By 'flowing along' the above vector–field, a *phase point* 'traces out' a solution $x_i(t)$, corresponding to a *trajectory* which is tangent to the vector–field. The entire phase plane is filled with trajectories (since each point can play the role of initial condition, depicting the so–called *phase portrait*. Every phase portrait has the following salient features (see [Str94]):

1. The fixed points, which satisfy: $f_i(x) = 0$, and correspond to the system's steady states or equilibria.

2. The closed orbits, corresponding to the *periodic solutions* (for which $x(t + T) = x(t)$, for all t, for some $T > 0$.

3. The specific *flow pattern*, i.e., the arrangement of trajectories near the fixed points and closed orbits.

4. The *stability* (attracting property) or *instability* (repelling property) of the fixed points and closed orbits.

Nothing more complicated than the fixed points and closed orbits can exist in the phase plane, according to the celebrated *Poincaré–Bendixson theorem*, which says that the dynamical possibilities in the phase plane are very limited. Specifically, *there cannot be chaotic behavior in the phase plane*. In other words, there is *no chaos in continuous 2D systems*.

However, there can exist chaotic behavior in *non–autonomous 2D continuous systems*, namely in the *forced nonlinear oscillators*, where explicit time–dependence actually represents the third dimension.

Free vs. Forced Nonlinear Oscillators

Here we give three examples of classical nonlinear oscillators, each in two modes: free (non–chaotic) and forced (possibly chaotic). For the simulation we use the technique called *time–phase plot*, combining an ordinary time plot with a phase–plane plot. We can see the considerable difference in complexity between unforced and forced oscillators (with all other parameters being the same). The reason for this is that *all forced 2D oscillators actually have dimension 3, although they are commonly written as a second–order ODE*. That is why for development of non–autonomous mechanics we use the *formalism of jet bundles*, see [II06b].

Spring

- Free (Rayleigh) spring (see Figure 1.3):

$$\dot{x} = y,$$
$$\dot{y} = -\frac{1}{m}(ax^3 + bx + cy),$$

where x is displacement, y is velocity, $m > 0$ is mass, $ax^3 + bx + cy$ is the restoring force of the spring, with $b > 0$; we have three possible cases: hard spring ($a > 0$), linear (Hooke) spring ($a = 0$), or soft spring ($a < 0$).[3]

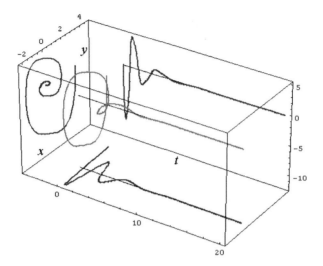

Fig. 1.3. Time–phase plot of the free hard spring with the following parameters: $m = 0.5\,kg$, $a = 1.3$, $b = 0.7$, $c = 0.5$, $x_0 = 3$, $y_0 = 0$, $t_{max} = 20\,s$. Simulated using $Mathematica^{TM}$.

- Forced (Duffing) spring (see Figure 1.4):

$$\dot{x} = y,$$
$$\dot{y} = -\frac{1}{m}(ax^3 + bx + cy) + F\cos(wt),$$
$$\dot{\theta} = w,$$

where F is the force amplitude, θ is the driving phase and w is the driving frequency; the rest is the same as above.

Self–Sustained Oscillator

- Free (Rayleigh) self–sustained oscillator (see Figure 1.5):

$$\dot{x} = y,$$
$$\dot{y} = -\frac{1}{CL}(x + By^3 - Ay),$$

---

[3] In his book *The Theory of Sound*, Lord Rayleigh introduced a series of methods that would prove quite general, such as the notion of a *limit cycle* – a periodic motion a system goes to regardless of the initial conditions.

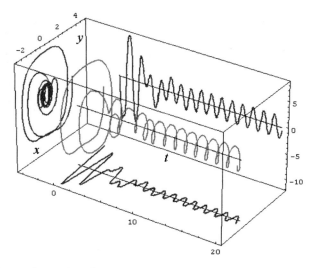

Fig. 1.4. Time–phase plot of the forced hard spring with the following parameters: $m = 0.5\,kg$, $a = 1.3$, $b = 0.7$, $c = 0.5$, $x_0 = 3$, $y_0 = 0$, $t_{max} = 20\,s$, $F = 10$, $w = 5$. Simulated using $Mathematica^{TM}$.

where x is current, y is voltage, $C > 0$ is capacitance and $L > 0$ is inductance; $By^3 - Ay$ (with $A, B > 0$) is the characteristic function of vacuum tube.

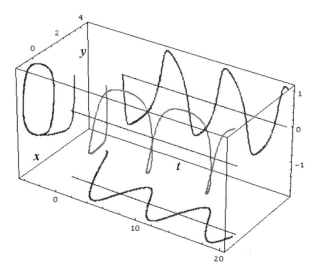

Fig. 1.5. Time–phase plot of the free Rayleigh's self–sustained oscillator with the following parameters: $A = 1.3$, $B = 1.5$, $C = 0.7$, $L = 1.5$, $x_0 = 3$, $y_0 = 0$, $t_{max} = 20\,s$. Simulated using $Mathematica^{TM}$.

14   1 Basics of Nonlinear and Chaotic Dynamics

- Forced (Rayleigh) self–sustained oscillator (see Figure 1.6):

$$\dot{x} = y,$$
$$\dot{y} = -\frac{1}{CL}(x + By^3 - Ay) + F\cos(wt),$$
$$\dot{\theta} = w.$$

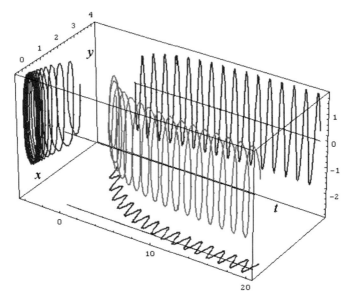

Fig. 1.6. Time–phase plot of the forced Rayleigh's self–sustained oscillator with the following parameters: $A = 1.3$, $B = 1.5$, $C = 0.7$, $L = 1.5$, $x_0 = 3$, $y_0 = 0$, $t_{max} = 20\,s$, $F = 10$, $w = 5$. Simulated using $Mathematica^{TM}$.

Van der Pol Oscillator

- Free Van der Pol oscillator (see Figure 1.7):

$$\dot{x} = y, \qquad (1.4)$$
$$\dot{y} = -\frac{1}{CL}[x + (Bx^2 - A)y].$$

- Forced Van der Pol oscillator (see Figure 1.8):

$$\dot{x} = y,$$
$$\dot{y} = -\frac{1}{CL}[x + (Bx^2 - A)y] + F\cos(wt),$$
$$\dot{\theta} = w.$$

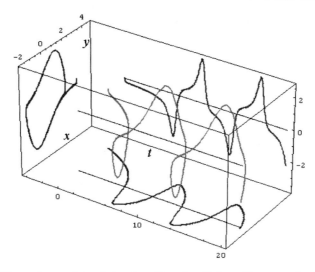

Fig. 1.7. Time–phase plot of the free Van der Pol oscillator with the following parameters: $A = 1.3$, $B = 1.5$, $C = 0.7$, $L = 1.5$, $x_0 = 3$, $y_0 = 0$, $t_{max} = 20\,s$. Simulated using $Mathematica^{TM}$.

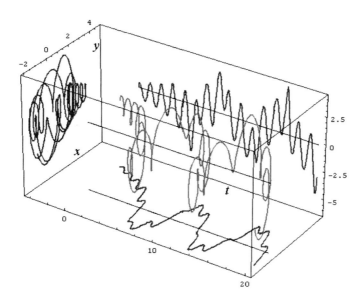

Fig. 1.8. Time–phase plot of the forced Van der Pol oscillator oscillator with the following parameters: $A = 1.3$, $B = 1.5$, $C = 0.7$, $L = 1.5$, $x_0 = 3$, $y_0 = 0$, $t_{max} = 20\,s$, $F = 10$, $w = 5$. Simulated using $Mathematica^{TM}$.

1.2 Basics of Attractor and Chaotic Dynamics

Recall from [II06b] that the concept of *dynamical system* has its origins in *Newtonian mechanics*. There, as in other natural sciences and engineering disciplines, the evolution rule of dynamical systems is given implicitly by a relation that gives the state of the system only a short time into the future. This relation is either a differential equation or difference equation. To determine the state for all future times requires iterating the relation many times–each advancing time a small step. The iteration procedure is referred to as solving the system or integrating the system. Once the system can be solved, given an initial point it is possible to determine all its future points, a collection known as a *trajectory* or *orbit*. All possible system trajectories comprise its *flow* in the phase–space.

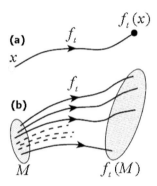

Fig. 1.9. Action of the *phase–flow* f_t in the phase–space manifold M: (a) Trajectory of a single initial point $x(t) \in M$, (b) Transporting the whole manifold M.

More precisely, recall from [II06b] that a *dynamical system* geometrically represents a *vector–field* (or, more generally, a *tensor–field*) in the system's phase–space manifold M, which upon *integration* (governed by the celebrated *existence & uniqueness theorems for ODEs*) defines a *phase–flow* in M (see Figure 1.9). This phase–flow $f_t \in M$, describing the complete behavior of a dynamical system at every time instant, can be either linear, nonlinear or chaotic.

Before the advent of fast computers, solving a dynamical system required sophisticated mathematical techniques and could only be accomplished for a small class of linear dynamical systems. Numerical methods executed on computers have simplified the task of determining the orbits of a dynamical system.

For simple dynamical systems, knowing the trajectory is often sufficient, but most dynamical systems are too complicated to be understood in terms of individual trajectories. The difficulties arise because:

1. The systems studied may only be known approximately–the parameters of the system may not be known precisely or terms may be missing from the equations. The approximations used bring into question the validity or relevance of numerical solutions. To address these questions several notions of stability have been introduced in the study of dynamical systems, such as *Lyapunov stability* or *structural stability*. The stability of the dynamical system implies that there is a class of models or initial conditions for which the trajectories would be equivalent. The operation for comparing orbits to establish their equivalence changes with the different notions of stability.

2. The type of trajectory may be more important than one particular trajectory. Some trajectories may be periodic, whereas others may wander through many different states of the system. Applications often require enumerating these classes or maintaining the system within one class. Classifying all possible trajectories has led to the qualitative study of dynamical systems, that is, properties that do not change under coordinate changes. Linear dynamical systems and systems that have two numbers describing a state are examples of dynamical systems where the possible classes of orbits are understood.

3. The behavior of trajectories as a function of a parameter may be what is needed for an application. As a parameter is varied, the dynamical systems may have *bifurcation points* where the qualitative behavior of the dynamical system changes. For example, it may go from having only periodic motions to apparently erratic behavior, as in the transition to *turbulence* of a fluid.

4. The trajectories of the system may appear erratic, as if random. In these cases it may be necessary to compute averages using one very long trajectory or many different trajectories. The averages are well defined for ergodic systems and a more detailed understanding has been worked out for *hyperbolic systems*. Understanding the probabilistic aspects of dynamical systems has helped establish the foundations of statistical mechanics and of chaos.

Now, let us start 'gently' with chaotic dynamics. Recall that a dynamical system may be defined as a deterministic rule for the time evolution of state observables. Well known examples are *ODEs* in which time is continuous,

$$\dot{\mathbf{x}}(t) = \mathbf{f}(\mathbf{x}(t)), \quad (\mathbf{x}, \mathbf{f} \in \mathbb{R}^n); \tag{1.5}$$

and *iterative maps* in which time is discrete:

$$\mathbf{x}(t+1) = \mathbf{g}(\mathbf{x}(t)), \quad (\mathbf{x}, \mathbf{g} \in \mathbb{R}^n). \tag{1.6}$$

In the case of maps, the evolution law is straightforward: from $\mathbf{x}(0)$ one computes $\mathbf{x}(1)$, and then $\mathbf{x}(2)$ and so on. For ODE's, under rather general assumptions on $\mathbf{f}$, from an initial condition $\mathbf{x}(0)$ one has a unique trajectory $\mathbf{x}(t)$ for $t > 0$ [Ott93]. Examples of regular behaviors (e.g., stable fixed–points, limit cycles) are well known, see Figure 1.10.

A rather natural question is the possible existence of less regular behaviors i.e., different from stable fixed–points, periodic or quasi-periodic motion.

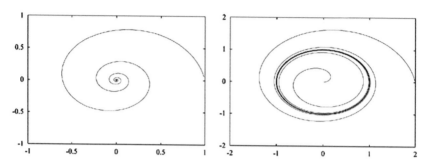

Fig. 1.10. Examples of regular attractors: fixed–point (left) and limit cycle (right). Note that limit cycles exist only in nonlinear dynamics.

After the seminal works of Poincaré, Lorenz, Smale, May, and Hénon (to cite only the most eminent ones) it is now well established that the so called chaotic behavior is ubiquitous. As a relevant system, originated in the geophysical context, we mention the celebrated *Lorenz system* [Lor63, Spa82]

$$\dot{x} = -\sigma(x - y)$$
$$\dot{y} = -xz + rx - y \quad (1.7)$$
$$\dot{z} = xy - bz$$

This system is related to the *Rayleigh–Bénard convection* under very crude approximations. The quantity x is proportional the circulatory fluid particle velocity; the quantities y and z are related to the temperature profile; σ, b and r are dimensionless parameters. Lorenz studied the case with $\sigma = 10$ and $b = 8/3$ at varying r (which is proportional to the Rayleigh number). It is easy to see by linear analysis that the fixed–point $(0, 0, 0)$ is stable for $r < 1$. For $r > 1$ it becomes unstable and two new fixed–points appear

$$C_{+,-} = (\pm\sqrt{b(r-1)}, \pm\sqrt{b(r-1)}, r-1), \quad (1.8)$$

these are stable for $r < r_c = 24.74$. A nontrivial behavior, i.e., non periodic, is present for $r > r_c$, as is shown in Figure 1.11.

In this 'strange', chaotic regime one has the so called sensitive dependence on initial conditions. Consider two trajectories, $\mathbf{x}(t)$ and $\mathbf{x}'(t)$, initially very close and denote with $\Delta(t) = ||\mathbf{x}'(t) - \mathbf{x}(t)||$ their separation. Chaotic behavior means that if $\Delta(0) \to 0$, then as $t \to \infty$ one has $\Delta(t) \sim \Delta(0) \exp \lambda_1 t$, with $\lambda_1 > 0$ [BLV01].

Let us notice that, because of its chaotic behavior and its dissipative nature, i.e.,

$$\frac{\partial \dot{x}}{\partial x} + \frac{\partial \dot{y}}{\partial y} + \frac{\partial \dot{z}}{\partial z} < 0, \quad (1.9)$$

the attractor of the Lorenz system cannot be a smooth surface. Indeed the attractor has a self–similar structure with a fractal dimension between 2 and

3. The Lorenz model (which had an important historical relevance in the development of chaos theory) is now considered a paradigmatic example of a chaotic system.

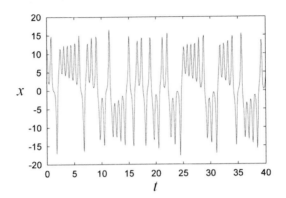

Fig. 1.11. Example of an aperiodic signal: the x variable of the Lorenz system (1.7) as function of time t, for $r = 28$.

Lyapunov Exponents

The sensitive dependence on the initial conditions can be formalized in order to give it a quantitative characterization. The main growth rate of trajectory separation is measured by the first (or maximum) *Lyapunov exponent*, defined as (see, e.g., [BLV01])

$$\lambda_1 = \lim_{t \to \infty} \lim_{\Delta(0) \to 0} \frac{1}{t} \ln \frac{\Delta(t)}{\Delta(0)}, \qquad (1.10)$$

As long as $\Delta(t)$ remains sufficiently small (i.e., infinitesimal, strictly speaking), one can regard the separation as a tangent vector $\mathbf{z}(t)$ whose time evolution is

$$\dot{z}_i = \frac{\partial f_i}{\partial x_j}|_{\mathbf{x}(t)} \cdot z_j, \qquad (1.11)$$

and, therefore,

$$\lambda_1 = \lim_{t \to \infty} \frac{1}{t} \ln \frac{||\mathbf{z}(t)||}{||\mathbf{z}(0)||}. \qquad (1.12)$$

In principle, λ_1 may depend on the initial condition $\mathbf{x}(0)$, but this dependence disappears for ergodic systems. In general there exist as many Lyapunov exponents, conventionally written in decreasing order $\lambda_1 \geq \lambda_2 \geq \lambda_3 \geq ...$, as the independent coordinates of the phase–space [BGG80]. Without entering the details, one can define the sum of the first k Lyapunov exponents as the growth rate of an infinitesimal kD volume in the phase–space. In particular, λ_1 is the growth rate of material lines, $\lambda_1 + \lambda_2$ is the growth rate of $2D$ surfaces, and so on. A numerical widely used efficient method is due to Benettin et al. [BGG80].

It must be observed that, after a transient, the growth rate of any generic small perturbation (i.e., distance between two initially close trajectories) is measured by the first (maximum) Lyapunov exponent λ_1, and $\lambda_1 > 0$ means chaos. In such a case, the state of the system is unpredictable on long times. Indeed, if we want to predict the state with a certain tolerance Δ then our forecast cannot be pushed over a certain time interval T_P, called *predictability time*, given by [BLV01]:

$$T_P \sim \frac{1}{\lambda_1} \ln \frac{\Delta}{\Delta(0)}. \tag{1.13}$$

The above relation shows that T_P is basically determined by $1/\lambda_1$, seen its weak dependence on the ratio $\Delta/\Delta(0)$. To be precise one must state that, for a series of reasons, relation (1.13) is too simple to be of actual relevance [BCF02].

Kolmogorov–Sinai Entropy

Deterministic chaotic systems, because of their irregular behavior, have many aspects in common with stochastic processes. The idea of using stochastic processes to mimic chaotic behavior, therefore, is rather natural [Chi79, Ben84]. One of the most relevant and successful approaches is symbolic dynamics [BS93]. For the sake of simplicity let us consider a discrete time dynamical system. One can introduce a partition $\mathcal{A}$ of the phase–space formed by N disjoint sets $A_1, ..., A_N$. From any initial condition one has a trajectory

$$\mathbf{x}(0) \to \mathbf{x}(1), \mathbf{x}(2), ..., \mathbf{x}(n), ... \tag{1.14}$$

dependently on the partition element visited, the trajectory (1.14), is associated to a symbolic sequence

$$\mathbf{x}(0) \to i_1, i_2, ..., i_n, ... \tag{1.15}$$

where i_n ($n = 1, 2, ..., N$) means that $\mathbf{x}(n) \in A_{i_n}$ at the step n, for $n = 1, 2, ...$. The coarse-grained properties of chaotic trajectories are therefore studied through the discrete time process (1.15).

An important characterization of symbolic dynamics is given by the *Kolmogorov–Sinai entropy* (KS), defined as follows. Let $C_n = (i_1, i_2, ..., i_n)$ be a generic 'word' of size n and $P(C_n)$ its occurrence probability, the quantity [BLV01]

$$H_n = \sup_{\mathcal{A}} [-\sum_{C_n} P(C_n) \ln P(C_n)], \tag{1.16}$$

is called *block entropy* of the n−sequences, and it is computed by taking the largest value over all possible partitions. In the limit of infinitely long sequences, the asymptotic entropy increment

$$h_{KS} = \lim_{n \to \infty} H_{n+1} - H_n, \qquad (1.17)$$

is the Kolmogorov–Sinai entropy. The difference $H_{n+1} - H_n$ has the intuitive meaning of average information gain supplied by the $(n+1)$−th symbol, provided that the previous n symbols are known. KS–entropy has an important connection with the positive Lyapunov exponents of the system [Ott93]:

$$h_{KS} = \sum_{\lambda_i > 0} \lambda_i. \qquad (1.18)$$

In particular, for low–dimensional chaotic systems for which only one Lyapunov exponent is positive, one has $h_{KS} = \lambda_1$.

We observe that in (1.16) there is a technical difficulty, i.e., taking the sup over all the possible partitions. However, sometimes there exits a special partition, called generating partition, for which one finds that H_n coincides with its superior bound. Unfortunately the generating partition is often hard to find, even admitting that it exist. Nevertheless, given a certain partition, chosen by physical intuition, the statistical properties of the related symbol sequences can give information on the dynamical system beneath. For example, if the probability of observing a symbol (state) depends only by the knowledge of the immediately preceding symbol, the symbolic process becomes a *Markov chain* (see [II06b]) and all the statistical properties are determined by the transition matrix elements W_{ij} giving the probability of observing a transition $i \to j$ in one time step. If the memory of the system extends far beyond the time step between two consecutive symbols, and the occurrence probability of a symbol depends on k preceding steps, the process is called *Markov process* of order k and, in principle, a k rank tensor would be required to describe the dynamical system with good accuracy. It is possible to demonstrate that if $H_{n+1} - H_n = h_{KS}$ for $n \geq k+1$, k is the (minimum) order of the required Markov process [Khi57]. It has to be pointed out, however, that to know the order of the suitable Markov process we need is of no practical utility if $k \gg 1$.

Second Motivating Example: Pinball Game and Periodic Orbits

Confronted with a potentially chaotic dynamical system, we analyze it through a sequence of three distinct stages: (i) diagnose, (ii) count, (iii) measure. First we determine the intrinsic dimension of the system – the minimum number of coordinates necessary to capture its essential dynamics. If the system is very turbulent we are, at present, out of luck. We know only how to deal with the transitional regime between regular motions and chaotic dynamics in a few dimensions. That is still something; even an infinite–dimensional system such as a burning flame front can turn out to have a very few chaotic degrees of freedom. In this regime the chaotic dynamics is restricted to a space of low dimension, the number of relevant parameters is small, and we can proceed to step (ii); we count and classify all possible topologically distinct trajectories of the system into a hierarchy whose successive layers require increased precision and patience on the part of the observer. If successful, we can proceed with step (iii): investigate the weights of the different pieces of the system [CAM05].

With the game of pinball we are lucky: it is only a 2D system, free motion in a plane. The motion of a point particle is such that after a collision with one disk it either continues to another disk or it escapes. If we label the three disks by 1, 2 and 3, we can associate every trajectory with an itinerary, a sequence of labels indicating the order in which the disks are visited. The itinerary is finite for a scattering trajectory, coming in from infinity and escaping after a finite number of collisions, infinite for a trapped trajectory, and infinitely repeating for a periodic orbit.[4] Such labelling is the simplest example of *symbolic dynamics*. As the particle cannot collide two times in succession with the same disk, any two consecutive symbols must differ. This is an example of *pruning*, a rule that forbids certain subsequences of symbols. Deriving pruning rules is in general a difficult problem, but with the game of pinball we are lucky, as there are no further pruning rules.[5]

Suppose you wanted to play a good game of pinball, that is, get the pinball to bounce as many times as you possibly can – what would be a winning strategy? The simplest thing would be to try to aim the pinball so it bounces many times between a pair of disks – if you managed to shoot it so it starts out in the periodic orbit bouncing along the line connecting two disk centers, it would stay there forever. Your game would be just as good if you managed to get it to keep bouncing between the three disks forever, or place it on any periodic orbit. The only rub is that any such orbit is unstable, so you have to aim very accurately in order to stay close to it for a while. So it is pretty clear that if one is interested in playing well, unstable periodic orbits are important – they form the skeleton onto which all trajectories trapped for long times cling.

Now, recall that a trajectory is *periodic* if it returns to its starting position and momentum. It is custom to refer to the set of periodic points that belong to a given periodic orbit as a *cycle*.

Short periodic orbits are easily drawn and enumerated, but it is rather hard to perceive the systematics of orbits from their shapes. In mechanics a trajectory is fully and uniquely specified by its position and momentum at

---

[4] The words *orbit* and *trajectory* here are synonymous.
[5] The choice of symbols is in no sense unique. For example, as at each bounce we can either proceed to the next disk or return to the previous disk, the above 3–letter alphabet can be replaced by a binary $\{0,1\}$ alphabet. A clever choice of an alphabet will incorporate important features of the dynamics, such as its symmetries.

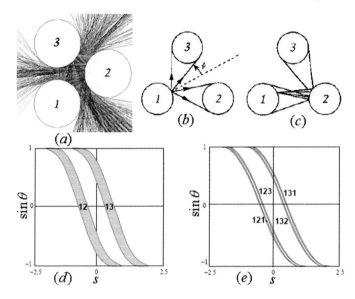

Fig. 1.12. A 3–disk pinball game. Up: (a) Elastic scattering around three hard disks (simulated in $Dynamics\ Solver^{TM}$); (b) A trajectory starting out from disk 1 can either hit another disk or escape; (c) Hitting two disks in a sequence requires a much sharper aim; the cones of initial conditions that hit more and more consecutive disks are nested within each other. Down: Poincaré section for the 3-disk pinball game, with trajectories emanating from the disk 1 with $x_0 = (arc-length, parallel\ momentum) = (s_0, p_0)$, disk radius: center separation ratio $a : R = 1 : 2.5$; (d) Strips of initial points M_{12}, M_{13} which reach disks 2, 3 in one bounce, respectively. (e) Strips of initial points M121, M131 M132 and M123 which reach disks 1, 2, 3 in two bounces, respectively; the Poincaré sections for trajectories originating on the other two disks are obtained by the appropriate relabelling of the strips (modified and adapted from [CAM05]).

a given instant, and no two distinct phase–space trajectories can intersect. Their projections on arbitrary subspaces, however, can and do intersect, in rather unilluminating ways. In the pinball example, the problem is that we are looking at the projections of a 4D phase–space trajectories onto its 2D subspace, the configuration space. A clearer picture of the dynamics is obtained by constructing a phase–space Poincaré section.

The position of the ball is described by a pair of numbers (the spatial coordinates on the plane), and the angle of its velocity vector. As far as a classical dynamist is concerned, this is a complete description. Now, suppose that the pinball has just bounced off disk 1. Depending on its position and outgoing angle, it could proceed to either disk 2 or 3. Not much happens in between the bounces – the ball just travels at constant velocity along a straight line – so we can reduce the 4D flow to a 2D map f that takes the coordinates of the pinball from one disk edge to another disk edge. Let us state this more precisely: the trajectory just after the moment of impact is defined by marking s_n, the arc–length position of the nth bounce along the billiard wall, and $p_n = p\sin\phi_n$ is the momentum component parallel to the billiard wall at the point of impact (see Figure 1.12). Such a section of a flow is called a *Poincaré section*, and the particular choice of coordinates (due to Birkhoff) is particularly smart, as it conserves the phase–space volume. In terms of the Poincaré section, the dynamics is reduced to the *return map*

$$P : (s_n, p_n) \to (s_{n+1}, p_{n+1}),$$

from the boundary of a disk to the boundary of the next disk.

Next, we mark in the Poincaré section those initial conditions which do not escape in one bounce. There are two strips of survivors, as the trajectories originating from one disk can hit either of the other two disks, or escape without further ado. We label the two strips M_0, M_1. Embedded within them there are four strips, $M_{00}, M_{10}, M_{01}, M_{11}$ of initial conditions that survive for two bounces, and so forth (see Figure 1.12). Provided that the disks are sufficiently separated, after n bounces the survivors are divided into 2^n distinct strips: the M_ith strip consists of all points with itinerary $i = s_1 s_2 s_3 ... s_n$, $s = \{0, 1\}$. The unstable cycles as a skeleton of chaos are almost visible here: each such patch contains a periodic point $\overline{s_1 s_2 s_3 ... s_n}$ with the basic block infinitely repeated. Periodic points are skeletal in the sense that as we look further and further, the strips shrink but the periodic points stay put forever.

We see now why it pays to utilize a symbolic dynamics; it provides a navigation chart through chaotic phase–space. There exists a unique trajectory for every admissible infinite length itinerary, and a unique itinerary labels every trapped trajectory. For example, the only trajectory labelled by $\overline{12}$ is the 2–cycle bouncing along the line connecting the centers of disks 1 and 2; any other trajectory starting out as $12\ldots$ either eventually escapes or hits the 3rd disk [CAM05].

Now we can ask what is a good physical quantity to compute for the game of pinball? Such system, for which almost any trajectory eventually leaves a finite region (the pinball table) never to return, is said to be open, or a *repeller*. The repeller escape rate is an eminently measurable quantity. An example of such a measurement would be an unstable molecular or nuclear state which can be well approximated by a classical potential with the possibility of escape in certain directions. In an experiment many projectiles are injected into such a non–confining potential and their mean escape rate is measured. The numerical experiment might consist of injecting the pinball between the disks in some random direction and asking how many times the pinball bounces on the average before it escapes the region between the disks. On the other hand, for a theorist a good game of pinball consists in predicting accurately the asymptotic lifetime (or the escape rate) of the pinball.

Here we briefly show how Cvitanovic's *periodic orbit theory* [Cvi91] accomplishes this for us. Each step will be so simple that you can follow even at the cursory pace of this overview, and still the result is surprisingly elegant.

Let us consider Figure 1.12 again. In each bounce, the initial conditions get thinned out, yielding twice as many thin strips as at the previous bounce. The total area that remains at a given time is the sum of the areas of the strips, so that the fraction of survivors after n bounces, or the *survival probability* is given by

$$\hat{\Gamma}_1 = \frac{|M_0|}{|M|} + \frac{|M_1|}{|M|}, \qquad (1.19)$$

$$\hat{\Gamma}_2 = \frac{|M_{00}|}{|M|} + \frac{|M_{10}|}{|M|} + \frac{|M_{01}|}{|M|} + \frac{|M_{11}|}{|M|},$$

$$\ldots$$

$$\hat{\Gamma}_n = \frac{1}{|M|} \sum_{i=1}^{(n)} |M_i|,$$

where $i = 01, 10, 11, \ldots$ is a label of the ith strip (not a binary number), $|M|$ is the initial area, and $|M_i|$ is the area of the ith strip of survivors. Since at each bounce one routinely loses about the same fraction of trajectories, one expects the sum (1.19) to fall off exponentially with n and tend to the limit

$$\Gamma_{n+1}/\hat{\Gamma}_n = e^{-\gamma n} \to e^{-\gamma},$$

where the quantity γ is called the *escape rate* from the repeller. In [Cvi91] and subsequent papers, Cvitanovic has showed that the escape rate γ can be extracted from a highly convergent exact expansion by reformulating the sum (1.19) in terms of *unstable periodic orbits*.

1.3 Brief History of Chaos Theory

Now, without pretending to give a complete history of chaos theory, in this section we present only its most prominent milestones (in our view). For a number of other important contributors, see [Gle87]). Before we embark on the quick historical journey of chaos theory, note that classical mechanics has not stood still since the foundational work of its father, *Sir Isaac Newton*. The mechanical formalism that we use today was developed mostly by the three giants: *Leonhard Euler, Joseph Louis Lagrange* and *Sir William Rowan Hamilton*. By the end of the 1800's the three problems that would lead to the notion of chaotic dynamics were already known: the *three–body problem* (see Figure 1.13), the *ergodic hypothesis*,[6] and *nonlinear oscillators* (see Figures 1.3–1.8).

---

[6] The second problem that played a key role in development of chaotic dynamics was the *ergodic hypothesis of Boltzmann*. Recall that *James Clerk Maxwell* and *Ludwig Boltzmann* had combined the mechanics of Newton with notions of probability in order to create statistical mechanics, deriving thermodynamics from the equations of mechanics. To evaluate the heat capacity of even a simple system,

1.3.1 Poincaré's Qualitative Dynamics, Topology and Chaos

Chaos theory really started with *Henry Jules Poincaré*, the last mathematical universalist, the father of both *dynamical systems* and *topology* (which he considered to be the two sides of the same coin). Together with the four dynamics giants mentioned above, Poincaré has been considered as one of the great scientific geniuses of all time.[7]

> Boltzmann had to make a great simplifying assumption of ergodicity: that the dynamical system would visit every part of the phase–space allowed by conservations law equally often. This hypothesis was extended to other averages used in statistical mechanics and was called the ergodic hypothesis. It was reformulated by Poincaré to say that a trajectory comes as close as desired to any phase–space point.
>
> Proving the ergodic hypothesis turned out to be very difficult. By the end of our own century it has only been shown true for a few systems and wrong for quite a few others. Early on, as a mathematical necessity, the proof of the hypothesis was broken down into two parts. First one would show that the mechanical system was ergodic (it would go near any point) and then one would show that it would go near each point equally often and regularly so that the computed averages made mathematical sense. Koopman took the first step in proving the ergodic hypothesis when he noticed that it was possible to reformulate it using the recently developed methods of *Hilbert spaces*. This was an important step that showed that it was possible to take a finite–dimensional nonlinear problem and reformulate it as a infinite–dimensional linear problem. This does not make the problem easier, but it does allow one to use a different set of mathematical tools on the problem. Shortly after Koopman started lecturing on his method, *John von Neumann* proved a version of the ergodic hypothesis, giving it the status of a theorem. He proved that if the mechanical system was ergodic, then the computed averages would make sense. Soon afterwards *George Birkhoff* published a much stronger version of the theorem (see [CAM05]).

[7] Recall that Henri Poincaré (April 29, 1854–July 17, 1912), was one of France's greatest mathematicians and theoretical physicists, and a philosopher of science. Poincaré is often described as the last 'universalist' (after Gauss), capable of understanding and contributing in virtually all parts of mathematics. As a mathematician and physicist, he made many original fundamental contributions to pure and applied mathematics, mathematical physics, and celestial mechanics. He was responsible for formulating the Poincaré conjecture, one of the most famous problems in mathematics. In his research on the three-body problem, Poincaré became the first person to discover a *deterministic chaotic system*. Besides, Poincaré introduced the modern principle of relativity and was the first to present the Lorentz transformations in their modern symmetrical form (Poincaré group).

Poincaré had the opposite philosophical views of Bertrand Russell and Gottlob Frege, who believed that mathematics were a branch of logic. Poincaré strongly disagreed, claiming that *intuition* was the *life of mathematics*. Poincaré gives an interesting point of view in his book 'Science and Hypothesis': "For a superficial observer, scientific truth is beyond the possibility of doubt; the logic of science is infallible, and if the scientists are sometimes mistaken, this is only from their mistaking its rule."

Poincaré conjectured and proved a number of theorems. Two of them related to chaotic dynamics are:

1. The *Poincaré–Bendixson theorem* says: Let F be a dynamical system on the real plane defined by

$$(\dot{x}, \dot{y}) = (f(x,y), g(x,y)),$$

where f and g are continuous differentiable functions of x and y. Let S be a closed bounded subset of the 2D phase–space of F that does not contain a stationary point of F and let C be a trajectory of F that never leaves S. Then C is either a limit–cycle or C converges to a limit–cycle. The Poincaré–Bendixson theorem limits the types of long term behavior that can be exhibited by continuous planar dynamical systems. One important implication is that a 2D continuous dynamical system cannot give rise to a *strange attractor*. If a strange attractor C did exist in such a system, then it could be enclosed in a closed and bounded subset of the phase–space. By making this subset small enough, any nearby stationary points could be excluded. But then the Poincaré–Bendixson theorem says that C is not a strange attractor at all – it is either a limit–cycle or it converges to a limit–cycle. The Poincaré–Bendixson theorem says that chaotic behavior can only arise in continuous dynamical systems whose phase–space has 3 or more dimensions. However, this restriction does not apply to discrete dynamical systems, where chaotic behavior can arise in two or even one–dimensional.

2. The *Poincaré–Hopf index theorem* says: Let M be a compact differentiable manifold and v be a vector–field on M with isolated zeroes. If M has boundary, then we insist that v be pointing in the outward normal direction along the boundary. Then we have the formula

$$\sum_i index_v = \chi(M),$$

where the sum is over all the isolated zeroes of v and $\chi(M)$ is the *Euler characteristic* of M. systems.

In 1887, in honor of his 60th birthday, Oscar II, King of Sweden offered a prize to the person who could answer the question "Is the Solar system stable?" Poincaré won the prize with his famous work on the *three–body problem*. He considered the Sun, Earth and Moon orbiting in a plane under their mutual gravitational attractions (see Figure 1.13). Like the pendulum, this system has some unstable solutions. Introducing a *Poincaré section*, he saw that *homoclinic tangles* must occur. These would then give rise to *chaos* and *unpredictability*.

Recall that trying to predict the motion of the Moon has preoccupied astronomers since antiquity. Accurate understanding of its motion was important for determining the longitude of ships while traversing open seas. The *Rudolphine Tables* of *Johannes Kepler* had been a great improvement over previous tables, and Kepler was justly proud of his achievements. Bernoulli

1 Basics of Nonlinear and Chaotic Dynamics

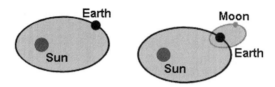

Fig. 1.13. The 2–body, problem solved by Newton (left), and the 3–body problem, first attacked by Poincaré, and still the point of active research (right).

used Newton's work on mechanics to derive the elliptic orbits of Kepler and set an example of how equations of motion could be solved by integrating. But the motion of the Moon is not well approximated by an ellipse with the Earth at a focus; at least the effects of the Sun have to be taken into account if one wants to reproduce the data the classical Greeks already possessed. To do that one has to consider the motion of three bodies: the Moon, the Earth, and the Sun. When the planets are replaced by point particles of arbitrary masses, the problem to be solved is known as the 3–body problem. The 3–body problem was also a model to another concern in astronomy. In the Newtonian model of the Solar system it is possible for one of the planets to go from an elliptic orbit around the Sun to an orbit that escaped its domain or that plunged right into it. Knowing if any of the planets would do so became the problem of the stability of the Solar system. A planet would not meet this terrible end if Solar system consisted of two celestial bodies, but whether such fate could befall in the 3–body case remained unclear.

After many failed attempts to solve the 3–body problem, natural philosophers started to suspect that it was impossible to integrate. The usual technique for integrating problems was to find the conserved quantities, quantities that do not change with time and allow one to relate the momenta and positions different times. The first sign on the impossibility of integrating the 3–body problem came from a result of Burns that showed that there were no conserved quantities that were polynomial in the momenta and positions. Burns' result did not preclude the possibility of more complicated conserved quantities. This problem was settled by Poincaré and Sundman in two very different ways.

In an attempt to promote the journal *Acta Mathematica*, *Gustaf Mittag–Leffler* got the permission of the King Oscar II of Sweden and Norway to establish a mathematical competition. Several questions were posed (although the king would have preferred only one), and the prize of 2500 kroner would go to the best submission. One of the questions was formulated by the 'father of modern analysis', *Karl Weierstrass*:

> "Given a system of arbitrary mass points that attract each other according to Newton's laws, under the assumption that no two points ever collide, try to find a representation of the coordinates of each

point as a series in a variable that is some known function of time and for all of whose values the series converges uniformly.
This problem, whose solution would *considerably extend our understanding of the Solar system...*"

Poincaré's submission won the prize. He showed that *conserved quantities that were analytic in the momenta and positions could not exist*. To show that he introduced methods that were very geometrical in spirit: the importance of phase flow, the role of *periodic orbits* and their cross sections, the *homoclinic points* (see [CAM05]).[8]

Poincaré pointed out that the problem was not correctly posed, and proved that a complete solution to it could not be found. His work was so impressive that in 1888 the jury recognized its value by awarding him the prize. He found that the evolution of such a system is often chaotic in the sense that a small perturbation in the initial state, such as a slight change in one body's initial position, might lead to a radically different later state. If the slight change is not detectable by our measuring instruments, then we will not be able to predict which final state will occur. One of the judges, the distinguished Karl Weierstrass, said, "This work cannot indeed be considered as furnishing the complete solution of the question proposed, but that it is nevertheless of such importance that its publication will inaugurate a new era in the history of celestial mechanics." Weierstrass did not know how accurate he was. In Poincaré's paper, he described new mathematical ideas such as homoclinic points. The memoir was about to be published in Acta Mathematica when an error was found by the editor. This error in fact led to further discoveries by Poincaré, which are now considered to be the beginning of *chaos theory*. The memoir was published later in 1890. Poincaré's research into orbits about Lagrange points and low-energy transfers was not utilized for more than a century afterwards.

---

[8] The interesting thing about Poincaré's work was that it did not solve the problem posed. He did not find a function that would give the coordinates as a function of time for all times. He did not show that it was impossible either, but rather that it could not be done with the Bernoulli technique of finding a conserved quantity and trying to integrate. Integration would seem unlikely from Poincaré's prize–winning memoir, but it was accomplished by the Finnish–born Swedish mathematician Sundman, who showed that to integrate the 3–body problem one had to confront the 2–body collisions. He did that by making them go away through a trick known as regularization of the collision manifold. The trick is not to expand the coordinates as a function of time t, but rather as a function of 3√t. To solve the problem for all times he used a conformal map into a strip. This allowed Sundman to obtain a series expansion for the coordinates valid for all times, solving the problem that was proposed by Weirstrass in the King Oscar II's competition. Though Sundman's work deserves better credit than it gets, it did not live up to Weirstrass's expectations, and the series solution did not 'considerably extend our understanding of the Solar system.' The work that followed from Poincaré did.

In 1889 Poincaré proved that for the restricted three body problem no integrals exist apart from the Jacobian. In 1890 Poincaré proved his famous recurrence theorem, namely that in any small region of phase–space trajectories exist which pass through the region infinitely often. Poincaré published 3 volumes of 'Les méthods nouvelle de la mécanique celeste' between 1892 and 1899. He discussed convergence and uniform convergence of the series solutions discussed by earlier mathematicians and proved them not to be uniformly convergent. The stability proofs of Lagrange and Laplace became inconclusive after this result.

Poincaré introduced further topological methods in 1912 for the theory of stability of orbits in the 3–body problem. It fact Poincaré essentially invented topology in his attempt to answer stability questions in the three body problem. He conjectured that there are infinitely many periodic solutions of the restricted problem, the conjecture being later proved by George *Birkhoff*. The stability of the orbits in the three body problem was also investigated by Levi–Civita, Birkhoff and others (see [II06b] for technical details).

To examine chaos, Poincaré used the idea of a section, today called the *Poincaré section*, which cuts across the orbits in phase–space. While the original dynamical system always *flows in continuous time*, on the Poincaré section we can observe *discrete–time steps*. More precisely, the original *phase–space flow* (see [II06b]) is replaced by an *iterated map*, which reduces the dimension of the phase–space by one (see Figure 1.14). Later, to show what

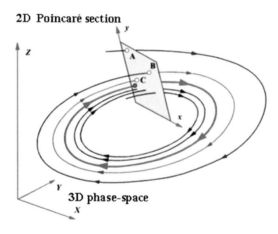

Fig. 1.14. The 2D Poincaré section, reducing the 3D phase–space, using the iterated map: $x_{new} = F(x,y)$, $y_{new} = G(x,y)$.

a Poincaré section would look like, Hénon devised a simple 2D–map, which is today called the *Hénon map*: $x_{new} = 1 - ax^2 + by$, $y_{new} = x$, with parameters $a = 1.4$, $b = 0.3$. Given any starting point, this map generates a sequence of points settling onto a chaotic attractor.

As an inheritance of Poincaré work, the chaos of the Solar system has been recently used for the SOHO project,[9] to minimize the fuel consumption need for the space flights. Namely, in a rotating frame, a spacecraft can remain stationary at 5 *Lagrange's points* (see Figure 1.15).

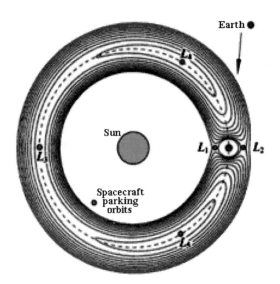

Fig. 1.15. The *Lagrange's points* $(L_1, ... L_5)$ used for the space flights. Points L_1, L_2, L_3 on the Sun–Earth axis are unstable. The SOHO spacecraft used a *halo orbit* around L_1 to observe the Sun. The triangular points L_4 and L_5 are often stable. A Japanese rescue mission used a chaotic Earth–Moon trajectory.

---

[9] The SOHO project is being carried out jointly by ESA (European Space Agency) and NASA (US National Aeronautics and Space Administration), as a cooperative effort between the two agencies in the framework of the Solar Terrestrial Science Program (STSP) comprising SOHO and CLUSTER, and the International Solar–Terrestrial Physics Program (ISTP), with Geotail (ISAS–Japan), Wind, and Polar. SOHO was launched on December 2, 1995. The SOHO spacecraft was built in Europe by an industry team led by Matra, and instruments were provided by European and American scientists. There are nine European Principal Investigators (PI's) and three American ones. Large engineering teams and more than 200 co-investigators from many institutions support the PI's in the development of the instruments and in the preparation of their operations and data analysis. NASA is responsible for the launch and mission operations. Large radio dishes around the world which form NASA's Deep Space Network are used to track the spacecraft beyond the Earth's orbit. Mission control is based at Goddard Space Flight Center in Maryland.

Poincaré had two protegés in the development of chaos theory in the new world: *George D. Birkhoff* (see [Bir15, Bir27, Bir17] for the *Birkhoff curve shortening flow*), and *Stephen Smale*.

In some detail, the theorems of John von Neumann and George Birkhoff on the *ergodic hypothesis* (see footnote 6 above) were published in 1912 and 1913. This line of enquiry developed in two directions. One direction took an abstract approach and considered *dynamical systems as transformations of measurable spaces into themselves*. Could we classify these transformations in a meaningful way? This lead *Andrey N. Kolmogorov* to the introduction of the fundamental concept of *entropy* for dynamical systems. With entropy as a *dynamical invariant* it became possible to classify a set of abstract dynamical systems known as the *Bernoulli systems*.

The other line that developed from the ergodic hypothesis was in trying to find mechanical systems that are ergodic. *An ergodic system could not have stable orbits, as these would break ergodicity.* So, in 1898 *Jacques S. Hadamard* published a paper on *billiards*, where he showed that the *motion of balls on surfaces of constant negative curvature is everywhere unstable*. This dynamical system was to prove very useful and it was taken up by Birkhoff.

Marston Morse in 1923 showed that it was possible to enumerate the orbits of a ball on a surface of constant negative curvature.[10] He did this by introducing a symbolic code to each orbit and showed that the number of possible codes grew exponentially with the length of the code. With contributions by E. Artin, G. Hedlund, and *Heinz Hopf* it was eventually proven that the motion of a ball on a surface of constant negative curvature was ergodic. The importance of this result escaped most physicists, one exception being N.M. *Krylov*, who understood that a physical billiard was a dynamical system on a surface of negative curvature, but with the curvature concentrated along the lines of collision. Sinai, who was the first to show that a physical billiard can be ergodic, knew Krylov's work well.

On the other hand, the work of Lord Rayleigh also received vigorous development. It prompted many experiments and some theoretical development by B. *Van der Pol*, G. *Duffing*, and D. *Hayashi*. They found other systems in which the nonlinear oscillator played a role and classified the possible motions

---

[10] Recall from [II06b] that in differential topology, the techniques of *Morse theory* give a very direct way of analyzing the topology of a manifold by studying differentiable functions on that manifold. According to the basic insights of Marston Morse, a differentiable function on a manifold will, in a typical case, reflect the topology quite directly. Morse theory allows one to find the so–called CW–structures and handle decompositions on manifolds and to obtain substantial information about their homology. Before Morse, Arthur Cayley and James Clerk Maxwell developed some of the ideas of Morse theory in the context of topography. Morse originally applied his theory to geodesics (critical points of the energy functional on paths). These techniques were later used by *Raoul Bott* in his proof of the celebrated Bott periodicity theorem.

of these systems. This concreteness of experiments, and the possibility of analysis was too much of temptation for M. L. *Cartwright* and J.E. *Littlewood*, who set out to prove that many of the structures conjectured by the experimentalists and theoretical physicists did indeed follow from the equations of motion.

Also, G. Birkhoff had found a 'remarkable curve' in a 2D map; it appeared to be non–differentiable and it would be nice to see if a smooth flow could generate such a curve. The work of Cartwright and Littlewood lead to the work of N. Levinson, which in turn provided the basis for the horseshoe construction of Steve Smale.

In Russia, *Aleksandr M. Lyapunov* paralleled the methods of Poincaré and initiated the strong Russian dynamical systems school. A. *Andronov*[11] carried on with the study of nonlinear oscillators and in 1937 introduced together with *Lev S. Pontryagin*[12] the notion of *coarse systems*. They were formalizing the understanding garnered from the study of nonlinear oscillators, the understanding that many of the details on how these oscillators work do not affect the overall picture of the phase–space: there will still be limit cycles if one changes the dissipation or spring force function by a little bit. And changing the system a little bit has the great advantage of eliminating exceptional cases in the mathematical analysis. Coarse systems were the concept that caught Smale's attention and enticed him to study dynamical systems (see [CAM05]).

The path traversed from ergodicity to entropy is a little more confusing. The general character of entropy was understood by *Norbert Wiener*,[13] who seemed to have spoken to *Claude E. Shannon*.[14] In 1948 Shannon published his results on *information theory*, where he discusses the entropy of the shift transformation.

In Russia, *Andrey N. Kolmogorov* went far beyond and suggested a definition of the metric entropy of an area preserving transformation in order to classify Bernoulli shifts. The suggestion was taken by his student Ya.G. *Sinai* and the results published in 1959. In 1967 D.V. Anosov[15] and Sinai

---

[11] Recall that both the *Andronov–Hopf bifurcation* and a crater on the Moon are named after Aleksandr Andronov.
[12] the father of modern optimal control theory (see [II06b])
[13] the father of cybernetics
[14] the father of information theory
[15] Recall that the *Anosov map* on a manifold M is a certain type of mapping, from M to itself, with rather clearly marked local directions of 'expansion' and 'contraction'. More precisely:

- If a differentiable map f on M has a hyperbolic structure on the tangent bundle, then it is called an *Anosov map*. Examples include the *Bernoulli map*, and *Arnold cat map*.
- If the Anosov map is a diffeomorphism, then it is called an *Anosov diffeomorphism*. Anosov proved that Anosov diffeomorphisms are *structurally stable*.
- If a flow on a manifold splits the tangent bundle into three invariant subbundles, with one subbundle that is exponentially contracting, and one that is exponen-

applied the notion of entropy to the study of dynamical systems. It was in the context of studying the entropy associated to a dynamical system that Sinai introduced *Markov partitions* (in 1968), which allow one *to relate dynamical systems and statistical mechanics*; this has been a very fruitful relationship. It adds measure notions to the topological framework laid down in Smale's dynamical systems paper. Markov partitions *divide the phase–space* of the dynamical system into nice little boxes that map into each other. Each box is labelled by a code and the dynamics on the phase–space maps the codes around, inducing a *symbolic dynamics*. From the number of boxes needed to cover all the space, Sinai was able to define the notion of entropy of a dynamical system. However, the relations with statistical mechanics became explicit in the work of *David Ruelle*.[16] Ruelle understood that the topology of the orbits could be specified by a symbolic code, and that one could associate an 'energy' to each orbit. The energies could be formally combined in a *partition function* (see [II06b]) to generate the invariant measure of the system.

1.3.2 Smale's Topological Horseshoe and Chaos of Stretching and Folding

The first deliberate, coordinated attempt to understand how global system's behavior might differ from its local behavior, came from topologist Steve Smale from the University of California at Berkeley. A young physicist, making a small talk, asked what Smale was working on. The answer stunned him: "Oscillators." It was absurd. Oscillators (pendulums, springs, or electric circuits) where the sort of problem that a physicist finished off early in his training. They were easy. Why would a great mathematician be studying elementary physics? However, Smale was looking at nonlinear oscillators, chaotic oscillators – and seing things that physicists had learned no to see [Gle87].

Smale's 1966 Fields Medal honored a famous piece of work in high–dimensional topology, proving *Poincaré conjecture* for all dimensions greater than 4; he later generalized the ideas in a 107 page paper that established the *H–cobordism theorem* (this seminal result provides algebraic algebraic topological criteria for establishing that higher–dimensional manifolds are diffeomorphic).

---

tially expanding, and a third, non–expanding, non–contracting 1D sub–bundle, then the flow is called an *Anosov flow*.

[16] David Ruelle is a mathematical physicist working on statistical physics and dynamical systems. Together with *Floris Takens*, he coined the term *strange attractor*, and founded a modern *theory of turbulence*. Namely, in a seminal paper [RT71] they argued that, as a function of an external parameter, the *route to chaos in a fluid flow* is a transition sequence leading from stationary (S) to single periodic (P), double periodic (QP_2), triple periodic (QP_3) and, possibly, quadruply periodic (QP_4) motions, before the flow becomes chaotic (C).

1.3 Brief History of Chaos Theory

After having made great strides in topology, Smale then turned to the study of nonlinear dynamical systems, where he made significant advances as well.[17] His first contribution is the famous *horseshoe map* [Sch81] that started–off significant research in dynamical systems and chaos theory.[18] Smale also outlined a mathematical research program carried out by many others. Smale is also known for injecting *Morse theory* into mathematical economics, as well as recent explorations of various theories of computation. In 1998 he compiled a list of 18 problems in mathematics to be solved in the 21st century. This list was compiled in the spirit of Hilbert's famous list of problems produced in 1900. In fact, Smale's list includes some of the original Hilbert problems. Smale's problems include the Jacobian conjecture and the Riemann hypothesis, both of which are still unsolved.

The *Smale horseshoe map* (see Figure 1.16) *is any member of a class of chaotic maps of the square into itself.* This topological transformation provided a basis for understanding the chaotic properties of dynamical systems. Its basis are simple: A space is stretched in one direction, squeezed in another, and then folded. When the process is repeated, it produces something like a many–layered pastry dough, in which a pair of points that end up close together may have begun far apart, while two initially nearby points can end completely far apart.[19]

---

[17] In the fall of 1961 Steven Smale was invited to Kiev where he met V.I. Arnol'd, (one of the fathers of modern geometrical mechanics [II06b]), D.V. Anosov, Sinai, and Novikov. He lectured there, and spent a lot of time with Anosov. He suggested a series of conjectures, most of which Anosov proved within a year. It was Anosov who showed that there are dynamical systems for which all points (as opposed to a nonwandering set) admit the hyperbolic structure, and it was in honor of this result that Smale named them *Axiom–A systems*. In Kiev Smale found a receptive audience that had been thinking about these problems. Smale's result catalyzed their thoughts and initiated a chain of developments that persisted into the 1970's.

[18] In his landmark 1967 Bulletin survey article entitled 'Differentiable dynamical systems' [Sch81], Smale presented his program for hyperbolic dynamical systems and stability, complete with a superb collection of problems. The major theorem of the paper was the Ω–Stability Theorem: the global foliation of invariant sets of the map into disjoint stable and unstable parts, whose proof was a tour de force in the new dynamical methods. Some other important ideas of this paper are the existence of a horseshoe and enumeration and ordering of all its orbits, as well as the use of zeta functions to study dynamical systems. The emphasis of the paper is on the global properties of the dynamical system, on how to understand the topology of the orbits. Smale's account takes us from a local differential equation (in the form of vector fields) to the global topological description in terms of horseshoes.

[19] Originally, Smale had hoped to explain all dynamical systems in terms of *stretching* and *squeezing* – with no folding, at least no folding that would drastically undermine a system's stability. But *folding* turned out to be necessary, and folding allowed sharp changes in dynamical behavior [Gle87].

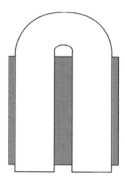

Fig. 1.16. The *Smale horseshoe map* consists of a sequence of operations on the unit square. First, stretch in the y−direction by more than a factor of two, then squeeze (compress) in the x−direction by more than a factor of two. Finally, fold the resulting rectangle and fit it back onto the square, overlapping at the top and bottom, and not quite reaching the ends to the left and right (and with a gap in the middle), as illustrated in the diagram. The shape of the stretched and folded map gives the horseshoe map its name. Note that it is vital to the construction process for the map to overlap and leave the middle and vertical edges of the initial unit square uncovered.

The horseshoe map was introduced by Smale while studying the behavior of the orbits of the *relaxation Van der Pol oscillator*. The action of the map is defined geometrically by squishing the square, then stretching the result into a long strip, and finally folding the strip into the shape of a horseshoe.

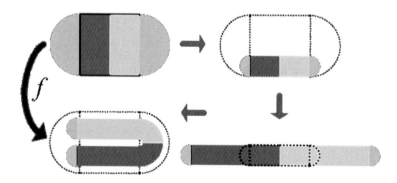

Fig. 1.17. The *Smale horseshoe map* f, defined by *stretching, folding* and *squeezing* of the system's phase–space.

Most points eventually leave the square under the action of the map f. They go to the side caps where they will, under iteration, converge to a *fixed–point* in one of the caps. The points that remain in the square under repeated iteration form a *fractal set* and are part of the *invariant set* of the map f (see Figure 1.17).

The *stretching, folding* and *squeezing* of the horseshoe map are the essential elements that must be present in any chaotic system. In the horseshoe map the squeezing and stretching are uniform. They compensate each other so that the area of the square does not change. The folding is done neatly, so that the orbits that remain forever in the square can be simply described.

Repeating this generates the horseshoe attractor. If one looks at a cross section of the final structure, it is seen to correspond to a *Cantor set*.

The Smale horseshoe map is the set of basic topological operations for constructing an attractor consist of stretching (which gives sensitivity to initial conditions) and folding (which gives the attraction). Since *trajectories in phase–space cannot cross*, the repeated stretching and folding operations result in an object of great topological complexity. For any horseshoe map we have:

- There is an infinite number of periodic orbits;
- Periodic orbits of arbitrarily long period exist;
- The number or periodic orbits grows exponentially with the period; and
- Close to any point of the fractal invariant set there is a point of a periodic orbit.

More precisely, the horseshoe map f is a *diffeomorphism* defined from a region S of the plane into itself. The region S is a square capped by two semi–disks. The action of f is defined through the composition of three geometrically defined transformations. First the square is contracted along the vertical direction by a factor $a < 1/2$. The caps are contracted so as to remain semi-disks attached to the resulting rectangle. Contracting by a factor smaller than one half assures that there will be a gap between the branches of the horseshoe. Next the rectangle is stretched by a factor of $1/a$; the caps remain unchanged. Finally the resulting strip is folded into a horseshoe–shape and placed back into S.

The interesting part of the dynamics is the image of the square into itself. Once that part is defined, the map can be extended to a diffeomorphism by defining its action on the caps. The caps are made to contract and eventually map inside one of the caps (the left one in the figure). The extension of f to the caps adds a fixed–point to the *non–wandering set* of the map. To keep the class of horseshoe maps simple, the curved region of the horseshoe should not map back into the square.

The horseshoe map is one–to–one (1–1, or injection): any point in the domain has a unique image, even though not all points of the domain are the image of a point. The inverse of the horseshoe map, denoted by f^{-1}, cannot have as its domain the entire region S, instead it must be restricted to the image of S under f, that is, the domain of f^{-1} is $f(S)$.

By folding the contracted and stretched square in different ways, other types of horseshoe maps are possible (see Figure 1.18). The contracted square

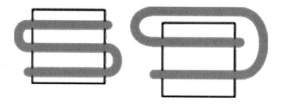

Fig. 1.18. Other types of horseshoe maps can be made by folding the contracted and stretched square in different ways.

cannot overlap itself to assure that it remains 1–1. When the action on the square is extended to a diffeomorphism, the extension cannot always be done on the plane. For example, the map on the right needs to be extended to a diffeomorphism of the sphere by using a 'cap' that wraps around the equator.

The horseshoe map is an Axiom A diffeomorphism that serves as a model for the general behavior at a transverse *homoclinic point*, where the *stable and unstable manifolds* of a periodic point intersect.

The horseshoe map was designed by Smale to reproduce the chaotic dynamics of a *flow* in the neighborhood of a given periodic *orbit*. The neighborhood is chosen to be a small disk perpendicular to the orbit. As the system evolves, points in this disk remain close to the given periodic orbit, tracing out orbits that eventually intersect the disk once again. Other orbits diverge.

The behavior of all the orbits in the disk can be determined by considering what happens to the disk. The intersection of the disk with the given periodic orbit comes back to itself every period of the orbit and so do points in its neighborhood. When this neighborhood returns, its shape is transformed. Among the points back inside the disk are some points that will leave the disk neighborhood and others that will continue to return. The set of points that never leaves the neighborhood of the given periodic orbit form a fractal.

A symbolic name can be given to all the orbits that remain in the neighborhood. The initial neighborhood disk can be divided into a small number of regions. Knowing the sequence in which the orbit visits these regions allows the orbit to be pinpointed exactly. The visitation sequence of the orbits provide the so–called *symbolic dynamics*[20]

---

[20] Symbolic dynamics is the practice of modelling a dynamical system by a space consisting of infinite sequences of abstract symbols, each sequence corresponding to a state of the system, and a shift operator corresponding to the dynamics. Symbolic dynamics originated as a method to study general dynamical systems, now though, its techniques and ideas have found significant applications in data storage and transmission, linear algebra, the motions of the planets and many other areas. The distinct feature in symbolic dynamics is that time is measured in discrete intervals. So at each time interval the system is in a particular state. Each state is associated with a symbol and the evolution of the system is described by an infinite sequence of symbols (see text below).

It is possible to describe the behavior of all initial conditions of the horseshoe map. An initial point $u_0 = x, y$ gets mapped into the point $u_1 = f(u_0)$. Its iterate is the point $u_2 = f(u_1) = f^2(u_0)$, and repeated iteration generates the orbit $u_0, u_1, u_2, \ldots$ Under repeated iteration of the horseshoe map, most orbits end up at the fixed–point in the left cap. This is because the horseshoe maps the left cap into itself by an *affine transformation*, which has exactly one fixed–point. Any orbit that lands on the left cap never leaves it and converges to the fixed–point in the left cap under iteration. Points in the right cap get mapped into the left cap on the next iteration, and most points in the square get mapped into the caps. Under iteration, most points will be part of orbits that converge to the fixed–point in the left cap, but some points of the square never leave.

Under forward iterations of the horseshoe map, the original square gets mapped into a series of horizontal strips. The points in these horizontal strips come from vertical strips in the original square. Let S_0 be the original square, map it forward n times, and consider only the points that fall back into the square S_0, which is a set of horizontal stripes $H_n = f^n(S_0) \cap S_0$. The points in the horizontal stripes came from the vertical stripes $V_n = f^{-n}(H_n)$, which are the horizontal strips H_n mapped backwards n times. That is, a point in V_n will, under n iterations of the horseshoe map, end up in the set H_n of vertical strips (see Figure 1.19).

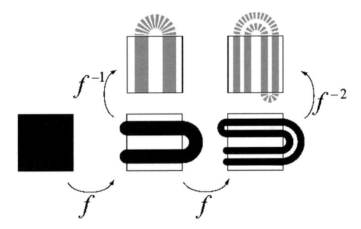

Fig. 1.19. Iterated horseshoe map: pre–images of the square region.

Now, if a point is to remain indefinitely in the square, then it must belong to an *invariant set* Λ that maps to itself. Whether this set is empty or not has to be determined. The vertical strips V_1 map into the horizontal strips H_1, but not all points of V_1 map back into V_1. Only the points in the intersection of V_1 and H_1 may belong to Λ, as can be checked by following points outside the intersection for one more iteration. The intersection of the horizontal and vertical stripes, $H_n \cap V_n$, are squares that converge in the limit $n \to \infty$ to the invariant set Λ (see Figure 1.20).

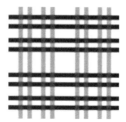

Fig. 1.20. Intersections that converge to the invariant set Λ.

The structure of invariant set Λ can be better understood by introducing a system of labels for all the intersections, namely a *symbolic dynamics*. The intersection $H_n \cap V_n$ is contained in V_1. So any point that is in Λ under iteration must land in the left vertical strip A of V_1, or on the right vertical strip B. The lower horizontal strip of H_1 is the image of A and the upper horizontal strip is the image of B, so $H_1 = f(A) \cap f(B)$. The strips A and B can be used to label the four squares in the intersection of V_1 and H_1 (see Figure 1.21) as:

$$\Lambda_{A \bullet A} = f(A) \cap A, \qquad \Lambda_{A \bullet B} = f(A) \cap B,$$
$$\Lambda_{B \bullet A} = f(B) \cap A, \qquad \Lambda_{B \bullet B} = f(B) \cap B.$$

The set $\Lambda_{B \bullet A}$ consist of points from strip A that were in strip B in the previous iteration. A dot is used to separate the region the point of an orbit is in from the region the point came from.

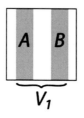

Fig. 1.21. The basic domains of the horseshoe map in symbolic dynamics.

This notation can be extended to higher iterates of the horseshoe map. The vertical strips can be named according to the sequence of visits to strip A or strip B. For example, the set $ABB \subset V_3$ consists of the points from A that will all land in B in one iteration and remain in B in the iteration after that:

$$ABB = \{x \in A | f(x) \in B \text{ and } f^2(x) \in B\}.$$

Working backwards from that trajectory determines a small region, the set ABB, within V_3.

The horizontal strips are named from their vertical strip pre–images. In this notation, the intersection of V_2 and H_2 consists of 16 squares, one of which is

$$\Lambda_{AB \bullet BB} = f^2(AB) \cap BB.$$

All the points in $\Lambda_{AB \bullet BB}$ are in B and will continue to be in B for at least one more iteration. Their previous trajectory before landing in BB was A followed by B.

Any one of the intersections $\Lambda_{P \bullet F}$ of a horizontal strip with a vertical strip, where P and F are sequences of As and Bs, is an affine transformation of a small region in V_1. If P has k symbols in it, and if $f^{-k}(\Lambda_{P \bullet F})$ and $\Lambda_{P \bullet F}$ intersect, then the region $\Lambda_{P \bullet F}$ will have a *fixed–point*. This happens when the sequence P is the same as F. For example, $\Lambda_{ABAB \bullet ABAB} \subset V_4 \cap H_4$ has at least one fixed–point. This point is also the same as the fixed–point in $\Lambda_{AB \bullet AB}$. By including more and more ABs in the P and F part of the label of intersection, the area of the intersection can be made as small as needed. It converges to a point that is part of a *periodic orbit of the horseshoe map*. The periodic orbit can be labelled by the simplest sequence of As and Bs that labels one of the regions the periodic orbit visits. For every sequence of As and Bs there is a periodic orbit.

The Smale horseshoe map is the same topological structure as the *homoclinic tangle*. To dynamically introduce homoclinic tangles, let us consider a classical engineering problem of *escape from a potential well*. Namely, if we have a motion, $x = x(t)$, of a damped particle in a well with potential energy $V = x^2/2 - x^3/3$ (see Figure 1.22) excited by a periodic driving force, $F\cos(wt)$ (with the period $T = 2\pi/w$), we are dealing with a nonlinear dynamical system given by [TS01]

$$\ddot{x} + a\dot{x} + x - x^2 = F\cos(wt). \qquad (1.20)$$

Now, if the driving is switched off, i.e., $F = 0$, we have an autonomous 2D–system with the phase–portrait (and the safe basin of attraction) given in Figure 1.22 (below). The grey area of escape starts over the hilltop to infinity. Once we start driving, the system (1.20) becomes 3–dimensional, with its 3D phase–space. We need to see the basin in a *stroboscopic section* (see Figure 1.23). The hill–top solution still has an inset and and outset. As the driving increases, the inset and outset get tangled. They intersect one another an infinite number of times. The boundary of the safe basin becomes fractal. As the driving increases even more, the so–called fractal–fingers created by the homoclinic tangling, make a sudden incursion into the safe basin. At that point, the integrity of the in–well motions is lost [TS01].

Now, topologically speaking (referring to the Figure 1.24), let X be the point of intersection, with X' ahead of X on one manifold and ahead of X''

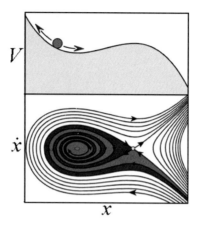

Fig. 1.22. Motion of a damped particle in a potential well, driven by a periodic force $F\cos(wt)$,. Up: potential $(x - V)$–plot, with $V = x^2/2 - x^3/3$; down: the corresponding phase $(x - \dot{x})$–portrait, showing the safe basin of attraction – if the driving is switched off $(F = 0)$.

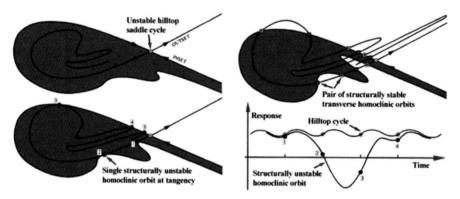

Fig. 1.23. Dynamics of a homoclinic tangle. The hill–top solution of a damped particle in a potential well driven by a periodic force. As the driving increases, the inset and outset get tangled.

of the other. The map of each of these points TX' and TX'' must be ahead of the map of X, TX. The only way this can happen is if the manifold loops back and crosses itself at a new *homoclinic point*, i.e., a point where a stable and an unstable separatrix (invariant manifold) from the same fixed–point or same family intersect. Another loop must be formed, with T^2X another homoclinic point. Since T^2X is closer to the hyperbolic point than TX, the distance between T^2X and TX is less than that between X and TX. Area preservation requires the area to remain the same, so each new curve (which is closer than the previous one) must extend further. In effect, the loops become longer and thinner. The network of curves leading to a dense area of homoclinic points is known as a homoclinic tangle or tendril. Homoclinic points appear where chaotic regions touch in a *hyperbolic fixed–point*.

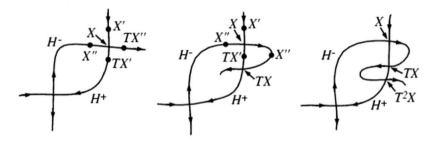

Fig. 1.24. More on homoclinic tangle (see text for explanation).

On the other hand, tangles are in general related to n−categories (see [II05, II06a, II06b]). Recall that in describing dynamical systems (processes) by means of n−categories, instead of classical starting with a *set of things:*

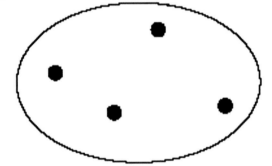

we can now start with a *category of things and processes between things:*

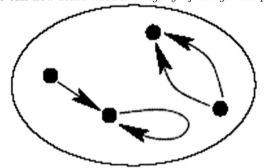

or, a 2−*category of things, processes, and processes between processes:*

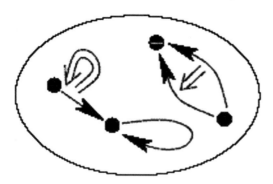

... and so on. In this way, topological $n-$categories form the natural framework for *high–dimensional chaos theory*.

1.3.3 Lorenz' Weather Prediction and Chaos

Recall that an *attractor* is a set of system's states (i.e., points in the system's phase–space), invariant under the dynamics, towards which neighboring states in a given *basin of attraction* asymptotically approach in the course of dynamic evolution.[21] An attractor is defined as the smallest unit which cannot be itself decomposed into two or more attractors with distinct basins of attraction. This restriction is necessary since a dynamical system may have multiple attractors, each with its own basin of attraction.

Conservative systems do not have attractors, since the motion is periodic. For dissipative dynamical systems, however, volumes shrink exponentially, so attractors have 0 volume in nD phase–space.

In particular, a stable *fixed–point* surrounded by a dissipative region is an attractor known as a *map sink*.[22] Regular attractors (corresponding to 0 *Lyapunov exponents*) act as *limit cycles*, in which trajectories circle around a limiting trajectory which they asymptotically approach, but never reach. The so–called *strange attractors*[23] are bounded regions of phase–space (corresponding to positive Lyapunov characteristic exponents) having zero measure in the embedding phase–space and a *fractal dimension*. Trajectories within a strange attractor appear to skip around randomly.

In 1963, Ed Lorenz from MIT was trying to improve weather forecasting. Using a primitive computer of those days, he discovered the first *chaotic attractor*. Lorenz used three Cartesian variables, (x, y, z), to define *atmospheric*

---

[21] A *basin of attraction* is a set of points in the system's phase–space, such that initial conditions chosen in this set dynamically evolve to a particular attractor.

[22] A *map sink* is a stable fixed–point of a map which, in a dissipative dynamical system, is an attractor.

[23] A strange attractor is an attracting set that has zero measure in the embedding phase–space and has fractal dimension. Trajectories within a strange attractor appear to skip around randomly.

1.3 Brief History of Chaos Theory 45

convection. Changing in time, these variables gave him a trajectory in a (Euclidean) 3D–space. From all starts, trajectories settle onto a chaotic, or *strange attractor*. [24]

More precisely, Lorenz reduced the *Navier–Stokes equations* for *convective Bénard fluid flow* (see section (1.120) below) into three first order coupled nonlinear differential equations, already introduced above as (1.7) and demonstrated with these the idea of sensitive dependence upon initial conditions and chaos (see [Lor63, Spa82]).

We rewrite the celebrated Lorenz equations here as

---

[24] Edward Lorenz is a professor of meteorology at MIT who wrote the first clear paper on *deterministic chaos*. The paper was called 'Deterministic Nonperiodic Flow' and it was published in the Journal of Atmospheric Sciences in 1963. Before that, in 1960, Lorenz began a project to simulate weather patterns on a computer system called the Royal McBee. Lacking much memory, the computer was unable to create complex patterns, but it was able to show the interaction between major meteorological events such as tornadoes, hurricanes, easterlies and westerlies. A variety of factors was represented by a number, and Lorenz could use computer printouts to analyze the results. After watching his systems develop on the computer, Lorenz began to see patterns emerge, and was able to predict with some degree of accuracy what would happen next. While carrying out an experiment, Lorenz made an accidental discovery. He had completed a run, and wanted to recreate the pattern. Using a printout, Lorenz entered some variables into the computer and expected the simulation to proceed the same as it had before. To his surprise, the pattern began to diverge from the previous run, and after a few 'months' of simulated time, the pattern was completely different. Lorenz eventually discovered why seemingly identical variables could produce such different results. When Lorenz entered the numbers to recreate the scenario, the printout provided him with numbers to the thousandth position (such as 0.617). However, the computer's internal memory held numbers up to the millionth position (such as 0.617395); these numbers were used to create the scenario for the initial run. This small deviation resulted in a completely divergent weather pattern in just a few months. This discovery creates the groundwork of chaos theory: In a system, small deviations can result in large changes. This concept is now known as a *butterfly effect*.

Lorenz definition of chaos is: "The property that characterizes a dynamical system in which most orbits exhibit sensitive dependence." Dynamical systems (like the weather) are all around us. They have recurrent behavior (it is always hotter in summer than winter) but are very difficult to pin down and predict apart from the very short term. 'What will the weather be tomorrow?' – can be anticipated, but 'What will the weather be in a months time?' is an impossible question to answer.

Lorenz showed that with a set of simple differential equations seemingly very complex turbulent behavior could be created that would previously have been considered as random. He further showed that accurate longer range forecasts in any chaotic system were impossible, thereby overturning the previous orthodoxy. It had been believed that the more equations you add to describe a system, the more accurate will be the eventual forecast.

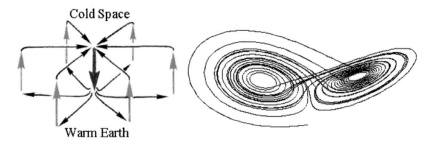

Fig. 1.25. Bénard cells, showing a typical vortex of a rolling air, with a warm air rising in a ring and a cool air descending in the center (left). A simple model of the Bénard cells provided by the celebrated 'Lorenz–butterfly' (or, 'Lorenz–mask') *strange attractor* (right).

$$\dot{x} = a(y-x), \qquad \dot{y} = bx - y - xz, \qquad \dot{z} = xy - cz, \qquad (1.21)$$

where x, y and z are dynamical variables, constituting the 3D *phase–space* of the *Lorenz system*; and a, b and c are the parameters of the system. Originally, Lorenz used this model to describe the unpredictable behavior of the weather, where x is the rate of convective overturning (convection is the process by which heat is transferred by a moving fluid), y is the horizontal temperature overturning, and z is the vertical temperature overturning; the parameters are: $a \equiv P$–proportional to the *Prandtl number* (ratio of the fluid viscosity of a substance to its thermal conductivity, usually set at 10), $b \equiv R$–proportional to the Rayleigh number (difference in temperature between the top and bottom of the system, usually set at 28), and $c \equiv K$–a number proportional to the physical proportions of the region under consideration (width to height ratio of the box which holds the system, usually set at 8/3). The Lorenz system (1.121) has the properties:

1. *Symmetry*: $(x, y, z) \to (-x, -y, z)$ for all values of the parameters, and
2. The z–axis ($x = y = 0$) is *invariant* (i.e., all trajectories that start on it also end on it).

Nowadays it is well–known that the Lorenz model is a paradigm for low–dimensional chaos in dynamical systems in synergetics and this model or its modifications are widely investigated in connection with modelling purposes in meteorology, hydrodynamics, laser physics, superconductivity, electronics, oil industry, chemical and biological kinetics, etc.

The 3D *phase–portrait* of the Lorenz system (1.52) shows the celebrated '*Lorenz mask*', a special type of *fractal attractor* (see Figure 1.52). It depicts the famous '*butterfly effect*', (i.e., sensitive dependence on initial conditions) – the popular idea in meteorology that 'the flapping of a butterfly's wings in Brazil can set off a tornado in Texas' (i.e., a tiny difference is amplified until two outcomes are totally different), so that the long term behavior becomes impossible to predict (e.g., long term weather forecasting). The Lorenz mask has the following characteristics:

1. Trajectory does not intersect itself in three dimensions;
2. Trajectory is not periodic or transient;
3. General form of the shape does not depend on initial conditions; and
4. Exact sequence of loops is very sensitive to the initial conditions.

1.3.4 Feigenbaum's Constant and Universality

Mitchell Jay Feigenbaum (born December 19, 1944; Philadelphia, USA) is a mathematical physicist whose pioneering studies in chaos theory led to the discovery of the *Feigenbaum constant*.

In 1964 he began graduate studies at the MIT. Enrolling to study electrical engineering, he changed to physics and was awarded a doctorate in 1970 for a thesis on dispersion relations under Francis Low. After short positions at Cornell University and Virginia Polytechnic Institute, he was offered a longer–term post at Los Alamos National Laboratory to study turbulence. Although the group was ultimately unable to unravel the intractable theory of turbulent fluids, his research led him to study chaotic maps.

Many mathematical maps involving a single linear parameter exhibit apparently random behavior known as chaos when the parameter lies in a certain range. As the parameter is increased towards this region, the map undergoes bifurcations at precise values of the parameter. At first there is one stable point, then bifurcating to oscillate between two points, then bifurcating again to oscillate between four points and so on. In 1975 Feigenbaum, using the HP-65 computer he was given, discovered that the ratio of the difference between the values at which such successive *period–doubling bifurcations* (called the *Feigenbaum cascade*) occur tends to a constant of around 4.6692. He was then able to provide a mathematical proof of the fact, and showed that the same behavior and the same constant would occur in a wide class of mathematical functions prior to the onset of chaos. For the first time this universal result enabled mathematicians to take their first huge step to unravelling the apparently intractable 'random' behavior of chaotic systems. This 'ratio of convergence' is now known as the Feigenbaum constant.

More precisely, the Feigenbaum constant δ is a universal constant for functions approaching chaos via successive period doubling bifurcations. It was discovered by Feigenbaum in 1975, while studying the fixed–points of the iterated function $f(x) = 1 - \mu|x|^r$, and characterizes the geometric approach of the bifurcation parameter to its limiting value (see Figure 1.26) as the parameter μ is increased for fixed x [Fei79].

The Logistic map is a well known example of the maps that Feigenbaum studied in his famous Universality paper [Fei78].

In 1986 Feigenbaum was awarded the Wolf Prize in Physics. He has been Toyota Professor at Rockefeller University since 1986.

For details on Feigenbaum universality, see [Gle87].

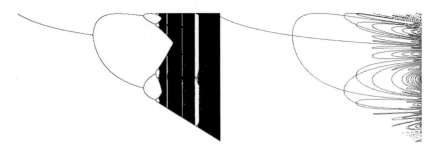

Fig. 1.26. Feigenbaum constant: approaching chaos via successive period doubling bifurcations. The plot on the left is made by iterating equation $f(x) = 1 - \mu|x|^r$ with $r = 2$ several hundred times for a series of discrete but closely spaced values of μ, discarding the first hundred or so points before the iteration has settled down to its fixed–points, and then plotting the points remaining. The plot on the right more directly shows the cycle may be constructed by plotting function $f^n(x) - x$ as a function of μ, showing the resulting curves for $n = 1, 2, 4$. Simulated in $Mathematica^{TM}$.

1.3.5 May's Population Modelling and Chaos

Let $x(t)$ be the population of the species at time t; then the *conservation law* for the population is conceptually given by (see [Mur02])

$$\dot{x} = births - deaths + migration, \qquad (1.22)$$

where $\dot{x} = dx/dt$. The above conceptual equation gave rise to a series of *population models*. The simplest continuous–time model, due to Thomas Malthus from 1798 [Mal798],[25] has no migration, while the birth and death terms are proportional to x,

---

[25] The Rev. Thomas Robert Malthus, FRS (February, 1766–December 23, 1834), was an English demographer and political economist best known for his pessimistic but highly influential views. Malthus's views were largely developed in reaction to the optimistic views of his father, Daniel Malthus and his associates, notably Jean-Jacques Rousseau and William Godwin. Malthus's essay was also in response to the views of the Marquis de Condorcet. In An Essay on the Principle of Population, first published in 1798, Malthus made the famous prediction that population would outrun food supply, leading to a decrease in food per person: "The power of population is so superior to the power of the earth to produce subsistence for man, that premature death must in some shape or other visit the human race. The vices of mankind are active and able ministers of depopulation. They are the precursors in the great army of destruction; and often finish the dreadful work themselves. But should they fail in this war of extermination, sickly seasons, epidemics, pestilence, and plague, advance in terrific array, and sweep off their thousands and tens of thousands. Should success be still incomplete, gigantic inevitable famine stalks in the rear, and with one mighty blow levels the population with the food of the world." This Principle of Population was based on the idea that population if unchecked increases at a geometric rate, whereas

$$\dot{x} = bx - dx \quad \Longrightarrow \quad x(t) = x_0 e^{(b-d)t}, \tag{1.23}$$

where b, d are positive constants and $x_0 = x(0)$ is the initial population. Thus, according to the *Malthus model* (1.23), if $b > d$, the population grows exponentially, while if $b < d$, it dies out. Clearly, this approach is fairly over-simplified and apparently fairly unrealistic. (However, if we consider the past and predicted growth estimates for the total world population from the 1900, we see that it has actually grown exponentially.)

This simple example shows that it is difficult to make long–term predictions (or, even relatively short–term ones), unless we know sufficient facts to incorporate in the model to make it a *reliable predictor*. In the long run, clearly, there must be some adjustment to such exponential growth. François Verhulst [Ver838, Ver845][26] proposed that a *self–limiting process* should operate when a population becomes too large. He proposed the so–called *logistic growth* population model,

$$\dot{x} = rx(1 - x/K), \tag{1.24}$$

where r, K are positive constants. In the Verhulst logistic model (2.17), the constant K is the *carrying capacity* of the environment (usually determined by the available sustaining resources), while the per capita birth rate $rx(1-x/K)$ is dependent on x. There are two steady states (where $\dot{x} = 0$) for (2.17): (i)

---

the food supply grows at an arithmetic rate. Only natural causes (eg. accidents and old age), misery (war, pestilence, and above all famine), moral restraint and vice (which for Malthus included infanticide, murder, contraception and homosexuality) could check excessive population growth. Thus, Malthus regarded his Principle of Population as an explanation of the past and the present situation of humanity, as well as a prediction of our future. The eight major points regarding evolution found in his 1798 *Essay* are: (i) Population level is severely limited by subsistence. (ii) When the means of subsistence increases, population increases. (iii) Population pressures stimulate increases in productivity. (iv) Increases in productivity stimulates further population growth. (v) Since this productivity can never keep up with the potential of population growth for long, there must be strong checks on population to keep it in line with carrying capacity. (vi) It is through individual cost/benefit decisions regarding sex, work, and children that population and production are expanded or contracted. (vii) Positive checks will come into operation as population exceeds subsistence level. (viii) The nature of these checks will have significant effect on the rest of the sociocultural system.

Evolutionists John Maynard Smith and Ronald Fisher were both critical of Malthus' theory, though it was Fisher who referred to the *growth rate r* (used in *logistic equation*) as the *Malthusian parameter*. Fisher referred to "...a relic of creationist philosophy..." in observing the fecundity of nature and deducing (as Darwin did) that this therefore drove natural selection. Smith doubted that famine was the great leveller that Malthus insisted it was.

[26] François Verhulst (October 28, 1804–February 15, 1849, Brussels, Belgium) was a mathematician and a doctor in number theory from the University of Ghent in 1825. Verhulst published in 1838 the logistic demographic model (2.17).

$x = 0$ (unstable, since linearization about it gives $\dot{x} \approx rx$); and (ii) $x = K$ (stable, since linearization about it gives $\frac{d}{dt}(x - K) \approx -r(x - K)$, so $\lim_{t \to \infty} x = K$). The carrying capacity K determines the size of the stable steady state population, while r is a measure of the rate at which it is reached (i.e., the measure of the dynamics) – thus $1/r$ is a representative timescale of the response of the model to any change in the population. The solution of (2.17) is

$$x(t) = \frac{x_0 K e^{rt}}{[K + x_0(e^{rt} - 1)]} \implies \lim_{t \to \infty} x(t) = K.$$

In general, if we consider a population to be governed by

$$\dot{x} = f(x), \qquad (1.25)$$

where typically $f(x)$ is a nonlinear function of x, then the equilibrium solutions x^* are solutions of $f(x) = 0$, and are linearly stable to small perturbations if $\dot{f}(x^*) < 0$, and unstable if $\dot{f}(x^*) > 0$ [Mur02].

In the mid 20th century, ecologists realised that many species had no overlap between successive generations and so population growth happens in discrete–time steps x_t, rather than in continuous–time $x(t)$ as suggested by the conservative law (1.22) and its Maltus–Verhulst derivations. This leads to study *discrete–time models* given by *difference equations*, or, *maps*, of the form

$$x_{t+1} = f(x_t), \qquad (1.26)$$

where $f(x_t)$ is some generic nonlinear function of x_t. Clearly, (1.26) is a discrete–time version of (1.25). However, instead of solving differential equations, if we know the particular form of $f(x_t)$, it is a straightforward matter to evaluate x_{t+1} and subsequent generations by simple recursion of (1.26). The skill in modelling a specific population's growth dynamics lies in determining the appropriate form of $f(x_t)$ to reflect known observations or facts about the species in question.

In 1970s, Robert May, a physicist by training, won the Crafoord Prize for 'pioneering ecological research in theoretical analysis of the dynamics of populations, communities and ecosystems', by proposing a simple *logistic map* model for the generic population growth (1.26).[27] May's model of population growth is the celebrated *logistic map* [May76, May73, May76],

---

[27] Lord Robert May received his Ph.D. in theoretical physics from University of Sydney in 1959. He then worked at Harvard University and the University of Sydney before developing an interest in animal population dynamics and the relationship between complexity and stability in natural communities. He moved to Princeton University in 1973 and to Oxford and the Imperial College in 1988. May was able to make major advances in the field of population biology through the application of mathematics. His work played a key role in the development of *theoretical ecology* through the 1970s and 1980s. He also applied these tools to the study of disease and to the study of *bio–diversity*.

1.3 Brief History of Chaos Theory

$$x_{t+1} = r\, x_t\, (1 - x_t), \tag{1.27}$$

where r is the *Malthusian parameter* that varies between 0 and 4, and the initial value of the population $x_0 = x(0)$ is restricted to be between 0 and 1. Therefore, in May's logistic map (1.27), the generic function $f(x_t)$ gets a specific quadratic form

$$f(x_t) = r\, x_t\, (1 - x_t).$$

For $r < 3$, the x_t have a single value. For $3 < r < 3.4$, the x_t oscillate between two values (see *bifurcation diagram*[28] on Figure 1.27). As r increases, bifurcations occur where the number of iterates doubles. These *period doubling bifurcations* continue to a limit point at $r_{lim} = 3.569944$ at which the period is 2^∞ and the dynamics become chaotic. The r values for the first two bifurcations can be found analytically, they are $r_1 = 3$ and $r_2 = 1 + \sqrt{6}$. We can label the successive values of r at which bifurcations occur as $r_1, r_2, \ldots$ The

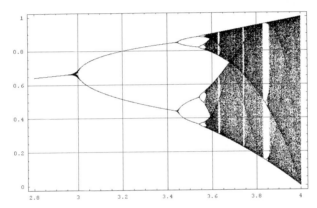

Fig. 1.27. Bifurcation diagram for the logistic map, simulated using *Mathematica*^{TM}.

universal number associated with such period doubling sequences is called the *Feigenbaum number*,

$$\delta = \lim_{k \to \infty} \frac{r_k - r_{k-1}}{r_{k+1} - r_k} \approx 4.669.$$

This series of period–doubling bifurcations says that close enough to r_{lim} the distance between bifurcation points decreases by a factor of δ for each bifurcation. The complex *fractal pattern* got in this way shrinks indefinitely.

---

[28] A bifurcation diagram shows the possible long–term values a variable of a system can get in function of a parameter of the system.

1.3.6 Hénon's Special 2D Map and Its Strange Attractor

Michel Hénon (born 1931 in Paris, France) is a mathematician and astronomer. He is currently at the Nice Observatory. In astronomy, Hénon is well known for his contributions to stellar dynamics, most notably the problem of *globular cluster* (see [Gle87]). In late 1960s and early 1970s he was involved in dynamical evolution of star clusters, in particular the globular clusters. He developed a numerical technique using *Monte Carlo methods*, to follow the dynamical evolution of a spherical star cluster much faster than the so–called n–body methods. In mathematics, he is well known for the Hénon map, a simple discrete dynamical system that exhibits chaotic behavior. Lately he has been involved in the restricted 3–body problem.

His celebrated *Hénon map* [Hen69] is a discrete–time dynamical system that is an extension of the *logistic map* (1.27) and exhibits a chaotic behavior. The map was introduced by Michel Hénon as a simplified model of the *Poincaré section* of the *Lorenz system* (1.121). This 2D–map takes a point (x, y) in the plane and maps it to a new point defined by equations

$$x_{n+1} = y_n + 1 - ax_n^2, \qquad y_{n+1} = bx_n,$$

The map depends on two parameters, a and b, which for the canonical Hénon map have values of $a = 1.4$ and $b = 0.3$ (see Figure 1.28). For the canonical values the Hénon map is chaotic. For other values of a and b the map may be chaotic, intermittent, or converge to a periodic orbit. An overview of the

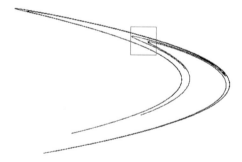

Fig. 1.28. *Hénon strange attractor* (see text for explanation), simulated using *Dynamics Solver*^{TM}.

type of behavior of the map at different parameter values may be obtained from its orbit (or, bifurcation) diagram (see Figure 1.29). For the canonical map, an initial point of the plane will either approach a set of points known as the *Hénon strange attractor*, or diverge to infinity. The Hénon attractor is a fractal, smooth in one direction and a Cantor set in another. Numerical estimates yield a correlation dimension of 1.42 ± 0.02 (Grassberger, 1983) and a Hausdorff dimension of 1.261 ± 0.003 (Russel 1980) for the Hénon attractor. As a dynamical system, the canonical Hénon map is interesting because, unlike the logistic map, its orbits defy a simple description. The Hénon map maps

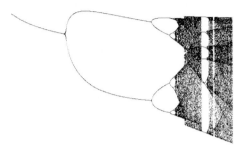

Fig. 1.29. Bifurcation diagram of the *Hénon strange attractor*, simulated using *Dynamics Solver*^{TM}.

two points into themselves: these are the invariant points. For the canonical values of a and b, one of these points is on the attractor: $x = 0.631354477...$ and $y = 0.189406343...$ This point is unstable. Points close to this fixed–point and along the slope 1.924 will approach the fixed–point and points along the slope -0.156 will move away from the fixed–point. These slopes arise from the linearizations of the *stable manifold* and *unstable manifold* of the fixed–point. The unstable manifold of the fixed–point in the attractor is contained in the strange attractor of the Hénon map. The Hénon map does not have a strange attractor for all values of the parameters a and b. For example, by keeping b fixed at 0.3 the bifurcation diagram shows that for $a = 1.25$ the Hénon map has a stable periodic orbit as an attractor. Cvitanovic et al. [CGP88] showed how the structure of the Hénon strange attractor could be understood in terms of unstable periodic orbits within the attractor.

For the (slightly modified) Hénon map: $x_{n+1} = ay_n + 1 - x_n^2$, $y_{n+1} = bx_n$, there are three *basins of attraction* (see Figure 1.30).

The *generalized Hénon map* is a 3D–system (see Figure 1.31)

$$x_{n+1} = a\,x_n - z\,(y_n - x_n^2)), \qquad y_{n+1} = z\,x_n + a\,(y_n - x_n^2)), \qquad z_{n+1} = z_n,$$

where $a = 0.24$ is a parameter. It is an *area–preserving map*, and simulates the *Poincaré map* of period orbits in *Hamiltonian systems*. Repeated random initial conditions are used in the simulation and their gray–scale color is selected at random.

Other Famous 2D Chaotic Maps

1. The *standard map*:

$$x_{n+1} = x_n + y_{n+1}/2\pi, \qquad y_{n+1} = y_n + a\sin(2\pi x_n).$$

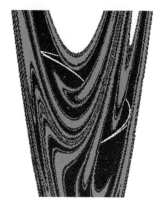

Fig. 1.30. Three basins of attraction for the Hénon map $x_{n+1} = ay_n + 1 - x_n^2$, $y_{n+1} = bx_n$, with $a = 0.475$.

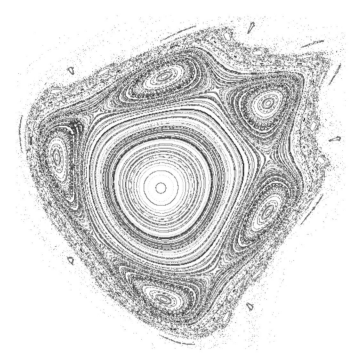

Fig. 1.31. Phase–plot of the *area–preserving generalized Hénon map*, simulated using $Dynamics\,Solver^{TM}$.

2. The *circle map*:

$$x_{n+1} = x_n + c + y_{n+1}/2\pi, \qquad y_{n+1} = by_n - a\sin(2\pi x_n).$$

3. The *Duffing map*:

$$x_{n+1} = y_n, \qquad y_{n+1} = -bx_n + ay_n - y_n^3.$$

4. The *Baker map*:

$$x_{n+1} = bx_n, \qquad y_{n+1} = y_n/a \qquad \text{if} \quad y_n \le a,$$
$$x_{n+1} = (1-c) + cx_n, \qquad y_{n+1} = (y_n - a)/(1-a) \qquad \text{if} \quad y_n > a.$$

5. The *Kaplan–Yorke map*:

$$x_{n+1} = ax_n \bmod 1, \qquad y_{n+1} = -by_n + \cos(2\pi x_n).$$

6. The *Ott–Grebogi–Yorke map*:

$$x_{n+1} = x_n + w_1 + aP_1(x_n, y_n) \bmod 1,$$
$$y_{n+1} = y_n + w_2 + aP_2(x_n, y_n) \bmod 1,$$

where the nonlinear functions P_1, P_2 are sums of sinusoidal functions $A_{rs}^{(i)} \sin[2\pi(rx + sy + B_{rs}^{(i)})]$, with $(r,s) = (0,1), (1,0), (1,1), (1,-1)$, while $A_{rs}^{(i)}, B_{rs}^{(i)}$ were selected randomly in the range $[0,1]$.

1.4 More Chaotic and Attractor Systems

Here we present numerical simulations of several popular chaotic systems (see, e.g., [Wig90, BCB92, Ach97]). Generally, to observe chaos in continuous time system, it is known that the dimension of the equation must be three or higher. That is, *there is no chaos in any phase plane* (see [Str94]), we need the third dimension for chaos in continuous dynamics. However, note that *all forced oscillators have actually dimension 3, although they are commonly written as second-order ODEs*.[29] On the other hand, in discrete–time systems like logistic map or Hénon map, we can see chaos even if the dimension is one.

Simple Pendulum

Recall (see [II05, II06a, II06b]) that a simple *undamped pendulum* (see Figure 1.32), given by equation

$$\ddot{\theta} + \frac{g}{l}\sin\theta = 0, \tag{1.28}$$

swings forever; it has closed orbits in a 2D phase–space (see Figure 1.33).

The conservative (un–damped) pendulum equation (1.28) does not take into account the effects of friction and dissipation. On the other hand, a simple

---

[29] Both Newtonian equation of motion and RLC circuit can generate chaos, provided they have a forcing term. This forcing (driving) term in second–order ODEs is the motivational reason for development of the jet–bundle formalism for non–autonomous dynamics (see [II06b]).

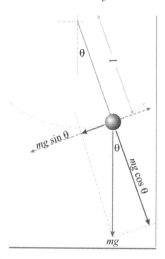

Fig. 1.32. Force diagram of a simple gravity pendulum.

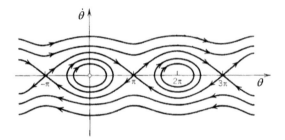

Fig. 1.33. Phase portrait of a simple gravity pendulum.

damped pendulum (see Figure 1.32) is given by modified equation, including a damping term proportional to the velocity,

$$\ddot{\theta} + \gamma\dot{\theta} + \frac{g}{l}\sin\theta = 0,$$

with the positive constant damping γ. This pendulum settles to rest (see Figure 1.34). Its spiralling orbits lead to a point attractor (focus) in a 2D phase–space. All closed trajectories for periodic solutions are destroyed, and the trajectories spiral around one of the critical points, corresponding to the vertical equilibrium of the pendulum. On the phase plane, these critical points are stable spiral points for the underdamped pendulum, and they are stable nodes for the overdamped pendulum. The unstable equilibrium at the inverted vertical position remains an unstable saddle point. It is clear physically that damping means loss of energy. The dynamical motion of the pendulum decays due to the friction and the pendulum relaxes to the equilibrium state in the vertical position.

1.4 More Chaotic and Attractor Systems

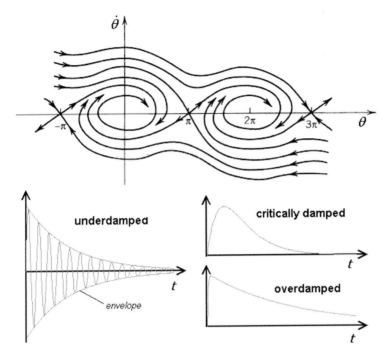

Fig. 1.34. A damped gravity pendulum settles to a rest: its phase portrait (up) shows spiralling orbits that lead to a focus attractor; its time plot (down) shows three common damping cases.

Finally, a *driven pendulum*, periodically forced by a force term $F\cos(w_D t)$, is given by equation (see our introductory example (1.1))

$$\ddot{\theta} + \gamma\dot{\theta} + \frac{g}{l}\sin\theta = F\cos(w_D t). \tag{1.29}$$

It has a 3D phase–space and can exhibit chaos (for certain values of its parameters, see Figure 1.35).

Van der Pol Oscillator

The unforced Van der Pol oscillator has the form of a second order ODE (compare with 1.4 above)

$$\ddot{x} = \alpha\left(1 - x^2\right)\dot{x} - \omega^2 x. \tag{1.30}$$

Its celebrated *limit cycle* is given in Figure 1.36. The simulation is performed with zero initial conditions and parameters $\alpha = \text{random}(0, 3)$, and $\omega = 1$. The Van der Pol oscillator was the first *relaxation oscillator*, used in 1928 as a model of human heartbeat (ω controls how much voltage is injected

58   1 Basics of Nonlinear and Chaotic Dynamics

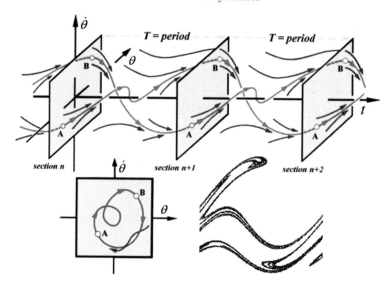

Fig. 1.35. A driven pendulum has a 3D phase–space with angle θ, angular velocity $\dot{\theta}$ and time t. Dashed lines denote steady states, while solid lines denote transients. Right–down we see a sample chaotic attractor (adapted and modified from [TS01]).

into the system, and α controls the way in which voltage flows through the system). The oscillator was also used as a model of an electronic circuit that appeared in very early radios in the days of vacuum tubes. The tube acts like a normal resistor when current is high, but acts like a negative resistor if the current is low. So this circuit pumps up small oscillations, but drags down large oscillations. α is a constant that affects how nonlinear the system is. For α equal to zero, the system is actually just a linear oscillator. As α grows the nonlinearity of the system becomes considerable.

The *sinusoidally–forced Van der Pol oscillator* is given by equation

$$\ddot{x} - \alpha\left(1 - x^2\right)\dot{x} + \omega^2\, x = \gamma \cos(\phi t), \tag{1.31}$$

where ϕ is the forcing frequency and γ is the amplitude of the forcing sinusoid.

Nerve Impulse Propagation

The nerve impulse propagation along the axon of a neuron can be studied by combining the equations for an excitable membrane with the differential equations for an electrical core conductor cable, assuming the axon to be an infinitely long cylinder. A well known approximation of FitzHugh [Fit61] and Nagumo [NAY60] to describe the propagation of voltage pulses $V(x,t)$ along the membranes of nerve cells is the set of coupled PDEs

$$V_{xx} - V_t = F(V) + R - I, \qquad R_t = c(V + a - bR), \tag{1.32}$$

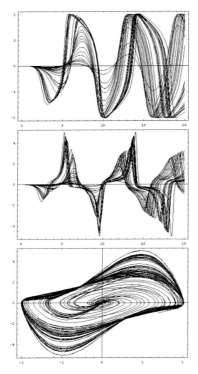

Fig. 1.36. Cascade of 30 unforced Van der Pol oscillators, simulated using $Mathematica^{TM}$; top–down: displacements, velocities and phase–plot (showing the celebrated limit cycle).

where $R(x,t)$ is the recovery variable, I the external stimulus and a, b, c are related to the membrane radius, specific resistivity of the fluid inside the membrane and temperature factor respectively.

When the spatial variation of V, namely V_{xx}, is negligible, (1.32) reduces to the Van der Pol oscillator,

$$\dot{V} = V - \frac{V^3}{3} - R + I, \qquad \dot{R} = c(V + a - bR),$$

with $F(V) = -V + \frac{V^3}{3}$. Normally the constants in (1.32) satisfy the inequalities $b < 1$ and $3a + 2b > 3$, though from a purely mathematical point of view this need not be insisted upon. Then with a periodic (ac) applied membrane current $A_1 \cos \omega t$ and a (dc) bias A_0, the Van der Pol equation becomes

$$\dot{V} = V - \frac{V^3}{3} - R + A_0 + A_1 \cos \omega t, \qquad \dot{R} = c(V + a - bR). \tag{1.33}$$

Further, (1.33) can be rewritten as a single second–order ODE by differentiating $\dot{V}$ with respect to time and using $\dot{R}$ for R,

$$\ddot{V} - (1-bc)\left\{1 - \frac{V^2}{1-bc}\right\}\dot{V} - c(b-1)V + \frac{bc}{3}V^3$$
$$= c(A_0 b - a) + A_1 \cos(\omega t + \phi), \qquad (1.34)$$

where $\phi = \tan^{-1}\frac{\omega}{bc}$. Using the transformation $x = (1-bc)^{-(1/2)}V$, $t \longrightarrow t' = t + \frac{\phi}{\omega}$, (1.34) can be rewritten as

$$\ddot{x} + p(x^2 - 1)\dot{x} + \omega_0^2 x + \beta x^3 = f_0 + f_1 \cos\omega t, \qquad \text{where} \qquad (1.35)$$

$$p = (1-bc), \qquad \omega_0^2 = c(1-b), \qquad \beta = bc\frac{(1-bc)}{3},$$

$$f_0 = c\frac{(A_0 b - a)}{\sqrt{1-bc}}, \qquad f_1 = \frac{A_1}{\sqrt{1-bc}}.$$

Note that (1.35), or its rescaled form

$$\ddot{x} + p(kx^2 + g)\dot{x} + \omega_0^2 x + \beta x^3 = f_0 + f_1 \cos\omega t, \qquad (1.36)$$

is the *Duffing–Van der Pol equation*. In the limit $k = 0$, we have the Duffing equation discussed below (with $f_0 = 0$), and in the case $\beta = 0$ ($g = -1$, $k = 1$) we have the forced van der Pol equation. Equation (1.36) exhibits a very rich variety of bifurcations and chaos phenomena, including quasi–periodicity, phase lockings and so on, depending on whether the potential $V = \frac{1}{2}\omega_0^2 x^2 + \frac{\beta x^4}{4}$ is i) a double well, ii) a single well or iii) a double hump [Lak97, Lak03].

Duffing Oscillator

The forced *Duffing oscillator* [Duf18] has the form similar to (1.31),

$$\ddot{x} + b\dot{x} - ax(1-x^2) = \gamma\cos(\phi t). \qquad (1.37)$$

Stroboscopic *Poincaré sections* of a *strange attractor* can be seen (Figure 1.37), with the *stretch–and–fold* action at work. The simulation is performed with parameters: $a = 1$, $b = 0.2$, and $\gamma = 0.3$, $\phi = 1$. The Duffing equation is used to model a double well oscillator such as the magneto–elastic mechanical system. This system consists of a beam positioned vertically between two magnets, with the top end fixed, and the bottom end free to swing. The beam will be attracted to one of the two magnets, and given some velocity will oscillate about that magnet until friction stops it. Each of the magnets creates a fixed–point where the beam may come to rest above that magnet and remain there in equilibrium. However, when this whole system is shaken by a periodic forcing term, the beam may jump back and forth from one magnet to the other in a seemingly random manner. Depending on how big the shaking term is, there may be no stable fixed–points and no stable fixed cycles in the system.

1.4 More Chaotic and Attractor Systems

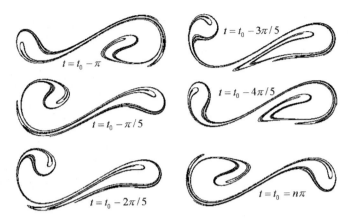

Fig. 1.37. Duffing strange attractor, showing stroboscopic Poincaré sections; simulated using *Dynamics Solver^{TM}*.

Rossler System

Classical *Rossler system* is given by equations

$$\dot{x} = -y - z, \qquad \dot{y} = x + by, \qquad \dot{z} = b + z(x - a). \qquad (1.38)$$

Using the parameter values $a = 4$ and $b = 0.2$, the phase–portrait is produced (see Figure 1.38), showing the celebrated attractor. The system is credited to O. Rossler and arose from work in chemical kinetics.

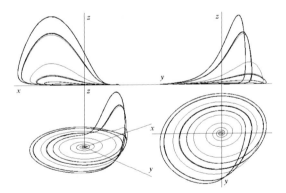

Fig. 1.38. The celebrated Rossler attractor, simulated using *Dynamics Solver^{TM}*.

Chua's Circuit

Chua's circuit is a simple electronic circuit that exhibits classic chaotic behavior. First introduced in 1983 by Leon O. Chua, its ease of construction has made it an ubiquitous real–world example of a chaotic system, leading some to declare it 'a paradigm for chaos'. It has been the subject of much study; hundreds of papers have been published on this topic (see [Chu94]).

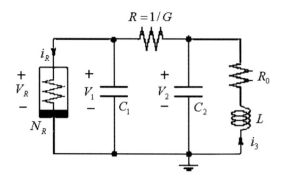

Fig. 1.39. Chua's circuit.

The *Chua's circuit* consists of two linear capacitors, two linear resistors, one linear inductor and a nonlinear resistor (see Figure 1.39). By varying the various circuit parameters, we can get complicated nonlinear and chaotic phenomena. Let us consider the case where we vary the conductance G of the resistor R and keep the other components fixed. In particular, we choose $L = 18\,mH, R_0 = 12.5\,Ohms, C_1 = 10\,nF, C_2 = 100\,nF$. The nonlinear resistor N_R (Chua's diode) is chosen to have a piecewise–linear $V - I$ characteristic of the form:
$$ i = -\begin{cases} G_b v + G_a - G_b & \text{if} \quad v > 1, \\ G_a v & \text{if} \quad |v| < 1, \\ G_b v + G_b - G_a & \text{if} \quad v < -1, \end{cases} $$
with $G_a = -0.75757\,mS$, and $G_b = -0.40909\,mS$.

Starting from low G–values, the circuit is stable and all trajectories converge towards one of the two stable equilibrium points. As G is increased, a limit cycle appears due to a *Hopf–like bifurcation*. In order to observe the period–doubling route to chaos, we need to further increase G. At the end of the period–doubling bifurcations, we observe a chaotic attractor. Because of symmetry, there exists a twin attractor lying in symmetrical position with respect the the origin. As G is further increased, these two chaotic attractors collide and form a 'double scroll' chaotic attractor.

After normalization, the state equations for the Chua's circuit read:
$$ \dot{x} = a(y - x - f(x)), \qquad \dot{y} = x - y + z, \qquad \dot{z} = -by - cz, \qquad (1.39) $$
where $f(x)$ is a nonlinear function to be manipulated to give various chaotic behaviors.

By using a specific form of the nonlinearity $f(x)$, a family of *multi–spiral strange attractors* have been generated in [Ala99] (see Figure 1.40).

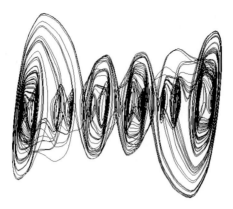

Fig. 1.40. A multi–spiral strange attractor of the Chua's circuit (modified from [Ala99]).

Inverted Pendulum

Stability of the *inverted driven pendulum* given by equation

$$\ddot{\theta} + k\dot{\theta} + (1 + a\sqrt{\phi}\cos(\phi t))\sin\theta = 0,$$

where θ is the angle, is simulated in Figure 1.41, using the parameter $a = 0.33$. It is possible to stabilize a mathematical pendulum around the upper vertical position by moving sinusoidally the suspension point in the vertical direction. Furthermore, the perturbed solution may be of two kinds: one goes to the vertical position while the other becomes periodic (see, e.g., [Ach97]).

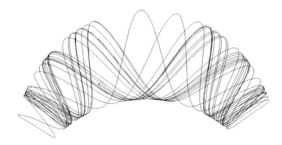

Fig. 1.41. Duffing strange attractor, showing stroboscopic Poincaré sections; simulated using *Dynamics Solver*^{TM}.

Elastic Pendulum

Elastic pendulum (Figure 1.42) of proper length l, mass m and elastic constant k is given by equation

$$\ddot{x} = x\sqrt{\dot{y}} + \cos y - a(x-1), \qquad \ddot{y} = -(2\dot{x}\dot{y} + \sin y)/x,$$

where the parameter $a = kl/mg = 0.4$. High values of a give raise to a simple pendulum.

Fig. 1.42. Phase–portrait of an elastic pendulum showing *Lissajous curves*; simulated using *Dynamics Solver*^{TM}.

Lorenz–Maxwell–Haken System

In 1975, H. Haken showed [Hak83] that the *Lorenz equations* (1.52) were isomorphic to the *Maxwell–Haken laser equations*

$$\dot{E} = \sigma(P-E), \qquad \dot{P} = \beta(ED-P), \qquad \dot{D} = \gamma(\sigma - 1 - D - \sigma EP),$$

Here, the variables in the Lorenz equations, namely x,y and z correspond to the slowly varying amplitudes of the electric field E and polarization P and the inversion D respectively in the Maxwell–Haken equations. The parameters are related via $c = \frac{\gamma}{\beta}$, $a = \frac{\sigma}{\beta}$ and $b = \sigma + 1$, where γ is the relaxation rate of the inversion, β is the relaxation rate of the polarization, σ is the field relaxation rate, and σ represents the normalized pump power.

Autocatalator System

This 4D *autocatalator* system from *chemical kinetics* (see Figure 1.43) is defined as (see, e.g., [BCB92])

$$\dot{x}_1 = -a\,x_1, \quad \dot{x}_2 = a\,x_1 - b\,x_2 - x_2\,x_3^2, \quad \dot{x}_3 = b\,x_2 - x_3 + x_2\,x_3^2, \quad \dot{x}_4 = x_3.$$

The simulation is performed with parameters: $a = 0.002$, and $b = 0.08$.

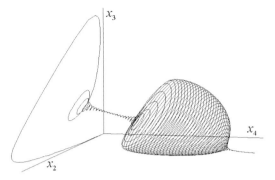

Fig. 1.43. 3D phase–portrait of the 4D autocatalator system, simulated using $Dynamics\,Solver^{TM}$.

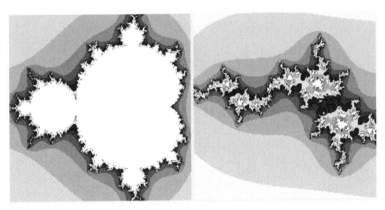

Fig. 1.44. The celebrated conformal Mandelbrot (left) and Julia (right) sets in the complex plane, simulated using $Dynamics\,Solver^{TM}$.

Mandelbrot and Julia Sets

Recall that *Mandelbrot and Julia sets* (see Figure 1.44) are celebrated *fractals*. Recall that fractals are sets with *fractional dimension* (see Figure 1.45). The Mandelbrot and Julia sets are defined either by a quadratic *conformal z–map* [Man80a, Man80b]

$$z_{n+1} = z_n^2 + c,$$

or by a real (x, y)–map

$$x_{n+1} = \sqrt{x_n} - \sqrt{y_n} + c_1, \qquad y_{n+1} = 2\,x_n\,y_n + c_2,$$

where c, c_1 and c_2 are parameters. For almost every c, this conformal transformation generates a fractal (probably, only for $c = -2$ it is not a fractal). Julia set J_c with $c \ll 1$, the *capacity dimension* is

$$d_{cap} = 1 + \frac{|c|^2}{4\ln 2} + O(|c|^3).$$

The set of all points for which J_c is connected is the Mandelbrot set.[30]

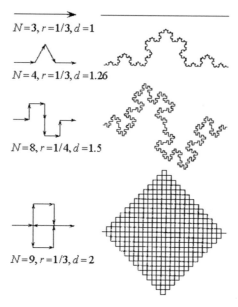

Fig. 1.45. Fractal dimension of curves in $\mathbb{R}^2$: $d = \frac{\log N}{\log 1/r}$.

Biomorphic Systems

Closely related to the Mandelbrot and Julia sets are *biomorphic systems*, which look like one–celled organisms. The term '*biomorph*' was proposed by C. Pickover from IBM [Pic86, Pic87]. Pickover's biomorphs inhabit the complex plane like the the Mandelbrot and Julia sets and exhibit a *protozoan morphology*. Biomorphs began for Pickover as a 'bug' in a program intended to probe the fractal properties of various formulas. He accidentally used an OR logical operator instead of an AND operator in the conditional test for the size of z's real and imaginary parts. The cilia that project from the biomorphs are a consequence of this 'error'. Each biomorph is generated by multiple iterations of a particular conformal map,

---

[30] The Mandelbrot set has its place in complex–valued dynamics, a field first investigated by the French mathematicians Pierre Fatou [Fat19, Fat22] and Gaston Julia [Jul18] at the beginning of the 20th century. For general families of holomorphic functions, the boundary of the Mandelbrot set generalizes to the bifurcation locus, which is a natural object to study even when the connectedness locus is not useful. A related *Mandelbar set* was encountered by mathematician John Milnor in his study of parameter slices of real cubic polynomials; it is not locally connected; this property is inherited by the connectedness locus of real cubic polynomials.

$$z_{n+1} = f(z_n, c),$$

where c is a parameter. Each iteration takes the output of the previous operations as the input of the next iteration. To generate a biomorph, one first needs to lay out a grid of points on a rectangle in the complex plane [And01]. The coordinate of each point constitutes the real and imaginary parts of an initial value, z_0, for the iterative process. Each point is also assigned a pixel on the computer screen. Depending on the outcome of a simple test on the 'size' of the real and imaginary parts of the final value, the pixel is colored either black or white. The biomorphs presented in Figure 1.46 are generated using the following conformal functions:

1. $f(z, c) = z^3$,
2. $f(z, c) = z^3 + c$, $\quad c = 10$,
3. $f(z, c) = z^3 + c$, $\quad c = 10 - 10i$,
4. $f(z, c) = z^5 + c$, $\quad c = 0.77 - 0.77i$,
5. $f(z, c) = z^3 + \sin z + c$, $\quad c = 1 - i$,
6. $f(z, c) = z^6 + \sin z + c$, $\quad c = 0.5 - 0.5i$,
7. $f(z, c) = z^2 \sin z + c$, $\quad c = 0.78 - 0.78i$,
8. $f(z, c) = z^c$, $\quad c = 5 - i$,
9. $f(z, c) = |z|^c \sin z$, $\quad c = 4$,
10. $f(z, c) = |z|^c \cos z + c$, $\quad c = 3 + 3i$,
11. $f(z, c) = |z|^c (\cos z + z) + c$, $\quad c = 3 + 2i$.

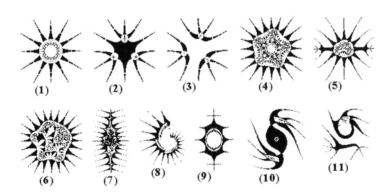

Fig. 1.46. *Pickover's biomorphs* (see text for details).

1.5 Continuous Chaotic Dynamics

The prediction of the behavior of a system when its evolution law is known, is a problem with an obvious interest in many fields of scientific research.

Roughly speaking, within this problem two main areas of investigation may be identified [CFL97]:

A) The definition of the 'predictability time'. If one knows the initial state of a system, with a precision $\delta_o = |\delta \mathbf{x}(0)|$, what is the maximum time T_p within which one is able to know the system future state with a given tolerance $\delta_{\max}$?

B) The understanding of the relaxation properties. What is the relation between the mean response of a system to an external perturbation and the features of its un–perturbed state [Lei78]? Using the terminology of statistical mechanics, one wants to reduce 'non–equilibrium' properties, such as relaxation and responses, to 'equilibrium' ones, such as correlation functions [KT85].

A remarkable example of type–A problem is the weather forecasting, where one has to estimate the maximum time for which the prediction is enough accurate. As an example of type–B problem, of geophysical interest, one can mention a volcanic eruption which induces a practically instantaneous change in the temperature. In this case it is relevant to understand how the difference between the atmospheric state after the eruption and the hypothetical un–perturbed state without the eruption evolves in time. In practice one wants to understand how a system absorbs, on average, the perturbation $\delta f(\tau)$ of a certain quantity f (e.g., the temperature), just looking at the statistical features, as correlations, of f in the un–perturbed regime. This is the so called fluctuation/relaxation problem [Lei75].

As far as problem–A is concerned, in the presence of deterministic chaos, a rather common situation, the distance between two initially close trajectories diverges exponentially:

$$|\delta \mathbf{x}(t)| \sim \delta_o \exp(\lambda t), \tag{1.40}$$

where λ is the *maximum Lyapunov exponent* of the system [BGS76]. From (2.23) it follows:

$$T_p \sim \frac{1}{\lambda} \ln\left(\frac{\delta_{max}}{\delta_o}\right). \tag{1.41}$$

Since the dependence on δ_{max} and δ_o is very weak, T_p appears to be proportional to the inverse of the Lyapunov exponent. We stress however that (2.24) is just a naive answer to the predictability problem, since it does not take into account the following relevant features of the chaotic systems:

i) The Lyapunov exponent is a global quantity, i.e., it measures the average exponential rate of divergence of nearby trajectories. In general there are finite-time fluctuations of this rate, described by means of the so called effective Lyapunov exponent $\gamma_t(\tau)$. This quantity depends on both the time delay τ and the time t at which the perturbation acted [PSV87]. Therefore, the predictability time T_p fluctuates, following the γ–variations [CJP93].

ii) In systems with many degrees of freedom one has to understand how a perturbation grows and propagates trough the different degrees of freedom. In fact, one can be interested in the prediction on certain variables, e.g., those associated with large scales in weather forecasting, while the perturbations act on a different set of variables, e.g., those associated to small scales.

iii) If one is interested into non–infinitesimal perturbations, and the system possess many characteristic times, such as the eddy turn–over times in fully developed turbulence, then T_p is determined by the detailed mechanism of propagation of the perturbations through different degrees of freedom, due to nonlinear effects. In particular, T_p may have no relation with λ [ABC96].

In addition to these points, when the evolution of a system is ruled by a set of ODEs,
$$\dot{\mathbf{x}} = \mathbf{f}(\mathbf{x}, t), \qquad (1.42)$$
which depend periodically on time,
$$\mathbf{f}(\mathbf{x}, t + T) = \mathbf{f}(\mathbf{x}, t), \qquad (1.43)$$
one can have a kind of 'seasonal' effect, for which the system shows an alternation, roughly periodic, of low and high predictability. This happens, for example, in the recently studied case of stochastic resonance in a chaotic deterministic system, where one observes a roughly periodic sequence of chaotic and regular evolution intervals [CFP94].

As far as problem–B is concerned, it is possible to show that in a chaotic system with an invariant measure $P(\mathbf{x})$, there exists a relation between the mean response $\langle \delta x_j(\tau) \rangle_P$ after a time τ from a perturbation $\delta x_i(0)$, and a suitable correlation function. Namely, one has the following equation [CFL97]

$$R_{ij}(\tau) \equiv \frac{\langle \delta x_j(\tau) \rangle_P}{\delta x_i(0)} = \left\langle x_j(\tau) \frac{\partial S(\mathbf{x}(0))}{\partial x_i} \right\rangle_P, \qquad (1.44)$$

where $S(\mathbf{x}) = -\ln P(\mathbf{x})$. Equation (1.44) ensures that the mean relaxation of the perturbed system is equal to some correlation of the un–perturbed system. As, in general, one does not know $P(\mathbf{x})$, (1.44) provides only a qualitative information.

1.5.1 Dynamics and Non–Equilibrium Statistical Mechanics

Recall that statistical mechanics, which was created at the end of the 19-th century by such people as Maxwell, Boltzmann, and Gibbs, consists of two rather different parts: equilibrium and non–equilibrium statistical mechanics. The success of equilibrium statistical mechanics has been spectacular. It has been developed to a high degree of mathematical sophistication, and applied with success to subtle physical problems like the study of critical phenomena. Equilibrium statistical mechanics also has highly nontrivial connections with the mathematical theory of smooth dynamical systems and the physical theory of quantum fields. By contrast, the progress of non–equilibrium statistical mechanics has been much slower. We still depend on the insights of Boltzmann for our basic understanding of irreversibility, and this understanding remains rather qualitative. Further progress has been mostly on dissipative phenomena close to equilibrium: Onsager reciprocity relations [Ons35], Green–Kubo formula [Gre51, Kub57], and related results. Yet, there is currently a strong revival of non–equilibrium statistical mechanics, based on taking seriously the complications of the underlying microscopic dynamics.

In this subsection, following [Rue78, Rue98, Rue99], we provide a general discussion of *irreversibility*. It is a fact of common observation that the behavior of bulk matter is often irreversible: if a warm and a cold body are put in contact, they will equalize their temperatures, but once they are at the same temperature they will not spontaneously go back to the warm and cold situation. Such facts have been organized into a body of knowledge named *thermodynamics*. According to thermodynamics, a number called *entropy* can be associated with macroscopic systems which are (roughly speaking) locally in equilibrium. The definition is such that, when the system is isolated, its entropy can only increase with time or stay constant (the celebrated *Second Law of thermodynamics*). A strict increases in entropy corresponds to an irreversible process. Such processes are also called *dissipative structures*, because they dissipate noble forms of energy (like mechanical energy) into heat. The flow of a viscous fluid, the passage of an electric current through a conductor, or a chemical reaction like a combustion are typical dissipative phenomena. The purpose of non–equilibrium statistical mechanics is to explain irreversibility on the basis of microscopic dynamics, and to give quantitative predictions for dissipative phenomena. In what follows we shall assume that the microscopic dynamics is classical. Unfortunately we shall have little to say about the quantum case.

Dynamics and Entropy

We shall begin with a naive discussion, and then see how we have to modify it to avoid the difficulties that will arise. The microscopic time evolution (f^t) which we want to consider is determined by an evolution equation

$$\dot{x} = F(x), \qquad (1.45)$$

in a phase–space M. More precisely, for an isolated system with Hamiltonian $H = H(q,p)$, we may rewrite (1.45) as

$$\frac{d}{dt}\begin{pmatrix} p \\ q \end{pmatrix} = \begin{pmatrix} -\partial_q H \\ \partial_p H \end{pmatrix}, \qquad (1.46)$$

1.5 Continuous Chaotic Dynamics

where p, and q are ND for a system with N DOF, so that the phase–space M is $2N$D manifold. Note that we are interested in a macroscopic description of a macroscopic system where N is large, and microscopic details cannot be observed. In fact many different points (p, q) in phase–space have the same macroscopic appearance. It appears therefore reasonable to describe a macroscopic state of our system by a probability measure $m(dx)$ on M. At equilibrium, $m(dx)$ should be an (f^t)-invariant measure. Remember now that the Hamiltonian time evolution determined by (1.46) preserves the energy and the volume element $dp\,dq$ in phase–space. This leads to the choice of the *micro–canonical ensemble* [Rue99]

$$m_K(dq\,dp) = \frac{1}{\Omega_K}\delta(H(q,p) - K)dq\,dp, \quad (1.47)$$

to describe the equilibrium state of energy K (Ω_K is a normalization constant, and the *ergodic hypothesis* asserts that m_K is ergodic). Note that the support of m_K is the energy shell

$$M_K = \{(q,p) : H(q,p) = K\}.$$

If the probability measure $m(dq\,dp)$ has density $\underline{m}(q,p)$ with respect to $dq\,dp$, we associate with the state described by this measure the entropy

$$S(\underline{m}) = -\int dq\,dp\,\underline{m}(q,p)\log\underline{m}(q,p). \quad (1.48)$$

This is what Lebowitz [Leb93] calls the *Gibbs entropy*, it is the accepted expression for the entropy of equilibrium states. There is a minor technical complication here. While (1.48) gives the right result for the canonical ensemble one should, for the micro–canonical ensemble, replace the reference measure $dp\,dq$ in (1.47) by $\delta(H(q,p) - K)dq\,dp$. This point is best appreciated in the light of the theory of equivalence of ensembles in equilibrium statistical mechanics. For our purposes, the easiest is to replace δ in (1.47) by the characteristic function of $[0, \epsilon]$ for small ϵ. The measure m_K is then still invariant, but no longer ergodic. Note also that the traditional definition of the entropy has a factor k (*Bolzmann constant*) in the right–hand side of (1.48), and it remains reasonable outside of equilibrium, as information theory for instance would indicate. Writing x instead of (q, p), we know that the density of $\widehat{m} = f^{t*}m$ is given by

$$\widehat{\underline{m}}(x) = \frac{\underline{m}(f^{-t}x)}{J_t(f^{-t}x)},$$

where J_t is the Jacobian determinant of f^t. We have thus [Rue99]

$$S(\underline{m}) = -\int dx\,\underline{m}(x)\log\underline{m}(x), \quad (1.49)$$

$$S(\widehat{m}) = -\int dx\, \widehat{m}(x) \log \widehat{m}(x) = -\int dx\, \frac{m(f^{-t}x)}{J_t(f^{-t}x)} \log \frac{m(f^{-t}x)}{J_t(f^{-t}x)},$$

$$= -\int dx\, \underline{m}(x) \log \frac{m(x)}{J_t(x)} = S(\underline{m}) + \int dx\, \underline{m}(x) \log J_t(x) \quad (1.50)$$

In the Hamiltonian situation that we are considering for the moment, since the volume element is preserved by f^t, $J_t = 1$, and therefore $S(\widehat{m}) = S(\underline{m})$. So, we have a problem: the entropy seems to remain constant in time. In fact, we could have expected that there would be a problem because the Hamiltonian evolution (1.46) has *time–reversal invariance* (to be discussed later), while we want to prove an increase of entropy which does not respect time reversal. We shall now present Boltzmann's way out of this difficulty.

Classical Boltzmann Theory

Let us cut M into cells c_i so that the coordinates p, q have roughly fixed values in a cell. In particular all points in c_i are macroscopically equivalent (but different c_i may be macroscopically indistinguishable). Instead of describing a macroscopic state by a probability measure m, we may thus give the weights $m(c_i)$. Now, time evolution will usually distort the cells, so that each $f^t c_i$ will now intersect a large number of cells. If the initial state m occupies N cells with weights $1/N$ (taken to be equal for simplicity), the state $f^{*t}m$ will occupy (thinly) N^t cells with weights $1/N^t$, where N^t may be much larger than N. If the c_i have side h, we have [Rue99]

$$\log N = -m(c_i) \log m(c_i) \approx -\int dx\, \underline{m}(x) \log(\underline{m}(x) h^{2N}) = S(\underline{m}) - 2N \log h,$$

i.e., the entropy associated with m is roughly $\log N + 2N \log h$. The apparent entropy associated with $f^{t*}m$ is similarly $\log N^t + 2N \log h$; it differs from $S(\widehat{m})$ because the density $\widehat{m}$, which may fluctuate rapidly in a cell, has been replaced by an average for the computation of the apparent entropy. The entropy increase $\log N^t - \log N$ is due to the fact that the initial state m is concentrated in a small region of phase–space, and becomes by time evolution spread (thinly) over a much larger region. The time evolved state, after a little smoothing (*coarse graining*), has a strictly larger entropy than the initial state. This gives a microscopic interpretation of the second law of thermodynamics. In specific physical examples (like that of two bodies in contact with initially different temperatures) one sees that the time evolved state has correlations (say between the microscopic states of the two bodies, after their temperatures have equalized) which are macroscopically unobservable. In the case of a macroscopic system locally close to equilibrium (small regions of space have a definite temperature, pressure,...) the above classical ideas of Boltzmann have been expressed in particularly clear and compelling manner by [Leb93]. He defines a local *Boltzmann entropy* to be the equilibrium entropy corresponding to the temperature, pressure,... which approximately describe locally a non–equilibrium state. The integral over space of the Boltzmann entropy is then what we have called apparent entropy, and is different from the Gibbs entropy defined by (1.48). These ideas would deserve a fuller discussion, but here we shall be content to refer to [Leb93], and to Boltzmann's original works. There is still some opposition to Boltzmann's ideas (notably by I. Prigogine and his school, see [Pri62]), but most workers in the area accept them, and so shall we. We shall however develop a formalism which is both rather different from and completely compatible with the ideas of Boltzmann and Lebowitz just discussed.

Description of States by Probability Measures

In the above discussion, we have chosen to describe states by probability measures. One may object that the state of a (classical) system is represented by a point in phase–space rather than by a probability measure. But for a many-particle system like those of interest in statistical mechanics it is practically impossible to know the position of all the particles, which changes rapidly with time anyway. The information that we have is macroscopic or statistical and it is convenient to take as description of our system a probability distribution ρ on phase–space compatible with the information available to us. Trying to define this ρ more precisely at this stage leads to the usual kind of difficulties that arise when one wants to get from a definition what should really come as a result of theory. For our later purposes it is technically important to work with states described by probability measures. Eventually, we shall get results about individual points of phase–space (true almost everywhere with respect to some measure). From a physical point of view, it would be desirable to make use here of points of phase–space which are typical for a macroscopic state of our system. Unfortunately, a serious discussion of this point seems out of reach in the present state of the theory.

Beyond Boltzmann

There are two reasons why one would want to go beyond the ideas presented above. One concerns explicit calculations, like that of a rate of entropy production; the other concerns situations far from equilibrium. We discuss these two points successively. If one follows Boltzmann and uses a decomposition of phase–space into cells to compute entropy changes, the result need not be monotone in time, and will in general depend on the particular decomposition used. Only after taking a limit $t \to \infty$ can one let the size of cells tend to 0, and get a result independent of the choice of coarse graining. This leaves open the problem of computing the entropy production per unit time. The idea of using local Boltzmann entropies works only for a macroscopic system locally close to equilibrium, and one may wonder what happens far from equilibrium. In fact one finds statements in the literature that biological processes are far from equilibrium (which is true), and that they may violate the second law of thermodynamics (which is not true). To see that life processes or other processes far from equilibrium cannot violate the second law, we can imagine a power plant fueled by sea water: it would produce electric current and reject colder water in the sea. Inside the plant there would be some life form or other physico-chemical system functioning far from equilibrium and violating the second law of thermodynamics. The evidence is that this is impossible, even though we cannot follow everything that happens inside the plant in terms of Boltzmann entropies of systems close to equilibrium. In fact, Boltzmann's explanation of irreversibility reproduced above applies also here, and the only unsatisfactory feature is that we do not have an effective definition of entropy far from equilibrium. The new formalism which we shall introduce below is quite in agreement with the ideas of Boltzmann which we have described. We shall however define the physical entropy by (1.48) (Gibbs entropy). To avoid the conclusion that this entropy does not change in time for a Hamiltonian time evolution, we shall idealize our physical setup differently, and in particular introduce a thermostat [Rue99].

Thermostats

Instead of investigating the approach to equilibrium as we have done above following Boltzmann, we can try to produce and study non–equilibrium steady states. To keep a finite system outside of equilibrium we subject it to non-Hamiltonian forces. We consider thus an evolution of the form (1.45), but not (1.46). Since we no longer have conservation of energy, $x(t)$ cannot be expected to stay in a bounded region of phase–space. This means that the system will heat up. Indeed, this is what is observed experimentally: dissipative systems produce heat. An experimentalist will eliminate excess heat by use of a thermostat, and if we want to study non–equilibrium steady states we have to introduce the mathematical equivalent of a thermostat. In the lab, the system in which we are interested (called *small system*) would be coupled with a *large system* constituting the thermostat. The obvious role of the large system is to take up the heat produced in the small system. At the same time, the thermostat allows entropy to be produced in the small system by a mechanism discussed above: microscopic correlations which exist between the small and the large system are rendered unobservable by the time evolution. An exact study of the pair small+large system would lead to the same problems that we have met above with Boltzmann's approach. For such an approach, see Jakšić and Pillet [JP98], where the large system has infinitely many noninteracting degrees of freedom. Studying the *diffusion and loss of correlations* in an infinite interacting system (say a system of particles with Hamiltonian interactions) appears to be very difficult in general, because the same particles may interact again and again, and it is hard to keep track of the correlations resulting from earlier interactions. This difficulty was bypassed by Lanford when he studied the Boltzmann equation in the Grad limit [Lan75], because in that limit, two particles that collide once will never see each other again.. Note however that the thermostats used in the lab are such that their

1.5 Continuous Chaotic Dynamics

state changes as little as possible under the influence of the small system. For instance the small system will consist of a small amount of fluid, surrounded by a big chunk of copper constituting the large system: because of the high thermal conductivity of copper, and the bulk of the large system, the temperature at the fluid-copper interface will remain constant to a good precision. In conclusion, the experimentalist tries to build an *ideal thermostat*, and we might as well do the same. Following [Rue99], we replace thus (1.45) by

$$\dot{x} = F(x) + \Theta(\omega(t), x),$$

where the effect $\Theta(\omega(t), x)$ of the thermostat depends on the state x of the small system, and on the state $\omega(t)$ of the thermostat, but the time evolution $t \to \omega(t)$ does not depend on the state x of the small system. We may think of $\omega(t)$ as random (corresponding to a *random thermostat*), but the simplest choice is to take ω constant, and use $\Theta(x) = \Theta(\omega, x)$ to keep $x(t)$ on a compact manifold M. For instance if M is the manifold $\{x : h(x) = K\}$, we may take

$$\Theta(x) = -\frac{(F(x) \cdot h(x))}{(\mathrm{grad}\,h(x) \cdot h(x))} \mathrm{grad}\,h(x).$$

This is the so-called Gaussian thermostat [Hoo86, EM90]. We shall be particularly interested later in the isokinetic thermostat, which is the special case where $x = (q, p)$ and $h(x) = p/2m$ (kinetic energy).

Non–Equilibrium Steady States

We assume for the moment that the phase–space of our system is reduced by the action of a thermostat to be a compact manifold M. The time evolution equation on M has the same form as (1.45):

$$\dot{x} = F(x), \tag{1.51}$$

where the vector–field F on M now describes both the effect of nonhamiltonian forces and of the thermostat. Note that (1.51) describes a general smooth evolution, and one may wonder if anything of physical interest is preserved at this level of generality. Perhaps surprisingly, the answer is yes, as we see when we ask what are the physical stationary states for (1.51). We start with a probability measure m on M such that $m(dx) = \underline{m}(x)\, dx$, where dx is the volume element for some Riemann metric on M (for simplicity, we shall say that m is absolutely continuous). At time t, m becomes $(f^t)^* m$, which still is *absolutely continuous*. If $(f^t)^* m$ has a limit ρ when $t \to \infty$, then ρ is invariant under time evolution, and in general singular with respect to the Riemann volume element dx (a time evolution of the form (1.51) does not in general have an absolutely continuous invariant measure). The probability measures

$$\rho = \lim_{t \to \infty} (f^t)^* m,$$

or more generally [Rue99]

$$\rho = \lim_{t\to\infty} \frac{1}{t} \int_0^t d\tau \, (f^\tau)^* m, \qquad (1.52)$$

(with m absolutely continuous) are natural candidates to describe non–equilibrium stationary states, or non–equilibrium steady states. Examples of such measures are the SRB states discussed later.

Entropy Production

We return now to our calculation (1.50) of the entropy production:

$$S(\widehat{m}) - S(m) = \int dx \, \underline{m}(x) \log J_t(x),$$

where $\underline{\widehat{m}}$ is the density of $\widehat{m} = f^{t*}m$. This is the amount of entropy gained by the system under the action of the external forces and thermostat in time t. The amount of entropy produced by the system and given to the external world in one unit of time is thus (with $J = J_1$)

$$e_f(m) = -\int m(dx) \log J(x).$$

Notice that this expression makes sense also when m is a singular measure. The average entropy production in t units of time is [Rue99]

$$\frac{1}{t} \sum_{k=0}^{t-1} e_f(f^k m) = \frac{1}{t}[S(\underline{m}) - S(\underline{\widehat{m}})]. \qquad (1.53)$$

When $t \to \infty$, this tends according to (1.52) towards

$$e_f(\rho) = -\int \rho(dx) \log J(x), \qquad (1.54)$$

which is thus the entropy production in the state ρ (more precisely, the entropy production per unit time in the non–equilibrium steady state ρ). Using (1.51) we can also write

$$e_f(\rho) = -\int \rho(dx) \, \text{div} F(x). \qquad (1.55)$$

Notice that the entropy S in (1.53) is bounded above (see the above definition (1.49)), so that $e_f(\rho) \geq 0$, and in many cases $e_f(\rho) > 0$ as we shall see later. Notice that for an arbitrary probability measure μ (invariant or not), $e_f(\mu)$ may be positive or negative, but the definition (1.52) of ρ makes a choice of the direction of time, and results in positive entropy production. It may appear paradoxical that the state ρ, which does not change in time, constantly gives entropy to the outside world. The solution of the paradox is that ρ is (normally) a singular measure and therefore has entropy $-\infty$: the non–equilibrium steady state ρ is thus a bottomless source of entropy.

1.5 Continuous Chaotic Dynamics

Recent Idealization of Non–Equilibrium Processes

We have now reached a new framework idealizing non–equilibrium processes. Instead of following Boltzmann in his study of *approach to equilibrium*, we try to understand the *non–equilibrium steady states* as given by (1.52). For the definition of the entropy production $e_f(\rho)$ we use the rate of phase–space contraction (1.54) or (1.55). Later we shall discuss the SRB states, which provide a mathematically precise *definition of the non–equilibrium steady states*. After that, the idea is to use SRB states to make interesting physical predictions, making hyperbolicity assumptions (this will be explained later) as strong as needed to get results. Such a program was advocated early by Ruelle, but only recently were interesting results actually obtained, the first one being the *fluctuation theorem* of Gallavotti and Cohen [GC95a, GC95b].

Diversity of Non–Equilibrium Regimes

Many important non–equilibrium systems are locally close to equilibrium, and the classical non–equilibrium studies have concentrated on that case, yielding such results as the Onsager reciprocity relations and the Green-Kubo formula. Note however that chemical reactions are often far from equilibrium. More exotic non–equilibrium systems of interest are provided by *metastable states*. Since quantum measurements (and the associated *collapse of wave packets*) typically involve metastable states, one would like to have a reasonable fundamental understanding of those states. Another class of exotic systems are *spin glasses*, which are almost part of equilibrium statistical mechanics, but evolve slowly, with extremely long relaxation times.

Further Discussion of Non–equilibrium steady states

Recall that above we have proposed a definition (1.52) for non–equilibrium steady states ρ. We now make this definition more precise and analyze it further. Write [Rue99]

$$\rho = \text{w.lim}_{t\to\infty} \frac{1}{t} \int_0^t d\tau \, (f^\tau)^* m, \tag{1.56}$$

$$m \quad \text{a.c. probability measure}, \tag{1.57}$$

$$\rho \quad \text{ergodic}. \tag{1.58}$$

In (1.56), w.lim is the weak or vague limit defined by $(\text{w.lim}\, m_t)(\Phi) = \lim(m_t(\Phi))$ for all continuous $\Phi : M \to \mathbf{C}$. The set of probability measures on M is compact and metrizable for the vague topology. There are thus always sequences (t_k) tending to ∞ such that

$$\rho = \text{w.lim}_{k\to\infty} \frac{1}{t_k} \int_0^{t_k} d\tau \, (f^\tau)^* m$$

exists; ρ is automatically $(f^\tau)^*$ invariant. By (1.57), we ask that the probability measure m be a.c. (absolutely continuous) with respect to the Riemann volume element dx (with respect to any metric) on M. The condition (1.58) is discussed below. Physically, we consider a system which in the distant past was in equilibrium with respect to a Hamiltonian time evolution

$$\dot{x} = F_0(x), \tag{1.59}$$

and described by the probability measure m on the energy surface M. According to conventional wisdom, m is the *micro–canonical ensemble*, which satisfies (1.57), and is ergodic with respect to the time evolution (1.59) when the *ergodic hypothesis* is satisfied. Even if the ergodic hypothesis is accepted, the physical justification of the micro–canonical ensemble remains a delicate problem, which we shall not further discuss here.. For our purposes, we might also suppose that m is an ergodic component of the micro–canonical ensemble or an integral over such ergodic components, provided (1.57) is satisfied. We assume now that, at some point in the distant past, (1.59) was replaced by the time evolution

$$\dot{x} = F(x), \tag{1.60}$$

representing nonhamiltonian forces plus a thermostat keeping x on M; we write the general solution of (1.60) as $x \mapsto f^t x$. We are interested in time averages of $f^t x$ for m–almost all x. Suppose therefore that

$$\rho_x = \text{w.lim}_{t \to \infty} \frac{1}{t} \int_0^t d\tau \, \delta_{f^\tau x} \tag{1.61}$$

exists for m–almost all x. In particular, with ρ defined by (1.56), we have

$$\rho = \int m(dx)\, \rho_x. \tag{1.62}$$

If (1.58) holds. Suppose that ρ is not ergodic but that $\rho = \alpha \rho' + (1-\alpha)\rho''$ with ρ' ergodic and $\alpha \neq 0$. Writing $S = \{x : \rho_x = \rho'\}$, we have $m(S) = \alpha$ and $\rho' = \int m'(dx)\, \rho_x$ with $m' = \alpha^{-1}\chi_S.m$. Therefore, (1.56–1.58) hold with ρ, m replaced by ρ', m'., (1.62) is equivalent to

$$\rho_x = \rho \quad \text{for } m\text{–almost all } x.$$

($\Leftarrow$ is obvious; if $\Rightarrow$ did not hold (1.62) would give a non-trivial decomposition $\rho = \alpha \rho' + (1-\alpha)\rho''$ in contradiction with ergodicity). As we have just seen, the ergodic assumption (1.58) allows us to replace the study of (1.61) by the study of (1.56), with the condition (1.57). This has interesting consequences, as we shall see, but note that (1.56–1.58) are not always satisfyable simultaneously (consider for instance the case $F = 0$). To study non–equilibrium steady states we shall modify or strengthen the conditions (1.56–1.58) in various ways. We may for simplicity replace the continuous time $t \in \mathbb{R}$ by a discrete time $t \in \mathbb{Z}$.

Uniform Hyperbolicity: Anosov Diffeomorphisms and Flows

Let M be a compact connected manifold. In what follows we shall be concerned with a time evolution (f^t) which either has discrete time $t \in \mathbb{Z}$, and is given by the iterates of a *diffeomorphism* f of M, or has continuous time $t \in \mathbb{R}$, and is the *flow* generated by some vector–field F on M. We assume that f or F are of class $C^{1+\alpha}$ (Hölder continuous first derivatives). The Anosov property is an assumption of strong chaoticity (in physical terms) or uniform hyperbolicity (in mathematical terms). For background see for instance Smale [Sch81] or Bowen [Bow75]). Choose a Riemann metric on M. The diffeomorphism f is Anosov if there are a continuous Tf–-invariant splitting of the tangent bundle: $TM = E^s \oplus E^u$ and constants $C > 0$, $\theta > 1$ such that [Rue99]

$$\text{for all } t \geq 0: \quad \begin{aligned} \|Tf^t \xi\| &\leq C\theta^{-t} \quad \text{if } \xi \in E^s \\ \|Tf^{-t}\xi\| &\leq C\theta^{-t} \quad \text{if } \xi \in E^u. \end{aligned} \quad (1.63)$$

One can show that $x \mapsto E^s_x$, E^u_x are Hölder continuous, but not C^1 in general. The flow (f^t) is Anosov if there are a continuous (Tf^t)–invariant splitting. Remember that F is the vector–field generating the flow: $F(x) = df^t x/dt|_{t=0}$. Therefore $\mathbb{R}.F$ is the 1D subbundle in the direction of the flow. $TM = \mathbb{R}.F \oplus E^s \oplus E^u$ and constants $C > 0$, $\theta > 1$ such that (1.63) again holds. In what follows we shall assume that the periodic orbits are dense in M. This is conjectured to hold automatically for Anosov diffeomorphisms, but there is a counterexample for flows (see [FW80]). Since M is connected, we are thus dealing with what is called technically a *mixing Anosov diffeomorphism* f, or a *transitive Anosov flow* (f^t). There is a powerful method, called *symbolic dynamics*, for the study of Anosov diffeomorphisms and flows. Recall that symbolic dynamics (see [Sin68a]) is based on the existence of *Markov partitions* [Sin68b, Rat69].

Uniform Hyperbolicity: Axiom A Diffeomorphisms and Flows

Smale [[Sch81]] has achieved an important generalization of Anosov dynamical systems by imposing hyperbolicity only on a subset Ω (the *non–wandering set*) of the manifold M. A point $x \in M$ is *wandering point* if there is an open set $O \ni x$ such that $O \cap f^t O \neq \emptyset$ for $|t| > 1$. The points of M which are not wandering constitute the non–wandering set Ω. A diffeomorphism or flow satisfies *Axiom A* if the following conditions hold (Aa) there is a continuous (Tf^t)–invariant splitting of $T_\Omega M$ (the tangent bundle restricted to the non–wandering set) verifying the above hyperbolicity conditions. (Ab) the periodic orbits are dense in Ω. Under these conditions, Ω is a finite union of disjoint compact (f^t)-invariant sets B (called *basic sets*) on which (f^t) is topologically transitive. (f^t) is topologically transitive on B if there is $x \in B$ such that the orbit $(f^t x)$ is dense in B.. This result is known as Smale's *spectral decomposition* theorem. If there is an open set $U \supset B$ such that $\cap_{t \geq 0} f^t U = B$, the basic set B is called an *attractor*. The set $\{x \in M : \lim_{t \to +\infty} d(f^t x, B) =$

$0\} = \uplus_{t\geq 0} f^{-t} U$ is the *basin of attraction* of the attractor B. Let B be a basic set. Given $x \in B$ and $\epsilon > 0$, write [Rue99]

$$W^s_{x,\epsilon} = \{y \in M : d(f^t y, f^t x) < \epsilon \text{ for } t \geq 0, \text{ and } \lim_{t \to +\infty} d(f^t y, f^t x) = 0\},$$
$$W^u_{x,\epsilon} = \{y \in M : d(f^t y, f^t x) < \epsilon \text{ for } t \leq 0, \text{ and } \lim_{t \to -\infty} d(f^t y, f^t x) = 0\}$$

Then for sufficienty small ϵ, $W^s_{x,\epsilon}$ and $W^u_{x,\epsilon}$ are pieces of smooth manifolds, called *locally stable* and *locally unstable* manifold respectively. There are also *globally* stable and unstable manifolds defined by

$$W^s_x = \{y \in M : \lim_{t \to +\infty} d(f^t y, f^t x) = 0\},$$
$$W^u_x = \{y \in M : \lim_{t \to -\infty} d(f^t y, f^t x) = 0\},$$

and tangent to E^s_x, E^u_x respectively. A basic set B is an attractor if and only if the stable manifolds $W^s_{x,\epsilon}$ for $x \in B$ cover a neighborhood of B. Also, a basic set B is an attractor if and only if the unstable manifolds $W^u_{x,\epsilon}$ for $x \in B$ are contained in B (see [BR75]). Markov partitions, symbolic dynamics (see [Bow70, Bow73]), and shadowing (see [Bow75]) are available on Axiom A basic sets as they were for Anosov dynamical systems.

SRB States on Axiom A Attractors

Let us cut an attractor B into a finite number of small cells such that the unstable manifolds are roughly parallel within a cell. Each cell is partitionned into a continuous family of pieces of local unstable manifolds, and we get thus a partition (Σ_α) of B into small pieces of unstable manifolds. If ρ is an invariant probability measure on B (for the dynamical system (f^t)), and if its conditional measures σ_α with respect to the partition (Σ_α) are a.c. with respect to the Riemann volume $d\sigma$ on the unstable manifolds, ρ is called an SRB measure. The study of SRB measures is transformed by use of symbolic dynamics into a problem of statistical mechanics: one can characterize SRB states as Gibbs states with respect to a suitable interaction. Such Gibbs states can in turn be characterized by a variational principle. In the end one has a variational principle for SRB measures, which we shall now describe. It is convenient to consider a general basic set B (not necessarily an attractor) and to define generalized SRB measures. First we need the concept of the *time entropy* $h_f(\mu)$, where $f = f^1$ is the time 1 map for our dynamical system, and μ an f-invariant probability measure on B. This entropy (or Kolmogorov–Sinai invariant) has the physical meaning of mean information production per unit time by the dynamical system (f^t) in the state μ (see for instance [6] for an exact definition). The time entropy $h_f(\mu)$ should not be confused with the *space entropy* S and the entropy production rate e_f which we have discussed above. We also need the concept of *expanding Jacobian* J^u. Since $T_x f$ maps E^u_x linearly to E^u_{fx}, and volume elements are defined in E^u_x, E^u_{fx} by a Riemann metric, we can define a volume expansion rate $J^u(x) > 0$.

It can be shown that the function $\log J^u : B \to \mathbb{R}$ is Hölder continuous. We say that the (f^t)-invariant probability measure ρ is a *generalized SRB measure* if it makes maximum the function

$$\mu \mapsto h_f(\mu) - \mu(\log J^u). \tag{1.64}$$

One can show that there is precisely one generalized SRB measure on each basic set B; it is ergodic and reduces to the unique SRB measure when B is an attractor. The value of the maximum of (9) is 0 precisely when B is an attractor, it is < 0 otherwise. If m is a measure absolutely continuous with respect to the Riemann volume element on M, and if its density $\underline{m}$ vanishes outside of the basin of attraction of an attractor B, then [Rue99]

$$\rho = \text{w.lim}_{t \to \infty} \frac{1}{t} \int_0^t d\tau \, (f^\tau)^* m \tag{1.65}$$

defines the unique SRB measure on B. We also have

$$\rho = \text{w.lim}_{t \to \infty} \frac{1}{t} \int_0^t d\tau \, \delta_{f^\tau x},$$

when x is in the domain of attraction of B and outside of a set of measure 0 for the Riemann volume. The conditions (1.56–1.58) and also (1.61) are thus satisfied, and the SRB state ρ is a natural non–equilibrium steady state. Note that if B is a *mixing*, i.e., for any two nonempty open sets $O_1, O_2 \subset B$, there is T such that $O_1 \cap f^t O_2 \neq \emptyset$ for $|t| \geq T$. attractor, (10) can be replaced by the stronger result

$$\rho = \text{w.lim}_{t \to \infty} (f^t)^* m.$$

See [Sin72, Rue76, BR75] for details.

SRB States Without Uniform Hyperbolicity

Remarkably, the theory of SRB states on Axiom A attractors extends to much more general situations. Consider a smooth dynamical system (f^t) on the compact manifold M, without any hyperbolicity condition, and let ρ be an ergodic measure for this system. Recall that the *Oseledec theorem* [Ose68, Rue79] permits the definition of *Lyapunov exponents* $\lambda_1 \leq \ldots \leq \lambda_{\dim M}$ which are the rates of expansion, ρ–almost everywhere, of vectors in TM. The λ_i are real numbers, positive (expansion), negative (contraction), or zero (neutral case). Pesin theory [Pes76, Pes77] allows the definition of stable and unstable manifolds ρ–almost everywhere. These are smooth manifolds; the dimension of the stable manifolds is the number of negative Lyapunov exponents while the dimension of the unstable manifold is the number of positive Lyapunov exponents. Consider now a family (Σ_α) constituted of pieces of (local) unstable manifolds, and forming, up to a set of ρ–measure 0, a partition of M. As in Section 4 above we define the conditional measures σ_α of ρ with respect to (Σ_α). If the measures σ_α are absolutely continuous with respect to the Riemann volumes of the corresponding unstable manifolds, we say that ρ is an SRB measure. For a C^2 diffeomorphism, the above definition of SRB measures is equivalent to the following condition (known as *Pesin formula*) for an ergodic measure ρ:

$$h(\rho) = \sum \text{positive Lyapunov exponents for } \rho,$$

(see [LS82, LY85] for the nontrivial proof). This is an extension of the result of Section 4, where the sum of the positive Lyapunov exponents is equal to $\rho(\log J^u)$. Note that in general $h(\mu)$ is $\leq$ the sum of the positive exponents for the ergodic measure μ (see [Rue78]). Suppose that the time t is discrete (diffeomorphism case), and that the Lyapunov exponents of the SRB state ρ are all different from zero: this is a weak (nonuniform) hyperbolicity condition. In this situation, there is a measurable set $S \subset M$ with positive Riemann volume such that [PS89]

$$\lim_{n\to\infty} \frac{1}{n} \sum_{k=0}^{n-1} \delta_{f^k x} = \rho,$$

for all $x \in S$. This result shows that ρ has the properties required of a non–equilibrium steady state. One expects that for continuous time (flow case), if supp ρ is not reduced to a point and if the Lyapunov exponents of the SRB state ρ except one. There is one zero exponent corresponding to the flow direction. are different from 0, there is a measurable set $S \subset M$ with positive Riemann volume such that [Rue99]

$$\rho = \text{w.lim}_{t\to\infty} \frac{1}{t} \int_0^t d\tau\, \delta_{f^\tau x}, \qquad \text{when } x \in S.$$

See [LS82, LY85, Via97] for details.

1.5.2 Statistical Mechanics of Nonlinear Oscillator Chains

Now, consider a model of a finite nonlinear chain of n d–dimensional oscillators, coupled to two Hamiltonian heat reservoirs initially at different temperatures T_L, T_R, each of which is described by a dD wave equation. A natural goal is to get a usable expression for the invariant (marginal) state of the chain analogous to the Boltzmann–Gibbs prescription $\mu = Z^{-1} \exp(-H/T)$ which one has in equilibrium statistical mechanics [BT00]. We assume that the Hamiltonian $H(p,q)$ of the isolated chain has the form

$$H(p,q) = \sum_{i=1}^{n} \frac{p_i^2}{2} + \sum_{i=1}^{n} U^{(1)}(q_i) + \sum_{i=1}^{n-1} U^{(2)}(q_i - q_{i+1}) \equiv \sum_{i=1}^{n} \frac{p_i^2}{2} + V(q), \quad (1.66)$$

where q_i and p_i are the coordinate and momentum of the ith particle, and where $U^{(1)}$ and $U^{(2)}$ are C^k confining potentials, i.e., $\lim_{|q|\to\infty} V(q) = +\infty$.

The coupling between the reservoirs and the chain is assumed to be of dipole approximation type and it occurs at the boundary only: the first particle of the chain is coupled to one reservoir and the nth particle to the other heat reservoir. At time $t = 0$ each reservoir is assumed to be in thermal equilibrium, i.e., the initial conditions of the reservoirs are distributed according to (Gaussian) Gibbs measure with temperature $T_1 = T_L$ and $T_n = T_R$ respectively. Projecting the dynamics onto the phase–space of the chain results in a set of integro–differential equations which differ from the Hamiltonian equations of motion by additional force terms in the equations for p_1 and p_n. Each of these terms consists of a deterministic integral part independent of temperature and a Gaussian random part with covariance proportional to the temperature. Due to the integral (memory) terms, the study of the long–time limit is a difficult mathematical problem. But by a further appropriate choice of couplings, the integral parts can be treated as auxiliary variables r_1 and r_n, the random parts become Markovian. Thus we get (see [EPR99] for details) the following system of Markovian stochastic differential equations (SDEs) on the extended phase–space $\mathbb{R}^{2dn+2d}$: For $x = (p, q, r)$ we have

$$\begin{aligned}
\dot{q}_1 &= p_1, & \dot{p}_1 &= -\nabla_{q_1} V(q) + r_1, \\
\dot{q}_j &= p_j, & \dot{p}_j &= -\nabla_{q_j} V(q), \quad (j = 2, \ldots, n-1) \\
\dot{q}_n &= p_n, & \dot{p}_n &= -\nabla_{q_n} V(q) + r_n, \\
dr_1 &= -\gamma(r_1 - \lambda^2 q_1)dt + (2\gamma\lambda^2 T_1)^{1/2} dw_1, \\
dr_n &= -\gamma(r_n - \lambda^2 q_1)dt + (2\gamma\lambda^2 T_n)^{1/2} dw_n.
\end{aligned} \quad (1.67)$$

In equation (1.67), $w_1(t)$ and $w_n(t)$ are independent dD Wiener processes, and λ^2 and γ are constants describing the couplings.

Now introduce a generalized Hamiltonian $G(p, q, r)$ on the extended phase–space, given by

$$G(p, q, r) = \sum_{i=1}^{n} \left(\frac{r_i^2}{2\lambda^2} - r_i q_i \right) + H(p, q), \quad (1.68)$$

where $H(p, q)$ is the Hamiltonian of the isolated systems of oscillators given by (1.66). We also introduce the parameters ε (the mean temperature of the reservoirs) and η (the relative temperature difference):

$$\varepsilon = \frac{T_1 + T_n}{2}, \qquad \eta = \frac{T_1 - T_n}{T_1 + T_n}. \quad (1.69)$$

Using (1.68), the equation (1.67) takes the form [BT00]

$$\begin{aligned}
\dot{q} &= \nabla_p G, \qquad \dot{p} = -\nabla_q G, \\
dr &= -\gamma\lambda^2 \nabla_r G dt + \varepsilon^{1/2}(2\gamma\lambda^2 D)^{1/2} dw,
\end{aligned} \quad (1.70)$$

where $p = (p_1, \ldots, p_n)$, $q = (q_1, \ldots, q_n)$, $r = (r_1, r_n)$ and where D is the $2d \times 2d$ matrix given by

$$D = \begin{pmatrix} 1+\eta & 0 \\ 0 & 1-\eta \end{pmatrix}.$$

The function G is a Lyapunov function, non–increasing in time, for the deterministic part of the flow (1.70). If the system is in equilibrium, i.e, if $T_1 = T_n = \varepsilon$ and $\eta = 0$, it is not difficult to check that the generalized Gibbs measure

$$\mu_\varepsilon = Z^{-1} \exp\left(-G(p, q, r)/\varepsilon\right),$$

is an invariant measure for the Markov process solving (1.70).

1.5.3 Geometrical Modelling of Continuous Dynamics

Here we give a paradigm of geometrical modelling and analysis of complex continuous dynamical systems (see [II06b] for technical details). This is essentially a *recipe* how to develop a *covariant formalism on smooth manifolds*, given a certain physical, or bio–physical, or psycho–physical, or socio–physical system, here labelled by a generic name: 'physical situation'. We present this recipe in the form of the following five–step algorithm.

(I) So let's start: given a certain physical situation, the first step in its predictive modelling and analysis, that is, in applying a powerful differential–geometric machinery to it, is to associate with this situation *two* independent coordinate systems, constituting two independent smooth Riemannian manifolds. Let us denote these two coordinate systems and their respective manifolds as:

- *Internal coordinates*: $x^i = x^i(t)$, $(i = 1, \ldots, m)$, constituting the mD *internal configuration manifold*: $M^m \equiv \{x^i\}$; and
- *External coordinates*: $y^e = y^e(t)$, $(e = 1, \ldots, n)$, constituting the nD *external configuration manifold*: $N^n \equiv \{y^e\}$.

The main example that we have in mind is a standard robotic or bio–dynamic (loco)motion system, in which x^i denote internal joint coordinates, while y^e denote external Cartesian coordinates of segmental centers of mass. However, we believe that such developed methodology can fit a generic physical situation.

Therefore, in this first, engineering step (I) of our differential–geometric modelling, we associate to the given natural system, not one but two different and independent smooth configuration manifolds, somewhat like viewing from two different satellites a certain place on Earth with a football game playing in it.

(II) Once that we have precisely defined two smooth manifolds, as two independent views on the given physical situation, we can apply our differential–geometric modelling to it and give it a natural physical interpretation. More precisely, once we have two smooth Riemannian manifolds, $M^m \equiv \{x^i\}$ and $N^n \equiv \{y^e\}$, we can formulate two smooth maps between them:[31]

$$f : N \to M, \text{ given by coordinate transformation: } x^i = f^i(y^e), \quad (1.71)$$

and

$$g : M \to N, \text{ given by coordinate transformation: } y^e = g^e(x^i). \quad (1.72)$$

If the Jacobian matrices of these two maps are nonsingular (regular), that is if their Jacobian determinants are nonzero, then these two maps are mutually inverse, $f = g^{-1}$, and they represent standard *forward and inverse kinematics*.

(III) Although, maps f and g define some completely general nonlinear coordinate (functional) transformations, which are even unknown at the moment, there is something linear and simple that we know about them (from calculus). Namely, the corresponding infinitesimal transformations are linear and homogenous: from (1.71) we have (applying everywhere Einstein's summation convention over repeated indices)

$$dx^i = \frac{\partial f^i}{\partial y^e} dy^e, \quad (1.73)$$

while from (1.72) we have

$$dy^e = \frac{\partial g^e}{\partial x^i} dx^i. \quad (1.74)$$

Furthermore, (1.73) implies the linear and homogenous transformation of *internal velocities*,

$$v^i \equiv \dot{x}^i = \frac{\partial f^i}{\partial y^e} \dot{y}^e, \quad (1.75)$$

while (1.74) implies the linear and homogenous transformation of *external velocities*,

$$u^e \equiv \dot{y}^e = \frac{\partial g^e}{\partial x^i} \dot{x}^i. \quad (1.76)$$

In this way, we have defined *two velocity vector–fields*, the internal one: $v^i = v^i(x^i, t)$ and the external one: $u^e = u^e(y^e, t)$, given respectively by the two nonlinear systems of ODEs, (1.75) and (1.76).[32]

(IV) The next step in our differential–geometrical modelling/analysis is to define second derivatives of the manifold maps f and g, that is the two *acceleration vector–fields*, which we will denote by $a^i = a^i(x^i, \dot{x}^i, t)$ and $w^e = w^e(y^e, \dot{y}^e, t)$, respectively. However, unlike simple physics in linear Euclidean spaces, these two acceleration vector–fields on manifolds M and N are not the simple time derivatives of the corresponding velocity vector–fields ($a^i \neq \dot{v}^i$

---

[31] This obviously means that we are working in the category of smooth manifolds.
[32] Although transformations of differentials and associated velocities are linear and homogeneous, the systems of ODE's define nonlinear vector–fields, as they include Jacobian (functional) matrices.

86   1 Basics of Nonlinear and Chaotic Dynamics

and $w^e \neq \dot{u}^e$), due to the existence of the *Levi–Civita connections* ∇_M and ∇_N on both M and N. Properly defined, these two acceleration vector–fields respectively read:

$$a^i = \dot{v}^i + \Gamma^i_{jk} v^j v^k = \ddot{x}^i + \Gamma^i_{jk} \dot{x}^j \dot{x}^k, \quad \text{and} \tag{1.77}$$

$$w^e = \dot{u}^e + \Gamma^e_{hl} u^h u^l = \ddot{y}^e + \Gamma^e_{hl} \dot{y}^h \dot{y}^l, \tag{1.78}$$

where Γ^i_{jk} and Γ^e_{hl} denote the (second–order) *Christoffel symbols* of the connections ∇_M and ∇_N.

Therefore, in the step (III) we gave the first–level model of our physical situation in the form of two ordinary vector–fields, the first–order vector–fields (1.75) and (1.76). For some simple situations (e.g., modelling ecological systems), we could stop at this modelling level. Using physical terminology we call them velocity vector–fields. Following this, in the step (IV) we have defined the two second–order vector–fields (1.77) and (1.78), as a connection–base derivations of the previously defined first–order vector–fields. Using physical terminology, we call them 'acceleration vector–fields'.

(V) Finally, following our generic physical terminology, as a natural next step we would expect to define some kind of generic Newton–Maxwell force–fields. And we can actually do this, with a little surprise that individual forces involved in the two force–fields will not be vectors, but rather the dual objects called 1–forms (or, 1D differential forms). Formally, we define the two *covariant force–fields* as

$$F_i = \mathfrak{m} g_{ij} a^j = \mathfrak{m} g_{ij}(\dot{v}^j + \Gamma^j_{ik} v^i v^k) = \mathfrak{m} g_{ij}(\ddot{x}^j + \Gamma^j_{ik} \dot{x}^i \dot{x}^k), \quad \text{and} \tag{1.79}$$

$$G_e = \mathfrak{m} g_{eh} w^h = \mathfrak{m} g_{eh}(\dot{u}^h + \Gamma^h_{el} u^e u^l) = \mathfrak{m} g_{eh}(\ddot{y}^h + \Gamma^h_{el} \dot{y}^e \dot{y}^l), \tag{1.80}$$

where $\mathfrak{m}$ is the mass of each single segment (unique, for simplicity), while $g_{ij} = g^M_{ij}$ and $g_{eh} = g^N_{eh}$ are the two *Riemannian metric tensors* corresponding to the manifolds M and N. The two force–fields, F_i defined by (1.79) and G_e defined by (1.80), are generic force–fields corresponding to the manifolds M and N, which represent the *material cause* for the given physical situation. Recall that they can be physical, bio–physical, psycho–physical or socio–physical force–fields. Physically speaking, they are the *generators* of the corresponding dynamics and kinematics.

Main geometrical relations behind this fundamental paradigm, forming the *covariant force functor*[33] [II06b], are depicted in Figure 1.47.

1.5.4 Lagrangian Chaos

A problem of great interest concerns the study of the spatial and temporal structure of the so–called passive fields, indicating by this term passively

---

[33] A functor is a generalized mapping from a domain category (a generalized group) to a codomain category, that preserves the structure of the domain category (see [II07b]).

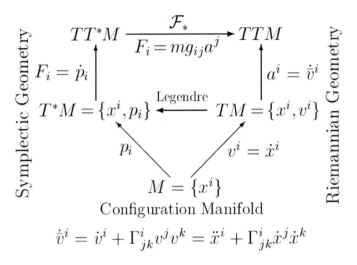

Fig. 1.47. The covariant force functor, including the main relations used by differential–geometric modelling of complex continuous systems.

quantities driven by the flow, such as the temperature under certain conditions [Mof83]. The equation for the evolution of a passive scalar field $\theta(\mathbf{x}, t)$, advected by a velocity field $\mathbf{v}(\mathbf{x}, t)$, is

$$\partial_t \theta + \nabla \cdot (\mathbf{v}\,\theta) = \chi \nabla^2 \theta, \tag{1.81}$$

where $\mathbf{v}(\mathbf{x}, t)$ is a given velocity field and χ is the molecular diffusion coefficient.

The problem (1.81) can be studied through two different approaches. Either one deals at any time with the field θ in the space domain covered by the fluid, or one deals with the trajectory of each fluid particle. The two approaches are usually designed as 'Eulerian' and 'Lagrangian', although both of them are due to Euler [Lam45]. The two points of view are in principle equivalent.

The motion of a fluid particle is determined by the differential equation

$$\dot{\mathbf{x}} = \mathbf{v}(\mathbf{x}, t), \tag{1.82}$$

which also describes the motion of test particles, for example a powder embedded in the fluid, provided that the particles are neutral and small enough not to perturb the velocity field, although large enough not to perform a Brownian motion. Particles of this type are commonly used for flow visualization in fluid mechanics experiments, see [Tri88]. Let us note that the true equation for the motion of a material particle in a fluid can be rather complicated [CFP90].

It is now well established that even in regular velocity field the motion of fluid particles can be very irregular [Hen66, Are84]. In this case initially nearby trajectories diverge exponentially and one speaks of *Lagrangian chaos*.

In general, chaotic behaviors can arise in 2D flow only for time dependent velocity fields in 2D, while it can be present even for stationary velocity fields in 3D.

If $\chi = 0$, it is easy to realize that (1.81) is equivalent to (1.82). In fact, we can write

$$\theta(\mathbf{x},t) = \theta_o(T^{-t}\mathbf{x}), \tag{1.83}$$

where $\theta_o(\mathbf{x}) = \theta(\mathbf{x}, t=0)$ and T is the formal evolution operator of (1.82),

$$\mathbf{x}(t) = T^t \mathbf{x}(0). \tag{1.84}$$

Taking into account the molecular diffusion χ, (1.81) is the Fokker–Planck equation of the Langevin equation [Cha43]

$$\dot{\mathbf{x}} = \mathbf{v}(\mathbf{x},t) + \eta(t), \tag{1.85}$$

where η is a Gaussian process with zero mean and variance

$$\langle \eta_i(t)\, \eta_j(t') \rangle = 2\chi \delta_{ij}\, \delta(t-t'). \tag{1.86}$$

In the following we will consider only incompressible flow

$$\nabla \cdot \mathbf{v} = 0, \tag{1.87}$$

for which the dynamical system (1.82) is conservative. In 2D, the constraint (2.29) is automatically satisfied assuming

$$v_1 = \frac{\partial \psi}{\partial x_2}, \quad v_2 = -\frac{\partial \psi}{\partial x_1}, \tag{1.88}$$

where $\psi(\mathbf{x},t)$ is the *stream function*. Inserting (2.30) into (1.82) the evolution equations become

$$\dot{x}_1 = \frac{\partial \psi}{\partial x_2}, \quad \dot{x}_2 = -\frac{\partial \psi}{\partial x_1}. \tag{1.89}$$

Formally (2.31) is a Hamiltonian system with the Hamiltonian given by the stream function ψ.

Examples of Lagrangian Chaos

As a first example we consider a $3n$ stationary velocity field, the so-called ABC flow

$$\mathbf{v} = (A \sin z + C \cos y,\ B \sin x + A \cos z,\ C \sin y + B \cos x), \tag{1.90}$$

where A, B and C are non zero real parameters. Because of the incompressibility condition, the evolution $\mathbf{x}(0) \to \mathbf{x}(t)$ defines a volume preserving, dynamics.

1.5 Continuous Chaotic Dynamics

Arnold [Arn65] argued that (2.33) is a good candidate for chaotic motion. Let us briefly repeat his elegant argument. For a steady state solution of the $3n$ Euler equation one has:

$$\nabla \cdot \mathbf{v} = 0, \qquad \mathbf{v} \times (\nabla \times \mathbf{v}) = \nabla \alpha, \qquad \alpha = \frac{P}{\rho} + \frac{\mathbf{v}^2}{2}, \qquad (1.91)$$

where P is the pressure and ρ the density. As a consequence of the Bernoulli theorem [LL59], $\alpha(\mathbf{x})$ is constant along a streamline – that is a Lagrangian trajectory $\mathbf{x}(t)$. One can easily verify that chaotic motion can appear only if $\alpha(\mathbf{x})$ is constant (i.e., $\nabla \alpha(\mathbf{x}) = 0$) in a part of the space. Otherwise the trajectory would be confined on a $2n$ surface $\alpha(\mathbf{x}) = $ constant, where the motion must be regular as a consequence of general arguments [Ott93]. In order to satisfy such a constraint, from (1.91) one has the Beltrami condition:

$$\nabla \times \mathbf{v} = \gamma(\mathbf{x}) \, \mathbf{v}. \qquad (1.92)$$

The reader can easily verify that the field $\mathbf{v}$ given by (2.33) satisfy (2.35) (in this case $\gamma(\mathbf{x}) = $ constant). Indeed, numerical experiments by Hénon [Hen66] provided evidence that Lagrangian motion under velocity (2.33) is chaotic for typical values of the parameters A, B, and C.

In a 2D incompressible stationary flows the motion of fluid particles is given by a time independent Hamiltonian system with one degree of freedom and, since trajectories follow iso-ψ lines, it is impossible to have chaos. However, for explicit time dependent stream function ψ the system (2.33) can exhibit chaotic motion [Ott93].

In the particular case of time periodic velocity fields, $\mathbf{v}(\mathbf{x}, t+T) = \mathbf{v}(\mathbf{x}, t)$, the trajectory of (1.82) can be studied in terms of discrete dynamical systems: the position $\mathbf{x}(t+T)$ is determined in terms of $\mathbf{x}(t)$. The map $\mathbf{x}(t) \to \mathbf{x}(t+T)$ will not depend on t thus (1.82) can be written in the form

$$\mathbf{x}(n+1) = \mathbf{F}[\mathbf{x}(n)], \qquad (1.93)$$

where now the time is measured in units of the period T. Because of incompressibility, the map (1.93) is conservative:

$$|\det A[\mathbf{x}]| = 1, \qquad \text{where} \qquad A_{ij}[\mathbf{x}] = \frac{\partial F_i[\mathbf{x}]}{\partial x_j}. \qquad (1.94)$$

An explicit deduction of the form of $\mathbf{F}$ for a general $2n$ or $3n$ flow is usually very difficult. However, in some simple model of can be deduced on the basis of physical features [AB86, CCT87].

Eulerian Properties and Lagrangian Chaos

In principle, the evolution of the velocity field $\mathbf{v}$ is described by partial differential equations, e.g., Navier–Stokes or Boussinesq equations. However, often in weakly turbulent situations, a good approximation of the flow can be geted by using a Galerkin approach, and reducing the Eulerian problem to a (small) system of F ordinary differential equations [BF79]. The motion of a fluid particle is then described by the $(n+F)$D dynamical system [BLV01]

$$\dot{\mathbf{Q}} = \mathbf{f}(\mathbf{Q}, t), \qquad \text{with } (\mathbf{Q}, \mathbf{f} \in \mathbb{R}^F), \tag{1.95}$$

$$\dot{\mathbf{x}} = \mathbf{v}(\mathbf{x}, \mathbf{Q}), \qquad \text{with } (\mathbf{x}, \mathbf{v} \in \mathbb{R}^n), \tag{1.96}$$

where n is the space dimensionality and $\mathbf{Q} = (Q_1, ... Q_F)$ are the F variables, usually normal modes, which are a representation of the velocity field $\mathbf{v}$. Note that the Eulerian equations (1.95) do not depend on the Lagrangian part (1.96) and can be solved independently.

In order to characterize the degree of chaos, three different Lyapunov exponents can be defined [FPV88]:

- a) λ_E for the Eulerian part (1.95);
- b) λ_L for the Lagrangian part (1.96), where the evolution of the velocity field is assumed to be known;
- c) λ_T per for the total system of the $n+F$ equations.

These Lyapunov exponents are defined as [BLV01]:

$$\lambda_{E,L,T} = \lim_{t \to \infty} \frac{1}{t} \ln \frac{|\mathbf{z}(t)^{(E,L,T)}|}{|\mathbf{z}(0)^{(E,L,T)}|}, \tag{1.97}$$

where the evolution of the three tangent vectors $\mathbf{z}$ are given by the linearized stability equations for the Eulerian part, for the Lagrangian part and for the total system, respectively:

$$\dot{z}_i^{(E)} = \sum_{j=1}^{F} \left. \frac{\partial f_i}{\partial Q_j} \right|_{\mathbf{Q}(t)} z_j^{(E)}, \qquad (\mathbf{z}^{(E)} \in \mathbb{R}^F), \tag{1.98}$$

$$\dot{z}_i^{(L)} = \sum_{j=1}^{n} \left. \frac{\partial v_i}{\partial x_j} \right|_{\mathbf{x}(t)} z_j^{(L)}, \qquad (\mathbf{z}^{(L)} \in \mathbb{R}^n), \tag{1.99}$$

$$\dot{z}_i^{(T)} = \sum_{j=1}^{n+F} \left. \frac{\partial G_i}{\partial y_j} \right|_{\mathbf{y}(t)} z_j^{(T)}, \qquad (\mathbf{z}^{(T)} \in \mathbb{R}^{F+n}), \tag{1.100}$$

and $\mathbf{y} = (Q_1, \ldots, Q_F, x_1, \ldots, x_n)$ and $\mathbf{G} = (f_1, \ldots, f_F, v_1, \ldots, v_n)$. The meaning of these Lyapunov exponents is evident:

- a) λ_E is the mean exponential rate of the increasing of the uncertainty in the knowledge of the velocity field (which is, by definition, independent on the Lagrangian motion);
- b) λ_L estimates the rate at which the distance $\delta x(t)$ between two fluid particles initially close increases with time, when the velocity field is given, i.e., a particle pair in the same Eulerian realization;

- c) λ_T is the rate of growth of the distance between initially close particle pairs, when the velocity field is not known with infinite precision.

There is no general relation between λ_E and λ_L. One could expect that in presence of a chaotic velocity field the particle motion has to be chaotic. However, the inequality $\lambda_L \geq \lambda_E$ – even if generic – sometimes does not hold, e.g., in some systems like the Lorenz model [FPV88] and in generic $2n$ flows when the Lagrangian motion happens around well defined vortex structures [BBP94] as discussed in the following. On the contrary, one has [CFP91]

$$\lambda_T = \max(\lambda_E, \lambda_L). \tag{1.101}$$

Lagrangian Chaos in 2D–Flows

Let us now consider the 2D *Navier–Stokes equations* with periodic boundary conditions at low Reynolds numbers, for which we can expand the stream function ψ in Fourier series and takes into account only the first F terms [BF79],

$$\psi = -i \sum_{j=1}^{F} k_j^{-1} Q_j e^{i\mathbf{k}_j \mathbf{x}} + \text{c.c.}, \tag{1.102}$$

where c.c. indicates the complex conjugate term and $\mathbf{Q} = (Q_1, \ldots, Q_F)$ are the Fourier coefficients. Inserting (1.102) into the Navier–Stokes equations and by an appropriate time rescaling, we get the system of F ordinary differential equations

$$\dot{Q}_j = -k_j^2 Q_j + \sum_{l,m} A_{jlm} Q_l Q_m + f_j, \tag{1.103}$$

in which f_j represents an external forcing.

Franceschini and coworkers have studied this truncated model with $F = 5$ and $F = 7$ [BF79]. The forcing were restricted to the 3th mode $f_j = \text{Re}\,\delta_{j,3}$. For $F = 5$ and $\text{Re} < \text{Re}_1 = 22.85\ldots$, there are four stable stationary solutions, say $\hat{\mathbf{Q}}$, and $\lambda_E < 0$. At $\text{Re} = \text{Re}_1$, these solutions become unstable, via a Hopf bifurcation [MM75], and four stable periodic orbits appear, still implying $\lambda_E = 0$. For $\text{Re}_1 < \text{Re} < \text{Re}_2 = 28.41\ldots$, one thus finds the stable limit cycles:

$$\mathbf{Q}(t) = \hat{\mathbf{Q}} + (\text{Re} - \text{Re}_1)^{1/2} \delta \mathbf{Q}(t) + O(\text{Re} - \text{Re}_1), \tag{1.104}$$

where $\delta \mathbf{Q}(t)$ is periodic with period

$$T(\text{Re}) = T_0 + O(\text{Re} - \text{Re}_1) \qquad T_0 = 0.7328\ldots \tag{1.105}$$

At $\text{Re} = \text{Re}_2$, these limit cycles lose stability and there is a period doubling cascade toward Eulerian chaos.

Let us now discuss the Lagrangian behavior of a fluid particle. For $\text{Re} < \text{Re}_1$, the stream function is asymptotically stationary, $\psi(\mathbf{x}, t) \to \widehat{\psi}(\mathbf{x})$, and the corresponding 1D Hamiltonian is time-independent, therefore Lagrangian trajectories are regular. For Re = $\text{Re}_1 + \epsilon$ the stream function becomes time dependent [BLV01]

$$\psi(\mathbf{x}, t) = \widehat{\psi}(\mathbf{x}) + \sqrt{\epsilon}\,\delta\psi(\mathbf{x}, t) + O(\epsilon), \qquad (1.106)$$

where $\widehat{\psi}(\mathbf{x})$ is given by $\widehat{\mathbf{Q}}$ and $\delta\psi$ is periodic in $\mathbf{x}$ and in t with period T. The region of phase–space, here the real 2D space, adjacent to a separatrix is very sensitive to perturbations, even of very weak intensity. Figure 1.48 shows the structure of the separatrices, i.e., the orbits of infinite periods at Re = $\text{Re}_1 - 0.05$.

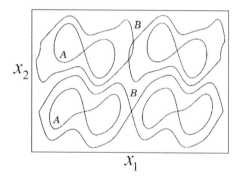

Fig. 1.48. Structure of the separatrices in the 5-mode model (1.102) with Re = $\text{Re}_1 - 0.05$ (adapted from [BLV01]).

Indeed, generically in 1D Hamiltonian systems, a periodic perturbation gives origin to stochastic layers around the separatrices where the motion is chaotic, as consequence of unfolding and crossing of the stable and unstable manifolds in domains centered at the hyperbolic fixed–points [Chi79, Ott93].

Chaotic and regular motion for small $\epsilon = \text{Re}_1 - \text{Re}$ can be studied by the Poincaré map

$$\mathbf{x}(nT) \to \mathbf{x}(nT + T). \qquad (1.107)$$

The period $T(\epsilon)$ is computed numerically. The size of the stochastic layers rapidly increase with ϵ. At $\epsilon = \epsilon_c \approx 0.7$ they overlap and it is practically impossible to distinguish between regular and chaotic zones. At $\epsilon > \epsilon_c$ there is always diffusive motion.

We stress that this scenario for the onset of Lagrangian chaos in 2D fluids is generic and does not depend on the particular truncated model. In fact, it is only related to the appearance of stochastic layers under the effects of small time–dependent perturbations in 1D integrable Hamiltonian systems. As consequence of a general features of 1D Hamiltonian systems we expect that a stationary stream function becomes time periodic through a Hopf bifurcation as occurs for all known truncated models of Navier–Stokes equations.

We have seen that there is no simple relation between Eulerian and Lagrangian behaviors. In the following, we shall discuss two important points [BLV01]:

- (i) what are the effects on the Lagrangian chaos of the transition to Eulerian chaos, i.e., from $\lambda_E = 0$ to $\lambda_E > 0$.
- (ii) whether a chaotic velocity field ($\lambda_E > 0$) always implies an erratic motion of fluid particles.

The first point can be studied again within the $F = 5$ modes model (1.103). Increasing Re, the limit cycles bifurcate to new double period orbits followed by a period doubling transition to chaos and a strange attractor appears at $\text{Re}_c \approx 28.73$, where λ_E becomes positive. These transitions have no signature on Lagrangian behavior, i.e., the onset of Eulerian chaos has no influence on Lagrangian properties.

This feature should be valid in most situations, since it is natural to expect that in generic cases there is a strong separation of the characteristic times for Eulerian and Lagrangian behaviors.

The second point – the conjecture that a chaotic velocity field always implies chaotic motion of particles – looks very reasonable. Indeed, it appears to hold in many systems [CFP91]. Nevertheless, one can find a class of systems where it is false, e.g., the equations (1.95), (1.96) may exhibit Eulerian chaoticity $\lambda_E > 0$, even if $\lambda_L = 0$ [BBP94].

Consider for example the motion of N point vortices in the plane with circulations Γ_i and positions $(x_i(t), y_i(t))$ $(i = 1, ..N)$ [Are83]:

$$\Gamma_i \dot{x}_i = \frac{\partial H}{\partial y_i}, \qquad \Gamma_i \dot{y}_i = -\frac{\partial H}{\partial x_i}, \qquad \text{where} \qquad (1.108)$$

$$H = -\frac{1}{4\pi} \Gamma_i \Gamma_j \ln r_{ij}, \qquad \text{and} \qquad r_{ij}^2 = (x_i - x_j)^2 + (y_i - y_j)^2. \qquad (1.109)$$

The motion of N point vortices is described in an Eulerian phase–space with $2ND$. Because of the presence of global conserved quantities, a system of three vortices is integrable and there is no exponential divergence of nearby trajectories in phase–space. For $N \geq 4$, apart from non generic initial conditions and/or values of the parameters Γ_i, the system is chaotic [Are83].

The motion of a passively advected particle located in $(x(t), y(t))$ in the velocity field defined by (2.41) is given

$$\dot{x} = -\frac{\Gamma_i}{2\pi} \frac{y - y_i}{R_i^2}, \qquad \dot{y} = \frac{\Gamma_i}{2\pi} \frac{x - x_i}{R_i^2}, \qquad (1.110)$$

where $R_i^2 = (x - x_i)^2 + (y - y_i)^2$.

Let us first consider the motion of advected particles in a three-vortices (integrable) system in which $\lambda_E = 0$. In this case, the stream function for the advected particle is periodic in time and the expectation is that the advected particles may display chaotic behavior. The typical trajectories of passive particles which have initially been placed respectively in close proximity of a vortex center or in the background field between the vortices display a very different behavior. The particle seeded close to the vortex center displays a regular motion around the vortex and thus $\lambda_L = 0$; by contrast, the particle in the background field undergoes an irregular and aperiodic trajectory, and λ_L is positive.

We now discuss a case where the Eulerian flow is chaotic i.e., $N = 4$ point vortices [BLV01]. Let us consider again the trajectory of a passive particle deployed in proximity of a vortex center. As before, the particle rotates around the moving vortex. The vortex motion is chaotic; consequently, the particle position is unpredictable on large times as is the vortex position. Nevertheless, the Lagrangian Lyapunov exponent for this trajectory is zero (i.e., two initially close particles around the vortex remain close), even if the Eulerian Lyapunov exponent is positive, see Figure 1.49.

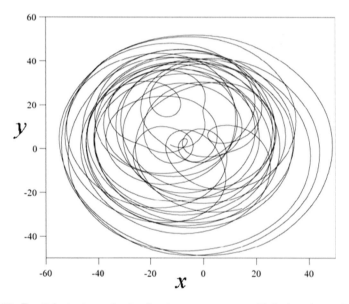

Fig. 1.49. Particle trajectories in the 4–vortex system. Eulerian dynamics in this case is chaotic. The left panel shows a regular Lagrangian trajectory while the right panel shows a chaotic Lagrangian trajectory. The different behavior of the two particles is due to different initial conditions (adapted from [BLV01]).

This result indicates once more that there is no strict link between Eulerian and Lagrangian chaoticity.

One may wonder whether a much more complex Eulerian flow, such as $2n$ turbulence, may give the same scenario for particle advection: i.e., regular trajectories close to the vortices and chaotic behavior between the vortices.

It has been shown that this is indeed the case [BBP94] and that the chaotic nature of the trajectories of advected particles is not strictly determined by the complex time evolution of the turbulent flow.

We have seen that there is no general relation between Lagrangian and Eulerian chaos. In the typical situation Lagrangian chaos may appear also for regular velocity fields. However, it is also possible to have the opposite situation, with $\lambda_L = 0$ in presence of Eulerian chaos, as in the example of Lagrangian motion inside vortex structures. As an important consequence of this discussion we remark that it is not possible to separate Lagrangian and Eulerian properties in a measured trajectory. Indeed, using the standard methods for data analysis [GP83a], from Lagrangian trajectories one extracts the total Lyapunov exponent λ_T and not λ_L or λ_E.

1.6 Standard Map and Hamiltonian Chaos

Chaos as defined in the theory of dynamical systems has been recognized in recent years as a common structural feature of phase–space flows. Its characterization involves geometric, dynamic, and symbolic aspects. Concepts developed in the framework of Poincaré–Birkhoff and Kolmogorov–Arnold–Moser theories describe the transition from regular to chaotic behavior in simple cases, or locally in phase–space. Their global application to real systems tends to interfere with non–trivial bifurcation schemes, and computer experimentation becomes an essential tool for comprehensive insight [Ric01, RSW90].

The prototypic example for *Hamiltonian chaos* is the standard map M, introduced by the plasma physicist B.V. Chirikov [Chi79]. It is an area–preserving map of a cylinder (or annulus) with coordinates (φ, r) onto itself,

$$M : \begin{pmatrix} \varphi \\ r \end{pmatrix} \mapsto \begin{pmatrix} \varphi' \\ r' \end{pmatrix} = \begin{pmatrix} \varphi + 2\pi(1 - r') \\ r + \mu \sin \varphi \end{pmatrix},$$

where φ is understood to be taken modulo 2π. The map has been celebrated for catching the essence of chaos in conservative systems, at least locally in the neighborhood of resonances. The number μ plays the role of a perturbation parameter. For $\mu = 0$, the map is integrable: all lines $r = const$ are invariant sets, with angular increment $\Delta\varphi = 2\pi(1 - r) = 2\pi W$. For obvious reasons, W is called *winding number*. For integer r, it is 0 mod 1; each point of the line is a fixed–point (see Figure 1.50). For rational $r = p/q$ with p, q coprime, the line carries a continuum of periodic orbits with period q, winding around the cylinder p/q times per period. However, when r is irrational, the orbit of an initial point never returns to it but fills the line densely. This makes for an important qualitative difference of rational and irrational invariant lines. The dependence of W on r is called the *twist* τ of the (un–perturbed) map. Here, $\tau = W/r = -1$. This has an intuitive interpretation: between any two integer values of r the map is a twist of the cylinder by one full turn. The behavior of such *twist maps* under small perturbations μ was the central issue of the stability discussion of which Poincaré's 'Méthodes Nouvelles' in 1892 and the discovery of the Kolmogorov–Arnold–Moser (KAM) theorem (see [AA68]) were the major highlights.

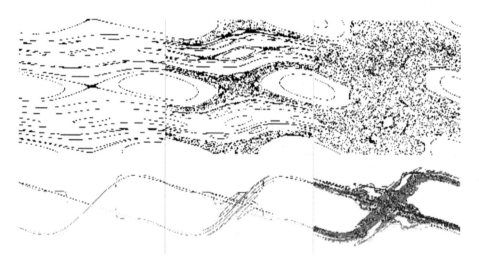

Fig. 1.50. Standard Map: Upper left: 20 trajectories with $\mu = 0.08$; the angle $\varphi \in [0, 2\pi]$ is plotted along the horizontal, the variable $r \in [0, 2]$ – along the vertical axis. Upper middle: the same for $\mu = 0.15$. Upper right: the same for $\mu = 0.25$. Lower left: beginning of the stable manifold (going from lower–left to upper–right) and unstable manifold (going from upper–left to lower–right) of the hyperbolic point $(\varphi, r) = 9\pi, 1)$ for $\mu = 0.25$. Lower middle: the same continued for two more iterations. Lower right: the same for additional 9 iterations (adapted from [Ric01]).

Consider the upper part of Figure 1.50. For perturbations $\mu = 0.08, 0.15, 0.25$, it shows 20 orbits generated from (almost) randomly chosen initial points. Three kinds of structures can be discerned: (i) islands, or chains of islands, surrounding elliptic periodic orbits; (ii) clouds of points centered at hyperbolic periodic orbits and surrounding the islands; (iii) the two left pictures contain invariant lines extending from left to right, i.e., surrounding the cylinder, whereas the right picture exhibits 'global chaos' in the sense that a single trajectory can wander along the vertical cylinder axis. These features illustrate three fundamental theorems that govern the *behavior of invariant sets under small perturbations*:

- *Poincaré–Birkhoff Theorem*: Poincaré's last theorem, proved by Birkhoff in 1913, states that of the uncountable infinity of periodic orbits on a given rational invariant set, only a finite number survives, half of them elliptic, the other half hyperbolic. More precisely, lines with rational winding number p/q decay into chains of q islands (sometimes, for reasons of symmetry, a finite multiple of q). Their centers are elliptic resonances E^k, $(k = 1, ..., q)$, i.e., fixed–points of the qth iterate M^q of the map. The Jacobian of M^q at E^k has eigenvalues of the form $\cos\rho \pm \mathrm{i}\sin\rho$, where $\cos\rho = \frac{1}{2}\mathrm{trace}\,M^q$ and $|\mathrm{trace}\,M^q| < 2$. The number ρ is a winding number of higher order; when points on small islands are parametrized by an angle, ρ is the angular increment between a point and its qth iterate. The islands of a given chain are separated by hyperbolic fixed–points H^k of M^q, $(k = 1, ..., q)$; the Jacobian at H^k has eigenvalues λ_u and λ_s with $|\lambda_u| > 1$ and $\lambda_s = 1/\lambda_u$. The corresponding eigenvectors $\mathbf{e}_u$ and $\mathbf{e}_s$ are, respectively, the *unstable and stable directions* of the hyperbolic orbits. Poincaré and Birkhoff found that for arbitrarily small perturbations, this breakup of rational lines is generic.

- *Kolmogorov–Arnold–Moser (KAM) Theorem*: It took some 70 years after Poincaré before the survival of invariant lines under small perturbations could be established with mathematical certainty [AA68].[34] For systems with n degrees of freedom and sufficiently small and smooth (but otherwise

---

[34] KAM–theorem is a result in dynamical systems about the persistence of quasi–periodic motions under small perturbations. The theorem partly resolves the small-divisor problem that arises in the perturbation theory of classical mechanics. The problem is whether or not a small perturbation of a conservative dynamical system results in a lasting quasi–periodic orbit. The original breakthrough to this problem was given by Kolmogorov in 1954. This was rigorously proved and extended by Arnold (in 1963 for analytic Hamiltonian systems) and Moser (in 1962 for smooth twist maps), and the general result is known as the KAM–theorem. The KAM–theorem, as it was originally stated, could not be applied to the motions of the solar system, although Arnold used the methods of KAM to prove the stability of elliptical orbits in the planar 3–body problem. The KAM–theorem is usually stated in terms of trajectories in phase–space of an integrable Hamiltonian system. The motion of an integrable system is confined to a doughnut–shaped surface, an *invariant torus*. Different initial conditions of the integrable Hamiltonian system will trace different invariant tori in phase–space. Plotting any of the coordinates of an integrable system would show that they are quasi–periodic. The KAM–theorem states that if the system is subjected to a weak nonlinear perturbation, some of the invariant tori are deformed and others are destroyed. The ones that survive are those that have 'sufficiently irrational' frequencies (this is known as the non–resonance condition). This implies that the motion continues to be quasi–periodic, with the independent periods changed (as a consequence of the non–degeneracy condition). The KAM–theorem specifies quantitatively what level of perturbation can be applied for this to be true. An important consequence of the KAM–theorem is that for a large set of initial conditions the motion remains perpetually quasi–periodic. The methods introduced by Kolmogorov, Arnold, and Moser have developed into a large body of results related to quasi–periodic motions. Notably, it has been extended to non–Hamiltonian systems (starting with Moser), to non–perturbative situations (as in the work of Michael Herman) and to systems with fast and slow frequencies. The non–resonance and non–degeneracy conditions of the KAM–theorem become increasingly difficult to satisfy for systems with more degrees of freedom. As the number of dimensions of the system

arbitrary) perturbations, the KAM–theorem guarantees the existence of fD *Liouville tori* in the $(2f-1)$D *energy surface*. For $f=2$, this is an extremely important stability result because orbits between two invariant tori are confined: if there is chaos, it cannot be global. For larger f, the situation is more delicate and still not fully explored.

The KAM–theorem also gives a hint as to which irrational tori are the most robust. The point is that one may distinguish various degrees of irrationality. A number W is called 'more irrational' than a number W' if rational approximations p/q with q smaller than a given $q_{\max}$ tend to be worse for W than for W':

$$\min_{p,q}\left|W-\frac{p}{q}\right| > \min_{p,q}\left|W'-\frac{p}{q}\right|.$$

The numbers W which qualify in the KAM–theorem for being sufficiently irrational, are characterized by constants $c>0$ and $\nu \geq 2$ in an estimate $\left|W-\frac{p}{q}\right| \geq \frac{c}{q^\nu}$, for arbitrary integers p and q. The set of W for which c and ν exist has positive measure. ν cannot be smaller than 2 because the *Liouville theorem* asserts that if p_k/q_k is a continued fraction approximation to W, then there exists a number C (independent of k) such that $|W - p_k/q_k| < C/q_k^2$. On the other hand, it is known that quadratic irrationals, i.e., solutions of quadratic equations with integer coefficients, have periodic continued fraction expansions which implies that the corresponding approximations behave asymptotically as [RSW90]

$$\left|W-\frac{p_k}{q_k}\right| = \frac{c_1}{q_k^2} + (-1)^k \frac{c_2}{q_k^4} + O(q_k^{-6}).$$

It is also known that there are no better rational approximations to a given W than its continued fraction approximations. In this sense, the quadratic irrationals are the 'most irrational' numbers because they have the smallest possible $\nu = 2$. Furthermore, among the quadratic irrationals, the so–called *noble numbers* have the largest possible $c_1 = 1/\sqrt{5}$; they are defined by continued fractions

$$W = w_0 + \cfrac{1}{w_1 + \cfrac{1}{w_2 + \cfrac{1}{w_3 + \ldots}}} = [w_0, w_1, w_2, w_3, \ldots],$$

where $w_k = 1$ for k greater than some K. Hence, invariant lines (or tori) with noble winding numbers tend to be the most robust. And finally, within the class of noble numbers, the so–called *golden number* $g = [0,1,1,1,\ldots] = \frac{1}{2}(\sqrt{5}-1) = 0.618\,034$, as well as $1-g = g^2$ and

---

increase the volume occupied by the tori decreases. The *KAM–tori* that are not destroyed by perturbation become invariant Cantor sets, termed *Cantori*.

1.6 Standard Map and Hamiltonian Chaos

$1 + g = 1/g$ are the 'champions of irrationality' because they have the smallest possible $c_2 = 1/(5\sqrt{5})$. More generally, the most irrational noble number between neighboring resonances p_1/q_1 and p_2/q_2 (for example, 2/5 and 1/2) is obtained with the so-called *Farey construction* [Sch91] $p_{n+1}/q_{n+1} = (p_{n-1} + p_n)/(q_{n-1} + q_n)$.

It has been found in numerous studies based on the discovery of the KAM–theorem, that when a Hamiltonian system can in some way be described as a perturbed twist map, the following scenario holds with growing perturbation. First, many small resonances (chains of q islands) and many irrational tori (invariant lines) coexist. Then the islands with small q grow; the chaos bands associated with their hyperbolic orbits swallow islands with large q and the less irrational tori. At one point, only one noble *KAM–torus* and few conspicuous resonances survive, before with further increasing perturbation the last torus decays and gives way to global chaos. More detailed analysis shows that a decaying KAM–torus leaves traces in the chaotic sea: invariant Cantor sets which act as 'permeable barriers' for the phase–space flow [Ric01, RSW90].

- *Smale–Zehnder Theorem*: E. Zehnder (1973) proved that each hyperbolic point H^k of the Poincaré–Birkhoff scenario is the center of a chaotic band, in the sense that it contains a *Smale horseshoe* (see Figure 1.51). To see this, consider a point P close to H^k on its *unstable eigen–direction* $\mathbf{e}_u$ and let P' be its image under the standard map M^q. Now iterate the line PP' with M^q. Step by step, this generates the *unstable manifold* of H^k, shown in the lower part of Figure 1.51: starting at the center, the two parts evolve towards the lower right and the upper left. It is seen from the succession of the three pictures that the manifold develops a folded structure of increasing complexity as the iteration goes on.[35] Eventually it densely fills a region of full measure. In the last picture shown here, the barriers associated with the former golden KAM–torus have not yet been penetrated, but it is believed that the closure of the unstable manifold is the entire chaos band connected to H^k.

Similarly, the *stable manifold* can be generated by backward iteration along the *stable eigen–direction* $\mathbf{e}_s$. The decisive point in Zehnder's theorem is the demonstration that the stable and unstable manifolds have transverse intersections, called homoclinic points (or heteroclinic if the two manifolds belong to different hyperbolic orbits, but this difference is not important). Once this is established, it may be concluded that part of the map has the character of a horseshoe map, hence it contains an invariant set on which the dynamics is chaotic. Two strips in the neighborhood of the central hyperbolic point, one red the other green, each stretched along a folded piece of the stable manifold and crossing the unstable manifold at a homoclinic point, are mapped a number of times and returned near the hyperbolic point in a transverse orientation. In that process they are contracted to-

---

[35] The manifold may not intersect itself because the map is injective.

wards and expanded along the unstable manifold. The intersection pattern of the strips contains a thin invariant Cantor set on which an iterate of the standard map is conjugate to the shift map on bi–infinite sequences of two symbols. This implies the existence of chaos. Zehnder showed that

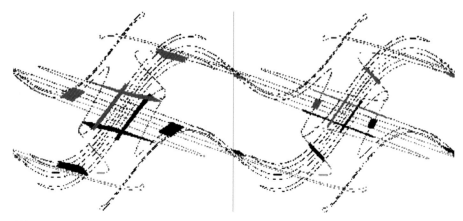

Fig. 1.51. Smale horseshoe in the standard map for $\mu = 0.4$; the angle $\varphi \in [0, 2\pi]$ is plotted along the horizontal, the variable $r \in [0, 2]$ – along the vertical axis. Left: the intersection of the two pairs of strips near the center contains an invariant set for M^3. Right: the same for M^4 (adapted from [Ric01]).

this situation is generic even for arbitrarily small perturbations μ, where it is impossible to give a graphical illustration because the homoclinic tangle is exponentially thin, and the number of iterations needed to return to the hyperbolic point very large. The message for our subsequent discussion of the three body problem is that when rational invariant sets of a twist map are perturbed, they break up into elliptic centers of order and hyperbolic centers of chaos. The invariant sets constructed on the basis of Figure 1.51 are of course not the whole story for the chaos band connected with the central hyperbolic point. We have shown the two perhaps most obvious horseshoes for this case, but an infinity of other horseshoes may be identified in the same chaotic region, using the same strategy with other homoclinic intersections. Each of them has a fractal dimension not much larger than 1, whereas the entire chaos band – the closure of the stable or unstable manifolds – is believed to have dimension 2. This has not yet been proven. Moreover, the computation of Lyapunov exponents is easy for a single horseshoe, but averaging them over the chaotic region is numerically delicate, analytically hopeless. So there remain important open questions [Ric01, RSW90].

1.7 Chaotic Dynamics of Binary Systems

Recall that *binary systems*, like cellular automata and neural networks, are described, in general, by a set of N binary variables $S_i = 0, 1$, $(i = 1, \ldots, N)$, or in short $\mathbf{S}$, that evolve according to dynamical rules. The natural metric for these systems is the *Hamming distance*

$$d_H(S - S') \equiv \sum_{i=1}^{N} |S_i - S'_i|.$$

The space $\{\mathbf{S}\}$ has 2^N possible states and so the topology constructed from d_H is discrete. Generally one is interested in studying these dynamical systems in the limit $N \to \infty$ since that is where interesting statistical properties appear, such as phase transitions, and it is possible to use powerful techniques like *mean–field theory*. Furthermore, numerical simulations which need to be done for finite, but large N, are understood as approximations of a system with infinite variables, much in the same way as floating point variables in computers are finite approximations of real numbers which generally have an infinite number of digits. Nevertheless, for $N \to \infty$, d_H is no longer a distance and the topology is ill defined in that limit. That makes our understanding of binary systems quite different from that of dynamical systems in $\mathbb{R}^d$ or in differentiable manifolds where one works with the usual topology of the real numbers. Here we will overcome this situation by extending the phase–space $\{\mathbf{S}\}$ to have an infinite number of states while preserving the equal status that the Hamming distance confers to each of the variables. That is to say, all the variables S_i give the same contribution to the distance for any i [WZ98].

Let us consider the Cartesian product of infinite copies of $\{\mathbf{S}\}$ and call this space Ω. We denote the elements of Ω by

$$S = (\mathbf{S}(0), \mathbf{S}(1), \mathbf{S}(2), \ldots).$$

We make Ω a topological space by introducing the following base:

$$\mathcal{N}_n(S) = \{S' \in \Omega | \mathbf{S}'(m) = \mathbf{S}(m), \text{ for all } m < n\}, \tag{1.111}$$

with $n = 1, 2, \ldots$. These base sets are closely related to the cylinders in one–sided shift spaces and Ω is homeomorphic to the space of symbols of the *symbolic dynamics* with 2^N symbols (see, e.g., [Wig90]). It follows that Ω is a cantor set. In symbolic dynamics the topology is usually derived from the metric

$$d(S, S') = \sum_{n=0}^{\infty} \frac{1}{2^n} d_n(S - S'), \quad \text{where} \tag{1.112}$$

$$d_n(S - S') \equiv \sum_{i=1}^{N} |S_i(n) - S'_i(n)| \tag{1.113}$$

is the *Hamming distance* of the nth copy of $\{\mathbf{S}\}$. One can check that if $\mathbf{S}(m) = \mathbf{S}'(m)$ for all $m < n$, then $d(S, S') < \frac{N+1}{2^{n-1}}$, so that (1.111) and (1.112) define the same topology.

Here and in the following our purpose is to study dynamical systems in Ω generated by a function $F : \Omega \longrightarrow \Omega$. This function may be continuous or discontinuous, unless explicitly stated below. Allowing discontinuous functions in principle opens the door to a richer variety of systems, which include neural networks and cellular automata.

Following [WZ98], we begin by generalizing in a natural way the definitions of chaos in subsets of $\mathbb{R}^N$ to Ω.

F has *sensitive dependence on initial conditions* on $\mathcal{A} \subset \Omega$ if there is $n \in \mathbb{N}$ for all $S \in \mathcal{A}$ and for all $\mathcal{N}_m(S)$ there is $S' \in \mathcal{N}_m(S) \cap \mathcal{A}$ and $k \in \mathbb{N}$ such that $F^k(S') \notin \mathcal{N}_n(F^k(S))$.

Let $\mathcal{A} \subset \Omega$ be a closed invariant set. Function $F : \Omega \longrightarrow \Omega$ is *topologically transitive* on $\mathcal{A} \subset \Omega$ if for any open sets $U, V \subset \mathcal{A}$ there is $n \in \mathbb{Z} \ni F^n(U) \cap V \neq \emptyset$. In the last expression, if F is non invertible we understand $F^{-k}(U)$ with $k > 0$, as the set of all points $S \in \Omega$ such that $F^k(S) \in U$.

Let $\mathcal{A} \subset \Omega$ be a compact set. Function $F : \mathcal{A} \longrightarrow \mathcal{A}$ is *chaotic* on $\mathcal{A}$ if F has sensitive dependence on initial conditions and is topologically transitive on $\mathcal{A}$.

A closed subset $\mathcal{M} \subset \Omega$ is called a *trapping region* if $F(\mathcal{M}) \subset \mathcal{M}$.

If F is a continuous function in Ω, $F^n(\mathcal{M})$ is compact and closed for all $n \in \mathbb{N}$. Since every closed subset of a compact set is compact, it follows that $\mathcal{M}$ is compact and since F is continuous $F^n(\mathcal{M})$ is compact. Since Ω is Hausdorff every compact subset of it is closed, so $F^n(\mathcal{M})$ is closed (see, e.g., [Mun75]).

The map $F : \Omega \longrightarrow \Omega$ has an *attractor* if it admits an asymptotically stable transitive set, i.e., if there exists a trapping region $\mathcal{M}$ such that

$$\Lambda \equiv \bigcap_{n \geq 0} F^n(\mathcal{M})$$

and F is topologically transitive on Λ.[36]

If F is a continuous function in Ω, Λ is compact and closed. If F is continuous, Λ is an intersection of closed sets, so it is closed. Since every closed subset of a compact space Ω is compact, it follows that Λ is compact.

Λ is called a *chaotic attractor* if function F is chaotic on Λ.

Let F be a continuous function in Ω, if Λ is a chaotic attractor then it is perfect. As Λ is closed, it remains to prove that every point in Λ is an accumulation point of Λ. By contradiction, let $S_0 \in \Lambda$ be an isolated point, then there exists $n \in \mathbb{N} \ni \mathcal{N}_n(S_0) \cap \Lambda = \{S_0\}$. Then, by topological transitivity Λ has an isolated orbit (the orbit of S_0) which implies that it is not sensitive to initial conditions on Λ.

---

[36] The trapping region is defined in the Ω space while in the theory of dynamical systems on manifolds, it is defined on the manifold

If F is a continuous function in Ω, and Λ is a chaotic attractor then it is a *Cantor set*.

1.7.1 Examples of Dynamical Maps

Here we consider some examples of dynamical maps $f : \Omega \longrightarrow \Omega$. The first one is the one–side shift map σ of symbolic dynamics which we introduce to familiarize the reader with the notation.

One–Sided Shift Map σ

The continuous map σ defined by

$$\sigma\left(\mathbf{S}\left(0\right), \mathbf{S}\left(1\right), ...\right) = \left(\mathbf{S}\left(1\right), \mathbf{S}\left(2\right), ...\right),$$

is chaotic in Ω (see [Dev89, KH95, Rob83]).

Note that σ is non–invertible and its action loses the information carried by the binary state $\mathbf{S}(0)$. The meaning and usefulness of this map is quite clear in the context of symbolic dynamics when the Conley–Moser conditions are satisfied [Mos73]. There one studies, in general, a non–invertible function $f : \Xi \longrightarrow \Xi$ where Ξ is a Cantor set embedded in $\mathbb{R}^N$. The set Ξ is divided in 2^N sectors I_α $\alpha = 0, 1, ..., 2^N$. Then it is possible to establish a topological conjugation between f and σ through a homeomorphism ψ, so that the following diagram commutes [Wig90]:

$$\begin{array}{ccc} \Xi & \stackrel{f}{\longrightarrow} & \Xi \\ \psi \downarrow & & \downarrow \psi \\ \Omega & \stackrel{\sigma}{\longrightarrow} & \Omega \end{array}$$

Moreover, let $S = \psi(x)$, then $\mathbf{S}(n)$ is the binary decomposition of the label α, such that $f^n(x) \in I_\alpha$.

Chaotic Maps with Non–Trivial Attractors in Ω

The shift map can be modified to create maps which are homeomorphic to the shift map on an asymptotically stable transitive subset of the space of symbols. In the following, we introduce two very simple examples. Firstly, take the space of symbols Ω with $N = 2$, homeomorphic to $\Xi \times \Xi$ where Ξ is the space of symbols with $N = 1$, that is the space of semi–infinite sequences $S = (S_0, S_1, S_2, ...)$. Then consider the function $f_c : \Xi \times \Xi \to \Xi \times \Xi$ given by $f_c = \sigma \times \zeta$. Where σ is the usual shift function and ζ is a right inverse of the shift function defined as follows [WZ98]:

$$\zeta(S_0, S_1, S_2, ...) = (0, S_0, S_1, S_2, ...).$$

It is easy to check that ζ is a continuous function, and of course so is the shift: so f_c is continuous. The set $\Xi \times \{0\}$ is an asymptotically stable transitive set, on which the restriction of f_c is the shift map σ.

As another example, consider the space Ω with $N = 1$. It can be split into the *disjoint union* of two Cantor sets $\Omega = \Lambda_0 \uplus \Lambda_1$. Where Λ_0 is the set of sequences such that $S_0 = 0$ and an analogous fashion for Λ_1. Take the continuous function $f_\pi = \pi \circ \sigma$, where σ is the shift map and π projects Ω in Λ_0 such that:

$$\pi(S_0, S_1, S_2, ...) = (0, S_1, S_2, ...).$$

Then the action of f_π is given by,

$$f_\pi(S_0, S_1, S_2, ...) = (0, S_2, S_3, ...).$$

It is easy to check that Λ_0 is a chaotic attractor of f_π.

Chaotic Maps in Ω Induced Through Chaotic Maps in Cantor Subsets of $\mathbb{R}^N$

Here, we consider a homeomorphism which relates a Cantor set $\chi \subset \mathbb{R}^N$ to the space Ω and allows one to construct chaotic maps in Ω from chaotic maps in χ through topological conjugation. Let $\chi \subset \mathbb{R}^N$ be the Cantor set that results from taking the Cartesian product of N Cantor sets χ_i;

$$\chi = \bigotimes_{i=1}^{N} \chi_i,$$

where the ith component χ_i is constructed by suppressing from the interval $[0, 1]$ the open middle $1/a_i$ part, $i = 1, \ldots, N$, $a_i > 1$, and repeating this procedure iteratively with the sub–intervals. Now, we define a map $\phi : \Omega \longrightarrow \chi$ by [WZ98]:

$$\phi_i(S) = \sum_{n=1}^{\infty} (l_{n-1} - l_n) S_i(n-1), \quad \text{where}$$

$$l_n = \frac{1}{2^n}\left(1 - \frac{1}{a_i}\right)^n,$$

is the length of each of the remaining 2^n intervals at the nth step of the construction of χ_i. If Ω is endowed with the metric (1.112) and $\chi \subset \mathbb{R}^N$ with the standard Euclidean metric, is easy to show that ϕ is a homeomorphism.

Now, if we have a map $f : \mathbb{R}^N \longrightarrow \mathbb{R}^N$ which is chaotic in χ we can construct a map $F : \Omega \longrightarrow \Omega$ which is chaotic in Ω, and is defined through the commutation of the following diagram:

$$\begin{array}{ccc} \chi & \xrightarrow{f} & \chi \\ \phi \uparrow & & \uparrow \phi \\ \Omega & \xrightarrow{F} & \Omega \end{array}$$

This leads to an interesting practical application of the homeomorphism ϕ, to realize computer simulations of chaotic systems on Cantor sets. If, for example, one iterates the *logistic map*

$$f(x) = \mu x (1-x), \quad \text{for } \mu \geq 4,$$

with a floating-point variable, the truncation errors nudge the trajectory away from the Cantor set and eventually $x \to -\infty$. The homeomorphism ϕ suggests a natural solution to this, which is to iterate the truncated binary states rather than the floating–point variable. To iterate the dynamics, one computes $x_i = \phi_i(S)$ for all $i = 1, \ldots, N$ by assuming that the truncated bits are all equal to zero, then applies f to get $x' = f(x)$. Since x' generally does *not* belong to the Cantor set (because of truncation errors), in the process of constructing $S' = \phi^{-1}(x')$, at some n one will find that this point does not belong to either the interval corresponding to $S_i(n) = 0$ or to $S_i(n) = 1$. This truncation error can be corrected by moving to the extremity of the interval which lies closest to x'_i. In this way, truncation errors are not allowed to draw the trajectory away from the Cantor set $\chi \subset \mathbb{R}^N$.

Binary Systems with Memory

Now we will define a map $\Gamma : \Omega \longrightarrow \Omega$ which is very useful to analyze binary systems with causal deterministic dynamics on N bits, such as neural networks, cellular automata, and neural networks with memory (see, e.g., [SK86]). Let

$$\gamma_i : \Omega \longrightarrow \{0,1\}, \quad (i = 1, \ldots, N), \tag{1.114}$$

be a set of continuous or discontinuous functions. The map $\Gamma : \Omega \longrightarrow \Omega$ is then defined by:

$$\Gamma_i(S) = (\gamma_i(S), S_i(0), S_i(1), \ldots),$$
$$\text{or, short–hand,} \quad \Gamma(S) = (\gamma(S), S). \tag{1.115}$$

Such maps have the following properties.

The shift map (1.114) is a left inverse of Γ since from (1.115) $\sigma \circ \Gamma(S) = S$. If Ω has an attracting set $\Lambda \subset \Omega$, then σ is also a right inverse in the restriction of Γ to Λ, so that, $\Gamma|_\Lambda^{-1} = \sigma$. For all $S \in \Lambda$ there is $S' \in \Lambda$ such that $\Gamma(S') = S$. Since

$$\Gamma(S') = (\gamma(S'), S') = S, \quad \text{and} \quad S = (\mathbf{S}(0), S_1),$$

where $S_1 \equiv (\mathbf{S}(1), \mathbf{S}(2), \ldots)$, one sees that $S' = S_1$. Thus,

$$\Gamma \circ \sigma(S) = \Gamma(S_1) = \Gamma(S') = S.$$

Γ has an attracting set Λ contained properly in Ω. Given S there are 2^N states $S' = (\mathbf{S}'(0), S)$ of which only one, $\Gamma(S) = (\gamma(S), S)$, belongs to $\Gamma(\Omega)$. Therefore the set

$$\Lambda \equiv \bigcap_{n \geq 0} \Gamma^n(\Omega)$$

is a proper subset of Ω [WZ98].

Classical Chaos in Lorenz and Laser ODEs

Before we focus on the turbulent geometry of the Navier–Stokes PDEs (1.120), let us briefly review the Lorenz reduced system of nonlinear ODEs

$$\dot{x} = a(y - x), \qquad \dot{y} = bx - y - xz, \qquad \dot{z} = xy - cz, \tag{1.121}$$

where x, y and z are dynamical variables, constituting the 3D *phase–space* of the *Lorenz flow*; and a, b and c are the parameters of the system. Originally, Lorenz used this model to describe the unpredictable behavior of the weather, where x is the rate of convective overturning (convection is the process by which heat is transferred by a moving fluid), y is the horizontal temperature overturning, and z is the vertical temperature overturning; the parameters are: $a \equiv P$–proportional to the Prandtl number (ratio of the fluid viscosity of a substance to its thermal conductivity, usually set at 10), $b \equiv R$–proportional to the Rayleigh number (difference in temperature between the top and bottom of the system, usually set at 28), and $c \equiv K$–a number proportional to the physical proportions of the region under consideration (width to height ratio of the box which holds the system, usually set at 8/3). The Lorenz system (1.121) has the properties:

1. *symmetry*: $(x, y, z) \rightarrow (-x, -y, z)$ for all values of the parameters, and
2. the z–axis ($x = y = 0$) is *invariant* (i.e., all trajectories that start on it also end on it).

Nowadays, it is well–known that the Lorenz model is a paradigm for low–dimensional chaos in dynamical systems in synergetics and this model or its modifications are widely investigated in connection with modelling purposes in meteorology, hydrodynamics, laser physics, superconductivity, electronics, oil industry, chemical and biological kinetics, etc.

The 3D *phase–portrait* of the Lorenz system (1.52) shows the celebrated 'Lorenz mask', a special type of *strange attractor* (see Figure 1.52), depicting the famous *butterfly effect*, (i.e., sensitive dependence on initial conditions). The Lorenz mask has the following characteristics:

1. Trajectory does not intersect itself in three dimensions,

---

high *Res* and is dominated by inertial forces, producing random eddies, vortices and other flow fluctuations. The transition between laminar and turbulent flow is often indicated by a critical Reynolds number (Re_{crit}), which depends on the exact flow configuration and must be determined experimentally. Within a certain range around this point there is a region of gradual transition where the flow is neither fully laminar nor fully turbulent, and predictions of fluid behavior can be difficult.

1.8 Spatio–Temporal Chaos and Turbulence in PDEs

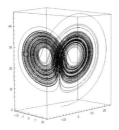

Fig. 1.52. The celebrated 'Lorenz–mask' strange attractor, obtained by simulating the equations (1.121) in $Mathematica^{TM}$.

2. Trajectory is not periodic or transient,
3. General form of the shape does not depend on initial conditions, and
4. Exact sequence of loops is very sensitive to the initial conditions.

In 1975, H. Haken showed in [Hak83, Hak93] that the Lorenz equations (1.52) were isomorphic to the *Maxwell–Haken laser equations*, that were the starting point for *Haken's synergetics*,

$$\dot{E} = \sigma(P - E), \qquad \dot{P} = \beta(ED - P), \qquad \dot{D} = \gamma(\sigma - 1 - D - \sigma EP),$$

Here, the variables in the Lorenz equations, namely x, y and z correspond to the slowly varying amplitudes of the electric field E and polarization P and the inversion D respectively in the Maxwell–Haken equations. The parameters are related via $c = \frac{\gamma}{\beta}$, $a = \frac{\sigma}{\beta}$ and $b = \sigma + 1$, where γ is the relaxation rate of the inversion, β is the relaxation rate of the polarization, σ is the field relaxation rate, and σ represents the normalized pump power.

Turbulent Flow

Recall that in fluid dynamics, *turbulent flow* is a flow regime characterized by low momentum diffusion, high momentum convection, and rapid variation of pressure and velocity in space and time. Flow that is not turbulent is called *laminar flow*. Also, recall that the *Reynolds number Re* characterizes whether flow conditions lead to laminar or turbulent flow. The structure of turbulent flow was first described by A. Kolmogorov. Consider the flow of water over a simple smooth object, such as a sphere. At very low speeds the flow is laminar, i.e., the flow is locally smooth (though it may involve vortices on a large scale). As the speed increases, at some point the transition is made to turbulent (or, chaotic) flow. In turbulent flow, unsteady vortices[39] appear on many scales and interact with each other. Drag due to boundary

---

[39] Recall that a *vortex* can be any circular or rotary flow that possesses vorticity. Vortex represents a spiral whirling motion (i.e., a spinning turbulent flow) with closed streamlines. The shape of media or mass rotating rapidly around a center forms a vortex. It is a flow involving rotation about an arbitrary axis.

layer skin friction increases. The structure and location of boundary layer separation often changes, sometimes resulting in a reduction of overall drag. Because laminar–turbulent transition is governed by Reynolds number, the same transition occurs if the size of the object is gradually increased, or the viscosity of the fluid is decreased, or if the density of the fluid is increased.

Vorticity Dynamics

Vorticity $\omega = \omega(x^i, t)$, $(i = 1, 2, 3)$ is a geometrical concept used in fluid dynamics, which is related to the amount of 'circulation' or 'rotation' in a fluid. More precisely, *vorticity* is the circulation per unit area at a point in the flow field, or formally, $\omega = \nabla \times \mathbf{u}$, where $\mathbf{u} = \mathbf{u}(x^i, t)$ is the fluid velocity. It is a vector quantity, whose direction is (roughly speaking) along the axis of the swirl. The movement of a fluid can be said to be vortical if the fluid moves around in a circle, or in a helix, or if it tends to spin around some axis. Such motion can also be called *solenoidal*. In the atmospheric sciences, vorticity is a property that characterizes large–scale rotation of air masses. Since the atmospheric circulation is nearly horizontal, the 3D vorticity is nearly vertical, and it is common to use the vertical component as a scalar vorticity.

A vortex can be seen in the spiraling motion of air or liquid around a center of rotation. Circular current of water of conflicting tides form vortex shapes. Turbulent flow makes many vortices. A good example of a vortex is the atmospheric phenomenon of a whirlwind or a *tornado*. This whirling air mass mostly takes the form of a helix, column, or spiral. Tornadoes develop from severe thunderstorms, usually spawned from squall lines and *supercell thunderstorms*, though they sometimes happen as a result of a *hurricane*.[40] Another example is a meso-vortex on the scale of a few miles (smaller than a hurricane but larger than a tornado). On a much smaller scale, a vortex is usually formed as water goes down a drain, as in a sink or a toilet. This occurs in water as the revolving mass forms a whirlpool.[41] This whirlpool is caused by water flowing out of a small opening in the bottom of a basin or reservoir. This swirling flow structure within a region of fluid flow opens downward from the water surface. In the hydrodynamic interpretation of the behavior of electromagnetic fields, the acceleration of electric fluid in a particular direction creates a positive vortex of magnetic fluid. This in turn creates around itself a corresponding negative vortex of electric fluid.

---

[40] Recall that a hurricane is a much larger, swirling body of clouds produced by evaporating warm ocean water and influenced by the Earth's rotation. In particular, polar vortex is a persistent, large–scale cyclone centered near the Earth's poles, in the middle and upper troposphere and the stratosphere. Similar, but far greater, vortices are also seen on other planets, such as the permanent Great Red Spot on Jupiter and the intermittent Great Dark Spot on Neptune.

[41] Recall that a whirlpool is a swirling body of water produced by ocean tides or by a hole underneath the vortex, where water drains out, as in a bathtub.

1.8 Spatio–Temporal Chaos and Turbulence in PDEs

Dynamical Similarity and Eddies

In order for two flows to be similar they must have the same geometry and equal Reynolds numbers. When comparing fluid behavior at homologous points in a model and a full–scale flow, we have $Re^* = Re$, where quantities marked with * concern the flow around the model and the other the real flow. This allows us to perform experiments with reduced models in water channels or wind tunnels, and correlate the data to the real flows. Note that true dynamic similarity may require matching other dimensionless numbers as well, such as the Mach number used in compressible flows, or the Froude number that governs free-surface flows.

In a turbulent flow, there is a range of scales of the fluid motions, sometimes called *eddies*. A single packet of fluid moving with a bulk velocity is called an *eddy*. The size of the largest scales (eddies) are set by the overall geometry of the flow. For instance, in an industrial smoke–stack, the largest scales of fluid motion are as big as the diameter of the stack itself. The size of the smallest scales is set by Re. As Re increases, smaller and smaller scales of the flow are visible. In the smoke–stack, the smoke may appear to have many very small bumps or eddies, in addition to large bulky eddies. In this sense, Re is an indicator of the range of scales in the flow. The higher the Reynolds number, the greater the range of scales.

In their first edition of Fluid Mechanics [LL59], Landau and Lifschitz proposed a *route to turbulence* in spatio–temporal fluid systems. Since then, much work, in dynamical systems, experimental fluid dynamics, and many other fields has been done concerning the routes to turbulence. Ever since the discovery of chaos in low–dimensional systems, researchers have been trying to use the concept of chaos to understand turbulence [RT71]. recall that there are two types of fluid motions: laminar flows and turbulent flows. Laminar flows look regular, and turbulent flows are non–laminar and look irregular. Chaos is more precise, for example, in terms of the so–called *Bernoulli shift dynamics*. On the other hand, even in low–dimensional systems, there are solutions which look irregular for a while, and then look regular again. Such a dynamics is often called a *transient chaos*.

Low–dimensional chaos is the starting point of a long journey toward understanding turbulence. To have a better connection between chaos and turbulence, one has to study chaos in PDEs [Li04].

1.8.2 Sine–Gordon Equation

Consider the simple perturbed *sine-Gordon equation* [Li04c]

$$u_{tt} = c^2 u_{xx} + \sin u + \epsilon[-au_t + \cos t \; \sin^3 u], \qquad (1.122)$$

subject to periodic boundary condition

$$u(t, x + 2\pi) = u(t, x) \; ,$$

as well as even or odd constraint,

$$u(t, -x) = u(t, x), \quad \text{or} \quad u(t, -x) = -u(t, x),$$

where u is a real–valued function of two real variables (t, x), c is a real constant, $\epsilon \geq 0$ is a small perturbation parameter, and $a > 0$ is an external parameter. One can view (1.122) as a *flow* (u, u_t) defined in the phase–space manifold $M \equiv H^1 \times L^2$, where H^1 and L^2 are the Sobolev spaces on $[0, 2\pi]$. A point in the phase–space manifold M corresponds to two profiles, $(u(x), u_t(x))$. [Li04c] has proved that there exists a homoclinic orbit $(u, u_t) = h(t, x)$ asymptotic to $(u, u_t) = (0, 0)$. Let us define two orbits segments

$$\eta_0 : (u, u_t) = (0, 0), \quad \text{and} \quad \eta_1 : (u, u_t) = h(t, x), \quad (t \in [-T, T]).$$

When T is large enough, η_1 is almost the entire homoclinic orbit (chopped off in a small neighborhood of $(u, u_t) = (0, 0)$). To any binary sequence

$$a = \{\cdots a_{-2} a_{-1} a_0, a_1 a_2 \cdots\}, \quad (a_k \in \{0, 1\}), \quad (1.123)$$

one can associate a pseudo–orbit

$$\eta_a = \{\cdots \eta_{a_{-2}} \eta_{a_{-1}} \eta_{a_0}, \eta_{a_1} \eta_{a_2} \cdots\}.$$

The pseudo–orbit η_a is not a true orbit but rather 'almost an orbit'. One can prove that for any such pseudo–orbit η_a, there is a unique true orbit in its neighborhood [Li04c]. Therefore, each binary sequence labels a true orbit. All these true orbits together form a chaos. In order to talk about sensitive dependence on initial data, one can introduce the *product topology* by defining the neighborhood basis of a binary sequence

$$a^* = \{\cdots a^*_{-2} a^*_{-1} a^*_0, a^*_1 a^*_2 \cdots\} \quad \text{as} \quad \Omega_N = \{a : \ a_n = a^*_n, \ |n| \leq N\}.$$

The Bernoulli shift on the binary sequence (1.123) moves the comma one step to the right. Two binary sequences in the neighborhood Ω_N will be of order Ω_1 away after N iterations of the Bernoulli shift. Since the binary sequences label the orbits, the orbits will exhibit the same feature. In fact, the Bernoulli shift is topologically conjugate to the perturbed sine–Gordon flow.

Replacing a homoclinic orbit by its fattened version – a homoclinic tube, or by a heteroclinic cycle, or by a heteroclinically tubular cycle; one can still obtain the same Bernoulli shift dynamics. Also, adding diffusive perturbation $\epsilon b u_{txx}$ to (1.122), one can still prove the existence of homoclinics or heteroclinics, but the Bernoulli shift result has not been established [Li04c].

1.8.3 Complex Ginzburg–Landau Equation

Consider the complex–valued *Ginzburg–Landau equation* [Li04a, Li04b],

$$iq_t = q_{xx} + 2[|q|^2 - \omega^2] + i\epsilon[q_{xx} - \alpha q + \beta], \quad (1.124)$$

which is subject to periodic boundary condition and even constraint

$$q(t, x + 2\pi) = q(t, x) , \quad q(t, -x) = q(t, x) ,$$

where q is a complex-valued function of two real variables (t, x), (ω, α, β) are positive constants, and $\epsilon \geq 0$ is a small perturbation parameter. In this case, one can prove the existence of homoclinic orbits [Li04a]. But the Bernoulli shift dynamics was established under generic assumptions [Li04b].

A real fluid example is the amplitude equation of Faraday water wave, which is also a complex Ginzburg–Landau equation [Li04d],

$$iq_t = q_{xx} + 2[|q|^2 - \omega^2] + i\epsilon[q_{xx} - \alpha q + \beta \bar{q}] , \quad (1.125)$$

subject to the same boundary condition as (1.124). For the first time, one can prove the existence of homoclinic orbits for a water wave equation (1.125). The Bernoulli shift dynamics was also established under generic assumptions [Li04d]. That is, one can prove the existence of chaos in water waves under generic assumptions.

The nature of the complex Ginzburg–Landau equation is a parabolic equation which is near a hyperbolic equation. The same is true for the perturbed sine–Gordon equation with the diffusive term $\epsilon b u_{t xx}$ added. They contain effects of diffusion, dispersion, and nonlinearity. The *Navier–Stokes equations* 1.120 are diffusion–advection equations. The advective term is missing from the perturbed sine–Gordon equation and the complex Ginzburg–Landau equation. Turbulence happens when the diffusion is weak, i.e., in the near hyperbolic regime. One should hope that turbulence should share some of the features of chaos in the perturbed sine–Gordon equation. There is a popular myth that turbulence is fundamentally different from chaos because turbulence contains many unstable modes. In both the perturbed sine–Gordon equation and the complex Ginzburg–Landau equation, one can incorporate as many unstable modes as one likes, the resulting Bernoulli shift dynamics is still the same. On a computer, the solution with more unstable modes may look rougher, but it is still chaos [Li04].

In a word, dynamics of strongly nonlinear classical fields is 'turbulent', not 'laminar'. On the other hand, field theories such as 4-dimensional QCD or gravity have many dimensions, symmetries, tensorial indices. They are far too complicated for exploratory forays into this forbidding terrain. Instead, we consider a simple spatio–temporally chaotic nonlinear system of physical interest [CCP96].

1.8.4 Kuramoto–Sivashinsky System

One of the simplest and extensively studied spatially extended dynamical systems is the Kuramoto–Sivashinsky (KS) system [Kur76, Siv77]

$$u_t = (u^2)_x - u_{xx} - \nu u_{xxxx}, \quad (1.126)$$

which arises as an amplitude equation for interfacial instabilities in a variety of contexts. The so–called *flame front* $u(x,t)$ has compact support, with $x \in [0, 2\pi]$ a periodic space coordinate. The u^2 term makes this a nonlinear system, $t \geq 0$ is the time, and ν is a 4–order 'viscosity' damping parameter that irons out any sharp features. Numerical simulations demonstrate that as the viscosity decreases (or the size of the system increases), the *flame front* becomes increasingly unstable and turbulent. The task of the theory is to describe this spatio-temporal turbulence and yield quantitative predictions for its measurable consequences.

For any finite spatial resolution, the KS system (1.126) follows approximately for a finite time a pattern belonging to a finite alphabet of admissible patterns, and the long term dynamics can be thought of as a walk through the space of such patterns, just as chaotic dynamics with a low dimensional attractor can be thought of as a succession of nearly periodic (but unstable) motions. The periodic orbit gives the machinery that converts this intuitive picture into precise calculation scheme that extracts asymptotic time predictions from the short time dynamics. For extended systems the theory gives a description of the asymptotics of partial differential equations in terms of recurrent spatio–temporal patterns.

The KS periodic orbit calculations of Lyapunov exponents and escape rates [CCP96] demonstrate that the *periodic orbit theory* predicts observable averages for deterministic but classically chaotic spatio–temporal systems. The main problem today is not how to compute such averages – periodic orbit theory as well as direct numerical simulations can handle that – but rather that there is no consensus on what the sensible experimental observables worth are predicting [Cvi00].

1.8.5 Burgers Dynamical System

Consider the following *Burgers dynamical system* on a *functional manifold* $M \subset C^k(\mathbb{R}; \mathbb{R})$:

$$u_t = u u_x + u_{xx}, \tag{1.127}$$

where $u \in M$, $t \in \mathbb{R}$ is an evolution parameter. The flow of (1.127) on M can be recast into a set of 2–forms $\{\alpha\} \subset \Lambda^2(J(\mathbb{R}^2; \mathbb{R}))$ upon the adjoint jet–manifold $J(\mathbb{R}^2; \mathbb{R})$ (see [II07b]) as follows [BPS98]:

$$\{\alpha\} = \{du^{(0)} \wedge dt - u^{(1)} dx \wedge dt = \alpha^1,\ du^{(0)} \wedge dx + u^{(0)} du^{(0)} \wedge dt$$
$$+ du^{(1)} \wedge dt = \alpha^2 : \left(x, t; u^{(0)}, u^{(1)}\right)^\top \in M^4 \subset J^1(\mathbb{R}^2; \mathbb{R})\}, \tag{1.128}$$

where M^4 is some finite–dimensional submanifold in $J^1(\mathbb{R}^2; \mathbb{R}))$ with coordinates $(x, t, u^{(0)} = u, u^{(1)} = u_x)$. The set of 2–forms (1.128) generates the closed ideal $\mathfrak{I}(\alpha)$, since

$$d\alpha^1 = dx \wedge \alpha^2 - u^{(0)} dx \wedge \alpha^1, \qquad d\alpha^2 = 0, \tag{1.129}$$

the integral submanifold $\bar{M} = \{x, t \in \mathbb{R}\} \subset M^4$ being defined by the condition $\mathfrak{J}(\alpha) = 0$. We now look for a reduced 'curvature' 1–form $\Gamma \in \Lambda^1(M^4) \otimes \mathcal{G}$, belonging to some not yet determined Lie algebra $\mathcal{G}$. This 1–form can be represented using (1.128), as follows:

$$\Gamma = b^{(x)}(u^{(0)}, u^{(1)})dx + b^{(t)}(u^{(0)}, u^{(1)})dt, \tag{1.130}$$

where elements $b^{(x)}, b^{(t)} \in \mathcal{G}$ satisfy such determining equations [BPS98]

$$\begin{array}{ll} \frac{\partial b^{(x)}}{\partial u^{(0)}} = g_2, & \frac{\partial b^{(x)}}{\partial u^{(1)}} = 0, \quad \frac{\partial b^{(t)}}{\partial u^{(0)}} = g_1 + g_2 u^{(0)}, \\ \frac{\partial b^{(t)}}{\partial u^{(1)}} = g_2, & [b^{(x)}, b^{(t)}] = -u^{(1)} g_1. \end{array} \tag{1.131}$$

The set (1.131) has the following unique solution

$$\begin{aligned} b^{(x)} &= A_0 + A_1 u^{(0)}, \\ b^{(t)} &= u^{(1)} A_1 + \frac{u^{(0)2}}{2} A_1 + [A_1, A_0] u^{(0)} + A_2, \end{aligned} \tag{1.132}$$

where $A_j \in \mathcal{G}$, $j = \overline{0, 2}$, are some constant elements on M of a Lie algebra $\mathcal{G}$ under search, obeying the *Lie structure equations* (see [I07b]):

$$\begin{aligned} &[A_0, A_2] = 0, \\ &[A_0, [A_1, A_0]] + [A_1, A_2] = 0, \\ &[A_1, [A_1, A_0]] + \tfrac{1}{2}[A_0, A_1] = 0. \end{aligned} \tag{1.133}$$

From (1.131) one can see that the curvature 2–form $\Omega \in span_\mathbb{R}\{A_1, [A_0, A_1] : A_j \in \mathcal{G}, j = 0, 1\}$. Therefore, reducing via the *Ambrose–Singer theorem* the associated principal fibred frame space $P(M; G = GL(n))$ to the principal fibre bundle $P(M; G(h))$, where $G(h) \subset G$ is the corresponding holonomy Lie group of the connection Γ on P, we need to satisfy the following conditions for the set $\mathcal{G}(h) \subset \mathcal{G}$ to be a Lie subalgebra in $\mathcal{G}$: $\nabla_x^m \nabla_t^n \Omega \in \mathcal{G}(h)$ for all $m, n \in \mathbb{Z}_+$.

Let us try now to close the above transfinitive procedure requiring that [BPS98]

$$\mathcal{G}(h) = \mathcal{G}(h)_0 = span_\mathbb{R}\{\nabla_x^m \nabla_x^n \Omega \in \mathcal{G} : m + n = 0\} \tag{1.134}$$

This means that

$$\mathcal{G}(h)_0 = span_\mathbb{R}\{A_1, A_3 = [A_0, A_1]\}. \tag{1.135}$$

To enjoy the set of relations (1.133) we need to use expansions over the basis (1.135) of the external elements $A_0, A_2 \in \mathcal{G}(h)$:

$$A_0 = q_{01} A_1 + q_{13} A_3, \qquad A_2 = q_{21} A_1 + q_{23} A_3. \tag{1.136}$$

Substituting expansions (1.136) into (1.133), we get that $q_{01} = q_{23} = \lambda$, $q_{21} = -\lambda^2/2$ and $q_{03} = -2$ for some arbitrary real parameter $\lambda \in \mathbb{R}$, that is $\mathcal{G}(h) = \text{span}_\mathbb{R}\{A_1, A_3\}$, where

$$[A_1, A_3] = A_3/2; \quad A_0 = \lambda A_1 - 2A_3, \quad A_2 = -\lambda^2 A_1/2 + \lambda A_3. \quad (1.137)$$

As a result of (1.137) we can state that the holonomy Lie algebra $\mathcal{G}(h)$ is a real 2D one, assuming the following (2×2)–matrix representation [BPS98]:

$$A_1 = \begin{pmatrix} 1/4 & 0 \\ 0 & -1/4 \end{pmatrix}, \quad A_3 = \begin{pmatrix} 0 & 1 \\ 0 & 0 \end{pmatrix},$$
$$A_0 = \begin{pmatrix} \lambda/4 & -2 \\ 0 & -\lambda/4 \end{pmatrix}, \quad A_2 = \begin{pmatrix} -\lambda^2/8 & \lambda \\ 0 & \lambda^2/8 \end{pmatrix}. \quad (1.138)$$

Thereby from (1.130), (1.132) and (1.138) we obtain the *reduced curvature 1–form* $\Gamma \in \Lambda^1(M) \otimes \mathcal{G}$,

$$\Gamma = (A_0 + uA_1)dx + ((u_x + u^2/2)A_1 - uA_3 + A_2)dt, \quad (1.139)$$

generating *parallel transport* of vectors from the representation space Y of the holonomy Lie algebra $\mathcal{G}(h)$:

$$dy + \Gamma y = 0 \quad (1.140)$$

upon the integral submanifold $\bar{M} \subset M^4$ of the ideal $\mathcal{I}(\alpha)$, generated by the set of 2–forms (1.128). The result (1.140) means also that the Burgers dynamical system (1.127) is endowed with the standard Lax type representation, having the spectral parameter $\lambda \in \mathbb{R}$ necessary for its integrability in quadratures.

1.8.6 2D Kuramoto–Sivashinsky Equation

A major goal in the study of spatio-temporal chaos (STC) [CH93] is to obtain quantitative connections between the chaotic dynamics of a system at small scales and the apparent stochastic behavior at large scales. The Kuramoto–Sivashinsky (KS) PDE [SM80]

$$\partial_t h = -\nabla^2 h - \nabla^4 h + (\nabla h)^2 \quad (1.141)$$

has been used as a paradigm in efforts to elucidate the micro–macro connections [Yak81, Zal89].

The qualitative behavior of the KS equation is quite simple. Cellular structures are generated at scales of the order $\ell_0 = 2\sqrt{2}\pi$ due to the linear instability. These cells then interact chaotically with each other via the nonlinear spatial coupling to form the STC steady state at scales much larger than ℓ_0. The characterization of the STC state has been studied extensively in one spatial dimension [Yak81, Zal89, SKJ92, CH95]. It was conjectured early by Yakhot [Yak81], based partially on symmetry grounds, that the large scale behavior of the 1D KS equation is equivalent to that of the 1D noisy

Burgers equation, also known as the *Karder–Parisi–Zhang equation* (KPZ) [FNS77]. This conjecture has since been validated by detailed numerical studies [Zal89, SKJ92]. More recently, an explicit coarse-graining procedure was used by Chow and Hwa [CH95] to derive a set of coupled effective equations describing the interaction between the chaotic cellular dynamics and the long wavelength fluctuations of the h–field. From this description, the large scale (KPZ–like) behavior of the 1D-KS system can be predicted quantitatively from the knowledge of various response functions at the 'mesoscopic scale' of several ℓ_0's.

The behavior of the 2D-KS equation is not as well understood. The simplest scenario is the generalization of Yakhot's conjecture to 2D, with the large scale behavior described by the 2D-KPZ equation, [BCH99]

$$\partial_t h = \nu \nabla^2 h + \frac{\lambda}{2}(\nabla h)^2 + \eta(r,t), \qquad (1.142)$$

where $\nu > 0$ can be interpreted as a stabilizing 'surface tension' for the height profile h, and η a stochastic noise with

$$\langle \eta(r,t)\eta(r',t')\rangle = 2D\delta^2(r-r')\delta(t-t').$$

For $\nu > 0$, the asymptotic scaling properties of (1.142) are described by 'strong–coupling' behavior with algebraic (rather than logarithmic) scaling in the roughness of h and super–diffusive dynamics, summarized by the scaling form

$$\langle [h(r,t) - h(0,t)]^2 \rangle = |r|^{2\alpha} f(|r|^z/t),$$

with anomalous exponents $\alpha \approx 0.4$ and $z = 2 - \alpha \approx 1.6$. The length scale at which the asymptotic regime is reached is given by $\ell_\times \sim e^{8\pi/g}$, where $g \equiv \lambda^2 D/\nu^3$. At scales below $\ell_\times$, the effect of the nonlinear term in (1.142) can be accounted for adequately via perturbation theory. The system behaves in this 'weak–coupling' regime as a linear stochastic diffusion equation with additive logarithmic corrections [NT92].

Previous studies of the 2D-KS equation [PJL92, LP92, JHP93] found behavior consistent with linear diffusion with logarithmic corrections but had different interpretations. [JHP93] performed a numerical analysis akin to Zaleski's on the 1D-KS [Zal89], and concluded that their results were consistent with the weak-coupling regime of the 2D-KPZ equation, with (in principle) a crossover to strong–coupling beyond a length of $\ell_\times \approx 10^{26}\ell_0$, for $g = 0.4$. Procaccia *et al.* [PJL92, LP92] used a comparative *Dyson–Wyld diagrammatic analysis* of the two equations to argue that 2D-KS and 2D-KPZ *cannot* belong to the same universality class.[42] They maintained instead that the asymptotic

---

[42] Their argument is contingent upon an equality (Eq. (23) of [PJL92]) relating the difference of two integration constants (called C_1 and C_2) and the bare coefficient of the diffusion term. The nonlocal solution is tenable if the equality is satisfied. They claim that the constants, which come from Wyld diagrammatic calculations,

behavior of the 2D-KS equation is described by a 'nonlocal' solution, consisting of diffusion with *multiplicative* logarithmic corrections.[43] We feel that the ensuing debate [LP94] failed to rule out either interpretation.

It is very difficult to distinguish between the above two scenarios numerically, as one must resolve different forms of logarithmic corrections to the (already logarithmic) correlation function of the linear diffusion equation. Theoretically, there is no *a priori* reason why simple symmetry considerations such as Yakhot's should be valid in two and higher dimensions. Unlike in 1D where there are only scalar density fluctuations, the 2D case is complicated because three or more large–k modes can couple and contribute to low–k fluctuations. Such nonlocal interactions in k may not be adequately accounted for in the type of analysis performed in [Zal89, JHP93], which numerically impose KPZ dynamics and then test for self-consistency.

In this section, following [BCH99], we perform a systematic symmetry analysis, taking into account the possibility of large–k coupling. Specifically, we extend the coarse graining procedure of [CH95] to two dimensions to derive a set of coupled equations describing the local arrangement of cells, and study their effect on the macroscopic dynamics of the h−field. The resulting behavior depends crucially on the small scale arrangement of the cells. In the simplest case, the strong–coupling 2D-KPZ behavior is recovered. Nevertheless, more complicated behaviors are allowed if the microscopic cellular arrangement exhibits *spontaneous rotational symmetry breaking*. A number of possible scenarios are listed for this case. To determine which of the allowed scenarios is selected by the 2D-KS equation, we performed numerical measurements of the cellular dynamics at the mesoscopic scale of 4 to 16 ℓ_0's. Our results disfavor the occurrence of the more exotic scenarios, leaving the strong–coupling 2D-KPZ behavior as the most likely possibility.

As in 1D, we coarse grain over a region of size $L \times L$, where L is several times the typical cellular size ℓ_0. $h(r,t)$ is separated into fast cellular modes $h_>$ and slow long wavelength modes $h_<$. Inserting

$$h(r,t) = h_<(r,t) + h_>(r,t)$$

---

are determined *uniquely* by the $(\nabla h)^2$ nonlinearity and therefore the *same* for both the KS and KPZ equations. However, the calculation involved integration of the 'full' response and correlation functions over the *entire* range of k. We note that these functions should be different for large values of k's where the microscopic dynamics matter. Consequently the constants C_1 and C_2 need *not* be the same for the KS and KPZ equations.

[43] The nonlocal solution of Ref. [PJL92] has the same form as that of the $g = 0$ fixed point of the 2D-KPZ equation. For the latter, multiplicative logarithmic correction to diffusion arises from the 'asymptotic freedom' of the system as $g \to 0^-$. The conjectured exponents of the logarithmic correction factor follow naturally from the non-renormalization of ν of the KPZ equation to all orders in λ.

1.8 Spatio–Temporal Chaos and Turbulence in PDEs

into (1.141), we get the following equations for the fast and slow modes [BCH99]:

$$\partial_t h_> = -\nabla^2 h_> - \nabla^4 h_> + (\nabla h_>)^2_> + 2(\nabla h_> \cdot \nabla h_<) \quad (1.143)$$
$$\partial_t h_< = -\nabla^2 h_< + (\nabla h_<)^2 + w(r,t) + O(\nabla^4 h_<). \quad (1.144)$$

where $w(r,t) \equiv (\nabla h_>)^2_<$ is the only contribution of the fast modes on the dynamics of $h_<$. It can be interpreted as the 'drift rate' of $h_<$ over a regime of $L \times L$ centered at r. To specify the dynamics of $h_<$, it is necessary to obtain the dynamics of w from the fast mode equation (1.143). Due to the structure of the nonlinear term, we must consider the tensor W, with elements $W_{ij} = 2\,\overline{\partial_i h_> \cdot \partial_j h_>}$, where the over–line denotes a spatial average over the coarse–graining scale L. It is convenient to introduce the curvature tensor K, with elements $K_{ij} = 2\partial_i \partial_j h_<$. In this notation, $w = \frac{1}{2}\mathrm{Tr}\,\mathsf{W}$ and $\kappa \equiv \nabla^2 h_< = \frac{1}{2}\mathrm{Tr}\,\mathsf{K}$. Taking the time derivative of W_{ij} and using (1.143), we get

$$\partial_t \mathsf{W} = \mathsf{F}\,[\mathsf{W}] + \mathsf{W}\cdot\mathsf{K} + \mathsf{K}\cdot\mathsf{W} \quad (1.145)$$

where F [W] contains purely fast mode dynamics and will be described shortly. The forms of the last two terms in Eq. (1.145) are fixed by the Galilean invariance of the KS equation and are exact.

Equation (1.145) can be made more transparent by rewriting the two tensors as

$$\mathsf{W} = w\cdot\mathbf{1} + \tilde{w}\cdot\mathsf{Q}(\phi) \quad \text{and} \quad \mathsf{K} = \kappa\cdot\mathbf{1} + \tilde{\kappa}\cdot\mathsf{Q}(\theta),$$

where $\mathbf{1}$ is the identity matrix and $\mathsf{Q}(\alpha)$ is a unit *traceless* matrix, represented by an angle α, e.g.,

$$Q_{12}(\alpha) = Q_{21}(\alpha) = \sin(2\alpha) \quad and \quad Q_{11}(\alpha) = -Q_{22}(\alpha) = \cos(2\alpha).$$

Adopting vector notation $\boldsymbol{\psi} = (\tilde{w}\cos 2\phi, \tilde{w}\sin 2\phi)$ and $\boldsymbol{\chi} = (\tilde{\kappa}\cos 2\theta, \tilde{\kappa}\sin 2\theta)$, equation (1.145) can be rewritten as [BCH99]

$$\partial_t w = f[w] + 2\kappa w + 2\boldsymbol{\chi}\cdot\boldsymbol{\psi} \quad (1.146)$$
$$\partial_t \boldsymbol{\psi} = \boldsymbol{\varphi}\,[\boldsymbol{\psi}] + \kappa\boldsymbol{\psi} + w\boldsymbol{\chi}, \quad (1.147)$$

to leading order, with f and $\boldsymbol{\varphi}$ obtained from the appropriate decomposition of F.

Equations (1.146) and (1.147), together with the slow mode equation (1.144), form a closed set of coarse grained equations which specifies the dynamics of $h_<$ once the effective forms of the small scale dynamics, i.e., f and $\boldsymbol{\varphi}$ are given. These equations are constructed from symmetry considerations, and can be regarded as the more complete generalization of Yakhot's conjecture for two dimensions. We first discuss the physical meaning of the coarse grained variables appearing in W and K.

The tensor K describes the local curvature of the slow modes $h_<$. With $\tilde{\kappa} = 0$, we have a symmetric paraboloid — a 'valley' if $\kappa > 0$ or a 'hill' if $\kappa < 0$.

With $\kappa = 0$, we have a 'saddle', with $\tilde{\kappa}$ and θ specifying the strength and the orientation. The tensor W characterizes the local *packing* of the cells. As in the 1D case [CH95], w gives the local cell density. The traceless component of W describes the local *anisotropy* in cell packing. Fluctuations in the anisotropic part of the curvature K will affect the local cell packing. For example, the cell density at the bottom of a valley will be higher and a saddle configuration in $h_<$ will induce anisotropy. Equations (1.146) and (1.147) describe these effects of curvature quantitatively, much like the relation between stress and strain in elastic systems. Cell packing in turn influences the slow mode dynamics via the w term in (1.144). The anisotropic parts of K and W are invariant upon a rotation by 180°. Thus, we can view the vector field $\psi(r,t)$ as a 'nematic' order parameter describing the local cellular orientation, and χ as an applied field biasing ψ towards a specific orientation.

If we turn off the applied field κ and χ in (1.146) and (1.147), we have $\partial_t w = f[w]$ and $\partial_t \psi = \varphi[\psi]$ for each coarse–grained region; thus f and φ describe the small scale dynamics. Even for a coarse–grained region of a few ℓ_0's, the small scale dynamics of h are already *chaotic*. The fields $w(t)$ and $\psi(t)$ are 'projections' of this small scale chaotic dynamics. They can be quantitatively characterized numerically as we will present shortly. Before doing so, we first construct some possible scenarios.

We expect that the h–field has on average a finite drift rate, i.e., a finite time–averaged value of w. The simplest dynamics of w is then [BCH99]

$$\partial_t w(r,t) = f[w] = -\alpha(w - w_0) + \xi(r,t),$$

where ξ is a stochastic forcing mimicking the chaotic small scale dynamics, and w_0 is a constant. This yields $w(r, t \to \infty) \to w_0$. The behavior of ψ is less straightforward. On symmetry grounds, the dynamics can take on the following form

$$\partial_t \psi = \varphi[\psi] = -\tilde{\alpha}\psi + \tilde{\beta}\ |\psi|^2\ \psi + \zeta(r,t), \tag{1.148}$$

where ζ is a vector stochastic forcing.

In the simplest scenario (where $\tilde{\alpha} > 0$, the $O(\psi^3)$ term need not be included), we have

$$\partial_t \psi = -\tilde{\alpha}\psi + \zeta(r,t)$$

to leading order, with ζ being a vectorial stochastic forcing. Equation (1.147) then yields (in the hydrodynamic limit) $\psi \simeq (2w_0/\tilde{\alpha})\chi$, where we took the asymptotic result $w = w_0$ and assumed that the typical curvature κ is small. Note that in this scenario, the cellular orientation *passively* follows the curvature. In particular, there is no orientational anisotropy on average if there is no external forcing. Inserting this result and $f[w]$ into (1.146), we find in the hydrodynamic limit

$$w \simeq w_0 + \frac{2w_0}{\alpha}\nabla^2 h_< + \frac{\xi}{\alpha} + O((\partial_i\partial_j h_<)^2). \tag{1.149}$$

1.8 Spatio–Temporal Chaos and Turbulence in PDEs

Substituting (1.149) into (1.144) yields an equation for $h_<$ of the KPZ form (1.142) to leading order, with $\nu = (2w_0/\alpha) - 1$ and $\eta(r,t) = \xi(r,t)/\alpha$, we have

$$\partial_t h_< \simeq \nu \nabla^2 h_< + (\nabla h_<)^2 + \eta(r,t), \qquad (1.150)$$

where

$$\nu = \frac{w_0}{\alpha} - 1, \qquad \eta(r,t) = \frac{\xi(r,t)}{\alpha} \qquad (1.151)$$

Dynamics of the KPZ universality class will be obtained if $\nu > 0$ and the noise ξ is uncorrelated between different coarse–graining regions. Unlike the constant α however, there is no a priori reason why the constant $\tilde{\alpha}$ cannot be negative. This would be the case if the microscopic chaotic dynamics has a preference for the *spontaneous* breaking of local isotropy. If $\tilde{\alpha} \leq 0$, then the dynamics of ψ would be more complicated. Higher order terms, e.g., $|\psi|^2\psi$, will be needed for stability. The minimal equation for (1.147) becomes [BCH99]

$$\partial_t \psi = -\tilde{\alpha}\psi - \tilde{\beta}|\psi|^2\psi + \gamma \nabla^2 \psi + w_0 \chi + \zeta(r,t) \qquad (1.152)$$

where $\tilde{\beta}$ is a positive constant, and the γ term describes the coupling of neighboring coarse grained regions. Equation (1.152) describes the relaxational dynamics of a nematic liquid crystal under an applied 'field' χ. Its behavior depends crucially on the dynamics of the phase field, ϕ, which is the Goldstone mode associated with symmetry breaking. The latter in turn depends on the parameters of (1.152), particularly the coupling constant γ and the amplitude of the noise ζ. The possibilities along with the effects on $h_<$ are:

Case (i). If the noise ζ dominates over the spatial coupling γ, then the local anisotropy will be destroyed at large scales due to the proliferation of topological defects (disclinations) in ϕ. Isotropy is restored and the KPZ universality class is recovered.

Case (ii). If the spatial coupling is large, then the direction of ψ may 'phase–lock' with the direction of χ, as manifested by $\langle(\theta - \phi)^2\rangle \ll 1$. Solving for the steady state of w in this case gives: $w \simeq w_0 + \frac{w_0}{\alpha}\kappa + \frac{\tilde{w}_0}{\alpha}\tilde{\kappa}$, leading to a slow mode equation which is explicitly not KPZ–like since $\tilde{\kappa} = \left[(h_{xx} - h_{yy})^2 + 4h_{xy}^2\right]^{1/2}$. (In the KPZ case, $\tilde{\kappa}$ comes in at second order and is presumed irrelevant; see (1.149)).

Case (iii). For intermediate parameters, there may exist a 'spin wave' phase characterized by $\langle\phi(r)\phi(0)\rangle = a\log|r|$. Here, 'spin wave' fluctuations would add a long range component to the effective KPZ noise, since $\langle\cos(\phi(r) - \phi(0))\rangle \sim r^{-a}$. For sufficiently small a, it would yield dynamics that are not in the KPZ universality class.

For more details, see [BCH99].

1.9 Basics of Chaos Control

1.9.1 Feedback and Non–Feedback Algorithms for Chaos Control

Although the presence of chaotic behavior is generic and robust for suitable nonlinearities, ranges of parameters and external forces, there are practical situations where one wishes to avoid or control chaos so as to improve the performance of the dynamical system. Also, although chaos is sometimes useful as in a mixing process or in heat transfer, it is often unwanted or undesirable. For example, increased drag in flow systems, erratic fibrillations of heart beats, extreme weather patterns and complicated circuit oscillations are situations where chaos is harmful. Clearly, the ability to control chaos, that is to convert chaotic oscillations into desired regular ones with a periodic time dependence would be beneficial in working with a particular system. The possibility of purposeful selection and stabilization of particular orbits in a normally chaotic system, using minimal, predetermined efforts, provides a unique opportunity to maximize the output of a dynamical system. It is thus of great practical importance to develop suitable control methods and to analyze their efficacy.

Let us consider a general nD nonlinear dynamical system,

$$\dot{x} = F(x, p, t), \qquad (1.153)$$

where $x = (x_1, x_2, x_3, ..., x_n)$ represents the n state variables and p is a control or external parameter. Let $x(t)$ be a chaotic solution of (1.153). Different control algorithms are essentially based on the fact that one would like to effect the most minimal changes to the original system so that it will not be grossly deformed. From this point of view, controlling methods or algorithms can be broadly classified into two categories:

(i) feedback methods, and
(ii) non–feedback algorithms.

Feedback methods essentially make use of the intrinsic properties of chaotic systems, including their sensitivity to initial conditions, to stabilize orbits already existing in the systems. Some of the prominent methods are the following (see, [Lak97, Lak03, Sch88, II06b]):

1. Adaptive control algorithm;
2. Nonlinear control algorithm;
3. Ott–Grebogi–Yorke (OGY) method of stabilizing unstable periodic orbits;
4. Singer's method of stabilizing unstable periodic orbits; and
5. Various control engineering approaches.

In contrast to feedback control techniques, non–feedback methods make use of a small perturbing external force such as a small driving force, a small noise term, a small constant bias or a weak modulation to some system parameter. These methods modify the underlying chaotic dynamical system weakly so that stable solutions appear. Some of the important controlling methods of this type are the following.

1. Parametric perturbation method
2. Addition of a weak periodic signal, constant bias or noise
3. Entrainment–open loop control
4. Oscillator absorber method.

Here is a typical example of adaptive control algorithm. We can control the chaotic orbit $X_s = (x_s, y_s)$ of the *Van der Pol oscillator* (1.35) by introducing the following dynamics on the parameter A_1:

$$\dot{x} = x - \frac{x^3}{3} - y + A_0 + A_1 \cos \omega t, \qquad \dot{y} = c(x + a - by),$$

$$\dot{A}_1 = -\epsilon[(x - x_s) - (y - y_s)], \qquad \epsilon \ll 1.$$

On the other hand, recall from [II06b] that a generic SISO nonlinear system

$$\dot{x} = f(x) + g(x)\,u \qquad y = h(x) \tag{1.154}$$

is said to have *relative degree* r at a point x^o if
 (i) $L_g L_f^k h(x) = 0$ for all x in a neighborhood of x^o and all $k < r - 1$
 (ii) $L_g L_f^{r-1} h(x^o) \neq 0$, where L_g denotes the *Lie derivative* in the direction of the vector–field g.

Now, the Van der Pol oscillator (1.30) has the state space form

$$\dot{x} = f(x) + g(x)\,u = \begin{bmatrix} x_2 \\ 2\omega\zeta\,(1 - \mu x_1^2)\,x_2 - \omega^2 x_1 \end{bmatrix} + \begin{bmatrix} 0 \\ 1 \end{bmatrix} u. \tag{1.155}$$

Suppose the output function is chosen as

$$y = h(x) = x_1. \tag{1.156}$$

In this case we have

$$L_g h(x) = \frac{\partial h}{\partial x} g(x) = \begin{bmatrix} 1 & 0 \end{bmatrix} \begin{bmatrix} 0 \\ 1 \end{bmatrix} = 0, \qquad \text{and} \tag{1.157}$$

$$L_f h(x) = \frac{\partial h}{\partial x} f(x) = \begin{bmatrix} 1 & 0 \end{bmatrix} \begin{bmatrix} x_2 \\ 2\omega\zeta\,(1 - \mu x_1^2)\,x_2 - \omega^2 x_1 \end{bmatrix} = x_2. \tag{1.158}$$

Moreover

$$L_g L_f h(x) = \frac{\partial (L_f h)}{\partial x} g(x) = \begin{bmatrix} 0 & 1 \end{bmatrix} \begin{bmatrix} 0 \\ 1 \end{bmatrix} = 1 \tag{1.159}$$

and thus we see that the Van der Pol oscillator system has relative degree 2 at any point x^o.

However, if the output function is, for instance

$$y = h(x) = \sin x_2 \tag{1.160}$$

then $L_g h(x) = \cos x_2$. The system has relative degree 1 at any point x^o, provided that $(x^o)_2 \neq (2k+1)\,\pi/2$. If the point x^o is such that this condition is violated, no relative degree can be defined.

Both adaptive and nonlinear control methods can be naturally extended to other chaotic systems, e.g., *Lorenz attractor* (see Figure 1.53).

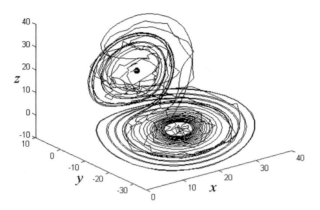

Fig. 1.53. Nonlinear control of the Lorenz system: targeting of unstable upper and lower states in the Lorenz attractor (after applying random perturbations, see [Pet96]), using a MIMO nonlinear controller (see [II06b]); simulated using $Matlab^{TM}$.

Hybrid Systems and Homotopy ODEs

Consider a *hybrid dynamical system of variable structure*, given by an nD ODE–system (see [MWH01])

$$\dot{x} = f(t, x), \tag{1.161}$$

where $x = x(t) \in \mathbb{R}^n$ and $f = f(t,x) : \mathbb{R}^+ \times \mathbb{R}^n \to \mathbb{R}^n$. Let the domain $G \subset \mathbb{R}^+ \times \mathbb{R}^n$, on which the vector–field $f(t,x)$ is defined, be divided into two subdomains, G^+ and G^-, by means of a smooth $(n-1)$–manifold M. In $G^+ \cup M$, let there be given a vector–field $f^+(t,x)$, and in $G^- \cup M$, let there be given a vector–field $f^-(t,x)$. Assume that both $f^+ = f^+(t,x)$ and $f^- = f^-(t,x)$ are continuous in t and smooth in x. For the system (1.161), let

$$f = \begin{cases} f^+ & \text{when } x \in G^+ \\ f^- & \text{when } x \in G^- \end{cases}.$$

Under these conditions, a solution $x(t)$ of ODE (1.161) is well–defined while passing through G until the manifold M is reached.

Upon reaching the manifold M, in physical systems with inertia, the transition

$$\text{from} \quad \dot{x} = f^-(t,x) \quad \text{to} \quad \dot{x} = f^+(t,x)$$

does not take place instantly on reaching M, but after some delay. Due to this delay, the solution $x(t)$ oscillates about M, $x(t)$ being displaced along M with some mean velocity.

As the delay tends to zero, the limiting motion and velocity along M are determined by the *linear homotopy ODE*

$$\dot{x} = f^0(t,x) \equiv (1-\alpha)\, f^-(t,x) + \alpha\, f^+(t,x), \tag{1.162}$$

where $x \in M$ and $\alpha \in [0,1]$ is such that the *linear homotopy segment* $f^0(t,x)$ is tangential to M at the point x, i.e., $f^0(t,x) \in T_x M$, where $T_x M$ is the tangent space to the manifold M at the point x.

The vector–field $f^0(t,x)$ of the system (1.162) can be constructed as follows: at the point $x \in M$, $f^-(t,x)$ and $f^+(t,x)$ are given and their ends are joined by the linear homotopy segment. The point of intersection between this segment and $T_x M$ is the end of the required vector–field $f^0(t,x)$. The vector function $x(t)$ which satisfies (1.161) in G^- and G^+, and (1.162) when $x \in M$, can be considered as a *solution* of (1.161) in a *general sense*.

However, there are cases in which the solution $x(t)$ cannot consist of a finite or even countable number of arcs, each of which passes through G^- or G^+ satisfying (1.161), or moves along the manifold M and satisfies the homotopic ODE (1.162). To cover such cases, assume that the vector–field $f = f(t,x)$ in ODE (1.161) is a Lebesgue–measurable function in a domain $G \subset \mathbb{R}^+ \times \mathbb{R}^n$, and that for any closed bounded domain $D \subset G$ there exists a summable function $K(t)$ such that almost everywhere in D we have $|f(t,x)| \leq K(t)$. Then the absolutely continuous vector function $x(t)$ is called the *generalized solution* of the ODE (1.161) *in the sense of Filippov* (see [MWH01]) if for almost all t, the vector $\dot{x} = \dot{x}(t)$ belongs to the least convex closed set containing all the limiting values of the vector–field $f(t,x^*)$, where x^* tends towards x in an arbitrary manner, and the values of the function $f(t,x^*)$ on a set of measure zero in $\mathbb{R}^n$ are ignored.

Such *hybrid systems* of variable structure occur in the study of nonlinear electric networks (endowed with electronic switches, relays, diodes, rectifiers, etc.), in models of both natural and artificial neural networks, as well as in feedback control systems (usually with continuous–time plants and digital controllers/filters).

1.9.2 Exploiting Critical Sensitivity

The fact that some dynamical systems showing the necessary conditions for chaotic behavior possess such a critical dependence on the initial conditions was known since the end of the last century. However, only in the last thirty years, experimental observations have pointed out that, in fact, chaotic systems are common in nature. They can be found, e.g., in chemistry (*Belouzov–Zhabotinski reaction*), in nonlinear optics (lasers), in electronics (*Chua–Matsumoto circuit*), in fluid dynamics (*Rayleigh–Bénard convection*), etc. Many natural phenomena can also be characterized as being chaotic. They can be found in meteorology, solar system, heart and brain of living organisms and so on.

Due to their critical dependence on the initial conditions, and due to the fact that, in general, experimental initial conditions are never known perfectly, these systems are intrinsically unpredictable. Indeed, the prediction trajectory emerging from an initial condition and the real trajectory emerging from the real initial condition diverge exponentially in course of time, so that the error in the prediction (the distance between prediction and real trajectories) grows exponentially in time, until making the system's real trajectory completely different from the predicted one at long times.

For many years, this feature made chaos undesirable, and most experimentalists considered such characteristic as something to be strongly avoided. Besides their critical sensitivity to initial conditions, chaotic systems exhibit two other important properties. Firstly, there is an infinite number of unstable periodic orbits embedded in the underlying chaotic set. In other words, the skeleton of a chaotic attractor is a collection of an infinite number of periodic orbits, each one being unstable. Secondly, the dynamics in the chaotic attractor is *ergodic*, which implies that during its temporal evolution the system ergodically visits small neighborhood of every point in each one of the unstable periodic orbits embedded within the chaotic attractor.

A relevant consequence of these properties is that a chaotic dynamics can be seen as shadowing some periodic behavior at a given time, and erratically jumping from one to another periodic orbit. The idea of controlling chaos is then when a trajectory approaches ergodically a desired periodic orbit embedded in the attractor, one applies small perturbations to stabilize such an orbit. If one switches on the stabilizing perturbations, the trajectory moves to the neighborhood of the desired periodic orbit that can now be stabilized. This fact has suggested the idea that the critical sensitivity of a chaotic system to changes (perturbations) in its initial conditions may be, in fact, very desirable in practical experimental situations. Indeed, if it is true that a small perturbation can give rise to a very large response in the course of time, it is also true that a judicious choice of such a perturbation can direct the trajectory to wherever one wants in the attractor, and to produce a series of desired dynamical states. This is exactly the idea of targeting [BGL00].

The important point here is that, because of chaos, one is able to produce an infinite number of desired dynamical behaviors (either periodic and not periodic) using the same chaotic system, with the only help of tiny perturbations chosen properly. We stress that this is not the case for a non–chaotic dynamics, wherein the perturbations to be done for producing a desired behavior must, in general, be of the same order of magnitude as the un–perturbed evolution of the dynamical variables.

The *idea* of *chaos control* was enunciated in 1990 at the University of Maryland, by E. Ott, C. Grebogi and J.A. Yorke [OGY90], widely referred to as Ott–Grebogi–Yorke (OGY, for short). In OGY–paper [OGY90], the ideas for controlling chaos were outlined and a method for stabilizing an unstable periodic orbit was suggested, as a proof of principle. The main idea consisted in waiting for a natural passage of the chaotic orbit close to the desired periodic behavior, and then applying a small judiciously chosen perturbation, in order to stabilize such periodic dynamics (which would be, in fact, unstable for the un–perturbed system). Through this mechanism, one can use a given laboratory system for producing an infinite number of different periodic behavior (the infinite number of its unstable periodic orbits), with a great flexibility in switching from one to another behavior. Much more, by constructing appropriate goal dynamics, compatible with the chaotic attractor, an operator may apply small perturbations to produce any kind of desired dynamics, even not periodic, with practical application in the coding process of signals.

A branch of the theory of dynamical systems has been developed with the aim of formalizing and quantitatively characterizing the sensitivity to initial conditions. The *largest Lyapunov exponent* λ (together with the related *Kaplan–Yorke dimension* d_{KY}) and the *Kolmogorov–Sinai entropy* h_{KS} are the two indicators for measuring the *rate of error growth* and *information* produced by the dynamical system [ER85].

1.9.3 Lyapunov Exponents and Kaplan–Yorke Dimension

The characteristic Lyapunov exponents are somehow an extension of the linear stability analysis to the case of aperiodic motions. Roughly speaking, they measure the typical rate of exponential divergence of nearby trajectories. In this sense they give information on the rate of growth of a very small error on the initial state of a system [BCF02].

Consider an nD dynamical system given by the set of ODEs of the form

$$\dot{x} = f(x), \tag{1.163}$$

where $x = (x_1, \ldots, x_n) \in \mathbb{R}^n$ and $f : \mathbb{R}^n \to \mathbb{R}^n$. Recall that since the r.h.s of equation (1.163) does not depend on t explicitly, the system is called *autonomous*. We assume that f is smooth enough that the evolution is well–defined for time intervals of arbitrary extension, and that the motion occurs in a bounded region R of the system phase–space M. We intend to study the separation between two trajectories in M, $x(t)$ and $x'(t)$, starting from two close initial conditions, $x(0)$ and $x'(0) = x(0) + \delta x(0)$ in $R_0 \subset M$, respectively.

As long as the difference between the trajectories, $\delta x(t) = x'(t) - x(t)$, remains infinitesimal, it can be regarded as a vector, $z(t)$, in the tangent space $T_x M$ of M. The time evolution of $z(t)$ is given by the linearized differential equations:

$$\dot{z}_i(t) = \left.\frac{\partial f_i}{\partial x_j}\right|_{x(t)} z_j(t).$$

Under rather general hypothesis, Oseledets [Ose68] proved that for almost all initial conditions $x(0) \in R$, there exists an orthonormal basis $\{e_i\}$ in the tangent space $T_x M$ such that, for large times,

$$z(t) = c_i e_i \exp(\lambda_i t), \tag{1.164}$$

where the coefficients $\{c_i\}$ depend on $z(0)$. The exponents $\lambda_1 \geq \lambda_2 \geq \cdots \geq \lambda_d$ are called *characteristic Lyapunov exponents*. If the dynamical system has an ergodic invariant measure on M, the spectrum of LEs $\{\lambda_i\}$ does not depend on the initial conditions, except for a set of measure zero with respect to the natural invariant measure.

Equation (1.164) describes how a dD spherical region $R = S^n \subset M$, with radius ϵ centered in $x(0)$, deforms, with time, into an ellipsoid of semi–axes $\epsilon_i(t) = \epsilon \exp(\lambda_i t)$, directed along the e_i vectors. Furthermore, for a generic small perturbation $\delta x(0)$, the distance between the reference and the perturbed trajectory behaves as

$$|\delta x(t)| \sim |\delta x(0)| \exp(\lambda_1 t) \left[1 + O\left(\exp{-(\lambda_1 - \lambda_2)t}\right)\right].$$

If $\lambda_1 > 0$ we have a rapid (exponential) amplification of an error on the initial condition. In such a case, the system is chaotic and, unpredictable on the long times. Indeed, if the initial error amounts to $\delta_0 = |\delta x(0)|$, and we purpose to predict the states of the system with a certain tolerance Δ, then the prediction is reliable just up to a *predictability time* given by

$$T_p \sim \frac{1}{\lambda_1} \ln\left(\frac{\Delta}{\delta_0}\right).$$

This equation shows that T_p is basically determined by the *positive leading Lyapunov exponent*, since its dependence on δ_0 and Δ is logarithmically weak. Because of its preeminent role, λ_1 is often referred as 'the leading positive Lyapunov exponent', and denoted by λ.

Therefore, Lyapunov exponents are average rates of expansion or contraction along the principal axes. For the ith principal axis, the corresponding Lyapunov exponent is defined as

$$\lambda_i = \lim_{t \to \infty} \{(1/t) \ln[L_i(t)/L_i(0)]\}, \qquad (1.165)$$

where $L_i(t)$ is the radius of the ellipsoid along the ith principal axis at time t. For technical details on calculating Lyapunov exponents from any time series data, see [Wol98].

An initial volume V_0 of the phase–space region R_0 evolves on average as

$$V(t) = V_0 e^{(\lambda_1 + \lambda_2 + \cdots + \lambda_{2n})t}, \qquad (1.166)$$

and therefore the rate of change of $V(t)$ is simply

$$\dot{V}(t) = \sum_{i=1}^{2n} \lambda_i V(t).$$

In the case of a 2D phase area A, evolving as $A(t) = A_0 e^{(\lambda_1 + \lambda_2)t}$, a *Lyapunov dimension* d_L is defined as

$$d_L = \lim_{\epsilon \to 0} \left[\frac{d(\ln(N(\epsilon)))}{d(\ln(1/\epsilon))}\right],$$

where $N(\epsilon)$ is the number of squares with sides of length ϵ required to cover $A(t)$, and d represents an ordinary *capacity dimension*,

$$d_c = \lim_{\epsilon \to 0} \left(\frac{\ln N}{\ln(1/\epsilon)} \right).$$

Lyapunov dimension can be extended to the case of nD phase–space by means of the *Kaplan–Yorke dimension* [Kap00, YAS96, OGY90]) as

$$d_{KY} = j + \frac{\lambda_1 + \lambda_2 + \cdots + \lambda_j}{|\lambda_{j+1}|},$$

where the λ_i are ordered (λ_1 being the largest) and j is the index of the smallest nonnegative Lyapunov exponent.

1.9.4 Kolmogorov–Sinai Entropy

The LE, λ, gives a first quantitative information on how rapidly we loose the ability of predicting the evolution of a system [BCF02]. A state, initially determined with an error $\delta x(0)$, after a time enough larger than $1/\lambda$, may be found almost everywhere in the region of motion $R \in M$. In this respect, the *Kolmogorov–Sinai* (KS) *entropy*, h_{KS}, supplies a more refined information. The error on the initial state is due to the maximal resolution we use for observing the system. For simplicity, let us assume the same resolution ϵ for each degree of freedom. We build a partition of the phase–space M with cells of volume ϵ^d, so that the state of the system at $t = t_0$ is found in a region R_0 of volume $V_0 = \epsilon^d$ around $x(t_0)$. Now we consider the trajectories starting from V_0 at t_0 and sampled at discrete times $t_j = j\tau$ ($j = 1, 2, 3, \ldots, t$). Since we are considering motions that evolve in a bounded region $R \subset M$, all the trajectories visit a finite number of different cells, each one identified by a symbol. In this way a unique sequence of symbols $\{s(0), s(1), s(2), \ldots\}$ is associated with a given trajectory $x(t)$. In a chaotic system, although each evolution $x(t)$ is univocally determined by $x(t_0)$, a great number of different symbolic sequences originates by the same initial cell, because of the divergence of nearby trajectories. The total number of the admissible symbolic sequences, $\widetilde{N}(\epsilon, t)$, increases exponentially with a rate given by the topological entropy

$$h_T = \lim_{\epsilon \to 0} \lim_{t \to \infty} \frac{1}{t} \ln \widetilde{N}(\epsilon, t).$$

However, if we consider only the number of sequences $N_{eff}(\epsilon, t) \leq \widetilde{N}(\epsilon, t)$ which appear with very high probability in the long time limit – those that can be numerically or experimentally detected and that are associated with the natural measure – we arrive at a more physical quantity, namely the *Kolmogorov–Sinai entropy* [ER85]:

$$h_{KS} = \lim_{\epsilon \to 0} \lim_{t \to \infty} \frac{1}{t} \ln N_{eff}(\epsilon, t) \leq h_T. \tag{1.167}$$

h_{KS} quantifies the long time exponential rate of growth of the number of the effective coarse-grained trajectories of a system. This suggests a link with information theory where the Shannon entropy measures the mean asymptotic growth of the number of the typical sequences – the ensemble of which has probability almost one – emitted by a source.

We may wonder what is the number of cells where, at a time $t > t_0$, the points that evolved from R_0 can be found, i.e., we wish to know how big is the coarse–grained volume $V(\epsilon, t)$, occupied by the states evolved from the volume V_0 of the region R_0, if the minimum volume we can observe is $V_{min} = \epsilon^d$. As stated above (4.45), we have

$$V(t) \sim V_0 \exp(t \sum_{i=1}^{d} \lambda_i).$$

However, this is true only in the limit $\epsilon \to 0$. In this (unrealistic) limit, $V(t) = V_0$ for a conservative system (where $\sum_{i=1}^{d} \lambda_i = 0$) and $V(t) < V_0$ for a dissipative system (where $\sum_{i=1}^{d} \lambda_i < 0$). As a consequence of limited resolution power, in the evolution of the volume $V_0 = \epsilon^d$ the effect of the contracting directions (associated with the negative Lyapunov exponents) is completely lost. We can experience only the effect of the expanding directions, associated with the positive Lyapunov exponents. As a consequence, in the typical case, the coarse grained volume behaves as

$$V(\epsilon, t) \sim V_0 \, e^{(\sum_{\lambda_i > 0} \lambda_i) t},$$

when V_0 is small enough. Since $N_{eff}(\epsilon, t) \propto V(\epsilon, t)/V_0$, one has

$$h_{KS} = \sum_{\lambda_i > 0} \lambda_i.$$

This argument can be made more rigorous with a proper mathematical definition of the metric entropy. In this case one derives the Pesin relation [Pes77, ER85]

$$h_{KS} \leq \sum_{\lambda_i > 0} \lambda_i. \qquad (1.168)$$

Because of its relation with the Lyapunov exponents – or by the definition (1.167) – it is clear that also h_{KS} is a fine-grained and global characterization of a dynamical system.

The metric entropy is an invariant characteristic quantity of a dynamical system, i.e., given two systems with invariant measures, their KS–entropies exist and they are equal iff the systems are isomorphic [Bil65].

1.9.5 Chaos Control by Ott, Grebogi and Yorke (OGY)

Besides the occurrence of chaos in a large variety of natural processes, chaos may also occur because one may wish to design a physical, biological or chemical experiment, or to project an industrial plant to behave in a chaotic manner.

1.9 Basics of Chaos Control

The OGY–idea is that chaos may indeed be desirable since it can be controlled by using small perturbation to some accessible parameter.

The major key ingredient for the OGY–control of chaos is the observation that a chaotic set, on which the trajectory of the chaotic process lives, has embedded within it a large number of unstable low–period periodic orbits. In addition, because of ergodicity, the trajectory visits or accesses the neighborhood of each one of these periodic orbits. Some of these periodic orbits may correspond to a desired system's performance according to some criterion. The second ingredient is the realization that chaos, while signifying sensitive dependence on small changes to the current state and henceforth rendering unpredictable the system state in the long time, also implies that the system's behavior can be altered by using small perturbations. Then, the accessibility of the chaotic systems to many different periodic orbits combined with its sensitivity to small perturbations allows for the control and the manipulation of the chaotic process. Specifically, the OGY approach is then as follows. One first determines some of the unstable low–period periodic orbits that are embedded in the chaotic set. One then examines the location and the stability of these orbits and chooses one which yields the desired system performance. Finally, one applies small control to stabilize this desired periodic orbit. However, all this can be done from data by using nonlinear time series analysis for the observation, understanding and control of the system. This is particularly important since chaotic systems are rather complicated and the detailed knowledge of the equations of the process is often unknown [BGL00].

Simple Example of Chaos Control: a 1D Map. The basic idea of controlling chaos can be understood [Lai94] by considering May's classical *logistic map* [May76] (1.27)

$$x_{n+1} = f(x_n, r) = rx_n(1 - x_n),$$

where x is restricted to the unit interval $[0, 1]$, and r is a control parameter. It is known that this map develops chaos via the *period–doubling bifurcation* route. For $0 < r < 1$, the asymptotic state of the map (or the attractor of the map) is $x = 0$; for $1 < r < 3$, the attractor is a nonzero fixed–point $x_F = 1 - 1/r$; for $3 < r < 1 + \sqrt{6}$, this fixed–point is unstable and the attractor is a stable period-2 orbit. As r is increased further, a sequence of period–doubling bifurcations occurs in which successive period–doubled orbits become stable. The period–doubling cascade accumulates at $r = r_\infty \approx 3.57$, after which chaos can arise.

Consider the case $r = 3.8$ for which the system is apparently chaotic. An important characteristic of a chaotic attractor is that there exists an infinite number of unstable periodic orbits embedded within it. For example, there are a fixed–point $x_F \approx 0.7368$ and a period-2 orbit with components $x(1) \approx 0.3737$ and $x(2) = 0.8894$, where $x(1) = f(x(2))$ and $x(2) = f(x(1))$.

Now suppose we want to avoid chaos at $r = 3.8$. In particular, we want trajectories resulting from a randomly chosen initial condition x_0 to be as close as possible to the period-2 orbit, assuming that this period−2 orbit gives the

Type of FP	λ_u	λ_s	x^* Locat.
Flip saddle	$\lambda_u < -1$	$-1 < \lambda_s < 1$	$x^* < s^*$
Saddle	$\lambda_u > 1$	$-1 < \lambda_s < 1$	$x^* > s^*$
Single–flip repeller	$\lambda_u > 1$	$\lambda_s < -1$	$x^* > s^*$
Double–flip repeller	$\lambda_u < -1$	$\lambda_s < -1$	$x^* < s^*$
Spiral (complex λ)	$|\lambda_u| > 1$	$|\lambda_s| > 1$	$x^* < s^*$

Table 1.1. Cases which lead to a stable fixed–point for the controlled dynamics. In all cases, it is assumed that $|\mathcal{A}| < 1$. (For the cases where $\lambda_s < -1$, the subscript s in λ_s is misleading in that the corresponding manifold is unstable. For the spiral, there is no stable manifold (adapted from [Kap00]).)

Although the local stability of the fixed–point is guaranteed for the cases in Table 1.1 for $-1 < A < 1$, the basin of attraction of this fixed–point may be small or large depending on A, C, s^*, λ_u and λ_s.

The endpoints of the basin of attraction can be derived analytically [Kap00]. The size of the basin of attraction will often be zero when A and C are chosen to match the stable manifold of the natural system. Therefore, in order to make the basin large, it is advantageous intentionally to misplace the control line and to put x^* in the direction indicated in Table 1.1. In addition, control may be enhanced by setting $A \neq \lambda_s$, for instance $A = 0$.

If the relationship between x^* and s^* is reversed from that given in Table 1.1, the controlled dynamics will not have a stable fixed points. To some extent, these can also be studied using 1D maps. The flip saddle and double–flip repeller can display stable period–2 orbits and chaos. For the non–flip saddle and single–flip repeller, control is unstable when $x^* < s^*$.

The fact that control may be successful or even enhanced when A and C are not matched to λ_s and s^* suggests that it may be useful to reverse the experimental procedure often followed in chaos control. Rather than first identifying the parameters of the natural unstable fixed points and then applying the control, one can blindly attempt control and then deduce the natural dynamics from the behavior of the controlled system. This use of PPF control is reminiscent of pioneering studies that used periodic stimulation to demonstrate the complex dynamics of biological preparations [GGS81].

As an example, consider the *Hénon map*:

$$s_{n+1} = 1.4 + 0.3 s_{n-1} - s_n^2.$$

This system has two distinct fixed points. There is a flip–saddle at $s^* = 0.884$ with $\lambda_u = -1.924$ and $\lambda_s = 0.156$ and a non–flip saddle at $s^* = -1.584$ with $\lambda_u = 3.26$ and $\lambda_s = -0.092$. In addition, there is an unstable flip–saddle orbit of period 2 following the sequence $1.366 \to -0.666 \to 1.366$. There are no real orbits of period 3, but there is an unstable orbit of period 4 following the sequence $.893 \to .305 \to 1.575 \to -.989 \to .893$. These facts can be deduced by algebraic analysis of the equations.

In an experiment using the controlled system, the control parameter $x^* = C/(1-A)$ can be varied. The theory presented above indicates that the controlled system should undergo a bifurcation as x^* passes through s^*. For each value of x^*, the controlled system was iterated from a random initial condition and the values of s_n plotted after allowing a transient to decay. A bifurcation from a stable fixed–point to a stable period 2 as x^* passes through the flip–saddle value of $s^* = 0.884$. A different type bifurcation occurs at the non–flip saddle fixed–point at $s^* = -1.584$. To the left of the bifurcation point, the iterates are diverging to $-\infty$ and are not plotted.

Adding gaussian dynamical noise (of standard deviation 0.05) does not substantially alter the bifurcation diagram, suggesting that examination of the truncation control bifurcation diagram may be a practical way to read off the location of the unstable fixed points in an experimental preparation.

Unstable periodic orbits can be difficult to find in uncontrolled dynamics because there is typically little data near such orbits. Application of PPF control, even blindly, can stabilize such orbits and dramatically improve the ability to locate them. This, and the robustness of the control, may prove particularly useful in biological experiments where orbits may drift in time as the properties of the system change [Kap00].

1.9.8 Jerk Functions of Simple Chaotic Flows

Recall that the celebrated *Lorenz equations* (1.121) can be rewritten as

$$\dot{x} = -ax + ay, \qquad \dot{y} = -xz + bx - y, \qquad \dot{z} = xy - cz. \qquad (1.177)$$

Note that there are seven terms in the phase–flow of these equations, two of which are nonlinear (xz and xy); also, there are three parameters, for which Lorenz found chaos with $a = 10$, $b = 28$, and $c = 8/3$. The number of independent parameters is generally $d+1$ less than the number of terms for a $d-$dimensional system, since each of the variables (x, y, and z in this case) and time (t) can be arbitrarily rescaled [SL00]. The Lorenz system has been extensively studied, and there is an entire book [Spa82] devoted to it.

Although the Lorenz system is often taken as the prototypical chaotic flow, it is not the algebraically simplest such system [SL00]. Recall that in 1976, *Rössler* [Ros76] proposed his equations (1.38), rewritten here as

$$\dot{x} = -y - z, \qquad \dot{y} = x + ay, \qquad \dot{z} = b + xz - cz. \qquad (1.178)$$

Rössler phase–flow also has seven terms and two parameters, which Rössler took as $a = b = 0.2$ and $b = 5.7$, but only a single quadratic nonlinearity (xz).

In 1994, Sprott [Spr94] embarked on an extensive search for autonomous three–dimensional chaotic systems with fewer than seven terms and a single quadratic nonlinearity and systems with fewer than six terms and two quadratic nonlinearities. The *brute-force* method [Spr93a, Spr93b] involved the numerical solution of a huge number (about 10^8) systems of autonomous

ODEs with randomly chosen real coefficients and initial conditions. The criterion for chaos was the existence of a leading Lyapunov exponent. He found fourteen algebraically distinct cases with six terms and one nonlinearity, and five cases with five terms and two nonlinearities. One case was volume conserving (conservative), and all the others were volume–contracting (dissipative), implying the existence of a strange attractor. Sprott provided a table of the spectrum of Lyapunov exponents, the related Kaplan–Yorke dimension, and the types and eigenvalues of the unstable fixed–points for each of the nineteen cases [SL00].

Subsequently, Hoover [Hoo95] pointed out that the conservative case found by Sprott in [Spr94]

$$\dot{x} = y, \qquad \dot{y} = -x + yz, \qquad \dot{z} = 1 - y^2, \qquad (1.179)$$

is a special case of the Nosé–Hoover thermostated dynamic system that had earlier been shown [PHV86] to exhibit time–reversible *Hamiltonian chaos*.

In response to Sprott's work, Gottlieb [Got96] pointed out that the conservative system (1.179) could be recast in the *explicit third–order form*

$$\dddot{x} = -\dot{x} + \ddot{x}(x + \ddot{x})/\dot{x},$$

which he called a '*jerk function*' since it involves a third derivative of $\ddot{x}$, which in a mechanical system is the time rate of change of the acceleration, also called the 'jerk' [Sch78]. It is known that any explicit ODE can be cast in the form of a system of coupled first–order ODEs, but the converse does not hold in general. Even if one can reduce the dynamical system to a jerk form for each of the phase–space variables, the resulting differential equations may look quite different. Gottlieb asked the provocative question 'What is the simplest jerk function that gives chaos?'

One response was provided by Linz [Lin97] who showed that both the original Rössler model and the Lorenz model can be reduced to jerk forms. The Rössler model (1.178) can be rewritten (in a slightly modified form) as

$$\dddot{x} + (c - \varepsilon + \varepsilon x - \dot{x})\ddot{x} + [1 - \varepsilon c - (1 + \varepsilon^2)x + \varepsilon \dot{x}]\dot{x} + (\varepsilon x + c)x + \varepsilon = 0,$$

where $\varepsilon = 0.2$ and $c = 5.7$ gives chaos. Note that the jerk form of the Rössler equation is a rather complicated quadratic polynomial with 10 terms.

The Lorenz model in (1.177) can be written as

$$\dddot{x} + (1 + \sigma + b - \dot{x}/x)\ddot{x} + [b(1 + \sigma + x^2) - (1 + \sigma)\dot{x}/x]\dot{x} - b\sigma(r - 1 - x^2)x = 0.$$

The jerk form of the Lorenz equation is not a polynomial since it contains terms proportional to $\dot{x}/x$ as is typical of dynamical systems with multiple nonlinearities. Its jerk form contains eight terms.

Linz [Lin97] showed that Sprott's case R model (see [Spr94]) can be written as a polynomial with only five terms and a single quadratic nonlinearity

1.9 Basics of Chaos Control

$$\dddot{x} + \ddot{x} - x\dot{x} + ax + b = 0,$$

with chaos for $a = 0.9$ and $b = 0.4$.

Sprott [Spr97] also took up Gottlieb's challenge and embarked on an extensive numerical search for chaos in systems of the explicit form

$$\dddot{x} = J(\ddot{x}, \dot{x}, x),$$

where the jerk function J is a simple quadratic or cubic polynomial. He found a variety of cases, including two with three terms and two quadratic nonlinearities in their jerk function,

$$\dddot{x} + ax\ddot{x} - \dot{x}^2 + x = 0,$$

with $a = 0.645$ and

$$\dddot{x} + ax\ddot{x} - x\dot{x} + x = 0,$$

with $a = -0.113$, and a particularly simple case with three terms and a single quadratic nonlinearity

$$\dddot{x} + a\ddot{x} \pm \dot{x}^2 + x = 0, \tag{1.180}$$

with $a = 2.017$. For this value of a, the Lyapunov exponents are $(0.0550, 0, -2.0720)$ and the Kaplan–Yorke dimension is $d_{KY} = 2.0265$.

Equation (1.180) is simpler than any previously discovered case. The range of the parameter a over which chaos occurs is quite narrow ($2.0168\ldots < a < 2.0577\ldots$). It also has a relatively small basin of attraction, so that initial conditions must be chosen carefully. One choice of initial conditions that lies in the basin of attraction is $(x, y, z) = (0, 0, \pm 1)$, where the sign is chosen according to the sign of the $\pm\dot{x}^2$ term.

All above systems share a common *route to chaos*. The control parameter a can be considered a *damping rate* for the nonlinear oscillator. For large values of a, there are one or more stable equilibrium points. As a decreases, a *Hopf bifurcation* (see [CD98]) occurs in which the equilibrium becomes unstable, and a stable limit cycle is born. The limit cycle grows in size until it bifurcates into a more complicated limit cycle with two loops, which then bifurcates into four loops, and so forth, in a *sequence of period doublings* until chaos finally onsets. A further decrease in a causes the *chaotic attractor* to grow in size, passing through infinitely many periodic windows, and finally becoming unbounded when the attractor grows to touch the boundary of its *basin of attraction* (a 'crisis').

Recently, Malasoma [Mal00] joined the search for simple chaotic jerk functions and found a cubic case as simple as (1.180) but of a different form

$$\dddot{x} + a\ddot{x} - x\dot{x}^2 + x = 0,$$

which exhibits chaos for $a = 2.05$. For this value of a, the Lyapunov exponents are $(0.0541, 0, -2.1041)$, and the Kaplan–Yorke dimension is $d_{KY} = 2.0257$.

This case follows the usual period–doubling route to chaos, culminating in a boundary crisis and unbounded solutions as a is lowered. The range of a over which chaos occurs is very narrow, $(2.0278\ldots < a < 2.0840\ldots)$. There is also a second extraordinarily small window of chaos for $(0.0753514\ldots < a < 0.0753624\ldots)$, which is five thousand times smaller than the previous case. Malasoma points out that this system is invariant under the parity transformation $x \to -x$ and speculates that this system is the simplest such example.

Both Linz and Sprott pointed out that if the jerk function is considered the time derivative of an acceleration of a particle of mass m, then *Newton's second law* implies a *force* F whose time derivative is $\dot{F} = mJ$. If the force has an explicit dependence on only $\dot{x}$, x, and time, it is considered to be 'Newtonian jerky'. The condition for $F = F(\dot{x}, x, t)$ is that J depends only linearly on $\ddot{x}$. In such a case the force in general includes a *memory term* of the form

$$M = \int_0^t G(x(\tau))\,d\tau,$$

which depends on the dynamical history of the motion.

The jerk papers by Linz [Lin97] and Sprott [Spr97] appeared in the same issue of the American Journal of Physics and prompted von Baeyer [Bae98] to comment: "The articles with those funny titles are not only perfectly serious, but they also illustrate in a particularly vivid way the revolution that is transforming the ancient study of mechanics into a new science – one that is not just narrowly concerned with the motion of physical bodies, but that deals with changes of all kinds." He goes on to say that the method of searching for chaos in a large class of systems "is not just empty mathematical formalism. Rather it illustrates the arrival of a new level of abstraction in physical science... At that higher level of abstraction, dynamics has returned to the classical Aristotelian goal of trying to understand all change."

1.9.9 Example: Chaos Control in Molecular Dynamics

Recall that classically modelled molecular dynamics are often characterized by the presence of chaotic behavior. A central issue in these studies is the role of a control variable. A small variation of this variable can cause a transition from the periodic or quasi–periodic regime to the chaotic regime. For example, the molecular energy can serve as a control variable. It was shown in [BRR95] that the isotopic mass could be viewed as a discrete control variable and related effects were also evident in quantum calculations.

In this approach the variation of the control variable or parameters changes the route taken by the dynamics of the system. It was shown that a small time–dependent perturbation of the control parameters could convert a chaotic attractor to any of a large number of possible attracting periodic orbits [Ott93]. The control parameter could stabilize the chaotic dynamics about some periodic orbit.

In [BRR95], a general method to control molecular systems has been proposed by employing optimally designed laser pulses. This approach is capable of designing the time–dependent laser pulse shapes to permit the steering of intramolecular dynamics to desired physical objectives. Such objectives include selective bond breaking through infrared excitation [SR91, SR90, SR91] and through electronic excitation between multiple surfaces [TR85, KRG89], control of curve-crossing reactions [GNR92, CBS91], selective electronic excitation in condensed media [GNR91, GNR93, SB89], control of the electric susceptibility of a molecular gas [SSR93] and selective rotational excitation [JLR90]. The pulse shape design is obtained by minimization of a design cost functional with respect to the control field pulse which corresponds to the optimization of the particular objective. Also, multiple solutions of the optimal control equations exist, which gives flexibility for adding further terms in the cost functional or changing certain design parameters so that the particular objective is better reached.

In this section, following [BRR95, BRR95], we will explore optimal control properties of nonlinear classical Hamiltonians under certain constraints in order to suppress their chaotic behavior. This approach to molecular dynamics simulations reduces the computational cost and instabilities especially associated with the treatment of the Lagrange multipliers.

Conventional optimal control with an ensemble of N trajectories would call for the introduction of rN Lagrange constraint functions where r is the dimension of the phase–space. In contrast the new method only requires the same number r of Lagrange multipliers as the dimension of the phase–space of a single molecule. The reduction of Lagrange multipliers is achieved by only following the control of the average trajectory. Here we present the application of these ideas to the classical control of chaotic Hamiltonian dynamics, where control is realized by a small external interaction. An important application of this approach is to finding regular orbits amongst a dense set of irregular trajectories.

Hamiltonian Chaotic Dynamics

The methodology introduced below is quit general, but to aid its presentation we develop it in the context of controlling the motion of a particular Hamiltonian [BRR95]

$$H = \frac{1}{2}(P_1^2 + P_2^2 + R_1^4 + R_2^4) - KR_1^2 R_2^2, \qquad (1.181)$$

where K is a coupling constant between R_1 and R_2. The Hamiltonian in (5.1) presents certain characteristics such as scaling, where the motion of the system at any energy can be determined from $E = 1$ by simple scaling relations.

This latter scaling property can be shown through Poincaré sections and from this approach it is also possible to see the transition to chaos. The *Poincaré section* is constructed as follows. The Hamiltonian is associated with a 4D phase–space (R_1, R_2, P_1, P_2). We can reduce this to a two dimensional phase–space by fixing the energy and one variable, for instance $R_1 = 0$. With this method, the Poincaré sections permit us to establish the character of the system motion at each desired energy and value of the parameter K. When we increase K, the cross term in this Hamiltonian increases its impact over the un–coupled quartic terms. Thus, for a fixed energy, K can be used as a perturbation control parameter, showing the scaling properties of the energy at different values of K.

Optimal Control Algorithm

The dynamics equation for the Hamiltonian at very strong coupling constant, in the chaotic regime, can be modified by the a small perturbation from an external interaction field. This external perturbation can guide the chaotic motion towards the average trajectory of an ensemble or draw the system towards a periodic or quasi–periodic orbit(if it exists). Based on this control idea, we introduce an interaction term for controlling the chaotic motion as [BRR95]
$$H_{int}(R(t), \epsilon(t)) = (R_1(t) - \eta R_2(t))\epsilon(t) \quad (1.182)$$
where η is a parameter. This term is similar to previous work in the control of chaos, where the parameter η is zero [SR91]. It is physically desirable to keep this external interaction small; an intense external interaction could itself lead to chaotic motion.

Let us define a set of N trajectories corresponding to the initial conditions as a vector $w(t) = [w_1(t), w_2(t)...w_{4N}(t)]$ where the first N values are the positions for $R_1(t)$, $N + 1$ to $2N$ the positions of $R_2(t)$, $2N + 1$ to $3N$ the values for $P_1(t)$ and the last values $3N+1$ to $4N$ for $P_2(t)$. With this definition the classical dynamics is governed by Hamiltonian's equations,

$$\begin{aligned} \dot{w}_1(t) &= g_1(w(t), \epsilon(t)), \\ \dot{w}_2(t) &= g_2(w(t), \epsilon(t)), \\ &\cdots \qquad \cdots \\ \dot{w}_{4N}(t) &= g_{4N}(\mathbf{w(t)}, \epsilon(t)), \end{aligned} \quad (1.183)$$

which depend on the dynamical observables and the time dependent field $\epsilon(t)$. The functions $g_i(w(t), \epsilon(t))$ may be readily identified as momentum or coordinate derivatives of the Hamiltonian. Alteration of the external control interaction $\epsilon(t)$ can steer about the classical dynamics and change its characteristic behavior.

The classical average of the coordinates and momenta are [BRR95]

$$\langle z_i(t) \rangle = \frac{1}{N} \sum_{j=(i-1)N+1}^{iN} w_j(t), \quad (i = 1..4), \quad (1.184)$$

1.9 Basics of Chaos Control

and the average approximate energy as a function of time is

$$E_l(t) = \frac{1}{N} \sum_{j=(l+1)N+1}^{(l+2)N} \frac{w_j^2(t)}{2} + \frac{1}{N} \sum_{j=(l-1)N+1}^{lN} V(w_j(t)), \quad (1.185)$$

where $l = 1$ and 2 labels the two degrees of freedom and $V(w_j(t))$ represents the un–coupled quartic term in the (5.1). This energy is approximate because it does not account for the nonlinear coupling term. With these definitions, we can proceed to define the control cost functional.

In the formulation of the classical optimal control problem, we first identify the physical objective and constraints. The objective at time T is to control the ensemble of trajectories by requiring that they are confined to a small region in the phase–space. In addition to the final time goal, another physical objective is to require that the chaotic trajectories evolve to either be close to the average trajectory $Z_i(t) = \langle z_i(t) \rangle$ or be confined around an imposed fiducial trajectory $Z_i(t)$. This fiducial trajectory may be a periodic or quasi–periodic trajectory. The cost functional, in quadratic form, that represents the objective and cost is [BRR95]

$$J[w,\epsilon] = \sum_{i=1}^{4} \sigma_i(\langle z_i(T) \rangle - \gamma_i)^2 + \frac{1}{N} \sum_{i=1}^{4} \sum_{j=(i-1)N+1}^{iN} \varrho_i(w_j(T) - Z_i(T))^2$$

$$+ \frac{1}{N} \int_0^T dt \sum_{i=1}^{4} \sum_{j=(i-1)N+1}^{iN} W_i(w_j(t) - Z_i(t))^2 + w_e \int_0^T dt\epsilon(t)^2, \quad (1.186)$$

where W_i, w_e, σ_i and ϱ_i are positive weights, which balance the importance of each term in the cost functional. The notation $J[w,\epsilon]$ indicated the functional dependence on $w(t), \epsilon(t)$ for $0 \leq t \leq T$. γ_i is the a specified constant target for each average variable in the phase–space. This cost functional can be decomposed into a sum of four terms where each of them represents the component average in the phase–space. $J[w,\epsilon]$ is to be minimized with respect to the external interaction subject to satisfying the equations of motion. This latter constraint can be satisfied by introducing Lagrange multipliers, $\lambda_i(t)$, to get an unconstrained functional

$$\bar{J}[w,\epsilon] = J[w,\epsilon] - \frac{1}{N} \int_0^T dt \sum_{i=1}^{4} \lambda_i(t) [\sum_{j=(i-1)N+1}^{iN} \dot{w}_j(t) - g_j(w(t),\epsilon(t))]. \quad (1.187)$$

Note that there are four Lagrange multipliers corresponding to constraints on the average equation of motion. These four Lagrange multipliers are introduced to be consistent with the cost functional (1.186) that represent the sum of four terms over each average component on the phase–space. The classical variational problem for the N initial conditions is given by

$$\epsilon(t) \to \epsilon(t) + \delta\epsilon(t), \qquad w_j(t) \to w_j(t) + \delta w_j(t), \qquad \lambda_i(t) \to \lambda_i(t) + \delta\lambda_i(t).$$

The variation of $\lambda_i(t)$ yields the average Hamiltonian equations of motion. The variation of $w_j(t)$ and $\epsilon(t)$, gives the following equations

$$\delta \bar{J}[w,\epsilon] = \sum_{i=1}^{4} \lambda_i(T)\delta\langle z_i(T)\rangle - 2\sum_{i=1}^{4} \sigma_i[\langle z_i(T)\rangle - \gamma_i]\delta\langle z_i(T)\rangle \qquad (1.188)$$

$$-\frac{2}{N}\sum_{i=1}^{4}\sum_{j=(i-1)N+1}^{iN} \varrho_i[w_j(T) - Z_i(T)][\delta w_j(T) - \delta Z_i(T)]$$

$$+ \int_0^T dt \sum_{i=1}^{4}\sum_{j=(i-1)N+1}^{iN} \{\dot{\lambda}_i(t)\delta\langle z_i(t)\rangle + \frac{2W_i}{N}[w_j(t) - Z_i(t)][\delta w_j(t) - \delta Z_i(t)]$$

$$+ \frac{\lambda_i(t)}{N}\sum_{k=1}^{4N} \frac{\partial g_j(w(t),\epsilon(t))}{\partial w_k(t)} \delta w_k(t)\}$$

$$+ \int_0^T dt[2\omega_\epsilon\epsilon(t) + \frac{1}{N}\sum_{i=1}^{4}\sum_{j=(i-1)N+1}^{iN} \lambda_i(t)\frac{\partial g_j(w(t),\epsilon(t))}{\partial \epsilon(t)}]\delta\epsilon(t),$$

where $\delta\langle z_i(t)\rangle$ is given by

$$\delta\langle z_i(t)\rangle = \frac{1}{N}\sum_{j=(i-1)N+1}^{iN} \delta w_j(t),$$

and take account all the variations for each initial condition at each instant of time t. The variation of $Z_i(t)$ is equal to zero if the target trajectory is a fiducial trajectory (constant trajectory in the control process) and $\delta Z_i(t) = \delta\langle z_i(t)\rangle$ if we consider the control evolution over its average trajectory. These two methods to control the chaotic behavior will be explained below.

The final conditions at time T,

$$\lambda_i(T)\delta\langle z_i(T)\rangle = 2\sigma_i[\langle z_i(T)\rangle - \gamma_i]\delta\langle z_i(T)\rangle \qquad (1.189)$$

$$+ \frac{2}{N}\sum_{j=(i-1)N+1}^{iN} \varrho_i[w_j(T) - Z_i(T)][\delta w_j(T) - \delta Z_i(T)]$$

are obtained through (1.188). The second variational equation for the Lagrange multipliers is derived from the second part of (1.188),

$$\int_0^T dt \sum_{i=1}^{4}\sum_{j=(i-1)N+1}^{iN} \{\dot{\lambda}_i(t)\delta\langle z_i(t)\rangle + \frac{2W_i}{N}[w_j(t) - Z_i(t)][\delta w_j(t) - \delta Z_i(t)]$$

$$+ \frac{\lambda_i(t)}{N}\sum_{k=1}^{4N} \frac{\partial g_j(w(t),\epsilon(t))}{\partial w_k(t)} \delta w_k(t)\} = 0. \qquad (1.190)$$

The gradient with respect to the field is given by

$$\frac{\delta \bar{J}[w,\epsilon]}{\delta \epsilon(t)} = 2w_e \epsilon(t) + \frac{1}{N} \sum_{i=1}^{4} \sum_{j=(i-1)N+1}^{iN} \lambda_i(t) \frac{\partial g_j(w(t), \epsilon(t))}{\partial \epsilon(t)}, \qquad (1.191)$$

and the minimum field solution is

$$\epsilon(t) = -\frac{1}{2Nw_e} \sum_{i=1}^{4} \sum_{j=(i-1)N+1}^{iN} \lambda_i(t) \frac{\partial g_j(w(t), \epsilon(t))}{\partial \epsilon(t)}. \qquad (1.192)$$

This minimum solution links Hamilton's equations with (1.190) for the Lagrange multipliers. In order to integrate the equations for the Lagrange multipliers we need to know the classical evolution for each component $w_j(t)$ and the final condition of the Lagrange multipliers at time T. These final conditions for the Lagrange multipliers in (1.189) are given by the objective in the control dynamics. The solution (if it exists) of these coupled equations prescribes the external control field interaction $\epsilon(t)$.

In order to find a solution of the (1.190), the Lagrange multipliers were chosen to satisfy the following equation

$$\sum_{i=1}^{4} \dot{\lambda}_i(t) \delta \langle z_i(t) \rangle = -\frac{1}{N} \sum_{i=1}^{4} \sum_{j=(i-1)N+1}^{iN} \{2W_i[w_j(t) - Z_i(t)][\delta w_j(t) - \delta Z_i(t)]$$

$$+ \lambda_i(t) \sum_{k=1}^{4N} \frac{\partial g_j(w(t), \epsilon(t))}{\partial w_k(t)} \delta w_k(t)\}, \qquad (1.193)$$

where each Lagrange multiplier gives information about the status of the average or fiducial trajectory $Z_i(t)$. The values of the Lagrange multipliers at each instant of time directs how the external field must evolve. The generalization of these equations is obvious when we increase the dimensionality of the system. Equations (1.193) are over–specified, and a physically motivated assumption on the nature of the solutions must be introduced to close the equations.

There is considerable flexibility in the choice of Lagrange multipliers and within this control formalism we will address two different ways to control the chaotic trajectories. The basic role of the Lagrange multipliers is to guide the physical system towards a desirable solution. We shall take advantage of these observations to develop the two approaches towards control of chaotic trajectories. The first method represents the desire to achieve control around the mean trajectory. In this case the control dynamics searches the phase–space, using the variational principle, finding the mean trajectory and the control field works to keep the ensemble around this trajectory. The second method directs the formerly chaotic trajectories towards a periodic or quasi–periodic orbit or other specified fiducial trajectory.

Following the Mean Trajectory

In this case the fiducial trajectory $Z_i(t)$ is not fixed, and can change at each cycle of cost functional minimization. Here the choice is the average trajectory, $Z_i(t) = \langle z_i(t) \rangle$. Substituting the equations of motion into (1.193) and (1.191), we have [BRR95]

$$\dot{\lambda}_1(t)\delta\langle z_1(t)\rangle = \frac{1}{N}\sum_{j=1}^{N}\delta w_j(t)\{2\lambda_3(t)[3w_j(t)^2 - Kw_{j+N}^2(t)]$$
$$-4K\lambda_4(t)w_j(t)w_{j+N}(t)\}$$
$$-\frac{2W_1}{N}\sum_{j=1}^{N}[w_j(t) - \langle z_1(t)\rangle][\delta w_j(t) - \delta\langle z_1(t)\rangle], \quad (1.194)$$

$$\dot{\lambda}_2(t)\delta\langle z_2(t)\rangle = \frac{1}{N}\sum_{j=1}^{N}\delta w_{j+N}(t)\{2\lambda_4(t)[3w_{j+N}^2(t) - Kw_j^2(t)]$$
$$-4K\lambda_3(t)w_j(t)w_{j+N}(t)\}$$
$$-\frac{2W_2}{N}\sum_{j=1}^{N}[w_{j+N}(t) - \langle z_2(t)\rangle][\delta w_{j+N}(t) - \delta\langle z_2(t)\rangle],$$

$$\dot{\lambda}_3(t)\delta\langle z_3(t)\rangle = -\lambda_1(t)\delta\langle z_3(t)\rangle - \frac{2W_3}{N}\sum_{j=1}^{N}[w_{j+2N}(t)$$
$$-\langle z_3(t)\rangle][\delta w_{j+2N}(t) - \delta\langle z_3(t)\rangle],$$

$$\dot{\lambda}_4(t)\delta\langle z_4(t)\rangle = -\lambda_2(t)\delta\langle z_4(t)\rangle - \frac{2W_4}{N}\sum_{j=1}^{N}[w_{j+3N}(t)$$
$$-\langle z_4(t)\rangle][\delta w_{j+3N}(t) - \delta\langle z_4(t)\rangle],$$

and the gradient of the cost functional with respect of the external field is

$$\frac{\delta \bar{J}[w(t),\epsilon(t)]}{\delta\epsilon(t)} = 2w_e\epsilon(t) - \lambda_3(t) + \eta\lambda_4(t). \quad (1.195)$$

This equation only depends on the values of the Lagrange multipliers, $\lambda_3(t)$, $\lambda_4(t)$, that represent the driving force. In this approach the control dynamics equations to be solved result from (1.195), (1.194) and the final conditions for the Lagrange multipliers are given by (1.189).

When the classical trajectories are very close or are attracted to a regular orbit, then these control equations for the Lagrange multipliers can be closed and written as

$$\dot{\lambda}_1(t) = 2\lambda_3(t)[3\langle z_1(t)\rangle^2 - K\langle z_2(t)\rangle^2] - 4K\lambda_4(t)\langle z_1(t)\rangle\langle z_2(t)\rangle,$$
$$\dot{\lambda}_2(t) = 2\lambda_4(t)[3\langle z_2(t)\rangle^2 - K\langle z_1(t)\rangle^2] - 4K\lambda_3(t)\langle z_1(t)\rangle\langle z_2(t)\rangle,$$
$$\dot{\lambda}_3(t) = -\lambda_1(t), \qquad \dot{\lambda}_4(t) = -\lambda_2(t). \quad (1.196)$$

This gives the control dynamics equations for one effective classical trajectory, in this case the average trajectory. Thus, the system to solve is (1.195) and (1.196) with boundary conditions (1.189).

Following a Fixed Trajectory

This case treats the desire to draw the chaotic trajectories around a specified fiducial trajectory. This trajectory is the target trajectory, $Z_i(t)$. In the present case the choice of this trajectory is the average trajectory produced from the original ensemble evolving from the initial conditions. The control dynamics equations are [BRR95]

$$\dot{\lambda}_1(t)\delta\langle z_1(t)\rangle = \frac{1}{N}\sum_{j=1}^{N}\delta w_j(t)\{2\lambda_3(t)[3w_j(t)^2 - Kw_{j+N}^2(t)]$$
$$-4K\lambda_4(t)w_j(t)w_{j+N}(t)\}$$
$$-\frac{2W_1}{N}\sum_{j=1}^{N}[w_j(t) - Z_1(t)]\delta w_j(t),$$

$$\dot{\lambda}_2(t)\delta\langle z_2(t)\rangle = \frac{1}{N}\sum_{j=1}^{N}\delta w_{j+N}(t)\{2\lambda_4(t)[3w_{j+N}^2(t) - Kw_j^2(t)]$$
$$-4K\lambda_3(t)w_j(t)w_{j+N}(t)\}$$
$$-\frac{2W_2}{N}\sum_{j=1}^{N}[w_{j+N}(t) - Z_2(t)]\delta w_{j+N}(t),$$

$$\dot{\lambda}_3(t)\delta\langle z_3(t)\rangle = -\lambda_1(t)\delta\langle z_3(t)\rangle - \frac{2W_3}{N}\sum_{j=1}^{N}[w_{j+2N}(t) - Z_3(t)]\delta w_{j+2N},\quad(1.197)$$

$$\dot{\lambda}_4(t)\delta\langle z_4(t)\rangle = -\lambda_2(t)\delta\langle z_4(t)\rangle - \frac{2W_4}{N}\sum_{j=1}^{N}[w_{j+3N}(t) - Z_4(t)]\delta w_{j+3N},$$

with the final conditions for the Lagrange multipliers as

$$\lambda_i(T)\delta\langle z_i(T)\rangle = 2\sigma_i[\langle z_i(T)\rangle - \gamma_i]\delta\langle z_i(T)\rangle + \frac{2}{N}\sum_{j=(i-1)N+1}^{iN}\varrho_i[w_j(T) - Z_i(T)]\delta w_j(T).$$
(1.198)

In this case the system of equations to solve are (1.195), (1.197) and the boundary conditions (1.198).

Once again, the self consistency of (1.197) needs to be addressed and here we seek the non-chaotic result of each trajectory closely following the corresponding initially specified mean $Z_i(t)$. Then for this one classical trajectory, we get the following equations

$$\dot\lambda_1(t) = 2\lambda_3(t)[3\langle z_1(t)\rangle^2 - K\langle z_2(t)\rangle^2]$$
$$-4K\lambda_4(t)\langle z_1(t)\rangle\langle z_2(t)\rangle - 2W_1[\langle z_1(t)\rangle - Z_1(t)],$$
$$\dot\lambda_2(t) = 2\lambda_4(t)[3\langle z_2(t)\rangle^2 - K\langle z_1(t)\rangle^2]$$
$$-4K\lambda_3(t)\langle z_1(t)\rangle\langle z_2(t)\rangle - 2W_2[\langle z_2(t)\rangle - Z_2(t)], \qquad (1.199)$$
$$\dot\lambda_3(t) = -\lambda_1(t) - 2W_3[\langle z_3(t)\rangle - Z_3(t)],$$
$$\dot\lambda_4(t) = -\lambda_2(t) - 2W_4[\langle z_4(t)\rangle - Z_4(t)].$$

This case expresses the desire and assumption that the formally chaotic trajectories are tightly drawn around the fiducial trajectory. The system to solve is (1.199), (1.195) and (1.198).

Utilizing these two methods to control the irregular motion using optimal control theory, we present two different algorithms in order to find the optimal solution $\epsilon(t)$. The first algorithm returns to the variational principle over all classical trajectories as in (1.193). This evaluation of the Lagrange multipliers takes account the single variation of each component in the phase–space. The procedure is implemented by evaluating the trajectory variations as

$$\delta w_j(t) \simeq w_j(t) - w_j^{old}(t) \qquad (1.200)$$

at each instant of time from two successive steps of the minimization process ($w_j^{old}(t)$ is the value in the previous minimization step). This first algorithm is a rigorous test (accepting the finite difference nature of (1.200)) of the arguments leading to the approximate Lagrange equations (1.196) and (1.199).

The second algorithm accepts the approximate form of (1.196) and (1.199) to ultimately yield the field $\epsilon(t)$. Although the average trajectory is used in this approach, the full dynamical equations for $w_j(t)$ are followed and used to evaluate the cost functional $J[\mathbf{w}, \epsilon]$ attesting to the quality of the results.

Computational Method

The optimal control of N classical trajectories is given by the solution of the (1.194), (1.189) and (1.195) for following the average trajectory, and (1.197), (1.198) and (1.195) for a fixed fiducial trajectory.

The following iterative scheme is adopted to find the external control field $\epsilon(t)$ that meets the physical objectives, for N classical trajectories:

a) Make an initial guess for the external interaction $\epsilon(t)$.

b) Integrate the equation of motion (1.183) forward in time for all initial conditions.

c) Calculate the cost functional and boundary conditions for the Lagrange multipliers $\lambda(T)$.

d) Integrate the Lagrange multiplier $\lambda(t)$ equation backwards in time.

f) Calculate the gradient and save all the classical trajectories as the 'old' configuration.

g) Upgrade the new external interaction as [BRR95]

$$\epsilon^{new}(t) = \epsilon(t) - \alpha \frac{\delta \bar{J}}{\delta \epsilon(t)},$$

where α is a suitable small positive constant. This process is repeated until the cost functional and field converge. In one of the algorithmic approaches we need to calculate the variation of $w_j(t)$ as

$$\delta w_j(t) \simeq w_j(t) - w_j^{old}(t),$$

where $w_j^{old}(t)$ is the previous value. In this case at the beginning of the minimization procedure, we choose one classical trajectory at random from the set of initial conditions as the 'old' configuration.

The procedure starts by examining a Poincaré section and choosing one point and giving a random distribution around this point (e.g., $R_2 \pm |\delta|, P_2 \pm |\delta|$), where δ is a random number between 0 and 1×10^{-3} (or another small number). After two cycles (one cycle is defined as the apparent quasi–period of the $R_1(t)$ motion at the beginning of the simulation), the system starts its chaotic behavior and the motion is bounded. The approximate energy for each degree of freedom reflects the irregular behavior with strong coupling between them. In five cycles the system is completely irregular, and the objective is to control this irregular motion by keeping the classical trajectories close during a time interval of 5 cycles; no demand is made on the behavior greater than 5 cycles, although, it is also possible to apply this method for more that 5 cycles [BRR95].

1.10 Spatio-Temporal Chaos Control

Excitable media denotes a class of systems that share a set of features which make their dynamical behavior qualitatively similar. These features include (i) the existence of two characteristic dynamical states, comprising a stable *resting state* and a meta-stable *excited state*, (ii) a *threshold* value associated with one of the dynamical variables characterizing the system, on exceeding which, the system switches from the resting state to the excited state, and (iii) a *recovery period* following an excitation, during which the response of the system to a supra-threshold stimulus is diminished, if not completely absent [KS98]. Natural systems which exhibit such features include, in biology, cells such as neurons, cardiac myocytes and pancreatic beta cells, all of which are vital to the function of a complex living organism. Other examples of dynamical phenomena associated with excitable media include cAMP waves observed during aggregation of slime mold, calcium waves observed in Xenopus oocytes, muscle contractions during childbirth in uterine tissue, chemical waves observed in the *Belusov–Zhabotinsky reaction* and concentration patterns in CO–oxidation reaction on Pt(110) surface. Excitation in such systems is observed as the characteristic *action potential*, where a variable associated with the system (e.g., *membrane potential*, in the case of biological cells) increases very fast from its resting value to the peak value corresponding to the excited state, followed by a slower process during which it gradually returns to the resting state [SS07].

The simplest model system capable of exhibiting all these features is the generic *Fitzhugh–Nagumo model*:

$$\dot{e} = e(1-e)(e-b) - g, \qquad \dot{g} = \epsilon(ke - g), \qquad (1.201)$$

which, having only two variables, is obviously incapable of exhibiting chaos. However, when several such sets are coupled together diffusively to simulate a spatially extended media (e.g., a piece of biological tissue made up of a large number of cells), the resulting high–dimensional dynamical system can display chaotic behavior. The genesis of this *spatio-temporal chaos* lies in the distinct property of interacting waves in excitable media, which mutually annihilate on colliding. This is a result of the fact that an excitation wavefront is followed by a region whose cells are all in the recovery period, and which, therefore, cannot be stimulated by another excitation wavefront, as for example when two waves cross each other.[46] Interaction between such waves result in the creation of spatial patterns, referred to variously as *reentrant excitations* (in 1D), *vortices* or *spiral waves* (in 2D) and *scroll waves* (in 3D), which form when an excitation wavefront is broken as the wave propagates across partially recovered tissue or encounters an inexcitable obstacle [JAD99]. The free ends of the wavefront gradually curl around to form spiral waves. Once formed, such waves become self-sustained sources of high–frequency excitation in the medium, and usually can only be terminated through external intervention. The existence of nonlinear properties of wave propagation in several excitable media (e.g., the dependence of the action potential duration, as well as the velocity of the excitation wave, on the distance of a wave from the preceding excitation wave) can lead to complex non–chaotic spatio-temporal rhythms, which are important targets of control as they are often associated with clinical arrhythmias (i.e., disturbances in the natural rhythm of the heart). Thus, spiral waves are associated with periodic as well as quasi-periodic patterns of temporal activity.

In this section, following [SS07], we shall not be discussing the many schemes proposed to terminate single spiral waves, but instead, focus on the control of spatiotemporally chaotic patterns seen in excitable media (in 2 or 3 dimensions), that occur when under certain conditions, spiral or scroll waves become unstable and break up. Various mechanisms of such breakup have been identified,[47] including meandering of the spiral focus. If the meandering is sufficiently high, the spiral wave can collide with itself and break up spontaneously, resulting in the creation of multiple smaller spirals (see Figure 1.54).

---

[46] Unlike waves in ordinary diffusive media which dissipate as they propagate further, excitation waves are self–regenerating.

[47] For a discussion of the multiple scenarios of spiral wave breakup, see [FCH02].

The process continues until the spatial extent of the system is spanned by several coexisting spiral waves that activate different regions without any degree of coherence. This state of *spiral turbulence* marks the onset of spatio-temporal chaos, as indicated by the *Lyapunov spectrum* and *Kaplan–Yorke dimension* [PPS02].

Fig. 1.54. Onset of spatio-temporal chaos in the 2D Panfilov model. The initial condition is a broken plane wave that is allowed to curl around into a spiral wave (left). Meandering of the spiral focus causes wave-breaks to occur (center) that eventually result in spiral turbulence, with multiple independent sources of high-frequency excitation (right) (adapted and modified from [SS07]).

Controlling spatio-temporal chaos in excitable media has certain special features. Unlike other chaotic systems, here the response to a control signal is not proportional to the signal strength because of the existence of a threshold. As a result, an excitable system shows discontinuous response to control. For instance, regions that have not yet recovered from a previous excitation or where the control signal is below the threshold, will not be affected by the control stimulus at all. Also, the focus of control in excitable media is to eliminate all activity rather than to stabilize unstable periodic behavior. This is because the problem of chaos termination has great practical importance in the clinical context, as the spatiotemporally chaotic state has been associated with the cardiac problem of ventricular fibrillation (VF). VF involves incoherent activation of the heart that results in the cessation of pumping of blood, and is fatal within minutes in the absence of external intervention. At present, the only effective treatment is electrical defibrillation, which involves applying very strong electrical shocks across the heart muscles, either externally using a defibrillator or internally through implanted devices. The principle of operation for such devices is to overwhelm the natural cardiac dynamics, so as to drive all the different regions of the heart to rest simultaneously, at which time the cardiac pacemaker can take over once again. Although the exact mechanism by which this is achieved is still not completely understood, the danger of using such large amplitude control (involving $\sim kV$ externally and $\sim 100V$ internally) is that, not only is it excruciatingly painful to the patient, but by causing damage to portions of cardiac tissue which subsequently result

1 Basics of Nonlinear and Chaotic Dynamics

in scars, it can potentially increase the likelihood of future arrhythmias. (i.e., abnormalities in the heart's natural rhythm). Therefore, devising a low–power control method for spatio-temporal chaos in excitable media promises a safer treatment for people at risk from potentially fatal cardiac arrhythmias.

In this section, we present most of the recent control methods that have been proposed for terminating spatio-temporal chaos in excitable media.[48]. These methods are also often applicable to the related class of systems known as oscillatory media, described by complex *Landau–Ginzburg equation* [AK02], which also exhibit spiral waves and spatio-temporal chaos through spiral breakup. We have broadly classified all control schemes into three types, depending on the nature of application of the control signal. If every region of the media is subjected to the signal (which, in general, can differ from region to region) it is termed as *global chaos control*; on the other hand, if the control signal is applied only at a small, localized region from which its effects spread throughout the media, this is called *local chaos control*. Between these two extremes lie control schemes where perturbations are applied simultaneously to a number of spatially distant regions. We have termed these methods as *non-global, spatially extended chaos control*. While global control may be the easiest to understand, involving as it does the principle of synchronizing the activity of all regions, it is also the most difficult to implement in any practical situation. On the other hand, local control will be the easiest to implement (requiring a single control point) but hardest to achieve [SS07].

1.10.1 Models of Spatio-Temporal Chaos in Excitable Media

The generic Fitzhugh-Nagumo model for excitable media (1.201) exhibits a structure that is common to most models used in the papers discussed here. Typically, the dynamics is described by a fast variable, $e(\mathbf{x}, t)$, and a slow variable, $g(\mathbf{x}, t)$, the ratio of timescales being given by ϵ. For biological cells, the fast variable is often associated with the transmembrane potential, while the slow (recovery) variable represents an effective membrane conductance that replaces the complexity of several different types of ion channels. For the spatially extended system, the fast variable of neighboring cells are coupled diffusively. There are several models belonging to this general class of excitable media which display breakup of spiral waves (in 2D) and scroll waves (in 3D), including the one proposed by Panfilov [PH93, Pan98]

$$\partial_t e = \nabla^2 e - f(e) - g, \qquad \partial_t g = \epsilon(e,g)(ke - g). \tag{1.202}$$

Here, $f(e)$ is the function specifying the initiation of the action potential and is piecewise linear: $f(e) = C_1 e$, for $e < e_1$, $f(e) = -C_2 e + a$, for $e_1 \le e \le e_2$, and $f(e) = C_3(e-1)$, for $e > e_2$. The physically appropriate parameters given in [Pan98] are $e_1 = 0.0026$, $e_2 = 0.837$, $C_1 = 20$, $C_2 = 3$, $C_3 = 15$, $a = 0.06$ and $k = 3$. The function $\epsilon(e, g)$ determines the time scale for the dynamics

---

[48] An earlier review, discussing methods proposed till 2002, can be found in [GHO02]

of the recovery variable: $\epsilon(e, g) = \epsilon_1$ for $e < e_2$, $\epsilon(e, g) = \epsilon_2$ for $e > e_2$, and $\epsilon(e, g) = \epsilon_3$ for $e < e_1$ and $g < g_1$ with $g_1 = 1.8$, $\epsilon_1 = 1/75$, $\epsilon_2 = 1.0$, and $\epsilon_3 = 0.3$ For details of the functional form of $f(e)$ and relevant parameter values, see [Pan98].

Simpler variants that also display spiral wave breakup in 2D include (i) the *Barkley model* [BKT90]:

$$\partial_t e = \nabla^2 e + \epsilon^{-1} e(1-e)(e - \frac{g+b}{a}), \qquad \partial_t g = e - g, \qquad (1.203)$$

the appropriate parameter values being given in [ASM03], are $a = 1.1$, $b = 0.19$ and $\epsilon = 0.02$, for which the spatially extended system exhibits expanding spiral waves (see Figure 1 in [ASM03]); and (ii) the *Bär–Eiswirth model* [BE93], which differs from (1.203) only in having $\partial g/\partial t = f(e) - g$. Here $f(e) = 0$ for $e < 1/3$, $f(e) = 1 - 6.75e(e-1)^2$ for $1/3 \le e \le 1$, and $f(e) = 1$ for $e > 1$. The parameters chosen are $a = 0.84$, $b = 0.12$ and $\epsilon = 0.074$, in which the system shows *turbulence* [BE93]. The *Aliev–Panfilov model* [AP96] is a modified form of the *Panfilov model*, that takes into account nonlinear effects such as the dependence of the action potential duration on the distance of the wavefront to the preceding wave-back. It has been used for control in [SF03, SK05] and is given by

$$\partial_t e = \nabla^2 e - K(e-a)(e-1) - eg, \qquad (1.204)$$
$$\partial_t g = (\epsilon + \frac{\mu_1 g}{\mu_2 + e})(-ke[e-b-1] - g).$$

The parameters are chosen to be $K = 8$, $\epsilon = 0.01$, $\mu_1 = 0.11$, $\mu_2 = 0.3$, $b = 0.1$, for which spiral chaos is observed at $a = 0.1$ [Pan99].

The *Karma model* [SS07] is defined as

$$\partial_t e = \tau_e^{-1}[-e + (e^* - g^M)\{1 - \tanh(e - e_h)\}\frac{e^2}{2}] + D\nabla^2 e, \qquad (1.205)$$
$$\partial_t g = \tau_g^{-1}[\frac{1 - g(1 - e^{-Re})}{1 - e^{-Re}} \Theta(e - e_n) - g\{1 - \Theta(e - e_n)\}],$$

$\Theta(x) = 0$ for $x \le 0$, $\Theta(x) = 1$ otherwise, is the Heaviside step function, and the parameters Re and M control the restitution and dispersion effects, respectively. Increasing Re makes the restitution curve steeper and makes alternans more likely, while increasing M weakens dispersion. The diffusion coefficient parameter can be chosen to be $D = 1$ cm^2 s^{-1}. The other parameters can be taken to be $\tau_e = 2.5$ ms, $\tau_g = 250$ ms, e$^* = 1.5415$, $e_h = 3$, $e_n = 1$, $M = 4$ and $Re = 1.5$, the last two values chosen to make both restitution and dispersion significant.

All the preceding models tend to disregard several complex features of actual biological cells, e.g., the different types of ion channels that allow passage of electrically charged ions across the cellular membrane. There exists a class of models inspired by the *Hodgkin–Huxley* formulation [HH52] describing action potential generation in the squid giant axon, which explicitly takes such details into account. While the simple models described above do reproduce generic features of several excitable media seen in nature, the more realistic models describe many properties of specific systems, e.g., cardiac tissue. The general form of such models are described by a PDE for the transmembrane potential V,

$$\partial_t V + \frac{I_{ion}}{C} = D\nabla^2 V,$$

where C is the membrane capacitance density and D is the diffusion constant, which, if the medium is isotropic, is a scalar. I_{ion} is the instantaneous total ionic-current density, and different realistic models essentially differ in its formulation. For example, in the *Luo–Rudy I model* [LR91] of guinea pig ventricular cells, I_{ion} is assumed to be composed of six different ionic current densities, which are themselves determined by several time–dependent ion–channel gating variables whose time–evolution is governed by ODEs of the form:

$$\dot{\xi} = \frac{\xi_\infty - \xi}{\tau_\xi}.$$

Here, $\xi_\infty = \alpha_\xi/(\alpha_\xi + \beta_\xi)$ is the steady state value of ξ and $\tau_\xi = \frac{1}{\alpha_\xi + \beta_\xi}$ is its time constant. The voltage–dependent rate constants, α_ξ and β_ξ, are complicated functions of V obtained by fitting experimental data.

1.10.2 Global Chaos Control

The first attempt at controlling chaotic activity in excitable media dates back almost to the beginning of the field of chaos control itself, when proportional perturbation feedback (PPF) control was used to stabilize cardiac arrhythmia in a piece of tissue from rabbit heart [GSD92]. This method applied small electrical stimuli, at intervals calculated using a feedback protocol, to stabilize an unstable periodic rhythm. Unlike in the original proposal for controlling chaos [OGY90], where the location of the stable manifold of the desired unstable periodic orbit (UPO) was moved using small perturbations, in the PPF method it is the state of the system that is moved onto the stable manifold. However, it has been later pointed out that PPF does not necessarily require the existence of UPOs (and, by extension, deterministic chaos) and can be used even in systems with stochastic dynamics [CC95]. Later, PPF method was used to control atrial fibrillation in human heart [DSS90]. However, the effectiveness of such control in suppressing spatio-temporal chaos, when applied only at a local region, has been questioned, especially as other experimental attempts in feedback control have not been able to terminate fibrillation by applying control stimuli at a single spatial location [GHO02].

More successful, at least in numerical simulations, have been schemes where control stimuli is applied throughout the system. Such global control

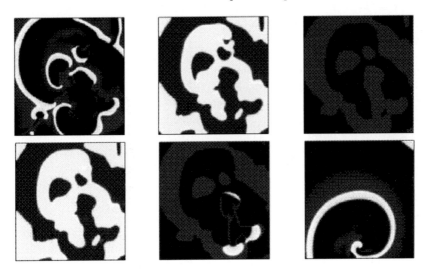

Fig. 1.55. Global control of the 2D Panfilov model starting from a spatiotemporally chaotic state (top left). Pseudo-gray-scale plots of excitability e show the result of applying a pulse of amplitude $A = 0.833$ between $t = 11$ ms and 27.5 ms (top centre) that eventually leads to elimination of all activity (top right). Applying the pulse between $t = 11$ ms and 33 ms (bottom left) results in some regions becoming active again after the control pulse ends (bottom center) eventually re-initiating spiral waves (bottom right) (adapted and modified from [SS07]).

schemes either apply small perturbations to the dynamical variables (e or g) or one of the parameters (usually the excitation threshold). The general scheme involves introducing an external control signal A into the model equations, e.g., in the Panfilov model (1.202),

$$\partial_t e = \nabla^2 e - f(e) - g + A,$$

for a control duration τ. If A is a small, positive perturbation, added to the fast variable, the result is an effective reduction of the threshold, thereby making simultaneous excitation of different regions more likely. In general, A can be periodic, consisting of a sequence of pulses. While in general, increasing the amplitude, or the duration, increases the likelihood of suppressing spatio-temporal chaos, it is not a simple, monotonic relationship. Depending on the initial state at which the control signal is applied, even a high amplitude (or long duration) control signal may not be able to uniformly excite all regions simultaneously. As a result, when the control signal is withdrawn, the inhomogeneous activation results in a few regions becoming active again and restarting the spatio-temporal chaotic behavior.

Most global control schemes are variations or modifications of the above scheme. [OC99] have shown that a low–amplitude signal used to change the value of the slow variable at the front and back of an excitation wave can result in different wavefront and wave-back velocities which destabilizes the travelling wave, eventually terminating all activity, and, hence, spatio-temporal chaos. [GPJ98] have investigated the termination of spiral wave breakup by using both short and long-duration pulses applied on the fast variable, in 2D and 3D systems. This study concluded that while short duration pulses affected only the fast variable, long duration pulses affected both fast and slow variables and that the latter is more efficient, i.e., uses less power, in terminating spatio-temporal chaos. The external control signal can also be periodic ($A = F sin(\omega t)$), in which case the critical amplitude F_c required for terminating activity has been found to be a function of the signal frequency ω [SF03]. In the Aliev–Panfilov model, the smallest value of F_c has been found to occur at a frequency close to that corresponding to the oscillatory instability due to nonlinear nature of wave propagation in this model.

Other schemes have proposed applying perturbations to the parameter controlling the excitation threshold, b. Applying a control pulse on this parameter ($b = b_f$, during duration of control pulse; $b = b_0$, otherwise) has been shown to cause splitting of an excitation wave into a pair of forward and backward moving waves [WM02]. Splitting of a spiral wave causes the two newly created spirals to annihilate each other on collision. For a spatiotemporally chaotic state, a sequence of such pulses may cause termination of all excitation, there being an optimal time interval between pulses that results in fastest control. Another control scheme that also applies perturbation to the threshold parameter is the uniform periodic forcing method suggested by [ASM03, ASM06] for controlling scroll wave turbulence in 3D excitable media. Such turbulence results from negative tension between scroll wave filaments, i.e., the line joining the phase singularities about which the scroll wave rotates. In this control method, the threshold is varied in periodic manner ($b = b_0 + b_f cos(\omega t)$) and the result depends on the relation between the control frequency ω and the spiral rotation frequency. If the former is higher than the latter, sufficiently strong forcing is seen to eliminate turbulence; otherwise, turbulence suppression is not achieved. The mechanism underlying termination has been suggested to be the effective increase of filament tension due to rapid forcing, such that, the originally negative tension between scroll wave filaments is changed to positive tension. This results in expanding scroll wave filaments to instead shrink and collapse, eliminating spatio-temporal chaotic activity. In a variant method, the threshold parameter has been perturbed by spatially uncorrelated Gaussian noise, rather than a periodic signal, which also results in suppression of scroll wave turbulence [ASS04].

As already mentioned, global control, although easy to understand, is difficult to achieve in experimental systems. A few cases in which such control could be implemented include the case of eliminating spiral wave patterns in populations of the Dictyostelium amoebae by spraying a fine mist of cAMP onto the agar surface over which the amoebae cells grow [LGC01]. Another experimental system where global control has been implemented is the photosensitive Belusov–Zhabotinsky reaction, where a light pulse shining over the entire system is used as a control signal [MP97]. Indeed, conventional defibrillation can be thought of as a kind of global control, where a large amplitude control signal is used to synchronize the phase of activity at all points by either exciting a previously unexcited region (advancing the phase) or slowing the recovery of an already excited region (delaying the phase).

1.10.3 Non-Global Spatially Extended Control

The control methods discussed so far apply control signal to all points in the system. As the chaotic activity is spatially extended, one may naively expect that any control scheme also has to be global. However, we will now discuss some schemes that, while being spatially extended, do not require the application of control stimuli at all points of the system.

Applying Control over a Mesh

The control method of [SPP01] involving supra-threshold stimulation along a grid of points is based on the observation that spatio-temporal chaos in excitable media is a long–lived transient that lasts long enough to establish a non–equilibrium statistical steady state displaying spiral turbulence. The lifetime of this transient, τ_L, increases rapidly with linear size of the system, L, e.g., increasing from 850 ms to 3200 ms as L increases from 100 to 128 in the 2D Panfilov model. This accords with the well-known observation that small mammals do not get life–threatening VF spontaneously whereas large mammals do [Win87] and has been experimentally verified by trying to initiate VF in swine ventricular tissue while gradually reducing its mass [KGI97]. A related observation is that non–conducting boundaries tend to absorb spiral excitations, which results in spiral waves not lasting for appreciable periods in small systems.

The essential idea of the control scheme is that a domain can be divided into electrically disconnected regions by creating boundaries composed of re-covering cells between them. These boundaries can be created by triggering excitation across a thin strip. For 2D media, the simulation domain (of size $L \times L$) is divided into K^2 smaller blocks by a network of lines with the block size $(L/K \times L/K)$ small enough so that spiral waves cannot form. For control in a 3D system, the mesh is used only on one of the faces of the simulation box. Control is achieved by applying a supra-threshold stimulation via the mesh for a duration τ. A network of excited and subsequently recovering cells then divides the simulation domain into square blocks whose length in each direction is fixed at a constant value L/K for the duration of control. The network effectively simulates non-conducting boundary conditions (for the block bounded by the mesh) for the duration of its recovery period, in so far as it absorbs spirals formed inside this block. Note that τ need not be large at all because the individual blocks into which the mesh divides the system (of linear size L/K) are so small that they do not sustain long spatiotemporally chaotic transients. Nor does K, which is related to the mesh density, have to be very large since the transient lifetime, τ_L, decreases rapidly with decreasing L. The method has been applied to multiple excitable models, including the Panfilov and Luo–Rudy models [SS07].

An alternative method [SK05] for controlling spiral turbulence that also uses a grid of control points has been demonstrated for the Aliev–Panfilov model. Two layers of excitable media are considered, where the first layer represents the 2D excitable media exhibiting spatio-temporal chaos that is to be controlled, and the second layer is a grid structure also made up of excitable media. The two layers are coupled using the fast variable but with asymmetric coupling constants, (i.e., $d_{12} \neq d_{21}$):

$$\partial_t e_1 = \nabla^2 e_1 - K(e_1 - a)(e_1 - 1) - e_1 g_1 + d_{12}(e_2 - e_1),$$
$$\partial_t e_2 = D\nabla^2 e_2 - K(e_2 - a)(e_2 - 1) - e_2 g_2 + d_{21}(e_1 - e_2),$$

with excitation pulses travelling $\sqrt{D}$ times faster in the second layer compared to the first. As the second layer consists only of grid lines, it is incapable of exhibiting chaotic behavior in the uncoupled state. For large d_{21} it can be driven into a state that reflects the spatio-temporal chaos of the first layer. However, if $d_{12} >> d_{21}$, If the coupling from the second layer to the first layer is sufficiently stronger than the other way round, the stable dynamics of the second layer (manifested as a single rotating spiral) overcomes the spiral chaos in the first layer, and drives it to an ordered state characterized by mutually synchronized spiral waves.

Applying Control over an Array of Points

An alternative method of spatially extended control is to apply perturbations at a series of points arranged in a regular array. [RFK99] had proposed using such an arrangement for applying a time–delayed feedback control scheme. However, this scheme does not control spatio-temporal chaos, that is oscillatory instability of wave propagation in excitable media from breaking up spiral waves. It should therefore be applied before the onset of spatio-temporal chaos.

Note that simulating a travelling wave using the array is found to be more effective at controlling spatio-temporal chaos than the simultaneous activation of all control points. The latter results in only incomplete excitation of the system, and after the control is stopped, regions with inhomogeneous activity remain that eventually re-initiate the spatio-temporal chaos. On the other hand, using a travelling wave allows the control signal to engage all high–frequency sources of excitation in the spiral turbulence regime, ultimately resulting in complete elimination of chaos. If, however, the control had only been applied locally the resulting wave could only have interacted with neighboring spiral waves and the effects of such control would not have been felt throughout the system. The efficacy of the control scheme depends upon the spacing between the control points, as well as the number of simulated travelling waves. Travelling waves have been used to control spatio-temporal chaos, although in the global control context with a spatiotemporally periodic signal being applied continuously for a certain duration, over the entire system [SS07].

1.10.4 Local Chaos Control

We now turn to the possibility of controlling spatio-temporal chaos by applying control at only a small localized region of the spatially extended system. Virtually all the proposed local control methods use *overdrive pacing*, generating a series of waves with frequency higher than any of the existing excitations in the spiral turbulent state. As low–frequency activity is progressively invaded by faster excitation, the waves generated by the control stimulation gradually sweep the chaotic activity to the system boundary where they are absorbed. Although we cannot speak of a single frequency source in the case of chaos, the relevant timescale is that of the spiral waves and is related to the recovery period of the medium. Control is manifested as a gradually growing region in which the waves generated by the control signal dominate, until the region expands to encompass the entire system. The time required to achieve termination depends on the frequency difference between the control stimulation and that of the chaotic activity, with control being achieved faster when this difference is greater.

Stamp [SOC02] has looked at the possibility of using low–amplitude, high–frequency pacing using a series of pulses to terminate spiral turbulence. However, using a series of pulses (having various waveform shapes) has met with only limited success in suppressing spatio-temporal chaos. By contrast, a periodic stimulation protocol [ZHH03] has successfully controlled chaos in the 2D Panfilov model. The key mechanism underlying such control is the periodic alternation between positive and negative stimulation. A more general control scheme uses *biphasic pacing*, i.e., applying a series of positive and negative pulses, that shortens the recovery period around the region of control stimulation, and thus allows the generation of very high–frequency waves than would have been possible using positive stimulation alone. A simple argument shows why a negative rectangular pulse decreases the recovery period for an excitable system. The stimulation vertically displaces the $e-$nullcline and therefore, the maximum value of g that can be attained is reduced. Consequently, the system will recover faster from the recovery period.

To understand how negative stimulation affects the response behavior of the spatially extended system, the so–called *pacing response diagrams* [SS07] can be used, indicating the relation between the control stimulation frequency f and the effective frequency f_{eff}, measured by applying a series of pulses at one site and then recording the number of pulses that reach another site located at a distance without being blocked by a region in the recovery period. Depending on the relative value of f^{-1} and the recovery period, we observe instances of $n:m$ response, i.e., m responses evoked by n stimuli. If, for any range of f, the corresponding f_{eff} is significantly higher than the effective frequency of spatio-temporal chaos, then termination of spiral turbulence is possible. It has been shown that there are ranges of stimulation frequencies that give rise to effective frequencies that dominate chaotic activity. As a result, the periodic waves emerging from the stimulation region gradually impose control over the regions exhibiting chaos. Note that, there is a tradeoff involved here. If f_{eff} is only slightly higher than the chaos frequency, control takes too long. On the other hand, if it is too high the waves suffer conduction block at inhomogeneities produced by chaotic activity which reduces the effective frequency, and therefore, control fails.

Recently, another local control scheme has been proposed [ZCW05] that periodically perturbs the model parameter governing the threshold. In fact, it is the local control analog of the global control scheme proposed by [ASM03] discussed above. As in the other methods discussed here, the local stimulation generates high-frequency waves that propagate into the medium and suppress spiral or scroll waves. Unlike the global control scheme, $b_f \gg b_0$, so that the threshold can be negative for a part of the time. This means that the regions in resting state can become spontaneously excited, which allow very high-frequency waves to be generated.

For more details on spatio-temporal chaos control, see [SS07].

1.10.5 Spatio-Temporal Chaos–Control in the Heart

Recall that a characteristic feature of *excitable media* is the formation of spiral waves and their subsequent breakup into spatio-temporal chaos. An example of obvious importance is the propagation of waves of electrical excitation along the heart wall, initiating the muscular contractions that enable the heart to pump blood. In fact, spiral turbulence has been identified by several investigators as the underlying cause of certain arrhythmias, i.e., abnormal cardiac rhythms, including *ventricular fibrillation* (VF) [GPJ98], a potentially fatal condition in which different regions of the heart are no longer activated coherently. As a result, the heart effectively stops pumping blood, resulting in death within a few minutes, if untreated. Current methods of defibrillatory treatment involve applying large electrical shocks to the entire heart in an attempt to drive it to the normal state. However, this is not only painful but also dangerous, as the resulting damage to heart tissue can form scars that act as substrates for future cardiac arrhythmias. Devising a low–amplitude control mechanism for spatio-temporal chaos in excitable media is therefore not only an exciting theoretical challenge but of potential significance for the treatment of VF. In this section. Following [BS04], we present a robust chaos–control method, using low–amplitude biphasic pacing.

Most of the methods proposed for controlling spatio-temporal chaos in excitable media involve applying perturbations either globally or over a spatially extended system of control points covering a significant proportion of the entire system [OC99, RFK99, SPP01]. However, in most real situations this may not be a feasible option. Further, in the specific context of controlling VF, a local control scheme has the advantage that it can be readily implemented with existing hardware of the *Implantable Cardioverter–Defibrillator* (ICD). This is a device implanted into patients at high risk from VF that monitors the heart rhythm and applies electrical treatment, when necessary, through electrodes laced on the heart wall. A low–energy control method involving ICDs should therefore aim towards achieving control of spatio-temporal chaos by applying small perturbations from a few local sources.

As a model for ventricular activation the modified Fitzhugh–Nagumo equations (1.202) proposed by Panfilov [PH93] are used. For simplicity we assume an isotropic medium; in this case the model is defined by the two equations governing the excitability e and recovery g variables [BS04],

$$\partial_t e = \nabla^2 e - f(e) - g, \qquad \partial_t g = \epsilon(e,g)(ke - g). \tag{1.206}$$

The function $f(e)$, which specifies fast processes (e.g., the initiation of excitation, i.e., the action potential) is piecewise linear,

$$f(e) = C_1 e, \quad \text{for} \quad e < e_1,$$
$$f(e) = -C_2 e + a, \quad \text{for} \quad e_1 \le e \le e_2,$$
$$f(e) = C_3(e - 1), \quad \text{for} \quad e > e_2.$$

The function $\epsilon(e,g)$, which determines the dynamics of the recovery variable, is

$$\epsilon(e,g) = \epsilon_1 \quad \text{for} \quad e < e_2,$$
$$\epsilon(e,g) = \epsilon_2 \quad \text{for} \quad e > e_2, \quad \text{and}$$
$$\epsilon(e,g) = \epsilon_3 \quad \text{for} \quad e < e_1 \quad \text{and} \quad g < g_1.$$

We use the physically appropriate parameter values given in [SPP01].

The model (1.206) has been solved in [BS04] using a *forward–Euler integration scheme*. The system was discretized on a 2D grid of points with spacing $\delta x = 0.5$ dimensionless units. The standard five-point difference stencil was used for the 2D Laplacian. The spatial grid consisted of $L \times L$ points. On the edges of the simulation region *no–flux Neumann boundary conditions* were used.

To achieve control of spatio-temporal chaos, a periodic perturbation, $AF(2\pi ft)$, was locally applied of amplitude A and frequency f. F can represent any periodic function, e.g., a series of pulses having a fixed amplitude and duration, applied at periodic intervals defined by the stimulation frequency f. The control mechanism can be understood as a process of overdriving the chaos by a source of periodic excitation having a significantly higher frequency. As noted in [Lee97, XQW99], in a competition between two sources of high frequency stimulation, the outcome is independent of the nature of the wave generation at either source, and is decided solely on the basis of their relative frequencies. This follows from the property of an excitable medium that waves annihilate when they collide with each other [KA83]. The lower frequency source is eventually entrained by the other source and will no longer generate waves when the higher frequency source is withdrawn. Although we cannot speak of a single frequency source in the case of chaos, the relevant timescale is that of spiral waves which is limited by the refractory period of the medium, τ_{ref}, the time interval during which an excited cell cannot be stimulated until it has recovered its resting state properties. To achieve control, one must use a periodic source with frequency $f > \tau_{ref}^{-1}$. This is almost impossible to achieve by using with purely excitatory stimuli as reported in [SPP01]; the effect of locally applying such perturbations is essentially limited by refractoriness to the immediate neighborhood of the stimulation point. As the effective frequency of a train of purely excitatory pulses cannot be higher than that of the spiral waves, and the area over which control is imposed is inversely related to the absolute difference of the two frequencies.

A simple argument shows why a negative rectangular pulse decreases the refractory period for the Panfilov model in the absence of the diffusion term. The stimulation vertically displaces the e−nullcline of (1.206) and therefore, the maximum value of g that can be attained is reduced. Consequently, the system will recover faster from the refractory state. To illustrate this, let us assume that the stimulation is applied when $e > e_2$. Then, the dynamics reduces to [BS04]

$$\dot{e} = -C_3(e-1) - g, \qquad \dot{g} = \epsilon_2(ke - g).$$

In this region of the (e, g)−plane, for sufficiently high g, the trajectory will be along the e−nullcline, i.e., $\dot{e} \simeq 0$. If a pulse stimulation of amplitude A is initiated at $t = 0$ (say), when $e = e(0), g = g(0)$, at a subsequent time t,

$$e(t) = 1 + \frac{A - g(t)}{C_3}, \qquad \text{and} \qquad g(t) = \frac{a}{b} - [\frac{a}{b} - g(0)]\exp(-bt),$$

where $\quad a = \epsilon_2 k(1 + \frac{A}{C_3}), \qquad b = \epsilon_2[1 + (k/C_3)].$

The negative stimulation has to be kept on till the system crosses into the region where $\dot{e} < 0$, after which no further increase of g can occur, as dictated by the dynamics of (1.206). Now, the time required by the system to enter this region in phase–space where $\dot{e} < 0$ is

$$\frac{1}{b}\ln\frac{a/b - g(0)}{a/b - \phi}, \qquad \text{where} \qquad \phi = C_3(1 - e_2) + A.$$

Therefore, this time is reduced when $A < 0$ and contributes to the decrease of the refractory period. Note that $g(0) > \phi$, as the stimulation is applied when the cell is not yet excitable. By looking at the trajectory, the significant decrease of the refractory period if self–evident. If the negative stimulation is kept on till the system crosses into the region where $\dot{e} < 0$, the system dynamics drives it to the $e < 0$ portion. As a result, the maximum value of g attained will be much smaller than normal, and consequently, the system will recover from the refractory state faster than normal. Besides, note that, the above discussion also indicates that a rectangular pulse will be more effective than a gradually increasing waveform, e.g., a sinusoidal wave (as used in [ZHH03]), provided the energy of stimulation is same in both cases, as the former allows a much smaller maximum value of g. Therefore, phase plane analysis of the response to negative stimulation allows us to design waveshapes for maximum efficiency in controlling spiral turbulence.

To understand how negative stimulation affects the response behavior of the spatially extended system, we first look at a 1D system, specifically into the relation between the stimulation frequency f and effective frequency f_{eff}, measured by applying a series of pulses at one site and then recording the number of pulses that reach another site located at a distance without being blocked by any refractory region. Depending on the relative value of f and τ_{ref}, we observe instances of $n : m$ response, i.e., m responses evoked by n stimuli. From the resulting effective frequencies f_{eff}, we can see that for purely excitatory stimulation, the relative refractory period can be reduced by increasing the amplitude A. However, this reduction is far more pronounced when the positive stimulation is alternated with negative stimulation, i.e., a negative stimulation is applied between every pair of positive pulses. There is an optimal time interval between applying the positive and negative pulses that maximally decreases the refractory period by as much as 50%. The highest effective frequencies correspond to a stimulation frequency in the range $0.1 - 0.25$, agreeing with the optimal time period of 2–5 time units between positive and negative stimulation.

A response diagram similar to the 1D case is also seen for stimulation in a 2D medium. A small region consisting of $n \times n$ points at the center of the simulation domain is chosen as the stimulation point. For the simulations reported here $n = 6$; for a smaller n, one requires a perturbation of larger amplitude to achieve a similar response. To understand control in two dimensions, we find out the characteristic time scale of spatio-temporal chaotic activity by obtaining its power spectral density. There is a peak at a frequency $f_c \simeq 0.0427$ [BS04] and there are ranges of stimulation frequencies that give rise to effective frequencies higher than this value. As a result, the periodic waves emerging from the stimulation point will gradually impose control over the regions exhibiting chaos. If f is only slightly higher than f_c control takes very long; if it is too high i.e., close to the refractory limit, the waves suffer conduction block at inhomogeneities produced by chaotic activity that reduces the effective frequency, and control fails. Note that, at lower frequencies the range of stimulation frequencies for which $f_{eff} > f_c$, is smaller than at higher frequencies. We also compare the performance of sinusoidal waves with rectangular pulses, adjusting the amplitudes so that they have the same energy. The former is much less effective than the latter at lower stimulation frequencies, which is the preferred operating region for the control method.

The effectiveness of overdrive control is limited by the size of the system sought to be controlled. As shown in [BS04], away from the control site, the generated waves are blocked by refractory regions, with the probability of block increasing as a function of distance from the site of stimulation.[49] To see whether the control method is effective in reasonably large systems, we used it to terminate chaos in the 2D *Panfilov model*, with $L = 500$.[50] Figure 1.10.5 shows a sequence of images illustrating the evolution of chaos control when a sequence of biphasic rectangular pulses are applied at the center. The time necessary to achieve the controlled state, when the waves from the stimulation point pervade the entire system, depends slightly on the initial state of the system when the control is switched on. Not surprisingly, we find that the stimulation frequency used to impose control in Figure 1.10.5 belongs to a range for which $f_{eff} > f_c$ [BS04].

Fig. 1.56. Control of spatio-temporal chaos in the 2D Panfilov model ($L = 500$) by applying biphasic pulses with amplitude $A = 18.9$ and frequency $f = 0.13$ at the center of the simulation domain (adapted and modified from [BS04]).

Although most of the simulations were performed with the Panfilov model, the arguments involving phase plane analysis apply in general to excitable media having a cubic–type nonlinearity. To ensure that our explanation is not

---

[49] The profile of the stimulated wave changes as it propagates along the medium, from biphasic at the stimulation source to gradually becoming indistinguishable from a purely excitatory stimulation. As a result, far away from the source of stimulation, the response cannot have a frequency higher than $\sim \tau_{ref}^{-1}$.

[50] The initial condition used for this purpose is a broken plane wave which is allowed to evolve for 5000 time units into a state displaying spiral turbulence.

sensitively model dependent we obtained similar stimulation response diagrams for the *Karma model* [Kar93].

Some local control schemes envisage stimulating at special locations, e.g., close to the tip of the spiral wave, thereby driving the spiral wave towards the edges of the system where they are absorbed [KPV95]. However, aside from the fact that spatio-temporal chaos involves a large number of coexisting spirals, in a practical situation it may not be possible to have a choice regarding the location of the stimulation point. We should therefore look for a robust control method which is not critically sensitive to the position of the control point in the medium. There have been some proposals to use periodic stimulation for controlling spatio-temporal chaos. For example, recently [ZHH03] have controlled some excitable media models by applying sinusoidal stimulation at the center of the simulation domain. Looking in detail into the mechanism of this type of control, we have come to the conclusion that the key feature is the alternation between positive and negative stimulation, i.e., biphasic pacing, and it is, therefore, a special case of the general scheme presented in [BS04].

Previous explanations of why biphasic stimulation is better than purely excitatory stimulation (that use only positive pulses), have concentrated on the response to very large amplitude electrical shocks typically used in conventional defibrillation [KL99, AT01] and have involved details of cardiac cell ion channels [JT00]. The present section gives the simplest and most general picture for understanding the efficacy of the biphasic scheme using very low amplitude perturbation, as it does not depend on the details of ion channels responsible for cellular excitation. This also allows us to provide guidelines about the optimal wave–shape for controlling VF through periodic stimulation.

For more details on spatio-temporal chaos control, see [BS04].

2
Phase Transitions and Synergetics

The *dynamical edge of chaos* physically corresponds to the popular *phase transitions*. In this Chapter we present the basic principles of phase transitions, which are used in *synergetics*. After the introduction and classification of equilibrium phase transitions, a brief on Landau's theory is given (providing a background for order–parameters and synergetics). The concept is subsequently generalized into non–equilibrium phase transitions, together with important examples of oscillatory, fractal and noise–induced transitions. This core Chapter of the book also introduces the concept of partition function, together with its general, path–integral description. After that the basic elements of Haken's synergetics are presented, and subsequently developed into synergetics of *attractor neural networks*.

2.1 Introduction to Phase Transitions

Recall that in thermodynamics, a phase transition (or phase change) is the transformation of a thermodynamic system from one phase to another (see Figure 2.1). The distinguishing characteristic of a phase transition is an abrupt change in one or more physical properties, in particular the heat capacity, with a small change in a thermodynamic variable such as the temperature.

2.1.1 Equilibrium Phase Transitions

In *thermodynamics*, a *phase transition* represents the transformation of a system from one phase to another. Here the therm *phase* denotes a set of states of a macroscopic physical system that have relatively uniform chemical composition and physical properties (i.e., density, crystal structure, index of refraction, and so forth.) The most familiar examples of phases are solids, liquids, and gases. Less familiar phases include plasmas, Bose–Einstein condensates and fermionic condensates and the paramagnetic and ferromagnetic phases of magnetic materials.

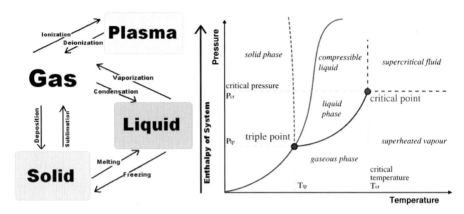

Fig. 2.1. Thermodynamic (or, equilibrium) phase transitions. Left: the term *phase transition* is most commonly used to describe transitions between solid, liquid and gaseous states of matter, in rare cases including plasma. Right: the transitions between the solid, liquid, and gaseous phases of a single component, due to the effects of temperature and/or pressure (adapted and modified from [Wik07].)

In essence, all thermodynamic properties of a system (entropy, heat capacity, magnetization, compressibility, and so forth) may be expressed in terms of the *free energy potential* $\mathcal{F}$ and its partial derivatives. For example, the *entropy* S is the first derivative of the free energy $\mathcal{F}$ with respect to the temperature T, i.e., $S = -\partial \mathcal{F}/\partial T$, while the *specific heat capacity* C is the second derivative, $C = T\,\partial S/\partial T$. As long as the free energy $\mathcal{F}$ remains analytic, all the thermodynamic properties will be well–behaved.

Now, the distinguishing characteristic of a phase transition is an *abrupt sudden change in one or more physical properties*, in particular the specific heat c, with a small change in a thermodynamic variable such as the temperature T. Standard examples of phase transitions are (see e.g., [LL78, Wik07]):

1. The transitions between the solid, liquid, and gaseous phases (boiling, melting, sublimation, etc.)
2. The transition between the ferromagnetic and paramagnetic phases of magnetic materials at the Curie point.
3. The emergence of superconductivity in certain metals when cooled below a critical temperature.
4. Quantum condensation of bosonic fluids, such as Bose–Einstein condensation and the superfluid transition in liquid helium.
5. The breaking of symmetries in the laws of physics during the early history of the universe as its temperature cooled.

When a system goes from one phase to another, there will generally be a stage where the free energy is non–analytic. This is known as a phase transition. Familiar examples of phase transitions are melting (solid to liquid), freezing (liquid to solid), boiling (liquid to gas), and condensation (gas to liquid). Due to this *non–analyticity*, the free energies on either side of the transition are two different functions, so one or more thermodynamic properties will behave very differently after the transition. The property most commonly examined in this context is the heat capacity. During a transition, the heat capacity may become infinite, jump abruptly to a different value, or exhibit a 'kink' or *discontinuity* in its derivative[1].

Therefore, phase transitions come about when the free energy of a system is non–analytic for some choice of thermodynamic variables. This non–analyticity generally stems from the interactions of an extremely large number of particles in a system, and does not appear in systems that are too small.

2.1.2 Classification of Phase Transitions

Ehrenfest Classification

The first attempt at classifying phase transitions was the *Ehrenfest classification scheme*, which grouped phase transitions based on the degree of non-analyticity involved. Though useful, Ehrenfest's classification is flawed, as we will discuss in the next section.

Under this scheme, phase transitions were labelled by the lowest partial derivative of the free energy that is discontinuous at the transition. *First–order phase transitions* exhibit a discontinuity in the first derivative of the free energy with respect to a thermodynamic variable. The various solid⇒liquid⇒gas transitions are classified as first–order transitions, as the density, which is the first partial derivative of the free energy with respect to chemical potential, changes discontinuously across the transitions[2]. *Second–order phase transitions* have a discontinuity in a second derivative of the free energy. These include the ferromagnetic phase transition in materials such as iron, where the magnetization, which is the first derivative of the free energy with the applied magnetic field strength, increases continuously from zero as the temperature is lowered below the Curie temperature. The magnetic susceptibility, the second derivative of the free energy with the field, changes discontinuously. Under the *Ehrenfest classification scheme*, there could in principle be third, fourth, and higher–order phase transitions.

---

[1] In practice, each type of phase is distinguished by a handful of relevant thermodynamic properties. For example, the distinguishing feature of a solid is its rigidity; unlike a liquid or a gas, a solid does not easily change its shape. Liquids are distinct from gases because they have much lower compressibility: a gas in a large container fills the container, whereas a liquid forms a puddle in the bottom. Not all the properties of solids, liquids, and gases are distinct; for example, it is not useful to compare their magnetic properties. On the other hand, the ferromagnetic phase of a magnetic material is distinguished from the paramagnetic phase by the presence of bulk magnetization without an applied magnetic field.

[2] The pressure must be continuous across the phase boundary in equilibrium.

Modern Classification

The Ehrenfest scheme is an inaccurate method of classifying phase transitions, for it is based on the *mean–field theory* of phases, which is inaccurate in the vicinity of phase transitions, as it neglects the role of thermodynamic fluctuations. For instance, it predicts a finite discontinuity in the heat capacity at the ferromagnetic transition, which is implied by Ehrenfest's definition of second–order transitions. In real ferromagnets, the heat capacity diverges to infinity at the transition.

In the modern classification scheme, phase transitions are divided into two broad categories, named similarly to the Ehrenfest classes:

- The *first–order phase transitions*, or, *discontinuous phase transitions*, are those that involve a latent heat. During such a transition, a system either absorbs or releases a fixed (and typically large) amount of energy. Because energy cannot be instantaneously transferred between the system and its environment, first–order transitions are associated with *mixed–phase regimes* in which some parts of the system have completed the transition and others have not. This phenomenon is familiar to anyone who has boiled a pot of water: the water does not instantly turn into gas, but forms a turbulent mixture of water and water vapor bubbles. Mixed–phase systems are difficult to study, because their dynamics are violent and hard to control. However, many important phase transitions fall in this category, including the solid⇒liquid⇒gas transitions.
- The *second–order phase transitions* are the *continuous phase transitions*. These have no associated latent heat. Examples of second–order phase transitions are the ferromagnetic transition, the superfluid transition, and Bose–Einstein condensation.

2.1.3 Basic Properties of Phase Transitions

Critical Points

In systems containing liquid and gaseous phases, there exist a special combination of pressure and temperature, known as the *critical point*, at which the transition between liquid and gas becomes a second–order transition. Near the critical point, the fluid is sufficiently hot and compressed that the distinction between the liquid and gaseous phases is almost non–existent.

This is associated with the phenomenon of critical opalescence, a milky appearance of the liquid, due to density fluctuations at all possible wavelengths (including those of visible light).

Symmetry

Phase transitions often (but not always) take place between phases with different symmetry. Consider, for example, the transition between a fluid (i.e., liquid or gas) and a crystalline solid. A fluid, which is composed of atoms arranged in a disordered but homogenous manner, possesses continuous translational symmetry: each point inside the fluid has the same properties as any other point. A crystalline solid, on the other hand, is made up of atoms arranged in a regular lattice. Each point in the solid is *not* similar to other points, unless those points are displaced by an amount equal to some lattice spacing.

Generally, we may speak of one phase in a phase transition as being more symmetrical than the other. The transition from the more symmetrical phase to the less symmetrical one is a *symmetry–breaking* process. In the fluid–solid transition, for example, we say that continuous translation symmetry is broken.

The ferromagnetic transition is another example of a symmetry–breaking transition, in this case the symmetry under reversal of the direction of electric currents and magnetic field lines. This symmetry is referred to as 'up–down symmetry' or 'time–reversal symmetry'. It is broken in the ferromagnetic phase due to the formation of magnetic domains containing aligned magnetic moments. Inside each domain, there is a magnetic field pointing in a fixed direction chosen spontaneously during the phase transition. The name *time–reversal symmetry* comes from the fact that electric currents reverse direction when the time coordinate is reversed.

The presence of symmetry–breaking (or nonbreaking) is important to the behavior of phase transitions. It was pointed out by Landau that, given any state of a system, one may unequivocally say whether or not it possesses a given symmetry [LL78]. Therefore, it cannot be possible to analytically deform a state in one phase into a phase possessing a different symmetry. This means, for example, that it is impossible for the solid–liquid phase boundary to end in a critical point like the liquid–gas boundary. However, symmetry–breaking transitions can still be either first or second order.

Typically, the more symmetrical phase is on the high–temperature side of a phase transition, and the less symmetrical phase on the low–temperature side. This is certainly the case for the solid–fluid and ferromagnetic transitions. This happens because the Hamiltonian of a system usually exhibits all the possible symmetries of the system, whereas the low–energy states lack some of these symmetries (this phenomenon is known as spontaneous symmetry breaking.) At low temperatures, the system tends to be confined to the low–energy states. At higher temperatures, thermal fluctuations allow the system to access states in a broader range of energy, and thus more of the symmetries of the Hamiltonian.

When symmetry is broken, one needs to introduce one or more extra variables to describe the state of the system. For example, in the ferromagnetic phase one must provide the net magnetization, whose direction was spontaneously chosen when the system cooled below the Curie point. Such variables are instances of *order parameters*. However, note that order parameters can also be defined for symmetry–nonbreaking transitions.

Symmetry–breaking phase transitions play an important role in cosmology. It has been speculated that, in the hot early universe, the vacuum (i.e., the various quantum fields that fill space) possessed a large number of symmetries. As the universe expanded and cooled, the vacuum underwent a series of symmetry–breaking phase transitions. For example, the electroweak transition broke the $SU(2) \times U(1)$ symmetry of the electroweak field into the $U(1)$ symmetry of the present–day electromagnetic field. This transition is important to understanding the asymmetry between the amount of matter and antimatter in the present–day universe.

Critical Exponents and Universality Classes

Continuous phase transitions are easier to study than first–order transitions due to the absence of latent heat, and they have been discovered to have many interesting properties. The phenomena associated with continuous phase transitions are called *critical phenomena*, due to their association with critical points.

It turns out that continuous phase transitions can be characterized by parameters known as critical exponents. For instance, let us examine the behavior of the heat capacity near such a transition. We vary the temperature T of the system while keeping all the other thermodynamic variables fixed, and find that the transition occurs at some critical temperature T_c. When T is near T_c, the heat capacity C typically has a *power law* behavior: $C \sim |T_c - T|^{-\alpha}$. Here, the constant α is the critical exponent associated with the heat capacity. It is not difficult to see that it must be less than 1 in order for the transition to have no latent heat. Its actual value depends on the type of phase transition we are considering. For $-1 < \alpha < 0$, the heat capacity has a 'kink' at the transition temperature. This is the behavior of liquid helium at the 'lambda transition' from a normal state to the superfluid state, for which experiments have found $\alpha = -0.013 \pm 0.003$. For $0 < \alpha < 1$, the heat capacity diverges at the transition temperature (though, since $\alpha < 1$, the divergence is not strong enough to produce a latent heat.) An example of such behavior is the 3D ferromagnetic phase transition. In the 3D Ising model for uniaxial magnets, detailed theoretical studies have yielded the exponent $\alpha \sim 0.110$.

Some model systems do not obey this power law behavior. For example, mean–field theory predicts a finite discontinuity of the heat capacity at the transition temperature, and the 2D Ising model has a logarithmic divergence. However, these systems are an exception to the rule. Real phase transitions exhibit power law behavior.

Several other critical exponents – β, γ, δ, ν, and η – are defined, examining the power law behavior of a measurable physical quantity near the phase transition.

It is a remarkable fact that phase transitions arising in different systems often possess the same set of critical exponents. This phenomenon is known as *universality*. For example, the critical exponents at the liquid–gas critical point have been found to be independent of the chemical composition of the fluid. More amazingly, they are an exact match for the critical exponents of the ferromagnetic phase transition in uniaxial magnets. Such systems are said to be in the same *universality class*. Universality is a prediction of the renormalization group theory of phase transitions, which states that the thermodynamic properties of a system near a phase transition depend only on a small number of features, such as dimensionality and symmetry, and is insensitive to the underlying microscopic properties of the system.

2.1.4 Landau's Theory of Phase Transitions

Landau's theory of phase transitions is a simple but powerful empirical thermodynamic theory by which the behavior of crystals at phase transitions can be described. It is based simply on a power series expansion of the free energy of the crystal with respect to one or a few prominent parameters distorting the symmetry of the crystal. The symmetry of the distortion decides which terms may be present, and which not. For example, odd terms on the power series expansion often are not allowed because the energy of the system is symmetric with respect to positive or negative distortion. With Landau's theory, the thermodynamics of the crystal (free energy, entropy, heat capacity) can be directly linked to it's structural state (volume, deviation from high symmetry, etc.), and both can be described as they change as a function of temperature or pressure.

More precisely, in Landau's theory, the probability (density) distribution function f is exponentially related to the potential $\mathcal{F}$,

$$f \approx e^{-\mathcal{F}(T)}, \tag{2.1}$$

if $\mathcal{F}$ is considered as a function of the order parameter o. Therefore, the most probable order parameter is determined by the requirement $\mathcal{F} = \min$.

When $M_\uparrow$ elementary magnets point upwards and $M_\downarrow$ elementary magnets point downwards, the magnetization order parameter o is given by

$$o = (M_\uparrow - M_\downarrow)\,m, \tag{2.2}$$

where m is the magnetic moment of a single elementary magnet.

We expand the potential $\mathcal{F} = \mathcal{F}(o, T)$ into a power series of o,

$$\mathcal{F}(o, T) = \mathcal{F}(0, T) + \mathcal{F}'(0, T)o + \cdots + \frac{1}{4!}\mathcal{F}''''(0, T)o^4 + \cdots, \tag{2.3}$$

and discus $\mathcal{F}$ as a function of o. In a number of cases $\mathcal{F}' = \mathcal{F}''' = 0$, due to inversion symmetry. In this case, $\mathcal{F}$ has the form

$$\mathcal{F}(o, T) = \mathcal{F}(0, T) + \frac{\sigma}{2}o^2 + \frac{\beta}{4}o^4, \tag{2.4}$$

where $\beta > 0$, and $\sigma = a(T - T_c)$, $(a > 0)$, i.e., it changes its sign at the critical temperature $T = T_c$.

Recall that the (negative) first partial derivative of the free energy potential $\mathcal{F}$ with respect to the *control parameter* – temperature T is the entropy

$$S = -\frac{\partial \mathcal{F}(q, T)}{\partial T}. \qquad (2.5)$$

For $T > T_c$, $\sigma > 0$, and the minimum of $\mathcal{F}$ lies at $o = o_0 = 0$, and

$$S = S_0 = -\frac{\partial \mathcal{F}(0, T)}{\partial T}.$$

Also recall that the second partial derivative of $\mathcal{F}$ with respect to T is the specific heat capacity (besides the factor T)

$$C = T \frac{\partial S}{\partial T}. \qquad (2.6)$$

One may readily check that S is continuous at $T = T_c$ for $\sigma = 0$. However, when we calculate the specific heat we get two different expressions above and below the critical temperature and thus a discontinuity at $T = T_c$.

Closely related to the Landau's theory of phase transitions is *Ginzburg–Landau model* of *superconductivity* (named after *Nobel Laureates Vitaly L. Ginzburg and Lev D. Landau*). It does not purport to explain the microscopic mechanisms giving rise to superconductivity. Instead, it examines the macroscopic properties of a superconductor with the aid of general thermodynamic arguments. Based on Landau's previously–established theory of second–order phase transitions, Landau and Ginzburg argued that the free energy F of a superconductor near the superconducting transition can be expressed in terms of a complex *order parameter* ψ, which describes how deep into the superconducting phase the system is. By minimizing the free energy with respect to fluctuations in the order parameter and the vector potential, one arrives at the *Ginzburg–Landau equation*, which is a generalization of the *nonlinear Schrödinger equation*.

2.1.5 Example: Order Parameters in Magnetite Phase Transition

The mechanism of the phase transition in magnetite (Fe_3O_4) at $T_V = 122$ K, discovered by Verwey [Ver39], has remained a big puzzle in the condensed matter physics for almost 70 years. Developments in experimental and theoretical methods during last years enabled to reveal subtle changes in the crystal and electronic structure below T_V [RAP01, LYA04]. A simple charge ordering picture in which metal–insulator transition is induced by electrostatic interactions was replaced by a highly complex scenario in which lattice, charge, spin and orbital degrees of freedom are involved. Recent theoretical studies revealed the important role of the local electron interactions and orbital correlations in the t_{2g} states on iron ions [LYA04].

The electronic interactions are complemented by the lattice deformation, which breaks the cubic symmetry and induces a low–temperature (LT) monoclinic phase driven by the electron–phonon interactions. In the work of [PPO06, PPO07], the phonon spectrum of magnetite has been analyzed using the *ab initio* computational technique [PLK97]. We have identified two primary *order parameters* (OPs) at $\mathbf{k}_X = \frac{2\pi}{a}(0,0,1)$ and $\mathbf{k}_\Delta = \frac{2\pi}{a}(0,0,\frac{1}{2})$ with the X_3 and Δ_5 symmetry, respectively, which both play important role in the Verwey transition (VT): (i) the Δ_5 mode is responsible for the doubling of the unit cell along the c direction in the monoclinic phase, while (ii) the X_3 phonon induces the *metal–insulator transition* by its coupling to the electronic states near the Fermi energy [PPO06]. Due to the *electron–phonon interaction* the above OPs are combinations of the electron (charge–orbital) and lattice components. This explains why the phonon soft mode has not been observed. Instead, low–energy critical fluctuations of OPs were found by the diffuse neutron scattering [FSY75]. The condensation of the OPs below T_V explains the crystal symmetry change as well as the charge-orbital ordering.

The group theory predicts also secondary OPs, which do not effect the symmetry below T_V but modify the properties of magnetite close to a transition point. At the Γ point, the T_{2g} mode can be classified as the secondary OP, and its coupling to the shear strain explains the softening of the C_{44} elastic constant [SNM05]. The lowest T_{2g} optic mode, marked in Fig. 1, could contribute quantitatively to the free energy, but it does not play any significant role for the VT.

In this section, following [PPO07], we analyze the *Landau free energy* for the VT, and discuss a solution corresponding to the LT monoclinic phase. We derive also the temperature dependence of C_{44}.

Free Energy

The Landau free energy can be expanded into a series of the components of the OPs. The invariant terms describing couplings between the OPs were derived using the group theory methods [SH02]. The only nonzero components of the primary OPs X_3 and Δ_5 are denoted by g and q, respectively. We include also the secondary OP with the T_{2g} symmetry (η) and shear–strain (ϵ). The free energy can be written in the form [PPO07]

$$\mathcal{F} = \mathcal{F}_0 + \frac{\alpha_1}{2}g^2 + \frac{\beta_1}{4}g^4 + \frac{\gamma_1}{6}g^6 + \frac{\alpha_2}{2}q^2 + \frac{\beta_2}{4}q^4 + \frac{\delta_1}{2}g^2q^2 \qquad (2.7)$$
$$+ \frac{\alpha_3}{2}\eta^2 + \frac{\alpha_4}{2}\epsilon^2 + \frac{\delta_2}{2}\eta g^2 + \frac{\delta_3}{2}\epsilon g^2 + \delta_4 \eta \epsilon,$$

were $\mathcal{F}_0$ is the part of the potential, which does not change through the transition. We assume that $\beta_1 > 0$, $\beta_2 > 0$ and $\gamma_1 > 0$ to ensure the stability of the potential at high temperatures. For the second–order terms we assume standard temperature behavior $\alpha_i = a_i(T - T_{ci})$ near the critical temperature T_{ci} for $i = 1, 2, 3$, which would correspond to a continuous phase transition.

The coefficient α_4 is the shear elastic constant at high temperatures (C_{44}^0). The coupling between the primary OPs is biquadratic, between the secondary and primary OPs has the linear–quadratic form, and the coupling between the components of the secondary OP is of the bilinear type. Taking first derivatives of $\mathcal{F}$ over all OPs we get [PPO07]

$$\frac{\partial \mathcal{F}}{\partial g} = g(\alpha_1 + \beta_1 g^2 + \gamma_1 g^4 + \delta_1 q^2 + \delta_2 \eta + \delta_3 \epsilon) = 0,$$

$$\frac{\partial \mathcal{F}}{\partial q} = q(\alpha_2 + \beta_2 q^2 + \delta_1 g^2) = 0,$$

$$\frac{\partial \mathcal{F}}{\partial \eta} = \alpha_3 \eta + \delta_4 \epsilon + \frac{\delta_2}{2} g^2 = 0,$$

$$\frac{\partial \mathcal{F}}{\partial \epsilon} = \alpha_4 \epsilon + \delta_4 \eta + \frac{\delta_3}{2} g^2 = 0.$$

The solution $g = q = \eta = \epsilon = 0$ corresponds to the high–temperature cubic symmetry ($Fd\bar{3}m$). We obtain the dependence between g and q,

$$q^2 = -\frac{\delta_1 g^2 + \alpha_2}{\beta_2}, \tag{2.8}$$

which has three possible solutions: (i) $g = 0$ and $q^2 = -\frac{\alpha_2}{\beta_2}$ if $\alpha_2 < 0$ (*Pbcm*), (ii) $q = 0$ and $g^2 = -\frac{\alpha_2}{\delta_1}$ if $\alpha_2 > 0$ and $\delta_1 > 0$ or $\alpha_2 < 0$ and $\delta_1 > 0$ (*Pmna*), (iii) $g \neq 0$ and $q \neq 0$ (*P2/c*). In the brackets we put the space group symbols, which characterize the low–symmetry phases. The solution (iii) which corresponds to the experimentally observed LT monoclinic phase requires simultaneous condensation of both primary OPs. The necessary condition for this is a negative value of δ_1. Indeed, it has been established by the *ab initio* studies that the total energy is lowered when the crystal is distorted by both X_3 and Δ_5 modes [PPO07]. For $\delta_1 < 0$, (2.8) has a non–zero solution provided that $|\delta_1|g^2 > \alpha_2$. It implies that for $\alpha_2 > 0$ ($T > T_{c2}$), the phase transition occurs when the OP g exceeds a critical value $\frac{\alpha_2}{|\delta_1|}$, so it has a discontinuous (first–order) character.

From above relations we get [PPO07]

$$\eta = \frac{\delta_3 \delta_4 - \delta_2 \alpha_4}{2\alpha_3 \alpha_4 - 2\delta_4^2} g^2 \equiv \lambda_1 g^2, \qquad \epsilon = \frac{\delta_2 \delta_4 - \delta_3 \alpha_3}{2\alpha_3 \alpha_4 - 2\delta_4^2} g^2 \equiv \lambda_2 g^2,$$

which shows that $\eta \neq 0$ and $\epsilon \neq 0$ only if $g \neq 0$. Eliminating q, η and ϵ, the potential $\mathcal{F}$ can be written as a function of g

$$\mathcal{F} = \mathcal{F}'_0 + \frac{\alpha}{2} g^2 + \frac{\beta}{4} g^4 + \frac{\gamma_1}{6} g^6,$$

where the renormalized coefficients are

$$\mathcal{F}'_0 = \mathcal{F}_0 - \frac{\alpha_2^2}{4\beta_2}, \qquad \alpha = \alpha_1 - \frac{\alpha_2 \delta_1}{\beta_2},$$

$$\beta = \beta_1 - \frac{\delta_1^2}{\beta_2} + 2\alpha_3 \lambda_1^2 + 2\alpha_4 \lambda_2^2 + 2\delta_2 \lambda_1 + 2\delta_3 \lambda_2 + 4\delta_4 \lambda_1 \lambda_2.$$

The zero–order and second–order terms depend on the parameters belonging to the primary OPs. The secondary OPs modify only the forth–order term. In this notation, the solution which minimizes the potential $\mathcal{F}$ reads

$$g_o^2 = \frac{-\beta + \sqrt{\beta^2 - 4\gamma\alpha}}{2\gamma}, \qquad q_o^2 = -\frac{\delta_1 g_o^2 + \alpha_2}{\beta_2},$$

$$\eta_o = \lambda_1 g_o^2, \qquad \epsilon_o = \lambda_2 g_o^2.$$

To study the softening of C_{44}, we have expressed the free energy as a function of ϵ only. In these calculations we have omitted the sixth–order term, which usually has a small contribution near the transition point. The elastic constant C_{44} is obtained using the standard definition

$$C_{44}(T) = \frac{\partial^2 \mathcal{F}}{\partial \epsilon^2} = C_{44}^0 - \frac{\delta^2}{\alpha'_3} - \frac{\delta_3^2}{\beta'_1}, \qquad \text{where} \qquad (2.9)$$

$$\delta = \delta_4 - \frac{\delta_2 \delta_3}{2\beta'_1}, \qquad \alpha'_3 = \alpha_3 - \frac{\delta_2^2}{2\beta'_1} = a_3(T - T'_{c3}), \qquad \beta'_1 = \beta_1 - \frac{\delta_1^2}{\beta_2},$$

with $T'_{c3} = T_{c3} + \delta_2^2/2\beta'_1$. The second and third term in (2.9) are negative at high temperatures, so both contribute to the softening of C_{44}. It means that all couplings included in (2.7) are involved in this behavior. The main temperature dependence is caused by the second term, but also the last term in (2.9) may depend on temperature. Omitting the last term, (2.9) can be written in the form [PPO07]

$$C_{44} = C_{44}^0 \frac{T - T_0}{T - T'_{c3}}, \qquad (2.10)$$

where $T_0 = T'_{c3} + \delta^2/C_{44}^0 a_3$.

For more technical details, see [PPO07] and references therein.

2.1.6 Universal Mandelbrot Set as a Phase–Transition Model

The problem of stability of time evolution is one of the most important in physics. Usually one can make the motion stable or unstable by changing some parameters which characterize Hamiltonian of the system. Stability regions can be represented on the phase diagram and transitions between them are described by catastrophe theory [Tho89, Arn78]. It can seem that a physical system or a mechanism can be taken from one domain of stability to any other by continuous and quasi–static variation of these parameters, i.e., that the phase diagram is connected. However sometimes this expectation is wrong, because domains of stability can be separated by points where our system is getting totally destroyed. Unfortunately today it is too difficult to explore the full phase diagram for generic physical system with many parameters. Therefore, following [LL03], it was proposed in [DM06] to consider as a simpler model the discrete dynamics of one complex variable. The phase diagram in this case is known as Universal Mandelbrot Set (UMS). MS is a well–known object in mathematics (see [Wik07, Fer23], as well as Figure 1.44 above), but its theory is too formal and not well adjusted to the use in the phase transition theory. In this section, following [Mor07], we will make MS more practical for physical applications.

Structure of MS

First of all we remind the definition of MS and UMS from [DM06], which different from conventional definition in mathematical literature, see below. Mandelbrot Set (MS) is a set of points in the complex c plane. MS includes a point c if the map $x \to f(x,c)$ has stable periodic orbits. As shown in Figure 2.2 MS consists of many clusters connected by trails, which in turn consist of smaller clusters and so on. Each cluster is linear connected and can be divided into elementary domains where only one periodic orbit is stable. Different elementary domains can merge and even overlap. Boundary of elementary domain of nth order, i.e., of a domain where an nth order orbit is stable, is a real curve $c(\alpha)$ given by the system:

$$\begin{cases} G_n(x,c) = 0 \\ F'_n(x,c) + 1 = e^{i\alpha} \end{cases}, \quad \text{with}$$

$$F_n(x,c) = f^{\circ n}(x,c) - x,$$
$$G_n(x,c) = \frac{F_n(x,c)}{\prod_m G_m(x,c)}, \quad (n = 1, ..., m).$$

Indeed, when $G_n(x,c)$ vanishes then x belongs to the orbit of exactly the nth order. This orbit is stable if $|\frac{\partial}{\partial x} f^{\circ n}(x,c)| < 1$, what implies the above equation. The solution of this system may give us more than a single nth order domain. Domains of different orders merge at the points c where

$$Resultant_x\,(G_n(x,c), G_k(x,c)) = 0. \qquad (2.11)$$

Clearly, two orbits and thereafter domains merge only if n is divisor of k, so it is reasonable to consider only $Resultant_x(G_n, G_{mn}) = 0$, with $m = 2, 3 \ldots$ and $Discriminant_x(G_n)$ for $k = n$.

Physically MS is a phase diagram of discrete dynamic of one complex variable. It is clear from Figure 2.2 that one should distinguish between three types of connectivity in different places of phase diagram. The first is linear

2.1 Introduction to Phase Transitions

connectivity: the possibility to connect any two points with a continuous line. The second type is weak connectivity: it means that only a closure of our set has linear connectivity. The third type we call strong connectivity: it means that any two interior points are connected with a thick tube.

Entire MS on Figure 2.2 is weakly[3] but not linearly connected and its clusters are linearly, but not strongly connected. Universal Mandelbrot Set (UMS) is unification of MS of different 1_c-parametric families. When we rise from MS to UMS we add more parameters to the base function. Thus entire UMS could become strongly connected, but it is unclear whether this really happens.

Fig. 2.2. The simplest examples of Mandelbrot sets $MS(x^2 + c)$ and $MS(x^3 + c)$, constructed by Fractal Explorer [Fer23]. The picture explains the terms 'clusters','elementary domain' and 'trails'. It is difficult to see any clusters except for the central one in the main figure, therefore one of the smaller clusters is shown in a separate enlarged picture to the left (adapted and modified from [Mor07]).

To explore MS of different functions we need to draw it. The method of MS construction which we can derive from the definition of MS is following. We are constructing the domains where different orbits are stable. We can build them using the fact that if orbit is stable then absolute value of derivative $\frac{df}{dx}$ is less than one.

Simplification of the Resultant Condition

In this section we prove that the resultant condition (2.11) can be substituted a by much simpler one [Mor07]:

$$Resultant_x\left(G_n(x), (F_n' + 1 - e^{i\alpha})\right) = 0, \qquad \alpha = \frac{2\pi}{m}k.$$

To prove this equation, it is enough to find the points where

$$Resultant_x(F_n, \frac{F_{nm}(x)}{F_n(x)}) = 0.$$

---

[3] MS is usually claimed to be locally connected [Wik07], i.e., any arbitrary small vicinity of a point of MS contains a piece of some cluster. In our opinion weak connectivity is another feature, especially important for physical applications.

Then:

$$\text{Resultant}_x(F_n, \frac{F_{nm}(x)}{F_n(x)}) = \text{Resultant}_x(F_n, \frac{F'_{nm}(x)}{F'_n(x)}) = 0.$$

By definition of $F(x)$,

$$F_{nm}(x) = F_{n(m-1)}(F_n(x) + x) + F_n(x).$$

Then:

$$\frac{F'_{nm}(x)}{F'_n(x)} = \frac{F'_{n(m-1)}(x)}{F'_n(x)}(F'_n(x) + 1) + 1 = \frac{F'_{n(m-1)}(x)}{F'_n(x)}e^{i\alpha} + 1 =$$

$$= (\ldots((e^{i\alpha} + 1)e^{i\alpha} + 1)e^{i\alpha} + 1)\ldots) + 1 = \sum_{l=0}^{m-1} e^{li\alpha} = \frac{e^{mi\alpha} - 1}{e^{i\alpha} - 1}.$$

Thus (2.11) implies that $e^{mi\alpha} - 1 = 0$ and therefore $\alpha = \frac{2\pi k}{m}$.

This theorem is a generalization of a well known fact for the central cardioid domain of $MS(x^2 + c)$ (see for example [Wik07]). This also provides a convenient parametrization of generic MS.

A Fast Method for MS simulation

Historically MS was introduced in a different way from our formulation. We call it $\widetilde{MS}$. It depends not only on the family of functions, but also on a point x_0. If c belongs to the $\widetilde{MS}(f, x_0)$ then

$$\lim_{n \to \infty} f^{on}(x_0) \neq \infty$$

In the literature one usually puts $x_0 = 0$ independently of the shape of $f(x)$. Such $\widetilde{MS}(f, 0) \neq MS(f)$, except for the families like $f = x^a + c$. Existing computer programs [Fer23] generate $\widetilde{MS}(f, 0)$, and can not be used to draw the proper $MS(f)$. Fortunately there is a simple relation [Mor07]:

$$MS(f) = \bigcup_{x_{cr}} \widetilde{MS}(f, x_{cr}).$$

where union is over all critical points of $f(x)$, $f'(x_{cr}) = 0$. Equation (2.1.6) is closely related to hyperbolic and local connectivity conjectures [Wik07]. It is also equivalent to the following two statements about the phase portrait in the complex x plane:

(I) If $\lim_{l \to \infty} f^{ol}(x_{cr}) \neq \infty$ then there is a stable periodic orbit O of finite order which attracts x_{cr}. It implies that

$$MS(f) \supseteq \bigcup_{x_{cr}} \widetilde{MS}(f, x_{cr}).$$

(II) If O is a stable periodic orbit, then a critical point x_{cr} exists, which is attracted to O. This implies that

$$MS(f) \subseteq \bigcup_{x_{cr}} \widetilde{MS}(f, x_{cr}).$$

The statement (I) says that if $\lim_{l \to \infty} f^{ol}(x_{cr}) \neq \infty$ then this limit exists and is a stable orbit with finite period, i.e., that there are no such things as strange attractors in discrete dynamics of one complex variable. This statement is unproved but we have no counter–examples.

The statement (II) is much easier. If x_0 is a stable fixed point, then it is surrounded by a disk–like domain, where $|f'(x)| < 1$. Its boundary is parametrized by $f'(x) = e^{i\alpha}$ and inside this area there is a point where $f'(x) = 0$, i.e., some critical point x_{cr} of f. It is important that this entire surrounding of x_0 – and thus this x_{cr} – lie inside the attraction domain of x_0:

$$|f(x_{cr}) - f(x)| < |x_{cr} - x|,$$

i.e., we found x_{cr} which is attracted to x_0. This argument can be easily extended to higher order orbits and can be used to prove (II). Equation (2.1.6) leads to a simple upgrade of programs, which construct MS.

Reducing $MS(x^3 + c)$ to $MS(x^2 + c)$

As application of our results we consider the 2_C–parametric section of UMS for the family

$$f(x) = a \cdot x^3 + (1-a) \cdot x^2 + c,$$

which interpolates between $MS(x^3+c)$ at $a = 1$ and $MS(x^2+c)$ at $a = 0$. We extend consideration of [DM07a] to non–trivial second order clusters which were beyond the reach of the methods used in that paper.

For more details on MS, see [Mor07].

2.1.7 Oscillatory Phase Transition

Coherent oscillations are observed in neural systems such as the *visual cortex* and the *hippocampus*. The synchronization of the oscillators is considered to play important roles in *neural information processing* [GKE89]. There are mainly two viewpoints in the research of the oscillatory activity in neural systems. In the first viewpoint, the activity of each neuron is expressed by the firing rate, and the coherent oscillation appears owing to the interaction of the excitatory and inhibitory neurons. Wilson–Cowan and Amari found first oscillatory behavior theoretically in interacting neurons [WC72, Ama72]. Recently, [RRR02] proposed a more elaborate model to explain various EEG rhythms and epileptic seizures. If the spatial freedom is taken into consideration, the excitation wave can propagate. Wilson–Cowan performed numerical simulations of two layers of excitable neurons and inhibitory neurons [WC72].

In the second viewpoint, each neuron is regarded as an oscillator. Coherent oscillation appears as the global synchronization of the coupled oscillators. The global synchronization in general coupled oscillators was first studied by Winfree [Win67]. Kuramoto proposed a globally coupled phase oscillator model as a solvable model for the global synchronization [Kur84]. The leaky–integrate–fire model is one of the simplest models for a single neuron and often used to study dynamical behaviors of neural networks. Each neuron receives an input via synaptic connections from other neurons and it fires when the input goes over a threshold and sends out impulses to other neurons. In that sense, the coupling is instantaneous, and then the model is called *pulse–coupled oscillators*. Mirollo and Strogatz studied a globally coupled system of the *integrate–and–fire neurons*, and showed that perfect synchronization occurs in a finite time [MS90]. The synchronization of pulse coupled oscillators has been studied in deterministic systems by many researchers [TMS93, GR93, VAE94]. If each oscillator's behavior is stochastic, the model is generalized to a noisy phase oscillator model and a noisy integrate–and–fire model. In the stochastic system, the coherent oscillation appears as an analogue of the phase transition in the statistical mechanics. Globally coupled noisy phase oscillators were studied in [SK86, Kur91, Sak02, KS03], and globally coupled noisy integrate–and–fire model were studied in [BH99, Bru00, HNT01]. The globally coupled system is a useful model for the detailed analyzes, however, local or non–local interactions are more plausible, since neurons interact with other neurons via long axons or gap junctions. The non–locally coupled system of the deterministic integrate–and–fire neurons was also studied [GE01]. In this section, following [Sak04], we study a non–locally coupled noisy integrate–and–fire model with the direct numerical simulation of the *Fokker–Planck equation*.

The equation for a noisy integrate–and–fire neuron is written as

$$\dot{x} = 1 - bx + I_0 + \xi(t), \qquad (2.12)$$

where x is a variable corresponding to the membrane potential, b is a positive parameter, I_0 denotes an external input, and $\xi(t)$ is the Gaussian white noise satisfying $\langle \xi(t)\xi(t') \rangle = 2D\delta(t-t')$. If x reaches a threshold 1, x jumps back to 0. If $b < 1 + I_0$, each neuron fires spontaneously. The Fokker–Planck equation for the Langevin equation (2.12) is [Sak04]

$$\partial_t P = -\frac{\partial}{\partial x}(1 - bx + I_0)P(x) + D\frac{\partial^2 P}{\partial x^2} + \delta(x)J_0(t), \qquad (2.13)$$

where $J_0(t) = -D(\partial P/\partial x)_{x=1}$ is the firing rate. The stationary distribution $P_0(x)$ for the Fokker–Planck equation (2.13) is written as [Ric77]

$$P_0(x) = P_0(0)e^{\{(ax-(1/2)bx^2\}/D}, \qquad \text{for } x < 0, \qquad (2.14)$$

$$= P_0(0)e^{\{(ax-(1/2)bx^2\}/D}\left[1 - \frac{\int_0^x e^{\{-az+(1/2)bz^2\}/D}dz}{\int_0^1 e^{\{-az+(1/2)bz^2\}/D}dz}\right], \quad \text{for } 0 < x < 1,$$

where $a = 1 + I_0$ and $P(0)$ is determined from the normalization condition $\int_{-\infty}^{1} P_0(x)dx = 1$. The firing rate J_0 is determined as

$$J_0 = DP_0(0)/\int_0^1 e^{\{-az+(1/2)bz^2\}/D} dz.$$

We have performed direct numerical simulation of (2.13) with the finite difference method with $\Delta x = 0.0002$ and $\Delta t = 2.5 \times 10^{-5}$, and checked that the stationary probability distribution (2.14) is successfully obtained.

We assume a non–locally coupled system composed of the noisy integrate–and–fire neurons. Each neuron interacts with other neurons via synaptic connections. Time delay exists generally for the synaptic connections. A model equation of the interacting noisy integrate–and–fire neurons is written as [Sak04]

$$\dot{x}_i = 1 - bx_i + I_i + \xi_i(t),$$

where x_i denotes the dimensionless membrane potential for the ith neuron, $\xi_i(t)$ denotes the noise term which is assumed to be mutually independent, i.e., $\langle \xi_i(t)\xi_j(t')\rangle = 2D\delta_{i,j}\delta(t-t')$, and I_i is the input to the ith neuron by the mutual interaction. The input I_i to the ith neuron from the other neurons is given by

$$I_i = \sum_j \sum_k g_{i,j}\frac{1}{\tau}e^{-(t-t_k^j)/\tau}, \qquad (2.15)$$

where t_k^j is the time of the kth firing for the jth neuron, $g_{i,j}$ denotes the interaction strength from the jth neuron to the ith neuron, and τ denotes a decay constant. The sum is taken only for $t > t_k^j$. The effect of the firing of the jth neuron to the ith neuron decays continuously with τ. If $\tau \to 0$, the coupling becomes instantaneous. Equation (2.15) is equivalent to

$$\dot{I}_i = -\{I_i - \sum_j \sum_k g_{i,j}\delta(t-t_k^j)\}/\tau.$$

If there are infinitely many neurons at each position y, we can define the number density of neurons with membrane potential x clearly at each position. The number density is expressed as $n(x,y,t)$ at position y and time t. The non–locally coupled system can be studied with a mean–field approach. In the mean–field approach, the number density is proportional to the probability distribution $P(x,y,t)$ for the probability variable x. The average value of $\delta(t-t_k^j)$ expresses the average firing rate at time t at the position y. It is expressed as $J_0(y,t) = -D(\partial n/\partial x)_{x=1}$. The number density $n(x,y,t)$ therefore obeys the Fokker–Planck type equation [Sak04]

$$\partial_t n(x,y) = -\frac{\partial}{\partial x}(1-bx+I(y,t))n(x,y) + D\frac{\partial^2 n}{\partial x^2} + \delta(x)J_0(y,t),$$

$$\dot{I}(y,t) = -\{I(y,t)-J(y,t)\}/\tau,$$

$$J(y,t) = \int g(y,y')J_0(y',t)dy',$$

where $g(y,y')$ is the coupling strength from the neuron located at y' to the one at y, and $I(y)$, $J_0(y)$ are respectively the input and the firing rate for the neuron at y.

We have assumed that the time delay for the signal to transmit between y' and y can be neglected and $g(y,y')$ depends only on the distance $|y-y'|$, i.e., $g(y,y') = g(|y-y'|)$. As two simple examples of the non-local coupling, we use

$$g_1(y,y') = c\exp(-\kappa|y-y'|) - d, \quad g_2(y,y') = c\exp(-\kappa|y-y'|) - d\exp(-\kappa'|y-y'|).$$

These forms of the coupling imply that the interaction is excitable locally, but the interaction strength decreases with the distance $|y-y'|$, and it becomes inhibitory when $|y-y'|$ is large. This *Mexican–hat coupling* was used in several neural models [Ama77], especially to study the competitive dynamics in neural systems. Although two layer models of excitatory neuron layer and inhibitory neuron layer may be more realistic, we consider the above simpler one–layer model. The inhibitory interaction approaches a constant value $-d$ for the coupling g_1, and 0 for the coupling g_2. The system size is assumed to be $L = 10$ as a simple example, and the periodic boundary conditions for the space variable y are imposed. We choose the damping constants κ and κ', as the exponential function decays to almost 0 for the distance $|y-y'| \sim L$. Therefore, the dynamical behaviors do not depend on the system size L qualitatively in the second model. But the dynamical behaviors depend on the system size L in the first model, because the range of the inhibitory interaction is infinite in the model.

There is a stationary and uniform solution $n(x,y,t) = n_0(x)$ and $I(y,t) = I_0$ in the non–locally coupled equation. The uniform solution satisfies

$$n_0(x) = n_0(0)e^{\{ax-(1/2)bx^2\}/D}, \quad \text{for } x < 0,$$

$$= n_0(0)e^{\{ax-(1/2)bx^2\}/D}\left[1 - \frac{\int_0^x e^{\{-az+(1/2)bz^2\}/D}dz}{\int_0^1 e^{\{-az+(1/2)bz^2\}/D}dz}\right], \quad \text{for } 0 < x < 1,$$

where the parameter a is determined by the self-consistent condition

$$a = 1 - g_0 D(\partial n_0(x)/\partial x)_{x=1}, \quad \text{where} \quad g_0 = \int g(y,y')dy'.$$

To study the linear stability of the stationary and uniform solution, we consider small deviations $\delta n(x,y,t) = n(x,y,t) - n_0(x)$ and $\delta I(y,t) = I - I_0$ from the uniform solution. The small deviations can be expressed with the Fourier series as

$$\delta n(x,y,t) = \sum \delta n_k(x,t)\exp(iky) \quad \text{and} \quad \delta I(y,t) = \sum \delta I_k \exp(iky)$$

under the periodic boundary conditions, where $k = 2\pi m/L$. The perturbations δn_k and δI_k obey coupled linear equations [Sak04]

$$\frac{\partial \delta n_k(x,t)}{\partial t} = -\frac{\partial}{\partial x}\{(1 - bx + I_0)\delta n_k(x,t) + \delta I_k(t)n_0(x)\} + D\frac{\partial^2 \delta n_k}{\partial x^2} + \delta(x)\delta J_0(t),$$

$$\frac{d\delta I_k(t)}{dt} = -\{\delta I_k(t) - g'\delta J_{0k}(t)\}/\tau, \quad \text{where} \tag{2.16}$$

$$\delta J_{0k}(t) = -D(\partial n_k/\partial x)_{x=1} \text{ and } g' = \int g(y,y')e^{ik(y'-y)}dy'.$$

For L is sufficiently large, $g' = 2c\kappa/(\kappa^2 + k^2) - dL\delta_{k,0}$ for the coupling g_1 and $g' = 2c\kappa/(\kappa^2 + k^2) - 2d\kappa'/(\kappa'^2 + k^2)$ for the coupling g_2. The stability of the stationary state is determined by the real part of the eigenvalues of the linear equation (10). But, the stationary solution $n_0(x)$ is a nontrivial function of x, and it is not so easy to obtain the eigenvalues. Here we have evaluated the real part of the largest eigenvalue of the linear equation for various k by direct numerical simulations of (2.16). The dynamical behavior in the long time evolution of (2.16) is approximately expressed with the largest eigenvalue λ, that is, δn_k and $\delta I_k \sim e^{\lambda t}$ for $t \gg 1$. We have numerically calculated the linear growth rate of the norm $\{\int (\delta n_k)^2 dx + (\delta I_k)^2\}^{1/2}$ (which grows as $e^{(\text{Re}\lambda)t}$ for $t \gg 1$) every time–interval 0.001. Since the norm grows to infinity or decays to zero in the natural time evolution of the linear equation, we have renormalized the variables every time–interval 0.001, as the norm is 1 by the rescaling $c\delta n_k \to \delta n_k$ and $c\delta I_k \to \delta I_k$ with a constant c.

Since the pulse propagates one round L with period $T = 3.23$, the velocity of the travelling pulse is $L/T \sim 3.1$. A regular limit cycle oscillation with period T is observed at each point. The directions depend on the initial conditions. The travelling pulse state is an ordered state in the non-locally coupled system. The locally excitable interaction facilitates the local synchronization of the firing, but the global inhibition suppresses the complete synchronization. As a result of the frustration, a travelling pulse appears. The pulse state is different form the travelling pulse observed in an excitable system, since the uniform state is unstable in our system and the pulse state is spontaneously generated from the stationary asynchronous state. The input $I(y,t)$ to the neuron at position y exhibits regular limit cycle oscillation.

As a second example, we consider a non–locally coupled system with the coupling function

$$g_2(y) = 1.8\exp(-4|y|) - 0.48\exp(-|y|).$$

A supercritical phase transition occurs at $D \sim 0.0155$, which is also consistent with the linear stability analysis. Near the critical value, the amplitude of the oscillation is small and the wavy state seems to be sinusoidal. As D is decreased, the oscillation amplitude increases and the sinusoidal waves change into pulse trains gradually. Pulses are created periodically near $x \sim 6$ and they are propagating alternatively in different directions. The inversely–propagating pulses collide at $x \sim 1$ and they disappear. Namely, there are a pacemaker region (a source region) and a sink region of travelling pulses in this solution. This type of wavy state including a pacemaker region and the simple pulse-train state are bistable.

In summary, we have studied the non–locally noisy integrate–and–fire model with the Fokker–Planck equation. We have found that a travelling pulse appears as a result of oscillatory phase transitions. We found also a pulse-train state by changing the form of the interaction. The wavy states appear as a phase transition from a asynchronous state when the noise strength is decreased. We have investigated a 1D system for the sake of simplicity of numerical simulations, but we can generalize the model equation to a 2D system easily. Our non–locally coupled integrate–and–fire model might be too simple, however, the wavy state is one of the typical dissipative structures far from equilibrium. Therefore, the spontaneously generated waves might be observed as some kind of brain waves also in real neural systems. For more details, see [Sak04].

2.1.8 Partition Function and Its Path–Integral Description

Recall that in *statistical mechanics*, the *partition function* Z is used for statistical description of a system in thermodynamic equilibrium. Z depends on the physical system under consideration and is a function of temperature T as well as other parameters (such as volume V enclosing a gas etc.). The partition function forms the basis for most calculations in statistical mechanics. It is most easily formulated in *quantum statistical mechanics* (see e.g., [Fey72]).

Classical Partition Function

A system subdivided into N subsystems, where each subsystem (e.g., a particle) can attain any of the energies ϵ_j $(j = 1, ..., N)$, has the partition function given by the sum of its Boltzmann factors,

$$\zeta = \sum_{j=0}^{\infty} e^{-\beta \epsilon_j},$$

where $\beta = \frac{1}{k_B T}$ and k_B is *Boltzmann constant*. The interpretation of ζ is that the probability that the subsystem will have energy ϵ_j is $e^{-\beta \epsilon_j}/\zeta$. When the number of energies ϵ_j is definite (e.g., particles with spin in a crystal lattice under an external magnetic field), then the indefinite sum is replaced with a definite sum. However, the total partition function for the system containing N subsystems is of the form

$$Z = \prod_{j=1}^{N} \zeta_j = \zeta_1 \zeta_2 \zeta_3 \cdot ...,$$

where ζ_j is the partition function for the jth subsystem. Another approach is to sum over all system's total energy states,

$$Z = \sum_{r=1}^{N} e^{-\beta E_r}, \quad \text{where} \quad E_j = n_1^{(j)}\epsilon_1 + n_2^{(j)}\epsilon_2 + \ldots$$

In case of a system containing N non–interacting subsystems (e.g., a real gas), the system's partition function is given by

$$Z = \frac{1}{N!}\zeta^N.$$

This equation also has the more general form

$$Z = \frac{1}{N!h^{3N}} \int \prod_{i=1}^{N} d^3q^i d^3p_i \sum_{i=1}^{N} e^{-\beta H_i},$$

where $H_i = H_i(q^i, p_i)$ is the ith subsystem's Hamiltonian, while h^{3N} is a normalization factor.

Given the partition function Z, the system's *free energy* F is defined as

$$F = -k_B T \ln Z,$$

while the *average energy* U is given by

$$U = \frac{1}{Z} E_i e^{-\frac{E_i}{k_B T}} = -\frac{d}{d\beta}(\ln Z).$$

Liner Harmonic Oscillators in Thermal Equilibrium. The partition function Z, free energy F, and average energy U of the system of M oscillators can be found as follows: The oscillators do not interact with each other, but only with the *heat bath*. Since each oscillator is independent, one can find F_i of the ith oscillator and then $F = \sum_{i=1}^{M} F_i$. For each ith oscillator (that can be in one of N states) we have [Fey72]

$$Z_i = \sum_{n=1}^{N} e^{-\frac{E_n^i}{k_B T}}, \quad F_i = -k_B T \ln Z_i, \quad U_i = \frac{1}{Z} \sum_{n=1}^{N} E_n^i e^{-\frac{E_n^i}{k_B T}}.$$

Quantum Partition Function

Partition function Z of a quantum–mechanical system may be written as a trace over all states (which may be carried out in any basis, as the trace is basis–independent),

$$Z = \text{Tr}(e^{-\beta \hat{H}}),$$

where $\hat{H}$ is the system's Hamiltonian operator. If $\hat{H}$ contains a dependence on a parameter λ, as in $\hat{H} = \hat{H}_0 + \lambda \hat{A}$, then the statistical average over $\hat{A}$ may be found from the dependence of the partition function on the parameter, by differentiation,

$$<\hat{A}> = -\beta^{-1}\frac{d}{d\lambda}\ln Z(\beta, \lambda).$$

However, if one is interested in the average of an operator that does not appear in the Hamiltonian, one often adds it artificially to the Hamiltonian, calculates Z as a function of the extra new parameter and sets the parameter equal to zero after differentiation.

More general, in *quantum field theory*, we have a *generating functional J* of the *field* $\phi(q)$ and the partition function is usually expressed by the *Feynman path integral* [Fey72] (see Chapter 4)

$$Z[J] = \int \mathcal{D}[\phi] e^{i(S[\phi] + \int d^N q\, J(q)\phi(q))},$$

where $S = S[\phi]$ is the *field action functional*.

Vibrations of Coupled Oscillators

In this subsection, following [Fey72], we give both classical and quantum analysis of vibrations of coupled oscillators. R. Feynman used this method as a generic model for the crystal lattice.

Consider a *crystal lattice* with A atoms per unit cell, such that $3A$ coordinates α must be given to locate each atom. Also let $Q_{\alpha,N}$ denote the *displacement* from equilibrium of the coordinate α in the Nth cell. $Q_{\alpha,N+M}$ is the displacement of an atom in a cell close to N.

The *kinetic energy* of the lattice is given by

$$T = \frac{1}{2} Q_{\alpha,N} Q_{\alpha,N},$$

while its *potential energy* (in linear approximation) is given by

$$V = \frac{1}{2} C^M_{\alpha\beta} Q_{\alpha,N} Q_{\beta,N+M}.$$

Classical Problem

The so–called *original Hamiltonian* is given by

$$H = \sum_i \frac{p_i\prime^2}{2m_i} + \frac{1}{2} C_{ij}\prime q^i\prime q^j\prime,$$

where the $q^i\prime$ are the coordinates of the amount of the lattice displacement from its equilibrium, $p_i\prime = m_i \dot{q}^i\prime$ are the canonical momenta, and $C_{ij}\prime = C_{ji}\prime$ are constants. To eliminate the mass constants m_i, let

$$q^i = q^i\prime \sqrt{m_i} \quad \text{and} \quad C_{ij} = \frac{C_{ij}\prime}{\sqrt{m_i m_j}}.$$

Then
$$p_i = \frac{\partial L}{\partial \dot{q}^i} = \frac{p_{i'}}{\sqrt{m_i}}, \qquad (L \text{ is the Lagrangian of the system})$$
and we get the *simplified Hamiltonian*
$$H = \frac{1}{2}\sum_i p_i^2 + \frac{1}{2}C_{ij}q^i q^j.$$

The Hamilton's equations of motion now read
$$\dot{q}^i = \partial_{p_i} H = p_i, \qquad \dot{p}_i = -\partial_{q^i} H = -C_{ij}q^j = \ddot{q}^i.$$

We now break the motion of the system into *modes*, each of which has its own frequency ω. The total motion of the system is a sum of the motions of the modes. Let the αth mode have frequency ω_α so that
$$q^i_{(\alpha)} = e^{-i\omega_\alpha t} a_i^{(\alpha)}$$
for the motion of the αth mode, with $a_i^{(\alpha)}$ independent of time. Then
$$\omega_\alpha^2 a_i^{(\alpha)} = C_{ij} a_j^{(\alpha)}.$$

In this way, the classical *problem of vibrations of coupled oscillators* has been reduced to the *problem of finding eigenvalues* and *eigenvectors* of the real, symmetric matrix $\|C_{ij}\|$. In order to get the ω_α we must solve the *characteristic equation*
$$\det \|C_{ij} - \omega^2 \delta_{ij}\| = 0.$$
Then the eigenvectors $a_i^{(\alpha)}$ can be found. It is possible to choose the $a_i^{(\alpha)}$ so that
$$a_i^{(\alpha)} a_i^{(\beta)} = \delta_{\alpha\beta}.$$
The general solution for q^i is
$$q^i = C_\alpha q^i_{(\alpha)},$$
where the C_α are arbitrary constants. If we take
$$Q_\alpha = C_\alpha e^{-i\omega_\alpha t},$$
we get
$$q^i = a_i^{(\alpha)} Q_\alpha.$$
From this it follows that
$$a_i^{(j)} q^i = a_i^{(j)} a_i^{(\alpha)} Q_\alpha = \delta_{\alpha j} Q_\alpha = Q_j.$$
Making the change of variables, $Q_j = a_i^{(j)} q^i$, we get $H = \sum_\alpha H_\alpha$, where
$$H_\alpha = \frac{1}{2} p_\alpha^2 + \frac{1}{2} \omega_\alpha^2 Q_\alpha.$$
This has the expected solutions: $Q_\alpha = C_\alpha e^{-i\omega_\alpha t}.$

Quantum–Mechanical Problem

Again we have the original Hamiltonian

$$H = \sum_i \frac{p_i'^2}{2m_i} + \frac{1}{2} C'_{ij} q'^i q'^j,$$

where this time

$$p_i' = \frac{1}{i} \frac{\partial}{\partial q'^i} \qquad \text{(in normal units } \hbar = 1\text{)}.$$

Making the same change of variables as before, we get

$$Q_\alpha = a_i^{(\alpha)} q^i = a_i^{(\alpha)} \sqrt{m_i} q'^i,$$

$$H = \sum_\alpha H_\alpha, \quad \text{where} \quad H_\alpha = -\frac{1}{2} \frac{\partial^2}{\partial Q_\alpha^2} + \frac{1}{2} \omega_\alpha^2 Q_\alpha.$$

It follows immediately that the eigenvalues of our original Hamiltonian are

$$E = \sum_\alpha (N_\alpha + \frac{1}{2}) \omega_\alpha.$$

The solution of a quantum–mechanical system of coupled oscillators is trivial once we have solved the characteristic equation

$$0 = \det \|C_{ij} - \omega^2 \delta_{ij}\| = \det \left\| \frac{C_{ij}'}{\sqrt{m_i m_j}} - \omega^2 \delta_{ij} \right\|.$$

If we have a solid with $\frac{1}{3}(10^{23})$ atoms we must apparently find the eigenvalues of a 10^{23} by 10^{23} matrix. But if the solid is *crystal*, the problem is enormously simplified. The *classical Hamiltonian for a crystal* is

$$H = \frac{1}{2} \sum_{\alpha,N} \dot{Q}_{\alpha,N}^2 + \frac{1}{2} \sum_{\alpha,\beta,N,M} C_{\alpha\beta}^M Q_{\alpha,N} Q_{\beta,N+M},$$

and the *classical equation of motion for a crystal lattice* is (using $C_{\alpha\beta}^M = C_{\beta\alpha}^{-M}$)

$$\ddot{Q}_{\alpha,N} = -\sum_{M,\beta} C_{\alpha\beta}^M Q_{\beta,N+M}.$$

In a given mode, if one cell of the crystal is vibrating in a certain manner, it is reasonable to expect all cells to vibrate the same way, but with different phases. So we try

$$Q_{\alpha,N} = a_\alpha(K) e^{-i\omega t} e^{iK \cdot N},$$

where K expresses the relative phase between cells. The $e^{iK \cdot N}$ factor allows for wave motion. We now want to find the dispersion relations, or $\omega = \omega(K)$.

2.1 Introduction to Phase Transitions

$$\omega^2 a_\alpha e^{iK\cdot N} = \sum_{M,\beta}(C^M_{\alpha\beta}a_\beta e^{iK\cdot M})e^{iK\cdot N}.$$

Let

$$\gamma_{\alpha\beta}(K) = \sum_M C^M_{\alpha\beta}e^{iK\cdot M}$$

(note that $\gamma_{\alpha\beta}(K)$ is Hermitian).

Then $\omega^2 a_\alpha = \sum_\beta \gamma_{\alpha\beta}a_\beta$, and we must solve the characteristic equation of a $3A$−by−$3A$ matrix:

$$\det|\gamma_{\alpha\beta} - \omega^2\delta_{\alpha\beta}| = 0.$$

The solutions of the characteristic equation are

$$\omega^{(r)}(K) = \omega\binom{r}{K},$$

where r runs from 1 to $3A$. The motion of a particular mode can be written

$$Q^{(r)}_{\alpha,N}(K) = a^r_\alpha(K)e^{-i\omega^{(r)}(K)^r}e^{iK\cdot N},$$

where

$$a^r_\alpha a^{*r'}_\alpha = \delta_{rr'}.$$

Then the general motion can be described by

$$Q_{\alpha,N} = \sum_{K,r} C_r(K)a^r_\alpha(K)e^{-i\omega^{(r)}(K)^r}e^{iK\cdot N},$$

where $C_r(K)$ are arbitrary constants.

Let $Q_r(K) = C_r(K)e^{-i\omega^{(r)}(K)^r}$. $Q_r(K)$ describe the motion of a particular mode. Then we have

$$Q_{\alpha,N} = \sum_{K,r} Q_r(K)a^r_\alpha(K)e^{iK\cdot N}.$$

It follows that

$$Q_r(K) \propto \sum_{\alpha,N} Q_{\alpha,N}a^{*r}_\alpha(K)e^{-iK\cdot N}, \qquad (\propto \text{ means 'proportional to'}),$$

and the Hamiltonian for the system is

$$\begin{aligned}H &= \frac{1}{2}\sum_{\alpha,N}\left[\dot{Q}^2_{\alpha,N} + \sum_{\beta,M}C^M_{\alpha\beta}Q_{\alpha,N}Q_{\beta,N+M}\right]\\&= \frac{1}{2}\sum_{K,r}\left[|\dot{Q}_r(K)|^2 + \omega^{2(r)}(K)|Q_r(K)|^2\right].\end{aligned}$$

A Cubic Lattice of Harmonic Oscillators

Assume the unit cell to be a cubic lattice with one atom per cell. Each atom behaves as an harmonic oscillator, with spring constants k_A (nearest neighbors), and k_B (diagonal–, or next–nearest neighbors). This case is fairly simple, and we can simplify the notation: $\alpha = 1, 2, 3$.

$$Q_{1,N} = X_N, \qquad Q_{2,N} = Y_N, \qquad Q_{3,N} = Q_N.$$

We wish to find the 3 natural frequencies associated with each k of the crystal. To do this, we must find $C_{\alpha\beta}^M$ and then $\gamma_{\alpha\beta}$. In complex coordinates,

$$V = \sum_{\alpha,\beta} V_{\alpha\beta}, \qquad \text{where} \qquad V_{\alpha\beta} = \sum_{N,M} C_{\alpha\beta}^M Q_{\alpha,N}^* Q_{\beta,N+M},$$

where $*$ denotes complex conjugation. For example,

$$V_{11} = \sum_{N,M} C_{11}^M X_N^* X_{N+M}.$$

If we express the displacement of atom N from its normal position as X_N, then the *potential energy* from the distortion of the spring between atoms N and M is

$$\frac{1}{2} k_M \left[(X_N - X_{N+M}) \cdot \frac{M}{|M|} \right]^2,$$

where $k_M = k_A$ for $N + M$ a nearest neighbor to N, $k_M = k_B$ for $N + M$ a next–nearest neighbor.

In summing over N and M to get the total potential energy we must *divide V by two*, for we count each spring twice. If we use complex coordinates, however, we *multiply V by two* to get the correct equations of motion:

$$V = \frac{1}{2} \sum_{N,M} k_M \left[(X_N - X_{N+M}) \cdot \frac{M}{|M|} \right]^2,$$

$$V_{11} = \frac{1}{2} \sum_{N,M} k_M \left(\frac{M_X}{|M|} \right)^2 (X_N^* - X_{N+M}^*)(X_N - X_{N+M})$$

$$= \frac{1}{2} \sum_{N,M} k_M \left(\frac{M_X}{|M|} \right)^2 [(X_N^* X_N + X_{N+M}^* X_{N+M}) - (X_N^* X_{N+M} + X_N X_{N+M}^*)]$$

$$= \sum_{N,M} k_M \left(\frac{M_X}{|M|} \right)^2 [X_N^* X_N - X_N X_{N+M}^*].$$

Comparing the above expressions we see that

$$C_{11}^0 = 2k_A + 4k_B, \qquad C_{11}^{\pm(1,0,0)} = -k_A, \qquad \text{and so on.}$$

In this way, all the $C_{\alpha\beta}^M$ can be found. We can then calculate

$$\gamma_{\alpha\beta}(K) = \sum_M C_{\alpha\beta}^M e^{iK \cdot M}.$$

We wish to solve

$$\det|\gamma_{\alpha\beta} - \omega^2 \delta_{\alpha\beta}| = 0.$$

For each relative phase K, there are 3 solutions for ω. Thus we get $3N$ values of ω and $\omega^{(r)}(K)$.

2.1.9 Noise–Induced Non–Equilibrium Phase Transitions

Noise is usually thought of as a phenomenon which perturbs the observation and creates disorder (see section 4.6.1). This idea is based mainly on our day to day experience and, in the context of physical theories, on the study of equilibrium systems. The effect of noise can, however, be quite different in *nonlinear non–equilibrium systems*. Several situations have been documented in the literature, in which the noise actually participates in the creation of ordered states or is responsible for surprising phenomena through its interaction with the nonlinearities of the system [HL84]. Recently, a quite spectacular phenomenon was discovered in a specific model of a spatially distributed system with multiplicative noise, white in space and time. It was found that the noise generates an *ordered symmetry–breaking state* through a genuine *second–order phase transition*, whereas no such transition is observed in the absence of noise [BPT94, BPT97].

Recently it has been shown that a white and *Gaussian multiplicative noise* can lead an *extended* dynamical system (fulfilling appropriate conditions) to undergo a *phase* transition towards an *ordered* state, characterized by a nonzero order parameter and by the breakdown of ergodicity [BPT94]. This result–first got within a Curie–Weiss–like *mean–field approximation*, and further extended to consider the simplest correlation function approach–has been confirmed through extensive numerical simulations [BPT97]. In addition to its *critical* nature as a function of the noise intensity σ, the newly found noise–induced phase transition has the noteworthy feature of being *reentrant*: for each value of D above a threshold one, the ordered state exists only inside a window $[\sigma_1, \sigma_2]$. At variance with the known case of *equilibrium* order$\Rightarrow$disorder transitions that are induced (in the simplest lattice models) by the nearest–neighbor coupling constant D and rely on the bi-stability of the local potential, the transition in the case at hand is led by the *combined effects* of D and σ through the nonlinearities of the system. Neither the zero–dimensional system (corresponding to the $D = 0$ limit) nor the deterministic one ($\sigma = 0$) show any transition.

General Zero–Dimensional System

To smoothly introduce the subject, we will start from the well–known *logistic equation*, and add to it a multiplicative white noise.

Noisy Logistic Equation

Recall that the logistic equation (also called the *Verhulst model* or *logistic growth curve*) is a model of population growth first published by P. Verhulst in 1845 (see [Wei05]). The model is continuous in time, but a modification of the continuous equation to a discrete quadratic recurrence equation known as the *logistic map* is widely used in *chaos theory*. The standard logistic equation

$$\dot{x} = \lambda x - x^2, \tag{2.17}$$

where the parameter λ is usually constrained to be positive, has a solution

$$x(t) = \frac{1}{1 + \left(\frac{1}{x_0} - 1\right) e^{-\lambda t}}.$$

Now, if we add a *multiplicative zero–mean Gaussian white noise* $\xi = \xi(t)$ with *noise intensity* σ to (2.17), we get the *Langevin SDE* (stochastic differential equation)

$$\dot{x} = \lambda x - x^2 + x\,\xi. \tag{2.18}$$

If we apply the *Stratonovitch interpretation* to the Langevin equation (2.18), we get the corresponding *Fokker–Planck equation*

$$\partial_t P(x,t) = -\partial_t \left(\lambda x - x^2\right) P(x,t) + \frac{\sigma^2}{2} \partial_x x\, \partial_x P(x,t) \tag{2.19}$$

dererminining the *probability density* $P(x,t)$ for the variable $x(t)$. The equation (2.19) has the *stationary probability density*

$$P_{st}(x) = \frac{1}{Z} x^{\frac{2\lambda}{\sigma^2} - 1} \exp\left(-\frac{2x}{\sigma^2}\right)$$

(where Z is a normalization constant), with *two extrema*:

$$x_1 = 0, \qquad x_2 = \lambda - \frac{\sigma^2}{2}.$$

General Zero–Dimensional Model

Now, following [BPT94, BPT97], we consider the following SDE that generalizes noisy logistic equation (2.18),

$$\dot{x} = f(x) + g(x)\,\xi, \tag{2.20}$$

where, as above, $\xi = \xi(t)$ denotes the Gaussian white noise with first two moments
$$\langle \xi(t) \rangle = 0, \qquad \langle \xi(t)\,\xi(t') \rangle = \sigma^2 \delta(t-t').$$

If we interpret equation (2.20) according to the Stratonovitch interpretation, we get the corresponding Fokker–Planck equation

$$\partial_t P(x,t) = -\partial_x \left[f(x) + P(x,t) \right] + \frac{\sigma^2}{2} \partial_x \left(g(x)\, \partial_x \left[g(x)\, P(x,t) \right] \right),$$

with the *steady-state solution*

$$P_{st}(x) = \frac{1}{Z} \exp\left(\int_0^x \frac{f(y) - \frac{\sigma^2}{2} g(y) g'(y)}{\frac{\sigma^2}{2} g^2(y)} dy \right), \qquad (2.21)$$

where $g'(x)$ stands for the derivative of $g(x)$ with respect to its argument. The extrema of the steady-state probability density obey the following equation

$$f(x) - \frac{\sigma^2}{2} g(x) g'(x) = 0. \qquad (2.22)$$

Note that this equation is not identical to the equation $f(x) = 0$ for the steady states in the absence of multiplicative noise. As a result, the most probable states need not coincide with the deterministic stationary states. More importantly, solutions can appear or existing solutions can be destabilized by the noise. These changes in the asymptotic behavior of the system have been generally named noise–induced phase transitions [HL84].

To illustrate this phenomenon, consider the case of a deterministically stable steady state at $x = 0$, e.g.,

$$f(x) = -x + o(x),$$

perturbed by a multiplicative noise. As is clear from equations (2.6–2.22), a noise term of the form
$$g(x) = 1 + x^2 + o(x^2)$$
will have a stabilizing effect, since
$$-(\sigma^2/2) g(x) g'(x) = -\sigma^2 x + o(x^2),$$

and it makes the coefficient of x more negative. On the other hand, noise of the form
$$g(x) = 1 - x^2 + o(x^2)$$
i.e., with maximal amplitude at the reference state $x = 0$, has the tendency to 'destabilize' the reference state. In fact, above a critical intensity $\sigma^2 > \sigma_c^2 = 1$, the stationary probability density will no longer have a maximum at $x = 0$, and 'noise–induced' maxima can appear. This phenomenon remains possible even if the deterministic steady-state equation, got by fixing the random value

of the noise to a constant value λ, namely, $f(x) + \lambda g(x) = 0$, has a unique solution for all λ [BPT94, BPT97].

Following the formalism for equilibrium states, it is tempting to introduce the notion of a 'stochastic potential' $U_{st}(x)$ by writing:

$$P_{st}(x) \sim \exp\left[-U_{st}(x)\right].$$

One concludes that for a system undergoing a noise–induced transition, e.g., for $g(x) = 1 - x^2 + o(x^2)$, and for $\sigma^2 > \sigma_c^2$, the stochastic potential has two minima. Consider now a spatially extended system got by coupling such units. The coupling is such that it favors the nearest–neighbor units, to stay at the same maximum of the probability density (minimum of the stochastic potential). In analogy to what happens for equilibrium models (such as the Landau–Ginzburg model), one expects that this system will undergo a phase transition for some critical value of the 'temperature' (noise intensity) σ^2. However, it turns out that this is not the case. It was shown in [BPT94, BPT97] that one needs a noise of precisely the other type, namely $g(x) = 1 + x^2 + o(x^2)$, to generate a genuine phase transition.

General d–Dimensional System

The general model has been introduced in [BPT94, BPT97]: a dD extended system of typical linear size L is restricted to a hypercubic lattice of $N = L^d$ points, whereas time is still regarded as a continuous variable. The state of the system at time t is given by the set of stochastic variables $\{x_i(t)\}$ ($i = 1, \ldots, N$) defined at the sites $\mathbf{r}_i$ of this lattice, which obey a system of coupled ordinary SDEs (with implemented Stratonovich interpretation)

$$\dot{x}_i = f(x_i) + g(x_i)\eta_i + \frac{D}{2d}\sum_{j \in n(i)}(x_j - x_i), \qquad (2.23)$$

Equations (2.23) are the discrete version of the *partial* SDE which in the continuum would determine the state of the extended system: we recognize in the first two terms the generalization of Langevin's equation for site i to the case of multiplicative noise (η_i is the *colored multiplicative noise* acting on site $\mathbf{r}_i$). For the specific example, perhaps the simplest one exhibiting the transition under analysis,

$$f(x) = -x(1+x^2)^2, \qquad g(x) = 1 + x^2. \qquad (2.24)$$

The last term in (2.23) is the lattice version of the Laplacian $\nabla^2 x$ of the extended stochastic variable $x(\mathbf{r},t)$ in a reaction–diffusion scheme. $n(i)$ stands for the set of $2d$ sites which form the immediate neighborhood of the site $\mathbf{r}_i$, and the coupling constant D between neighboring lattice sites is the diffusion coefficient.

Here we want to investigate the effects of the self–correlation time τ of the multiplicative noise on the model system just described [MDT00]. To that end we must assume a specific form for the noises $\{\eta_i = \eta_i(t)\}$: we choose *Ornstein–Uhlenbeck noise*, i.e., Gaussian distributed stochastic variables with zero mean and exponentially decaying correlations,

$$\langle \eta_i(t)\,\eta_j(t') \rangle = \delta_{ij}(\sigma^2/2\tau)\exp(-|t-t'|/\tau). \qquad (2.25)$$

They arise as solutions of an *un–coupled* set of Langevin SDEs,

$$\tau\dot{\eta}_i = -\eta_i + \sigma\xi_i \qquad (2.26)$$

where the $\{\xi_i = \xi_i(t)\}$ are white noises–namely, Gaussian stochastic variables with zero mean and δ–correlated:

$$\langle \xi_i(t)\,\xi_j(t') \rangle = \delta_{ij}\delta(t-t').$$

For $\tau \to 0$, the Ornstein–Uhlenbeck noise $\eta_i(t)$ approaches the white–noise limit $\xi_i^W(t)$ with correlations

$$\langle \xi_i^W(t)\,\xi_j^W(t') \rangle = \sigma^2\delta_{ij}\delta(t-t').$$

Mean–Field Approximation

The mean–field approximation here follows closely Curie–Weiss' mean–field approach to magnetism (see [MDT00]), and consists in replacing the last term in (2.23)

$$\Delta_i \equiv \frac{D}{2d}\sum_{j\in n(i)}(x_j - x_i), \qquad (2.27)$$

by

$$\bar{\Delta}_i \equiv D(\bar{x} - x_i), \qquad (2.28)$$

where $\bar{x}$ is the *order parameter* that will be determined self–consistently. In other words, the (short–ranged) interactions are substituted by a time– and space–independent 'external' field whose value *depends on the state* of the system. Since in this approximation equations (2.23) get immediately decoupled, there is no use in keeping the subindex i and we may refer to the systems in (2.23) and (2.26) as if they were single equations (Hereafter, the primes will indicate derivatives with respect to x (clearly $\bar{\Delta}' = -D$)).

If we take the time derivative of (2.23), replace first $\dot{\eta}$ in terms of η and ξ from (2.26) and then η in terms of $\dot{x}$ and x from (2.23), we get the following *non–Markovian* SDE:

$$\tau\left(\ddot{x} - \frac{g'}{g}\dot{x}^2\right) = -\left(1 - \tau\left[(f+\bar{\Delta})' - \frac{g'}{g}(f+\bar{\Delta})\right]\right)\dot{x} + (f+\bar{\Delta}) + \sigma g\xi. \qquad (2.29)$$

Now, following [MDT00], we perform an *adiabatic elimination* of variables, namely, neglecting $\ddot{x}$ and $\dot{x}^2$, so that the system's dynamics becomes governed by a Fokker–Planck equation. The resulting equation, being *linear* in $\dot{x}$ (but not in x), can be immediately solved for $\dot{x}$, giving

$$\dot{x} = Q(x;\bar{x}) + S(x;\bar{x})\xi, \tag{2.30}$$

with

$$Q(x;\bar{x}) \equiv (f + \bar{\Delta})\theta, \tag{2.31}$$
$$S(x;\bar{x}) \equiv \sigma g \theta, \tag{2.32}$$
$$\theta(x;\bar{x}) \equiv \{1 - \tau g[(f + \bar{\Delta})/g]'\}^{-1}. \tag{2.33}$$

The Fokker–Planck equation associated to the SDE (2.30) is

$$\partial_t P(x,t;\bar{x}) = -\partial_x [R_1(x;\bar{x})P(x,t;\bar{x})] + \frac{1}{2}\partial_x^2 [R_2(x;\bar{x})P(x,t;\bar{x})], \tag{2.34}$$

with *drift* and *diffusion* coefficients given by

$$R_1(x;\bar{x}) = Q + \frac{1}{4}(S^2)' \tag{2.35}$$
$$R_2(x;\bar{x}) = S^2. \tag{2.36}$$

The solution of the time-independent Fokker–Planck equation leads to the stationary probability density

$$P_{st}(x;\bar{x}) = \frac{1}{Z} \exp\left[\int_0^x dx' \frac{2R_1(x';\bar{x}) - \partial_{x'} R_2(x';\bar{x})}{R_2(x';\bar{x})}\right]. \tag{2.37}$$

The value of $\bar{x}$ arises from a *self-consistency relation*, once we equate it to the average value of the random variable x_i in the stationary state

$$\bar{x} = \langle x \rangle \equiv \int_{-\infty}^{\infty} dx\, x\, P_{st}(x;\bar{x}) \equiv F(\bar{x}). \tag{2.38}$$

Now, the condition

$$\left.\frac{dF}{d\bar{x}}\right|_{\bar{x}=0} = 1 \tag{2.39}$$

allows us to find the *transition line* between the *ordered* and the *disordered* phases.

Results

The mean–field approximation of the general dD extended system are the following (see [MDT00]):

A. As in the white–noise case $\tau = 0$, the ordering phase transition is *reentrant with respect to* σ: for a range of values of D that depends on τ, ordered states can only exist within a window $[\sigma_1, \sigma_2]$. The fact that this window shifts to the right *for small* τ means that, for fixed D, color *destroys* order just above σ_1 but *creates* it just above σ_2.

B. For fixed $\sigma > 1$ and $\tau \neq 0$, ordered states exist *only within a window* of values for D. Thus the ordering phase transition is *also reentrant with respect to D*. For τ small enough the maximum value of D compatible with the ordered phase increases rather steeply with σ, reaching a maximum around $\sigma \sim 5$ and then decreases gently. For $\tau \geq 0.1$ it becomes evident (in the ranges of D and σ analyzed) that the region sustaining the ordered phase is *closed*, and shrinks to a point for a value slightly larger than $\tau = 0.123$.

C. For fixed values of $\sigma > 1$ and D larger than its minimum for $\tau = 0$, the system *always* becomes disordered for τ large enough. The maximum value of τ consistent with order altogether corresponds to $\sigma \sim 5$ and $D \sim 32$. In other words, ordering is possible *only* if the multiplicative noise inducing it has short memory.

D. The fact that the region sustaining the ordered phase finally shrinks to a point means that even for that small region in the σ–D plane for which order is induced by color, a further increase in τ destroys it. In other words, the phase transition is *also reentrant with respect to τ*. For D large enough there may exist even *two* such windows.

Order Parameter

As already mentioned above, the *order parameter* in this system is $m \equiv |\bar{x}|$, namely, the positive solution of the consistency equation (2.38). Consistently with what has been discussed in (A) and (C), we see that as τ increases the window of σ values where ordering occurs shrinks until it disappears. One also notices that at least for this D, the value of σ corresponding to the maximum order parameter varies very little with τ.

The *short–time evolution* of $\langle x \rangle$ can be obtained multiplying (2.34) by x and integrating:

$$\frac{d\langle x \rangle}{dt} = \int_{-\infty}^{\infty} dx\, R_1(x; \bar{x}) P(x, t; \bar{x}). \qquad (2.40)$$

Let us assume an initial condition such that at early times $P(x, t \sim 0; \bar{x}) = \delta(x - \bar{x})$. Equating $\bar{x} = \langle x \rangle$ as before, we get the *order parameter equation*

$$\frac{d\langle x \rangle}{dt} = R_1(\bar{x}, \bar{x}). \qquad (2.41)$$

The solution of (2.41) has an initial *rising* period (it is initially *unstable*) reaching very soon a maximum and tending to zero afterwards.

For $D/\sigma^2 \to \infty$, equation (2.41) is valid also in the *asymptotic regime* since $P_{st}(x) = \delta(x - \bar{x})$ [BPT97]. According to this criterion, in the $D/\sigma^2 \to \infty$ limit the system undergoes a second–order phase transition *if* the corresponding zero-dimensional model presents *a linear instability in its short–time dynamics*, i.e., if after linearizing (2.41):

$$\langle \dot{x} \rangle = -\alpha \langle x \rangle, \qquad (2.42)$$

one finds that $\alpha < 0$. We then see that the trivial (disordered) solution $\langle x \rangle = 0$ is stable only for $\alpha > 0$. For $\alpha < 0$ other stable solutions with $\langle x \rangle \neq 0$ appear, and the system develops order through a genuine *phase* transition. In this case, $\langle x \rangle$ can be regarded as the *order parameter*. In the white noise limit $\tau = 0$ this is known to be the case for sufficiently large values of the coupling D and for a window of values for the noise amplitude $\sigma \in [\sigma_1, \sigma_2]$.

In summary, we have:

A. Multiplicative noise can shift or induce phase transitions in 0D systems.
B. Multiplicative noise can induce phase transitions in spatially extended systems.
C. Mean–field approximation predicts a minimal coupling strength for the appearance of noise induced phase transitions.
D. Mean–field approximation predicts, that the phase transition is reentrant, i.e., the ordered phase is destroyed by even higher noise intensity.
E. Appearance of an ordered phase results from a nontrivial cooperative effect between multiplicative noise, nonlinearity and diffusion.

2.1.10 Noise–Driven Ferromagnetic Phase Transition

Over the last decade, the dynamics of ferromagnetic systems below their critical temperatures in a periodically oscillating magnetic field have been studied both theoretically [TO90, LP90, Ach97, SRM98, KWR01, FTR01, YTY02] and experimentally [JY95]. The systems exhibit two qualitatively different behaviors referred to as *symmetry–restoring oscillation* (SRO) and *symmetry–breaking oscillation* (SBO), depending on the frequency Ω and the amplitude h of the applied magnetic field. It has been established that there exists a sharp transition line between SRO and SBO on the (Ω, h) plane, which is called the *dynamical phase transition* (DPT). The DPT was first observed numerically in the deterministic mean–field system for a ferromagnet in a periodically oscillating field [TO90], and has subsequently been studied in numerous *Monte Carlo simulations* of the kinetic Ising system below critical temperature [LP90, Ach97, SRM98, KWR01]. It has also been observed experimentally in [JY95].

Recently, the DPT was investigated by introducing the model equation [FTR01]
$$\dot{s}(t) = (T_c - T)s - s^3 + h \cos \Omega t.$$

This equation is a simplified model for the Ising spin system at the temperature T below its critical value T_c in an external periodic magnetic field. By appropriately scaling the magnetization s, time t, and the applied field, this equation is written as
$$\dot{s}(t) = s - s^3 + h \cos \Omega t. \tag{2.43}$$

The SBO and SRO are observed in (2.43) and the transition line between them on the (Ω, h) plane is determined analytically [FTR01].

It is quite interesting to ask whether DPT is observed under another kind of applied field, especially random field with bounded amplitude. The fundamental aim of the present section is to study the dynamics of $s(t)$ with a dichotomous Markov noise (DMN) $F(t)$ instead of periodically oscillating external field $h\cos\Omega t$ (see, e.g., [Kam92]).

The equation of motion [OHF06]

$$\dot{s} = f(s) + F(t), \qquad (2.44)$$

with a nonlinear function $f(s)$ and the DMN $F(t)$ has been extensively studied by many authors (see, e.g., [KHL79, BBK02, HR83]). It is well known that the master equation for the system can be derived, and then transition phenomena of stationary probability densities concerning the intensity of $F(t)$, for example, are studied, which are referred to as the noise-induced phase transition [KHL79, HL84]. The asymptotic drift velocity $\langle \dot{s} \rangle$ in the case of $f(s)$ being periodic functions are also discussed as a specific dynamic property [BBK02]. Furthermore, the *mean first-passage time* (MFPT) and transition rates are investigated as another important dynamic property when $f(s)$ is the force associated with the bistable potential given by (2.44).

Symmetry–Breaking Phase Transition

Model Equation and Noise–Induced Phase Transition

We consider the equation of motion driven by the external field $F(t)$ [OHF06],

$$\frac{ds(t)}{dt} = f(s) + F(t), \qquad (f(s) = s - s^3) \qquad (2.45)$$

where $F(t)$ is a symmetric DMN with taking the values $\pm H_0$. Here the probability $p(\tau)$ that $F(t)$ continues to take the identical value $+H_0$ or $-H_0$ longer than time τ is given by

$$p(\tau) = e^{-\tau/\tau_f}. \qquad (2.46)$$

This implies that the correlation time of $F(t)$ is equal to $\tau_f/2$. Throughout this section, numerical integrations of (2.45) are carried out by using the Euler difference scheme with the time increment $\Delta t = 1/100$.

Without DMN, $s(t)$ eventually approaches either of the stationary fixed points ± 1, one of which is achieved according to the initial condition $s(0)$ as shown in Figure 2.3. In the presence of DMN, if $H_0 < H_c$, H_c being defined by

$$H_c \equiv 2(1/3)^{3/2} = 0.3849 \cdots, \qquad (2.47)$$

then $f(s) + H_0 = 0 (f(s) - H_0 = 0)$ has three real roots $s_{j+}(s_{j-})$, ($j = 1, 2,$ and 3). Each value of $s_{j\pm}$ is graphically shown in Figure 2.3(a). On the other hand, if $H_0 > H_c$, then $f(s) + H_0 = 0$ ($f(s) - H_0 = 0$) has only one real root s_+ (s_-) given by

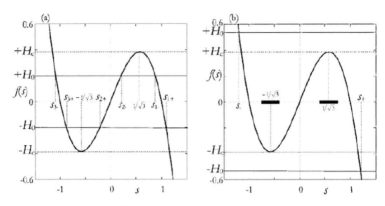

Fig. 2.3. The function $f(s) = s - s^3$ is shown by solid lines. (a) Real roots $s_{j\pm}$, ($j = 1, 2, 3$) for $H_0 < H_c$ and (b) real roots $s_\pm$ for $H_0 > H_c$ of the algebraic equation $s - s^3 \pm H_0 = 0$. The definition of H_c is graphically represented (adapted and modified from [OHF06]).

$$s_\pm = \left[\frac{1}{2}\left(\pm H_0 + \sqrt{H_0^2 - H_c^2}\right)\right]^{1/3} + \left[\frac{1}{2}\left(\pm H_0 - \sqrt{H_0^2 - H_c^2}\right)\right]^{1/3}, \quad (2.48)$$

which are indicated in Figure 2.3(b). Next let us consider the dynamics described by (2.45) for $H_0 < H_c$ and for $H_0 > H_c$, and discuss similarity and difference between the dynamics in the periodically oscillating field case and those in the present DMN case. A part of our results belongs to the context of the noise–induced phase transition and MFPT in [KHL79, HL84, BBK02, HR83]. In the case of $H_0 < H_c$, three motions numerically integrated are shown in Figures 2.4(a) and (b). Two motions confined in the ranges $s_{1-} < s(t) < s_{1+}$ and $s_{3-} < s(t) < s_{3+}$ are both stable. The long time average $\langle s(t) \rangle$ of each motion does not vanish, and the motion is called SBM in relation to DPT in the oscillating external field case. On the other hand, the motion $s_u(t)$ confined in the range $s_{2+} < s_u(t) < s_{2-}$ is unstable. The long time average of $s_u(t)$ vanishes, and in this sense the motion is called SRM. It should be noted that this unstable SRM is located between two stable SBM, which has a similar characteristic to SBO of DPT [FTR01].

The motion of $s(t)$ for $H_0 > H_c$ is shown in Figures 2.4(c) and (d). One observes that there exists a stable SRM confined in the range $s_- < s(t) < s_+$. For SRM, the time average of $s(t)$ vanishes, i.e., $\langle s(t) \rangle = 0$. The comparison between Figures 2.4(b) and (d) suggests that the SRM for $H_0 > H_c$ is generated via the "attractor merging crisis" [Ott93] of the two SBM's and one unstable SRM, i.e., the two SBM's and one unstable SRM disappear and then one stable SRM takes place at $H_0 = H_c$. This situation is similar to that in the DPT case. However, in contrast to the DPT case, the transition line on the (τ_f^{-1}, H_0) plane is independent of the correlation time τ_f of $F(t)$ and the average $\langle s(t) \rangle$ depends discontinuously on H_0.

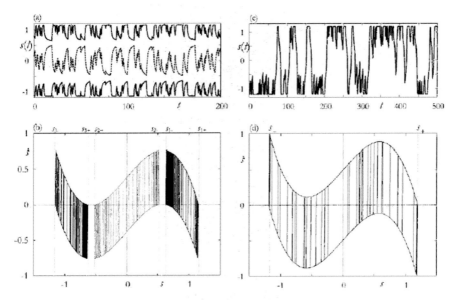

Fig. 2.4. Figures (a) and (b) show the motions obtained by numerically integrating (2.45) for $H_0 = 0.38(< H_c)$ and $\tau_f = 5$, where two SBM's (solid line) and an unstable SRM (dashed line) are drawn. The unstable SRM is evaluated by replacing $t \to -t$. On the other hand, Figures (c) and (d) show the motions for $H_0 = 0.5(> H_c)$ and $\tau_f = 10$ (adapted and modified from [OHF06]).

Stationary Distribution Functions

In this subsection, we discuss the stationary distribution functions for SBM and SRM. To this aim, we first consider a slightly general nonlinear Langevin equation of motion driven by DMN [OHF06],

$$\dot{x}(t) = f(x) + g(x)F(t), \quad (2.49)$$

where $f(x)$ and $g(x)$ are generally nonlinear functions of x and $F(t)$ is DMN [Ris84]. The temporal evolution of the distribution function $P(x, F, t)$ that $x(t)$ and $F(t)$ respectively take the values x and $F(= \pm H_0)$ is determined by [KHL79, HL84]

$$\frac{\partial}{\partial t}P(x,t) = -\frac{\partial}{\partial x}\left[f(x)P(x,t) + H_0 g(x)q(x,t)\right],$$
$$\frac{\partial}{\partial t}q(x,t) = -\frac{2}{\tau_f}q(x,t) - \frac{\partial}{\partial x}\left[f(x)q(x,t) + H_0 g(x)P(x,t)\right], \quad (2.50)$$

where we put $P(x,t) \equiv P(x,+H_0,t) + P(x,-H_0,t)$ and $q(x,t) \equiv P(x,+H_0,t) - P(x,-H_0,t)$. The stationary distribution $P^{st}(x) \equiv P(x,\infty)$ is solved to yield

$$P^{st}(x) = \qquad (2.51)$$

$$N \frac{g(x)}{H_0^2 g(x)^2 - f(x)^2} \exp\left\{-\frac{1}{\tau_f}\int^x dx' \left[\frac{1}{f(x') - H_0 g(x')} + \frac{1}{f(x') + H_0 g(x')}\right]\right\},$$

provided that each of the equations

$$\dot{x} = f(x) + H_0 g(x), \qquad \dot{x} = f(x) - H_0 g(x)$$

has at least one stable fixed point, where N is the normalization constant.

By substituting $f(x) = x - x^3$ and $g(x) = 1$, ((2.45)), into (2.51), the stationary distribution function $P^{st}_{SBM}(s)$ for SBM ($H_0 < H_c$) for $s_{3-} < s < s_{3+}$ or $s_{1-} < s < s_{1+}$ is written as

$$P^{st}_{SBM}(s) \propto |s^2 - s_{1+}^2|^{-\beta_{1+}} |s^2 - s_{1-}^2|^{-\beta_{1-}} |s^2 - s_{2+}^2|^{-\beta_{2+}}, \qquad (2.52)$$

$$\beta_{j\pm} = 1 - \tau_f^{-1}|(s_{j\pm} - s_{k\pm})(s_{j\pm} - s_{l\pm})|,$$

where $(j, k, l) = (1, 2, 3)$, $(2, 3, 1)$, and $(3, 1, 2)$.

On the other hand, the stationary distribution function $P^{st}_{SRM}(s)$ for the SRM ($H_0 > H_c$) for $s_- < s < s_+$ is obtained as

$$P^{st}_{SRM}(s) \propto |s^2 - s_+^2|^{\frac{\tau_f^{-1}}{3s_+^2 - 1} - 1} \left[(s^2 + s_+^2 - 1)^2 - s_+^2 s^2\right]^{-\frac{\tau_f^{-1}}{3s_+^2 - 1} - 1}$$

$$\times \exp\left\{\frac{\tau_f^{-1} s_+}{(s_+^2 - 1/3)\sqrt{3s_+^2 - 4}}\left[\arctan\left(\frac{2s - s_+}{\sqrt{3s_+^2 - 4}}\right)\right.\right.$$

$$\left.\left. - \arctan\left(\frac{2s + s_+}{\sqrt{3s_+^2 - 4}}\right)\right]\right\}. \qquad (2.53)$$

As H_0 is increased, the form of the stationary distribution function changes drastically from the forms in (2.52) to (2.53) at $H_0 = H_c$. This phenomenon which is induced by the disappearance of two pairs of stable and unstable fixed points [BBK02] is an example of the *noise–induced phase transitions* [HL84].

It turns out that the transition line between SRM and SBM on the (τ_f^{-1}, H_0) plane is given by $H_0 = H_c$. Furthermore, the long time average of $s(t)$, $\langle s(t) \rangle$, depends discontinuously on H_0 at $H_0 = H_c$. These behaviors are quite different from those of the DPT case driven by periodically oscillating field, $F(t) = h\cos(\Omega t)$ [Ach97, FTR01]. The transition point h_c for a fixed Ω between SRO and SBO depends on the frequency Ω, and $\langle s(t) \rangle$ is a continuous function of h.

MFPT Through the Channels

We hereafter discuss the dynamics for H_0 slightly above H_c. Let us first consider the behavior obeying the equations [OHF06]

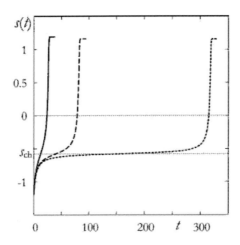

Fig. 2.5. Three orbits of the equation of motion (2.54) with $\epsilon = +$ for $H_0 > H_c$. The values of H_0 are set to be $H_0 = 0.386$ (dotted line), 0.4 (dashed line), and 0.5 (solid line), where all the initial conditions are chosen as $s_0 = s_-$ (adapted and modified from [OHF06]).

$$\dot{s} = s - s^3 + \epsilon H_0, \qquad (\epsilon = + \text{ or } -) \qquad (2.54)$$

for $H_0 > H_c$, i.e., $F(t)$ is fixed to be either $+H_0$ or $-H_0$. Equation (2.54) for $H_0 > H_c$ is integrated to yield

$$t = -\frac{1}{2(3s_\epsilon^2 - 1)} \ln \frac{(s - s_\epsilon)^2}{s^2 + s_\epsilon s + s_\epsilon^2 - 1} \frac{s_0^2 + s_\epsilon s_0 + s_\epsilon^2 - 1}{(s_0 - s_\epsilon)^2} \qquad (2.55)$$

$$+ \frac{6s_\epsilon}{2(3s_\epsilon^2 - 1)\sqrt{3s_\epsilon^2 - 4}} \left[\arctan\left(\frac{2s + s_\epsilon}{\sqrt{3s_\epsilon^2 - 4}}\right) - \arctan\left(\frac{2s_0 + s_\epsilon}{\sqrt{3s_\epsilon^2 - 4}}\right) \right],$$

where $s_0 = s(0)$ and s_ϵ has been defined in (2.48). Figure 2.5 displays three orbits given by (2.55) with $s_0 = s_-$ and $\epsilon = +$, which shows that $s(t)$ approaches s_+ in the limit $t \to \infty$. One observes that $s(t)$ stays for a long time in the vicinity of $s = -1/\sqrt{3}$ for H_0 slightly above H_c. The small region including the position $s = -1/\sqrt{3}$ is called the 'channel'. From the symmetry of the system, there also exists the channel near $s = 1/\sqrt{3}$ for $F(t) = -H_0$, as shown in Figure 2.3(b). Let us express the positions s_{ch} of the channels as

$$s_{ch} = \begin{cases} -1/\sqrt{3}, & \text{if } F(t) = +H_0 \\ +1/\sqrt{3}, & \text{if } F(t) = -H_0 \end{cases}.$$

The characteristic time τ_{ch} is then defined as the time span that the state point $s(t)$ passes through one of the channels for a constant $F(t)$, either $+H_0$ or $-H_0$. τ_{ch} can be estimated by integrating (2.54) around $s \simeq s_{ch}$ as follows. First, consider the case $F(t) = -H_0$. By setting $u(t) = s(t) - s_{ch}$ and assuming $|u| \ll s_{ch}$, (2.54) is approximated as [OHF06]

$$\dot{u} = -3s_{ch}u^2 - (H_0 - H_c).$$

This can be integrated to give

$$u(t) = -\sqrt{\frac{H_0 - H_c}{3s_{ch}}} \tan\left[\sqrt{3s_{ch}(H_0 - H_c)}\,t\right] \quad (2.56)$$

with the initial condition $u(0) = 0$. τ_{ch} is estimated by the condition $u(\tau_{ch}) = \infty$ and thus

$$\tau_{ch} = \frac{C}{(H_0 - H_c)^{1/2}}, \quad C = \frac{\pi}{2\sqrt{3s_{ch}}}. \quad (2.57)$$

Let us next consider the process that the state point $s(t)$ passes through the channels under DMN. One finds that the time of passing through channels increases as H_0 approaches H_c. The MFPT $\bar{\tau}$ through channels was calculated in [HR83] by analyzing the master equation. In the present subsection, we will derive MFPT in terms of the time scales τ_f and τ_{ch} from a phenomenological viewpoint without use of the analysis made in [HR83].

The condition for passing through a channel is that $F(t)$ continues to take the identical value either $+H_0$ or $-H_0$ for time longer than τ_{ch}. For H_0 satisfying $\tau_f > \tau_{ch}$, we obtain $p(\tau_{ch}) = e^{-\tau_{ch}/\tau_f} \simeq 1$, which implies that $F(t)$ almost always satisfies the condition for passing through the channel. Therefore, $\bar{\tau}$ in the case of $\tau_f > \tau_{ch}$ is nearly equivalent to τ_{ch}, i.e.,

$$\bar{\tau} \simeq \frac{C}{(H_0 - H_c)^{1/2}}. \quad (2.58)$$

In the case of $\tau_f \ll \tau_{ch}$, on the other hand, (2.46) gives $p(\tau_{ch}) \ll 1$. This fact implies that the probability that $F(t)$ continues to take the identical value for time longer than τ_{ch} is quite small and hence that $\bar{\tau}$ is much longer than τ_{ch} because it needs a long time to satisfy the condition for the state point to pass through the channel. $\bar{\tau}$ in the case of $\tau_f \ll \tau_{ch}$ is explicitly determined as follows. For a long $\bar{\tau}$, let us divide $\bar{\tau}$ into subintervals each of which has the time span τ_f. The divided individual time series are approximately independent of each other. Therefore, $\tau_f/\bar{\tau}$ is the probability that the state point passes through a channel once because $\bar{\tau}$ is MFPT through the channel. On the other hand, $p(\tau_{ch})$ is identical to the probability for $s(t)$ to pass through the channel once by definition of the probability. Therefore we get the relation $p(\tau_{ch}) \simeq \tau_f/\bar{\tau}$, which leads to

$$\bar{\tau}^{-1} \simeq \tau_f^{-1} e^{-\tau_{ch}/\tau_f} = \tau_f^{-1} \exp\left[-\frac{C}{\tau_f(H_0 - H_c)^{1/2}}\right] \quad (2.59)$$

with the constant C defined in (2.57). This expression agrees with the result obtained in [HR83].

Equation (2.59) reveals that MFPT through the channel depends on $H_0 - H_c$ in a stretched exponential form for $\tau_f \ll \tau_{ch}$, and is quite different from the asymptotic form (2.58).

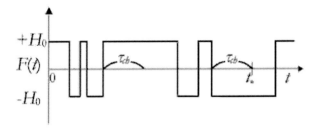

Fig. 2.6. Time series of $F(t)$ schematically indicating the time that $s(t)$ jumps from s_+ to s_-. $s(t)$ jumps at the time t_* in this case, where $F(t)$ takes the value $-H_0$ longer than τ_{ch} for the first time in the time series of $F(t)$ (adapted and modified from [OHF06]).

Phenomenological Analysis

In order to discuss statistical characteristics of the dynamics passing through the channels for $\tau_f \ll \tau_{ch}$, we here develop a phenomenological approach. The behaviors of $s(t)$ for which we attempt to model are first summarized. The initial condition of $s(t)$ is set to be in the vicinity of s_+. If a time interval of $F(t)$ satisfying the condition $F(t) = -H_0$ becomes longer than τ_{ch} for the first time, then $s(t)$ passes through s_{ch} and approaches s_- in the time interval. See Figure 2.6. The event in which $s(t)$ jumps from s_+ to s_- occurs only in this case. It should be noted that the jumps from $s(t) > 0$ ($s(t) < 0$) to $s(t) < 0$ ($s(t) > 0$) are approximately independent of subsequent jumps.

Let us discretize the time t in the form $t = k\Delta t$, ($k = 1, 2, 3, \cdots$) as a simple approach to develop the phenomenological analysis according to the process noted above, where Δt is a certain small time step. Then, τ_{ch} is discretized as $\tau_{ch} \equiv n_{ch}\Delta t$ with the corresponding integer n_{ch}. $F(t)$ is assumed to keep the same value for the interval Δt, which is denoted as $F_k = F(k\Delta t)$. The conditional probability p that F_{j+1} takes the same value as F_j is given by

$$p = e^{-\Delta t/\tau_f}, \tag{2.60}$$

and the probability q that F_{k+1} is different from F_k is therefore given by

$$q = 1 - p. \tag{2.61}$$

The system is analyzed phenomenologically as follows:

- We introduce the variable s_k at a discretized time $k\Delta t$ which takes two values ± 1.
- s_k and F_k are initially set to $s_0 = +1$ and $F_0 = +H_0$, respectively.
- s_k jumps from $+1$ (-1) to -1 $(+1)$ only if F_k continues to take the identical value $-H_0$ $(+H_0)$ for a time interval longer than $n_{ch}\Delta t$.
- s_k does not jump from $+1$ (-1) to -1 $(+1)$ even though F_k continues to take $+H_0$ $(-H_0)$ for any time interval longer than $n_{ch}\Delta t$.

MFPT $\bar{\tau}$ through the Channel

We first derive the exact expression for the MFPT $\bar{\tau}$ through the channel with the phenomenological approach. In considering the time series having F_k, $\bar{\tau}$ is evaluated as

$$\bar{\tau} = \sum_l l \Delta t \sum_{0 \leq k \leq l} g_{k,l}^{(n_{ch})} q^k p^{l-k} \bigg|_{q=1-p}. \tag{2.62}$$

Here $g_{k,l}^{(n_{ch})}$ is the number of the time sequences $\{F_j\}$ for $0 \leq j < l$ satisfying that F_j changed its value k times in each $\{F_j\}$ and s_j jumps from $+1$ to -1 for the first time at $t = l\Delta t$.

Equation (2.62) is, furthermore, rewritten as

$$\bar{\tau} = \hat{T} Q_{n_{ch}}(q,p) \bigg|_{q=1-p}$$

with the differential operator $\hat{T}$ and the quantity $Q_{n_{ch}}(q,p)$ defined by

$$\hat{T} \equiv \Delta t \left(q \frac{\partial}{\partial q} + p \frac{\partial}{\partial p} \right) \tag{2.63}$$

and

$$Q_{n_{ch}}(q,p) \equiv \sum_l \sum_{0 \leq k \leq l} g_{k,l}^{(n_{ch})} q^k p^{l-k}. \tag{2.64}$$

One should note that the q- and p-dependences in $Q_{n_{ch}}$ are crucial and that q and p are considered to be independent in (2.64).

The explicit form of $Q_{n_{ch}}(q,p)$ is then determined so as to satisfy the following conditions:

1. In considering any length of time series giving F_k, there exists a time interval of length n_{ch} in the last of the time series, where all the F_k take the same value $-H_0$, i.e., the condition that s_k jumps from $+1$ to -1 is satisfied.
2. The condition for s_k to jump from $+1$ to -1 is not satisfied before the last time interval.

One should note that the equality $Q_n(1-p,p) = 1$ holds for any n, because the time interval described above always exists somewhere in a long time series. Particularly, for $n = n_{ch}$, $Q_{n_{ch}}(1-p,p)$ is obviously equal to the probability that s_j changes its sign, which must be unity for $H_0 > H_c$.

The explicit form of $Q_n(q,p)$ is given by [OHF06]

$$Q_n(q,p) = \frac{(1-p)qp^{n-1}}{(1-p)^2 - q^2(1-p^{n-1})}, \tag{2.65}$$

where the condition $Q_n(1-p,p) = 1$ is easily confined. Applying the operator (2.63) to the explicit form (2.65) with $n = n_{ch}$ yields the relation

2.1 Introduction to Phase Transitions   215

$$\bar{\tau} = \hat{T}Q_{n_{ch}}(q,p)\Big|_{q=1-p} = \Delta t \frac{2 - p^{n_{ch}-1}}{(1-p)p^{n_{ch}-1}}$$

$$= \Delta t \frac{2 - e^{\Delta t/\tau_f}e^{-\tau_{ch}/\tau_f}}{(1 - e^{-\Delta t/\tau_f})e^{\Delta t/\tau_f}e^{-\tau_{ch}/\tau_f}},$$

where the last equality is obtained by using Eqs. (2.60) and (2.61) with the relation $\tau_{ch} = n_{ch}\Delta t$. The exact expression of $\bar{\tau}$ is finally given by

$$\bar{\tau} = \tau_f \left(2e^{\tau_{ch}/\tau_f} - 1\right) \tag{2.66}$$

in the limit of $\Delta t \to 0$ by keeping τ_{ch} constant. Equation (2.66) qualitatively agrees for $\tau_{ch}/\tau_f \gg 1$ with the result (2.59).

Distribution Function $P(\tau)$ for the Passage Time τ

The distribution function $P(\tau)$ for the passage time τ through the channel s_{ch} is determined by solving the equation

$$P(\tau) = \delta(\tau - \hat{T})Q_{n_{ch}}(q,p)\Big|_{q=1-p}, \tag{2.67}$$

where $\delta(x)$ is the delta function. The Laplace transform $\mathcal{L}[P](z)$ should be calculated in order to solve (2.67). By using the series expansion of $Q_{n_{ch}}(q,p)$ given by (2.64), the Laplace transform of $P(\tau)$ is obtained as

$$\mathcal{L}[P](z) \equiv \int_0^\infty e^{-\tau z} P(\tau) d\tau = e^{-z\hat{T}} Q_{n_{ch}}(q,p)\Big|_{q=1-p}$$

$$= \sum_l \sum_{0 \le k \le l} g_{k,l}^{(n_{ch})} (e^{-z\Delta t}q)^k (e^{-z\Delta t}p)^{l-k} \Big|_{q=1-p}. \tag{2.68}$$

Equation (2.68) implies that $\mathcal{L}[P](z)$ can be obtained by replacing q and p by $e^{-z\Delta t}q$ and $e^{-z\Delta t}p$ in $Q_{n_{ch}}(q,p)$, respectively, i.e.,

$$\mathcal{L}[P(\tau)](z) = Q_{n_{ch}}(e^{-z\Delta t}q, e^{-z\Delta t}p)\Big|_{q=1-p}. \tag{2.69}$$

Substituting the explicit form (2.65) for $n = n_{ch}$ into (2.69) yields the equation

$$\mathcal{L}[P(\tau)](z) = \frac{(\tau_f z + 1)e^{-(z+\tau_f^{-1})\tau_{ch}}}{\tau_f^2 z^2 + 2\tau_f z + e^{-(z+\tau_f^{-1})\tau_{ch}}} \tag{2.70}$$

in the limit of $\Delta t \to 0$ by keeping τ_{ch} constant. By applying the inverse Laplace transform to (2.70), the distribution function $P(\tau)$ is analytically evaluated in the series expansion as

$$P(\tau) = \tau_f^{-1} e^{-\tau/\tau_f} \sum_{k=0}^{\infty} \theta(t_{k+1}) \frac{(-x)^k}{k!} \frac{d^k}{dx^k} \cosh\sqrt{x}\bigg|_{x=(t_{k+1})^2} =$$

$$\tau_f^{-1} e^{-\tau/\tau_f} \bigg[\theta(t_1)\cosh(t_1) - \theta(t_2)\frac{t_2 \sinh(t_2)}{2} + \theta(t_3)\frac{t_3^2 \cosh(t_3) - t_3 \sinh(t_3)}{8}$$

$$- \theta(t_4)\frac{t_4^3 \sinh(t_4) - 3t_4^2 \cosh(t_4) + 3t_4 \sinh(t_4)}{48} + \cdots \bigg], \quad (2.71)$$

where $t_k(\tau) \equiv (\tau - k\tau_{ch})/\tau_f$ and $\theta(t)$ is the Heaviside function defined by

$$\theta(t) = \begin{cases} 1 \text{ for } t \geq 0 \\ 0 \text{ for } t < 0 \end{cases}. \quad (2.72)$$

For details of the derivation of (2.71), see [OHF06].

Let us suppose to truncate the expansion (2.71) at $k = k_c$ for an arbitrary k_c. It should be noted that (2.71) gives the exact distribution for $0 < \tau < k_c\tau_{ch}$ even though the truncation is executed, since all the terms individually include $\theta(t_k)$ and so the terms for $k > k_c$ do not contribute to $P(\tau)$ for $\tau < k_c\tau_{ch}$. The analytical result (2.71) is compared with the numerically evaluated distribution. One observes that the phenomenological analysis quantitatively explains the statistical property of passing through the channels. The characteristics obtained from the figure are summarized as follows [OHF06]:

1. There exists a region where $P(\tau) = 0$ for $\tau < \tau_{ch}$, which presents the minimal time of passing through the channels.

2. $P(\tau)$ decreases exponentially for $\tau \gg \tau_{ch}$, $P(\tau) \propto e^{-\alpha\tau}$ with a constant α.

3. The rate α increases as τ_f is increased. This tendency is consistent with the fact that the probability of passing through channels increases as τ_f is increased since DMN will often continue to take an identical value longer than τ_{ch}.

On the other hand, the expansion (2.71) disagrees with the correct value in an exponential way for $\tau > k_c\tau_{ch}$. Let us try to obtain the asymptotic solution of $P(\tau)$ for $\tau \gg \tau_{ch}$. We have [OHF06]

$$\langle e^{-z(\tau - \tau_{ch})} \rangle \simeq \frac{1}{1 + (\bar{\tau} - \tau_{ch})z} \quad \text{for } |z| \ll \tau_{ch}^{-1}, \quad (2.73)$$

where $\bar{\tau}$ is MFPT given in (2.66). The inverse Laplace transform of (2.73) is straightforwardly calculated to give

$$P(\tau) \simeq \frac{1}{\bar{\tau} - \tau_{ch}} \exp\left(-\frac{\tau - \tau_{ch}}{\bar{\tau} - \tau_{ch}}\right),$$

which reveals that $P(\tau)$ decreases exponentially with the damping rate $\alpha = (\bar{\tau} - \tau_{ch})^{-1}$ for $\tau \gg \tau_{ch}$.

Phenomenological Fourier Spectrum

We derive the Fourier spectrum of a time series $s(t)$ by the phenomenological analysis to focus on the dynamical characteristics in the SRM phase. The Fourier spectrum $I_x(\omega)$ is defined by

$$I_x(\omega) = \lim_{T \to \infty} \frac{1}{T} \left\langle \left| \int_0^T x(t) e^{-i\omega t} dt \right|^2 \right\rangle, \tag{2.74}$$

i.e., the ensemble average of the Fourier transform of a time series $x(t)$.

Let us first consider $s_0(t) \equiv \text{sgn}[s(t)]$. Then the time series $s_0(t)$ is expressed as

$$s_0(t) = (-1)^{n-1}, \quad \text{for} \quad t_{n-1} \le t < t_n \tag{2.75}$$

with $n \ge 1$, where t_n denotes the nth time to cross zero for $s(t)$. Hereafter, t_0 is set to be zero without loss of generality. By identifying that $\tau_n \equiv t_n - t_{n-1}$ is independently distributed according to (2.71), one obtains the Fourier spectrum of $s_0(t)$ by the phenomenological analysis shown (see [OHF06]), in the form

$$I_{s_0}(\omega) = \frac{4}{\bar{\tau}\omega^2} \Re \left(\frac{1 - \langle e^{-i\omega\tau_n} \rangle}{1 + \langle e^{-i\omega\tau_n} \rangle} \right) = \frac{4}{\bar{\tau}\omega^2} \frac{1 - |\langle e^{-i\omega\tau_n} \rangle|^2}{|1 + \langle e^{-i\omega\tau_n} \rangle|^2} \tag{2.76}$$

where $\Re(X)$ represents the real part of X, and $\lim_{N\to\infty} \frac{t_N}{N} = \langle \tau_n \rangle = \bar{\tau}$ is used.

Substituting the explicit form of $\langle e^{-i\omega\tau_n} \rangle$ given in (2.70) with $z = i\omega$ into (2.76) yields

$$I_{s_0}(\omega) = \left(\frac{4\tau_f}{\bar{\tau}\omega} \right) \frac{\omega^3 \tau_f^3 + (4 - e^{-2\tau_{ch}/\tau_f})\omega\tau_f - 2e^{-\tau_{ch}/\tau_f}(\omega\tau_f \cos\omega\tau_{ch} + 2\sin\omega\tau_{ch})}{(4 + \omega^2\tau_f^2)(\omega^2\tau_f^2 - 2\omega\tau_f e^{-\tau_{ch}/\tau_f}\sin\omega\tau_{ch} + e^{-2\tau_{ch}/\tau_f})}. \tag{2.77}$$

The above result is confirmed by comparing with the numerically evaluated Fourier spectrum for the normalized time series $s_0(t)$ (see [OHF06]).

Let us finally modify the phenomenological analysis which is compatible with the numerically evaluated spectrum of the original time series $s(t)$ without normalization. Instead of (2.75), let us define

$$\tilde{s}(t) = (-1)^{n-1}[1 - a(t - t_{n-1})] \quad \text{for} \quad t_{n-1} \le t < t_n$$

with $n \ge 1$, where $a(\Delta t)$ incorporates the wave form of the time series passing through the channel and is assumed to be $a(\Delta t) = 0$ for $\Delta t > \tau_{ch}$. Note that by setting $a(\Delta t) = 0$ also for $\Delta t \le \tau_{ch}$ the result of original phenomenological analysis is recovered. As shown in [OHF06], the Fourier spectrum $I_{\tilde{s}}(\omega)$ for $\tilde{s}(t)$ as a modification to $I_{s_0}(\omega)$ is obtained in the form

$$I_{\tilde{s}}(\omega) = I_{s_0}(\omega) \frac{1 + |\hat{a}(\omega)|^2 + 2\Re[\hat{a}(\omega)]}{4}, \tag{2.78}$$

where
$$\hat{a}(\omega) \equiv 1 - i\omega \int_0^{\tau_{ch}} a(t)e^{-i\omega t} dt.$$
By approximating as $a(\Delta t) = 1 + |s_{ch}|$ for $0 < \Delta t < \tau_{ch}$, (2.78) reduces to
$$I_{\tilde{s}}(\omega) = I_{s_0}(\omega) \left(\frac{1 + s_{ch}^2}{2} + \frac{1 - s_{ch}^2}{2} \cos \omega \tau_{ch} \right). \quad (2.79)$$

Fore more technical details, see [OHF06].

2.1.11 Phase Transition in a Reaction–Diffusion System

Recently 1D *reaction–diffusion systems* have received much attention because they show a variety of interesting critical phenomena such as *non–equilibrium phase transitions* [Sch01]. A simple system of this type, which has been studied widely in related literatures, is the Asymmetric Simple Exclusion Process (ASEP) [DEH93]. In this 2–states model the particles are injected from the left site of an open discreet lattice of length L. They diffuse in the system and at the end of the lattice are extracted from the system. It is known that depending on the injection and the extraction rates the ASEP shows different boundary induced phase transitions. Non–equilibrium phase transition may also happen in the systems with *non–conserving dynamics*. For instance in [EKL02] the authors investigate a 3–states model consists of two species of particles besides vacancies on a lattice with ring geometry. The dynamics of this model consists of diffusion, creation and annihilation of both species of the particles. They have found that the phase diagram of the model highly depends on the annihilation rate of the particles. By changing the annihilation rate of the particles, the system transfers from a maximal current phase to a fluid phase. The density of the vacancies changes discontinuously from one phase to the other phase.

In this section, following [JG07], we study a reaction–diffusion model on a discrete lattice of length L with periodic boundary condition. Besides the vacancies there are two different types of particles in the system. Throughout this paper the vacancies and the particles are denoted by E, A and B. The dynamics of the system is not conserving. The particles of type A and type B hop to the left and to the right respectively. The total number of particles of type B is a conserved quantity and assumed to be equal to M. The density of these particles is defined as $\rho_B = \frac{M}{L}$. In contrast, the total number of particles of type A is not a conserved quantity due to the creation and annihilation of them. Only the nearest neighbors interactions are allowed and the model evolves through the following processes

$$\begin{aligned}
A\emptyset &\longrightarrow \emptyset A \quad \text{with rate } 1, \\
\emptyset B &\longrightarrow B\emptyset \quad \text{with rate } 1, \\
AB &\longrightarrow BA \quad \text{with rate } 1, \\
A\emptyset &\longrightarrow \emptyset\emptyset \quad \text{with rate } \omega, \\
\emptyset\emptyset &\longrightarrow A\emptyset \quad \text{with rate } 1.
\end{aligned} \quad (2.80)$$

As can be seen the parameter ω determines the annihilation rate for the particles of type A which besides the number of the particles of type B i.e., ρ_B are the free parameters of the model. One should note that the annihilation in our model only takes place for one species of particles. Our main aim in the present work is to study the phase diagram of the model in terms of ω and the density of the B particles.

In our model if one starts with a lattice without any vacancies the dynamics of the model prevents it from evolving into other configurations consisting of vacancies. In this case the system remains in its initial configuration and the steady state of the system is trivial. In order to study the non-trivial case we consider those configurations which have at least one vacancy. In order to find the stationary probability distribution function of the system we apply the *Matrix Product Formalism* (MPF) first introduced in [DEH93] and then generalized in [KS97]. According to this formalism the stationary probability for any configuration of a system with periodic boundary condition is proportional to the trace of product of non–commuting operators which satisfy a quadratic algebra. In our model we have three different states at each site of the lattice associated with the presence of vacancies, the A particles and the B particles. We assign three different operators $\mathbf{E}$, $\mathbf{A}$ and $\mathbf{B}$ to each state. Now the unnormalized steady state probability of a configuration $\mathcal{C}$ is given by

$$P(\mathcal{C}) = \frac{1}{\mathcal{Z}_L} Tr[\prod_{i=1}^{L} \mathbf{X}_i], \qquad (2.81)$$

in which $\mathbf{X}_i = \mathbf{E}$ if the site i is empty, $\mathbf{X}_i = \mathbf{A}$ if the site i is occupied by a particle of type A and $\mathbf{X}_i = \mathbf{B}$ if it is occupied by a particle of type B. The normalization factor $\mathcal{Z}_L$ in the denominator of (2.81) is called the partition function of the system and is given by the sum of unnormalized weights of all accessible configurations. By applying the MPF one finds the following quadratic algebra for our model [JG07]

$$\begin{aligned}\mathbf{AB} &= \mathbf{A} + \mathbf{B}, \\ \mathbf{AE} &= \mathbf{E}, \\ \mathbf{EB} &= \mathbf{E}, \\ \mathbf{E}^2 &= \omega \mathbf{E}.\end{aligned} \qquad (2.82)$$

Now by defining

$$\mathbf{E} = \omega |V\rangle\langle W| \quad \text{in which} \quad \langle W|V\rangle = 1,$$

one can simply find

$$\begin{aligned}\mathbf{AB} &= \mathbf{A} + \mathbf{B}, \\ \mathbf{A}|V\rangle &= |V\rangle, \\ \langle W|\mathbf{B} &= \langle W|, \\ \mathbf{E} &= \omega|V\rangle\langle W|.\end{aligned} \qquad (2.83)$$

The first three relations in (2.83) make a quadratic algebra which is well known in the related literatures. It is the quadratic algebra of the ASEP when the boundary rates are equal to one.[4] This algebra has an infinite dimensional representation given by the following matrices and vectors

$$\mathbf{A} = \begin{pmatrix} 1 & 1 & 0 & 0 & \cdots \\ 0 & 1 & 1 & 0 & \\ 0 & 0 & 1 & 1 & \\ 0 & 0 & 0 & 1 & \\ \vdots & & & & \ddots \end{pmatrix}, \quad \mathbf{B} = \mathbf{A}^T,$$

$$|V\rangle = \begin{pmatrix} 1 \\ 0 \\ 0 \\ 0 \\ \vdots \end{pmatrix}, \quad \langle W| = |V\rangle^T,$$

in which T stands for transpose. In order to find the phase structure of the system one can calculate the generating function of the partition function of the system and study its singularities. In what follows we first calculate the grandcanonical partition function of the system $\mathcal{Z}_L(\xi)$ by introducing a fugacity ξ for particles of type B. We then fix the fugacity of the B particles using the following relation

$$\rho_B = \lim_{L \to \infty} \frac{\xi}{L} \frac{\partial \ln \mathcal{Z}_L(\xi)}{\partial \xi}. \tag{2.84}$$

The grandcanonical *partition function* of the system $\mathcal{Z}(\xi)$ can now be calculated from (2.81) and is given by [JG07]

$$\mathcal{Z}_L(\xi) = \sum_{\mathcal{C}} Tr[\prod_{i=1}^{L} \mathbf{X}_i] = Tr[(\mathbf{A} + \xi\mathbf{B} + \mathbf{E})^L - (\mathbf{A} + \xi\mathbf{B})^L]. \tag{2.85}$$

One should note that the operator **B** in (2.81) is replaced with the operator $\xi\mathbf{B}$ in which ξ should be fixed using (2.84). As we mentioned above the stationary state of the system without vacancies is a trivial one, therefore in (2.85) we have considered those configurations with at least one vacancy. The generating function for $\mathcal{Z}_L(\xi)$ can now be calculated using (2.83). Using the same procedure introduced in [RSS00] and after some straightforward algebra one finds

$$\mathcal{G}(\xi, \lambda) = \sum_{L=0}^{\infty} \lambda^{L-1} \mathcal{Z}_L(\xi) = \frac{\frac{d}{d\lambda}\omega U(\xi, \lambda)}{1 - \omega U(\xi, \lambda)}, \tag{2.86}$$

in which

---

[4] The operator **A** and **B** should be regarded as the operators associated with the particles and vacancies respectively.

2.1 Introduction to Phase Transitions

$$U(\xi, \lambda) = \sum_{L=0}^{\infty} \lambda^{L+1} \langle W|(\mathbf{A} + \xi \mathbf{B})^L |V\rangle.$$

The convergence radius of the formal series (2.86) which is the absolute value of its nearest singularity to the origin can be written as

$$R(\xi) = \lim_{L \to \infty} \mathcal{Z}_L(\xi)^{\frac{-1}{L}}.$$

This is also the inverse of the largest eigenvalue of $\mathcal{Z}_L(\xi)$. In the large L limit using (2.84) this results in the following relation

$$\rho_B = \xi \frac{\partial}{\partial \xi} \ln \frac{1}{R(\xi)}. \tag{2.87}$$

Therefore one should only find the singularities of (2.86) and decide in which region of the phase diagram which singularity is the smallest one. In order to find the singularities of (2.86) one should first calculate $U(\xi, \lambda)$. This can easily be done by noting that the matrix $\mathbf{A} + \xi \mathbf{B}$ defined as

$$\mathbf{A} + \xi \mathbf{B} = \begin{pmatrix} 1+\xi & 1 & 0 & 0 & \cdots \\ \xi & 1+\xi & 1 & 0 & \\ 0 & \xi & 1+\xi & 1 & \\ 0 & 0 & \xi & 1+\xi & \\ \vdots & & & & \ddots \end{pmatrix},$$

satisfy the following *eigenvalue relation*

$$(\mathbf{A} + \xi \mathbf{B})|\theta\rangle = (1 + \xi + 2\sqrt{\xi} Cos(\theta))|\theta\rangle, \tag{2.88}$$

for $-\pi \leq \theta \leq \pi$ in which we have defined [JG07]

$$|\theta\rangle = \begin{pmatrix} Sin(\theta) \\ \xi^{1/2} Sin(2\theta) \\ \xi Sin(3\theta) \\ \xi^{3/2} Sin(4\theta) \\ \xi^2 Sin(5\theta) \\ \vdots \end{pmatrix}.$$

Considering the fact that

$$\int_{-\pi}^{\pi} \frac{d\theta}{\pi} Sin(\theta) Sin(n\theta) = \delta_{1,n},$$

one can easily see that

$$|V\rangle = \int_{-\pi}^{\pi} \frac{d\theta}{\pi} Sin(\theta)|\theta\rangle. \tag{2.89}$$

Now using (2.88)–(2.89) and the fact that $\langle W|\theta\rangle = Sin(\theta)$ one can easily show that $U(\xi, \lambda)$ is given by

$$U(\xi, \lambda) = \int_{-\pi}^{\pi} \frac{d\theta}{\pi} \frac{\lambda Sin(\theta)^2}{1 - \lambda(1 + \xi + 2\sqrt{\xi}Cos(\theta))} \tag{2.90}$$

which is valid for $\lambda < \frac{1}{(1+\sqrt{\xi})^2}$. The integral (2.90) can easily be calculated using the Cauchy residue theorem by noting that it has three poles in the complex plane. Two of them are inside the contour which is a circle of unite radius around the origin and one is outside it. After some calculations one finds

$$U(\xi, \lambda) = \frac{1 - \lambda - \lambda\xi - \sqrt{(\lambda + \lambda\xi - 1)^2 - 4\lambda^2\xi}}{2\lambda\xi}.$$

Now that $U(\xi, \lambda)$ is calculated one can simply find the singularities of (2.86). It turns out that (2.86) has two different kinds of singularities: a simple root singularity $R_1 = \frac{\omega}{(1+\omega)(\omega+\xi)}$ which come from the denominator of (2.86) and a square root singularity $R_2 = \frac{1}{(1+\sqrt{\xi})^2}$. Therefore the model has two different phases. The relation between the density of B particles and their fugacity in each phase should be obtained from (2.87). In terms of the density of the B particles ρ_B we find

$$R_1 = \frac{1 - \rho_B}{1 + \omega} \quad \text{and} \quad R_2 = (1 - \rho_B)^2.$$

Two different scenarios might happen: defining $\omega_c = \frac{\rho_B}{1-\rho_B}$ we find that for $\omega > \omega_c$ the nearest singularity to the origin is R_1 and for $\omega < \omega_c$ it is R_2. The density of the vacancies can be calculated quit similar to that of the B particles. It is given by

$$\rho_E = \omega \frac{\partial}{\partial \omega} \ln \frac{1}{R(\xi)}, \tag{2.91}$$

in which $R(\xi)$ is again the nearest singularity to the origin in each phase. The density of the A particles is in turn $\rho_A = 1 - \rho_B - \rho_E$. Let us now investigate the current of the particles in each phase. Noting that the configurations without vacancies are inaccessible, the particle current for each species is obtained to be [JG07]

$$J_\mathbf{A} = \frac{Tr[(\xi\mathbf{AB} + \mathbf{AE})(\mathbf{A} + \xi\mathbf{B} + \mathbf{E})^{L-2} - (\xi\mathbf{AB})(\mathbf{A} + \xi\mathbf{B})^{L-2}]}{Tr[(\mathbf{A} + \xi\mathbf{B} + \mathbf{E})^L - (\mathbf{A} + \xi\mathbf{B})^L]},$$

$$J_\mathbf{B} = \frac{Tr[(\xi\mathbf{AB} + \xi\mathbf{EB})(\mathbf{A} + \xi\mathbf{B} + \mathbf{E})^{L-2} - (\xi\mathbf{AB})(\mathbf{A} + \xi\mathbf{B})^{L-2}]}{Tr[(\mathbf{A} + \xi\mathbf{B} + \mathbf{E})^L - (\mathbf{A} + \xi\mathbf{B})^L]}.$$

These relations can be simplified in the thermodynamic limit and one finds

$$J_A = R(\xi)(1 + (\xi - 1)\rho_A),$$
$$J_B = R(\xi)(\xi + (1 - \xi)\rho_B).$$

As we mentioned above for $w < w_c$ the nearest singularity to the origin is always R_2. The particle currents in this case are equal and we find

$$J_A = J_B = \rho_B(1 - \rho_B).$$

In contrast for $w > w_c$ the nearest singularity to the origin is always R_1 and it turns out that the currents are not equal. We find $J_A = \frac{w}{(1+w)^2}$ and $J_B = \rho_B(1 - \rho_B)$. In what follows we bring the summery of the results concerning the phase structure of the system

for $w < w_c$
$$\begin{cases} \rho_A = 1 - \rho_B & J_A = \rho_B(1 - \rho_B) \\ \rho_B = \rho_B & J_B = \rho_B(1 - \rho_B) \\ \rho_E = 0 \end{cases} \quad \text{and}$$

for $w > w_c$
$$\begin{cases} \rho_A = \frac{1}{1+w} & J_A = \frac{w}{(1+w)^2} \\ \rho_B = \rho_B & J_B = \rho_B(1 - \rho_B) \\ \rho_E = \frac{w}{1+w} - \rho_B \end{cases}.$$

As can be seen for $w < w_c$ the density of vacancies is equal to zero which means there are only A and B particles on the lattice. Since the density of the B particles is fixed and equal to ρ_B the density of A particles should be $1 - \rho_B$. In this case, according to (2.80), both A and B particles have simply ASEP dynamics and therefore their currents should be of the form $\rho(1 - \rho)$ and that $J_A = J_B$. This is in quite agreement with our calculations for J_A and J_B. On the other hand for $w > w_c$ the density of vacancies on the lattice is no longer zero. At the transition point w_c the density of the vacancies is zero but it increases linearly in this phase. In terms of the density of the vacancies the phase transition is a continuous transition. In this phase for $w < 1$ we always have $J_A > J_B$ while for $w > 1$ we have

$$J_A > J_B \quad \text{for} \quad \rho_B < \frac{1}{1+w}$$
$$\text{and} \quad J_A < J_B \quad \text{for} \quad \frac{1}{1+w} < \rho_B < \frac{w}{1+w}.$$

In this paper we studied a 3-states model consists of A and B particles besides the vacancies. The A particles are created and annihilated which is controlled by w while the B particles only diffuse on the lattice and have a fixed density ρ_B. We found that the system have two phases depending on ρ_B and w. The current of the B particles is always a constant $\rho_B(1 - \rho_B)$ throughout the phase diagram while for the A particles it is given by different expressions in each phase. As a generalization one could also consider a more general process

$$A\emptyset \longrightarrow \emptyset A \quad \text{with rate } \alpha,$$
$$\emptyset B \longrightarrow B\emptyset \quad \text{with rate } \beta,$$
$$AB \longrightarrow BA \quad \text{with rate } 1,$$
$$A\emptyset \longrightarrow \emptyset\emptyset \quad \text{with rate } \lambda,$$
$$\emptyset\emptyset \longrightarrow A\emptyset \quad \text{with rate } \lambda',$$

with the quadratic algebra given by [JG07]

$$\mathbf{AB} = \mathbf{A} + \mathbf{B},$$
$$\alpha \, \mathbf{A}|V\rangle = |V\rangle,$$
$$\beta \, \langle W|\mathbf{B} = \langle W|,$$
$$\mathbf{E} = \tfrac{\lambda}{\lambda'\alpha}|V\rangle\langle W|.$$

and apply the same approach used in present section to study its phase diagram.

2.1.12 Phase Transition in Negotiation Dynamics

Statistical physics has recently proved to be a powerful framework to address issues related to the characterization of the collective social behavior of individuals, such as culture dissemination, the spreading of linguistic conventions, and the dynamics of opinion formation [Wei00].

According to the 'herding behavior' described in sociology [Cha03], processes of opinion formation are usually modelled as simple collective dynamics in which the agents update their opinions following local majority [Gla63] or imitation rules [Lig85]. Starting from random initial conditions, the system self–organizes through an ordering process eventually leading to the emergence of a global consensus, in which all agents share the same opinion. Deviations from purely herding behavior are considered by introducing a certain level of noise. In analogy with kinetic Ising models and contact processes [Odo04], the presence of noise can induce non–equilibrium phase transitions from the *consensus* state to disordered configurations, in which more than one opinion is present.

The principle of 'bounded confidence' [DNA00], on the other hand, consists in enabling interactions only between agents that share already some cultural features (defined as discrete objects) [Axe97] or with not too different opinions (in a continuous space) [DNA00, BKR03]. The overall behavior of the system depends on the method used to discriminate 'different' and 'similar' opinions. By tuning some threshold parameter, transitions are observed concerning the number of opinions surviving in the (frozen) final state. This can be a situation of *consensus*, in which all agents share the same opinion, *polarization*, in which a finite number of groups with different opinions survive, or *fragmentation*, with a final number of opinions scaling with the system size. For instance, in the *Axelrod model* [Axe97] a consensus–to–fragmentation transition occurs as the variability of cultural features is increased [CMV00].

2.1 Introduction to Phase Transitions

In this section, following [BDB07], we present a model of *opinion dynamics* in which a consensus–polarization–fragmentation non–equilibrium phase transition is driven by external noise, intended as an 'irresolute attitude' of the agents in making decisions. The primary attribute of the model is that it is based on a negotiation process, in which memory and feedback play a central role. Moreover, apart from the consensus state, no configuration is frozen: the stationary states with several coexisting opinions are still dynamical, in the sense that the agents are still able to evolve, in contrast to the Axelrod model [Axe97].

Let us consider a population of N agents, each one endowed with a memory, in which an a priori undefined number of opinions can be stored. In the initial state, agents memories are empty. At each time step, an ordered pair of neighboring agents is randomly selected. This choice is consistent with the idea of *directed attachment* in the socio–psychological literature (see for instance [Fri90]). The negotiation process is described by a local pairwise interaction rule: a) the first agent selects randomly one of its opinions (or creates a new opinion if its memory is empty) and conveys it to the second agent; b) if the memory of the latter contains such an opinion, with probability β the two agents update their memories erasing all opinions except the one involved in the *interaction (agreement)*, while with probability $1 - \beta$ nothing happens; c) if the memory of the second agent does not contain the uttered opinion, it adds such an opinion to those already stored in its *memory (learning)*. Note that, in the special case $\beta = 1$, the negotiation rule reduces to the Naming Game rule [BFC06], a model used to describe the emergence of a communication system or a set of linguistic conventions in a population of individuals. In our modelling the parameter β plays roughly the same role as the *probability of acknowledged influence* in the socio–psychological literature [Fri90]. Furthermore, as already stated for other models [Cas05], when the system is embedded in heterogeneous topologies, different pair selection criteria influence the dynamics. In the *direct strategy*, the first agent is picked up randomly in the population, and the second agent is randomly selected among its neighbors. The opposite choice is called *reverse strategy*; while the *neutral strategy* consists in randomly choosing a link, assigning it an order with equal probability.

At the beginning of the dynamics, a large number of opinions is created, the total number of different opinions growing rapidly up to $\mathcal{O}(N)$. Then, if β is sufficiently large, the number of opinions decreases until only one is left and the consensus state is reached (as for the Naming Game in the case $\beta = 1$). In the opposite limit, when $\beta = 0$, opinions are never eliminated, therefore the only possible stationary state is the trivial state in which every agent possesses all opinions. Thus, a non–equilibrium phase transition is expected for some critical value β_c of the parameter β governing the update efficiency. In order to find β_c, we exploit the following general stability argument. Let us consider the consensus state, in which all agents possess the same unique opinion, say A. Its stability may be tested by considering a situation in which

A and another opinion, say B, are present in the system: each agent can have either only opinion A or B, or both (AB state). The critical value β_c is provided by the threshold value at which the perturbed configuration with these three possible states does not converge back to consensus.

The simplest assumption in modelling a population of agents is the homogeneous mixing (i.e., mean–field (MF) approximation), where the behavior of the system is completely described by the following evolution equations for the densities n_i of agents with the opinion i [BDB07]

$$dn_A/dt = -n_A n_B + \beta n_{AB}^2 + \frac{3\beta - 1}{2} n_A n_{AB},$$
$$dn_B/dt = -n_A n_B + \beta n_{AB}^2 + \frac{3\beta - 1}{2} n_B n_{AB}, \qquad (2.92)$$

and $n_{AB} = 1 - n_A - n_B$. Imposing the steady state condition $\dot{n}_A = \dot{n}_B = 0$, we get three possible solutions: 1) $n_A = 1$, $n_B = 0$, $n_{AB} = 0$; 2) $n_A = 0$, $n_B = 1$, $n_{AB} = 0$; and 3) $n_A = n_B = b(\beta)$, $n_{AB} = 1 - 2b(\beta)$ with $b(\beta) = \frac{1+5\beta-\sqrt{1+10\beta+17\beta^2}}{4\beta}$ (and $b(0) = 0$). The study of the solutions' stability predicts a phase transition at $\beta_c = 1/3$. The maximum non–zero eigenvalue of the linearized system around the consensus solution becomes indeed positive for $\beta < 1/3$, i.e., the consensus becomes unstable, and the population polarizes in the $n_A = n_B$ state, with a finite density of undecided agents n_{AB}. The model therefore displays a first order non–equilibrium transition between the *frozen* absorbing consensus state and an *active polarized state*, in which global observables are stationary on average, but not frozen, i.e., the population is split in three dynamically evolving parts (with opinions A, B, and AB), whose densities fluctuate around the average values $b(\beta)$ and $1 - 2b(\beta)$.

We have checked the predictions of (2.92) by numerical simulations of N agents interacting on a complete graph [BDB07]. The convergence time t_{conv} required by the system to reach the consensus state indeed diverges at $\beta_c = 1/3$, with a power–law behavior $(\beta - \beta_c)^{-a}$, $a \simeq 0.3$. [5] Very interestingly however, the analytical and numerical analysis of (2.92) predicts that the relaxation time diverges instead as $(\beta - \beta_c)^{-1}$. This apparent discrepancy arises in fact because (2.92) consider that the agents have at most two different opinions at the same time, while this number is unlimited in the original model (and in fact diverges with N). Numerical simulations reproducing the two opinions case allow to recover the behavior of t_{conv} predicted from (2.92). We have also investigated the case of a finite number m of opinions available to the agents. The analytical result $a = 1$ holds also for $m = 3$ (but analytical analysis for larger m becomes out of reach), whereas preliminary numerical simulations performed for $m = 3, 10$ with the largest reachable population size

---

[5] The low value of a, which moreover slightly decreases as the system size increases, does not allow to exclude a logarithmic divergence. This issue deserves more investigations that we leave for future work.

($N = 10^6$) lead to an exponent $a \simeq 0.74 \div 0.8$. More extensive and systematic simulations are in order to determine the possible existence of a series of universality classes varying the memory size for the agents. In any case, the models with finite (m opinions) or unlimited memory define at least two clearly different universality classes for this non–equilibrium phase transition between consensus and polarized states (see [CG94] for similar findings in the framework of non–equilibrium q-state systems).

Also, the transition at β_c is only the first of a series of transitions: when decreasing $\beta < \beta_c$, a system starting from empty initial conditions self-organizes into a fragmented state with an increasing number of opinions. In principle, this can be shown analytically considering the mean-field evolution equations for the partial densities when $m > 2$ opinions are present, and studying, as a function of β, the sign of the eigenvalues of a $(2^m - 1) \times (2^m - 1)$ stability matrix for the stationary state with m opinions. For increasing values of m, such a calculation becomes rapidly very demanding, thus we limit our analysis to the numerical insights, from which we also get that the number of residual opinions in the fragmented state follows the *exponential law* [BDB07]

$$m(\beta) \propto \exp\left[(\beta_c - \beta)/C\right],$$

where C is a constant depending on the initial conditions (not shown).

We now extend the present analysis to more general interactions topologies, in which agents are placed on the vertices of a network, and the edges define the possible interaction patterns [BDB07]. When the network is a homogeneous random one (Erdös–Rényi (ER) graph [ER59]), the degree distribution is peaked around a typical value $\langle k \rangle$, and the evolution equations for the densities when only two opinions are present provide the same transition value $\beta_c = 1/3$ and the same exponent -1 for the divergence of t_{conv} as in MF.

Since any real negotiation process takes place on social groups, whose topology is generally far from being homogeneous, we have simulated the model on various uncorrelated heterogeneous networks (using the *uncorrelated configuration model* (UCM) [CBP05]), with power–law degree distributions $P(k) \sim k^{-\gamma}$ with exponents $\gamma = 2.5$ and $\gamma = 3$.

Very interestingly, the model still presents a consensus-polarization transition, in contrast with other *opinion–dynamics* models, like for instance the Axelrod model [KET03], for which the transition disappears for heterogeneous networks in the thermodynamic limit.

To understand these numerical results, we analyze, as for the fully connected case, the evolution equations for the case of two possible opinions. Such equations can be written for general correlated complex networks whose topology is completely defined by the degree distribution $P(k)$, i.e., the probability that a node has degree k, and by the degree–degree conditional probability $P(k'|k)$ that a node of degree k' is connected to a node of degree k (*Markovian networks*). Using partial densities

$$n_A^k = N_A^k/N_k, n_B^k = N_B^k/N_k \quad \text{and} \quad n_{AB}^k = N_{AB}^k/N_k,$$

i.e., the densities on classes of degree k, one derives mean–field type equations in analogy with epidemic models. Let us consider for definiteness the neutral pair selection strategy, the equation for n_A^k is in this case [BDB07]

$$\frac{dn_A^k}{dt} = -\frac{1}{\langle k \rangle} n_A^k k \sum_{k'} P(k'|k) n_B^{k'} - \frac{1}{2\langle k \rangle} n_A^k k \sum_{k'} P(k'|k) n_{AB}^{k'} \qquad (2.93)$$
$$+ \frac{3\beta}{2\langle k \rangle} k n_{AB}^k \sum_{k'} P(k'|k) n_A^{k'} + 2\frac{\beta}{2\langle k \rangle} k n_{AB}^k \sum_{k'} P(k'|k) n_{AB}^{k'},$$

$$\frac{dn_A^k}{dt} = -\frac{k n_A^k}{\langle k \rangle} \sum_{k'} P(k'|k) n_B^{k'} - \frac{k n_A^k}{2\langle k \rangle} \sum_{k'} P(k'|k) n_{AB}^{k'}$$
$$+ \frac{3\beta k n_{AB}^k}{2\langle k \rangle} \sum_{k'} P(k'|k) n_A^{k'} + \frac{\beta k n_{AB}^k}{\langle k \rangle} \sum_{k'} P(k'|k) n_{AB}^{k'},$$

and similar equations hold for n_B^k and n_{AB}^k. The first term corresponds to the situation in which an agent of degree k' and opinion B chooses as second actor an agent of degree k with opinion A. The second term corresponds to the case in which an agent of degree k' with opinions A and B chooses the opinion B, interacting with an agent of degree k and opinion A. The third term is the sum of two contributions coming from the complementary interaction; while the last term accounts for the increase of agents of degree k and opinion A due to the interaction of pairs of agents with AB opinion in which the first agent chooses the opinion A.

Now, let us define [BDB07]

$$\Theta_i = \sum_{k'} P(k'|k) n_i^{k'}, \quad (\text{for } i = A, B, AB).$$

Under the un-correlation hypothesis for the degrees of neighboring nodes, i.e., $P(k'|k) = k' P(k')/\langle k \rangle$, we get the following relation for the total densities $n_i = \sum_k P(k) n_i^k$,

$$\frac{d(n_A - n_B)}{dt} = \frac{3\beta - 1}{2} \Theta_{AB}(\Theta_A - \Theta_B). \qquad (2.94)$$

If we consider a small perturbation around the consensus state $n_A = 1$, with $n_A^k \gg n_B^k$ for all k, we can argue that

$$\Theta_A - \Theta_B = \sum_k k P(k)(n_A^k - n_B^k)/\langle k \rangle$$

is still positive, i.e., the consensus state is stable only for $\beta > 1/3$. In other words, the transition point does not change in heterogeneous topologies when the neutral strategy is assumed. This is in agreement with our numerical simulations, and in contrast with the other selection strategies.

For more details on negotiation dynamics and its non–equilibrium phase transitions, see [BDB07].

2.2 Elements of Haken's Synergetics

In this section we present the basics of the most powerful tool for high–dimensional chaos control, which is the *synergetics*. This powerful scientific tool to *extract order from chaos* has been developed outside of chaos theory, with intention to deal with much more complex, high–dimensional, hierarchical systems, in the realm of *synergetics*. Synergetics is an interdisciplinary field of research that was founded by H. Haken in 1969 (see [Hak83, Hak93, Hak96, Hak00]). Synergetics deals with complex systems that are composed of many individual parts (components, elements) that interact with each other and are able to produce spatial, temporal or functional structures by self–organization. In particular, synergetics searches for general principles governing self–organization irrespective of the nature of the individual parts of the systems that may belong to a variety of disciplines such as physics (lasers, fluids, plasmas), meteorology, chemistry (pattern formation by chemical reactions, including flames), biology (morphogenesis, evolution theory) movement science, brain activities, computer sciences (synergetic computer), sociology (e.g., city growth) psychology and psychiatry (including Gestalt psychology).

The aim of synergetics has been to describe processes of *spontaneous self–organization and cooperation* in complex systems built from many subsystems which themselves can be complicated nonlinear objects (like many individual neuro–muscular components of the human motion system, having their own excitation and contraction dynamics, embedded in a synergistic way to produce coordinated human movement). General properties of the subsystems are their own nonlinear/chaotic dynamics as well as mutual nonlinear/chaotic interactions. Furthermore, the systems of synergetics are *open*. The influence from outside is measured by a certain set of *control parameters* $\{\sigma\}$ (like amplitudes, frequencies and time characteristics of neuro–muscular driving forces). Processes of self-organization in synergetics, (like musculo–skeletal coordination in human motion dynamics) are observed as temporal macroscopic patterns. They are described by a small set of *order parameters* $\{o\}$, similar to those in Landau's *phase–transition theory* (named after *Nobel Laureate Lev D. Landau*) of physical systems in *thermal equilibrium* [Hak83].

Now, recall that the *measure for the degree of disorder* in any isolated, or conservative, system (such a system that does not interact with its surrounding, i.e., does neither dissipate nor gain energy) is entropy. The *second law of thermodynamics*[6] states that in every conservative irreversible system the entropy ever increases to its maximal value, i.e., to the total disorder of the system (or remains constant for a reversible system).

Example of such a system is conservative Hamiltonian dynamics of human skeleton in the phase–space Γ defined by all joint angles q^i and momenta p_i[7],

---

[6] This is the only physical law that implies the arrow of time.
[7] If we neglect joints dissipation and muscular driving forces, we are dealing with pure skeleton conservative dynamics.

defined by ordinary (conservative) Hamilton's equations

$$\dot{q}^i = \partial_{p_i} H, \qquad \dot{p}_i = -\partial_{q^i} H. \tag{2.95}$$

The basic fact of the conservative Hamiltonian system is that its phase–flow, the time evolution of equations (5.1), preserves the phase–space volume (the so–called *Liouville measure*), as proposed by the *Liouville theorem*. This might look fine at first sight, however, the preservation of phase–space volume causes *structural instability* of the conservative Hamiltonian system, i.e., the phase–space spreading effect, by which small phase regions R_t will tend to get distorted from the initial one R_0 during the system evolution. The problem is much more serious in higher dimensions than in lower dimensions, since there are so many 'directions' in which the region can locally spread. Here we see the work of the second law of thermodynamics on an irreversible process: the increase of entropy towards the total disorder/chaos [Pen89]. In this way, the conservative Hamiltonian systems of the form (5.1) cover the wide range of dynamics, from completely integrable, to completely ergodic. Biodynamics of human–like movement is probably somewhere in the middle of this range, the more DOF included in the model, the closer to the ergodic case. One can easily imagine that the conservative skeleton–like system with 300 DOF, which means 600–D system of the form (5.1), which is full of trigonometry (coming from its noncommutative rotational matrices), is probably closer to the ergodic than to the completely integrable case.

On the other hand, when we manipulate a system from the outside, by the use of certain *control parameters* $\{\sigma\}$, we can change its *degree of order* (see [Hak83, Hak93]). Consider for example *water vapor*. At elevated temperature its molecules move freely without mutual correlation. When temperature is lowered, a liquid drop is formed, the molecules now keep a mean distance between each other. Their motion is thus highly correlated. Finally, at still lower temperature, at the freezing point, water is transformed into ice crystals. The transitions between the different aggregate states, also called phases, are quite abrupt. Though the same kind of molecules are involved all the time, the macroscopic features of the three phases differ drastically.

Similar type of ordering, but not related to the thermal equilibrium conditions, occurs in *lasers*, mathematically given by Lorenz–like attractor equations. Lasers are certain types of lamps which are capable of emitting coherent light. A typical laser consists of a crystal rod filled with gas, with the following features important from the synergetics point of view: when the atoms the laser material consists of are excited or 'pumped' from the outside, they emit light waves. So, the pump power, or pump rate represents the control parameter σ. At low pump power, the waves are entirely uncorrelated as in a usual lamp. Could we hear light, it would sound like noise to us [Hak83].

When we increase the pump rate to a critical value σ_c, the noise disappears and is replaced by a pure tone. This means that the atoms emit a pure sinusoidal light wave which in turn means that the individual atoms act in a perfectly correlated way – they become self–organized. When the pump rate

2.2 Elements of Haken's Synergetics

is increased beyond a second critical value, the laser may periodically emit very intense and short pulses. In this way the following *instability sequence* occurs [Hak83]:

noise $\mapsto$ {coherent oscillation at frequency ω_1} $\mapsto$
periodic pulses at frequency ω_2 which modulate oscillation at frequency ω_1
i.e., no oscillation $\mapsto$ first frequency $\mapsto$ second frequency.

Under different conditions the light emission may become *chaotic* or even *turbulent*. The frequency spectrum becomes broadened.

The laser played a crucial role in the development of synergetics for various reasons [Hak83]. In particular, it allowed detailed theoretical and experimental study of the phenomena occurring within the transition region: *lamp $\leftrightarrow$ laser*, where a surprising and far–reaching analogy with phase transitions of systems in thermal equilibrium was discovered. This analogy includes all basic *phase–transition effects*: a *symmetry breaking instability*, *critical slowing down* and *hysteresis effect*.

2.2.1 Phase Transitions and Synergetics

Besides water vapor, a typical example is a *ferromagnet* [Hak83]. When a ferromagnet is heated, it suddenly loses its magnetization. When temperature is lowered, the magnet suddenly regains its magnetization. What happens on a microscopic, atomic level, is this: We may visualize the magnet as being composed of many, elementary (atomic) magnets (called spins). At elevated temperature, the elementary magnets point in random directions. Their magnetic moments, when added up, cancel each other and no macroscopic magnetization results. Below a critical value of temperature T_c, the elementary magnets are lined up, giving rise to a macroscopic magnetization. Thus the *order on the microscopic level* is a cause of a *new feature* of the material *on the macroscopic level*. The change of one phase to the other one is called *phase transition*.

A thermodynamical description of a ferromagnet is based on analysis of its *free energy potential* (in thermal equilibrium conditions). The free energy $\mathcal{F}$, depends on the *control parameter* $\sigma = T$, the temperature. We seek the minimum of the potential $\mathcal{F}$ for a fixed value of magnetization o, which is called *order parameter* in Landau's theory of phase transitions (see Appendix).

This phenomenon is called a *phase transition of second order* because the second derivative (specific heat) of the free energy potential $\mathcal{F}$ is discontinuous. On the other hand, the entropy S (the first derivative of $\mathcal{F}$) itself is continuous so that this transition is also referred to as a *continuous phase transition*.

In statistical physics one also investigates the temporal change of the order parameter – magnetization o. Usually, in a more or less phenomenological manner, one assumes that o obeys an equation of the form

$$\dot{o} = -\frac{\partial \mathcal{F}}{\partial o} = -\sigma o - \beta o^3. \qquad (2.96)$$

For $\sigma \to 0$ we observe a phenomenon called *critical slowing down*, because the 'particle' with coordinate o falls down the slope of the 'potential well' more and more slowly. Simple relation (2.96) is called *order parameter equation*.

We now turn to the case where the free energy potential has the form

$$\mathcal{F}(o,T) = \frac{\sigma}{2}o^2 + \frac{\gamma}{3}o^3 + \frac{\beta}{4}o^4, \qquad (2.97)$$

(β and γ – positive but σ may change its sign according to $\sigma = a(T-T_c)$, ($a > 0$)). When we change the control parameter – temperature T, i.e., the parameter σ, we pass through a sequence of deformations of the potential curve.

When lowering temperature, the local minimum first remains at $o_0 = 0$. When lowering temperature, the 'particle' may fall down from o_0 to the new (global) minimum of $\mathcal{F}$ at o_1. The entropies of the two states, o_0 and o_1, differ. This phenomenon is called a *phase transition of first order* because the first derivative of the potential $\mathcal{F}$ with respect to the control parameter T is discontinuous. Since the entropy S is discontinuous this transition is also referred to as a *discontinuous phase transition*. When we now increase the temperature, is apparent that the system stays at o_1 longer than it had been before when lowering the control parameter. This represents *hysteresis effect*.

In the case of the potential (2.97) the order parameter equation gets the form

$$\dot{o} = -\sigma o - \gamma o^2 - \beta o^3.$$

Similar *disorder* $\Rightarrow$ *order* transitions occur also in various non–equilibrium systems of physics, chemistry, biology, psychology, sociology, as well as in human motion dynamics. The analogy is subsumed in Table 1.

Table 1. Phase transition analogy

System in thermal equilibrium	Non–equilibrium system
Free energy potential $\mathcal{F}$	Generalized potential V
Order parameters o_i	Order parameters o_i
$\dot{o}_i = -\frac{\partial \mathcal{F}}{\partial o_i}$	$\dot{o}_i = -\frac{\partial V}{\partial o_i}$
Temperature T	Control input u
Entropy S	System output y
Specific Heat c	System efficiency e

In the case of human motion dynamics, natural control inputs u_i are muscular torques F_i, natural system outputs y_i are joint coordinates q^i and momenta p_i, while the system efficiencies e_i represent the changes of coordinates and momenta with changes of corresponding muscular torques for the ith joint,

$$e_i^q = \frac{\partial q^i}{\partial F_i}, \qquad e_i^p = \frac{\partial p_i}{\partial F_i}.$$

Order parameters o_i represent certain important qualities of the human motion system, depending on muscular torques as control inputs, similar to *magnetization*, and usually defined by equations similar to (2.96) or

$$\dot{o}_i = -\sigma o - \gamma o^2 - \beta o^3,$$

with nonnegative parameters σ, β, γ, and corresponding to the second and first order phase transitions, respectively. The choice of actual order parameters is a matter of *expert knowledge* and *purpose of macroscopic system modelling* [Hak83].

2.2.2 Order Parameters in Human/Humanoid Biodynamics

Basic Hamiltonian Model of Biodynamics

To describe the biodynamics of human–like movement, namely our *covariant force law* [II05, II06a, II06b]:

$$F_i = mg_{ij}a^j, \qquad \text{that 'in plain English' reads :}$$

force 1–form–field = mass distribution × acceleration vector–field,

we can also start from generalized Hamiltonian vector–field X_H describing the behavior of the *human–like locomotor system*

$$\dot{q}^i = \frac{\partial H}{\partial p_i} + \frac{\partial R}{\partial p_i}, \qquad (2.98)$$

$$\dot{p}_i = F_i - \frac{\partial H}{\partial q^i} + \frac{\partial R}{\partial q^i}, \qquad (2.99)$$

where the vector–field X_H is generating time evolution, or *phase–flow*, of $2n$ *system variables*: n generalized coordinates (joint angles q^i) and n generalized momenta (joint angular momenta p_i), $H = H(q,p)$ represents the system's conservative energy: kinetic energy + various mechano–chemical potentials, $R = R(q,p)$ denotes the nonlinear dissipation of energy, and $F_i = F_i(t,q,p,\sigma)$ are external control forces (biochemical energy inputs). The *system parameters* include inertia tensor with mass distribution of all body segments, stiffness and damping tensors for all joints (labelled by index i, which is, for geometric reasons, written as a subscript on angle variables, and as a superscript on momentum variables), as well as amplitudes, frequencies and time characteristics of all active muscular forces (supposed to be acting in all the joints; if some of the joints are inactive, we have the affine Hamiltonian control system, see chapter 6).

The equation (2.98) is called the *velocity equation*, representing the *flow* of the system (analogous to current in electrodynamics), while the equation (2.99) is a Newton–like *force equation*, representing the *effort* of the system (analogous to voltage). Together, these two functions represent Hamiltonian formulation of the *biomechanical force–velocity relation* of A.V. Hill

[Hil38]. From engineering perspective, their (inner) product, *flow · effort*, represents the total system's *power*, equal to the time–rate–of–change of the total system's energy (included in H, R and F_i functions). And energy itself is transformed into the *work* done by the system.

Now, the reasonably accurate musculo–skeletal biodynamics would include say a hundred DOF, which means a hundred of joint angles and a hundred of joint momenta, which further means a hundred of coupled equations of the form of (2.98–2.99). And the *full coupling* means that each angle (and momentum) includes the information of all the other angles (and momenta), the *chain coupling* means that each angle (and momentum) includes the information of all the previous (i.e., children) angles (and momenta), the *nearest neighbor coupling* includes the information of the nearest neighbors, etc.

No matter which coupling we use for modelling the dynamics of human motion, one thing is certain: the *coupling is nonlinear*. And we obviously have to fight chaos within several hundreds of variables.

Wouldn't it be better if we could somehow be able to obtain a synthetic information about the whole musculo–skeletal dynamics, synthesizing the hundreds of equations of motion of type (2.98–2.99) into a small number of equations describing the time evolution of the so–called *order parameters*? If we could do something similar to principal component analysis in multivariate statistics and neural networks, to get something like 'nonlinear factor dynamics'?

Starting from the basic system (2.98–2.99), on the lowest, *microscopic level of human movement organization*, the *order parameter equations of macroscopic synergetics* can be (at least theoretically), either exactly derived along the lines of *mezoscopic synergetics*, or phenomenologically stated by the use of the certain biophysical analogies and nonlinear identification and control techniques (a highly complex nonlinear system like human locomotor apparatus could be neither identified nor controlled by means of standard linear engineering techniques).

Mezoscopic Derivation of Order Parameters

Basic Hamiltonian equations (2.98–2.99) are in general quite complicated and can hardly be solved completely in the whole locomotor phase–space Γ, spanned by the set of possible joint vectors $\{q^i(t), p_i(t)\}$. We therefore have to restrict ourselves to local concepts for analyzing the behavior of our locomotor system. To this end we shall consider a reference musculo–skeletal state $\{q_0, p_0\}$ and its neighborhood. Following the procedures of the mezoscopic synergetics (see [Hak83, Hak93]), we assume that the reference state has the properties of an attractor and is a comparably low–dimensional object in Γ. In order to explore the behavior of our locomotor system (dependent on the set of control parameters σ) in the neighborhood of $\{q_0, p_0\}$ we look for the time development of small deviations from the reference state (to make the formalism as simple as possible, we drop the joint index in this section)

2.2 Elements of Haken's Synergetics

$$q(t) = q_0 + \delta q(t), \qquad p(t) = p_0 + \delta p(t),$$

and consider $\delta q(t)$ and $\delta p(t)$ as small entities. As a result we may linearize the equations of δq and δp in the vicinity of the reference state $\{q_0, p_0\}$. We get

$$\partial_t \delta q(t) = L[q_0, p_0, \sigma] \delta q(t), \qquad \partial_t \delta p(t) = K[q_0, p_0, \sigma] \delta p(t),$$

where $L[.]$ and $K[.]$ are linear matrices independent of $\delta q(t)$ and $\delta p(t)$, which can be derived from the basic Hamiltonian vector–field (2.98–2.99) by standard synergetics methods [Hak83, Hak93, Hak96, Hak00]. We now assume that we can construct a complete set of eigenvectors $\{l^{(j)}(t), k^{(j)}(t)\}$ corresponding to (2.96). These eigenvectors allow us to decompose arbitrary deviations $\delta q(t)$ and $\delta p(t)$ into elementary collective deviations along the directions of the eigenvectors

$$\delta q(t) = \xi_j(t)\, l^j(t), \qquad \delta p(t) = \zeta_j(t)\, k^j(t), \tag{2.100}$$

where $\xi_j(t)$ and $\zeta_j(t)$ represent the excitations of the system along the directions in the phase–space Γ prescribed by the eigenvectors $l^j(t)$ and $k^j(t)$, respectively. These amplitudes are still dependent on the set of control parameters $\{\sigma\}$. We note that the introduction of the eigenvectors $\{l^j(t), k^j(t)\}$ is of crucial importance. In the realm of synergetics they are considered as the *collective modes or patterns* of the system. Whereas the basic Hamiltonian equation (2.98–2.99) is formulated on the basis of the human locomotor–system variables (coordinates and momenta) of the single subsystems (joints), we can now give a new formulation which is based on these collective patterns and describes the dynamical behavior of the locomotor system in terms of these different collective patterns. Inserting relations (2.100) into the basic system (2.98–2.99) we get equations for the amplitudes $\xi_j(t)$ and $\zeta_j(t)$,

$$\dot{\xi}_i(t) = A_{ij} \cdot \xi_j(t) + \text{nonlinear terms}, \qquad \dot{\zeta}_j(t) = B_{ij} \cdot \zeta_j(t) + \text{nonlinear terms},$$

where $\cdot$ denotes the scalar product, and it is assumed that the time dependence of the linear matrices L and K is carried out by the eigenvectors leaving us with constant matrices A and B.

We now summarize the results by discussing the following time–evolution formulas for joint coordinates $q(t)$ and momenta $p(t)$,

$$q(t) = q_0 + \xi_j(t)\, l^j(t), \qquad p(t) = p_0 + \zeta_j(t)\, k^j(t), \tag{2.101}$$

which describes the time dependence of the phase vectors $q(t)$ and $p(t)$ through the evolution of the collective patterns. Obviously, the reference musculo–skeletal state $\{q_0(t), p_0(t)\}$ can be called stable when all the possible excitations $\{\xi_j(t), \zeta_j(t)\}$ decay during the curse of time. When we now change the control parameters $\{\sigma\}$ some of the $\{\xi_j(t), \zeta_j(t)\}$ can become unstable and start to grow in time. The border between decay and growth in parameter space is called a Tablecritical region. Haken has shown that the few well known (see, e.g., [Att71, Hak91]). When we view the *Necker cube*, which is a classic example of perspective alternation, a part of the figure is perceived either as front or back of a cube and our perception switches between the two different interpretations (see Figure 2.7). In this circumstance the external stimulus is kept constant, but perception undergoes involuntary and random–like change. The measurements have been quantified in psychophysical experiments and it becomes evident that the times between such changes are approximately Gamma distributed [BMA72, BCR82, Hak91].

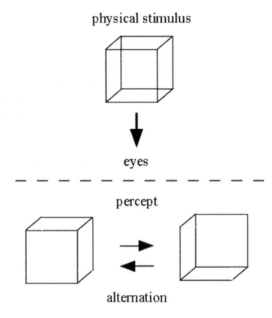

Fig. 2.7. Perception of the Necker cube with its two alternative interpretations (modified and adapted from [NNM00]).

Mathematical model approaches to explaining the facts have been made mainly from three situations based on the *synergetics* [DT89, DT90, CA93], the BSB (brain–state–in–a–box) neural network model [KA85, RMS90, MM95], and the PDP (parallel distributed processing) schema model [RM86, SKW95, IN96]. Common to these approaches is that top–down designs are applied so that the model can be manipulable by a few parameters and upon this basis fluctuating sources are brought in. The major interests seem to be not in the relation between the whole function and its element (neuron), but in the model building at the phenomenological level.

So far diverse types of chaotic dynamics have been confirmed at several hierarchical levels in the real neural systems from single cells to cortical networks (e.g. ionic channels, spike trains from cells, EEG) [Arb95].

Following [NNM00], in this section we present a perception model of ambiguous patterns based on the chaotic neural network from the viewpoint of bottom–up approach [NNM97], aiming at the functioning of chaos in dynamic perceptual processes.

The chaotic neural network (CNN) composed of N chaotic neurons is described as [ATT90, NKF97] (summation upon repeated indices is always understood)

$$X_i(t+1) = f(\eta_i(t+1) + \zeta_i(t+1)), \qquad (2.104)$$

$$\eta_i(t+1) = w_{ij} \sum_{d=0}^{t} k_f^d X_j(t-d), \qquad (2.105)$$

$$\zeta_i(t+1) = -\alpha \sum_{d=0}^{t} k_r^d X_i(t-d) - \theta_i, \qquad (2.106)$$

where X_i : output of neuron $i(-1 \leq X_i \leq 1)$, w_{ij} : synaptic weight from neuron j to neuron i, θ_i : threshold of neuron i, $k_f(k_r)$: decay factor for the feedback (refractoriness) $(0 \leq k_f, k_r < 1)$, α : refractory scaling parameter, f : output function defined by $f(y) = \tanh(y/2\varepsilon)$ with the steepness parameter ε. Owing to the exponentially decaying form of the past influence, (2.105) and (2.106) can be reduced to

$$\eta_i(t+1) = k_f \eta_i(t) + w_{ij} X_j(t), \qquad (2.107)$$
$$\zeta_i(t+1) = k_r \zeta_i(t) - \alpha X_i(t) + a, \qquad (2.108)$$

where a is temporally constant $a \equiv -\theta_i(1-k_r)$. All neurons are updated in parallel, that is, synchronously. The network corresponds to the conventional *Hopfield discrete–time network*:

$$X_i(t+1) = f\left[w_{ij} X_j(t) - \theta_i\right], \qquad (2.109)$$

when $\alpha = k_f = k_r = 0$ (Hopfield network point (HNP)). The asymptotical stability and chaos in discrete–time neural networks are theoretically investigated in [MW89, CA97]. The stochastic fluctuation $\{F_i\}$ is attached to (2.109) of HNP together with the external stimulus $\{\sigma_i\}$:

$$X_i(t+1) = f\left[w_{ij} X_j(t) + \sigma_i + F_i(t)\right],$$
$$\text{where} \quad \begin{cases} < F_i(t) > = 0 \\ < F_i(t) F_j(t') > = D^2 \delta_{tt'} \delta_{ij}. \end{cases}$$

Under external stimuli, (2.104) is influenced as

$$X_i(t+1) = f\left[\eta_i(t+1) + \zeta_i(t+1) + \sigma_i\right], \qquad (2.110)$$

where $\{\sigma_i\}$ is the effective term by external stimuli. This is a simple and un–artificial incorporation of stimuli as the changes of neural active potentials.

The two competitive interpretations are embedded in the network as minima of the energy map (see Figure 2.8):

$$E = -\frac{1}{2} w_{ij} X_i X_j,$$

at HNP. This is done by using a iterative perception learning rule for $p(< N)$ patterns $\{\xi_i^\mu\} \equiv (\xi_1^\mu, \cdots, \xi_N^\mu), (\mu = 1, \cdots, p; \xi_i^\mu = +1 \text{ or} -1)$ in the form:

$$w_{ij}^{new} = w_{ij}^{old} + \sum_\mu \delta w_{ij}^\mu, \quad \text{with} \quad \delta w_{ij}^\mu = \frac{1}{N} \theta(1 - \gamma_i^\mu) \xi_i^\mu \xi_j^\mu,$$

where $\quad \gamma_i^\mu \equiv \xi_i^\mu w_{ij} \xi_j^\mu,$

and $\theta(h)$ is the unit step function. The learning mode is separated from the performance mode by (2.110).

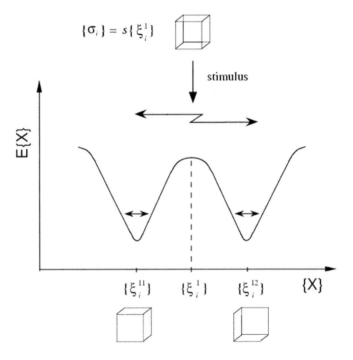

Fig. 2.8. Conceptual psychodynamic model of [NNM00], illustrating state transitions induced by chaotic activity.

Simulations of the CNN have shown that the neural chaos leads to perceptual alternations as responses to ambiguous stimuli in the chaotic neural network. Its emergence is based on the simple process in a realistic bottom–up framework. In the same stage, similar results can not be obtained by the stochastic activity. This simulation suggests functional usefulness of the chaotic activity in perceptual systems even at higher cognitive levels. The perceptual alternation appears to be an inherent feature built in the chaotic neuron assembly. It may be interesting to study the brain with the experimental technique (e.g., functional MRI) under the circumstance where the perceptual alternation is running [NNM00].

2.2.5 Kick Dynamics and Dissipation–Fluctuation Theorem

Deterministic Delayed Kicks

Following [Hak02], we consider the mechanical example of a soccer ball that is kicked by a soccer player and rolls over grass, whereby its motion will be slowed down. We start with the Newton's (second) law of motion, $m\dot{v} = force$, and in order to get rid of superfluous constants, we put temporarily $m = 1$. The $force$ on the r.h.s. consists of the damping force $-\gamma v(t)$ of the grass (where γ is the damping constant) and the sharp force $F(t) = s\delta(t - \sigma)$ of the individual kick occurring at time $t = \sigma$ (where s is the strength of the kick, and δ is the Dirac's 'delta' function). In this way, the (single) *kick equation* of the ball motion becomes

$$\dot{v} = -\gamma v(t) + s\delta(t - \sigma), \qquad (2.111)$$

with the general solution

$$v(t) = sG(t - \sigma),$$

where $G(t - \sigma)$ is the *Green's function*[8]

$$G(t - \sigma) = \begin{cases} 0 & \text{for } t < \sigma \\ e^{-\gamma(t-\sigma)} & \text{for } t \geq \sigma \end{cases}.$$

Now, we can generalize the above to N kicks with individual strengths s_j, occurring at a sequence of times $\{\sigma_j\}$, so that the total kicking force becomes

$$F(t) = \sum_{j=1}^{N} s_j \delta(t - \sigma_j).$$

In this way, we get the *multi-kick equation* of the ball motion

---

[8] This is the Green's function of the first order system (2.111). Similarly, the Green's function

$$G(t - \sigma) = \begin{cases} 0 & \text{for } t < \sigma \\ (t - \sigma)e^{-\gamma(t-\sigma)} & \text{for } t \geq \sigma \end{cases}$$

corresponds to the second order system

$$\left(\frac{d}{dt} + \gamma\right)^2 G(t - \sigma) = \delta(t - \sigma).$$

$$\dot{v} = -\gamma v(t) + \sum_{j=1}^{N} s_j \delta(t - \sigma_j),$$

with the general solution

$$v(t) = \sum_{j=1}^{N} s_j G(t - \sigma_j). \tag{2.112}$$

As a final generalization, we would imagine that the kicks are continuously exerted on the ball, so that kicking force becomes

$$F(t) = \int_{t_0}^{T} s(\sigma)\delta(t - \sigma)d\sigma \equiv \int_{t_0}^{T} d\sigma F(\sigma)\delta(t - \sigma),$$

so that the continuous multi–kick equation of the ball motion becomes

$$\dot{v} = -\gamma v(t) + \int_{t_0}^{T} s(\sigma)\delta(t - \sigma)d\sigma \equiv -\gamma v(t) + \int_{t_0}^{T} d\sigma F(\sigma)\delta(t - \sigma),$$

with the general solution

$$v(t) = \int_{t_0}^{T} d\sigma F(\sigma)G(t - \sigma) = \int_{t_0}^{T} d\sigma F(\sigma)e^{-\gamma(t-\sigma)}. \tag{2.113}$$

Random Kicks and Langevin Equation

We now denote the times at which kicks occur by t_j and indicate their direction in a one–dimensional game by $(\pm 1)_j$, where the choice of the plus or minus sign is random (e.g., throwing a coin). Thus the kicking force can be written in the form [Hak02]

$$F(t) = s \sum_{j=1}^{N} \delta(t - t_j)(\pm 1)_j, \tag{2.114}$$

where for simplicity we assume that all kicks have the same strength s. When we observe many games, then we may perform an average $< \ldots >$ over all these different *performances*,

$$< F(t) > = s < \sum_{j=1}^{N} \delta(t - t_j)(\pm 1)_j > . \tag{2.115}$$

Since the direction of the kicks is assumed to be independent of the time at which the kicks happen, we may split (2.115) into the product

$$< F(t) > = s < \sum_{j=1}^{N} \delta(t - t_j) > < (\pm 1)_j > .$$

2.2 Elements of Haken's Synergetics

As the kicks are assumed to happen with equal frequency in both directions, we get the cancellation

$$< (\pm 1)_j > = 0,$$

which implies that the average kicking force also vanishes,

$$< F(t) > = 0.$$

In order to characterize the strength of the force (2.114), we consider a quadratic expression in F, e.g., by calculating the *correlation function* for two times t, t',

$$< F(t)F(t') > = s^2 < \sum_j \delta(t - t_j)(\pm 1)_j \sum_k \delta(t' - t_k)(\pm 1)_k > .$$

As the ones for $j \neq k$ will cancel each other and for $j = k$ will become 1, the correlation function becomes a single sum

$$< F(t)F(t') > = s^2 < \sum_j \delta(t - t_j)\delta(t' - t_k) >, \quad (2.116)$$

which is usually evaluated by assuming the *Poisson process* for the times of the kicks.

Now, proper description of random motion is given by *Langevin rate equation*, which describes the *Brownian motion*: when a particle is immersed in a fluid, the velocity of this particle is slowed down by a force proportional to its velocity and the particle undergoes a zig–zag motion (the particle is steadily pushed by much smaller particles of the liquid in a random way). In physical terminology, we deal with the behavior of a system (particle) which is coupled to a *heat bath* or reservoir (namely the liquid). The heat bath has two effects [Hak02]:

A. It decelerates the mean motion of the particle; and
B. It causes statistical fluctuation.

The standard Langevin equation has the form

$$\dot{v} = -\gamma v(t) + F(t), \quad (2.117)$$

where $F(t)$ is a *fluctuating force* with the following properties:

A. Its statistical average (2.115) vanishes; and
B. Its correlation function (2.116) is given by

$$< F(t)F(t') > = Q\delta(t - t_0), \quad (2.118)$$

where $t_0 = T/N$ denotes the mean free time between kicks, and $Q = s^2/t_0$ is the *random fluctuation*.

The general solution of the Langevin equation (2.117) is given by (2.113).

The average velocity vanishes, $<v(t)>=0$, as both directions are possible and cancel each other. Using the integral solution (2.113) we get

$$<v(t)v(t')>=<\int_{t_0}^{t}d\sigma\int_{t_0}^{t'}d\sigma' F(\sigma)F(\sigma')e^{-\gamma(t-\sigma)}e^{-\gamma(t'-\sigma')}>,$$

which, in the steady-state, reduces to

$$<v(t)v(t')>=\frac{Q}{2\gamma}e^{-\gamma(t-\sigma)},$$

and for equal times

$$<v(t)^2>=\frac{Q}{2\gamma}.$$

If we now repeat all the steps performed so far with $m \neq 1$, the final result reads

$$<v(t)^2>=\frac{Q}{2\gamma m}. \quad (2.119)$$

Now, according to thermodynamics, the *mean kinetic energy* of a particle is given by

$$\frac{m}{2}<v(t)^2>=\frac{1}{2}k_B T, \quad (2.120)$$

where T is the (absolute) temperature, and k_B is the Boltzman's constant. Comparing (2.119) and (2.120), we get the important Einstein's result

$$Q = 2\gamma k_B T,$$

which says that whenever there is damping, i.e., $\gamma \neq 0$, then there are random fluctuations (or noise) Q. In other words, fluctuations or noise are inevitable in any physical system. For example, in a resistor (with the resistance R) the electric field E fluctuates with a correlation function (similar to (2.118))

$$<E(t)E(t')>=2Rk_B T\delta(t-t_0).$$

This is the simplest example of the *fluctuation–dissipation theorem*.

2.3 Synergetics of Recurrent and Attractor Neural Networks

Recall that *recurrent neural networks* are neural networks with synaptic feedback loops. Provided that we restrict ourselves to large neural systems, we can apply to their analysis tools from statistical mechanics. Here, we have two possibilities. Under the common conditions of synaptic symmetry, the stochastic process of evolving neuron states leads towards an equilibrium situation

2.3 Synergetics of Recurrent and Attractor Neural Networks

where the microscopic state probabilities are known, where the classical techniques of *equilibrium statistical mechanics* can be applied. On the other hand, for non–symmetric networks, where the asymptotic (stationary) statistics are not known, synergetic techniques from *non–equilibrium statistical mechanics* are the only tools available for analysis. Here, the 'natural' set of macroscopic *order parameters* to be calculated can be defined in practice as the smallest set which will obey closed deterministic equations in the limit of an infinitely large network.

Being high–dimensional nonlinear systems with extensive feedback, the dynamics of recurrent neural networks are generally dominated by a wealth of different attractors, and the practical use of recurrent neural networks (in both biology and engineering) lies in the potential for creation and manipulation of these attractors through adaptation of the network parameters (synapses and thresholds). Input fed into a recurrent neural network usually serves to induce a specific initial configuration (or firing pattern) of the neurons, which serves as a cue, and the 'output' is given by the (static or dynamic) attractor which has been triggered by this cue. The most familiar types of recurrent neural network models, where the idea of creating and manipulating attractors has been worked out and applied explicitly, are the so–called *attractor neural networks* for associative memory, designed to store and retrieve information in the form of neuronal firing patterns and/or sequences of neuronal firing patterns. Each pattern to be stored is represented as a microscopic state vector. One then constructs synapses and thresholds such that the dominant attractors of the network are precisely the pattern vectors (in the case of static recall), or where, alternatively, they are trajectories in which the patterns are successively generated microscopic system states. From an initial configuration (the 'cue', or input pattern to be recognized) the system is allowed to evolve in time autonomously, and the final state (or trajectory) reached can be interpreted as the pattern (or pattern sequence) recognized by network from the input. For such programmes to work one clearly needs recurrent neural networks with extensive *ergodicity breaking*: the state vector will during the course of the dynamics (at least on finite time–scales) have to be confined to a restricted region of state space (an 'ergodic component'), the location of which is to depend strongly on the initial conditions. Hence our interest will mainly be in systems with many attractors. This, in turn, has implications at a theoretical/mathematical level: solving models of recurrent neural networks with extensively many attractors requires advanced tools from disordered systems theory, such as statical replica theory and dynamical partition function analysis.

The *equilibrium* statistical mechanical techniques can provide much detailed quantitative information on the behavior of recurrent neural networks, but they obviously have serious restrictions. The first one is that, by definition, they will only provide information on network properties in the stationary state. For associative memories, for instance, it is not clear how one can calculate quantities like sizes of domains of attraction without solving the dynamics. The second, and more serious, restriction is that for equilibrium statistical mechanics to apply the dynamics of the network under study must obey detailed balance, i.e., absence of microscopic probability currents in the stationary state. For recurrent networks in which the dynamics take the form of a stochastic alignment of neuronal firing rates to post–synaptic potentials which, in turn, depend linearly on the firing rates, this requirement of detailed balance usually implies symmetry of the synaptic matrix. From a physiological point of view this requirement is clearly unacceptable, since it is violated in any network that obeys Dale's law as soon as an excitatory neuron is connected to an inhibitory one. Worse still, in any network of graded–response neurons detailed balance will always be violated, even when the synapses are symmetric. The situation will become even worse when we turn to networks of yet more realistic (spike–based) neurons, such as integrate-and-fire ones. In contrast to this, *non–equilibrium* statistical mechanical techniques, it will turn out, do not impose such biologically non–realistic restrictions on neuron types and synaptic symmetry, and they are consequently the more appropriate avenue for future theoretical research aimed at solving biologically more realistic models (for details, see [Coo01, SC00, SC01, CKS05]).

2.3.1 Stochastic Dynamics of Neuronal Firing States

Recall that the simplest non–trivial definition of a recurrent neural network is that where N binary neurons $\sigma_i \in \{-1, 1\}$ (in which the states '1' and '-1' represent firing and rest, respectively) respond iteratively and synchronously to post–synaptic potentials (or local fields) $h_i(\boldsymbol{\sigma})$, with $\boldsymbol{\sigma} = (\sigma_1, \ldots, \sigma_N)$. The fields are assumed to depend linearly on the instantaneous neuron states (summation convention upon repeated indices is always used):

Parallel Dynamics: $\quad \sigma_i(\ell+1) = \mathrm{sgn}\left[h_i(\boldsymbol{\sigma}(\ell)) + T\eta_i(\ell)\right], \quad h_i(\sigma) = J_{ij}\sigma_j + \theta_i.$ \hfill (2.121)

The stochasticity is in the independent random numbers $\eta_i(\ell) \in \mathbb{R}$ (representing threshold noise), which are all drawn according to some distribution $w(\eta)$. The parameter T is introduced to control the amount of noise. For $T = 0$ the process (2.121) is deterministic: $\sigma_i(\ell+1) = \mathrm{sgn}[h_i(\sigma(\ell))]$. The opposite extreme is choosing $T = \infty$, here the system evolution is fully random. The external fields θ_i represent neural thresholds and/or external stimuli, J_{ij} represents the synaptic efficacy at the junction $j \to i$ ($J_{ij} > 0$ implies excitation, $J_{ij} < 0$ inhibition). Alternatively we could decide that at each iteration step ℓ only a single randomly drawn neuron σ_{i_ℓ} is to undergo an update of the type (2.121):

Sequential Dynamics: $\quad \begin{array}{l} i \neq i_\ell : \sigma_i(\ell+1) = \sigma_i(\ell), \\ i = i_\ell : \sigma_i(\ell+1) = \mathrm{sgn}\left[h_i(\boldsymbol{\sigma}(\ell)) + T\eta_i(\ell)\right], \end{array}$ \hfill (2.122)

with the local fields as in (2.121). The stochasticity is now both in the independent random numbers $\eta_i(\ell)$ (the threshold noise) and in the site i_ℓ to be

2.3 Synergetics of Recurrent and Attractor Neural Networks

updated, drawn randomly from the set $\{1,\ldots,N\}$. For simplicity we assume $w(-\eta)=w(\eta)$, and define

$$g(z)=2\int_0^z d\eta\, w(\eta): \qquad g(-z)=-g(z), \qquad \lim_{z\to\pm\infty}g(z)=\pm 1, \qquad \partial_z g(z)\geq 0.$$

Popular choices for the threshold noise distributions are

$$w(\eta)=(2\pi)^{-\frac{1}{2}}e^{-\frac{1}{2}\eta^2}: \qquad g(z)=\mathrm{Erf}[z/\sqrt{2}],$$

$$w(\eta)=\frac{1}{2}[1-\tanh^2(\eta)]: \qquad g(z)=\tanh(z).$$

Now, from the microscopic equations (2.121,2.122), which are suitable for numerical simulations, we can derive an equivalent but mathematically more convenient description in terms of microscopic state probabilities $p_\ell(\boldsymbol{\sigma})$. Equations (2.121,2.122) state that, if the system state $\boldsymbol{\sigma}(\ell)$ is given, a neuron i to be updated will obey

$$\mathrm{Prob}\,[\sigma_i(\ell+1)]=\frac{1}{2}\,[1+\sigma_i(\ell+1)\,g[\beta h_i(\boldsymbol{\sigma}(\ell))]]\,, \qquad (2.123)$$

with $\beta=T^{-1}$. In the case (2.121) this rule applies to all neurons, and thus we simply get $p_{\ell+1}(\boldsymbol{\sigma})=\prod_{i=1}^N\frac{1}{2}[1+\sigma_i\,g[\beta h_i(\boldsymbol{\sigma}(\ell))]]$. If, on the other hand, instead of $\boldsymbol{\sigma}(\ell)$ only the probability distribution $p_\ell(\boldsymbol{\sigma})$ is given, this expression for $p_{\ell+1}(\boldsymbol{\sigma})$ is to be averaged over the possible states at time ℓ:

Parallel Dynamics : $\qquad p_{\ell+1}(\boldsymbol{\sigma})=\sum_{\boldsymbol{\sigma}'}W\,[\boldsymbol{\sigma};\boldsymbol{\sigma}']\,p_\ell(\boldsymbol{\sigma}'), \qquad (2.124)$

$$W\,[\boldsymbol{\sigma};\boldsymbol{\sigma}']=\prod_{i=1}^N\frac{1}{2}\,[1+\sigma_i\,g[\beta h_i(\boldsymbol{\sigma}')]]\,.$$

This is the standard representation of a *Markov chain*. Also the sequential process (2.122) can be formulated in terms of probabilities, but here expression (2.123) applies only to the randomly drawn candidate i_ℓ. After averaging over all possible realisations of the sites i_ℓ we get

$$p_{\ell+1}(\boldsymbol{\sigma})=\frac{1}{N}\left\{[\prod_{j\neq i}\delta_{\sigma_j,\sigma_j(\ell)}]\,\frac{1}{2}\,[1+\sigma_i\,g[\beta h_i(\boldsymbol{\sigma}(\ell))]]\right\},$$

with the Kronecker symbol: $\delta_{ij}=1$, if $i=j$ and $\delta_{ij}=0$, otherwise. If, instead of $\boldsymbol{\sigma}(\ell)$, the probabilities $p_\ell(\boldsymbol{\sigma})$ are given, this expression is to be averaged over the possible states at time ℓ, with the result:

$$p_{\ell+1}(\boldsymbol{\sigma})=\frac{1}{2N}\,[1+\sigma_i\,g[\beta h_i(\boldsymbol{\sigma})]]\,p_\ell(\boldsymbol{\sigma})$$
$$+\frac{1}{2N}\,[1+\sigma_i\,g[\beta h_i(F_i\boldsymbol{\sigma})]]\,p_\ell(F_i\boldsymbol{\sigma}),$$

with the state-flip operators $F_i\Phi(\boldsymbol{\sigma}) = \Phi(\sigma_1,\ldots,\sigma_{i-1},-\sigma_i,\sigma_{i+1},\ldots,\sigma_N)$. This equation can again be written in the standard form

$$p_{\ell+1}(\boldsymbol{\sigma}) = \sum_{\boldsymbol{\sigma}'} W[\boldsymbol{\sigma};\boldsymbol{\sigma}']\, p_\ell(\boldsymbol{\sigma}'),$$

but now with the transition matrix

$$\text{Sequential Dynamics:}\quad W[\boldsymbol{\sigma};\boldsymbol{\sigma}'] = \delta_{\boldsymbol{\sigma},\boldsymbol{\sigma}'} + \frac{1}{N}\{w_i(F_i\boldsymbol{\sigma})\delta_{\boldsymbol{\sigma},F_i\boldsymbol{\sigma}'} - w_i(\boldsymbol{\sigma})\delta_{\boldsymbol{\sigma},\boldsymbol{\sigma}'}\},\tag{2.125}$$

where $\quad\delta_{\boldsymbol{\sigma},\boldsymbol{\sigma}'} = \prod_i \delta_{\sigma_i,\sigma'_i}\quad$ and $\quad w_i(\boldsymbol{\sigma}) = \frac{1}{2}[1 - \sigma_i \tanh[\beta h_i(\boldsymbol{\sigma})]].$
$$\tag{2.126}$$

Note that, as soon as $T > 0$, the two transition matrices $W[\boldsymbol{\sigma};\boldsymbol{\sigma}']$ in (2.124,2.125) both describe *ergodic systems*: from any initial state $\boldsymbol{\sigma}'$ one can reach any final state $\boldsymbol{\sigma}$ with nonzero probability in a finite number of steps (being one in the parallel case, and N in the sequential case). It now follows from the standard theory of stochastic processes (see e.g., [Kam92, Gar85]) that in both cases the system evolves towards a unique stationary distribution $p_\infty(\boldsymbol{\sigma})$, where all probabilities $p_\infty(\boldsymbol{\sigma})$ are non–zero [Coo01, SC00, SC01, CKS05].

The above processes have the (mathematically and biologically) less appealing property that time is measured in discrete units. For the sequential case we will now assume that the *duration* of each of the iteration steps is a continuous random number (for parallel dynamics this would make little sense, since all updates would still be made in full synchrony). The statistics of the durations are described by a function $\pi_\ell(t)$, defined as the probability that at time t precisely ℓ updates have been made. Upon denoting the previous discrete-time probabilities as $\hat{p}_\ell(\boldsymbol{\sigma})$, our new process (which now includes the randomness in step duration) will be described by

$$p_t(\boldsymbol{\sigma}) = \sum_{\ell \geq 0} \pi_\ell(t)\hat{p}_\ell(\boldsymbol{\sigma}) = \sum_{\ell \geq 0} \pi_\ell(t) \sum_{\boldsymbol{\sigma}'} W^\ell[\boldsymbol{\sigma};\boldsymbol{\sigma}']\, p_0(\boldsymbol{\sigma}'),$$

and time has become a continuous variable. For $\pi_\ell(t)$ we make the Poisson choice,

$$\pi_\ell(t) = \frac{1}{\ell+} \left(\frac{t}{\Delta}\right)^\ell e^{-t/\Delta}.$$

From $\langle \ell \rangle_\pi = t/\Delta$ and $\langle \ell^2 \rangle_\pi = t/\Delta + t^2/\Delta^2$ it follows that Δ is the average duration of an iteration step, and that the relative deviation in ℓ at a given t vanishes for $\Delta \to 0$ as $\sqrt{\langle \ell^2 \rangle_\pi - \langle \ell \rangle_\pi^2}/\langle \ell \rangle_\pi = \sqrt{\Delta/t}$. The nice properties of the Poisson distribution under temporal derivation allow us to derive:

$$\Delta \dot{p}_t(\boldsymbol{\sigma}) = \sum_{\boldsymbol{\sigma}'} W[\boldsymbol{\sigma};\boldsymbol{\sigma}']\, p_t(\boldsymbol{\sigma}') - p_t(\boldsymbol{\sigma}).$$

For sequential dynamics, we choose $\Delta = \frac{1}{N}$ so that, as in the parallel case, in one time unit each neuron will on average be updated once. The master equation corresponding to (2.125) acquires the form $w_i(\boldsymbol{\sigma})$; (2.126) now play the role of *transition rates*. The choice $\Delta = \frac{1}{N}$ implies $\sqrt{\langle \ell^2 \rangle_\pi - \langle \ell \rangle_\pi^2 / \langle \ell \rangle_\pi} = \sqrt{1/Nt}$, so we will still for $N \to \infty$ no longer have uncertainty in where we are on the t axis.

Alternatively, we could start with continuous neuronal variables σ_i (representing e.g., firing frequencies or oscillator phases), where $i = 1, \ldots, N$, and with stochastic equations of the form

$$\sigma_i(t + \Delta) = \sigma_i(t) + \Delta f_i(\boldsymbol{\sigma}(t)) + \sqrt{2T\Delta}\xi_i(t). \qquad (2.127)$$

Here, we have introduced (as yet unspecified) deterministic state–dependent forces $f_i(\boldsymbol{\sigma})$, and uncorrelated Gaussian distributed random forces $\xi_i(t)$ (the noise), with

$$\langle \xi_i(t) \rangle = 0 \quad \text{and} \quad \langle \xi_i(t)\xi_j(t') \rangle = \delta_{ij}\delta_{t,t'}.$$

As before, the parameter T controls the amount of noise in the system, ranging from $T = 0$ (deterministic dynamics) to $T = \infty$ (completely random dynamics). If we take the limit $\Delta \to 0$ in (2.127) we find a Langevin equation (with a continuous time variable) [Coo01, SC00, SC01, CKS05]:

$$\dot{\sigma}_i(t) = f_i(\boldsymbol{\sigma}(t)) + \eta_i(t). \qquad (2.128)$$

This equation acquires its meaning only as the limit $\Delta \to 0$ of (2.127). The moments of the new noise variables $\eta_i(t) = \xi_i(t)\sqrt{2T/\Delta}$ in (2.128) are given by

$$\langle \eta_i(t) \rangle = 0 \quad \text{and} \quad \langle \eta_i(t)\eta_j(t') \rangle = 2T\delta_{ij}\delta(t - t').$$

This can be derived from the moments of the $\xi_i(t)$. For instance:

$$\langle \eta_i(t)\eta_j(t') \rangle = \lim_{\Delta \to 0} \frac{2T}{\Delta} \langle \xi_i(t)\xi_j(t') \rangle$$

$$= 2T\delta_{ij} \lim_{\Delta \to 0} \frac{1}{\Delta}\delta_{t,t'} = 2TC\delta_{ij}\delta(t - t').$$

The constant C is found by summing over t', before taking the limit $\Delta \to 0$, in the above equation:

$$\int dt' \, \langle \eta_i(t)\eta_j(t') \rangle = \lim_{\Delta \to 0} 2T \sum_{t'=-\infty}^{\infty} \langle \xi_i(t)\xi_j(t') \rangle$$

$$= 2T\delta_{ij} \lim_{\Delta \to 0} \sum_{t'=-\infty}^{\infty} \delta_{t,t'} = 2T\delta_{ij}.$$

Thus $C = 1$, which indeed implies $\langle \eta_i(t)\eta_j(t') \rangle = 2T\delta_{ij}\delta(t-t')$. More directly, one can also calculate the moment partition function:

$$\langle e^{i\int dt\,\psi_i(t)\eta_i(t)}\rangle = \lim_{\Delta\to 0}\prod_{i,t}\int\frac{dz}{\sqrt{2\pi}}\,e^{-\frac{1}{2}z^2+iz\psi_i(t)\sqrt{2T\Delta}} \quad (2.129)$$

$$= \lim_{\Delta\to 0}\prod_{i,t} e^{-T\Delta\psi_i^2(t)} = e^{-T\int dt\,\sum_i \psi_i^2(t)}. \quad (2.130)$$

On the other hand, a mathematically more convenient description of the process (2.128) is provided by the Fokker–Planck equation for the microscopic state probability density $p_t(\boldsymbol{\sigma}) = \langle\delta[\boldsymbol{\sigma}-\boldsymbol{\sigma}(t)]\rangle$, which we will now derive. For the discrete–time process (2.127) we expand the δ–distribution in the definition of $p_{t+\Delta}(\boldsymbol{\sigma})$ (in a distributional sense) [Coo01, SC00, SC01, CKS05]:

$$p_{t+\Delta}(\boldsymbol{\sigma}) - p_t(\boldsymbol{\sigma}) = \langle\delta\big[\boldsymbol{\sigma}-\boldsymbol{\sigma}(t)-\Delta\mathbf{f}(\boldsymbol{\sigma}(t))-\sqrt{2T\Delta}\boldsymbol{\xi}(t)\big]\rangle - \langle\delta[\boldsymbol{\sigma}-\boldsymbol{\sigma}(t)]\rangle$$

$$= -\frac{\partial}{\partial\sigma_i}\langle\delta[\boldsymbol{\sigma}-\boldsymbol{\sigma}(t)]\big[\Delta f_i(\boldsymbol{\sigma}(t))+\sqrt{2T\Delta}\xi_i(t)\big]\rangle$$

$$+ T\Delta\frac{\partial^2}{\partial\sigma_i\partial\sigma_j}\langle\delta[\boldsymbol{\sigma}-\boldsymbol{\sigma}(t)]\xi_i(t)\xi_j(t)\rangle + \mathcal{O}(\Delta^{\frac{3}{2}}).$$

The variables $\boldsymbol{\sigma}(t)$ depend only on noise variables $\xi_j(t')$ with $t' < t$, so that for any function A,

$$\langle A[\boldsymbol{\sigma}(t)]\xi_i(t)\rangle = \langle A[\boldsymbol{\sigma}(t)]\rangle\langle\xi_i(t)\rangle = 0, \quad \text{and}$$
$$\langle A[\boldsymbol{\sigma}(t)]\xi_i(t)\xi_j(t)\rangle = \delta_{ij}\langle A[\boldsymbol{\sigma}(t)]\rangle.$$

As a consequence, we have:

$$\frac{1}{\Delta}[p_{t+\Delta}(\boldsymbol{\sigma})-p_t(\boldsymbol{\sigma})] = -\frac{\partial}{\partial\sigma_i}\langle\delta[\boldsymbol{\sigma}-\boldsymbol{\sigma}(t)]f_i(\boldsymbol{\sigma}(t))\rangle$$

$$+ T\frac{\partial^2}{\partial\sigma_i^2}\langle\delta[\boldsymbol{\sigma}-\boldsymbol{\sigma}(t)]\rangle + \mathcal{O}(\Delta^{\frac{1}{2}})$$

$$= -\frac{\partial}{\partial\sigma_i}[p_t(\boldsymbol{\sigma})f_i(\boldsymbol{\sigma})] + T\frac{\partial^2}{\partial\sigma_i^2}p_t(\boldsymbol{\sigma}) + \mathcal{O}(\Delta^{\frac{1}{2}}).$$

By taking the limit $\Delta \to 0$ we then arrive at the Fokker–Planck equation:

$$\dot{p}_t(\boldsymbol{\sigma}) = -\frac{\partial}{\partial\sigma_i}[p_t(\boldsymbol{\sigma})f_i(\boldsymbol{\sigma})] + T\frac{\partial^2}{\partial\sigma_i^2}p_t(\boldsymbol{\sigma}). \quad (2.131)$$

In the case of graded–response neurons, the continuous variable σ_i represents the membrane potential of neuron i, and (in their simplest form) the deterministic forces are given by

$$f_i(\boldsymbol{\sigma}) = J_{ij}\tanh[\gamma\sigma_j] - \sigma_i + \theta_i, \quad \text{with} \quad \gamma > 0,$$

and with the θ_i representing injected currents. Conventional notation is restored by putting $\sigma_i \to u_i$. Thus equation (2.128) specializes to

2.3 Synergetics of Recurrent and Attractor Neural Networks

$$\dot{u}_i(t) = J_{ij}\tanh[\gamma u_j(t)] - u_i(t) + \theta_i + \eta_i(t). \tag{2.132}$$

One often chooses $T = 0$ (i.e., $\eta_i(t) = 0$), the rationale being that threshold noise is already assumed to have been incorporated via the nonlinearity in (2.132).

In our second example the variables σ_i represent the phases of coupled neural oscillators, with forces of the form

$$f_i(\boldsymbol{\sigma}) = J_{ij}\sin(\sigma_j - \sigma_i) + w_i.$$

Individual synapses J_{ij} now try to enforce either pair–wise synchronization ($J_{ij} > 0$), or pair–wise anti-synchronization ($J_{ij} < 0$), and the w_i represent the natural frequencies of the individual oscillators. Conventional notation dictates $\sigma_i \to \xi_i$, giving [Coo01, SC00, SC01, CKS05]

$$\dot{\xi}_i(t) = w_i + J_{ij}\sin[\xi_j(t) - \xi_i(t)] + \eta_i(t). \tag{2.133}$$

2.3.2 Synaptic Symmetry and Lyapunov Functions

In the deterministic limit $T \to 0$ the rules (2.121) for networks of synchronously evolving binary neurons reduce to the deterministic map

$$\sigma_i(\ell + 1) = \text{sgn}\left[h_i(\boldsymbol{\sigma}(\ell))\right]. \tag{2.134}$$

It turns out that for systems with symmetric interactions, $J_{ij} = J_{ji}$ for all (ij), one can construct a Lyapunov function, i.e., a function of $\boldsymbol{\sigma}$ which during the dynamics decreases monotonically and is bounded from below (see e.g., [Kha92]):

Binary & Parallel Dynamics: $\quad L[\boldsymbol{\sigma}] = -\sum_i |h_i(\boldsymbol{\sigma})| - \sigma_i\theta_i. \tag{2.135}$

Clearly, $L \geq -\sum_i[\sum_j |J_{ij}| + |\theta_i|] - \sum_i |\theta_i|$. During iteration of (2.134) we find:

$$L[\boldsymbol{\sigma}(\ell+1)] - L[\boldsymbol{\sigma}(\ell)] = -\sum_i |h_i(\boldsymbol{\sigma}(\ell+1))|$$
$$+ \sigma_i(\ell+1)[J_{ij}\sigma_j(\ell) + \theta_i] - \theta_i[\sigma_i(\ell+1) - \sigma_i(\ell)]$$
$$= -\sum_i |h_i(\boldsymbol{\sigma}(\ell+1))| + \sigma_i(\ell)h_i(\boldsymbol{\sigma}(\ell+1))$$
$$= -\sum_i |h_i(\boldsymbol{\sigma}(\ell+1))|\left[1 - \sigma_i(\ell+2)\sigma_i(\ell)\right] \leq 0,$$

where we have used (2.134) and $J_{ij} = J_{ji}$. So L decreases monotonically until a stage is reached where $\sigma_i(\ell+2) = \sigma_i(\ell)$ for all i. Thus, with symmetric interactions this system will in the deterministic limit always end up in a limit cycle with period ≤ 2. A similar result is found for networks with binary neurons and sequential dynamics. In the limit $T \to 0$ the rules (2.122) reduce to the map

$$\sigma_i(\ell+1) = \delta_{i,i_\ell} \,\mathrm{sgn}\,[h_i(\boldsymbol{\sigma}(\ell))] + [1 - \delta_{i,i_\ell}]\sigma_i(\ell). \tag{2.136}$$

(in which we still have randomness in the choice of site to be updated). For systems with symmetric interactions and without self–interactions, i.e., $J_{ii} = 0$ for all i, we again find a Lyapunov function:

Binary & Sequential Dynamics: $\quad L[\boldsymbol{\sigma}] = -\frac{1}{2}\sigma_i J_{ij}\sigma_j - \sigma_i\theta_i. \tag{2.137}$

This quantity is bounded from below, $L \geq -\frac{1}{2}\sum_{ij}|J_{ij}| - \sum_i |\theta_i|$. Upon calling the site i_ℓ selected for update at step ℓ simply i, the change in L during iteration of (2.136) can be written as [Coo01, SC00, SC01, CKS05]:

$$\begin{aligned} L[\boldsymbol{\sigma}(\ell+1)] - L[\boldsymbol{\sigma}(\ell)] &= -\theta_i[\sigma_i(\ell+1) - \sigma_i(\ell)] \\ &\quad - \frac{1}{2}J_{ik}[\sigma_i(\ell+1)\sigma_k(\ell+1) - \sigma_i(\ell)\sigma_k(\ell)] \\ &\quad - \frac{1}{2}J_{ji}[\sigma_j(\ell+1)\sigma_i(\ell+1) - \sigma_j(\ell)\sigma_i(\ell)] \\ &= [\sigma_i(\ell) - \sigma_i(\ell+1)][J_{ij}\sigma_j(\ell) + \theta_i] \\ &= -|h_i(\boldsymbol{\sigma}(\ell))|\,[1 - \sigma_i(\ell)\sigma_i(\ell+1)] \;\leq\; 0. \end{aligned}$$

Here we used (2.136), $J_{ij} = J_{ji}$, and absence of self–interactions. Thus L decreases monotonically until $\sigma_i(t+1) = \sigma_i(t)$ for all i. With symmetric synapses, but without diagonal terms, the sequentially evolving binary neurons system will in the deterministic limit always end up in a stationary state.

Now, one can derive similar results for models with continuous variables. Firstly, in the deterministic limit the graded–response equations (2.132) simplify to

$$\dot{u}_i(t) = J_{ij}\tanh[\gamma u_j(t)] - u_i(t) + \theta_i. \tag{2.138}$$

Symmetric networks again admit a Lyapunov function (without a need to eliminate self-interactions):

Graded–Response Dynamics : $\quad L[\mathbf{u}] = -\frac{1}{2}J_{ij}\tanh(\gamma u_i)\tanh(\gamma u_j) +$

$$\sum_i \left[\gamma \int_0^{u_i} dv\, v[1 - \tanh^2(\gamma v)] - \theta_i \tanh(\gamma u_i)\right].$$

Clearly, $L \geq -\frac{1}{2}\sum_{ij}|J_{ij}| - \sum_i |\theta_i|$; the term in $L[\mathbf{u}]$ with the integral is non–negative. During the noise–free dynamics (2.138) one can use the identity

$$\frac{\partial L}{\partial u_i} = -\gamma \dot{u}_i[1 - \tanh^2(\gamma u_i)],$$

valid only when $J_{ij} = J_{ji}$, to derive [Coo01, SC00, SC01, CKS05]

$$\dot{L} = \frac{\partial L}{\partial u_i}\dot{u}_i = -\gamma \sum_i [1 - \tanh^2(\gamma u_i)]\,\dot{u}_i^2 \leq 0.$$

Again L is found to decrease monotonically, until $\dot{u}_i = 0$ for all i, i.e., until we are at a fixed–point.

The coupled oscillator equations (2.133) reduce in the *noise–free limit* to

$$\dot{\xi}_i(t) = \omega_i + J_{ij}\sin[\xi_j(t) - \xi_i(t)]. \qquad (2.139)$$

Note that self-interactions J_{ii} always drop out automatically. For symmetric oscillator networks, a construction of the type followed for the graded–response equations would lead us to propose

Coupled Oscillators Dynamics: $\qquad L[\boldsymbol{\xi}] = -\frac{1}{2}J_{ij}\cos[\xi_i - \xi_j] - \omega_i\xi_i. \quad (2.140)$

This function decreases monotonically, due to $\partial L/\partial \xi_i = -\dot{\xi}_i$:

$$\dot{L} = \frac{\partial L}{\partial \xi_i}\dot{\xi}_i = -\sum_i \dot{\xi}_i^2 \leq 0.$$

Actually, (2.139) describes gradient descent on the surface $L[\boldsymbol{\xi}]$. However, due to the term with the natural frequencies ω_i the function $L[\boldsymbol{\xi}]$ is not bounded, so it cannot be a Lyapunov function. This could have been expected; when $J_{ij} = 0$ for all (i,j), for instance, one finds continually increasing phases, $\xi_i(t) = \xi_i(0) + \omega_i t$. Removing the ω_i, in contrast, gives the bound $L \geq -\sum_j |J_{ij}|$. Now the system must go to a fixed-point. In the special case $\omega_i = \omega$ (N identical natural frequencies) we can transform away the ω_i by putting $\xi(t) = \tilde{\xi}_i(t) + \omega t$, and find the relative phases $\tilde{\xi}_i$ to go to a fixed-point.

2.3.3 Detailed Balance and Equilibrium Statistical Mechanics

The results got above indicate that networks with symmetric synapses are a special class [Coo01, SC00, SC01, CKS05]. We now show how synaptic symmetry is closely related to the detailed balance property, and derive a number of consequences. An ergodic Markov chain of the form (2.124,2.125), i.e.,

$$p_{\ell+1}(\boldsymbol{\sigma}) = \sum_{\boldsymbol{\sigma}'} W[\boldsymbol{\sigma};\boldsymbol{\sigma}']\,p_\ell(\boldsymbol{\sigma}'), \qquad (2.141)$$

is said to obey detailed balance if its (unique) stationary solution $p_\infty(\boldsymbol{\sigma})$ has the property

$$W[\boldsymbol{\sigma};\boldsymbol{\sigma}']\,p_\infty(\boldsymbol{\sigma}') = W[\boldsymbol{\sigma}';\boldsymbol{\sigma}]\,p_\infty(\boldsymbol{\sigma}), \qquad \text{(for all } \boldsymbol{\sigma},\boldsymbol{\sigma}'\text{)}. \qquad (2.142)$$

All $p_\infty(\boldsymbol{\sigma})$ which satisfy (2.142) are stationary solutions of (2.141), this is easily verified by substitution. The converse is not true. Detailed balance states that, in addition to $p_\infty(\boldsymbol{\sigma})$ being stationary, one has *equilibrium*: there is no net probability current between any two microscopic system states.

It is not a trivial matter to investigate systematically for which choices of the threshold noise distribution $w(\eta)$ and the synaptic matrix $\{J_{ij}\}$ detailed balance holds. It can be shown that, apart from trivial cases (e.g., systems with self–interactions only) a Gaussian distribution $w(\eta)$ will not support detailed balance. Here we will work out details only for the choice $w(\eta) = \frac{1}{2}[1 - \tanh^2(\eta)]$, and for $T > 0$ (where both discrete systems are ergodic). For parallel dynamics the transition matrix is given in (2.124), now with $g[z] = \tanh[z]$, and the detailed balance condition (2.142) becomes

$$\frac{e^{\beta \sigma_i h_i(\boldsymbol{\sigma}')} p_\infty(\boldsymbol{\sigma}')}{\prod_i \cosh[\beta h_i(\boldsymbol{\sigma}')]} = \frac{e^{\beta \sigma_i' h_i(\boldsymbol{\sigma})} p_\infty(\boldsymbol{\sigma})}{\prod_i \cosh[\beta h_i(\boldsymbol{\sigma})]}, \qquad \text{(for all } \boldsymbol{\sigma}, \boldsymbol{\sigma}'\text{)}. \qquad (2.143)$$

All $p_\infty(\boldsymbol{\sigma})$ are non-zero (ergodicity), so we may safely put

$$p_\infty(\boldsymbol{\sigma}) = e^{\beta[\theta_i \sigma_i + K(\boldsymbol{\sigma})]} \prod_i \cosh[\beta h_i(\boldsymbol{\sigma})],$$

which, in combination with definition (2.121), simplifies the detailed balance condition to:

$$K(\boldsymbol{\sigma}) - K(\boldsymbol{\sigma}') = \sigma_i \left[J_{ij} - J_{ji}\right] \sigma_j', \qquad \text{(for all } \boldsymbol{\sigma}, \boldsymbol{\sigma}'\text{)}. \qquad (2.144)$$

Averaging (2.144) over all possible $\boldsymbol{\sigma}'$ gives $K(\boldsymbol{\sigma}) = \langle K(\boldsymbol{\sigma}') \rangle_{\boldsymbol{\sigma}'}$ for all $\boldsymbol{\sigma}$, i.e., K is a constant, whose value follows from normalizing $p_\infty(\boldsymbol{\sigma})$. So, if detailed balance holds the equilibrium distribution must be [Coo01, SC00, SC01, CKS05]:

$$p_{\text{eq}}(\boldsymbol{\sigma}) \sim e^{\beta \theta_i \sigma_i} \prod_i \cosh[\beta h_i(\boldsymbol{\sigma})]. \qquad (2.145)$$

For symmetric systems detailed balance indeed holds: (2.145) solves (2.143), since $K(\boldsymbol{\sigma}) = K$ solves the reduced problem (2.144). For non-symmetric systems, however, there can be no equilibrium. For $K(\boldsymbol{\sigma}) = K$ the condition (2.144) becomes $\sum_{ij} \sigma_i [J_{ij} - J_{ji}] \sigma_j' = 0$ for all $\boldsymbol{\sigma}, \boldsymbol{\sigma}' \in \{-1,1\}^N$. For $N \geq 2$ the vector pairs $(\boldsymbol{\sigma}, \boldsymbol{\sigma}')$ span the space of all $N \times N$ matrices, so $J_{ij} - J_{ji}$ must be zero. For $N = 1$ there simply exists no non-symmetric synaptic matrix. In conclusion: for binary networks with parallel dynamics, interaction symmetry implies detailed balance, and vice versa.

For sequential dynamics, with $w(\eta) = \frac{1}{2}[1 - \tanh^2(\eta)]$, the transition matrix is given by (2.125) and the detailed balance condition (2.142) simplifies to

$$\frac{e^{\beta \sigma_i h_i(F_i \boldsymbol{\sigma})} p_\infty(F_i \boldsymbol{\sigma})}{\cosh[\beta h_i(F_i \boldsymbol{\sigma})]} = \frac{e^{-\beta \sigma_i h_i(\boldsymbol{\sigma})} p_\infty(\boldsymbol{\sigma})}{\cosh[\beta h_i(\boldsymbol{\sigma})]}, \qquad \text{(for all } \boldsymbol{\sigma}, i\text{)}.$$

2.3 Synergetics of Recurrent and Attractor Neural Networks

Self–interactions J_{ii}, inducing $h_i(F_i\boldsymbol{\sigma}) \neq h_i(\boldsymbol{\sigma})$, complicate matters. Therefore we first consider systems where all $J_{ii} = 0$. All stationary probabilities $p_\infty(\boldsymbol{\sigma})$ being non–zero (ergodicity), we may write:

$$p_\infty(\boldsymbol{\sigma}) = e^{\beta[\theta_i\sigma_i + \frac{1}{2}\sigma_i J_{ij}\sigma_j + K(\boldsymbol{\sigma})]}. \tag{2.146}$$

Using relations like

$$J_{kl} F_i(\sigma_k \sigma_l) = J_{kl}\sigma_k\sigma_l - 2\sigma_i [J_{ik} + J_{ki}]\sigma_k,$$

we can simplify the detailed balance condition to

$$K(F_i\boldsymbol{\sigma}) - K(\boldsymbol{\sigma}) = \sigma_i [J_{ik} - J_{ki}]\sigma_k, \quad \text{(for all } \boldsymbol{\sigma}, i\text{)}.$$

If to this expression we apply the general identity

$$[1 - F_i] f(\boldsymbol{\sigma}) = 2\sigma_i \langle \sigma_i f(\boldsymbol{\sigma}) \rangle_{\sigma_i},$$

we find for $i \neq j$ [Coo01, SC00, SC01, CKS05]:

$$K(\boldsymbol{\sigma}) = -2\sigma_i\sigma_j [J_{ij} - J_{ji}], \quad \text{(for all } \boldsymbol{\sigma} \text{ and all } i \neq j\text{)}.$$

The left–hand side is symmetric under permutation of the pair (i, j), which implies that the interaction matrix must also be symmetric: $J_{ij} = J_{ji}$ for all (i, j). We now find the trivial solution $K(\boldsymbol{\sigma}) = K$ (constant), detailed balance holds and the corresponding equilibrium distribution is

$$p_{\text{eq}}(\boldsymbol{\sigma}) \sim e^{-\beta H(\boldsymbol{\sigma})}, \quad H(\boldsymbol{\sigma}) = -\frac{1}{2}\sigma_i J_{ij}\sigma_j - \theta_i\sigma_i.$$

In conclusion: for binary networks with sequential dynamics, but without self–interactions, interaction symmetry implies detailed balance, and vice versa. In the case of self–interactions the situation is more complicated. However, here one can still show that non-symmetric models with detailed balance must be pathological, since the requirements can be met only for very specific choices for the $\{J_{ij}\}$.

Now, let us turn to the question of when we find microscopic equilibrium (stationarity without probability currents) in continuous models described by a Fokker–Planck equation (2.131). Note that (2.131) can be seen as a continuity equation for the density of a conserved quantity:

$$\dot{p}_t(\boldsymbol{\sigma}) + \frac{\partial}{\partial \sigma_i} J_i(\boldsymbol{\sigma}, t) = 0.$$

The components $J_i(\boldsymbol{\sigma}, t)$ of the current density are given by

$$J_i(\boldsymbol{\sigma}, t) = [f_i(\boldsymbol{\sigma}) - T\frac{\partial}{\partial \sigma_i}] p_t(\boldsymbol{\sigma}).$$

Stationary distributions $p_\infty(\boldsymbol{\sigma})$ are those which give $\sum_i \frac{\partial}{\partial \sigma_i} J_i(\boldsymbol{\sigma}, \infty) = 0$ (divergence-free currents). Detailed balance implies the stronger statement $J_i(\boldsymbol{\sigma}, \infty) = 0$ for all i (zero currents), so

$$f_i(\boldsymbol{\sigma}) = T \frac{\partial \log p_\infty(\boldsymbol{\sigma})}{\partial \sigma_i}, \qquad \text{or}$$

$$f_i(\boldsymbol{\sigma}) = -\frac{\partial H(\boldsymbol{\sigma})}{\partial \sigma_i}, \qquad p_\infty(\boldsymbol{\sigma}) \sim e^{-\beta H(\boldsymbol{\sigma})}, \qquad (2.147)$$

for some $H(\boldsymbol{\sigma})$, i.e., the forces $f_i(\boldsymbol{\sigma})$ must be conservative. However, one can have conservative forces without a normalizable equilibrium distribution. Just take $H(\boldsymbol{\sigma}) = 0$, i.e., $f_i(\boldsymbol{\sigma}, t) = 0$: here we have $p_{\text{eq}}(\boldsymbol{\sigma}) = C$, which is not normalizable for $\boldsymbol{\sigma} \in \mathbb{R}^N$. For this particular case equation (2.131) is solved easily:

$$p_t(\boldsymbol{\sigma}) = [4\pi T t]^{-N/2} \int d\boldsymbol{\sigma}'\, p_0(\boldsymbol{\sigma}') e^{-[\boldsymbol{\sigma}-\boldsymbol{\sigma}']^2/4Tt},$$

so the limit $\lim_{t \to \infty} p_t(\boldsymbol{\sigma})$ does not exist. One can prove the following (see e.g., [Zin93]). If the forces are conservative and if $p_\infty(\boldsymbol{\sigma}) \sim e^{-\beta H(\boldsymbol{\sigma})}$ is normalizable, then it is the unique stationary solution of the Fokker–Planck equation, to which the system converges for all initial distributions $p_0 \in L^1[\mathbb{R}^N]$ which obey $\int_{\mathbb{R}^N} d\boldsymbol{\sigma}\, e^{\beta H(\boldsymbol{\sigma})} p_0^2(\boldsymbol{\sigma}) < \infty$.

Note that conservative forces must obey [Coo01, SC00, SC01, CKS05]

$$\frac{\partial f_i(\boldsymbol{\sigma})}{\partial \sigma_j} - \frac{\partial f_j(\boldsymbol{\sigma})}{\partial \sigma_i} = 0, \qquad (\text{for all } \boldsymbol{\sigma} \text{ and all } i \neq j). \qquad (2.148)$$

In the graded–response equations (2.138) the deterministic forces are

$$f_i(\mathbf{u}) = J_{ij} \tanh[\gamma u_j] - u_i + \theta_i, \qquad \text{where}$$

$$\frac{\partial f_i(\mathbf{u})}{\partial u_j} - \frac{\partial f_j(\mathbf{u})}{\partial u_i} = \gamma\{J_{ij}[1 - \tanh^2[\gamma u_j] - J_{ji}[1 - \tanh^2[\gamma u_i]\}.$$

At $\mathbf{u} = \mathbf{0}$ this reduces to $J_{ij} - J_{ji}$, i.e., the interaction matrix must be symmetric. For symmetric matrices we find away from $\mathbf{u} = \mathbf{0}$:

$$\frac{\partial f_i(\mathbf{u})}{\partial u_j} - \frac{\partial f_j(\mathbf{u})}{\partial u_i} = \gamma J_{ij}\{\tanh^2[\gamma u_i] - \tanh^2[\gamma u_j]\}.$$

The only way for this to be zero for any $\mathbf{u}$ is by having $J_{ij} = 0$ for all $i \neq j$, i.e., all neurons are disconnected (in this trivial case the system (2.138) does indeed obey detailed balance). Network models of interacting graded–response neurons of the type (2.138) apparently never reach equilibrium, they will always violate detailed balance and exhibit microscopic probability currents. In the case of coupled oscillators (2.133), where the deterministic forces are

$$f_i(\boldsymbol{\xi}) = J_{ij} \sin[\xi_j - \xi_i] + \omega_i,$$

one finds the left-hand side of condition (2.148) to give

$$\frac{\partial f_i(\boldsymbol{\xi})}{\partial \xi_j} - \frac{\partial f_j(\boldsymbol{\xi})}{\partial \xi_i} = [J_{ij} - J_{ji}] \cos[\xi_j - \xi_i].$$

Requiring this to be zero for any $\boldsymbol{\xi}$ gives the condition $J_{ij} = J_{ji}$ for any $i \neq j$. We have already seen that symmetric oscillator networks indeed have conservative forces:

$$f_i(\boldsymbol{\xi}) = -\partial H(\boldsymbol{\xi})/\partial \xi_i, \qquad \text{with} \qquad H(\boldsymbol{\xi}) = -\frac{1}{2} J_{ij} \cos[\xi_i - \xi_j] - \omega_i \xi_i.$$

If in addition we choose all $\omega_i = 0$ the function $H(\boldsymbol{\sigma})$ will also be bounded from below, and, although $p_\infty(\boldsymbol{\xi}) \sim e^{-\beta H(\boldsymbol{\xi})}$ is still not normalizable on $\boldsymbol{\xi} \in \mathbb{R}^N$, the full 2π–periodicity of the function $H(\boldsymbol{\sigma})$ now allows us to identify $\xi_i + 2\pi \equiv \xi_i$ for all i, so that now $\boldsymbol{\xi} \in [-\pi, \pi]^N$ and $\int d\boldsymbol{\xi}\, e^{-\beta H(\boldsymbol{\xi})}$ does exist. Thus symmetric coupled oscillator networks with zero natural frequencies obey detailed balance. In the case of non-zero natural frequencies, in contrast, detailed balance does not hold.

The above results establish the link with *equilibrium statistical mechanics* (see e.g., [Yeo92, PB94]). For binary systems with symmetric synapses (in the sequential case: without self-interactions) and with threshold noise distributions of the form

$$w(\eta) = \frac{1}{2}[1 - \tanh^2(\eta)],$$

detailed balance holds and we know the equilibrium distributions. For sequential dynamics it has the Boltzmann form (2.147) and we can apply standard equilibrium statistical mechanics. The parameter β can formally be identified with the inverse 'temperature' in equilibrium, $\beta = T^{-1}$, and the function $H(\boldsymbol{\sigma})$ is the usual *Ising–spin Hamiltonian*. In particular we can define the *partition function* Z and the *free energy* F [Coo01, SC00, SC01, CKS05]:

$$p_{\text{eq}}(\boldsymbol{\sigma}) = \frac{1}{Z} e^{-\beta H(\boldsymbol{\sigma})}, \qquad H(\boldsymbol{\sigma}) = -\frac{1}{2} \sigma_i J_{ij} \sigma_j - \theta_i \sigma_i, \qquad (2.149)$$

$$Z = \sum_{\boldsymbol{\sigma}} e^{-\beta H(\boldsymbol{\sigma})}, \qquad F = -\beta^{-1} \log Z. \qquad (2.150)$$

The free energy can be used as the partition function for equilibrium averages. Taking derivatives with respect to external fields θ_i and interactions J_{ij}, for instance, produces $\langle \sigma_i \rangle = -\partial F/\partial \theta_i$ and $\langle \sigma_i \sigma_j \rangle = -\partial F/\partial J_{ij}$, whereas equilibrium averages of arbitrary state variable $f(\boldsymbol{\sigma})$ can be obtained by adding suitable partition terms to the Hamiltonian:

$$H(\boldsymbol{\sigma}) \rightarrow H(\boldsymbol{\sigma}) + \lambda f(\boldsymbol{\sigma}), \qquad \langle f \rangle = \lim_{\lambda \to 0} \frac{\partial F}{\partial \lambda}.$$

In the parallel case (2.145) we can again formally write the equilibrium probability distribution in the Boltzmann form [Per84] and define a corresponding partition function $\tilde{Z}$ and a free energy $\tilde{F}$:

$$p_{\text{eq}}(\boldsymbol{\sigma}) = \frac{1}{\tilde{Z}} e^{-\beta \tilde{H}(\boldsymbol{\sigma})}, \qquad \tilde{H}(\boldsymbol{\sigma}) = -\theta_i \sigma_i - \frac{1}{\beta} \sum_i \log 2 \cosh[\beta h_i(\boldsymbol{\sigma})], \quad (2.151)$$

$$\tilde{Z} = \sum_{\boldsymbol{\sigma}} e^{-\beta \tilde{H}(\boldsymbol{\sigma})}, \qquad \tilde{F} = -\beta^{-1} \log \tilde{Z}, \quad (2.152)$$

which again serve to generate averages: $\tilde{H}(\boldsymbol{\sigma}) \to \tilde{H}(\boldsymbol{\sigma}) + \lambda f(\boldsymbol{\sigma})$, $\langle f \rangle = \lim_{\lambda \to 0} \partial \tilde{F}/\partial \lambda$. However, standard thermodynamic relations involving derivation with respect to β need no longer be valid, and derivation with respect to fields or interactions generates different types of averages, such as [Coo01, SC00, SC01, CKS05]

$$-\frac{\partial \tilde{F}}{\partial \theta_i} = \langle \sigma_i \rangle + \langle \tanh[\beta h_i(\boldsymbol{\sigma})] \rangle, \qquad -\frac{\partial \tilde{F}}{\partial J_{ii}} = \langle \sigma_i \tanh[\beta h_i(\boldsymbol{\sigma})] \rangle,$$

$$i \neq j: \quad \frac{\partial \tilde{F}}{\partial J_{ij}} = \langle \sigma_i \tanh[\beta h_j(\boldsymbol{\sigma})] \rangle + \langle \sigma_j \tanh[\beta h_i(\boldsymbol{\sigma})] \rangle.$$

One can use $\langle \sigma_i \rangle = \langle \tanh[\beta h_i(\boldsymbol{\sigma})] \rangle$, which can be derived directly from the equilibrium equation $p_{\text{eq}}(\boldsymbol{\sigma}) = \sum_{\boldsymbol{\sigma}'} W[\boldsymbol{\sigma}; \boldsymbol{\sigma}'] p_{\text{eq}}(\boldsymbol{\sigma})$, to simplify the first of these identities.

A connected network of graded–response neurons can never be in an equilibrium state, so our only model example with continuous neuronal variables for which we can set up the equilibrium statistical mechanics formalism is the system of coupled oscillators (2.133) with symmetric synapses and absent (or uniform) natural frequencies ω_i. If we define the phases as $\xi_i \in [-\pi, \pi]$ we have again an equilibrium distribution of the Boltzmann form, and we can define the standard thermodynamic quantities:

$$p_{\text{eq}}(\boldsymbol{\xi}) = \frac{1}{Z} e^{-\beta H(\boldsymbol{\xi})}, \qquad H(\boldsymbol{\xi}) = -\frac{1}{2} J_{ij} \cos[\xi_i - \xi_j], \quad (2.153)$$

$$Z = \int_{-\pi}^{\pi} \cdots \int_{-\pi}^{\pi} d\boldsymbol{\xi} \, e^{-\beta H(\boldsymbol{\xi})}, \qquad F = -\beta^{-1} \log Z. \quad (2.154)$$

These generate equilibrium averages in the usual manner. For instance

$$\langle \cos[\xi_i - \xi_j] \rangle = -\frac{\partial F}{\partial J_{ij}},$$

whereas averages of arbitrary state variables $f(\boldsymbol{\xi})$ follow, as before, upon introducing suitable partition terms:

$$H(\boldsymbol{\xi}) \to H(\boldsymbol{\xi}) + \lambda f(\boldsymbol{\xi}), \qquad \langle f \rangle = \lim_{\lambda \to 0} \frac{\partial F}{\partial \lambda}.$$

2.3.4 Simple Recurrent Networks with Binary Neurons

Networks with Uniform Synapses

We now turn to a simple toy model to show how equilibrium statistical mechanics is used for solving neural network models, and to illustrate similarities and differences between the different dynamics types [Coo01, SC00, SC01, CKS05]. We choose uniform infinite-range synapses and zero external fields, and calculate the free energy for the binary systems (2.121,2.122), parallel and sequential, and with threshold–noise distribution $w(\eta) = \frac{1}{2}[1 - \tanh^2(\eta)]$:

$$J_{ij} = J_{ji} = J/N, \qquad (i \neq j), \qquad J_{ii} = \theta_i = 0, \qquad \text{(for all } i\text{)}.$$

The free energy is an extensive object, $\lim_{N \to \infty} F/N$ is finite. For the models (2.121,2.122) we now get:

Binary & Sequential Dynamics:

$$\lim_{N \to \infty} F/N = -\lim_{N \to \infty} (\beta N)^{-1} \log \sum_\sigma e^{\beta N[\frac{1}{2}Jm^2(\sigma)]},$$

Binary & Parallel Dynamics:

$$\lim_{N \to \infty} \tilde{F}/N = -\lim_{N \to \infty} (\beta N)^{-1} \log \sum_\sigma e^{N[\log 2 \cosh[\beta Jm(\sigma)]]},$$

with the average activity $m(\sigma) = \frac{1}{N} \sum_k \sigma_k$. We have to count the number of states σ with a prescribed average activity $m = 2n/N - 1$ (n is the number of neurons i with $\sigma_i = 1$), in expressions of the form

$$\frac{1}{N} \log \sum_\sigma e^{NU[m(\sigma)]} = \frac{1}{N} \log \sum_{n=0}^N \binom{N}{n} e^{NU[2n/N-1]}$$

$$= \frac{1}{N} \log \int_{-1}^1 dm \, e^{N[\log 2 - c^*(m) + U[m]]},$$

$$\lim_{N \to \infty} \frac{1}{N} \log \sum_\sigma e^{NU[m(\sigma)]} = \log 2 + \max_{m \in [-1,1]} \{U[m] - c^*(m)\},$$

with the *entropic function*

$$c^*(m) = \frac{1}{2}(1+m)\log(1+m) + \frac{1}{2}(1-m)\log(1-m).$$

In order to get there we used Stirling's formula to get the leading term of the factorials (only terms which are exponential in N survive the limit $N \to \infty$), we converted (for $N \to \infty$) the summation over n into an integration over $m = 2n/N - 1 \in [-1, 1]$, and we carried out the integral over m via *saddle–point integration* (see e.g., [Per92]). This leads to a saddle–point problem whose solution gives the free energies [Coo01, SC00, SC01, CKS05]:

$$\lim_{N\to\infty} F/N = \min_{m\in[-1,1]} f_{\text{seq}}(m), \qquad \beta f_{\text{seq}}(m) = c^*(m) - \log 2 - \frac{1}{2}\beta J m^2.$$
(2.155)

$$\lim_{N\to\infty} \tilde{F}/N = \min_{m\in[-1,1]} f_{\text{par}}(m), \qquad \beta f_{\text{par}}(m) = c^*(m) - 2\log 2 - \log\cosh[\beta J m].$$
(2.156)

The equations from which to solve the minima are easily got by differentiation, using $\frac{d}{dm}c^*(m) = \tanh^{-1}(m)$. For sequential dynamics we find

Binary & Sequential Dynamics: $\qquad m = \tanh[\beta J m],$ (2.157)

which is the so-called *Curie–Weiss law*. For parallel dynamics we find

$$m = \tanh\left[\beta J \tanh[\beta J m]\right].$$

One finds that the solutions of the latter equation again obey a Curie–Weiss law. The definition $\hat{m} = \tanh[\beta|J|m]$ transforms it into the coupled equations $m = \tanh[\beta|J|\hat{m}]$ and $\hat{m} = \tanh[\beta|J|m]$, from which we derive

$$0 \leq [m - \hat{m}]^2 = [m - \hat{m}]\left[\tanh[\beta|J|\hat{m}] - \tanh[\beta|J|m]\right] \leq 0.$$

Since $\tanh[\beta|J|m]$ is a monotonically increasing function of m, this implies $\hat{m} = m$, so

Binary & Parallel Dynamics: $\qquad m = \tanh[\beta|J|m].$ (2.158)

Our study of the toy models has thus been reduced to analyzing the nonlinear equations (2.157) and (2.158). If $J \geq 0$ (excitation) the two types of dynamics lead to the same behavior. At high noise levels, $T > J$, both minimisation problems are solved by $m = 0$, describing a disorganized (paramagnetic) state. This can be seen upon writing the right–hand side of (2.157) in integral form [Coo01, SC00, SC01, CKS05]:

$$m^2 = m\tanh[\beta J m] = \beta J m^2 \int_0^1 dz\, [1 - \tanh^2[\beta J m z]] \leq \beta J m^2.$$

So $m^2[1 - \beta J] \leq 0$, which gives $m = 0$ as soon as $\beta J < 1$. A phase transition occurs at $T = J$ (a bifurcation of non–trivial solutions of (2.157)), and for $T < J$ the equations for m are solved by the two non-zero solutions of (2.157), describing a state where either all neurons tend to be firing ($m > 0$) or where they tend to be quiet ($m < 0$). This becomes clear when we expand (2.157) for small m: $m = \beta J m + \mathcal{O}(m^3)$, so precisely at $\beta J = 1$ one finds a de–stabilization of the trivial solution $m = 0$, together with the creation of (two) stable non–trivial ones. Furthermore, using the identity $c^*(\tanh x) = x\tanh x - \log\cosh x$, we get from (2.155,2.156) the relation $\lim_{N\to\infty} \tilde{F}/N = 2\lim_{N\to\infty} F/N$. For $J < 0$ (inhibition), however, the two types of dynamics give quite different results. For sequential dynamics the relevant minimum is located at $m = 0$ (the paramagnetic state). For parallel dynamics, the minimization problem is invariant under $J \to -J$, so the behavior is again of the Curie-Weiss type, with a paramagnetic state for $T > |J|$, a phase transition at $T = |J|$, and order for $T < |J|$. This difference between the two types of dynamics for $J < 0$ is explained by studying dynamics. For the present (toy) model in the limit $N \to \infty$ the average activity evolves in time according to the deterministic laws [Coo01, SC00, SC01, CKS05]

$$\dot{m} = \tanh[\beta Jm] - m, \qquad m(t+1) = \tanh[\beta Jm(t)],$$

for sequential and parallel dynamics, respectively. For $J < 0$ the sequential system always decays towards the trivial state $m = 0$, whereas for sufficiently large β the parallel system enters the stable limit–cycle $m(t) = M_\beta(-1)^t$, where M_β is the non-zero solution of (2.158). The concepts of 'distance' and 'local minima' are quite different for the two dynamics types; in contrast to the sequential case, parallel dynamics allows the system to make the transition $m \to -m$ in equilibrium.

Phenomenology of Hopfield Models

Recall that the Hopfield model [Hop82] represents a network of binary neurons of the type (2.121,2.122), with threshold noise $w(\eta) = \frac{1}{2}[1 - \tanh^2(\eta)]$, and with a specific recipe for the synapses J_{ij} aimed at storing patterns, motivated by suggestions made in the late nineteen-forties [Heb49]. The original model was in fact defined more narrowly, as the zero noise limit of the system (2.122), but the term has since then been accepted to cover a larger network class. Let us first consider the simplest case and try to store a single pattern $\boldsymbol{\xi} \in \{-1,1\}^N$ in noise–less infinite–range binary networks. Appealing candidates for interactions and thresholds would be $J_{ij} = \xi_i \xi_j$ and $\theta_i = 0$ (for sequential dynamics we put $J_{ii} = 0$ for all i). With this choice the Lyapunov function (2.137) becomes:

$$L_{\text{seq}}[\boldsymbol{\sigma}] = \frac{1}{2}N - \frac{1}{2}[\xi_i \sigma_i]^2.$$

This system indeed reconstructs dynamically the original pattern $\boldsymbol{\xi}$ from an input vector $\boldsymbol{\sigma}(0)$, at least for sequential dynamics. However, *en passant* we have created an additional attractor: the state $-\boldsymbol{\xi}$. This property is shared by all binary models in which the external fields are zero, where the Hamiltonians $H(\boldsymbol{\sigma})$ (2.149) and $\tilde{H}(\boldsymbol{\sigma})$ (2.151) are invariant under an overall sign change $\boldsymbol{\sigma} \to -\boldsymbol{\sigma}$. A second feature common to several (but not all) attractor neural networks is that *each* initial state will lead to pattern reconstruction, even nonsensical (random) ones.

The Hopfield model is got by generalizing the previous simple one-pattern recipe to the case of an arbitrary number p of binary patterns $\boldsymbol{\xi}^\mu = (\xi_1^\mu, \dots, \xi_N^\mu) \in \{-1,1\}^N$ [Coo01, SC00, SC01, CKS05]:

$$J_{ij} = \frac{1}{N}\xi_i^\mu \xi_j^\mu, \qquad \theta_i = 0 \quad (\text{for all } i; \ \mu = 1, \dots, p), \qquad (2.159)$$

$$(\text{sequential dynamics} : \ J_{ii} \to 0, \ \text{for all } i).$$

The prefactor N^{-1} has been inserted to ensure that the limit $N \to \infty$ will exist in future expressions. The process of interest is that where, triggered by correlation between the initial state and a stored pattern $\boldsymbol{\xi}^\lambda$, the state vector $\boldsymbol{\sigma}$ evolves towards $\boldsymbol{\xi}^\lambda$. If this happens, pattern $\boldsymbol{\xi}^\lambda$ is said to be recalled. The similarity between a state vector and the stored patterns is measured by so–called *Hopfield overlaps*

$$m_\mu(\boldsymbol{\sigma}) = \frac{1}{N} \xi_i^\mu \sigma_i. \tag{2.160}$$

The Hopfield model represents as an associative memory, in which the recall process is described in terms of overlaps.

Analysis of Hopfield Models Away From Saturation

A binary Hopfield network with parameters given by (2.159) obeys detailed balance, and the Hamiltonian $H(\boldsymbol{\sigma})$ (2.149) (corresponding to sequential dynamics) and the pseudo-Hamiltonian $\tilde{H}(\boldsymbol{\sigma})$ (2.151) (corresponding to parallel dynamics) become [Coo01, SC00, SC01, CKS05]

$$H(\boldsymbol{\sigma}) = -\frac{1}{2} N \sum_{\mu=1}^p m_\mu^2(\boldsymbol{\sigma}) + \frac{1}{2} p, \tag{2.161}$$

$$\tilde{H}(\boldsymbol{\sigma}) = -\frac{1}{\beta} \sum_i \log 2 \cosh[\beta \xi_i^\mu m_\mu(\boldsymbol{\sigma})],$$

with the overlaps (2.160). Solving the statics implies calculating the free energies F and $\tilde{F}$:

$$F = -\frac{1}{\beta} \log \sum_{\boldsymbol{\sigma}} e^{-\beta H(\boldsymbol{\sigma})}, \qquad \tilde{F} = -\frac{1}{\beta} \log \sum_{\boldsymbol{\sigma}} e^{-\beta \tilde{H}(\boldsymbol{\sigma})}.$$

Upon introducing the short-hand notation $\mathbf{m} = (m_1, \ldots, m_p)$ and $\boldsymbol{\xi}_i = (\xi_i^1, \ldots, \xi_i^p)$, both free energies can be expressed in terms of the density of states $\mathcal{D}(\mathbf{m}) = 2^{-N} \sum_{\boldsymbol{\sigma}} \delta[\mathbf{m} - \mathbf{m}(\boldsymbol{\sigma})]$:

$$F/N = -\frac{1}{\beta} \log 2 - \frac{1}{\beta N} \log \int d\mathbf{m}\, \mathcal{D}(\mathbf{m})\, e^{\frac{1}{2}\beta N \mathbf{m}^2} + \frac{p}{2N}, \tag{2.162}$$

$$\tilde{F}/N = -\frac{1}{\beta} \log 2 - \frac{1}{\beta N} \log \int d\mathbf{m}\, \mathcal{D}(\mathbf{m})\, e^{\sum_{i=1}^N \log 2 \cosh[\beta \boldsymbol{\xi}_i \cdot \mathbf{m}]}, \tag{2.163}$$

using $\int d\mathbf{m}\, \delta[\mathbf{m} - \mathbf{m}(\boldsymbol{\sigma})] = 1$. In order to proceed, we need to specify how the number of patterns p scales with the system size N. In this section we will follow [AGS85] (equilibrium analysis following sequential dynamics) and [FK88] (equilibrium analysis following parallel dynamics), and assume p to be finite. One can now easily calculate the leading contribution to the density

2.3 Synergetics of Recurrent and Attractor Neural Networks

of states, using the integral representation of the δ-function and keeping in mind that according to (2.162,2.163) only terms exponential in N will retain statistical relevance for $N \to \infty$:

$$\lim_{N \to \infty} \frac{1}{N} \log \mathcal{D}(\mathbf{m}) = \lim_{N \to \infty} \frac{1}{N} \log \int d\mathbf{x}\, e^{iN\mathbf{x} \cdot \mathbf{m}} \langle e^{-i\sigma_i \boldsymbol{\xi}_i \cdot \mathbf{x}} \rangle_\sigma$$

$$= \lim_{N \to \infty} \frac{1}{N} \log \int d\mathbf{x}\, e^{N[i\mathbf{x} \cdot \mathbf{m} + \langle \log \cos[\boldsymbol{\xi} \cdot \mathbf{x}]\rangle_{\boldsymbol{\xi}}]},$$

with the abbreviation $\langle \Phi(\boldsymbol{\xi}) \rangle_{\boldsymbol{\xi}} = \lim_{N \to \infty} \frac{1}{N} \sum_{i=1}^{N} \Phi(\boldsymbol{\xi}_i)$. The leading contribution to both free energies can be expressed as a finite-dimensional integral, for large N dominated by that saddle–point (extremum) for which the extensive exponent is real and maximal [Coo01, SC00, SC01, CKS05]:

$$\lim_{N \to \infty} F/N = -\frac{1}{\beta N} \log \int d\mathbf{m}d\mathbf{x}\, e^{-N\beta f(\mathbf{m},\mathbf{x})} = \mathrm{extr}_{\mathbf{x},\mathbf{m}}\, f(\mathbf{m},\mathbf{x}),$$

$$\lim_{N \to \infty} \tilde{F}/N = -\frac{1}{\beta N} \log \int d\mathbf{m}d\mathbf{x}\, e^{-N\beta \tilde{f}(\mathbf{m},\mathbf{x})} = \mathrm{extr}_{\mathbf{x},\mathbf{m}}\, \tilde{f}(\mathbf{m},\mathbf{x}), \qquad \text{with}$$

$$f(\mathbf{m},\mathbf{x}) = -\tfrac{1}{2}\mathbf{m}^2 - i\mathbf{x} \cdot \mathbf{m} - \beta^{-1}\langle \log 2\cos[\beta \boldsymbol{\xi} \cdot \mathbf{x}]\rangle_{\boldsymbol{\xi}},$$

$$\tilde{f}(\mathbf{m},\mathbf{x}) = -\beta^{-1}\langle \log 2\cosh[\beta \boldsymbol{\xi} \cdot \mathbf{m}]\rangle_{\boldsymbol{\xi}} - i\mathbf{x} \cdot \mathbf{m} - \beta^{-1}\langle \log 2\cos[\beta \boldsymbol{\xi} \cdot \mathbf{x}]\rangle_{\boldsymbol{\xi}}.$$

The saddle–point equations for f and $\tilde{f}$ are given by:

$$f : \mathbf{x} = i\mathbf{m}, \qquad\qquad i\mathbf{m} = \langle \boldsymbol{\xi} \tan[\beta \boldsymbol{\xi} \cdot \mathbf{x}]\rangle_{\boldsymbol{\xi}},$$

$$\tilde{f} : \mathbf{x} = i\langle \boldsymbol{\xi} \tanh[\beta \boldsymbol{\xi} \cdot \mathbf{m}]\rangle_{\boldsymbol{\xi}},\ i\mathbf{m} = \langle \boldsymbol{\xi} \tan[\beta \boldsymbol{\xi} \cdot \mathbf{x}]\rangle_{\boldsymbol{\xi}}.$$

In saddle-points $\mathbf{x}$ turns out to be purely imaginary. However, after a shift of the integration contours, putting $\mathbf{x} = i\mathbf{x}^\star(\mathbf{m}) + \mathbf{y}$ (where $i\mathbf{x}^\star(\mathbf{m})$ is the imaginary saddle–point, and where $\mathbf{y} \in \mathbb{R}^p$) we can eliminate $\mathbf{x}$ in favor of $\mathbf{y} \in \mathbb{R}^p$ which does have a real saddle–point, by construction.(Our functions to be integrated have no poles, but strictly speaking we still have to verify that the integration segments linking the original integration regime to the shifted one will not contribute to the integrals. This is generally a tedious and distracting task, which is often skipped. For simple models, however (e.g., networks with uniform synapses), the verification can be carried out properly, and all is found to be safe.) We then get

Sequential Dynamics: $\mathbf{m} = \langle \boldsymbol{\xi} \tanh[\beta \boldsymbol{\xi} \cdot \mathbf{m}]\rangle_{\boldsymbol{\xi}},$

Parallel Dynamics: $\mathbf{m} = \langle \boldsymbol{\xi} \tanh[\beta \boldsymbol{\xi} \cdot [\langle \boldsymbol{\xi}' \tanh[\beta \boldsymbol{\xi}' \cdot \mathbf{m}]\rangle_{\boldsymbol{\xi}'}]]\rangle_{\boldsymbol{\xi}},$

(compare to e.g., (2.157,2.158)). The solutions of the above two equations will in general be identical. To see this, let us denote $\hat{\mathbf{m}} = \langle \boldsymbol{\xi} \tanh[\beta \boldsymbol{\xi} \cdot \mathbf{m}]\rangle_{\boldsymbol{\xi}}$, with which the saddle point equation for $\tilde{f}$ decouples into:

$$\mathbf{m} = \langle \boldsymbol{\xi} \tanh[\beta \boldsymbol{\xi} \cdot \hat{\mathbf{m}}]\rangle_{\boldsymbol{\xi}}, \qquad \hat{\mathbf{m}} = \langle \boldsymbol{\xi} \tanh[\beta \boldsymbol{\xi} \cdot \mathbf{m}]\rangle_{\boldsymbol{\xi}}, \qquad \text{so}$$

$$[\mathbf{m} - \hat{\mathbf{m}}]^2 = \langle [(\boldsymbol{\xi} \cdot \mathbf{m}) - (\boldsymbol{\xi} \cdot \hat{\mathbf{m}})][\tanh(\beta\boldsymbol{\xi}\cdot\hat{\mathbf{m}}) - \tanh(\beta\boldsymbol{\xi}\cdot\mathbf{m})]\rangle_{\boldsymbol{\xi}}.$$

Since tanh is a monotonicaly–increasing function, we must have $[\mathbf{m} - \hat{\mathbf{m}}] \cdot \boldsymbol{\xi} = 0$ for each $\boldsymbol{\xi}$ that contributes to the averages $\langle \ldots \rangle_{\boldsymbol{\xi}}$. For all choices of patterns where the covariance matrix $C_{\mu\nu} = \langle \xi_\mu \xi_\nu \rangle_{\boldsymbol{\xi}}$ is positive definite, we thus get $\mathbf{m} = \hat{\mathbf{m}}$. The final result is: for both types of dynamics (sequential and parallel) the overlap order parameters in equilibrium are given by the solution $\mathbf{m}^*$ of

$$\mathbf{m} = \langle \boldsymbol{\xi} \tanh[\beta \boldsymbol{\xi} \cdot \mathbf{m}]\rangle_{\boldsymbol{\xi}}, \qquad \text{which minimises} \qquad (2.164)$$

$$f(\mathbf{m}) = \frac{1}{2}\mathbf{m}^2 - \frac{1}{\beta}\langle \log 2\cosh[\beta\boldsymbol{\xi}\cdot\mathbf{m}]\rangle_{\boldsymbol{\xi}}. \qquad (2.165)$$

The free energies of the ergodic components are $\lim_{N\to\infty} F/N = f(\mathbf{m}^*)$ and $\lim_{N\to\infty} \tilde{F}/N = 2f(\mathbf{m}^*)$. Adding partition terms of the form $H \to H + \lambda g[\mathbf{m}(\boldsymbol{\sigma})]$ to the Hamiltonians allows us identify $\langle g[\mathbf{m}(\boldsymbol{\sigma})]\rangle_{\text{eq}} = \lim_{\lambda \to 0} \partial F/\partial \lambda = g[\mathbf{m}^*]$. Thus, in equilibrium the fluctuations in the overlap order parameters $\mathbf{m}(\boldsymbol{\sigma})$ (2.160) vanish for $N \to \infty$. Their deterministic values are simply given by $\mathbf{m}^*$. Note that in the case of sequential dynamics we could also have used linearization with Gaussian integrals (as used previously for coupled oscillators with uniform synapses) to arrive at this solution, with p auxiliary integrations, but that for parallel dynamics this would not have been possible.

Now, in analysis of *order parameter equations*, we will restrict our further discussion to the case of randomly drawn patterns, so [Coo01, SC00, SC01, CKS05]

$$\langle \Phi(\boldsymbol{\xi})\rangle_{\boldsymbol{\xi}} = 2^{-p} \sum_{\boldsymbol{\xi}\in\{-1,1\}^p} \Phi(\boldsymbol{\xi}), \qquad \langle \xi_\mu\rangle_{\boldsymbol{\xi}} = 0, \qquad \langle \xi_\mu\xi_\nu\rangle_{\boldsymbol{\xi}} = \delta_{\mu\nu}.$$

We first establish an upper bound for the temperature for where non–trivial solutions $\mathbf{m}^*$ could exist, by writing (2.164) in integral form:

$$m_\mu = \beta\langle \xi_\mu (\boldsymbol{\xi}\cdot\mathbf{m}) \int_0^1 d\lambda [1 - \tanh^2[\beta\lambda\boldsymbol{\xi}\cdot\mathbf{m}]]\rangle_{\boldsymbol{\xi}},$$

from which we deduce

$$0 = \mathbf{m}^2 - \beta\langle (\boldsymbol{\xi}\cdot\mathbf{m})^2 \int_0^1 d\lambda[1 - \tanh^2[\beta\lambda\boldsymbol{\xi}\cdot\mathbf{m}]]\rangle_{\boldsymbol{\xi}}$$

$$\geq \mathbf{m}^2 - \beta\langle (\boldsymbol{\xi}\cdot\mathbf{m})^2\rangle_{\boldsymbol{\xi}} = \mathbf{m}^2(1-\beta),$$

For $T > 1$ the only solution of (2.164) is the paramagnetic state $\mathbf{m} = 0$, which gives for the free energy per neuron $-T\log 2$ and $-2T\log 2$ (for sequential and parallel dynamics, respectively). At $T = 1$ a phase transition occurs, which follows from expanding (2.164) for small $|\mathbf{m}|$ in powers of $\tau = \beta - 1$:

2.3 Synergetics of Recurrent and Attractor Neural Networks

$$m_\mu = (1+\tau)m_\mu - \frac{1}{3}m_\nu m_\rho m_\lambda \langle \xi_\mu \xi_\nu \xi_\rho \xi_\lambda \rangle_\xi$$
$$+ \mathcal{O}(\mathbf{m}^5, \tau\mathbf{m}^3) = m_\mu[1+\tau - \mathbf{m}^2 + \frac{2}{3}m_\mu^2] + \mathcal{O}(\mathbf{m}^5, \tau\mathbf{m}^3).$$

The new saddle–point scales as $m_\mu = \tilde{m}_\mu \tau^{1/2} + \mathcal{O}(\tau^{3/2})$, with for each μ: $\tilde{m}_\mu = 0$ or $0 = 1 - \tilde{\mathbf{m}}^2 + \frac{2}{3}\tilde{m}_\mu^2$.

The solutions are of the form $\tilde{m}_\mu \in \{-\tilde{m}, 0, \tilde{m}\}$. If we denote with n the number of non-zero components in the vector $\tilde{\mathbf{m}}$, we derive from the above identities: $\tilde{m}_\mu = 0$ or $\tilde{m}_\mu = \pm\sqrt{3}/\sqrt{3n-2}$. These saddle-points are called *mixture states*, since they correspond to microscopic configurations correlated equally with a finite number n of the stored patterns (or their negatives). Without loss of generality we can always perform gauge transformations on the set of stored patterns (permutations and reflections), such that the mixture states acquire the form [Coo01, SC00, SC01, CKS05]

$$\mathbf{m} = m_n(\overbrace{1,\ldots,1}^{n \text{ times}}, \overbrace{0,\ldots,0}^{p-n \text{ times}}), \qquad (2.166)$$
$$m_n = [\frac{3}{3n-2}]^{\frac{1}{2}}(\beta-1)^{1/2} + \ldots$$

These states are in fact saddle–points of the surface $f(\mathbf{m})$ (2.165) for any finite temperature, as can be verified by substituting (2.166) as an *ansatz* into (2.164):

$$\mu \leq n: \qquad m_n = \langle \xi_\mu \tanh[\beta m_n \sum_{\nu \leq n} \xi_\nu]\rangle_\xi,$$

$$\mu > n: \qquad 0 = \langle \xi_\mu \tanh[\beta m_n \sum_{\nu \leq n} \xi_\nu]\rangle_\xi.$$

The second equation is automatically satisfied since the average factorizes. The first equation leads to a condition determining the amplitude m_n of the mixture states:

$$m_n = \langle [\frac{1}{n}\sum_{\mu \leq n}\xi_\mu] \tanh[\beta m_n \sum_{\nu \leq n}\xi_\nu]\rangle_\xi. \qquad (2.167)$$

The corresponding values of $f(\mathbf{m})$, to be denoted by f_n, are

$$f_n = \frac{1}{2}nm_n^2 - \frac{1}{\beta}\langle \log 2\cosh[\beta m_n \sum_{\nu \leq n}\xi_\nu]\rangle_\xi. \qquad (2.168)$$

The relevant question at this stage is whether or not these saddle-points correspond to local minima of the surface $f(\mathbf{m})$ (2.165). The second derivative of $f(\mathbf{m})$ is given by

$$\frac{\partial^2 f(\mathbf{m})}{\partial m_\mu \partial m_\nu} = \delta_{\mu\nu} - \beta\langle \xi_\mu \xi_\nu \left[1 - \tanh^2[\beta\boldsymbol{\xi} \cdot \mathbf{m}]\right]\rangle_\xi, \qquad (2.169)$$

where a local minimum corresponds to a positive definite second derivative. In the trivial saddle–point $\mathbf{m} = 0$ this gives simply $\delta_{\mu\nu}(1-\beta)$, so at $T = 1$ this state destabilizes. In a mixture state of the type (2.166) the second derivative becomes:

$$D^{(n)}_{\mu\nu} = \delta_{\mu\nu} - \beta \langle \xi_\mu \xi_\nu [1 - \tanh^2[\beta m_n \sum_{\rho \leq n} \xi_\rho]] \rangle_\xi.$$

Due to the symmetries in the problem the spectrum of the matrix $D^{(n)}$ can be calculated. One finds the following eigen–spaces, with

$$Q = \langle \tanh^2[\beta m_n \sum_{\rho \leq n} \xi_\rho] \rangle_\xi \quad \text{and} \quad R = \langle \xi_1 \xi_2 \tanh^2[\beta m_n \sum_{\rho \leq n} \xi_\rho] \rangle_\xi,$$

Eigenspace : Eigenvalue :
I : $\mathbf{x} = (0, \ldots, 0, x_{n+1}, \ldots, x_p)$, $1 - \beta[1 - Q]$,
II : $\mathbf{x} = (1, \ldots, 1, 0, \ldots, 0)$, $1 - \beta[1 - Q + (1-n)R]$,
III : $\mathbf{x} = (x_1, \ldots, x_n, 0, \ldots, 0)$, $\sum_\mu x_\mu = 0$, $1 - \beta[1 - Q + R]$.

The eigen–space III and the quantity R only come into play for $n > 1$. To find the smallest eigenvalue we need to know the sign of R. With the abbreviation $M_\xi = \sum_{\rho \leq n} \xi_\rho$ we find [Coo01, SC00, SC01, CKS05]:

$$n(n-1)R = \langle M_\xi^2 \tanh^2[\beta m_n M_\xi] \rangle_\xi - n \langle \tanh^2[\beta m_n M_\xi] \rangle_\xi$$
$$= \langle [M_\xi^2 - \langle M_{\xi'}^2 \rangle_{\xi'}] \tanh^2[\beta m_n | M_\xi |] \rangle_\xi$$
$$= \langle [M_\xi^2 - \langle M_{\xi'}^2 \rangle_{\xi'}] \left\{ \tanh^2[\beta m_n \sqrt{M_\xi^2}] - \tanh^2[\beta m_n \sqrt{\langle M_{\xi'}^2 \rangle_{\xi'}}] \right\} \rangle_\xi \geq 0.$$

We may now identify the conditions for an n–mixture state to be a local minimum of $f(\mathbf{m})$. For $n = 1$ the relevant eigenvalue is I, now the quantity Q simplifies considerably. For $n > 1$ the relevant eigenvalue is III, here we can combine Q and R into one single average:

$$n = 1 : 1 - \beta[1 - \tanh^2[\beta m_1]] > 0$$
$$n = 2 : 1 - \beta > 0$$
$$n \geq 3 : 1 - \beta[1 - \langle \tanh^2[\beta m_n \sum_{\rho=3}^n \xi_\rho] \rangle_\xi] > 0$$

The $n = 1$ states, correlated with one pattern only, are the desired solutions. They are stable for all $T < 1$, since partial differentiation with respect to β of the $n = 1$ amplitude equation (2.167) gives

$$m_1 = \tanh[\beta m_1] \quad \to \quad 1 - \beta[1 - \tanh^2[\beta m_1]]$$
$$= m_1 [1 - \tanh^2[\beta m_1]](\partial m_1 / \partial \beta)^{-1},$$

so that clearly $\text{sgn}[m_1] = \text{sgn}[\partial m_1/\partial \beta]$. The $n = 2$ mixtures are always unstable. For $n \geq 3$ we have to solve the amplitude equations (2.167) numerically to evaluate their stability. It turns out that only for odd n will there be a critical temperature below which the n−mixture states are local minima of $f(\mathbf{m})$.

We have now solved the model in equilibrium for finite p and $N \to \infty$. For non–random patterns one simply has to study the bifurcation properties of equation (2.164) for the new pattern statistics at hand; this is only qualitatively different from the random pattern analysis explained above. The occurrence of multiple saddle–points corresponding to local minima of the free energy signals ergodicity breaking. Although among these only the *global minimum* will correspond to the thermodynamic equilibrium state, the non–global minima correspond to true ergodic components, i.e., on finite time–scales they will be just as relevant as the global minimum.

2.3.5 Simple Recurrent Networks of Coupled Oscillators

Coupled Oscillators with Uniform Synapses

Models with continuous variables involve integration over states, rather than summation. For a coupled oscillator network (2.133) with uniform synapses $J_{ij} = J/N$ and zero frequencies $\omega_i = 0$ (which is a simple version of the model in [Kur84]) we get for the free energy per oscillator [Coo01, SC00, SC01, CKS05]:

$$\lim_{N \to \infty} F/N = - \lim_{N \to \infty} \frac{1}{\beta N} \log \int_{-\pi}^{\pi} \cdots \int_{-\pi}^{\pi} d\boldsymbol{\xi} \times$$
$$\times e^{(\beta J/2N)\left[\left[\sum_i \cos(\xi_i)\right]^2 + \left[\sum_i \sin(\xi_i)\right]^2\right]}.$$

We would now have to 'count' microscopic states with prescribed average cosines and sines. A faster route exploits auxiliary Gaussian integrals, via the identity

$$e^{\frac{1}{2}y^2} = \int Dz\, e^{yz}, \qquad (2.170)$$

with the short–hand $Dx = (2\pi)^{-\frac{1}{2}} e^{-\frac{1}{2}x^2} dx$ (this alternative would also have been open to us in the binary case; my aim in this section is to explain both methods):

$$\lim_{N \to \infty} F/N = - \lim_{N \to \infty} \frac{1}{\beta N} \log \int_{-\pi}^{\pi} \cdots \int_{-\pi}^{\pi} d\boldsymbol{\xi} \int DxDy \times$$
$$\times e^{\sqrt{\beta J/N}\left[x \sum_i \cos(\xi_i) + y \sum_i \sin(\xi_i)\right]}$$
$$= - \lim_{N \to \infty} \frac{1}{\beta N} \log \int DxDy \left[\int_{-\pi}^{\pi} d\xi\, e^{\cos(\xi)\sqrt{\beta J(x^2+y^2)/N}}\right]^N$$
$$= - \lim_{N \to \infty} \frac{1}{\beta N} \log \int_0^\infty dq\, q e^{-\frac{1}{2}N\beta|J|q^2} \times$$
$$\times \left[\int_{-\pi}^{\pi} d\xi\, e^{\beta|J|q \cos(\xi) \sqrt{\mathrm{rmsgn}(J)}}\right]^N,$$

where we have transformed to polar coordinates, $(x, y) = q\sqrt{\beta|J|N}(\cos\theta, \sin\theta)$, and where we have already eliminated (constant) terms which will not survive the limit $N \to \infty$. Thus, saddle–point integration gives us, quite similar to the previous cases (2.155,2.156):

$$\lim_{N\to\infty} F/N = \min_{q\geq 0} f(q), \quad \begin{array}{l} J > 0 : \beta f(q) = \frac{1}{2}\beta|J|q^2 - \log[2\pi I_0(\beta|J|q)] \\ J < 0 : \beta f(q) = \frac{1}{2}\beta|J|q^2 - \log[2\pi I_0(i\beta|J|q)] \end{array}, \quad (2.171)$$

in which the $I_n(z)$ are the Bessel functions (see e.g., [AS72]). The equations from which to solve the minima are got by differentiation, using $\frac{d}{dz}I_0(z) = I_1(z)$:in which the $I_n(z)$ are the Bessel functions (see e.g., [AS72]). The equations from which to solve the minima are got by differentiation, using $\frac{d}{dz}I_0(z) = I_1(z)$:

$$J > 0: \quad q = \frac{I_1(\beta|J|q)}{I_0(\beta|J|q)}, \quad J < 0: \quad q = i\frac{I_1(i\beta|J|q)}{I_0(i\beta|J|q)}. \quad (2.172)$$

Again, in both cases the problem has been reduced to studying a single non-linear equation. The physical meaning of the solution follows from the identity $-2\partial F/\partial J = \langle N^{-1}\sum_{i\neq j}\cos(\xi_i - \xi_j)\rangle$:

$$\lim_{N\to\infty}\langle[\frac{1}{N}\sum_i \cos(\xi_i)]^2\rangle + \lim_{N\to\infty}\langle[\frac{1}{N}\sum_i \sin(\xi_i)]^2\rangle = \text{sgn}(J)\, q^2.$$

From this equation it also follows that $q \leq 1$. Note: since $\partial f(q)/\partial q = 0$ at the minimum, one only needs to consider the explicit derivative of $f(q)$ with respect to J. If the synapses induce anti-synchronization, $J < 0$, the only solution of (2.172) (and the minimum in (2.171)) is the trivial state $q = 0$. This also follows immediately from the equation which gave the physical meaning of q. For synchronizing forces, $J > 0$, on the other hand, we again find the trivial solution at high noise levels, but a globally synchronized state with $q > 0$ at low noise levels. Here a phase transition occurs at $T = \frac{1}{2}J$ (a bifurcation of non–trivial solutions of (2.172)), and for $T < \frac{1}{2}J$ the minimum of (2.171) is found at two non-zero values for q. The critical noise level is again found upon expanding the saddle–point equation, using

$$I_0(z) = 1 + \mathcal{O}(z^2) \quad \text{and} \quad I_1(z) = \frac{1}{2}z + \mathcal{O}(z^3) : q = \frac{1}{2}\beta J q + \mathcal{O}(q^3).$$

Precisely at $\beta J = 2$ one finds a de-stabilization of the trivial solution $q = 0$, together with the creation of (two) stable non–trivial ones. Note that, in view of (2.171), we are only interested in non–negative values of q. One can prove, using the properties of the Bessel functions, that there are no other (discontinuous) bifurcations of non–trivial solutions of the saddle–point equation. Note, finally, that the absence of a state with global anti-synchronization for

2.3 Synergetics of Recurrent and Attractor Neural Networks

$J < 0$ has the same origin as the absence of an anti-ferromagnetic state for $J < 0$ in the previous models with binary neurons. Due to the long-range nature of the synapses $J_{ij} = J/N$ such states simply cannot exist: whereas any set of oscillators can be in a fully synchronized state, if two oscillators are in anti-synchrony it is already impossible for a third to be simultaneously in anti-synchrony with the first two (since anti-synchrony with one implies synchrony with the other) [Coo01, SC00, SC01, CKS05].

Coupled Oscillator Attractor Networks

Let us now turn to an alternative realisation of information storage in a recurrent network based upon the creation of attractors [Coo01, SC00, SC01, CKS05]. We will solve models of coupled neural oscillators of the type (2.133), with zero natural frequencies (since we wish to use equilibrium techniques), in which real-valued patterns are stored as stable configurations of oscillator phases, following [Coo89]. Let us, however, first find out how to store a single pattern $\boldsymbol{\xi} \in [-\pi, \pi]^N$ in a noise-less infinite-range oscillator network. For simplicity we will draw each component ξ_i independently at random from $[-\pi, \pi]$, with uniform probability density. This allows us to use asymptotic properties such as $|N^{-1} \sum_j e^{i\ell\xi_j}| = \mathcal{O}(N^{-\frac{1}{2}})$ for any integer ℓ. A sensible choice for the synapses would be $J_{ij} = \cos[\xi_i - \xi_j]$. To see this we work out the corresponding Lyapunov function (2.140):

$$L[\boldsymbol{\xi}] = -\frac{1}{2N^2} \cos[\xi_i - \xi_j] \cos[\xi_i - \xi_j],$$

$$L[\boldsymbol{\xi}] = -\frac{1}{2N^2} \cos^2[\xi_i - \xi_j] = -\frac{1}{4} + \mathcal{O}(N^{-\frac{1}{2}}),$$

where the factors of N have been inserted to achieve appropriate scaling in the $N \to \infty$ limit. The function $L[\boldsymbol{\xi}]$, which is obviously bounded from below, must decrease monotonically during the dynamics. To find out whether the state $\boldsymbol{\xi}$ is a stable fixed-point of the dynamics we have to calculate L and derivatives of L at $\boldsymbol{\xi} = \boldsymbol{\xi}$:

$$\left.\frac{\partial L}{\partial \xi_i}\right|_{\boldsymbol{\xi}} = \frac{1}{2N^2} \sum_j \sin[2(\xi_i - \xi_j)],$$

$$\left.\frac{\partial^2 L}{\partial \xi_i^2}\right|_{\boldsymbol{\xi}} = \frac{1}{N^2} \sum_j \cos^2[\xi_i - \xi_j],$$

$$i \neq j : \quad \left.\frac{\partial^2 L}{\partial \xi_i \partial \xi_j}\right|_{\boldsymbol{\xi}} = -\frac{1}{N^2} \cos^2[\xi_i - \xi_j].$$

Clearly $\lim_{N\to\infty} L[\boldsymbol{\xi}] = -\frac{1}{4}$. Putting $\boldsymbol{\xi} = \boldsymbol{\xi} + \Delta\boldsymbol{\xi}$, with $\Delta\xi_i = \mathcal{O}(N^0)$, we find

$$L[\boldsymbol{\xi} + \Delta\boldsymbol{\xi}] - L[\boldsymbol{\xi}] = \Delta\xi_i \frac{\partial L}{\partial \xi_i}|_{\boldsymbol{\xi}} \qquad (2.173)$$

$$+ \frac{1}{2}\Delta\xi_i \Delta\xi_j \frac{\partial^2 L}{\partial \xi_i \partial \xi_j}|_{\boldsymbol{\xi}} + \mathcal{O}(\Delta\boldsymbol{\xi}^3)$$

$$= \frac{1}{4N}\sum_i \Delta\xi_i^2 - \frac{1}{2N^2}\Delta\xi_i \Delta\xi_j \cos^2[\xi_i - \xi_j] + \mathcal{O}(N^{-\frac{1}{2}}, \Delta\boldsymbol{\xi}^3)$$

$$= \frac{1}{4}\{\frac{1}{N}\sum_i \Delta\xi_i^2 - [\frac{1}{N}\sum_i \Delta\xi_i]^2 - [\frac{1}{N}\Delta\xi_i \cos(2\xi_i)]^2$$

$$- [\frac{1}{N}\sum_i \Delta\xi_i \sin(2\xi_i)]^2\} + \mathcal{O}(N^{-\frac{1}{2}}, \Delta\boldsymbol{\xi}^3).$$

In leading order in N the following three vectors in $\mathbb{R}^N$ are normalized and orthogonal:

$$\mathbf{e}_1 = \frac{1}{\sqrt{N}}(1, 1, \ldots, 1), \qquad \mathbf{e}_2 = \frac{\sqrt{2}}{\sqrt{N}}(\cos(2\xi_1), \ldots, \cos(2\xi_N)),$$

$$\mathbf{e}_2 = \frac{\sqrt{2}}{\sqrt{N}}(\sin(2\xi_1), \ldots, \sin(2\xi_N)).$$

We may therefore use

$$\Delta\boldsymbol{\xi}^2 \geq (\Delta\boldsymbol{\xi}\cdot_1)^2 + (\Delta\boldsymbol{\xi}\cdot_2)^2 + (\Delta\boldsymbol{\xi}\cdot_3)^2,$$

insertion of which into (2.173) leads to

$$L[\boldsymbol{\xi} + \Delta\boldsymbol{\xi}] - L[\boldsymbol{\xi}] \geq [\frac{1}{2N}\sum_i \Delta\xi_i \cos(2\xi_i)]^2$$

$$+ [\frac{1}{2N}\sum_i \Delta\xi_i \sin(2\xi_i)]^2 + \mathcal{O}(N^{-\frac{1}{2}}, \Delta\boldsymbol{\xi}^3).$$

Thus for large N the second derivative of L is non-negative at $\boldsymbol{\xi} = \boldsymbol{\xi}$, and the phase pattern $\boldsymbol{\xi}$ has indeed become a fixed–point attractor of the dynamics of the noise-free coupled oscillator network. The same is found to be true for the states $\boldsymbol{\xi} = \pm\boldsymbol{\xi} + \alpha(1, \ldots, 1)$ (for any α).

We next follow the strategy of the Hopfield model and attempt to simply extend the above recipe for the synapses to the case of having a finite number p of phase patterns $\boldsymbol{\xi}^\mu = (\xi_1^\mu, \ldots, \xi_N^\mu) \in [-\pi, \pi]^N$, giving

$$J_{ij} = \frac{1}{N}\sum_{\mu=1}^p \cos[\xi_i^\mu - \xi_j^\mu], \qquad (2.174)$$

where the factor N, as before, ensures a proper limit $N \to \infty$ later. In analogy with our solution of the Hopfield model we define the following averages over pattern variables:

2.3 Synergetics of Recurrent and Attractor Neural Networks

$$\langle g[\boldsymbol{\xi}]\rangle_{\boldsymbol{\xi}} = \lim_{N\to\infty} \sum_i g[\boldsymbol{\xi}_i], \quad \boldsymbol{\xi}_i = (\xi_i^1,\ldots,\xi_i^p) \in [-\pi,\pi]^p.$$

We can write the Hamiltonian $H(\boldsymbol{\xi})$ of (2.153) in the form [Coo01, SC00, SC01, CKS05]

$$H(\boldsymbol{\xi}) = -\frac{1}{2N}\sum_{\mu=1}^p \cos[\xi_i^\mu - \xi_j^\mu]\cos[\xi_i - \xi_j]$$

$$= -\frac{N}{2}\sum_{\mu=1}^p \{m_{cc}^\mu(\boldsymbol{\xi})^2 + m_{cs}^\mu(\boldsymbol{\xi})^2 + m_{sc}^\mu(\boldsymbol{\xi})^2 + m_{ss}^\mu(\boldsymbol{\xi})^2\},$$

in which

$$m_{cc}^\mu(\boldsymbol{\xi}) = \frac{1}{N}\cos(\xi_i^\mu)\cos(\xi_i), \qquad (2.175)$$

$$m_{cs}^\mu(\boldsymbol{\xi}) = \frac{1}{N}\cos(\xi_i^\mu)\sin(\xi_i),$$

$$m_{sc}^\mu(\boldsymbol{\xi}) = \frac{1}{N}\sin(\xi_i^\mu)\cos(\xi_i), \qquad (2.176)$$

$$m_{ss}^\mu(\boldsymbol{\xi}) = \frac{1}{N}\sin(\xi_i^\mu)\sin(\xi_i).$$

The free energy per oscillator can now be written as

$$F/N = -\frac{1}{\beta N}\log\int\cdots\int d\boldsymbol{\xi}\, e^{-\beta H(\boldsymbol{\xi})} =$$

$$-\frac{1}{\beta N}\log\int\cdots\int d\boldsymbol{\xi}\, e^{\frac{1}{2}\beta N\sum_\mu\sum_{\star\star} m_{\star\star}^\mu(\boldsymbol{\xi})^2},$$

with $\star\star \in \{cc, ss, cs, sc\}$. Upon introducing the notation $\mathbf{m}_{\star\star} = (m_{\star\star}^1,\ldots,m_{\star\star}^p)$ we can again express the free energy in terms of the density of states $\mathcal{D}(\{\mathbf{m}_{\star\star}\}) = (2\pi)^{-N}\int\cdots\int d\boldsymbol{\xi}\prod_{\star\star}\delta[\mathbf{m}_{\star\star} - \mathbf{m}_{\star\star}(\boldsymbol{\sigma})]$:

$$F/N = -\frac{1}{\beta}\log(2\pi) - \frac{1}{\beta N}\log\int\prod_{\star\star}d\mathbf{m}_{\star\star}\,\mathcal{D}(\{\mathbf{m}_{\star\star}\})e^{\frac{1}{2}\beta N\sum_{\star\star}\mathbf{m}_{\star\star}^2}. \qquad (2.177)$$

Since p is finite, the leading contribution to the density of states (as $N\to\infty$), which will give us the entropy, can be calculated by writing the δ−functions in integral representation:

$$\lim_{N\to\infty} \frac{1}{N} \log \mathcal{D}(\{\mathbf{m}_{\star\star}\}) = \lim_{N\to\infty} \frac{1}{N} \log \int \prod_{\star\star} [d\mathbf{x}_{\star\star}\ e^{iN\mathbf{x}_{\star\star}\cdot\mathbf{m}_{\star\star}}] \ \times$$

$$\int\cdots\int \frac{d\boldsymbol{\xi}}{(2\pi)^N} \ \times$$

$$e^{-i[x_{cc}^\mu \cos(\xi_i^\mu)\cos(\xi_i)+x_{cs}^\mu \cos(\xi_i^\mu)\sin(\xi_i)+x_{sc}^\mu \sin(\xi_i^\mu)\cos(\xi_i)+x_{ss}^\mu \sin(\xi_i^\mu)\sin(\xi_i)]}$$

$$= \operatorname{extr}_{\{\mathbf{x}_{\star\star}\}}\{i\sum_{\star\star}\mathbf{x}_{\star\star}\cdot\mathbf{m}_{\star\star} + \langle\log\int\frac{d\xi}{2\pi}\ \times$$

$$e^{-i[x_{cc}^\mu \cos(\xi_\mu)\cos(\xi)+x_{cs}^\mu \cos(\xi_\mu)\sin(\xi)+x_{sc}^\mu \sin(\xi_\mu)\cos(\xi)+x_{ss}^\mu \sin(\xi_\mu)\sin(\xi)]}\rangle_\xi\}.$$

The relevant extremum is purely imaginary so we put $\mathbf{x}_{\star\star} = i\beta\mathbf{y}_{\star\star}$ (see our previous discussion for the Hopfield model) and, upon inserting the density of states into our original expression for the free energy per oscillator, arrive at

$$\lim_{N\to\infty} F/N = \operatorname{extr}_{\{\mathbf{m}_{\star\star},\mathbf{y}_{\star\star}\}} f(\{\mathbf{m}_{\star\star},\mathbf{y}_{\star\star}\}),$$

$$f(\{\mathbf{m}_{\star\star},\mathbf{y}_{\star\star}\}) = -\frac{1}{\beta}\log(2\pi) - \frac{1}{2}\sum_{\star\star}\mathbf{m}_{\star\star}^2 + \sum_{\star\star}\mathbf{y}_{\star\star}\cdot\mathbf{m}_{\star\star}$$

$$-\frac{1}{\beta}\langle\log\int\frac{d\xi}{2\pi} e^{\beta[y_{cc}^\mu \cos(\xi_\mu)\cos(\xi)+y_{cs}^\mu \cos(\xi_\mu)\sin(\xi)+y_{sc}^\mu \sin(\xi_\mu)\cos(\xi)+y_{ss}^\mu \sin(\xi_\mu)\sin(\xi)]}\rangle_\xi$$

Taking derivatives with respect to the order parameters $\mathbf{m}_{\star\star}$ gives us $\mathbf{y}_{\star\star} = \mathbf{m}_{\star\star}$, with which we can eliminate the $\mathbf{y}_{\star\star}$. Derivation with respect to the $\mathbf{m}_{\star\star}$ subsequently gives the saddle–point equations

$$m_{cc}^\mu = \tag{2.178}$$

$$\langle\cos[\xi_\mu]\frac{\int d\xi \cos[\xi] e^{\beta\cos[\xi][m_{cc}^\nu\cos[\xi_\nu]+m_{sc}^\nu\sin[\xi_\nu]]+\beta\sin[\xi][m_{cs}^\nu\cos[\xi_\nu]+m_{ss}^\nu\sin[\xi_\nu]]}}{\int d\xi\ e^{\beta\cos[\xi][m_{cc}^\nu\cos[\xi_\nu]+m_{sc}^\nu\sin[\xi_\nu]]+\beta\sin[\xi][m_{cs}^\nu\cos[\xi_\nu]+m_{ss}^\nu\sin[\xi_\nu]]}}\rangle_\xi,$$

$$m_{cs}^\mu = \tag{2.179}$$

$$\langle\cos[\xi_\mu]\frac{\int d\xi \sin[\xi] e^{\beta\cos[\xi][m_{cc}^\nu\cos[\xi_\nu]+m_{sc}^\nu\sin[\xi_\nu]]+\beta\sin[\xi][m_{cs}^\nu\cos[\xi_\nu]+m_{ss}^\nu\sin[\xi_\nu]]}}{\int d\xi\ e^{\beta\cos[\xi][m_{cc}^\nu\cos[\xi_\nu]+m_{sc}^\nu\sin[\xi_\nu]]+\beta\sin[\xi][m_{cs}^\nu\cos[\xi_\nu]+m_{ss}^\nu\sin[\xi_\nu]]}}\rangle_\xi,$$

$$m_{sc}^\mu = \tag{2.180}$$

$$\langle\sin[\xi_\mu]\frac{\int d\xi \cos[\xi] e^{\beta\cos[\xi][m_{cc}^\nu\cos[\xi_\nu]+m_{sc}^\nu\sin[\xi_\nu]]+\beta\sin[\xi][m_{cs}^\nu\cos[\xi_\nu]+m_{ss}^\nu\sin[\xi_\nu]]}}{\int d\xi\ e^{\beta\cos[\xi][m_{cc}^\nu\cos[\xi_\nu]+m_{sc}^\nu\sin[\xi_\nu]]+\beta\sin[\xi][m_{cs}^\nu\cos[\xi_\nu]+m_{ss}^\nu\sin[\xi_\nu]]}}\rangle_\xi,$$

$$m_{ss}^\mu = \tag{2.181}$$

$$\langle\sin[\xi_\mu]\frac{\int d\xi \sin[\xi] e^{\beta\cos[\xi][m_{cc}^\nu\cos[\xi_\nu]+m_{sc}^\nu\sin[\xi_\nu]]+\beta\sin[\xi][m_{cs}^\nu\cos[\xi_\nu]+m_{ss}^\nu\sin[\xi_\nu]]}}{\int d\xi\ e^{\beta\cos[\xi][m_{cc}^\nu\cos[\xi_\nu]+m_{sc}^\nu\sin[\xi_\nu]]+\beta\sin[\xi][m_{cs}^\nu\cos[\xi_\nu]+m_{ss}^\nu\sin[\xi_\nu]]}}\rangle_\xi,$$

The equilibrium values of the observables $\mathbf{m}_{\star\star}$, as defined in (2.175,2.176), are now given by the solution of the coupled equations (2.178-2.181) which minimizes

2.3 Synergetics of Recurrent and Attractor Neural Networks

$$f(\{\mathbf{m}_{\star\star}\}) = \frac{1}{2}\sum_{\star\star} \mathbf{m}_{\star\star}^2 - \frac{1}{\beta}\langle \log \int d\xi \times \quad (2.182)$$

$$e^{\beta \cos[\xi][m^\nu_{cc}\cos[\xi_\nu]+m^\nu_{sc}\sin[\xi_\nu]]+\beta \sin[\xi][m^\nu_{cs}\cos[\xi_\nu]+m^\nu_{ss}\sin[\xi_\nu]]}\rangle_{\boldsymbol{\xi}}.$$

We can confirm that the relevant saddle–point must be a minimum by inspecting the $\beta = 0$ limit (infinite noise levels):

$$\lim_{\beta \to 0} f(\{\mathbf{m}_{\star\star}\}) = \frac{1}{2}\sum_{\star\star} \mathbf{m}_{\star\star}^2 - \frac{1}{\beta}\log(2\pi).$$

From now on we will restrict our analysis to phase pattern components ξ_i^μ which have all been drawn independently at random from $[-\pi, \pi]$, with uniform probability density, so that $\langle g[\boldsymbol{\xi}]\rangle_{\boldsymbol{\xi}} = (2\pi)^{-p}\int_{-\pi}^{\pi}\ldots\int_{-\pi}^{\pi}d\boldsymbol{\xi}\, g[\boldsymbol{\xi}]$. At $\beta = 0$ ($T = \infty$) one finds only the trivial state $m_{\star\star}^\mu = 0$. It can be shown that there will be no discontinuous transitions to a non–trivial state as the noise level (temperature) is reduced. The continuous ones follow upon expansion of the equations (2.178-2.181) for small $\{\mathbf{m}_{\star\star}\}$, which is found to give (for each μ and each combination $\star\star$):

$$m_{\star\star}^\mu = \frac{1}{4}\beta m_{\star\star}^\mu + \mathcal{O}(\{\mathbf{m}_{\star\star}^2\}).$$

Thus a continuous transition to recall states occurs at $T = \frac{1}{4}$. Full classification of all solutions of (2.178-2.181) is ruled out. Here we will restrict ourselves to the most relevant ones, such as the pure states, where $m_{\star\star}^\mu = m_{\star\star}\delta_{\mu\lambda}$ (for some pattern label λ). Here the oscillator phases are correlated with only one of the stored phase patterns (if at all). Insertion into the above expression for $f(\{\mathbf{m}_{\star\star}\})$ shows that for such solutions we have to minimize [Coo01, SC00, SC01, CKS05]

$$f(\{m_{\star\star}\}) = \frac{1}{2}\sum_{\star\star} m_{\star\star}^2 - \frac{1}{\beta}\int \frac{d\xi}{2\pi}\log\int d\xi \times \quad (2.183)$$

$$e^{\beta\cos[\xi][m_{cc}\cos[\xi]+m_{sc}\sin[\xi]]+\beta\sin[\xi][m_{cs}\cos[\xi]+m_{ss}\sin[\xi]]}.$$

We anticipate solutions corresponding to the (partial) recall of the stored phase pattern $\boldsymbol{\xi}^\lambda$ or its mirror image (modulo overall phase shifts $\xi_i \to \xi_i + \delta$, under which the synapses are obviously invariant). Insertion into (2.178-2.181) of the state

$$\xi_i = \xi_i^\lambda + \delta \quad \text{gives} \quad (m_{cc}, m_{sc}, m_{cs}, m_{ss}) = \frac{1}{2}(\cos\delta, -\sin\delta, \sin\delta, \cos\delta).$$

Similarly, insertion into (2.178-2.181) of

$$\xi_i = -\xi_i^\lambda + \delta \quad \text{gives} \quad (m_{cc}, m_{sc}, m_{cs}, m_{ss}) = \frac{1}{2}(\cos\delta, \sin\delta, \sin\delta, -\cos\delta).$$

Thus we can identify retrieval states as those solutions which are of the form

(i) retrieval of $\boldsymbol{\xi}^\lambda$: $(m_{cc}, m_{sc}, m_{cs}, m_{ss}) = m(\cos\delta, -\sin\delta, \sin\delta, \cos\delta)$,
(ii) retrieval of $-\boldsymbol{\xi}^\lambda$: $(m_{cc}, m_{sc}, m_{cs}, m_{ss}) = m(\cos\delta, \sin\delta, \sin\delta, -\cos\delta)$,

with full recall corresponding to $m = \frac{1}{2}$. Insertion into the saddle–point equations and into (2.183), followed by an appropriate shift of the integration variable ξ, shows that the free energy is independent of δ (so the above two ansätzes solve the saddle–point equations for any δ) and that

$$m = \frac{1}{2} \frac{\int d\xi \ \cos[\xi] e^{\beta m \cos[\xi]}}{\int d\xi \ e^{\beta m \cos[\xi]}}, \qquad f(m) = m^2 - \frac{1}{\beta} \log \int d\xi \ e^{\beta m \cos[\xi]}.$$

Expansion in powers of m, using $\log(1+z) = z - \frac{1}{2}z^2 + \mathcal{O}(z^3)$, reveals that non-zero minima m indeed bifurcate continuously at $T = \beta^{-1} = \frac{1}{4}$:

$$f(m) + \frac{1}{\beta}\log[2\pi] = (1 - \frac{1}{4}\beta)m^2 + \frac{1}{64}\beta^3 m^4 + \mathcal{O}(m^6). \tag{2.184}$$

Retrieval states are obviously not the only pure states that solve the saddle–point equations. The function (2.183) is invariant under the following discrete (non-commuting) transformations:

$$\text{I}: (m_{cc}, m_{sc}, m_{cs}, m_{ss}) \to (m_{cc}, m_{sc}, -m_{cs}, -m_{ss}),$$
$$\text{II}: (m_{cc}, m_{sc}, m_{cs}, m_{ss}) \to (m_{cs}, m_{ss}, m_{cc}, m_{sc}).$$

We expect these to induce solutions with specific symmetries. In particular we anticipate the following symmetric and antisymmetric states:

symmetric under I : $(m_{cc}, m_{sc}, m_{cs}, m_{ss}) = \sqrt{2}m(\cos\delta, \sin\delta, 0, 0)$,
antisymmetric under I : $(m_{cc}, m_{sc}, m_{cs}, m_{ss}) = \sqrt{2}m(0, 0, \cos\delta, \sin\delta)$,
symmetric under II : $(m_{cc}, m_{sc}, m_{cs}, m_{ss}) = m(\cos\delta, \sin\delta, \cos\delta, \sin\delta)$,
antisymmetric under II : $(m_{cc}, m_{sc}, m_{cs}, m_{ss}) = m(\cos\delta, \sin\delta, -\cos\delta, -\sin\delta)$.

Insertion into the saddle–point equations and into (2.183) shows in all four cases the parameter δ is arbitrary and that always

$$m = \frac{1}{\sqrt{2}} \int \frac{d\xi}{2\pi} \cos[\xi] \frac{\int d\xi \ \cos[\xi] e^{\beta m \sqrt{2}\cos[\xi]\cos[\xi]}}{\int d\xi \ e^{\beta m \sqrt{2}\cos[\xi]\cos[\xi]}},$$
$$f(m) = m^2 - \frac{1}{\beta} \int \frac{d\xi}{2\pi} \log \int d\xi \ e^{\beta m \sqrt{2}\cos[\xi]\cos[\xi]}.$$

Expansion in powers of m reveals that non-zero solutions m here again bifurcate continuously at $T = \frac{1}{4}$:

$$f(m) + \frac{1}{\beta}\log[2\pi] = (1 - \frac{1}{4}\beta)m^2 + \frac{3}{2} \cdot \frac{1}{64}\beta^3 m^4 + \mathcal{O}(m^6). \tag{2.185}$$

However, comparison with (2.184) shows that the free energy of the pure recall states is lower. Thus the system will prefer the recall states over the above solutions with specific symmetries.

Note, that the free energy and the order parameter equation for the pure recall states can be written in terms of *Bessel functions* as follows:

$$m = \frac{1}{2}\frac{I_1(\beta m)}{I_0(\beta m)}, \qquad f(m) = m^2 - \frac{1}{\beta}\log[2\pi I_0(\beta m)]$$

2.3.6 Attractor Neural Networks with Binary Neurons

The simplest non–trivial recurrent neural networks consist of N binary neurons $\sigma_i \in \{-1, 1\}$, which respond stochastically to post-synaptic potentials (or local fields) $h_i(\boldsymbol{\sigma})$, with $\boldsymbol{\sigma} = (\sigma_1, \ldots, \sigma_N)$. The fields depend linearly on the instantaneous neuron states,

$$h_i(\boldsymbol{\sigma}) = J_{ij}\sigma_j + \theta_i,$$

with the J_{ij} representing synaptic efficacies, and the θ_i representing external stimuli and/or neural thresholds [Coo01, SC00, SC01, CKS05].

Closed Macroscopic Laws for Sequential Dynamics

Here, we will first show how for sequential dynamics (where neurons are updated one after the other) one can calculate, from the microscopic stochastic laws, differential equations for the probability distribution of suitably defined macroscopic observables. For mathematical convenience our starting point will be the continuous-time master equation for the microscopic probability distribution $p_t(\boldsymbol{\sigma})$

$$\dot{p}_t(\boldsymbol{\sigma}) = \{w_i(F_i\boldsymbol{\sigma})p_t(F_i\boldsymbol{\sigma}) - w_i(\boldsymbol{\sigma})p_t(\boldsymbol{\sigma})\}, \qquad (2.186)$$

$$w_i(\boldsymbol{\sigma}) = \frac{1}{2}[1 - \sigma_i \tanh[\beta h_i(\boldsymbol{\sigma})]],$$

with $F_i\Phi(\boldsymbol{\sigma}) = \Phi(\sigma_1, \ldots, \sigma_{i-1}, -\sigma_i, \sigma_{i+1}, \ldots, \sigma_N)$.

Let us illustrate the basic ideas with the help of a simple (infinite range) toy model: $J_{ij} = (J/N)\eta_i\xi_j$ and $\theta_i = 0$ (the variables η_i and ξ_i are arbitrary, but may not depend on N). For $\eta_i = \xi_i = 1$ we get a network with uniform synapses. For $\eta_i = \xi_i \in \{-1, 1\}$ and $J > 0$ we recover the Hopfield [Hop82] model with one stored pattern. Note: the synaptic matrix is non-symmetric as soon as a pair (ij) exists such that $\eta_i\xi_j \neq \eta_j\xi_i$, so in general equilibrium statistical mechanics will not apply. The local fields become $h_i(\boldsymbol{\sigma}) = J\eta_i m(\boldsymbol{\sigma})$ with $m(\boldsymbol{\sigma}) = \frac{1}{N}\sum_k \xi_k\sigma_k$. Since they depend on the microscopic state $\boldsymbol{\sigma}$ only through the value of m, the latter quantity appears to constitute a natural macroscopic level of description. The probability density of finding the macroscopic state $m(\boldsymbol{\sigma}) = m$ is given by $\mathcal{P}_t[m] = \sum_{\boldsymbol{\sigma}} p_t(\boldsymbol{\sigma})\delta[m - m(\boldsymbol{\sigma})]$. Its time derivative follows upon inserting (2.186):

$$\dot{\mathcal{P}}_t[m] = \sum_\sigma p_t(\boldsymbol{\sigma})w_k(\boldsymbol{\sigma})\left\{\delta[m - m(\boldsymbol{\sigma}) + \frac{2}{N}\xi_k\sigma_k] - \delta[m - m(\boldsymbol{\sigma})]\right\}$$

$$= \frac{\partial}{\partial m}\left\{\sum_\sigma p_t(\boldsymbol{\sigma})\delta[m - m(\boldsymbol{\sigma})] \frac{2}{N}\xi_k\sigma_k w_k(\boldsymbol{\sigma})\right\} + \mathcal{O}(\frac{1}{N}).$$

Inserting our expressions for the transition rates $w_i(\boldsymbol{\sigma})$ and the local fields $h_i(\boldsymbol{\sigma})$ gives:

$$\dot{\mathcal{P}}_t[m] = \frac{\partial}{\partial m}\left\{\mathcal{P}_t[m]\left[m - \frac{1}{N}\xi_k\tanh[\eta_k\beta Jm]\right]\right\} + \mathcal{O}(N^{-1}).$$

In the limit $N \to \infty$ only the first term survives. The general solution of the resulting Liouville equation is

$$\mathcal{P}_t[m] = \int dm_0\, \mathcal{P}_0[m_0]\delta[m - m(t|m_0)],$$

where $m(t|m_0)$ is the solution of

$$\dot{m} = \lim_{N\to\infty} \frac{1}{N}\xi_k\tanh[\eta_k\beta Jm] - m, \qquad m(0) = m_0. \tag{2.187}$$

This describes deterministic evolution; the only uncertainty in the value of m is due to uncertainty in initial conditions. If at $t = 0$ the quantity m is known exactly, this will remain the case for finite time–scales; m turns out to evolve in time according to (2.187).

Let us now allow for less trivial choices of the synaptic matrix $\{J_{ij}\}$ and try to calculate the evolution in time of a given set of macroscopic observables $\boldsymbol{\Omega}(\boldsymbol{\sigma}) = (\Omega_1(\boldsymbol{\sigma}), \ldots, \Omega_n(\boldsymbol{\sigma}))$ in the limit $N \to \infty$. There are no restrictions yet on the form or the number n of these state variables; these will, however, arise naturally if we require the observables $\boldsymbol{\Omega}$ to obey a closed set of deterministic laws, as we will see. The probability density of finding the system in macroscopic state $\boldsymbol{\Omega}$ is given by [Coo01, SC00, SC01, CKS05]:

$$\mathcal{P}_t[\boldsymbol{\Omega}] = \sum_\sigma p_t(\boldsymbol{\sigma})\delta[\boldsymbol{\Omega} - \boldsymbol{\Omega}(\boldsymbol{\sigma})]. \tag{2.188}$$

Its time derivative is got by inserting (2.186). If in those parts of the resulting expression which contain the operators F_i we perform the transformations $\boldsymbol{\sigma} \to F_i\boldsymbol{\sigma}$, we arrive at

$$\dot{\mathcal{P}}_t[\boldsymbol{\Omega}] = \sum_\sigma p_t(\boldsymbol{\sigma})w_i(\boldsymbol{\sigma})\{\delta[\boldsymbol{\Omega} - \boldsymbol{\Omega}(F_i\boldsymbol{\sigma})] - \delta[\boldsymbol{\Omega} - \boldsymbol{\Omega}(\boldsymbol{\sigma})]\}.$$

If we define

$$\Omega_\mu(F_i\boldsymbol{\sigma}) = \Omega_\mu(\boldsymbol{\sigma}) + \Delta_{i\mu}(\boldsymbol{\sigma})$$

and make a Taylor expansion in powers of $\{\Delta_{i\mu}(\boldsymbol{\sigma})\}$, we finally get the so-called *Kramers–Moyal expansion*:[9]

$$\dot{\mathcal{P}}_t[\boldsymbol{\Omega}] = \sum_{\ell \geq 1} \frac{(-1)^\ell}{\ell+} \sum_{\mu_1=1}^n \cdots \sum_{\mu_\ell=1}^n \frac{\partial^\ell}{\partial \Omega_{\mu_1} \cdots \partial \Omega_{\mu_\ell}} \left\{ \mathcal{P}_t[\boldsymbol{\Omega}] F^{(\ell)}_{\mu_1 \cdots \mu_\ell}[\boldsymbol{\Omega}; t] \right\}. \tag{2.189}$$

It involves conditional averages $\langle f(\boldsymbol{\sigma}) \rangle_{\boldsymbol{\Omega}; t}$ and the 'discrete derivatives'

$$\Delta_{j\mu}(\boldsymbol{\sigma}) = \Omega_\mu(F_j \boldsymbol{\sigma}) - \Omega_\mu(\boldsymbol{\sigma}),$$

$$F^{(\ell)}_{\mu_1 \cdots \mu_\ell}[\boldsymbol{\Omega}; t] = \langle w_j(\boldsymbol{\sigma}) \Delta_{j\mu_1}(\boldsymbol{\sigma}) \cdots \Delta_{j\mu_\ell}(\boldsymbol{\sigma}) \rangle_{\boldsymbol{\Omega}; t},$$

$$\langle f(\boldsymbol{\sigma}) \rangle_{\boldsymbol{\Omega}; t} = \frac{\sum_{\boldsymbol{\sigma}} p_t(\boldsymbol{\sigma}) \delta[\boldsymbol{\Omega} - \boldsymbol{\Omega}(\boldsymbol{\sigma})] f(\boldsymbol{\sigma})}{\sum_{\boldsymbol{\sigma}} p_t(\boldsymbol{\sigma}) \delta[\boldsymbol{\Omega} - \boldsymbol{\Omega}(\boldsymbol{\sigma})]}. \tag{2.190}$$

Retaining only the $\ell = 1$ term in (2.189) would lead us to a Liouville equation, which describes deterministic flow in $\boldsymbol{\Omega}$ space. Including also the $\ell = 2$ term leads us to a Fokker–Planck equation which, in addition to flow, describes diffusion of the macroscopic probability density. Thus a sufficient condition for the observables $\boldsymbol{\Omega}(\boldsymbol{\sigma})$ to evolve in time deterministically in the limit $N \to \infty$ is:

$$\lim_{N \to \infty} \sum_{\ell \geq 2} \frac{1}{\ell+} \sum_{\mu_1=1}^n \cdots \sum_{\mu_\ell=1}^n \langle |\Delta_{j\mu_1}(\boldsymbol{\sigma}) \cdots \Delta_{j\mu_\ell}(\boldsymbol{\sigma})| \rangle_{\boldsymbol{\Omega}; t} = 0. \tag{2.191}$$

In the simple case where all observables Ω_μ scale similarly in the sense that all 'derivatives' $\Delta_{j\mu} = \Omega_\mu(F_i\boldsymbol{\sigma}) - \Omega_\mu(\boldsymbol{\sigma})$ are of the same order in N (i.e., there is a monotonic function $\tilde{\Delta}_N$ such that $\Delta_{j\mu} = \mathcal{O}(\tilde{\Delta}_N)$ for all $j\mu$), for instance, criterion (2.191) becomes:

$$\lim_{N \to \infty} n \tilde{\Delta}_N \sqrt{N} = 0. \tag{2.192}$$

If for a given set of observables condition (2.191) is satisfied, we can for large N describe the evolution of the macroscopic probability density by a Liouville equation:

$$\dot{\mathcal{P}}_t[\boldsymbol{\Omega}] = -\frac{\partial}{\partial \Omega_\mu} \left\{ \mathcal{P}_t[\boldsymbol{\Omega}] F^{(1)}_\mu[\boldsymbol{\Omega}; t] \right\},$$

whose solution describes deterministic flow,

---

[9] The Kramers–Moyal expansion (2.189) is to be interpreted in a distributional sense, i.e., only to be used in expressions of the form $\int d\boldsymbol{\Omega}_t(\boldsymbol{\Omega}) G(\boldsymbol{\Omega})$ with smooth functions $G(\boldsymbol{\Omega})$, so that all derivatives are well-defined and finite. Furthermore, (2.189) will only be useful if the $\Delta_{j\mu}$, which measure the sensitivity of the macroscopic quantities to single neuron state changes, are sufficiently small. This is to be expected: for finite N any observable can only assume a finite number of possible values; only for $N \to \infty$ may we expect smooth probability distributions for our macroscopic quantities.

$$\mathcal{P}_t[\boldsymbol{\Omega}] = \int d\boldsymbol{\Omega}_0 \mathcal{P}_0[\boldsymbol{\Omega}_0]\delta[\boldsymbol{\Omega} - \boldsymbol{\Omega}(t|\boldsymbol{\Omega}_0)],$$

with $\boldsymbol{\Omega}(t|\boldsymbol{\Omega}_0)$ given, in turn, as the solution of

$$\dot{\boldsymbol{\Omega}}(t) = \mathbf{F}^{(1)}\left[\boldsymbol{\Omega}(t); t\right], \qquad \boldsymbol{\Omega}(0) = \boldsymbol{\Omega}_0. \tag{2.193}$$

In taking the limit $N \to \infty$, however, we have to keep in mind that the resulting deterministic theory is got by taking this limit for *finite* t. According to (2.189) the $\ell > 1$ terms do come into play for sufficiently large times t; for $N \to \infty$, however, these times diverge by virtue of (2.191).

Now, equation (2.193) will in general not be autonomous; tracing back the origin of the explicit time dependence in the right–hand side of (2.193) one finds that to calculate $\mathbf{F}^{(1)}$ one needs to know the microscopic probability density $p_t(\boldsymbol{\sigma})$. This, in turn, requires solving equation (2.186) (which is exactly what one tries to avoid). We will now discuss a mechanism via which to eliminate the offending explicit time dependence, and to turn the observables $\boldsymbol{\Omega}(\boldsymbol{\sigma})$ into an autonomous level of description, governed by *closed* dynamic laws. The idea is to choose the observables $\boldsymbol{\Omega}(\boldsymbol{\sigma})$ in such a way that there is no explicit time dependence in the flow field $\mathbf{F}^{(1)}[\boldsymbol{\Omega}; t]$ (if possible). According to (2.190) this implies making sure that there exist functions $\Phi_\mu[\boldsymbol{\Omega}]$ such that

$$\lim_{N \to \infty} w_j(\boldsymbol{\sigma})\Delta_{j\mu}(\boldsymbol{\sigma}) = \Phi_\mu\left[\boldsymbol{\Omega}(\boldsymbol{\sigma})\right], \tag{2.194}$$

in which case the time dependence of $\mathbf{F}^{(1)}$ indeed drops out and the macroscopic state vector simply evolves in time according to [Coo01, SC00, SC01, CKS05]:

$$\dot{\boldsymbol{\Omega}} = \boldsymbol{\Phi}\left[\boldsymbol{\Omega}\right], \qquad \boldsymbol{\Phi} = (\Phi_1[\boldsymbol{\Omega}], \ldots, \Phi_n[\boldsymbol{\Omega}]).$$

Clearly, for this closure method to apply, a suitable separable structure of the synaptic matrix is required. If, for instance, the macroscopic observables Ω_μ depend linearly on the microscopic state variables $\boldsymbol{\sigma}$ (i.e., $\Omega_\mu(\boldsymbol{\sigma}) = \frac{1}{N}\sum_{j=1}^N \omega_{\mu j}\sigma_j$), we get with the transition rates defined in (2.186):

$$\dot{\Omega}_\mu = \lim_{N \to \infty} \frac{1}{N}\omega_{\mu j}\tanh(\beta h_j(\boldsymbol{\sigma})) - \Omega_\mu, \tag{2.195}$$

in which case the only further condition for (2.194) to hold is that all local fields $h_k(\boldsymbol{\sigma})$ must (in leading order in N) depend on the microscopic state $\boldsymbol{\sigma}$ only through the values of the observables $\boldsymbol{\Omega}$; since the local fields depend linearly on $\boldsymbol{\sigma}$ this, in turn, implies that the synaptic matrix must be separable: if $J_{ij} = \sum_\mu K_{i\mu}\omega_{\mu j}$ then indeed $h_i(\boldsymbol{\sigma}) = \sum_\mu K_{i\mu}\Omega_\mu(\boldsymbol{\sigma}) + \theta_i$. Next we will show how this approach can be applied to networks for which the matrix of synapses has a separable form (which includes most symmetric and non–symmetric Hebbian type attractor models). We will restrict ourselves to models with $\theta_i = 0$; introducing non–zero thresholds is straightforward and does not pose new problems.

Application to Separable Attractor Networks

We consider the following class of models, in which the interaction matrices have the form

$$J_{ij} = \frac{1}{N} Q(\boldsymbol{\xi}_i; \boldsymbol{\xi}_j), \qquad \boldsymbol{\xi}_i = (\xi_i^1, \ldots, \xi_i^p). \tag{2.196}$$

The components ξ_i^μ, representing the information ('patterns') to be stored or processed, are assumed to be drawn from a finite discrete set Λ, containing n_Λ elements (they are not allowed to depend on N). The Hopfield model [Hop82] corresponds to choosing $Q(\mathbf{x}; \mathbf{y}) = \mathbf{x} \cdot \mathbf{y}$ and $\Lambda \equiv \{-1, 1\}$. One now introduces a partition of the system $\{1, \ldots, N\}$ into n_Λ^p so–called sublattices I_η:

$$I_\eta = \{i | \ \boldsymbol{\xi}_i = \boldsymbol{\eta}\}, \qquad \{1, \ldots, N\} = \bigcup_\eta I_\eta, \qquad (\boldsymbol{\eta} \in \Lambda^p).$$

The number of neurons in sublattice I_η is denoted by $|I_\eta|$ (this number will have to be large). If we choose as our macroscopic observables the average activities ('magnetizations') within these sublattices, we are able to express the local fields h_k solely in terms of macroscopic quantities:

$$m_{\boldsymbol{\eta}}(\boldsymbol{\sigma}) = \frac{1}{|I_\eta|} \sum_{i \in I_\eta} \sigma_i, \qquad h_k(\boldsymbol{\sigma}) = p_{\boldsymbol{\eta}} Q(\boldsymbol{\xi}_k; \boldsymbol{\eta}) \, m_{\boldsymbol{\eta}}, \tag{2.197}$$

with the relative sublattice sizes $p_{\boldsymbol{\eta}} = |I_\eta|/N$. If all $p_{\boldsymbol{\eta}}$ are of the same order in N (which, for example, is the case if the vectors $\boldsymbol{\xi}_i$ have been drawn at random from the set Λ^p) we may write $\Delta_{j\boldsymbol{\eta}} = \mathcal{O}(n_\Lambda^p N^{-1})$ and use (2.192). The evolution in time of the sublattice activities is then found to be deterministic in the $N \to \infty$ limit if $\lim_{N \to \infty} p/\log N = 0$. Furthermore, condition (2.194) holds, since

$$w_j(\boldsymbol{\sigma}) \Delta_{j\boldsymbol{\eta}}(\boldsymbol{\sigma}) = \tanh[\beta p_{\boldsymbol{\eta}'} Q(\boldsymbol{\eta}; \boldsymbol{\eta}') \, m_{\boldsymbol{\eta}'}] - m_{\boldsymbol{\eta}}.$$

This situation is described by (2.195), and that the evolution in time of the sublattice activities is governed by the following autonomous set of differential equations [RKH88]:

$$\dot{m}_{\boldsymbol{\eta}} = \tanh[\beta p_{\boldsymbol{\eta}'} Q(\boldsymbol{\eta}; \boldsymbol{\eta}') \, m_{\boldsymbol{\eta}'}] - m_{\boldsymbol{\eta}}. \tag{2.198}$$

We see that, in contrast to the equilibrium techniques as described above, here there is no need at all to require symmetry of the interaction matrix or absence of self-interactions. In the symmetric case $Q(\mathbf{x}; \mathbf{y}) = Q(\mathbf{y}; \mathbf{x})$ the system will approach equilibrium; if the kernel Q is positive definite this can be shown, for instance, by inspection of the *Lyapunov function* $\mathcal{L}\{m_{\boldsymbol{\eta}}\}$:[10]

---

[10] Recall that *Lyapunov function* is a function of the state variables which is bounded from below and whose value decreases monotonically during the dynamics, see e.g., [Kha92]. Its existence guarantees evolution towards a stationary state (under some weak conditions).

$$\mathcal{L}\{m_\eta\} = \frac{1}{2} p_\eta m_\eta Q(\eta; \eta') m_{\eta'} p_{\eta'}$$
$$- \frac{1}{\beta} \sum_\eta p_\eta \log \cosh[\beta Q(\eta; \eta') m_{\eta'} p_{\eta'}],$$

which is bounded from below and obeys:

$$\dot{\mathcal{L}} = -[p_\eta \dot{m}_\eta] Q(\eta; \eta') [p_{\eta'} \dot{m}_{\eta'}] \leq 0. \tag{2.199}$$

Note that from the sublattice activities, in turn, follow the *Hopfield overlaps* $m_\mu(\sigma)$,

$$m_\mu(\sigma) = \frac{1}{N} \xi_i^\mu \sigma_i = p_\eta \eta_\mu m_\eta. \tag{2.200}$$

Simple examples of relevant models of the type (2.196), the dynamics of which are for large N described by equation (2.198), are for instance the ones where one applies a nonlinear operation Φ to the standard Hopfield–type [Hop82] (or Hebbian) interactions. This nonlinearity could result from e.g., a clipping procedure or from retaining only the *sign* of the Hebbian values:

$$J_{ij} = \frac{1}{N} \Phi(\xi_i^\mu \xi_j^\mu) : \quad \text{e.g.,}$$

$$\Phi(x) = \begin{cases} -K & \text{for} \quad x \leq K \\ x & \text{for} \quad -K < x < K \\ K & \text{for} \quad x \geq K \end{cases}, \quad \text{or} \quad \Phi(x) = \text{sgn}(x).$$

The effect of introducing such nonlinearities is found to be of a quantitative nature, giving rise to little more than a re-scaling of critical noise levels and storage capacities. We will illustrate this statement by working out the $p = 2$ equations for randomly drawn pattern bits $\xi_i^\mu \in \{-1, 1\}$, where there are only four sub-lattices, and where $p_\eta = \frac{1}{4}$ for all η (details can be found in e.g., [DHS91]). Using $\Phi(0) = 0$ and $\Phi(-x) = -\Phi(x)$ (as with the above examples) we get from (2.198):

$$\dot{m}_\eta = \tanh[\frac{1}{4}\beta \Phi(2)(m_\eta - m_{-\eta})] - m_\eta. \tag{2.201}$$

Here the choice made for $\Phi(x)$ shows up only as a re-scaling of the temperature. From (2.201) we further get $\frac{d}{dt}(m_\eta + m_{-\eta}) = -(m_\eta + m_{-\eta})$. The system decays exponentially towards a state where, according to (2.200), $m_\eta = -m_{-\eta}$ for all η. If at $t = 0$ this is already the case, we find (at least for $p = 2$) decoupled equations for the sub-lattice activities.

Now, equations (2.198,2.200) suggest that at the level of overlaps there will be, in turn, closed laws if the kernel Q is bilinear, $Q(\mathbf{x}; \mathbf{y}) = \sum_{\mu\nu} x_\mu A_{\mu\nu} y_\nu$, or [Coo01, SC00, SC01, CKS05]:

$$J_{ij} = \frac{1}{N} \xi_i^\mu A_{\mu\nu} \xi_j^\nu, \quad \boldsymbol{\xi}_i = (\xi_i^1, \ldots, \xi_i^p). \tag{2.202}$$

2.3 Synergetics of Recurrent and Attractor Neural Networks

We will see that now the ξ_i^μ need not be drawn from a finite discrete set (as long as they do not depend on N). The Hopfield model corresponds to $A_{\mu\nu} = \delta_{\mu\nu}$ and $\xi_i^\mu \in \{-1, 1\}$. The fields h_k can now be written in terms of the overlaps m_μ:

$$h_k(\boldsymbol{\sigma}) = \boldsymbol{\xi}_k \cdot A\mathbf{m}(\boldsymbol{\sigma}), \qquad \mathbf{m} = (m_1, \ldots, m_p), \qquad m_\mu(\boldsymbol{\sigma}) = \frac{1}{N}\xi_i^\mu \sigma_i.$$

For this choice of macroscopic variables we find $\Delta_{j\mu} = \mathcal{O}(N^{-1})$, so the evolution of the vector $\mathbf{m}$ becomes deterministic for $N \to \infty$ if, according to (2.192), $\lim_{N\to\infty} p/\sqrt{N} = 0$. Again (2.194) holds, since

$$w_j(\boldsymbol{\sigma})\Delta_{j\mu}(\boldsymbol{\sigma}) = \frac{1}{N}\boldsymbol{\xi}_k \tanh[\beta\boldsymbol{\xi}_k \cdot A\mathbf{m}] - \mathbf{m}.$$

Thus the evolution in time of the overlap vector $\mathbf{m}$ is governed by a closed set of differential equations:

$$\dot{\mathbf{m}} = \langle \boldsymbol{\xi} \tanh[\beta\boldsymbol{\xi} \cdot A\mathbf{m}]\rangle_{\boldsymbol{\xi}} - \mathbf{m}, \qquad \langle \Phi(\boldsymbol{\xi})\rangle_{\boldsymbol{\xi}} = \int d\boldsymbol{\xi}\, \rho(\boldsymbol{\xi})\Phi(\boldsymbol{\xi}), \qquad (2.203)$$

with $\rho(\boldsymbol{\xi}) = \lim_{N\to\infty} N^{-1}\sum_i \delta[\boldsymbol{\xi} - \boldsymbol{\xi}_i]$. Symmetry of the synapses is not required. For certain non–symmetric matrices A one finds stable limit–cycle solutions of (2.203). In the symmetric case $A_{\mu\nu} = A_{\nu\mu}$ the system will approach equilibrium; the Lyapunov function (2.199) for positive definite matrices A now becomes:

$$\mathcal{L}\{\mathbf{m}\} = \frac{1}{2}\mathbf{m} \cdot A\mathbf{m} - \frac{1}{\beta}\langle \log \cosh[\beta\boldsymbol{\xi} \cdot A\mathbf{m}]\rangle_{\boldsymbol{\xi}}.$$

As a second simple application of the flow equations (2.203) we turn to the relaxation times corresponding to the attractors of the Hopfield model (where $A_{\mu\nu} = \delta_{\mu\nu}$). Expanding (2.203) near a stable fixed–point $\mathbf{m}^*$, i.e., $\mathbf{m}(t) = \mathbf{m}^* + \mathbf{x}(t)$ with $|\mathbf{x}(t)| \ll 1$, gives the linearized equation

$$\dot{x}_\mu = [\beta\langle \xi_\mu \xi_\nu \tanh[\beta\boldsymbol{\xi} \cdot \mathbf{m}^*]\rangle_{\boldsymbol{\xi}} - \delta_{\mu\nu}]x_\nu + \mathcal{O}(\mathbf{x}^2). \qquad (2.204)$$

The Jacobian of (2.203), which determines the linearized equation (2.204), turns out to be *minus* the curvature matrix of the free energy surface at the fixed-point. The asymptotic relaxation towards any stable attractor is generally exponential, with a characteristic time τ given by the inverse of the smallest eigenvalue of the curvature matrix. If, in particular, for the fixed–point $\mathbf{m}^*$ we substitute an n–mixture state, i.e., $m_\mu = m_n$ ($\mu \leq n$) and $m_\mu = 0$ ($\mu > n$), and transform (2.204) to the basis where the corresponding curvature matrix $\mathbf{D}^{(n)}$ (with eigenvalues D_λ^n) is diagonal, $\mathbf{x} \to \tilde{\mathbf{x}}$, we get $\tilde{x}_\lambda(t) = \tilde{x}_\lambda(0)e^{-tD_\lambda^n} + \ldots$ so $\tau^{-1} = \min_\lambda D_\lambda^n$, which we have already calculated in determining the character of the saddle–points of the free-energy surface. The relaxation time for the n–mixture attractors decreases monotonically with the degree of mixing n, for any noise level. At the transition where a macroscopic state $\mathbf{m}^*$ ceases to correspond to a local minimum of the free energy surface, it also de–stabilizes in terms of the linearized dynamic equation (2.204) (as it should). The Jacobian develops a zero eigenvalue, the relaxation time diverges, and the long–time behavior is no longer got from the linearized equation. This gives rise to critical slowing down (i.e., power law relaxation as opposed to exponential relaxation). For instance, at the transition temperature $T_c = 1$ for the $n = 1$ (pure) state, we find by expanding (2.203):

$$\dot{m}_\mu = m_\mu [\frac{2}{3} m_\mu^2 - \mathbf{m}^2] + \mathcal{O}(\mathbf{m}^5),$$

which gives rise to a relaxation towards the trivial fixed–point of the form $\mathbf{m} \sim t^{-\frac{1}{2}}$.

If one is willing to restrict oneself to the limited class of models (2.202) (as opposed to the more general class (2.196)) and to the more global level of description in terms of p overlap parameters m_μ instead of n_A^p sublattice activities m_η, then there are two rewards. Firstly there will be no restrictions on the stored pattern components ξ_i^μ (for instance, they are allowed to be real-valued); secondly the number p of patterns stored can be much larger for the deterministic autonomous dynamical laws to hold ($p \ll \sqrt{N}$ instead of $p \ll \log N$, which from a biological point of view is not impressive [Coo01, SC00, SC01, CKS05].

Closed Macroscopic Laws for Parallel Dynamics

We now turn to the parallel dynamics counterpart of (2.186), i.e., the Markov chain

$$p_{\ell+1}(\boldsymbol{\sigma}) = \sum_{\boldsymbol{\sigma}'} W[\boldsymbol{\sigma}; \boldsymbol{\sigma}'] p_\ell(\boldsymbol{\sigma}'), \tag{2.205}$$

$$W[\boldsymbol{\sigma}; \boldsymbol{\sigma}'] = \prod_{i=1}^N \frac{1}{2}[1 + \sigma_i \tanh[\beta h_i(\boldsymbol{\sigma}')]],$$

with $\sigma_i \in \{-1, 1\}$, and with local fields $h_i(\boldsymbol{\sigma})$ defined in the usual way. The evolution of macroscopic probability densities will here be described by discrete maps, in stead of differential equations.

Let us first see what happens to our previous toy model: $J_{ij} = (J/N)\eta_i \xi_j$ and $\theta_i = 0$. As before we try to describe the dynamics at the (macroscopic) level of the quantity $m(\boldsymbol{\sigma}) = \frac{1}{N} \sum_k \xi_k \sigma_k$. The evolution of the macroscopic probability density $\mathcal{P}_t[m]$ is got by inserting (2.205):

2.3 Synergetics of Recurrent and Attractor Neural Networks

$$\mathcal{P}_{t+1}[m] = \sum_{\sigma\sigma'} \delta[m - m(\sigma)] W[\sigma;\sigma'] p_t(\sigma')$$

$$= \int dm' \, \tilde{W}_t[m, m'] \mathcal{P}_t[m'], \quad \text{with} \qquad (2.206)$$

$$\tilde{W}_t[m, m'] = \frac{\sum_{\sigma\sigma'} \delta[m - m(\sigma)] \delta[m' - m(\sigma')] W[\sigma;\sigma'] p_t(\sigma')}{\sum_{\sigma'} \delta[m' - m(\sigma')] p_t(\sigma')}.$$

We now insert our expression for the transition probabilities $W[\sigma;\sigma']$ and for the local fields. Since the fields depend on the microscopic state σ only through $m(\sigma)$, the distribution $p_t(\sigma)$ drops out of the above expression for $\tilde{W}_t$ which thereby loses its explicit time–dependence, $\tilde{W}_t[m, m'] \to \tilde{W}[m, m']$:

$$\tilde{W}[m, m'] = e^{-\sum_i \log \cosh(\beta J m' \eta_i)} \langle \delta[m - m(\sigma)] e^{\beta J m' \eta_i \sigma_i} \rangle_\sigma$$

with $\quad \langle \ldots \rangle_\sigma = 2^{-N} \sum_\sigma \ldots$

Inserting the integral representation for the δ–function allows us to perform the average:

$$\tilde{W}[m, m'] = \left[\frac{\beta N}{2\pi}\right] \int dk \, e^{N\Psi(m, m', k)},$$

$$\Psi = i\beta km + \langle \log \cosh \beta [J\eta m' - ik\xi] \rangle_{\eta,\xi} - \langle \log \cosh \beta [J\eta m'] \rangle_\eta.$$

Since $\tilde{W}[m, m']$ is (by construction) normalized, $\int dm \, \tilde{W}[m, m'] = 1$, we find that for $N \to \infty$ the expectation value with respect to $\tilde{W}[m, m']$ of any sufficiently smooth function $f(m)$ will be determined only by the value $m^*(m')$ of m in the relevant saddle–point of Ψ:

$$\int dm \, f(m) \tilde{W}[m, m'] = \frac{\int dm dk \, f(m) e^{N\Psi(m, m', k)}}{\int dm dk \, e^{N\Psi(m, m', k)}} \to f(m^*(m')), \quad (N \to \infty).$$

Variation of Ψ with respect to k and m gives the two saddle–point equations:

$$m = \langle \xi \tanh \beta [J\eta m' - \xi k] \rangle_{\eta,\xi}, \qquad k = 0.$$

We may now conclude that $\lim_{N\to\infty} \tilde{W}[m, m'] = \delta[m - m^*(m')]$ with $m^*(m') = \langle \xi \tanh(\beta J\eta m') \rangle_{\eta,\xi}$, and that the macroscopic equation (2.206) becomes [Coo01, SC00, SC01, CKS05]:

$$\mathcal{P}_{t+1}[m] = \int dm' \, \delta\left[m - \langle \xi \tanh(\beta J\eta m') \rangle_{\eta\xi}\right] \mathcal{P}_t[m'], \quad (N \to \infty).$$

This describes deterministic evolution. If at $t = 0$ we know m exactly, this will remain the case for finite time-scales, and m will evolve according to a discrete version of the sequential dynamics law (2.187):

$$m_{t+1} = \langle \xi \tanh[\beta J\eta m_t] \rangle_{\eta,\xi}.$$

We now try to generalize the above approach to less trivial classes of models. As for the sequential case we will find in the limit $N \to \infty$ closed deterministic evolution equations for a more general set of intensive macroscopic state variables $\boldsymbol{\Omega}(\boldsymbol{\sigma}) = (\Omega_1(\boldsymbol{\sigma}), \ldots, \Omega_n(\boldsymbol{\sigma}))$ if the local fields $h_i(\boldsymbol{\sigma})$ depend on the microscopic state $\boldsymbol{\sigma}$ only through the values of $\boldsymbol{\Omega}(\boldsymbol{\sigma})$, and if the number n of these state variables necessary to do so is not too large. The evolution of the ensemble probability density (2.188) is now got by inserting the Markov equation (2.205):

$$\mathcal{P}_{t+1}[\boldsymbol{\Omega}] = \int d\boldsymbol{\Omega}'\, \tilde{W}_t[\boldsymbol{\Omega}, \boldsymbol{\Omega}']\, \mathcal{P}_t[\boldsymbol{\Omega}'], \tag{2.207}$$

$$\tilde{W}_t[\boldsymbol{\Omega}, \boldsymbol{\Omega}'] = \frac{\sum_{\boldsymbol{\sigma}\boldsymbol{\sigma}'} \delta[\boldsymbol{\Omega} - \boldsymbol{\Omega}(\boldsymbol{\sigma})]\,\delta[\boldsymbol{\Omega}' - \boldsymbol{\Omega}(\boldsymbol{\sigma}')]\,W[\boldsymbol{\sigma}; \boldsymbol{\sigma}']\,p_t(\boldsymbol{\sigma}')}{\sum_{\boldsymbol{\sigma}'} \delta[\boldsymbol{\Omega}' - \boldsymbol{\Omega}(\boldsymbol{\sigma}')]\,p_t(\boldsymbol{\sigma}')} \tag{2.208}$$

$$= \langle \delta[\boldsymbol{\Omega} - \boldsymbol{\Omega}(\boldsymbol{\sigma})]\,\langle e^{[\beta\sigma_i h_i(\boldsymbol{\sigma}') - \log\cosh(\beta h_i(\boldsymbol{\sigma}'))]}\rangle_{\boldsymbol{\Omega}';t}\rangle_{\boldsymbol{\sigma}},$$

with $\langle \ldots \rangle_{\boldsymbol{\sigma}} = 2^{-N} \sum_{\boldsymbol{\sigma}} \ldots$, and with the conditional (or sub-shell) average defined as in (2.190). It is clear from (2.208) that in order to find autonomous macroscopic laws, i.e., for the distribution $p_t(\boldsymbol{\sigma})$ to drop out, the local fields must depend on the microscopic state $\boldsymbol{\sigma}$ only through the macroscopic quantities $\boldsymbol{\Omega}(\boldsymbol{\sigma})$: $h_i(\boldsymbol{\sigma}) = h_i[\boldsymbol{\Omega}(\boldsymbol{\sigma})]$. In this case $\tilde{W}_t$ loses its explicit time–dependence, $\tilde{W}_t[\boldsymbol{\Omega}, \boldsymbol{\Omega}'] \to \tilde{W}[\boldsymbol{\Omega}, \boldsymbol{\Omega}']$. Inserting integral representations for the δ–functions leads to:

$$\tilde{W}[\boldsymbol{\Omega}, \boldsymbol{\Omega}'] = \left[\frac{\beta N}{2\pi}\right]^n \int d\mathbf{K}\, e^{N\Psi(\boldsymbol{\Omega}, \boldsymbol{\Omega}', \mathbf{K})},$$

$$\Psi = i\beta \mathbf{K} \cdot \boldsymbol{\Omega} + \frac{1}{N} \log \langle e^{\beta[\sum_i \sigma_i h_i[\boldsymbol{\Omega}']] - iN\mathbf{K}\cdot\boldsymbol{\Omega}(\boldsymbol{\sigma})]} \rangle_{\boldsymbol{\sigma}}$$

$$- \frac{1}{N} \sum_i \log \cosh[\beta h_i[\boldsymbol{\Omega}']].$$

Using the normalization $\int d\boldsymbol{\Omega}\, \tilde{W}[\boldsymbol{\Omega}, \boldsymbol{\Omega}'] = 1$, we can write expectation values with respect to $\tilde{W}[\boldsymbol{\Omega}, \boldsymbol{\Omega}']$ of macroscopic quantities $f[\boldsymbol{\Omega}]$ as

$$\int d\boldsymbol{\Omega}\, f[\boldsymbol{\Omega}]\tilde{W}[\boldsymbol{\Omega}, \boldsymbol{\Omega}'] = \frac{\int d\boldsymbol{\Omega} d\mathbf{K}\, f[\boldsymbol{\Omega}]e^{N\Psi(\boldsymbol{\Omega}, \boldsymbol{\Omega}', \mathbf{K})}}{\int d\boldsymbol{\Omega} d\mathbf{K}\, e^{N\Psi(\boldsymbol{\Omega}, \boldsymbol{\Omega}', \mathbf{K})}}. \tag{2.209}$$

For saddle–point arguments to apply in determining the leading order in N of (2.209), we encounter restrictions on the number n of our macroscopic quantities (as expected), since n determines the dimension of the integrations in (2.209). The restrictions can be found by expanding Ψ around its maximum Ψ^*. After defining $\mathbf{x} = (\boldsymbol{\Omega}, \mathbf{K})$, of dimension $2n$, and after translating the location of the maximum to the origin, one has [Coo01, SC00, SC01, CKS05]

$$\Psi(\mathbf{x}) = \Psi^* - \frac{1}{2}x_\mu x_\nu H_{\mu\nu} + x_\mu x_\nu x_\rho L_{\mu\nu\rho} + \mathcal{O}(\mathbf{x}^4), \quad \text{giving}$$

2.3 Synergetics of Recurrent and Attractor Neural Networks

$$\frac{\int d\mathbf{x}\ g(\mathbf{x})e^{N\Psi(\mathbf{x})}}{\int d\mathbf{x}\ g(\mathbf{x})e^{N\Psi(\mathbf{x})}} - g(\mathbf{0}) = \frac{\int d\mathbf{x}\ [g(\mathbf{x}) - g(\mathbf{0})]e^{-\frac{1}{2}N\mathbf{x}\cdot\mathbf{H}\mathbf{x} + Nx_\mu x_\nu x_\rho L_{\mu\nu\rho} + \mathcal{O}(N\mathbf{x}^4)}}{\int d\mathbf{x}\ e^{-\frac{1}{2}N\mathbf{x}\cdot\mathbf{H}\mathbf{x} + Nx_\mu x_\nu x_\rho L_{\mu\nu\rho} + \mathcal{O}(N\mathbf{x}^4)}}$$

$$= \frac{\int d\mathbf{y}\ [g(\mathbf{y}/\sqrt{N}) - g(\mathbf{0})]e^{-\frac{1}{2}\mathbf{y}\cdot\mathbf{H}\mathbf{y} + y_\mu y_\nu y_\rho L_{\mu\nu\rho}/\sqrt{N} + \mathcal{O}(\mathbf{y}^4/N)}}{\int d\mathbf{y}\ e^{-\frac{1}{2}\mathbf{y}\cdot\mathbf{H}\mathbf{y} + y_\mu y_\nu y_\rho L_{\mu\nu\rho}/\sqrt{N} + \mathcal{O}(\mathbf{y}^4/N)}} =$$

$$\frac{\int d\mathbf{y}\ \left[N^{-\frac{1}{2}}\mathbf{y}\cdot\nabla g(\mathbf{0}) + \mathcal{O}(\mathbf{y}^2/N)\right]e^{-\frac{1}{2}\mathbf{y}\cdot\mathbf{H}\mathbf{y}}\left[1 + y_\mu y_\nu y_\rho L_{\mu\nu\rho}/\sqrt{N} + \mathcal{O}(\mathbf{y}^6/N)\right]}{\int d\mathbf{y}\ e^{-\frac{1}{2}\mathbf{y}\cdot\mathbf{H}\mathbf{y}}\left[1 + y_\mu y_\nu y_\rho L_{\mu\nu\rho}/\sqrt{N} + \mathcal{O}(\mathbf{y}^6/N)\right]}$$

$$= \mathcal{O}(n^2/N) + \mathcal{O}(n^4/N^2) + \text{non-dominant terms}, \qquad (N, n \to \infty),$$

with $\mathbf{H}$ denoting the Hessian (curvature) matrix of the surface Ψ at the minimum Ψ^*. We thus find

$$\lim_{N\to\infty} n/\sqrt{N} = 0: \qquad \lim_{N\to\infty} \int d\mathbf{\Omega}\ f[\mathbf{\Omega}]\tilde{W}[\mathbf{\Omega}, \mathbf{\Omega}'] = f[\mathbf{\Omega}^*(\mathbf{\Omega}')],$$

where $\mathbf{\Omega}^*(\mathbf{\Omega}')$ denotes the value of $\mathbf{\Omega}$ in the saddle–point where Ψ is minimized. Variation of Ψ with respect to $\mathbf{\Omega}$ and $\mathbf{K}$ gives the saddle–point equations:

$$\mathbf{\Omega} = \frac{\langle\mathbf{\Omega}(\boldsymbol{\sigma})e^{\beta[\sigma_i h_i[\mathbf{\Omega}'] - iN\mathbf{K}\cdot\mathbf{\Omega}(\boldsymbol{\sigma})]}\rangle_{\boldsymbol{\sigma}}}{\langle e^{\beta[\sigma_i h_i[\mathbf{\Omega}'] - iN\mathbf{K}\cdot\mathbf{\Omega}(\boldsymbol{\sigma})]}\rangle_{\boldsymbol{\sigma}}}, \qquad \mathbf{K} = 0.$$

We may now conclude that $\lim_{N\to\infty} \tilde{W}[\mathbf{\Omega}, \mathbf{\Omega}'] = \delta[\mathbf{\Omega} - \mathbf{\Omega}^*(\mathbf{\Omega}')]$, with

$$\mathbf{\Omega}^*(\mathbf{\Omega}') = \frac{\langle\mathbf{\Omega}(\boldsymbol{\sigma})e^{\beta\sigma_i h_i[\mathbf{\Omega}']}\rangle_{\boldsymbol{\sigma}}}{\langle e^{\beta\sigma_i h_i[\mathbf{\Omega}']}\rangle_{\boldsymbol{\sigma}}},$$

and that for $N \to \infty$ the macroscopic equation (2.207) becomes $\mathcal{P}_{t+1}[\mathbf{\Omega}] = \int d\mathbf{\Omega}'\ \delta[\mathbf{\Omega} - \mathbf{\Omega}^*(\mathbf{\Omega}')]\mathcal{P}_t[\mathbf{\Omega}']$. This relation again describes deterministic evolution. If at $t = 0$ we know $\mathbf{\Omega}$ exactly, this will remain the case for finite time–scales and $\mathbf{\Omega}$ will evolve according to

$$\mathbf{\Omega}(t+1) = \frac{\langle\mathbf{\Omega}(\boldsymbol{\sigma})e^{\beta\sigma_i h_i[\mathbf{\Omega}(t)]}\rangle_{\boldsymbol{\sigma}}}{\langle e^{\beta\sigma_i h_i[\mathbf{\Omega}(t)]}\rangle_{\boldsymbol{\sigma}}}. \tag{2.210}$$

As with the sequential case, in taking the limit $N \to \infty$ we have to keep in mind that the resulting laws apply to finite t, and that for sufficiently large times terms of higher order in N do come into play. As for the sequential case, a more rigorous and tedious analysis shows that the restriction $n/\sqrt{N} \to 0$ can in fact be weakened to $n/N \to 0$. Finally, for macroscopic quantities $\mathbf{\Omega}(\boldsymbol{\sigma})$ which are linear in $\boldsymbol{\sigma}$, the remaining $\boldsymbol{\sigma}$–averages become trivial, so that [Ber91]:

$$\Omega_\mu(\boldsymbol{\sigma}) = \frac{1}{N}\omega_{\mu i}\sigma_i: \qquad \Omega_\mu(t+1) = \lim_{N\to\infty}\frac{1}{N}\omega_{\mu i}\tanh\left[\beta h_i[\mathbf{\Omega}(t)]\right].$$

Application to Separable Attractor Networks

The separable attractor models (2.196), described at the level of sublattice activities (2.197), indeed have the property that all local fields can be written in terms of the macroscopic observables. What remains to ensure deterministic evolution is meeting the condition on the number of sublattices. If all relative sublattice sizes p_η are of the same order in N (as for randomly drawn patterns) this condition again translates into $\lim_{N\to\infty} p/\log N = 0$ (as for sequential dynamics). Since the sublattice activities are linear functions of the σ_i, their evolution in time is governed by equation (2.210), which acquires the form:

$$m_\eta(t+1) = \tanh[\beta p_{\eta'} Q(\eta;\eta') m_{\eta'}(t)]. \tag{2.211}$$

As for sequential dynamics, symmetry of the interaction matrix does not play a role.

At the more global level of overlaps $m_\mu(\boldsymbol{\sigma}) = N^{-1}\sum_i \xi_i^\mu \sigma_i$ we, in turn, get autonomous deterministic laws if the local fields $h_i(\boldsymbol{\sigma})$ can be expressed in terms if $\mathbf{m}(\boldsymbol{\sigma})$ only, as for the models (2.202) (or, more generally, for all models in which the interactions are of the form $J_{ij} = \sum_{\mu\leq p} f_{i\mu}\xi_j^\mu$), and with the following restriction on the number p of embedded patterns: $\lim_{N\to\infty} p/\sqrt{N} = 0$ (as with sequential dynamics). For the bilinear models (2.202), the evolution in time of the overlap vector $\mathbf{m}$ (which depends linearly on the σ_i) is governed by (2.210), which now translates into the iterative map:

$$\mathbf{m}(t+1) = \langle \boldsymbol{\xi} \tanh[\beta \boldsymbol{\xi} \cdot \mathbf{A}\mathbf{m}(t)]\rangle_{\boldsymbol{\xi}}, \tag{2.212}$$

with $\rho(\boldsymbol{\xi})$ as defined in (2.203). Again symmetry of the synapses is not required. For parallel dynamics it is far more difficult than for sequential dynamics to construct Lyapunov functions, and prove that the macroscopic laws (2.212) for symmetric systems evolve towards a stable fixed–point (as one would expect), but it can still be done. For non–symmetric systems the macroscopic laws (2.212) can in principle display all the interesting, but complicated, phenomena of non–conservative nonlinear systems. Nevertheless, it is also not uncommon that the equations (2.212) for non–symmetric systems can be mapped by a time–dependent transformation onto the equations for related symmetric systems (mostly variants of the original Hopfield model).

Note that the fixed–points of the macroscopic equations (2.198) and (2.203) (derived for sequential dynamics) are identical to those of (2.211) and (2.212) (derived for parallel dynamics). The stability properties of these fixed–points, however, need not be the same, and have to be assessed on a case–by–case basis. For the Hopfield model, i.e., equations (2.203,2.212) with $A_{\mu\nu} = \delta_{\mu\nu}$, they are found to be the same, but already for $A_{\mu\nu} = -\delta_{\mu\nu}$ the two types of dynamics would behave differently [Coo01, SC00, SC01, CKS05].

2.3.7 Attractor Neural Networks with Continuous Neurons

Closed Macroscopic Laws

We have seen above that models of recurrent neural networks with continuous neural variables (e.g., graded–response neurons or coupled oscillators) can often be described by a Fokker–Planck equation for the microscopic state probability density $p_t(\boldsymbol{\sigma})$:

$$\dot{p}_t(\boldsymbol{\sigma}) = -\frac{\partial}{\partial \sigma_i}[p_t(\boldsymbol{\sigma})f_i(\boldsymbol{\sigma})] + T\sum_i \frac{\partial^2}{\partial \sigma_i^2}p_t(\boldsymbol{\sigma}).$$

Averages over $p_t(\boldsymbol{\sigma})$ are denoted by $\langle G \rangle = \int d\boldsymbol{\sigma}\, p_t(\boldsymbol{\sigma})G(\boldsymbol{\sigma},t)$. From (2.131) one gets directly (through integration by parts) an equation for the time derivative of averages [Coo01, SC00, SC01, CKS05]:

$$\frac{d}{dt}\langle G \rangle = \langle \frac{\partial G}{\partial t} \rangle + \langle \left[f_i(\boldsymbol{\sigma}) + T\frac{\partial}{\partial \sigma_i}\right]\frac{\partial G}{\partial \sigma_i}\rangle. \qquad (2.213)$$

In particular, if we apply (2.213) to $G(\boldsymbol{\sigma},t) = \delta[\boldsymbol{\Omega} - \boldsymbol{\Omega}(\boldsymbol{\sigma})]$, for any set of macroscopic observables $\boldsymbol{\Omega}(\boldsymbol{\sigma}) = (\Omega_1(\boldsymbol{\sigma}),\ldots,\Omega_n(\boldsymbol{\sigma}))$ (in the spirit of the previous section), we get a dynamic equation for the macroscopic probability density $P_t(\boldsymbol{\Omega}) = \langle \delta[\boldsymbol{\Omega} - \boldsymbol{\Omega}(\boldsymbol{\sigma})]\rangle$, which is again of the Fokker–Planck form:

$$\dot{P}_t(\boldsymbol{\Omega}) = -\frac{\partial}{\partial \Omega_\mu}\left\{P_t(\boldsymbol{\Omega})\langle\left[f_i(\boldsymbol{\sigma}) + T\frac{\partial}{\partial \sigma_i}\right]\frac{\partial}{\partial \sigma_i}\Omega_\mu(\boldsymbol{\sigma})\rangle_{\boldsymbol{\Omega};t}\right\} \qquad (2.214)$$
$$+ T\frac{\partial^2}{\partial \Omega_\mu \partial \Omega_\nu}\left\{P_t(\boldsymbol{\Omega})\langle\left[\frac{\partial}{\partial \sigma_i}\Omega_\mu(\boldsymbol{\sigma})\right]\left[\frac{\partial}{\partial \sigma_i}\Omega_\nu(\boldsymbol{\sigma})\right]\rangle_{\boldsymbol{\Omega};t}\right\},$$

with the conditional (or sub-shell) averages:

$$\langle G(\boldsymbol{\sigma})\rangle_{\boldsymbol{\Omega},t} = \frac{\int d\boldsymbol{\sigma}\, p_t(\boldsymbol{\sigma})\delta[\boldsymbol{\Omega} - \boldsymbol{\Omega}(\boldsymbol{\sigma})]G(\boldsymbol{\sigma})}{\int d\boldsymbol{\sigma}\, p_t(\boldsymbol{\sigma})\delta[\boldsymbol{\Omega} - \boldsymbol{\Omega}(\boldsymbol{\sigma})]}.$$

From (2.214) we infer that a sufficient condition for the observables $\boldsymbol{\Omega}(\boldsymbol{\sigma})$ to evolve in time deterministically (i.e., for having vanishing diffusion matrix elements in (2.214)) in the limit $N \to \infty$ is

$$\lim_{N \to \infty} \langle \sum_i \left[\sum_\mu \left|\frac{\partial}{\partial \sigma_i}\Omega_\mu(\boldsymbol{\sigma})\right|\right]^2 \rangle_{\boldsymbol{\Omega};t} = 0. \qquad (2.215)$$

If (2.215) holds, the macroscopic Fokker–Planck equation (2.214) reduces for $N \to \infty$ to a Liouville equation, and the observables $\boldsymbol{\Omega}(\boldsymbol{\sigma})$ will evolve in time according to the coupled deterministic equations:

$$\dot{\Omega}_\mu = \lim_{N \to \infty} \langle \left[f_i(\boldsymbol{\sigma}) + T\frac{\partial}{\partial \sigma_i}\right]\frac{\partial}{\partial \sigma_i}\Omega_\mu(\boldsymbol{\sigma})\rangle_{\boldsymbol{\Omega};t}. \qquad (2.216)$$

The deterministic macroscopic equation (2.216), together with its associated condition for validity (2.215) will form the basis for the subsequent analysis.

The general derivation given above went smoothly. However, the equations (2.216) are not yet closed. It turns out that to achieve closure even for simple continuous networks we can no longer get away with just a finite (small) number of macroscopic observables (as with binary neurons). This we will now illustrate with a simple toy network of graded–response neurons:

$$\dot{u}_i(t) = J_{ij}\, g[u_j(t)] - u_i(t) + \eta_i(t),$$

with $g[z] = \frac{1}{2}[\tanh(\gamma z) + 1]$ and with the standard Gaussian white noise $\eta_i(t)$. In the language of (2.131) this means

$$f_i(\mathbf{u}) = J_{ij}\, g[u_j] - u_i.$$

We choose uniform synapses $J_{ij} = J/N$, so $f_i(\mathbf{u}) \to (J/N)\sum_j g[u_j] - u_i$. If (2.215) were to hold, we would find the deterministic macroscopic laws

$$\dot{\Omega}_\mu = \lim_{N\to\infty} \langle \sum_i [\frac{J}{N}\sum_j g[u_j] - u_i + T\frac{\partial}{\partial u_i}]\frac{\partial}{\partial u_i}\Omega_\mu(\mathbf{u})\rangle_{\Omega;t}. \qquad (2.217)$$

In contrast to similar models with binary neurons, choosing as our macroscopic level of description $\Omega(\mathbf{u})$ again simply the average $m(\mathbf{u}) = N^{-1}\sum_i u_i$ now leads to an equation which fails to close [Coo01, SC00, SC01, CKS05]:

$$\dot{m} = \lim_{N\to\infty} J\, \langle \frac{1}{N}\sum_j g[u_j]\rangle_{m;t} - m.$$

The term $N^{-1}\sum_j g[u_j]$ cannot be written as a function of $N^{-1}\sum_i u_i$. We might be tempted to try dealing with this problem by just including the offending term in our macroscopic set, and choose $\Omega(\mathbf{u}) = (N^{-1}\sum_i u_i, N^{-1}\sum_i g[u_i])$. This would indeed solve our closure problem for the m–equation, but we would now find a new closure problem in the equation for the newly introduced observable. The only way out is to choose an observable *function*, namely the distribution of potentials

$$\rho(u;\mathbf{u}) = \frac{1}{N}\sum_i \delta[u - u_i], \qquad (2.218)$$

$$\rho(u) = \langle \rho(u;\mathbf{u})\rangle = \langle \frac{1}{N}\sum_i \delta[u - u_i]\rangle.$$

This is to be done with care, in view of our restriction on the number of observables: we evaluate (2.218) at first only for n specific values u_μ and take the limit $n \to \infty$ only after the limit $N \to \infty$. Thus we define $\Omega_\mu(\mathbf{u}) = \frac{1}{N}\sum_i \delta[u_\mu - u_i]$, condition (2.215) reduces to the familiar expression $\lim_{N\to\infty} n/\sqrt{N} = 0$, and we get for $N \to \infty$ and $n \to \infty$ (taken in that order) from (2.217) a diffusion equation for the distribution of membrane potentials (describing a so-called 'time-dependent Ornstein–Uhlenbeck process' [Kam92, Gar85]):

$$\dot{p}(u) = -\frac{\partial}{\partial u}\left\{ p(u)\left[J\int du'\, p(u')g[u'] - u\right]\right\} + T\frac{\partial^2}{\partial u^2}p(u). \qquad (2.219)$$

The natural solution of (2.219) [11] is the Gaussian distribution

$$p_t(u) = [2\pi \Sigma^2(t)]^{-\frac{1}{2}} e^{-\frac{1}{2}[u-\bar{u}(t)]^2/\Sigma^2(t)}, \qquad (2.220)$$

in which $\Sigma = [T + (\Sigma_0^2 - T)e^{-2t}]^{\frac{1}{2}}$, and $\bar{u}$ evolves in time according to

$$\frac{d}{dt}\bar{u} = J\int Dz\, g[\bar{u} + \Sigma z] - \bar{u},$$

with $Dz = (2\pi)^{-\frac{1}{2}} e^{-\frac{1}{2}z^2} dz$. We can now also calculate the distribution $p(s)$ of neuronal firing activities $s_i = g[u_i]$ at any time,

$$p(s) = \int du\, p(u)\, \delta[s - g[u]] = \frac{p(g^{\text{inv}}[s])}{\int_0^1 ds'\, p(g^{\text{inv}}[s'])}.$$

For our choice $g[z] = \frac{1}{2} + \frac{1}{2}\tanh[\gamma z]$ we have $g^{\text{inv}}[s] = \frac{1}{2\gamma}\log[s/(1-s)]$, so in combination with (2.220)[12]

$$0 < s < 1: \qquad p(s) = \frac{e^{-\frac{1}{2}[(2\gamma)^{-1}\log[s/(1-s)] - \bar{u}]^2/\Sigma^2}}{\int_0^1 ds'\, e^{-\frac{1}{2}[(2\gamma)^{-1}\log[s'/(1-s')] - \bar{u}]^2/\Sigma^2}}.$$

Application to Graded–Response Attractor Networks

We will now turn to attractor networks with graded–response neurons of the type (2.132), in which p binary patterns $\xi^\mu = (\xi_1^\mu, \ldots, \xi_N^\mu) \in \{-1,1\}^N$ have been stored via separable Hebbian synapses (2.202): $J_{ij} = (2/N)\xi_i^\mu A_{\mu\nu}\xi_j^\nu$ (the extra factor 2 is inserted for future convenience). Adding suitable thresholds $\theta_i = -\frac{1}{2}\sum_j J_{ij}$ to the right-hand sides of (2.132), and choosing the nonlinearity $g(z) = \frac{1}{2}(1 + \tanh[\gamma z])$ would then give us [Coo01, SC00, SC01, CKS05]

---

[11] For non-Gaussian initial conditions $p_0(u)$ the solution of (2.219) would in time converge towards the Gaussian solution.

[12] None of the above results (not even those on the stationary state) could have been got within equilibrium statistical mechanics, since any network of connected graded-response neurons will violate detailed balance. Secondly, there appears to be a qualitative difference between simple networks (e.g., $J_{ij} = J/N$) of binary neurons versus those of continuous neurons, in terms of the types of macroscopic observables needed for deriving closed deterministic laws: a single number $m = N^{-1}\sum_i \sigma_i$ versus a distribution $\rho(\sigma) = N^{-1}\sum_i \delta[\sigma - \sigma_i]$.

$$\dot{u}_i(t) = \sum_{\mu\nu} \xi_i^\mu A_{\mu\nu} \frac{1}{N} \sum_j \xi_j^\nu \tanh[\gamma u_j(t)] - u_i(t) + \eta_i(t),$$

so the deterministic forces are $f_i(\mathbf{u}) = N^{-1} \sum_{\mu\nu} \xi_i^\mu A_{\mu\nu} \sum_j \xi_j^\nu \tanh[\gamma u_j] - u_i$. Choosing our macroscopic observables $\Omega(\mathbf{u})$ such that (2.215) holds, would lead to the deterministic macroscopic laws

$$\dot{\Omega}_\mu = \lim_{N\to\infty} \sum_{\mu\nu} A_{\mu\nu} \langle \left[\frac{1}{N}\xi_j^\nu \tanh[\gamma u_j]\right] \left[\xi_i^\mu \frac{\partial}{\partial u_i}\Omega_\mu(\mathbf{u})\right]\rangle_{\Omega;t} \quad (2.221)$$
$$+ \lim_{N\to\infty} \langle \left[T\frac{\partial}{\partial u_i} - u_i\right] \frac{\partial}{\partial u_i}\Omega_\mu(\mathbf{u})\rangle_{\Omega;t}.$$

As with the uniform synapses case, the main problem to be dealt with is how to choose the $\Omega_\mu(\mathbf{u})$ such that (2.221) closes. It turns out that the canonical choice is to turn to the distributions of membrane potentials within each of the 2^p sub–lattices, as introduced above :

$$I_\eta = \{i|\ \xi_i = \eta\}: \qquad \rho_\eta(u;\mathbf{u}) = \frac{1}{|I_\eta|}\sum_{i\in I_\eta} \delta[u - u_i], \quad (2.222)$$
$$\rho_\eta(u) = \langle \rho_\eta(u;\mathbf{u})\rangle,$$

with $\eta \in \{-1,1\}^p$ and $\lim_{N\to\infty}|I_\eta|/N = p_\eta$. Again we evaluate the distributions in (2.222) at first only for n specific values u_μ and send $n \to \infty$ after $N \to \infty$. Now condition (2.215) reduces to $\lim_{N\to\infty} 2^p/\sqrt{N} = 0$. We will keep p finite, for simplicity. Using identities such as $\sum_i \ldots = \sum_\eta \sum_{i\in I_\eta} \ldots$ and $i \in I_\eta$:

$$\frac{\partial}{\partial u_i}\rho_\eta(u;\mathbf{u}) = -|I_\eta|^{-1}\frac{\partial}{\partial u}\delta[u-u_i], \qquad \frac{\partial^2}{\partial u_i^2}\rho_\eta(u;\mathbf{u}) = |I_\eta|^{-1}\frac{\partial^2}{\partial u^2}\delta[u-u_i],$$

we then get for $N \to \infty$ and $n \to \infty$ (taken in that order) from equation (2.221) 2^p coupled diffusion equations for the distributions $\rho_\eta(u)$ of membrane potentials in each of the 2^p sub–lattices I_η:

$$\dot{\rho}_\eta(u) = -\frac{\partial}{\partial u}\{\rho_\eta(u)[\eta_\mu A_{\mu\nu} p_{\eta'}\eta'_\nu \int du'\ \rho_{\eta'}(u')\tanh[\gamma u'] - u]\} + T\frac{\partial^2}{\partial u^2}\rho_\eta(u). \quad (2.223)$$

Equation (2.223) is the basis for our further analysis. It can be simplified only if we make additional assumptions on the system's initial conditions, such as δ–distributed or Gaussian distributed $\rho_\eta(u)$ at $t=0$ (see below); otherwise it will have to be solved numerically.

It is clear that (2.223) is again of the time–dependent Ornstein–Uhlenbeck form, and will thus again have Gaussian solutions as the natural ones [Coo01, SC00, SC01, CKS05]:

$$\rho_{t,\eta}(u) = [2\pi\Sigma_\eta^2(t)]^{-\frac{1}{2}} e^{-\frac{1}{2}[u-\bar{u}_\eta(t)]^2/\Sigma_\eta^2(t)},$$

in which $\Sigma_\eta(t) = [T + (\Sigma_\eta^2(0) - T)e^{-2t}]^{\frac{1}{2}}$, and with the $\bar{u}_\eta(t)$ evolving in time according to

$$\frac{d}{dt}\bar{u}_\eta = p_{\eta'}(\eta \cdot \mathbf{A}\eta') \int Dz\ \tanh[\gamma(\bar{u}_{\eta'} + \Sigma_{\eta'} z)] - \bar{u}_\eta. \qquad (2.224)$$

Our problem has thus been reduced successfully to the study of the 2^p coupled scalar equations (2.224). We can also measure the correlation between the firing activities $s_i(u_i) = \frac{1}{2}[1 + \tanh(\gamma u_i)]$ and the pattern components (similar to the overlaps in the case of binary neurons). If the pattern bits are drawn at random, i.e., $\lim_{N\to\infty} |I_\eta|/N = p_\eta = 2^{-p}$ for all η, we can define a 'graded–response' equivalent $m_\mu(\mathbf{u}) = 2N^{-1}\xi_i^\mu s_i(u_i) \in [-1, 1]$ of the Hopfield pattern overlaps:

$$m_\mu(\mathbf{u}) = \frac{2}{N}\xi_i^\mu s_i(\mathbf{u}) = \frac{1}{N}\xi_i^\mu \tanh(\gamma u_i) + \mathcal{O}(N^{-\frac{1}{2}})$$

$$= p_\eta\ \eta_\mu \int du\ \rho_\eta(u; \mathbf{u})\tanh(\gamma u) + \mathcal{O}(N^{-\frac{1}{2}}).$$

Full recall of pattern μ implies $s_i(u_i) = \frac{1}{2}[\xi_i^\mu + 1]$, giving $m_\mu(\mathbf{u}) = 1$. Since the distributions $\rho_\eta(u)$ obey deterministic laws for $N \to \infty$, the same will be true for the overlaps $\mathbf{m} = (m_1, \ldots, m_p)$. For the Gaussian solutions (2.224) of (2.223) we can now proceed to replace the 2^p macroscopic laws (2.224), which reduce to $\frac{d}{dt}\bar{u}_\eta = \eta \cdot \mathbf{A}\mathbf{m} - \bar{u}_\eta$ and give $\bar{u}_\eta = \bar{u}_\eta(0)e^{-t} + \eta \cdot \mathbf{A}\int_0^t ds\ e^{s-t}\mathbf{m}(s)$, by p integral equations in terms of overlaps only:

$$m_\mu(t) = p_\eta\ \eta_\mu \int Dz\ \tanh[\gamma(\bar{u}_\eta(0)e^{-t} \qquad (2.225)$$

$$+ \eta \cdot \mathbf{A}\int_0^t ds\ e^{s-t}\mathbf{m}(s) + z\sqrt{T + (\Sigma_\eta^2(0) - T)e^{-2t}})],$$

with $Dz = (2\pi)^{-\frac{1}{2}}e^{-\frac{1}{2}z^2}dz$. Here the sub–lattices only come in via the initial conditions.

The equations describing the asymptotic (stationary) state can be written entirely without sub-lattices,[13]

$$m_\mu = \langle \xi_\mu \int Dz\ \tanh[\gamma(\boldsymbol{\xi} \cdot \mathbf{A}\mathbf{m} + z\sqrt{T})]\rangle_{\boldsymbol{\xi}}, \qquad (2.226)$$

$$\rho_\eta(u) = [2\pi T]^{-\frac{1}{2}}e^{-\frac{1}{2}[u - \eta \cdot \mathbf{A}\mathbf{m}]^2/T},$$

by taking the $t \to \infty$ limit in (2.225), using $\bar{u}_\eta \to \eta \cdot \mathbf{A}\mathbf{m}$, $\Sigma_\eta \to \sqrt{T}$, and the familiar notation

---

[13] Note the appealing similarity with previous results on networks with binary neurons in equilibrium. For $T = 0$ the overlap equations (2.226) become identical to those found for attractor networks with binary neurons and finite p (hence our choice to insert an extra factor 2 in defining the synapses), with γ replacing the inverse noise level β in the former.

$$\langle g(\boldsymbol{\xi})\rangle_{\boldsymbol{\xi}} = \lim_{N\to\infty} \frac{1}{N}\sum_i g(\boldsymbol{\xi}_i) = 2^{-p}\sum_{\boldsymbol{\xi}\in\{-1,1\}^p} g(\boldsymbol{\xi}).$$

For the simplest non–trivial choice, $A_{\mu\nu} = \delta_{\mu\nu}$ (i.e., $J_{ij} = (2/N)\sum_\mu \xi_i^\mu \xi_j^\mu$, as in the Hopfield [Hop82] model) equation (2.226) yields the familiar pure and mixture state solutions. For $T = 0$ we find a continuous phase transition from non–recall to pure states of the form $m_\mu = m\delta_{\mu\nu}$ (for some ν) at $\gamma_c = 1$. For $T > 0$ we have in (2.226) an additional Gaussian noise, absent in the models with binary neurons. Again the pure states are the first non–trivial solutions to enter the stage. Substituting $m_\mu = m\delta_{\mu\nu}$ into (2.226) gives

$$m = \int Dz \ \tanh[\gamma(m + z\sqrt{T})]. \tag{2.227}$$

Writing (2.227) as $m^2 = \gamma m \int_0^m dk[1 - \int Dz \ \tanh^2[\gamma(k + z\sqrt{T})]] \le \gamma m^2$, reveals that $m = 0$ as soon as $\gamma < 1$. A continuous transition to an $m > 0$ state occurs when

$$\gamma^{-1} = 1 - \int Dz \ \tanh^2[\gamma z\sqrt{T}].$$

A parametrization of this transition line in the (γ, T)–plane is given by

$$\gamma^{-1}(x) = 1 - \int Dz \ \tanh^2(zx),$$
$$T(x) = x^2/\gamma^2(x), \qquad x \ge 0.$$

Discontinuous transitions away from $m = 0$ (for which there is no evidence) would have to be calculated numerically. For $\gamma = \infty$ we get the equation $m = \mathrm{erf}[m/\sqrt{2T}]$, giving a continuous transition to $m > 0$ at $T_c = 2/\pi \approx 0.637$. Alternatively the latter number can also be found by taking $\lim_{x\to\infty} T(x)$ in the above parametrization:

$$T_c(\gamma = \infty) = \lim_{x\to\infty} x^2[1 - \int Dz \ \tanh^2(zx)]^2$$
$$= \lim_{x\to\infty} [\int Dz \ \frac{\partial}{\partial z}\tanh(zx)]^2 = [2\int Dz \ \delta(z)]^2 = 2/\pi.$$

Let us now turn to dynamics. It follows from (2.226) that the 'natural' initial conditions for $\bar{u}_\eta$ and Σ_η are of the form: $\bar{u}_\eta(0) = \boldsymbol{\eta}\cdot\mathbf{k}_0$ and $\Sigma_\eta(0) = \Sigma_0$ for all $\boldsymbol{\eta}$. Equivalently:

$$t = 0: \qquad p_\eta(u) = [2\pi\Sigma_0^2]^{-\frac{1}{2}} e^{-\frac{1}{2}[u-\boldsymbol{\eta}\cdot\mathbf{k}_0]^2/\Sigma_0^2},$$
$$\mathbf{k}_0 \in \mathbb{R}^p, \ \Sigma_0 \in \mathbb{R}.$$

These would also be the typical and natural statistics if we were to prepare an initial firing state $\{s_i\}$ by hand, via manipulation of the potentials $\{u_i\}$.

2.3 Synergetics of Recurrent and Attractor Neural Networks

For such initial conditions we can simplify the dynamical equation (2.225) to [Coo01, SC00, SC01, CKS05]

$$m_\mu(t) = \langle\, \xi_\mu \int Dz\ \tanh[\gamma(\boldsymbol{\xi}\cdot[\mathbf{k}_0 e^{-t} \tag{2.228}$$

$$+ \mathbf{A}\int_0^t ds\ e^{s-t}\mathbf{m}(s)] + z\sqrt{T + (\Sigma_0^2 - T)e^{-2t}})]\rangle_{\boldsymbol{\xi}}. \tag{2.229}$$

For the special case of the Hopfield synapses, i.e., $A_{\mu\nu} = \delta_{\mu\nu}$, it follows from (2.228) that recall of a given pattern ν is triggered upon choosing $k_{0,\mu} = k_0\delta_{\mu\nu}$ (with $k_0 > 0$), since then equation (2.228) generates $m_\mu(t) = m(t)\delta_{\mu\nu}$ at any time, with the amplitude $m(t)$ following from

$$m(t) = \int Dz\ \tanh[\gamma[k_0 e^{-t} + \int_0^t ds\ e^{s-t}m(s) + z\sqrt{T+(\Sigma_0^2-T)e^{-2t}}]], \tag{2.230}$$

which is the dynamical counterpart of equation (2.227) (to which indeed it reduces for $t \to \infty$).

We finally specialize further to the case where our Gaussian initial conditions are not only chosen to trigger recall of a single pattern $\boldsymbol{\xi}^\nu$, but in addition describe uniform membrane potentials within the sub-lattices, i.e., $k_{0,\mu} = k_0 \delta_{\mu\nu}$ and $\Sigma_0 = 0$, so $\rho_\eta(u) = \delta[u - k_0\eta_\nu]$. Here we can derive from (2.230) at $t=0$ the identity $m_0 = \tanh[\gamma k_0]$, which enables us to express k_0 as $k_0 = (2\gamma)^{-1}\log[(1+m_0)/(1-m_0)]$, and find (2.230) reducing to

$$m(t) = \int Dz\ \tanh[e^{-t}\log[\frac{1+m_0}{1-m_0}]^{\frac{1}{2}} + \gamma[\int_0^t ds\ e^{s-t}m(s) + z\sqrt{T(1-e^{-2t})}]].$$

Compared to the overlap evolution in large networks of binary networks (away from saturation) one can see richer behavior, e.g., non–monotonicity [Coo01, SC00, SC01, CKS05].

2.3.8 Correlation– and Response–Functions

We now turn to correlation functions $C_{ij}(t,t')$ and response functions $G_{ij}(t,t')$. These will become the language in which the partition function methods are formulated, which will enable us to solve the dynamics of recurrent networks in the (complex) regime near saturation (we take $t > t'$):

$$C_{ij}(t,t') = \langle \sigma_i(t)\sigma_j(t')\rangle, \qquad G_{ij}(t,t') = \partial\langle\sigma_i(t)\rangle/\partial\theta_j(t') \tag{2.231}$$

The $\{\sigma_i\}$ evolve in time according to equations of the form (2.186) (binary neurons, sequential updates), (2.205) (binary neurons, parallel updates) or (2.131) (continuous neurons). The θ_i represent thresholds and/or external stimuli, which are added to the local fields in the cases (2.186,2.205), or added to the deterministic forces in the case of a Fokker–Planck equation (2.131).

We retain $\theta_i(t) = \theta_i$, except for a perturbation $\delta\theta_j(t')$ applied at time t' in defining the response function. Calculating averages such as (2.231) requires determining joint probability distributions involving neuron states at different times.

Fluctuation–Dissipation Theorems

For networks of binary neurons with discrete time dynamics of the form $p_{\ell+1}(\boldsymbol{\sigma}) = \sum_{\boldsymbol{\sigma}'} W[\boldsymbol{\sigma};\boldsymbol{\sigma}']p_\ell(\boldsymbol{\sigma}')$, the probability of observing a given 'path' $\boldsymbol{\sigma}(\ell') \to \boldsymbol{\sigma}(\ell'+1) \to \ldots \to \boldsymbol{\sigma}(\ell-1) \to \boldsymbol{\sigma}(\ell)$ of successive configurations between step ℓ' and step ℓ is given by the product of the corresponding transition matrix elements (without summation):

$$\mathrm{Prob}[\boldsymbol{\sigma}(\ell'),\ldots,\boldsymbol{\sigma}(\ell)] = W[\boldsymbol{\sigma}(\ell);\boldsymbol{\sigma}(\ell-1)]W[\boldsymbol{\sigma}(\ell-1);\boldsymbol{\sigma}(\ell-2)]\ldots W[\boldsymbol{\sigma}(\ell'+1);\boldsymbol{\sigma}(\ell')]p_{\ell'}(\boldsymbol{\sigma}(\ell')).$$

This allows us to write [Coo01, SC00, SC01, CKS05]

$$C_{ij}(\ell,\ell') = \sum_{\boldsymbol{\sigma}(\ell')}\cdots\sum_{\boldsymbol{\sigma}(\ell)} \mathrm{Prob}[\boldsymbol{\sigma}(\ell'),\ldots,\boldsymbol{\sigma}(\ell)]\sigma_i(\ell)\sigma_j(\ell') \quad (2.232)$$

$$= \sum_{\boldsymbol{\sigma\sigma'}} \sigma_i \sigma_j' W^{\ell-\ell'}[\boldsymbol{\sigma};\boldsymbol{\sigma}']p_{\ell'}(\boldsymbol{\sigma}'),$$

$$G_{ij}(\ell,\ell') = \sum_{\boldsymbol{\sigma\sigma'\sigma''}} \sigma_i W^{\ell-\ell'-1}[\boldsymbol{\sigma};\boldsymbol{\sigma}'']\left[\frac{\partial}{\partial\theta_j}W[\boldsymbol{\sigma}'';\boldsymbol{\sigma}']\right]p_{\ell'}(\boldsymbol{\sigma}'). \quad (2.233)$$

From (2.232) and (2.233) it follows that both $C_{ij}(\ell,\ell')$ and $G_{ij}(\ell,\ell')$ will in the stationary state, i.e., upon substituting $p_{\ell'}(\boldsymbol{\sigma}') = p_\infty(\boldsymbol{\sigma}')$, only depend on $\ell - \ell'$: $C_{ij}(\ell,\ell') \to C_{ij}(\ell-\ell')$ and $G_{ij}(\ell,\ell') \to G_{ij}(\ell-\ell')$. For this we do not require detailed balance. Detailed balance, however, leads to a simple relation between the response function $G_{ij}(\tau)$ and the temporal derivative of the correlation function $C_{ij}(\tau)$.

We now turn to equilibrium systems, i.e., networks with symmetric synapses (and with all $J_{ii} = 0$ in the case of sequential dynamics). We calculate the derivative of the transition matrix that occurs in (2.233) by differentiating the equilibrium condition $p_{\mathrm{eq}}(\boldsymbol{\sigma}) = \sum_{\boldsymbol{\sigma}'} W[\boldsymbol{\sigma};\boldsymbol{\sigma}']p_{\mathrm{eq}}(\boldsymbol{\sigma}')$ with respect to external fields:

$$\frac{\partial}{\partial\theta_j}p_{\mathrm{eq}}(\boldsymbol{\sigma}) = \sum_{\boldsymbol{\sigma}'}\{\frac{\partial W[\boldsymbol{\sigma};\boldsymbol{\sigma}']}{\partial\theta_j}p_{\mathrm{eq}}(\boldsymbol{\sigma}') + W[\boldsymbol{\sigma};\boldsymbol{\sigma}']\frac{\partial}{\partial\theta_j}p_{\mathrm{eq}}(\boldsymbol{\sigma}')\}.$$

Detailed balance implies $p_{\mathrm{eq}}(\boldsymbol{\sigma}) = Z^{-1}e^{-\beta H(\boldsymbol{\sigma})}$ (in the parallel case we simply substitute the appropriate Hamiltonian $H \to \tilde{H}$), giving $\partial p_{\mathrm{eq}}(\boldsymbol{\sigma})/\partial\theta_j = -[Z^{-1}\partial Z/\partial\theta_j + \beta\partial H(\boldsymbol{\sigma})/\partial\theta_j]p_{\mathrm{eq}}(\boldsymbol{\sigma})$, so that

2.3 Synergetics of Recurrent and Attractor Neural Networks

$$\sum_{\sigma'} \frac{\partial W[\sigma;\sigma']}{\partial \theta_j} p_{\text{eq}}(\sigma') = \beta \{ \sum_{\sigma'} W[\sigma;\sigma'] \frac{\partial H(\sigma')}{\partial \theta_j} p_{\text{eq}}(\sigma') - \frac{\partial H(\sigma)}{\partial \theta_j} p_{\text{eq}}(\sigma) \},$$

in which the term containing Z drops out. We now get for the response function (2.233) in equilibrium the following result:

$$G_{ij}(\ell) = \qquad (2.234)$$

$$\beta \sum_{\sigma \sigma'} \sigma_i W^{\ell-1}[\sigma;\sigma'] \left\{ \sum_{\sigma''} W[\sigma';\sigma''] \frac{\partial H(\sigma'')}{\partial \theta_j} p_{\text{eq}}(\sigma'') - \frac{\partial H(\sigma')}{\partial \theta_j} p_{\text{eq}}(\sigma') \right\}.$$

The structure of (2.234) is similar to what follows upon calculating the evolution of the equilibrium correlation function (2.232) in a single iteration step:

$$C_{ij}(\ell) - C_{ij}(\ell-1) = \sum_{\sigma \sigma'} \sigma_i W^{\ell-1}[\sigma;\sigma'] \times \qquad (2.235)$$

$$\times \{ \sum_{\sigma''} W[\sigma';\sigma''] \sigma_j'' p_{\text{eq}}(\sigma'') - \sigma_j' p_{\text{eq}}(\sigma') \}.$$

Finally we calculate the relevant derivatives of the two Hamiltonians

$$H(\sigma) = -J_{ij}\sigma_i \sigma_j + \theta_i \sigma_i, \quad \text{and}$$
$$\tilde{H}(\sigma) = -\theta_i \sigma_i - \beta^{-1} \sum_i \log 2 \cosh[\beta h_i(\sigma)]$$

(with $h_i(\sigma) = J_{ij}\sigma_j + \theta_i$),

$$\frac{\partial H(\sigma)}{\partial \theta_j} = -\sigma_j, \qquad \frac{\partial \tilde{H}(\sigma)}{\partial \theta_j} = -\sigma_j - \tanh[\beta h_j(\sigma)].$$

For sequential dynamics we hereby arrive directly at a *fluctuation–dissipation theorem*. For parallel dynamics we need one more identity (which follows from the definition of the transition matrix in (2.205) and the detailed balance property) to transform the *tanh* occurring in the derivative of $\tilde{H}$:

$$\tanh[\beta h_j(\sigma')] p_{\text{eq}}(\sigma') = \sum_{\sigma''} \sigma_j'' W[\sigma'';\sigma'] p_{\text{eq}}(\sigma')$$
$$= \sum_{\sigma''} W[\sigma';\sigma''] \sigma_j'' p_{\text{eq}}(\sigma'').$$

For parallel dynamics ℓ and ℓ' are the real time labels t and t', and we get, with $\tau = t - t'$:

$$G_{ij}(\tau > 0) = -\beta[C_{ij}(\tau+1) - C_{ij}(\tau-1)], \qquad G_{ij}(\tau \leq 0) = 0. \qquad (2.236)$$

For the continuous-time version (2.186) of sequential dynamics the time t is defined as $t = \ell/N$, and the difference equation (2.235) becomes a differential equation. For perturbations at time t' in the definition of the response function (2.233) to retain a non-vanishing effect at (re-scaled) time t in the limit $N \to \infty$, they will have to be re-scaled as well: $\delta\theta_j(t') \to N\delta\theta_j(t')$. As a result:

$$G_{ij}(\tau) = -\beta\theta(\tau)\frac{d}{d\tau}C_{ij}(\tau). \qquad (2.237)$$

The need to re-scale perturbations in making the transition from discrete to continuous times has the same origin as the need to re-scale the random forces in the derivation of the continuous-time Langevin equation from a discrete-time process. Going from ordinary derivatives to function derivatives (which is what happens in the continuous–time limit), implies replacing Kronecker delta's $\delta_{t,t'}$ by Dirac delta-functions according to $\delta_{t,t'} \to \Delta\delta(t-t')$, where Δ is the average duration of an iteration step. Equations (2.236) and (2.237) are examples of so–called *fluctuation–dissipation theorems* (FDT).

For systems described by a Fokker–Planck equation (2.131) the simplest way to calculate correlation- and response-functions is by first returning to the underlying discrete-time system and leaving the continuous time limit $\Delta \to 0$ until the end. We saw above that for small but finite time-steps Δ the underlying discrete-time process is described by [Coo01, SC00, SC01, CKS05]

$$t = \ell\Delta, \qquad p_{\ell\Delta+\Delta}(\boldsymbol{\sigma}) = [1 + \Delta\mathcal{L}_{\boldsymbol{\sigma}} + \mathcal{O}(\Delta^{\frac{3}{2}})]p_{\ell\Delta}(\boldsymbol{\sigma}),$$

with $\ell = 0, 1, 2, \ldots$ and with the differential operator

$$\mathcal{L}_{\boldsymbol{\sigma}} = -\frac{\partial}{\partial\sigma_i}[f_i(\boldsymbol{\sigma}) - T\frac{\partial}{\partial\sigma_i}].$$

From this it follows that the conditional probability density $p_{\ell\Delta}(\boldsymbol{\sigma}|\boldsymbol{\sigma}', \ell'\Delta)$ for finding state $\boldsymbol{\sigma}$ at time $\ell\Delta$, given the system was in state $\boldsymbol{\sigma}'$ at time $\ell'\Delta$, must be

$$p_{\ell\Delta}(\boldsymbol{\sigma}|\boldsymbol{\sigma}', \ell'\Delta) = [1 + \Delta\mathcal{L}_{\boldsymbol{\sigma}} + \mathcal{O}(\Delta^{\frac{3}{2}})]^{\ell-\ell'}\delta[\boldsymbol{\sigma} - \boldsymbol{\sigma}']. \qquad (2.238)$$

Equation (2.238) will be our main building block. Firstly, we will calculate the correlations:

$$C_{ij}(\ell\Delta, \ell'\Delta) = \langle \sigma_i(\ell\Delta)\sigma_j(\ell'\Delta) \rangle$$

$$= \int d\boldsymbol{\sigma} d\boldsymbol{\sigma}' \ \sigma_i \sigma_j' \ p_{\ell\Delta}(\boldsymbol{\sigma}|\boldsymbol{\sigma}', \ell'\Delta)p_{\ell'\Delta}(\boldsymbol{\sigma}')$$

$$= \int d\boldsymbol{\sigma} \ \sigma_i[1 + \Delta\mathcal{L}_{\boldsymbol{\sigma}} + \mathcal{O}(\Delta^{\frac{3}{2}})]^{\ell-\ell'} \int d\boldsymbol{\sigma}' \ \sigma_j' \delta[\boldsymbol{\sigma} - \boldsymbol{\sigma}']p_{\ell'\Delta}(\boldsymbol{\sigma}')$$

$$= \int d\boldsymbol{\sigma} \ \sigma_i[1 + \Delta\mathcal{L}_{\boldsymbol{\sigma}} + \mathcal{O}(\Delta^{\frac{3}{2}})]^{\ell-\ell'}[\sigma_j \ p_{\ell'\Delta}(\boldsymbol{\sigma})].$$

At this stage, we can take the limits $\Delta \to 0$ and $\ell, \ell' \to \infty$, with $t = \ell\Delta$ and $t' = \ell'\Delta$ finite, using $\lim_{\Delta\to 0}[1 + \Delta A]^{k/\Delta} = e^{kA}$:

2.3 Synergetics of Recurrent and Attractor Neural Networks

$$C_{ij}(t,t') = \int d\boldsymbol{\sigma}\; \sigma_i\; e^{(t-t')\mathcal{L}_\sigma}\left[\sigma_j\; p_{t'}(\boldsymbol{\sigma})\right]. \tag{2.239}$$

Next we turn to the response function. A perturbation applied at time $t' = \ell'\Delta$ to the Langevin forces $f_i(\boldsymbol{\sigma})$ comes in at the transition $\boldsymbol{\sigma}(\ell'\Delta) \to \boldsymbol{\sigma}(\ell'\Delta + \Delta)$. As with sequential dynamics binary networks, the perturbation is re-scaled with the step size Δ to retain significance as $\Delta \to 0$:

$$G_{ij}(\ell\Delta, \ell'\Delta) = \frac{\partial\langle\sigma_i(\ell\Delta)\rangle}{\Delta\partial\theta_j(\ell'\Delta)} = \frac{\partial}{\Delta\partial\theta_j(\ell'\Delta)}\int d\boldsymbol{\sigma} d\boldsymbol{\sigma}'\; \sigma_i\; p_{\ell\Delta}(\boldsymbol{\sigma}|\boldsymbol{\sigma}',\ell'\Delta)p_{\ell'\Delta}(\boldsymbol{\sigma}')$$

$$= \int d\boldsymbol{\sigma} d\boldsymbol{\sigma}' d\boldsymbol{\sigma}''\; \sigma_i\; p_{\ell\Delta}(\boldsymbol{\sigma}|\boldsymbol{\sigma}'',\ell'\Delta + \Delta) \left[\frac{\partial p_{\ell''\Delta+\Delta}(\boldsymbol{\sigma}''|\boldsymbol{\sigma}',\ell'\Delta)}{\Delta\partial\theta_j}\right] p_{\ell'\Delta}(\boldsymbol{\sigma}')$$

$$= \int d\boldsymbol{\sigma} d\boldsymbol{\sigma}' d\boldsymbol{\sigma}''\; \sigma_i[1+\Delta\mathcal{L}_\sigma+\mathcal{O}(\Delta^{\frac{3}{2}})]^{\ell-\ell'-1}\delta[\boldsymbol{\sigma}-\boldsymbol{\sigma}''] \times$$

$$\left[\frac{1}{\Delta}\frac{\partial}{\partial\theta_j}[1+\Delta\mathcal{L}_{\sigma''}+\mathcal{O}(\Delta^{\frac{3}{2}})]\delta[\boldsymbol{\sigma}''-\boldsymbol{\sigma}']\right]p_{\ell'\Delta}(\boldsymbol{\sigma}')$$

$$= -\int d\boldsymbol{\sigma} d\boldsymbol{\sigma}' d\boldsymbol{\sigma}''\; \sigma_i[1+\Delta\mathcal{L}_\sigma+\mathcal{O}(\Delta^{\frac{3}{2}})]^{\ell-\ell'-1} \times$$

$$\delta[\boldsymbol{\sigma}-\boldsymbol{\sigma}'']\delta[\boldsymbol{\sigma}''-\boldsymbol{\sigma}'][\frac{\partial}{\partial\sigma'_j}+\mathcal{O}(\Delta^{\frac{1}{2}})]\; p_{\ell'\Delta}(\boldsymbol{\sigma}')$$

$$= -\int d\boldsymbol{\sigma}\; \sigma_i[1+\Delta\mathcal{L}_\sigma+\mathcal{O}(\Delta^{\frac{3}{2}})]^{\ell-\ell'-1}[\frac{\partial}{\partial\sigma_j}+\mathcal{O}(\Delta^{\frac{1}{2}})]\; p_{\ell'\Delta}(\boldsymbol{\sigma}).$$

We take the limits $\Delta \to 0$ and $\ell, \ell' \to \infty$, with $t = \ell\Delta$ and $t' = \ell'\Delta$ finite:

$$G_{ij}(t,t') = -\int d\boldsymbol{\sigma}\; \sigma_i\; e^{(t-t')\mathcal{L}_\sigma}\frac{\partial}{\partial\sigma_j}p_{t'}(\boldsymbol{\sigma}). \tag{2.240}$$

Equations (2.239) and (2.240) apply to arbitrary systems described by Fokker–Planck equations. In the case of conservative forces, i.e., $f_i(\boldsymbol{\sigma}) = -\partial H(\boldsymbol{\sigma})/\partial\sigma_i$, and when the system is in an equilibrium state at time t' so that $C_{ij}(t,t') = C_{ij}(t-t')$ and $G_{ij}(t,t') = G_{ij}(t-t')$, we can take a further step using $p_{t'}(\boldsymbol{\sigma}) = p_{eq}(\boldsymbol{\sigma}) = Z^{-1}e^{-\beta H(\boldsymbol{\sigma})}$. In that case, taking the time derivative of expression (2.239) gives

$$\frac{\partial}{\partial\tau}C_{ij}(\tau) = \int d\boldsymbol{\sigma}\; \sigma_i\; e^{\tau\mathcal{L}_\sigma}\mathcal{L}_\sigma\left[\sigma_j\; p_{eq}(\boldsymbol{\sigma})\right].$$

Working out the key term in this expression gives

$$\mathcal{L}_\sigma[\sigma_j\; p_{eq}(\boldsymbol{\sigma})] = -\sum_i \frac{\partial}{\partial\sigma_i}[f_i(\boldsymbol{\sigma}) - T\frac{\partial}{\partial\sigma_i}][\sigma_j\; p_{eq}(\boldsymbol{\sigma})]$$

$$= T\frac{\partial}{\partial\sigma_j}p_{eq}(\boldsymbol{\sigma}) - \sum_i \frac{\partial}{\partial\sigma_i}[\sigma_j J_i(\boldsymbol{\sigma})],$$

with the components of the probability current density $J_i(\boldsymbol{\sigma}) = [f_i(\boldsymbol{\sigma}) - T\frac{\partial}{\partial \sigma_i}]p_{\text{eq}}(\boldsymbol{\sigma})$. In equilibrium, however, the current is zero by definition, so only the first term in the above expression survives. Insertion into our previous equation for $\partial C_{ij}(\tau)/\partial \tau$, and comparison with (2.240) leads to the FDT for continuous systems [Coo01, SC00, SC01, CKS05]:

$$\text{Continuous Dynamics:} \qquad G_{ij}(\tau) = -\beta\theta(\tau)\frac{d}{d\tau}C_{ij}(\tau).$$

We will now calculate the correlation and response functions explicitly, and verify the validity or otherwise of the FDT relations, for attractor networks away from saturation.

Simple Attractor Networks with Binary Neurons

We will consider the continuous time version (2.186) of the sequential dynamics, with the local fields $h_i(\boldsymbol{\sigma}) = J_{ij}\sigma_j + \theta_i$, and the separable interaction matrix (2.202). We already solved the dynamics of this model for the case with zero external fields and away from saturation (i.e., $p \ll \sqrt{N}$). Having non–zero, or even time–dependent, external fields does not affect the calculation much; one adds the external fields to the internal ones and finds the macroscopic laws (2.203) for the overlaps with the stored patterns being replaced by [Coo01, SC00, SC01, CKS05]

$$\dot{\mathbf{m}}(t) = \lim_{N\to\infty} \frac{1}{N}\boldsymbol{\xi}_i \tanh[\beta\boldsymbol{\xi}_i \cdot \mathbf{Am}(t) + \theta_i(t)] - \mathbf{m}(t), \qquad (2.241)$$

Fluctuations in the local fields are of vanishing order in N (since the fluctuations in $\mathbf{m}$ are), so that one can easily derive from the master equation (2.186) the following expressions for spin averages:

$$\frac{d}{dt}\langle\sigma_i(t)\rangle = \tanh\beta[\boldsymbol{\xi}_i \cdot \mathbf{Am}(t) + \theta_i(t)] - \langle\sigma_i(t)\rangle, \qquad (2.242)$$

$$i \neq j: \quad \frac{d}{dt}\langle\sigma_i(t)\sigma_j(t)\rangle = \tanh\beta[\boldsymbol{\xi}_i \cdot \mathbf{Am}(t) + \theta_i(t)]\langle\sigma_j(t)\rangle$$
$$+ \tanh\beta[\boldsymbol{\xi}_j \cdot \mathbf{Am}(t) + \theta_j(t)]\langle\sigma_i(t)\rangle - 2\langle\sigma_i(t)\sigma_j(t)\rangle.$$

Correlations at different times are calculated by applying (2.242) to situations where the microscopic state at time t' is known exactly, i.e., where $p_{t'}(\boldsymbol{\sigma}) = \delta_{\boldsymbol{\sigma},\boldsymbol{\sigma}'}$ for some $\boldsymbol{\sigma}'$:

$$\langle\sigma_i(t)\rangle|_{\boldsymbol{\sigma}(t')=\boldsymbol{\sigma}'} = \sigma'_i e^{-(t-t')} + \int_{t'}^{t} ds\ e^{s-t} \tanh\beta \times \qquad (2.243)$$
$$\times [\boldsymbol{\xi}_i \cdot \mathbf{Am}(s;\boldsymbol{\sigma}',t') + \theta_i(s)],$$

with $\mathbf{m}(s;\boldsymbol{\sigma}',t')$ denoting the solution of (2.241) following initial condition $\mathbf{m}(t') = \frac{1}{N}\sigma'_i\boldsymbol{\xi}_i$. If we multiply both sides of (2.243) by σ'_j and average over all possible states $\boldsymbol{\sigma}'$ at time t' we get in leading order in N:

$$\langle\sigma_i(t)\sigma_j(t')\rangle = \langle\sigma_i(t')\sigma_j(t')\rangle e^{-(t-t')} + \int_{t'}^t ds\ e^{s-t}\langle \tanh\beta[\boldsymbol{\xi}_i\cdot\mathbf{Am}(s;\boldsymbol{\sigma}(t'),t') + \theta_i(s)]\sigma_j(t')\rangle.$$

Because of the existence of deterministic laws for the overlaps $\mathbf{m}$ in the $N\to\infty$ limit, we know with probability one that during the stochastic process the actual value $\mathbf{m}(\boldsymbol{\sigma}(t'))$ must be given by the solution of (2.241), evaluated at time t'. As a result we get, with $C_{ij}(t,t') = \langle\sigma_i(t)\sigma_j(t')\rangle$:

$$C_{ij}(t,t') = C_{ij}(t',t')e^{-(t-t')} + \qquad (2.244)$$
$$\int_{t'}^t ds\ e^{s-t}\tanh\beta[\boldsymbol{\xi}_i\cdot\mathbf{Am}(s) + \theta_i(s)]\langle\sigma_j(t')\rangle.$$

Similarly we get from the solution of (2.242) an equation for the leading order in N of the response functions, by derivation with respect to external fields:

$$\frac{\partial\langle\sigma_i(t)\rangle}{\partial\theta_j(t')} = \beta\theta(t-t')\int_{-\infty}^t ds\ e^{s-t}\left[1 - \tanh^2\beta[\boldsymbol{\xi}_i\cdot\mathbf{Am}(s) + \theta_i(s)]\right]\times$$
$$\times\left[\frac{1}{N}(\boldsymbol{\xi}_i\cdot\mathbf{A}\boldsymbol{\xi}_k)\frac{\partial\langle\sigma_k(s)\rangle}{\partial\theta_j(t')} + \delta_{ij}\delta(s-t')\right],\qquad\text{or}$$

$$G_{ij}(t,t') = \beta\delta_{ij}\theta(t-t')e^{-(t-t')}\left[1 - \tanh^2\beta[\boldsymbol{\xi}_i\cdot\mathbf{Am}(t') + \theta_i(t')]\right](2.245)$$
$$+\ \beta\theta(t-t')\int_{t'}^t ds\ e^{s-t}\times$$
$$\times\left[1 - \tanh^2\beta[\boldsymbol{\xi}_i\cdot\mathbf{Am}(s) + \theta_i(s)]\right]\frac{1}{N}(\boldsymbol{\xi}_i\cdot\mathbf{A}\boldsymbol{\xi}_k)G_{kj}(s,t').$$

For $t = t'$ we retain in leading order in N only the instantaneous single-site contribution

$$\lim_{t'\uparrow t} G_{ij}(t,t') = \beta\delta_{ij}\left[1 - \tanh^2\beta[\boldsymbol{\xi}_i\cdot\mathbf{Am}(t) + \theta_i(t)]\right].\qquad (2.246)$$

This leads to the following ansatz for the scaling with N of the $G_{ij}(t,t')$, which can be shown to be correct by insertion into (2.245), in combination with the correctness at $t = t'$ following from (2.246):

$$i = j:\quad G_{ii}(t,t') = \mathcal{O}(1),\qquad i\neq j:\quad G_{ij}(t,t') = \mathcal{O}(N^{-1})$$

Note that this implies $\frac{1}{N}(\boldsymbol{\xi}_i\cdot\mathbf{A}\boldsymbol{\xi}_k)G_{kj}(s,t') = \mathcal{O}(\frac{1}{N})$. In leading order in N we now find

$$G_{ij}(t,t') = \beta\delta_{ij}\theta(t-t')e^{-(t-t')}\left[1 - \tanh^2 \beta[\boldsymbol{\xi}_i \cdot \mathbf{Am}(t') + \theta_i(t')]\right]. \quad (2.247)$$

For those cases where the macroscopic laws (2.241) describe evolution to a stationary state $\mathbf{m}$, obviously requiring stationary external fields $\theta_i(t) = \theta_i$, we can take the limit $t \to \infty$, with $t-t' = \tau$ fixed, in (2.244) and (2.247). Using the $t \to \infty$ limits of (2.242) we subsequently find time translation invariant expressions: $\lim_{t\to\infty} C_{ij}(t, t-\tau) = C_{ij}(\tau)$ and $\lim_{t\to\infty} G_{ij}(t, t-\tau) = G_{ij}(\tau)$, with in leading order in N

$$C_{ij}(\tau) = \tanh\beta[\boldsymbol{\xi}_i \cdot \mathbf{Am} + \theta_i]\tanh\beta[\boldsymbol{\xi}_j \cdot \mathbf{Am} + \theta_j]$$
$$+ \delta_{ij}e^{-\tau}\left[1 - \tanh^2\beta[\boldsymbol{\xi}_i \cdot \mathbf{Am} + \theta_i]\right],$$

$$G_{ij}(\tau) = \beta\delta_{ij}\theta(\tau)e^{-\tau}\left[1 - \tanh^2\beta[\boldsymbol{\xi}_i \cdot \mathbf{Am} + \theta_i]\right],$$

for which the fluctuation–dissipation theorem (2.237) holds [Coo01, SC00, SC01, CKS05]:

$$G_{ij}(\tau) = -\beta\theta(\tau)\frac{d}{d\tau}C_{ij}(\tau).$$

We now turn to the parallel dynamical rules (2.205), with the local fields $h_i(\boldsymbol{\sigma}) = J_{ij}\sigma_j + \theta_i$, and the interaction matrix (2.202). As before, having time-dependent external fields amounts simply to adding these fields to the internal ones, and the dynamic laws (2.212) are found to be replaced by

$$\mathbf{m}(t+1) = \lim_{N\to\infty}\frac{1}{N}\boldsymbol{\xi}_i\tanh\left[\beta\boldsymbol{\xi}_i \cdot \mathbf{Am}(t) + \theta_i(t)\right]. \quad (2.248)$$

Fluctuations in the local fields are again of vanishing order in N, and the parallel dynamics versions of equations (2.242), to be derived from (2.205), are found to be

$$\langle\sigma_i(t+1)\rangle = \tanh\beta[\boldsymbol{\xi}_i \cdot \mathbf{Am}(t) + \theta_i(t)], \quad (2.249)$$

$$i \neq j: \quad \langle\sigma_i(t+1)\sigma_j(t+1)\rangle = \quad (2.250)$$
$$\tanh\beta[\boldsymbol{\xi}_i \cdot \mathbf{Am}(t) + \theta_i(t)]\tanh\beta[\boldsymbol{\xi}_j \cdot \mathbf{Am}(t) + \theta_j(t)].$$

With $\mathbf{m}(t; \boldsymbol{\sigma}', t')$ denoting the solution of the map (2.248) following initial condition $\mathbf{m}(t') = \frac{1}{N}\sigma'_i\boldsymbol{\xi}_i$, we immediately get from equations (2.249,2.250) the correlation functions:

$$C_{ij}(t,t) = \delta_{ij} + [1 - \delta_{ij}]\tanh\beta[\boldsymbol{\xi}_i \cdot \mathbf{Am}(t-1) + \theta_i(t-1)] \times$$
$$\tanh\beta[\boldsymbol{\xi}_j \cdot \mathbf{Am}(t-1) + \theta_j(t-1)],$$

$$t > t': \quad C_{ij}(t,t') = \langle\tanh\beta[\boldsymbol{\xi}_i \cdot \mathbf{Am}(t-1; \boldsymbol{\sigma}(t'), t') + \theta_i(t-1)]\sigma_j(t')\rangle$$
$$= \tanh\beta[\boldsymbol{\xi}_i \cdot \mathbf{Am}(t-1) + \theta_i(t-1)]\tanh\beta[\boldsymbol{\xi}_j \cdot \mathbf{Am}(t'-1) + \theta_j(t'-1)].$$

From (2.249) also follow equations determining the leading order in N of the response functions $G_{ij}(t,t')$, by derivation with respect to the external fields $\theta_j(t')$:

$t' > t-1: G_{ij}(t,t') = 0,$
$t' = t-1: G_{ij}(t,t') = \beta\delta_{ij}\left[1 - \tanh^2\beta[\boldsymbol{\xi}_i \cdot \mathbf{A}\mathbf{m}(t-1) + \theta_i(t-1)]\right],$
$t' < t-1: G_{ij}(t,t') = \beta\left[1 - \tanh^2\beta[\boldsymbol{\xi}_i \cdot \mathbf{A}\mathbf{m}(t-1) + \theta_i(t-1)]\right] \times$
$\times \frac{1}{N}(\boldsymbol{\xi}_i \cdot \mathbf{A}\boldsymbol{\xi}_k)G_{kj}(t-1,t').$

It now follows iteratively that all off–diagonal elements must be of vanishing order in N: $G_{ij}(t,t-1) = \delta_{ij}G_{ii}(t,t-1) \rightarrow G_{ij}(t,t-2) = \delta_{ij}G_{ii}(t,t-2) \rightarrow \ldots$, so that in leading order

$$G_{ij}(t,t') = \beta\delta_{ij}\delta_{t,t'+1}\left[1 - \tanh^2\beta[\boldsymbol{\xi}_i \cdot \mathbf{A}\mathbf{m}(t') + \theta_i(t')]\right].$$

For those cases where the macroscopic laws (2.248) describe evolution to a stationary state $\mathbf{m}$, with stationary external fields, we can take the limit $t \to \infty$, with $t - t' = \tau$ fixed above. We find time translation invariant expressions: $\lim_{t\to\infty} C_{ij}(t,t-\tau) = C_{ij}(\tau)$ and $\lim_{t\to\infty} G_{ij}(t,t-\tau) = G_{ij}(\tau)$, with in leading order in N:

$$C_{ij}(\tau) = \tanh\beta[\boldsymbol{\xi}_i \cdot \mathbf{A}\mathbf{m} + \theta_i]\tanh\beta[\boldsymbol{\xi}_j \cdot \mathbf{A}\mathbf{m} + \theta_j]$$
$$+ \delta_{ij}\delta_{\tau,0}\left[1 - \tanh^2\beta[\boldsymbol{\xi}_i \cdot \mathbf{A}\mathbf{m} + \theta_i]\right],$$
$$G_{ij}(\tau) = \beta\delta_{ij}\delta_{\tau,1}\left[1 - \tanh^2\beta[\boldsymbol{\xi}_i \cdot \mathbf{A}\mathbf{m} + \theta_i]\right],$$

obeying the Fluctuation-Dissipation Theorem (2.236):

$$G_{ij}(\tau > 0) = -\beta[C_{ij}(\tau+1) - C_{ij}(\tau-1)].$$

Graded–Response Neurons with Uniform Synapses

Let us finally find out how to calculate correlation and response function for the simple network (2.132) of graded–response neurons, with (possibly time–dependent) external forces $\theta_i(t)$, and with uniform synapses $J_{ij} = J/N$ [Coo01, SC00, SC01, CKS05]:

$$\dot{u}_i(t) = \frac{J}{N}\sum_j g[\gamma u_j(t)] - u_i(t) + \theta_i(t) + \eta_i(t). \tag{2.251}$$

For a given realisation of the external forces and the Gaussian noise variables $\{\eta_i(t)\}$ we can formally integrate (2.251) and find

$$u_i(t) = u_i(0)e^{-t} + \int_0^t ds\, e^{s-t}\left[J\int du\, \rho(u;\mathbf{u}(s))\, g[\gamma u] + \theta_i(s) + \eta_i(s)\right], \tag{2.252}$$

with the distribution of membrane potentials $\rho(u;\mathbf{u}) = N^{-1}\sum_i \delta[u-u_i]$. The correlation function $C_{ij}(t,t') = \langle u_i(t)u_j(t')\rangle$ immediately follows from (2.252). Without loss of generality we can define $t \geq t'$. For absent external forces (which were only needed in order to define the response function), and upon using $\langle \eta_i(s)\rangle = 0$ and $\langle \eta_i(s)\eta_j(s')\rangle = 2T\delta_{ij}\delta(s-s')$, we arrive at

$$C_{ij}(t,t') = T\delta_{ij}(e^{t'-t} - e^{-t'-t}) +$$
$$\left\langle \left[u_i(0)e^{-t} + J\int du\, g[\gamma u]\int_0^t ds\, e^{s-t}\rho(u;\mathbf{u}(s))\right] \times \right.$$
$$\left. \times \left[u_j(0)e^{-t'} + J\int du\, g[\gamma u]\int_0^{t'} ds'\, e^{s'-t'}\rho(u;\mathbf{u}(s'))\right] \right\rangle.$$

For $N \to \infty$, however, we know the distribution of potentials to evolve deterministically: $\rho(u;\mathbf{u}(s)) \to \rho_s(u)$ where $\rho_s(u)$ is the solution of (2.219). This allows us to simplify the above expression to

$$N \to \infty: \quad C_{ij}(t,t') = T\delta_{ij}(e^{t'-t} - e^{-t'-t}) \tag{2.253}$$
$$+ \left\langle \left[u_i(0)e^{-t} + J\int du\, g[\gamma u]\int_0^t ds\, e^{s-t}\rho_s(u)\right] \times \right.$$
$$\left. \times \left[u_j(0)e^{-t'} + J\int du\, g[\gamma u]\int_0^{t'} ds'\, e^{s'-t'}\rho_{s'}(u)\right] \right\rangle.$$

Next we turn to the response function $G_{ij}(t,t') = \delta\langle u_i(t)\rangle/\delta\xi_j(t')$ (its definition involves functional rather than scalar differentiation, since time is continuous). After this differentiation the forces $\{\theta_i(s)\}$ can be put to zero. Functional differentiation of (2.252), followed by averaging, then leads us to

$$G_{ij}(t,t') = \theta(t-t')\,\delta_{ij}\, e^{t'-t} - J\int du\, g[\gamma u]\frac{\partial}{\partial u} \times$$
$$\int_0^t ds\, e^{s-t}\frac{1}{N}\sum_k \lim_{\theta \to 0}\langle \delta[u-u_k(s)]\frac{\delta u_k(s)}{\delta\theta_j(t')}\rangle.$$

In view of (2.252) we make the self-consistent ansatz $\delta u_k(s)/\delta\xi_j(s') = \mathcal{O}(N^{-1})$ for $k \neq j$. This produces

$$N \to \infty: \quad G_{ij}(t,t') = \theta(t-t')\,\delta_{ij}\, e^{t'-t}.$$

Since equation (2.219) evolves towards a stationary state, we can also take the limit $t \to \infty$, with $t-t' = \tau$ fixed, in (2.253). Assuming non–pathological decay of the distribution of potentials allows us to put $\lim_{t\to\infty}\int_0^t ds\, e^{s-t}\rho_s(u) = \rho(u)$ (the stationary solution of (2.219)), with which we find also (2.253) reducing to time translation invariant expressions for $N \to \infty$, $\lim_{t\to\infty} C_{ij}(t,t-\tau) = C_{ij}(\tau)$ and $\lim_{t\to\infty} G_{ij}(t,t-\tau) = G_{ij}(\tau)$, in which

$$C_{ij}(\tau) = T\delta_{ij}\mathrm{e}^{-\tau} + J^2\left\{\int du\, \rho(u) g[\gamma u]\right\}^2,$$

$$G_{ij}(\tau) = \theta(\tau)\delta_{ij}\mathrm{e}^{-\tau}.$$

Clearly the leading orders in N of these two functions obey the fluctuation-dissipation theorem:

$$G_{ij}(\tau) = -\beta\theta(\tau)\frac{d}{d\tau}C_{ij}(\tau).$$

As with the binary neuron attractor networks for which we calculated the correlation and response functions earlier, the impact of detailed balance violation (occurring when $A_{\mu\nu} \neq A_{\nu\mu}$ in networks with binary neurons and synapses (2.202), and in all networks with graded–response neurons on the validity of the fluctuation-dissipation theorems, vanishes for $N \to \infty$, provided our networks are relatively simple and evolve to a stationary state in terms of the macroscopic observables (the latter need not necessarily happen. Detailed balance violation, however, would be noticed in the finite size effects [CCV98].

3

Geometry and Topology Change in Complex Systems

While the natural stage for linear dynamics comprises of flat, Euclidean geometry (with the corresponding calculation tools from linear algebra and analysis), the natural stage for nonlinear dynamics is curved, *Riemannian geometry* (with the corresponding tools from tensor algebra and analysis). In both cases, the system's (kinetic) energy is defined by the metric form, either Euclidean or Riemannian. The extreme nonlinearity – chaos – corresponds to the *topology change* of this curved geometrical stage, usually called configuration manifold. This Chapter elaborates on geometry and topology change in relation with complex nonlinearity and chaos.

3.1 Riemannian Geometry of Smooth Manifolds

3.1.1 Riemannian Manifolds: an Intuitive Picture

Smooth Manifolds

The core of modern geometrical dynamics represents the concept of a *manifold*. A manifold is an abstract mathematical space, which locally (i.e., in a close–up view) resembles the spaces described by *Euclidean geometry*, but which globally (i.e., when viewed as a whole) may have a more complicated structure. For example, the *surface of Earth* is a manifold; locally it seems to be flat, but viewed as a whole from the outer space (globally) it is actually round. A manifold can be constructed by 'gluing' separate *Euclidean spaces* together; for example, a world map can be made by gluing many maps of local regions together, and accounting for the resulting distortions.[1]

---

[1] On a sphere, the sum of the angles of a triangle is not equal to $180°$. A sphere is not a Euclidean space, but locally the laws of the Euclidean geometry are good approximations. In a small triangle on the face of the earth, the sum of the angles is very nearly $180°$. A sphere can be represented by a collection of two dimensional maps, therefore a sphere is a manifold.

3 Geometry and Topology Change in Complex Systems

Another example of a manifold is a *circle* S^1. A small piece of a circle appears to be like a slightly–bent part of a straight line segment, but overall the circle and the segment are different 1D manifolds (see Figure 3.1). A circle can be formed by bending a straight line segment and gluing the ends together.[2]

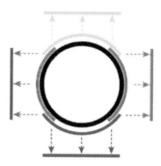

Fig. 3.1. The four charts each map part of the circle to an open interval, and together cover the whole circle.

---

[2] Locally, the circle looks like a line. It is 1D, that is, only one coordinate is needed to say where a point is on the circle locally. Consider, for instance, the top part of the circle (Figure 3.1), where the y–coordinate is positive. Any point in this part can be described by the x–coordinate. So, there is a continuous *bijection* χ_{top} (a mapping which is 1–1 both ways), which maps the top part of the circle to the open interval $(-1, 1)$, by simply projecting onto the first coordinate: $\chi_{top}(x, y) = x$. Such a function is called a *chart*. Similarly, there are charts for the bottom, left, and right parts of the circle. Together, these parts *cover* the whole circle and the four charts form an *atlas* (see the next subsection) for the circle. The top and right charts overlap: their intersection lies in the quarter of the circle where both the x– and the y–coordinates are positive. The two charts χ_{top} and χ_{right} map this part bijectively to the interval $(0, 1)$. Thus a function T from $(0, 1)$ to itself can be constructed, which first inverts the top chart to reach the circle and then follows the right chart back to the interval:

$$T(a) = \chi_{right}\left(\chi_{top}^{-1}(a)\right) = \chi_{right}\left(a, \sqrt{1-a^2}\right) = \sqrt{1-a^2}.$$

Such a function is called a *transition map*. The top, bottom, left, and right charts show that the circle is a manifold, but they do not form the only possible atlas. Charts need not be geometric projections, and the number of charts is a matter of choice. T and the other transition functions in Figure 3.1 are differentiable on the interval $(0, 1)$. Therefore, with this atlas the circle is a *differentiable*, or *smooth manifold*.

The surfaces of a *sphere*[3] and a *torus*[4] are examples of 2D manifolds. Manifolds are important objects in mathematics, physics and control theory, because they allow more complicated structures to be expressed and understood in terms of the well–understood properties of simpler Euclidean spaces.

The *Cartesian product* of manifolds is also a manifold (note that not every manifold can be written as a product). The dimension of the product manifold is the sum of the dimensions of its factors. Its topology is the product topology, and a Cartesian product of charts is a chart for the product manifold. Thus, an atlas for the product manifold can be constructed using atlases for its factors.

---

[3] The surface of the sphere S^2 can be treated in almost the same way as the circle S^1. It can be viewed as a subset of $\mathbb{R}^3$, defined by: $S = \{(x,y,z) \in \mathbb{R}^3 | x^2 + y^2 + z^2 = 1\}$. The sphere is 2D, so each chart will map part of the sphere to an open subset of $\mathbb{R}^2$. Consider the northern hemisphere, which is the part with positive z coordinate. The function χ defined by $\chi(x,y,z) = (x,y)$, maps the northern hemisphere to the open unit disc by projecting it on the (x,y)–plane. A similar chart exists for the southern hemisphere. Together with two charts projecting on the (x,z)–plane and two charts projecting on the (y,z)–plane, an atlas of six charts is obtained which covers the entire sphere. This can be easily generalized to an nD sphere $S^n = \{(x_1, x_2, ..., x_n) \in \mathbb{R}^n | x_1^2 + x_2^2 + ... + x_n^2 = 1\}$.

An n–sphere S^n can be also constructed by gluing together two copies of $\mathbb{R}^n$. The transition map between them is defined as $\mathbb{R}^n \setminus \{0\} \to \mathbb{R}^n \setminus \{0\} : x \mapsto x/\|x\|^2$. This function is its own inverse, so it can be used in both directions. As the transition map is a (C^∞)–*smooth function*, this atlas defines a *smooth manifold*.

[4] A torus (pl. tori), denoted by T^2, is a doughnut–shaped surface of revolution generated by revolving a circle about an axis coplanar with the circle. The sphere S^2 is a special case of the torus obtained when the axis of rotation is a diameter of the circle. If the axis of rotation does not intersect the circle, the torus has a hole in the middle and resembles a ring doughnut, a hula hoop or an inflated tire. The other case, when the axis of rotation is a chord of the circle, produces a sort of squashed sphere resembling a round cushion.

A torus can be defined parametrically by:

$$x(u,v) = (R + r\cos v)\cos u, \qquad y(u,v) = (R + r\cos v)\sin u, \qquad z(u,v) = r\sin v,$$

where $u, v \in [0, 2\pi]$, R is the distance from the center of the tube to the center of the torus, and r is the radius of the tube. According to a broader definition, the generator of a torus need not be a circle but could also be an ellipse or any other conic section.

Topologically, a torus is a closed surface defined as product of two circles: $T^2 = S^1 \times S^1$. The surface described above, given the relative topology from $\mathbb{R}^3$, is *homeomorphic* to a topological torus as long as it does not intersect its own axis.

One can easily generalize the torus to arbitrary dimensions. An n–torus T^n is defined as a product of n circles: $T^n = S^1 \times S^1 \times \cdots \times S^1$. Equivalently, the n–torus is obtained from the n–cube (the $\mathbb{R}^n$–generalization of the ordinary cube in $\mathbb{R}^3$) by gluing the opposite faces together.

An n–torus T^n is an example of an nD *compact manifold*. It is also an important example of a *Lie group* (see below).

If these atlases define a differential structure on the factors, the corresponding atlas defines a differential structure on the product manifold. The same is true for any other structure defined on the factors. If one of the factors has a boundary, the product manifold also has a boundary. Cartesian products may be used to construct tori and cylinders, for example, as $S^1 \times S^1$ and $S^1 \times [0,1]$, respectively.

Manifolds need not be *connected* (all in 'one piece'): a pair of separate circles is also a *topological manifold* (see below). Manifolds need not be *closed*: a line segment without its ends is a manifold. Manifolds need not be *finite*: a parabola is a topological manifold.

Manifolds can be viewed using either extrinsic or intrinsic view. In the *extrinsic view*, usually used in geometry and topology of surfaces, an nD manifold M is seen as embedded in an $(n+1)$D Euclidean space $\mathbb{R}^{n+1}$. Such a manifold is called a 'codimension 1 space'. With this view it is easy to use intuition from Euclidean spaces to define additional structure. For example, in a Euclidean space it is always clear whether a vector at some point is tangential or normal to some surface through that point. On the other hand, the *intrinsic view* of an nD manifold M is an abstract way of considering M as a topological space by itself, without any need for surrounding $(n+1)$D Euclidean space. This view is more flexible and thus it is usually used in high–dimensional mechanics and physics (where manifolds used represent configuration and phase spaces of dynamical systems), can make it harder to imagine what a tangent vector might be.

Additional structures are often defined on manifolds. Examples of manifolds with additional structure include:

- *differentiable* (or, *smooth manifolds*, on which one can do calculus;
- *Riemannian manifolds*, on which *distances* and *angles* can be defined;
- *symplectic manifolds*, which serve as the *phase space* in mechanics and physics;
- 4D pseudo–Riemannian manifolds which model *space–time* in general relativity.

The study of manifolds combines many important areas of mathematics: it generalizes concepts such as curves and surfaces as well as ideas from linear algebra and topology. Certain special classes of manifolds also have additional algebraic structure; they may behave like groups, for instance.

Historically, before the modern concept of a manifold there were several important results:

A. *Carl Friedrich Gauss* was arguably the first to consider abstract spaces as mathematical objects in their own right. His 'Theorema Egregium' gives a method for computing the curvature of a surface S without considering the ambient *Euclidean 3D space* $\mathbb{R}^3$ in which the surface lies. Such a surface would, in modern terminology, be called a manifold.

3.1 Riemannian Geometry of Smooth Manifolds

B. *Non–Euclidean geometry* considers spaces where Euclid's 'Parallel Postulate' fails. Saccheri first studied them in 1733. Lobachevsky, Bolyai, and Riemann developed them 100 years later. Their research uncovered two more types of spaces whose geometric structures differ from that of classical *Euclidean nD space* $\mathbb{R}^n$; these gave rise to *hyperbolic geometry* and *elliptic geometry*. In the modern theory of manifolds, these notions correspond to manifolds with negative and positive *curvature*, respectively.

C. The *Euler characteristic* is an example of a *topological property* (or *topological invariant*) of a manifold. For a convex polyhedron in Euclidean 3D space $\mathbb{R}^3$, with V vertices, E edges and F faces, Euler showed that $V - E + F = 2$. Thus the number 2 is called the Euler characteristic of the space $\mathbb{R}^3$. The Euler characteristic of other 3D spaces is a useful topological invariant, which can be extended to higher dimensions using the so–called *Betti numbers*. The study of other topological invariants of manifolds is one of the central themes of topology.

D. *Bernhard Riemann* was the first to do extensive work generalizing the idea of a surface to higher dimensions. The name manifold comes from Riemann's original German term, 'Mannigfaltigkeit', which W.K. *Clifford* translated as *'manifoldness'*. In his famous Göttingen inaugural lecture entitled 'On the Hypotheses which lie at the Bases of Geometry', Riemann described the set of all possible values of a variable with certain constraints as a 'manifoldness', because the variable can have many values. He distinguishes between continuous manifoldness and discontinuous manifoldness, depending on whether the value changes continuously or not. As continuous examples, Riemann refers to not only colors and the locations of objects in space, but also the possible shapes of a spatial figure. Using *mathematical induction*, Riemann constructs an n times extended manifoldness, or nD manifoldness, as a continuous stack of $(n-1)$D manifoldnesses. Riemann's intuitive notion of a 'manifoldness' evolved into what is today formalized as a manifold.

E. *Henri Poincaré* studied 3D manifolds at the end of the 19th Century, and raised a question, today known as the *Poincaré conjecture*. Hermann *Weyl* gave an *intrinsic definition for differentiable manifolds* in 1912. During the 1930s, H. *Whitney* and others clarified the foundational aspects of the subject, and thus intuitions dating back to the latter half of the 19th Century became precise, and developed through differential geometry (in particular, by the *Lie group theory* introduced by *Sophus Lie* in 1870, see below).

Riemannian Manifolds

Now, to measure distances and angles on manifolds, the manifold must be Riemannian. A *Riemannian manifold* is an analytic manifold in which each *tangent space* is equipped with an *inner product* $g = \langle \cdot, \cdot \rangle$, in a manner which varies smoothly from point to point. Given two tangent vectors X and Y, the inner product $\langle X, Y \rangle$ gives a real number. The dot (or scalar) product is a typical example of an inner product. This allows one to define various notions such as length, angles, areas (or, volumes), curvature, gradients of functions and divergence of vector–fields. Most familiar curves and surfaces, including $n-$spheres and Euclidean space, can be given the structure of a Riemannian manifold.

Any smooth manifold admits a Riemannian metric, which often helps to solve problems of differential topology. It also serves as an entry level for the more complicated structure of pseudo–Riemannian manifolds, which (in four dimensions) are the main objects of the general relativity theory.[5]

Every smooth submanifold of $\mathbb{R}^n$ (see extrinsic view above) has an induced Riemannian metric g: the inner product on each tangent space is the restriction of the inner product on $\mathbb{R}^n$. Therefore, one could define a Riemannian manifold as a *metric space* which is *isometric* to a smooth submanifold of $\mathbb{R}^n$ with the induced intrinsic metric, where isometry here is meant in the sense of preserving the length of curves.

Usually a Riemannian manifold M is defined as a smooth manifold with a smooth section of positive–definite quadratic forms on the associated *tangent bundle TM*. Then one has to work to show that it can be turned to a metric space.

Even though Riemannian manifolds are usually 'curved' (e.g., the space–time of general relativity), there is still a notion of 'straight line' on them: the *geodesics*. These are curves which locally join their points along shortest paths.

In Riemannian manifolds, the notions of geodesic completeness, topological completeness and metric completeness are the same: that each implies the other is the content of the *Hopf–Rinow Theorem* [II07b].

Riemann Surfaces

A *Riemann surface*, is a 1D complex manifold. Riemann surfaces can be thought of as 'deformed versions' of the complex plane: locally near every point they look like patches of the complex plane, but the global topology can be quite different. For example, they can look like a sphere, or a torus, or a couple of sheets glued together.

The main point of Riemann surfaces is that holomorphic (analytic complex) functions may be defined between them. Riemann surfaces are nowadays considered the natural setting for studying the global behavior of these functions, especially multi–valued functions such as the square root or the logarithm.

Every Riemann surface is a 2D real analytic manifold (i.e., a surface), but it contains more structure (specifically, a complex structure) which is needed

---

[5] A pseudo–Riemannian manifold is a variant of Riemannian manifold where the metric tensor is allowed to have an indefinite signature (as opposed to a positive–definite one).

3.1 Riemannian Geometry of Smooth Manifolds

for the unambiguous definition of holomorphic functions. A 2D real manifold can be turned into a Riemann surface (usually in several inequivalent ways) iff it is *orientable*. So the sphere and torus admit complex structures, but the Möbius strip, Klein bottle and projective plane do not.

Geometrical facts about Riemann surfaces are as 'nice' as possible, and they often provide the intuition and motivation for generalizations to other curves and manifolds. The Riemann–Roch Theorem is a prime example of this influence.[6]

Examples of Riemann surfaces include: the complex plane[7], open subsets of the complex plane[8], *Riemann sphere*[9], and many others.

Riemann surfaces naturally arise in string theory as models of string interactions [II07b].

Riemannian Geometry

Riemannian geometry is the study of smooth manifolds with Riemannian metrics g, i.e., a choice of positive–definite quadratic form $g = \langle \cdot, \cdot \rangle$ on a manifold's tangent spaces which varies smoothly from point to point. This gives in particular local ideas of angle, length of curves, and volume. From those

---

[6] Formally, let X be a Hausdorff space. A homeomorphism from an open subset $U \subset X$ to a subset of $\mathbb{C}$ is a chart. Two charts f and g whose domains intersect are said to be compatible if the maps $f \circ g^{-1}$ and $g \circ f^{-1}$ are holomorphic over their domains. If A is a collection of compatible charts and if any $x \in X$ is in the domain of some $f \in A$, then we say that A is an atlas. When we endow X with an atlas A, we say that (X, A) is a Riemann surface.

Different atlases can give rise to essentially the same Riemann surface structure on X; to avoid this ambiguity, one sometimes demands that the given atlas on X be maximal, in the sense that it is not contained in any other atlas. Every atlas A is contained in a unique maximal one by *Zorn's lemma*.

[7] The complex plane $\mathbb{C}$ is perhaps the most trivial Riemann surface. The map $f(z) = z$ (the identity map) defines a chart for $\mathbb{C}$, and f is an atlas for $\mathbb{C}$. The map $g(z) = z^*$ (the conjugate map) also defines a chart on $\mathbb{C}$ and g is an atlas for $\mathbb{C}$. The charts f and g are not compatible, so this endows $\mathbb{C}$ with two distinct Riemann surface structures.

[8] In a fashion analogous to the complex plane, every open subset of the complex plane can be viewed as a Riemann surface in a natural way. More generally, every open subset of a Riemann surface is a Riemann surface.

[9] The Riemann sphere is a useful visualization of the extended complex plane, which is the complex plane plus a point at infinity. It is obtained by imagining that all the rays emanating from the origin of the complex plane eventually meet again at a point called the point at infinity, in the same way that all the meridians from the south pole of a sphere get to meet each other at the north pole.

Formally, the Riemann sphere is obtained via a one–point compactification of the complex plane. This gives it the topology of a 2–sphere. The sphere admits a unique complex structure turning it into a Riemann surface. The Riemann sphere can be characterized as the unique simply–connected, compact Riemann surface.

some other global quantities can be derived by integrating local contributions [II07b].

The manifold may also be given an *affine connection*,[10] which is roughly an idea of change from one point to another. If the metric does not 'vary from point to point' under this connection, we say that the metric and connection are compatible, and we have a Riemann–Cartan manifold. If this connection is also self–commuting when acting on a scalar function, we say that it is torsion–free, and the manifold is a Riemannian manifold.

Levi–Civita Connection

In Riemannian geometry, the *Levi–Civita connection* (named after Tullio Levi–Civita) is the torsion–free Riemannian connection, i.e., a torsion–free connection of the tangent bundle, preserving a given Riemannian metric (or, pseudo–Riemannian metric).[11] The fundamental Theorem of Riemannian geometry states that there is a unique connection which satisfies these properties.[12]

In the theory of Riemannian and pseudo–Riemannian manifolds the term covariant derivative is often used for the Levi–Civita connection. The coordi-

---

[10] Connection (or, covariant derivative) is a way of specifying a derivative of a vector–field along another vector–field on a manifold. That is an application to tangent bundles; there are more general connections, used to formulate intrinsic differential equations. Connections give rise to parallel transport along a curve on a manifold. A connection also leads to invariants of curvature, and the so–called *torsion*.

An affine connection is a connection on the tangent bundle TM of a smooth manifold M. In general, it might have a non–vanishing torsion.

The curvature of a connected manifold can be characterized intrinsically by taking a vector at some point and *parallel transporting* it along a curve on the manifold. Although comparing vectors at different points is generally not a well–defined process, an affine connection ∇ is a rule which describes how to legitimately move a vector along a curve on the manifold without changing its direction ('keeping the vector parallel').

[11] Formally, let (M, g) be a Riemannian manifold (or pseudo–Riemannian manifold); then an affine connection is the Levi–Civita connection if it satisfies the following conditions:

A. *Preserves metric* g, i.e., for any three vector–fields $X, Y, Z \in M$ we have $Xg(Y, Z) = g(\nabla_X Y, Z) + g(Y, \nabla_X Z)$, where $Xg(Y, Z)$ denotes the derivative of a function $g(Y, Z)$ along a vector–field X.
B. *Torsion–free*, i.e., for any two vector–fields $X, Y, Z \in M$ we have $\nabla_X Y - \nabla_Y X = [X, Y]$, where $[X, Y]$ is the *Lie bracket* for vector–fields X and Y.

[12] The Levi–Civita connection defines also a derivative along curves, usually denoted by D. Given a smooth curve (a path) $\gamma = \gamma(t) : \mathbb{R} \to M$ and a vector–field $X = X(t)$ on γ, its derivative along γ is defined by: $D_t X = \nabla_{\dot\gamma(t)} X$. This equation defines the *parallel transport* for a vector–field X.

nate expression of the connection is given by *Christoffel symbols*.[13] Note that connection is not a tensor, except on jet bundles.

Fundamental Riemannian Tensors

The two basic objects in Riemannian geometry are the metric tensor and the curvature tensor. The *metric tensor* $g = \langle \cdot, \cdot \rangle$ is a symmetric second–order (i.e., $(0,2)$) tensor that is used to measure distance in a space. In other words, given a Riemannian manifold, we make a choice of a $(0,2)$–tensor on the manifold's tangent spaces.[14] At a given point in the manifold, this tensor takes a pair of vectors in the tangent space to that point, and gives a real number. This concept is just like a dot product, or inner product. This function from vectors into the real numbers is required to vary smoothly from point to point.

On any Riemannian manifold, from its second–order metric tensor $g = \langle \cdot, \cdot \rangle$, one can derive the associated fourth–order *Riemann curvature tensor* [II07b]. This tensor is the most standard way to express curvature of Rieman-

---

[13] The Christoffel symbols, named for Elwin Bruno Christoffel (1829–1900), are coordinate expressions for the Levi–Civita connection derived from the metric tensor. The Christoffel symbols are used whenever practical calculations involving geometry must be performed, as they allow very complex calculations to be performed without confusion. In particular, if we denote the unit vectors on M as $e_i = \partial/\partial x_i$, then the Christoffel symbols of the second kind are defined by $\Gamma^k_{ij} = \langle \nabla_{e_i}; e_j, e_k \rangle$. Alternatively, using the metric tensor g_{ik} (see below) we get the explicit expression for the Christoffel symbols in a holonomic coordinate basis:

$$\Gamma^i_{kl} = \frac{1}{2} g^{im} \left(\frac{\partial g_{mk}}{\partial x^l} + \frac{\partial g_{ml}}{\partial x^k} - \frac{\partial g_{kl}}{\partial x^m} \right).$$

In a general, nonholonomic coordinates they include the additional commutation coefficients. The Christoffel symbols are used to define the covariant derivative of various tensor–fields, as well as the Riemannian curvature. Also, they figure in the *geodesic equation*:

$$\frac{d^2 x^i}{dt^2} + \Gamma^i_{jk} \frac{dx^j}{dt} \frac{dx^k}{dt} = 0$$

for the curve $x^i = x^i(t)$ on the smooth manifold M.

[14] The most familiar example is that of basic high–school geometry: the 2D Euclidean metric tensor, in the usual $x - y$ coordinates, reads: $g = \begin{bmatrix} 1 & 0 \\ 0 & 1 \end{bmatrix}$. The associated length of a curve is given by the familiar calculus formula: $L = \int_a^b \sqrt{(dx)^2 + (dy)^2}$.

The unit sphere in $\mathbb{R}^3$ comes equipped with a natural metric induced from the ambient Euclidean metric. In standard spherical coordinates (θ, ϕ) the metric takes the form: $g = \begin{bmatrix} 1 & 0 \\ 0 & \sin^2 \theta \end{bmatrix}$, which is usually written as: $g = d\theta^2 + \sin^2 \theta \, d\phi^2$.

nian manifolds, or more generally, any manifold with an affine connection, torsionless or with torsion.[15]

---

[15] The Riemann curvature tensor is given in terms of a Levi–Civita connection ∇ (more generally, an affine connection, or covariant differentiation, see below) by the following formula:

$$R(u,v)w = \nabla_u \nabla_v w - \nabla_v \nabla_u w - \nabla_{[u,v]} w,$$

where u, v, w are tangent vector–fields and $R(u,v)$ is a linear transformation of the tangent space of the manifold; it is linear in each argument. If $u = \partial/\partial x_i$ and $v = \partial/\partial x_j$ are coordinate vector–fields then $[u,v] = 0$ and therefore the above formula simplifies to

$$R(u,v)w = \nabla_u \nabla_v w - \nabla_v \nabla_u w,$$

i.e., the curvature tensor measures non–commutativity of the covariant derivative. The linear transformation $w \mapsto R(u,v)w$ is also called the curvature transformation or endomorphism.

In local coordinates x^μ (e.g., in general relativity) the Riemann curvature tensor can be written using the Christoffel symbols of the manifold's Levi–Civita connection:

$$R^\rho{}_{\sigma\mu\nu} = \partial_\mu \Gamma^\rho_{\nu\sigma} - \partial_\nu \Gamma^\rho_{\mu\sigma} + \Gamma^\rho_{\mu\lambda} \Gamma^\lambda_{\nu\sigma} - \Gamma^\rho_{\nu\lambda} \Gamma^\lambda_{\mu\sigma}.$$

The Riemann curvature tensor has the following symmetries:

$$R(u,v) = -R(v,u), \qquad \langle R(u,v)w, z \rangle = -\langle R(u,v)z, w \rangle,$$
$$R(u,v)w + R(v,w)u + R(w,u)v = 0.$$

The last identity was discovered by Ricci, but is often called the first Bianchi identity or algebraic Bianchi identity, because it looks similar to the Bianchi identity below. These three identities form a complete list of symmetries of the curvature tensor, i.e. given any tensor which satisfies the identities above, one can find a Riemannian manifold with such a curvature tensor at some point. Simple calculations show that such a tensor has $n^2(n^2-1)/12$ independent components.

The *Bianchi identity* involves the covariant derivatives:

$$\nabla_u R(v,w) + \nabla_v R(w,u) + \nabla_w R(u,v) = 0.$$

A contracted curvature tensor is called the *Ricci tensor*. It is a symmetric second–order tensor given by:

$$R_{ik} = \frac{\partial \Gamma^l{}_{ik}}{\partial x^l} - \frac{\partial \Gamma^l{}_{il}}{\partial x^k} + \Gamma^l{}_{ik} \Gamma^m{}_{lm} - \Gamma^m{}_{il} \Gamma^l{}_{km}.$$

Its further contraction gives the *Ricci scalar curvature*, $R = g^{ik} R_{ik}$. The *Einstein tensor* G_{ik} is defined in terms of the Ricci tensor R_{ik} and the Ricci scalar R,

$$G_{ik} = R_{ik} - \frac{1}{2} g_{ik} R.$$

Example: Lagrangian Mechanics

Riemannian manifolds are natural stage for the *Lagrangian mechanics*, which is a re–formulation of classical mechanics introduced by Joseph Louis Lagrange in 1788. In Lagrangian mechanics, the trajectory of an object is derived by finding the path which minimizes the action, a quantity which is the integral of the Lagrangian over time. The Lagrangian for classical mechanics L is taken to be the difference between the kinetic energy T and the potential energy V, so $L = T - V$. This considerably simplifies many physical problems.

For example, consider a bead on a hoop. If one were to calculate the motion of the bead using *Newtonian mechanics*, one would have a complicated set of equations which would take into account the forces that the hoop exerts on the bead at each moment. The same problem using Lagrangian mechanics is much simpler. One looks at all the possible motions that the bead could take on the hoop and mathematically finds the one which *minimizes the action*. There are fewer equations since one is not directly calculating the influence of the hoop on the bead at a given moment.

Lagrange's Equations

The equations of motion in Lagrangian mechanics are *Lagrange's equations*, also known as *Euler–Lagrange equations*. Below, we sketch out the derivation of Lagrange's equation from Newton's laws of motion (see next chapter for details).

Consider a single mechanical particle with mass m and position vector $\mathbf{r}$. The applied force, $\mathbf{F}$, can be expressed as the gradient (denoted ∇) of a scalar potential energy function $V(\mathbf{r}, t)$:

$$\mathbf{F} = -\nabla V.$$

Such a force is independent of third– or higher–order derivatives of $\mathbf{r}$, so *Newton's Second Law* forms a set of 3 second–order ODEs. Therefore, the motion of the particle can be completely described by 6 independent variables, or degrees of freedom (DOF). An obvious set of variables is the Cartesian components of $\mathbf{r}$ and their time derivatives, at a given instant of time, that is position (x, y, z) and velocity (v_x, v_y, v_z).

More generally, we can work with a set of generalized coordinates, q^i, ($i = 1, ..., n$), and their time derivatives, the generalized velocities, $\dot{q}^i$. The position vector $\mathbf{r}$ is related to the generalized coordinates by some transformation equation: $\mathbf{r} = \mathbf{r}(q^i, t)$. The term 'generalized coordinates' is really a leftover from the period when Cartesian coordinates were the default coordinate system. In the q^i–coordinates the Lagrange's equations read:

$$\frac{\partial L}{\partial q^i} = \frac{d}{dt} \frac{\partial L}{\partial \dot{q}^i},$$

where $L = T - V$ is the system's Lagrangian.

The time integral of the Lagrangian L, denoted S is called the *action*:[16]

$$S = \int L\, dt.$$

Let q_0 and q_1 be the coordinates at respective initial and final times t_0 and t_1. Using the *calculus of variations*, it can be shown the Lagrange's equations are equivalent to the *Hamilton's principle*: "The system undergoes the trajectory between t_0 and t_1 whose action has a stationary value." This is formally written:

$$\delta S = 0,$$

where by 'stationary', we mean that the action does not vary to first–order for infinitesimal deformations of the trajectory, with the end–points (q_0, t_0) and (q_1, t_1) fixed.[17]

---

[16] The action principle is an assertion about the nature of motion, from which the trajectory of a dynamical system subject to some forces can be determined. The path of an object is the one that yields a stationary value for a quantity called the action. Thus, instead of thinking about an object accelerating in response to applied forces, one might think of them picking out the path with a stationary action. The action is a scalar (a number) with the unit of measure for Action as Energy × Time. Although equivalent in classical mechanics with Newton's laws, the action principle is better suited for generalizations and plays an important role in modern physics. Indeed, this principle is one of the great generalizations in physical science. In particular, it is fully appreciated and best understood within quantum mechanics. Richard Feynman's path integral formulation of quantum mechanics is based on a stationary–action principle, using path integrals. Maxwell's equations can be derived as conditions of stationary action.

[17] More generally, a Lagrangian $\mathcal{L}[\varphi^i]$ of a dynamical system is a function of the dynamical variables $\varphi^i(x)$ and concisely describes the equations of motion of the system in coordinates x^i, $(i = 1, ..., n)$. The equations of motion are obtained by means of an *action principle*, written as

$$\frac{\delta \mathcal{S}}{\delta \varphi_i} = 0,$$

where the *action* is a functional

$$\mathcal{S}[\varphi^i] = \int \mathcal{L}[\varphi^i(s)]\, d^n x,$$

$(d^n x = dx^1...dx^n)$.

The equations of motion obtained by means of the functional derivative are identical to the usual Euler–Lagrange equations. Dynamical system whose equations of motion are obtainable by means of an action principle on a suitably chosen Lagrangian are known as Lagrangian dynamical systems. Examples of Lagrangian dynamical systems range from the (classical version of the) Standard Model, to Newton's equations, to purely mathematical problems such as geodesic equations and the Plateau's problem.

The total energy function called *Hamiltonian*, denoted by H, is obtained by performing a *Legendre transformation* on the Lagrangian.[18] The Hamiltonian is the basis for an alternative formulation of classical mechanics known as Hamiltonian mechanics (see below).

In 1948, R.P. Feynman invented the *path–integral formulation* extending the principle of least action to quantum mechanics for electrons and photons. In this formulation, particles travel every possible path between the initial and final states; the probability of a specific final state is obtained by summing over all possible trajectories leading to it. In the classical regime, the path integral formulation cleanly reproduces the Hamilton's principle, as well as the Fermat's principle in optics.

3.1.2 Smooth Manifolds and Their (Co)Tangent Bundles

As we have already got the initial feeling, in the heart of geometrical dynamics is the concept of a *manifold* (see [Rha84]). To get some *dynamical intuition* behind this concept, let us consider a simple 3DOF mechanical system determined by three generalized coordinates, $q^i = \{q^1, q^2, q^3\}$. There is a unique way to represent this system as a 3D manifold, such that to each point of the manifold there corresponds a definite configuration of the mechanical system with coordinates q^i; therefore, we have a geometrical representation of the configurations of our mechanical system, called the *configuration manifold*.

The Lagrangian mechanics is important not just for its broad applications, but also for its role in advancing deep understanding of physics. Although Lagrange sought to describe classical mechanics, the action principle that is used to derive the Lagrange's equation is now recognized to be deeply tied to quantum mechanics: physical action and quantum–mechanical phase (waves) are related via Planck's constant, and the *Principle of stationary action* can be understood in terms of constructive interference of wave functions. The same principle, and the Lagrangian formalism, are tied closely to *Noether Theorem*, which relates physical conserved quantities to continuous symmetries of a physical system; and Lagrangian mechanics and Noether's Theorem together yield a natural formalism for first quantization by including commutators between certain terms of the Lagrange's equations of motion for a physical system.

More specifically, in field theory, occasionally a distinction is made between the Lagrangian L, of which the action is the time integral $S = \int L dt$ and the Lagrangian density $\mathcal{L}$, which one integrates over all space–time to get the 4D action:
$$S[\varphi^i] = \int \mathcal{L}[\varphi^i(x)] \, d^4x.$$
The Lagrangian is then the spatial integral of the Lagrangian density.

[18] The Hamiltonian is the Legendre transform of the Lagrangian:
$$H(q, p, t) = \sum_i \dot{q}^i p_i - L(q, \dot{q}, t).$$

If the mechanical system moves in any way, its coordinates are given as the functions of the time. Thus, the motion is given by equations of the form: $q^i = q^i(t)$. As t varies (i.e., $t \in \mathbb{R}$), we observe that the system's *representative point* in the configuration manifold describes a *curve* and $q^i = q^i(t)$ are the equations of this curve.

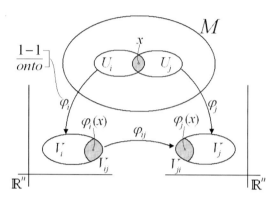

Fig. 3.2. An intuitive geometrical picture behind the manifold concept (see text).

On the other hand, to get some *geometrical intuition* behind the concept of a manifold, consider a set M (see Figure 3.2) which is a *candidate* for a manifold. Any point $x \in M$[19] has its *Euclidean chart*, given by a 1–1 and *onto* map $\varphi_i : M \to \mathbb{R}^n$, with its *Euclidean image* $V_i = \varphi_i(U_i)$. More precisely, a chart φ_i is defined by

$$\varphi_i : M \supset U_i \ni x \mapsto \varphi_i(x) \in V_i \subset \mathbb{R}^n,$$

where $U_i \subset M$ and $V_i \subset \mathbb{R}^n$ are open sets (see [Arn78, Rha84]).

Clearly, any point $x \in M$ can have several different charts (see Figure 3.2). Consider a case of two charts, $\varphi_i, \varphi_j : M \to \mathbb{R}^n$, having in their images two open sets, $V_{ij} = \varphi_i(U_i \cap U_j)$ and $V_{ji} = \varphi_j(U_i \cap U_j)$. Then we have *transition functions* φ_{ij} between them,

$$\varphi_{ij} = \varphi_j \circ \varphi_i^{-1} : V_{ij} \to V_{ji}, \qquad \text{locally given by} \qquad \varphi_{ij}(x) = \varphi_j(\varphi_i^{-1}(x)).$$

If transition functions φ_{ij} exist, then we say that two charts, φ_i and φ_j are *compatible*. Transition functions represent a general (nonlinear) *transformations of coordinates*, which are the core of classical *tensor calculus* (Appendix).

A set of compatible charts $\varphi_i : M \to \mathbb{R}^n$, such that each point $x \in M$ has its Euclidean image in at least one chart, is called an *atlas*. Two atlases are *equivalent* iff all their charts are compatible (i.e., transition functions exist

---

[19] Note that sometimes we will denote the point in a manifold M by m, and sometimes by x (thus implicitly assuming the existence of coordinates $x = (x^i)$).

between them), so their union is also an atlas. A *manifold structure* is a class of equivalent atlases.

Finally, as charts $\varphi_i : M \to \mathbb{R}^n$ were supposed to be 1-1 and onto maps, they can be either *homeomorphisms*, in which case we have a *topological* (C^0) manifold, or *diffeomorphisms*, in which case we have a *smooth* (C^∞) manifold.

Slightly more precisely, a topological (respectively smooth) manifold is a separable space M which is locally homeomorphic (resp. diffeomorphic) to Euclidean space $\mathbb{R}^n$, having the following properties (reflected in Figure 3.2):

A. M is a *Hausdorff space*: For every pair of points $x_1, x_2 \in M$, there are disjoint open subsets $U_1, U_2 \subset M$ such that $x_1 \in U_1$ and $x_2 \in U_2$.
B. M is *second–countable space*: There exists a countable basis for the topology of M.
C. *M is locally Euclidean of dimension n*: Every point of M has a neighborhood that is homeomorphic (resp. diffeomorphic) to an open subset of $\mathbb{R}^n$.

This implies that for any point $x \in M$ there is a homeomorphism (resp. diffeomorphism) $\varphi : U \to \varphi(U) \subseteq \mathbb{R}^n$, where U is an open neighborhood of x in M and $\varphi(U)$ is an open subset in $\mathbb{R}^n$. The pair (U, φ) is called a *coordinate chart* at a point $x \in M$, etc.

Definition of a Smooth Manifold

Given a chart (U, φ), we call the set U a *coordinate domain*, or a coordinate neighborhood of each of its points. If in addition $\varphi(U)$ is an open ball in $\mathbb{R}^n$, then U is called a *coordinate ball*. The map φ is called a (*local*) *coordinate map*, and the component functions $(x^1, ..., x^n)$ of φ, defined by $\varphi(m) = (x^1(m), ..., x^n(m))$, are called *local coordinates* on U.

Two charts (U_1, φ_1) and (U_2, φ_2) such that $U_1 \cap U_2 \neq \emptyset$ are called *compatible* if $\varphi_1(U_1 \cap U_2)$ and $\varphi_2(U_2 \cap U_1)$ are open subsets of $\mathbb{R}^n$. A family $(U_\alpha, \varphi_\alpha)_{\alpha \in A}$ of compatible charts on M such that the U_α form a *covering* of M is called an *atlas*. The maps $\varphi_{\alpha\beta} = \varphi_\beta \circ \varphi_\alpha^{-1} : \varphi_\alpha(U_{\alpha\beta}) \to \varphi_\beta(U_{\alpha\beta})$ are called the *transition maps*, for the atlas $(U_\alpha, \varphi_\alpha)_{\alpha \in A}$, where $U_{\alpha\beta} = U_\alpha \cap U_\beta$.

An atlas $(U_\alpha, \varphi_\alpha)_{\alpha \in A}$ for a manifold M is said to be a C^∞–*atlas*, if all transition maps $\varphi_{\alpha\beta} : \varphi_\alpha(U_{\alpha\beta}) \to \varphi_\beta(U_{\alpha\beta})$ are of class C^∞. Two C^∞ atlases are called C^∞–*equivalent*, if their union is again a C^∞–atlas for M. An equivalence class of C^∞–atlases is called a C^∞–*structure* on M. In other words, a smooth structure on M is a *maximal* smooth atlas on M, i.e., such an atlas that is not contained in any strictly larger smooth atlas. By a C^∞–*manifold* M, we mean a topological manifold together with a C^∞–structure and a chart on M will be a chart belonging to some atlas of the C^∞–structure. Smooth manifold means C^∞–manifold, and the word '*smooth*' is used synonymously for C^∞ [Rha84].

Sometimes the terms 'local coordinate system' or 'parametrization' are used instead of charts. That M is not defined with any particular atlas, but with an equivalence class of atlases, is a mathematical formulation of the *general covariance* principle. Every suitable coordinate system is equally good. A Euclidean chart may well suffice for an open subset of $\mathbb{R}^n$, but this coordinate system is not to be preferred to the others, which may require many charts (as with polar coordinates), but are more convenient in other respects.

For example, the atlas of an n−sphere S^n has two charts. If $N = (1, 0, ..., 0)$ and $S = (-1, ..., 0, 0)$ are the north and south poles of S^n respectively, then the two charts are given by the stereographic projections from N and S:

$$\varphi_1 : S^n \setminus \{N\} \to \mathbb{R}^n, \varphi_1(x^1, ..., x^{n+1}) = (x^2/(1-x^1), ..., x^{n+1}/(1-x^1)), \text{ and}$$
$$\varphi_2 : S^n \setminus \{S\} \to \mathbb{R}^n, \varphi_2(x^1, ..., x^{n+1}) = (x^2/(1+x^1), ..., x^{n+1}/(1+x^1)),$$

while the overlap map $\varphi_2 \circ \varphi_1^{-1} : \mathbb{R}^n \setminus \{0\} \to \mathbb{R}^n \setminus \{0\}$ is given by the diffeomorphism $(\varphi_2 \circ \varphi_1^{-1})(z) = z/||z||^2$, for z in $\mathbb{R}^n \setminus \{0\}$, from $\mathbb{R}^n \setminus \{0\}$ to itself.

Various *additional structures* can be imposed on $\mathbb{R}^n$, and the corresponding manifold M will inherit them through its covering by charts. For example, if a covering by charts takes their values in a *Banach space* E, then E is called the *model space* and M is referred to as a C^∞−*Banach manifold* modelled on E. Similarly, if a covering by charts takes their values in a *Hilbert space* $\mathcal{H}$, then $\mathcal{H}$ is called the *model space* and M is referred to as a C^∞−*Hilbert manifold* modelled on $\mathcal{H}$. If not otherwise specified, we will consider M to be an Euclidean manifold, with its covering by charts taking their values in $\mathbb{R}^n$.

For a Hausdorff C^∞−manifold the following properties are equivalent [II07b]: (i) it is paracompact; (ii) it is metrizable; (iii) it admits a Riemannian metric;[20] (iv) each connected component is separable.

Smooth Maps Between Manifolds

A map $\varphi : M \to N$ between two manifolds M and N, with $M \ni m \mapsto \varphi(m) \in N$, is called a *smooth map*, or C^∞−map, if for each $m \in M$ and each chart (V, ψ) on N with $\varphi(m) \in V$ there is a chart (U, ϕ) on M with $m \in U, \varphi(U) \subseteq V$, and $\Phi = \psi \circ \varphi \circ \phi^{-1}$ is C^∞.

Let M and N be smooth manifolds and let $\varphi : M \to N$ be a smooth map. The map φ is called a *covering*, or equivalently, M is said to *cover* N, if φ is surjective and each point $n \in N$ admits an open neighborhood V such that $\varphi^{-1}(V)$ is a union of disjoint open sets, each diffeomorphic via φ to V.

A C^∞−map $\varphi : M \to N$ is called a C^∞−*diffeomorphism* if φ is a bijection, $\varphi^{-1} : N \to M$ exists and is also C^∞. Two manifolds are called diffeomorphic

---
[20] Recall the corresponding properties of a Euclidean metric d. For any three points $x, y, z \in \mathbb{R}^n$, the following axioms are valid:

$$M_1 : d(x, y) > 0, \text{ for } x \neq y; \quad \text{and} \quad d(x, y) = 0, \text{ for } x = y;$$
$$M_2 : d(x, y) = d(y, x); \quad M_3 : d(x, y) \leq d(x, z) + d(z, y).$$

if there exists a diffeomorphism between them. All smooth manifolds and smooth maps between them form the category $\mathcal{M}$.

(Co)Tangent Bundles of a Smooth Manifold

Intuition Behind a Tangent Bundle

In mechanics, to each nD *configuration manifold* M there is associated its $2nD$ *velocity phase–space manifold*, denoted by TM and called the tangent bundle of M (see Figure 3.3). The original smooth manifold M is called the *base* of TM. There is an onto map $\pi : TM \to M$, called the *projection*. Above each point $x \in M$ there is a *tangent space* $T_x M = \pi^{-1}(x)$ to M at x, which is called a *fibre*. The fibre $T_x M \subset TM$ is the subset of TM, such that the total tangent bundle, $TM = \bigsqcup_{m \in M} T_x M$, is a *disjoint union* of tangent spaces $T_x M$ to M for all points $x \in M$. From dynamical perspective, the most important quantity in the tangent bundle concept is the smooth map $v : M \to TM$, which is an inverse to the projection π, i.e, $\pi \circ v = \mathrm{Id}_M$, $\pi(v(x)) = x$. It is called the *velocity vector–field*. Its graph $(x, v(x))$ represents the *cross–section* of the tangent bundle TM. This explains the dynamical term velocity phase–space, given to the tangent bundle TM of the manifold M.

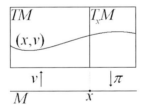

Fig. 3.3. A sketch of a tangent bundle TM of a smooth manifold M (see text for explanation).

Definition of a Tangent Bundle

Recall that if $[a, b]$ is a closed interval, a C^0–map $\gamma : [a, b] \to M$ is said to be *differentiable* at the endpoint a if there is a chart (U, ϕ) at $\gamma(a)$ such that the following limit exists and is finite [II07b]:

$$\frac{d}{dt}(\phi \circ \gamma)(a) \equiv (\phi \circ \gamma)'(a) = \lim_{t \to a} \frac{(\phi \circ \gamma)(t) - (\phi \circ \gamma)(a)}{t - a}. \tag{3.1}$$

Generalizing (3.1), we get the notion of the *curve on a manifold*. For a smooth manifold M and a point $m \in M$ a curve at m is a C^0–map $\gamma : I \to M$ from an interval $I \subset \mathbb{R}$ into M with $0 \in I$ and $\gamma(0) = m$.

Two curves γ_1 and γ_2 passing though a point $m \in U$ are *tangent at m with respect to the chart* (U, ϕ) if $(\phi \circ \gamma_1)'(0) = (\phi \circ \gamma_2)'(0)$. Thus, two curves are tangent if they have identical tangent vectors (same direction and speed) in a local chart on a manifold.

For a smooth manifold M and a point $m \in M$, the *tangent space* $T_m M$ to M at m is the *set of equivalence classes* of curves at m:

$$T_m M = \{[\gamma]_m : \gamma \text{ is a curve at a point } m \in M\}.$$

A C^∞−map $\varphi : M \ni m \mapsto \varphi(m) \in N$ between two manifolds M and N induces a linear map $T_m\varphi : T_m M \to T_{\varphi(m)} N$ for each point $m \in M$, called a *tangent map*, with the *natural projection* $\pi_M : TM \to M$, given by $\pi_M(T_m M) = m$, that takes a tangent vector v to the point $m \in M$ at which the vector v is attached i.e., $v \in T_m M$.

For an nD smooth manifold M, its nD *tangent bundle* TM is the disjoint union of all its tangent spaces $T_m M$ at all points $m \in M$, $TM = \bigsqcup_{m \in M} T_m M$.

To define the smooth structure on TM, we need to specify how to construct local coordinates on TM. To do this, let $(x^1(m), ..., x^n(m))$ be local coordinates of a point m on M and let $(v^1(m), ..., v^n(m))$ be components of a tangent vector in this coordinate system. Then the $2n$ numbers $(x^1(m), ..., x^n(m), v^1(m), ..., v^n(m))$ give a *local coordinate system* on TM.

$TM = \bigsqcup_{m \in M} T_m M$ defines a family of vector spaces parameterized by M.

The inverse image $\pi_M^{-1}(m)$ of a point $m \in M$ under the natural projection π_M is the tangent space $T_m M$. This space is called the *fibre* of the tangent bundle over the point $m \in M$ [Ste72].

Definition of a Cotangent Bundle

A *dual* notion to the tangent space $T_m M$ to a smooth manifold M at a point m is its *cotangent space* $T_m^* M$ at the same point m. Similarly to the tangent bundle, for a smooth manifold M of dimension n, its *cotangent bundle* T^*M is the disjoint union of all its cotangent spaces $T_m^* M$ at all points $m \in M$, i.e., $T^*M = \bigsqcup_{m \in M} T_m^* M$. Therefore, the cotangent bundle of an n−manifold M is the vector bundle $T^*M = (TM)^*$, the (real) dual of the tangent bundle TM.

If M is an n−manifold, then T^*M is a $2n$−manifold. To define the smooth structure on T^*M, we need to specify how to construct local coordinates on T^*M. To do this, let $(x^1(m), ..., x^n(m))$ be local coordinates of a point m on M and let $(p_1(m), ..., p_n(m))$ be components of a covector in this coordinate system. Then the $2n$ numbers $(x^1(m), ..., x^n(m), p_1(m), ..., p_n(m))$ give a local coordinate system on T^*M. This is the basic idea one uses to prove that indeed T^*M is a $2n$−manifold.

$T^*M = \bigsqcup_{m \in M} T_m^* M$ defines a family of vector spaces parameterized by M, with the *conatural projection*, $\pi_M^* : T^*M \to M$, given by $\pi_M^*(T_m^* M) = m$, that takes a covector p to the point $m \in M$ at which the covector p is attached i.e., $p \in T_m^* M$. The inverse image $\pi_M^{-1}(m)$ of a point $m \in M$ under the conatural projection π_M^* is the cotangent space $T_m^* M$. This space is called the *fibre* of the cotangent bundle over the point $m \in M$.

In a similar way, a C^∞−map $\varphi : M \to N$ between two manifolds M and N induces a linear *cotangent map* $T^*\varphi : T^*M \to T^*N$ between their cotangent bundles.

All cotangent bundles and their cotangent maps form the category $\mathcal{T}^*\mathcal{B}$. The category $\mathcal{T}^*\mathcal{B}$ is the natural stage for *Hamiltonian dynamics*.

Now, we can formulate the *dual version of the global chain rule*. If $\varphi : M \to N$ and $\psi : N \to P$ are two smooth maps, then we have $T^*(\psi \circ \varphi) = T^*\psi \circ T^*\varphi$.

Vector–Fields and Their Flows on Manifolds

Vector–Fields on M

A *vector–field* X on U, where U is an open chart in n−manifold M, is a smooth function from U to M assigning to each point $m \in U$ a vector at that point, i.e., $X(m) = (m, X(m))$. If $X(m)$ is tangent to M for each $m \in M$, X is said to be a *tangent vector–field* on M. If $X(m)$ is orthogonal to M (i.e., $X(p) \in M_m^\perp$) for each $X(m) \in M$, X is said to be a *normal vector–field* on M.

In other words, let M be a C^∞−manifold. A C^∞−vector–field on M is a C^∞−section of the tangent bundle TM of M. Thus a vector–field X on a manifold M is a C^∞−map $X : M \to TM$ such that $X(m) \in T_m M$ for all points $m \in M$, and $\pi_M \circ X = Id_M$. Therefore, a vector–field assigns to each point m of M a vector based (i.e., bound) at that point. The set of all C^∞ vector–fields on M is denoted by $\mathcal{X}^k(M)$.

A vector–field $X \in \mathcal{X}^k(M)$ represents a field of direction indicators [Thi79]: to every point m of M it assigns a vector in the tangent space $T_m M$ at that point. If X is a vector–field on M and (U, ϕ) is a chart on M and $m \in U$, then we have $X(m) = X(m)\phi^i \frac{\partial}{\partial \phi^i}$. Following [II07b], we write $X|_U = X\phi^i \frac{\partial}{\partial \phi^i}$.

Let M be a connected n−manifold, and let $f : U \to \mathbb{R}$ (U an open set in M) and $c \in \mathbb{R}$ be such that $M = f^{-1}(c)$ (i.e., M is the *level set* of the function f at *height* c) and $\nabla f(m) \neq 0$ for all $m \in M$. Then there exist on M exactly two smooth unit normal vector–fields $N_{1,2}(m) = \pm \frac{\nabla f(m)}{|\nabla f(m)|}$ (here $|X| = (X \cdot X)^{1/2}$ denotes the norm or length of a vector X, and $(\cdot)$ denotes the scalar product on M) for all $m \in M$, called *orientations* on M.

Let $\varphi : M \to N$ be a smooth map. Recall that two vector–fields $X \in \mathcal{X}^k(M)$ and $Y \in \mathcal{X}(N)$ are called φ−related, if $T\varphi \circ X = Y \circ \varphi$ holds. In particular, a diffeomorphism $\varphi : M \to N$ induces a linear map between vector–fields on two manifolds, $\varphi^* : \mathcal{X}^k(M) \to \mathcal{X}(N)$, such that $\varphi^* X = T\varphi \circ X \circ \varphi^{-1} : N \to TN$.

A C^∞ *time–dependent vector–field* is a C^∞–map $X : \mathbb{R} \times M \to TM$ such that $X(t,m) \in T_m M$ for all $(t,m) \in \mathbb{R} \times M$, i.e., $X_t(m) = X(t,m)$.

Integral Curves as Dynamical Trajectories

Recall (3.1.2) that a curve γ at a point m of an n–manifold M is a C^0–map from an open interval $I \subset \mathbb{R}$ into M such that $0 \in I$ and $\gamma(0) = m$. For such a curve we may assign a tangent vector at each point $\gamma(t)$, $t \in I$, by $\dot{\gamma}(t) = T_t\gamma(1)$.

Let X be a smooth tangent vector–field on the smooth n–manifold M, and let $m \in M$. Then there exists an open interval $I \subset \mathbb{R}$ containing 0 and a parameterized curve $\gamma : I \to M$ such that:

A. $\gamma(0) = m$;
B. $\dot{\gamma}(t) = X(\gamma(t))$ for all $t \in I$; and
C. If $\beta : \tilde{I} \to M$ is any other parameterized curve in M satisfying (1) and (2), then $\tilde{I} \subset I$ and $\beta(t) = \gamma(t)$ for all $t \in \tilde{I}$.

A parameterized curve $\gamma : I \to M$ satisfying condition (2) is called an *integral curve* of the tangent vector–field X. The unique γ satisfying conditions (1)–(3) is the *maximal integral curve* of X through $m \in M$.

In other words, let $\gamma : I \to M$, $t \mapsto \gamma(t)$ be a smooth curve in a manifold M defined on an interval $I \subseteq \mathbb{R}$. $\dot{\gamma}(t) = \frac{d}{dt}\gamma(t)$ defines a smooth vector–field along γ since we have $\pi_M \circ \dot{\gamma} = \gamma$. Curve γ is called an *integral curve* or *flow line* of a vector–field $X \in \mathcal{X}^k(M)$ if the tangent vector determined by γ equals X at every point $m \in M$, i.e.,

$$\dot{\gamma} = X \circ \gamma.$$

On a chart (U,ϕ) with coordinates $\phi(m) = (x^1(m),...,x^n(m))$, for which $\varphi \circ \gamma : t \mapsto \gamma_i(t)$ and $T\varphi \circ X \circ \varphi^{-1} : x^i \mapsto (x^i, X_i(m))$, this is written

$$\dot{\gamma}_i(t) = X_i(\gamma(t)), \text{ for all } t \in I \subseteq \mathbb{R}, \qquad (3.2)$$

which is an ordinary differential equation of first–order in n dimensions.

The *velocity* $\dot{\gamma}$ of the parameterized curve $\gamma(t)$ is a vector–field along γ defined by

$$\dot{\gamma}(t) = (\gamma(t), \dot{x}^1(t), \ldots \dot{x}^n(t)).$$

Its length $|\dot{\gamma}| : I \to \mathbb{R}$, defined by $|\dot{\gamma}|(t) = |\dot{\gamma}(t)|$ for all $t \in I$, is a function along α. $|\dot{\gamma}|$ is called *speed* of γ [Arn78].

Each vector–field X along γ is of the form $X(t) = (\gamma(t), X_1(t), \ldots, X_n(t))$, where each component X_i is a function along γ. X is *smooth* if each $X_i : I \to M$ is smooth. The *derivative* of a smooth vector–field X along a curve $\gamma(t)$ is the vector–field $\dot{X}$ along γ defined by

$$\dot{X}(t) = (\gamma(t), \dot{X}_1(t), \ldots \dot{X}_n(t)).$$

$\dot{X}(t)$ measures the *rate of change of the vector part* $(X_1(t), \ldots X_n(t))$ of $X(t)$ *along* γ. Thus, the *acceleration* $\ddot{\gamma}(t)$ of a parameterized curve $\gamma(t)$ is the vector–field along γ get by differentiating the velocity field $\dot{\gamma}(t)$.

Differentiation of vector–fields along parameterized curves has the following properties. For X and Y smooth vector–fields on M along the parameterized curve $\gamma : I \to M$ and f a smooth function along γ, we have:

A. $\frac{d}{dt}(X + Y) = \dot{X} + \dot{Y}$;
B. $\frac{d}{dt}(fX) = \dot{f}X + f\dot{X}$; and
C. $\frac{d}{dt}(X \cdot Y) = \dot{X}Y + X\dot{Y}$.

A *geodesic* in M is a parameterized curve $\gamma : I \to M$ whose acceleration $\ddot{\gamma}$ is everywhere orthogonal to M; that is, $\ddot{\gamma}(t) \in M^{\perp}_{\alpha(t)}$ for all $t \in I \subset \mathbb{R}$. Thus a geodesic is a curve in M which always goes 'straight ahead' in the surface. Its acceleration serves only to keep it in the surface. It has no component of acceleration tangent to the surface. Therefore, it also has a constant speed $\dot{\gamma}(t)$.

Let $v \in M_m$ be a vector on M. Then there exists an open interval $I \subset \mathbb{R}$ containing 0 and a geodesic $\gamma : I \to M$ such that:

A. $\gamma(0) = m$ and $\dot{\gamma}(0) = v$; and
B. If $\beta : \tilde{I} \to M$ is any other geodesic in M with $\beta(0) = m$ and $\dot{\beta}(0) = v$, then $\tilde{I} \subset I$ and $\beta(t) = \gamma(t)$ for all $t \in \tilde{I}$.

The geodesic γ is now called the *maximal geodesic* in M passing through m with initial velocity v.

By definition, a parameterized curve $\gamma : I \to M$ is a geodesic of M iff its acceleration is everywhere perpendicular to M, i.e., iff $\ddot{\gamma}(t)$ is a multiple of the orientation $N(\gamma(t))$ for all $t \in I$, i.e., $\ddot{\gamma}(t) = g(t) N(\gamma(t))$, where $g : I \to \mathbb{R}$. Taking the scalar product of both sides of this equation with $N(\gamma(t))$ we find $g = -\dot{\gamma}\dot{N}(\gamma(t))$. Thus $\gamma : I \to M$ is geodesic iff it satisfies the differential equation

$$\ddot{\gamma}(t) + \dot{N}(\gamma(t))\, N(\gamma(t)) = 0.$$

This vector equation represents the system of second–order component ODEs

$$\ddot{x}^i + N_i(x+1, \ldots, x^n) \frac{\partial N_j}{\partial x^k}(x+1, \ldots, x^n)\, \dot{x}^j \dot{x}^k = 0.$$

The substitution $u^i = \dot{x}^i$ reduces this second–order differential system (in n variables x^i) to the first–order differential system

$$\dot{x}^i = u^i, \qquad \dot{u}^i = -N_i(x+1, \ldots, x^n) \frac{\partial N_j}{\partial x^k}(x+1, \ldots, x^n)\, \dot{x}^j \dot{x}^k$$

(in $2n$ variables x^i and u^i). This first–order system is just the differential equation for the integral curves of the vector–field X in $U \times \mathbb{R}$ (U open chart in M), in which case X is called a *geodesic spray*.

Now, when an integral curve $\gamma(t)$ is the path a mechanical system Ξ follows, i.e., the solution of the equations of motion, it is called a *trajectory*. In this case the parameter t represents time, so that (3.2) describes motion of the system Ξ on its configuration manifold M.

If $X_i(m)$ is C^0 the existence of a local solution is guaranteed, and a *Lipschitz condition* would imply that it is unique. Therefore, exactly one integral curve passes through every point, and different integral curves can never cross. As $X \in \mathcal{X}^k(M)$ is C^∞, the following statement about the solution with arbitrary initial conditions holds [Thi79, Arn78]:

Theorem. Given a vector–field $X \in \mathcal{X}(M)$, for all points $p \in M$, there exist $\eta > 0$, a neighborhood V of p, and a function $\gamma : (-\eta, \eta) \times V \to M$, $(t, x^i(0)) \mapsto \gamma(t, x^i(0))$ such that

$$\dot{\gamma} = X \circ \gamma, \qquad \gamma(0, x^i(0)) = x^i(0) \qquad \text{for all } x^i(0) \in V \subseteq M.$$

For all $|t| < \eta$, the map $x^i(0) \mapsto \gamma(t, x^i(0))$ is a diffeomorphism f_t^X between V and some open set of M. For proof, see [Die69], I, 10.7.4 and 10.8.

This theorem states that trajectories that are near neighbors cannot suddenly be separated. There is a well–known estimate (see [Die69], I, 10.5) according to which points cannot diverge faster than exponentially in time if the derivative of X is uniformly bounded.

An integral curve $\gamma(t)$ is said to be *maximal* if it is not a restriction of an integral curve defined on a larger interval $I \subseteq \mathbb{R}$. It follows from the existence and uniqueness theorems for ODEs with smooth r.h.s and from elementary properties of Hausdorff spaces that for any point $m \in M$ there exists a maximal integral curve γ_m of X, passing for $t = 0$ through point m, i.e., $\gamma(0) = m$.

Theorem (Local Existence, Uniqueness, and Smoothness) [AM78, II07b]. Let E be a Banach space, $U \subset E$ be open, and suppose $X : U \subset E \to E$ is of class C^∞, $k \geq 1$. Then

1. For each $x_0 \in U$, there is a curve $\gamma : I \to U$ at x_0 such that $\dot{\gamma}(t) = X(\gamma(t))$ for all $t \in I$.
2. Any two such curves are equal on the intersection of their domains.
3. There is a neighborhood U_0 of the point $x_0 \in U$, a real number $a > 0$, and a C^∞ map $F : U_0 \times I \to E$, where I is the open interval $]-a, a[$, such that the curve $\gamma_u : I \to E$, defined by $\gamma_u(t) = F(u, t)$ is a curve at $u \in E$ satisfying the ODEs $\dot{\gamma}_u(t) = X(\gamma_u(t))$ for all $t \in I$.

Proposition (Global Uniqueness). Suppose γ_1 and γ_2 are two integral curves of a vector–field X at a point $m \in M$. Then $\gamma_1 = \gamma_2$ on the intersection of their domains [AM78, II07b].

If for every point $m \in M$ the curve γ_m is defined on the entire real axis $\mathbb{R}$, then the vector–field X is said to be *complete*.

The *support of a vector–field* X defined on a manifold M is defined to be the closure of the set $\{m \in M | X(m) = 0\}$. A C^∞ vector–field with compact support on a manifold M is complete. In particular, a C^∞ vector–field on

3.1 Riemannian Geometry of Smooth Manifolds

a compact manifold is complete. Completeness corresponds to well–defined dynamics persisting eternally.

Now, following [AM78, II07b], for the *derivative* of a C^∞ function $f : E \to \mathbb{R}$ in the direction X we use the notation $X[f] = df \cdot X$, where df stands for the *derivative map*. In standard coordinates on $\mathbb{R}^n$ this is a standard *gradient*

$$df(x) = \nabla f = (\partial_{x^1} f, ..., \partial_{x^n} f), \quad \text{and} \quad X[f] = X^i \partial_{x^i} f.$$

Let F_t be the flow of X. Then $f(F_t(x)) = f(F_s(x))$ if $t \geq s$.

For example, Newtonian equations for a moving particle of mass m in a potential field V in $\mathbb{R}^n$ are given by $\ddot{q}^i(t) = -(1/m)\nabla V\left(q^i(t)\right)$, for a smooth function $V : \mathbb{R}^n \to \mathbb{R}$. If there are constants $a, b \in \mathbb{R}$, $b \geq 0$ such that $(1/m)V(q^i) \geq a - b \|q^i\|^2$, then *every solution exists for all time*. To show this, rewrite the second–order equations as a first–order system $\dot{q}^i = (1/m)\, p_i$, $\dot{p}_i = -V(q^i)$ and note that the energy $E(q^i, p_i) = (1/2m)\|p_i\|^2 + V(q)$ is a *first integral* of the motion. Thus, for any solution $(q^i(t), p_i(t))$ we have $E\left(q^i(t), p_i(t)\right) = E\left(q^i(0), p_i(0)\right) = V(q(0))$.

Let X_t be a C^∞ time–dependent vector–field on an $n-$manifold M, $k \geq 1$, and let m_0 be an *equilibrium* of X_t, that is, $X_t(m_0) = 0$ for all t. Then for any T there exists a neighborhood V of m_0 such that any $m \in V$ has integral curve existing for time $t \in [-T, T]$.

Dynamical Flows on M

The so–called *flow* F_t of a C^∞ vector–field $X \in \mathcal{X}^k(M)$ is the *one–parameter group of diffeomorphisms* $F_t : M \to M$ such that $t \mapsto F_t(m)$ is the integral curve of X with initial condition m for all $m \in M$ and $t \in I \subseteq \mathbb{R}$. The flow $F_t(m)$ is C^∞ by induction on k. It is defined as [AM78, II07b]:

$$\frac{d}{dt} F_t(x) = X(F_t(x)).$$

Existence and uniqueness theorems for ODEs guarantee that F_t is smooth in m and t. From uniqueness, we get the *flow property*:

$$F_{t+s} = F_t \circ F_s$$

along with the initial conditions $F_0 =$ identity. The flow property generalizes the situation where $M = V$ is a linear space, $X(x) = Ax$ for a (bounded) linear operator A, and where $F_t(x) = e^{tA}x$ – to the nonlinear case. Therefore, the flow $F_t(m)$ can be defined as a *formal exponential*

$$F_t(m) = \exp(t\, X) = (I + t\, X + \frac{t^2}{2} X^2 + ...) = \sum_{k=0}^{\infty} \frac{X^k t^k}{k!}.$$

A *time–dependent vector–field* is a map $X : M \times \mathbb{R} \to TM$ such that $X(m, t) \in T_m M$ for each point $m \in M$ and $t \in \mathbb{R}$. An *integral curve* of X is a curve $\gamma(t)$ in M such that

$$\dot{\gamma}(t) = X(\gamma(t), t), \qquad \text{for all } t \in I \subseteq \mathbb{R}.$$

In this case, the flow is the one–parameter group of diffeomorphisms $F_{t,s} : M \to M$ such that $t \mapsto F_{t,s}(m)$ is the integral curve $\gamma(t)$ with initial condition $\gamma(s) = m$ at $t = s$. Again, the existence and uniqueness theorem from ODE–theory applies here, and in particular, uniqueness gives the time–dependent flow property, i.e., the *Chapman–Kolmogorov law*

$$F_{t,r} = F_{t,s} \circ F_{s,r}.$$

If X happens to be time independent, the two notions of flows are related by $F_{t,s} = F_{t-s}$ (see [MR99]).

3.1.3 Local Riemannian Geometry

An important class of problems in Riemannian geometry is to understand the interaction between the curvature and topology on a smooth manifold [II06b, II07b]. A prime example of this interaction is the *Gauss–Bonnet formula* on a closed surface M^2, which says

$$\int_M K \, dA = 2\pi \chi(M), \tag{3.3}$$

where dA is the area element of a metric g on M, K is the Gaussian curvature of g, and $\chi(M)$ is the *Euler characteristic* of M.

To study the geometry of a smooth manifold we need an additional structure: the *Riemannian metric tensor*. The metric is an inner product on each of the tangent spaces and tells us how to measure angles and distances infinitesimally. In local coordinates $(x^1, x^2, \cdots, x^n)$, the metric g is given by $g_{ij}(x) \, dx^i \otimes dx^j$, where $(g_{ij}(x))$ is a positive definite symmetric matrix at each point x. For a smooth manifold one can differentiate functions. A Riemannian metric defines a natural way of differentiating vector–fields: *covariant differentiation*. In Euclidean space, one can change the order of differentiation. On a Riemannian manifold the commutator of twice covariant differentiating vector–fields is in general nonzero and is called the *Riemann curvature tensor*, which is a 4–tensor–field on the manifold.

For surfaces, the Riemann curvature tensor is equivalent to the *Gaussian curvature* K, a scalar function. In dimensions 3 or more, the Riemann curvature tensor is inherently a tensor–field. In local coordinates, it is denoted by R_{ijkl}, which is anti-symmetric in i and k and in j and l, and symmetric in the pairs $\{ij\}$ and $\{kl\}$. Thus, it can be considered as a bilinear form on 2–forms which is called the *curvature operator*. We now describe heuristically the various curvatures associated to the Riemann curvature tensor. Given a point $x \in M^n$ and 2–plane Π in the tangent space of M at x, we can define a surface S in M to be the union of all geodesics passing through x and tangent to Π. In a neighborhood of x, S is a smooth 2D submanifold of M. We define

the *sectional curvature* $K(\Pi)$ of the 2−plane to be the Gauss curvature of S at x:
$$K(\Pi) = K_S(x).$$

Thus the sectional curvature K of a Riemannian manifold associates to each 2−plane in a tangent space a real number. Given a line L in a tangent space, we can average the sectional curvatures of all planes through L to get the *Ricci tensor* $Rc(L)$. Likewise, given a point $x \in M$, we can average the Ricci curvatures of all lines in the tangent space of x to get the *scalar curvature* $R(x)$. In local coordinates, the Ricci tensor is given by $R_{ik} = g^{jl} R_{ijkl}$ and the scalar curvature is given by $R = g^{ik} R_{ik}$, where $(g^{ij}) = (g_{ij})^{-1}$ is the inverse of the metric tensor (g_{ij}).

Riemannian Metric on M

Riemann in 1854 observed that around each point $m \in M$ one can pick a *special* coordinate system $(x^1, \ldots, x^n)$ such that there is a symmetric $(0,2)$−tensor−field $g_{ij}(m)$ called the *metric tensor* defined as (see [Pet99, Pet98, II06b, II07b])

$$g_{ij}(m) = g(\partial_{x^i}, \partial_{x^j}) = \delta_{ij}, \qquad \partial_{x^k} g_{ij}(m) = 0.$$

Thus the metric, at the specified point $m \in M$, in the coordinates $(x^1, \ldots, x^n)$ looks like the Euclidean metric on $\mathbb{R}^n$. We emphasize that these conditions only hold at the specified point $m \in M$. When passing to different points it is necessary to pick different coordinates. If a curve γ passes through m, say, $\gamma(0) = m$, then the acceleration at 0 is defined by firstly, writing the curve out in our special coordinates

$$\gamma(t) = (\gamma^1(t), \ldots, \gamma^n(t)),$$

secondly, defining the tangent, *velocity* vector−field, as

$$\dot{\gamma} = \dot{\gamma}^i(t) \cdot \partial_{x^i},$$

and finally, the *acceleration* vector−field as

$$\ddot{\gamma}(0) = \ddot{\gamma}^i(0) \cdot \partial_{x^i}.$$

Here, the background idea is that we have a *connection*.

Recall that a *connection* on a smooth manifold M tells us how to *parallel transport* a vector at a point $x \in M$ to a vector at a point $x' \in M$ along a curve $\gamma \in M$. Roughly, to parallel transport vectors along curves, it is enough if we can define parallel transport under an infinitesimal displacement: given a vector X at x, we would like to define its parallel transported version $\tilde{X}$ after an infinitesimal displacement by ϵv, where v is a tangent vector to M at x.

This length is independent of our parametrization of the curve γ. Thus the curve γ can be reparameterized, in such a way that it has unit velocity. The *distance between two points* m_1 *and* m_2 *on* M, $d(m_1, m_2)$, can now be defined as the infimum of the lengths of all curves from m_1 to m_2, i.e.,

$$L(\gamma, I) \to \min.$$

This means that the distance measures the shortest way one can travel from m_1 to m_2.

If we take a variation $V(s,t): (-\varepsilon, \varepsilon) \times [0, \ell] \to M$ of a smooth curve $\gamma(t) = V(0,t)$ parameterized by arc–length L and of length ℓ, then the first derivative of the arc–length function

$$L(s) = \int_0^\ell |\dot{V}| \, dt, \quad \text{is given by}$$

$$\frac{dL(0)}{ds} \equiv \dot{L}(0) = g(\dot{\gamma}, X)\big|_0^\ell - \int_0^\ell g(\gamma, X)\, dt, \quad (3.8)$$

where $X(t) = \frac{\partial V}{\partial s}(0,t)$ is the so–called *variation vector–field*. Equation (3.8) is called the *first variation formula*. Given any vector–field X along γ, one can produce a variation whose variational field is X. If the variation fixes the endpoints, $X(a) = X(b) = 0$, then the second term in the formula drops out, and we note that the length of γ can always be decreased as long as the acceleration of γ is not everywhere zero. Thus the *Euler–Lagrangian equations* for the arc–length functional are the equations for a curve to be a *geodesic*.

Recall that in local coordinates $x^i \in U$, where U is an open subset in the Riemannian manifold M, the geodesics are defined by the *geodesic equation*

$$\ddot{x}^i + \Gamma^i_{jk} \dot{x}^j \dot{x}^k = 0, \quad (3.9)$$

where overdot means derivative upon the line parameter s, while Γ^i_{jk} are Christoffel symbols of the affine Levi–Civita connection ∇ on M. From (3.9) it follows that the linear *connection homotopy*,

$$\bar{\Gamma}^i_{jk} = s\Gamma^i_{jk} + (1-s)\Gamma^i_{jk}, \quad (0 \le s \le 1),$$

determines the same geodesics as the original Γ^i_{jk}.

Riemannian Curvature on M

The *Riemann curvature tensor* is a rather ominous tensor of type $(1,3)$; i.e., it has three vector variables and its value is a vector as well. It is defined through the *Lie bracket*[21] as [Pet99, Pet98, II06b, II07b]

$$R(X,Y)Z = \left(\nabla_{[X,Y]} - [\nabla_X, \nabla_Y]\right)Z = \nabla_{[X,Y]}Z - \nabla_X \nabla_Y Z + \nabla_Y \nabla_X Z.$$

---

[21] If $X, Y \in \mathcal{X}^k(M)$, $k \ge 1$ are two vector–fields on M, then

3.1 Riemannian Geometry of Smooth Manifolds

This turns out to be a vector valued $(1,3)$−tensor−field in the three variables $X, Y, Z \in \mathcal{X}^k(M)$. We can then create a $(0,4)$−tensor,

$$R(X, Y, Z, W) = g\left(\nabla_{[X,Y]}Z - \nabla_X \nabla_Y Z + \nabla_Y \nabla_X Z, W\right).$$

$$[\mathcal{L}_X, \mathcal{L}_Y] = \mathcal{L}_X \circ \mathcal{L}_Y - \mathcal{L}_Y \circ \mathcal{L}_X$$

is a derivation map from $C^{k+1}(M, \mathbb{R})$ to $C^{k-1}(M, \mathbb{R})$. Then there is a unique vector−field, $[X, Y] \in \mathcal{X}^k(M)$ of X and Y such that $\mathcal{L}_{[X,Y]} = [\mathcal{L}_X, \mathcal{L}_Y]$ and $[X, Y](f) = X(Y(f)) - Y(X(f))$ holds for all functions $f \in C^\infty(M, \mathbb{R})$. This vector−field is also denoted $\mathcal{L}_X Y$ and is called the Lie derivative of Y with respect to X, or the *Lie bracket* of X and Y. In a local chart (U, ϕ) at a point $m \in M$ with coordinates $(x^1, ..., x^n)$, for $X|_U = X^i \partial_{x^i}$ and $Y|_U = Y^i \partial_{x^i}$ we have

$$\left[X^i \partial_{x^i}, Y^j \partial_{x^j}\right] = \left(X^i \left(\partial_{x^i} Y^j\right) - Y^i \left(\partial_{x^i} X^j\right)\right) \partial_{x^j},$$

since second partials commute. If, also X has flow F_t, then [AM78, II07b]

$$\frac{d}{dt}(F_t^* Y) = F_t^* (\mathcal{L}_X Y).$$

In particular, if $t = 0$, this formula becomes

$$\frac{d}{dt}|_{t=0} (F_t^* Y) = \mathcal{L}_X Y.$$

Then the unique C^{k-1} vector−field $\mathcal{L}_X Y = [X, Y]$ on M defined by

$$[X, Y] = \frac{d}{dt}|_{t=0} (F_t^* Y),$$

is called the Lie derivative of Y with respect to X, or the Lie bracket of X and Y, and can be interpreted as the leading order term that results from the sequence of flows

$$F_t^{-Y} \circ F_t^{-X} \circ F_t^{Y} \circ F_t^{X}(m) = \epsilon^2 [X, Y](m) + \mathcal{O}(\epsilon^3), \qquad (3.10)$$

for some real $\epsilon > 0$. Therefore a Lie bracket can be interpreted as a 'new direction' in which the system can flow, by executing the sequence of flows (3.10).

Lie bracket satisfies the following property:

$$[X, Y][f] = X[Y[f]] - Y[X[f]],$$

for all $f \in C^{k+1}(U, \mathbb{R})$, where U is open in M.

An important relationship between flows of vector−fields is given by the *Campbell–Baker–Hausdorff formula*:

$$F_t^Y \circ F_t^X = F_t^{X+Y+\frac{1}{2}[X,Y]+\frac{1}{12}([X,[X,Y]]-[Y,[X,Y]])+...} \qquad (3.11)$$

Essentially, if given the composition of multiple flows along multiple vector−fields, this formula gives the one flow along one vector−field which results in the same net flow. One way to prove the Campbell–Baker–Hausdorff formula (3.11) is to expand the product of two formal exponentials and equate terms in the resulting formal power series [II06b, II07b].

Clearly this tensor is skew–symmetric in X and Y, and also in Z and $W \in \mathcal{X}^k(M)$. This was already known to Riemann, but there are some further, more subtle properties that were discovered a little later by Bianchi. The *Bianchi symmetry condition* reads

$$R(X, Y, Z, W) = R(Z, W, X, Y).$$

The *Ricci tensor* is the $(1,1)-$ or $(0,2)-$tensor defined by

$$\mathrm{Ric}(X) = R(\partial_{x^i}, X)\partial_{x^i}, \qquad \mathrm{Ric}(X, Y) = g(R(\partial_{x^i}, X)\partial_{x^i}, Y),$$

for any orthonormal basis (∂_{x^i}). In other words, the Ricci curvature is a trace of the curvature tensor. Similarly one can define the *scalar curvature* as the trace

$$\mathrm{scal}(m) = \mathrm{Tr}\,(\mathrm{Ric}) = \mathrm{Ric}(\partial_{x^i}, \partial_{x^i}).$$

When the Riemannian manifold has dimension 2, all of these curvatures are essentially the same. Since $\dim \Lambda^2 TM = 1$ and is spanned by $X \wedge Y$ where $X, Y \in \mathcal{X}^k(M)$ form an orthonormal basis for $T_m M$, we see that the curvature tensor depends only on the scalar value

$$K(m) = R(X, Y, X, Y),$$

which also turns out to be the *Gaussian curvature*. The Ricci tensor is a homothety

$$\mathrm{Ric}(X) = K(m)X, \qquad \mathrm{Ric}(Y) = K(m)Y,$$

and the scalar curvature is twice the Gauss curvature. In dimension 3 there are also some redundancies as $\dim TM = \dim \Lambda^2 TM = 3$. In particular, the Ricci tensor and the curvature tensor contain the same amount of information.

The *sectional curvature* is a kind of generalization of the Gauss curvature whose importance Riemann was already aware of. Given a 2–plane $\pi \subset T_m M$ spanned by an orthonormal basis $X, Y \in \mathcal{X}^k(M)$ it is defined as

$$\sec(\pi) = R(X, Y, X, Y).$$

The remarkable observation by Riemann was that the *curvature operator is a homothety*, i.e., looks like $\mathfrak{R} = kI$ on $\Lambda^2 T_m M$ iff all sectional curvatures of planes in $T_m M$ are equal to k. This result is not completely trivial, as the sectional curvature is not the entire quadratic form associated to the symmetric operator $\mathfrak{R}$. In fact, it is not true that $\sec \geq 0$ implies that the curvature operator is nonnegative in the sense that all its eigenvalues are nonnegative. What Riemann did was to show that our special coordinates $(x^1, \ldots, x^n)$ at m can be chosen to be *normal* at m, i.e., satisfy the condition

$$x^i = \delta^i_j x^j, \qquad (\delta^i_j x^j = g_{ij})$$

on a neighborhood of m. One can show that such coordinates are actually exponential coordinates together with a choice of an orthonormal basis for

T_mM so as to identify T_mM with $\mathbb{R}^n$. In these coordinates one can then expand the metric as follows:

$$g_{ij} = \delta_{ij} - \frac{1}{3} R_{ikjl} x^k x^l + O(r^3).$$

Now the equations $x^i = g_{ij} x^j$ evidently give conditions on the curvatures R_{ijkl} at m.

If $\Gamma^i_{jk}(m) = 0$, the manifold M is flat at the point m. This means that the $(1,3)$ curvature tensor, defined locally at $m \in M$ as [Pet99, Pet98, II06b, II07b]

$$R^l_{ijk} = \partial_{x^j} \Gamma^l_{ik} - \partial_{x^k} \Gamma^l_{ij} + \Gamma^l_{rj} \Gamma^r_{ik} - \Gamma^l_{rk} \Gamma^r_{ij},$$

also vanishes at that point, i.e., $R^l_{ijk}(m) = 0$.

Now, the rate of change of a vector–field A^k on the manifold M along the curve $x^i(s)$ is properly defined by the *absolute covariant derivative*

$$\frac{D}{ds} A^k = \dot{x}^i \nabla_i A^k = \dot{x}^i \left(\partial_{x^i} A^k + \Gamma^k_{ij} A^j \right) = \dot{A}^k + \Gamma^k_{ij} \dot{x}^i A^j.$$

By applying this result to itself, we can get an expression for the second covariant derivative of the vector–field A^k along the curve $x^i(s)$:

$$\frac{D^2}{ds^2} A^k = \frac{d}{ds} \left(\dot{A}^k + \Gamma^k_{ij} \dot{x}^i A^j \right) + \Gamma^k_{ij} \dot{x}^i (\dot{A}^j + \Gamma^j_{mn} \dot{x}^m A^n).$$

In the local coordinates $(x^1(s), ..., x^n(s))$ at a point $m \in M$, if $\delta x^i = \delta x^i(s)$ denotes the *geodesic deviation*, i.e., the infinitesimal vector describing perpendicular separation between the two neighboring geodesics, passing through two neighboring points $m, n \in M$, then the *Jacobi equation of geodesic deviation* on the manifold M holds [Pet99, Pet98, II06b, II07b]

$$\frac{D^2 \delta x^i}{ds^2} + R^i_{jkl} \dot{x}^j \delta x^k \dot{x}^l = 0. \tag{3.12}$$

This equation describes the *relative acceleration* between two infinitesimally close facial geodesics, which is proportional to the facial curvature (measured by the Riemann tensor R^i_{jkl} at a point $m \in M$), and to the geodesic deviation δx^i. Solutions of equation (3.12) are called *Jacobi fields*.

In particular, if the manifold M is a 2D–surface in $\mathbb{R}^3$, the Riemann curvature tensor simplifies into

$$R^i_{jmn} = \frac{1}{2} R g^{ik} (g_{km} g_{jn} - g_{kn} g_{jm}),$$

where R denotes the *scalar Gaussian curvature*. Consequently the equation of geodesic deviation (3.12) also simplifies into

$$\frac{D^2}{ds^2} \delta x^i + \frac{R}{2} \delta x^i - \frac{R}{2} \dot{x}^i (g_{jk} \dot{x}^j \delta x^k) = 0. \tag{3.13}$$

This simplifies even more if we work in a locally Cartesian coordinate system; in this case the covariant derivative $\frac{D^2}{Ds^2}$ reduces to an ordinary derivative $\frac{d^2}{ds^2}$ and the metric tensor g_{ij} reduces to identity matrix I_{ij}, so our 2D equation of geodesic deviation (3.13) reduces into a simple second–order ODE in just two coordinates x^i $(i=1,2)$

$$\ddot{x}^i + \frac{R}{2}\delta x^i - \frac{R}{2}\dot{x}^i(I_{jk}\,\dot{x}^j\,\delta x^k) = 0.$$

For more technical details on local Riemannian geometry, see [II07b].

3.1.4 Global Riemannian Geometry

In this subsection we briefly describe the global Riemann geometry. For more technical details, see [II07b].

The Second Variation Formula

Cartan also establishes another important property of manifolds with nonpositive curvature. First he observes that all spaces of constant zero curvature have torsion–free fundamental groups. This is because any isometry of finite order on Euclidean space must have a fixed point (the center of mass of any orbit is necessarily a fixed point). Then he notices that one can geometrically describe the L^∞ center of mass of finitely many points $\{m_1,\ldots,m_k\}$ in Euclidean space as the unique minimum for the strictly convex function [Pet99, Pet98, II06b, II07b]

$$x \to \max_{i=1,\cdots,k} \frac{1}{2}\left\{(d(m_i,x))^2\right\}.$$

In other words, the center of mass is the center of the ball of smallest radius containing $\{m_1,\ldots,m_k\}$. Now Cartan's observation from above was that the exponential map is expanding and globally distance nondecreasing as a map:

$$(T_mM,\ \text{Euclidean metric}) \to (T_mM,\ \text{with pull–back metric}).$$

Thus distance functions are convex in nonpositive curvature as well as in Euclidean space. Hence the above argument can in fact be used to conclude that any Riemannian manifold of nonpositive curvature must also have torsion free fundamental group.

Now, let us set up the *second variation formula* and explain how it is used. We have already seen the first variation formula and how it can be used to characterize geodesics. Now suppose that we have a unit speed geodesic $\gamma(t)$ parameterized on $[0,\ell]$ and consider a variation $V(s,t)$, where $V(0,t)=\gamma(t)$. Synge then shows that $(\ddot{L}\equiv\frac{d^2L}{ds^2})$

$$\ddot{L}(0) = \int_0^\ell \{g(\dot{X}, \dot{X}) - (g(\dot{X}, \dot{\gamma}))^2 - g(R(X, \dot{\gamma})X, \dot{\gamma})\}dt + g(\dot{\gamma}, A)|_0^\ell,$$

where $X(t) = \frac{\partial V}{\partial s}(0, t)$ is the variational vector–field, $\dot{X} = \nabla_{\dot{\gamma}} X$, and $A(t) = \nabla_{\frac{\partial V}{\partial s}} X$. In the special case where the variation fixes the endpoints, i.e., $s \to V(s, a)$ and $s \to V(s, b)$ are constant, the term with A in it falls out. We can also assume that the variation is perpendicular to the geodesic and then drop the term $g\left(\dot{X}, \dot{\gamma}\right)$. Thus, we arrive at the following simple form [Pet99, Pet98, II06b, II07b]

$$\ddot{L}(0) = \int_0^\ell \{g(\dot{X}, \dot{X}) - g(R(X, \dot{\gamma})X, \dot{\gamma})\}dt = \int_0^\ell \{|\dot{X}|^2 - \sec(\dot{\gamma}, X)|X|^2\}dt.$$

Therefore, if the sectional curvature is nonpositive, we immediately observe that any geodesic locally minimizes length (that is, among close–by curves), even if it does not minimize globally (for instance γ could be a closed geodesic). On the other hand, in positive curvature we can see that if a geodesic is too long, then it cannot minimize even locally. The motivation for this result comes from the unit sphere, where we can consider geodesics of length $> \pi$. Globally, we know that it would be shorter to go in the opposite direction. However, if we consider a variation of γ where the variational field looks like $X = \sin\left(t \cdot \frac{\pi}{\ell}\right) E$ and E is a unit length parallel field along γ which is also perpendicular to γ, then we get

$$\begin{aligned}\ddot{L}(0) &= \int_0^\ell \left\{ \left|\dot{X}\right|^2 - \sec(\dot{\gamma}, X)|X|^2 \right\} dt \\ &= \int_0^\ell \left\{ \left(\frac{\pi}{\ell}\right)^2 \cdot \cos^2\left(t \cdot \frac{\pi}{\ell}\right) - \sec(\dot{\gamma}, X)\sin^2\left(t \cdot \frac{\pi}{\ell}\right) \right\} dt \\ &= \int_0^\ell \left(\left(\frac{\pi}{\ell}\right)^2 \cdot \cos^2\left(t \cdot \frac{\pi}{\ell}\right) - \sin^2\left(t \cdot \frac{\pi}{\ell}\right) \right) dt = -\frac{1}{2\ell}\left(\ell^2 - \pi^2\right),\end{aligned}$$

which is negative if the length ℓ of the geodesic is greater than π. Therefore, the variation gives a family of curves that are both close to and shorter than γ. In the general case, we can then observe that if $\sec \geq 1$, then for the same type of variation we get

$$\ddot{L}(0) \leq -\frac{1}{2\ell}\left(\ell^2 - \pi^2\right).$$

Thus we can conclude that, if the space is complete, then the diameter must be $\leq \pi$ because in this case any two points are joined by a segment, which cannot minimize if it has length $> \pi$. With some minor modifications one can now conclude that any complete Riemannian manifold (M, g) with $\sec \geq k^2 > 0$ must satisfy $\mathrm{diam}(M, g) \leq \pi \cdot k^{-1}$. In particular, M must be compact. Since the universal covering of M satisfies the same curvature hypothesis, the conclusion must also hold for this space; hence M must have compact universal covering space and finite fundamental group.

In odd dimensions all spaces of constant positive curvature must be orientable, as orientation reversing orthogonal transformation on odd–dimensional spheres have fixed points. This can now be generalized to manifolds of varying positive curvature. Synge did it in the following way: Suppose M is not simply–connected (or not orientable), and use this to find a shortest closed geodesic in a free homotopy class of curves (that reverses orientation). Now consider parallel translation around this geodesic. As the tangent field to the geodesic is itself a parallel field, we see that parallel translation preserves the orthogonal complement to the geodesic. This complement is now odd dimensional (even dimensional), and by assumption parallel translation preserves (reverses) the orientation; thus it must have a fixed point. In other words, there must exist a closed parallel field X perpendicular to the closed geodesic γ. We can now use the above second variation formula [Pet99, Pet98, II06b, II07b]

$$\ddot{L}(0) = \int_0^\ell \{|\dot{X}|^2 - |X|^2 \sec(\dot\gamma, X)\} dt + g(\dot\gamma, A)|_0^\ell = -\int_0^\ell |X|^2 \sec(\dot\gamma, X) \, dt.$$

Here the boundary term drops out because the variation closes up at the endpoints, and $\dot{X} = 0$ since we used a parallel field. In case the sectional curvature is always positive we then see that the above quantity is negative. But this means that the closed geodesic has nearby closed curves which are shorter. However, this is in contradiction with the fact that the geodesic was constructed as a length minimizing curve in a free homotopy class.

In 1941 Myers generalized the diameter bound to the situation where one only has a lower bound for the Ricci curvature. The idea is that $\mathrm{Ric}(\dot\gamma, \dot\gamma) = \sum_{i=1}^{n-1} \sec(E_i, \dot\gamma)$ for any set of vector–fields E_i along γ such that $\dot\gamma, E_1, \ldots, E_{n-1}$ forms an orthonormal frame. Now assume that the fields are parallel and consider the $n-1$ variations coming from the variational vector–fields $\sin\left(t \cdot \frac{\pi}{\ell}\right) E_i$. Adding up the contributions from the variational formula applied to these fields then induces

$$\sum_{i=1}^{n-1} \ddot{L}(0) = \sum_{i=1}^{n-1} \int_0^\ell \left\{\left(\frac{\pi}{\ell}\right)^2 \cdot \cos^2\left(t \cdot \frac{\pi}{\ell}\right) - \sec(\dot\gamma, E_i)\sin^2\left(t \cdot \frac{\pi}{\ell}\right)\right\} dt$$

$$= \int_0^\ell \left\{(n-1)\left(\frac{\pi}{\ell}\right)^2 \cdot \cos^2\left(t \cdot \frac{\pi}{\ell}\right) - \mathrm{Ric}(\dot\gamma, \dot\gamma)\sin^2\left(t \cdot \frac{\pi}{\ell}\right)\right\} dt.$$

Therefore, if $\mathrm{Ric}(\dot\gamma, \dot\gamma) \geq (n-1)k^2$ (this is the Ricci curvature of S_k^n), then

$$\sum_{i=1}^{n-1} \ddot{L}(0) \leq (n-1) \int_0^\ell \left\{\left(\frac{\pi}{\ell}\right)^2 \cdot \cos^2\left(t \cdot \frac{\pi}{\ell}\right) - k^2 \sin^2\left(t \cdot \frac{\pi}{\ell}\right)\right\} dt$$

$$= -(n-1)\frac{1}{2\ell}\left(\ell^2 k^2 - \pi^2\right),$$

which is negative when $\ell > \pi \cdot k^{-1}$ (the diameter of S_k^n). Thus at least one of the contributions $\frac{d^2 L_i}{ds^2}(0)$ must be negative as well, implying that the geodesic cannot be a segment in this situation.

Gauss–Bonnet Formula

In 1926 Hopf proved that in fact there is a *Gauss–Bonnet formula* for all even–dimensional hypersurfaces $H^{2n} \subset \mathbb{R}^{2n+1}$. The idea is that the determinant of the differential of the Gauss map $G : H^{2n} \to S^{2n}$ is the Gaussian curvature of the hypersurface. Moreover, this is an intrinsically computable quantity. If we integrate this over the hypersurface, we get [Pet99, Pet98, II06b, II07b]

$$\frac{1}{\operatorname{Vol} S^{2n}} \int_H \det(DG) = \deg(G),$$

where $\deg(G)$ is the *Brouwer degree* of the *Gauss map*. Note that this can also be done for odd–dimensional surfaces, in particular curves, but in this case the degree of the Gauss map will depend on the embedding or immersion of the hypersurface. Instead one gets the so–called winding number. Hopf then showed, as Dyck had earlier done for surfaces, that $\deg(G)$ is always half the *Euler characteristic* of H, thus yielding

$$\frac{2}{\operatorname{Vol} S^{2n}} \int_H \det(DG) = \chi(H). \tag{3.14}$$

Since the l.h.s of this formula is in fact intrinsic, it is natural to conjecture that such a formula should hold for all manifolds.

Ricci Flow on M

Ricci flow, or the *parabolic Einstein equation*, was introduced by R. Hamilton in 1982 [Ham82] in the form

$$\partial_t g_{ij} = -2R_{ij}. \tag{3.15}$$

Now, because of the minus sign in the front of the Ricci tensor R_{ij} in this equation, the solution metric g_{ij} to the Ricci flow shrinks in positive Ricci curvature direction while it expands in the negative Ricci curvature direction. For example, on the 2–sphere S^2, any metric of positive Gaussian curvature will shrink to a point in finite time. Since the Ricci flow (3.15) does not preserve volume in general, one often considers the *normalized* Ricci flow defined by

$$\partial_t g_{ij} = -2R_{ij} + \frac{2}{n} r g_{ij}, \tag{3.16}$$

where $r = \int R dV / \int dV$ is the average scalar curvature. Under this normalized flow, which is equivalent to the (unnormalized) Ricci flow (3.15) by re-parameterizing in time t and scaling the metric in space by a function of t, the volume of the solution metric is constant in time. Also that Einstein metrics (i.e., $R_{ij} = cg_{ij}$) are fixed points of (3.16).

Hamilton [Ham82] showed that on a closed Riemannian 3−manifold M^3 with initial metric of positive Ricci curvature, the solution $g(t)$ to the normalized Ricci flow (3.16) exists for all time and the metrics $g(t)$ converge exponentially fast, as time t tends to the infinity, to a constant positive sectional curvature metric g_∞ on M^3.

Since the Ricci flow lies in the realm of parabolic partial differential equations, where the prototype is the heat equation, here is a brief review of the *heat equation* [CC99].

Let (M^n, g) be a Riemannian manifold. Given a C^2 function $u : M \to \mathbb{R}$, its Laplacian is defined in local coordinates $\{x^i\}$ to be

$$\Delta u = \mathrm{Tr}\left(\nabla^2 u\right) = g^{ij} \nabla_i \nabla_j u,$$

where $\nabla_i = \nabla_{\partial_{x^i}}$ is its associated covariant derivative (Levi–Civita connection). We say that a C^2 function $u : M^n \times [0,T) \to \mathbb{R}$, where $T \in (0, \infty]$, is a solution to the heat equation if

$$\partial_t u = \Delta u.$$

One of the most important properties satisfied by the heat equation is the maximum principle, which says that for any smooth solution to the heat equation, whatever pointwise bounds hold at $t = 0$ also hold for $t > 0$. Let $u : M^n \times [0,T) \to \mathbb{R}$ be a C^2 solution to the heat equation on a complete Riemannian manifold. If $C_1 \leq u(x, 0) \leq C_2$ for all $x \in M$, for some constants $C_1, C_2 \in \mathbb{R}$, then $C_1 \leq u(x,t) \leq C_2$ for all $x \in M$ and $t \in [0,T)$ [CC99].

Now, given a smooth manifold M, a one–parameter family of metrics $g(t)$, where $t \in [0,T)$ for some $T > 0$, is a solution to the Ricci flow if (3.15) is valid at all $x \in M$ and $t \in [0,T)$. The minus sign in the equation (3.15) makes the Ricci flow a *forward* heat equation (with the normalization factor 2). In local geodesic coordinates $\{x^i\}$, we have [CC99]

$$g_{ij}(x) = \delta_{ij} - \frac{1}{3} R_{ipjq} x^p x^q + O\left(|x|^3\right), \qquad \text{therefore,} \qquad \Delta g_{ij}(0) = -\frac{1}{3} R_{ij},$$

where Δ is the standard Euclidean Laplacian. Hence the Ricci flow is like the heat equation for a Riemannian metric

$$\partial_t g_{ij} = 6 \Delta g_{ij}.$$

The practical study of the Ricci flow is made possible by the following short–time existence result: Given any smooth compact Riemannian manifold (M, g_o), there exists a unique smooth solution $g(t)$ to the Ricci flow defined on some time interval $t \in [0, \epsilon)$ such that $g(0) = g_o$ [CC99].

Now, given that short–time existence holds for any smooth initial metric, one of the main problems concerning the Ricci flow is to determine under what conditions the solution to the normalized equation exists for all time and converges to a constant curvature metric. Results in this direction have been established under various curvature assumptions, most of them being some sort of positive curvature. Since the Ricci flow (3.15) does not preserve volume in general, one often considers, as we mentioned in the Introduction, the normalized Ricci flow (3.16). Under this flow, the volume of the solution $g(t)$ is independent of time.

To study the long–time existence of the normalized Ricci flow, it is important to know what kind of curvature conditions are preserved under the equation. In general, the Ricci flow tends to preserve some kind of positivity of curvatures. For example, positive scalar curvature is preserved in all dimensions. This follows from applying the maximum principle to the evolution equation for scalar curvature R, which is

$$\partial_t R = \Delta R + 2\,|R_{ij}|^2\,.$$

In dimension 3, positive Ricci curvature is preserved under the Ricci flow. This is a special feature of dimension 3 and is related to the fact that the Riemann curvature tensor may be recovered algebraically from the Ricci tensor and the metric in dimension 3. Positivity of sectional curvature is not preserved in general. However, the stronger condition of positive curvature operator is preserved under the Ricci flow [II07b].

3.2 Riemannian Approach to Chaos

In this section, following [CC96], we present the Riemannian approach to chaos.

During the last two decades or so, there has been a growing evidence of the independence of the two properties of *determinism* and *predictability* of classical dynamics. In fact, predictability for arbitrary long times requires also the *stability* of the motions with respect to variations, however small, of the initial conditions.

With the exception of integrable systems, the generic situation of classical dynamical systems describing, say, N particles interacting through physical potentials, is *instability* of the trajectories in the Lyapunov sense. Nowadays such an instability is called intrinsic stochasticity, or *chaoticity*, of the dynamics and is a consequence of nonlinearity of the equations of motion.

Likewise any other kind of instability, dynamical instability brings about the exponential growth of an initial perturbation, in this case it is the distance between a reference trajectory and any other trajectory originating in its close vicinity that locally grows exponentially in time. Quantitatively, the degree of chaoticity of a dynamical system is characterized by the *largest Lyapunov exponent* λ_1 that, if positive, measures the mean instability rate of nearby trajectories averaged along a sufficiently long reference trajectory. The exponent λ_1 also measures the typical time scale of memory loss of the initial conditions.

Recall that if
$$\dot{x}^i = X^i(x^1 \ldots x^N) \tag{3.17}$$
is a given dynamical system, i.e., a realisation in local coordinates of a one-parameter group of diffeomorphisms of a manifold M, that is of $\phi^t : M \to M$, and if we denote by
$$\dot{\xi}^i = \mathcal{J}^i_k[x(t)]\,\xi^k \tag{3.18}$$
the usual tangent dynamics equation, i.e., the realisation of the mapping $d\phi^t : TM \to TM$, where TM is the tangent bundle of M and $[\mathcal{J}^i_k]$ is the Jacobian matrix of $[X^i]$, then the largest Lyapunov exponent λ_1 is defined by
$$\lambda_1 = \lim_{t \to \infty} \frac{1}{t} \ln \frac{\|\xi(t)\|}{\|\xi(0)\|} \tag{3.19}$$
and, by setting
$$\Lambda[x(t), \xi(t)] = \xi^T\,\mathcal{J}[x(t)]\,\xi / \xi^T \xi \equiv \xi^T \dot{\xi}/\xi^T \xi = \frac{1}{2}\frac{d}{dt}\ln(\xi^T\xi),$$
this can be formally expressed as a time average
$$\lambda_1 = \lim_{t \to \infty} \frac{1}{2t} \int_0^t d\tau\,\Lambda[x(\tau), \xi(\tau)]. \tag{3.20}$$

Even though λ_1 is the most important indicator of chaos of classical dynamical systems, it is used only as a diagnostic tool in numerical simulations. With the exception of a few simple discrete-time systems (maps of the interval), no theoretical method exists to compute λ_1.[22] This situation reveals that a satisfactory theory of deterministic chaos is still lacking, at least for systems of physical relevance.

In the *conventional chaos theory*, dynamical instability is caused by homoclinic intersections of perturbed separatrices, however this theory has many problems: (i) it needs *action–angle coordinates*, (ii) it works in conditions of weak perturbation of an integrable system, (iii) to compute quantities like *Mel'nikov integrals* one needs the analytic expressions of the unperturbed separatrices: at large N this is hopeless, moreover the generalization of *Poincaré–Birkhoff theorem* is still problematic at $N > 2$; (iv) finally, there is no computational relationship between homoclinic intersections and Lyapunov exponents. Therefore this theory seems not adequate to treat chaos in Hamiltonian

---

[22] In the paper [Goz83], the authors show the connection between Lyapunov exponents and supersymmetry through a path-integral formulation of both stochastic and deterministic dynamics. This is a very interesting approach because Lyapunov exponents are represented as expectation values of some observables, thus in principle they can be computed by means of standard field-theoretic methods.

systems with many degrees of freedom at arbitrary degree of nonlinearity, with potentials that can be hardly transformed in action–angle coordinates, not to speak of accounting for phenomena like the transition from weak to strong chaos in Hamiltonian systems [PL90, PC91]. Motivated by the need of understanding this transition from weak to strong chaos, recently has been proposed in [Pet93, CP93, CLP95, CP95, CP96, PV95] to tackle Hamiltonian chaos in a theoretical framework different from that of homoclinic intersections. This new method makes use of the well–known possibility of formulating Hamiltonian dynamics in the language of Riemannian geometry (see [II07b]) so that the stability or instability of a geodesic flow depends on curvature properties of some suitably defined manifold.

In the early 1940s, N. S. Krylov already got a hold of the potential interest of this differential–geometric framework to account for dynamical instability and hence for phase space mixing [Kry79]. The follow–up of his intuition can be found in abstract ergodic theory [Sin89] and in a very few mathematical works concerning the ergodicity of geodesic flows of physical interest [Kna87, Gut77]. However, Krylov's work did not entail anything useful for a more general understanding of chaos in nonlinear newtonian dynamics because one soon hits against unsurmountable mathematical obstacles. By filling certain mathematical gaps with numerical investigations, these obstacles have been overcome and a rich scenario emerged about the relationship between stability and curvature

Based on the so–obtained information, this section aims at bringing a substantial contribution to the development of a Riemannian theory of Hamiltonian chaos. The new contribution consists of a method to analytically compute the largest Lyapunov exponent λ_1 for physically meaningful Hamiltonian systems of arbitrary large number of degrees of freedom. A preliminary and limited account of the results presented here can be found in [CLP95].

3.2.1 Geometrization of Newtonian Dynamics

Let us briefly recall how Newtonian dynamics can be rephrased in the language of Riemannian geometry. We shall deal with standard autonomous systems, i.e., described by the Lagrangian function

$$L = T - V = \frac{1}{2} a_{ij} \dot{q}_i \dot{q}_j - V(q_1, \ldots, q_N) , \qquad (3.21)$$

so that the Hamiltonian function $H = T + V \equiv E$ is a constant of motion.

According to the principle of stationary action – in the form of Maupertuis – among all the possible iso-energetic paths $\gamma(t)$ with fixed end points, the paths that make vanish the first variation of the action functional [CC96]

$$A = \int_{\gamma(t)} p_i \, dq_i = \int_{\gamma(t)} \frac{\partial L}{\partial \dot{q}_i} \dot{q}_i \, dt \qquad (3.22)$$

are natural motions.

As the kinetic energy T is a homogeneous function of degree two, we have $2T = \dot{q}_i \partial L/\partial \dot{q}_i$, and *Maupertuis' principle* reads

$$\delta A = \delta \int_{\gamma(t)} 2T \, dt = 0 . \tag{3.23}$$

The configuration space M of a system with N degrees of freedom is an ND differentiable manifold and the lagrangian coordinates $(q_1, \ldots, q_N)$ can be used as local coordinates on M. The manifold M is naturally given a proper Riemannian structure. In fact, let us consider the matrix

$$g_{ij} = 2[E - V(q)]a_{ij},$$

so that (3.23) becomes

$$\delta \int_{\gamma(t)} 2T \, dt = \delta \int_{\gamma(t)} \left(g_{ij}\dot{q}^i\dot{q}^j\right)^{1/2} dt = \delta \int_{\gamma(s)} ds = 0,$$

thus natural motions are geodesics of M, provided we define ds as its arc-length. The metric tensor g_J of M is then defined by

$$g_J = g_{ij} \, dq^i \otimes dq^j,$$

where $(dq^1, \ldots, dq^N)$ is a natural base of T_q^*M, the cotangent space at the point q, in the local chart $(q^1, \ldots, q^N)$. This is known as Jacobi (or kinetic energy) metric. Denoting by ∇ the canonical Levi–Civita connection, the geodesic equation

$$\nabla_{\dot\gamma}\dot\gamma = 0$$

becomes, in the local chart $(q^1, \ldots, q^N)$ [II07b]

$$\frac{d^2q^i}{ds^2} + \Gamma^i_{jk}\frac{dq^j}{ds}\frac{dq^k}{ds} = 0, \tag{3.24}$$

where the Christoffel symbols are the components of ∇ defined by

$$\Gamma^i_{jk} = \langle dq^i, \nabla_j e_k\rangle = \frac{1}{2}g^{im}\left(\partial_j g_{km} + \partial_k g_{mj} - \partial_m g_{jk}\right), \tag{3.25}$$

where $\partial_i = \partial/\partial q^i$. Without loss of generality consider $g_{ij} = 2[E - V(q)]\delta_{ij}$, from (3.24) we get

$$\frac{d^2q^i}{ds^2} + \frac{1}{2(E-V)}\left[2\frac{\partial(E-V)}{\partial q_j}\frac{dq^j}{ds}\frac{dq^i}{ds} - g^{ij}\frac{\partial(E-V)}{\partial q_j}g_{km}\frac{dq^k}{ds}\frac{dq^m}{ds}\right] = 0,$$

and, using $ds^2 = 2(E-V)^2 dt^2$, we can easily verify that these equations yield

$$\frac{d^2q^i}{dt^2} = -\frac{\partial V}{\partial q_i}, \quad (i = 1, \ldots, N).$$

which are Newton equations.

As discussed in [Pet93, CP93], there are other possibilities to associate a Riemannian manifold to a standard Hamiltonian system. Among the others we mention a structure, defined by Eisenhart [Eis29], that will be used in the following for computational reasons. In this case the ambient space is an enlarged configuration space-time $M \times \mathbb{R}^2$, with local coordinates $(q^0, q^1, \ldots, q^N, q^{N+1})$, with $(q^1, \ldots, q^N) \in M$, $q^0 \in \mathbb{R}$ is the time coordinate, $q^{N+1} \in \mathbb{R}$ is a coordinate closely related to the action; Eisenhart defines a pseudo–Riemannian non–degenerate metric g_E on $M \times \mathbb{R}^2$ as

$$ds_E^2 = g_{\mu\nu} \, dq^\mu \otimes dq^\nu = a_{ij} \, dq^i \otimes dq^j - 2V(q) \, dq^0 \otimes dq^0 + dq^0 \otimes dq^{N+1} + dq^{N+1} \otimes dq^0. \tag{3.26}$$

Natural motions are now given by the canonical projection π of the geodesics of $(M \times \mathbb{R}^2, g_E)$ on configuration space-time: $\pi : M \times \mathbb{R}^2 \to M \times \mathbb{R}$. However, among all the geodesics of g_E we must consider only those for which the arc-length is positive definite and given by [CC96]

$$ds^2 = g_{\mu\nu} dq^\mu dq^\nu = 2C^2 dt^2, \tag{3.27}$$

or, equivalently, we have to consider only those geodesics such that the coordinate q^{N+1} evolves according to

$$q^{N+1} = C^2 t + C_1^2 - \int_0^t L \, d\tau, \tag{3.28}$$

where C and C_1 are real constants. Since the values of these constants are arbitrary, we fix $C^2 = 1/2$ in order that $ds^2 = dt^2$ along a physical geodesic. For a diagonal kinetic energy matrix $a_{ij} = \delta_{ij}$, the non vanishing components of the connection ∇ are simply

$$\Gamma_{00}^i = -\Gamma_{0i}^{N+1} = \partial_i V,$$

therefore it is easy to check that also the geodesics of g_E yield Newton equations together with the differential versions of (3.28) and of $q^0 = t$ (details can be found in [Pet93, CP93]).

3.2.2 Geometric Description of Dynamical Instability

The actual interest of the Riemannian formulation of dynamics stems from the possibility of studying the instability of natural motions through the instability of geodesics of a suitable manifold, a circumstance that has several advantages. First of all a powerful mathematical tool exists to investigate the stability or instability of a geodesic flow: the *Jacobi–Levi-Civita equation* (JLC) for geodesic spread. The JLC equation describes covariantly how nearby geodesics locally scatter and it is a familiar object both in Riemannian geometry and theoretical physics (it is of fundamental interest in experimental General Relativity). Moreover the JLC equation relates the stability or instability of a geodesic flow with *curvature* properties of the ambient manifold, thus opening a wide and largely unexplored field of investigation of the connections among geometry, topology and geodesic instability, hence chaos.

Jacobi–Levi Civita Equation for Geodesic Spread

A *congruence of geodesics* is defined as a family of geodesics $\{\gamma_\tau(s) = \gamma(s,\tau) \,|\, \tau \in \mathbb{R}\}$ that, originating in some neighbourhood $\mathcal{I}$ of any given point of a manifold, are differentiably parametrized by some parameter τ. Choose a reference geodesic $\bar{\gamma}(s,\tau_0)$, denote by $\dot{\gamma}(s)$ the field of vectors tangent at s to $\bar{\gamma}$ and denote by $J(s)$ the field of vectors tangent at τ_0 to the curves $\gamma_s(\tau)$ at fixed s. The field $J = (\partial\gamma/\partial\tau)_{\tau_0}$ is known as *geodetic separation field* and it has the property: $\mathcal{L}_{\dot{\gamma}} J = 0$, where $\mathcal{L}$ is the *Lie derivative* (see [II07b]). Locally we can measure the distance between two nearby geodesics by means of J.

The evolution of the geodetic separation field J conveys information about stability or instability of the reference geodesic $\bar{\gamma}$, in fact, if $\|J\|$ exponentially grows with s then the geodesic is unstable in the sense of Lyapunov, otherwise it is stable [CC96].

The evolution of J is described by [Car92]

$$\frac{\nabla^2 J(s)}{ds^2} + R(\dot{\gamma}(s), J(s))\,\dot{\gamma}(s) = 0, \qquad (3.29)$$

known as Jacobi–Levi–Civita (JLC) equation. Here $J(s) \in T_{\gamma(s)}M$;

$$R(X,Y) = \nabla_X \nabla_Y - \nabla_Y \nabla_X - \nabla_{[X,Y]}$$

is the Riemann–Christoffel curvature tensor; $\dot{\gamma} = d\gamma/ds$; ∇/ds is the covariant derivative and $\gamma(s)$ is a normal geodesic, i.e., such that s is the length. In the following we assume that $J(s)$ is normal, i.e., $\langle J, \dot{\gamma}\rangle = 0$ [II07b]. This equation relates the stability or instability of nearby geodesics to the curvature properties of the ambient manifold. If the ambient manifold is endowed with a metric (e.g. Jacobi or Eisenhart) derived from the Lagrangian of a physical system, then stable or unstable (chaotic) motions will depend on the curvature properties of the manifold. Therefore it is reasonable to guess that some *average* global geometric property will provide information, at least, about an *average degree of chaoticity* of the dynamics independently of the knowledge of the trajectories, that is independently of the numerical integration of the equations of motion.

In local coordinates the JLC equation (3.29) reads as [CC96]

$$\frac{\nabla^2 J^i}{ds^2} + R^i{}_{jkl}\frac{dq^j}{ds} J^k \frac{dq^l}{ds} = 0, \qquad (3.30)$$

where

$$R^i{}_{jkl} = \langle dq^i, R(e_{(k)}, e_{(l)})e_{(j)}\rangle$$

are the components of the curvature tensor, and the covariant derivative is

$$(\nabla J^i/ds) = dJ^i/ds + \Gamma^i_{jk} J^k dq^j/ds.$$

There are $\mathcal{O}(N^4)$ of such components, $N = \dim(M)$, therefore – even if this number can be considerably reduced by symmetry considerations – equation (3.30) appears untractable already at rather small N. It is worth mentioning that some exception exists. Such is the case of *isotropic* manifolds for which (3.30) can be reduced to the simple form

$$\frac{\nabla^2 J^i}{ds^2} + K J^i = 0, \qquad (i = 1, \ldots, N), \tag{3.31}$$

where K is the constant value assumed throughout the manifold by the sectional curvature.

The sectional curvature of a manifold is the ND generalization of the gaussian curvature of 2D surfaces of $\mathbb{R}^3$. Consider two arbitrary vectors $X, Y \in T_x M$, where $x \in M$ is an arbitrary point of M, and define

$$\|X \wedge Y\| = (\|X\|^2 \|Y\|^2 - \langle X, Y \rangle)^{1/2}, \tag{3.32}$$

if $\|X \wedge Y\| \neq 0$ the vectors X, Y span a two-dimensional plane $\pi \subset T_x M$, then the sectional curvature at x relative to the plane π is defined by

$$K(X, Y) = K(x, \pi) = \frac{\langle R(Y, X) X, Y \rangle}{\|X \wedge Y\|^2}, \tag{3.33}$$

which is only a property of M at x independently of $X, Y \in \pi$ (*Gauss' Theorema Egregium*). For an isotropic manifold $K(x, \pi)$ is also independent of the choice of π and thus, according to Schur's theorem, K turns out also independent of $x \in M$.

Unstable solutions of the equation (3.31) are of the form

$$J(s) = w(0)(-K)^{-1/2} \sinh\left(\sqrt{-K}\,s\right), \tag{3.34}$$

once the initial conditions are assigned as $J(0) = 0$ and $dJ(0)/ds = w(0)$ and $K < 0$. In abstract ergodic theory geodesic flows on compact manifolds of constant negative curvature have been considered in classical works [Ano67]. In this case the quantity $\sqrt{-K}$, uniform on the manifold, measures the degree of instability of nearby geodesics.

While (3.31) holds true only for constant curvature manifolds, a similar form of general validity can be obtained for JLC equation at $N = 2$.

In this low-dimensional case (3.30) is exactly rewritten as

$$\frac{d^2 J}{ds^2} + \frac{1}{2} \mathcal{R}(s) J = 0, \tag{3.35}$$

where a parallely transported frame is used and $\mathcal{R}(s)$ is the scalar curvature. Using Jacobi metric one finds ($N = 2$): $\mathcal{R} = \Delta V / W^2 + (\nabla V)^2 / W^3$, with

$W = E - V$, so that for smooth and binding potentials $\mathcal{R}$ can be negative only where $\Delta V < 0$, i.e., nowhere for nonlinearly coupled oscillators as described by the Hénon-Heiles model [CP96] or for quartic oscillators [PV95]. $\Delta V < 0$ is only possible if the potential V has inflection points.

Recent detailed analyses of two-degrees of freedom systems [CP96, PV95] have shown that chaos can be produced by *parametric instability* due to a fluctuating positive curvature along the geodesics.

Let us remember that parametric instability is a generic property of dynamical systems with parameters that are periodically or quasi-periodically varying in time, even if for each value of the varying parameter the system has stable solutions [NM79]. A harmonic oscillator with periodically modulated frequency, described by the Mathieu equation, is perhaps the prototype of such a parametric instability mechanism.

Numerical simulations have shown that all the informations about order and chaos obtained by standard means (Lyapunov exponent and Poincaré sections) are fully retrieved by using (3.35). As in the case of tangent dynamics, (3.35) has to be computed along a reference geodesic (trajectory).

Let us now cope with the large N case. It is convenient to rewrite the JLC equation (3.30) in the following form [CC96]

$$\frac{\nabla^2 J(s)}{ds^2} + \frac{1}{N-1} [\text{Ric}(\dot{\gamma}(s), \dot{\gamma}(s)) J(s) - \text{Ric}(\dot{\gamma}(s), J(s)) \dot{\gamma}(s)] \quad (3.36)$$
$$+ W(\dot{\gamma}(s), J(s)) \dot{\gamma}(s) = 0,$$

where W is the Weyl projective curvature tensor whose components W^i_{jkl} are given by [Gol65]

$$W^i_{jkl} = R^i_{jkl} - \frac{1}{N-1}(R_{jl}\delta^i_k - R_{jk}\delta^i_l), \quad (3.37)$$

and Ric is the Ricci curvature tensor of components $R_{ij} = R^m_{imj}$. Weyl's projective tensor W (not to be confused with Weyl's *conformal* curvature tensor) measures the deviation from isotropy of a given manifold. For an isotropic manifold $W^i_{jkl} = 0$, and we recognize in (3.36) equation (3.31), in fact in this case $R_{jl}\dot{q}^j\dot{q}^l/(N-1)$ is just the constant value of sectional curvature. Remind that the Ricci curvature at $x \in M$ is $K_R(X_{(b)}) = R_{jl}X^j_{(b)}X^l_{(b)} = \sum_{a=1}^{N-1} K(X_{(b)}, X_{(a)})$ where $X_{(1)}, \ldots, X_{(N)}$ form an orthonormal basis of T_xM. Hence we understand that (3.36) retains the structure of (3.31) up to its second term that now has the meaning of a mean sectional curvature averaged, at any given point, over the independent orientations of the planes spanned by $X_{(a)}$ and $X_{(b)}$; this mean sectional curvature is no longer constant along $\gamma(s)$. The last term of (3.36) accounts for the local degree of anisotropy of the ambient manifold.

Let us now consider the following decomposition for the Jacobi field J

$$J(s) = \sum_i J_i(s) e_{(i)}(s),$$

where $\{e_{(1)} \ldots e_{(N)}\}$ is an orthonormal system of parallely transported vectors. In this reference frame it is

$$\frac{\nabla^2 J}{ds^2} = \sum_i \frac{d^2 J_i}{ds^2} e_{(i)}(s)$$

and the last term of (3.36) is

$$W(\dot{\gamma}, J)\dot{\gamma} = \sum_j \langle W(\dot{\gamma}, J)\dot{\gamma}, e_{(j)}\rangle e_{(j)}$$
$$= \sum_j \langle W(\dot{\gamma}, \sum_i J_i e_{(i)})\dot{\gamma}, e_{(j)}\rangle e_{(j)}$$
$$= \sum_{ij} \langle W(\dot{\gamma}, e_{(i)})\dot{\gamma}, e_{(j)}\rangle J_i e_{(j)} ,$$

the same decomposition applies to the third term of (3.36) which is finally rewritten as

$$\frac{d^2 J_j}{ds^2} + k_R(s)\, J_j + \sum_i (w_{ij} + r_{ij})\, J_i = 0, \qquad (3.38)$$

where

$$k_R = K_R/(N-1),\; w_{ij} = \langle W(\dot{\gamma}, e_{(i)})\dot{\gamma}, e_{(j)}\rangle \quad \text{and}$$
$$r_{ij} = \langle \text{Ric}(\dot{\gamma}, e_{(i)})\dot{\gamma}, e_{(j)}\rangle/(N-1).$$

Clearly, k_R is independent of the coordinate system. The elements w_{ij} still depend on the dynamics and on the behavior of the vectors $e_{(k)}(s)$, thus, in order to obtain a stability equation, for the geodesic flow, that depends only on average curvature properties of the ambient manifold, we try to conveniently approximate the w_{ij}. To this purpose define at any point $x \in M$ the trilinear mapping $R' : T_x M \times T_x M \times T_x M \to T_x M$ by

$$\langle R'(X, Y, U), Z\rangle = \langle X, U\rangle\langle Y, Z\rangle - \langle Y, U\rangle\langle X, Z\rangle, \qquad (3.39)$$

for all $X, Y, U, Z \in T_x M$. It is well known [Car92] that, if and only if M is isotropic then $R = K_0 R'$, where R is the Riemann curvature tensor of M and K_0 is the constant sectional curvature.

Let us now assume that the ambient manifold is *quasi–isotropic manifold*, i.e., that it looks like an isotropic manifold after a coarse-graining that smears out all the metric fluctuations, and let us formulate this assumption by putting $R \approx K(s)R'$ and $\text{Ric} \approx K(s)g$, although $K(s)$ is no longer a constant. Now we use (3.39) to find

$$w_{ij} \approx \delta K(s)[\langle \dot{\gamma}, \dot{\gamma}\rangle\langle e_{(i)}, e_{(j)}\rangle - \langle e_{(i)}, \dot{\gamma}\rangle\langle \dot{\gamma}, e_{(j)}\rangle],$$

then we use $\text{Ric} \propto g$ and $g(\dot{\gamma}, J) = 0$ to find $r_{ij} = 0$ thus (3.38) becomes [CC96]

$$\frac{d^2 J_j}{ds^2} + k_R(s)\, J_j + \delta K(s)\, J_j = 0, \tag{3.40}$$

by $\delta K(s) = K(s) - \overline{K}$ we denote the local deviation of sectional curvature from its coarse–grained value $\overline{K}$, thus $\delta K(s)$ measures the fluctuation of sectional curvature along a geodesic due to the local deviation from isotropy. The problem is that $\delta K(s)$ still depends on a moving plane $\pi(s)$ determined by $\dot\gamma(s)$ and $J(s)$. In order to get rid of this dependence, consider that if $x \in M$ is an isotropic point then the components of the Ricci tensor are $R_{lh} = (N-1)K(x)g_{lh}$ and the scalar curvature is $R = N(N-1)K(x)$; with these quantities one constructs the Einstein tensor $G_{lh} = R_{lh} - \frac{1}{2} g_{lh} R$ whose divergence vanishes identically $(G_{lh|l} = 0)$ so that it is immediately found that, if a manifold consists entirely of isotropic points, then $\partial K(x)/\partial x^l = 0$ and so $\partial K_R(x)/\partial x^l = 0$, i.e., the manifold is a space of constant curvature (Schur's theorem [Car92]). Conversely, the local variation of Ricci curvature detects the local loss of isotropy, thus a reasonable approximation of the *average* variation $\delta K(s)$ along a geodesic may be given by the variation of Ricci curvature.

Next let us model $\delta K(s)$ along a geodesic by a stochastic process. In fact $K(s)$ is obtained by summing a large number of terms, each one depending on different combinations of the components of J and on the the coordinates q^i, moreover, unless we tackle an integrable model, the dynamics is always chaotic and the functions $q^i(s)$ behave irregularly. By invoking a *Central–limit–theorem* argument, at large N, $\delta K(s)$ is expected to behave, in first approximation, as a gaussian stochastic process. More generally, the probability distribution $\mathcal{P}(\delta K)$ may be other than gaussian and in practice it could be determined by computing its cumulants along a geodesic $\gamma(s)$.

Now we make quantitative the previous statement – about using the variation of Ricci curvature along a geodesic to estimate $\delta K(s)$ – by putting

$$\mathcal{P}(\delta K) \simeq \mathcal{P}(\delta K_R). \tag{3.41}$$

Both δK and δK_R are zero mean variations, so the first moments vanish; according to (3.41) the following relation for the second moments will hold [CC96]

$$\langle [K(s) - \overline{K}]^2 \rangle_s \simeq \frac{1}{N-1} \langle [K_R(s) - \langle K_R \rangle_s]^2 \rangle_s, \tag{3.42}$$

where $\langle \cdot \rangle_s$ stands for proper-time average along a geodesic $\gamma(s)$. Let us comment about the numerical factor in the r.h.s. of (3.42) where a factor $\frac{1}{N^2}$ might be expected. At increasing N the mean square fluctuations of k_R drop to zero as $\frac{1}{N}$ because k_R is the mean of independent quantities, however this cannot be the case of the mean square fluctuations of K, in fact out of the sum K_R of all the sectional curvatures, in (3.40) only one sectional curvature is 'picked-up' from point to point by δK so that δK remains finite with increasing N. Therefore, as the second cumulant of δK does not vanish with N, we have to keep finite the second cumulant of δK_R, what is simply achieved by properly adjusting the numerical factor in (3.42).

The lowest order approximation of a cumulant expansion of the stochastic process $\delta K(s)$ is the gaussian approximation

$$\delta K(s) \simeq \frac{1}{\sqrt{N-1}} \langle \delta^2 K_R \rangle_s^{1/2} \eta(s), \qquad (3.43)$$

where $\eta(s)$ is a random gaussian process with zero mean and unit variance. Finally, in order to decouple the stability equation from the dynamics, we replace time averages with static averages computed with a suitable ergodic invariant measure μ. As we deal with autonomous Hamiltonian systems, a natural choice is the micro-canonical measure on the constant energy surface of phase space[23]

$$\mu \propto \delta(\mathcal{H} - E), \qquad (3.44)$$

so that (3.43) becomes

$$\delta K(s) \simeq \frac{1}{\sqrt{N-1}} \langle \delta^2 K_R \rangle_\mu^{1/2} \eta(s). \qquad (3.45)$$

Similarly, $k_R(s)$ in (3.40) is replaced by $\langle k_R \rangle_\mu$, in fact at large N the fluctuations of k_R – as already noticed above – vanish as $\frac{1}{N}$ because the coarse-grained manifold is isotropic, so that we finally have [CC96]

$$\frac{d^2\psi}{ds^2} + \langle k_R \rangle_\mu \psi + \frac{1}{\sqrt{N-1}} \langle \delta^2 K_R \rangle_\mu^{1/2} \eta(s) \psi = 0, \qquad (3.46)$$

where ψ stands for any of the components J^j, since all of them now obey the same effective equation of motion. The instability growth-rate of ψ measures the instability growth-rate of $\|J\|^2$ and thus provides the dynamical instability exponent in our Riemannian framework. Equation (3.46) is a scalar equation which, *independently of the knowledge of dynamics*, provides a measure of the average degree of instability of the dynamics itself through the behavior of $\psi(s)$. The peculiar properties of a given Hamiltonian system enter (3.46) through the global geometric properties $\langle k_R \rangle_\mu$ and $\langle \delta^2 K_R \rangle_\mu$ of the ambient Riemannian manifold (whose geodesics are natural motions) and are sufficient to determine the average degree of chaoticity of the dynamics. Moreover, according to (3.44), $\langle k_R \rangle_\mu$ and $\langle \delta^2 K_R \rangle_\mu$ are functions of the energy E of the system – or of the energy density $\varepsilon = E/N$ which is the relevant parameter as $N \to \infty$ – so that from (3.46) we can obtain the energy dependence of the geometric instability exponent.

---

[23] At $N \geq 3$, after the *Poincaré–Fermi theorem*, generic non-integrable systems have no smooth invariant besides energy, thus the whole constant energy surface is topologically accessible. One might think that troubles with ergodicity could be raised by the KAM theorem, but at large N this is not possible because a positive measure of tori survives a perturbation of amplitude smaller than some threshold value, and the threshold typically drops to zero as $\exp(-cN \ln N)$, c is a constant.

Stochastic Oscillator Equation

In this subsection we briefly describe how to cope with the *stochastic oscillator problem* (which we will enconuter in the following text). For more details, see [Kam92].

A stochastic differential equation can be put in the general form

$$F(x, \Omega) = 0, \qquad (3.47)$$

where F is an assigned function and the variable Ω is a random process, defined by a mean, a standard deviation and an autocorrelation function. A function $\xi(\Omega)$ is a solution of this equation if $F(\xi(\Omega), \Omega) = 0$ for all Ω. If equation (3.47) is linear of order n, it is written as

$$\dot{\mathbf{u}} = \mathbf{A}(t, \Omega)\mathbf{u}, \qquad (3.48)$$

where $\mathbf{u} \in \mathbb{R}^n$ and $\mathbf{A}$ is a $n \times n$ matrix whose elements are randomly dependent on time.

For the purposes of our work we are interested in studying the evolution of the average carried over all the realizations of the process, $\langle \mathbf{u}(t) \rangle$. Let us consider the matrix $\mathbf{A}$ as the sum

$$\mathbf{A}(t, \Omega) = \mathbf{A}_0(t) + \alpha \mathbf{A}_1(t, \Omega), \qquad (3.49)$$

where the first term is Ω-independent and the second one is randomly fluctuating with zero mean. Let us also assume that $\mathbf{A}_0$ is time-independent. If the parameter α – that determines the fluctuation amplitude – is small we can treat (3.48) by means of a perturbation expansion. It is convenient to use the interaction picture, thus we put

$$\mathbf{u}(t) = \exp(\mathbf{A}_0 t) \mathbf{v}(t), \qquad \mathbf{A}_1(t) = \exp(\mathbf{A}_0 t) \mathbf{v}(t) \exp(-\mathbf{A}_0 t).$$

Formally one is led to a Dyson expansion for the solution $\mathbf{v}(t)$. Then, going back to the previous variables and averaging, the second order approximation gives

$$\frac{d}{dt} \langle \mathbf{u}(t) \rangle = \{\mathbf{A}_0 + \alpha^2 \int_{-\infty}^{+\infty} \langle \mathbf{A}_1(t) \exp(\mathbf{A}_0 \tau) \mathbf{A}_1(t - \tau) \rangle \exp(-\mathbf{A}_0 \tau) d\tau \} \langle \mathbf{u}(t) \rangle. \qquad (3.50)$$

Following the same procedure one can find also the evolution of the second moments (and by iterating also the evolution of higher moments). In fact, with the components of $\mathbf{u} \in \mathbb{R}^n$ we can make n^2 quantities $u_\nu u_\mu$ that obey the differential equation

$$\frac{d}{dt}(u_\nu u_\mu) = \sum_{k,\lambda} \tilde{A}_{\nu\mu,k\lambda}(t)(u_k u_\lambda),$$

where

3.2 Riemannian Approach to Chaos

$$\tilde{A}_{\nu\mu,k\lambda} = A_{\nu k}\delta_{\mu\lambda} + \delta_{\nu k}A_{\mu\lambda}. \tag{3.51}$$

The above presented averaging method can be now applied to this new equation.

Now, if we consider a random harmonic oscillator, (3.48) has the form

$$\frac{d}{dt}\begin{pmatrix} x \\ \dot{x} \end{pmatrix} = \begin{pmatrix} 0 & 1 \\ -\Omega & 0 \end{pmatrix}\begin{pmatrix} x \\ \dot{x} \end{pmatrix},$$

with the random squared frequency $\Omega = \Omega_0 + \sigma_\Omega \eta(t)$. In particular, we are interested in working out the second moments equation when the process $\eta(t)$ is gaussian and δ-correlated. Using (3.51) one finds that

$$\frac{d}{dt}\begin{pmatrix} x^2 \\ \dot{x}^2 \\ x\dot{x} \end{pmatrix} = \begin{pmatrix} 0 & 0 & 2 \\ 0 & 0 & -2\Omega \\ -\Omega & 1 & 0 \end{pmatrix}\begin{pmatrix} x^2 \\ \dot{x}^2 \\ x\dot{x} \end{pmatrix} = \mathbf{A}\begin{pmatrix} x^2 \\ \dot{x}^2 \\ x\dot{x} \end{pmatrix}.$$

Because of our assumptions for this system, (3.50) is more than a second order approximation, it is exact. In fact, the Dyson series can be written in compact form as

$$\begin{pmatrix} \langle x^2(t) \rangle \\ \langle \dot{x}^2(t) \rangle \\ \langle x(t)\dot{x}(t) \rangle \end{pmatrix} = \lceil \langle \exp\left(\int_0^t \mathbf{A}(t')dt'\right)\rangle \rceil \begin{pmatrix} \langle x^2(0) \rangle \\ \langle \dot{x}^2(0) \rangle \\ \langle x(0)\dot{x}(0) \rangle \end{pmatrix}, \tag{3.52}$$

where the brackets $\lceil ... \rceil$ stand for a chronological product. According to Wick's procedure we can rewrite (3.52) as a cumulant expansion, and when the cumulants of order higher than the second vanish (as is the case of interest to us) one can easily show that (3.50) is exact.

Likewise in (3.49), the matrix $\mathbf{A}$ splits as

$$\mathbf{A}(t) = \mathbf{A}_0 + \sigma_\Omega \eta(t)\mathbf{A}_1 = \begin{pmatrix} 0 & 0 & 2 \\ 0 & 0 & -2\Omega_0 \\ -\Omega_0 & 1 & 0 \end{pmatrix} + \sigma_\Omega \eta(t)\begin{pmatrix} 0 & 0 & 0 \\ 0 & 0 & -2 \\ -1 & 0 & 0 \end{pmatrix},$$

therefore the equation for the averages becomes

$$\frac{d}{dt}\begin{pmatrix} \langle x^2 \rangle \\ \langle \dot{x}^2 \rangle \\ \langle x\dot{x} \rangle \end{pmatrix} = \{\mathbf{A}_0 + \sigma_\Omega^2 \int_{-\infty}^{+\infty} \langle \eta(t)\eta(t-\tau)\rangle \mathbf{B}(\tau)d\tau\}\begin{pmatrix} \langle x^2 \rangle \\ \langle \dot{x}^2 \rangle \\ \langle x\dot{x} \rangle \end{pmatrix},$$

where $\mathbf{B}(\tau) = \mathbf{A}_1 \exp(\mathbf{A}_0\tau)\mathbf{A}_1 \exp(-\mathbf{A}_0\tau)$. As $\langle \eta(t)\eta(t-\tau)\rangle = \tau\delta(\tau)$, with τ a characteristic time scale of the process, we obtain

$$\frac{d}{dt}\begin{pmatrix} \langle x^2 \rangle \\ \langle \dot{x}^2 \rangle \\ \langle x\dot{x} \rangle \end{pmatrix} = \{\mathbf{A}_0 + \sigma_\Omega^2 \tau \mathbf{B}(0)\}\begin{pmatrix} \langle x^2 \rangle \\ \langle \dot{x}^2 \rangle \\ \langle x\dot{x} \rangle \end{pmatrix}.$$

From the definition of $\mathbf{B}(\tau)$ it follows that $\mathbf{B}(0) = \mathbf{A}_1^2$, then by easy calculations we find

$$\mathbf{A}_0 + \sigma_\Omega^2 \tau \mathbf{A}_1^2 = \begin{pmatrix} 0 & 0 & 2 \\ 2\sigma_\Omega^2 \tau & 0 & -2\Omega_0 \\ -\Omega_0 & 1 & 0 \end{pmatrix}.$$

Largest Lyapunov Exponent

By transforming (3.29) into (3.46) the original complexity of the JLC equation has been considerably reduced: from a tensor equation we have worked out an effective scalar equation formally representing a stochastic oscillator. In fact (3.46), with a self-evident notation, is in the form [CC96]

$$\frac{d^2\psi}{ds^2} + \Omega(s)\,\psi = 0, \tag{3.53}$$

where $\Omega(s)$ is a gaussian stochastic process.

Now, passing from proper time s to physical time t, (3.53) simply reads

$$\frac{d^2\psi}{dt^2} + \Omega(t)\,\psi = 0, \tag{3.54}$$

where

$$\Omega(t) = \langle k_R \rangle_\mu + \frac{1}{\sqrt{N}} \langle \delta^2 K_R \rangle_\mu^{1/2} \eta(t), \tag{3.55}$$

if the Eisenhart metric is used (because of the affine parametrization of the arclength with time, (3.27)); if Jacobi metric is used, we have [CC96]

$$\Omega(t) = \langle k_R \rangle_\mu + \left\langle -\frac{1}{4}\left(\frac{\dot W}{W}\right)^2 + \frac{1}{2}\frac{d}{dt}\left(\frac{\dot W}{W}\right)\right\rangle_\mu + \frac{1}{\sqrt{N}} \langle \delta^2 K_R \rangle_\mu^{1/2} \eta(t) \tag{3.56}$$

(see [Pet93, CP96]), note that $d/dt = \dot q^j (\partial/\partial q^j)$. Being interested in the large N limit, we replaced $N-1$ with N in (3.55) and (3.56). Obviously, Ricci curvature has different expressions according to the metric used.

The stochastic process $\Omega(t)$ is not completely determined unless its time correlation function $\Gamma_\Omega(t_1, t_2)$ is given. We consider a stationary and δ−correlated process $\Omega(t)$ so that $\Gamma_\Omega(t_1, t_2) = \Gamma_\Omega(|t_2 - t_1|)$ and

$$\Gamma_\Omega(t) = \tau \sigma_\Omega^2 \delta(t), \tag{3.57}$$

where τ is a characteristic time scale of the process. In order to estimate τ, let us notice that for a geodesic flow on a smooth manifold the assumption of δ−correlation of $\Omega(t)$ will be reasonable only down to some time scale below which the differentiable character of the geodesics will be felt. In other words, we have to think that in reality the power spectrum of $\Omega(t)$ is flat up to some high frequency cutoff, let us denote it by $\nu_\star$; therefore, by representing the δ function as the limit for $\nu \to \infty$ of $\delta_\nu(t) = \frac{\sin(\nu t)}{\pi t}$, a more realistic representation of the autocorrelation function $\Gamma_\Omega(t)$ in (3.57) could be

$$\Gamma_\Omega^\star(t) = \sigma_\Omega^2 \frac{1}{\pi}\frac{\sin(\nu_\star t)}{\nu_\star t} \equiv \tau_\star \sigma_\Omega^2 \delta_{\nu_\star}(t),$$

whence $\tau_\star = 1/\nu_\star$. Notice that

$$\int_{0^-}^\infty \Gamma_\Omega(t)dt = \tau\sigma_\Omega^2 \quad \text{and} \quad \int_{0^-}^\infty \Gamma_\Omega^\star(t)dt = \frac{1}{2}\tau_\star\sigma_\Omega^2$$

thus $\tau = \tau_\star/2$. For practical computational reasons it is convenient to use $\Gamma_\Omega(t)$ in the form given by (3.57) (with the implicit assumption that $\nu_\star$ is sufficiently large), however, being $\nu_\star$ finite, the definition $\tau = \tau_\star/2$ will be kept. To estimate $\tau_\star$ we proceed as follows. A first time scale, which we will refer to as τ_1, is associated to the time needed to cover the average distance between two successive conjugate points along a geodesic.[24] In fact, at distances smaller than this one the geodesics are minimal and far from looking like random walks, whereas at each crossing of a conjugate point the separation vector field increases as if the geodesics in the local congruence were kicked (this is what happens when parametric instability is active). From Rauch's comparison theorem [Car92] we know that if sectional curvature K is bounded as follows: $0 < L \leq K \leq H$, then the distance d between two successive conjugate points is bounded by $\frac{\pi}{\sqrt{H}} < d < \frac{\pi}{\sqrt{L}}$. We need the lower bound estimate that, for strongly convex domains,[25] is slightly modified to $d > \frac{\pi}{2\sqrt{H}}$.

Hence we define τ_1 through

$$\tau_1 = \left\langle \frac{dt}{ds} \right\rangle d_\star = \left\langle \frac{dt}{ds} \right\rangle \frac{\pi}{2\sqrt{\Omega_0 + \sigma_\Omega}}, \tag{3.58}$$

where $\left\langle \frac{dt}{ds} \right\rangle$ is the average of the ratio between proper and physical time ($\left\langle \frac{dt}{ds} \right\rangle = 1$ if Eisenhart metric is used) and the upper bound H of K is replaced by the N-th fraction of a typical peak value of Ricci curvature, which is in turn estimated as its average Ω_0 plus the typical value δK of the (positive) fluctuation, i.e., in a gaussian approximation $\delta K = \sigma_\Omega$. This time scale is expected to be the most relevant only as long as curvature is positive and the fluctuations, compared to the average, are small.

Another time scale, referred to as τ_2, is related to local curvature fluctuations. These will be felt on a length scale of the order of, at least, $l = 1/\sqrt{\sigma_\Omega}$ (the average fluctuation of curvature radius). The scale l is expected to be relevant one when the fluctuations are of the same order of magnitude as the average curvature. When the sectional curvature is positive (resp. negative), lengths and time intervals – on a scale l – are enlarged (resp. shortened) by a factor $(l^2 K/6)$,[26] so that the period $\frac{2\pi}{\sqrt{\Omega_0}}$ has a fluctuation amplitude d_2 given

---

[24] The conjugate points are defined by the vanishing of Jacobi field of geodesic separation.

[25] A subset $\mathcal{M}$ of a Riemannian manifold is said to be strongly convex if every minimal geodesic joining two of its points always lies in $\mathcal{M}$. The lowering of the bound follows from a theorem due to Whitehead, see [CE75].

[26] This is a consequence of the possibility of approximating locally the metric of a manifold by $g_{ik} \simeq \delta_{ik} - \frac{1}{6} R_{ikjl} u^i u^k$ obtained by using normal coordinates and displacements u^i (see, e.g., [Car92]).

by $d_2 = \frac{l^2 K}{6} \frac{2\pi}{\sqrt{\Omega_0}}$; replacing K by its most probable value Ω_0 one gets [CC96]

$$\tau_2 = \left\langle \frac{dt}{ds} \right\rangle d_2 = \left\langle \frac{dt}{ds} \right\rangle \frac{l^2 \Omega_0}{6} \frac{2\pi}{\sqrt{\Omega_0}} \simeq \left\langle \frac{dt}{ds} \right\rangle \frac{\Omega_0^{1/2}}{\sigma_\Omega}. \qquad (3.59)$$

Finally τ in (3.57) is obtained by combining τ_1 with τ_2 as follows

$$\tau^{-1} = 2\tau_\star^{-1} = 2\left(\tau_1^{-1} + \tau_2^{-1}\right). \qquad (3.60)$$

The present estimate of τ is very close, though not equal, to the one of [CLP95].

Whenever $\Omega(t)$ in (3.54) has a non–vanishing stochastic component the solution $\psi(t)$ has an exponentially growing envelope [Kam92] whose growth-rate provides a measure of the degree of chaoticity. Let us call this quantity Lyapunov exponent and denote it by λ.

Our exponent λ is defined as

$$\lambda = \lim_{t \to \infty} \frac{1}{2t} \log \frac{\psi^2(t) + \dot{\psi}^2(t)}{\psi^2(0) + \dot{\psi}^2(0)}, \qquad (3.61)$$

where $\psi(t)$ is solution of (3.54).

The ratio $(\psi^2(t) + \dot{\psi}^2(t))/(\psi^2(0) + \dot{\psi}^2(0))$ is computed by means of a technique, developed by Van Kampen and sketched in Appendix A, which is based on the possibility of computing analytically the evolution of the second moments of ψ and $\dot{\psi}$, averaged over the realizations of the stochastic process, from

$$\frac{d}{dt}\begin{pmatrix} \langle \psi^2 \rangle \\ \langle \dot{\psi}^2 \rangle \\ \langle \psi\dot{\psi} \rangle \end{pmatrix} = \begin{pmatrix} 0 & 0 & 2 \\ 2\sigma_\Omega^2 \tau & 0 & -2\Omega_0 \\ -\Omega_0 & 1 & 0 \end{pmatrix} \begin{pmatrix} \langle \psi^2 \rangle \\ \langle \dot{\psi}^2 \rangle \\ \langle \psi\dot{\psi} \rangle \end{pmatrix} \qquad (3.62)$$

where Ω_0 and σ_Ω are respectively the mean and the variance of $\Omega(t)$ above defined. By diagonalizing the matrix in the r.h.s. of (3.62) one finds two complex conjugate eigenvalues, and one real eigenvalue related to the evolution of $\frac{1}{2}\left(\langle \psi^2 \rangle + \langle \dot{\psi}^2 \rangle\right)$. According to (3.61) the exponent λ is half the real eigenvalue. Simple algebra leads to the final expression [CC96]

$$\lambda(\Omega_0, \sigma_\Omega, \tau) = \frac{1}{2}\left(\Lambda - \frac{4\Omega_0}{3\Lambda}\right),$$

$$\Lambda = \left(2\sigma_\Omega^2 \tau + \sqrt{\left(\frac{4\Omega_0}{3}\right)^3 + (2\sigma_\Omega^2 \tau)^2}\right)^{1/3}.$$

All the quantities Ω_0, σ_Ω and τ can be computed as *static* averages, therefore – within the validity limits of the assumptions made above – Eqs. (3.67) provide an analytic formula to compute the largest Lyapunov exponent independently of the numerical integration of the dynamics and of the tangent dynamics.

Lyapunov exponent and Eisenhart metric

Let us consider dynamical systems described by the Lagrangian function (5.62) with a diagonal kinetic energy matrix, i.e., $a_{ij} = \delta_{ij}$, and let us choose as ambient manifold the enlarged configuration space-time equipped with the Eisenhart metric (3.26).

Trivial algebra gives $\Gamma^i_{00} = (\partial V/\partial q_i)$ and $\Gamma^{N+1}_{0i} = (-\partial V/\partial q^i)$ as the only non-vanishing Christoffel coefficients and hence the Riemann curvature tensor has only the following non-vanishing components

$$R_{0i0j} = \frac{\partial^2 V}{\partial q^i \partial q^j}. \tag{3.64}$$

The JLC equation (3.29) is thus rewritten in local coordinates as [CC96]

$$\frac{\nabla}{ds}\frac{\nabla}{ds}J^0 + R^0_{i0j}\frac{dq^i}{ds}J^0\frac{dq^j}{ds} + R^0_{0ij}\frac{dq^0}{ds}J^i\frac{dq^j}{ds} = 0,$$

$$\frac{\nabla}{ds}\frac{\nabla}{ds}J^i + R^i_{0j0}\left(\frac{dq^0}{ds}\right)^2 J^j + R^i_{00j}\frac{dq^0}{ds}J^0\frac{dq^j}{ds} + R^i_{j00}\frac{dq^j}{ds}J^0\frac{dq^0}{ds} = 0,$$

$$\frac{\nabla}{ds}\frac{\nabla}{ds}J^{N+1} + R^{N+1}_{i0j}\frac{dq^i}{ds}J^0\frac{dq^j}{ds} + R^{N+1}_{ij0}\frac{dq^i}{ds}J^j\frac{dq^0}{ds} = 0. \tag{3.65}$$

As $\Gamma^0_{ij} = 0$ implies $\nabla J^0/ds = dJ^0/ds$ and as $R^0_{ijk} = 0$, we find that the first of these equations reads

$$\frac{d^2 J^0}{ds^2} = 0,$$

hence J^0 does not accelerate and, without loss of generality, we can set $\dot{J}^0(0) = J^0(0) = 0$, this yields (using $\nabla J^i/ds = dJ^i/ds + \Gamma^i_{0k}\dot{q}^0 J^k + \Gamma^i_{k0}\dot{q}^k J^0$)

$$\frac{\nabla^2 J^i}{ds^2} = \frac{d^2 J^i}{ds^2}$$

and the second equation in (3.65) gives, for the projection in configuration space of the separation vector,

$$\frac{d^2 J^i}{ds^2} + \frac{\partial^2 V}{\partial q_i \partial q_k}\left(\frac{dq^0}{ds}\right)^2 J_k = 0, \quad (i = 1, ..., N); \tag{3.66}$$

the third of equations (3.65) describes the passive evolution of J^{N+1} which does not contribute the norm of J because $g_{N+1\,N+1} = 0$, so we can disregard it.

As already mentioned in the previous Section, along the physical geodesics of g_E it is $ds^2 = (dq^0)^2 = dt^2$ therefore (3.66) is exactly the usual tangent dynamics equation reported in the Introduction, provided that the obvious identification $\xi = (\xi_q, \xi_p) \equiv (J, \dot{J})$ is made. This clarifies the relationship between the geometric description of the instability of a geodesic flow and the conventional description of dynamical instability. It has been recently shown [CP96, PV95] that the solutions of the equations (3.66) and (3.35) (where $\mathcal{R}$ is computed with Jacobi metric) are strikingly close one another in the case of two degrees of freedom systems. This result is reasonable because the geodesics of $(M \times \mathbb{R}^2, g_E)$ – that are natural motions – project themselves onto the geodesics of (M, g_J), and as the extra coordinates q^0 and q^{N+1} do not contribute to the instability of the geodesic flow, both local and global instability properties must be the same with either Jacobi or Eisenhart metrics, independently of N.

With Eisenhart metric the only non-vanishing component of the Ricci tensor is $R_{00} = \triangle V$, where $\triangle$ is the Euclidean Laplacian in configuration space. Hence Ricci curvature is $k_R(q) = \triangle V/(N-1)$ (remember that we choose the constant C such that $ds^2 = dt^2$ along a physical geodesic) and the stochastic process $\Omega(t)$ in (3.54) is specified by [CC96]

$$\Omega_0 = \langle k_R \rangle_\mu = \frac{1}{N} \langle \triangle V \rangle_\mu, \tag{3.67a}$$

$$\sigma_\Omega^2 = \frac{1}{N} \langle \delta^2 K_R \rangle_\mu = \frac{1}{N} \left(\langle (\triangle V)^2 \rangle_\mu - \langle \triangle V \rangle_\mu^2 \right), \tag{3.67b}$$

$$2\tau = \frac{\pi \sqrt{\Omega_0}}{2\sqrt{\Omega_0(\Omega_0 + \sigma_\Omega)} + \pi \sigma_\Omega}. \tag{3.67c}$$

Averages of geometric quantities

Let us now sketch how to compute the mean and the variance of any observable function $f(q)$, a geometric quantity of the chosen ambient manifold, by means of the micro-canonical measure (3.44), i.e., [CC96]

$$\langle f(q) \rangle_\mu = \frac{1}{\omega_E} \int f(q)\, \delta(\mathcal{H}(q,p) - E)\, dq\, dp, \tag{3.68}$$

where

$$\omega_E = \int \delta(\mathcal{H}(q,p) - E)\, dq\, dp$$

and $q = (q_1 \dots q_N)$, $p = (p_1 \dots p_N)$. By using the configurational partition function $Z_C(\beta)$, given by

$$Z_C(\beta) = \int dq\, e^{-\beta V(q)},$$

where $dq = \prod_{i=1}^N dq_i$, we can compute the Gibbsian average $\langle f \rangle^G$ of the observable f as

$$\langle f \rangle^G = [Z_C(\beta)]^{-1} \int dq\, f(q)\, e^{-\beta V(q)}.$$

Whenever this average is known, we can obtain the micro-canonical average of f [LPV67] in the following parametric form [CC96]

$$\langle f\rangle_\mu(\varepsilon) \to \begin{cases} \langle f\rangle_\mu(\beta) = \langle f\rangle^G(\beta) \\ \varepsilon(\beta) = \frac{1}{2\beta} - \frac{1}{N}\frac{\partial}{\partial\beta}[\log Z_C(\beta)] \end{cases} \quad (3.69)$$

By replacing f with the explicit expression for Ricci curvature $k_R = \frac{1}{N}K_R$ we can work out Ω_0. Notice that (3.69) is strictly valid in the thermodynamic limit; at finite N it is

$$\langle f\rangle_\mu(\beta) = \langle f\rangle^G(\beta) + \mathcal{O}(\frac{1}{N}).$$

At variance with the computation of $\langle f\rangle$, which is insensitive to the choice of the probability measure in the $N \to \infty$ limit, computing the fluctuations of f, i.e., of $\langle \delta^2 f\rangle = \frac{1}{N}\langle (f - \langle f\rangle)^2\rangle$, by means of the canonical or micro-canonical measures yields different results. The relationship between the canonical – i.e., computed with the Gibbsian weight $e^{-\beta\mathcal{H}}$ – and the micro-canonical fluctuations is given by the well known formula [LPV67]

$$\langle \delta^2 f\rangle_\mu(\varepsilon) = \langle \delta^2 f\rangle^G(\beta) - \frac{\beta^2}{C_V}\left[\frac{\partial\langle f\rangle^G(\beta)}{\partial\beta}\right]^2, \quad (3.70)$$

where

$$C_V = -\frac{\beta^2}{N}\frac{\partial\langle E\rangle}{\partial\beta}$$

is the specific heat at constant volume and $\beta = \beta(\varepsilon)$ is given in implicit form by the second equation in (3.69).

By replacing f with k_R we can work out σ_Ω^2.

3.2.3 Examples

The Fermi–Pasta–Ulam β–Model

The FPU β-model is defined by the Hamiltonian [FPU55]

$$\mathcal{H}(p,q) = \sum_{i=1}^{N}\frac{1}{2}p_i^2 + \sum_{i=1}^{N}\left[\frac{1}{2}(q_{i+1} - q_i)^2 + \frac{\mu}{4}(q_{i+1} - q_i)^4\right]. \quad (3.71)$$

This is a paradigmatic model of nonlinear classical many-body systems that has been extensively studied over the last decades and that stimulated remarkable developments in nonlinear dynamics, one example: the discovery of solitons. For a recent review we refer to [For92]. Also the transition between weak and strong chaos has been first discovered in this model [PL90, PC91] and then, the effort of understanding the origin of such a threshold has stimulated the development of the geometric theory presented here.

Let us now compute the average Ricci curvature Ω_0 and its fluctuations σ_Ω. We have seen above that, using Eisenhart metric, k_R is given by

$$k_R = \frac{1}{N} \sum_{i=1}^{N} \frac{\partial^2 V(q)}{\partial q_i^2},$$

for the FPU β-model this reads

$$k_R = 2 + \frac{6\mu}{N} \sum_{i=1}^{N} (q_{i+1} - q_i)^2, \qquad (3.72)$$

note that k_R is always positive.

In order to compute the Gibbssian average of k_R and its fluctuations, we rewrite the *configurational partition function* as [CC96]

$$\tilde{Z}_C(\alpha) = \int_{-\infty}^{+\infty} \prod_{i=1}^{N} dq_i \exp\left\{-\beta \sum_{i=1}^{N} \left[\frac{\alpha}{2}(q_{i+1} - q_i)^2 + \frac{\mu}{4}(q_{i+1} - q_i)^4\right]\right\}, \qquad (3.73)$$

which, in terms of the arbitrary parameter α and of Z_C, is expressed as $\tilde{Z}_C(\alpha) = Z_C(\alpha\beta, \mu/\alpha)$ and leads to the following identity

$$\langle k_R \rangle(\beta) = 2 - \frac{12\mu}{\beta N} \frac{1}{Z_C} \left[\frac{\partial}{\partial \alpha} \tilde{Z}_C(\alpha)\right]_{\alpha=1}. \qquad (3.74)$$

Thus we have to compute

$$\frac{1}{N Z_C} \left[\frac{\partial}{\partial \alpha} \tilde{Z}_C(\alpha)\right]_{\alpha=1} = \frac{1}{N} \left[\frac{\partial}{\partial \alpha} \log \tilde{Z}_C(\alpha)\right]_{\alpha=1}, \qquad (3.75)$$

using

$$\tilde{Z}_C(\alpha) = [\tilde{z}_C(\alpha)]^N f(\alpha),$$

where $f(\alpha)$ is a quantity $\mathcal{O}(1)$, $\tilde{z}_C(\alpha)$ is the single particle partition function

$$\tilde{z}_C(\alpha) = \Gamma\left(\frac{1}{2}\right) \left(\frac{\beta\mu}{2}\right)^{-1/4} \exp(\frac{1}{4}\alpha^2\theta^2) D_{-1/2}(\alpha\theta), \qquad (3.76)$$

Γ is the Euler function, $D_{-1/2}$ is a parabolic cylinder function and

$$\theta = \left(\frac{\beta}{2\mu}\right)^{1/2}. \qquad (3.77)$$

The final result in parametric form of the average Ricci curvature of $(M \times \mathbb{R}^2, g_E)$, with the constant energy constraint, is (details can be found in [CP93])

$$\Omega_0(\varepsilon) \to \begin{cases} \langle k_R \rangle(\theta) = 2 + \frac{3}{\theta} \frac{D_{-3/2}(\theta)}{D_{-1/2}(\theta)} \\ \varepsilon(\theta) = \frac{1}{8\sigma} \left[\frac{3}{\theta^2} + \frac{1}{\theta} \frac{D_{-3/2}(\theta)}{D_{-1/2}(\theta)}\right] \end{cases}. \qquad (3.78)$$

3.2 Riemannian Approach to Chaos

Let us now compute

$$\sigma_\Omega^2(\varepsilon) = \frac{1}{N}\langle \delta^2 K_R\rangle_\mu(\varepsilon) = \frac{1}{N}\langle (K_R - \langle K_R\rangle)^2\rangle_\mu.$$

According to (3.70), first the Gibbsian average of this quantity, $\langle \delta^2 k_R\rangle^G(\beta) = \frac{1}{N}\langle (K_R - \langle K_R\rangle)^2\rangle^G(\beta)$, has to be computed and then the correction term must be added. Now define

$$Q = \sum_{i=1}^{N}(q_{i+1} - q_i)^2;$$

after (3.72),

$$\frac{1}{N}\langle \delta^2 K_R\rangle^G(\beta) = \frac{1}{N}\langle (K_R - \langle K_R\rangle)^2\rangle^G = \frac{36\mu^2}{N}\langle (Q - \langle Q\rangle)^2\rangle^G, \qquad (3.79)$$

hence using (3.73)

$$\langle (Q - \langle Q\rangle)^2\rangle^G = \frac{4}{\beta^2}\left[\frac{\partial^2}{\partial\alpha^2}\log \tilde{Z}_C(\alpha)\right]_{\alpha=1}, \qquad (3.80)$$

and finally

$$\frac{1}{N}\langle \delta^2 K_R\rangle^G = \frac{144\mu^2}{\beta^2}\left[\frac{\partial^2}{\partial\alpha^2}\log \tilde{z}_C(\alpha)\right]_{\alpha=1}. \qquad (3.81)$$

Simple algebra gives

$$\left[\frac{\partial^2}{\partial\alpha^2}\log \tilde{z}_C(\alpha)\right]_{\alpha=1} = \frac{\theta^2}{4}\left\{2 - 2\theta\frac{D_{-3/2}(\theta)}{D_{-1/2}(\theta)} - \left[\frac{D_{-3/2}(\theta)}{D_{-1/2}(\theta)}\right]^2\right\}, \qquad (3.82)$$

so that from (3.81) we obtain

$$\frac{1}{N}\langle \delta^2 K_R\rangle^G(\theta) = \frac{9}{\theta^2}\left\{2 - 2\theta\frac{D_{-3/2}(\theta)}{D_{-1/2}(\theta)} - \left[\frac{D_{-3/2}(\theta)}{D_{-1/2}(\theta)}\right]^2\right\}. \qquad (3.83)$$

According to the prescription of (3.70), the final result for the fluctuations of Ricci curvature is [CC96]

$$\sigma_\Omega^2(\varepsilon) \to \begin{cases} \frac{1}{N}\langle \delta^2 K_R\rangle_\mu(\theta) = \frac{1}{N}\langle \delta^2 K_R\rangle^G(\theta) - \frac{\beta^2}{c_V(\theta)}\left(\frac{\partial \langle k_R\rangle(\theta)}{\partial \beta}\right)^2 \\ \varepsilon(\theta) = \frac{1}{8\mu}\left[\frac{3}{\theta^2} + \frac{1}{\theta}\frac{D_{-3/2}(\theta)}{D_{-1/2}(\theta)}\right] \end{cases} \qquad (3.84)$$

where $\langle \delta^2 K_R\rangle^G(\theta)$ is given by (3.83), the derivative part of the correction term is

$$\frac{\partial \langle k_R \rangle (\theta)}{\partial \beta} = \frac{3}{8\mu \theta^3} \frac{\theta D_{-3/2}^2(\theta) + 2(\theta^2-1)D_{-1/2}(\theta)D_{-3/2}(\theta) - 2\theta D_{-1/2}^2(\theta)}{D_{-1/2}^2(\theta)}, \tag{3.85}$$

and the specific heat per particle c_V is found to be

$$c_V(\theta) = \frac{1}{16 D_{-1/2}^2(\theta)} \left\{ (12 + 2\theta^2) D_{-1/2}^2(\theta) + 2\theta D_{-1/2}(\theta) D_{-3/2}(\theta) \right.$$
$$\left. - \theta^2 D_{-3/2}(\theta) \left[2\theta D_{-1/2}(\theta) + D_{-3/2}(\theta) \right] \right\}. \tag{3.86}$$

Analytic result for $\lambda_1(\varepsilon)$ and its comparison with numeric results

Now we use (3.78) and (3.84) to compute τ according to its definition in (3.67c), then we substitute $\Omega_0(\varepsilon)$, $\sigma_\Omega^2(\varepsilon)$ and $\tau(\varepsilon)$ into (3.67) to obtain the analytic prediction for $\lambda_1(\varepsilon)$ in the limit $N \to \infty$.

A Chain of Coupled Rotators

Let us now consider the system described by the Hamiltonian [CC96]

$$\mathcal{H}(p,q) = \sum_{i=1}^{N} \left\{ \frac{p_i^2}{2} + J[1 - \cos(q_{i+1} - q_i)] \right\}. \tag{3.87}$$

If the canonical coordinates q_i and p_i are given the meaning of angular coordinates and momenta, this Hamiltonian describes a linear chain of N rotators constrained to rotate on a plane and coupled by a nearest-neighbor interaction.

This model can be formally obtained by restricting to one spatial dimension the classical Heisenberg model whose potential energy is $V = -J \sum_{\langle i,j \rangle} \mathbf{S}_i \cdot \mathbf{S}_j$, where the sum is extended only over nearest-neighbor pairs, J is the coupling constant and each $\mathbf{S}_i$ has unit module and rotates on a plane. To each "spin" $\mathbf{S}_i = (\cos q_i, \sin q_i)$ the velocity $\frac{d}{dt}\mathbf{S}_i = (-\frac{dq_i}{dt}\sin q_i, \frac{dq_i}{dt}\cos q_i)$ is associated so that (3.87) follows from $\mathcal{H} = \sum_{i=1}^{N} \frac{1}{2}\dot{\mathbf{S}}_i^2 - J\sum_{\langle i,j \rangle} \mathbf{S}_i \cdot \mathbf{S}_j$.

The Hamiltonian (3.87) has two integrable limits. In the limit of vanishing energy it represents a chain of harmonic oscillators

$$\mathcal{H}(p,q) \simeq \sum_{i=1}^{N} \left\{ \frac{p_i^2}{2} + J(q_{i+1} - q_i)^2 \right\}, \tag{3.88}$$

whereas in the limit of indefinitely growing energy a system of freely rotating objects is found because of potential boundedness.

The expression of Ricci curvature K_R, computed with Eisenhart metric, is

$$K_R = \sum_{i=1}^{N} \frac{\partial^2 V(q)}{\partial q_i^2} = 2J \sum_{i=1}^{N} \cos(q_{i+1} - q_i).$$

Let us observe that for this model a relation exists between potential energy V and Ricci curvature K_R:

$$V(q) = JN - \frac{K_R}{2}. \qquad (3.89)$$

This relation binds the fluctuating quantity that enters the analytic formula for λ_1. This constraint does not exist for the sectional curvature thus a-priori it may be expected that some problem will arise.

The configurational partition function for a chain of coupled rotators is [CC96]

$$\begin{aligned} Z_C(\beta) &= \int_{-\pi}^{\pi} \prod_{i=1}^{N} dq_i \exp\left\{-\beta \sum_{i=1}^{N} J[1 - \cos(q_{i+1} - q_i)]\right\} \\ &= \exp(-\beta JN) \int_{-\pi}^{\pi} \prod_{i=1}^{N} d\omega_i \exp(\beta J \sum_{i=1}^{N} \cos \omega_i) \qquad (3.90) \\ &= \exp(-\beta JN)[I_0(\beta J)]^N (2\pi)^N g(\overline{\omega}) \,. \end{aligned}$$

where $I_0(x) = \frac{1}{\pi} \int_0^{+\pi} e^{x \cos \theta} d\theta$ is the modified Bessel function of index zero; $\omega_i = q_{i+1} - q_i$, $i \in (1, \ldots, N-1)$, $\omega_N = \overline{q} - q_N$, $\overline{q} = \overline{\omega}$ depend on the initial conditions. The function $g(\overline{\omega})$ contributes with a term of $\mathcal{O}(\frac{1}{N})$ thus vanishing in the thermodynamic limit.

In order to compute Ω_0 and σ_Ω^2 we follow the same procedure adopted for the FPU model, i.e., we define

$$\begin{aligned} \tilde{Z}_C(\alpha) &= \int_{-\pi}^{+\pi} \prod_{i=1}^{N} dq_i \exp\left\{-\beta \sum_{i=1}^{N} [1 - \alpha \cos(q_{i+1} - q_i)]\right\} \\ &= \exp(-\beta JN) \left[I_0(\beta J \alpha)\right]^N g(\overline{\omega})(2\pi)^N. \end{aligned}$$

and by observing that

$$\langle k_R \rangle_\mu (\beta) = \frac{2}{N\beta} \left[\frac{\partial}{\partial \alpha} \log \tilde{Z}_C(\alpha)\right]_{\alpha=1} . \qquad (3.91)$$

we find $\Omega_0(\varepsilon)$ in parametric form

$$\Omega_0(\varepsilon) \to \begin{cases} \langle k_R \rangle_\mu (\beta) = 2J \frac{I_0(\beta J)}{I_1(\beta J)} \\ \varepsilon(\beta) = \frac{1}{2\beta} + J \left(1 - \frac{I_1(\beta J)}{I_0(\beta J)}\right). \end{cases} \qquad (3.92)$$

In order to work out the average of the square fluctuations of Ricci curvature we use the following identity

$$\frac{1}{N} \langle \delta^2 K_R \rangle^G = \frac{4}{\beta^2 N} \left[\frac{\partial^2}{\partial \alpha^2} \log \tilde{Z}_C(\alpha)\right]_{\alpha=1},$$

whence

$$\frac{1}{N}\langle\delta^2 K_R\rangle^G = 4J^2 \frac{\beta J I_0^2(\beta J) - I_1(\beta J)I_0(\beta J) - \beta J I_1^2(\beta J)}{\beta J I_0^2(\beta J)}.$$

The computation of the correction term $\left[\frac{\partial\langle k_R\rangle(\beta)}{\partial\beta}\right]^2 / \frac{\partial\varepsilon(\beta)}{\partial\beta}$ involves the following derivatives

$$\frac{\partial\varepsilon(\beta)}{\partial\beta} = -\frac{1}{2\beta} - J^2\left\{1 - \frac{1}{\beta J}\frac{I_1(\beta J)}{I_0(\beta J)} - \left[\frac{I_1(\beta J)}{I_0(\beta J)}\right]^2\right\},$$

$$\frac{\partial\langle k_R\rangle(\beta)}{\partial\beta} = 2J^2\left\{1 - \frac{1}{\beta J}\frac{I_1(\beta J)}{I_0(\beta J)} - \left[\frac{I_1(\beta J)}{I_0(\beta J)}\right]^2\right\}.$$

Finally, gluing together the different terms, we obtain

$$\sigma_\Omega^2(\varepsilon) \to \begin{cases} \frac{1}{N}\langle\delta^2 K_R\rangle(\beta) = \frac{4J}{\beta}\frac{\beta J I_0^2(\beta J) - I_0(\beta J)I_1(\beta J) - \beta J I_1^2(\beta J)}{I_0^2(\beta J)[1+2(\beta J)^2] - 2\beta J I_1(\beta J)I_0(\beta J) - 2[\beta J I_1(\beta J)]^2} \\ \varepsilon(\beta) = \frac{1}{2\beta} + J\left[1 - \frac{I_1(\beta J)}{I_0(\beta J)}\right]. \end{cases}$$
(3.93)

Analytic result for $\lambda_1(\varepsilon)$ and its comparison with numeric results

By inserting into (3.67) the analytic expressions of $\Omega_0(\varepsilon)$ and $\sigma_\Omega^2(\varepsilon)$ given in (3.92) and (3.93) – and also $\tau(\varepsilon)$ which is a function of the latter quantities – we find $\lambda_1(\varepsilon)$.

Using Eisenhart metric, the explicit expression of the sectional curvature $K(v,\xi)$, relative to the plane spanned by the velocity vector v and a generic vector $\xi \perp v$ (here we use ξ to denote the geodesic separation vector in order to avoid confusion with J which is the notation for the coupling constant), is [CC96]

$$K(v,\xi) = R_{0i0k}\frac{dq^0}{dt}\frac{\xi^i}{\|\xi\|}\frac{dq^0}{dt}\frac{\xi^k}{\|\xi\|} \equiv \frac{\partial^2 V}{\partial q^i \partial q^k}\frac{\xi^i\xi^k}{\|\xi\|^2},$$

hence we get

$$K(v,\xi) = \frac{J}{\|\xi\|^2}\sum_{i=1}^{N}\cos(q_{i+1} - q_i)\left[\xi^{i+1} - \xi^i\right]^2 \qquad (3.94)$$

for the coupled rotators model. We realize, by simple inspection of (3.94), that K can take negative values with non-vanishing probability regardless of the value of ε, whereas – as long as $\varepsilon < J$ – this possibility is lost in the replacement of K by Ricci curvature that we adopted in our theory. In fact, because of the constraint (3.89), at each point of the manifold it is $k_R(\varepsilon) \geq 2(J - \varepsilon)$,

3.2 Riemannian Approach to Chaos

thus our approximation fails in accounting for the presence of negative sectional curvatures at small values of ε. In (3.94) the cosines have different and variable weights, $[\xi^{i+1}-\xi^i]^2$, that in principle make possible to find somewhere along a geodesic $K < 0$ also with only one negative cosine. This is not the case of k_R where all the cosines have the same weight. Therefore the probability of finding $K < 0$ along a geodesic must be related to the probability of finding an angular difference greater than $\frac{\pi}{2}$ between two nearest-neighboring rotators. If the energy is sufficiently low this event will be very unlikely, but we can guess that it will become considerable where the theoretical prediction is not satisfactory, i.e., when chaos is strong. Notice that the frequent occurrence of $K < 0$ along a geodesic adds to parametric instability another instability mechanism that enforces chaos [(3.34)].

Our strategy is to modify the model for $K(s)$ in some *effective* way that takes into account the mentioned difficulty of $k_R(s)$ to adequately model $K(s)$. This will be achieved by suitably 're-normalizing' Ω_0 or σ_Ω to obtain an *effective gaussian process* for the behavior of the sectional curvature.

From (3.94) we see that N directions of the vector ξ exist such that the sectional curvatures – relative to the N planes spanned by these vectors together with v – are just $\cos(q_{i+1} - q_i)$. Hence the probability $P(\varepsilon)$ of occurrence of a negative value of the cosine is used to estimate the probability of occurrence of negative sectional curvatures along the geodesics. This probability function has the following simple expression [CC96]

$$P(\varepsilon) = \frac{\int_{-\pi}^{\pi} \Theta(-\cos x) e^{\beta J \cos x} dx}{\int_{-\pi}^{\pi} e^{\beta J \cos x} dx} = \frac{\int_{\frac{\pi}{2}}^{\frac{3\pi}{2}} e^{\beta J \cos x} dx}{2\pi I_0(\beta J)}, \qquad (3.95)$$

where $\Theta(x)$ is the Heaviside unit step function.

The function $P(\varepsilon)$ begins to increase at $\varepsilon \simeq 0.2$, just where the analytic prediction begins to fail, and when it approaches its asymptotic value of $\frac{1}{2}$, around the end of the knee, a good agreement is again found between theory and numeric results. The simplest way to account for the existence of negative sectional curvatures is to shift the peak of the distribution $\mathcal{P}(\delta K_R)$ toward the negative axis. This is achieved by the replacement

$$\langle k_R(\varepsilon) \rangle \to \frac{\langle k_R(\varepsilon) \rangle}{1 + \alpha P(\varepsilon)}. \qquad (3.96)$$

This correction neither has influence when $P(\varepsilon) \simeq 0$ (below $\varepsilon \simeq 0.2$) nor when $P(\varepsilon) \simeq 1/2$ (because in this case $\langle k_R(\varepsilon) \rangle \to 0$). The value of the parameter α in (3.96) must be estimated *a posteriori* in order to obtain the best agreement between numerical and theoretical data over the whole range of energies.

3.3 Morse Topology of Smooth Manifolds

Recall that *topology* is a kind of *abstraction of Euclidean geometry*, and also a natural framework for the study of *continuity*.[27] Euclidean geometry is abstracted by regarding triangles, circles, and squares as being the same basic object. Continuity enters because in saying this one has in mind a *continuous deformation* of a triangle into a square or a circle, or any arbitrary shape. On the other hand, a disk with a hole in the center is topologically different from a circle or a square because one cannot create or destroy holes by continuous deformations. Thus using topological methods one does not expect to be able to identify a geometrical figure as being a triangle or a square. However, one does expect to be able to detect the presence of gross features such as holes or the fact that the figure is made up of two disjoint pieces etc. In this way topology produces theorems that are usually qualitative in nature – they may assert, for example, the existence or non–existence of an object. They will not, in general, give the means for its construction [Nas83, II07b].

3.3.1 Intro to Euler Characteristic and Morse Topology

The Euler Characteristic

The so–called *Euler characteristic* is a *topological invariant*, a number that describes one aspect of a topological space's shape or structure. It is commonly denoted by the Greek letter χ. The Euler characteristic was originally defined for polyhedra and used to prove various theorems about them, including the classification of the Platonic solids. Leonhard Euler, for whom the concept is named, was responsible for much of this early work. In modern mathematics, the Euler characteristic arises from *homology theory* and connects to many other topological invariants.

The Euler characteristic χ was classically defined for polyhedra, according to the formula
$$\chi = V - E + F,$$
where $V, E,$ and F are respectively the numbers of *vertices* (corners), *edges* and *faces* in the given polyhedron. Any convex polyhedron is *homeomorphic* to a sphere S^2, so its Euler characteristic is
$$\chi = V - E + F = 2.$$

---

[27] Intuitively speaking, a function $f : \mathbb{R} \to \mathbb{R}$ is continuous near a point x in its domain if its value does not jump there. That is, if we just take δx to be small enough, the two function values $f(x)$ and $f(x + \delta x)$ should approach each other arbitrarily closely. In more rigorous terms, this leads to the following definition: A function $f : \mathbb{R} \to \mathbb{R}$ is continuous at $x \in \mathbb{R}$ if for all $\epsilon > 0$, there exists a $\delta > 0$ such that for all $y \in \mathbb{R}$ with $|y - x| < \delta$, we have that $|f(y) - f(x)| < \epsilon$. The whole function is called continuous if it is continuous at every point x.

3.3 Morse Topology of Smooth Manifolds

This result is known as *Euler's formula*, and can be applied not only to polyhedra but also to embedded planar *graphs*. A proof is given inductively below.

For a tetrahedron (with $V = 4$, $E = 6$, $F = 4$) we have $\chi = 2$.
For a cube (or, hexahedron, with $V = 8$, $E = 12$, $F = 6$) we have $\chi = 2$.
For a octahedron (with $V = 6$, $E = 12$, $F = 8$) we have $\chi = 2$.
For a dodecahedron (with $V = 20$, $E = 30$, $F = 12$) we have $\chi = 2$.
For a icosahedron (with $V = 12$, $E = 30$, $F = 20$) we have $\chi = 2$.

The first rigorous proof of Euler's formula, was given by a 20-year-old A. Cauchy.

If M and N are any two topological spaces, then the Euler characteristic of their *disjoint union* (denoted by $\uplus$) is the sum of their Euler characteristics:

$$\chi(M \uplus N) = \chi(M) + \chi(N).$$

Also, the Euler characteristic of any product space $M \times N$ is

$$\chi(M \times N) = \chi(M) \cdot \chi(N).$$

These addition and multiplication properties are also enjoyed by cardinality of sets. In this way, the Euler characteristic can be viewed as a generalization of cardinality.

As a corollary of the so–called *Poincaré duality*, the Euler characteristic of any closed odd–dimensional manifold is zero.

The Euler characteristic of a closed orientable surface can be calculated from its *genus* g[28] as

$$\chi = 2 - 2g.$$

For closed Riemannian manifolds, the Euler characteristic can be found by integrating the curvature, via the *Gauss–Bonnet theorem* (3.3). A discrete analog of the Gauss–Bonnet theorem is Descartes' theorem that the 'total defect' of a polyhedron, measured in full circles, is the Euler characteristic of the polyhedron.

For any contractible space its Euler characteristic is 1. This case includes Euclidean space $\mathbb{R}^n$ of any dimension, as well as the solid unit ball in any Euclidean space: the 1D interval, the 2D disk, the 3D ball, etc.

As was said before, for a sphere S^2, $\chi = 2$.
For a torus T^2, $\chi = 0$.
For a double torus, $\chi = -2$.
For a triple torus, $\chi = -4$.
For both a Möbius strip and a Klein bottle, $\chi = 0$.

---

[28] Genus g of a closed orientable surface is the number of tori in a connected sum decomposition of the surface. Intuitively, it can be depicted as the number of handles.

Morse Topology: an Intuitive Picture

On the other hand, Morse theory gives a very direct way of analyzing the topology of a manifold by studying smooth functions on it. Consider, for purposes of illustration, a mountainous landscape M. If f is the function $M \to \mathbb{R}$ sending each point to its elevation, then the inverse image of a point in $\mathbb{R}$ (a level set) is simply a contour line. Each connected component of a contour line is either a point, a simple closed curve, or a closed curve with a double point. Contour lines may also have points of higher order (triple points, etc.), but these are unstable and may be removed by a slight deformation of the landscape. Double points in contour lines occur at saddle points, or passes. Saddle points are points where the surrounding landscape curves up in one direction and down in the other.

Now, imagine flooding this landscape with water. Then, assuming the ground is porous, the region covered by water when the water reaches an elevation of a is $f^{-1}(-\infty, a]$, or the points with elevation less than or equal to a. Consider how the topology of this region changes as the water rises. It appears, intuitively, that it does not change except when a passes the height of a critical point; that is, a point where the gradient of f is 0. In other words, it does not change except when the water either (i) starts filling a basin, (ii) covers a saddle (a mountain pass), or (iii) submerges a peak.

To each of these three types of *critical points*: basins, passes, and peaks (also called minima, saddles, and maxima), one associates a number called the *index*. Intuitively speaking, the *index* of a critical point b is the number of independent directions around b in which f decreases. Therefore, we have natural indices of basins, passes, and peaks as 0, 1, and 2, respectively.

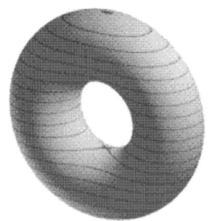

Define M^a as $f^{-1}(-\infty, a]$. Leaving the context of topography, one can make a similar analysis of how the topology of M^a changes as a increases when M is a torus oriented as in the image above and f is projection on a vertical axis, taking a point to its height above the plane.

3.3 Morse Topology of Smooth Manifolds

When a is less than 0, M^a is the empty set. After a passes the level of p (a critical point of index 0), when $0 < a < f(q)$, then M^a is a disk, which is *homotopy equivalent* to a point, (or, a 0–cell) which has been 'attached' to the empty set. Next, when a exceeds the level of q (a critical point of index 1), and $f(q) < a < f(r)$, then M^a is a cylinder, and is homotopy equivalent to a disk with a 1–cell attached. Once a passes the level of r (a critical point of index 1), and $f(r) < a < f(s)$, then M^a is a torus with a disk removed, which is homotopy equivalent to a cylinder with a 1–cell attached. Finally, when a is greater than the critical level of s (a critical point of index 2) M^a is our original torus.

We therefore appear to have the following rule: the topology of M^a does not change except when a passes the height of a critical point, and when a passes the height of a critical point of index γ, a γ–cell is attached to M^a. This does not address the question of what happens when two critical points are at the same height. That situation can be resolved by a slight perturbation of f. In the case of a landscape (or, a manifold embedded in Euclidean space), this perturbation might simply be tilting the landscape slightly, or rotating the coordinate system.[29]

The *Euler characteristic* $\chi(M)$ of a manifold M can be defined as

$$\chi(M) = \sum (-1)^\gamma C^\gamma,$$

where C^γ is the number of critical points of index γ.

3.3.2 Sets and Topological Spaces

Sets and Maps between Them

Given a map (or, a function) $f : A \to B$, the set A is called the *domain* of f, and denoted $\mathrm{Dom}\, f$. The set B is called the *codomain* of f, and denoted $\mathrm{Cod}\, f$. The codomain is not to be confused with the *range* of $f(A)$, which is in general only a subset of B.

A map $f : X \to Y$ is called *injective* or 1–1 or an *injection* if for every y in the codomain Y there is at most one x in the domain X with $f(x) = y$. Put another way, given x and x' in X, if $f(x) = f(x')$, then it follows that $x = x'$. A map $f : X \to Y$ is called *surjective* or *onto* or a *surjection* if for every y in the codomain $\mathrm{Cod}\, f$ there is at least one x in the *domain* X with $f(x) = y$. Put another way, the *range* $f(X)$ is equal to the codomain Y. A map

---

[29] This rule, however, is false as stated. To see this, let M equal $\mathbb{R}$ and let $f(x) = x^3$. Then 0 is a critical point of f, but the topology of M^a does not change when a passes 0. In fact, the concept of index does not make sense. The problem is that the second derivative is also 0 at 0. This kind of situation is called a *degenerate critical point*. Note that this situation is unstable: by rotating the coordinate system under the graph, the degenerate critical point either is removed or breaks up into two non-degenerate critical points.

is *bijective* iff it is both injective and surjective. Injective functions are called the *monomorphisms*, and surjective functions are called the *epimorphisms* in the *category of sets* (see below).

Two main classes of maps (or, functions) that we will use int this book are: (i) continuous maps (denoted as C^0−class), and (ii) smooth or differentiable maps (denoted as C^∞−class). The former class is the core of topology, the later of differential geometry. They are both used in the core concept of manifold.

A *relation* is any subset of a *Cartesian product* (see below). By definition, an *equivalence relation* α on a set X is a relation which is *reflexive, symmetrical* and *transitive*, i.e., relation that satisfies the following three conditions:

A. *Reflexivity*: each element $x \in X$ is equivalent to itself, i.e., $x\alpha x$,
B. *Symmetry*: for any two elements $x, x' \in X$, $x\alpha x'$ implies $x'\alpha x$, and
C. *Transitivity*: $a \leq b$ and $b \leq c$ implies $a \leq c$.

Similarly, a relation $\leq$ defines a *partial order* on a set S if it has the following properties:

A. *Reflexivity*: $a \leq a$ for all $a \in S$,
B. *Antisymmetry*: $a \leq b$ and $b \leq a$ implies $a = b$, and
C. *Transitivity*: $a \leq b$ and $b \leq c$ implies $a \leq c$.

A *partially ordered set* (or *poset*) is a set taken together with a partial order on it. Formally, a partially ordered set is defined as an ordered pair $P = (X, \leq)$, where X is called the *ground set* of P and $\leq$ is the partial order of P.

Recall that a *map* (or, *function*) f is a *rule* that assigns to each element x in a set A exactly one element, called $f(x)$, in a set B. A map could be thought of as a *machine* $[[f]]$ with x−input (the *domain* of f is the set of all possible inputs) and $f(x)$−output (the *range* of f is the set of all possible outputs) [Stu99]

$$x \to [[f]] \to f(x)$$

There are four possible ways to represent a function (or map): (i) verbally (by a description in words); (ii) numerically (by a table of values); (iii) visually (by a graph); and (iv) algebraically (by an explicit formula). The most common method for visualizing a function is its *graph*. If f is a function with domain A, then its graph is the set of ordered input–output pairs

$$\{(x, f(x)) : x \in A\}.$$

A generalization of the graph concept is a concept of a *cross–section of a fibre bundle*, which is one of the core geometrical objects for dynamics of complex systems.

Let f and g be maps with domains A and B. Then the maps $f + g$, $f - g$, fg, and f/g are defined as follows:

3.3 Morse Topology of Smooth Manifolds

$$(f+g)(x) = f(x) + g(x) \quad \text{domain} = A \cap B,$$
$$(f-g)(x) = f(x) - g(x) \quad \text{domain} = A \cap B,$$
$$(fg)(x) = f(x)g(x) \quad \text{domain} = A \cap B,$$
$$\left(\frac{f}{g}\right)(x) = \frac{f(x)}{g(x)} \quad \text{domain} = \{x \in A \cap B : g(x) \neq 0\}.$$

Given two maps f and g, the composite map $f \circ g$ (also called the *composition* of f and g) is defined by

$$(f \circ g)(x) = f(g(x)).$$

The $(f \circ g)$–machine is composed of the g–machine (first) and then the f–machine:

$$x \to [[g]] \to g(x) \to [[f]] \to f(g(x))$$

For example, suppose that $y = f(u) = \sqrt{u}$ and $u = g(x) = x^2 + 1$. Since y is a function of u and u is a function of x, it follows that y is ultimately a function of x. We calculate this by substitution

$$y = f(u) = f \circ g = f(g(x)) = f(x^2 + 1) = \sqrt{x^2 + 1}.$$

If f and g are both differentiable (or smooth, i.e., C^∞) maps and $h = f \circ g$ is the composite map defined by $h(x) = f(g(x))$, then h is differentiable and h' is given by the product:

$$h'(x) = f'(g(x)) g'(x).$$

In Leibniz notation, if $y = f(u)$ and $u = g(x)$ are both differentiable maps, then

$$\frac{dy}{dx} = \frac{dy}{du} \frac{du}{dx}.$$

The reason for the name *chain rule* becomes clear if we add another link to the chain. Suppose that we have one more differentiable map $x = h(t)$. Then, to calculate the derivative of y with respect to t, we use the chain rule twice,

$$\frac{dy}{dt} = \frac{dy}{du} \frac{du}{dx} \frac{dx}{dt}.$$

1–1 continuous (i.e., C^0) map T with a nonzero *Jacobian* $\left|\frac{\partial(x,...)}{\partial(u,...)}\right|$ that maps a region S onto a region R, (see [Stu99]) we have the following substitution formulas:

1. for a single integral,

$$\int_R f(x)\,dx = \int_S f(x(u)) \frac{\partial x}{\partial u} du,$$

2. for a double integral,
$$\iint_R f(x,y)\,dA = \iint_S f(x(u,v),y(u,v)) \left|\frac{\partial(x,y)}{\partial(u,v)}\right| du\,dv,$$

3. for a triple integral,
$$\iiint_R f(x,y,z)\,dV = \iiint_S f(x(u,v,w),y(u,v,w),z(u,v,w)) \left|\frac{\partial(x,y,z)}{\partial(u,v,w)}\right| du\,dv\,dw$$

4. similarly for n-tuple integrals.

Topological Spaces

Study of topology starts with the fundamental notion of *topological space*. Let X be any *set* and $Y = \{X_\alpha\}$ denote a collection, finite or infinite of subsets of X. Then X and Y form a topological space provided the X_α and Y satisfy:

A. Any finite or infinite subcollection $\{Z_\alpha\} \subset X_\alpha$ has the property that $\cup Z_\alpha \in Y$, and
B. Any *finite subcollection* $\{Z_{\alpha_1}, ..., Z_{\alpha_n}\} \subset X_\alpha$ has the property that $\cap Z_{\alpha_i} \in Y$.

The set X is then called a topological space and the X_α are called *open sets*. The choice of Y satisfying (2) is said to give a topology to X.

Given two topological spaces X and Y, a *function* (or, a *map*) $f : X \to Y$ is *continuous* if the inverse image of an open set in Y is an open set in X.

The main general idea in topology is to study spaces which can be continuously deformed into one another, namely the idea of *homeomorphism*. If we have two topological spaces X and Y, then a map $f : X \to Y$ is called a homeomorphism iff

A. f is continuous (C^0), and
B. There exists an inverse of f, denoted f^{-1}, which is also continuous.

Definition (2) implies that if f is a homeomorphism then so is f^{-1}. Homeomorphism is the main topological example of *reflexive, symmetrical* and *transitive relation*, i.e., *equivalence relation*. Homeomorphism divides all topological spaces up into *equivalence classes*. In other words, a pair of topological spaces, X and Y, belong to the same equivalence class if they are homeomorphic.

The second example of topological equivalence relation is *homotopy*. While homeomorphism generates equivalence classes whose members are topological spaces, homotopy generates equivalence classes whose members are continuous (C^0) maps. Consider two continuous maps $f, g : X \to Y$ between topological spaces X and Y. Then the map f is said to be *homotopic* to the map g if f can be continuously deformed into g (see below for the precise definition of homotopy). Homotopy is an equivalence relation which divides the space of continuous maps between two topological spaces into equivalence classes [Nas83].

Another important notions in topology are *covering, compactness* and *connectedness*. Given a family of sets $\{X_\alpha\} = X$ say, then X is a *covering* of another set Y if $\cup X_\alpha$ contains Y. If all the X_α happen to be open sets the covering is called an *open covering*. Now consider the set Y and all its possible open coverings. The set Y is *compact* if for every open covering $\{X_\alpha\}$ with $\cup X_\alpha \supset Y$ there always exists a finite sub-covering $\{X_1, ..., X_n\}$ of Y with $X_1 \cup ... \cup X_n \supset Y$. Again, we define a set Z to be *connected* if it cannot be written as $Z = Z_1 \cup Z_2$, where Z_1 and Z_2 are both open and $Z_1 \cap Z_2$ is an empty set.

Let $A_1, A_2, ..., A_n$ be closed subspaces of a topological space X such that $X = \cup_{i=1}^n A_i$. Suppose $f_i : A_i \to Y$ is a function, $1 \leq i \leq n$, iff

$$f_i|A_i \cap A_j = f_j|A_i \cap A_j, 1 \leq i, j \leq n. \tag{3.97}$$

In this case f is continuous iff each f_i is. Using this procedure we can define a C^0–function $f : X \to Y$ by cutting up the space X into closed subsets A_i and defining f on each A_i separately in such a way that $f|A_i$ is obviously continuous; we then have only to check that the different definitions agree on the *overlaps* $A_i \cap A_j$.

The *universal property of the Cartesian product*: let $p_X : X \times Y \to X$, and $p_Y : X \times Y \to Y$ be the *projections* onto the first and second factors, respectively. Given any pair of functions $f : Z \to X$ and $g : Z \to Y$ there is a unique function $h : Z \to X \times Y$ such that $p_X \circ h = f$, and $p_Y \circ h = g$. Function h is continuous iff both f and g are. This property characterizes X/α up to homeomorphism. In particular, to check that a given function $h : Z \to X$ is continuous it will suffice to check that $p_X \circ h$ and $p_Y \circ h$ are continuous.

The *universal property of the quotient*: let α be an equivalence relation on a topological space X, let X/α denote the *space of equivalence classes* and $p_\alpha : X \to X/\alpha$ the *natural projection*. Given a function $f : X \to Y$, there is a function $f' : X/\alpha \to Y$ with $f' \circ p_\alpha = f$ iff $x\alpha x'$ implies $f(x) = f(x')$, for all $x \in X$. In this case f' is continuous iff f is. This property characterizes X/α up to homeomorphism.

Betti Numbers and the Euler Characteristic

The *Betti number* of a topological space is, in intuitive terms, a way of counting the maximum number of cuts that can be made without dividing the space into two pieces. This defines, in fact, what is called the first Betti number. There is a sequence of *Betti numbers* defined. Each Betti number is a natural number, or infinity. For the most reasonable spaces (such as *compact manifolds*, finite *simplicial complexes* or *CW complexes*), the sequence of Betti numbers is 0 from some points onwards, and consists of natural numbers. For example, (i) the Betti number sequence for a circle S^1 is 1, 1, 0, 0, 0, ...; (ii) Betti number sequence for a two-torus T^2 is 1, 2, 1, 0, 0, 0, ...; (iii) Betti number sequence for a three-torus is 1, 3, 3, 1, 0, 0, 0, ... (iv) for an $n-$torus T^n one should see the binomial coefficients; this is a case of the Künneth theorem.

For any topological space X, we can define the nth Betti number b_n as the rank of the nth *homology group* (see next paragraph). The Euler characteristic can then be defined as the alternating sum

$$\chi = b_0 - b_1 + b_2 - b_3 + \cdots .$$

This quantity is well-defined if the Betti numbers are all finite and if they are zero beyond a certain index n_0.

Simplicial Homology

Simplicial homology concerns topological spaces whose building blocks are $n-$simplexes, the nD analogs of triangles. By definition, such a space is homeomorphic to a simplicial complex S (see Figure 3.4); more precisely, the geometric realization of an abstract simplicial complex. Such a homeomorphism is referred to as a triangulation of the given space. Replacing $n-$simplexes by their continuous images in a given topological space gives singular homology. The *simplicial homology* of a simplicial complex is naturally isomorphic to the *singular homology* of its geometric realization. This implies, in particular, that the simplicial homology of a space does not depend on the triangulation chosen for the space. For technical details on simplicial homology and related (co)homology theories, see [II07b]. Here we give a very short brief, necessary for the comprehensive reading of the following text.

Fig. 3.4. An example of a simplicial complex with two 1–holes.

Let S be a simplicial complex. A simplicial $k-chain$ is a formal sum of $k-$simplices:

$$\sum_{i=1}^{N} c_i \sigma^i.$$

The group of $k-$chains on S, the *free Abelian group* (see [II07b]) defined on the set of $k-$simplices in S, is denoted C_k. Consider a basis element of C_k, a $k-$simplex,

$$\sigma = \left\langle v^0, v^1, ..., v^k \right\rangle.$$

The so-called *boundary operator*

$$\partial_k : C_k \to C_{k-1}$$

is a *homomorphism* defined by:

$$\partial_k(\sigma) = \sum_{i=0}^{K}(-1)^i \left\langle v^0, ..., \hat{v}^i, ..., v^k \right\rangle,$$

where the simplex

$$\left\langle v^0, ..., \hat{v}^i, ..., v^k \right\rangle$$

is the ith face of σ obtained by deleting its ith vertex. In C_k, elements of the subgroup

$$Z_k = \ker \partial_k$$

are called *cycles*, and the subgroup

$$B_k = \operatorname{im} \partial_{k+1}$$

is said to consist of *boundaries*. Direct computation shows that B_k lies in Z_k. The *boundary of a boundary* must be a cycle. In other words, (C_k, ∂_k) form a simplicial *chain complex*.

The kth homology group H_k of S is now defined to be the *quotient space*

$$H_k(S) = Z_k / B_k.$$

A homology group H_k is not trivial if the complex at hand contains k–cycles which are not boundaries. This indicates that there are k–dimensional holes in the complex. For example consider the complex obtained by gluing together two triangles (with no interior) along one edge (see Figure 3.4). This is a triangulation of the figure eight. The edges of each triangle form a cycle. These two cycles are by construction not boundaries (there are no 2–chains). Therefore the figure has two '1–holes'.

Holes can be of different dimensions. The rank of the homology groups, that is, the numbers

$$\beta_k = \operatorname{rank}(H_k(S))$$

are called the *Betti numbers* of the space S, and gives a measure of the number of k–dimensional holes in S.

Topological Invariants of Manifolds

Now, restricting to the topology of nD compact (i.e., closed and bounded) and connected manifolds, the only cases in which we have a complete understanding of topology are $n = 0, 1, 2$. The only compact and connected 0D manifold is a point. A 1D compact and connected manifold can either be a line element or a circle, and it is intuitively clear (and can easily be proven) that these two spaces are topologically different. In 2D, there is already an infinite number of different topologies: a 2D compact and connected surface can have an arbitrary number of handles and boundaries, and can either be orientable or non–orientable (see figure 3.5). Again, it is intuitively quite clear that two surfaces are not homeomorphic if they differ in one of these respects. On the other hand, it can be proven that any two surfaces for which these data are the same can be continuously mapped to one another, and hence this gives a complete classification of the possible topologies of such surfaces.

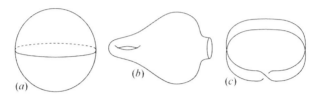

Fig. 3.5. Three examples of 2D manifolds: (*a*) The sphere S^2 is an orientable manifold without handles or boundaries. (*b*) An orientable manifold with one boundary and one handle. (*c*) The *Möbius strip* is an unorientable manifold with one boundary and no handles.

A quantity such as the number of boundaries of a surface is called a *topological invariant*. A topological invariant is a number, or more generally any type of structure, which one can associate to a topological space, and which does not change under continuous mappings. Topological invariants can be used to distinguish between topological spaces: if two surfaces have a different number of boundaries, they can certainly not be topologically equivalent. On the other hand, the knowledge of a topological invariant is in general not enough to decide whether two spaces are homeomorphic: a torus and a sphere have the same number of boundaries (zero), but are clearly not homeomorphic. Only when one has some *complete set* of topological invariants, such as the number of handles and boundaries in the 2D case, is it possible to determine whether or not two topological spaces are homeomorphic. In more than 2D, many topological invariants are known, but for no dimension larger than two has a complete set of topological invariants been found. In 3D, it is generally believed that a finite number of countable invariants would suffice for compact manifolds, but this is not rigorously proven, and in particular there is at present no generally accepted construction of a complete set. A very interesting and intimately related problem is the famous *Poincaré conjecture*, stating that if a 3D manifold has a certain set of topological invariants called its 'homotopy groups' equal to those of the 3–sphere S^3, it is actually homeomorphic to the three-sphere. In four or more dimensions, a complete set of topological invariants would consist of an uncountably infinite number of invariants! A general classification of topologies is therefore very hard to get, but even without such a general classification, each new invariant that can be constructed gives us a lot of interesting new information. For this reason, the construction of topological invariants of manifolds is one of the most important issues in topology.

Homotopy

Let I be a compact unit interval $I = [0,1]$. A *homotopy* from X to Y is a continuous function $F : X \times I \to Y$. For each $t \in I$ one has $F_t : X \to Y$ defined by $F_t(x) = F(x,t)$ for all $x \in X$. The functions F_t are called the 'stages' of the homotopy. If $f, g : X \to Y$ are two continuous maps, we say f is homotopic to g, and write $f \simeq g$, if there is a homotopy $F : X \times I \to Y$ such that $F_0 = f$ and $F_1 = g$. In other words, f can be continuously deformed into g through the stages F_t. If $A \subset X$ is a subspace, then F is a homotopy relative to A if $F(a,t) = F(a,0)$, for all $a \in A, t \in I$.

The homotopy relation $\simeq$ is an equivalence relation. To prove that we have $f \simeq f$ is obvious; take $F(x, t = f(x)$, for all $x \in X$, $t \in I$. If $f \simeq g$ and F is a homotopy from f to g, then $G : X \times I \to Y$ defined by $G(x,t) = F(x, 1-t)$, is a homotopy from g to f, i.e., $g \simeq f$. If $f \simeq g$ with homotopy F and $g \simeq f$ with homotopy G, then $f \simeq h$ with homotopy H defined by

$$H(x,t) = \begin{cases} F(x,t), & 0 \le t \le 1/2 \\ G(x, 2t-1), & 1/2 \le t \le 1 \end{cases}.$$

To show that H is continuous we use the relation (3.97).

In this way, the set of all C^0–functions $f : X \to Y$ between two topological spaces X and Y, called the *function space* and denoted by Y^X, is partitioned into equivalence classes under the relation $\simeq$. The equivalence classes are called *homotopy classes*, the homotopy class of f is denoted by $[f]$, and the set of all homotopy classes is denoted by $[X;Y]$.

If α is an equivalence relation on a topological space X and $F : X \times I \to Y$ is a homotopy such that each stage F_t factors through X/α, i.e., $x\alpha x'$ implies $F_t(x) = F_t(x')$, then F induces a homotopy $F' : (X/\alpha) \times I \to Y$ such that $F' \circ (p_\alpha \times 1) = F$.

Homotopy theory has a range of applications of its own, outside topology and geometry, as for example in proving Cauchy theorem in complex variable theory, or in solving nonlinear equations of artificial neural networks.

A *pointed set* (S, s_0) is a set S together with a distinguished point $s_0 \in S$. Similarly, a *pointed topological space* (X, x_0) is a space X together with a distinguished point $x_0 \in X$. When we are concerned with pointed spaces $(X, x_0), (Y, y_0)$, etc., we always require that all functions $f : X \to Y$ shell preserve base points, i.e., $f(x_0) = y_0$, and that all homotopies $F : X \times I \to Y$ be relative to the base point, i.e., $F(x_0, t) = y_0$, for all $t \in I$. We denote the homotopy classes of base point–preserving functions by $[X, x_0; Y, y_0]$ (where homotopies are relative to x_0). $[X, x_0; Y, y_0]$ is a pointed set with base point f_0, the constant function: $f_0(x) = y_0$, for all $x \in X$.

A *path* $\gamma(t)$ from x_0 to x_1 in a topological space X is a continuous map $\gamma : I \to X$ with $\gamma(0) = x_0$ and $\gamma(1) = x_1$. Thus X^I is the space of all paths in X with the compact–open topology. We introduce a relation $\sim$ on X by saying $x_0 \sim x_1$ iff there is a path $\gamma : I \to X$ from x_0 to x_1. $\sim$ is clearly an equivalence relation, and the set of equivalence classes is denoted by $\pi_0(X)$. The elements of $\pi_0(X)$ are called the *path components*, or $0-$*components* of X. If $\pi_0(X)$ contains just one element, then X is called *path connected*, or $0-$*connected*. A *closed path*, or *loop* in X at the point x_0 is a path $\gamma(t)$ for which $\gamma(0) = \gamma(1) = x_0$. The *inverse loop* $\gamma^{-1}(t)$ based at $x_0 \in X$ is defined by $\gamma^{-1}(t) = \gamma(1-t)$, for $0 \le t \le 1$. The *homotopy of loops* is the particular case of the above defined homotopy of continuous maps.

Functors

In *algebraic topology*, one attempts to assign to every topological space X some algebraic object $\mathcal{F}(X)$ in such a way that to every C^0-function $f : X \to Y$ there is assigned a homomorphism $\mathcal{F}(f) : \mathcal{F}(X) \to \mathcal{F}(Y)$. One advantage of this procedure is, e.g., that if one is trying to prove the non–existence of a C^0-function $f : X \to Y$ with certain properties, one may find it relatively easy to prove the non–existence of the corresponding algebraic function $\mathcal{F}(f)$ and hence deduce that f could not exist. In other words, $\mathcal{F}$ is to be a 'homomorphism' from one category (e.g., $\mathcal{T}$) to another (e.g., $\mathcal{G}$ or $\mathcal{A}$). Formalization of this notion is a *functor*, a generic *picture* projecting one category into another. See [II07b] for technical details on categories and functors.

For example, if (X, x_0) is a pointed space, then we may regard $\pi_0(X)$ as a pointed set with the $0-$component of x_0 as a base point. We use the notation $\pi_0(X, x_0)$ to denote $p_0(X, x_0)$ thought of as a pointed set. If $f : X \to Y$ is a map then f sends $0-$components of X into $0-$components of Y and hence defines a function $\pi_0(f) : \pi_0(X) \to \pi_0(Y)$. Similarly, a base point–preserving map $f : (X, x_0) \to (Y, y_0)$ induces a map of pointed sets $\pi_0(f) : \pi_0(X, x_0) \to \pi_0(Y, y_0)$. In this way defined π_0 represents a 'functor' from the 'category' of topological (point) spaces to the underlying category of (point) sets.

Combination of topology and calculus gives differential topology, or differential geometry (see [II07b]).

3.3.3 A Brief Intro to Morse Theory

At the same time the variational calculus was discovered, a related technique, called *Morse theory*, was introduced into *Riemannian geometry*. This theory was developed by M. Morse, first for functions on manifolds in 1925, and then in 1934, for the loop space. The latter theory, as we shall see, sets up a very nice connection between the first and second variation formulae from the previous section and the topology of M.

3.3 Morse Topology of Smooth Manifolds

The essence of *Morse theory* is to study the topology of a manifold M by analyzing the critical points of a smooth, real function $f : M \to \mathbb{R}$ (see [Mil63, Mil65a, Mil65]).

Let $f : M \to \mathbb{R}$ be a smooth function and $dim(M) = n$. A point $p \in M$ is a *critical point* of f if, in some coordinate system we have

$$\left.\frac{\partial f}{\partial x_1}\right|_p = \ldots = \left.\frac{\partial f}{\partial x_n}\right|_p = 0$$

and p is a *non-degenerate* critical point if

$$det\left(\left.\frac{\partial^2 f}{\partial x_i \, \partial x_j}\right|_p\right) \neq 0$$

A *critical value* $c_i \in \mathbb{R}$ is the image of a critical point, $c_i = f(p_i)$. For the rest of this article all critical points will be assumed to be nondegenerate.

The key result is the *Morse lemma*: If $p \in M$ is a nondegenerate critical point of f, then

$$f(x_1 \ldots x_n) = f(p) - x_1^2 - \ldots - x_\lambda^2 + x_{\lambda+1}^2 + \ldots + x_n^2$$

in some coordinate system in a neighborhood $U(p)$. λ is called the *index* of the (nondegenerate) critical point $p \in M$.

Let $(M; M_0, M_1)$ be a cobordism. A smooth, real function $f : M \to [a, b]$ is a *Morse function* if:
1) $f^{-1}(a) = M_0$;
2) $f^{-1}(b) = M_1$; and
3) all the critical points p_i of f are interior (i.e., $p_i \notin \partial M$) and nondegenerate.

Without loss of generality, we can assume that the critical values of f are distinct, $p_i \neq p_j \Rightarrow c_i = f(p_i) \neq f(p_j) = c_j$ (i.e., the Morse function is called *proper*).

The *Morse number* of a cobordism $\mu(M)$ is the minimum (over all the Morse functions defined on M) of the number of critical points,

$$\mu(M) = min_f\{\#\text{of critical points of}| \ f - Morse\}.$$

Thus, the sphere has $\mu(S^3) = 2$, the cylinder $\mu(\Sigma \times I) = 0$ and the torus $\mu(T^2) = 4$.

We have the following general theorem [Mil63, Mil65]:
(i) Every cobordism has a Morse function.
(ii) A Morse function has a finite number of critical points.
(iii) If $f : M \to \mathbb{R}$ is a Morse function with no critical points, then M is topologically trivial, $M \cong \Sigma \times [0, 1]$.

Note that the converse of (iii) is, however, not true. A topologically trivial manifold (e.g., $\Sigma \times I$) can have a nontrivial Morse function (i.e. with critical points). An example is an U-shaped cylinder which has 2 critical points (the

Morse function is the height function from a plane tangent to bottom of the cylinder). As a consequence of (iii), topology change can occur only at the critical points. Since the number of critical points is finite and the critical values are distinct, we can 'slice up' the manifold between the critical points. Thus, any cobordism can be expressed as a composition of cobordisms with Morse number $\mu = 1$, these being the *building blocks* [Ion97].

3.3.4 Morse Theory and Energy Functionals

In other words, if we have a *proper function* $f : M \to \mathbb{R}$, then its Hessian (as a quadratic form) is in fact well defined at its *critical points* without specifying an underlying Riemannian metric. The *nullity* of f at a *critical point* is defined as the dimension of the *kernel* of $\nabla^2 f$, while the *index* is the number of negative eigenvalues counted with multiplicity. A function is said to be a *Morse function* if the nullity at any of its critical points is zero. Note that this guarantees in particular that all critical points are isolated. The first fundamental Theorem of Morse theory is that one can determine the topological structure of a manifold from a Morse function. More specifically, if one can order the critical points $x_1, \ldots, x_k$ so that $f(x_1) < \cdots < f(x_k)$ and the index of x_i is denoted λ_i, then M has the structure of a CW complex with a cell of dimension λ_i for each i. Note that in case M is closed then x_1 must be a minimum and so $\lambda_1 = 0$, while x_k is a maximum and $\lambda_k = n$. The classical example of Milnor of this Theorem in action is a torus in 3–space and f the height function.

We are now left with the problem of trying to find appropriate Morse functions. While there are always plenty of such functions, there does not seem to be a natural way of finding one. However, there are natural choices for Morse functions on the loop space to a Riemannian manifold. This is, somewhat inconveniently, infinite–dimensional. Still, one can develop Morse theory as above for suitable functions, and moreover *the loop space of a manifold determines the topology of the underlying manifold*.

If $m, p \in M$, then we denote by Ω_{mp} the space of all C^k paths from m to p. The first observation about this space is that

$$\pi_{i+1}(M) = \pi_i(\Omega_{mp}).$$

To see this, just fix a path from m to q and then join this path to every curve in Ω_{mp}. In this way Ω_{mp} is identified with Ω_m, the space of loops fixed at m. For this space the above relationship between the homotopy groups is almost self-evident.

On the space Ω_{mp} we have two naturally defined functions, the *arc–length* and *energy functionals*:

$$L(\gamma, I) = \int_I |\dot{\gamma}|\, dt, \quad \text{and} \quad E(\gamma, I) = \frac{1}{2} \int_I |\dot{\gamma}|^2 dt.$$

While the energy functional is easier to work with, it is the arc–length functional that we are really interested in. In order to make things work out nicely for the arc–length functional, it is convenient to parameterize all curves on $[0,1]$ and proportionally to arc–length. We shall think of Ω_{mp} as an *infinite-dimensional manifold*. For each curve $\gamma \in \Omega_{mp}$ the natural choice for the tangent space consists of the vector–fields along γ which vanish at the endpoints of γ. This is because these vector–fields are exactly the variational fields for curves through γ in Ω_{mp}, i.e., fixed endpoint variations of γ. An inner product on the tangent space is then naturally defined by

$$(X,Y) = \int_0^1 g(X,Y)\, dt.$$

Now the first variation formula for arc–length tells us that the gradient for L at γ is $-\nabla_{\dot\gamma}\dot\gamma$. Actually this cannot be quite right, as $-\nabla_{\dot\gamma}\dot\gamma$ does not vanish at the endpoints. The real gradient is gotten in the same way we find the gradient for a function on a surface in space, namely, by projecting it down into the correct tangent space. In any case we note that the critical points for L are exactly the geodesics from m to p. The second variation formula tells us that the Hessian of L at these critical points is given by

$$\nabla^2 L(X) = \ddot{X} + R(X,\dot\gamma)\,\dot\gamma,$$

at least for vector–fields X which are perpendicular to γ. Again we ignore the fact that we have the same trouble with endpoint conditions as above. We now need to impose the Morse condition that this Hessian is not allowed to have any kernel. The vector–fields J for which $\ddot{J} + R(J,\dot\gamma)\,\dot\gamma = 0$ are called *Jacobi fields*. Thus we have to Figure out whether there are any Jacobi fields which vanish at the endpoints of γ. The first observation is that Jacobi fields must always come from geodesic variations. The Jacobi fields which vanish at m can therefore be found using the exponential map $\exp_m$. If the Jacobi field also has to vanish at p, then p must be a critical value for $\exp_m$. Now, the so-called *Sard's Theorem* asserts that the set of critical values has measure zero. For given $m \in M$ it will therefore be true that the arc–length functional on Ω_{mp} is a Morse function for almost all $p \in M$. Note that it may not be possible to choose $p = m$, the simplest example being the standard sphere. We are now left with trying to decide what the *index* should be. This is the dimension of the largest subspace on which the Hessian is negative definite. It turns out that this index can also be computed using Jacobi fields and is in fact always finite. Thus one can calculate the topology of Ω_{mp}, and hence M, by finding all the geodesics from m to p and then computing their index.

In geometrical situations it is often unrealistic to suppose that one can calculate the index precisely, but as we shall see it is often possible to given lower bounds for the index. As an example, note that if M is not simply-connected, then Ω_{mp} is not connected. Each curve of minimal length in the path components is a geodesic from m to p which is a local minimum for the arc–length functional. Such geodesics evidently have index zero. In particular, if one can show that all geodesics, except for the minimal ones from m to p, have index > 0, then the manifold must be simply–connected.

3.3.5 Morse Theory and Riemannian Geometry

Recall that on any smooth manifold M there exist many Riemannian metrics g. Each of these metrics is *locally defined* in a particular point $q \in M$ as a symmetric $(0, 2)$ tensor–field such that $g|_q : T_qM \times T_qM \to \mathbb{R}$ is a positively defined inner product for each point $q \in M$. In an open local chart $U \in M$ containing the point q, this metric is given as $g|_q \mapsto g_{ij}(q)\, dq^i dq^j$. With each metric $g|_q$ there is associated a *local geodesic* on M.

Now, two main *global geodesics problems* on the biodynamical configuration manifold M with the Riemannian metrics g, can be formulated as follows (compare with subsection 3.3.4 above):

A. *Is there a minimal geodesic $\gamma_0(t)$ between two points A and B on M?* In other words, does an arc of geodesic $\gamma_0(t)$ with extremities A, B actually have minimum length among all rectifiable curves $\gamma(t) = (q^i(t), p_i(t))$ joining A and B?

B. *How many geodesic arcs are there joining two points A and B on M?*

Locally these problems have a *complete answer*: each point of the biodynamics manifold M has an open neighborhood V such that for any two distinct points A, B of V *there is* exactly *one* arc of a geodesic *contained in* V and joining A and B, and it is the *unique minimal geodesic* between A and B.

Recall (see subsection (3.3.4) above), that seven decades ago, Morse considered the set $\Omega = \Omega(M; A, B)$ of *piecewise smooth paths* on a Riemannian manifold M having fixed extremities A, B, defined as continuous maps $\gamma : [0, 1] \to M$ such that $\gamma(0) = A$, $\gamma(1) = B$, and there were a finite number of points

$$t_0 = 0 < t_1 < t_2 < \cdots < t_{m-1} < t_m = 1, \qquad (3.98)$$

such that in every *closed interval* $[t_i, t_{i+1}]$, γ was a C^∞–function. The parametrization was always chosen such that for $t_j \leq t \leq t_{j+1}$,

$$t - t_j = \frac{t_{j+1} - t_j}{l_j} \int_{t_j}^{t} \left\| \frac{d\gamma}{du} \right\| du, \qquad \text{with} \qquad l_j = \int_{t_j}^{t_{j+1}} \left\| \frac{d\gamma}{du} \right\| du. \quad (3.99)$$

In other words, $t - t_j$ was proportional to the length of the image of $[t_j, t]$ by γ. Then

$$L(\gamma) = \sum_{j=0}^{m} l_j,$$

the length of γ, was a function of γ in Ω. A minimal arc from A to B should be a path γ for which $L(\gamma)$ is *minimum* in Ω, and a geodesic arc from A to

3.3 Morse Topology of Smooth Manifolds

B should be a path that is a 'critical point' for the function L. This at first has no meaning, since Ω is not a differential manifold; the whole of Morse's theory consists in showing that it is possible to substitute for Ω genuine differential manifolds to which his results on critical points can be applied ([Mor34, II07b]).

To study the geodesics joining two points A, B it is convenient, instead of working with the length $L(\gamma)$, to work with the *energy of a path* $\gamma : [A, B] \to M$, defined by ([Die88])

$$E_A^B(\gamma) = \int_A^B \left\| \frac{d\gamma}{du} \right\|^2 du. \tag{3.100}$$

With the chosen parametrization (3.99), $E(\gamma) = (B - A)L(\gamma)^2$, and the extremals of E are again the geodesics, but the computations are easier with E.

Morse theory can be divided into several steps (see [Mil63, II07b]).

Step 1 is essentially a presentation of the classical Lagrangian method that brings to light the analogy with the critical points of a C^∞− function on M. No topology is put on Ω; a *variation* of a path $\gamma \in \Omega$ is a continuous map α into M, defined in a product $]-\varepsilon, \varepsilon[\times [0, 1]$ with the following properties:

A. $\alpha(0, t) = \gamma(t)$;
B. $\alpha(u, 0) = A$, $\alpha(u, 1) = B$ for $-\varepsilon < u < \varepsilon$; and
C. There is a decomposition (3.98) such that α is C^∞ in each set

$$]-\varepsilon, \varepsilon[\times [t_i, t_{i+1}].$$

A *variation vector–field* $t \mapsto W(t)$ is associated to each variation α, where $W(t)$ is a tangent vector in the tangent space $T_{\gamma(t)}M$ to M, defined by

$$W(t) = \partial_u \alpha(0, t). \tag{3.101}$$

It is a continuous map of $[0, 1]$ into the tangent bundle TM, smooth in each interval $[t_i, t_{i+1}]$. These maps are the substitute for the *tangent vectors* at the point γ; they form an infinite–dimensional vector space written $T\Omega(\gamma)$.

More generally the interval $]-\varepsilon, \varepsilon[$ can be replaced in the definition of a variation by a neighborhood of 0 in some $\mathbb{R}^n$, defining an $n-$*parameter variation*.

A *critical path* $\gamma_0 \in \Omega$ for a function $F : \Omega \to \mathbb{R}$ is defined by the condition that for every variation α of γ_0 the function

$$u \mapsto F(\alpha(u, \cdot))$$

is derivable for $u = 0$ and its derivative is 0.

Step 2 is a modern presentation of the formulas of Riemannian geometry, giving the *first variation* and *second variation* of the energy (3.100) of a path $\gamma_0 \in \Omega$, which form the basis of Jacobi results.

First consider an arbitrary path $\omega_0 \in \Omega$, its *velocity* $\dot\omega(t) = d\omega/dt$, and its *acceleration* in the Riemannian sense

$$\ddot\omega(t) = \nabla_t \dot\omega(t),$$

where ∇_t denotes the Bianchi covariant derivative. They belong to $T_{\omega(t)}M$ for each $t \in [0,1]$, are defined and continuous in each interval $[t_i, t_{i+1}]$ in which ω is smooth, and have limits at both extremities. Now let α be a variation of ω and $t \mapsto W(t)$ be the corresponding variation vector–field (3.101). The *first variation formula* gives the first derivative

$$\frac{1}{2}\frac{d}{du}E(\alpha(u,\cdot))|_{u=0} = -\sum_i (W(t_i)|\dot\omega(t_i+) - \dot\omega(t_i-)) - \int_0^1 (W(t)|\ddot\omega(t))\, dt,$$

where $(x|y)$ denotes the scalar product of two vectors in a tangent space. It follows from this formula that $\gamma_0 \in \Omega$ is a critical path for E iff γ is a *geodesic*.

Next, fix such a geodesic γ and consider a two–parameter variation:

$$\alpha : U \times [0,1] \to M,$$

where U is a neighborhood of 0 in $\mathbb{R}^2$, so that

$$\alpha(0,0,t) = \gamma(t), \qquad \partial_{u_1}\alpha(0,0,t) = W_1(t), \qquad \partial_{u_2}\alpha(0,0,t) = W_2(t),$$

in which W_1 and W_2 are in $T\Omega(\gamma)$. The *second variation formula* gives the mixed second derivative

$$\frac{1}{2}\frac{\partial^2}{\partial u_1 \partial u_2}E(\alpha(u_1,u_2,\cdot)))|_{(0,0)} = -\sum_i (W_2(t_i)|\nabla_t W_1(t_i+) - \nabla_t W_1(t_i-))$$
$$- \int_0^1 (W_2(t)|\nabla_t^2 W_1(t) + R(V(t) \wedge W_1(t)) \cdot V(t))\, dt, \qquad (3.102)$$

where $Z \mapsto R(X \wedge Y) \cdot Z$ is the curvature of the Levi–Civita connection. The l.h.s of (3.102) is thus a *bilinear symmetric form*

$$(W_1, W_2) \mapsto E_{**}(W_1, W_2)$$

on the product $T\Omega(\gamma) \times T\Omega(\gamma)$. For a one–parameter variation α

$$E_{**}(W,W) = \frac{1}{2}\frac{d^2}{du^2}E(\alpha(u,\cdot))|_{u=0},$$

from which it follows that if γ is a *minimal* geodesic in Ω, $E_{**}(W,W) \geq 0$ in $T\Omega(\gamma)$. As usual, we shall speak of E_{**} indifferently as a symmetric bilinear form or as a quadratic form $W \mapsto E_{**}(W,W)$.

Formula (3.102) naturally leads to the junction with Jacobi work (see [Die88, II07b]): consider the smooth vector–fields $t \mapsto J(t)$ along $\gamma \in M$, satisfying the equation

3.3 Morse Topology of Smooth Manifolds

$$\nabla_t^2 J(t) + R(V(t) \wedge J(t)) \cdot V(t) = 0 \qquad \text{for } 0 \leq t \leq 1. \tag{3.103}$$

With respect to a frame along γ moving by parallel translation on M this relation is equivalent to a system of n linear homogeneous ODEs of order 2 with C^∞–coefficients; the solutions J of (3.103) are called the *Jacobi fields* along γ and form a vector space of dimension $2n$. If for a value $a \in]0,1]$ of the parameter t there exists a Jacobi field along γ that is not identically 0 but *vanishes for* $t = 0$ *and* $t = a$, then the points $A = \gamma(0)$ and $r = \gamma(a)$ are conjugate along γ with a *multiplicity* equal to the dimension of the vector space of Jacobi fields vanishing for $t = 0$ and $t = a$.

Jacobi fields on the biodynamical configuration manifold M may also be defined as variation vector–fields for *geodesic variations* of the path $\gamma \in M$: they are C^∞–maps

$$\alpha :]-\varepsilon, \varepsilon[\times [0,1] \to M,$$

such that for any $u \in]-\varepsilon, \varepsilon[, t \mapsto \alpha(u,t)$ is a geodesic and $\alpha(0,t) = \gamma(t)$.

It can be proved that the Jacobi fields along $\gamma \in M$ that vanish at A and B (hence belong to $T\Omega(\gamma)$) are exactly the vector–fields $J \in T\Omega(\gamma)$ such that

$$E_{**}(J,W) = 0$$

for every $W \in T\Omega(\gamma)$. Although $T\Omega(\gamma)$ is infinite–dimensional, the form E_{**} is again called *degenerate* if the vector space of the Jacobi fields vanishing at A and B is note reduced to 0 and the dimension of that vector space is called the *nullity* of E_{**}. Therefore, E_{**} is thus degenerate iff A and B are conjugate along γ and the nullity of E_{**} is the multiplicity of B.

Step 3 is the beginning of Morse's contributions (see [Mil63, II07b]). He first considered a *fixed* geodesic $\gamma : [0,1] \to M$ with extremities $A = \gamma(0)$, $B = \gamma(1)$ and the bilinear symmetric form $E_{**} : T\Omega(\gamma) \times T\Omega(\gamma) \to \mathbb{R}$. By analogy with the finite–dimensional quadratic form, the *index* of E_{**} is defined as the maximum dimension of a vector subspace of $T\Omega(\gamma)$ in which E_{**} is *strictly negative* (i.e., nondegenerate and taking values $E_{**}(W,W) < 0$ except for $W = 0$). Morse's central result gives the value of the index of E_{**} and is known as the *index Theorem*.

Suppose a subdivision (3.98) is chosen such that each arc $\gamma([t_{i-1}, t_i])$ is contained in an open set $U_i \subset M$ such that any two points of U_i are joined by a unique geodesic arc contained in U_i that is *minimal*; $\gamma([t_{i-1}, t_i])$ is such an arc. In the infinite–dimensional vector space $T\Omega(\gamma)$, consider the two vector subspaces:

A. $T\Omega(\gamma; t_0, t_1, \cdots, t_m)$ consisting of all continuous vector–fields $t \mapsto W(t)$ along γ, vanishing for $t = 0$ and $t = 1$, such that each restriction $W|[t_{i-1}, t_i]$ is a *Jacobi field* (hence smooth) along $\gamma([t_{i-1}, t_i])$; that subspace is finite–dimensional;

B. T' consisting of the vector–fields $t \mapsto W(t)$ along γ, such that $W(t_0) = 0, W(t_1) = 0, \cdots, W(t_m) = 0$.

$T\Omega(\gamma)$ is then the *direct sum* $T\Omega(\gamma; t_0, t_1, \cdots, t_m) \oplus T'$; these two subspaces are orthogonal for the bilinear form E_{**}, and E_{**} is *strictly positive* in T', so that the index of E_{**} is equal to the index of its restriction to the subspace $T\Omega(\gamma; t_0, t_1, \cdots, t_m)$.

To calculate the nullity and index of E_{**}, due to this decomposition, apply their definitions either to vector subspaces of $T\Omega(\gamma)$ or to vector subspaces of $T\Omega(\gamma; t_0, t_1, \cdots, t_m)$. The computation of the index of E_{**} is done by considering the geodesic arc $\gamma_\tau : [0, \tau] \to M$, the restriction of γ to $[0, \tau]$, and its energy

$$E(\gamma_\tau) = \tau \int_0^\tau \left\| \frac{d\gamma}{du} \right\|^2 du.$$

E_{**}^τ is the corresponding quadratic form on $T\Omega(\gamma_\tau)$, and $\lambda(\tau)$ is its index; one studies the variation of $\lambda(\tau)$ when *tau* varies from 0 to 1, and $\lambda(1)$ is the index of E_{**}.

The index Theorem says: *the index of E_{**} is the sum of the multiplicities of the points conjugate to A along B and distinct from B.*

We have seen that the dimension of $T\Omega(\gamma; t_0, t_1, \cdots, t_m)$ is finite; it follows that the index of E_{**} is always *finite*, and therefore the number of points conjugate to A along γ is also *finite*.

Step 4 of Morse theory introduces a *topology* on the set $\Omega = \Omega(M; A, B)$. On the biodynamical configuration manifold M the usual topology can be defined by a *distance* $\rho(A, B)$, the g.l.b. of the lengths of all piecewise smooth paths joining A and B. For any pair of paths ω_1, ω_2 in $\Omega(M; A, B)$, consider the function $d(\omega_1, \omega_2) \in M$

$$d(\omega_1, \omega_2) = \sup_{0 \leq t \leq 1} \rho(\omega_1(t), \omega_2(t)) + \sqrt{\int_0^1 (\dot{s}_1 - \dot{s}_2)^2 \, dt},$$

where $s_1(t)$ (resp. $s_2(t)$) is the length of the path $\tau \mapsto \omega_1(\tau)$ (resp. $\tau \mapsto \omega_2(\tau)$) defined in $[0, t]$. This distance on Ω such that the function $\omega \mapsto E_A^B(\omega)$ is *continuous* for that distance.

3.3.6 Morse Topology in Human/Humanoid Biodynamics

Morse Functions and Boundary Operators

Let $f : M \to \mathbb{R}$ represents a C^∞–function on the biodynamical configuration manifold M. Recall that $z = (q, p) \in M$ is the *critical point* of f if $df(z) \equiv df[(q, p)] = 0$. In local coordinates $(x^1, ..., x^n) = (q^1, ..., q^n, p_1, ..., p_n)$ in a neighborhood of z, this means $\frac{\partial f}{\partial x^i}(z) = 0$ for $i = 1, ..., n$. The Hessian of f at a critical point z defines a symmetric bilinear form $\nabla df(z) = d^2 f(z)$ on $T_z M$, in local coordinates $(x^1, ..., x^n)$ represented by the matrix $\left(\frac{\partial^2 f}{\partial x^i \partial x^j} \right)$. Index and nullity of this matrix are called index and nullity of the critical point z of f.

3.3 Morse Topology of Smooth Manifolds

Now, we assume that all critical points $z_1, ..., z_n$ of $f \in M$ are nondegenerate in the sense that the Hessians $d^2 f(z_i)$, $i = 1, ..., m$, have maximal rank. Let z be such a critical point of f of *Morse index* s (= number of negative eigenvalues of $d^2 f(z_i)$, counted with multiplicity). The eigenvectors corresponding to these negative eigenvalues then span a subspace $V_z \subset T_z M$ of dimension s. We choose an orthonormal basis $e_1, ..., e_s$ of V_z w.r.t. the Riemannian metric g on M (induced by the system's kinetic energy), with dual basis $dx^1, ..., dx^s$. This basis then defines an orientation of V_z which we may also represent by the s--form $dx^1 \wedge ... \wedge dx^s$. We now let z' be another critical point of f, of Morse index $s - 1$. We consider paths $\gamma(t)$ of the steepest descent of f from z to z', i.e., integral curves of the vector–field $-\nabla f(\gamma)$. Thus $\gamma(t)$ defines the *gradient flow* of f [II06b, II07b]

$$\dot{\gamma}(t) = -\nabla f(\gamma(t)), \quad \text{with} \quad \begin{cases} \lim_{t \to -\infty} \gamma(t) = z, \\ \lim_{t \to \infty} \gamma(t) = z' \end{cases}. \quad (3.104)$$

A path $\gamma(t)$ obviously depends on the Riemannian metric g on M as

$$\nabla f = g^{ij}\, \partial_{x^i} f\, \partial_{x^j} f.$$

From [Sma60, Sma67] it follows that for a generic metric g, the Hessian $\nabla df(y)$ has only nondegenerate eigenvalues. Having a metric g induced by the system's kinetic energy, we let $\tilde{V}_y \subset T_y M$ be the space spanned by the eigenvectors corresponding to the $s - 1$ lowest eigenvalues. Since z' has Morse index $s - 1$, $\nabla df(z') = d^2 f(z')$ has precisely $s - 1$ negative eigenvalues. Therefore, $\tilde{V}_{z'} \equiv \lim_{t \to \infty} \tilde{V}_{\gamma(t)} = V_{z'}$, while the unit tangent vector of γ at z', i.e., $\lim_{t \to \infty} \frac{\dot{\gamma}(t)}{\|\dot{\gamma}(t)\|}$, lies in the space of directions corresponding to positive eigenvalues and is thus orthogonal to $V_{z'}$. Likewise, the unit tangent vector v_z of γ at z, while contained in V_z, is orthogonal to $\tilde{V}_z$, because it corresponds to the largest one among the s negative eigenvalues of $d^2 f(z)$. Taking the interior product $i(v_z)\, dx^1 \wedge ... \wedge dx^s$ defines an orientation of $\tilde{V}_z$. Since $\tilde{V}_y$ depends smoothly on y, we may transport the orientation of $\tilde{V}_z$ to $\tilde{V}_{z'}$ along γ. We then define $n_\gamma = +1$ or -1, depending on whether this orientation of $\tilde{V}_{z'}$ coincides with the chosen orientation of $V_{z'}$ or not, and further define $n(z, z') = \sum_\gamma n_\gamma$, where the sum is taken over all such paths γ of the steepest descent from p to p'.

Now, let M^s be the *set of critical points* of f of *Morse index* s, and let H^s_f be the vector space over $\mathbb{R}$ spanned by the elements of M^s. We define a *boundary operator*

$$\delta : H^{s-1}_f \to H^s_f, \quad \text{by putting, for } z' \in M^{s-1},$$

$$\delta(z') = \sum_{n \in M^s} n(z', z)\, z, \quad \text{and extending } \delta \text{ by linearity.}$$

This operator satisfies $\delta^2 = 0$ and therefore defines a cohomology theory. Using *Conley's continuation principle*, Floer [Flo88] showed that the resulting cohomology theories resulting from different choices of f are canonically isomorphic.

In his QFT–based rewriting the Morse topology, Ed Witten [Wit82] considered also the operators:

$$d_t = e^{-tf} d e^{tf}, \quad \text{their adjoints}: \quad d_t^* = e^{tf} d e^{-tf},$$
$$\text{as well as their Laplacian}: \quad \Delta_t = d_t d_t^* + d_t^* d_t.$$

For $t = 0$, Δ_0 is the standard *Hodge–de Rham Laplacian*, whereas for $t \to \infty$, one has the following expansion

$$\Delta_t = dd^* + d^*d + t^2 \|df\|^2 + t \sum_{k,j} \frac{\partial^2 h}{\partial x^k \partial x^j} [i \partial_{x^k}, dx^j],$$

where $(\partial_{x^k})_{k=1,\ldots,n}$ is an orthonormal frame at the point under consideration. This becomes very large for $t \to \infty$, except at the critical points of f, i.e., where $df = 0$. Therefore, the eigenvalues of Δ_t will concentrate near the critical points of f for $t \to \infty$, and we get an *interpolation* between de Rham cohomology and Morse cohomology.

Morse Homology on M

Now, following [Mil99, IP05b, II07b], for any Morse function f on the configuration manifold M we denote by $\text{Crit}_p(f)$ the set of its critical points of index p and define $C_p(f)$ as a free Abelian group generated by $\text{Crit}_p(f)$. Consider the gradient flow generated by (3.104). Denote by $\mathcal{M}_{f,g}(M)$ the set of all $\gamma : \mathbb{R} \to M$ satisfying (3.104) such that

$$\int_{-\infty}^{+\infty} \left|\frac{d\gamma}{dt}\right|^2 dt < \infty.$$

The spaces

$$\mathcal{M}_{f,g}(x^-, x^+) = \{\gamma \in \mathcal{M}_{f,g}(M) \mid \gamma(t) \to x^{\pm} \text{ as } t \to \pm\infty\}$$

are smooth manifolds of dimension $m(x^+) - m(x^-)$, where $m(x)$ denotes the Morse index of a critical point x. Note that

$$\mathcal{M}_{f,g}(x, y) \cong W_g^u(x) \cap W_g^s(y),$$

where $W_g^s(y)$ and $W_g^u(x)$ are the stable and unstable manifolds of the gradient flow (3.104). For generic g the intersection above is transverse (Morse–Smale condition). The group $\mathbb{R}$ acts on $\mathcal{M}_{f,g}(x,y)$ by $\gamma \mapsto \gamma(\cdot + t)$. We denote

$$\widehat{\mathcal{M}}_{f,g}(x,y) = \mathcal{M}_{f,g}(x,y)/\mathbb{R}.$$

The manifolds $\widehat{\mathcal{M}}_{f,g}(x,y)$ can be given a coherent orientation σ (see [Sch93]). Now, we can define the boundary operator, as

$$\partial : C_p(f) \to C_{p-1}(f), \qquad \partial x = \sum_{y \in \mathrm{Crit}_{p-1}(f)} n(x,y) y,$$

where $n(x,y)$ is the number of points in 0D manifold $\widehat{\mathcal{M}}_{f,g}(x,y)$ counted with the sign with respect to the orientation σ. The proof of $\partial \circ \partial = 0$ is based on gluing and cobordism arguments [Sch93]. Now Morse homology groups are defined by

$$H_p^{\mathrm{Morse}}(f) = \mathrm{Ker}(\partial)/\mathrm{Im}(\partial).$$

For generic choices of Morse functions f_1 and f_2 the groups $H_p(f_1)$ and $H_p(f_2)$ are isomorphic. Furthermore, they are isomorphic to the singular homology group of M, i.e.,

$$H_p^{\mathrm{Morse}}(f) \cong H_p^{\mathrm{sing}}(M),$$

for generic f [Mil65].

The construction of isomorphism is given (see [Mil99, IP05b, II07b]) as

$$h_{\alpha\beta} : H_p(f^\alpha) \to H_p(f^\beta), \qquad (3.105)$$

for generic Morse functions f^α, f^β. Consider the 'connecting trajectories', i.e., the solutions of non–autonomous equation

$$\dot{\gamma} = -\nabla f_t^{\alpha\beta}, \qquad (3.106)$$

where $f_t^{\alpha\beta}$ is a homotopy connecting f^α and f^β such that for some $R > 0$

$$f_t^{\alpha\beta} \equiv \begin{cases} f^\alpha \text{ for } t \leq -R \\ f^\beta \text{ for } t \geq R \end{cases}.$$

For $x^\alpha \in \mathrm{Crit}_p(f^\alpha)$ and $x^\beta \in \mathrm{Crit}_p(f^\beta)$ denote

$$\mathcal{M}_{f^{\alpha\beta},g}(x^\alpha, x^\beta) = \{\gamma : \gamma \text{ satisfies (3.106) and } \lim_{t \to -\infty} \gamma = x^\alpha, \lim_{t \to \infty} \gamma = x^\beta\}.$$

As before, $\mathcal{M}_{f^{\alpha\beta},g}$ is a smooth finite–dimensional manifold. Now, define

$$(h_{\alpha\beta})_\sharp : C_p(f^\alpha) \to C_p(f^\beta), \qquad \text{by}$$
$$(h_{\alpha\beta})_\sharp x^\alpha = \sum_{x^\beta \in \mathrm{Crit}_p(f^\beta)} n(x^\alpha, x^\beta) x^\beta, \text{ for } x^\alpha \in \mathrm{Crit}_p(f^\alpha),$$

where $n(x^\alpha, x^\beta)$ is the algebraic number of points in 0D manifold $\mathcal{M}_{f^{\alpha\beta},g}(x^\alpha, x^\beta)$ counted with the signs defined by the orientation of $\mathcal{M}_{f^{\alpha\beta},g}$. Homomorphisms $(h_{\alpha\beta})_\sharp$ commute with ∂ and thus define the homomorphisms $h_{\alpha\beta}$ in homology which, in addition, satisfy $h_{\alpha\beta} \circ h_{\beta\gamma} = h_{\alpha\gamma}$.

Now, if we fix a Morse function $f : M \to \mathbb{R}$ instead of a metric g, we establish the isomorphism (see [Mil99, IP05b, II07b])

$$h_{\alpha\beta} : H_p(g^\alpha, f) \to H_p(g^\beta, f)$$

between the two Morse homology groups defined by means of two generic metrics g^α and g^β in a similar way, by considering the 'connecting trajectories',

$$\dot\gamma = -\nabla^{g_t^{\alpha\beta}} f. \tag{3.107}$$

Here $g_t^{\alpha\beta}$ is a homotopy connecting g^α and g^β such that for some $R > 0$

$$g_t^{\alpha\beta} \equiv \begin{cases} g^\alpha \text{ for } t \leq -R, \\ g^\beta \text{ for } t \geq R, \end{cases}$$

and ∇^g is a gradient defined by metric g.

Note that f is decreasing along the trajectories solving autonomous gradient equation (3.104). Therefore, the boundary operator ∂ preserves the downward filtration given by level sets of f. In other words, if we denote

$$\mathrm{Crit}_p^\lambda(f) = \mathrm{Crit}_p(f) \cap f^{-1}((-\infty,\lambda]), \quad \text{and}$$
$$C_p^\lambda(f) = \text{ free Abelian group generated by } \mathrm{Crit}_p^\lambda(f),$$

then the boundary operator ∂ restricts to $\partial^\lambda : C_p^\lambda(f) \to C_{p-1}^\lambda(f)$. Obviously, $\partial^\lambda \circ \partial^\lambda = 0$, thus we can define the relative Morse homology groups

$$H_p^\lambda(f) = \mathrm{Ker}(\partial^\lambda)/\mathrm{Im}(\partial^\lambda).$$

Following the standard algebraic construction, we define (relative) Morse cohomology. We set

$$C_\lambda^p(f) = \mathrm{Hom}(C_p^\lambda(f), \mathbb{Z}), \quad \text{and}$$
$$\delta^\lambda : C_\lambda^p(f) \to C_\lambda^{p+1}(f), \quad \langle \delta^\lambda a, x \rangle = \langle a, \partial^\lambda x \rangle$$

and define

$$H_\lambda^p(f) = \mathrm{Ker}(\delta^\lambda)/\mathrm{Im}(\delta^\lambda).$$

Since $\mathrm{Crit}_p(f)$ is finite, we have $H_p^\lambda(f) = H_p(f)$ and $H_\lambda^p(f) = H^p(f)$.

3.3.7 Cobordism Topology on Smooth Manifolds

Cobordism appeared as a revival of Poincaré's unsuccessful 1895 attempts to define homology using only manifolds. Smooth manifolds (without boundary) are again considered as 'negligible' when they are *boundaries* of smooth manifolds–with–boundary. But there is a big difference, which keeps definition of 'addition' of manifolds from running into the difficulties encountered by Poincaré; it is now the disjoint union. The (un-oriented) *cobordism relation* between two compact smooth manifolds M_1, M_2 of same dimension n means that their disjoint union $\partial W = M_1 \uplus M_2$ is the boundary ∂W of an $(n+1)$D smooth manifold–with–boundary W. This is an *equivalence relation*, and the classes for that relation of nD manifolds form a *commutative group* $\mathfrak{N}_n$ in

3.3 Morse Topology of Smooth Manifolds

which every element has order 2. The direct sum $\mathfrak{N}_\bullet = \oplus_{n\geq 0} \mathfrak{N}_n$ is a ring for the multiplication of classes deduced from the Cartesian product of manifolds.

More precisely, a manifold M is said to be a *cobordism* from A to B if there exists a diffeomorphism from a disjoint sum, $\varphi \in \text{diff}(A^* \uplus B, \partial M)$. Two cobordisms $M(\varphi)$ and $M'(\varphi')$ are equivalent if there is a $\Phi \in \text{diff}(M, M')$ such that $\varphi' = \Phi \circ \varphi$. The equivalence class of cobordisms is denoted by $M(A, B) \in Cob(A, B)$ [Sto68, II07b].

Composition c_{Cob} of cobordisms comes from gluing of manifolds [BD95]. Let $\varphi' \in \text{diff}(C^* \uplus D, \partial N)$. One can glue cobordism M with N by identifying B with C^*, $(\varphi')^{-1} \circ \varphi \in \text{diff}(B, C^*)$. We get the glued cobordism $(M \circ N)(A, D)$ and a semigroup operation,

$$c(A, B, D) : Cob(A, B) \times Cob(B, D) \longrightarrow Cob(A, D).$$

A *surgery* is an operation of cutting a manifold M and gluing to cylinders.[30] A surgery gives new cobordism: from $M(A, B)$ into $N(A, B)$. The disjoint sum of $M(A, B)$ with $N(C, D)$ is a cobordism $(M \uplus N)(A \uplus C, B \uplus D)$. We got a 2–graph of cobordism Cob with $Cob_0 = Man_d$, $Cob_1 = Man_{d+1}$, whose 2–cells from Cob_2 are surgery operations.

There is an n–category of cobordisms $\mathcal{BO}$ [Lei03, II07b] with:

- 0–cells: 0–manifolds, where 'manifold' means 'compact, smooth, oriented manifold'. A typical 0–cell is • • • • •.
- 1–cells: 1–manifolds with corners, i.e., cobordisms between 0–manifolds,

such as  (this being a 1–cell from the 4–point manifold to the 2–point 0–manifold).

---

[30] In geometry and topology, surgery theory is the name given to a collection of techniques used to produce one manifold from another in a 'controlled' way. Surgery refers to cutting out parts of the manifold and replacing it with a part of another manifold, matching up along the cut or boundary. More technically, the idea is to start with a well–understood manifold M and perform surgery on it to produce a manifold M' having some desired property, in such a way that the effects on the homology, homotopy groups, or other interesting topological invariants of the manifold are known.

- 2−cells: 2−manifolds with corners, such as the so-called '*trousers*'

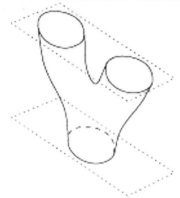

- 3−cells, 4−cells,... are defined similarly;
- Composition is gluing of manifolds.

The cobordisms theme was taken a step further by [BD95], when they started a programme to understand the subtle relations between certain TMFT models for manifolds of different dimensions, frequently referred to as the dimensional ladder. This programme is based on higher−dimensional algebra, a generalization of the theory of categories and functors to n−categories and n−functors. In this framework a topological quantum field theory (TMFT) becomes an n−functor from the n−category $\mathcal{BO}$ of n−cobordisms to the n−category of n−Hilbert spaces.

3.4 Topology Change in 3D

Recall that *topology change* has become recently a subject of increasing research interest [GH92b, GH92a, Sor97, MS97]. The image of fluctuating topology at the Planck scale is due to Wheeler, who was the first one to point out the dynamical topology inherent to that scale, the now famous *foamlike structure* of space-time [Whe62].

Although from a classical point of view topology change is excluded [Ger67], in the quantum case this is different due to fluctuations of the metric. There are many arguments in favor of topology change. In a sum over histories approach to quantum gravity, the sum over metrics is naturally extended to a sum over topologies. Another argument comes from the Big–Bang, which implies a topological transition $\emptyset \to S^3$. This implies that topology change becomes an essential ingredient of Planck scale physics.

In this article we investigate topology change by using Morse theory and handle decomposition. Applied to the case of 3−manifolds, this yields the set of *building blocks*, i.e., those elementary cobordisms from which any 3−fold can be built (up to a homeomorphism).

3.4 Topology Change in 3D

While finalizing this article, reference [DG98] came to our attention, in which a similar framework of handle decomposition was used for topology change.

From the beginning, it is important to make a distinction between a topological and a Lorentzian cobordism [Yod72, Rei63].

By a *topological cobordism* we understand a smooth, compact, nD manifold M whose boundary has 2 disjoint components $\partial M = M_0 \uplus M_1$, with M_0 and M_1 two smooth, closed, $(n-1)$D manifolds (possible empty or non-connected). Two manifolds are topologically cobordant if and only if they have the same Stiefel–Whitney and Pontrjagin numbers (the oriented case) or only the Stiefel–Whitney numbers (the non–oriented case) [MS74)].

A *Lorentzian cobordism* is a topological cobordism $(M; M_0, M_1)$ together with a nonsingular vector field $\mathbf{v}$ which is interior normal to M_0 and exterior normal to M_1. In this case we can define a nonsingular Lorentz metric $g^L_{\mu\nu}$ on M [Ion97]

$$g^L_{\mu\nu} = g^R_{\mu\nu} - \frac{2v_\mu v_\nu}{g^R_{\alpha\beta} v_\alpha v_\beta} \tag{3.108}$$

where $g^R_{\mu\nu}$ is a Riemann metric on M. This is always possible, since there is a 1–1 correspondence between nonsingular vector fields $\mathbf{v}$ and Lorentz metrics $g^L_{\mu\nu}$ on M. With respect to $g^L_{\mu\nu}$, M_0 and M_1 are space-like and we will denote them as the initial and final hyper-surfaces of the cobordism.

Following a celebrated theorem of Geroch [Ger67], topology change implies either *closed timelike curves* (CTCs) or *singularities* in the metric. For the rest of this article we assume there are no CTCs, and therefore we admit singularities in the metric, in order to have topology change. A consequence of the other choice (CTCs, no singularities) for topology change in *Kaluza–Klein theories* has been studied in [Ion97].

The first question we ask is: How serious are such singularities? First of all, we have to point out that these are not *curvature singularities* (like $r = 0$ in the Schwarzschild metric). In our case space-time is a smooth manifold and therefore the curvature is bounded. However, at the singular points the metric $g_{\mu\nu}$ fails to be invertible, this being related to the singularities in the vector field $\mathbf{v}$ which defines the *time–flow* in (3.108). As Horowitz pointed out in [Hor91], if we allow degenerate tetrads, the singularities can be very mild, since the curvature is bounded.

The viewpoint adopted here is the following: until we have a full theory of quantum gravity we shall leave all the options open, and therefore singular metrics are a legitimate object to study.

The important result is the following theorem [Sor86b]: Every topological cobordism admits a metric which is Lorentzian everywhere, except for a finite number of singularities.

3.4.1 Attaching Handles

Next, we focus on the structure of the elementary cobordisms (i.e., those with Morse number $\mu = 1$) [Ion97].

A cobordism $(M; M_0, M_1)$ which has a Morse function with a single critical point of index λ is called an *elementary cobordism of index λ* (or shortly, a λ–*cobordism*). Obviously, a λ–cobordism is an elementary building block.

In any dimension n, there can be only $n + 1$ types of (non-degenerate, or Morse) critical points, since $\lambda = 0 \ldots n$. Moreover, a λ–cobordism is homeomorphic to one of the $(n - \lambda)$–cobordisms (since for given λ, there can be several λ-cobordism which are not homeomorphic– see below the case of 1-cobordisms in 3 dimensions). This can be easily seen from the following argument. If f has a critical point p of index λ, then for the Morse function $g = -f$, p is a critical point of index $n - \lambda$. Thus the $(n - \lambda)$–cobordism represents the same cobordism as a λ–cobordism, but 'upside–down', and therefore it mediates the inverse topological transition $\Sigma_{final} \to \Sigma_{initial}$.

The following is a standard theorem in cobordism theory [Mil65, FR84]: Any cobordism $(M; M_0, M_1)$ can be obtained from the trivial cobordism $M_0 \times I$ by attaching a finite number of λ–handles.

By *attaching a handle* to a boundary M_0 we understand gluing an n–ball $D^n = D^\lambda \times D^{n-\lambda}$ via an arbitrary embedding $h : S^{\lambda-1} \times D^{n-\lambda} \to M_0$.

In order to find the λ–cobordisms we start with the cylinder $M_0 \times I$ over an arbitrary boundary and find all the embeddings $h : S^{\lambda-1} \times D^{n-\lambda} \to M_0$ for the boundary of a given λ–handle. The manifold obtained from the cylinder $M_0 \times I$ after gluing the λ–handle along this embedding will be a λ–cobordism.

Handles in 2D

We start with a 'warm–up' example in 2D [Ion97]. The possible boundary for a 2–manifold is S^1 (the only closed 1D manifold), or a disjoint sum of circles $S^1 \uplus \ldots \uplus S^1$. Attaching a handle in 2D is equivalent to gluing a disk (i.e. a 2-ball) D^2 along different parts of its boundary.

0–Cobordism

In this case we have to attach $D^2 = D^0 \times D^2$ via the empty set $S^{-1} \times D^2 = \emptyset$ (since $S^{-1} = \emptyset$).

Starting with a manifold M with boundary ∂M, attaching a 0–handle is equivalent to creating an S^1 boundary out of nothing and therefore the new boundary will be $\partial M \uplus S^1$. Thus, the 0–cobordism is simply a disk D^2 and the topology transition mediated is $\emptyset \to S^1$.

1–Cobordism

The same disk $D^2 = D^1 \times D^1$ is glued now along $S^0 \times D^1$ (two disjoint line segments). There are two different embeddings of the two segments $S^0 \times D^1$

in the boundary of a 2–manifold. The two segments can belong either to the same S^1 component, or to different S^1 components of the boundary. In the first case we have Figure 3.6, while the second case corresponds to Figure 3.7

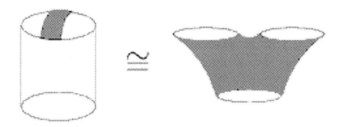

Fig. 3.6. The first case (adapted and modified from [Ion97]).

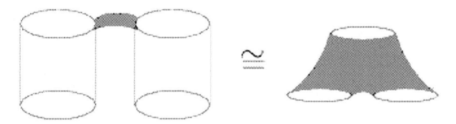

Fig. 3.7. The second case (adapted and modified from [Ion97]).

Another possibility is to twist the *ribbon* D^2 before gluing its two ends on the same S^1 component; the resulting manifold is the connected sum of a *Möbius band* and a disk. Since we are not interested in non-orientable boundaries, we do not consider this case here.

2–Cobordism

This is the reverse of a 0–cobordism. The disk $D^2 = D^2 \times D^0$ is glued along its whole $S^1 \times D^0 = S^1$ boundary (D^0 is just a point). Attaching a 2–handle reduces then to gluing a disk to one of the existing S^1 boundaries. Therefore, the only two elementary cobordisms in 2D are the *trousers* and the *Big–Bang* (or, *yarmulke* [LS97], see Figure 3.8.

Handles in 3D

Now we can do the same analysis for the 3D case. The general boundary of a 3-manifold is homeomorphic to a genus g surface Σ_g (or a disjoint sum of such

λ					
0	+	+	⌣	:	$\emptyset \to S^1$
1	−	+	⋈	:	$\begin{cases} S^1 \to S^1 \uplus S^1 \\ S^1 \uplus S^1 \to S^1 \end{cases}$
2	−	−	⌢	:	$S^1 \to \emptyset$

Fig. 3.8. the only two elementary cobordisms in 2D are the *trousers* and the *Big–Bang* (or, *yarmulke* (adapted and modified from [Ion97]).

surfaces). We consider only manifolds with orientable boundaries, therefore we exclude from the possible boundaries closed 2–manifolds which have the projective plane $\mathbb{RP}^2$ as a factor.

In order to construct the λ–cobordisms we start with the cylinder $\Sigma_g \times I$ and attach to one of the Σ_g boundaries a λ–handle. The cylinder $\Sigma_g \times I$ can be viewed as a hollow genus g handle-body. As a 3–manifold, it has two Σ_g boundaries: the exterior one and the interior one, which is shaded in Figure 3.9.

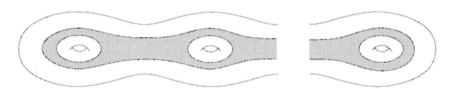

Fig. 3.9. Two Σ_g boundaries: the exterior one and the interior one (adapted and modified from [Ion97]).

0–Cobordism

This is similar to the 2D case and represents the creation of an S^2 boundary out of the vacuum (we glue $D^0 \times D^3$ via the empty set $S^{-1} \times D^3$). The cobordism is just a three–ball D^3 which mediates the transition $\emptyset \to S^2$.

1–Cobordism

Attach the three–ball $D^1 \times D^2$ along $S^0 \times D^2$ (two disks) on an arbitrary Σ_g boundary. It is equivalent to gluing a solid tube along its two opposite ends. There are two possible embeddings.

$$\Sigma_g \to \Sigma_{g+1}:$$

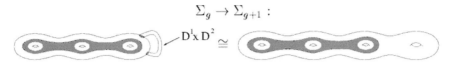

Fig. 3.10. The wormhole creation: both ends on the same boundary (adapted and modified from [Ion97]).

(i) Both ends on the same boundary, see Figure 3.10.
(ii) The two disks glued on disjoint boundaries, see Figure 3.11.

$$\Sigma_{g_1} \uplus \Sigma_{g_2} \to \Sigma_{g_1+g_2}:$$

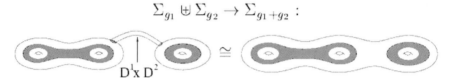

Fig. 3.11. The creation of an Einstein–Rosen bridge (connecting two disjoint 'universes') (adapted and modified from [Ion97]).

2–Cobordism

This should be equivalent to one of the 1–cobordisms, as we can check. Attach $D^2 \times D^1$ (viewed as a solid cylinder) along its lateral surface $S^1 \times D^1$. We have to find different embeddings of this surface in the Σ_g boundary. This can be seen in Figure 3.11, where the gluings are done on the interior Σ_g of the cylinder $\Sigma_g \times I$. The gluings of type (i) sever one of the handles, and thus they are equivalent to $\Sigma_g \to \Sigma_{g-1}$. Type (ii) gluings separate the inner boundary into two disjoint boundaries, $\Sigma_g \to \Sigma_k \uplus \Sigma_{g-k}$, with $k = 0 \ldots g$. Type (iii) reduces to type (i) after a homeomorphism of the Σ_g boundary.

3–Cobordism

Simply cap an S^2 boundary, $S^2 \to \emptyset$; the cobordism is again a 3–ball D^3, see Figure 3.12.

The summary is given in Figure 3.13.
The physical interpretation of these elementary building blocks in 3D is:
i) Big–Bang/Big–Crunch: $\emptyset \longleftrightarrow S^2$
ii) wormhole creation/annihilation: $\Sigma_g \longleftrightarrow \Sigma_{g+1}$
iii) Einstein–Rosen bridge creation/annihilation: $\Sigma_{g_1} \uplus \Sigma_{g_2} \longleftrightarrow \Sigma_{g_1+g_2}$

A similar approach, but using spherical modifications instead of handle decomposition was used in [Yod72, Yod73]; however, the authors omitted the 2–cobordism representing the Einstein–Rosen bridge, $\Sigma_{g_1} \uplus \Sigma_{g_2} \to \Sigma_{g_1+g_2}$. Note that at first sight, it seems that the cobordism $\Sigma_{g_1} \uplus \Sigma_{g_2} \to \Sigma_{g_1+g_2}$ is a

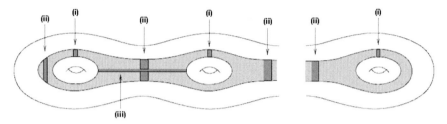

Fig. 3.12. Different ways of attaching a 2–handle and the resulting 2–cobordisms: (i) $\Sigma_g \to \Sigma_{g-1}$; (ii) $\Sigma_g \to \Sigma_k \uplus \Sigma_{g-k}$, ($k = 0\ldots g$); (iii) this reduces, after a homeomorphism, to case (i), $\Sigma_g \to \Sigma_{g-1}$ (adapted and modified from [Ion97]).

$$
\begin{array}{llll}
\lambda & & & \\
0 & +\ +\ + & \emptyset \to S^2 & \\
1 & -\ +\ + & \left\{ \begin{array}{l} \Sigma_g \to \Sigma_{g+1} \\ \Sigma_{g_1} \uplus \Sigma_{g_2} \to \Sigma_{g_1+g_2} \end{array} \right. \\
2 & -\ -\ + & \left\{ \begin{array}{l} \Sigma_g \to \Sigma_{g-1} \\ \Sigma_{g_1+g_2} \to \Sigma_{g_1} \uplus \Sigma_{g_2} \end{array} \right. \\
3 & -\ -\ - & S^2 \to \emptyset &
\end{array}
$$

Fig. 3.13. Summary of the 3–cobordism (adapted and modified from [Ion97]).

composite one. Thus, we could try to obtain it from $S^2 \uplus S^2 \to S^2$ by applying on each S^2 boundary an appropriate number of *wormhole creation* cobordisms $\Sigma_g \to \Sigma_{g+1}$. However, this is not so, and a counterexample is given in Figure 3.14.

Consider the building block $T^2 \uplus S^2 \to T^2$. It is easy to see that this is just the connected sum of the cylinder $T^2 \times I$ and the 3–ball D^3 (any S^2 boundary can be obtained by taking the connected sum with D^3), namely $(T^2 \times I)\#D^3$. On the other hand, we can start with the simplest *trousers* $S^2 \uplus S^2 \to S^2$ and glue on two of the boundaries the 1–cobordism $S^2 \to T^2$. The resulting manifold is the connected sum of two solid tori and a three–ball, i.e. $(D^2 \times S^1)\#(D^2 \times S^1)\#D^3$. The two cobordisms are not homeomorphic, since $T^2 \times I \not\cong (D^2 \times S^1)\#(D^2 \times S^1)$. This can be checked by computing the Euler characteristics of the two cobordisms, $\chi(T^2 \times I) = 0$, whereas $\chi[(D^2 \times S^1)\#(D^2 \times S^1)] = 2\chi(D^2 \times S^1) - 2 = -2$. The difference between these two cobordisms is depicted schematically in Figure 3.14 (both are constructed

3.4 Topology Change in 3D   401

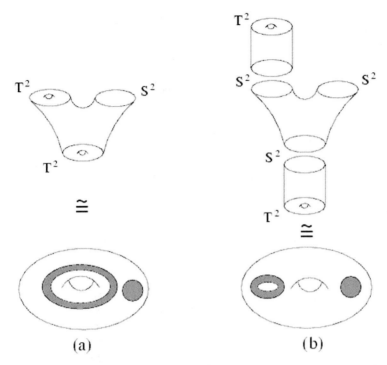

Fig. 3.14. Non-homeomorphic cobordisms (the shaded regions are holes in the solid tori): (a) $(T^2 \times I) \# D^3$; (b) $(D^2 \times S^1) \# (D^2 \times S^1) \# D^3$ (adapted and modified from [Ion97]).

from a solid torus by removing from its interior a ball D^3 and another solid torus $D^2 \times S^1$, the difference being in the way the $D^2 \times S^1$ is removed).

Handles in nD

In general, finding the elementary cobordisms in n dimensions is difficult, since it requires at least a classification of closed $(n-1)$D manifolds, which are the boundaries. The only manageable cases are the 0 and 1–cobordisms (and their duals, the $n-$ and $(n-1)$–cobordisms). Index 0–cobordism is just a creation of an S^{n-1} boundary, whereas the n–cobordism is the 'capping' of an S^{n-1} boundary. Both are equivalent to an n–ball D^n [Ion97].

The 1–handle is more complicated. Intuitively, it is similar to the 1–handle of the 3D case. We have to attach the $D^n = D^1 \times D^{n-1}$ along an arbitrary embedding of $S^0 \times D^{n-1}$ (two $(n-1)$–balls). This 1–handle is the higher dimensional analog to the solid tube from the 3D case. Again, the two ends can be on the same $(n-1)$ boundary, say V^{n-1} (wormhole creation), or they can be on two different boundaries (Einstein–Rosen bridge creation). In the first case, this is equivalent to taking the connected sum of the initial boundary with the nD wormhole $S^{n-2} \times S^1$, see Figure 3.15, with V^{n-1}, W^{n-1} arbitrary closed $(n-1)$–manifolds.

$$
\begin{array}{cl}
\lambda & \\
0 & \emptyset \to S^{n-1} \\
\\
1 & \begin{cases} V^{n-1} \to V^{n-1} \# (S^{n-2} \times S^1) \\ V^{n-1} \uplus W^{n-1} \to V^{n-1} \# W^{n-1} \end{cases} \\
\vdots & \\
n-1 & \begin{cases} V^{n-1} \# (S^{n-2} \times S^1) \to V^{n-1} \\ V^{n-1} \# W^{n-1} \to V^{n-1} \uplus W^{n-1} \end{cases} \\
\\
n & S^{n-1} \to \emptyset
\end{array}
$$

Fig. 3.15. Summary on handles in nD (adapted and modified from [Ion97]).

3.4.2 Oriented Cobordism and Surgery Theory

Recall that an oriented manifold X is an *oriented cobordism* between the oriented manifolds M and M' if ∂X, with the induced orientation, is diffeomorphic to the disjoint union of M and $-M'$. Here, $-$ denotes orientation reversal. Cobordism defines an equivalence relation on the space of oriented manifolds. Thus the manifold X has connected boundary $\partial X = M'$, neglecting the change in orientation [Har03].

Given a cobordism X, it is possible to obtain a different cobordism X', with $\partial X = \partial X'$, through *surgery* on X. Suppose X is nD. Intuitively, *surgery*, also known as *spherical modification*, should be thought of as removing an embedded kD sphere S^k and replacing it with an embedded sphere S^{n-k-1} of dimension $n-k-1$. A more precise description is as follows [Wal60, Mil65].

Start with an embedding of $\phi : S^k \times D^{n-k} \to X$. The boundary of the embedding is $S^k \times S^{n-k-1}$, which is also the boundary of $D^{k+1} \times S^{n-k-1}$. We may thus remove the interior of the embedding and replace it with the interior of $D^{k+1} \times S^{n-k-1}$. The result is the manifold

$$X' = \left(X - \phi(S^k \times 0)\right) + \left(D^{k+1} \times S^{n-k-1}\right),$$

where $-$ denotes removal and $+$ denotes an identification of $\phi(u, \theta v)$ with $(\theta u, v)$ for each $u \in S^k, v \in S^{n-k-1}$ and $0 < \theta \leq 1$. This is usually called a *type $(k, n-k-1)$ surgery*.[31] The process is illustrated in Figure 3.16.

Fig. 3.16. Surgery between X and X', both with boundary S^1. An $S^1 \times D^1$ is removed and replaced with a $D^2 \times S^0$. The change in topology is evident (adapted and modified from [Har03]).

Handle Decomposition and Causal Continuity

Once we have obtained an interesting cobordism X, it will be useful to consider its *handle decomposition* [RS72, Mil65, DG98]. A handle of index k on an nD manifold X is an n–disc D^n such that $X \cap D^n \subset \partial X$, and there is a homeomorphism $h : D^k \times D^{n-k} \to D^n$, such that $h(S^{k-1} \times D^{n-k}) = X \cap D^n$, where $\partial D^k = S^{k-1}$. Two simple examples are shown in Figure 3.17. Adding a handle is closely related to performing a surgery, as we shall see below.

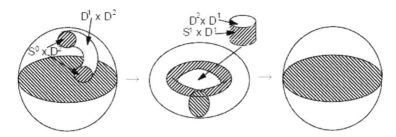

Fig. 3.17. Adding a 1-handle to D^2 to obtain a solid torus. Adding a 2-handle to a solid torus to reobtain the ball (adapted and modified from [Har03]).

A *handle decomposition* of a cobordism X from M to M' is a presentation

$$X = C_0 \cup H_1 \cup \cdots \cup H_t,$$

where $C_0 = M \times [0, 1]$ and H_k is a handle on the cobordism

$$X_{k-1} = C_0 \cup \{\cup H_l \mid l \leq k-1\}.$$

---

[31] Note that we are using surgery to modify the cobordism itself. This should not be confused with the use of surgery to construct cobordisms by modifying manifolds without boundaries.

This gives a procedure for constructing X from the trivial cobordism. If ∂X has a single connected component, M', then one may start from the disc $C_0 = D^n$. The handle decomposition of X is not unique. For example, Figure 2 shows a handle decomposition of a 2–disc as a 2–disc with a 1–handle and a 2–handle added. Handle decompositions are generic by the following theorem [RS72]: Every cobordism admits a handle decomposition.

Handle decomposition is also closely related to *Morse theory* [Mil65, Mil63]. Morse theory is used to define an almost Lorentzian metric on the cobordism with certain causal properties. Recall that a function $f : X \to \mathbb{R}$ has a *critical point* at $p \in X$ if $\partial_i f(p) = 0$. The critical point is non-degenerate if $\det[\partial_i \partial_j f(p)] \neq 0$. A *Morse function* on a cobordism X is a function $f : X \to \mathbb{R}$ that is constant on each connected component of ∂X and whose critical points are in the interior of X and non-degenerate. Every cobordism admits a Morse function, a result that follows from the previous theorem [Har03].

The *index* of a non-degenerate critical point p is the number of negative eigenvalues of the *Hessian* $\partial_i \partial_j f(p)$. The number of critical points with index k will be denoted $m_k(f)$. The following result is important (this is theorem 3.12 of [Mil65] translated into the language of handles): Given a handle decomposition of the cobordism X, then X admits a Morse function with exactly one critical point of index k for each k–handle in the decomposition.

The power of this result is that it gives us an equality for the number of critical points of a Morse function. This should be contrasted with the well-known *weak Morse inequalities* [Mil63] $b_k \leq m_k(f)$, where b_k are the Betti numbers of the manifold, X.

Given a Morse function f on X and a Riemannian metric G on X, which always exists, one may then construct an *almost Lorentzian metric* [LS97]

$$g_{\mu\nu} = G^{\rho\sigma} \partial_\rho f \partial_\sigma f G_{\mu\nu} - \zeta \partial_\mu f \partial_\nu f,$$

where $\zeta > 1$ is a real number. This metric is Lorentzian everywhere except at the critical points and has a well-defined causal structure because f acts as a time function. The timelike direction is $G^{\mu\nu} \partial_\nu f$. This almost Lorentzian metric is said to define a Morse space-time.

The final idea we need is that of *causal continuity* [HS74]. Intuitively, a space-time is causally discontinuous if the volume of the causal past or future of some point changes discontinuously under a continuous change in the point.

It is conjectured [DS98] that causally discontinuous space-times do not contribute to the Lorentzian sum over histories. It was further conjectured [DS98] that causal continuity should be associated with critical points of index 1 and $n - 1$ of Morse functions. It was later proven the following theorem [DGS99, Sor97]: If all Morse functions on a cobordism X contain critical points of index 1 or $(n-1)$, then the cobordism supports only causally discontinuous Morse space-times. Conversely, if X admits a Morse function with no critical points of index 1 or $(n-1)$, then it does support causally continuous Morse space-times. Thus, there is a selection rule for topology change. Topology change requires a cobordism with a handle decomposition with no 1–handles or $(n-1)$–handles [Har03].

3.5 Topology Change in Quantum Gravity

In this subsection, following [Dow02], we present the current understanding of topology change in quantum gravity.

3.5.1 A Top–Down Framework for Topology Change

What is meant by a topology change in quantum gravity is a space-time based on an nD manifold M, with an initial space-like $(n-1)$D hypersurface Σ_0, and a final space-like hypersurface Σ_1, not diffeomorphic to Σ_0. For simplicity in what follows we take Σ_0 and Σ_1 to be closed and M compact so that the boundary of M is the disjoint union of Σ_0 and Σ_1. This restriction means effectively that we're studying topology changes that are localized. For example, a topology change from $\mathbb{R}^3$ to $\mathbb{R}^3$ with a *handle* – an $S^2 \times S^1$ – attached can be reduced to the compact case because infinity in both cases is topologically the same.

Following S. Hawking in [Haw78, Haw78b], we will use the *path–integral approach* (see Chapter 4) to quantum gravity which can be summarized in the following formula for the *transition amplitude* between the Riemannian metric h_0 on $(n-1)$–manifold Σ_0 and the Riemannian metric h_1 on $(n-1)$–manifold Σ_1 [Dow02]

$$\langle h_1 \Sigma_1 | h_0 \Sigma_0 \rangle = \sum_M \int_g [dg]\omega(g) \ . \qquad (3.109)$$

The sum is over all n–manifolds M, called cobordisms, whose boundary is the disjoint union of Σ_0 and Σ_1, and the functional integral is over all metrics on M which restrict to h_0 on Σ_0 and h_1 on Σ_1. Each metric contributes a weight, $\omega(g)$, to the amplitude. It is clear from this that the path–integral framework lends itself to the study of topology change as it readily accommodates the inclusion of topology changing manifolds in the sum. Despite the fact that we may not be able to turn (3.109) into a mathematically well–defined object within the top down approach, if even the basic form of this transition amplitude is correct then we can already draw some conclusions. We can say that a topology change from Σ_0 to Σ_1 can only occur if there is at least one manifold which interpolates between them, in other words if they are cobordant. This does not place any restriction on topology change in 3+1 space-time dimensions since all closed three–manifolds are cobordant, but it does in all higher dimensions: not all closed four–manifolds are cobordant, for example. We can also say that even if cobordisms exist, there must also exist appropriate metrics on at least one cobordism and so we come to the question of what the metrics should be. There are many possibilities and just three are listed here [Dow02]:

A. Euclidean (i.e., positive definite signature) metrics. This choice is of course closely associated with Hawking and the whole programme of Euclidean quantum gravity [GH93]. It is to this tradition and to Hawking's influence that we attribute my enduring belief that topology change does occur in quantum gravity. Indeed, Euclidean (equivalently, Riemannian) metrics exist on any cobordism and it would seem perverse to exclude different topologies from the path–integral.
B. Lorentzian (i.e., $(-, +, +, \cdots +)$ signature) metrics. With this choice we are forced, by a theorem of Geroch [Ger67], to contemplate closed time-like curves (CTC's or time machines). Geroch proved that if a Lorentzian metric exists on a topology changing cobordism then it must contain CTC's or be time non-orientable. Hawking has been at the forefront of the study of these causal pathologies, formulating his famous Chronology Protection Conjecture [Haw92]. Hawking and Gibbons also proved that requiring an $SL(2,\mathbb{C})$ spin structure for fundamental fermi fields on a Lorentzian cobordism produces a further restriction on allowed topology changing transitions [GH92a, GH92b].
C. Causal metrics. By this we mean metrics which give rise to a well–defined 'partial order' on the set of space-time events. A partial order is a binary relation, $\prec$, on a set P, with the properties:
- (i) transitivity: $(\forall x, y, z \in P)(x \prec y \prec z \Rightarrow x \prec z)$
- (ii) irreflexivity: $(\forall x \in P)(x \not\prec x)$.

A Lorentzian metric provides a partial order via the identification $x \prec y \Leftrightarrow x \in J^-(y)$, where the latter condition means that there's a future directed curve from x to y whose tangent vector is nowhere space-like (a 'causal curve'), so long as the metric contains no closed causal curves. The information contained in the order $\prec$ is called the 'causal structure' of the space-time. By Geroch's theorem, we know there are no Lorentzian metrics on a topology changing cobordism that give rise to a well–defined causal structure. But, there are metrics on any cobordism which are Lorentzian almost everywhere which do [Sor89]. These metrics avoid Geroch's theorem by being degenerate at a finite number of points but the causal structure at the degenerate points is nevertheless meaningful.

3.5.2 Morse Metrics and Elementary Topology Changes

Morse theory gives us a way of breaking a cobordism into a sequence of elementary topology changes [Yod72, Sor89]. On any cobordism M there exists a Morse function, $f: M \to [0, 1]$, with $f|_{\Sigma_0} = 0$, $f|_{\Sigma_1} = 1$ such that f possesses a set of critical points $\{p_k\}$ where $\partial_a f|_{p_k} = 0$ and the Hessian, $\partial_a \partial_b f|_{p_k}$, is invertible. These critical points, or Morse points, of f are isolated and, because M is compact, there are finitely many of them. The index, λ_k, of each Morse point, p_k is the number of negative eigenvalues of the Hessian at p_k. It is the number of maxima in the generalized saddle point at p_k if f is interpreted as a height function. For space-time dimension n, there are $n+1$ possible values for the indices, $(0, 1, \ldots n)$. A cobordism with a single Morse point is called an elementary cobordism [Dow02]. Three elementary cobordisms for $n = 2$ are shown in Figure 3.18. They are the $\lambda = 2$ yarmulke in which a circle is

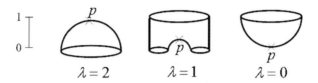

Fig. 3.18. Three elementary cobordisms for $n = 2$ and $\lambda = 2, 1, 0$: the 'yarmulke', 'trousers' and time–reverse of the 'yarmulke' (adapted and modified from [Dow02]).

destroyed, the $\lambda = 1$ trousers in which two circles join to form a single circle and the $\lambda = 0$ time reverse of the yarmulke in which a circle is created from nothing. (The upside–down trousers is in fact also a $\lambda = 1$ elementary cobordism: locally the Morse point looks the same as the regular trousers with one maximum and one minimum.) For higher space-time dimensions, n, the generalizations of these are easy to visualize: the index n yarmulke (or its time reverse of index 0) is half an n-sphere, the index 1 trousers (or its time reverse of index $n - 1$) is an n-sphere with three balls deleted creating three S^{n-1} boundaries. For $n > 3$ qualitatively different types of Morse point exist with at least two maxima and two minima, i.e. $\lambda \neq 0, 1, n - 1, n$.

Using a Morse function, f, on M we can construct *Morse metrics*, which are Lorentzian everywhere except at the Morse points where they are zero. The precise form is not important here, but roughly the Morse function is used as a time function as you'd expect. These Morse metrics are our candidates for inclusion in the path–integral for quantum gravity.

Now, there is important counter-evidence to the claim that topology change occurs in quantum gravity. This is work which shows that the expectation value of the energy–momentum tensor of a massless scalar field propagating on a $(1+1)$ Morse trousers is singular along the future light cone of the Morse point [MCD88]. In addition, one can look instead at the in-out matrix element of the energy momentum tensor and one finds a singularity along both the future and past light cones of the Morse point (calculation described in [Sor89]). This last result in particular, if it can be extended to all Morse metrics on the $(1+1)$–trousers, can be taken to suggest that in the full path–integral expression for the transition amplitude, integrating out over the scalar field first will leave an expression for the effective action for g that is infinitely sensitive to fluctuations in g. Thus, destructive interference between nearby metrics will suppress the contribution of any metric on the trousers. To be sure, this is a heuristic argument that would need to be strengthened but suppose it is valid. Would this mean, as DeWitt has argued, that all topology change is suppressed? The answer is not necessarily, especially if the following two conjectures hold.

The first conjecture is based on the idea that it is a certain property, called 'causal discontinuity', of the causal structure of the $(1+1)$–trousers that is the origin of the bad behavior of quantum fields on it. So the conjecture (Sorkin) is that quantum fields will be singular on causally discontinuous space-times but well–behaved on causally continuous space-times. The second conjecture (Borde and Sorkin) is that only Morse metrics containing index 1 or $n-1$ points (trousers type) are causally discontinuous.

So what is *causal discontinuity*? S. Hawking invented this concept in work with R.K. Sachs [HS74]. That paper is a piece of hard mathematical physics but there is a physically intuitive way of understanding the concept. Roughly, a space-time is causally discontinuous if the causal past, or future, of a point changes discontinuously as the point is moved continuously in space-time. We can see from Figure 3.19 that it is very plausible that the $(1+1)$–trousers is causally discontinuous: an observer down in one of the legs will have a causal past that is contained only in that leg, but as the observer moves up into the waist region, as they pass the future light cone of the Morse point, their causal past will suddenly get bigger and include a whole new region contained in the other leg. Hawking and Sachs conclude their paper by saying, "There is some reason, but no fully convincing argument, for regarding causal continuity as a basic macro-physical property."

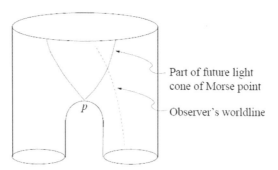

Fig. 3.19. The $(1+1)$–trousers with part of the future light cone of the Morse point, p. The other part goes up the back. The past light cone of the Morse point also has two parts, one down each leg (adapted and modified from [Dow02]).

3.5.3 'Good' and 'Bad' Topology Change

If we assume the two conjectures of the previous section hold, and that the argument about the consequent suppression of causally discontinuous metrics in the path–integral is valid, the implication is that cobordisms which admit

3.5 Topology Change in Quantum Gravity

only Morse metrics containing index 1 and/or $n-1$ points will be suppressed in the sum over manifolds. We can use this to draw conclusions about many interesting topology changing processes in quantum gravity. They divide into 'good' ones which can occur and 'bad' ones which do not. The results in this section and the next are taken from a series of papers on topology change [DG98, DS98, BDG99, DGS99].

Good processes include the pair production of black holes of different sorts in a variety of scenarios worked on by many people including Hawking (see the contribution by Simon Ross in this volume). For example, in 4 space-time dimensions, the manifold of the instanton that is used to calculate the non-extremal black hole pair production rate [GS91] admits a Morse function with a single index 2 Morse point. The pair production of Kaluza–Klein monopoles [DGG94] is also good, as is the nucleation of spherical bubbles of Kaluza-Klein $(n-5)$-branes in magnetic fluxbrane backgrounds [DGG96]. The decay of the Kaluza–Klein vacuum [Wit82] is good, which is slightly disappointing: one might have hoped that it would be stabilized by these considerations. We know, however, that the cobordism for KK vacuum decay is the same as that for pair production of KK monopoles [DGG95] and so if the latter is a good process so must the former be.

The Big Bang, or creation of an $(n-1)$-sphere from nothing via the yarmulke, is good. Notice that in this way of treating topology change as a sequence of elementary changes, the universe, if created from nothing, must start off as a sphere. No other topology is cobordant to the empty set via an elementary cobordism. The conifold transition in string theory [GMS95] where a three–cycle shrinks down to a point and blows up again as a two–cycle is good. Indeed, the shrinking and blowing up process traces out the 7D cobordism (each stage of the process is a level surface of a corresponding Morse function) and the fact that it is a three–cycle that degenerates and a two–cycle that blows up tells us that the index of the cobordism is three [Dow02].

Bad topology changes include space-time wormholes where an S^3 baby universe is born by branching off a parent universe, the epitome of a trousers cobordism. Hawking founded the study of baby universes and space-time wormholes [Haw88] within the Euclidean quantum gravity framework where our present considerations do not apply. However, if one takes the view that Euclidean solutions, instantons, are to be thought of as devices for calculating transition amplitudes which are nevertheless defined as sums over real, causal, space-times, then the badness of the trousers would be counter-evidence for the relevance of space-time wormholes.

Another bad topology change is the pair production or annihilation of topological geons, particles made from non-trivial spatial topology. This deals a serious blow to the hope that the processes of pair production and annihilation of geons could restore to geons the spin–statistics correlation that they lack if their number is fixed [DS98b].

In (1+1) and (2+1) space-time dimensions, all topology changes except for the yarmulkes and their time-reverses are bad ones. This raises the question, is this not in conflict with string theory and the finiteness of topology changing amplitudes in $(2+1)$–quantum gravity [Wit89]? It would seem that the first order formalism and the metric formalism are genuinely different theories of gravity and distinguishing between them might be an observational issue [Dow02].

3.5.4 Borde–Sorkin Conjecture

Having looked at some of the consequences of the conjectures, we can ask how plausible they are. There is fragmentary evidence for the conjecture that causal discontinuity leads to badly behaved quantum fields but causally continuous topology changes allow regular quantum field behavior [Sor89]. A key investigation that needs to be done is of quantum field theory on a four dimensional space-time with an index two point, which is conjectured to be regular.

On the other hand we are well on the way to proving the Borde–Sorkin conjecture that Morse space-times are causally continuous if and only if they contain no index 1 or $n-1$ points [Dow02]. We sketch here the basic ideas involved in the progress made to date. If we think about the causal structure around the Morse point, p, of the $(1+1)$–trousers, it seems intuitive that the causal past of p should contain two separate parts, one down each 'leg' of the trousers. And the causal future of the Morse point also divides into two lobes, one up the front and one up the back of the trousers. It's also intuitive that the causal discontinuities of the $(1+1)$–trousers should be related to the disconnectedness of the causal past and future of p in the neighborhood of p [Dow02]. Flattening out the crotch region, we should obtain a causal structure that looks like that shown in Figure 3.20. There is a special metric for the $(1+1)$–trousers in which the causal structure can be proved to be exactly as shown: the past (future) of the Morse point p consists of the two regions P_1 and P_2 (F_1 and F_2).

There are two types of causal discontinuity here. The first type is when an observer starting in P_1, say, crosses the past light cone of p into S_1, say. As it does so the causal future of the observer, which at first contains regions in both F_1 and F_2, jumps so that it no longer contains any points in F_2. The second type is when an observer in S_1, say, crosses the future light cone of p into F_1. As this happens, the causal past of the observer which contained no points in P_2 suddenly grows to contain a whole new region in P_2.

The special metric in which this behavior can be demonstrated exactly generalizes to higher dimensions and all Morse indices. For dimension n and index $\lambda \neq 0, n$ (no yarmulkes for now), the causal past and future of p are obtained from figure 3.20 by rotating it around the x-axis by $SO(n-\lambda)$ and around the y-axis by $SO(\lambda)$. We see that when $\lambda = 1$ the past of p remains in two disconnected pieces and when $\lambda = n-1$ the future of p remains in

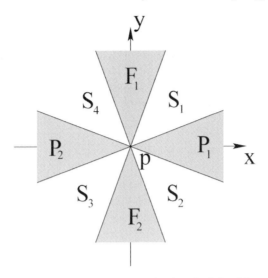

Fig. 3.20. The causal structure in the neighborhood of the Morse point, p, of the $(1+1)$-trousers. The past (future) of p consists of the two regions P_1 and P_2 (F_1 and F_2). The 'elsewhere' of p divides into four regions $S_1, \ldots S_4$ (adapted and modified from [Dow02]).

two pieces. But when $\lambda \neq 1, n-1$ then both the future and past of p become connected sets.

These suggestive pictures can be turned into a proof that, for these special metrics, index 1 and $n-1$ Morse points produce causal discontinuities and the other indices do not. We can further show that, not just these special metrics, but any index 1 and $n-1$ Morse metric is causally discontinuous. It is also true that any Morse metric on the yarmulke is causally continuous. It remains to be proved that any Morse metric on a $\lambda \neq 0, 1, n-1, n$ elementary cobordism is causally continuous.

For more technical details on topology change in quantum gravity, see [Dow02].

3.6 A Handle-Body Calculus for Topology Change

In this section, following [DG98], we a handle-body calculus for generic topology change in smooth manifolds.

The question of whether the topology of space can change is a basic one in the search for a theory of quantum gravity. The theorems of Geroch [Ger67] and Tipler [Tip70] are widely understood to show that there is no topology change in classical general relativity, so that we should look to the quantum theory to see it, if it occurs. Though the definitive statement about the occurrence of topology change may well have to wait until we have a fully developed theory of quantum gravity it is nevertheless generally believed that topology change does happen. A general calculus for topology change within the path–integral approach, based on *Morse theory*, has been suggested by R. Sorkin [Sor90]. The paper [DG98] has reviewed this picture, and used it to investigate certain interesting physical topology–changing processes.

Let an n–geometry (M, g) consist of an nD manifold M and a metric g on M–strictly, a geometry is an equivalence class of such pairs under diffeomorphism. A *topology change* in n space-time dimensions is a transition from a Riemannian $(n-1)$–geometry (W_0, h_0) to another Riemannian $(n-1)$–geometry (W_1, h_1) in which W_0 and W_1 are non-diffeomorphic.

We call (M, V_0, V_1) a *smooth manifold triad* if M is a compact smooth n–manifold whose boundary is the disjoint union of the two closed submanifolds V_0 and V_1, $\partial M = V_0 \uplus V_1$. Given two closed smooth $(n-1)$–manifolds, W_0 and W_1, a *topological cobordism* from W_0 to W_1 is a 5-tuple (M, V_0, V_1, d_0, d_1) where (M, V_0, V_1) is a smooth manifold triad and $d_i : V_i \to W_i$ is a diffeomorphism, $i = 0, 1$. cobordism gives rise to an equivalence relation on the set of $(n-1)$ manifolds. We say that W_0 and W_1 are in the same cobordism class if a topological cobordism exists between them. A *Lorentzian cobordism* from geometry (W_0, h_0) to (W_1, h_1) is a 6-tuple $(M, V_0, V_1, d_0, d_1, g)$ where (M, V_0, V_1, d_0, d_1) is a topological cobordism and g is a Lorentzian metric on M such that $(d_i^{-1})^*(g_{|V_i}) = h_i$, $i = 0, 1$ and such that V_0 is a past space-like boundary and V_1 is a future space-like boundary. We will often drop the explicit mention of the diffeomorphisms in what follows and unless otherwise stated all cobordisms M will be compact [DG98].

A necessary and sufficient condition for a topological cobordism to exist between a given pair of manifolds is that their *Steifel–Whitney and Pontrjagin numbers* coincide when both are oriented or just their Steifel–Whitney numbers in the non-oriented case [MS74], Sto68]. Hence the number of cobordism classes equals that of distinct combinations of Steifel–Whitney and Pontrjagin numbers, which has a finite value, depending on the dimension. As it happens, all 3–manifolds are cobordant, while 4–manifolds divide into four cobordism classes.

Now that we have explained what we mean by topology changing transitions between two space-like hyper-surfaces, we must decide how to investigate them. Among the different approaches to quantum gravity, the path–integral (see Chapter 4) affords the most natural expression for topology changing transition amplitudes [DG98]

$$\langle W_1, h_1; W_0, h_0 \rangle = \sum_{(M, V_0, V_1, d_0, d_1)} \omega(M, d_0, d_1) \int_{\mathcal{C}} Dg \, e^{iS[g]}, \qquad (3.110)$$

where the sum is over topological cobordisms and $\mathcal{C}$ is a class of metrics, g, on M such that $(d_i^{-1})^*(g|_{V_i}) = h_i$, $i = 0, 1$. The weight $\omega(M, d_0, d_1)$ will not concern us here but is discussed in [Sor97]. Although this formal expression is far from being defined, and indeed may never be so without recourse to

3.6 A Handle-Body Calculus for Topology Change

a possibly discrete underlying theory, we can already draw some conclusions from its general form. For example, if W_0 and W_1 are not (co)bordant then the amplitude for the topology change is zero.

There are various proposals for the type of metrics over which the functional integral runs for each topological cobordism M. Following Sorkin [Sor97] we start with the view that the integral should be over all Lorentzian metrics but this immediately raises a problem. In the event of topology change, the geometry (M, g) cannot be both Lorentzian and causally ordered. This follows from the following theorem of Geroch [Ger67]: If a smooth triad (M, V_0, V_1), with V_0 and V_1 closed, admits a time–orientable Lorentzian metric g without closed time-like curves and such that V_0 and V_1 are space-like with respect to g, then $V_0 \cong V_1$ and $M \cong V_0 \times I$ where I is the unit interval, i.e., there is no topology change.

So which do we choose to keep: causal order or the equivalence principle? Following Sorkin [Sor86b] we plump for casual order. For one thing, if we were instead to insist on globally time–orientable Lorentzian metrics this would rule out the production of Kaluza–Klein monopole–antimonopole pairs since there does not exist such a metric on any topological cobordism for this process [Sor86b, Sor86c]. Also, if causal sets are the correct description of the discrete substructure of space-time then causal order is more fundamental than metric [Sor97]. Pursuing this route, however, means we must allow singularities of some sort in the geometries (M, g) that contribute to the amplitude for a topology changing process. So what singularities are allowed? Sorkin has suggested that *Morse theory* (see section 3.3 above) furnishes the appropriate metrics that are Lorentzian almost everywhere and exist on all topological cobordisms.

A *Morse function* on a manifold M is a smooth function $f : M \to \mathbb{R}$ such that $\partial_\mu f$ vanishes only at a finite number of points p_k where the Hessian $\partial_\mu \partial_\nu f|_{p_k}$ is a non degenerate matrix. The *Morse index* λ_k of each critical point p_k is the number of negative eigenvalues of the Hessian matrix evaluated at p_k. The critical values of f are the values it takes at the critical points; we will often denote them $c_k = f(p_k)$. The abundance of Morse functions on a manifold is enough to ensure the following theorem [DFN95, MS74)]: For any smooth triad, (M, V_0, V_1), there exists a Morse function $f : M \to [0, 1]$ such that we have the following theorem [DG98]:

A. $f^{-1}(0) = V_0$ and $f^{-1}(1) = V_1$;
B. f has no critical points on $\partial M = V_0 \uplus V_1$

Then given any Riemannian metric G on M and a real number $\zeta > 1$, we can construct an almost Lorentzian metric g associated with f as follows:

$$g_{\mu\nu} = \partial_\rho f \partial_\sigma f G^{\rho\sigma} G_{\mu\nu} - \zeta \partial_\mu f \partial_\nu f \qquad (3.111)$$

and we call this a *Morse metric*. It is Lorentzian everywhere except for the Morse points and $G^{\mu\nu} \partial_\nu f$ defines a time-like direction. If moreover Riemannian metrics are given on V_0 and V_1, we can demand that g has the correct restrictions by choosing G appropriately. We summarize these statements as the following lemma: Let (M, V_0, V_1, d_0, d_1) be a topological cobordism between W_0 and W_1 and let h_0 and h_1 be Riemannian metrics on W_0 and W_1. Then there exists a Morse metric g on M such that $(d_i^{-1})^*(g_{|V_i}) = h_i$, $i = 0, 1$. For the proof see [DG98].

Any topological cobordism has an infinite number of Morse metrics associated with it. We call each such geometry (M, g) an Almost Lorentzian (AL) cobordism and restrict the functional integral to be over such cobordisms. Since these geometries are singular, it will be necessary to extend the definition of the action S to these cases. In this view the critical points are not to be sent to infinity as in [Yod72] but rather remain part of the space-time and indeed the causal order is well defined with the Morse points present.

Now, Sorkin suggested that it might be necessary to impose a stronger condition on the set of contributing metrics. This observation is motivated by a very simple example in (1+1) dimensions: quantum field theory on the trousers cobordism. The $(1+1)$D trousers admits an everywhere flat AL metric with a single index one Morse point at the crotch which singularity is the source for an infinite burst of energy of a scalar quantum field propagating on the trousers [AD86]. Anderson and DeWitt have argued that this provides evidence against topology change. But the regular propagation of a quantum field on the $(1+1)$ 'yarmulke' topology, a hemisphere mediating the transition $\emptyset \to S^1$, suggests that it might be a particular feature of the trousers topology, and not a general flaw of all nontrivial cobordisms, that causes the un-physical energy burst. A crucial difference between the trousers and the yarmulke topologies is that the former has a causal discontinuity whereas the latter does not (roughly speaking a causal discontinuity is a discontinuous change in the volume of the causal past or future of a continuously varied point [HS74]). Generalizing this idea Sorkin conceived the following conjectures:

(i) A quantum field propagating on an AL cobordism (M, g) has an unphysically singular behavior if and only if (M, g) is causally discontinuous; and

(ii) An nD AL cobordism (M, g) is causally discontinuous if and only if the Morse function from which g is constructed has either an index 1 or index $(n-1)$ critical point.

3.6.1 Handle-body Decompositions

Define the *Morse structure* of a Morse function f on M to be a complete ordered list $\{(p_k, \lambda_k) : k = 1, \ldots r\}$ of its Morse points and corresponding Morse indices. As we shall see, a handle-body decomposition of a manifold, M, implies the existence of Morse functions on M with totally determined Morse structure. The following definitions follow very closely the first pages of Kirby's book [Kir89]. They make extensive use of the concepts of closed or open n-balls, n−spheres and their respective boundaries, which are listed here [DG98]:

3.6 A Handle-Body Calculus for Topology Change

$$B^n = \{x \in \mathbb{R}^n : |x|^2 \le 1\} \qquad S^n = \{x \in \mathbb{R}^{n+1} : |x|^2 = 1\}$$
$$\partial B^n = S^{(n-1)} \qquad \partial S^n = \emptyset$$
$$\dot{B}^n = \{x \in \mathbb{R}^n : |x|^2 < 1\}$$

By $\dot{A}$ we mean the interior of the set A, i.e., the largest open set contained in A. Note for future reference that when A is a subset of the manifold with boundary M and ∂M is not empty $\dot{A}$ may contain part of it.

A *handle-body decomposition* of an nD compact manifold M is a nested sequence of manifolds $\emptyset = M_{-1} \subset M_0 \subset M_1 \subset \cdots \subset M_r = M$ where M_k is obtained by adjoining a λ_k handle to M_{k-1}, i.e., $M_k = M_{k-1} +_{h_k} B^{\lambda_k} \times B^{n-\lambda_k}$ via an embedding, $h_k : \partial B^{\lambda_k} \times B^{n-\lambda_k} \hookrightarrow \partial M_{k-1}$ of the boundary of the λ_k handle into the boundary of M_{k-1}. Note that $M_0 = B^n$ in any such handle-body sequence. This definition involves two operations whereby a pair of manifolds with boundary can be combined: adjunction ($+$) and product ($\times$).

Associated with any smooth handle-body decomposition is a Morse function $f : M \to [0,1]$ with as many critical points as handles being attached. For the $r+1$-handled-body in the definition, the function f would have $r+1$ non-degenerate critical points, $\{p_k\}$, $k = 0, 1, \ldots r$, which can be taken to lie in different level surfaces, i.e., $f(p_0) < f(p_1) < \cdots < f(p_r)$. Each critical point may be located at the center $(0,0)$ of $B^{\lambda_k} \times B^{n-\lambda_k}$; then $B^{\lambda_k} \times \{\mathbf{0}\}$ is the descending manifold and $\{\mathbf{0}\} \times B^{n-\lambda_k}$ the ascending manifold. By this we mean that around p_k the function f admits an expansion (Morse lemma, see [MS74]):

$$f(q) = f(p_k) - x_1^2 - x_2^2 - x_{\lambda_k}^2 + x_{\lambda_k + 1}^2 + \cdots + x_n^2.$$

The first λ_k local coordinates parameterize B^{λ_k}, the last $n - \lambda_k$ local coordinates parameterize $B^{n-\lambda_k}$ and p_k is identified as a Morse point of index λ_k. In other words, we can define f following the sequence of manifolds. It is zero at some point of M_0, which is the index 0 critical point p_0; it then increases in a regular way except for the critical point associated with each handle attachment.

The Morse function, f, associated with the handle-body decomposition given above is 1 on the boundary of M. As such, it is appropriate for the case of the topology change from the empty set to ∂M. We are interested in the more general case of topology change from V_0 to V_1. In that case we have a manifold, M, whose boundary is the disjoint union of V_0 and V_1. A generalized handle-body decomposition of M is a nested sequence $V_0 \times B^1 = M_0 \subset M_1 \subset \ldots M_r = M$ where M_k is obtained by attaching a λ_k handle to M_{k-1}. But now there is a restriction on each embedding h_k: its image must not intersect the initial V_0 component of the boundary of M_k. In this handle-body calculus, the addition of each handle can be thought of as an elementary topological transition from ∂M_k to ∂M_{k+1}.

A useful property of handle-body decompositions is 'right distributivity' of a B^m. It allows us to deduce from a handle-body decomposition for a manifold

M a whole series of higher dimensional handle-bodies for the manifolds $M \times B^m$. The crucial point is that the new B^m factor does not actively partake in the induced imbedding $\partial B^\lambda \times B^{n+m-\lambda} \hookrightarrow \partial(M \times B^m)$. More explicitly we have the following lemma: Let M be an nD manifold; then if

$$M = B^n + \sum_{k=1}^{r} B^{\lambda_k} \times B^{n-\lambda_k}$$

it follows that

$$M \times B^m = B^{n+m} + \sum_{k=1}^{r} B^{\lambda_k} \times B^{n+m-\lambda_k}.$$

For the proof, see [DG98].

Given that $M \cong L +_h B^\lambda \times B^{n-\lambda}$ through the embedding $h : \partial B^\lambda \times B^{n-\lambda} \hookrightarrow \partial L$, we define

$$\tilde{h} : \partial B^\lambda \times (B^{n-\lambda} \times B^1) \hookrightarrow \partial L \times B^1 \subset \partial(L \times B^1)$$

by

$$\tilde{h}(x,t) = (h(x), t)$$

where $x \in \partial B^\lambda \times B^{n-\lambda}$ and $t \in B^1$. This new embedding induces a map $f : (L +_h B^\lambda \times B^{n-\lambda}) \times B^1 \to L \times B^1 +_{\tilde{h}} B^\lambda \times B^{n+1-\lambda}$ given by [DG98]:

$$f(([z]_h, t)) = [(z,t)]_{\tilde{h}} \tag{3.112}$$

The map f is well defined, independent of class representative, and since the same holds for its obvious inverse, f is a bijection. It is also a homeomorphism of topological spaces. It can be shown that f has differentiable local representatives even at the smoothed corner set [KS77] once it has been composed with the relevant smoothing maps. Thus f is a diffeomorphism; it expresses distributivity between adjunction and product of manifolds with boundary. Figure 3.21 illustrates a simple case of B^1 distributivity: the handle-body for the annulus $S^1 \times B^1 \cong B^2 + B^1 \times B^1$ gives rise to the handle-body $S^1 \times B^2 \cong B^3 + B^1 \times B^2$.

With the machinery of Morse theory and handle-bodies in hand we can investigate *topology changing processes*. First of all the content of the conjectures translates into the following statements. If a smooth triad (M, V_0, V_1) has a handle-body decomposition which does not include a $B^1 \times B^{n-1}$ nor a $B^{n-1} \times B^1$ handle, then it admits a CCAL metric and according to our premises M is to be included in the path–integral for the process. On the other hand, if a smooth triad has a handle-body decomposition which does contain a 1-handle or an $(n-1)$-handle then we cannot draw the contrary conclusion. For example consider two decompositions of B^3 (Figure 3.22):

$$B^3 = B^3 + B^1 \times B^2 + B^2 \times B^1 \tag{3.113}$$

3.6 A Handle-Body Calculus for Topology Change

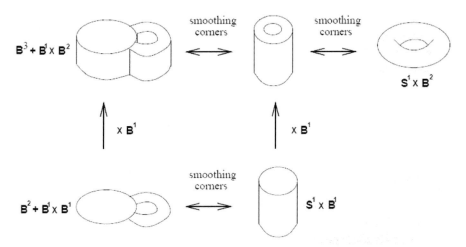

Fig. 3.21. Right B^1 distributivity lifting a 2D handle-body, the hollow cylinder (bottom-right corner), to a 2D handle-body, the solid torus (top-right corner) (adapted and modified from [DG98]).

$$B^3 = B^3 \tag{3.114}$$

In view of (3.113) alone we would be wrong to conclude that B^3 supports no CCAL cobordisms for the creation of S^2 since (3.114) shows that B^3 does support causally continuous cobordisms.

However, the Morse inequalities [MS74] do furnish a sufficient, but not necessary, criterion for automatically discarding certain cobordisms. Consider the triad (M, V_0, V_1). Let $\beta_\lambda(M, V_0)$ be the λth Betti number of M relative to V_0 and let μ_λ denote the number of critical points of index λ of a Morse function $f : M \to [0, 1]$ with $f^{-1}(0) = V_0$ and $f^{-1}(1) = V_1$. Then a weak version of the Morse inequalities establishes that:

$$\mu_\lambda \geq \beta_\lambda(M, V_0) \tag{3.115}$$

So if the first or $(n-1)^{\text{th}}$ homology of M relative to V_0 has non-trivial torsion free part, any Morse function on M must have index 1 or index $(n-1)$ points.

As an example consider the cobordism $B^4 \times S^1$ for creation of an $S^3 \times S^1$ (V_0 is empty here). We can compute its homology using the *Kunneth formula* [GH81] for the homology groups of the product of two spaces when both have torsion–free homologies, namely $H_q(X \times Y) = \sum_{p=0}^{q} H_p(X) \otimes H_{q-p}(Y)$. Applying this to $B^4 \times S^1$ gives [DG98]:

$$H_1(B^4 \times S^1) = H_0(B^4) \otimes H_1(S^1) + H_1(B^4) \otimes H_0(S^1)$$
$$= \mathbb{Z} \otimes \mathbb{Z} + 0 \otimes \mathbb{Z} = \mathbb{Z}$$

The same, applied to $B^2 \times S^3$ gives:

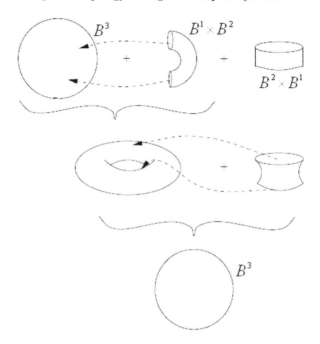

Fig. 3.22. 'Redundant' decomposition of the 3–ball (adapted and modified from [DG98]).

$$H_1(B^2 \times S^3) = H_0(B^2) \otimes H_1(S^3) + H_1(B^2) \otimes H_0(S^3)$$
$$= Z \otimes 0 + 0 \otimes Z = 0$$

Thus $\beta_1(B^4 \times S^1) = 1$, while $\beta_1(B^2 \times S^3) = 0$. In conjunction with (3.115) $\beta_1(B^4 \times S^1) = 1$ tells us that there are no Morse functions on $B^4 \times S^1$ without index 1 critical points and so this cobordism does not admit CCAL metrics. But for general M a vanishing β_1 does not guarantee that there is an allowed Morse function on M, since there is no reason why Morse functions should exist that saturate the inequalities. In particular, from $\beta_1(B^2 \times S^3) = 0$ alone we could not infer that $B^2 \times S^3$ admits CCAL metrics. It is only in view of the handle-body decomposition given earlier that we can so conclude.

3.6.2 Instantons in Quantum Gravity

Instantons in quantum gravity are the analogues of tunnelling solutions in quantum mechanics. When we consider tunnelling of a point particle from an unstable minimum x_∞ to a position x_0 of zero momentum, we calculate the transition amplitude $< x_0, 0 | x_{-\infty}, -\infty >$. One can show that the path–integral is well approximated by Ae^{-S} where A is a prefactor and S is the action of the classical Euclidean solution. By analogy, an instanton in gravity is a solution of the Euclidean Einstein equations that interpolates between an initial unstable state U_0, approached asymptotically, and a zero–momentum hyper-surface U_1 which is initial data for the post–decay Lorentzian evolution. The existence of such an instanton is usually taken as strong evidence that the transition takes place and the amplitude is approximately given by Ae^{-S} where S is the action of the instanton. We are investigating the suggestion that the path–integral in quantum gravity be defined fundamentally as a sum over CCAL cobordisms – or over AL cobordisms with causal continuity enforced dynamically. That means first of all that there must be some CCAL cobordisms for the transition under consideration. Secondly, it seems reasonable that there would only be an instanton approximation if the instanton had a background topology that was included in the sum over manifolds in equation (4.14), i.e., one which admits CCAL metrics. Thus we want to check that when instantons are invoked as evidence that topology changing processes occur, the instanton manifolds admit CCAL metrics.

Localised Topology Change

Before turning to our specific examples, we first prove some results necessary because the processes to be considered are embedded in an ambient asymptotically flat region. We could think of this as the topology change taking place within a lab with fixed walls say. Clearly our Morse and handle-body technology will have to be adapted to apply to these non-compact manifolds. This will not be difficult because, with the assumption that the topology change is localised in space, we can reduce the questions to the closed case by, roughly speaking, closing off space. Once we demonstrate the existence of CCAL metrics in the compact cobordism, we open back to the physical manifolds. That this can be done without disrupting the Morse structure of the metric is the content of the 'decompactifying' lemmas stated below. Their proof is given in [DG98].

We use the concept of a gradient-like vector–field for a Morse function f on a manifold M. Defining such a vector–field amounts to covering M with a congruence of curves, along which f increases, without reference to any particular Riemannian metric on M. We borrow the definition from [MS74], while our construction of a concrete vector–field is a simple generalization to non-elementary cobordisms of the one given therein. Let f be a Morse function on the nD manifold M with a set of r Morse points $P = \{p_k\}$. For simplicity we assume that each Morse point occurs on a distinct level surface of f though this assumption can easily be dropped.

A *gradient-like vector–field* ξ for f is a smooth vector–field on M with properties:

(i) $\xi(f) > 0 \ \forall q \notin P$

(ii) ξ has coordinates $(-2x_1, \cdots, -2x_{\lambda_k}, 2x_{\lambda_k+1}, \cdots, 2x_n)$ in a neighborhood of p_k where f admits expansion $f(q) = f(p_k) - \sum_{1 \leq i \leq \lambda_k} x_i^2 + \sum_{\lambda_k < j \leq n} x_j^2$

A vector–field satisfying these two conditions can always be found in M. Indeed, pick an atlas $\mathcal{A} = (U_\alpha, \phi_\alpha)$ $\alpha = 1, \cdots N$ so that a single chart U_k contains the critical point p_k and so that, dividing the range $\{\alpha\}$ as $\{k, a\}$ $k = 1, \cdots, r$ $a = r+1, \cdots, N$, the following hold:

A. For each k, there is a smaller neighborhood $U'_k \subset U_k$ satisfying

$$\bar{U}'_k \cap U_\alpha = \begin{cases} \bar{U}'_k & \text{if } k = \alpha \\ \emptyset & \text{otherwise} \end{cases}$$

and $\phi_k(U'_k) = \{\mathbf{x} \in \mathbb{R}^n : |\mathbf{x}|^2 < \varepsilon\}$ for some small ε

where $\bar{U}'_k$ means the closure of U'_k.

B. In U_k f has local representative $f_k(\mathbf{x}) \equiv f \circ \phi_k^{-1}(\mathbf{x}) = c_k - \sum_1^{\lambda_k} x_i^2 + \sum_{\lambda_k+1}^n x_i^2$

C. In U_a f has local representative $f_a(\mathbf{x}) \equiv f \circ \phi_a^{-1}(\mathbf{x}) = \text{const} + x_1^{(a)}$

We now define ξ chart by chart. In U_k we give it the components $\xi^{(k)} = (-2x_1, \cdots, -2x_{\lambda_k}, 2x_{\lambda_k+1}, \cdots, 2x_n)$ and in U_a $\xi^{(a)} = (1, 0, \cdots, 0)$. Then we combine the local representatives $\xi^{(\alpha)}$, through a partition of unity $\{\theta_\alpha\}$ for $\mathcal{A}$ to obtain a vector–field, $\xi = \sum_\alpha \theta_\alpha \xi^{(\alpha)}$, which clearly satisfies condition (i) and condition (ii) in the neighborhood U'_k of p_k.

Covering the case of ordinary asymptotic flatness we have the following two lemmas [DG98]:

1. Consider two non-compact asymptotically flat (n-1)-geometries (U_0, h_0) and (U_1, h_1). Suppose that the closed manifolds V_0 and V_1 are one-point compactifications of U_0 and U_1, in the sense that there are points $\tilde{q}_i \in V_i$ and diffeomorphisms $\tilde{d}_i : V_i - \tilde{q}_i \to U_i$, $i = 0, 1$. Further suppose that there is a triad (M, V_0, V_1) with a Morse function $f : M \to [0, 1]$ with no index n critical points. Then we have:

(i) there is an integral curve $\mathbf{C}$ of a gradient-like vector–field for f which traverses M, from V_0 to V_1 without intercepting any critical point; and

(ii) the manifold $L \equiv M - \mathbf{C}$ is a cobordism between U_0 and U_1 and there is an AL metric on L which has the same Morse structure as f, is asymptotically flat and has the correct restrictions, the pull-backs of h_0 and h_1, on the boundary $\partial L = (V_0 - q_0) \uplus (V_1 - q_1)$ where $q_i = V_i \cap \mathbf{C}$.

For asymptotic Kaluza–Klein boundary conditions consider compactifying $\mathbb{R}^3 \times S^1$, the topology of a spatial section in the 5D Kaluza–Klein vacuum: we add a whole circle, one point at infinity of $\mathbb{R}^3$ for each point of S^1. In the reverse process an S^1 must be removed to recover the physical boundaries from the closed manifold. While all points in a manifold are equivalent, in general not all embedded circles are: given a manifold V, the manifolds $V - C$ and $V - \tilde{C}$ may not be diffeomorphic if C and $\tilde{C}$ are different embedded circles. In order to decompactify to Kaluza–Klein boundary conditions, we enlarge our list of hypotheses with a further condition which guarantees the equivalence of all subtracted circles in the closed boundary V_1.

2. Consider two asymptotically Kaluza–Klein flat (n-1)-geometries (U_0, h_0) and (U_1, h_1). Suppose that there exist closed manifolds V_0 and V_1, with V_1 connected and simply connected, and diffeomorphisms $\tilde{d}_i : V_i - \tilde{C}_i \to U_i$ with $\tilde{C}_i \subset V_i$ diffeomorphic to S^1, $(i = 0, 1)$. Further suppose that there is a triad (M, V_0, V_1) with a Morse function $f : M \to [0, 1]$ with no index n or $(n-1)$ critical points. Then we have [DG98]:

(i) There is an 'integral annulus' **A** for the gradient-like vector-field ξ — by this we mean an S^1 worth of integral curves of ξ, i.e., an imbedding $i : B^1 \times S^1 \hookrightarrow M$ such that for each point in the circle $\psi \in [0, 2\pi)$ the segment $i(B^1 \times \{\psi\})$ is an integral curve of ξ — which traverses M, from V_0 to V_1 without intercepting any critical point.

(ii) The manifold $L \equiv M - \mathbf{A}$ is a cobordism between U_0 and U_1 and there is an AL metric on L which has the same Morse structure as f, is asymptotically flat and has correct restrictions, the pull-backs of h_0 and h_1, on the boundary $\partial L = (V_0 - C_0) \uplus (V_1 - C_1)$ where $C_i = V_i \cap \mathbf{A}$, i=0,1.

Pair–Production of Black Holes

Due to the positive energy theorems, the Minkowski vacuum M^4 is stable with respect to semi-classical decay. However a cylindrically symmetric magnetic field described by the Melvin solution can decay into a pair of oppositely charged black holes thanks to the extra energy contained in the field [GS91]. The instanton that governs the decay is the Euclideanised Ernst solution. To see the topologies associated with the metrics involved, the reader is encouraged to consult [GS91]. We take them as the starting point for analyzing the cobordism. They are a space-like hyper-surface of Melvin, $\mathbb{R}^3$, a post-tunnelling space-like hyper-surface containing a pair of black holes, $S^2 \times S^1 - \{point\}$ and the doubled instanton, or 'bounce' topology, $S^2 \times S^2 - \{point\}$. Removing a point from a 4D closed manifold is equivalent to removing a closed ball B^4: it gives a non-compact manifold. We compactify by adding the point back in and cut the bounce in half to obtain $\underline{M} \cong S^2 \times B^2$. We then delete an open 4-ball to create the initial boundary. The manifold $\underline{M} - \dot{B}^4$ is M in Lemma 1, $V_0 \cong S^3$ is the initial boundary and $V_1 \cong S^2 \times S^1$ is the final boundary.

Combining Figure 3.23 with right–distributivity gives the following handle-body decomposition for $\underline{M}$:

$$\underline{M} = S^2 \times B^2 \cong (S^2 \times B^1) \times B^1 \cong (B^3 + B^2 \times B^1) \times B^1$$
$$= B^4 + B^2 \times B^2$$

Thus there exists a Morse function on M which contains only a Morse point of index 2. Dowker and Surya gave an earlier proof by explicitly constructing an allowed Morse function on $\underline{M}$ that can, in fact, be regarded as associated with the handle-body decomposition given above. There is an asymptotically

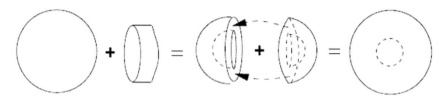

Fig. 3.23. Handle-body for $S^2 \times B^1$: a ball with a hole at the center (adapted and modified from [DG98]).

flat CCAL metric on the non-compact cobordism $L = M - \mathbf{C} \cong S^2 \times B^2 - \dot{B}^4 - B^1$ where $\mathbf{C}$ is an integral curve of a gradient-like vector–field of the Morse function. L is diffeomorphic to the original cobordism, half of $(S^2 \times S^2 - \{point\})$, up to the observation that the initial boundary in L is at some finite time in the past whereas in the original cobordism it is in the infinite past.

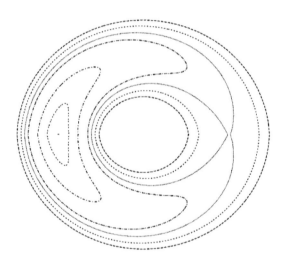

Fig. 3.24. Levels of a Morse function f in the cobordism $S^2 \to S^1 \times S^1$. On the left, an index 0 point accounts for the creation of S^2; on the right an index 1 point marks the transition to $S^1 \times S^1$. The function f increases in the direction of the expanding spheres and then in the direction of the expanding tori. Critical character of the Morse points is reflected in a same behavior of f along a Cartesian direction and its opposite (adapted and modified from [DG98]).

We can illustrate the location of the critical points and the critical levels in a lower dimension, $n = 3$. The equivalent process would be $S^2 \to S^1 \times S^1$, which is mediated by part of the handle-body $B^2 \times S^1 = B^3 + B^1 \times B^2$, and hence contains an unwanted index 1 point. The two critical points lie in the interior of the solid torus $B^2 \times S^1$ as depicted in Figure 3.24, a 2D section.

3.6 A Handle-Body Calculus for Topology Change

This construction generalizes to higher dimensions, so that black hole pair creation is feasible whenever $n \geq 4$. Indeed, applying right distributivity of the B^2 ball to $S^{n-2} = B^{n-2} + B^{n-2}$ we obtain:

$$S^{n-2} \times B^2 = B^n + B^{n-2} \times B^2$$

Thus the cobordism $S^{n-1} \to S^{n-2} \times S^1$ contains only an index $(n-2)$ critical point, which respects causal continuity whenever $n \geq 4$.

Kaluza–Klein Gravity

In 5D Kaluza–Klein gravity a fifth compact dimension is added to ordinary 4D space-time. The corresponding metric has fifteen degrees of freedom, which can be interpreted as one dilaton scalar, four components of the electromagnetic field and ten components of the space-time 4–metric. The 4D space-time associated with a given 5–geometry is obtained by reduction along a *Killing vector–field* of closed orbits. Both the Kaluza–Klein vacuum and the Kaluza–Klein version of the Melvin solution have a background topology $\mathbb{R}^4 \times S^1$ and are semi-classically unstable [DGG95].

The topology change is the same in both cases, from the unstable space-like hyper-surface $\mathbb{R}^3 \times S^1$ to the starting hyper-surface for post–decay $\mathbb{R}^2 \times S^2 \cong S^4 - S^1$. The double instanton has topology $\mathbb{R}^2 \times S^3 \cong S^5 - S^1$. Once more we compactify by replacing the circle and then halve the closed S^5 bounce to get $M \cong B^5$, with $\partial B^5 = S^4$. Finally we delete an open thickened circle $S^1 \times \dot{B}^4$ from M to create the initial boundary. This yields M with $\partial M = S^1 \times S^3 \uplus S^4$. That is, we have the triple $(M, V_0, V_1) = (B^5 - S^1 \times \dot{B}^4, S^3 \times S^1, S^4)$. We seek a handle-body decomposition for B^5 which truncates into a cobordism from $S^3 \times S^1$ to S^4. The 'redundant' B^3 decomposition equation (3.113) and right-distributivity imply the identity [DG98]:

$$\begin{aligned}
B^5 &= B^3 \times B^2 \\
&= (\underbrace{B^3 + B^1 \times B^2}_{B^2 \times S^1} + B^2 \times B^1) \times B^2 \\
&= \underbrace{B^5 + B^1 \times B^4}_{B^4 \times S^1} + B^2 \times B^3.
\end{aligned}$$

The first term, B^5, corresponds to the creation of S^4 from $\emptyset$ and the first handle addition corresponds to the transition from S^4 to $\partial(B^4 \times S^1) = S^3 \times S^1$, i.e., the (closed) KK vacuum space. The second handle addition is therefore the one that corresponds to the process we are investigating, $S^3 \times S^1 \to S^4$. This means that in the cobordism between $S^1 \times S^3$ and S^4, which involves only the handle $B^2 \times B^3$, there is a Morse function with exactly one critical point of index 2, i.e., no index 1 or 4 points.

The next figure represents a section of the 3D analogue of the cobordism M. The whole cobordism between $S^1 \times S^1$ and S^2 is generated by revolution around the z–axis. The reader can try and picture a cylinder between the inner boundary and the outer boundary that is orthogonal to the contours and does not touch the critical point at the center.

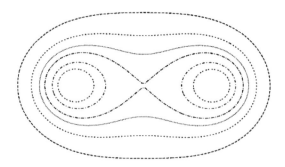

Fig. 3.25. Levels of the Morse function that represents the change $S^1 \times S^1 \to S^2$. Notice that this is in fact Figure 3.24 turned inside out: the time reversed cobordism if time progresses from inner to outer surfaces . Thus the central point, which was there an index 1 point is here an index 2 point: the Morse function increases in the z-direction and decreases in the other two (adapted and modified from [DG98]).

This result also generalizes to a countable family of higher–dimensional cobordisms that mediate the nucleation of various p–branes [DGG96]. In nD Kaluza–Klein theory the Kerr instanton manifold $\mathbb{R}^2 \times S^{n-2} \cong S^n - S^1$ is the double of the cobordism that mediates the transition $\mathbb{R}^{n-2} \times S^1 \to \mathbb{R}^2 \times S^{n-3}$, which in the closed case reads $S^{n-2} \times S^1 \to S^{n-1}$. These are respectively the boundaries of $B^{n-1} \times S^1$ and B^n; since

$$B^n = B^3 \times B^{n-3} = B^n + B^1 \times B^{n-1} + B^2 \times B^{n-2}.$$

Again, it is the second handle addition that corresponds to the process of interest and we see that there exists a Morse function with only one critical point of index 2, which respects causal continuity when $n \geq 4$.

For more technical details, see [DG98].

4
Nonlinear Dynamics of Path Integrals

In this Chapter we develop the *action–amplitude formalism* of *Feynman path integrals*, the essential tool in highly–nonlinear high–dimensional dynamics, both continuous and discrete, deterministic and stochastic. We start from the basic facts of the quantum probability concept. Note that the advanced, sum–over–geometries version of the path integral, has already been used in the previous Chapter.

4.1 Sum over Histories

The pivot–point of theoretical physics in the last half of a Century has been the celebrated *path integral*, a powerful conceptual and computational tool, first conceived by Richard (Dick) Feynman, and later generalized by Ed Witten, Stephen Hawking and other pioneers of physical science. Recall that in the path–integral formalism, we first formulate the specific classical *action of a new theory*, and subsequently perform its quantization by means of the associated *transition amplitude*. This *action–amplitude picture* is the core structure in any new physical theory. Its *virtual paths* are in general neither deterministic nor smooth, although they include bundles and jets of deterministic and smooth paths, as well as Markov chains. Yet, it is essentially a (broader) geometrical dynamics, with its Riemannian and symplectic versions, among many others. At the beginning, it worked only for *conservative* physical systems. Today it includes also *dissipative structures*, as well as various *sources and sinks, geometries* and *topologies*. Its smooth part reveals all celebrated equations of physics, both classical and quantum. It is the core of modern quantum gravity and superstring theory. It is arguably the most important construct of mathematical physics. At the edge of a new millennium, if you asked a typical theoretical physicist: what will be your main research tool in the new millennium, they would most probably say: path integral. And today, we see it moving out from physics, into the realm of social sciences. Finally, since Feynman's fairly intuitive invention of the path integral [Fey51],

4.1.1 Intuition Behind a Path Integral

Classical Probability Concept

Recall that a *random variable* X is defined by its *distribution function* $f(x)$. Its *probabilistic description* is based on the following rules: (i) $P(X = x_i)$ is the probability that $X = x_i$; and (ii) $P(a \leq X \leq b)$ is the probability that X lies in a closed interval $[a, b]$. Its statistical description is based on: (i) μ_X or $E(X)$ is the mean or expectation of X; and (ii) σ_X is the standard deviation of X. There are two cases of random variables: discrete and continuous, each having its own probability (and statistics) theory.

Discrete Random Variable

A discrete random variable X has only a countable number of values $\{x_i\}$. Its distribution function $f(x_i)$ has the following properties:

$$P(X = x_i) = f(x_i), \qquad f(x_i) \geq 0, \qquad \sum_i f(x_i)\,dx = 1.$$

Statistical description of X is based on its discrete mean value μ_X and standard deviation σ_X, given respectively by

$$\mu_X = E(X) = \sum_i x_i f(x_i), \qquad \sigma_X = \sqrt{E(X^2) - \mu_X^2}.$$

Continuous Random Variable

Here $f(x)$ is a piecewise continuous function such that:

$$P(a \leq X \leq b) = \int_a^b f(x)\,dx, \qquad f(x) \geq 0, \qquad \int_{-\infty}^{\infty} f(x)\,dx = \int_{\mathbb{R}} f(x)\,dx = 1.$$

Statistical description of X is based on its continuous mean μ_X and standard deviation σ_X, given respectively by

$$\mu_X = E(X) = \int_{-\infty}^{\infty} x f(x)\,dx, \qquad \sigma_X = \sqrt{E(X^2) - \mu_X^2}.$$

Now, let us observe the similarity between the two descriptions. The same kind of similarity between discrete and continuous quantum spectrum stroke Dirac when he suggested the combined integral approach, that he denoted by (see, e.g., [Dra06]): $\oint$ – meaning 'both integral and sum at once', that is, integration over the continuous spectrum and summing over the discrete spectrum.

To emphasize this similarity even further, as well as to set–up the stage for the path integral, recall the notion of a *cumulative distribution function* of a random variable X, that is a function $F : \mathbb{R} \to \mathbb{R}$, defined by

$$F(a) = P(X) \leq a.$$

In particular, suppose that $f(x)$ is the distribution function of X. Then

$$F(x) = \sum_{x_i \leq x} f(x_i), \quad \text{or} \quad F(x) = \int_{-\infty}^{\infty} f(t)\, dt,$$

according to as x is a discrete or continuous random variable. In either case, $F(a) \leq F(b)$ whenever $a \leq b$. Also,

$$\lim_{x \to -\infty} F(x) = 0 \quad \text{and} \quad \lim_{x \to \infty} F(x) = 1,$$

that is, $F(x)$ is monotonic and its limit to the left is 0 and the limit to the right is 1. Furthermore, its cumulative probability is given by

$$P(a \leq X \leq b) = F(b) - F(a),$$

and the Fundamental Theorem of Calculus tells us that, in the continuum case,

$$f(x) = \partial_x F(x).$$

General Markov Stochastic Dynamics

Recall that *Markov stochastic process* is a random process characterized by a *lack of memory*, i.e., the statistical properties of the immediate future are uniquely determined by the present, regardless of the past [Gar85].

For example, a *random walk* is an example of the *Markov chain*, i.e., a discrete–time Markov process, such that the motion of the system in consideration is viewed as a sequence of states, in which the transition from one state to another depends only on the preceding one, or the probability of the system being in state k depends only on the previous state $k-1$. The property of a Markov chain of prime importance in biomechanics is the existence of an *invariant distribution of states*: we start with an initial state x_0 whose absolute probability is 1. Ultimately the states should be distributed according to a specified distribution.

Between the pure deterministic dynamics, in which all DOF of the system in consideration are explicitly taken into account, leading to classical dynamical equations, for example in Hamiltonian form,

$$\dot{q}^i = \partial_{p_i} H, \qquad \dot{p}_i = -\partial_{q^i} H$$

428   4 Nonlinear Dynamics of Path Integrals

– and pure stochastic dynamics (Markov process), there is so–called *hybrid dynamics*, particularly *Brownian dynamics*, in which some of DOF are represented only through their *stochastic influence* on others. As an example, suppose a system of particles interacts with a viscous medium. Instead of specifying a detailed interaction of each particle with the particles of the viscous medium, we represent the medium as a *stochastic force* acting on the particle. The stochastic force *reduces the dimensionality* of the dynamics.

Recall that the Brownian dynamics represents the phase–space trajectories of a collection of particles that individually obey *Langevin rate equations* in the field of force (i.e., the particles interact with each other via some deterministic force). For a free particle, the Langevin equation reads [Gar85]:

$$m\dot{v} = R(t) - \beta v,$$

where m denotes the mass of the particle and v its velocity. The right–hand side represent the coupling to a *heat bath*; the effect of the random force $R(t)$ is to heat the particle. To balance overheating (on the average), the particle is subjected to *friction* β. In humanoid dynamics this is performed with the Rayleigh–Van der Pol's *dissipation*. Formally, the solution to the Langevin equation can be written as

$$v(t) = v(0) \exp\left(-\frac{\beta}{m}t\right) + \frac{1}{m}\int_0^t \exp[-(t-\tau)\beta/m]\,R(\tau)\,d\tau,$$

where the integral on the right–hand side is a *stochastic integral* and the solution $v(t)$ is a random variable. The stochastic properties of the solution depend significantly on the stochastic properties of the random force $R(t)$. In the Brownian dynamics the random force $R(t)$ is Gaussian distributed. Then the problem boils down to finding the solution to the Langevin stochastic differential equation with the supplementary condition (mean zero and variance)

$$<R(t)> = 0, \qquad <R(t)\,R(0)> = 2\beta k_B T \delta(t),$$

where $<.>$ denotes the mean value, T is temperature, k_B–*equipartition* (i.e., uniform distribution of energy) coefficient, Dirac $\delta(t)$–function.

Algorithm for computer simulation of the Brownian dynamics (for a single particle) can be written as [Hee90]:

A. Assign an initial position and velocity.
B. Draw a random number from a Gaussian distribution with mean zero and variance.
C. Integrate the velocity to get v^{n+1}.
D. Add the random component to the velocity.

Another approach to taking account the coupling of the system to a heat bath is to subject the particles to collisions with *virtual particles* [Hee90]. Such collisions are imagined to affect only momenta of the particles, hence

4.1 Sum over Histories

they affect the kinetic energy and introduce fluctuations in the total energy. Each stochastic collision is assumed to be an instantaneous event affecting only one particle.

The collision–coupling idea is incorporated into the Hamiltonian dynamics by adding a stochastic force $R_i = R_i(t)$ to the $\dot{p}$ equation

$$\dot{q}^i = \partial_{p_i} H, \qquad \dot{p}_i = -\partial_{q^i} H + R_i(t).$$

On the other hand, the so–called *Ito stochastic integral* represents a kind of classical Riemann–Stieltjes integral from linear functional analysis, which is (in 1D case) for an arbitrary time–function $G(t)$ defined as the *mean square limit*

$$\int_{t_0}^{t} G(t) dW(t) = ms \lim_{n \to \infty} \{\sum_{i=1}^{n} G(t_{i-1}[W(t_i) - W(t_{i-1})]\}.$$

Now, the general ND Markov process can be defined by *Ito stochastic differential equation* (SDE),

$$dx_i(t) = A_i[x^i(t), t] dt + B_{ij}[x^i(t), t] dW^j(t),$$
$$x^i(0) = x_{i0}, \qquad (i, j = 1, \ldots, N)$$

or corresponding *Ito stochastic integral equation*

$$x^i(t) = x^i(0) + \int_0^t ds\, A_i[x^i(s), s] + \int_0^t dW^j(s)\, B_{ij}[x^i(s), s],$$

in which $x^i(t)$ is the variable of interest, the vector $A_i[x(t), t]$ denotes deterministic *drift*, the matrix $B_{ij}[x(t), t]$ represents continuous stochastic *diffusion* fluctuations, and $W^j(t)$ is an N-variable *Wiener process* (i.e., generalized Brownian motion) [Wie61], and $dW^j(t) = W^j(t+dt) - W^j(t)$.

Now, there are three well–known special cases of the *Chapman–Kolmogorov equation* (see [Gar85]):

A. When both $B_{ij}[x(t), t]$ and $W(t)$ are zero, i.e., in the case of pure deterministic motion, it reduces to the *Liouville equation*

$$\partial_t P(x', t'|x'', t'') = -\sum_i \frac{\partial}{\partial x^i} \{A_i[x(t), t]\, P(x', t'|x'', t'')\}.$$

B. When only $W(t)$ is zero, it reduces to the *Fokker–Planck equation*

$$\partial_t P(x', t'|x'', t'') = -\sum_i \frac{\partial}{\partial x^i} \{A_i[x(t), t]\, P(x', t'|x'', t'')\}$$
$$+ \frac{1}{2} \sum_{ij} \frac{\partial^2}{\partial x^i \partial x^j} \{B_{ij}[x(t), t]\, P(x', t'|x'', t'')\}.$$

C. When both $A_i[x(t), t]$ and $B_{ij}[x(t), t]$ are zero, i.e., the state–space consists of integers only, it reduces to the *Master equation* of discontinuous jumps

$$\partial_t P(x', t'|x'', t'') = \int dx \{W(x'|x'', t) P(x', t'|x'', t'') - W(x''|x', t) P(x', t'|x'', t'')\}.$$

The *Markov assumption* can now be formulated in terms of the conditional probabilities $P(x^i, t_i)$: if the times t_i increase from right to left, the conditional probability is determined entirely by the knowledge of the most recent condition. Markov process is generated by a set of conditional probabilities whose probability–density $P = P(x', t'|x'', t'')$ evolution obeys the general *Chapman–Kolmogorov integro–differential equation*

$$\partial_t P = -\sum_i \frac{\partial}{\partial x^i} \{A_i[x(t), t] P\}$$
$$+ \frac{1}{2} \sum_{ij} \frac{\partial^2}{\partial x^i \partial x^j} \{B_{ij}[x(t), t] P\} + \int dx \{W(x'|x'', t) P - W(x''|x', t) P\}$$

including *deterministic drift, diffusion fluctuations* and *discontinuous jumps* (given respectively in the first, second and third terms on the r.h.s.).

It is this general Chapman–Kolmogorov integro–differential equation, with its conditional probability density evolution, $P = P(x', t'|x'', t'')$, that we are going to model by various forms of the Feynman path integral, providing us with the physical insight behind the abstract (conditional) probability densities.

Quantum Probability Concept

An alternative concept of probability, the so–called *quantum probability*, is based on the following physical facts (elaborated in detail in this section):

A. The *time–dependent Schrödinger equation* represents a *complex–valued generalization* of the real–valued *Fokker–Planck equation* for describing the spatio–temporal *probability density function* for the system exhibiting continuous–time Markov stochastic process.
B. The *Feynman path integral* $\oint$ is a generalization of the time–dependent Schrödinger equation, including both continuous–time and discrete–time Markov stochastic processes.
C. Both Schrödinger equation and path integral give 'physical description' of any system they are modelling in terms of its physical energy, instead of an abstract probabilistic description of the Fokker–Planck equation.

Therefore, the *Feynman path integral* $\oint$, as a generalization of the time–dependent Schrödinger equation, gives a unique physical description for the general Markov stochastic process, in terms of the physically based generalized probability density functions, valid both for continuous–time and discrete–time Markov systems.

Basic consequence: a different way for calculating probabilities. The difference is rooted in the fact that *sum of squares is different from the square of sums*, as is explained in the following text.

Namely, in Dirac–Feynman quantum formalism, each possible route from the initial system state A to the final system state B is called a *history*. This history comprises any kind of a route (see Figure 4.1), ranging from continuous and smooth deterministic (mechanical–like) paths to completely discontinues and random Markov chains (see, e.g., [Gar85]). Each history (labelled by index i) is quantitatively described by a *complex number*[1] z_i called the 'individual transition amplitude'. Its absolute square, $|z_i|^2$, is called the *individual transition probability*. Now, the *total transition amplitude* is the sum of all individual transition amplitudes, $\sum_i z_i$, called the *sum–over–histories*. The absolute square of this sum–over–histories, $|\sum_i z_i|^2$, is the *total transition probability*.

In this way, the overall probability of the system's transition from some initial state A to some final state B is given *not* by adding up the probabilities for each history–route, but by 'head–to–tail' adding up the sequence of amplitudes making–up each route first (i.e., performing the sum–over–histories) – to get the total amplitude as a 'resultant vector', and then squaring the total amplitude to get the overall transition probability.

Quantum Coherent States

Recall that a *quantum coherent state* is a specific kind of quantum state of the quantum harmonic oscillator whose dynamics most closely resemble the oscillating behavior of a classical harmonic oscillator. It was the first example of quantum dynamics when Erwin Schrödinger derived it in 1926 while searching for solutions of the *Schrödinger equation* that satisfy the *correspondence principle*. The quantum harmonic oscillator and hence, the coherent state, arise in the quantum theory of a wide range of physical systems. For instance, a coherent state describes the oscillating motion of the particle in a quadratic potential well. In the quantum electrodynamics and other bosonic quantum field theories they were introduced by the 2005 Nobel Prize winning work of R. Glauber in 1963 [Gla63a, Gla63b]. Here the coherent state of a field describes an oscillating field, the closest quantum state to a classical sinusoidal wave such as a continuous laser wave.

---

[1] Recall that a *complex number* $z = x + iy$, where $i = \sqrt{-1}$ is the *imaginary unit*, x is the *real part* and y is the *imaginary part*, can be represented also in its polar form, $z = r(\cos\theta + i\sin\theta)$, where the radius vector in the complex–plane, $r = |z| = \sqrt{x^2 + y^2}$, is the *modulus* or *amplitude*, and angle θ is the *phase*; as well as in its exponential form $z = re^{i\theta}$. In this way, complex numbers actually represent 2D vectors with usual vector 'head–to–tail' addition rule.

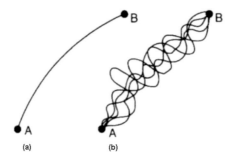

Fig. 4.1. Two ways of physical *transition* from an *initial state A* to the corresponding *final state B*. (a) Classical physics proposes *a single deterministic trajectory*, minimizing the total system's energy. (b) Quantum physics proposes *a family of Markov stochastic histories*, namely *all possible routes* from A to B, both continuous–time and discrete–time Markov chains, each giving an equal contribution to the total *transition probability*.

In classical optics, light is thought of as electromagnetic waves radiating from a source. Specifically, coherent light is thought of as light that is emitted by many such sources that are in phase. For instance, a light bulb radiates light that is the result of waves being emitted at all the points along the filament. Such light is incoherent because the process is highly random in space and time. On the other hand, in a laser, light is emitted by a carefully controlled system in processes that are not random but interconnected by stimulation and the resulting light is highly ordered, or coherent. Therefore a coherent state corresponds closely to the quantum state of light emitted by an ideal laser. Semi–classically we describe such a state by an electric field oscillating as a stable wave. Contrary to the coherent state, which is the most wave–like quantum state, the *Fock state* (e.g., a single photon) is the most particle–like state. It is indivisible and contains only one quanta of energy. These two states are examples of the opposite extremes in the concept of *wave–particle duality*. A coherent state distributes its quantum–mechanical uncertainty equally, which means that the phase and amplitude uncertainty are approximately equal. Conversely, in a single–particle state the phase is completely uncertain.

Formally, the coherent state $|\alpha\rangle$ is defined to be the eigenstate of the annihilation operator a, i.e., $a|\alpha\rangle = \alpha|\alpha\rangle$. Note that since a is not Hermitian, $\alpha = |\alpha|e^{i\theta}$ is complex. $|\alpha|$ and θ are called the *amplitude* and *phase* of the state.

Physically, $a|\alpha\rangle = \alpha|\alpha\rangle$ means that a coherent state is left unchanged by the detection (or annihilation) of a particle. Consequently, in a coherent state, one has exactly the same probability to detect a second particle. Note, this condition is necessary for the coherent state's *Poisson detection statistics*.

Compare this to a single–particle's Fock state: Once one particle is detected, we have zero probability of detecting another.

Now, recall that a *Bose–Einstein condensate* (BEC) is a collection of boson atoms that are all in the same quantum state. An approximate theoretical description of its properties can be derived by assuming the BEC is in a coherent state. However, unlike photons, atoms interact with each other so it now appears that it is more likely to be one of the *squeezed coherent states* (see [BSM97]). In quantum field theory and string theory, a generalization of coherent states to the case of infinitely many degrees of freedom is used to define a *vacuum state* with a different vacuum expectation value from the original vacuum.

Dirac's $< bra\,|\,ket >$ Transition Amplitude

Now, we are ready to move–on into the realm of quantum mechanics. Recall that P. Dirac [Dir49] described behavior of quantum systems in terms of complex–valued *ket–vectors* $|A>$ living in the Hilbert space $\mathcal{H}$, and their duals, *bra–covectors* $<B|$ (i.e., 1–forms) living in the *dual* Hilbert space $\mathcal{H}^*$. The *Hermitian inner product* of kets and bras, the *bra–ket* $<B|A>$, is a *complex number*, which is the evaluation of the ket $|A>$ by the bra $<B|$. This complex number, say $re^{i\theta}$ represents the system's *transition amplitude*[2] from its *initial state* A to its *final state* B[3], i.e.,

$$Transition\,Amplitude\; =<B|A>=re^{i\theta}.$$

That is, there is a process that can mediate a transition of a system from initial state A to the final state B and the amplitude for this transition equals $<B|A>=re^{i\theta}$. The absolute square of the amplitude, $|<B|A>|^2$ represents the *transition probability*. Therefore, the probability of a transition event equals the absolute square of a complex number, i.e.,

$$Transition\,Probability\; =|<B|A>|^2 = |re^{i\theta}|^2.$$

These complex amplitudes obey the usual *laws of probability*: when a transition event can happen in alternative ways then we add the complex numbers,

$$<B_1|A_1> + <B_2|A_2> = r_1e^{i\theta_1} + r_2e^{i\theta_2},$$

and when it can happen only as a succession of intermediate steps then we multiply the complex numbers,

$$<B|A> = <B|c><c|A> = (r_1e^{i\theta_1})(r_2e^{i\theta_2}) = r_1r_2e^{i(\theta_1+\theta_2)}.$$

In general,

---

[2] Transition amplitude is otherwise called *probability amplitude*, or just *amplitude*.
[3] Recall that in quantum mechanics, complex numbers are regarded as the *vacuum–state*, or the *ground–state*, and the entire amplitude $<b|a>$ is a *vacuum–to–vacuum amplitude* for a process that includes the *creation* of the state a, its *transition* to b, and the *annihilation* of b to the vacuum once more.

A. The amplitude for *n mutually alternative processes* equals the *sum* $\sum_{k=1}^{n} r_k e^{i\theta_k}$ of the amplitudes for the alternatives; and
B. If transition from A to B occurs in a *sequence of m steps*, then the total transition amplitude equals the *product* $\prod_{j=1}^{m} r_j e^{i\theta_j}$ of the amplitudes of the steps.

Formally, we have the so–called *expansion principle*, including both products and sums,[4]

$$<B|A> = \sum_{i=1}^{n} <B|c_i><c_i|A>. \qquad (4.1)$$

Feynman's Sum–over–Histories

Now, *iterating* the Dirac's expansion principle (4.1) over a *complete set of all possible states* of the system, leads to the simplest form of the *Feynman path integral*, or, *sum–over–histories*. Imagine that the *initial* and *final* states, A and B, are points on the vertical lines $x = 0$ and $x = n + 1$, respectively, in the $x - y$ plane, and that $(c(k)_{i(k)}, k)$ is a given point on the line $x = k$ for $0 < i(k) < m$ (see Figure 4.2). Suppose that the sum of projectors for each intermediate state is complete[5] Applying the completeness iteratively, we get the following expression for the transition amplitude:

$$<B|A> = \sum\sum\ldots\sum <B|c(1)_{i(1)}><c(1)_{i(1)}|c(2)_{i(2)}> \ldots <c(n)_{i(n)}|A>,$$

where the sum is taken over all $i(k)$ ranging between 1 and m, and k ranging between 1 and n. Each term in this sum can be construed as a *combinatorial route* from A to B in the 2D space of the $x - y$ plane. Thus the transition amplitude for the system going from some initial state A to some final state B is seen as a summation of contributions from *all the routes* connecting A to B.

Feynman used this description to produce his celebrated *path integral* expression for a transition amplitude (see, e.g., [GS98, Sch81]). His path integral takes the form

---

[4] In Dirac's language, the *completeness* of intermediate states becomes the statement that a certain sum of projectors is equal to the identity. Namely, suppose that $\sum_i |c_i><c_i| = 1$ with $<c_i|c_i> = 1$ for each i. Then

$$<b|a> = <b||a> = <b|\sum_i |c_i><c_i||a> = \sum_i <b|c_i><c_i|a>.$$

[5] We assume that following sum is equal to one, for each k from 1 to $n-1$:

$$|c(k)_1><c(k)_1| + \ldots + |c(k)_m><c(k)_m| = 1.$$

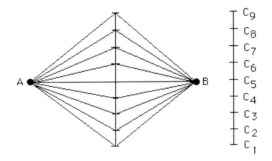

Fig. 4.2. Analysis of all possible routes from the source A to the detector B is simplified to include only double straight lines (in a plane).

$$Transition\ Amplitude = <B|A> = \oint \mathcal{D}[x]\, e^{i\mathcal{S}[x]}, \qquad (4.2)$$

where the sum–integral $\oint$ is taken over all possible routes $x = x(t)$ from the initial point $A = A(t_{ini})$ to the final point $B = B(t_{fin})$, and $\mathcal{S} = \mathcal{S}[x]$ is the classical *action* for a particle to travel from A to B along a given extremal path x. In this way, Feynman took seriously Dirac's conjecture interpreting the exponential of the classical *action functional* ($\mathcal{D}e^{i\mathcal{S}}$), resembling a complex number ($re^{i\theta}$), as an *elementary amplitude*. By integrating this elementary amplitude, $\mathcal{D}e^{i\mathcal{S}}$, over the infinitude of all possible histories, we get the total system's transition amplitude.[6]

---

[6] For the quantum physics associated with a classical (Newtonian) particle the action S is given by the integral along the given route from a to b of the difference $T - V$ where T is the classical kinetic energy and V is the classical potential energy of the particle.

The beauty of Feynman's approach to quantum physics is that it shows the relationship between the classical and the quantum in a particularly transparent manner. Classical motion corresponds to those regions where all nearby routes contribute constructively to the summation. This classical path occurs when the *variation of the action* is null. To ask for those paths where the variation of the action is zero is a problem in the calculus of variations, and it leads directly to Newton's equations of motion (derived using the Euler–Lagrangian equations). Thus with the appropriate choice of action, classical and quantum points of view are unified.

Also, a discretization of the Schrodinger equation

$$i\hbar \frac{d\psi}{dt} = -\frac{\hbar^2}{2m}\frac{d^2\psi}{dx^2} + V\psi,$$

leads to a sum–over–histories that has a discrete path integral as its solution. Therefore, the transition amplitude is equivalent to the wave ψ. The particle travelling on the x–axis is executing a one–step *random walk*, see Figure 4.3.

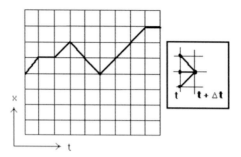

Fig. 4.3. Random walk (a particular case of Markov chain) on the x−axis.

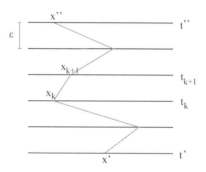

Fig. 4.4. A piecewise linear particle path contributing to the discrete Feynman propagator.

The Basic Form of a Path Integral

In Feynman's version of non–relativistic quantum mechanics, the time evolution $\psi(x', t') \mapsto \psi(x'', t'')$ of the wave function $\psi = \psi(x, t)$ of the elementary 1D particle may be described by the integral equation [GS98]

$$\psi(x'', t'') = \int K(x'', x'; t'', t') \, \psi(x', t'), \qquad (4.3)$$

where the *propagator* or *Feynman kernel* $K = K(x'', x'; t'', t')$ is defined through a limiting procedure,

$$K(x'', x'; t'', t') = \lim_{\epsilon \to 0} A^{-N} \prod_{k=1}^{N-1} \int dx_k \, e^{i \sum_{j=0}^{N-1} \epsilon L(x_{j+1}, (x_{j+1}-x_j)/\epsilon)}. \qquad (4.4)$$

The time interval $t'' - t'$ has been discretized into N steps of length $\epsilon = (t'' - t')/N$, and the r.h.s. of (4.4) represents an integral over all piecewise linear paths $x(t)$ of a 'virtual' particle propagating from x' to x'', illustrated in Figure 4.4.

The prefactor A^{-N} is a normalization and L denotes the Lagrangian function of the particle. Knowing the propagator G is tantamount to having solved the quantum dynamics. This is the simplest instance of a path integral, and is often written schematically as

$$K(x', t'; x'', t'') = \int \mathcal{D}[x(t)] \, e^{iS[x(t)]},$$

where $\mathcal{D}[x(t)]$ is a functional measure on the 'space of all paths', and the exponential weight depends on the classical action $S[x(t)]$ of a path. Recall also that this procedure can be defined in a mathematically clean way if we Wick–rotate the time variable t to imaginary values $t \mapsto \tau = it$, thereby making all integrals real [RS75].

Adaptive Path Integral

Now, we can extend the Feynman sum–over–histories (4.2), by adding the synaptic–like weights $w^i = w^i(t)$ into the measure $\mathcal{D}[x]$, to get the *adaptive path integral*:

$$Adaptive\, Transition\, Amplitude\, = <B|A>_w = \int \mathcal{D}[w, x] \, e^{iS[x]}, \quad (4.5)$$

where the *adaptive measure* $\mathcal{D}[w, x]$ is defined by the weighted product (of discrete time steps)

$$\mathcal{D}[w, x] = \lim_{n \to \infty} \prod_{t=1}^{n} w^i(t) \, dx^i(t). \quad (4.6)$$

In (4.298) the *synaptic weights* $w^i = w^i(t)$ are updated by the unsupervised Hebbian–like learning rule [Heb49]:

$$w^i(t+1) = w^i(t) + \frac{\sigma}{\eta}(w^i_d(t) - w^i_a(t)), \quad (4.7)$$

where $\sigma = \sigma(t)$, $\eta = \eta(t)$ represent local *signal* and *noise* amplitudes, respectively, while superscripts d and a denote *desired* and *achieved* system states, respectively. Theoretically, equations (4.57–4.290) define an ∞–*dimensional complex–valued neural network*.[7] Practically, in a computer simulation we can use $10^7 \leq n \leq 10^8$, approaching the number of neurons in the brain. Such equations are usually solved using Markov–Chain Monte–Carlo methods on parallel (cluster) computers (see, e.g., [WW83a, WW83b]).

4.1.2 Basic Path–Integral Calculations

Consider a particle moving in one dimension, the Hamiltonian being of the usual form:

---

[7] For details on complex–valued neural networks, see e.g., complex–domain extension of the standard backpropagation learning algorithm [GK92, BP02].

The propagator becomes

$$K = \int \prod_{j=1}^{N-1} dq_j \exp\left[-i\delta \sum_{j=0}^{N-1} V(\bar{q}_j) \prod_{j=0}^{N-1}\left(\sqrt{\frac{m}{2\pi i\delta}} \exp i\delta \frac{m\dot{q}_j^2}{2}\right)\right]$$

$$= \left(\frac{m}{2\pi i\delta}\right)^{N/2} \int \prod_{j=1}^{N-1} dq_j \exp\left[i\delta \sum_{j=0}^{N-1}\left(\frac{m\dot{q}_j^2}{2} - V(\bar{q}_j)\right)\right]. \quad (4.16)$$

The argument of the exponential is a discrete approximation of the action of a path passing through the points $q_0 = q, q_1, \cdots, q_{N-1}, q_N = q'$. As above, we can write this in the more compact form

$$K = \int \mathcal{D}[q(t)] e^{iS[q(t)]}. \quad (4.17)$$

This is our final result, and is known as the *configuration–space path integral*. Again, (4.17) should be viewed as a notation for the more precise expression (4.16), as $N \to \infty$.

Elementary Path–Integral Examples

As a first example, let us compute the propagator $K(q', T; q, 0)$ for a *free particle*, described by the Hamiltonian $H = p^2/2m$. The propagator can be computed straightforwardly using ordinary quantum mechanics. To this end, we write

$$K = \langle q'| e^{-iHT} |q\rangle = \langle q'| e^{-iT\hat{p}^2/2m} \int \frac{dp}{2\pi} |p\rangle \langle p| q\rangle \quad (4.18)$$

$$= \int \frac{dp}{2\pi} e^{-iTp^2/2m} \langle q'| p\rangle \langle p| q\rangle = \int \frac{dp}{2\pi} e^{-iT(p^2/2m)+i(q'-q)p}.$$

The integral is Gaussian; we obtain

$$K = \left(\frac{m}{2\pi iT}\right)^{1/2} e^{im(q'-q)^2/2T}. \quad (4.19)$$

Let us now see how the same result can be attained using path–integrals. The configuration space path–integral (4.17) is

$$K = \lim_{N\to\infty} \left(\frac{m}{2\pi i\delta}\right)^{N/2} \int \prod_{j=1}^{N-1} dq_j \exp\left[i\frac{m\delta}{2} \sum_{j=0}^{N-1}\left(\frac{q_{j+1}-q_j}{\delta}\right)^2\right] =$$

$$\lim_{N\to\infty} \left(\frac{m}{2\pi i\delta}\right)^{N/2} \int \prod_{j=1}^{N-1} dq_j \exp\{i\frac{m}{2\delta}\big[(q_N - q_{N-1})^2 + (q_{N-1} - q_{N-2})^2$$

$$+ \cdots + (q_2 - q_1)^2 + (q_1 - q_0)^2\big]\},$$

where $q_0 = q$ and $q_N = q'$ are the initial and final points. The integrals are Gaussian, and can be evaluated exactly, although the fact that they are coupled complicates matters significantly. The result is

$$K = \lim_{N \to \infty} \left(\frac{m}{2\pi i \delta}\right)^{N/2} \frac{1}{\sqrt{N}} \left(\frac{2\pi i \delta}{m}\right)^{(N-1)/2} e^{im(q'-q)^2/2N\delta}$$

$$= \lim_{N \to \infty} \left(\frac{m}{2\pi i N \delta}\right)^{1/2} e^{im(q'-q)^2/2N\delta}.$$

But $N\delta$ is the total time interval T, resulting in

$$K = \left(\frac{m}{2\pi i T}\right)^{1/2} e^{im(q'-q)^2/2T},$$

in agreement with (4.19).

A couple of remarks are in order. First, we can write the argument of the exponential as $T \cdot \frac{1}{2}m((q'-q)/T)^2$, which is just the action $S[q_c]$ for a particle moving along the classical path (a straight line in this case) between the initial and final points.

Secondly, we can restore the factors of $\hbar$ if we want, by ensuring correct dimensions. The argument of the exponential is the action, so in order to make it a pure number we must divide by $\hbar$; furthermore, the propagator has the dimension of the inner product of two position eigenstates, which is inverse length; in order that the coefficient have this dimension we must multiply by $\hbar^{-1/2}$. The final result is

$$K = \left(\frac{m}{2\pi i \hbar T}\right)^{1/2} e^{iS[q_c]/\hbar}. \tag{4.20}$$

This result typifies a couple of important features of calculations in this subject, which we will see repeatedly in these lectures. First, the propagator separates into two factors, one of which is the phase $\exp iS[q_c]/\hbar$. Second, calculations in the path–integral formalism are typically quite a bit more lengthy than using standard techniques of quantum mechanics.

As a second example, let us compute the propagator for the *harmonic oscillator* using the path–integral method. Let us start with the somewhat-formal version of the configuration–space path–integral, (4.17):

$$K(q', T; q, 0) = \int \mathcal{D}[q(t)] e^{iS[q(t)]}.$$

For the harmonic oscillator,

$$S[q(t)] = \int_0^T dt \left(\frac{1}{2}m\dot{q}^2 - \frac{1}{2}m\omega^2 q^2\right).$$

The paths over which the integral is to be performed go from $q(0) = q$ to $q(T) = q'$. To do this path–integral, suppose we know the solution of the classical problem, $q_c(t)$:

$$\ddot{q}_c + \omega^2 q_c = 0, \qquad q_c(0) = q, \qquad q_c(T) = q'.$$

We can write $q(t) = q_c(t) + y(t)$, and perform a change of variables in the path–integral to $y(t)$, since integrating over all deviations from the classical path is equivalent to integrating over all possible paths. Since at each time q and y differ by a constant, the Jacobian of the transformation is 1. Furthermore, since q_c obeys the correct boundary conditions, the paths $y(t)$ over which we integrate go from $y(0) = 0$ to $y(T) = 0$. The action for the path $q_c(t) + y(t)$ can be written as a power series in y:

$$S[q_c(t) + y(t)] = \int_0^T dt \left(\frac{1}{2} m \dot{q}_c^2 - \frac{1}{2} m \omega^2 q_c^2 \right)$$

$$+ \underbrace{(\text{linear in } y)}_{=0} + \int_0^T dt \left(\frac{1}{2} m \dot{y}^2 - \frac{1}{2} m \omega^2 y^2 \right).$$

The term linear in y vanishes by construction: q_c, being the classical path, is that path for which the action is stationary! So we may write $S[q_c(t) + y(t)] = S[q_c(t)] + S[y(t)]$. We substitute this into (4.17), yielding

$$K(q', T; q, 0) = e^{iS[q_c(t)]} \int \mathcal{D}[y(t)] e^{iS[y(t)]}. \tag{4.21}$$

As mentioned above, the paths $y(t)$ over which we integrate go from $y(0) = 0$ to $y(T) = 0$: the only appearance of the initial and final positions is in the classical path, i.e., in the classical action. Once again, the path–integral separates into two factors. The first is written in terms of the action of the classical path, and the second is a path–integral over deviations from this classical path. The second factor is independent of the initial and final points.

This separation into a factor depending on the action of the classical path and a second one, a path–integral which is independent of the details of the classical path, is a recurring theme, and an important one. Indeed, it is often the first factor which contains most of the useful information contained in the propagator, and it can be deduced without even performing a path–integral. It can be said that much of the work in the game of path integrals consists in avoiding having to actually compute one!

As for the evaluation of (4.21), a number of fairly standard techniques are available. One can calculate the path–integral directly in position space, as was done above for the harmonic oscillator (see [Sch81], Chapter 6). Alternatively, one can compute it in Fourier space (writing $y(t) = \sum_k a_k \sin(k\pi t/T)$ and integrating over the coefficients $\{a_k\}$). This latter approach is outlined in [FH65], Section 3.11. The result is

$$K(q', T; q, 0) = \left(\frac{m\omega}{2\pi i \sin \omega T} \right)^{1/2} e^{iS[q_c(t)]}. \tag{4.22}$$

The classical action can be evaluated straightforwardly (note that this is not a path–integral problem, nor even a quantum mechanics problem!); the result is

$$S[q_c(t)] = \frac{m\omega}{2\sin\omega T}\left((q'^2 + q^2)\cos\omega T - 2q'q\right).$$

We close this section with two remarks. First, the path–integral for any quadratic action can be evaluated exactly, essentially since such a path–integral consists of Gaussian integrals; the general result is given in [Sch81]. Second, the following fact is not difficult to prove: $K(q', T; q, 0)$ (whether computed via path–integrals or not) is the amplitude to propagate from one point to another in a given time interval. But this is the response to the following question: If a particle is initially at position q, what is its wave function after the elapse of a time T? Thus, if we consider K as a function of the final position and time, it is none other than the wave function for a particle with a specific initial condition. As such, the propagator satisfies the Schrödinger equation at its final point.

4.1.3 Brief History of Feynman's Path Integral

Extract from Feynman's Nobel Lecture

In his Nobel Lecture, December 11, 1965, Richard (Dick) Feynman said that he and his PhD supervisor, John Wheeler, had found the *action* $A = A[x; t_i, t_j]$, directly involving the *motions of the charges only*,[14]

$$A[x; t_i, t_j] = m_i \int (\dot{x}^i_\mu \dot{x}^i_\mu)^{\frac{1}{2}} dt_i + \frac{1}{2} e_i e_j \int\int \delta(I_{ij}^2) \dot{x}^i_\mu(t_i)\dot{x}^j_\mu(t_j)\, dt_i dt_j$$
$$\text{with } (i \neq j) \qquad (4.23)$$
$$I_{ij}^2 = \left[x^i_\mu(t_i) - x^j_\mu(t_j)\right]\left[x^i_\mu(t_i) - x^j_\mu(t_j)\right],$$

where $x^i_\mu = x^i_\mu(t_i)$ is the four–vector *position* of the ith particle as a function of the proper time t_i, while $\dot{x}^i_\mu(t_i) = dx^i_\mu(t_i)/dt_i$ is the *velocity* four–vector.

The first term in the action $A[x; t_i, t_j]$ (4.294) is the integral of the proper time t_i, the *ordinary action of relativistic mechanics of free particles of mass m_i* (summation over μ). The second term in the action $A[x; t_i, t_j]$ (4.294) represents the *electrical interaction of the charges*. It is summed over each pair of charges (the factor $\frac{1}{2}$ is to count each pair once, the term $i = j$ is omitted to avoid self–action). The *interaction is a double integral over a delta function of the square of space–time interval I^2 between two points on the paths*. Thus, interaction occurs only when this interval vanishes, that is, *along light cones* (see [WF49]).

Feynman comments here: "The fact that the interaction is exactly one-half advanced and half–retarded meant that we could write such a principle of least action, whereas interaction via retarded waves alone cannot be written

---

[14] *Wheeler–Feynman Idea* [WF49] "The energy tensor can be regarded only as a provisional means of representing matter. In reality, *matter consists of electrically charged particles*."

in such a way. So, all of classical electrodynamics was contained in this very simple form."

"...The problem is only to make a quantum theory, which has as its classical analog, this expression (4.294). Now, there is no unique way to make a quantum theory from classical mechanics, although all the textbooks make believe there is. What they would tell you to do, was find the momentum variables and replace them by $(\hbar/i)(\partial/\partial x)$, but I couldn't find a momentum variable, as there wasn't any."

"The character of quantum mechanics of the day was to write things in the famous *Hamiltonian way* (in the form of Schrödinger equation), which described how the wave function changes from instant to instant, and in terms of the Hamiltonian operator H. If the classical physics could be reduced to a Hamiltonian form, everything was all right. Now, least action does not imply a Hamiltonian form if the action is a function of anything more than positions and velocities at the same moment. If the action is of the form of the integral of the Lagrangian $L = L(\dot{x}, x)$, a function of the velocities and positions at the same time t,

$$S[x] = \int L(\dot{x}, x)\, dt, \qquad (4.24)$$

then you can start with the Lagrangian L and then create a Hamiltonian H and work out the quantum mechanics, more or less uniquely. But the action $A[x; t_i, t_j]$ (4.294) involves the key variables, positions (and velocities), at two different times t_i and t_j and therefore, it was not obvious what to do to make the quantum–mechanical analogue..."

So, Feynman was looking for the action integral in quantum mechanics. He says: "...I simply turned to Professor Jehle and said, 'Listen, do you know any way of doing quantum mechanics, starting with action – where the action integral comes into the quantum mechanics?" "No", he said, "but Dirac has a paper in which the Lagrangian, at least, comes into quantum mechanics." What Dirac said was the following: There is in quantum mechanics a very important quantity which carries the wave function from one time to another, besides the differential equation but equivalent to it, a kind of a kernel, which we might call $K(x', x)$, which carries the wave function $\psi(x)$ known at time t, to the wave function $\psi(x')$ at time $t + \varepsilon$,

$$\psi(x', t + \varepsilon) = \int K(x', x)\, \psi(x, t)\, dx.$$

Dirac points out that this function K was analogous to the quantity in classical mechanics that you would calculate if you took the exponential of [$i\varepsilon$ multiplied by the Lagrangian $L(\dot{x}, x)$], imagining that these two positions x, x' corresponded to t and $t + \varepsilon$. In other words,

$$K(x', x) \quad \text{is analogous to} \quad e^{i\varepsilon L(\frac{x'-x}{\varepsilon}, x)/\hbar}.$$

So, Feynman continues: "What does he mean, they are analogous; what does that mean, *analogous*? What is the use of that?" Professor Jehle said, "You

Americans! You always want to find a use for everything!" I said that I thought that Dirac must mean that they were *equal*. "No", he explained, "he doesn't mean they are equal." "Well", I said, "Let's see what happens if we make them equal."

"So, I simply put them equal, taking the simplest example where the Lagrangian is

$$L = \frac{1}{2} M \dot{x}^2 - V(x),$$

but soon found I had to put a constant of proportionality N in, suitably adjusted. When I substituted for K to get

$$\psi(x', t + \varepsilon) = \int N \exp\left[\frac{i\varepsilon}{\hbar} L\left(\frac{x' - x}{\varepsilon}, x\right)\right] \psi(x, t) \, dx \quad (4.25)$$

and just calculated things out by Taylor series expansion, *out came the Schrödinger equation*. So, I turned to Professor Jehle, not really understanding, and said, "Well, you see, Dirac meant that they were proportional." Professor Jehle's eyes were bugging out – he had taken out a little notebook and was rapidly copying it down from the blackboard, and said, "No, no, this is an important discovery. You Americans are always trying to find out how something can be used. That's a good way to discover things!" So, I thought I was finding out what Dirac meant, but, as a matter of fact, had made the discovery that what Dirac thought was analogous, was, in fact, equal. I had then, at least, the connection between the Lagrangian and quantum mechanics, but still with wave functions and infinitesimal times."

"It must have been a day or so later when I was lying in bed thinking about these things, that I imagined what would happen if I wanted to calculate the wave function at a finite interval later. I would put one of these factors $e^{i\varepsilon L}$ in here, and that would give me the wave functions the next moment, $t + \varepsilon$, and then I could substitute that back into (4.296) to get another factor of $e^{i\varepsilon L}$ and give me the wave function the next moment, $t + 2\varepsilon$, and so on and so on. In that way I found myself thinking of a large number of integrals, one after the other in sequence. In the integrand was the product of the exponentials, which was the exponential of the sum of terms like εL. Now, L is the Lagrangian and ε is like the time interval dt, so that if you took a sum of such terms, that's exactly like an integral. That's like Riemann's formula for the integral $\int L dt$, you just take the value at each point and add them together. We are to take the limit as $\varepsilon \to 0$. Therefore, the connection between the wave function of one instant and the wave function of another instant a finite time later could be get by an infinite number of integrals (because ε goes to zero), of exponential where S is the action expression (4.295). At last, I had succeeded in representing quantum mechanics directly in terms of the action $S[x]$."

Fully satisfied, Feynman comments: "This led later on to the idea of the *transition amplitude* for a path: that for each possible way that the particle can go from one point to another in space–time, there's an amplitude. That amplitude is **e** to the power of [i/ℏ times the action $S[x]$ for the path], i.e.,

$e^{iS[x]/\hbar}$. Amplitudes from various paths superpose by addition. This then is another, a third way, of describing quantum mechanics, which looks quite different from that of Schrödinger or Heisenberg, but which is equivalent to them."

"...Now immediately after making a few checks on this thing, what we wanted to do, was to substitute the action $A[x; t_i, t_j]$ (4.294) for the other $S[x]$ (4.295). The first trouble was that I could not get the thing to work with the relativistic case of spin one–half. However, although I could deal with the matter only nonrelativistically, I could deal with the light or the photon interactions perfectly well by just putting the interaction terms of (4.294) into any action, replacing the mass terms by the non–relativistic $L dt = \frac{1}{2} M \dot{x}^2 dt$,

$$A[x; t_i, t_j] = \frac{1}{2} \sum_i m_i \int (\dot{x}^i_\mu)^2 dt_i + \frac{1}{2} \sum_{i,j(i\neq j)} e_i e_j \int \int \delta(I^2_{ij}) \, \dot{x}^i_\mu(t_i) \dot{x}^j_\mu(t_j) \, dt_i dt_j.$$

When the action has a delay, as it now had, and involved more than one time, I had to lose the idea of a wave function. That is, I could no longer describe the program as: given the amplitude for all positions at a certain time to calculate the amplitude at another time. However, that didn't cause very much trouble. It just meant developing a new idea. *Instead of wave functions we could talk about this: that if a source of a certain kind emits a particle, and a detector is there to receive it, we can give the amplitude that the source will emit and the detector receive*, $e^{iA[x;t_i,t_j]/\hbar}$. We do this without specifying the exact instant that the *source* emits or the exact instant that any *detector* receives, without trying to specify the state of anything at any particular time in between, but by just finding the *amplitude for the complete experiment*. And, then we could discuss how that amplitude would change if you had a scattering sample in between, as you rotated and changed angles, and so on, without really having any wave functions...It was also possible to discover what the old concepts of energy and momentum would mean with this generalized action. And, so I believed that I had a quantum theory of classical electrodynamics – or rather of this new classical electrodynamics described by the action $A[x; t_i, t_j]$ (4.294)..."

Lagrangian Path Integral

Dirac and Feynman first developed the lagrangian approach to functional integration. To review this approach, we start with the time–dependent *Schrödinger equation*

$$i\hbar \, \partial_t \psi(x,t) = -\partial_{x^2} \psi(x,t) + V(x) \, \psi(x,t)$$

appropriate to a particle of mass m moving in a potential $V(x)$, $x \in$. A solution to this equation can be written as an integral (see e.g., [Kla97, Kla00]),

$$\psi(x'', t'') = \int K(x'', t''; x', t') \, \psi(x', t') \, dx',$$

which represents the wave function $\psi(x'', t'')$ at time t'' as a linear superposition over the wave function $\psi(x', t')$ at the initial time t', $t' < t''$. The integral kernel $K(x'', t''; x', t')$ is known as the *propagator*, and according to Feynman [Fey48] it may be given by

$$K(x'', t''; x', t') = \mathcal{N} \int \mathcal{D}[x] \, e^{(i/\hbar) \int [(m/2)\dot{x}^2(t) - V(x(t))] \, dt},$$

which is a formal expression symbolizing an integral over a suitable set of paths. This integral is supposed to run over all continuous paths $x(t)$, $t' \leq t \leq t''$, where $x(t'') = x''$ and $x(t') = x'$ are fixed end points for all paths. Note that the integrand involves the *classical Lagrangian* for the system.

To overcome the convergence problems, Feynman adopted a *lattice regularization* as a procedure to yield well–defined integrals which was then followed by a limit as the lattice spacing goes to zero called the continuum limit. With $\varepsilon > 0$ denoting the lattice spacing, the details regarding the lattice regularization procedure are given by

$$K(x'', t''; x', t') = \lim_{\varepsilon \to 0} (m/2\pi i \hbar \varepsilon)^{(N+1)/2} \int \cdots$$

$$\cdots \int \exp\{(i/\hbar) \sum_{l=0}^{N} [(m/2\varepsilon)(x_{l+1} - x_l)^2 - \varepsilon \, V(x_l)]\} \prod_{l=1}^{N} dx_l,$$

where $x_{N+1} = x''$, $x_0 = x'$, and $\varepsilon \equiv (t'' - t')/(N+1)$, $N \in \{1, 2, 3, \dots\}$. In this version, at least, we have an expression that has a reasonable chance of being well defined, provided, that one interprets the conditionally convergent integrals involved in an appropriate manner. One common and fully acceptable interpretation adds a convergence factor to the exponent of the preceding integral in the form $-(\varepsilon^2/2\hbar) \sum_{l=1}^{N} x_l^2$, which is a term that formally makes no contribution to the final result in the continuum limit save for ensuring that the integrals involved are now rendered absolutely convergent.

Hamiltonian Path Integral

It is necessary to retrace history at this point to recall the introduction of the *phase–space path integral* by Feynman [Fey51, GS98]. In Appendix B to this article, Feynman introduced a formal expression for the configuration or q–space propagator given by (see e.g., [Kla97, Kla00])

$$K(q'', t''; q', t') = \mathcal{M} \int \mathcal{D}[p] \, \mathcal{D}[q] \, \exp\{(i/\hbar) \int [p\dot{q} - H(p, q)] \, dt\}.$$

In this equation one is instructed to integrate over all paths $q(t)$, $t' \leq t \leq t''$, with $q(t'') \equiv q''$ and $q(t') \equiv q'$ held fixed, as well as to integrate over all paths $p(t)$, $t' \leq t \leq t''$, without restriction.

It is widely appreciated that the phase–space path integral is more generally applicable than the original, Lagrangian, version of the path integral. For example, the original configuration space path integral is satisfactory for Lagrangians of the general form

$$L(x) = \frac{1}{2} m\dot{x}^2 + A(x)\dot{x} - V(x) ,$$

but it is unsuitable, for example, for the case of a relativistic particle with the Lagrangian

$$L(x) = -m\,qrt{1 - \dot{x}^2}$$

expressed in units where the speed of light is unity. For such a system – as well as many more general expressions – the phase–space form of the path integral is to be preferred. In particular, for the relativistic free particle, the phase–space path integral

$$\mathcal{M} \int \mathcal{D}[p]\,\mathcal{D}[q]\ \exp\{(i/\hbar) \int [p\dot{q} - qrt{p^2 + m^2}\,]\,dt\},$$

is readily evaluated and induces the correct propagator.

Feynman–Kac Formula

Through his own research, M. Kac was fully aware of *Wiener's theory of Brownian motion* and the *associated diffusion equation* that describes the corresponding *distribution function*. Therefore, it is not surprising that he was well prepared to give a path integral expression in the sense of Feynman for an equation similar to the time–dependent Schrödinger equation save for a rotation of the time variable by $-\pi/2$ in the complex–plane, namely, by the change $t \to -it$ (see e.g., [Kla97, Kla00]). In particular, Kac [Kac51] considered the equation

$$\partial_t \rho(x,t) = \partial_{x^2} \rho(x,t) - V(x)\,\rho(x,t). \tag{4.26}$$

This equation is analogous to Schrödinger equation but differs from it in certain details. Besides certain constants which are different, and the change $t \to -it$, the nature of the dependent variable function $\rho(x,t)$ is quite different from the normal quantum mechanical wave function. For one thing, if the function ρ is initially real it will remain real as time proceeds. Less obvious is the fact that if $\rho(x,t) \geq 0$ for all x at some time t, then the function will continue to be nonnegative for all time t. Thus we can interpret $\rho(x,t)$ more like a probability density; in fact in the special case that $V(x) = 0$, then $\rho(x,t)$ is the probability density for a Brownian particle which underlies the *Wiener measure*. In this regard, ν is called the diffusion constant.

The fundamental solution of (4.26) with $V(x) = 0$ is readily given as

$$W(x,T;y,0) = \frac{1}{qrt{2\pi\nu T}}\,\exp\left(-\frac{(x-y)^2}{2\nu T}\right),$$

4.1 Sum over Histories

which describes the solution to the diffusion equation subject to the initial condition

$$\lim_{T \to 0+} W(x, T; y, 0) = \delta(x - y).$$

Moreover, it follows that the solution of the diffusion equation for a general initial condition is given by

$$\rho(x'', t'') = \int W(x'', t''; x', t') \, \rho(x', t') \, dx'.$$

Iteration of this equation N times, with $\epsilon = (t'' - t')/(N+1)$, leads to the equation

$$\rho(x'', t'') = N' \int \cdots \int e^{-(1/2\nu\epsilon) \sum_{l=0}^{N}(x_{l+1}-x_l)^2} \prod_{l=1}^{N} dx_l \, \rho(x', t') \, dx',$$

where $x_{N+1} \equiv x''$ and $x_0 \equiv x'$. This equation features the imaginary time propagator for a free particle of unit mass as given formally as

$$W(x'', t''; x', t') = \mathcal{N} \int \mathcal{D}[x] \, e^{-(1/2\nu) \int \dot{x}^2 \, dt},$$

where $\mathcal{N}$ denotes a formal normalization factor.

The similarity of this expression with the Feynman path integral [for $V(x) = 0$] is clear, but there is a profound difference between these equations. In the former (Feynman) case the underlying measure is only *finitely additive*, while in the latter (Wiener) case the continuum limit actually defines a genuine measure, i.e., a *countably additive measure* on paths, which is a version of the famous *Wiener measure*. In particular,

$$W(x'', t''; x', t') = \int d\mu_W^\nu(x),$$

where μ_W^ν denotes a measure on continuous paths $x(t)$, $t' \leq t \leq t''$, for which $x(t'') \equiv x''$ and $x(t') \equiv x'$. Such a measure is said to be a *pinned Wiener measure*, since it specifies its path values at two time points, i.e., at $t = t'$ and at $t = t'' > t'$.

We note that Brownian motion paths have the property that with probability one they are concentrated on continuous paths. However, it is also true that the time derivative of a Brownian path is almost nowhere defined, which means that, with probability one, $\dot{x}(t) = \pm\infty$ for all t.

When the potential $V(x) \neq 0$ the propagator associated with (4.26) is formally given by

$$W(x'', t''; x', t') = \mathcal{N} \int \mathcal{D}[x] e^{-(1/2\nu) \int \dot{x}^2 \, dt - \int V(x) \, dt},$$

an expression which is well defined if $V(x) \geq c$, $-\infty < c < \infty$. A mathematically improved expression makes use of the Wiener measure and reads

$$W(x'', t''; x', t') = \int e^{-\int V(x(t))\, dt} \, d\mu_W^\nu(x).$$

This is an elegant relation in that it represents a solution to the differential equation (4.26) in the form of an integral over Brownian motion paths suitably weighted by the potential V. Incidentally, since the propagator is evidently a strictly positive function, it follows that the solution of the differential equation (4.26) is nonnegative for all time t provided it is nonnegative for any particular time value.

Itô Formula

Itô [Ito60] proposed another version of a *continuous–time regularization* that resolved some of the troublesome issues. In essence, the proposal of Itô takes the form given by

$$\lim_{\nu \to \infty} \mathcal{N}_\nu \int \mathcal{D}[x]\, \exp\{(i/\hbar) \int [\tfrac{1}{2} m\dot{x}^2 - V(x)]\, dt\} \exp\{-(1/2\nu) \int [\ddot{x}^2 + \dot{x}^2]\, dt\}.$$

Note well the alternative form of the auxiliary factor introduced as a regulator. The additional term $\ddot{x}^2$, the square of the second derivative of x, acts to smooth out the paths sufficiently well so that in the case of (21) both $x(t)$ and $\dot{x}(t)$ are continuous functions, leaving $\ddot{x}(t)$ as the term which does not exist. However, since only x and $\dot{x}$ appear in the rest of the integrand, the indicated path integral can be well defined; this is already a positive contribution all by itself (see e.g., [Kla97, Kla00]).

4.1.4 Path–Integral Quantization

Canonical versus Path–Integral Quantization

Recall that in the usual, *canonical formulation* of quantum mechanics, the system's phase–space coordinates, q, and momenta, p, are replaced by the corresponding Hermitian operators in the Hilbert space, with real measurable eigenvalues, which obey *Heisenberg commutation relations*.

The *path–integral quantization* is instead based directly on the notion of a propagator $K(q_f, t_f; q_i, t_i)$ which is defined such that (see [Ryd96, CL84, Gun03])

$$\psi(q_f, t_f) = \int K(q_f, t_f; q_i, t_i)\, \psi(q_i, t_i)\, dq_i, \tag{4.27}$$

i.e., the wave function $\psi(q_f, t_f)$ at final time t_f is given by a Huygens principle in terms of the wave function $\psi(q_i, t_i)$ at an initial time t_i, where we have to integrate over all the points q_i since all can, in principle, send out little wavelets that would influence the value of the wave function at q_f at the later time t_f. This equation is very general and is an expression of causality. We use the normal units with $\hbar = 1$.

According to the usual interpretation of quantum mechanics, $\psi(q_f, t_f)$ is the *probability amplitude* that the particle is at the point q_f and the time t_f, which means that $K(q_f, t_f; q_i, t_i)$ is the probability amplitude for a transition from q_i and t_i to q_f and t_f. The probability that the particle is observed at q_f at time t_f if it began at q_i at time t_i is

$$P(q_f, t_f; q_i, t_i) = |K(q_f, t_f; q_i, t_i)|^2.$$

Let us now divide the time interval between t_i and t_f into two, with t as the intermediate time, and q the intermediate point in space. Repeated application of (4.27) gives

$$\psi(q_f, t_f) = \int \int K(q_f, t_f; q, t)\, dq\, K(q, t; q_i, t_i)\, \psi(q_i, t_i)\, dq_i,$$

from which it follows that

$$K(q_f, t_f; q_i, t_i) = \int dq\, K(q_f, t_f; q, t)\, K(q, t; q_i, t_i).$$

This equation says that the transition from (q_i, t_i) to (q_f, t_f) may be regarded as the result of the transition from (q_i, t_i) to all available intermediate points q followed by a transition from (q, t) to (q_f, t_f). This notion of *all possible paths* is crucial in the *path–integral formulation* of quantum mechanics.

Now, recall that the *state vector* $|\psi, t\rangle_S$ in the *Schrödinger picture* is related to that in the *Heisenberg picture* $|\psi\rangle_H$ by

$$|\psi, t\rangle_S = e^{-iHt} |\psi\rangle_H,$$

or, equivalently,

$$|\psi\rangle_H = e^{iHt} |\psi, t\rangle_S.$$

We also define the vector

$$|q, t\rangle_H = e^{iHt} |q\rangle_S,$$

which is the Heisenberg version of the Schrödinger state $|q\rangle$. Then, we can equally well write

$$\psi(q, t) = \langle q, t | \psi \rangle_H. \tag{4.28}$$

By completeness of states we can now write

$$\langle q_f, t_f | \psi \rangle_H = \int \langle q_f, t_f | q_i, t_i \rangle_H \langle q_i, t_i | \psi \rangle_H\, dq_i,$$

which with the definition of (4.28) becomes

$$\psi(q_f, t_f) = \int \langle q_f, t_f | q_i, t_i \rangle_H\, \psi(q_i, t_i)\, dq_i.$$

Comparing with (4.27), we get

$$K(q_f, t_f; q_i, t_i) = \langle q_f, t_f \, | q_i, t_i \rangle_H.$$

Now, let us calculate the *quantum–mechanics propagator*

$$\langle q', t' \, | q, t \rangle_H = \langle q' | e^{-iH(t-t')} | q \rangle$$

using the *path–integral formalism* that will incorporate the direct quantization of the coordinates, without Hilbert space and Hermitian operators.

The first step is to divide up the time interval into $n+1$ tiny pieces: $t_l = l\varepsilon + t$ with $t' = (n+1)\varepsilon + t$. Then, by completeness, we can write (dropping the Heisenberg picture index H from now on)

$$\langle q', t' \, | q, t \rangle = \int dq_1(t_1)... \int dq_n(t_n) \, \langle q', t' \, | q_n, t_n \rangle \times$$
$$\times \langle q_n, t_n \, | q_{n-1}, t_{n-1} \rangle ... \langle q_1, t_1 \, | q, t \rangle. \qquad (4.29)$$

The integral $\int dq_1(t_1)...dq_n(t_n)$ is an *integral over all possible paths*, which are not trajectories in the normal sense, since there is no requirement of continuity, but rather *Markov chains*.

Now, for small ε we can write

$$\langle q', \varepsilon \, | q, 0 \rangle = \langle q' | e^{-i\varepsilon H(P,Q)} | q \rangle = \delta(q'-q) - i\varepsilon \, \langle q' | H(P,Q) | q \rangle,$$

where $H(P,Q)$ is the Hamiltonian (e.g., $H(P,Q) = \frac{1}{2}P^2 + V(Q)$, where P, Q are the momentum and coordinate operators). Then we have (see [Ryd96, CL84, Gun03])

$$\langle q' | H(P,Q) | q \rangle = \int \frac{dp}{2\pi} \, e^{ip(q'-q)} H\left(p, \frac{1}{2}(q'+q)\right).$$

Putting this into our earlier form we get

$$\langle q', \varepsilon \, | q, 0 \rangle \simeq \int \frac{dp}{2\pi} \exp\left[i\left\{p(q'-q) - \varepsilon H\left(p, \frac{1}{2}(q'+q)\right)\right\}\right],$$

where the 0th order in $\varepsilon \to \delta(q'-q)$ and the 1st order in $\varepsilon \to -i\varepsilon \, \langle q' | H(P,Q) | q \rangle$. If we now substitute many such forms into (4.29) we finally get

$$\langle q', t' \, | q, t \rangle = \lim_{n\to\infty} \int \prod_{i=1}^{n} dq_i \prod_{k=1}^{n+1} \frac{dp_k}{2\pi} \times \qquad (4.30)$$
$$\times \exp\left\{i \sum_{j=1}^{n+1} [p_j(q_j - q_{j-1})] - H\left(p_j, \frac{1}{2}(q_j + q_{j+1})\right)(t_j - t_{j-1})]\right\},$$

with $q_0 = q$ and $q_{n+1} = q'$. Roughly, the above formula says to *integrate over all possible momenta and coordinate values associated with a small interval*, weighted by something that is going to turn into the *exponential of the action* e^{iS} in the limit where $\varepsilon \to 0$. It should be stressed that the different q_i and p_k integrals are independent, which implies that p_k for one interval can be completely different from the $p_{k'}$ for some other interval (including the neighboring intervals). In principle, the integral (4.30) should be defined by *analytic continuation into the complex–plane* of, for example, the p_k integrals.

Now, if we go to the differential limit where we call $t_j - t_{j-1} \equiv d\tau$ and write $\frac{(q_j - q_{j-1})}{(t_j - t_{j-1})} \equiv \dot{q}$, then the above formula takes the form

$$\langle q', t' | q, t \rangle = \int \mathcal{D}[p]\mathcal{D}[q] \exp\left\{ i \int_t^{t'} [p\dot{q} - H(p,q)] d\tau \right\},$$

where we have used the shorthand notation

$$\int \mathcal{D}[p]\mathcal{D}[q] \equiv \int \prod_\tau \frac{dq(\tau)dp(\tau)}{2\pi}.$$

Note that the above integration is an integration over the p and q values at every time τ. This is what we call a *functional integral*. We can think of a given set of choices for all the $p(\tau)$ and $q(\tau)$ as defining a *path in the 6D phase–space*. The most important point of the above result is that we have get an expression for a *quantum–mechanical transition amplitude* in terms of an integral involving only pure complex numbers, without operators.

We can actually perform the above integral for Hamiltonians of the type $H = H(P, Q)$. We use square completion in the exponential for this, defining the integral in the complex p plane and continuing to the physical situation. In particular, we have

$$\int_{-\infty}^{\infty} \frac{dp}{2\pi} \exp\left\{ i\varepsilon(p\dot{q} - \frac{1}{2}p^2) \right\} = \frac{1}{\sqrt{2\pi i\varepsilon}} \exp\left[\frac{1}{2} i\varepsilon \dot{q}^2 \right],$$

(see [Ryd96, CL84, Gun03]) which, substituting into (4.30) gives

$$\langle q', t' | q, t \rangle = \lim_{n \to \infty} \int \prod_i \frac{dq_i}{\sqrt{2\pi i \varepsilon}} \exp\{i\varepsilon \sum_{j=1}^{n+1} [\frac{1}{2}(\frac{q_j - q_{j-1}}{\varepsilon})^2 - V(\frac{q_j + q_{j+1}}{2})]\}.$$

This can be formally written as

$$\langle q', t' | q, t \rangle = \int \mathcal{D}[q] \, e^{iS[q]},$$

where

$$\int \mathcal{D}[q] \equiv \int \prod_i \frac{dq_i}{\sqrt{2\pi i\varepsilon}},$$

while
$$S[q] = \int_t^{t'} L(q, \dot{q})\, d\tau$$
is the *standard action* with the *Lagrangian*
$$L = \frac{1}{2}\dot{q}^2 - V(q).$$

Generalization to many degrees of freedom is straightforward:
$$\langle q_1'...q_N', t'|q_1...q_N, t\rangle = \int \mathcal{D}[p]\mathcal{D}[q]\, \exp\left\{i\int_t^{t'}\left[\sum_{n=1}^N p_n \dot{q}_n - H(p_n, q_n)\right] d\tau\right\},$$
with
$$\int \mathcal{D}[p]\mathcal{D}[q] = \int \prod_{n=1}^N \frac{dq_n dp_n}{2\pi}.$$

Here, $q_n(t) = q_n$ and $q_n(t') = q_n'$ for all $n = 1, ..., N$, and we are allowing for the full Hamiltonian of the system to depend upon all the N momenta and coordinates collectively.

Elementary Applications

(i) Consider first
$$\langle q', t'|Q(t_0)|q, t\rangle$$
$$= \int \prod dq_i(t_i)\, \langle q', t'|q_n, t_n\rangle \ldots \langle q_{i0}, t_{i0}|Q(t_0)|q_{i-1}, t_{i-1}\rangle \ldots \langle q_1, t_1|q, t\rangle,$$
where we choose one of the time interval ends to coincide with t_0, i.e., $t_{i0} = t_0$. If we operate $Q(t_0)$ to the left, then it is replaced by its eigenvalue $q_{i0} = q(t_0)$. Aside from this one addition, everything else is evaluated just as before and we will obviously get
$$\langle q', t'|Q(t_0)|q, t\rangle = \int \mathcal{D}[p]\mathcal{D}[q]\, q(t_0)\, \exp\left\{i\int_t^{t'}[p\dot{q} - H(p,q)]d\tau\right\}.$$

(ii) Next, suppose we want a *path–integral expression* for $\langle q', t'|Q(t_1)Q(t_2)|q, t\rangle$ in the case where $t_1 > t_2$. For this, we have to insert as intermediate states $|q_{i1}, t_{i1}\rangle \langle q_{i1}, t_{i1}|$ with $t_{i1} = t_1$ and $|q_{i2}, t_{i2}\rangle \langle q_{i2}, t_{i2}|$ with $t_{i2} = t_2$ and since we have ordered the times at which we do the insertions we must have the first insertion to the left of the 2nd insertion when $t_1 > t_2$. Once these insertions are done, we evaluate $\langle q_{i1}, t_{i1}|Q(t_1) = \langle q_{i1}, t_{i1}|q(t_1)$ and $\langle q_{i2}, t_{i2}|Q(t_2) = \langle q_{i2}, t_{i2}|q(t_2)$ and then proceed as before and get
$$\langle q', t'|Q(t_1)Q(t_2)|q, t\rangle = \int \mathcal{D}[p]\mathcal{D}[q]\, q(t_1)\, q(t_2)\, \exp\left\{i\int_t^{t'}[p\dot{q} - H(p,q)]d\tau\right\}.$$

Now, let us ask what the above integral is equal to if $t_2 > t_1$? It is obvious that what we get for the above integral is $\langle q', t'|Q(t_2)Q(t_1)|q, t\rangle$. Clearly, this generalizes to an arbitrary number of Q operators.

(iii) When we enter into quantum field theory, the Q's will be replaced by fields, since it is the fields that play the role of coordinates in the 2nd quantization conditions.

Sources

The *source* is represented by modifying the Lagrangian:

$$L \to L + J(t)q(t).$$

Let us define $|0,t\rangle^J$ as the ground state (vacuum) vector (in the moving frame, i.e., with the e^{iHt} included) in the presence of the source. The required *transition amplitude* is

$$Z[J] \propto \langle 0, +\infty|0, -\infty\rangle^J,$$

where the source $J = J(t)$ plays a role analogous to that of an electromagnetic current, which acts as a source of the electromagnetic field. In other words, we can think of the scalar product $J_\mu A^\mu$, where J_μ is the current from a scalar (or Dirac) field acting as a source of the potential A^μ. In the same way, we can always define a current J that acts as the source for some arbitrary field ϕ. $Z[J]$ (otherwise denoted by $W[J]$) is a functional of the current J, defined as (see [Ryd96, CL84, Gun03])

$$Z[J] \propto \int \mathcal{D}[p]\mathcal{D}[q] \exp\left\{i\int_t^{t'} [p(\tau)\dot{q}(\tau) - H(p,q) + J(\tau)q(\tau)]d\tau\right\},$$

with the *normalization condition* $Z[J = 0] = 1$. Here, the argument of the exponential depends upon the functions $q(\tau)$ and $p(\tau)$ and we then integrate over all possible forms of these two functions. So the exponential is a functional that maps a choice for these two functions into a number. For example, for a quadratically completable $H(p,q)$, the p integral can be performed as a q integral

$$Z[J] \propto \int \mathcal{D}[q] \exp\left\{i\int_{-\infty}^{+\infty}\left(L + Jq + \frac{1}{2}i\varepsilon q^2\right)d\tau\right\},$$

where the addittion to H was chosen in the form of a *convergence factor* $-\frac{1}{2}i\varepsilon q^2$.

Fields

Let us now treat the *abstract scalar field* $\phi(x)$ as a coordinate in the sense that we imagine dividing space up into many little cubes and the average value of the field $\phi(x)$ in that cube is treated as a coordinate for that little cube. Then, we go through the multi–coordinate analogue of the procedure we just considered above and take the continuum limit. The final result is

$$Z[J] \propto \int \mathcal{D}[\phi] \exp\left\{i \int d^4x \left(\mathcal{L}(\phi(x)) + J(x)\phi(x) + \frac{1}{2}i\varepsilon\phi^2\right)\right\},$$

where for $\mathcal{L}$ we would employ the *Klein–Gordon Lagrangian* form. In the above, the dx_0 integral is the same as $d\tau$, while the $d^3\mathbf{x}$ integral is summing over the sub–Lagrangians of all the different little cubes of space and then taking the continuum limit. $\mathcal{L}$ is the *Lagrangian density* describing the Lagrangian for each little cube after taking the many–cube limit (see [Ryd96, CL84, Gun03]) for the full derivation).

We can now introduce *interactions*, $\mathcal{L}_I$. Assuming the simple form of the Hamiltonian, we have

$$Z[J] \propto \int \mathcal{D}[\phi] \exp\left\{i \int d^4x \left(\mathcal{L}(\phi(x)) + \mathcal{L}_I(\phi(x)) + J(x)\phi(x)\right)\right\},$$

again using the normalization factor required for $Z[J=0]=1$.

For example of Klein Gordon theory, we would use

$$\mathcal{L} = \mathcal{L}_0 + \mathcal{L}_I, \qquad \mathcal{L}_0 \frac{1}{2}[\partial_\mu \phi \partial^\mu \phi - \mu^2 \phi^2], \qquad \mathcal{L}_I = \mathcal{L}_I(\phi),$$

where $\partial_\mu \equiv \partial_{x^\mu}$ and we can freely manipulate indices, as we are working in Euclidean space [3]. In order to define the above $Z[J]$, we have to include a convergence factor $i\varepsilon\phi^2$,

$$\mathcal{L}_0 \to \frac{1}{2}[\partial_\mu \phi \partial^\mu \phi - \mu^2\phi^2 + i\varepsilon\phi^2], \qquad \text{so that}$$

$$Z[J] \propto \int \mathcal{D}[\phi] \exp\{i \int d^4x (\frac{1}{2}[\partial_\mu \phi \partial^\mu \phi - \mu^2\phi^2 + i\varepsilon\phi^2] + \mathcal{L}_I(\phi(x)) + J(x)\phi(x))\}$$

is the appropriate *generating function* in the free field theory case.

Gauges

In the path integral approach to quantization of the *gauge theory*, we implement *gauge fixing* by restricting in some manner or other the path integral over gauge fields $\int \mathcal{D}[A_\mu]$. In other words we will write instead

$$Z[J] \propto \int \mathcal{D}[A_\mu]\, \delta\, (\text{some gauge fixing condition}) \exp\{i \int d^4x \mathcal{L}(A_\mu)\}.$$

A common approach would be to start with the *gauge condition*

$$\mathcal{L} = -\frac{1}{4}F_{\mu\nu}F^{\mu\nu} - \frac{1}{2}(\partial^\mu A_\mu)^2$$

where the electrodynamic field tensor is given by $F_{\mu\nu} = \partial_\mu A_\nu - \partial_\nu A_\mu$, and calculate

$$Z[J] \propto \int \mathcal{D}[A_\mu] \exp\left\{ i \int d^4x \left[\mathcal{L}(A_\mu(x)) + J_\mu(x) A^\mu(x) \right] \right\}$$

as the *generating function* for the *vacuum expectation* values of *time ordered products* of the A_μ fields. Note that J_μ should be conserved ($\partial^\mu J_\mu = 0$) in order for the full expression $\mathcal{L}(A_\mu) + J_\mu A^\mu$ to be *gauge–invariant* under the integral sign when $A_\mu \to A_\mu + \partial^\mu \Lambda$. For a proper approach, see [Ryd96, CL84, Gun03].

Riemannian–Symplectic Geometries

In this subsection, following [SK98b], we describe path integral quantization on Riemannian–symplectic manifolds. Let $\hat{q}^j$ be a set of Cartesian coordinate canonical operators satisfying the Heisenberg commutation relations $[\hat{q}^j, \hat{q}^k] = i\omega^{jk}$. Here $\omega^{jk} = -\omega^{kj}$ is the canonical symplectic structure. We introduce the canonical coherent states as $|q\rangle \equiv e^{iq^j \omega_{jk} \hat{q}^k} |0\rangle$, where $\omega_{jn}\omega^{nk} = \delta_j^k$, and $|0\rangle$ is the ground state of a harmonic oscillator with unit angular frequency. Any state $|\psi\rangle$ is given as a function on phase–space in this representation by $\langle q|\psi\rangle = \psi(q)$. A general operator $\hat{A}$ can be represented in the form $\hat{A} = \int dq\, a(q) |q\rangle\langle q|$, where $a(q)$ is the lower symbol of the operator and dq is a properly normalized form of the Liouville measure. The function $A(q, q') = \langle q|\hat{A}|q'\rangle$ is the *kernel of the operator*.

The main object of the path integral formalism is the integral kernel of the evolution operator

$$K_t(q, q') = \langle q| e^{-it\hat{H}} |q'\rangle = \int_{q(0)=q'}^{q(t)=q} \mathcal{D}[q]\, e^{i \int_0^t d\tau \left(\frac{1}{2} q^j \omega_{jk} \dot{q}^k - h \right)}. \tag{4.31}$$

Here $\hat{H}$ is the Hamiltonian, and $h(q)$ its symbol. The measure formally implies a sum over all phase-space paths pinned at the initial and final points, and a Wiener measure regularization implies the following replacement

$$\mathcal{D}[q] \to \mathcal{D}[\mu_\nu(q)] = \mathcal{D}[q]\, e^{-\frac{1}{2\nu} \int_0^t d\tau\, \dot{q}^2} = N_\nu(t)\, d\mu_W^\nu(q). \tag{4.32}$$

The factor $N_\nu(t)$ equals $2\pi e^{\nu t/2}$ for every degree of freedom, $d\mu_W^\nu(q)$ stands for the Wiener measure, and ν denotes the diffusion constant. We denote by $K_t^\nu(q, q')$ the integral kernel of the evolution operator for a finite ν. The Wiener measure determines a stochastic process on the *flat* phase–space. The integral of the symplectic 1–form $\int q\omega dq$ is a stochastic integral that is interpreted in the Stratonovich sense. Under general coordinate transformations $q = q(\bar{q})$, the Wiener measure describes *the same* stochastic process on *flat* space in the curvilinear coordinates $dq^2 = d\sigma(\bar{q})^2$, so that the value of the integral is not changed apart from a possible phase term. After the calculation of the integral, the evolution operator kernel is get by taking the limit $\nu \to \infty$. The existence of this limit, and also the covariance under general phase-space coordinate transformations, can be proved through the *operator* formalism for the regularized kernel $K_t^\nu(q, q')$.

Note that the integral (4.31) with the Wiener measure inserted can be regarded as an ordinary Lagrangian path integral with a complex action, where the configuration space is the original phase–space and the Hamiltonian $h(q)$ serves as a potential. Making use of this observation it is not hard to derive the corresponding Schrödinger–like equation

$$\partial_t K_t^\nu(q, q') = \left[\frac{\nu}{2} \left(\partial_{q^j} + \frac{i}{2} \omega_{jk} q^k \right)^2 - i h(q) \right] K_t^\nu(q, q'), \qquad (4.33)$$

subject to the initial condition $K_{t=0}^\nu(q, q') = \delta(q - q'), 0 < \nu < \infty$. One can show that $\hat{K}_t^\nu \to \hat{K}_t$ as $\nu \to \infty$ for all $t > 0$. The covariance under general coordinate transformations follows from the covariance of the "kinetic" energy of the Schrödinger operator in (4.33): The Laplace operator is replaced by the Laplace–Beltrami operator in the new curvilinear coordinates $q = q(\bar{q})$, so the solution is not changed, but written in the new coordinates. This is similar to the covariance of the ordinary Schrödinger equation and the corresponding *Lagrangian* path integral relative to general coordinate transformations on the configuration space: The kinetic energy operator (the Laplace operator) in the ordinary Schrödinger equation gives a term *quadratic* in time derivatives in the path integral measure which is sufficient for the general coordinate covariance. We remark that the regularization procedure based on the modified Schrödinger equation (4.33) applies to far more general Hamiltonians than those quadratic in canonical momenta and leading to the conventional *Lagrangian* path integral.

4.1.5 Statistical Mechanics via Path Integrals

The Feynman path integral turns out to provide an elegant way of doing statistical mechanics, as the *partition function* can be written as a path–integral.

Recall that the standard partition function is defined as

$$Z = \sum_j e^{-\beta E_j}, \qquad (4.34)$$

where $\beta = 1/k_B T$ and E_j is the energy of the state $|j\rangle$. We can write

$$Z = \sum_j \langle j| e^{-\beta H} |j\rangle = \text{Tr}\, e^{-\beta H}.$$

But recall the definition of the *propagator* [Mac00]

4.1 Sum over Histories

$$K(q', T; q, 0) = \langle q' | e^{-iHT} | q \rangle.$$

Suppose we consider T to be a complex parameter, and consider it to be pure imaginary, so that we can write $T = -i\beta$, where β is real. Then we have

$$K(q', -i\beta; q, 0) = \langle q' | e^{-iH(-i\beta)} | q \rangle = \langle q' | e^{-\beta H} \underbrace{\sum_j |j\rangle \langle j|}_{=1} | q \rangle$$

$$= \sum_j e^{-\beta E_j} \langle q' | j \rangle \langle j | q \rangle = \sum_j e^{-\beta E_j} \langle j | q \rangle \langle q' | j \rangle.$$

Putting $q' = q$ and integrating over q, we get

$$\int dq\, K(q, -i\beta; q, 0) = \sum_j e^{-\beta E_j} \langle j | \underbrace{\int dq |q\rangle \langle q|}_{=1} |j\rangle = Z. \tag{4.35}$$

This is the central observation of this section: that the propagator evaluated at negative imaginary time is related to the partition function [Mac00].

We can easily work out an elementary example such as the harmonic oscillator. Recall the path integral for it, (4.22):

$$K(q', T; q, 0) = \left(\frac{m\omega}{2\pi i \sin \omega T}\right)^{1/2} \exp\left\{i\frac{m\omega}{2\sin \omega T}\left((q'^2 + q^2)\cos \omega T - 2q'q\right)\right\}.$$

We can put $q' = q$ and $T = -i\beta$:

$$K(q, -i\beta; q, 0) = \left(\frac{m\omega}{2\pi \sinh(\beta\omega)}\right)^{1/2} \exp\left\{-\frac{m\omega q^2}{\sinh(\beta\omega)}(\cosh(\beta\omega) - 1)\right\}.$$

The partition function is thus

$$Z = \int dq\, K(q, -i\beta; q, 0) = \left(\frac{m\omega}{2\pi \sinh(\beta\omega)}\right)^{1/2} \sqrt{\frac{\pi}{\frac{m\omega}{\sinh(\beta\omega)}(\cosh(\beta\omega) - 1)}}$$

$$= [2(\cosh(\beta\omega) - 1)]^{-1/2} = \left[e^{\beta\omega/2}(1 - e^{-\beta\omega})\right]^{-1} = \frac{e^{-\beta\omega/2}}{1 - e^{-\beta\omega}} = \sum_{j=0}^{\infty} e^{-\beta(j+1/2)\omega}.$$

Putting $\hbar$ back in, we get the standard result:

$$Z = \sum_{j=0}^{\infty} e^{-\beta(j+1/2)\hbar\omega}.$$

We can rewrite the partition function in terms of a path–integral. In ordinary (real) time,

$$K(q', T; q, 0) = \int \mathcal{D}q(t) \exp\left[i\int_0^T dt \left(\frac{m\dot{q}^2}{2} - V(q)\right)\right],$$

where the integral is over all paths from $(q, 0)$ to (q', T). With $q' = q, T \to -i\beta$,

$$K(q, -i\beta; q, 0) = \int \mathcal{D}q(t) \exp\left[i \int_0^{-i\beta} dt \left(\frac{m\dot{q}^2}{2} - V(q)\right)\right],$$

where we now integrate along the negative imaginary time axis (Figure 4.6).

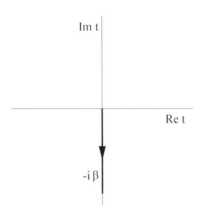

Fig. 4.6. Path in the complex time plane (adapted from [Mac00]).

Let us now define a real variable for this integration, $\tau = it$. τ is called the imaginary time, since when the time t is imaginary, τ is real. Then the integral over τ is along *its* real axis: when $t : 0 \to -i\beta$, then $\tau : 0 \to \beta$. We can write q as a function of the variable τ: $q(t) \to q(\tau)$; then $\dot{q} = i dq/d\tau$. The propagator becomes

$$K(q, -i\beta; q, 0) = \int \mathcal{D}q(\tau) \exp - \int_0^\beta d\tau \left(\frac{m}{2}\left(\frac{dq}{d\tau}\right)^2 + V(q)\right). \tag{4.36}$$

The integral is over all functions $q(\tau)$ such that $q(0) = q(\beta) = q$.

The result (4.36) is an 'imaginary–time' or *Euclidean path integral*, defined by associating to each path an amplitude (statistical weight) $\exp -S_E$, where S_E is the so–called *Euclidean action*, obtained from the usual *Minkowski action* by changing the sign of the potential energy term.

4.1.6 Path Integrals and Green's Functions

In quantum field theory we are interested in objects such as [Mac00]

$$\langle 0 | T \hat{\phi}(x_1) \hat{\phi}(x_2) \cdots \hat{\phi}(x_n) | 0 \rangle,$$

the vacuum expectation value of a *time–ordered product* of *Heisenberg field operators*. This object is known as a *Green's function*,[15] or as a *correlation*

---

[15] Recall that a *Green's function*, $G(x, s)$, of a linear operator L acting on distributions over a manifold M, at a point x_0, is any solution of *linear operator equation*

$$L\,G(x, s) = \delta(x - s),$$

where $\delta(x)$ is the Dirac delta function at a point $x \in M$. This technique can be used to solve differential equations of the form

$$Lu(x) = f(x).$$

If the kernel of the operator L is nontrivial, then the Green's function is not unique. However, in practice, some combination of *symmetry*, *boundary conditions* and/or other externally imposed criteria would give us a unique Green's function. Also, Green's functions in general are distributions (or generalized functions like the Dirac delta function $\delta(x)$), not necessarily proper functions.

Green's functions are a useful tool in condensed matter theory, where they allow the resolution of the *diffusion equation*, as well as in quantum mechanics, where the Green's function of the Hamiltonian is a key concept, with important links to the concept of density of states. The Green's functions used in those two domains are highly similar, due to the analogy in the mathematical structure of the real–valued *diffusion equation* and complex–valued Schrödinger equation. Briefly, if such a function $G(x, s)$ can be found for the operator L, then if we multiply the equation $L\,G(x, s) = \delta(x - s)$ for the Green's function by $f(s)$, and then perform an integration in the s variable, we get

$$\int L G(x, s) f(s) ds = \int \delta(x - s) f(s) ds = f(x).$$

As the right hand side is actually $Lu(x) = f(x)$, we get

$$Lu(x) = \int L G(x, s) f(s) ds.$$

Further, because the operator L is linear and acts on the variable x alone (not on the variable of integration s), we can take the operator L outside of the integration on the right hand side, getting

$$Lu(x) = L \left(\int G(x, s) f(s) ds \right),$$

which implies

$$u(x) = \int G(x, s) f(s) ds.$$

Thus, we can obtain the function $u(x)$ through knowledge of the Green's function in the first equation, and the source term on the right hand side in second equation. This process has resulted from the linearity of the operator L.

A *convolution* with a Green's function gives solutions to inhomogeneous differential–integral equations, most commonly a *Sturm–Liouville problem*. If G is the Green's function of an operator L, then the solution for u of the equation $Lu = f$ is given by *function*. The order of the operators is such that the earliest field is written last (right-most), the second earliest second last, etc. For example,

$$T\hat{\phi}(x_1)\hat{\phi}(x_2) = \begin{cases} \hat{\phi}(x_1)\hat{\phi}(x_2), & x_1^0 > x_2^0, \\ \hat{\phi}(x_2)\hat{\phi}(x_1), & x_2^0 > x_1^0, \end{cases}$$

Green's functions are related to amplitudes for physical processes such as scattering and decay processes.

Let us look at the analogous object in quantum mechanics,

$$G^{(n)}(t_1, t_2, \cdots, t_n) = \langle 0| T\hat{q}(t_1)\hat{q}(t_2)\cdots \hat{q}(t_n) |0\rangle.$$

We will develop a path–integral expression for this.

First, we must recast the path–integral in terms of Heisenberg representation objects. The operator $\hat{q}(t)$ is the usual Heisenberg operator, defined in terms of the Schrödinger operator $\hat{q}$ by

$$\hat{q}(t) = e^{iHt}\hat{q}e^{-iHt}.$$

The eigenstates of the Heisenberg operator are $|q,t\rangle$: $\hat{q}(t)|q,t\rangle = q|q,t\rangle$. The relation with the time–independent eigenstates is $|q,t\rangle = e^{iHt}|q\rangle$ [Mac00]. Then we can write the path–integral,

$$K = \langle q'| e^{-iHT} |q\rangle = \langle q', T| q, 0\rangle = \int \mathcal{D}[q]e^{iS}.$$

We can now calculate the *two–point function* $G(t_1, t_2)$, via the path–integral. We will proceed in two steps. First, we will calculate the following expression:

$$\langle q', T| T\hat{q}(t_1)\hat{q}(t_2) |q, 0\rangle.$$

We will then devise a method for extracting the vacuum contribution to the initial and final states.

Suppose first that $t_1 > t_2$. Then

$$\langle q', T| T\hat{q}(t_1)\hat{q}(t_2) |q, 0\rangle = \langle q', T| \hat{q}(t_1)\hat{q}(t_2) |q, 0\rangle$$
$$= \int dq_1 dq_2 \langle q', T|q_1, t_1\rangle \underbrace{\langle q_1, t_1| \hat{q}(t_1)}_{\langle q_1, t_1|q_1} \underbrace{\hat{q}(t_2)|q_2, t_2\rangle}_{q_2|q_2, t_2\rangle} \langle q_2, t_2| q, 0\rangle$$
$$= \int dq_1 dq_2\, q_1 q_2 \langle q', T|q_1, t_1\rangle \langle q_1, t_1| q_2, t_2\rangle \langle q_2, t_2| q, 0\rangle.$$

$$u(x) = \int f(s)G(x, s)ds.$$

This can be thought of as an expansion of the function f according to a Dirac delta function basis (projecting f over $\delta(x-s)$) and a superposition of the solution on each projection. Such an integral is known as a *Fredholm integral equation*, the study of which constitutes *Fredholm theory*.

4.1 Sum over Histories

Each of these matrix elements is a path–integral:

$$\langle q', T| T\hat{q}(t_1)\hat{q}(t_2) |q,0\rangle =$$
$$\int dq_1 dq_2\, q_1 q_2 \int_{q_1,t_1}^{q',T} \mathcal{D}[q] e^{iS} \int_{q_2,t_2}^{q_1,t_1} \mathcal{D}[q] e^{iS} \int_{q,0}^{q_2,t_2} \mathcal{D}[q] e^{iS}.$$

This expression consists of a first path–integral from the initial position q to an arbitrary position q_2, a second one from there to a second arbitrary position q_1, and a third one from there to the final position q'. So we are integrating over all paths from q to q', subject to the restriction that the paths pass through the intermediate points q_1 and q_2. We then integrate over the two arbitrary positions, so that in fact we are integrating over *all* paths: we can combine these three path integrals plus the integrations over q_1 and q_2 into one path–integral. The factors q_1 and q_2 in the above integral can be incorporated into this path–integral by simply including a factor $q(t_1)q(t_2)$ in the path–integral. So

$$\langle q', T| \hat{q}(t_1)\hat{q}(t_2) |q,0\rangle = \int_{q,0}^{q',T} \mathcal{D}[q]\, q(t_1)q(t_2) e^{iS}, \qquad (t_1 > t_2).$$

An identical calculation shows that exactly this same final expression is also valid for $t_2 < t_1$: magically, the path–integral does the time ordering automatically. Thus for all times

$$\langle q', T| T\hat{q}(t_1)\hat{q}(t_2) |q,0\rangle = \int_{q,0}^{q',T} \mathcal{D}[q]\, q(t_1)q(t_2) e^{iS}.$$

As for how to obtain vacuum-to-vacuum matrix elements, our work on statistical mechanics provides us with a clue. We can expand the states $\langle q', T|$ and $|q,0\rangle$ in terms of eigenstates of the Hamiltonian. If we evolve towards a *negative imaginary time*, the contribution of all other states will decay away relative to that of the ground state. We have (resetting the initial time to $-T$ for convenience)

$$\langle q', T| q, -T\rangle \propto \langle 0, T| 0, -T\rangle,$$

where on the right the '0' denotes the ground state. The proportionality involves the ground state wave function and an exponential factor

$$\exp[2iE_0 T] = \exp[-2E_0|T|].$$

We could perform all calculations in a Euclidean theory and analytically continue to real time when computing physical quantities (many books do this), but to be closer to physics we can also consider T not to be pure imaginary and negative, but to have a small negative imaginary phase: $T = |T|e^{-i\epsilon}$ ($\epsilon > 0$). With this,

$$\langle 0, T| 0, -T\rangle \propto \langle q', T| q, -T\rangle = \int \mathcal{D}[q]\, e^{iS}.$$

4 Nonlinear Dynamics of Path Integrals

To compute the Green's functions, we must simply add $T\hat{q}(t_1)\hat{q}(t_2)\cdots\hat{q}(t_n)$ to the matrix element, and the corresponding factor $q(t_1)q(t_2)\cdots q(t_n)$ inside the path–integral:

$$\langle 0,T|\,T\hat{q}(t_1)\hat{q}(t_2)\cdots\hat{q}(t_n)\,|0,-T\rangle \propto \int \mathcal{D}[q]\, q(t_1)q(t_2)\cdots q(t_n) e^{iS}.$$

The proportionality sign is a bit awkward; fortunately, we can rid ourselves of it. To do this, we note that the left hand expression is not exactly what we want: the vacua $|0,\pm T\rangle$ differ by a phase. We wish to eliminate this phase; to this end, the Green's functions are defined

$$G^{(n)}(t_1,t_2,\cdots,t_n) = \langle 0|\,T\hat{q}(t_1)\hat{q}(t_2)\cdots\hat{q}(t_n)\,|0\rangle \equiv$$

$$\frac{\langle 0,T|\,T\hat{q}(t_1)\hat{q}(t_2)\cdots\hat{q}(t_n)\,|0,-T\rangle}{\langle 0,T|0,-T\rangle} = \frac{\int \mathcal{D}[q]\, q(t_1)q(t_2)\cdots q(t_n)e^{iS}}{\int \mathcal{D}[q]\, e^{iS}},$$

with no proportionality sign. The wave functions and exponential factors in the numerator and denominator cancel.

To compute the numerator, we can once again use the trick we used in perturbation theory in quantum mechanics, namely, adding a *source* to the *action*. We define

$$Z[J] = \frac{\int \mathcal{D}[q]\, e^{i(S+\int dt\, J(t)q(t))}}{\int \mathcal{D}[q]\, e^{iS}} = \frac{\langle 0|0\rangle_J}{\langle 0|0\rangle_{J=0}}.$$

If we operate on $Z[J]$ with $i^{-1}\delta/\delta J(t_1)$, this gives

$$\left(\frac{1}{i}\frac{\delta}{\delta J(t_1)}Z[J]\right)\bigg|_{J=0} = \left(\frac{\int \mathcal{D}[q]\, q(t_1)e^{i(S+\int dt\, J(t)q(t))}}{\int \mathcal{D}[q]\, e^{iS}}\right)\bigg|_{J=0}$$

$$= \frac{\int \mathcal{D}[q]\, q(t_1)e^{iS}}{\int \mathcal{D}q\, e^{iS}} = \frac{\langle 0,T|\,\hat{q}(t_1)\,|0,-T\rangle}{\langle 0,T|0,-T\rangle} = \langle 0|\,\hat{q}(t_1)\,|0\rangle.$$

(The expectation values are evaluated in the *absence* of J.)

Repeating this procedure, we obtain a path–integral with several q's in the numerator. This ordinary product of q's in the path–integral corresponds, as discussed earlier in this section, to a time-ordered product in the matrix element. So we make the following conclusion:

$$\left(\frac{1}{i}\frac{\delta}{\delta J(t_1)}\cdots\frac{1}{i}\frac{\delta}{\delta J(t_n)}Z[J]\right)\bigg|_{J=0} =$$

$$\frac{\int \mathcal{D}[q]\, q(t_1)\cdots q(t_n)e^{iS}}{\int \mathcal{D}[q]e^{iS}} = \langle 0|\,T\hat{q}(t_1)\cdots\hat{q}(t_1)\,|0\rangle.$$

For obvious reasons, the functional $Z[J]$ is called the *generating functional* for Green's functions; it is a very handy tool in quantum field theory and in statistical mechanics.

4.1 Sum over Histories

To be able to calculate $Z[J]$, let us examine the numerator,

$$N \equiv \int \mathcal{D}[q] e^{i(S + \int dt\, J(t)q(t))}.$$

Suppose initially that S is the harmonic oscillator action (denoted S_0):

$$S_0 = \int dt \left(\frac{1}{2} m\dot{q}^2 - \frac{1}{2} m\omega^2 q^2 \right),$$

Then the corresponding numerator, N_0, is the non–Euclidean, or real-time, version of the propagator $K_E^0[J]$ that we used before. We can calculate $N_0[J]$ in the same way as $K_E^0[J]$. Since the calculation repeats much of that of $K_E^0[J]$, we will be succinct.

By definition,

$$N_0 = \int \mathcal{D}q(t) \exp\left[i \int dt \left(\frac{1}{2} m\dot{q}^2 - \frac{1}{2} m\omega^2 q^2 + Jq \right) \right].$$

We do the path integral over a new variable y, defined by $q(t) = q_c(t) + y(t)$, where q_c is the classical solution. Then the path–integral over y is a constant (independent of J) and we can avoid calculating it. (It will cancel against the denominator in $Z[J]$.) Calling it C, we have

$$N_0 = C e^{i S_{0J}[q_c]}, \quad \text{where}$$

$$S_{0J}[q_c] = \int dt \left(\frac{1}{2} m\dot{q}_c^2 - \frac{1}{2} m\omega^2 q_c^2 + J q_c \right) = \frac{1}{2} \int dt\, J(t) q_c(t),$$

using the fact that q_c satisfies the equation of motion. We can write the classical path in terms of the Green's function (to be determined shortly), defined by

$$\left(\frac{d^2}{dt^2} + \omega^2 \right) G(t, t') = -i\delta(t - t'). \tag{4.37}$$

Then

$$q_c(t) = -i \int dt'\, G(t, t') J(t').$$

We can now write

$$N_0 = C \exp \frac{1}{2} \int dt\, dt'\, J(t) G(t, t') J(t').$$

Dividing by the denominator merely cancels the factor C, giving our final result [Mac00]:

$$Z[J] = \exp \frac{1}{2} \int dt\, dt'\, J(t) G(t, t') J(t').$$

We can solve (4.37) for the Green's function by going into momentum space; the result is

$$G(t,t') = G(t-t') = \int \frac{dk}{2\pi} \frac{i}{k^2 - \omega^2} e^{-ik(t-t')}.$$

However, there are poles on the axis of integration. The Green's function is ambiguous until we give it a 'pole prescription', i.e., a *boundary condition*. But remember that our time T has a small, negative imaginary part. We require that G go to zero as $T \to \infty$. The correct pole prescription then turns out to be

$$G(t-t') = \int \frac{dk}{2\pi} \frac{i}{k^2 - \omega^2 + i\epsilon} e^{-ik(t-t')}. \tag{4.38}$$

We could at this point do a couple of practice calculations to get used to this formalism. Examples would be to compute perturbatively the generating functional for an action which has terms beyond quadratic (for example, a q^4 term), or to compute some Green's function in either the quadratic or quartic theory. But since these objects aren't really useful in quantum mechanics, without further delay we will go directly to the case of interest: quantum field theory.

4.1.7 Monte Carlo Simulation of the Path Integral

The *Monte Carlo method* came into being roughly around the same time as the Feynman path integral. Anecdotally, the idea of gaining insight into a complex phenomenon by making various trials and studying the proportions of the respective outcomes occurred to Stanislaw Ulam while playing solitaire during an illness in 1946. The immediate application was, the problem of neutron diffusion studied in Los Alamos at that time. The name of the procedure first appeared in print in a classic paper by Metropolis and Ulam in 1949 [MU49], where the authors explicitly mentioned that the method they presented as a statistical approach to the study of integro–differential equations would sometimes be referred to as the Monte Carlo method. In classical statistical mechanics it quickly became a standard calculational tool.

The object of interest in Monte Carlo evaluations of Feynman's path integral is the quantum statistical partition function Z, given, in operator language, as the trace of the density operator $\exp(-\beta \hat{H})$ of the canonical ensemble ($\beta = 1/k_B T$) associated with a Hamilton operator describing N particles of mass m_i moving under the influence of a potential V,

$$\hat{H} = \sum_{i=1}^{N} \frac{\hat{\mathbf{p}}_i^2}{2m_i} + V(\hat{\mathbf{r}}_1, \ldots, \hat{\mathbf{r}}_N).$$

Expressed as a Feynman integral, the density matrix elements read

$$\langle \mathbf{r} | \exp(-\beta \hat{H}) | \mathbf{r}' \rangle = \int_{\mathbf{r}(0)=\mathbf{r}}^{\mathbf{r}(\hbar\beta)=\mathbf{r}'} \mathcal{D}\mathbf{r}(\tau) \exp\left\{ -\frac{1}{\hbar} \int_0^{\hbar\beta} L(\{\dot{\mathbf{r}}_i(\tau), \mathbf{r}_i(\tau)\}) d\tau \right\}, \tag{4.39}$$

where $\mathbf{r} \equiv \{\mathbf{r}_1, \ldots, \mathbf{r}_N\}$, L denotes the classical Lagrangian

$$L(\{\dot{\mathbf{r}}_i(\tau), \mathbf{r}_i(\tau)\}) = \sum_{i=1}^{N} \frac{m_i}{2} \dot{r}_i^2 + V(\mathbf{r}_1, \ldots, \mathbf{r}_N(\tau))$$

expressed in imaginary time τ.[16] The particles are assumed to be distinguishable. To evaluate the trace, we only need to set $\mathbf{r} = \mathbf{r}'$ and integrate over $\mathbf{r}$. To take into account Bose or Fermi statistics for indistinguishable particles, the partition function splits into a sum of the direct Boltzmann part and parts with permuted endpoints.

The right hand side of (4.39) is a path integral for the $3N$ functions $\mathbf{r}$. The idea of a Monte Carlo evaluation of this quantity is to sample these paths stochastically and to get (approximate) information about the quantum statistics of the system by averaging over the finite set of paths generated in the sampling process.

Monte Carlo data always come with error bars and, in general, the errors associated with numerical Monte Carlo data stem from two distinct sources. A *systematic error* of Monte Carlo evaluations of the path integral follows from the need to identify the paths by a finite amount of computer information. This can be done by discretizing the paths at some set of points in the interval $(0, \hbar\beta)$. For a single particle moving in one dimension, the simplest discrete time approximation for L time slices reads ($\epsilon = \hbar\beta/L$)

$$\langle x | \exp(-\beta \hat{H}) | x' \rangle =$$

$$\lim_{L \to \infty} \frac{1}{A} \prod_{j=1}^{L-1} \left[\int \frac{dx_j}{A} \right] \exp \left\{ -\frac{1}{\hbar} \sum_{j=1}^{L} \left[\frac{m(x_j - x_{j-1})^2}{2\epsilon} + \epsilon V(x_{j-1}) \right] \right\} \quad (4.40)$$

where $A = (2\pi\hbar\epsilon/m)^{1/2}$ and $x_0 = x$ and $x_L = x'$. Alternatively, one may expand the individual paths in terms of an orthogonal function basis, e.g. by the *Fourier decomposition*,

$$x(\tau) = x + \frac{(x' - x)\tau}{\hbar\beta} + \sum_{k=1}^{\infty} a_k \sin \frac{k\pi\tau}{\hbar\beta},$$

and express the density matrix as

$$\langle x | \exp(-\beta \hat{H}) | x' \rangle = \lim_{L' \to \infty} J \exp \left\{ -\frac{m}{2\hbar^2 \beta}(x - x')^2 \right\} \times$$

$$\times \int \prod_{k=1}^{L'} da_k \exp \left\{ -\frac{a_k^2}{2\sigma_k^2} \right\} \times \exp \left\{ -\frac{1}{\hbar} \int_0^{\hbar\beta} V(x(\tau)) d\tau \right\},$$

---

[16] There have been attempts to apply the Monte Carlo method to path integrals also for real time. However, due to the oscillating exponential one then has to deal with problems of numerical cancellation, and it is much harder to obtain results of some numerical accuracy. Therefore, we will here restrict myself to Monte Carlo work in imaginary time.

where $\sigma_k = [2\hbar^2\beta/m(\pi k)^2]^{1/2}$ and J is the Jacobian of the transformation from the integral over all paths to the integral over all Fourier coefficients. A systematic error then arises from the loss of information by the finite number L of points x_i on the discretized time axis or by the finite number L' of Fourier components a_k that are taken into account in the Monte Carlo sampling of the paths.

The other error source of Monte Carlo data is the *statistical error* due to the finite number N_m of paths that form the sample used for evaluating the statistical averages. To make matters worse, the probability of configurations is, in general highly peaked, making an independent sampling of paths highly inefficient in most cases. The remedy is to introduce some way of 'importance sampling' where configurations are generated according to their probability given by the exponential in (4.39). Statistical averages may then be computed as simple arithmetic means. A way to achieve this is by constructing Markov chains where transition probabilities between configuration are constructed that allow to generate a new configuration from a given one such that in the limit of infinitely many configurations the correct probability distribution of paths results. A very simple and universally applicable algorithm to set up such a *Markov chain* is the *Metropolis algorithm* introduced in 1953 [MRR53]. Here a new configuration is obtained by looking at some configuration with only one variable changed and accepting or rejecting it for the sample on the basis of a simple rule that depends only on the respective energies of the two configurations. The advantages of importance sampling on the basis of Markov chains are obtained on the cost that, in general, successive configurations are not statistically independent but autocorrelated. The crucial quantity is the integrated autocorrelation time $\tau_{\mathcal{O}}^{\text{int}}$ of a quantity of interest $\mathcal{O} = \langle \overline{\mathcal{O}} \rangle$ with $\overline{\mathcal{O}} = (1/N_m)\sum_{i=1}^{N_m} \mathcal{O}_i$ and $\mathcal{O}_i$ computed for each path i in the sample. It enters the statistical error estimate $\Delta_{\mathcal{O}}$ for expectation values of $\mathcal{O}$ computed from a Monte Carlo sample of N_m autocorrelated configurations as

$$\Delta\overline{\mathcal{O}} = \sqrt{\frac{\sigma_{\mathcal{O}_i}^2}{N_m}}\sqrt{2\tau_{\mathcal{O}}^{\text{int}}},$$

where $\sigma_{\mathcal{O}_i}^2$ is the variance of $\mathcal{O}_i$.

With Monte Carlo generated samples of Feynman paths one can thus 'measure' thermodynamic properties of quantum systems like the internal energy and the specific heat, but also gain more detailed information about correlation functions, probability distributions and the like. In the low–temperature limit, $\beta \to \infty$, quantum mechanical ground state properties are recovered.

The feasibility of evaluating the quantum statistical partition function of many–particle systems by Monte Carlo sampling of paths was well established by the early eighties and the method began to be applied to concrete problems, in particular in the chemical physics literature. It had also become clear that the method had severe restrictions if numerical accuracy was called for. In addition to the statistical error inherent to the Monte Carlo method, a systematic error was unavoidably introduced by the necessary discretization of the paths. Attempts to improve the accuracy by algorithmic improvements to reduce both the systematic and the statistical errors were reported in subsequent years. The literature is abundant and rather than trying to review the field we will only indicate some pertinent paths of development.

In Fourier PIMC methods, introduced in 1983 in the chemical physics context by Doll and Freeman [DF84, FD84], the systematic error arises from the fact that only a finite number of Fourier components are taken into account. Here the systematic error could be reduced by the method of partial averaging [CFD86].

In discrete time approximations arising from the short–time propagator or, equivalently, the high–temperature Green's function various attempts have been made to find more rapidly converging formulations. Among these are attempts to include higher terms in an expansion of the Wigner–Kirkwood type, i.e. an expansion in terms of $\hbar^2/2m$. Taking into account the first term of such an expansion would imply to replace the potential term $\epsilon V(x_{j-1})$ in (4.40) by [RR83, LB87, KTL88]

$$\epsilon V(x_{j-1}) \to \frac{\epsilon}{x-x'} \int_x^{x'} dy V(y).$$

This improves the convergence of the density matrix (4.40) (from even less than $\mathcal{O}(1/L)$) to $\mathcal{O}(1/L^2)$. For the full partition function, the convergence of the simple discretization scheme is already of order $\mathcal{O}(1/L^2)$ since due to the cyclic property of the trace, the discretization $\epsilon V(x_{j-1})$ is then equivalent to a symmetrized potential term $\epsilon(V(x_{j-1})+V(x_j))/2$. The convergence behavior of these formulations follows from the *Trotter decomposition formula*,

$$e^{-(A+B)} = \left[e^{-\frac{A}{L}}e^{-\frac{B}{L}}\right]^L + \mathcal{O}(\frac{1}{L}) = \left[e^{-\frac{A}{2L}}e^{-\frac{B}{L}}e^{-\frac{A}{2L}}\right]^L + \mathcal{O}(\frac{1}{L^2}),$$

valid for non-commuting operators A and B in a *Banach space*, identifying A with the kinetic energy $\beta \sum \hat{\mathbf{p}}_i^2/2m_i$ and B with the potential energy $\beta V(\{\hat{x}_i\})$. More rapidly converging discretization schemes were investigated on the basis of higher-order decompositions. Unfortunately, a direct, 'fractal' decomposition [Suz90] of the form

$$e^{-(A+B)} = \lim_{L\to\infty} \left[e^{\alpha_1 \frac{A}{L}}e^{\beta_1 \frac{B}{L}}e^{\alpha_2 \frac{A}{L}}e^{\beta_2 \frac{B}{L}}\ldots\right]^L, \quad \sum \alpha_i = \sum \beta_i = 1,$$

inevitably leads to negative coefficients for higher decompositions [Suz91] and is therefore not amenable to Monte Carlo sampling of paths [JS92]. Higher–order Trotter decomposition schemes involving commutators have proven to be more successful [RR83, LB87, KTL88]. In particular, a decomposition of the form

$$Z = \lim_{L\to\infty} \mathrm{Tr} \left[e^{-\frac{A}{2L}}e^{-\frac{B}{2L}}e^{-\frac{[[B,A],B]}{24L^3}}e^{-\frac{B}{2L}}e^{-\frac{A}{2L}}\right]^L,$$

derivable by making use of the cyclic property of the trace, is convergent of order $\mathcal{O}(1/L^4)$ and amounts to simply replacing the potential ϵV in (4.40) by an effective potential

$$V_{\text{eff}} = V + \frac{(\beta\hbar)^2}{24mL^2}(V')^2.$$

Another problem for the numerical accuracy of PIMC simulations arises from the analog of the *critical slowing down* problem well–known for local update algorithms at second–order *phase transitions* in the simulation of spin systems and lattice field theory. Since the correlations $\langle x_j x_{j+k} \rangle$ between variables x_j and x_{j+k} in the discrete time approximation only depend on the temperature and on the gaps between the energy levels and not, or at least not appreciably, on the discretization parameter ϵ, the correlation length ζ along the discretized time axis always diverges linearly with L when measured in units of the lattice spacing ϵ. Hence in the continuum limit of $\epsilon \to 0$ with β fixed or, equivalently, of $L \to \infty$ for local, importance sampling update algorithms, like the standard Metropolis algorithm, a slowing down occurs because paths generated in the Monte Carlo process become highly correlated. Since for simulations using the Metropolis algorithm autocorrelation times diverge as $\tau_\mathcal{O}^{\text{int}} \propto L^z$ with $z \approx 2$ the computational effort (CPU time) to achieve comparable numerical accuracy in the continuum limit $L \to \infty$ diverges as $L \times L^z = L^{z+1}$.

To overcome this drawback, ad hoc algorithmic modifications like introducing collective moves of the path as a whole between local Metropolis updates were introduced then and again. One of the earliest more systematic and successful attempts to reduce autocorrelations between successive path configurations was introduced by [PC84]. Rewriting the discretized path integral, their method essentially amounts to a recursive transformation of the variables x_i in such a way that the kinetic part of the energy can be taken care of by sampling direct Gaussian random variables and a Metropolis choice is made for the potential part. The recursive transformation can be done between some fixed points of the discretized paths, and the method has been applied in such a way that successively finer discretizations of the path were introduced between neighbouring points. Invoking the polymer analog of the discretized path this method was christened the *staging algorithm* by [SKC85].

The staging algorithm decorrelates successive paths very effectively because the whole staging section of the path is essentially sampled independently. In 1993, another explicitly non–local update was applied to PIMC simulations [JS93] by transferring the so–called *multigrid method* known from the simulation of spin systems. Originating in the theory of numerical solutions of partial differential equations, the idea of the multigrid method is to introduce a hierarchy of successively coarser grids in order to take into account long wavelength fluctuations more effectively. Moving variables of the coarser grids then amounts to a collective move of neighbouring variables of the finer grids, and the formulation allows to give a recursive description of how to cycle most effectively through the various levels of the multigrid. Particularly successful is the so–called *W–cycle*. Both the staging algorithm and the multigrid W–cycle have been shown to beat the slowing down problem in the continuum limit completely by reducing the exponent z to $z \approx 0$ [JS96].

Another cause of severe correlations between paths arises if the probability density of configurations is sharply peaked with high maxima separated by regions of very low probability density. In the statistical mechanics of spin systems this is the case at a first-order phase transition. In PIMC simulations the problem arises for tunneling situations like, e.g., for a double well potential with a high potential barrier between the two wells. In these cases, an unbiased probing of the configuration space becomes difficult because the system tends to get stuck around one of the probability maxima. A remedy to this problem is to simulate an auxiliary distribution that is flat between the maxima and to recover the correct Boltzmann distribution by an appropriate reweighting of the sample. The procedure is known under the name of *umbrella sampling* or *multicanonical sampling*. It was shown to reduce autocorrelations for PIMC simulations of a single particle in a 1D double well, and it can also be combined with multigrid acceleration [JS94].

The statistical error associated with a Monte Carlo estimate of an observable $\mathcal{O}$ cannot only be reduced by reducing autocorrelation times $\tau_{\mathcal{O}}^{\text{int}}$. If the observable can be measured with two different estimators U_i that yield the same mean $U_i^{(L)} = \langle U_i \rangle$ with $\mathcal{O} = \lim_{L\to\infty} U_i^{(L)}$, the estimator with the smaller variance $\sigma_{U_i}^2$ is to be preferred. Straightforward differentiation of the discretized path integral (4.40) leads to an estimator of the energy that explicitly measures the kinetic and potential parts of the energy by

$$U_{\mathrm{k}} = \frac{L}{2\beta} - \frac{m}{2L} \sum \left(\frac{x_j - x_{j-1}}{\epsilon}\right)^2 + \frac{1}{L} \sum_{i=1}^{L} V(x_i).$$

The variance of this so-called *kinetic–energy estimator* diverges with L. Another estimator can be derived by invoking the path analog of the virial theorem

$$\frac{L}{2\beta} - \frac{m}{2} \left\langle \left(\frac{x_j - x_{j-1}}{\epsilon}\right)^2 \right\rangle = \frac{1}{2} \langle x_j V'(x_j) \rangle,$$

and the variance of the *virial estimator*

$$U_{\mathrm{v}} = \frac{1}{2L} \sum_{i=1}^{L} x_i V'(x_i) + \frac{1}{L} \sum_{i=1}^{L} V(x_i)$$

does not depend on L. In the early eighties, investigations of the 'kinetic' and the 'virial' estimators focussed on their variances [PR84]. Some years later, it was pointed out that a correct assessment of the accuracy also has to take into account the autocorrelations, and it was demonstrated that for a standard Metropolis simulation of the harmonic oscillator the allegedly less successful 'kinetic' estimator gave smaller errors than the 'virial' estimator. In 1989 it was shown [CB89] that conclusions about the accuracy also depend on the particular Monte Carlo update algorithm at hand since modifications of the update scheme such as inclusion of collective moves of the whole path affect the autocorrelations of the two estimators in a different way. A careful comparison of the two estimators which disentangles the various factors involved was given in [JS97]. Here it was also shown that a further reduction of the error may be achieved by a proper combination of both estimators without extra cost.

Application of the Monte Carlo method to quantum systems is not restricted to direct sampling of Feynman paths but this method has attractive features. It is not only conceptually suggestive but also allows for algorithmic improvements that help to make the method useful even when the problems at hand requires considerable numerical accuracy. However, algorithmic improvements like the ones alluded to above have tended to be proposed and tested mainly for simple, one–particle systems. On the other hand, the power of the Monte Carlo method is, of course, most welcome in those cases where analytical methods fail. For more complicated systems, however, evaluation of the algorithms and control of numerical accuracy is also more difficult. Only recently, a comparison of the efficiency of Fourier– and discrete–time path integral Monte Carlo for a cluster of 22 hydrogen molecules was presented [CGC98]—and debated [DF99, CGC99]. Nevertheless, path integral Monte Carlo simulations have become an essential tool for the treatment of strongly interacting quantum systems, like, e.g., the theory of condensed helium [Cep95].

For more details on path–integral Monte carlo techniques, see [Sau01].

4.2 Sum over Geometries and Topologies

Recall that the term *quantum gravity* (or *quantum geometrodynamics*, or *quantum geometry*), is usually understood as a consistent fundamental quantum description of gravitational space–time geometry whose classical limit is Einstein's general relativity. Among the possible ramifications of such a theory are a model for the structure of space–time near the Planck scale, a consistent calculational scheme to calculate gravitational effects at all energies, a description of quantum geometry near space–time singularities and a non–perturbative quantum description of 4D black holes. It might also help us in understanding cosmological issues about the beginning and end of the universe, i.e., the so–called *'big bang'* and *'Big–Crunch'* (see e.g., [Pen89, Pen94, Pen97]).

From what we know about the quantum dynamics of other fundamental interactions it seems eminently plausible that also the gravitational excitations should at very short scales be governed by quantum laws. Now, conventional *perturbative path integral expansions of gravity*, as well as perturbative expansion in the string coupling in the case of unified approaches, both have difficulty in finding any direct or indirect evidence for quantum gravitational effects, be they experimental or observational, which could give a feedback for model building. The outstanding problems mentioned above require a non–perturbative treatment; it is not sufficient to know the first few terms of a perturbation series. The real goal is to search for a *non–perturbative* definition of such a theory, where the initial input of any fixed 'background metric' is inessential (or even undesirable), and where 'space–time' is determined *dynamically*. Whether or not such an approach necessarily requires the inclusion of higher dimensions and fundamental supersymmetry is currently unknown (see [AK93, AL98, AJL00a, AJL00b, AJL01a, AJL01b, AJL01d, DL01]).

Such a non–perturbative viewpoint is very much in line with how one proceeds in classical geometrodynamics, where a metric space–time $(M, g_{\mu\nu})$ (+ matter) emerges only as a *solution* to the familiar Einstein equation

$$G_{\mu\nu}[g] \equiv R_{\mu\nu}[g] - \frac{1}{2} g_{\mu\nu} R[g] = -8\pi T_{\mu\nu}[\Phi], \qquad (4.41)$$

which define the classical dynamics of fields $\Phi = \Phi^{\mu\nu}$ on the space $\mathcal{M}(M)$, the space of all metrics $g = g_{\mu\nu}$ on a given smooth manifold M. The analogous question we want to address in the quantum theory is: Can we get 'quantum space–time' as a solution to a set of non–perturbative quantum equations of motion on a suitable quantum analogue of $\mathcal{M}(M)$ or rather, of the space of geometries, $Geom(M) = \mathcal{M}(M)/Diff(M)$?

Now, this is not a completely straightforward task. Whichever way we want to proceed non–perturbatively, if we give up the privileged role of a flat, Minkowskian background space–time on which the quantization is to take place, we also have to abandon the central role usually played by the Poincaré group, and with it most standard quantum field–theoretic tools for regularization and renormalization. If one works in a continuum metric formulation of gravity, the symmetry group of the *Einstein–Hilbert action* is instead the group $Diff(M)$ of diffeomorphisms on M, which in terms of local charts are the smooth invertible coordinate transformations $x^\mu \mapsto y^\mu(x^\mu)$.

In the following, we will describe a non–perturbative path integral approach to quantum gravity, defined on the *space of all geometries*, without distinguishing any background metric structure [Lol01]. This is closely related in spirit with the canonical approach of loop quantum gravity [Rov98] and its more recent incarnations using so–called *spin networks* (see, e.g., [Ori01]). 'Non–perturbative' here means in a covariant context that the path sum or integral will have to be performed explicitly, and not just evaluated around its stationary points, which can only be achieved in an appropriate regularization. The method we will employ uses a discrete lattice regularization as an intermediate step in the construction of the quantum theory.

4.2.1 Simplicial Quantum Geometry

In this section we will explain how one may construct a theory of quantum gravity from a non–perturbative path integral, using the method of Lorentzian dynamical triangulations. The method is minimal in the sense of employing standard tools from quantum field theory and the theory of critical phenomena and adapting them to the case of *generally covariant systems*, without invoking any symmetries beyond those of the classical theory. At an intermediate stage of the construction, we use a regularization in terms of simplicial *Regge geometries*, that is, piecewise linear manifolds. In this approach, 'computing the path integral' amounts to a conceptually simple and geometrically transparent 'counting of geometries', with additional weight factors which are determined by the EH action. This is done first of all at a regularized level. Subsequently, one searches for interesting continuum limits of these discrete models which are possible candidates for theories of quantum gravity, a step that will always involve a renormalization. From the point of view of statistical mechanics, one may think of Lorentzian dynamical triangulations as a new class of statistical models of Lorentzian random surfaces in various dimensions, whose building blocks are flat simplices which carry a 'time arrow', and whose dynamics is entirely governed by their intrinsic geometric properties.

Before describing the details of the construction, it may be helpful to recall the path integral representation for a 1D non–relativistic particle (see previous subsection). The time evolution of the particle's wave function ψ may be described by the integral equation (4.3) above, where the *propagator*, or the *Feynman kernel* G, is defined through a limiting procedure (4.4). The time interval $t'' - t'$ has been discretized into N steps of length $\epsilon = (t'' - t')/N$, and the r.h.s. of (4.4) represents an integral over all piecewise linear paths $x(t)$ of a 'virtual' particle propagating from x' to x'', illustrated in Figure 4.4 above.

The prefactor A^{-N} is a normalization and L denotes the Lagrangian function of the particle. Knowing the propagator G is tantamount to having solved the quantum dynamics. This is the simplest instance of a path integral, and is often written schematically as

$$G(x',t';x'',t'') = \int \mathcal{D}[x(t)] \, e^{\mathrm{i}S[x(t)]}, \tag{4.42}$$

where $\mathcal{D}[x(t)]$ is a functional measure on the 'space of all paths', and the exponential weight depends on the classical action $S[x(t)]$ of a path. Recall also that this procedure can be defined in a mathematically clean way if we Wick–rotate the time variable t to imaginary values $t \mapsto \tau = \mathrm{i}t$, thereby making all integrals real [RS75].

Can a similar strategy work for the case of Einstein geometrodynamics? As an analogue of the particle's position we can take the geometry $[g_{ij}(x)]$ (i.e., an equivalence class of spatial metrics) of a constant–time slice. Can one then define a gravitational propagator

$$G([g'_{ij}],[g''_{ij}]) = \int_{Geom(M)} \mathcal{D}[g_{\mu\nu}] \, e^{\mathrm{i}S^{\mathrm{EH}}[g_{\mu\nu}]} \tag{4.43}$$

from an initial geometry $[g']$ to a final geometry $[g'']$ (Figure 4.7) as a limit of some discrete construction analogous to that of the non-relativistic particle (4.4)? And crucially, what would be a suitable class of 'paths', that is, space–times $[g_{\mu\nu}]$ to sum over?

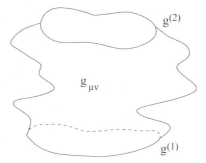

Fig. 4.7. The time–honoured way [HE79] of illustrating the gravitational path integral as the propagator from an initial to a final spatial boundary geometry.

Now, to be able to perform the integration $\oint \mathcal{D}[g_{\mu\nu}]$ in a meaningful way, the strategy we will be following starts from a regularized version of the space $Geom(M)$ of all geometries. A regularized path integral $G(a)$ can be defined which depends on an ultraviolet cutoff a and is *convergent* in a non–trivial region of the space of coupling constants. Taking the continuum limit corresponds to letting $a \to 0$. The resulting continuum theory – if it can be shown to exist – is then investigated with regard to its geometric properties and in particular its semiclassical limit.

4.2.2 Discrete Gravitational Path Integrals

Trying to construct non–perturbative path integrals for gravity from sums over discretized geometries, using approach of *Lorentzian dynamical triangulations*, is not a new idea. Inspired by the successes of lattice gauge theory, attempts to describe quantum gravity by similar methods have been popular on and off since the late 70's. Initially the emphasis was on gauge–theoretic, first–order formulations of gravity, usually based on (compactified versions of) the Lorentz group, followed in the 80's by 'quantum Regge calculus', an attempt to represent the gravitational path integral as an integral over certain piecewise linear geometries (see [Wil97] and references therein), which had first made an appearance in approximate descriptions of *classical* solutions of the Einstein equations. A variant of this approach by the name of 'dynamical triangulation(s)' attracted a lot of interest during the 90's, partly because it had proved a powerful tool in describing 2D quantum gravity (see the textbook [ADJ97] and lecture notes [AJL00a] for more details).

The problem is that none of these attempts have so far come up with convincing evidence for the existence of an underlying continuum theory of 4D quantum gravity. This conclusion is drawn largely on the basis of numerical simulations, so it is by no means water–tight, although one can make an argument that the 'symptoms' of failure are related in the various approaches [Lol98]. What goes wrong generically seems to be a dominance in the continuum limit of highly degenerate geometries, whose precise form depends on the approach chosen. One would expect that non–smooth geometries play a decisive role, in the same way as it can be shown in the particle case that the support of the measure in the continuum limit is on a set of nowhere differentiable paths. However, what seems to happen in the case of the path integral for 4–geometries is that the structures get are *too* wild, in the sense of not generating, even at coarse–grained scales, an effective geometry whose dimension is anywhere near four.

The schematic phase diagram of Euclidean dynamical triangulations shown in Figure 4.8 gives an example of what can happen. The picture turns out to be essentially the same in both three and four dimensions: the model possesses infinite-volume limits everywhere along the critical line $k_3^{\text{crit}}(k_0)$, which fixes the bare cosmological constant as a function of the inverse Newton constant $k_0 \sim G_N^{-1}$. Along this line, there is a critical point k_0^{crit} (which we now know to be of first–order in $d = 3, 4$) below which geometries generically have a very large effective or Hausdorff dimension.[17] Above k_0^{crit} we find the opposite phenomenon of 'polymerization': a typical element contributing to the state sum is a thin branched polymer, with one or more dimensions 'curled up' such that its effective dimension is around two.

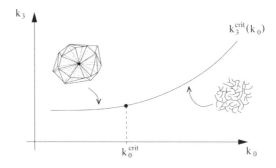

Fig. 4.8. The phase diagram of 3D and 4D Euclidean dynamical triangulations (adapted from [AJL00b, AJL01a]).

This problem has to do with the fact that the gravitational action is unbounded below, causing potential havoc in Euclidean versions of the path integral. Namely, what all the above-mentioned approaches have in common is that they work from the outset with *Euclidean* geometries, and associated Boltzmann-type weights $\exp(-S^{\text{eu}})$ in the path integral. In other words, they integrate over 'space–times' which know nothing about time, light cones and

---

[17] In terms of geometry, this means that there are a few vertices at which the entire space–time 'condenses' in the sense that almost every other vertex in the simplicial space–time is about one link-distance away from them.

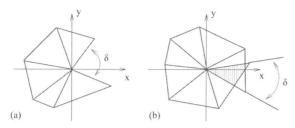

Fig. 4.9. Positive (a) and negative (b) space–like deficit angles δ (adapted from [Lol01, Lol98]).

causality. This is done mainly for technical reasons, since it is difficult to set up simulations with complex weights and since until recently a suitable Wick rotation was not known.

'Lorentzian dynamical triangulations', first proposed in [AL98] and further elaborated in [AJL00b, AJL01a] tries to establish a logical connection between the fact that non–perturbative path integrals were constructed for Euclidean instead of Lorentzian geometries and their apparent failure to lead to an interesting continuum theory.

4.2.3 Regge Calculus

The use of simplicial methods in general relativity goes back to the pioneering work of Regge [Reg61]. In classical applications one tries to approximate a classical space–time geometry by a triangulation, that is, a piecewise linear space get by gluing together flat simplicial building blocks, which in dimension d are dD generalizations of triangles. By 'flat' we mean that they are isometric to a subspace of dD Euclidean or Minkowski space. We will only be interested in gluings leading to genuine manifolds, which therefore look locally like an R^d. A nice feature of such simplicial manifolds is that their geometric properties are completely described by the discrete set $\{l_i^2\}$ of the squared lengths of their edges. Note that this amounts to a description of geometry *without the use of coordinates*. There is nothing to prevent us from re–introducing coordinate patches covering the piecewise linear manifold, for example, on each individual simplex, with suitable transition functions between patches. In such a coordinate system the metric tensor will then assume a definite form. However, for the purposes of formulating the path integral we will not be interested in doing this, but rather work with the edge lengths, which constitute a direct, regularized parametrization of the space $Geom(M)$ of geometries.

How precisely is the intrinsic geometry of a simplicial space, most importantly, its curvature, encoded in its edge lengths? A useful example to keep in mind is the case of dimension two, which can easily be visualized. A 2d piecewise linear space is a triangulation, and its scalar curvature $R(x)$ coincides with the *Gaussian curvature* (see [II07b]). One way of measuring this curvature is by parallel–transporting a vector around closed curves in the manifold. In our piecewise–flat manifold such a vector will always return to its original orientation *unless* it has surrounded lattice vertices v at which the surrounding angles did not add up to 2π, but $\sum_{i \supset v} \alpha_i = 2\pi - \delta$, for $\delta \neq 0$, see Figure 4.9. The so–called deficit angle δ is precisely the rotation angle picked up by the vector and is a direct measure for the scalar curvature at the vertex. The operational description to get the scalar curvature in higher dimensions is very similar, one basically has to sum in each point over the Gaussian curvatures of all 2D submanifolds. This explains why in Regge calculus the curvature part of the EH action is given by a sum over building blocks of dimension $(d - 2)$ which are the objects dual to those local 2d submanifolds. More precisely, the continuum curvature and volume terms of the action become

$$\frac{1}{2} \int_{\mathcal{R}} d^d x \sqrt{|\det g|} \,^{(d)} R \longrightarrow \sum_{i \in \mathcal{R}} Vol(i^{th} \, (d-2)\text{–simplex}) \, \delta_i \quad (4.44)$$

$$\int_{\mathcal{R}} d^d x \sqrt{|\det g|} \longrightarrow \sum_{i \in \mathcal{R}} Vol(i^{th} \, d\text{–simplex}) \quad (4.45)$$

in the simplicial discretization. It is then a simple exercise in trigonometry to express the volumes and angles appearing in these formulas as functions of the edge lengths l_i, both in the Euclidean and the Minkowskian case.

The approach of dynamical triangulations uses a certain class of such simplicial space–times as an explicit, regularized realization of the space $Geom(M)$. For a given volume N_d, this class consists of all gluings of manifold–type of a set of N_d simplicial building blocks of top–dimension d whose edge lengths are restricted to take either one or one out of two values. In the Euclidean case we set $l_i^2 = a^2$ for all i, and in the Lorentzian case we allow for both space- and time–like links with $l_i^2 \in \{-a^2, a^2\}$, where the geodesic distance a serves as a short-distance cutoff, which will be taken to zero later. Coming from the classical theory this may seem a grave restriction at first, but this is indeed not the case. Firstly, keep in mind that for the purposes of the quantum theory we want to sample the space of geometries 'ergodically' at a coarse-grained scale of order a. This should be contrasted with the classical theory where the objective is usually to approximate a given, *fixed* space–time to within a length scale a. In the latter case one typically requires a much finer topology on the space of metrics or geometries. It is also straightforward to see that no local curvature degrees of freedom are suppressed by fixing the edge lengths; deficit angles in all directions are still present, although they take on only a discretized set of values. In this sense, in dynamical triangulations all geometry is in the gluing of the fundamental building blocks. This is dual to how quantum Regge calculus is set up, where one usually fixes a triangulation T and then 'scans' the space of geometries by letting the l_i's run continuously over all values compatible with the triangular inequalities.

In a nutshell, Lorentzian dynamical triangulations give a definite meaning to the 'integral over geometries', namely, as a sum over inequivalent Lorentzian gluings T over any number N_d of d–simplices,

$$\oint_{Geom(M)} \mathcal{D}[g_{\mu\nu}] \, e^{iS[g_{\mu\nu}]} \xrightarrow{\text{LDT}} \sum_{T \in \mathcal{T}} \frac{1}{C_T} e^{iS^{\text{Reg}}(T)}, \qquad (4.46)$$

where the symmetry factor $C_T = |\text{Aut}(T)|$ on the r.h.s. is the order of the automorphism group of the triangulation, consisting of all maps of T onto itself which preserve the connectivity of the simplicial lattice. We will specify below what precise class $\mathcal{T}$ of triangulations should appear in the summation.

It follows from the above that in this formulation all curvatures and volumes contributing to the *Regge simplicial action* come in discrete units. This can be illustrated by the case of a 2D triangulation with Euclidean signature, which according to the prescription of dynamical triangulations consists of *equilateral* triangles with squared edge lengths $+a^2$. All interior angles of such a triangle are equal to $\pi/3$, which implies that the deficit angle at any vertex v can take the values $2\pi - k_v \pi/3$, where k_v is the number of triangles meeting at v. As a consequence, the Einstein–Regge action S^{Reg} assumes the simple form

$$S^{\text{Reg}}(T) = \kappa_{d-2} N_{d-2} - \kappa_d N_d, \qquad (4.47)$$

where the coupling constants $\kappa_i = \kappa_i(\lambda, G_N)$ are simple functions of the bare cosmological and Newton constants in d dimensions. Substituting this into the path sum in (4.46) leads to

$$Z(\kappa_{d-2}, \kappa_d) = \sum_{N_d} e^{-i\kappa_d N_d} \sum_{N_{d-2}} e^{i\kappa_{d-2} N_{d-2}} \sum_{T|_{N_d, N_{d-2}}} \frac{1}{C_T}, \qquad (4.48)$$

The point of taking separate sums over the numbers of $d-$ and $(d-2)$–simplices in (4.48) is to make explicit that 'doing the sum' is tantamount to the combinatorial problem of *counting* triangulations of a given volume and number of simplices of codimension 2 (corresponding to the last summation in (4.48)).[18] It turns out that at least in two space–time dimensions the counting of geometries can be done completely explicitly, turning both Lorentzian and Euclidean quantum gravity into exactly soluble statistical models.

4.2.4 Lorentzian Path Integral

Now, the simplicial building blocks of the models are taken to be pieces of Minkowski space, and their edges have squared lengths $+a^2$ or $-a^2$. For example, the two types of 4–simplices that are used in Lorentzian dynamical triangulations in dimension four are shown in Figure 4.10. The first of them has four time–like and six space–like links (and therefore contains 4 time–like

---

[18] The symmetry factor C_T is almost always equal to 1 for large triangulations.

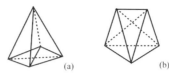

Fig. 4.10. Two types of Minkowskian 4–simplices in 4D (adapted from [Lol01, Lol98]).

and 1 space–like tetrahedron), whereas the second one has six time–like and four space–like links (and contains 5 time–like tetrahedra). Since both are subspaces of flat space with signature $(-+++)$, they possess well–defined light–cone structures everywhere [Lol01, Lol98].

In general, gluings between pairs of $d-$simplices are only possible when the metric properties of their $(d-1)-$faces match. Having local light cones implies causal relations between pairs of points in local neighborhoods. Creating closed time–like curves will be avoided by requiring that all space–times contributing to the path sum possess a global 'time' function t. In terms of the triangulation this means that the $d-$simplices are arranged such that their space–like links all lie in slices of constant integer t, and their time–like links interpolate between adjacent spatial slices t and $t+1$. Moreover, with respect to this time, we will not allow for any *spatial* topology changes[19].

This latter condition is always satisfied in classical applications, where 'trouser points' like the one depicted in Figure 4.14 (see previous Chapter) are ruled out by the requirement of having a non–degenerate Lorentzian metric defined everywhere on M (it is geometrically obvious that the light cone and hence $g_{\mu\nu}$ must degenerate in at least one point along the 'crotch'). Another way of thinking about such configurations (and their time–reversed counterparts) is that the causal past (future) of an observer changes discontinuously as her world–line passes near the singular point. See [Dow02] and references therein for related discussions about the issue of topology change in quantum gravity.

There is no *a priori* reason in the quantum theory to not relax some of these classical causality constraints. After all, as we stressed right at the outset, path integral histories are not in general classical solutions, nor can we attribute any other direct physical meaning to them individually. It might well be that one can construct models whose path integral configurations violate causality in this strict sense, but where this notion is somehow recovered in the resulting continuum theory. What the approach of Lorentzian dynamical triangulations has demonstrated is that *imposing causality constraints will in general lead to a different continuum theory*. This is in contrast with the intuition one may have that 'including a few isolated singular points will not

---

[19] Note that if we were in the continuum and had introduced coordinates on space–time, such a statement would actually be diffeomorphism–invariant.

make any difference'. On the contrary, tampering with causality in this way is not innocent at all, as was already anticipated by Teitelboim many years ago [Tei83].

We want to point out that one cannot conclude from the above that spatial topology changes or even *fluctuations in the space–time topology* cannot be treated in the formulation of dynamical triangulations. However, if one insists on including geometries of variable topology in a Lorentzian discrete context, one has to come up with a prescription of how to weigh these singular points in the path integral, both before and after the Wick rotation [Das02].

Having said this, we next have to address the question of the *Wick rotation*, in other words, of how to get rid of the factor of i in the exponent of (4.48). Without it, this expression is an infinite sum (since the volume can become arbitrarily large) of complex terms whose convergence properties will be very difficult to establish. In this situation, a Wick rotation is simply a technical tool which – in the best of all worlds – enables us to perform the state sum and determine its continuum limit. The end result will have to be Wick–rotated back to Lorentzian signature.

Fortunately, Lorentzian dynamical triangulations come with a natural notion of Wick rotation, and the strategy we just outlined can be carried out explicitly in two space–time dimensions, leading to a unitary theory. In higher dimensions we do not yet have sufficient analytical control of the continuum theories to make specific statements about the *inverse* Wick rotation. Since we use the Wick rotation at an intermediate step, one can ask whether other Wick rotations would lead to the same result. Currently this is a somewhat academic question, since it is in practice difficult to find such alternatives. In fact, it is quite miraculous we have found a single prescription for Wick–rotating in our regularized setting, and it does not seem to have a direct continuum analogue (for more comments on this issue, see [DL01, Das02]).

Our Wick rotation W in any dimension is an injective map from Lorentzian– to Euclidean–signature simplicial space–times. Using the notation T for a simplicial manifold together with length assignments l_s^2 and l_t^2 to its space– and time–like links, it is defined by

$$\mathrm{T}^{\mathrm{lor}} = (T, \{l_s^2 = a^2, l_t^2 = -a^2\}) \xmapsto{W} \mathrm{T}^{\mathrm{eu}} = (T, \{l_s^2 = a^2, l_t^2 = a^2\}). \quad (4.49)$$

Note that we have not touched the connectivity of the simplicial manifold T, but only its metric properties, by mapping all time–like links of T into space–like ones, resulting in a Euclidean 'space–time' of equilateral building blocks. It can be shown [AJL01a] that at the level of the corresponding weight factors in the path integral this Wick rotation[20] has precisely the desired effect of rotating to the exponentiated Regge action of the 'Euclideanized' geometry,

---

[20] To get a genuine Wick rotation and not just a discrete map, one introduces a complex parameter α in $l_t^2 = -\alpha a^2$. The proper prescription leading to (4.50) is then an analytic continuation of α from 1 to -1 through the lower–half complex–plane.

$$e^{iS(T^{\mathrm{lor}})} \xmapsto{W} e^{-S(T^{\mathrm{eu}})}. \tag{4.50}$$

The Euclideanized path sum after the Wick rotation has the form

$$\begin{aligned} Z^{\mathrm{eu}}(\kappa_{d-2},\kappa_d) &= \sum_T \frac{1}{C_T}\, e^{-\kappa_d N_d(T)+\kappa_{d-2}N_{d-2}(T)} \\ &= \sum_{N_d} e^{-\kappa_d N_d} \sum_{T|N_d} \frac{1}{C_T}\, e^{\kappa_{d-2}N_{d-2}(T)} \\ &= \sum_{N_d} e^{-\kappa_d N_d}\, e^{\kappa_d^{\mathrm{crit}}(\kappa_{d-2})N_d} \times \mathrm{subleading}(N_d). \end{aligned} \tag{4.51}$$

In the last equality we have used that the number of Lorentzian triangulations of discrete volume N_d to leading order scales exponentially with N_d for large volumes. This can be shown explicitly in space–time dimension 2 and 3. For $d=4$, there is strong (numerical) evidence for such an exponential bound for *Euclidean triangulations*, from which the desired result for the Lorentzian case follows (since W maps to a strict subset of all Euclidean simplicial manifolds).

From the functional form of the last line of (4.51) one can immediately read off some qualitative features of the phase diagram, an example of which appeared already earlier in Figure 4.8. Namely, the sum over geometries Z^{eu} converges for values $\kappa_d > \kappa_d^{\mathrm{crit}}$ of the bare cosmological constant, and diverges (ie. is not defined) below this critical line. Generically, for all models of dynamical triangulations the infinite–volume limit is attained by approaching the critical line $\kappa_d^{\mathrm{crit}}(\kappa_{d-2})$ from above, ie. from inside the region of convergence of Z^{eu}. In the process of taking $N_d \to \infty$ and the cutoff $a \to 0$, one gets a renormalized cosmological constant Λ through

$$(\kappa_d - \kappa_d^{\mathrm{crit}}) = a^\mu \Lambda + O(a^{\mu+1}). \tag{4.52}$$

If the scaling is canonical (which means that the dimensionality of the renormalized coupling constant is the one expected from the classical theory), the exponent is given by $\mu = d$. Note that this construction requires a positive *bare* cosmological constant in order to make the state sum converge. Moreover, by virtue of relation (4.52) also the *renormalized* cosmological constant must be positive. Other than that, its numerical value is not determined by this argument, but by comparing observables of the theory which depend on Λ with actual physical measurements.[21] Another interesting observation is that the inclusion of a sum over topologies in the discretized sum (4.51) would lead to a super–exponential growth of at least $\propto N_d!$ of the number of triangulations with the volume N_d. Such a divergence of the path integral cannot be compensated by an additive renormalization of the cosmological constant of the kind outlined above.

---

[21] The non–negativity of the renormalized cosmological coupling may be taken as a first 'prediction' of our construction, which in the physical case of four dimensions is indeed in agreement with current observations.

4.2 Sum over Geometries and Topologies

There are ways in which one can sum divergent series of this type, for example, by performing a Borel sum. The problem with these stems from the fact that two different functions can share the same asymptotic expansion. Therefore, the series in itself is *not* sufficient to define the underlying theory uniquely. The non–uniqueness arises because of non–perturbative contributions to the path integral which are not represented in the perturbative expansion.[22] In order to fix these uniquely, an independent, non–perturbative definition of the theory is necessary. Unfortunately, for dynamically triangulated models of quantum gravity, no such definitions have been found so far. In the context of 2D (Euclidean) quantum gravity this difficulty is known as the 'absence of a physically motivated double-scaling limit' [AK93].

Lastly, getting an interesting continuum limit may or may not require an additional fine–tuning of the inverse gravitational coupling κ_{d-2}, depending on the dimension d. In four dimensions, one would expect to find a second-order transition along the critical line, corresponding to local gravitonic excitations. The situation in $d = 3$ is less clear, but results get so far indicate that no fine–tuning of Newton's constant is necessary [AJL01b, AJL01d].

Before delving into the details, let me summarize briefly the results that have been get so far in the approach of Lorentzian dynamical triangulations. At the regularized level, that is, in the presence of a finite cutoff a for the edge lengths and an infrared cutoff for large space–time volume, they are well–defined statistical models of Lorentzian random geometries in $d = 2, 3, 4$. In particular, they obey a suitable notion of reflection-positivity and possess self–adjoint Hamiltonians.

The crucial questions are then to what extent the underlying combinatorial problems of counting all dD geometries with certain causal properties can be solved, whether continuum theories with non–trivial dynamics exist and how their bare coupling constants get renormalized in the process. What we know about Lorentzian dynamical triangulations so far is that they lead to continuum theories of quantum gravity in dimension 2 and 3. In $d = 2$, there is a complete analytic solution, which is distinct from the continuum theory produced by Euclidean dynamical triangulations. Also the matter–coupled model has been studied. In $d = 3$, there are numerical and partial analytical results which show that both a continuum theory exists and that it again differs from its Euclidean counterpart. Work on a more complete analytic solution which would give details about the geometric properties of the quantum theory is under way. In $d = 4$, the first numerical simulations are currently being set up. The challenge here is to do this for sufficiently large lattices, to be able to perform meaningful measurements. So far, we cannot make any statements about the existence and properties of a continuum theory in this physically most interesting case.

---

[22] A field–theoretic example would be instantons and renormalons in QCD.

4.2.5 Non-Perturbative Quantum Gravity

A fascinating and deep question about nature is what one would see if one could probe space and time at smaller and smaller distances. Already the 19th–century founders of modern geometry contemplated the possibility that a piece of empty space that looks completely smooth and structure-less to the naked eye might have an intricate microstructure at a much smaller scale. Our vastly increased understanding of the physical world acquired during the 20th century has made this a certainty. Two pillars of contemporary physics support the expectation that as we resolve the fabric of space-time with an imaginary microscope at ever smaller scales, space-time will turn from an immutable stage into the actor itself. First, due to Heisenberg's uncertainty relations, probing space-time at very short distances is necessarily accompanied by large quantum fluctuations in energy and momentum – the shorter the distance, the larger the energy–momentum uncertainty. Second, according to Einstein's theory of general relativity, the presence of these energy fluctuations, like that of any form of energy, will deform the geometry of the space-time in which it resides, imparting curvature which is detectable through the bending of light rays and particle trajectories. Taking these two things together leads to the prediction that the quantum structure of space and time at the Planck scale must be highly curved and dynamical [AJL06].

A long held ambition of theoretical physicists is to find a consistent description of this dynamical microstructure within a *theory of quantum gravity*, which unifies quantum theory and general relativity, and to determine its ramifications for high–energy physics and cosmology. Given the extraordinary smallness of the Planck length, how can we achieve progress in describing a physical situation that cannot be directly probed by experiment in the foreseeable future? The way this is usually done is by first postulating additional dynamical principles or fundamental symmetries at small distances, which are not accessible to direct experimental verification, second, verifying that these do not conflict with standard quantum physics or general relativity as one goes to larger scales, and third, predicting new physical phenomena that can (at least in principle) be tested, or confirmed indirectly by astrophysical observations. Examples of fundamental building principles are that the universe is made up of tiny vibrating strings, or that space-time at the Planck scale is not a continuum, but consists of tiny discrete grains.

Research into quantum gravity falls broadly into two categories [Kie04, Smo00]: non-perturbative approaches to quantum gravity, whose primary aim is to quantize the gravitational degrees of freedom per se, introducing little or no additional structure such as supersymmetry or extra dimensions, and string–theoretic approaches, where the quantization of gravity appears almost as a by-product of a unified higher–dimensional and supersymmetric 'theory of everything', whose fundamental objects are strings and (mem)branes [Wit02, Zwi04].

Quantum gravity is quite unlike any other fundamental quantum interaction in that it describes the dynamics of an entity that in most physical situations is considered as fixed and given, namely, that of space-time itself. Recall that the degrees of freedom of a space-time in classical general relativity can be described by the space-time metric $g_{\mu\nu}(x)$, a local field variable which determines the values of distance and angle measurements in space-time, or, equivalently, how space-time is bent and curved locally. Space-time is classically determined by solving the Einstein equations for $g_{\mu\nu}(x)$, subject to boundary conditions and a particular matter content of the universe or a piece thereof. In the same manner, in order to determine what space-time *is* from a quantum–theoretical point of view, one would like to formulate a quantum analogue of Einstein's equations, from which *quantum space-time* should then emerge as a *solution*. This should be contrasted with usual quantum field theory, which describes the dynamics of elementary particles and their interactions on a *fixed* space-time background, usually that of the flat, 4D Minkowski space of special relativity. Since at short distances the gravitational forces are so much weaker than the electromagnetic ones, say, it is usually an excellent approximation to treat the gravitational degrees of freedom as 'frozen in' and non-dynamical. The trivial geometric structure of the Minkowski metric forms merely part of the immutable background structure of how quantum field theories are formulated. On the other hand, the physical situations that quantum gravity aims to explain are not in general describable in terms of linear fluctuations of the metric field around Minkowski space or some other fixed background metric. These include the quantitative description of 'empty' space-time at very short distances of the order of the Planck scale, 10^{-35} m, and of the extreme and ultradense state our universe presumably was in when it was very young. From a technical point of view this implies that in quantum gravity one has to modify standard quantization techniques which rely (sometimes implicitly) on the presence of a fixed metric background structure. This is often phrased by saying that gravity must ultimately be quantized in a way that is both *background–independent* (i.e., does not distinguish any particular background metric at the outset) and non-perturbative (i.e., does not simply describe the dynamics of linear perturbations around some fixed background space-time). The most promising approach to constructing such a theory of causal non-perturbative quantum gravity, is the method of causal yynamical triangulations (CDT) [AJL06].

CDT Quantum Gravity

The CDT approach to quantum gravity is based on the space-time geometrical generalization of the Feynman path integral. Recall from the previous section that the basic idea of the quantum–mechanical path integral is to obtain a solution to the quantum dynamics of a physical system by taking a superposition of 'all possible' configurations of the system, where each configuration contributes a complex weight $\exp(iS)$ to the path integral, which depends on the classical action $S = \int L(t)\,dt$ of the configuration, where L denotes the system's Lagrangian. For the case of a nonrelativistic particle moving in a potential, the configurations are literally paths in space, i.e., continuous trajectories $\mathbf{x}(\tau)$ describing the particle's position as a function of time τ, which runs through an interval $\tau \in [0, t]$. Superposing (that is, adding or integrating up) the associated quantum amplitudes $\exp i S^{\text{part}}[\mathbf{x}(\tau)]$ as in (4.53) below, one obtains a solution to the Schródinger equation of the particle.

It is important to realize that the *individual* paths $\mathbf{x}(\tau)$ appearing in the path integral are *not* themselves physical trajectories the particle could move on, and even less solutions to the particle's classical equations of motion. Instead, they are so–called *virtual paths*, that is, a bunch of curves one can draw between fixed initial and final points $\mathbf{x}_i$ and $\mathbf{x}_f$. The magic of the path integral

$$G(\mathbf{x}_i, \mathbf{x}_f, t) = \oint_{\text{paths: } \mathbf{x}_i \to \mathbf{x}_f} \mathcal{D}[x] e^{i S^{\text{part}}[\mathbf{x}(\tau)]} \tag{4.53}$$

is that the true quantum physics of the particle is encoded precisely in the *superposition* of all these virtual paths. In order to extract these physical properties, one has to evaluate suitable quantum operators $\hat{O}$ on the ensemble of paths contributing to (4.53). For example, one may be interested in computing expectation values for the position or the energy of the particle, together with their quantum fluctuations. Of course, the path integral or *propagator* (4.53) also allows us to retrieve the classical behavior of the particle in a particular limit (in this case, when its mass becomes big), but it contains more information, describing the full quantum dynamics of the system.

Analogously, a path integral for gravity is a superposition of all virtual 'paths' our universe (or a part thereof) can follow as time unfolds. These paths are simply the different configurations for the metric field variables $g_{\mu\nu}(x)$ mentioned earlier. It is important to realize that a single path is now no longer an assignment of just three numbers (the coordinates x_i of the particle) to every moment τ in time, but rather the assignment to every τ of a whole array of numbers (the space-space components $\mathbf{g}_{ij}(x) \equiv \mathbf{g}_{ij}(\mathbf{x}, \tau)$ of the metric tensor $g_{\mu\nu}(x)$) for each spatial point $\mathbf{x}$. This is simply a consequence of gravity being a field theory with infinitely many degrees of freedom. The path integral for gravity can thus be written as [AJL06]

$$G(\mathbf{g}_i, \mathbf{g}_f, t) = \oint_{\text{space-times: } \mathbf{g}_i \to \mathbf{g}_f} \mathcal{D}[g] e^{i S^{\text{grav}}[g_{\mu\nu}(\mathbf{x}, \tau)]}, \tag{4.54}$$

where S^{grav} now denotes the classical gravitational action associated with a space-time metric $g_{\mu\nu}$ with initial and final boundary condition $\mathbf{g}_i$ and $\mathbf{g}_f$, separated by a time distance t. Like in the particle case, the individual space-time configurations interpolating between the initial and final spatial geometries have nothing a priori to do with classical space-times, and are much more general objects. Again, one would expect to be able to retrieve the full quantum dynamics of space-time from the path integral (4.54), which is a superposition of all possible ways in which an empty space-time can be curved. In other words, the collective behavior of the virtual space-times contributing to the gravitational propagator (4.54) should tell us what quantum space-time *is*. To extract this geometric information, we will again have to evaluate suitable quantum operators $\hat{O}$ on the ensemble of geometries contributing to (4.54). Suffice it to say that making the gravitational path integral well–defined and extracting the desired physical information is very much more difficult than in the case of the quantum particle. The way in which CDT proceeds is by giving a precise prescription of how the path integral (4.54) should be computed, and in particular how the class of virtual paths should be chosen. In addition, it provides a set of technical tools to extract concrete physical information about the quantum geometry thus created by the principle of quantum superposition.

There are a number of ways in which the path integral of CDT differs from that of previous approaches. In the first instance, it is genuinely non-perturbative, in that the contributing geometries can have very large curvature fluctuations at very small scales and thus be arbitrarily far away from any classical space-time. Our summation is 'democratic' in that no particular space-time geometry is distinguished at the outset. In fact, path integral histories which have any geometric resemblance to a classical space-time are so rare that their contribution to the path integral is effectively negligible.[23] Secondly, as we will see in the following, the causal structure of the geometries contributing to the path sum plays an important role in the method of causal dynamical triangulations, and is a key new element in comparison with previous, so–called *Euclidean path–integral* approaches to quantum gravity [AJL06].

Space-Time Geometry in CDT

What we need to do next in order to make sense of the expression (4.54) for the non-perturbative quantum–gravitational propagator is to define the precise class of space-time geometries (labelled above by $g_{\mu\nu}$) over which the sum or integral is to be taken. As elsewhere in quantum field theory, one is immediately confronted with the fact that unless one chooses a careful regularization for the path integral, it will be wildly divergent and simply not exist in any meaningful mathematical sense (and thus be useless for extracting physical information). 'Regularizing' means making the path integral finite by introducing certain cutoff parameters for the contributing configurations, which at a later stage will be removed in a controlled manner.

---

[23] This is completely analogous to the particle case, where it can be shown rigorously that classical paths 'form a set of measure zero' with respect to the Wiener measure of the path integral [RS75]. Maybe surprisingly, the paths which contribute non-trivially are nowhere differentiable, and thus 'consist only of corners'. One expects a similarly nonclassical behavior for the dominant configurations of the gravitational path integral.

We start with explaining the nature of the regularized space-times used by CDT, which are called 'piecewise flat geometries'. Recall that the dynamical degrees of freedom of a geometry are the ways in which it is locally curved. Piecewise flat geometries are simply spaces that are flat (the same as straight or uncurved, that is, structureless from a geometric point of view) everywhere apart from small subspaces where curvature is said to be concentrated. This in a way discretizes curvature and vastly reduces the different number of ways space-time can be curved. The type of geometry we will use is a triangulated space, also sometimes called a *Regge geometry*, after the physicist who first introduced it into (classical) general relativity [Reg61]. It can be thought of as a space glued together from elementary building blocks which are (higher–dimensional generalizations of) triangles, so–called 'simplices'. The geometric structure of each simplex is trivial, since it is by itself flat by definition and therefore carries no curvature. Local curvature only appears along lower–dimensional interfaces when one starts gluing the simplices together [AJL06].

This can be visualized most easily in the 2D case. Consider a set of identical equilateral 2D triangles cut out from a piece of cardboard which is perfectly straight and unbendable (and hence flat). To obtain a larger surface, start gluing these triangles together by identifying their 1D sides or edges pairwise. Points where several edges meet are also called vertices. One can obtain a piece of flat space by arranging the triangles in a regular pattern so that exactly six triangles and edges meet at each vertex. However, there are many more ways to create *curved spaces* by the same gluing procedure. Namely, whenever the number of triangles meeting at a vertex is smaller or larger than six, this vertex will carry a positive or negative curvature. By 'curvature' we mean the *intrinsic curvature* of the 2D surface, i.e. the curvature that can be detected from within the surface – for example, by studying the trajectories of particles or light rays –, and is independent of any higher-dimensional space in which it may be imbedded. This mirrors a property of the physical theory of general relativity in four dimensions, which likewise depends only on the intrinsic geometry of space-time. The set-up in higher dimensions is identical, with the 2D triangles (or 'two–simplices') substituted by the corresponding flat higher–dimensional simplices (three–simplices (or tetrahedra) in dimension 3, four–simplices in dimension 4, etc.). Generally speaking, the fundamental building blocks in dimension d are glued together pairwise along their $(d-1)$–dimensional faces, and their intrinsic curvature is concentrated on the $(d-2)$–dimensional intersections of these faces.

The so–called *Regge calculus* [Wil97] was originally designed to approximate smooth classical space-times, or, more precisely, solutions to the Einstein equations, by these piecewise flat, triangulated spaces. There are two reasons for why this is a very economical way of describing a space-time. Firstly, only a finite amount of data is necessary to completely characterize a finite piece of space-time, namely, the geodesic invariant lengths of all the 1D edges of all the simplices involved, and the way in which the d–dimensional simplices are glued together. Secondly, because no coordinate system need ever be introduced on the simplices, this formulation does not share the usual coordinate redundancy of Einstein gravity described in terms of the field variables $g_{\mu\nu}(x)$.

The use of Regge geometries in the quantum theory is not new, and CDT builds on previous attempts of both 'Quantum Regge Calculus' [Ham00] and 'Dynamical Triangulations' [AJ92] to define a theory of quantum gravity from a non-perturbative Euclidean path integral[24]. To avoid misunderstandings, it should be emphasized that the use of triangulated space-times differs in classical and quantum applications. The objective in the former is to approximate a single, smooth space-time (which may or may not be known exactly by some other method) as well as possible. This can be achieved by choosing a sequence of triangulations, where in each step of the sequence the triangulation is chosen finer than in the previous step (i.e., the typical edge length is decreased) and therefore can converge to the smooth manifold in a point-wise sense. In the 2D example, it is quite clear that such an approximation can be very good when the edge lengths become much smaller than the scale at which the smooth space-time is curved.

By contrast, the objective of the quantum theory, and that of CDT in particular, is to approximate the integral (4.54) as well as possible, or, more precisely, to *define* it since there is currently no alternative, independent way of doing the computation. This is a completely different task, since the integral does not represent a single classical geometry, but a quantum superposition, where each single contributing space-time is a highly nonclassical object, as we pointed out earlier. There is no accurate mathematical statement to guide this construction, but one would expect that the path integral should provide an 'ergodic sampling' of the space of geometries. This may seem like a very vague characterization, but one is in practice very much constrained by the requirement of making the regularized path integral mathematically well–defined and obtaining a sensible classical limit.

The short–distance cutoff a is an important part of our regularization of the space-time geometries in the gravitational propagator. We will take the limit $a \to 0$ as part of the search for a so-called continuum limit of the path integral over the regularized geometries. This has to be done in order to obtain a final theory which does not depend on many of the arbitrary details which have gone into constructing the regularized model, which itself constitutes only an intermediate step in the construction of the theory. Using a finite 'lattice spacing' a and taking $a \to 0$ (while renormalizing the coupling constants of the theory as a function of a) is a method borrowed from the theory of critical phenomena and virtually ensures that the end result does not depend on a

---

[24] The essential difference between the two approaches is that in Quantum Regge Calculus one fixes a triangulation or 'gluing', so that the path integral takes the form of a (multiple) integral over the lengths of the edges of that triangulation, whereas in Dynamical Triangulations one fixes *all* edge lengths to a common value a, in which case the path integral takes the form of a discrete sum over all inequivalent ways to glue the (then identical-looking) simplicial building blocks together.

variety of regularization details. This latter property of 'universality' is only a necessary condition and does by no means guarantee that this construction leads to a viable theory of quantum gravity, as opposed to e.g., describing the dynamics of certain 1D polymers.

Ensemble of Virtual Space-Time Geometries

Now that we have introduced the regularized triangulated geometries the question still remains as to exactly what ensemble of such objects should be included in the sum over geometries in (4.54). Here is where CDT differs in a crucial way from previous approaches, and where the notion of 'causality' comes into play. We mentioned above that precursors of CDT's non-perturbative path integral are 'Euclidean' in nature. What this means is that the integration is not performed over so-called *Lorentzian space-times* (which have one time- and three space-directions) but over Euclidean *spaces* (which have four spatial directions, and thus no notion of time, light rays or causality). Classically, Euclidean 'space-times' are bizarre and unphysical entities, in which moving back and forth in time is just as easy as moving back and forth in space. Their use in the (mainly perturbative) gravitational path integral was made popular in the late 1970s by the influential work of S. Hawking and collaborators on black holes and quantum cosmology in the context of Euclidean quantum gravity [GH93]. The reason for using them instead of Lorentzian space-times of the correct physical signature[25] is mainly technical: in the Euclidean case, the weights $\exp(iS^{\text{grav}})$ are no longer complex but real numbers, which simplifies a discussion of the convergence properties of the path integral, and also makes *Monte–Carlo simulations* possible.[26] The potential catch is that in gravity, unlike in other quantum field theories on a flat background, there is no obvious relation between a non-perturbative path integral for Lorentzian and one for Euclidean geometries. In fact, causal dynamically triangulated gravity in dimensions two [AL98], three [AJL01b] and four [AJL00b] has for the first time provided concrete evidence that the two path integrals are genuinely inequivalent and possess completely different properties.

It would seem straightforward to write down a regularized version of the gravitational propagator as [AJL06]

$$G^{\text{reg}}(\mathbf{T}_i, \mathbf{T}_f, t) = \sum_{\text{triangulations T}: \mathbf{T}_i \to \mathbf{T}_f} e^{iS^{\text{reg}}[T]}, \qquad (4.55)$$

---

[25] The *signature* refers to the signs of the four eigenvalues of the symmetric matrix $g_{\mu\nu}(x)$; it is (+,+,+,+) in the Euclidean case and (-,+,+,+) in the Lorentzian case.

[26] In order to simplify notation, we will always use the notation $\exp(iS)$ to denote Boltzmann weights, with the implicit understanding that S is a real action when we talk about Lorentzian signature (and the weight thus a complex phase factor), and a purely imaginary one in a Euclidean context (and $\exp(iS)$ therefore a *real* quantity).

where T denotes a triangulated space-time, glued from four–simplices, and with two spatial triangulated boundary geometries $\mathbf{T}_i$ and $\mathbf{T}_f$ (glued from three–simplices), between which it interpolates. The gravitational action for a piecewise flat space-time T schematically takes the form

$$S^{\text{reg}}(T) = -1/G_N \, Curvature(T) + \lambda \, Volume(T), \qquad (4.56)$$

and there is a definite prescription for how to compute the curvature and volume of a given triangulation T in terms of the lengths of its edges and its connectivity (that is, the way the four–simplices are glued together). The two coupling constants of the theory appearing in (4.56) are Newton's constant G_N, governing the strength of gravitational interactions, and the cosmological constant λ, another constant of nature, which may be responsible for the 'dark energy' pervading our universe [Sah04].

As mentioned in footnote 24, all simplices used in DT are equilateral, and the path integral assumes the form of a discrete sum over inequivalent ways in which the simplicial building blocks can be glued together. The only thing that remains then to be specified in (4.55) is whether any gluing of the building blocks is to be allowed, or whether further restrictions need to be imposed. One condition turns out to be essential for making the path integral construction well–defined. Call $\mathcal{N}(N_4)$ the number of distinct gluings of N_4 four–simplices, for a particular set of gluing rules. Clearly, this number will grow with N_4, but the important question is whether it will grow exponentially as a function of N_4 or faster, namely, 'super-exponentially', for example, like $\exp(cN_4^\nu)$, with $c > 0$ and $\nu > 1$. In the latter case, and noting that $N_4(T)$ is proportional to $Volume(T)$, there is no way in which the exponential weights $\exp i S^{\text{reg}}[T]$ could ever counterbalance the growth of the number of contributing geometries (the *growth of the entropy* of the system). The path integral would then be too divergent to lead to a fundamental theory of gravity.

These considerations preclude the inclusion in the path integral (4.55) of a so–called *sum–over–topologies*.[27] Therefore, the topology of the space-times contributing to the non-perturbative path integral has to be fixed. It is typically chosen to be a 4D sphere or torus. This state of affairs is somewhat ironic, because the possibility of including a sum over topologies has often been praised as an advantage of the path integral formulation over canonical quantization methods, which employ a 3+1 split of space-time into space and time. As we have argued, this is only true at a formal level, that is, as long as one does not perform concrete computations (and therefore has to worry about

---

[27] The *topology* of a space-time describes the way in which it hangs together. For example, the topology of a two-dimensional compact and closed surface is completely characterized by the number of its 'holes' or 'handles'. It could have the form of a surface of a ball (no holes), of a surface of a torus or bicycle inner tube (one hole), of a surface of a double-torus or two-hole doughnut (two holes), and so on. In four dimensions, the labelling of different topologies is much more involved.

the convergence or otherwise of an expression like (4.55)). At least from a Euclidean point of view, there are now no further natural restrictions one may impose on the geometries, and it is from this starting point that the original approach of Dynamical Triangulations proceeded [AJ92, AJ95], in order to study the properties of the theory (hopefully) defined by the continuum limit of (4.55).

This may be a convenient moment to make some non-technical remarks on how (C)DT evaluates the path integral and extracts physical information from it, such as the expectation values of certain geometric observables. A direct analytical evaluation of (4.55), although available in lower–dimensional models, is formidably difficult. However, unlike in a variety of other approaches to quantum gravity, DT possesses a set of powerful and well-developed numerical tools, whose value can hardly be overstated. They have been adapted from statistical mechanics and the theory of critical phenomena to the case where the individual configurations are curved geometries, rather than spin or field configurations on a fixed background space or lattice. The ensemble of space-times underlying the path integral is simulated by Monte Carlo methods [NB99], generating a random walk in the space of all configurations according to a probability distribution defined by (4.55).[28] The limitations of the computer imply that this procedure can only be implemented on a (possibly large but) finite space of geometric configurations. This is usually taken into account by performing the simulations on the ensemble of triangulations of a fixed discrete volume N_4. By repeating the numerical measurements for a variety of different N_4's, one then tries to extrapolate the results in a systematic way to the physically relevant limit $N_4 \to \infty$. This well-known technique is known as 'finite–size scaling' [AJL06].

Now, what kind of 'quantum geometry' does one expect to see with the help of these tools? If all goes well, the quantum superposition (4.55) of geometries should be able to reproduce a classical space-time at large scales L, that is, in the classical limit. However, at small scales l, with $a \ll l \ll L$, one expects quantum fluctuations to dominate, with a resulting highly nonclassical behavior of the geometry. To cut a long story short, this was unfortunately not what was found for the Euclidean dynamically triangulated path integral studied in the 1990s. This was not immediately realized, but emerged gradually as more numerical simulations were performed [BB96]. It turned out that the quantum geometry generated by Euclidean DT could be in either one of two 'phases'. In the first one the geometry was completely crumpled, and in the other totally polymerized, that is, degenerated into thin branching threads. These structures persist also at large scales, and as a result the DT path integral appears to have no meaningful classical limit, and therefore does

---

[28] For the Euclidean path integral, one can directly use the real weights $\exp iS^{\mathrm{reg}}[T]$. For the Lorentzian case of CDT, in order to obtain a probability distribution from (4.55), one first has to apply a so-called 'Wick rotation' which converts the complex to real weights [AJL01a].

not satisfy a necessary criterion for a theory of quantum gravity. (One can only wonder how long it may have taken to realize this, had one not been in a state to perform extensive simulations of the model.)

The starting point of CDT was the hypothesis that this failure may have to do with the un-physical *Euclidean* nature of the construction, and that one may be able to rectify the situation by encoding the causal structure of *Lorentzian space-times* explicitly in the choices of building blocks and gluing rules. Several years passed since this initial conjecture, in which CDT's causal quantization program was implemented and its viability tested in lower dimensions [AL98, AJL01b]. Namely, superpositions like (4.55) can be defined also by considering space-times glued from 2D or 3D building blocks. This gives rise to simplified toy models which share some, but by no means all properties of the true CDT path integral. On the plus side, they can be tackled both analytically and numerically, and compared with other quantization approaches to Einstein gravity in two and three space-time dimensions. These extensive investigations showed unequivocally that the causal, Lorentzian path integral in all cases gave different results from the corresponding Euclidean path integral [AJL06].

Sum over Topologies and Quantum Gravity

Many attempts of constructing a non-perturbative path integral for gravity start from the premise that this should also contain a sum over space-time topologies, formally written as

$$Z = \sum_{topol.} \int \mathcal{D}g_{\mu\nu} e^{iS[g_{\mu\nu}]}, \quad (4.57)$$

with the action

$$S = \int d^4x \sqrt{|\det g|}(\kappa R - \lambda), \quad (4.58)$$

where each term in the sum (4.57) is given by the functional integral over equivalence classes $[g_{\mu\nu}]$ of metrics on a space-time of a particular topology. This assertion is usually followed by immediately dropping the sum again, since no way can be found to enumerate the different topologies, let alone perform the sum explicitly.

Needless to say that this state of affairs is highly unsatisfactory. Whether or not a sum over topologies should be included is connected to the nature of the fundamental degrees of freedom governing quantum gravity at the very shortest scale, about which little is known. Topological excitations seem a natural enough candidate, and pictures of a multiply–connected space-time foam for a review and bibliography) may be suggestive to the imagination, but there is so far little direct or indirect evidence that such structures are realized in nature.

Is there then anything we can say about the issue of topology change[29], in the absence of a full–fledged non-perturbative theory of quantum gravity? If we managed to make sense, mathematically and physically, of the sum over topologies, how would the final theory be affected by the inclusion? Any theory predicting finite probabilities for *macroscopic topology changes* is likely to be already in contradiction with observational data.

There are to our mind strong indications – at least within the realm of *Euclidean quantum gravity* – that the topological sum cannot be made meaningful, simply because it results in too many configurations contributing to the path integral. This is true even in dimension two, where toy models of quantum gravity (in the form of generally covariant non-perturbative Euclidean path integrals over geometries) can be defined and solved exactly. In this case, no difficulty arises with the labelling of topologies, which amounts to a single parameter g, the genus or number of handles (holes) of the 2D geometry. The topological expansion in g was the subject of intense study in the early 1990s, because it is an example of the non-perturbative sum over world–sheets of a bosonic string, in a 0D target space. The problem in making the sum well–defined stems from the factorial growth in V of the number of inequivalent 2D surfaces of a given volume V. Moreover, the coefficients in the $g-$expansion are positive, obstructing Borel–summability, and no way has been found to define the non-perturbative sum unambiguously.

Given the recent successes in obtaining quantum gravity theories from state sums over Lorentzian geometries in 2D [AL98] and 3D [AJL01b, Lol01], the question arises of how a topological sum can be incorporated in these models and whether any progress can be made in performing the sum. For quantum gravity in two space-time dimensions the problem is indeed ameliorated by going to a Lorentzian signature: consideration of their causal properties leads to a natural restriction on the topology–changing geometries entering the regularized path integral, as will be explained below. The combined sum over topologies and geometries can be performed exactly, and possesses a well-defined double-scaling limit, involving both the cosmological and the gravitational coupling constants, Λ and G. For $G \to 0$, standard Lorentzian quantum gravity without holes is recovered, whereas for values larger than zero, the presence of holes leads to an observable and non-local scattering of light rays traversing the space-time. At $G = G_{max}$, the system undergoes a transition to a phase of 'handle condensation'. In addition to a further instance of how Lorentzian-ness and causality lead to path integrals that are better behaved than their Euclidean counterparts, this opens up a new playground for gravity–inspired 2D statistical models.

---

[29] Performing a sum over (space-time) topologies in a path integral with fixed initial and final boundary conditions implies configurations whose spatial topology changes in time. For reviews of the issue of topology change in gravity, see [Dow02].

Implementing Topology Change

There is some freedom in how to include topology–changing (1+1)D space-time configurations in the gravitational path integral. Our implementation will be minimal in the sense that each hole will be allowed to exist for an infinitesimal proper–time interval only. In our discrete, triangulated framework this will mean that a hole will come into existence at some integer time t and disappear again at $t+1$. The number of allowed holes per time step $\Delta t = 1$ (in the continuum limit) will be arbitrary. As in Lorentzian quantum gravity for fixed topology, all configurations possess a globally defined proper time variable. For the sake of definiteness, we will work with spatially compact slices. Therefore, by construction a spatial slice at some integer t will have the topology S^1 of a circle, whereas for all times in the open interval $]t, t+1[$ it will be split into a constant number $g+1$ of S^1–components [LW03].

Although this seems the very mildest form of topology change imaginable in two dimensions, we will see that generic space-times of this kind are extremely ill–behaved in their geometric and causal properties, even if there is only a single hole in the entire space-time. The essential difference with the Euclidean case is that the presence (almost everywhere) of a Lorentzian structure allows us to quantify how badly causality is violated (as it necessarily must be in a topology–changing geometry). We will then argue for a restriction of the state sum to geometries whose causality violations are relatively mild. This is motivated by the search for continuum limits which do not necessarily exhibit macroscopic acausal and therefore physically unacceptable behavior (adopting a similar line of argument as one would in 4d).

What we find is an exactly soluble model of 2D gravity with dynamical topology, with a well–defined double–scaling limit involving both the gravitational and cosmological coupling constants. Its acausal properties can be probed by light rays, and get larger with increasing (renormalized) gravitational constant G. This means that for $G \neq 0$ there is always a non-trivial effect coming from the 'infinitesimal' holes, which may still be compatible with observation if the measuring instruments that could detect the acausality were not sufficiently sensitive. However, as we will see, for sufficiently large G any experimental detection threshold will eventually be exceeded. We interpret this behavior as an a posteriori justification for having restricted the allowed space-time histories in the first place, in the sense that it seems unlikely that a model with significantly more general types of holes will possess any physically acceptable ground state whatsoever.

We first give a qualitative description of space-time geometries with 'bad' and 'not-so-bad' topology changes, and then present a concrete realization of the latter within the framework of 2D Lorentzian dynamical triangulations. The generation of holes of either type is illustrated in Figure 4.11. At time t, an initial spatial slice S^1 splits into $g+1$ components, (a1), giving rise to g saddle points. The components evolve in time until $t+\Delta t$, where they re-unite to a single S^1. A difference now arises, depending on which pairs of points are

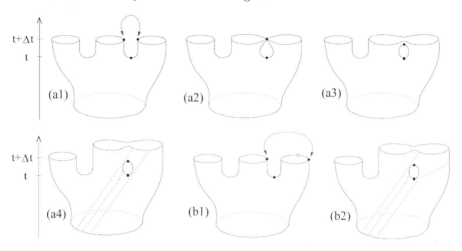

Fig. 4.11. A connected spatial slice splits into three components at time t, (a1). In the first example, two circles at time $t + \Delta t$ are merged into one by identifying two points which are time-like separated from the original branching point, (a2) and (a3). Parallel light rays passing between t and $t + \Delta t$ are unaffected unless they are scattered non-locally by the hole to another part of the manifold, as is the central light ray in (a4). – If one of the merging points is space-like separated from the branching point at t, a twist (indicated by the arrow in the embedded picture, (b1)) is required before the regluing. In the resulting geometry, the distance between two parallel light rays which pass on either side of the hole has jumped discontinuously after the merger, (b2) (adapted and modified from [LW03]).

identified in the process of merging. In the not–so–bad topology change, the upper saddle point of a hole is by definition time– or light–like related to the lower saddle point of the same hole, in either component, as indicated in (a2), (a3) (for simplicity of illustration, we perform the merger only for two of the components). A merger which is not of this type is illustrated in Figure 4.11, (b1). The marked point at time $t + \Delta t$ on the right-most cylinder component is supposed to lie outside the light cone of the lower saddle point.

To illustrate the qualitative difference between the two cases, we follow a set of parallel light rays through the resulting space-times, as indicated in Figure 4.11, (a4) and (b2). In both cases, a light ray which 'hits the hole' is scattered non-trivially to a different part of the manifold. However, in the case where an additional relative twist of at least one of the cylinder components is present (in Figure 4.11, (b1) and (b2), the right cylinder has been twisted by an angle π), there is another non-local effect, consisting in a permutation of different finite sections of the propagating light front, which will persist after the hole has disappeared. The effect on the two outer parallel light rays depicted in Figure 4.11, (b2), is that they are still parallel after time $t + \Delta t$, but their mutual distance will have jumped.

Note that while the effect of the direct scattering by a single hole will vanish in the limit as $\Delta t \to 0$ (corresponding to the continuum limit in the discretized model), the effect of globally rearranging parts of space-time with respect to each other for the 'bad' topology changes will persist in the same limit, and represents an observable, macroscopic violation of causality. We will discard such configurations from the path integral, since we do not think that these large-scale causality violations can cancel out in any superposition of such geometries. Moreover, they completely outnumber the geometries with 'not–so–bad' topology changes. The precise definition of the resulting 2D quantum gravity model, its continuum limit, and its physical properties will be the subject of the following subsection.

Lorentzian Quantum Gravity with Holes

We will now discuss how to implement 'baby holes' of the type introduced in the previous section explicitly in the framework of piecewise linear two–manifolds. Recall that space-time geometries in Lorentzian dynamical triangulations are constructed by gluing together strips of height $\Delta t = 1$, where t is a discrete analogue of proper time. A given strip between integer times t and $t+1$ consists of N_t Minkowskian triangles (each with one space-like and two time-like edges), and is periodically identified in the spatial direction.

We create a hole of minimal time extension $\Delta t = 1$, and associated with a 'not–so–bad' topology change, by identifying two time-like links in the same strip $[t, t+1]$ (these are links interpolating between the slices of constant time t and $t + 1$), and cutting them open in the perpendicular direction, thereby creating two cylinders and a minimal hole in between (see Figure 4.12). This process generates two curvature singularities, at the beginning and end of the hole, which after the Wick rotation will be of the standard conical type, and we will choose their Boltzmann weights accordingly. As anticipated earlier, the number of possible strip geometries of this type as a function of the total strip volume scales exponentially, and both the state sum and its continuum limit are completely well defined [LW03].

As in the original Lorentzian model [AL98], it suffices to examine the combinatorics of a single strip to determine the bulk behavior of the model in the continuum limit (as well as the associated quantum Hamiltonian). After the Wick rotation, the relevant partition function is

$$Z(\lambda, \kappa) = \sum_{l_{in}} \sum_{l_{out}} G_{\lambda, \kappa}(l_{in}, l_{out}; t = 1), \qquad (4.59)$$

where we have performed a sum over both the initial and final boundary geometries of length l_{in} and l_{out}, and where the propagator $G_{\lambda, \kappa}$ is given by

$$G_{\lambda, \kappa}(l_{in}, l_{out}; t = 1) = e^{-\lambda(l_{in} + l_{out})} \sum_{T|_{l_{in}, l_{out}}} e^{-\kappa g(T)}. \qquad (4.60)$$

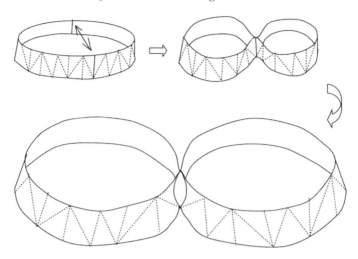

Fig. 4.12. Constructing a strip with one hole by identifying two of the time-like edges between times t and $t+1$ of a regular Lorentzian strip and separating them perpendicularly as indicated, thereby creating a hole between the two integer times (adapted and modified from [LW03]).

The sum in (4.277) is taken over all triangulated strip geometries with boundary lengths l_{in} and l_{out}, and $g \geq 0$ denotes the number of holes of the strip. In writing (4.277) we have used that the discrete volume of a strip is given by $N \equiv N_t = l_{in} + l_{out}$, as in Lorentzian gravity without topology change. Fixing l_{in} and l_{out} (and for convenience putting a mark on the entrance loop), the number

$$\tilde{G}(l_{in}, l_{out}) = \binom{l_{in} + l_{out} - 1}{l_{in} - 1} \tag{4.61}$$

of distinct (marked) interpolating strip triangulations without holes gives rise to an overall factor 2^{N-1}. For a given triangulated strip of volume $N = l_{in} + l_{out}$, holes are created according to the prescription given above and as shown in Figure 4.12.

An alternative, planar representation of the creation of a single hole is given in Figure 4.13, which shows a cut through the strip halfway between times t and $t+1$. The N time-like links of the strip appear as dots on the circle (Figure 4.13a). The procedure for several holes is completely analogous. The only restriction one needs to impose in order to obtain a well-defined two–geometry with $g+1$ cylindrical components is that the g arrows identifying pairs of points in the corresponding planar diagram should not cross each other. Also, we will impose the regularity condition that there should be at most one arrow per point. This avoids some double counting of identical geometries and eliminates a few geometries with cylinders of the size of the cutoff, and is not expected to affect the continuum limit in any way. Note also that we are including some geometries where one or more cylinders degenerate to a point either at time t or $t+1$; this is merely to simplify some of the combinatorial formulas and again will not have any consequences for the continuum limit.

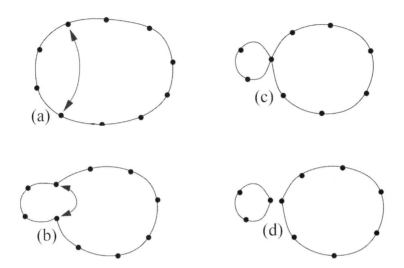

Fig. 4.13. Inserting a single hole into a regular strip of length $N = 10$ (a), as it appears in a slice of constant time half–way between the two boundaries. The strip is pinched along a pair of time-like links (appearing as dots), as indicated by the arrows (b), until a figure eight is obtained (c), after which the strip is separated into two cylinders as indicated in (d) (adapted and modified from [LW03]).

Our next task is therefore to count the number of ways of inserting g holes into a strip of volume N, which is equivalent to counting graphs with N points and g arrows. This is readily done by noting that the number of ways to pick $2g$ out of N points, $2g \leq N$, is given by $\binom{N}{2g}$, since the N points can be regarded as distinguishable (at large volumes N, a generic triangulated strip without holes will not have any symmetries). For a given set of $2g$ points, we then have to count the number of ways of connecting them by non-intersecting arches. Fortunately, this is a well-known combinatorial problem whose resolution is given by the so-called *Catalan numbers*

$$A(2g) = \frac{(2g)!}{g!(g+1)!}. \tag{4.62}$$

The complete formula for the partition function (4.59) is therefore

$$Z(\lambda, \kappa) = \frac{1}{2} \sum_{N=0}^{\infty} \sum_{g=0}^{[N/2]} \binom{N}{2g} \frac{(2g)!}{g!(g+1)!} e^{-2\kappa g} e^{-(\lambda - \log 2)N}. \tag{4.63}$$

After an exchange of the two sums, both of them can be performed explicitly, leading to

$$Z(\lambda, \kappa) = \frac{1}{2(1 - e^{-(\lambda - \log 2)})} \frac{1 - \sqrt{1 - 4z}}{2z}, \qquad (4.64)$$

where the second term on the right-hand side[30] depends only on the specific combination

$$z = e^{-2\kappa}(e^{\lambda - \log 2} - 1)^{-2} \qquad (4.65)$$

of the bare coupling constants κ and λ. The partition function (4.63) is convergent for (real) $\lambda > \log 2$ and $z < 1/4$. We are now interested in constructing a continuum limit of Z. This will necessarily involve an infinite–volume limit $N \to \infty$. It is straightforward to compute the expectation value of the discrete volume,

$$\langle N \rangle = -\frac{1}{Z} \frac{\partial Z}{\partial \lambda} = \frac{e^\lambda}{(e^\lambda - 2)\sqrt{1 - 4z}} - 1, \qquad (4.66)$$

from which we deduce that the infinite–volume limit can be obtained by letting λ approach $\log 2$ from above, just like in standard 2D Lorentzian gravity. However, this is only consistent if one stays inside the combined region of convergence of both λ and z. From the explicit form (4.65) of z this is only possible if one scales the bare inverse gravitational coupling κ in such a way as to counterbalance the divergence coming from the inverse powers of $(e^{\lambda - \log 2} - 1)$. More specifically, if we make a standard ansatz of canonical scaling for the cosmological coupling constant,

$$\lambda = \lambda^{crit} + a^2 \Lambda + O(a^3) \equiv \log 2 + a^2 \Lambda + O(a^3), \qquad (4.67)$$

where Λ denotes the renormalized, dimensionful cosmological constant, we obtain for any fixed value $z = c < 1/4$ of z an equation for κ, namely,

$$\kappa = -\frac{1}{2} \log \left(c \, (a^2 \Lambda)^2 \right) + O(a), \qquad (4.68)$$

which determines the leading–order behavior of κ as a function of the cutoff a. This relation can now be read as the defining equation for the renormalized inverse gravitational coupling K,

$$K = \kappa - 2 \log \frac{1}{a\sqrt{\Lambda}} + O(a), \quad \text{with} \quad K = \frac{1}{2} \log \frac{1}{c}. \qquad (4.69)$$

The logarithmic subtraction is what one would expect for the renormalization of a dimensionless coupling constant. Introducing the renormalized Newton's constant $G = 1/K$, and substituting the expansions (4.67) and (4.69) into (4.64), one obtains to lowest order in a

---

[30] This term is recognized as the generating function for the Catalan numbers, and has previously appeared in a statistical model of certain 'decorated' 2d Lorentzian geometries without topology changes.

$$Z = \frac{1}{a^2}\frac{e^{2/G}}{4\Lambda}\left(1 - \sqrt{1 - 4e^{-2/G}}\right) = \frac{1}{a^2}Z^R(\Lambda, G), \qquad (4.70)$$

where we have performed a wave function renormalization to arrive at a finite renormalized partition function $Z^R(\Lambda, G)$. In summary, we have been led to (4.70) by taking a well-defined *double-scaling limit* of both the gravitational and the cosmological coupling constants, very similar to what has always been hoped for in 2D *Euclidean quantum gravity* [LW03].

The physical properties of the resulting continuum theory are governed by the values of these two couplings. The expectation value of the space-time volume $V = a^2 N$, computed in analogy with (4.66), is given by the inverse cosmological constant,

$$\langle V \rangle = -\frac{1}{Z^R}\frac{\partial Z^R}{\partial \Lambda} = \frac{1}{\Lambda}, \qquad (4.71)$$

as one may have anticipated. The role of the gravitational constant is exhibited by computing the expectation value of the number of holes per time interval,

$$\langle g \rangle = \frac{1}{2}G^2\frac{1}{Z^R}\frac{\partial Z^R}{\partial G} = -\frac{1}{2}\left(1 - \frac{1}{\sqrt{1 - 4e^{-2/G}}}\right). \qquad (4.72)$$

For small coupling $G \approx 0$, there are basically no holes, up to $G \approx 1.33$, where there is an average of a single hole in the entire strip. Beyond this value, the number of holes increases rapidly and diverges at the boundary $G = 2/\log 4$ of the allowed interval.

The average genus is an interesting quantity because it relates in a direct way to an 'observable', namely, the part of a light beam that undergoes scattering when passing through a Lorentzian space-time with baby holes. In first approximation, this is given by

$$\text{scattered portion of light beam} \propto \langle g \rangle \frac{T}{L}, \qquad (4.73)$$

where L is the characteristic linear spatial extension of the *quantum universe*, and T is the (continuum) time of propagation of the light beam.

Causality Implies 4D

We will now describe the first piece of evidence which showed that CDT can reproduce at least *some* aspects of classical geometry correctly. This concerns a point where previous related quantization attempts have failed, namely, to generate a geometric object that can be said to be 4D on sufficiently large scales. It may come as a surprise that a superposition of locally 4D geometries can give anything that is *not* again 4D. After all, we have obtained our geometric building blocks by cutting out small pieces from a 4D flat space. However, as is illustrated by Euclidean dynamically triangulated models, it is perfectly possible that the dimension comes out not as four, and this is indeed what seems to happen *generically*. The crumpled and polymeric phases of the Euclidean model mentioned above are characterized by a so-called *Hausdorff dimension*, which takes the values ∞ and 2, respectively.

How can one obtain spaces with such strange dimensionalities? Roughly speaking, the Hausdorff dimension is obtained by comparing the typical linear size r of a convex subspace of a given space (e.g. its diameter) with its volume $V(r)$. If the leading behavior is $V(r) \sim r^{d_H}$, the space is said to have the Hausdorff dimension d_H. To obtain an effectively infinite-dimensional space from gluing N_4 4D simplices with edge length a, one may consider a sequence of triangulations whose volume goes to infinity, $N_4 \to \infty$, where the gluing for each fixed N_4 is chosen such that every single building block shares a given vertex. That is, no matter how large N_4 gets, all building blocks of the triangulated space crowd around a single point. This is a procedure which is compatible with the gluing rules, but gives rise to a space whose dimensionality diverges, simply because its linear size always stays at the cutoff length a. Conversely, one can get an effectively 1D space by gluing the 4D building blocks into a long and thin tube. That is, as $N_4 \to \infty$ and $a \to 0$, one keeps three out of the four directions at a size of the order of the cutoff a, and only grows the geometry along the fourth direction [AJL06].

This argument shows that there are spaces with 'exotic' dimensionality which can be obtained as limiting cases of regular simplicial manifolds. Of course, the relevant question for the gravitational path integral is whether geometries of this nature indeed *dominate the path integral in the continuum limit*. This is a genuinely dynamical question which cannot be decided a priori. It depends on the relative weight of 'energy' and 'entropy', that is to say, the Boltzmann weight of a given geometry (which in turn is a function of the values of the bare coupling constants) and the number of geometries with a given, fixed Boltzmann weight. Thus it may happen that an exotic geometry (for example, one of the highly crumpled objects above) has a very large Boltzmann weight and is therefore 'energetically favored', but that there are relatively speaking far fewer of such objects in the ensemble than there are of the more 'normal' geometries, such that the contribution of the former will in the end play no role in the path integral in the continuum limit. As we have seen, this is not what happens in Euclidean dynamically triangulated models for quantum gravity whose state sums, depending on the values of the coupling constants, are dominated by exotic geometries which are either maximally crumpled ($d_H = \infty$) or of the form of so-called branched polymers (with $d_H = 2$).

The finding that 'dimensionality' is turned into a dynamical quantity is a consequence of the fact that the non-perturbative gravitational path integral contains highly nonclassical geometries which are curved (and even highly curved) at the cutoff scale a. It can and indeed does happen that geometries with such an unruly short–scale behavior dominate the path integral as $a \to 0$. As already remarked earlier, this is exactly what one would expect in analogy with the path integral for the particle, which in the continuum limit is dominated by totally nonclassical paths with 'infinitely many corners'. It is important to emphasize that the short-scale picture of geometry that arises in CDT is completely different from that of the classical theory. If one looks at a piece of classical space-time, no matter how curved, with an ever finer resolution, it will *always* eventually start looking like a piece of flat space-time, namely, when the observed scale becomes much smaller than the characteristic scale at which the space is curved. By contrast, a typical 'quantum space-time' generated by our non-perturbative path integral construction will *never* resemble a flat space, no matter how fine we choose the resolution of our virtual magnifying glass [AJL06].

Having understood that quantum geometry will necessarily look very nonclassical at short scales, we presumably are still left with many possibilities for the precise microstructure that is generated by various prescriptions for setting up the gravitational path integral. Can we formulate criteria for recognizing when a particular prescription stands a chance of leading to the correct theory of quantum gravity? Fortunately, the answer is yes, and the criteria in question have to do with reproducing features of classical geometry at sufficiently large scales. As alluded to above, the simplest such test is whether the quantum geometry has the correct dimension four at large distances. A path integral which does not pass this test simply does not qualify as a candidate for a theory of quantum gravity.

The art is then to come up with a path integral which allows for large short-scale fluctuations in curvature, but in such a way that the resulting large-scale geometry nevertheless does not degenerate completely, so that a sensible classical limit may exist. The method of causal dynamical triangulations has for the first time in the history of the non-perturbative gravitational path integral given us an explicit example of such a geometry. What has been found to be crucial in its derivation are certain causal rules one imposes on the triangular building blocks, which make explicit reference to the Lorentzian structure of the individual geometries contributing to the path integral. The new and intriguing physical insight that can therefore be deduced from this result is that causality at sub-Planckian scales may be responsible for the fact that our universe is 4D [AJL04, AJL05C]. A related lesson that has been made explicit by the dynamical triangulations approach in general is the fact that once geometric excitations are 'let loose' in a non-perturbative formulation of quantum gravity, just about anything can happen. Not even the dimensionality of (what we thought of as) the space-time emerging from the quantum superposition has to come out right. At the same time one could therefore also worry that other non-perturbative quantum gravity approaches may suffer related pathologies, which have only gone undetected because one has not been able to determine expectation values like that of the Hausdorff dimension $\langle d_H \rangle$ explicitly.

The reader may by now be curious about the precise nature of the causality conditions present in the CDT approach. They are simply that each space-time appearing in the sum over geometries should have a specific form. Namely, it should be a geometric object which can be obtained by evolving a purely spatial geometry in time, in such a way that its spatial topology (the way in which space hangs together) is unchanged as a function of time. An example of a forbidden space-time is one where an initially connected space splits into two or several components, or the converse process where several components of a space reunite into a single one [Dow02]. Space-times with so-called wormholes also fall into this category and will therefore not be included in the sum over geometries. So, what is wrong with these geometries? Why do they violate causality? Let us start by explaining why these geometries are pathological *from a classical point of view*. Imagine a 3D space that undergoes a branching process as time progresses (see Figure 4.14).

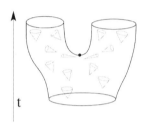

Fig. 4.14. At a branching point associated with a spatial topology change, light-cones get 'squeezed'.

Initially the space consists of a single piece (or component), which simply means that any point in the space can be reached from any other point along a continuous path. At some moment in time, the space splits into two components which then remain cut off from each other. Classically this represents a highly singular process, with nothing to suggest such processes actually occur in nature.[31] From a space-time point of view, something in these processes goes wrong with the light cone structure. The assignment of light cones to space-time points cannot be smooth, since there must be at least one point in space-time (precisely the branching point) where it cannot be decided whether a light ray arriving from the past should be continued to the future in one or the other of the two spatial components. Since the light cones define the causal structure of space-time, this is an example of a geometry where causality is violated. The classical Einstein equations simply cannot describe such topology–changing space-times. Two more things should be noted: firstly, the absence of branching points (and their time inverses, the joining points) from a Lorentzian geometry is invariant under diffeomorphisms because different

---

[31] If this were the case, we would see whole chunks of space (together with their contents) suddenly disappear.

notions of time always share the same overall direction of their 'time flow'.[32] In order to introduce branching points and their associated 'baby universes' (those parts of the universe that branch out from the 'mother universe', never to return), one would need to reverse the time flow in entire open regions of space-time, which cannot be done by an allowed coordinate transformation. Secondly, in the Euclidean theory, which has no distinguished (class of) time direction(s), one simply cannot talk about the absence or presence of analogous branching points in a meaningful (i.e., coordinate–invariant) manner.

Returning to our discussion of the *quantum theory*, the premise of CDT is to use the Lorentzian structure of its contributing geometries explicitly and exclude all space-times with topology changes and therefore acausal behavior.[33] Although classical considerations of causality have motivated a similar implementation of causality in CDT, it should be emphasized that such constraints on the path integral histories can never be derived conclusively from the classical theory. After all, the individual path integral geometries are never going to be smooth classical objects (let alone solutions to the equations of motion), nor even close to classical geometries. There is hence no obvious reason to forbid any particular quantum fluctuations of the geometry, including those that include topology changes. In principle, a quantum superposition of acausal space-times could lead to a quantum space-time where causality by some mechanism is restored dynamically, at least macroscopically. However, although this is a theoretical possibility, it is not what one has observed in the Euclidean version of DT which does not have such causality restrictions, and which goes wrong already in trying to reconstruct a 4D space. By the same token, the fact that individual path integral geometries in CDT are causal is also not by itself sufficient to guarantee that the quantum geometry it generates has again the same property. Whether this is indeed the case is not yet known, and requires a more detailed knowledge about the local geometric structure than is currently available. For example, one would like to ascertain that at a sufficiently coarse–grained level the quantum geometry possesses a well–defined light cone structure by defining and measuring suitable quantum observables.

Quantum Space-Time Generated by CDT?

The dimensionality of space-time is only one of many quantum observables one may try to evaluate in order to determine the properties of the ground state geometry generated by CDT at various length scales. It is the coarsest

---

[32] We are not considering the possibility of a complete reversion of the time flow, which exchanges past and future globally.

[33] It is rather straightforward to implement the causality conditions on the triangulated geometries of CDT. Each space-time is built from layers of fixed duration (one 'length step' in proper time), and one implements gluing rules for the simplices which ensure that no change of spatial topology can occur during the step [AJL01a].

such variable, because the dimensionality of a space-time, at least in classical differential geometry, precedes that of specifying a metric structure. Here, one must keep in mind that an innocent-looking question like 'what is the value of the metric tensor $g_{\mu\nu}$ at point x?' is among the most difficult to answer in a non-perturbative approach like ours. Firstly, although CDT histories come with a notion of proper time, they do not otherwise carry any natural coordinate system. Even if we introduced coordinate systems on the individual triangulated space-times, there is no way to mark 'the same point' simultaneously in all of them. This is a consequence of the fact that individual points do not have any physical significance in empty space; in the absence of matter there is simply nothing we could 'mark' the point x with. We are thus forced to phrase any question about local curvature properties, say, in terms of quantities that are meaningful in the context of a diffeomorphism-invariant theory, for example, $n-$point correlation functions where the location of each of the n points has been averaged over space-time.[34]

The correlation function that has been studied up to now in CDT measures the correlation between the volumes $V_{\text{space}}(\tau)$ of spatial slices (slices of constant time τ) some fixed proper–time distance $\Delta\tau$ apart, that is, a suitably normalized version of the expectation value [AJL06]

$$\langle V_{\text{space}}(0)V_{\text{space}}(\Delta\tau)\rangle = \sum_{\tau=0}^{t}\langle V_{\text{space}}(\tau)V_{\text{space}}(\tau+\Delta\tau)\rangle, \qquad (4.74)$$

where the ensemble average is taken over simplicial space-times with time extension t and with fixed four-volume [AJL04, AJL05a, AJL05C]. One piece of evidence for the four-dimensionality of space-time at large distances is the fact that in order to map the functions $\langle V_{\text{space}}(0)V_{\text{space}}(\Delta\tau)\rangle$ on top of each other for different values of the space-time volume N_4, the time distance $\Delta\tau$ has to be rescaled by the power N_4^{1/D_H}, where the 'cosmological Hausdorff dimension' is $D_H = 4$ within measuring accuracy [AJL04, AJL05C]. This means that what we would like to call a continuum 'time' really scales with the correct fraction of the total space-time volume. Such a 'canonical scaling' is what one would have expected naïvely, but is absolutely not ensured a priori in the presence of large geometric quantum fluctuations, even though the individual building blocks at the cutoff scale are 4D.[35]

Before looking at another striking result on dimensionality to have come out of CDT, let us review what else we know about the large-scale geometry of the quantum space-time dynamically generated by CDT. This concerns a result which enables us to make contact with (quantum) cosmology. Recall

---

[34] Two-point functions of this type have been measured previously in Euclidean DT [AJK93].

[35] Further, independent evidence that the volumes $V_{\text{space}}(\tau)$ of the spatial slices also scale canonically as N_4 is increased, $V_{\text{space}} \sim N_4^{3/D_H}$, with $D_H = 4$, can be found in [AJL05C].

the remarkable fact that almost every aspect of today's standard model of cosmology, describing the large–scale structure of our universe, is based on a radical truncation of (the geometric sector of) Einstein's theory to a *single* global degree of freedom, the so–called *scale factor* $a(\tau)$. It describes the linear size (or 'scale') of the universe as a function of time τ.[36] This truncation is justified if the universe is homogeneous and isotropic at the largest scales, which means that it looks the same everywhere and in all spatial directions, something that is usually assumed to be true. An entirely different question is whether one can extract information about the *quantum behavior* of the universe (for example, very close to the big bang where quantum effects should come into play) by quantizing the classically truncated system of just a single geometric variable $a(\tau)$. One may wonder whether in this way one is not missing important physics contained in the quantum fluctuations of all the local gravitational degrees of freedom which the cosmological description ignores [AJL06].

Having in hand an explicit construction of quantum geometry à la CDT where no such truncation is present, one can ask what predictions it makes for the dynamics of the scale factor, and compare those to standard quantum cosmology. The answer obtained is intriguing: it is indeed possible to extract an effective action for the scale factor from CDT by integrating out all other degrees of freedom in the full quantum theory. The resulting action takes the *same* functional form as the standard action of a 'minisuperspace' cosmology for a closed universe, *up to an overall sign* [AJL05a]. The collective effect of the local gravitational excitations seems to result in the same kind of contribution as that coming from the scale factor itself, but with the opposite sign. One way to understand this from an analogous continuum point of view may be in terms of so–called Faddeev–Popov determinants, which contribute to the effective action as a result of gauge–fixing [Lol01]. The potentially far–reaching consequences of this result for quantum cosmology are currently being explored. What has already been established is that the computer–generated quantum geometry can in the semiclassical approximation be understood as a so–called 'bounce', a particular type of solution to the Euclidean equation of motion. On the basis of this, the infamous 'wave function of the universe' $\Psi_0(a)$ [HH83, Vil03] has been computed in CDT as a function of the scale factor a [AJL05a].

However, what is also clear from the computer simulations is that the semiclassical approximation is no longer an adequate description of the observed behavior of the scale factor when the latter becomes small. This is of course to be expected and is an indicator for new quantum-gravitational effects appearing at short distances. Having gathered some nontrivial evidence that CDT's quantum geometry reproduces well-known aspects of classical general

---

[36] Recall that our present universe not only expands, but expands at an ever increasing rate, that is, both $\dot{a}(\tau) > 0$ and $\ddot{a}(\tau) > 0$. A 'big bang' or 'big crunch' corresponds classically to a singular point where $a = 0$.

relativity on sufficiently large scales, the main focus of research has to be on what the actual *quantum modifications* of this structure are. This is the place where new quantum physics will appear, and our effort will go into describing it in both a qualitative and quantitative manner.

CDT has already given us first insights into what the microstructure of quantum space-time may be. The evidence comes from yet another way of probing the effective dimensionality of space-time. The idea is to define a diffusion process (equivalently, a random walk) on the triangulated geometries in the path integral over space-times, and to deduce geometric information of the underlying quantum space-time from the behavior of the diffusion as a function of the diffusion time σ inherent to the process. The beauty of this procedure is its wide applicability, since diffusion processes cannot just be defined on smooth manifolds, but on much more general spaces, such as our triangulations and even on fractal structures [BH00]. The quantity we are interested in is the so-called *spectral dimension*, which is really the effective dimension of the carrier space 'seen' by the diffusion process. It can be extracted from the return probability $P(\sigma)$ which measures the probability of a random walk to have returned to its origin after diffusion time σ (or σ evolution steps if the diffusion is implemented discretely). For diffusion on a flat d−dimensional manifold, we have the exact relation $P(\sigma) = 1/(4\pi\sigma)^{d/2}$. For general spaces we define the spectral dimension $D_S(\sigma)$ as the logarithmic derivative[37]

$$D_S(\sigma) = -2 \frac{d \log P(\sigma)}{d \log \sigma}. \tag{4.75}$$

Note that in general this dimension will depend on σ: small values of σ probe the small-distance properties of the underlying space, and large values its large–distance geometry.[38] The spectral dimension extracted for the quantum geometry of CDT is a twofold average over the starting point of the diffusion process (which is initially peaked at a given four–simplex) and over all geometries contributing to the path integral [AJL05b, AJL05C].

What one observes is indeed a scale-dependence of the space-time dimension! At large distances it approaches the value four asymptotically, in agreement with the dimension obtained previously from scaling arguments, and in agreement with our classical expectation. However, as we probe the geometry at ever shorter distances (and *before* we enter the region where the simulations become unreliable due to discretization effects), this dimension decreases continuously to an extrapolated value of two within measuring accuracy. Such a scale–dependence has never before been observed in statistical models of quantum gravity and is a clear indication that space-time behaves highly nonclassically at short distances close to the Planck length. Further

---

[37] The complete expression for the return probability has correction terms because of the finite size of the computer-generated geometries which we are suppressing for simplicity. A more detailed discussion can be found in [AJL05C].

[38] As usual in a random walk, the linear distance probed will be of the order of $\sqrt{\sigma}$.

investigations of a number of critical exponents and dimensions associated with the geometric structure of spatial slices and 'sandwiches' (of time extension $\Delta\tau = 1$) [AJL05C] suggest the presence of a *fractal microstructure* of quantum space-time, whose details are the subject of ongoing research.

In an independent development, a similar smooth running from four to two of the spectral dimension has been derived within a renormalization group approach to quantum gravity in the continuum [LR05], which posits (and provides some evidence for) the existence of a nontrivial fixed point in the ultraviolet (i.e. short-distance) regime of quantum gravity [Reu98]. Although this coincidence by no means proves that either formulation is correct, it is nevertheless remarkable that the same unexpected result has been obtained in two very different approaches to quantum gravity. If the result can indeed be shown to be part and parcel of a viable quantum gravity theory, its implications for how we view space-time and how we compute quantum processes of the other fundamental interactions on space-time may be profound. For example, it could provide a natural ultraviolet cutoff for scattering amplitudes in high–energy physics [AJL06].

4.3 Dynamics of Fields and Strings

4.3.1 Topological Quantum Field Theory

Before we come to (super)strings, we give a brief on *topological quantum field theory* (TQFT), as developed by Ed Witten, from his original path integral point of view (see [Wit88b, LL98]). TQFT originated in 1982, when Witten rewrote classical *Morse theory* (see [II07b]) in Dick Feynman's language of quantum field theory [Wit82]. Witten's arguments made use of Feynman's path integrals and consequently, at first, they were regarded as mathematically non–rigorous. However, a few years later, A. Floer reformulated a rigorous Morse–Witten theory [Flo87] (that won a Fields medal for Witten). This trend in which some mathematical structure is first constructed by quantum field theory methods and then reformulated in a rigorous mathematical ground constitutes one of the tendencies in modern physics.

In TQFT our basic topological space is an nD Riemannian manifold M with a metric $g_{\mu\nu}$. Let us consider on it a set of fields $\{\phi_i\}$, and let $S[\phi_i]$ be a real functional of these fields which is regarded as the *action* of the theory. We consider 'operators', $O_\alpha(\phi_i)$, which are in general arbitrary functionals of the fields. In TQFT these functionals are real functionals labelled by some set of indices α carrying topological or group–theoretical data. The *vacuum expectation value* (VEV) of a product of these operators is defined as

$$\langle O_{\alpha_1} O_{\alpha_2} \cdots O_{\alpha_p} \rangle = \int [D\phi_i] O_{\alpha_1}(\phi_i) O_{\alpha_2}(\phi_i) \cdots O_{\alpha_p}(\phi_i) \exp\left(-S[\phi_i]\right).$$

A quantum field theory is considered *topological* if the following relation is satisfied:

$$\frac{\delta}{\delta g^{\mu\nu}} \langle O_{\alpha_1} O_{\alpha_2} \cdots O_{\alpha_p} \rangle = 0, \qquad (4.76)$$

i.e., if the VEV of some set of selected operators is independent of the metric $g_{\mu\nu}$ on M. If such is the case those operators are called 'observables'.

There are two ways to guarantee, at least formally, that condition (4.76) is satisfied. The first one corresponds to the situation in which both, the action $S[\phi_i]$, as well as the operators O_{α_i} are *metric independent*. These TQFTs are called of *Schwarz* type. The most important representative is *Chern–Simons gauge theory*. The second one corresponds to the case in which there exist a *symmetry*, whose infinitesimal form is denoted by δ, satisfying the following properties:

$$\delta O_{\alpha_i} = 0, \qquad T_{\mu\nu} = \delta G_{\mu\nu}, \qquad (4.77)$$

where $T_{\mu\nu}$ is the SEM–tensor of the theory, i.e.,

$$T_{\mu\nu}(\phi_i) = \frac{\delta}{\delta g^{\mu\nu}} S[\phi_i]. \qquad (4.78)$$

The fact that δ in (4.77) is a symmetry of the theory implies that the transformations $\delta \phi_i$ of the fields are such that both $\delta A[\phi_i] = 0$ and $\delta O_{\alpha_i}(\phi_i) = 0$. Conditions (4.77) lead, at least formally, to the following relation for VEVs:

$$\frac{\delta}{\delta g^{\mu\nu}} \langle O_{\alpha_1} O_{\alpha_2} \cdots O_{\alpha_p} \rangle = -\int [D\phi_i] O_{\alpha_1}(\phi_i) O_{\alpha_2}(\phi_i) \cdots O_{\alpha_p}(\phi_i) T_{\mu\nu} e^{-S[\phi_i]}$$

$$= -\int [D\phi_i] \delta \left(O_{\alpha_1}(\phi_i) O_{\alpha_2}(\phi_i) \cdots O_{\alpha_p}(\phi_i) G_{\mu\nu} \exp(-S[\phi_i]) \right) = 0, \qquad (4.79)$$

which implies that the quantum field theory can be regarded as topological. This second type of TQFTs are called of *Witten type*. One of its main representatives is the theory related to Donaldson invariants, which is a twisted version of $N = 2$ supersymmetric *Yang–Mills gauge theory*. It is important to remark that the symmetry δ must be a scalar symmetry, i.e., that its symmetry parameter must be a scalar. The reason is that, being a global symmetry, this parameter must be covariantly constant and for arbitrary manifolds this property, if it is satisfied at all, implies strong restrictions unless the parameter is a scalar.

Most of the TQFTs of *cohomological type* satisfy the relation:

$$S[\phi_i] = \delta \Lambda(\phi_i), \qquad (4.80)$$

for some functional $\Lambda(\phi_i)$. This has far–reaching consequences, for it means that the topological observables of the theory, in particular the partition function, (path integral) itself are independent of the value of the coupling constant. Indeed, let us consider for example the VEV:

$$\langle O_{\alpha_1} O_{\alpha_2} \cdots O_{\alpha_p} \rangle = \int [D\phi_i] O_{\alpha_1}(\phi_i) O_{\alpha_2}(\phi_i) \cdots O_{\alpha_p}(\phi_i) e^{-\frac{1}{g^2} S[\phi_i]}. \qquad (4.81)$$

Under a change in the coupling constant, $1/g^2 \to 1/g^2 - \Delta$, one has (assuming that the observables do not depend on the coupling), up to first–order in Δ:

$$\langle O_{\alpha_1} O_{\alpha_2} \cdots O_{\alpha_p} \rangle \longrightarrow \langle O_{\alpha_1} O_{\alpha_2} \cdots O_{\alpha_p} \rangle$$
$$+ \Delta \int [D\phi_i] \delta \left[O_{\alpha_1}(\phi_i) O_{\alpha_2}(\phi_i) \cdots O_{\alpha_p}(\phi_i) \Lambda(\phi_i) \exp\left(-\frac{1}{g^2} S[\phi_i]\right) \right]$$
$$= \langle O_{\alpha_1} O_{\alpha_2} \cdots O_{\alpha_p} \rangle. \tag{4.82}$$

Hence, observables can be computed either in the *weak coupling limit*, $g \to 0$, or in the *strong coupling limit*, $g \to \infty$.

So far we have presented a rather general definition of TQFT and made a series of elementary remarks. Now we will analyze some aspects of its structure. We begin pointing out that given a theory in which (4.77) holds one can build *correlators* which correspond to *topological invariants* (in the sense that they are invariant under deformations of the metric $g_{\mu\nu}$) just by considering the operators of the theory which are *invariant under the symmetry*. We will call these operators *observables*. In virtue of (4.79), if one of these operators can be written as a symmetry transformation of another operator, its presence in a correlation function will make it vanish. Thus we may identify operators satisfying (4.77) which differ by an operator which corresponds to a symmetry transformation of another operator. Let us denote the set of the resulting classes by $\{\Phi\}$. By restricting the analysis to the appropriate set of operators, one has that in fact,

$$\delta^2 = 0. \tag{4.83}$$

Property (4.83) has consequences on the features of TQFT. First, the *symmetry must be odd* which implies the presence in the theory of *commuting* and *anticommuting fields*. For example, the tensor $G_{\mu\nu}$ in (4.77) must be anticommuting. This is the first appearance of an odd non–spinorial field in TQFT. Those kinds of objects are standard features of *cohomological* TQFTs. Second, if we denote by Q the operator which implements this symmetry, the observables of the theory can be described as the cohomology classes of Q:

$$\{\Phi\} = \frac{\operatorname{Ker} Q}{\operatorname{Im} Q}, \qquad Q^2 = 0. \tag{4.84}$$

Equation (4.77) means that in addition to the *Poincaré group* the theory possesses a symmetry generated by an odd version of the Poincaré group. The corresponding odd generators are constructed out of the tensor $G_{\mu\nu}$ in much the same way as the ordinary Poincaré generators are built out of $T_{\mu\nu}$. For example, if P_μ represents the ordinary momentum operator, there exists a corresponding odd one G_μ such that

$$P_\mu = \{Q, G_\mu\}. \tag{4.85}$$

Now, let us discuss the structure of the Hilbert space of the theory in virtue of the symmetries that we have just described. The states of this space must correspond to representations of the algebra generated by the operators in the Poincaré groups and by Q. Furthermore, as follows from our analysis of operators leading to (4.84), if one is interested only in states $|\Psi\rangle$ leading to topological invariants one must consider states which satisfy

$$Q|\Psi\rangle = 0, \tag{4.86}$$

and two states which differ by a Q–exact state must be identified. The odd Poincaré group can be used to generate descendant states out of a state satisfying (4.86). The operators G_μ act non–trivially on the states and in fact, out of a state satisfying (4.86) we can build additional states using this generator. The simplest case consists of

$$\int_{\gamma_1} G_\mu |\Psi\rangle,$$

where γ_1 is a 1–cycle. One can verify using (4.77) that this new state satisfies (4.86):

$$Q \int_{\gamma_1} G_\mu |\Psi\rangle = \int_{\gamma_1} \{Q, G_\mu\}|\Psi\rangle = \int_{\gamma_1} P_\mu |\Psi\rangle = 0.$$

Similarly, one may construct other invariants tensoring n operators G_μ and integrating over n–cycles γ_n:

$$\int_{\gamma_n} G_{\mu_1} G_{\mu_2} \ldots G_{\mu_n} |\Psi\rangle. \tag{4.87}$$

Notice that since the operator G_μ is odd and its algebra is Poincaré–like the integrand in this expression is an exterior differential n–form. These states also satisfy condition (4.86). Therefore, starting from a state $|\Psi\rangle \in \ker Q$ we have built a set of partners or descendants giving rise to a topological multiplet. The members of a multiplet have well defined *ghost* number. If one assigns ghost number -1 to the operator G_μ the state in (4.87) has ghost number $-n$ plus the ghost number of $|\Psi\rangle$. Now, n is bounded by the dimension of the manifold X. Among the states constructed in this way there may be many which are related via another state which is Q–exact, i.e., which can be written as Q acting on some other state. Let us try to single out representatives at each level of ghost number in a given topological multiplet.

Consider an $(n-1)$–cycle which is the boundary of an nD surface, $\gamma_{n-1} = \partial S_n$. If one builds a state taking such a cycle one finds ($P_\mu = -i\partial_\mu$),

$$\int_{\gamma_{n-1}} G_{\mu_1} G_{\mu_2} \ldots G_{\mu_{n-1}} |\Psi\rangle = i \int_{S_n} P_{[\mu_1} G_{\mu_2} G_{\mu_3} \ldots G_{\mu_n]} |\Psi\rangle \tag{4.88}$$

$$= iQ \int_{S_n} G_{\mu_1} G_{\mu_2} \ldots G_{\mu_n} |\Psi\rangle,$$

i.e., it is Q–exact. The square–bracketed subscripts in (4.88) denote that all indices between them must by antisymmetrized. In (4.88) use has been made of (4.85). This result tells us that the representatives we are looking for are built out of the homology cycles of the manifold X. Given a manifold X, the homology cycles are equivalence classes among cycles, the equivalence relation being that two n−cycles are equivalent if they differ by a cycle which is the boundary of an $n+1$ surface. Thus, knowledge on the homology of the manifold on which the TQFT is defined allows us to classify the representatives among the operators (4.87). Let us assume that X has dimension d and that its homology cycles are γ_{i_n}, $(i_n = 1, ..., d_n,\ n = 0, ..., d)$, where d_n is the dimension of the n−homology group, and d the dimension of X. Then, the non–trivial partners or descendants of a given $|\Psi\rangle$ highest–ghost–number state are labelled in the following way:

$$\int_{\gamma_{i_n}} G_{\mu_1} G_{\mu_2} ... G_{\mu_n} |\Psi\rangle, \qquad (i_n = 1, ..., d_n,\ n = 0, ..., d).$$

A similar construction to the one just described can be made for fields. Starting with a field $\phi(x)$ which satisfies,

$$[Q, \phi(x)] = 0, \qquad (4.89)$$

one can construct other fields using the operators G_μ. These fields, which we call *partners* are antisymmetric tensors defined as,

$$\phi^{(n)}_{\mu_1 \mu_2 ... \mu_n}(x) = \frac{1}{n!}[G_{\mu_1}, [G_{\mu_2}...[G_{\mu_n}, \phi(x)\}...\}\}, \qquad (n = 1, ..., d).$$

Using (4.85) and (4.89) one finds that these fields satisfy the so–called *topological descent equations*:

$$d\phi^{(n)} = i[Q, \phi^{(n+1)}\},$$

where the subindices of the forms have been suppressed for simplicity, and the highest–ghost–number field $\phi(x)$ has been denoted as $\phi^{(0)}(x)$. These equations enclose all the relevant properties of the observables which are constructed out of them. They constitute a very useful tool to build the observables of the theory.

4.3.2 TQFT and Seiberg–Witten Theory

Recall that the field of low–dimensional geometry and topology [Ati88b] has undergone a dramatic phase of progress in the last decade of the 20th Century, prompted, to a large extend, by new ideas and discoveries in mathematical physics. The discovery of quantum groups [Dri86] in the study of the *Yang–Baxter equation* [Bax82] has reshaped the theory of knots and links [Jon85, RT91, ZGD91]; the study of conformal field theory and quantum *Chern–Simons theory* [Wit89] in physics had a profound impact on the theory of 3–manifolds; and most importantly, investigations of the classical Yang–Mills

(YM) theory led to the creation of the *Donaldson theory* of 4–manifolds [FU84, Don87]. Witten [Wit94] discovered a new set of invariants of 4–manifolds in the study of the Seiberg–Witten (SW) monopole equations, which have their origin in supersymmetric gauge theory. The SW theory, while closely related to Donaldson theory, is much easier to handle. Using SW theory, proofs of many theorems in Donaldson theory have been simplified, and several important new results have also been obtained [Tau90, Tau94].

In [ZOC95] a topological quantum field theory was introduced which reproduces the SW invariants of 4–manifolds. A geometrical interpretation of the 3D quantum field theory was also given.

SW Invariants and Monopole Equations

Recall that the SW monopole equations are classical field theoretical equations involving a $U(1)$ gauge field and a complex Weyl spinor on a 4D manifold. Let X denote the 4–manifold, which is assumed to be oriented and closed. If X is spin, there exist positive and negative spin bundles $S^{\pm}$ of rank two. Introduce a complex line bundle $L \to X$. Let A be a connection on L and M be a section of the product bundle $S^+ \otimes L$. Recall that the SW monopole equations read

$$\mathcal{F}^+_{kl} = -\frac{i}{2}\bar{M}\Gamma_{kl}M, \qquad D_A M = 0, \qquad (4.90)$$

where D_A is the twisted Dirac operator, $\Gamma_{ij} = \frac{1}{2}[\gamma_i, \gamma_j]$, and $\mathcal{F}^+$ represents the self–dual part of the curvature of L with connection A.

If X is not a spin manifold, then spin bundles do not exist. However, it is always possible to introduce the so called $Spin_c$ bundles $S^{\pm} \otimes L$, with L^2 being a line bundle. Then in this more general setting, the SW monopoles equations look formally the same as (4.90), but the M should be interpreted as a section of the the $Spin_C$ bundle $S^+ \otimes L$.

Denote by $\mathcal{M}$ the moduli space of solutions of the SW monopole equations up to gauge transformations. Generically, this space is a manifold. Its virtual dimension is equal to the number of solutions of the following equations

$$(d\psi)^+_{kl} + \frac{i}{2}\left(\bar{M}\Gamma_{kl}N + \bar{N}\Gamma_{kl}M\right) = 0, \qquad D_A N + \psi M = 0,$$

$$\nabla_k \psi^k + \frac{i}{2}(\bar{N}M - \bar{M}N) = 0, \qquad (4.91)$$

where A and M are a given solution of (4.90), $\psi \in \Omega^1(X)$ is a one form, $(d\psi)^+ \in \Omega^{2,+}(X)$ is the self dual part of the two form $d\psi$, and $N \in S^+ \otimes L$. The first two of the equations in (4.91) are the linearization of the monopole equations (4.90), while the last one is a gauge fixing condition. Though with a rather unusual form, it arises naturally from the dual operator governing gauge transformations

$$C : \Omega^0(X) \to \Omega^1(X) \oplus (S^+ \otimes L), \qquad \phi \mapsto (-d\phi, i\phi M).$$
Let $T : \Omega^1(X) \oplus (S^+ \otimes L) \to \Omega^0(X) \oplus \Omega^{2,+}(X) \oplus (S^- \otimes L),$

be the operator governing equation (4.91), namely, the operator which allows us to rewrite (4.91) as $T(\psi, N) = 0$. Then T is an elliptic operator, the index $\text{Ind}(T)$ of which yields the virtual dimension of $\mathcal{M}$. A straightforward application of the Atiyah–Singer index Theorem gives

$$\text{Ind}(T) = -\frac{2\chi(X) + 3\sigma(X)}{4} + c_1(L)^2,$$

where $\chi(X)$ is the Euler character of X, $\sigma(X)$ its signature index and $c_1(L)^2$ is the square of the first Chern class of L evaluated on X in the standard way.

When $\text{Ind}(T)$ equals zero, the moduli space generically consists of a finite number of points, $\mathcal{M} = \{p_t : t = 1, 2, ..., I\}$. Let ϵ_t denote the sign of the determinant of the operator T at p_t, which can be defined with mathematical rigor. Then the SW invariant of the 4–manifold X is defined by $\sum_1^I \epsilon_t$.

The fact that this is indeed an invariant(i.e., independent of the metric) of X is not very difficult to prove, and we refer to [Wit94] for details. As a matter of fact, the number of solutions of a system of equations weighted by the sign of the operator governing the equations(i.e., the analog of T) is a topological invariant in general [Wit94]. This point of view has been extensively explored by Vafa and Witten [VW94] within the framework of topological quantum field theory in connection with the so called S duality. Here we wish to explore the SW invariants following a similar line as that taken in [Wit88b, VW94].

Topological Lagrangian

Introduce a *Lie super–algebra* with an odd generator Q and two even generators U and δ obeying the following (anti)commutation relations [ZOC95]

$$[U, Q] = Q, \qquad [Q, Q] = 2\delta, \qquad [Q, \delta] = 0. \tag{4.92}$$

We will call U the ghost number operator, and Q the *BRST–operator*.

Let A be a connection of L and $M \in S^+ \otimes L$. We define the action of the super–algebra on these fields by requiring that δ coincide with a gauge transformation with a gauge parameter $\phi \in \Omega^0(X)$. The field multiplets associated with A and M furnishing representations of the super–algebra are (A, ψ, ϕ), and (M, N), where $\psi \in \Omega^1(X)$, $\phi \in \Omega^0(X)$, and N is a section of $S^+ \otimes L$. They transform under the action of the super–algebra according to

$$[Q, A_i] = \psi_i, \qquad [Q, M] = N,$$
$$[Q, \psi_i] = -\partial_i \phi, \qquad [Q, N] = i\phi M, \qquad [Q, \phi] = 0.$$

We assume that both A and M have ghost number 0, and thus will be regarded as bosonic fields when we study their quantum field theory. The ghost numbers of other fields can be read off the above transformation rules. We have that ψ and N are of ghost number 1, thus are fermionic, and ϕ is of ghost number 2 and bosonic. Note that the multiplet (A, ψ, ϕ) is what one would get in the topological field theory for Donaldson invariants except that our gauge group is $U(1)$, while the existence of M and N is a new feature. Also note that both M and ψ have the wrong statistics.

In order to construct a quantum field theory which will reproduce the SW invariants as correlation functions, anti–ghosts and Lagrangian multipliers are also required. We introduce the anti–ghost multiplet $(\lambda, \eta) \in \Omega^0(X)$, such that

$$[U, \lambda] = -2\lambda, \qquad [Q, \lambda] = \eta, \qquad [Q, \eta] = 0,$$

and the Lagrangian multipliers $(\chi, H) \in \Omega^{2,+}(X)$, and $(\mu, \nu) \in S^- \otimes L$ such that

$$[U, \chi] = -\chi, \qquad [Q, \chi] = H, \qquad [Q, H] = 0;$$
$$[U, \mu] = -\mu, \qquad [Q, \mu] = \nu, \qquad [Q, \nu] = i\phi\mu.$$

With the given fields, we construct the following functional which has ghost number -1:

$$V = \int_X \left\{ [\nabla_k \psi^k + \frac{i}{2}(\overline{N}M - \overline{M}N)]\lambda - \chi^{kl}\left(H_{kl} - \mathcal{F}^+_{kl} - \frac{i}{2}\overline{M}\Gamma_{kl}M\right) \right.$$
$$\left. - \bar{\mu}(\nu - iD_A M) - \overline{(\nu - iD_A M)}\mu \right\}, \qquad (4.93)$$

where the indices of the tensorial fields are raised and lowered by a given metric g on X, and the integration measure is the standard $\sqrt{g}d^4x$. Also, $\overline{M}$ and $\bar{\mu}$ etc. represent the Hermitian conjugate of the spinorial fields. In a formal language, $\overline{M} \in S^+ \otimes L^{-1}$ and $\bar{\mu}, \bar{\nu}, \overline{D_A M} \in S^- \otimes L^{-1}$. Following the standard procedure in constructing topological quantum field theory, we take the classical action of our theory to be [ZOC95]: $S = [Q, V]$, which has ghost number 0. One can easily show that S is also BRST invariant, i.e., $[Q, S] = 0$, thus it is invariant under the full super–algebra (4.92).

The bosonic Lagrangian multiplier fields H and ν do not have any dynamics, and so can be eliminated from the action by using their equations of motion

$$H_{kl} = \frac{1}{2}\left(\mathcal{F}^+_{kl} + \frac{i}{2}\overline{M}\Gamma_{kl}M\right), \qquad \nu = \frac{1}{2}iD_A M. \qquad (4.94)$$

Then we arrive at the following expression for the action [ZOC95]

$$S = \int_X \left\{ [-\Delta\phi + \overline{M}M\phi - i\overline{N}N]\lambda - [\nabla_k\psi^k + \frac{i}{2}(\overline{N}M - \overline{M}N)]\eta + 2i\phi\bar{\mu}\mu \right.$$
$$+ \overline{(iD_A N - \gamma.\psi M)}\mu - \bar{\mu}(iD_A N - \gamma.\psi M)$$
$$\left. - \chi^{kl}\left[\left(\nabla_k\psi^l - \nabla_l\psi^k\right)^+ + \frac{i}{2}\left(\overline{M}\Gamma_{kl}N + \overline{N}\Gamma_{kl}M\right)\right] \right\} + S_0, \qquad (4.95)$$

where S_0 is given by

$$S_0 = \int_X \left\{ \frac{1}{4}|\mathcal{F}^+|^2 + \frac{i}{2}\bar{M}\Gamma M|^2 + \frac{1}{2}|D_A M|^2 \right\}.$$

It is interesting to observe that S_0 is nonnegative, and vanishes if and only if A and M satisfy the SW monopole equations. As pointed out in [Wit94], S_0 can be rewritten as

$$S_0 = \int_X \left\{ \frac{1}{4}|\mathcal{F}^+|^2 + \frac{1}{4}|M|^4 + \frac{1}{8}R|M|^2 + g^{ij}\overline{D_i M} D_j M \right\},$$

where R is the scalar curvature of X associated with the metric g. If R is nonnegative over the entire X, then the only square integrable solution of the monopole equations (4.90) is A is a anti-self-dual connection and $M = 0$.

Quantum Field Theory

We will now investigate the quantum field theory defined by the classical action (4.95) with the path integral method. Let $\mathcal{F}$ collectively denote all the fields. The partition function of the theory is defined by [ZOC95]

$$Z = \int \mathcal{DF} \exp(-\frac{1}{e^2} S),$$

where $e \in \mathbb{R}$ is the coupling constant. The integration measure $\mathcal{DF}$ is defined on the space of all the fields. However, since S is invariant under the gauge transformations, we assume the integration over the gauge field to be performed over the gauge orbits of A. In other words, we fix a gauge for the A field using, say, a *Faddeev–Popov procedure*. This can be carried out in the standard manner, thus there is no need for us to spell out the details here. The integration measure $\mathcal{DF}$ can be shown to be invariant under the super charge Q. Also, it does not explicitly involve the metric g of X.

Let W be any operator in the theory. Its correlation function is defined by

$$Z[W] = \int \mathcal{DF} \exp(-\frac{1}{e^2} S) W.$$

It follows from the Q invariance of both the action S and the path integration measure that for any operator W,

$$Z[[Q, W]] = \int \mathcal{DF} \exp(-\frac{1}{e^2} S)[Q, W] = 0.$$

For the purpose of constructing topological invariants of the 4-manifold X, we are particularly interested in operators W which are BRST–closed,

$$[Q, W] = 0, \tag{4.96}$$

but not BRST–exact, i.e., can not be expressed as the (anti)–commutators of Q with other operators. For such a W, if its variation with respect to the metric g is $BRST$ exact,

$$\delta_g W = [Q, W'], \tag{4.97}$$

then its correlation function $Z[W]$ is a topological invariant of X (by that we really mean that it does not depend on the metric g):

$$\delta_g Z[W] = \int \mathcal{DF} \exp(-\frac{1}{e^2} S)[Q, W' - \frac{1}{e^2} \delta_g V.W] = 0.$$

In particular, the partition function Z itself is a topological invariant.

Another important property of the partition function is that it does not depend on the coupling constant e:

$$\frac{\partial Z}{\partial e^2} = \int \mathcal{DF} \frac{1}{e^4} \exp(-\frac{1}{e^2} S)[Q, V] = 0.$$

Therefore, Z can be computed *exactly* in the limit when the coupling constant goes to zero. Such a computation can be carried out in the standard way: Let A^o, M^o be a solution of the equations of motion of A and M arising from the action S. We expand the fields A and M around this classical configuration,

$$A = A^o + ea, \qquad M = M^o + em,$$

where a and m are the quantum fluctuations of A and M respectively. All the other fields do not acquire background components, thus are purely quantum mechanical. We scale them by the coupling constant e, by setting N to eN, ϕ to $e\phi$ etc.. To the order $o(1)$ in e^2, we have

$$Z = \sum_p \exp(-\frac{1}{e^2} S_{cl}^{(p)}) \int \mathcal{DF}' \exp(-S_q^{(p)}),$$

where $S_q^{(p)}$ is the quadratic part of the action in the quantum fields and depends on the gauge orbit of the classical configuration A^o, M^o, which we label by p. Explicitly [ZOC95],

$$S_q^{(p)} = \int_X \left\{ [-\Delta\phi + \overline{M}^o M^o \phi - i\overline{N}N]\lambda - [\nabla_k \psi^k + \frac{i}{2}(\overline{N}M^o - \overline{M}^o N)]\eta + 2i\phi\bar{\mu}\mu \right.$$
$$+ \overline{(iD_{A^o} N - \gamma.\psi M^o)}\mu - \bar{\mu}(iD_{A^o} N - \gamma.\psi M^o)$$
$$- \chi^{kl}\left[\left(\nabla_k \psi^l - \nabla_l \psi^k\right)^+ + \frac{i}{2}\left(\overline{M}^o \Gamma_{kl} N + \overline{N} \Gamma_{kl} M^o\right)\right]$$
$$\left. + \frac{1}{4}|f^+ + \frac{i}{2}(\bar{m}\Gamma M^o + \overline{M}^o \Gamma m)|^2 + \frac{1}{2}|iD_{A^o} m + \gamma.aM^o|^2 \right\},$$

with f^+ the self–dual part of $f = da$. The classical part of the action is given by $S_{cl}^{(p)} = S_0|_{A=A^o, M=M^o}$. The integration measure $\mathcal{DF}'$ has exactly the same form as $\mathcal{DF}$ but with A replaced by a, and M by m, $\bar{M}$ by $\bar{m}$ respectively. Needless to say, the summation over p runs through all gauge classes of classical configurations.

Let us now examine further features of our quantum field theory. A gauge class of classical configurations may give a non–zero contribution to the partition function in the limit $e^2 \to 0$ only if $S_{cl}^{(p)}$ vanishes, and this happens if and only if A^o and M^o satisfy (4.90). Therefore, the SW monopole equations are recovered from the quantum field theory.

The equations of motion of the fields ψ and N in the semi–classical approximation can be easily derived from the quadratic action $S_q^{(p)}$, solutions of which are the zero modes of the quantum fields ψ and N. The equations of motion read

$$(d\psi)_{kl}^+ + \frac{i}{2}\left(\bar{M}^o \Gamma_{kl} N + \bar{N}\Gamma_{kl} M^o\right) = 0, \qquad D_{A^o} N + \gamma.\psi M^o = 0,$$

$$\nabla_k \psi^k + \frac{i}{2}(\bar{N}M - \bar{M}N) = 0. \qquad (4.98)$$

Note that they are exactly the same equations which we have already discussed in (4.91). The first two equations are the linearization of the monopole equations, while the last is a 'gauge fixing condition' for ψ. The dimension of the space of solutions of these equations is the virtual dimension of the moduli space $\mathcal{M}$. Thus, within the context of our quantum field theoretical model, the virtual dimension of $\mathcal{M}$ is identified with the number of the zero modes of the quantum fields ψ and N.

For simplicity we assume that there are no zero modes of ψ and N, i.e., the moduli space is zero–dimensional. Then no zero modes exist for the other two fermionic fields χ and μ. To compute the partition function in this case, we first observe that the quadratic action $S_q^{(p)}$ is invariant under the supersymmetry obtained by expanding Q to first order in the quantum fields around the monopole solution A^o, M^o (equations of motion for the nonpropagating fields H and ν should also be used.). This supersymmetry transforms the set of 8 real bosonic fields (each complex field is counted as two real ones; the a_i contribute 2 upon gauge fixing.) and the set of 16 fermionic fields to each other. Thus at a given monopole background we get [ZOC95]

$$\int \mathcal{DF}' \exp(-S_q^{(p)}) = \frac{\text{Pfaff}(\nabla_{\mathcal{F}})}{|\text{Pfaff}(\nabla_{\mathcal{F}})|} = \epsilon^{(p)},$$

where $\epsilon^{(p)}$ is $+1$ or -1. In the above equation, $\nabla_{\mathcal{F}}$ is the skew symmetric first order differential operator defining the fermionic part of the action $S_q^{(p)}$, which can be read off from $S_q^{(p)}$ to be $\nabla_{\mathcal{F}} = \begin{pmatrix} 0 & T \\ -T^* & 0 \end{pmatrix}$. Therefore, $\epsilon^{(p)}$ is the sign of the determinant of the elliptic operator T at the monopole background A^o, M^o, and the partition function $Z = \sum_p \epsilon^{(p)}$ coincides with the SW invariant of the 4–manifold X.

When the dimension of the moduli space $\mathcal{M}$ is greater than zero, the partition function Z vanishes identically, due to integration over zero modes of the fermionic fields. In order to get any non trivial topological invariants for the underlying manifold X, we need to examine correlations functions of operators satisfying equations (4.96) and (4.97). A class of such operators can be constructed following the standard procedure [Wit94]. We define the following set of operators

$$W_{k,0} = \frac{\phi^k}{k!}, \qquad W_{k,1} = \psi W_{k-1,0},$$
$$W_{k,2} = \mathcal{F} W_{k-1,0} - \frac{1}{2} \psi \wedge \psi W_{k-2,0}, \qquad (4.99)$$
$$W_{k,3} = \mathcal{F} \wedge \psi W_{k-2,0} - \frac{1}{3!} \psi \wedge \psi \wedge \psi W_{k-3,0},$$
$$W_{k,4} = \frac{1}{2} \mathcal{F} \wedge \mathcal{F} W_{k-2,0} - \frac{1}{2} \mathcal{F} \wedge \psi \wedge \psi W_{k-3,0} - \frac{1}{4!} \psi \wedge \psi \wedge \psi \wedge \psi W_{k-4,0}.$$

These operators are clearly independent of the metric g of X. Although they are not BRST invariant except for $W_{k,0}$, they obey the following equations [ZOC95]

$$dW_{k,0} = -[Q, W_{k,1}], \qquad dW_{k,1} = [Q, W_{k,2}],$$
$$dW_{k,2} = -[Q, W_{k,3}], \qquad dW_{k,3} = [Q, W_{k,4}], \qquad dW_{k,4} = 0,$$

which allow us to construct BRST invariant operators from the the W's in the following way: Let X_i, $i = 1, 2, 3$, $X_4 = X$, be compact manifolds without boundary embedded in X. We assume that these submanifolds are homologically nontrivial. Define

$$\widehat{O}_{k,0} = W_{k,0}, \qquad \widehat{O}_{k,i} = \int_{X_i} W_{k,i}, \qquad (i = 1, 2, 3, 4). \qquad (4.100)$$

As we have already pointed out, $\widehat{O}_{k,0}$ is BRST invariant. It follows from the descendent equations that

$$[Q, \widehat{O}_{k,i}] = \int_{X_i} [Q, W_{k,i}] = \int_{X_i} dW_{k,i-1} = 0.$$

Therefore the operators $\widehat{O}$ indeed have the properties (4.96) and (4.97). Also, for the boundary ∂K of an $i+1$D manifold K embedded in X, we have

$$\int_{\partial K} W_{k,i} = \int_K dW_{k,i} = [Q, \int_K W_{k,i+1}],$$

is BRST trivial. The correlation function of $\int_{\partial K} W_{k,i}$ with any BRST invariant operator is identically zero. This in particular shows that the $\widehat{O}$'s only depend on the homological classes of the submanifolds X_i.

Dimensional Reduction and 3D Field Theory

In this subsection we dimensionally reduce the quantum field theoretical model for the SW invariant from 4D to 3D, thus to get a new topological quantum field theory defined on 3–manifolds. Its partition function yields a 3–manifold invariant, which can be regarded as the SW version of Casson's invariant [AM90, Tau94].

We take the 4-manifold X to be of the form $Y \times [0, 1]$ with Y being a compact 3–manifold without boundary. The metric on X will be taken to be

$$(ds)^2 = (dt)^2 + g_{ij}(x)dx^i dx^j,$$

where the 'time' t–independent $g(x)$ is the Riemannian metric on Y. We assume that Y admits a spin structure which is compatible with the $Spin_c$ structure of X, i.e., if we think of Y as embedded in X, then this embedding induces maps from the $Spin_c$ bundles $S^\pm \otimes L$ of X to $\tilde{S} \otimes L$, where $\tilde{S}$ is a spin bundle and L is a line bundle over Y.

To perform the dimensional reduction, we impose the condition that all fields are t in dependent. This leads to the following action [ZOC95]

$$S = \int \sqrt{g} d^3x \left\{ [-\Delta\phi + \overline{M}M\phi - i\overline{N}N]\lambda - [\nabla_k \psi^k + \frac{i}{2}(\overline{N}M - \overline{M}N)]\eta + 2i\phi\bar{\mu}\mu \right.$$
$$+ \overline{[i(D_A + b)N - (\sigma.\psi - \tau)M]}\mu - \bar{\mu}[i(D_A + b)N - (\sigma.\psi - \tau)M]$$
$$- 2\chi^k \left[-\partial_k \tau + *(\nabla\psi)_k - \overline{M}\sigma_k N - \overline{N}\sigma_k M \right]$$
$$\left. + \frac{1}{4}|*\mathcal{F} - \partial b - \overline{M}\sigma M|^2 + \frac{1}{2}|(D_A + b)M|^2 \right\}, \qquad (4.101)$$

where the k is a 3D index, and σ_k are the Pauli matrices. The fields $b, \tau \in \Omega^0(Y)$ respectively arose from A_0 and ψ_0 of the 4D theory, while the meanings of the other fields are clear. The BRST symmetry in 4D carries over to the 3D theory. The BRST transformations rules for (A_i, ψ_i, ϕ), $i = 1, 2, 3$, (M, N), and (λ, η) are the same as before, but for the other fields, we have

$$[Q, b] = \tau, \qquad [Q, \tau] = 0,$$
$$[Q, \chi_k] = \frac{1}{2}\left(*\mathcal{F}_k - \partial_k b - \overline{M}\sigma_k M\right),$$
$$[Q, \mu] = \frac{1}{2}i(D_A + b)M.$$

The action S is cohomological in the sense that $S = [Q, V_3]$, with V_3 being the dimensionally reduced version of V defined by (4.93), and $[Q, S] = 0$. Thus it gives rise to a topological field theory upon quantization. The partition function of the theory

$$Z = \int \mathcal{D}\mathcal{F} \exp(-\frac{1}{e^2}S),$$

can be computed exactly in the limit $e^2 \to 0$, as it is coupling constant independent. We have, as before,

$$Z = \sum_p \exp(-\frac{1}{e^2} S_{cl}^{(p)}) \int \mathcal{DF}' \exp(-S_q^{(p)}),$$

where $S_q^{(p)}$ is the quadratic part of S expanded around a classical configuration with the classical parts for the fields A, M, b being A^o, M^o, b^o, while those for all the other fields being zero. The classical action $S_{cl}^{(p)}$ is given by

$$S_{cl}^{(p)} = \int_Y \left\{ \frac{1}{4} |*\mathcal{F}^o - db^o - \bar{M}^o \sigma M^o|^2 + \frac{1}{2} |(D_{A^o} + b^o) M^o|^2 \right\},$$

which can be rewritten as [ZOC95]

$$S_{cl}^{(p)} = \int_Y \left\{ \frac{1}{4} |*\mathcal{F}^o - \bar{M}^o \sigma M^o|^2 + \frac{1}{2} |D_{A^o} M^o|^2 + \frac{1}{2} |db^o|^2 + \frac{1}{2} |b^o M^o|^2 \right\}.$$

In order for the classical configuration to have non–vanishing contributions to the partition function, all the terms in $S_{cl}^{(p)}$ should vanish separately. Therefore,

$$*\mathcal{F}^o - \bar{M}^o \sigma M^o = 0, \qquad D_{A^o} M^o = 0, \qquad b^o = 0, \qquad (4.102)$$

where the last condition requires some explanation. When we have a trivial solution of the equations (4.102), it can be replaced by the less stringent condition $db^o = 0$. However, in a more rigorous treatment of the problem at hand, we in general perturb the equations (4.102), then the trivial solution does not arise.

Let us define an operator

$$\tilde{T}: \Omega^0(Y) \oplus \Omega^1(Y) \oplus (\tilde{S} \otimes L) \to \Omega^0(Y) \oplus \Omega^1(Y) \oplus (\tilde{S} \otimes L),$$
$$(\tau, \psi, N) \mapsto (-d^*\psi + \frac{i}{2}(\bar{N}M - \bar{M}N), \quad *(d\psi) - d\tau - \bar{N}\sigma M - \bar{M}\sigma N,$$
$$iD_A N - (\sigma.\psi - \tau) M), \qquad (4.103)$$

where the complex bundle $\tilde{S} \otimes L$ should be regarded as a real one with twice the rank. This operator is self–adjoint, and is also obviously elliptic. We will assume that it is Fredholm as well. In terms of $\tilde{T}$, the equations of motion of the fields χ^i and μ can be expressed as [ZOC95] $\tilde{T}^{(p)}(\tau, \psi, N) = 0$, where $\tilde{T}^{(p)}$ is the opeartor $\tilde{T}$ with the background fields (A^o, M^o) belonging to the gauge class p of classical configurations.

When the kernel of $\tilde{T}$ is zero, the partition function Z does not vanish identically. An easy computation leads to $Z = \sum_p \epsilon^{(p)}$, where the sum is over all gauge inequivalent solutions of (4.102), and $\epsilon^{(p)}$ is the sign of the determinant of $\tilde{T}^{(p)}$.

A rigorous definition of the sign of the $\det(\tilde{T})$ can be devised. However, if we are to compute only the absolute value of Z, then it is sufficient to know the sign of $\det(\tilde{T})$ relative to a fixed gauge class of classical configurations. This can be achieved using the $mod-2$ spectral flow of a family of Fredholm operators $\tilde{T}_t$ along a path of solutions of (4.102). More explicitly, let (A^o, M^o) belong to the gauge class of classical configurations p, and $(\tilde{A}^o, \tilde{M}^o)$ in $\tilde{p}$. We consider the solution of the SW equation on $X = Y \times [0, 1]$ with $A_0 = 0$ and also satisfying the following conditions

$$(A, M)|_{t=0} = (A^o, M^o), \qquad (A, M)|_{t=1} = (\tilde{A}^o, \tilde{M}^o).$$

Using this solution in $\tilde{T}$ results in a family of *Fredholm operators*, which has zero kernels at $t = 0$ and 1. The spectral flow of $\tilde{T}_t$, denoted by $q(p, \tilde{p})$, is defined to be the number of eigenvalues which cross zero with a positive slope minus the number which cross zero with a negative slope. This number is a well defined quantity, and is given by the index of the operator $\frac{\partial}{\partial t} - \tilde{T}_t$. In terms of the spectral flow, we have [ZOC95]

$$\frac{\det(\tilde{T}^{(p)})}{\det(\tilde{T}^{(\tilde{p})})} = (-1)^{q(p,\tilde{p})}.$$

Equations (4.102) can be derived from the functional

$$S_{c-s} = \frac{1}{2} \int_Y A \wedge \mathcal{F} + i \int_Y \sqrt{g} d^3 x \overline{M} D_A M.$$

(It is interesting to observe that this is almost the standard Lagrangian of a $U(1)$ Chern–Simons theory coupled to spinors, except that we have taken M to have bosonic statistics.) S_{c-s} is gauge invariant modulo a constant arising from the Chern–Simons term upon a gauge transformation. Therefore, $(\frac{\delta S_{c-s}}{\delta A}, \frac{\delta S_{c-s}}{\delta M})$ defines a vector field on the quotient space of all $U(1)$ connections $\mathcal{A}$ tensored with the $\tilde{S} \times L$ sections by the $U(1)$ gauge group G, i.e., $\mathcal{W} = (\mathcal{A} \times (\tilde{S} \otimes L))/G$. Solutions of (4.102) are zeros of this vector field, and $\tilde{T}^{(p)}$ is the Hessian at the point $p \in \mathcal{W}$. Thus the partition Z is nothing else but the Euler character of $\mathcal{W}$. This geometrical interpretation will be spelt out more explicitly in the next subsection by re–interpreting the theory using the *Mathai–Quillen formula* [MQ86].

Geometrical Interpretation

To elucidate the geometric meaning of the 3D theory obtained in the last section, we now cast it into the framework of Atiyah and Jeffrey [AJ90]. Let us briefly recall the geometric set up of the *Mathai–Quillen formula* as reformulated in [AJ90]. Let P be a Riemannian manifold of dimension $2m + \dim G$, and G be a compact Lie group acting on P by isometries. Then $P \to P/G$ is a principle bundle. Let V be a $2m$ dimensional real vector space, which furnishes a representation $G \to SO(2m)$. Form the associated vector bundle $P \times_G V$. Now the Thom form of $P \times_G V$ can be expressed [ZOC95]

$$U = \frac{\exp(-x^2)}{(2\pi)^{\dim G} \pi^m} \int \exp\left\{\frac{i\chi\phi\chi}{4} + i\chi dx - i\langle \delta\nu, \lambda\rangle \right. \\ \left. - \langle \phi, R\lambda\rangle + \langle \nu, \eta\rangle \right\} \mathcal{D}\eta \mathcal{D}\chi \mathcal{D}\phi \mathcal{D}\lambda, \tag{4.104}$$

where $x = (x^1, ..., x^{2m})$ is the coordinates of V, ϕ and λ are bosonic variables in the Lie algebra g of G, and η and χ are Grassmannian variables valued in the Lie algebra and the tangent space of the fiber respectively. In the above equation, C maps any $\eta \in g$ to the element of the vertical part of TP generated by η; ν is the g-valued one form on P defined by $\langle \nu(\alpha), \eta\rangle = \langle \alpha, C(\eta)\rangle$, for all vector fields α; and $R = C^*C$. Also, δ is the exterior derivative on P.

Now we choose a G invariant map $s: P \to V$, and pull back the Thom form U. Then the top form on P in s^*U is the Euler class. If $\{\delta p\}$ forms a basis of the cotangent space of P (note that ν and δs are one forms on P), we replace it by a set of Grassmannian variables $\{\psi\}$ in s^*U, then intergrate them away. We arrive at

$$\Upsilon = \frac{1}{(2\pi)^{\dim G} \pi^m} \int \exp\left\{-|s|^2 + \frac{i\chi\phi\chi}{4} + i\chi\delta s - i\langle \delta\nu, \lambda\rangle \right. \\ \left. - \langle \phi, R\lambda\rangle + \langle \psi, C\eta\rangle \right\} \mathcal{D}\eta \mathcal{D}\chi \mathcal{D}\phi \mathcal{D}\lambda \mathcal{D}\psi, \tag{4.105}$$

the precise relationship of which with the Euler character of $P \times_G V$ is

$$\int_P \Upsilon = \text{Vol}(G)\chi(P \times_G).$$

It is rather obvious that the action S defined by (4.95) for the 4D theory can be interpreted as the exponent in the integrand of (4.105), if we identify P with $\mathcal{A} \times \Gamma(W^+)$, and V with $\Omega^{2,+}(X) \times \Gamma(W^-)$, and set $s = (\mathcal{F}^+ + \frac{i}{2}\bar{M}\Gamma M, D_A M)$. Here $\mathcal{A}$ is the space of all $U(1)$ connections of $\det(W^+)$, and $\Gamma(W^\pm)$ are the sections of $S^\pm \otimes L$ respectively.

For the 3D theory, we wish to show that the partition function yields the Euler number of $\mathcal{W}$. However, the tangent bundle of $\mathcal{W}$ cannot be regarded as an associated bundle with the principal bundle, for which for the formulae (4.104) or (4.105) can readily apply, some further work is required.

Let P be the principal bundle over P/G, V, V' be two orthogonal representations of G. Suppose there is an embedding from $P \times_G V'$ to $P \times_G V$ via a G-map $\gamma(p): V' \to V$ for $p \in P$. Denote the resulting quotient bundle as E. In order to derive the Thom class for E, one needs to choose a section of E, or equivalently, a G-map $s: P \to V$ such that $s(p) \in (\text{Im}\gamma(p))^\perp$. Then the Euler class of E can be expressed as $\pi_*\rho^*U$, where U is the Thom class of $P \times_G V$, ρ is a G-map: $P \times V' \to P \times V$ defined by

$$\rho(p, \tau) = (p, \gamma(p)\tau + s(p)),$$

and π_* is the integration along the fiber for the projection $\pi : P \times V' \to P/G$. Explicitly, [ZOC95]

$$\pi_*\rho^*(U) = \int \exp\{-|\gamma(p)\tau + s(p)|^2 + i\chi\phi\chi + i\chi\delta(\gamma(p)\tau + s(p)) - i\langle\delta\nu,\lambda\rangle - \langle\phi, R\lambda\rangle + \langle\nu, C\eta\rangle\} \mathcal{D}\chi\mathcal{D}\phi\mathcal{D}\tau\mathcal{D}\eta\mathcal{D}\lambda. \quad (4.106)$$

Consider the exact sequence

$$0 \longrightarrow (\mathcal{A} \times \Gamma(W)) \times_G \Omega^0(Y) \xrightarrow{j} (\mathcal{A} \times \Gamma(W)) \times_G (\Omega^1(Y) \times \Gamma(W)),$$

where $j_{(A,M)} : b \mapsto (-db, bM)$ (assuming that $M \neq 0$). Then the tangent bundle of $\mathcal{A} \times_G \Gamma(W)$ can be Regarded as the quotient bundle

$$(\mathcal{A} \times \Gamma(W)) \times_G (\Omega^1(Y) \times \Gamma(W))/\mathrm{Im}(j).$$

We define a vector field on $\mathcal{A} \times_G \Gamma(W)$ by

$$s(A, M) = (*\mathcal{F}_A - \bar{M}\sigma M, D_A M),$$

which lies in $\mathrm{Im}(j)^\perp$:

$$\int_Y (*\mathcal{F}_A - \bar{M}\sigma M) \wedge *(-db) + \int_Y \sqrt{g}d^3x \langle D_A M, bM\rangle = 0, \quad (4.107)$$

where we have used the short hand notation $\langle M_1, M_2\rangle = \frac{1}{2}(\bar{M}_1 M_2 + \bar{M}_2 M_1)$.

Formally applying the formula (4.106) to the present infinite–dimensional situation, we get the *Euler class* $\pi_*\rho^*(U)$ for the tangent bundle $T(\mathcal{A} \times_G \Gamma(W))$, where ρ is the $\mathcal{G}$–invariant map ρ is defined by

$$\rho : \quad \Omega^0(Y) \longrightarrow \Omega^1(Y) \times \Gamma(W), \qquad \rho(b) = (-db + *\mathcal{F}_A - \bar{M}\sigma M, (D_A + b)M),$$

π is the projection $(\mathcal{A} \times \Gamma(W)) \times_G \Omega^0(Y) \longrightarrow \mathcal{A} \times_G \Gamma(W)$, and π_* signifies the integration along the fiber. Also U is the Thom form of the bundle

$$(\mathcal{A} \times \Gamma(W)) \times_G (\Omega^1(Y) \times \Gamma(W)) \longrightarrow \mathcal{A} \times_G \Gamma(W).$$

To get a concrete feel about U, we need to explain the geometry of this bundle. The metric on Y and the Hermitian metric $\langle .\,, .\rangle$ on $\Gamma(W)$ naturally define a connection. The *Maurer–Cartan connection* on $\mathcal{A} \longrightarrow \mathcal{A}/G$ is flat while the Hermitian connection on has the curvature $i\phi\mu \wedge \bar{\mu}$. This gives the expression of term $i(\chi,\mu)\phi(\chi,\mu)$ in (4.105) in our case.

In our infinite–dimensional setting, the map C is given by

$$C : \quad \Omega^0(Y) \longrightarrow T_{(A,M)}(\mathcal{A} \times \Gamma(W)), \qquad C(\eta) = (-d\eta, i\eta M),$$

and its dual is given by

$$C^*: \quad \Omega^1(Y) \times \Gamma(W) \longrightarrow \Omega^0(Y), \qquad C^*(\psi, N) = -d^*\psi + \langle N, \mathrm{i}M \rangle.$$

The one form $\langle \nu, \eta \rangle$ on $\mathcal{A} \times \Gamma(W)$ takes the value

$$\langle (\psi, N), C\eta \rangle = \langle -d^*\psi, \eta \rangle + \langle N, \mathrm{i}M \rangle \eta$$

on the vector field (ψ, N). We also easily get $R(\lambda) = -\Delta\lambda + \langle M, M \rangle \lambda$, where $\Delta = d^*d$. The $\langle \delta\nu, \lambda \rangle$ is a two form on $\mathcal{A} \times \Gamma(W)$ whose value on $(\psi_1, N_1), (\psi_2, N_2)$ is $-\langle N_1, N_2 \rangle \lambda$.

Combining all the information together, we arrive at the following formula,

$$\pi_* \rho^*(U) = \int \exp\left\{-\frac{1}{2}|\rho|^2 + \mathrm{i}(\chi, \mu)\delta\rho + 2\mathrm{i}\phi\mu\bar{\mu}\right.$$
$$+ \langle \Delta\phi, \lambda \rangle - \phi\lambda\langle M, M \rangle + \mathrm{i}\langle N, N \rangle \lambda$$
$$\left. + \langle \nu, \eta \rangle \right\} \mathcal{D}\chi\mathcal{D}\phi\mathcal{D}\lambda\mathcal{D}\eta\mathcal{D}b. \qquad (4.108)$$

Note that the 1–form $\mathrm{i}(\chi, \mu)\delta\rho$ on $\mathcal{A} \times \Gamma(W) \times \Omega^0(Y)$ contacted with the vector field (ϕ, N, b) leads to

$$2\chi^k \left[-\partial_k \tau + *(\nabla\psi)_k - \bar{M}\sigma_k N - \bar{N}\sigma_k M \right] + 2\langle \mu, [\mathrm{i}(D_A + b)N - (\sigma.\psi - \tau)M] \rangle;$$

and the relation (4.107) gives $|\rho|^2 = |*\mathcal{F} - \bar{M}\sigma M|^2 + |db|^2 + |D_A M|^2 + b^2|M|^2$. Finally we get the *Euler character*

$$\pi_* \rho^*(U) = \int \exp(-S) \mathcal{D}\chi \mathcal{D}\phi \mathcal{D}\lambda \mathcal{D}\eta \mathcal{D}b, \qquad (4.109)$$

where S is the action (4.101) of the 3D theory defined on the manifold Y.

Integrating (4.109) over $\mathcal{A} \times_G \Gamma(W)$ leads to the Euler number

$$\sum_{[(A,M)]:s(A,M)=0} \epsilon^{(A,M)},$$

which coincides with the partition function Z of our 3D theory.

4.3.3 Stringy Actions and Amplitudes

Now we give a brief review of modern path–integral methods in superstring theory (mainly following [DEF99]). Recall that the fundamental quantities in quantum field theory (QFT) are the transition amplitudes $Amp : IN \Longrightarrow OUT$, describing processes in which a number IN of incoming particles scatter to produce a number OUT of outgoing particles. The square modulus of the transition amplitude yields the *probability* for this process to take place.

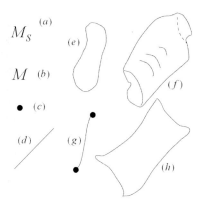

Fig. 4.15. Basic geometrical objects of string theory: (*a*) a space with fixed time; (*b*) a space–time picture; (*c*) a point–particle; (*d*) a world–line of a point–particle; (*e*) a closed string; (*f*) a world–sheet of a closed string; (*g*) an open string; (*h*) a world–sheet of an open string.

Strings

Recall that in string theory, *elementary particles* are not described as 0–dimensional *points*, but instead as 1D *strings*. If M_s and $M(\sim \mathbb{R} \times M_s)$ denote the 3D space and 4D space–time manifolds respectively, then we picture strings as in Figure 4.15.

While the point–particle sweeps out a 1D *world–line*, the string sweeps out a *world–sheet*, i.e., a 2D real surface. For a *free string*, the topology of the world–sheet is a *cylinder* (in the case of a closed string) or a *sheet* (for an open string).

Roughly, different elementary particles correspond to different *vibration modes* of the string just as different minimal notes correspond to different vibrational modes of musical string instruments.

It turns out that the physical size of strings is set by *gravity*, more precisely the *Planck length* $\ell_P \sim 10^{-33}$ cm. This scale is so small that we effectively only see point–particles at our distance scales. Thus, for length scales much larger than ℓ_P, we expect to recover a QFT–description of point–particles, plus typical *string corrections* that represent physics at the Planck scale.

Interactions

While the string itself is an extended 1D object, the fundamental string interactions are *local*, just as for point–particles. The interaction takes place when strings overlap in space at the same time. In case of *closed string theories* the interactions have a form depicted in Figure 4.16, while in case of *open string theories* the interactions have a form depicted in Figure 4.17. Other interactions result from combining the interactions defined above.

530   4 Nonlinear Dynamics of Path Integrals

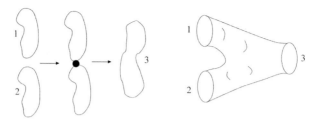

Fig. 4.16. Interactions in closed string theories (left 2D–picture and right 3D–picture).

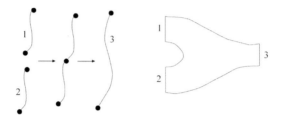

Fig. 4.17. Interactions in open string theories (left 2D–picture and right 3D–picture).

In point–particle theories, the fundamental interactions are read off from the QFT–Lagrangian. An interaction occurs at a *geometrical point*, where the world–lines join and cease to be a manifold. In *Lorentz–invariant theories* (where manifold M is a flat Minkowski space–time), the interaction point is Lorentz–invariant. To specify how the point–particles interact, additional data must be supplied at the interaction point, giving rise to *many possible* distinct quantum field theories.

In string theory, the interaction point *depends* upon the Lorentz frame chosen to observe the process. In the Figure above, equal time slices are indicated from the point of view of two different Lorentz frames, schematically indicated by t and t'. The closed string interaction, as seen from frames t and t', occurs at times t_2 and t'_2 and at (distinct) points P and P' respectively.

Lorentz invariance of interaction forbids that any point on the world–sheet be singled out as interaction point. Instead, the interaction results purely from the joining and splitting of strings. While free closed strings are characterized by their topology being that of a cylinder, interacting strings are characterized by the fact that their associated world–sheet is connected to at least 3 strings, incoming and/or outgoing.

As a result, the free string determines the nature of the interactions completely, leaving only the string coupling constant undetermined.

The *orientation* is an additional structure of closed strings, dividing them into two categories: (i) *oriented strings*, in which all world–sheets are assumed to be orientable; and (ii) non–oriented strings, in which world–sheets are non–orientable, such as the Möbius strip, Klein bottle, etc.

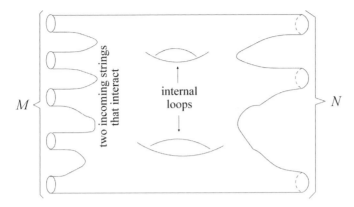

Fig. 4.18. Boundary components and handles of closed oriented system of M incoming strings, interacting through internal loops, to produce N outgoing strings. Note the striking similarity with MIMO–systems of *nonlinear control theory*, with M input processes and N output processes (see [II06b]).

Loop Expansion – Topology of Closed Surfaces

For simplicity, here we consider closed oriented strings only, so that the associated world–sheet is also oriented. A general string configuration describing the process in which M incoming strings interact and produce N outgoing strings looks at the topological level like a closed surface with $M + N = E$ *boundary components* and any number of *handles* (see Figure 4.18). This picture is a kind of topological generalization of nonlinear control MIMO–systems with M inputs, N outputs X states.

The internal loops may arise when virtual particle pairs are produced, just as in quantum field theory. For example, a *Feynman diagram* in quantum field theory that involves a loop is shown in Figure 4.19 together with the corresponding string diagram.

Surfaces associated with closed oriented strings have two topological invariants: (i) the *number of boundary components* $E = M + N$ (which may be shrunk to punctures, under certain conditions), and (ii) the number h of *handles* on the surface, which equals the surface *genus*.

When $E = 0$, we just have the topological classification of compact oriented surfaces without boundary. Rendering $E > 0$ is achieved by removing E discs from the surface.

Recall that in QFT, an expansion in powers of Planck's constant $\hbar$ yields an expansion in the number of loops of the associated Feynman diagram, for a given number of external states:

532   4 Nonlinear Dynamics of Path Integrals

Fig. 4.19. A QFT Feynman diagram that involves an internal loop (left), with the corresponding string diagram (right).

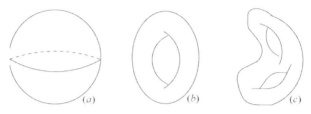

Fig. 4.20. Number h of handles on the surface of closed oriented strings, which equals the string–surface genus: (a) $h = 0$ for sphere S^2; (b) $h = 1$ for torus T^2; (c) $h = 2$ for string–surfaces with higher genus, etc.

$$\hbar^{E+h-1} = \begin{cases} \hbar & \text{for every propagator} \\ \hbar^{-1} & \text{for every vertex} \\ -1 & \text{for overall momentum conservation} \end{cases}$$

Thus, in string theory we expect that, for a given number of external strings E, the topological expansion genus by genus should correspond to a *loop* expansion as well.

Recall that in QFT, there are in general many Feynman diagrams that correspond to an amplitude with a given number of external particles and a given number of loops. For example, for $E = 4$ external particles and $h = 1$ loop in ϕ^3 theory are given in Figure 4.21, together with the same process in string theory (for closed oriented strings), where it is described by just a single diagram (right).

Much of recent interest has been focused on the so–called D–branes. A D–*brane* is a submanifold of space–time with the property that strings can *end* or *begin* on it.

4.3.4 Transition Amplitudes for Strings

The only way we have today to define string theory is by giving a *rule* for the evaluation of transition amplitudes, order by order in the loop expansion, i.e., genus by genus. The rule is to assign a relative weight to a given configuration and then to sum over all configurations [DEF99]. To make this more precise, we first describe the system's configuration manifold M (see Figure 4.22).

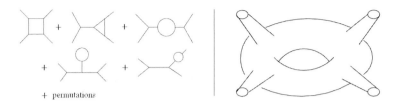

Fig. 4.21. Feynman QFT–diagrams for ϕ^3 theory with $E = 4$ external particles and $h = 1$ loop (left), and a single corresponding string diagram (right). In this way the usual Feynman diagrams of quantum field theory are generalized by arbitrary Riemannian surfaces.

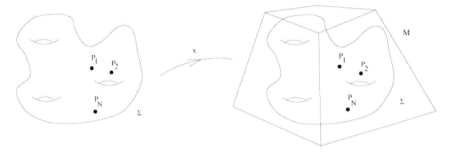

Fig. 4.22. The embedding map x from the reference surface Σ into the pseudo–Riemannian configuration manifold M (see text for explanation).

We assume that Σ and M are smooth manifolds, of dimensions 2 and n respectively, and that x is a continuous map from Σ to M. If ξ^m, (for $m = 1, 2$), are local coordinates on Σ and x^μ, ($\mu = 1, \ldots, n$), are local coordinates on M then the map x may be described by functions $x^\mu(\xi^m)$ which are continuous.

To each system configuration we can associate a weight $e^{-S[x, \Sigma, M]}$, (for $S \in \mathbb{C}$) and the transition amplitude Amp for specified external strings (incoming and outgoing) is get by summing over all surfaces Σ and all possible maps x,

$$Amp = \sum_{surfaces\ \Sigma} \sum_x e^{-S[x, \Sigma, M]}.$$

We now need to specify each of these ingredients:

(1) We assume M to be an nD Riemannian manifold, with metric g. A special case is flat Euclidean space–time $\mathbb{R}^n$. The space–time metric is assumed *fixed*.

$$ds^2 = (dx, dx)_g = g_{\mu\nu}(x) dx^\mu \otimes dx^\nu.$$

(2) The metric g on M induces a metric on Σ: $\gamma = x^*(g)$,

$$\gamma = \gamma_{mn} d\xi^m \otimes d\xi^n, \qquad \gamma_{mn} = g_{\mu\nu} \frac{\partial x^\mu}{\partial \xi^m} \frac{\partial x^\nu}{\partial \xi^n}.$$

This metric is non–negative, but depends upon x. It is advantageous to introduce an intrinsic Riemannian metric g on Σ, independently of x; in local coordinates, we have
$$g = g_{mn}(\xi) d\xi^m \otimes d\xi^n.$$

A natural intrinsic candidate for S is the area of $x(\Sigma)$, which gives the so–called *Nambu–Goto action*[39]
$$Area\,(x(\Sigma)) = \int_\Sigma d\mu_\gamma = \int_\Sigma n^2\xi \sqrt{\det \gamma_{mn}}, \tag{4.110}$$

which depends only upon g and x, but not on $\dot{g}$ [Got71]. However, the transition amplitudes derived from the Nambu–Goto action are *not* well–defined quantum–mechanically.

Otherwise, we can take as starting point the so–called *Polyakov action*[40]
$$S[x, g] = \kappa \int_\Sigma (dx, *dx)_g = \kappa \int_\Sigma d\mu_g\, g^{mn} \partial_m x^\mu \partial_n x^\nu g_{\mu\nu}(x), \tag{4.111}$$

where κ is the *string tension* (a positive constant with dimension of inverse length square). The stationary points of S with respect to g are at $g^0 = e^\phi \gamma$ for some function ϕ on Σ, and thus $S[x, g^0] \sim Area\,(x(\Sigma))$.

The Polyakov action leads to *well–defined* transition amplitudes, get by integration over the space $Met(\Sigma)$ of all positive metrics on Σ for a given topology, as well as over the space of all maps $Map(\Sigma, M)$. We can define the path integral
$$Amp = \sum_{\substack{topologies \\ \Sigma}} \int_{Met(\Sigma)} \frac{1}{N(g)} \int_{Map(\Sigma,M)} \mathcal{D}[x]\, e^{-S[x,g,g]},$$

where N is a normalization factor, while the measures $\mathcal{D}[g]$ and $\mathcal{D}[x]$ are constructed from $Diff^+(\Sigma)$ and $Diff(M)$ invariant L^2 norms on Σ and M. For fixed metric g, the action S is well–known: its stationary points are the harmonic maps $x\colon \Sigma \to M$ (see, e.g., [EL78]). However, g here varies and in fact is to be integrated over. For a general metric g, the action S defines a *nonlinear sigma model*, which is renormalizable because the dimension of Σ is 2. It would not in general be renormalizable in dimension higher than 2, which is usually regarded as an argument against the existence of fundamental membrane theories (see [DEF99]).

---

[39] Nambu–Goto action is the starting point of the analysis of string behavior, using the principles of ordinary Lagrangian mechanics. Just as the Lagrangian for a free point particle is proportional to its proper time-i.e., the 'length' of its world–line, a relativistic string's Lagrangian is proportional to the area of the sheet which the string traces as it travels through space–time.

[40] The Polyakov action is the 2D action from *conformal field theory*, used in string theory to describe the world–sheet of a moving string.

The Nambu–Goto action (4.110) and Polyakov action (4.111) represent the core of the so–called *bosonic string theory*, the original version of string theory, developed in the late 1960s. Although it has many attractive features, it also predicts a particle called the *tachyon* possessing some unsettling properties, and it has no fermions. All of its particles are *bosons*, the matter particles. The physicists have also calculated that bosonic string theory requires 26 space–time dimensions: 25 spatial dimensions and one dimension of time. In the early 1970s, *supersymmetry* was discovered in the context of string theory, and a new version of string theory called *superstring theory* (i.e., supersymmetric string theory) became the real focus, as it includes also *fermions*, the force particles. Nevertheless, bosonic string theory remains a very useful 'toy model' to understand many general features of *perturbative string theory*.

4.3.5 Weyl Invariance and Vertex Operator Formulation

The action S is also invariant under *Weyl rescalings* of the metric g by a positive function on σ: $\Sigma \to \mathbb{R}$, given by $g \to e^{2\sigma}g$. In general, Weyl invariance of the full amplitude may be spoiled by anomalies. Assuming Weyl invariance of the full amplitude, the integral defining Amp may be *simplified* in two ways.

1) The integration over $Met(\Sigma)$ effectively collapses to an integration over the *moduli space of surfaces*, which is finite dimensional, for each genus h.

2) The boundary components of Σ — characterizing external string states — may be mapped to regular points on an underlying compact surface without boundary by conformal transformations. The *data*, such as momenta and other quantum numbers of the external states, are mapped into *vertex operators*. The amplitudes are now given by the path integral

$$Amp = \sum_{h=0}^{\infty} \int_{Met(\Sigma)} \mathcal{D}[g] \frac{1}{N(g)} \int_{Map(\Sigma,M)} \mathcal{D}[x] \, V_1 \ldots V_N \, e^{-S},$$

for suitable vertex operators $V_1, \ldots V_N$.

4.3.6 More General Stringy Actions

Generalizations of the action S given above are possible when M carries extra structure.

1) M carries a 2–form $B \in \Omega^{(2)}(M)$. The resulting contribution to the action is also that of a 'nonlinear sigma model'

$$S_B[x, B] = \int_\Sigma x^*(B) = \int_\Sigma dx^\mu \wedge dx^\nu B_{\mu\nu}(x)$$

2) M may carry a *dilaton field* $\Phi \in \Omega^{(0)}(M)$ so that

$$S_\Phi[x,\Phi] = \int_\Sigma d\mu_g R_g \Phi(x).$$

where R_g is the Gaussian curvature of Σ for the metric g.

3) There may be a *tachyon field* $T \in \Omega^{(0)}(M)$ contributing

$$S_T[x,T] = \int_\Sigma d\mu_g T(x).$$

4.3.7 Transition Amplitude for a Single Point Particle

The transition amplitude for a single point–particle could in fact be get in a way analogous to how we prescribed string amplitudes. Let space–time be again a Riemannian manifold M, with metric g. The prescription for the transition amplitude of a particle travelling from a point $y \in M$ to a point y' to M is expressible in terms of a sum over all (continuous) paths connecting y to y':

$$Amp(y,y') = \sum_{\substack{paths \\ joining\ y\ and\ y'}} e^{-S[path]}.$$

Paths may be parametrized by maps from $C = [0,1]$ into M with $x(0) = y$, $x(1) = y'$. A simple world–line action for a massless particle is get by introducing a metric g on $[0,1]$

$$S[x,g] = \frac{1}{2}\int_C d\tau\, g(\tau)^{-1} \dot{x}^\mu \dot{x}^\nu g_{\mu\nu}(x),$$

which is invariant under $Diff^+(C)$ and $Diff(M)$.

Recall that the analogous prescription for the point–particle transition amplitude is the path integral

$$Amp(y,y') = \int_{Met(C)} \mathcal{D}[g] \frac{1}{N} \int_{Map(C,M)} \mathcal{D}[x]\, e^{-S[x,g]}.$$

Note that for string theory, we had a prescription for transition amplitudes valid for all topologies of the world–sheet. For point–particles, there is only the topology of the interval C, and we can only describe a single point–particle, but not interactions with other point–particles. To put those in, we would have to supply additional information.

Finally, it is very instructive to work out the amplitude Amp by carrying out the integrations. The only $Diff^+(C)$ invariant of g is the length $L = \int_0^1 d\tau\, g(\tau)$; all else is generated by $Diff^+(C)$. Defining the normalization factor to be the volume of $Diff(C)$: $N = \text{Vol}(Diff(C))$ we have $\mathcal{D}[g] = \mathcal{D}[v]\,dL$ and the transition amplitude becomes

$$Amp(y,y') = \int_0^\infty dL \int \mathcal{D}[x]\, e^{-\frac{1}{2L}\int_0^1 d\tau (\dot{x},\dot{x})_g} = \int_0^\infty dL \left\langle y'|e^{-L\Delta}|y\right\rangle = \left\langle y'|\frac{1}{\Delta}|y\right\rangle.$$

Thus, the amplitude is just the Green function at (y,y') for the Laplacian Δ and corresponds to the propagation of a massless particle (see [DEF99]).

4.3.8 Witten's Open String Field Theory

Noncommutative nature of space–time has often appeared in non–perturbative aspects of string theory. It has been used in a formulation of interacting open string field theory by Ed Witten [Wit88c, Wit86a]. Witten has written a classical action of open string field theory in terms of noncommutative geometry, where the noncommutativity appears in a product of string fields. Later, the *Dirichlet branes* (or, D–branes) have been recognized as solitonic objects in superstring theory [Pol95]. Further, it has been found that the low energy behavior of the D–branes are well described by supersymmetric Yang–Mills theory (SYM) [Wit96]. In the situation of some D–branes coinciding, the space–time coordinates are promoted to matrices which appear as the fields in SYM. Then the size of the matrices corresponds to the number of the D–branes, so noncommutativity of the matrices is related to the noncommutative nature of space–time.

In this subsection, mainly following [Sug00], we review some basic properties of Witten's bosonic open string field theory [Wit88c] and its explicit construction based on a *Fock space* representation of string field functional and δ-function overlap vertices [GJ87a, GJ87b, CST86].

Witten introduced a beautiful formulation of open string field theory in terms of a noncommutative extension of differential geometry, where string fields, the BRST operator Q and the integration over the string configurations $\int$ in string field theory are analogs of differential forms, the exterior derivative d and the integration over the manifold $\int_M$ in the differential geometry, respectively. The ghost number assigned to the string field corresponds to the degree of the differential form. Also the (noncommutative) product between the string fields $*$ is interpreted as an analog of the wedge product $\wedge$ between the differential forms.

The axioms obeyed by the system of $\int$, $*$ and Q are

$$\int QA = 0, \qquad Q(A*B) = (QA)*B + (-1)^{n_A} A*(QB),$$

$$(A*B)*C = A*(B*C), \qquad \int A*B = (-1)^{n_A n_B} \int B*A,$$

where A, B and C are arbitrary string fields, whose ghost number is half–integer valued: The *ghost number* of A is defined by the integer n_A as $n_A + \frac{1}{2}$.

Then Witten discussed the following *string–field–theory action*

$$S = \frac{1}{G_s} \int \left(\frac{1}{2} \psi * Q\psi + \frac{1}{3} \psi * \psi * \psi \right), \qquad (4.112)$$

where G_s is the open string coupling constant and ψ is the string field with the ghost number $-\frac{1}{2}$. The action is invariant under the gauge transformation

$$\delta\psi = Q\Lambda + \psi * \Lambda - \Lambda * \psi,$$

with the gauge parameter Λ of the ghost number $-\frac{3}{2}$.

Operator Formulation of String Field Theory

The objects defined above can be explicitly constructed by using the operator formulation, where the string field is represented as a Fock space, and the integration $\int$ as an inner product on the Fock space. It was considered by [GJ87a, GJ87b] in the case of the *Neumann boundary condition*. We will heavily use the notation of [GJ87a, GJ87b]. In the operator formulation, the action (4.112) is described as

$$S = \frac{1}{G_s}\left(\frac{1}{2}{}_{12}\langle V_2||\psi\rangle_1 Q|\psi\rangle_2 + \frac{1}{3}{}_{123}\langle V_3||\psi\rangle_1|\psi\rangle_2|\psi\rangle_3\right), \qquad (4.113)$$

where the structure of the product $*$ in the kinetic and potential terms is encoded to that of the overlap vertices $\langle V_2|$ and $\langle V_3|$ respectively (here, subscripts put to vectors in the Fock space label the strings concerning the vertices).

As a preparation for giving the explicit form of the overlaps, let us consider open strings in 26-dimensional space–time with the constant metric G_{ij} in the Neumann boundary condition. The world sheet action is given by

$$S_{WS} = \frac{1}{4\pi\alpha'}\int d\tau \int_0^\pi d\sigma G_{ij}(\partial_\tau X^i \partial_\tau X^j - \partial_\sigma X^i \partial_\sigma X^j) + S_{gh}, \qquad (4.114)$$

where S_{gh} is the action of the bc–ghosts:

$$S_{gh} = \frac{i}{2\pi}\int d\tau \int_0^\pi d\sigma [c_+(\partial_\tau - \partial_\sigma)b_+ + c_-(\partial_\tau + \partial_\sigma)b_-]. \qquad (4.115)$$

Under the Neumann boundary condition, the string coordinates have the standard mode expansions:

$$X^j(\tau,\sigma) = x^j + 2\alpha'\tau p^j + i\sqrt{2\alpha'}\sum_{n\neq 0}\frac{1}{n}\alpha_n^j e^{-in\tau}\cos(n\sigma), \qquad (4.116)$$

also the mode expansions of the ghosts are given by

$$c_\pm(\tau,\sigma) = \sum_{n\in\mathbb{Z}} c_n e^{-in(\tau\pm\sigma)} \equiv c(\tau,\sigma) \pm i\pi_b(\tau,\sigma),$$

$$b_\pm(\tau,\sigma) = \sum_{n\in\mathbb{Z}} b_n e^{-in(\tau\pm\sigma)} \equiv \pi_c(\tau,\sigma) \mp ib(\tau,\sigma).$$

As a result of the quantization, the modes obey the commutation relatons:

$$[x^i, p^j] = iG^{ij}, \qquad [\alpha_n^i, \alpha_m^j] = nG^{ij}\delta_{n+m,0}, \qquad \{b_n, c_m\} = \delta_{n+m,0},$$

while the other therms vanish.

The overlap $|V_N\rangle = |V_N\rangle^X|V_N\rangle^{gh}$, $(N = 1, 2, \cdots)$ is the state satisfying the continuity conditions for the string coordinates and the ghosts at the N–string vertex of the string field theory. The superscripts X and gh show the contribution of the sectors of the coordinates and the ghosts respectively. The continuity conditions for the coordinates are

$$(X^{(r)j}(\sigma) - X^{(r-1)j}(\pi - \sigma))|V_N\rangle^X = 0, \quad (P_i^{(r)}(\sigma) + P_i^{(r-1)}(\pi - \sigma))|V_N\rangle^X = 0, \tag{4.117}$$

for $0 \leq \sigma \leq \frac{\pi}{2}$ and $r = 1, \cdots, N$. Here $P_i(\sigma)$ is the momentum conjugate to the coordinate $X^j(\sigma)$ at $\tau = 0$, and the superscript (r) labels the string (r) meeting at the vertex. In the above formulas, we regard $r = 0$ as $r = N$ because of the cyclic property of the vertex. For the ghost sector, we impose the following conditions on the variables $c(\sigma)$, $b(\sigma)$ and their conjugate momenta $\pi_c(\sigma)$, $\pi_b(\sigma)$:

$$(\pi_c^{(r)}(\sigma) - \pi_c^{(r-1)}(\pi - \sigma))|V_N\rangle^{gh} = 0, \quad (b^{(r)}(\sigma) - b^{(r-1)}(\pi - \sigma))|V_N\rangle^{gh} = 0,$$
$$(c^{(r)}(\sigma) + c^{(r-1)}(\pi - \sigma))|V_N\rangle^{gh} = 0, \quad (\pi_b^{(r)}(\sigma) + \pi_b^{(r-1)}(\pi - \sigma))|V_N\rangle^{gh} = 0,$$

for $0 \leq \sigma \leq \frac{\pi}{2}$ and $r = 1, \cdots, N$.

Open Strings in Constant B–Field Background

We consider a constant background of the second–rank antisymmetric tensor field B_{ij} in addition to the constant metric g_{ij} where open strings propagate. Then the boundary condition at the end points of the open strings changes from the Neumann type, and thus the open string has a different mode expansion from the Neumann case (4.116). As a result, the end point is to be noncommutative, in the picture of the D–branes which implies noncommutativity of the world volume coordinates on the D–branes. Here we derive the mode–expanded form of the open string coordinates as a preparation for a calculation of the overlap vertices in the next section.

We start with the world sheet action

$$S_{WS}^B = \frac{1}{4\pi\alpha'} \int d\tau \int_0^\pi d\sigma [g_{ij}(\partial_\tau X^i \partial_\tau X^j - \partial_\sigma X^i \partial_\sigma X^j) \\ - 2\pi\alpha' B_{ij}(\partial_\tau X^i \partial_\sigma X^j - \partial_\sigma X^i \partial_\tau X^j)] + S_{gh}. \tag{4.118}$$

Because the term proportional to B_{ij} can be written as a total derivative term, it does not affect the equation of motion but does the boundary condition, which requires

$$g_{ij}\partial_\sigma X^j - 2\pi\alpha' B_{ij}\partial_\tau X^j = 0 \tag{4.119}$$

on $\sigma = 0, \pi$. This can be rewritten to the convenient form

$$E_{ij}\partial_- X^j = (E^T)_{ij}\partial_+ X^j, \tag{4.120}$$
$$\text{where} \quad E_{ij} \equiv g_{ij} + 2\pi\alpha' B_{ij},$$

and $\partial_\pm$ are derivatives with respect to the light cone variables $\sigma^\pm = \tau \pm \sigma$. We can easily see that $X^j(\tau, \sigma)$ satisfying the boundary condition (4.120) has the following mode expansion:

$$X^j(\tau,\sigma) = \tilde{x}^j + \alpha'\left[(E^{-1})^{jk}g_{kl}p^l\sigma^- + (E^{-1T})^{jk}g_{kl}p^l\sigma^+\right] \quad (4.121)$$
$$+ i\sqrt{\frac{\alpha'}{2}}\sum_{n\neq 0}\frac{1}{n}\left[(E^{-1})^{jk}g_{kl}\alpha_n^l e^{-in\sigma^-} + (E^{-1T})^{jk}g_{kl}\alpha_n^l e^{-in\sigma^+}\right].$$

We will get the commutators between the modes from the propagator of the open strings, which gives another derivation different from the method by [Ch99] based on the quantization via the Dirac bracket. When performing the *Wick rotation*: $\tau \to -i\tau$ and mapping the world sheet to the upper half plane

$$z = e^{\tau+i\sigma}, \qquad \bar{z} = e^{\tau-i\sigma} \ (0 \leq \sigma \leq \pi),$$

the boundary condition (4.120) becomes

$$E_{ij}\partial_{\bar{z}}X^j = (E^T)_{ij}\partial_z X^j, \quad (4.122)$$

which is imposed on the real axis $z = \bar{z}$. The propagator $\langle X^i(z,\bar{z})X^j(z',\bar{z}')\rangle$ satisfying the boundary condition (4.122) is determined as

$$\langle X^i(z,\bar{z})X^j(z',\bar{z}')\rangle = -\alpha'\left[g^{ij}\ln|z-z'| - g^{ij}\ln|z-\bar{z}'|\right.$$
$$\left. + G^{ij}\ln|z-\bar{z}'|^2 + \frac{1}{2\pi\alpha'}\theta^{ij}\ln\frac{z-\bar{z}'}{\bar{z}-z'} + D^{ij}\right],$$

where G^{ij} and θ^{ij} are given by

$$G^{ij} = \frac{1}{2}(E^{T-1} + E^{-1})^{ij} = (E^{T-1}gE^{-1})^{ij} = (E^{-1}gE^{T-1})^{ij}, \quad (4.123)$$

$$\theta^{ij} = 2\pi\alpha'\cdot\frac{1}{2}(E^{T-1} - E^{-1})^{ij} = (2\pi\alpha')^2(E^{T-1}BE^{-1})^{ij} \quad (4.124)$$
$$= -(2\pi\alpha')^2(E^{-1}BE^{T-1})^{ij}.$$

Also the constant D^{ij} remains unknown from the boundary condition alone. However it is an irrelevant parameter, so we can fix an appropriate value. The mode-expanded form (4.121) is mapped to

$$X^j(z,\bar{z}) = \tilde{x}^j - i\alpha'[(E^{-1})^{jk}p_k\ln\bar{z} + (E^{-1T})^{jk}p_k\ln z]$$
$$+ i\sqrt{\frac{\alpha'}{2}}\sum_{n\neq 0}\frac{1}{n}\left[(E^{-1})^{jk}\alpha_{n,k}\bar{z}^{-n} + (E^{-1T})^{jk}\alpha_{n,k}z^{-n}\right].$$

Note that the indices of p^l and α_n^l were lowered by the metric g_{ij} not G_{ij}. Recall the definition of the propagator

$$\langle X^i(z,\bar{z})X^j(z',\bar{z}')\rangle \equiv R(X^i(z,\bar{z})X^j(z',\bar{z}')) - N(X^i(z,\bar{z})X^j(z',\bar{z}')), \quad (4.125)$$

where R and N stand for the radial ordering and the normal ordering respectively. We take a prescription for the normal ordering which pushes p_i

to the right and $\tilde{x}_j$ to the left with respect to the zero–modes p_i and $\tilde{x}_j$. It corresponds to considering the vacuum satisfying

$$p_j|0\rangle = \alpha_{n,j}|0\rangle = 0 \quad (n>0), \qquad \langle 0|\alpha_{n,j} = 0 \quad (n<0), \tag{4.126}$$

which is the standard prescription for calculating the propagator of the massless scalar field in 2D conformal field theory from the operator formalism. Making use of (4.125), (4.126) and techniques of the contour integration, it is easy to get the commutators

$$[\alpha_{n,i}, \alpha_{m,j}] = n\delta_{n+m,0}G_{ij}, \qquad [\tilde{x}^i, p_j] = i\delta^i_j,$$

where the first equation holds for all integers with $\alpha_{0,i} \equiv \sqrt{2\alpha'}p_i$. The constant D^{ij} is written as $\alpha'D^{ij} = -\langle 0|\tilde{x}^i\tilde{x}^j|0\rangle$. Let us fix D^{ij} as $\alpha'D^{ij} = -\frac{i}{2}\theta^{ij}$, which is the convention taken in [SW99]. Then the coordinates $\tilde{x}^i$ become noncommutative:

$$[\tilde{x}^i, \tilde{x}^j] = i\theta^{ij},$$

but the center of mass coordinates $x^i \equiv \tilde{x}^i + \frac{1}{2}\theta^{ij}p_j$ can be seen to commute each other.

Now we have the mode–expanded form of the string coordinates and the commutation relations between the modes, which are

$$X^j(\tau, \sigma) = x^j + 2\alpha' \left(G^{jk}\tau + \frac{1}{2\pi\alpha'}\theta^{jk}\left(\sigma - \frac{\pi}{2}\right) \right) p_k$$
$$+ i\sqrt{2\alpha'} \sum_{n\neq 0} \frac{1}{n} e^{-in\tau} \left[G^{jk}\cos(n\sigma) - i\frac{1}{2\pi\alpha'}\theta^{jk}\sin(n\sigma) \right] \alpha_{n,k},$$

$$[\alpha_{n,i}, \alpha_{m,j}] = n\delta_{n+m,0}G_{ij}, \qquad [\tilde{x}^i, p_j] = i\delta^i_j,$$

with all the other commutators vanishing.

Also, due to the formula

$$\sum_{n=1}^{\infty} \frac{2}{n} \sin(n(\sigma+\sigma')) = \begin{cases} \pi - \sigma - \sigma', & (\sigma+\sigma' \neq 0, 2\pi) \\ 0, & (\sigma+\sigma' = 0, 2\pi), \end{cases}$$

we can see by a direct calculation that the end points of the string become noncommutative

$$[X^i(\tau,\sigma), X^j(\tau,\sigma')] = \begin{cases} i\theta^{ij}, & (\sigma=\sigma'=0) \\ -i\theta^{ij}, & (\sigma=\sigma'=\pi) \\ 0, & \text{(otherwise)}. \end{cases}$$

On the other hand, it is noted that the conjugate momenta have the mode expansion identical with that in the Neumann case:

$$P_i(\tau,\sigma) = \frac{1}{2\pi\alpha'}(g_{ij}\partial_\tau - 2\pi\alpha' B_{ij}\partial_\sigma)X^j(\tau,\sigma)$$
$$= \frac{1}{\pi}p_i + \frac{1}{\pi\sqrt{2\alpha'}}\sum_{n\neq 0} e^{-in\tau}\cos(n\sigma)\alpha_{n,i}.$$

Note that the relations (4.123) and (4.124) are in the same form as a T–duality transformation, although the correspondence is a formal sense, because we are not considering any compactification of space–time. The generalized T–duality transformation, namely $O(D,D)$–transformation, is defined by

$$E' = (aE + b)(cE + d)^{-1}, \qquad (4.127)$$

with a, b, c and d being $D \times D$ real matrices. (D is the dimension of space–time.) The matrix $h = \begin{pmatrix} a & b \\ c & d \end{pmatrix}$ is $O(D,D)$ matrix, which satisfies

$$h^T J h = J, \qquad \text{where} \qquad J = \begin{pmatrix} 0 & \mathbf{1}_D \\ \mathbf{1}_D & 0 \end{pmatrix}.$$

The relations (4.123) and (4.124) correspond to the case of the inversion $a = d = 0$, $b = c = \mathbf{1}_D$.

Construction of Overlap Vertices

Here we construct Witten's open string theory in the constant B–field background by obtaining the explicit formulas of the overlap vertices. As is understood from the fact that the action of the ghosts (4.115) contains no background fields, the ghost sector is not affected by turning on the B–field background. Thus we may consider the coordinate sector only. First, let us see the mode-expanded forms of the coordinates and the momenta at $\tau = 0$

$$X^j(\sigma) = G^{jk} y_k + \frac{1}{\pi} \theta^{jk} \left(\sigma - \frac{\pi}{2}\right) p_k$$

$$+ 2\sqrt{\alpha'} \sum_{n=1}^{\infty} \left[G^{jk} \cos(n\sigma) x_{n,k} + \frac{1}{2\pi\alpha'} \theta^{jk} \sin(n\sigma) \frac{1}{n} p_{n,k} \right],$$

$$P_i(\sigma) = \frac{1}{\pi} p_i + \frac{1}{\pi\sqrt{\alpha'}} \sum_{n=1}^{\infty} \cos(n\sigma) p_{n,i},$$

where $x^j = G^{jk} y_k$, the coordinates and the momenta for the oscillator modes are

$$x_{n,k} = \frac{i}{2}\sqrt{\frac{2}{n}}(a_{n,k} - a^\dagger_{n,k}) = \frac{i}{\sqrt{2n}}(\alpha_{n,k} - \alpha_{-n,k}),$$

$$p_{n,k} = \sqrt{\frac{n}{2}}(a_{n,k} + a^\dagger_{n,k}) = \frac{1}{\sqrt{2}}(\alpha_{n,k} + \alpha_{-n,k}).$$

The non-vanishing commutators are given by

$$[x_{n,k}, p_{m,l}] = i G_{kl} \delta_{n,m}, \qquad [y_k, p_l] = i G_{kl}. \qquad (4.128)$$

We should note that the metric appearing in eqs. (4.128) is G_{ij}, instead of g_{ij}. So it can be seen that if we employ the variables with the lowered space–time indices y_k, p_k, $x_{n,k}$ and $p_{n,k}$, the metric used in the expression of the overlaps is G^{ij} not g^{ij}.

The continuity condition (4.117) is universal for any background, and the mode expansion of the momenta $P_i(\sigma)$'s is of the same form as in the Neumann case, thus the continuity conditions for the momenta in terms of the modes $p_{n,i}$ are identical with those in the Neumann case. Also, since $p_{n,i}$'s mutually commute, it is natural to find a solution of the continuity condition, assuming the following form for the overlap vertices:

$$|\hat{V}_N\rangle^X_{1\ldots N} = \exp\left[\frac{i}{4\pi\alpha}\theta^{ij}\sum_{r,s=1}^N p^{(r)}_{n,i} Z^{rs}_{nm} p^{(s)}_{m,j}\right]|V_N\rangle^X_{1\ldots N}, \quad (4.129)$$

where $|\hat{V}_N\rangle^X_{1\ldots N}$ and $|V_N\rangle^X_{1\ldots N}$ are the overlaps in the background corresponding to the world sheet actions (4.118) and (4.114) respectively, the explicit form of the latter is given in appendix A. Clearly the expression (4.129) satisfies the continuity conditions for the modes of the momenta, and the coefficients Z^{rs}_{nm} are determined so that the continuity conditions for the coordinates are satisfied [Sug00].

- $|\hat{I}\rangle^X \equiv |\hat{V}_1\rangle^X$

For the $N = 1$ case, we consider the identity overlap $|\hat{I}\rangle^X \equiv |\hat{V}_1\rangle^X$. The continuity conditions for the momenta require that

$$P_i(\sigma) + P_i(\pi - \sigma) = \frac{2}{\pi}p_i + \frac{2}{\pi\sqrt{\alpha'}}\sum_{n=2,4,6,\cdots}\cos(n\sigma)p_{n,i}$$

should vanish for $0 \leq \sigma \leq \frac{\pi}{2}$, namely,

$$p_i = 0, \qquad p_{n,i} = 0 \quad (n = 2, 4, 6, \cdots), \quad (4.130)$$

which is satisfied by the overlap in the Neumann case $|I\rangle$. In addition, the conditions for the coordinates are that

$$X^j(\sigma) - X^j(\pi - \sigma) = \frac{2}{\pi}\theta^{jk}(\sigma - \frac{\pi}{2})p_k + \quad (4.131)$$

$$4\sqrt{\alpha'}\sum_{n=1,3,5,\cdots} G^{jk}\cos(n\sigma)x_{n,k} + 4\sqrt{\alpha'}\sum_{n=2,4,6,\cdots}\frac{1}{2\pi\alpha'}\theta^{jk}\sin(n\sigma)\frac{1}{n}p_{n,k},$$

should vanish for $0 \leq \sigma \leq \frac{\pi}{2}$. The first and third lines in the r. h. s. can be put to zero by using (4.130). So what we have to consider is the remaining condition $x_{n,k} = 0$ for $n = 1, 3, 5, \cdots$, which however is nothing but the continuity condition for the coordinates in the Neumann case. It can be understood from the point that the second line in (4.131) does not depend on θ^{ij}. Thus it turns out that the continuity conditions in the case of the B-field turned on are satisfied by the identity overlap made in the Neumann case. The solution is [Sug00]

$$|\hat{I}\rangle^X = |I\rangle^X = \exp\left[-\frac{1}{2}G^{ij}\sum_{n=0}^{\infty}(-1)^n a_{n,i}^\dagger a_{n,j}^\dagger\right]|0\rangle, \quad (4.132)$$

where also the zero modes y_i and p_i are written by using the creation and annihilation operators $a_{0,i}^\dagger$ and $a_{0,i}$ as

$$y_i = \frac{i}{2}\sqrt{2\alpha'}(a_{0,i} - a_{0,i}^\dagger), \qquad p_i = \frac{1}{\sqrt{2\alpha'}}(a_{0,i} + a_{0,i}^\dagger).$$

- $|\hat{V}_2\rangle_{12}^X$

For the $N = 2$ case, we are to do the same argument as in the $N = 1$ case. The continuity conditions mean that

$$P_i^{(1)}(\sigma) + P_i^{(2)}(\pi - \sigma) = \frac{1}{\pi}(p_i^{(1)} + p_i^{(2)}) + \frac{1}{\pi\sqrt{\alpha'}}\sum_{n=1}^{\infty}\cos(n\sigma)(p_{n,i}^{(1)} + (-1)^n p_{n,i}^{(2)}),$$

$$X^{(1)j}(\sigma) - X^{(2)j}(\pi - \sigma) = G^{jk}(y_k^{(1)} - y_k^{(2)}) + \frac{1}{\pi}\theta^{jk}(\sigma - \frac{\pi}{2})(p_k^{(1)} + p_k^{(2)})$$

$$+ 2\sqrt{\alpha'}\sum_{n=1}^{\infty}\Big[G^{jk}\cos(n\sigma)(x_{n,k}^{(1)} - (-1)^n x_{n,k}^{(2)})$$

$$+ \frac{1}{2\pi\alpha'}\theta^{jk}\sin(n\sigma)\frac{1}{n}(p_{n,k}^{(1)} + (-1)^n p_{n,k}^{(2)})\Big]$$

should be zero for $0 \leq \sigma \leq \pi$. It turns out again that the conditions for the modes are identical with those in the Neumann case:

$$p_i^{(1)} + p_i^{(2)} = 0, \qquad p_{n,i}^{(1)} + (-1)^n p_{n,i}^{(2)} = 0,$$
$$y_i^{(1)} - y_i^{(2)} = 0, \qquad x_{n,i}^{(1)} - (-1)^n x_{n,i}^{(2)} = 0,$$

for $n \geq 1$. Thus we have the solution [Sug00]

$$|\hat{V}_2\rangle_{12}^X = |V_2\rangle_{12}^X = \exp\left[-G^{ij}\sum_{n=0}^{\infty}(-1)^n a_{n,i}^{(1)\dagger} a_{n,j}^{(2)\dagger}\right]|0\rangle_{12}. \quad (4.133)$$

- $|\hat{V}_4\rangle^X_{1234}$

We find a solution of the continuity conditions (4.117) in the $N = 4$ case assuming the form

$$|\hat{V}_4\rangle^X_{1234} = \exp\left[\frac{i}{4\pi\alpha}\theta^{ij}\sum_{r,s=1}^{4}P^{(r)}_{n,i}Z^{rs}_{nm}P^{(s)}_{m,j}\right]|V_4\rangle^X_{1234}. \quad (4.134)$$

When considering the continuity conditions, it is convenient to employ the Z_4-Fourier transformed variables:

$$Q^j_1(\sigma) = \frac{1}{2}[iX^{(1)j}(\sigma) - X^{(2)j}(\sigma) - iX^{(3)j}(\sigma) + X^{(4)j}(\sigma)] \equiv Q^j(\sigma),$$

$$Q^j_2(\sigma) = \frac{1}{2}[-X^{(1)j}(\sigma) + X^{(2)j}(\sigma) - X^{(3)j}(\sigma) + X^{(4)j}(\sigma)],$$

$$Q^j_3(\sigma) = \frac{1}{2}[-iX^{(1)j}(\sigma) - X^{(2)j}(\sigma) + iX^{(3)j}(\sigma) + X^{(4)j}(\sigma)] \equiv \bar{Q}^j(\sigma),$$

$$Q^j_4(\sigma) = \frac{1}{2}[X^{(1)j}(\sigma) + X^{(2)j}(\sigma) + X^{(3)j}(\sigma) + X^{(4)j}(\sigma)].$$

For the momentum variables we also define the Z_4-Fourier transformed variables $P_{1,i}(\sigma)(\equiv P_i(\sigma))$, $P_{2,i}(\sigma)$, $P_{3,i}(\sigma)(\equiv \bar{P}_i(\sigma))$ and $P_{4,i}(\sigma)$ by the same combinations of $P^{(r)}_i(\sigma)$'s as the above. These variables have the following mode expansions

$$P_{t,i}(\sigma) = \frac{1}{\pi\sqrt{2\alpha'}}P_{t,0,i} + \frac{1}{\pi\sqrt{\alpha'}}\sum_{n=1}^{\infty}\cos(n\sigma)P_{t,n,i},$$

$$Q^j_t(\sigma) = G^{jk}\sqrt{2\alpha'}Q_{t,0,k} + \frac{1}{\pi}\theta^{jk}(\sigma - \frac{\pi}{2})\frac{1}{\sqrt{2\alpha'}}P_{t,0,k} \quad (4.135)$$

$$+ \sqrt{2\alpha'}\sum_{n=1}^{\infty}\left[G^{jk}\cos(n\sigma)Q_{t,n,k} + \frac{1}{2\pi\alpha'}\theta^{jk}\sin(n\sigma)\frac{1}{n}P_{t,n,k}\right],$$

where $t = 1, 2, 3, 4$. From now on, we frequently omit the subscript t for the $t = 1$ case, and at the same time we employ the expression with a bar instead of putting the subscript t for the $t = 3$ case.

Using those variables, the continuity conditions are written as

$$Q^j_4(\sigma) - Q^j_4(\pi - \sigma) = 0, \quad P_{4,i}(\sigma) + P_{4,i}(\pi - \sigma) = 0,$$
$$Q^j_2(\sigma) + Q^j_2(\pi - \sigma) = 0, \quad P_{2,i}(\sigma) - P_{2,i}(\pi - \sigma) = 0,$$
$$Q^j(\sigma) - i Q^j(\pi - \sigma) = 0, \quad P_i(\sigma) + i P_i(\pi - \sigma) = 0,$$
$$\bar{Q}^j(\sigma) + i \bar{Q}^j(\pi - \sigma) = 0, \quad \bar{P}_i(\sigma) - i \bar{P}_i(\pi - \sigma) = 0 \quad (4.136)$$

for $0 \leq \sigma \leq \frac{\pi}{2}$. In terms of the modes, the conditions for the sectors of $t = 2$ and 4 are identical with the Neumann case

$$(1-C)|Q_{4,k}\rangle|\hat{V}_4\rangle^X = (1+C)|P_{4,k}\rangle|\hat{V}_4\rangle^X = 0,$$
$$(1+C)|Q_{2,k}\rangle|\hat{V}_4\rangle^X = (1-C)|P_{2,k}\rangle|\hat{V}_4\rangle^X = 0,$$

which can be seen from the point that the conditions (4.136) for the sectors of $t = 2$ and 4 lead the same relations between the modes as those without the terms containing θ^{jk}. Here we adopted the vector notation for the modes

$$|Q_{t,k}\rangle = \begin{bmatrix} Q_{t,0,k} \\ Q_{t,1,k} \\ \vdots \end{bmatrix}, \quad |P_{t,k}\rangle = \begin{bmatrix} P_{t,0,k} \\ P_{t,1,k} \\ \vdots \end{bmatrix},$$

and C is a matrix such that $(C)_{nm} = (-1)^n \delta_{nm}$ ($n, m \geq 0$). Thus there is needed no correction containing θ^{ij} for the sectors of $t = 2$ and 4, so it is natural to assume the form of the phase factor in (4.134) as

$$\frac{1}{2}\theta^{ij}\sum_{r,s=1}^{4}\langle p_i^{(r)}|Z^{rs}|p_j^{(s)}\rangle = \theta^{ij}\langle P_i|Z|\bar{P}_j\rangle \tag{4.137}$$

with Z being anti–Hermitian.

Next let us consider the conditions for the sectors of $t = 1$ and 3. We rewrite the mode expansions of $Q^j(\sigma)$ and $\bar{Q}^j(\sigma)$ as [Sug00]

$$Q^j(\sigma) = G^{jk}(\sqrt{2\alpha'}Q_{0,k} + 2\sqrt{\alpha'}\sum_{n=1}^{\infty}\cos(n\sigma)Q_{n,k})$$
$$+ \theta^{jk}\left[\int_{\pi/2}^{\sigma}d\sigma' P_i(\sigma') + \frac{1}{\pi\sqrt{\alpha'}}\sum_{n=1,3,5,\cdots}\frac{1}{n}(-1)^{(n-1)/2}P_{n,k}\right]$$
$$\equiv \theta^{jk}\int_{\pi/2}^{\sigma}d\sigma' P_i(\sigma') + \Delta Q^j(\sigma), \tag{4.138}$$

$$\bar{Q}^j(\sigma) = G^{jk}(\sqrt{2\alpha'}\bar{Q}_{0,k} + 2\sqrt{\alpha'}\sum_{n=1}^{\infty}\cos(n\sigma)\bar{Q}_{n,k})$$
$$+ \theta^{jk}\left[\int_{\pi/2}^{\sigma}d\sigma' \bar{P}_i(\sigma') + \frac{1}{\pi\sqrt{\alpha'}}\sum_{n=1,3,5,\cdots}\frac{1}{n}(-1)^{(n-1)/2}\bar{P}_{n,k}\right]$$
$$\equiv \theta^{jk}\int_{\pi/2}^{\sigma}d\sigma' \bar{P}_i(\sigma') + \Delta\bar{Q}^j(\sigma). \tag{4.139}$$

Using the conditions for $P_i(\sigma)$ and $\bar{P}_i(\sigma)$ in (4.136), we can reduce the conditions for $Q^j(\sigma)$ and $\bar{Q}^j(\sigma)$ to those for $\Delta Q^j(\sigma)$ and $\Delta\bar{Q}^j(\sigma)$:

$$\Delta Q^j(\sigma) = \begin{cases} i\Delta Q^j(\pi - \sigma) & (0 \leq \sigma \leq \frac{\pi}{2}) \\ -i\Delta Q^j(\pi - \sigma) & (\frac{\pi}{2} \leq \sigma \leq \pi), \end{cases}$$

$$\Delta\bar{Q}^j(\sigma) = \begin{cases} -i\Delta\bar{Q}^j(\pi - \sigma) & (0 \leq \sigma \leq \frac{\pi}{2}) \\ i\Delta\bar{Q}^j(\pi - \sigma) & (\frac{\pi}{2} \leq \sigma \leq \pi). \end{cases}$$

These formulas are translated to the relations between the modes via the Fourier transformation. The result is expressed in the vector notation as

$$(1-X)|\mathcal{Q}_i\rangle|\hat{V}_4\rangle^X = (1+X)|\overline{\mathcal{Q}}_i\rangle|\hat{V}_4\rangle^X = 0, \qquad (4.140)$$

where the vectors $|\mathcal{Q}_i\rangle$ and $|\overline{\mathcal{Q}}_i\rangle$ stand for

$$|\mathcal{Q}_i\rangle = \begin{bmatrix} Q_{0,i} + \frac{i}{4\alpha'} G_{ik}\theta^{kj} \sum_{n=0}^{\infty} X_{0n} P_{n,j} \\ Q_{1,i} \\ Q_{2,i} \\ \vdots \end{bmatrix},$$

$$|\overline{\mathcal{Q}}_i\rangle = \begin{bmatrix} \bar{Q}_{0,i} + \frac{i}{4\alpha'} G_{ik}\theta^{kj} \sum_{n=0}^{\infty} X_{0n} \bar{P}_{n,j} \\ \bar{Q}_{1,i} \\ \bar{Q}_{2,i} \\ \vdots \end{bmatrix}.$$

In (4.140), passing the vectors through the phase factor of the $|\hat{V}_4\rangle$ and using the continuity conditions in the Neumann case

$$\begin{aligned}(1+X)|P_i\rangle|V_4\rangle^X &= (1-X)|\bar{P}_i\rangle|V_4\rangle^X = 0, \\ (1-X)|Q_i\rangle|V_4\rangle^X &= (1+X)|\bar{Q}_i\rangle|V_4\rangle^X = 0,\end{aligned} \qquad (4.141)$$

we get the equations, which the coefficients Z_{nm}'s should satisfy,

$$[(1-X)_{m0}\sum_{n=0}^{\infty}(\bar{Z}_{0n}+i\frac{\pi}{2}\bar{X}_{0n})P_{n,j} + \sum_{n=1}^{\infty}(1-X)_{mn}\sum_{n'=0}^{\infty}\bar{Z}_{nn'}P_{n',j}]|V_4\rangle^X = 0$$

$$[(1+X)_{m0}\sum_{n=0}^{\infty}(Z_{0n}-i\frac{\pi}{2}X_{0n})\bar{P}_{n,j} + \sum_{n=1}^{\infty}(1+X)_{mn}\sum_{n'=0}^{\infty}Z_{nn'}\bar{P}_{n',j}]|V_4\rangle^X = 0$$

for $m \geq 0$. Now all our remaining task is to solve these equations. It is easy to see that a solution of them is given by [Sug00]

$$Z_{mn} = -i\frac{\pi}{2}(1-X)_{mn} + i\beta\frac{\pi}{2}C_{mn}, \quad (m,n \geq 0, \text{ except for } m=n=0),$$

$$Z_{00} = i\beta\frac{\pi}{2},$$

if we pay attention to (4.141). Here β is an unknown real constant, which is not fixed by the continuity conditions alone. This ambiguity of the solution comes from the property of the matrix X: $XC = -CX$. However it will become clear that the term containing the constant β does not contribute to the vertex $|\hat{V}_4\rangle^X$.

Therefore, we have the expression of the phase (4.137)

$$\theta^{ij}(P_i|Z|\bar{P}_j) = \theta^{ij}[\mathrm{i}\frac{\pi}{2}P_{0,i}\bar{P}_{0,j} + \mathrm{i}\beta\frac{\pi}{2}\sum_{n=0}^{\infty}(-1)^n P_{n,i}\bar{P}_{n,j}$$
$$-\mathrm{i}\frac{\pi}{2}\sum_{m,n=0}^{\infty}P_{m,i}(1-X)_{mn}\bar{P}_{n,j}].$$

Then recalling (4.141) again, the last term in the r. h. s. can be discarded. Also we can rewrite the term containing β

$$-\frac{\theta^{ij}}{4\alpha'}\beta(P_i|C|\bar{P}_j) = +\frac{\theta^{ij}}{4\alpha'}\beta(P_i|X^T C X|\bar{P}_j) = +\frac{\theta^{ij}}{4\alpha'}\beta(P_i|C|\bar{P}_j),$$

on $|V_4\rangle^X$. The above formula means that the term containing β can be set to zero on $|V_4\rangle^X$. After all, the form of the 4-string vertex becomes

$$|\hat{V}_4\rangle^X_{1234} = \exp\left[-\frac{\theta^{ij}}{4\alpha'}P_{0,i}\bar{P}_{0,j}\right]|V_4\rangle^X_{1234}.$$

Note that the phase factor has the cyclic symmetric form

$$-\frac{\theta^{ij}}{4\alpha'}P_{0,i}\bar{P}_{0,j} = \mathrm{i}\frac{\theta^{ij}}{8\alpha'}(p^{(1)}_{0,i}p^{(2)}_{0,j} + p^{(2)}_{0,i}p^{(3)}_{0,j} + p^{(3)}_{0,i}p^{(4)}_{0,j} + p^{(4)}_{0,i}p^{(1)}_{0,j}),$$

which is a property the vertices should have[41].

- $|\hat{V}_3\rangle^X_{123}$

We can get the 3-string overlap in the similar manner as in the 4-string case. First, we introduce the Z_3–Fourier transformed variables

$$Q^j_1(\sigma) = \frac{1}{\sqrt{3}}[eX^{(1)j}(\sigma) + \bar{e}X^{(2)j}(\sigma) + X^{(3)j}(\sigma)] \equiv Q^j(\sigma),$$
$$Q^j_2(\sigma) = \frac{1}{\sqrt{3}}[\bar{e}X^{(1)j}(\sigma) + eX^{(2)j}(\sigma) + X^{(3)j}(\sigma)] \equiv \bar{Q}^j(\sigma),$$
$$Q^j_3(\sigma) = \frac{1}{\sqrt{3}}[X^{(1)j}(\sigma) + X^{(2)j}(\sigma) + X^{(3)j}(\sigma)],$$

where $e \equiv e^{\mathrm{i}2\pi/3}$, $\bar{e} \equiv e^{-\mathrm{i}2\pi/3}$. The momenta $P_{1,i}(\sigma)(\equiv P_i(\sigma))$, $P_{2,i}(\sigma)(\equiv \bar{P}_i(\sigma))$ and $P_{3,i}(\sigma)$ are defined in the same way. The mode expansions take the same form as those in (4.135). In these variables, the continuity conditions require

$$Q^j(\sigma) - eQ^j(\pi-\sigma) = 0, \qquad P_i(\sigma) + eP_i(\pi-\sigma) = 0,$$
$$\bar{Q}^j(\sigma) - \bar{e}\bar{Q}^j(\pi-\sigma) = 0, \qquad \bar{P}_i(\sigma) + \bar{e}\bar{P}_i(\pi-\sigma) = 0,$$
$$Q^j_3(\sigma) - Q^j_3(\pi-\sigma) = 0, \qquad P_{3,i}(\sigma) + P_{3,i}(\pi-\sigma) = 0$$

---

[41] Here the momentum $p^{(r)}_{0,i}$ is given by $p^{(r)}_{0,i} = \sqrt{2\alpha'}p^{(r)}_i$.

for $0 \leq \sigma \leq \frac{\pi}{2}$. The conditions imposed to the modes with respect to the $t = 3$ component are identical with those in the Neumann case

$$(1+C)|P_{3,i}\rangle|\hat{V}_3\rangle^X = (1-C)|Q_{3,i}\rangle|\hat{V}_3\rangle^X = 0.$$

Thus the $t = 3$ component does not couple with θ^{ij}, so we can find the solution by determining the single anti–Hermitian matrix Z in the phase factor whose form is assumed as [Sug00]

$$\frac{1}{2}\theta^{ij} \sum_{r,s=1}^{3} (p_i^{(r)}|Z^{rs}|p_j^{(s)}) = \theta^{ij}(P_i|Z|\bar{P}_j). \tag{4.142}$$

For the sectors of $t = 1$ and 2, the same argument goes on as in the 4-string case. $Q^j(\sigma)$ and $\bar{Q}^j(\sigma)$ have the mode expansions same as in eqs. (4.138) and (4.139). The conditions we have to consider are

$$\Delta Q^j(\sigma) = \begin{cases} e \Delta Q^j(\pi - \sigma), & (0 \leq \sigma \leq \frac{\pi}{2}) \\ \bar{e} \Delta Q^j(\pi - \sigma), & (\frac{\pi}{2} \leq \sigma \leq \pi), \end{cases}$$

$$\Delta \bar{Q}^j(\sigma) = \begin{cases} \bar{e} \Delta \bar{Q}^j(\pi - \sigma), & (0 \leq \sigma \leq \frac{\pi}{2}) \\ e \Delta \bar{Q}^j(\pi - \sigma), & (\frac{\pi}{2} \leq \sigma \leq \pi), \end{cases}$$

which are rewritten as the relations between the modes

$$(1-Y)|\mathcal{Q}_i\rangle|\hat{V}_3\rangle^X = (1-Y^T)|\bar{\mathcal{Q}}_i\rangle|\hat{V}_3\rangle^X = 0. \tag{4.143}$$

Recalling the conditions in the Neumann case

$$(1+Y)|P_i\rangle|V_3\rangle^X = (1+Y^T)|\bar{P}_i\rangle|V_3\rangle^X = 0,$$
$$(1-Y)|Q_i\rangle|V_3\rangle^X = (1-Y^T)|\bar{Q}_i\rangle|V_3\rangle^X = 0, \tag{4.144}$$

we end up with the following equations

$$[(1-Y)_{m0} \sum_{n=0}^{\infty} (\bar{Z}_{0n} + \frac{\pi}{2}\bar{X}_{0n}) P_{n,j} + \sum_{n=1}^{\infty} (1-Y)_{mn} \sum_{n'=0}^{\infty} \bar{Z}_{nn'} P_{n',j}]|V_3\rangle^X = 0,$$

$$[(1-Y^T)_{m0} \sum_{n=0}^{\infty} (Z_{0n} - i\frac{\pi}{2} X_{0n}) \bar{P}_{n,j} + \sum_{n=1}^{\infty} (1-Y^T)_{mn} \sum_{n'=0}^{\infty} Z_{nn'} \bar{P}_{n',j}]|V_3\rangle^X = 0$$

for $m \geq 0$. It can be easily found out that the expression

$$Z_{mn} = -i\frac{\pi}{\sqrt{3}}(1+Y^T)_{mn} \quad (m, n \geq 0, \text{ except for } m = n = 0),$$
$$Z_{00} = 0,$$

satisfies the above equations. It should be noted that in this case, because of $CYC = \bar{Y} \neq -Y$, it does not contain any unknown constant differently from the 4–string case.

Owing to the condition (4.144) we can write the phase factor only in terms of the zero-modes. Finally we have [Sug00]

$$|\hat{V}_3\rangle_{123}^X = \exp\left[-\frac{\theta^{ij}}{4\sqrt{3}\alpha'}P_{0,i}\bar{P}_{0,j}\right]|V_3\rangle_{123}^X$$

$$= \exp\left[i\frac{\theta^{ij}}{12\alpha'}(p_{0,i}^{(1)}p_{0,j}^{(2)} + p_{0,i}^{(2)}p_{0,j}^{(3)} + p_{0,i}^{(3)}p_{0,j}^{(1)})\right]|V_3\rangle_{123}^X. \quad (4.145)$$

It is not clear whether the solutions we have obtained here are unique or not. However we can show that the phase factors are consistent with the relations between the overlaps which they should satisfy,

$$_3\langle\hat{I}|\hat{V}_3\rangle_{123} = |\hat{V}_2\rangle_{12}, \quad _4\langle\hat{I}|\hat{V}_4\rangle_{1234} = |\hat{V}_3\rangle_{123}, \quad _{34}\langle\hat{V}_2||\hat{V}_3\rangle_{123}|\hat{V}_3\rangle_{456} = |\hat{V}_4\rangle_{1256},$$

by using the momentum conservation on the vertices $(p_i^{(1)} + \cdots + p_i^{(N)})|\hat{V}_N\rangle_{1\cdots N}^X = 0$. Furthermore we can see that the phase factors successfully reproduce the Moyal product structures of the correlators among vertex operators obtained in the perturbative approach to open string theory in the constant B-field background [SW99]. These facts convince us that the solutions obtained here are physically meaningful.

Transformation of String Fields

In the previous section, we have explicitly constructed the overlap vertices in the operator formulation under the constant B-field background. Then we have obtained the vertices with a new noncommutative structure of the Moyal type originating from the constant B-field, in addition to the ordinary product $*$ of string fields. Denoting the product with the new structure by $\star$, the action of the string field theory is written as

$$S_B = \frac{1}{G_s}\int\left(\frac{1}{2}\psi \star Q\psi + \frac{1}{3}\psi \star \psi \star \psi\right)$$

$$= \frac{1}{G_s}\left(\frac{1}{2}{}_{12}\langle\hat{V}_2||\psi\rangle_1 Q|\psi\rangle_2 + \frac{1}{3}{}_{123}\langle\hat{V}_3||\psi\rangle_1|\psi\rangle_2|\psi\rangle_3\right), \quad (4.146)$$

where the BRST charge Q is constructed from the world sheet action (4.118). The theory (4.146) gives the noncommutative $U(1)$ Yang–Mills theory in the low energy region in the same sense as Witten's open string field theory in the case of the Neumann boundary condition leads to the ordinary $U(1)$ Yang–Mills theory in the low energy limit.[42]

---
[42] It can be explicitly seen by repeating a similar calculation as that carried out in [Dea90].

4.3 Dynamics of Fields and Strings

In [SW99] the authors argued that open string theory in the constant B–field background leads to either commutative or noncommutative Yang–Mills theories, corresponding to the different regularization scheme (the so-called Pauli–Villars regularization or the point–splitting regularization) in the world sheet formulation. They discussed a map between the gauge fields in the commutative and noncommutative Yang–Mills theories. In string field theory perspective, there also should be a certain transformation (hopefully simpler than the Yang–Mills case) from the string field ψ in (4.146) to a string field in a new string field theory which leads to the commutative Yang–Mills theory in the low energy limit.

Here we get the new string field theory by finding a unitary transformation which absorbs the noncommutative structure of the Moyal type in the product $\star$ into a redefinition of the string fields. There are used the two vertices $|\hat{V}_2\rangle$ and $|\hat{V}_3\rangle$ in the action (4.146). Recall that the 2–string vertex is in the same form as in the Neumann case and has no Moyal type noncommutative structure. First, we consider the phase factor of the 3–string vertex which multiplies in front of $|V_3\rangle$ (see (4.145)). Making use of the continuity conditions

$$P_{0,i} = -2\sum_{n=1}^{\infty} Y_{0n} P_{n,i}, \qquad \bar{P}_{0,i} = -2\sum_{n=1}^{\infty} \bar{Y}_{0n} \bar{P}_{n,i}, \qquad (4.147)$$

it can be rewritten as [Sug00]

$$-\frac{\theta^{ij}}{4\sqrt{3}\alpha'} P_{0,i}\bar{P}_{0,j} = \frac{\theta^{ij}}{4\sqrt{3}\alpha'} \sum_{n=1}^{\infty} (P_{0,i}\bar{Y}_{0n}\bar{P}_{n,j} + P_{n,i} Y_{0n}\bar{P}_{0,j})$$

$$= -\frac{\theta^{ij}}{24\alpha'} \sum_{n=1}^{\infty} X_{0n}[(-p_{0,i}^{(2)} - p_{0,i}^{(3)} + 2p_{0,i}^{(1)})p_{n,j}^{(1)}$$

$$+ (-p_{0,i}^{(3)} - p_{0,i}^{(1)} + 2p_{0,i}^{(2)})p_{n,j}^{(2)} + (-p_{0,i}^{(1)} - p_{0,i}^{(2)} + 2p_{0,i}^{(3)})p_{n,j}^{(3)}]$$

$$= -\frac{\theta^{ij}}{8\alpha'} \sum_{r=1}^{3}\sum_{n=1}^{\infty} X_{0n} p_{0,i}^{(r)} p_{n,j}^{(r)},$$

where we used the property of the matrix Y: $Y_{0n} = -\bar{Y}_{0n} = \frac{\sqrt{3}}{2} X_{0n}$ for $n \geq 1$ and the momentum conservation on $|V_3\rangle$: $p_{0,i}^{(1)} + p_{0,i}^{(2)} + p_{0,i}^{(3)} = 0$. We manage to represent the phase factor of the Moyal type as a form factorized into the product of the unitary operators

$$\mathcal{U}_r = \exp\left(\frac{\theta^{ij}}{8\alpha'} \sum_{n=1,3,5,\cdots} X_{0n} p_{0,i}^{(r)} p_{n,j}^{(r)}\right). \qquad (4.148)$$

Note that the unitary operator acts to a single string field. So the Moyal type noncommutativity can be absorbed by the unitary rotation of the string field

$$_{123}\langle \hat{V}_3 || \psi\rangle_1 |\psi\rangle_2 |\psi\rangle_3 =_{123}\langle V_3 | \mathcal{U}_1 \mathcal{U}_2 \mathcal{U}_3 | \psi\rangle_1 |\psi\rangle_2 |\psi\rangle_3 =_{123}\langle V_3 || \tilde{\psi}\rangle_1 |\tilde{\psi}\rangle_2 |\tilde{\psi}\rangle_3, \qquad (4.149)$$

with $|\tilde{\psi}\rangle_r = \mathcal{U}_r|\psi\rangle_r$. It should be remarked that this manipulation has been suceeded owing to the factorized expression of the phase factor, which originates from the continuity conditions relating the zero-modes to the nonzero-modes (4.147). It is a characteristic feature of string field theory that can not be found in any local field theories.

Next let us see the kinetic term. In doing so, it is judicious to write the kinetic term as follows:

$$_{12}\langle \hat{V}_2||\psi\rangle_1(Q|\psi\rangle_2) =_{123} \langle \hat{V}_3||\psi\rangle_1(Q_L|I\rangle_2|\psi\rangle_3 + |\psi\rangle_2 Q_L|I\rangle_3), \qquad (4.150)$$

where Q_L is defined by integrating the BRST current $j_{BRST}(\sigma)$ with respect to σ over the left half region

$$Q_L = \int_0^{\pi/2} d\sigma j_{BRST}(\sigma).$$

Equation (4.150) is also represented by the product $\star$ as

$$\psi \star (Q\psi) = \psi \star [(Q_L I) \star \psi + \psi \star (Q_L I)]. \qquad (4.151)$$

Here, I stands for the identity element with respect to the $\star$-product, carrying the ghost number $-\frac{3}{2}$, which corresponds to $|I\rangle$ in the operator formulation. As is discussed by [HLR86], in order to show the relation (4.151) we need the formulas

$$Q_R I = -Q_L I, \qquad (Q_R \psi) \star \xi = -(-1)^{n_\psi} \psi \star (Q_L \xi) \qquad (4.152)$$

for arbitrary string fields ψ and ξ, where Q_R is the integrated BRST current over the right half region of σ. n_ψ stands for the ghost number of the string field ψ minus $\frac{1}{2}$, and takes an integer value. The first formula means that the identity element is a physical quantity, also the second does the conservation of the BRST charge. By using these formulas, the first term in the bracket in r. h. s. of (4.151) becomes

$$(Q_L I) \star \psi = -(Q_R I) \star \psi = I \star (Q_L \psi) = Q_L \psi.$$

Also, it turns out that the second term is equal to $Q_R \psi$. Combining these, we can see that (4.151) holds.

Further, we should remark that because the BRST current does not contain the center of mass coordinate x^j, it commute with the momentum p_i. From the continuity condition $p_i|I\rangle = 0$, it can be seen that $p_i Q_L|I\rangle = 0$. Expanding the exponential in the expression of the unitary operator (4.148) and passing the momentum $p_{0,i}$ to the right, we get

$$\mathcal{U} Q_L|I\rangle = Q_L|I\rangle. \qquad (4.153)$$

Now we can write down the result of the kinetic term. As a result of the same manipulation as in eq. (4.149) and the use of eq. (4.153), we have[43]

$$_{12}\langle \hat{V}_2||\psi\rangle_1(Q|\psi\rangle_2) =_{123}\langle \hat{V}_3||\psi\rangle_1(Q_L|I\rangle_2|\psi\rangle_3 + |\psi\rangle_2 Q_L|I\rangle_3)$$
$$=_{123}\langle V_3||\tilde{\psi}\rangle_1(Q_L|I\rangle_2|\tilde{\psi}\rangle_3 + |\tilde{\psi}\rangle_2 Q_L|I\rangle_3) =_{12}\langle V_2||\tilde{\psi}\rangle_1(Q|\tilde{\psi}\rangle_2). \quad (4.154)$$

Here we have a comment [Sug00]. If we considered the kinetic term itself without using (4.150), what would be going on? Let us see this. From the continuity conditions for $|\hat{V}_2\rangle_{12}^X = |V_2\rangle_{12}^X$:

$$p_{0,i}^{(1)} + p_{0,i}^{(2)} = 0, \qquad p_{n,i}^{(1)} + (-1)^n p_{n,i}^{(2)} = 0 \qquad (n = 1, 2, \cdots),$$

it could be shown that the 2-string overlap is invariant under the unitary rotation

$$\mathcal{U}_1 \mathcal{U}_2 |V_2\rangle_{12} = |V_2\rangle_{12}.$$

So we would find the expression for the kinetic term after the rotation

$$_{12}\langle V_2||\psi\rangle_1 Q|\psi\rangle_2 =_{12}\langle V_2||\tilde{\psi}\rangle_1 \tilde{Q}|\tilde{\psi}\rangle_2,$$

where $\tilde{Q}$ is the BRST charge similarity transformed by $\mathcal{U}$

$$\tilde{Q} = \mathcal{U} Q \mathcal{U}^\dagger. \quad (4.155)$$

However, after some computations of the r. h. s. of (4.155), we could see that $\tilde{Q}$ has divergent term proportional to

$$\sum_{n=1,3,5,\cdots} 1$$

and thus it is not well-defined. It seems that this procedure is not correct and needs some suitable regularization, which preserves the conformal symmetry[44]. It is considered that the use of eq. (4.150) gives that kind of regularization, which will be justified at the end of the next section.

---

[43] Strictly speaking, in general this formula holds in the case that both of the string fields $|\psi\rangle$ and $|\tilde{\psi}\rangle$ belong to the Fock space which consists of states excited by *finite* number of creation operators. This point is subtle for giving a proof. However, for the infinitesimal θ case, by keeping arbitrary finite order terms in the expanded form of the exponential of $\mathcal{U}$, we can make the situation of both $|\psi\rangle$ and $|\tilde{\psi}\rangle$ being inside the Fock space, and thus clearly eq. (4.154) holds. From this fact, it is plausible to expect that eq. (4.154) is correct in the finite θ case.

[44] That divergence comes from the mid-point singularity of the string coordinates transformed by $\mathcal{U}$. In fact, after some calculations, we have

$$\mathcal{U} X^j(\sigma) \mathcal{U}^\dagger = X^j(\sigma) - i\frac{\theta^{jk}}{4\sqrt{2\alpha'}} \sum_{n=1,3,5,\cdots} X_{n0} p_{n,k} - \frac{\theta^{jk}}{4} p_k \operatorname{sgn}\left(\sigma - \frac{\pi}{2}\right). \quad (4.156)$$

The last term leads to the mid-point sigularity in the energy-momentum tensor and the BRST charge Q. It seems that the use of (4.150) corresponds to taking the

Therefore, the string field theory action (4.146) with the Moyal type noncommutativity added to the ordinary noncommutativity is equivalently rewritten as the one with the ordinary noncommutativity alone [Sug00]:

$$S_B = \frac{1}{G_s} \int \left(\frac{1}{2} \tilde{\psi} * Q\tilde{\psi} + \frac{1}{3} \tilde{\psi} * \tilde{\psi} * \tilde{\psi} \right)$$
$$= \frac{1}{G_s} \left(\frac{1}{2}{}_{12}\langle V_2 || \tilde{\psi}\rangle_1 Q |\tilde{\psi}\rangle_2 + \frac{1}{3}{}_{123}\langle V_3 || \tilde{\psi}\rangle_1 |\tilde{\psi}\rangle_2 |\tilde{\psi}\rangle_3 \right). \quad (4.157)$$

It is noted that the BRST charge Q, which is constructed from the world sheet action (4.118), has the same form as the one obtained from the action (4.114) with the relation (4.123). So all the B–dependence has been stuffed into the string fields except that existing in the metric G_{ij}. Furthermore, recalling that the relation between the metrics G^{ij} and g_{ij} is the same form as the T–duality inversion transformation, which was pointed out at the end of section 3, we can make the metric g_{ij} appear in the overlap vertices, instead of the metric G_{ij}. To do so, we consider the following transformation for the modes:

$$\hat{\alpha}_n^i = (E^{T-1})^{ik} \alpha_{n,k}, \qquad \hat{p}^i = (E^{T-1})^{ik} p_k, \qquad \hat{x}_i = E_{ik} x^k. \quad (4.158)$$

By this transformation, the commutators become

$$[\hat{\alpha}_n^i, \hat{\alpha}_m^j] = n g^{ij} \delta_{n+m,0}, \qquad [\hat{p}^i, \hat{x}_j] = -i\delta_j^i,$$

and the bilinear form of the modes

$$G^{ij} \alpha_{n,i} \alpha_{m,j} = g_{ij} \hat{\alpha}_n^i \hat{\alpha}_m^j, \qquad G^{ij} p_i \alpha_{m,j} = g_{ij} \hat{p}^i \hat{\alpha}_m^j, \qquad G^{ij} p_i p_j = g_{ij} \hat{p}^i \hat{p}^j. \quad (4.159)$$

4.3.9 Topological Strings

The 2D field theories we have constructed are already very similar to string theories. However, one ingredient from string theory is missing: in string theory, the world–sheet theory does not only involve a path integral over the maps ϕ^i to the target space and their fermionic partners, but also a path integral over the world–sheet metric $h_{\alpha\beta}$. So far, we have set this metric to a fixed background value.

We have also encountered a drawback of our construction. Even though the theories we have found can give us some interesting 'semi–topological' information about the target spaces, one would like to be able to define general nonzero n–point functions at genus g instead of just the partition function

---

point splitting regularization with respect to the mid–point. Because of the discontinuity of the last term in (4.156), it is considered that the transformed string coordinates have no longer a good picture as a string. It could be understood from the point that the transformation $\mathcal{U}$ drives states around a perturbative vacuum to those around highly non–perturbative one like coherent states.

at genus one and the particular correlation functions we calculated at genus zero.

It turns out that these two remarks are intimately related. In this section we will go from topological field theory to topological string theory by introducing integrals over all metrics, and in doing so we will find interesting nonzero correlation functions at any genus (see [Von05]).

Coupling to Topological Gravity

In coupling an ordinary field theory to gravity, one has to perform the following three steps.

- First of all, one rewrites the Lagrangian of the theory in a covariant way by replacing all the flat metrics by the dynamical ones, introducing covariant derivatives and multiplying the measure by a factor of $\sqrt{\det h}$.
- Secondly, one introduces an Einstein–Hilbert term as the 'kinetic' term for the metric field, plus possibly extra terms and fields to preserve the symmetries of the original Lagrangian.
- Finally, one has to integrate the resulting theory over the space of all metrics.

Here we will not discuss the first two steps in this procedure. As we have seen in our discussion of topological field theories, the precise form of the Lagrangian only plays a comparatively minor role in determining the properties of the theory, and we can derive many results without actually considering a Lagrangian. Therefore, let us just state that it is possible to carry out the analog of the first two steps mentioned above, and construct a Lagrangian with a 'dynamical' metric which still possesses the topological Q–symmetry we have constructed. The reader who is interested in the details of this construction is referred to the paper [Wit90] and to the lecture notes [DVV91].

The third step, integrating over the space of all metrics, is the one we will be most interested in here. Naively, by the metric independence of our theories, integrating their partition functions over the space of all metrics, and then dividing the results by the volume of the topological 'gauge group', would be equivalent to multiplication by a factor of 1,

$$Z[h_0] \stackrel{?}{=} \frac{1}{G_{top}} \int \mathcal{D}[h]\, Z[h], \qquad (4.160)$$

for any arbitrary background metric h_0. There are several reasons why this naive reasoning might go wrong:

- There may be metric configurations which cannot be reached from a given metric by continuous changes.
- There may be anomalies in the topological symmetry at the quantum level preventing the conclusion that all gauge fixed configurations are equivalent.

- The volume of G_{top} is infinite, so even if we could rigorously define a path integral the above multiplication and division would not be mathematically well–defined.

For these reasons, we should really be more careful and precisely define what we mean by the 'integral over the space of all metrics'. Let us note the important fact that just like in ordinary string theory (and even before twisting), the 2D sigma models become conformal field theories when we include the metric in the Lagrangian. This means that we can borrow the technology from string theory to integrate over all conformally equivalent metrics. As is well known, and as we will discuss in more detail later, the conformal symmetry group is a huge group, and integrating over conformally equivalent metrics leaves only a nD integral over a set of world–sheet moduli. Therefore, our strategy will be to use the analogy to ordinary string theory to first do this integral over all conformally equivalent metrics, and then perform the integral over the remaining nD moduli space.

In integrating over conformally equivalent metrics, one usually has to worry about conformal anomalies. However, here a very important fact becomes our help. To understand this fact, it is useful to rewrite our twisting procedure in a somewhat different language (see [Von05]).

Let us consider the SEM–tensor $T_{\alpha\beta}$, which is the conserved Noether current with respect to global translations on $\mathbb{C}$. From conformal field theory, it is known that $T_{z\bar{z}} = T_{\bar{z}z} = 0$, and the fact that T is a conserved current, $\partial_\alpha T^\alpha{}_\beta = 0$, means that $T_{zz} \equiv T(z)$ and $T_{\bar{z}\bar{z}} \equiv \bar{T}(\bar{z})$ are (anti–)holomorphic in z. One can now expand $T(z)$ in Laurent modes,

$$T(z) = \sum L_m z^{-m-2}. \tag{4.161}$$

The L_m are called the *Virasoro generators*, and it is a well–known result from conformal field theory that in the quantum theory their commutation relations are

$$[L_m, L_n] = (m-n)L_{m+n} + \frac{c}{12}m(m^2-1)\delta_{m+n}.$$

The number c depends on the details of the theory under consideration, and it is called the central charge. When this central charge is nonzero, one runs into a technical problem. The reason for this is that the equation of motion for the metric field reads

$$\frac{\delta S}{\delta h^{\alpha\beta}} = T_{\alpha\beta} = 0.$$

In conformal field theory, one imposes this equation as a constraint in the quantum theory. That is, one requires that for physical states $|\psi\rangle$,

$$L_m|\psi\rangle = 0 \quad \text{(for all } m \in \mathbb{Z}\text{)}.$$

However, this is clearly incompatible with the above commutation relation unless $c = 0$. In string theory, this value for c can be achieved by taking the target space of the theory to be 10D. If $c \neq 0$ the quantum theory is problematic to define, and we speak of a 'conformal anomaly' [Von05].

The whole above story repeats itself for $\bar{T}(\bar{z})$ and its modes $\bar{L}_m$. At this point there is a crucial difference between open and closed strings. On an open string, left–moving and right–moving vibrations are related in such a way that they combine into standing waves. In our complex notation, 'left–moving' translates into 'z–dependent' (i.e., holomorphic), and 'right–moving' into '$\bar{z}$–dependent' (i.e., anti–holomorphic). Thus, on an open string all holomorphic quantities are related to their anti–holomorphic counterparts. In particular, $T(z)$ and $\bar{T}(\bar{z})$, and their modes L_m and $\bar{L}_m$, turn out to be complex conjugates. There is therefore only one independent algebra of Virasoro generators L_m.

On a *closed string* on the other hand, which is the situation we have been studying so far, left– and right–moving waves are completely independent. This means that all holomorphic and anti–holomorphic quantities, and in particular $T(z)$ and $\bar{T}(\bar{z})$, are independent. One therefore has two sets of Virasoro generators, L_m and $\bar{L}_m$.

Let us now analyze the problem of central charge in the twisted theories. To twist the theory, we have used the $U(1)$–symmetries. Any global $U(1)$–symmetry of our theory has a conserved current J_α. The fact that it is conserved again means that $J_z \equiv J(z)$ is holomorphic and $J_{\bar{z}} \equiv \bar{J}(\bar{z})$ is anti–holomorphic. Once again, on an open string J and $\bar{J}$ will be related, but in the closed string theory we are studying they will be independent functions. In particular, this means that we can view a global $U(1)$–symmetry as really consisting of two independent, left– and right–moving, $U(1)$–symmetries, with generators F_L and F_R.

Note that the sum of $U(1)$–symmetries $F_V + F_A$ only acts on objects with a + index. That is, it acts purely on left–moving quantities. Similarly, $F_V - F_A$ acts purely on right–moving quantities. From our discussion above, it is therefore natural to identify these two symmetries with the two components of a single global $U(1)$ symmetry:

$$F_V = \frac{1}{2}(F_L + F_R) \qquad F_A = \frac{1}{2}(F_L - F_R).$$

A more detailed construction shows that this can indeed be done.

Let us expand the left–moving conserved $U(1)$–current into Laurent modes,

$$J(z) = \sum J_m z^{-m-1}. \qquad (4.162)$$

The commutation relations of these modes with one another and with the Virasoro modes can be calculated, either by writing down all of the modes in terms of the fields of the theory, or by using more abstract knowledge from the theory of superconformal symmetry algebras. In either case, one finds

$$[L_m, L_n] = (m-n)L_{m+n} + \frac{c}{12}m(m^2-1)\delta_{m+n}[L_m, J_n]$$
$$= -nJ_{m+n}[J_m, J_n] = \frac{c}{3}m\delta_{m+n}.$$

Note that the same central charge c appears in the $J-$ and in the $L-$commutators. This turns out to be crucial.

Following the standard Noether procedure, we can now construct a conserved charge by integrating the conserved current $J(z)$ over a space–like slice of the $z-$plane. In string theory, the physical time direction is the radial direction in the $z-$plane, so a space–like slice is just a curve around the origin. The integral is therefore calculated using the *Cauchy Theorem*,

$$F_L = \oint_{z=0} J(z) dz = 2\pi i J_0.$$

In the quantum theory, it will be this operator that generates the $U(1)_L$ − symmetry. Now recall that to twist the theory we want to introduce new Lorentz rotation generators,

$$M_A = M - F_V = M - \frac{1}{2}(F_L + F_R) M_B = M - F_A = M - \frac{1}{2}(F_L - F_R).$$

A well–known result from string theory (see [Von05]) is that the generator of Lorentz rotations is $M = 2\pi i (L_0 - \bar{L}_0)$. Therefore, we find that the twisting procedure in this new language amounts to

$$A: L_{0,A} = L_0 - \frac{1}{2}J_0, \quad \bar{L}_{0,A} = \bar{L}_0 + \frac{1}{2}\bar{J}_0,$$
$$B: L_{0,B} = L_0 - \frac{1}{2}J_0, \quad \bar{L}_{0,B} = \bar{L}_0 - \frac{1}{2}\bar{J}_0.$$

Let us now focus on the left–moving sector; we see that for both twistings the new Lorentz rotation generator is the difference of L_0 and $\frac{1}{2}J_0$. The new Lorentz generator should also correspond to a conserved 2–tensor, and from (4.161) and (4.162) there is a very natural way to get such a current:

$$\tilde{T}(z) = T(z) + \frac{1}{2}\partial J(z), \qquad (4.163)$$

which clearly satisfies $\bar{\partial}\tilde{T} = 0$ and

$$\tilde{L}_m = L_m - \frac{1}{2}(m+1)J_m, \qquad (4.164)$$

so in particular we find that $\tilde{L}_0$ can serve as $L_{0,A}$ or $L_{0,B}$. We should apply the same procedure (with a minus sign in the $A-$model case) in the right–moving sector. Equations (4.163) and (4.164) tell us how to implement the twisting procedure not only on the conserved charges, but on the whole $N=2$

superconformal algebra – or at least on the part consisting of the $J-$ and $L-$modes, but a further investigation shows that this is the only part that changes. We have motivated, but not rigorously derived (4.163); for a complete justification the reader is referred to the original papers [LVW89] and [CV91].

Now, we come to the crucial point. The algebra that the new modes $\tilde{L}_m$ satisfy can be directly calculated from (4.163), and we find

$$[\tilde{L}_m, \tilde{L}_n] = (m-n)\tilde{L}_{m+n}.$$

That is, there is no central charge left. This means that we do not have any restriction on the dimension of the theory, and topological strings will actually be well–defined in target spaces of any dimension.

From this result, we see that we can integrate our partition function over conformally equivalent metrics without having to worry about the conformal anomaly represented by the nonzero central charge. After having integrated over this large part of the space of all metrics, it turns out that there is a nD integral left to do. In particular, it is known that one can always find a conformal transformation which in the neighborhood of a chosen point puts the metric in the form $h_{\alpha\beta} = \eta_{\alpha\beta}$, with η the usual flat metric with diagonal entries ± 1. (Or, $+1$ in the Euclidean setting.) On the other hand, when one considers the global situation, it turns out that one cannot always enforce this gauge condition everywhere. For example, if the world–sheet is a torus, there is a left–over complex parameter τ that cannot be gauged away. The easiest way to visualize this parameter (see [Von05]) is by drawing the resulting torus in the complex–plane and rescaling it in such a way that one of its edges runs from 0 to 1; the other edge then runs from 0 to τ, see Figure 4.23. It seems intuitively clear that a conformal transformation – which should leave all angles fixed – will never deform τ, and even though intuition often fails when considering conformal mappings, in this case this can indeed be proven. Thus, τ is really a modular parameter which we need to integrate over. Another fairly intuitive result is that any locally flat torus can, after a rescaling, be drawn in this form, so τ indeed is the only modulus of the torus.

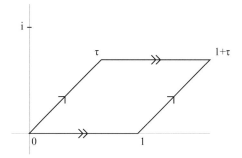

Fig. 4.23. The only modulus τ of a torus T^2.

More generally, one can show that a Riemann surface of genus g has $m_g = 3(g-1)$ complex modular parameters. As usual, this is the *virtual* dimension of the moduli space. If $g > 1$, one can show that this virtual dimension equals the actual dimension. For $g = 0$, the sphere, we have a negative virtual dimension $m_g = -3$, but the actual dimension is 0: there is always a flat metric on a surface which is topologically a sphere (just consider the sphere as a plane with a point added at infinity), and after having chosen this metric there are no remaining parameters such as τ in the torus case. For $g = 1$, the virtual dimension is $m_g = 0$, but as we have seen the actual dimension is 1.

We can explain these discrepancies using the fact that, after we have used the conformal invariance to fix the metric to be flat, the sphere and the torus have leftover symmetries. In the case of the sphere, it is well known in string theory that one can use these extra symmetries to fix the positions of three labelled points. In the case of the torus, after fixing the metric to be flat we still have rigid translations of the torus left, which we can use to fix the position of a single labelled point. To see how this leads to a difference between the virtual and the actual dimensions, let us for example consider tori with n labelled points on them. Since the virtual dimension of the moduli space of tori without labelled points is 0, the virtual dimension of the moduli space of tori with n labelled points is n. One may expect that at some point (and in fact, this happens already when $n = 1$), one reaches a sufficiently generic situation where the virtual dimension really is the actual dimension. However, even in this case we can fix one of the positions using the remaining conformal (translational) symmetry, so the positions of the points only represent $n-1$ moduli. Hence, there must be an nth modulus of a different kind, which is exactly the shape parameter τ that we have encountered above. In the limiting case where $n = 0$, this parameter survives, thus causing the difference between the virtual and the real dimension of the moduli space.

For the sphere, the reasoning is somewhat more formal: we analogously expect to have three 'extra' moduli when $n = 0$. In fact, three extra parameters are present, but they do not show up as moduli. They must be viewed as the three parameters which need to be added to the problem to find a 0D moduli space.

Since the cases $g = 0, 1$ are thus somewhat special, let us begin by studying the theory on a Riemann surface with $g > 1$. To arrive at the topological string correlation functions, after gauge fixing we have to integrate over the remaining moduli space of complex dimension $3(g-1)$. To do this, we need to fix a measure on this moduli space. That is, given a set of $6(g-1)$ tangent vectors to the moduli space, we want to produce a number which represents the size of the volume element spanned by these vectors, see Figure 4.24. We should do this in a way which is invariant under coordinate redefinitions of both the moduli space and the world–sheet. Is there a 'natural' way to do this?

To answer this question, let us first ask how we can describe the tangent vectors to the moduli space (see [Von05]). In two dimensions, conformal

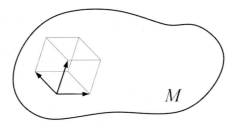

Fig. 4.24. A measure on the moduli space M assigns a number to every set of three tangent vectors. This number is interpreted as the volume of the element spanned by these vectors.

transformations are equivalent to holomorphic transformations: $z \mapsto f(z)$. It thus seems natural to assume that the moduli space we have left labels different complex structures on Σ, and indeed this can be shown to be the case. Therefore, a tangent vector to the moduli space is an infinitesimal change of complex structure, and these changes can be parameterized by holomorphic 1–forms with anti–holomorphic vector indices,

$$dz \mapsto dz + \epsilon \mu_{\bar z}^z(z) d\bar z.$$

The dimension counting above tells us that there are $3(g-1)$ independent $(\mu_i)_{\bar z}^z$, plus their $3(g-1)$ complex conjugates which change $d\bar z$. So the tangent space is spanned by these $\mu_i(z,\bar z), \bar\mu_i(z,\bar z)$. How do we get a number out of a set of these objects? Since μ_i has a z and a $\bar z$ index, it seems natural to integrate it over Σ. However, the z–index is an upper index, so we need to lower it first with some tensor with two z–indices. It turns out that a good choice is to use the Q–partner G_{zz} of the SEM–tensor component T_{zz}, and thus to define the integration over moduli space as

$$\int_{M_g} \prod_{i=1}^{3g=3} \left(dm^i d\bar m^i \int_\Sigma G_{zz}(\mu_i)_{\bar z}^z \int_\Sigma G_{\bar z\bar z}(\bar\mu_i)_z^{\bar z} \right). \quad (4.165)$$

Note that by construction, this integral is also invariant under a change of basis of the moduli space. There are several reasons why using G_{zz} is a natural choice. First of all, this choice is analogous to what one does in bosonic string theory. There, one integrates over the moduli space using exactly the same formula, but with G replaced by the conformal ghost b. This ghost is the BRST–partner of the SEM–tensor in exactly the same way as G is the Q–partner of T. Secondly, one can make the not unrelated observation that since $\{Q,G\} = T$, we can still use the standard arguments to show independence of the theory of the parameters in a Lagrangian of the form $L = \{Q,V\}$. The only difference is that now we also have to commute Q through G to make it act on the vacuum, but since $T_{\alpha\beta}$ itself is the derivative of the action with respect to the metric $h_{\alpha\beta}$, the terms we get in this way amount to integrating a total derivative over the moduli space. Therefore, apart from possible boundary terms these contributions vanish. Note that this reasoning also gives us an argument for using G_{zz} instead of T_{zz} (which is more or less the only other reasonable option) in (4.165): if we had chosen T_{zz} then all path integrals would have been over total derivatives on the moduli space, and apart from boundary contributions the whole theory would have become trivial.

If we consider the vector and axial charges of the full path integral measure, including the new path integral over the world–sheet metric h, we find a surprising result. Since the world–sheet metric does not transform under R–symmetry, naively one might expect that its measure does not either. However, this is clearly not correct since one should also take into account the explicit G–insertions in (4.165) that do transform under R–symmetry. From the $N=2$ superconformal algebra (or, more down–to–earth, from expressing the operators in terms of the fields), it follows that the product of G and $\bar{G}$ has vector charge zero and axial charge 2. Therefore, the total vector charge of the measure remains zero, and the axial charge gets an extra contribution of $6(g-1)$, so we find a total axial R–charge of $6(g-1)-2m(g-1)$. From this, we see that the case of complex target space dimension 3 is very special: here, the axial charge of the measure vanishes for any g, and hence the partition function is nonzero at every genus. If $m>3$ and $g>1$, the total axial charge of the measure is negative, and we have seen that we cannot cancel such a charge with local operators. Therefore, for these theories only the partition function at $g=1$ and a specific set of correlation functions at genus zero give nonzero results. Moreover, for $m=2$ and $m=1$, the results can be shown to be trivial by other arguments. Therefore, a Calabi–Yau threefold is by far the most interesting target space for a topological string theory. It is a 'happy coincidence' (see [Von05]) that this is exactly the dimension we are most interested in from the string theory perspective.

Finally, let us come back to the special cases of genus 0 and 1. At genus zero, the Riemann surface has a single point as its moduli space, so there are no extra integrals or G–insertions to worry about. Therefore, we can copy the topological field theory result saying that we have to introduce local operators with total degree (m,m) in the theory. The only remnant of the fact that we are integrating over metrics is that we should also somehow fix the remaining three symmetries of the sphere. The most straightforward way to do this is to consider 3–point functions with insertions on three labelled points. As a gauge choice, we can then for example require these points to be at the points 0, 1 and ∞ in the compactified complex–plane. For example, in the A–model on a Calabi–Yau threefold, the 3–point function of three operators corresponding to $(1,1)$–forms would thus give a nonzero result.

In the case of the torus, we have seen that there is one 'unexpected' modular parameter over which we have to integrate. This means we have to insert one G– and one $\bar{G}$–operator in the measure, which spoils the absence of the axial anomaly we had for $g=1$ in the topological field theory case. However, we also must fix the one remaining translational symmetry, which we can do by inserting a local operator at a labelled point. Thus, we can restore the axial R-charge to its zero value by choosing this to be an operator of degree $(1,1)$.

Summarizing, we have found that in topological string theory on a target Calabi–Yau 3–fold, we have a non–vanishing 3–point function of total degree (3,3) at genus zero; a non–vanishing 1–point function of degree (1,1) at genus one, and a non–vanishing partition ('zero–point') function at all genera $g > 1$.

Nonlocal Operators

In one respect, what we have achieved is great progress: we can now for any genus define a nonzero partition function (or for low genus a correlation function) of the topological string theory. On the other hand, we would also like to define correlation functions of an arbitrary number of operators at these genera. As we have seen, the insertion of extra local operators in the correlation functions is not possible, since any such insertion will spoil our carefully constructed absence of R–symmetry anomalies. Therefore, we have to introduce nonlocal operators.

There is one class of nonlocal operators which immediately becomes mind. Before we saw, using the descent equations, that for every local operator we can define a corresponding 1–form and a 2–form operator. If we check the axial and vector charges of these operators, we find that if we start with an operator of degree $(1,1)$, the 2–form operator we end up with actually has vanishing axial and vector charges. This has two important consequences. First of all, we can add the integral of this operator to our action [Von05],

$$S[t] = S_0 + t^a \int \mathcal{O}_a^{(2)},$$

without spoiling the axial and vector symmetry of the theory. Secondly, we can insert the integrated operator into correlation functions,

$$\langle \int \mathcal{O}_1^{(2)} \cdots \int \mathcal{O}_n^{(2)} \rangle$$

and still get a nonzero result by the vanishing of the axial and vector charges. These two statements are related: one obtains such correlators by differentiating $S[t]$ with respect to the appropriate t's, and then setting all $t^a = 0$.

A few remarks are in place here. First of all, recall that the integration over the insertion points of the operators can be viewed as part of the integration over the moduli space of Riemann surfaces, where now we label a certain number of points on the Riemann surface. From this point of view, the $g = 0, 1$ cases fit naturally into the same framework. We could unite the descendant fields into a world–sheet *super–field*,

$$\Phi_a = \mathcal{O}_a^{(0)} + \mathcal{O}_{a\alpha}^{(1)} \theta^\alpha + \mathcal{O}_{a\alpha\beta}^{(2)} \theta^\alpha \theta^\beta$$

where we formally replaced each dz and $d\bar{z}$ by corresponding fermionic coordinates θ^z and $\theta^{\bar{z}}$. Now, one can write the above correlators as integrals over n copies of this *super–space*,

$$\int \prod_{s=1}^{n} d^2 z_s d^2 \theta_s \, \langle \Phi_{a_1}(z_1, \theta_1) \cdots \Phi_{a_n}(z_n, \theta_n) \rangle$$

The integration prescription at genus 0 and 1 tells us to fix 3 and 1 points respectively, so we need to remove this number of super–space integrals. Then, integrating over the other super–space coordinates, the genus 0 correlators indeed become

$$\langle \mathcal{O}_{a_1}^{(0)} \mathcal{O}_{a_2}^{(0)} \mathcal{O}_{a_3}^{(0)} \int \mathcal{O}_{a_4}^{(2)} \cdots \int \mathcal{O}_{a_n}^{(2)} \rangle$$

From this prescription we note that these expressions are symmetric in the exchange of all a_i and a_j. In particular, this means that the genus zero 3–point functions at arbitrary t,

$$c_{abc}[t] = \langle \mathcal{O}_a^{(0)} \mathcal{O}_b^{(0)} \mathcal{O}_c^{(0)} \rangle [t]$$

have symmetric derivatives:

$$\frac{\partial c_{abc}}{\partial t^d} = \frac{\partial c_{abd}}{\partial t^c},$$

and similarly with permuted indices. These equations can be viewed as integrability conditions, and using the Poincaré lemma we see that they imply that

$$c_{ijk}[t] = \frac{\partial Z_0[t]}{\partial t^i \partial t^j \partial t^k}.$$

for some function $Z_0[t]$. Following the general philosophy that n–point functions are nth derivatives of the t–dependent partition function, we see that $Z_0[t]$ can be naturally thought of as the partition function at genus zero. Similarly, the partition function at genus 1 can be defined by integrating up the one-point functions once.

The quantities we have calculated above should be semi–topological invariants, meaning that they only depend on 'half' of the moduli (either the Kähler ones or the complex structure ones) of the target space. For example, in the A–model we find the Gromov–Witten invariants. In the B–model, it turns out that $F_0[t] = \ln Z_0[t]$ is actually a quantity we already knew: it is the prepotential of the Calabi–Yau manifold. A discussion of why this is the case can be found in the paper [BCO94]. The higher genus partition functions can be thought of as 'quantum corrections' to the prepotential.

Finally, there is a type of operator we have not discussed at all so far. Recall that in the topological string theory, the metric itself is now a dynamical field. We could not include the metric in our physical operators, since this would spoil the topological invariance. However, the metric is part of a

Q–multiplet, and the highest field in this multiplet is a scalar field which is usually labelled φ. (It should not be confused with the fields ϕ^i.) We can get more correlation functions by inserting operators φ^k and the operators related to them by the descent equations into the correlation functions. These operators are called 'gravitational descendants'. Even the case where the power is $k = 0$ is nontrivial; it does not insert any operator, but it does label a certain point, and hence changes the moduli space one integrates over. This operator is called the 'puncture operator'.

All of this seems to lead to an enormous amount of semi-topological target space invariants that can be calculated, but there are many recursion relations between the several correlators. This is similar to how we showed before that all correlators for the cohomological field theories follow from the 2–and 3–point functions on the sphere. Here, it turns out that the set of all correlators has a structure which is related to the theory of integrable hierarchies. Unfortunately, a discussion of this is outside the scope of both these lectures and the author's current knowledge.

The Holomorphic Anomaly

We have now defined the partition function and correlation functions of topological string theory, but even though the expressions we obtained are much simpler than the path integrals for ordinary quantum field or string theories, it would still be very hard to explicitly calculate them. Fortunately, it turns out that the t–dependent partition and correlation functions are actually 'nearly holomorphic' in t, and this is a great aid in exactly calculating these quantities.

Let us make this 'near holomorphy' more precise. As we have seen, calculating correlation functions of primary operators in topological string theories amounts to taking t–derivatives of the corresponding perturbed partition function $Z[t]$ and consequently setting $t = 0$. Recall that $Z[t]$ is defined through adding terms to the action of the form

$$t^a \int_\Sigma \mathcal{O}_a^{(2)}, \qquad (4.166)$$

Let us for definiteness consider the B–twisted model. We want to show that the above term is Q_B–exact. For simplicity, we assume that $\mathcal{O}_a^{(2)}$ is a bosonic operator, but what we are about to say can by inserting a few signs straightforwardly be generalized to the fermionic case. From the descent equations we studied above, we know that

$$(\mathcal{O}_a^{(2)})_{+-} = -\{G_+, [G_-, \mathcal{O}_a^{(0)}]\}, \qquad (4.167)$$

where G_+ is the charge corresponding to the current G_{zz}, and G_- the one corresponding to $G_{\bar z \bar z}$. We can in fact express $G_\pm$ in terms of the $N = (2,2)$ supercharges $\mathcal{Q}$. So, according to [Von05], we have

$$H = 2\pi i(L_0 + \bar{L}_0) = \frac{1}{2}\{Q_+, \bar{Q}_+\} - \frac{1}{2}\{Q_-, \bar{Q}_-\}P$$

$$= 2\pi i(L_0 - \bar{L}_0) = \frac{1}{2}\{Q_+, \bar{Q}_+\} + \frac{1}{2}\{Q_-, \bar{Q}_-\}.$$

Thus, we find that the left– and right–moving SEM charges satisfy

$$T_+ = 2\pi i L_0 = \frac{1}{2}\{Q_+, \bar{Q}_+\} \quad T_- = 2\pi i \bar{L}_0 = -\frac{1}{2}\{Q_-, \bar{Q}_-\}.$$

To find G in the B–model, we should write these charges as commutators with respect to $Q_B = \bar{Q}_+ + \bar{Q}_-$, which gives

$$T_+ = \frac{1}{2}\{Q_B, Q_+\} \quad T_- = -\frac{1}{2}\{Q_B, Q_-\},$$

so we arrive at the conclusion that for the B–model,

$$G_+ = \frac{1}{2}Q_+ \quad G_- = -\frac{1}{2}Q_-.$$

Now, we can rewrite (4.167) as

$$(\mathcal{O}_a^{(2)})_{+-} = -\{G_+, [G_-, \mathcal{O}_a^{(0)}]\} = \frac{1}{4}\{Q_+, [Q_-, \mathcal{O}_a^{(0)}]\} \qquad (4.168)$$

$$= \frac{1}{8}\{Q_B, [(Q_- - Q_+), \mathcal{O}_a^{(0)}]\},$$

which proves our claim that $\mathcal{O}_a^{(2)}$ is Q_B–exact.

An $N = (2,2)$ sigma model with a real action does, apart from the term (4.166), also contain a term

$$t^{\bar{a}} \int_\Sigma \bar{\mathcal{O}}_a^{(2)}, \qquad (4.169)$$

where $t^{\bar{a}}$ is the complex conjugate of t^a. It is not immediately clear that $\bar{\mathcal{O}}_a^{(2)}$ is a physical operator: we have seen that physical operators in the B–model correspond to forms that are $\bar{\partial}$–closed, but the complex conjugate of such a form is ∂–closed. However, by taking the complex conjugate of (4.168), we see that

$$(\bar{\mathcal{O}}_a^{(2)})_{+-} = \frac{1}{8}\{Q_B, [(\bar{Q}_- - \bar{Q}_+), \bar{\mathcal{O}}_a^{(0)}]\},$$

so not only is the operator Q_B–closed, it is even Q_B–exact. This means that we can add terms of the form (4.169) to the action, and taking $t^{\bar{a}}$–derivatives inserts Q_B–exact terms in the correlation functions. Naively, we would expect this to give a zero result, so all the physical quantities seem to be t–independent, and thus holomorphic in t. We will see in a moment that this naive expectation turns out to be almost right, but not quite.

However, before doing so, let us comment briefly on the generalization of the above argument in the case of the A–model. It seems that a straightforward generalization of the argument fails, since Q_A is its own complex conjugate, and the complex conjugate of the de Rham operator is also the same operator. However, note that the $N = (2,2)$–theory has a different kind of 'conjugation symmetry': we can exchange the two supersymmetries, or in other words, exchange θ^+ with $\bar\theta^+$ and θ^- with $\bar\theta^-$. This exchanges Q_A with an operator which we might denote as $Q_{\bar A} \equiv \bar Q_+ + \bar Q_-$. Using the above argument, we then find that the physical operators $\mathcal{O}_a^{(2)}$ are $Q_{\bar A}$–exact, and that their conjugates in the new sense are Q_A–exact. We can now add these conjugates to the action with parameters $t^{\bar a}$, and we again naively find independence of these parameters. In this case it is less natural to choose t^a and $t^{\bar a}$ to be complex conjugates, but we are free to choose this particular 'background point' and study how the theory behaves if we then vary t^a and $t^{\bar a}$ independently.

Now, let us see how the naive argument showing independence of the theory of $t^{\bar a}$ fails. In fact, the argument above would certainly hold for topological *field* theories. However, in topological string theories (see [Von05]), we have to worry about the insertions in the path integral of

$$G \cdot \mu_i \equiv \int d^2 z\, G_{zz}\,(\mu_i)^z_{\bar z},$$

and their complex conjugates, when commuting the Q_B towards the vacuum and making sure it gives a zero answer. Indeed, the Q_B–commutator of the above factor is not zero, but it gives

$$\{Q_B, G \cdot \mu_i\} = T \cdot \mu_i.$$

Now recall that $T_{\alpha\beta} = \partial_{h^{\alpha\beta}} S$. We did not give a very precise definition of μ_i above, but we know that it parameterizes the change in the metric under an infinitesimal change of the coordinates m_i on the moduli space. One can make this intuition precise, and then finds the following 'chain rule': $T \cdot \mu_i = \partial_{m^i} S$. Inserting this into the partition function, we find that

$$\frac{\partial F_g}{\partial t^{\bar a}} =$$

$$\int_{M_g} \prod_{i=1}^{3g-3} dm^i d\bar m^i \sum_{j,k} \frac{\partial^2}{\partial m^j \partial \bar m^k} \left\langle (\prod_{l\neq j}\int \mu_l \cdot G)(\prod_{l\neq k}\int \bar\mu_l \cdot \bar G)\int \bar{\mathcal{O}}_a^{(2)} \right\rangle,$$

where $F_g = \ln Z_g$ is the free energy at genus g, and the reason F_g appears in the above equation instead of Z_g is, as usual in quantum field theory, that the expectation values in the r.h.s. are normalized such that $\langle 1 \rangle = 1$, and so the l.h.s. should be normalized accordingly and equal $Z_g^{-1}\partial_{\bar a} Z_g = \partial_{\bar a} F_g$ [Von05].

Thus, as we have claimed before, we are integrating a total derivative over the moduli space of genus g surfaces. If the moduli space did not have a boundary, this would indeed give zero, but in fact the moduli space does have a boundary. It consists of the moduli which make the genus g surface degenerate. This can happen in two ways: an internal cycle of the genus g

surface can be pinched, leaving a single surface of genus $g-1$, as in Figure 4.25 (a), or the surface can split up into two surfaces of genus g_1 and $g_2 = g - g_1$, as depicted in Figure 4.25 (b). By carefully considering the boundary contributions to the integral for these two types of boundaries, it was shown in [BCO94] that

$$\frac{\partial F_g}{\partial t^{\bar{a}}} = \frac{1}{2} c_{\bar{a}\bar{b}\bar{c}} e^{2K} G^{\bar{b}d} G^{\bar{c}e} \left(D_d D_e F_{g-1} + \sum_{r=1}^{g-1} D_d F_r D_e F_{g-r} \right),$$

where G is the so-called *Zamolodchikov metric* on the space parameterized by the coupling constants $t^a, t^{\bar{a}}$; K is its Kähler potential, and the D_a are covariant derivatives on this space. The coefficients $c_{\bar{a}\bar{b}\bar{c}}$ are the 3–point functions on the sphere of the operators $\bar{\mathcal{O}}_a^{(0)}$. We will not derive the above formula in detail, but the reader should notice that the contributions from the two types of boundary are quite clear.

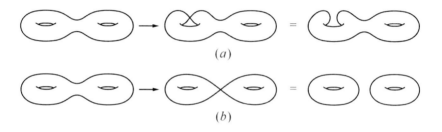

(a)

(b)

Fig. 4.25. At the boundary of the moduli space of genus g surfaces, the surfaces degenerate because certain cycles are pinched. This either lowers the genus of the surface (a) or breaks the surface into two lower genus ones (b) (see text for explanation).

Using this formula, one can inductively determine the $t^{\bar{a}}$ dependence on the partition functions if the holomorphic t^a–dependence is known. Holomorphic functions on complex spaces (or more generally holomorphic sections of complex vector bundles) are quite rare: usually, there is only a nD space of such functions. The same turns out to hold for our topological string partition functions: even though they are not quite holomorphic, their anti–holomorphic behavior is determined by the holomorphic dependence on the coordinates, and as a result there is a finite number of coefficients which determines them.

Thus, just from the above structure and without doing any path integrals, one can already determine the topological string partition functions up to a finite number of constants. This leads to a feasible program for completely determining the topological string partition function for a given target space and at given genus. From the holomorphic anomaly equation, one first has to find the general form of the partition function. Then, all one has left to do is

4.3 Dynamics of Fields and Strings

to fix the unknown constants. Here, the fact that in the A−model the partition function counts a number of points becomes our help: by requiring that the A−model partition functions are integral, one can often fix the unknown constants and completely determine the t−dependent partition function. In practice, the procedure is still quite elaborate, so we will not describe any examples here, but several have been worked out in detail in the literature. Once again, the pioneering work for this can be found in the paper [BCO94].

4.3.10 Geometrical Transitions

Conifolds

Recall that a *conifold* is a generalization of the notion of a manifold. Unlike manifolds, a conifold can (or, should) contain conical singularities i.e., points whose neighborhood looks like a cone with a certain base. The base is usually a 5D manifold.

In string theory, a *conifold transition* represents such an evolution of the Calabi–Yau manifold in which its fabric rips and repairs itself, yet with mild and acceptable physical consequences in the context of string theory. However, the tears involved are more severe than those in an 'weaker' *flop transition* (see [Gre00]). The geometrically singular conifolds were shown to lead to completely smooth physics of strings. The divergences are 'smeared out' by D3–branes wrapped on the shrinking 3–sphere S^3, as originally pointed out by A. Strominger, who, together with D. Morrison and B. Greene have also found that the topology near the conifold singularity can undergo a topological phase–transition. It is believed that nearly all Calabi–Yau manifolds can be connected via these 'critical transitions'.

More precisely, the conifold is the simplest example of a non–compact Calabi–Yau 3–fold: it is the set of solutions to the equation

$$x_1 x_2 - x_3 x_4 = 0$$

in $\mathbb{C}^4$. The resulting manifold is a cone, meaning in this case that any real multiple of a solution to this equation is again a solution. The point $(0,0,0,0)$ is the 'tip' of this cone, and it is a singular point of the solution space. Note that by writing

$$x_1 = z_1 + iz_2, \qquad x_2 = z_1 - iz_2, \qquad x_3 = z_3 + iz_4, \qquad x_4 = -z_3 + iz_4,$$

where the z_i are still *complex* numbers, one can also write the equation as

$$z_1^2 + z_2^2 + z_3^2 + z_4^2 = 0.$$

Writing each z_i as $a_i + ib_i$, with a_i and b_i real, we get the two equations

$$|a|^2 - |b|^2 = 0, \qquad a \cdot b = 0. \qquad (4.170)$$

Here $a \cdot b = \sum_i a_i b_i$ and $|a|^2 = a \cdot a$. Since the geometry is a cone, let us focus on a 'slice' of this cone given by

$$|a|^2 + |b|^2 = 2r^2,$$

for some $r \in \mathbb{R}$. On this slice, the first equation in (4.170) becomes

$$|a|^2 = r^2, \qquad (4.171)$$

which is the equation defining a 3–sphere S^3 of radius r. The same holds for b, so both a and b lie on 3–spheres. However, we also have to take the second equation in (4.170) into account. Let us suppose that we fix an a satisfying (4.171). Then b has to lie on a 3–sphere, but also on the plane through the origin defined by $a \cdot b = 0$. That is, b lies on a 2–sphere. This holds for every a, so the slice we are considering is a fibration of 2–spheres over the 3–sphere. With a little more work, one can show that this fibration is trivial, so the conifold is a cone over $S^2 \times S^3$.

Since the conifold is a singular geometry, we would like to find geometries which approximate it, but which are non–singular. There are two interesting ways in which this can be done. The simplest way is to replace the defining equation by

$$x_1 x_2 - x_3 x_4 = \mu^2. \qquad (4.172)$$

From the two equations constraining a and b, we now see that $|a|^2 \geq \mu^2$. In other words, the parameter r should be at least μ. At $r = \mu$, the a–sphere still has finite radius μ, but the b–sphere shrinks to zero size. This geometry is called the *deformed conifold*. Even though this is not clear from the picture, from the equation (4.172) one can straightforwardly show that it is nonsingular. One can also show that it is topologically equivalent to the cotangent bundle on the 3–sphere, T^*S^3. Here, the S^3 on which the cotangent bundle is defined is exactly the S^3 at the 'tip' of the deformed conifold.

The second way to change the conifold geometry arises from studying the two equations

$$x_1 A + x_3 B = 0, \qquad x_4 A + x_2 B = 0. \qquad (4.173)$$

Here, we require A and B to be homogeneous complex coordinates on a $\mathbb{CP}^1$, i.e.,

$$(A, B) \neq (0, 0), \qquad (A, B) \sim (\lambda A, \lambda B)$$

where λ is any nonzero complex number. If one of the x_i is nonzero, say x_1, one can solve for A or B, e.g., $A = -\frac{x_3 B}{x_1}$, and insert this in the other equation to get

$$x_1 x_2 - x_3 x_4 = 0$$

which is the conifold equation. However, if all x_i are zero, any A and B solve the system of equations (4.173). In other words, we have constructed a geometry which away from the former singularity is completely the same as the conifold, but the singularity itself is replaced by a $\mathbb{CP}^1$, which topologically is the same as an S^2. From the defining equations one can again show that the resulting geometry is nonsingular, so we have now replaced our conifold geometry by the so-called *resolved conifold*.

Topological D–branes

Since topological string theories are in many ways similar to an ordinary (bosonic) string theories, one natural question which arises is: are there also open topological strings which can end on D–branes? To answer the above question rigorously, we would have to study boundary conditions on world–sheets with boundaries which preserve the Q–symmetry.

In the A–model, one can only construct 3D–branes wrapping so-called 'Lagrangian' submanifolds of M. Here, 'Lagrangian' means that the Kähler form ω vanishes on this submanifold. In the B–model, one can construct D–branes of any even dimension, as long as these branes wrap *holomorphic* submanifolds of M.

Just like in ordinary string theory, when we consider open topological strings ending on a D–brane, there should be a field theory on the brane world–volume describing the low–energy physics of the open strings. Moreover, since we are studying topological theories, one may expect such a theory to inherit the property that it only depends on a restricted amount of data of the manifolds involved. A key example is the case of the A–model on the deformed conifold, $M = T^*S^3$, where we wrap ND–branes on the S^3 in the base. (One can show that this is indeed a Lagrangian submanifold.) In ordinary string theory, the world–volume theory on ND–branes has a $U(N)$ gauge symmetry, so putting the ingredients together we can make the guess that the world–volume theory is a 3D topological field theory with $U(N)$ gauge symmetry. There is really only one candidate for such a theory: the *Chern–Simons gauge theory*. Recall that it consists of a single $U(N)$ gauge field, and has the action

$$S = \frac{k}{4\pi} \int_{S^3} \text{Tr}\left(A \wedge dA + \frac{2}{3} A \wedge A \wedge A\right). \qquad (4.174)$$

Before the invention of D–branes, E. Witten showed that this is indeed the theory one gets. In fact, he showed even more: this theory actually describes the *full* topological string–field theory on the D–branes, even without going to a low–energy limit [Wit95].

Let us briefly outline the argument that gives this result. In his paper, Witten derived the open string–field theory action for the open A–model topological string; it reads

$$S = \int \text{Tr}\left(\mathcal{A} * Q_A \mathcal{A} + \frac{2}{3} \mathcal{A} * \mathcal{A} * \mathcal{A}\right).$$

The form of this action is very similar to Chern–Simons theory, but its interpretation is completely different: $\mathcal{A}$ is a *string–field* (a wave function on the space of all maps from an open string to the space–time manifold), Q_A is the topological symmetry generator, which has a natural action on the string–field, and $*$ is a certain noncommutative product. Witten shows that the topological properties of the theory imply that only the constant maps contribute, so $\mathcal{A}$ becomes a field on M – and since open strings can only end on D–branes, it actually becomes a field on S^3. Moreover, recall that Q_A can be interpreted as a de Rham differential. Using these observations and the precise definition of the star product one can indeed show that the string–field theory action reduces to Chern–Simons theory on S^3.

4.3.11 Topological Strings and Black Hole Attractors

Topological string theory is naturally related to *black hole dynamics*. Namely, critical string theory compactified on Calabi–Yau manifolds has played a central role in both the mathematical and physical development of modern string theory. The physical relevance of the data provided by the topological string $\hat{c} = 6$ (of A and B types) has been that it computes F–type terms in the corresponding four dimensional theory [BCO94, AGN94]. These higher–derivative F–type terms for Type II superstring on a Calabi–Yau manifold are of the general form

$$\int d^4x d^4\theta (W_{ab}W^{ab})^g F_g(X^\Lambda), \qquad (4.175)$$

where W_{ab} is the *graviphoton super–field* of the $N = 2$ *super–gravity* and X^Λ are the *vector multiplet fields*. The lowest component of W is F the graviphoton field strength and the highest one is the Riemann tensor. The lowest components of X^Λ are the complex scalars parameterizing Calabi–Yau moduli and their highest components are the associated $U(1)$ vector–fields. These terms contribute to multiple graviphoton–graviton scattering. (4.175) includes (after θ integrations) an $R^2 F^{2g-2}$ term. The topological string partition function Z_{top} represents the canonical ensemble for multi–particle spinning five dimensional black holes [BMP97, KKV99].

Recently, [OSV04] proposed a simple and direct relationship between the second–quantized topological string partition function Z_{top} and black hole partition function Z_{BH} in four dimensions of the form

$$Z_{BH}(p^\Lambda, \phi^\Lambda) = |Z_{\text{top}}(X^\Lambda)|^2, \qquad \text{where} \qquad X^\Lambda = p^\Lambda + \frac{i}{\pi}\phi^\Lambda$$

in a certain *Kähler gauge*. The l.h.s. here is evaluated as a function of integer magnetic charges p^Λ and continuous electric potentials ϕ^Λ, which are conjugate to integer electric charges q_Λ. The r.h.s. is the holomorphic square of the partition function for a gas of topological strings on a Calabi–Yau whose moduli are those associated to the charges/potentials $(p^\Lambda, \phi^\Lambda)$ via the attractor equations [OSV04]. Both sides of (4.176) are defined in a perturbation expansion in $1/Q$, where Q is the graviphoton charge carried by the black hole.[45]

---
[45] The string coupling g_s is in a hypermultiplet and decouples from the computation.

The non–perturbative completion of either side of (4.176) might in principle be defined as the partition function of the holographic CFT dual to the black hole, as in [SV96b]. Then we have the triple equality,

$$Z_{CFT} = Z_{BH} = |Z_{\text{top}}|^2.$$

The existence of fundamental connection between 4D black holes and the topological string might have been anticipated from the following observation. Calabi–Yau spaces have two types of moduli: Kähler and complex structure. The world–sheet twisting which produces the A (B) model topological string from the critical superstring eliminates all dependence on the complex structure (Kähler) moduli at the perturbative level. Hence the perturbative topological string depends on only half the moduli. Black hole entropy on the other hand, insofar as it is an intrinsic property of the black hole, cannot depend on any externally specified moduli. What happens at leading order is that the moduli in vector multiplets are driven to attractor values at the horizon which depend only on the black hole charges and not on their asymptotically specified values. Hypermultiplet vevs on the other hand are not fixed by an attractor mechanism but simply drop out of the entropy formula. It is natural to assume this is valid to all orders in a $1/Q$ expansion. Hence the perturbative topological string and the large black hole partition functions depend on only half the Calabi–Yau moduli. It would be surprising if string theory produced two functions on the same space that were not simply related. Indeed [OSV04] argued that they were simply related as in (4.176).

Supergravity Area–Entropy Formula

Recall that a well–known hypothesis by J. Bekenstein and S. Hawking states that *the entropy of a black hole is proportional to the area of its horizon* (see [HE79]). This *area* is a function of the black hole mass, or in the extremal case, of its charges. Here we review the leading *semiclassical area–entropy formula* for a general $N = 2$, $d = 4$ extremal black hole characterized by magnetic and electric charges (p^Λ, q_Λ), recently reviewed in [OSV04]. The asymptotic values of the moduli in vector multiplets, parameterized by complex projective coordinates X^Λ, $(\Lambda = 0, 1, \ldots, n_V)$ in the black hole solution, are arbitrary. These moduli couple to the electromagnetic fields and accordingly vary as a function of the radius. At the horizon they approach an attractor point whose location in the moduli space depends only on the charges. The locations of these attractor points can be found by looking for supersymmetric solutions with constant moduli. They are determined by the attractor equations,

$$p^\Lambda = \text{Re}[CX^\Lambda], \qquad q_\Lambda = \text{Re}[CF_{0\Lambda}], \qquad (4.176)$$

where $F_{0\Lambda} = \partial F_0/\partial X^\Lambda$ are the *holomorphic periods*, and the subscript 0 distinguishes these from the string loop corrected periods to appear in the next subsection. Both (p^Λ, q_Λ) and $(X^\Lambda, F_{0\Lambda})$ transform as vectors under the $Sp(2n + 2; Z)$ duality group.

4 Nonlinear Dynamics of Path Integrals

The $(2n_v+2)$ real equations (4.176) determine the (n_v+2) complex quantities (C, X^Λ) up to Kähler transformations, which act as

$$K \to K - f(X) - \bar{f}(\bar{X}), \quad X^\Lambda \to e^f X^\Lambda, \quad F_0 \to e^{2f} F_0, \quad C \to e^{-f} C,$$

where the *Kähler potential* K is given by

$$e^{-K} = i(\bar{X}^\Lambda F_{0\Lambda} - X^\Lambda \bar{F}_{0\Lambda}).$$

We could at this point set $C = 1$ and fix the Kähler gauge but later we shall find other gauges useful. It is easy to see that (as required) the charges (p^Λ, q_Λ) determined by the attractor equations (4.176) are invariant under Kähler transformations. Given the horizon attractor values of the moduli determined by (4.176) the *Bekenstein–Hawking entropy* S_{BH} may be written as

$$S_{BH} = \frac{1}{4}\text{Area} = \pi |Q|^2,$$

where $Q = Q_m + iQ_e$ is a complex combination of the magnetic and electric graviphoton charges and

$$|Q|^2 = \frac{i}{2}\left(q_\Lambda \bar{C} \bar{X}^\Lambda - p^\Lambda \bar{C} \bar{F}_{0\Lambda}\right) = \frac{C\bar{C}}{4} e^{-K}.$$

The normalization of Q here is chosen so that $|Q|$ equals the radius of the two sphere at the horizon.

It is useful to rephrase the above results in the context of type IIB superstrings in terms of geometry of Calabi–Yau. In this case the attractor equations fix the complex geometry of the Calabi–Yau. The electric/magentic charges correlate with three cycles of Calabi–Yau. Choosing a symplectic basis for the three cycles gives a choice of the splitting to electric and magnetic charges. Let A_Λ denote a basis for the electric three cycles, B^Σ the dual basis for the magnetic charges and Ω the holomorphic 3–form at the attractor point. Ω is fixed up to an overall multiplication by a complex number $\Omega \to \lambda \Omega$. There is a unique choice of λ such that the resulting Ω has the property that

$$p^\Lambda = \int_{A_\Lambda} \text{Re}\,\Omega = \text{Re}[CX^\Lambda], \qquad q_\Lambda = \int_{B^\Lambda} \text{Re}\,\Omega = \text{Re}[CF_{0\Lambda}],$$

where $\text{Re}\,\Omega = \frac{1}{2}(\Omega + \bar{\Omega})$.

In terms of this choice, the black hole entropy can be written as

$$S_{BH} = \frac{\pi}{4} \int_{CY} \Omega \wedge \bar{\Omega}.$$

Higher–Order Corrections

F–term corrections to the action are encoded in a string loop corrected holomorphic prepotential

$$F(X^\Lambda, W^2) = \sum_{h=0}^{\infty} F_h(X^\Lambda) W^{2h}, \qquad (4.177)$$

where F_h can be computed by topological string amplitudes (as we review in the next section) and W^2 involves the square of the anti–self dual graviphoton field strength. This obeys the homogeneity equation

$$X^\Lambda \partial_\Lambda F(X^\Lambda, W^2) + W \partial_W F(X^\Lambda, W^2) = 2 F(X^\Lambda, W^2). \qquad (4.178)$$

Near the black hole horizon, the attractor value of W^2 obeys $C^2 W^2 = 256$, and therefore the exact attractor equations read

$$p^\Lambda = \mathrm{Re}[C X^\Lambda], \qquad q_\Lambda = \mathrm{Re}\left[C F_\Lambda \left(X^\Lambda, \frac{256}{C^2} \right) \right]. \qquad (4.179)$$

This is essentially the only possibility consistent with *symplectic invariance*. It has been then argued that the entropy as a function of the charges is

$$S_{BH} = \frac{\pi i}{2} (q_\Lambda \bar{C} \bar{X}^\Lambda - p^\Lambda \bar{C} \bar{F}_\Lambda) + \frac{\pi}{2} \mathrm{Im}[C^3 \partial_C F], \qquad (4.180)$$

where F_Λ, X^Λ and C are expressed in terms of the charges using (4.179).

Topological Strings

Partition Functions for Black Hole and Topological Strings. The notion of *topological string* was introduced in [Wit90]. Subsequently a connection between them and superstring was discovered: It was shown in [BCO94, AGN94], that the superstring loop corrected F–terms (4.177) can be computed as topological string amplitudes. The purpose of this subsection is to translate the super–gravity notation of the previous section to the topological string notation.

The second quantized partition function for the topological string may be written

$$Z_{\mathrm{top}}(t^A, g_{\mathrm{top}}) = \exp\left[F_{\mathrm{top}}(t^A, g_{\mathrm{top}}) \right], \qquad \text{where}$$
$$F_{\mathrm{top}}(t^A, g_{\mathrm{top}}) = \sum_h g_{\mathrm{top}}^{2h-2} F_{\mathrm{top}, h}(t^A),$$

and $F_{\mathrm{top}, h}$ is the h–loop *topological string amplitude*. The Kähler moduli are expressed in the flat coordinates

$$t^A = \frac{X^A}{X^0} = \theta^A + ir^A,$$

where r^A are the Kähler classes of the Calabi–Yau M and θ^A are periodic $\theta^A \sim \theta^A + 1$.

We would like to determine relations between super–gravity quantities and topological string quantities. Using the homogeneity property (4.178) and the expansion (4.177), the holomorphic prepotential in super–gravity can be expressed as

$$F(CX^A, 256) = (CX^0)^2 F\left(\frac{X^A}{X^0}, \frac{256}{(CX^0)^2}\right)$$
$$= \sum_{h=0}^{\infty} (CX^0)^{2-2h} f_h(t^A), \qquad (4.181)$$

where $f_h(t^A)$ is related to $F_h(X^A)$ in (4.177) as

$$f_h(t^A) = 16^{2h} F_h\left(\frac{X^A}{X^0}\right).$$

This suggests an identification of the form $f_h(t^A) \sim F_{\text{top},h}(t^A)$ and $g_{\text{top}} \sim (CX^0)^{-1}$. For later purposes, we need precise relations between super–gravity and topological string quantities, including numerical coefficients. These can be determined by studying the limit of a large Calabi–Yau space.

In the *super-gravity notation*, the genus 0 and 1 terms in the large volume are given by

$$F(CX^A, 256) = C^2 D_{ABC} \frac{X^A X^B X^C}{X^0} - \frac{1}{6} c_{2A} \frac{X^A}{X^0} + \cdots$$
$$= (CX^0)^2 D_{ABC} t^A t^B t^C - \frac{1}{6} c_{2A} t^A + \cdots,$$

where $\quad c_{2A} = \int_M c_2 \wedge \alpha_A,$

with c_2 being the *second Chern class* of M, and $C_{ABC} = -6 D_{ABC}$ are the 4–cycle intersection numbers. These terms are normalized so that the mixed entropy S_{BH} is given by (4.180). On the other hand, the topological string amplitude in this limit is given by

$$F_{\text{top}} = -\frac{(2\pi)^3 i}{g_{\text{top}}^2} D_{ABC} t^A t^B t^C - \frac{\pi i}{12} c_{2A} t^A + \cdots \qquad (4.182)$$

The normalization here is fixed by the holomorphic anomaly equations in [BCO94], which are nonlinear equations for $F_{\text{top},h}$.

Comparing the one–loop terms in (4.181) and (4.182), which are independent of g_{top}, we find

4.3 Dynamics of Fields and Strings   577

$$F(CX^\Lambda, 256) = -\frac{2i}{\pi} F_{\text{top}}(t^\Lambda, g_{\text{top}}).$$

Given this, we can compare the genus 0 terms to find

$$g_{\text{top}} = \pm \frac{4\pi i}{CX^0}.$$

This implies

$$\ln Z_{BH} = -\pi \operatorname{Im}\left[F(CX^\Lambda, 256)\right] = F_{\text{top}} + \bar{F}_{\text{top}} \quad \text{and}$$
$$Z_{BH}(\phi^\Lambda, p^\Lambda) = |Z_{\text{top}}(t^\Lambda, g_{\text{top}})|^2, \quad \text{with}$$
$$t^\Lambda = \frac{p^\Lambda + i\phi^\Lambda/\pi}{p^0 + i\phi^0/\pi}, \quad g_{\text{top}} = \pm \frac{4\pi i}{p^0 + i\phi^0/\pi}.$$

Supergravity Approach to Z_{BH}. The above relation

$$Z_{BH} = |Z_{\text{top}}|^2 \tag{4.183}$$

can have a simpler super–gravity derivation [OSV04].

A main ingredient in this derivation is the observation that the $N = 2$ super–gravity coupled to vector multiplets can be written as the action

$$S = \int d^4x d^4\theta \,(\text{super} - -\text{volume form}) + h.c. = \int d^4x \sqrt{-g} R + ..., \tag{4.184}$$

where the super–volume form in the above depends non–trivially on curvature of the fields. This reproduces the ordinary action after integrating over $d^4\theta$ and picking up the θ^4 term in the super–volume. In the context of black holes the boundary terms accompanying (4.184) give the classical black hole entropy.

We now become the derivation of (4.183). As was observed in [BCO94, AGN94], topological string computes the terms

$$F = \sum_{h=0}^{\infty} \int d^4x d^4\theta F_h(X)(W^2)^g + c.c. \tag{4.185}$$

There are various terms one can get from the above action after integrating over $d^4\theta$. Let us concentrate on one of the terms which turns out to be the relevant one for us: Take the top components of X^Λ and W^2, and absorb the $d^4\theta$ integral from the super–volume measure as in (4.184). We will work in the gauge $X^0 \sim 1$ and thus $C \sim 1/g_{\text{top}}$. As noted before in the near–horizon black hole geometry in this gauge the top component $W^2 \sim 1/C^2 \sim g_{\text{top}}^2$ and the X^Λ are fixed by the attractor mechanism. Thus, we have the black hole free energy

$$\ln Z_{BH} = \sum_{h=0}^{\infty} g_{\text{top}}^{2h} F_{\text{top},h}(X^\Lambda/X^0) \int d^4x d^4\theta + \text{c.c.}$$

$$= \sum_{g=0}^{\infty} (g_{\text{top}})^{2h-2} F_{\text{top},h}(X^\Lambda/X^0) + \text{c.c.}$$

$$= 2 \text{ Re } F_{\text{top}}, \quad (\text{using } \int d^4x d^4\theta \sim 1/g_{\text{top}}^2).$$

Upon exponentiation this leads to (4.183).

Here we have shown that if we consider one absorption of θ^4 term in (4.185) upon $d^4\theta$ integral we get the desired result. That there be no other terms is not obvious. For example another way to absorb the θ's would have given the familiar term $R^2 F^{2g-2}$ where F is the graviphoton field. However, such terms do not contribute in the black hole background. It would be nice to find a simple way to argue why these terms do not contribute and that we are left with this simple absorption of the θ integrals.

4.4 Chaos Field Theory

In [Cvi00], Cvitanovic re–examined the path–integral formulation and the role that the classical solutions play in quantization of strongly nonlinear fields. In the path integral formulation of a field theory the dominant contributions come from saddle–points, the classical solutions of equations of motion. Usually one imagines *one dominant saddle point*, the 'vacuum' (see Figure 4.26, (a)).

The *Feynman diagrams* of quantum electrodynamics (QED) and quantum chromodynamics (QCD), associated to their path integrals, give us a visual and intuitive scheme to calculate the correction terms to this starting semiclassical, *Gaussian saddlepoint approximation*. But there might be other saddles (Figure 4.26, (b)). That field theories might have a rich repertoire of classical solutions became apparent with the discovery of *instantons* [BPS75], analytic solutions of the classical $SU(2)$ *Yang–Mills relation*, and the realization that the associated *instanton vacua* receive contributions from countable ∞'s of saddles. What is not clear is whether these are the important classical saddles. Cvitanovic asks the question: could it be that the strongly nonlinear theories are dominated by altogether different classical solutions?

The search for the classical solutions of nonlinear field theories such as the *Yang–Mills* and *gravity* has so far been neither very successful nor very systematic. In modern field theories the main emphasis has been on symmetries (compactly collected in action functionals that define the theories) as guiding principles in writing down the actions. But writing down a differential equation is only the start of the story; even for systems as simple as 3 coupled ordinary differential equations one in general has no clue what the nature of the long time solutions might be.

4.4 Chaos Field Theory

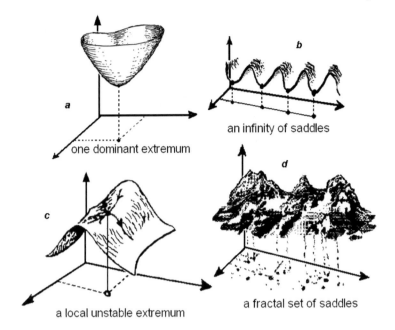

Fig. 4.26. Path integrals and chaos field theory (see text for explanation).

These are hard problems, and in explorations of modern field theories the dynamics tends to be is neglected, and understandably so, because the wealth of the classical solutions of nonlinear systems can be truly bewildering. If the classical behavior of these theories is anything like that of the field theories that describe the classical world – the hydrodynamics, the magneto–hydrodynamics, the *Burgers dynamical system* (1.127), *Ginzburg–Landau equation* (1.124), or *Kuramoto–Sivashinsky equation* (1.126), there should be very many solutions, with very few of the important ones analytical in form; the strongly nonlinear classical field theories are turbulent, after all. Furthermore, there is not a dimmest hope that such solutions are either beautiful or analytic, and there is not much enthusiasm for grinding out numerical solutions as long as one lacks ideas as what to do with them.

By late 1970's it was generally understood that even the simplest nonlinear systems exhibit chaos. Chaos is the norm also for generic Hamiltonian flows, and for path integrals that implies that instead of a few, or countably few saddles (Figure 4.26, (c)), classical solutions populate fractal sets of saddles (Figure 4.26, (d)). For the path–integral formulation of quantum mechanics such solutions were discovered and accounted for by [Gut90] in late 1960's. In this framework the spectrum of the theory is computed from a set of its unstable classical periodic solutions and quantum corrections. The new aspect is that the individual saddles for classically chaotic systems are nothing like the harmonic oscillator degrees of freedom, the quarks and gluons of QCD

– they are all unstable and highly nontrivial, accessible only by numerical techniques.

So, if one is to develop a semiclassical field theory of systems that are *classically chaotic* or *turbulent*, the problem one faces is twofold [Cvi00]

A. Determine, classify, and order by relative importance the classical solutions of nonlinear field theories.
B. Develop methods for calculating perturbative corrections to the corresponding classical saddles.

4.5 Non–Physical Applications of Path Integrals

4.5.1 Stochastic Optimal Control

A path–integral based optimal control model for nonlinear stochastic systems has recently been developed in [Kap05]. The author addressed the role of noise and the issue of efficient computation in stochastic optimal control problems. He considered a class of nonlinear control problems that can be formulated as a *path integral* and where the noise plays the role of temperature. The path integral displays symmetry breaking and there exist a critical noise value that separates regimes where optimal control yields qualitatively different solutions. The path integral can be computed efficiently by Monte Carlo integration or by Laplace approximation, and can therefore be used to solve high dimensional stochastic control problems.

Recall that *optimal control of nonlinear systems in the presence of noise* is a very general problem that occurs in many areas of science and engineering. It underlies *autonomous system behavior*, such as the control of movement and planning of actions of animals and robots, but also optimization of financial investment policies and control of chemical plants. The problem is stated as: *given that the system is in this configuration at this time, what is the optimal course of action to reach a goal state at some future time.* The *cost* of each time course of actions consists typically of a path contribution, that specifies the amount of work or other cost of the trajectory, and an end cost, that specifies to what extend the trajectory reaches the goal state.

Also recall that in the absence of noise, the optimal control problem can be solved in two ways: using (i) the *Pontryagin Maximum Principle* (PMP, see previous subsection), which represents a pair of ordinary differential equations that are similar to the Hamiltonian equations; or (ii) the *Hamilton–Jacobi–Bellman* (HJB) equation, which is a partial differential equation (PDE) [BK64].

In the presence of *Wiener noise*, the PMP formalism is replaced by a set of stochastic differential equations (SDEs), which become difficult to solve (compare with [YZ99]). The inclusion of noise in the HJB framework is mathematically quite straightforward, yielding the so–called *stochastic HJB equation* [Ste93]. However, its solution requires a *discretization of space and time* and the computation becomes intractable in both memory requirement and CPU time in high dimensions. As a result, deterministic control can be computed efficiently using the PMP approach, but stochastic control is intractable due to the curse of dimensionality.

For small noise, one expects that optimal stochastic control resembles optimal deterministic control, but for larger noise, the optimal stochastic control can be entirely different from the deterministic control [RN03]. However, there is currently no good understanding how *noise* affects optimal control.

In this subsection, we address both the issue of efficient computation and the role of noise in stochastic optimal control. We consider a class of nonlinear stochastic control problems, that can be formulated as a statistical mechanics problem. This class of control problems includes arbitrary dynamical systems, but with a limited control mechanism. It contains linear–quadratic [Ste93] control as a special case. We show that under certain conditions on the noise, the HJB equation can be written as a *linear* PDE

$$-\partial_t \psi = H\psi, \qquad (4.186)$$

with H a (non–Hermitian) operator. Equation (4.186) must be solved subject to a boundary condition at the end time. As a result of the linearity of (4.186), the solution can be obtained in terms of a diffusion process evolving forward in time, and can be written as a path integral. The path–integral has a direct interpretation as a free energy, where noise plays the role of temperature.

This link between stochastic optimal control and a free energy has an immediate consequence that phenomena that allow for a free energy description, typically display phase transitions. [Kap05] has argued that for stochastic optimal control one can identify a critical noise value that separates regimes where the optimal control has been qualitatively different. He showed how the Laplace approximation can be combined with Monte Carlo sampling to efficiently calculate the optimal control.

Path–Integral Formalism

Let x^i be an nD stochastic variable that is subject to the SDE

$$dx^i = (b^i(x^i, t) + u^i)dt + d\xi^i \qquad (4.187)$$

with $d\xi^i$ being an nD Wiener process with $\langle d\xi_i d\xi_j \rangle = \nu_{ij} dt$, and functions ν_{ij} independent of x^i, u^i and time t. The term $b^i(x^i, t)$ is an arbitrary nD function of x^i and t, and u^i represents an nD vector of control variables. Given the value of x^i at an initial time t, the *stochastic optimal control problem* is to find the control path $u^i(\cdot)$ that minimizes

$$C(x^i, t, u^i(\cdot)) = \left\langle \phi(x^i(t_f)) + \int_t^{t_f} d\tau (\frac{1}{2} u_i(\tau) R u^i(\tau) + V(x^i(\tau), \tau)) \right\rangle_{x^i}, \qquad (4.188)$$

with R a matrix, $V(x^i, t)$ a time–dependent potential, and $\phi(x^i)$ the *end cost*. The brackets $\langle \rangle_{x^i}$ denote expectation value with respect to the stochastic trajectories (4.187) that start at x^i.

One defines the *optimal cost–to–go function* from any time t and state x^i as

$$J(x^i, t) = \min_{u^i(\cdot)} C(x^i, t, u^i(\cdot)).$$

J satisfies the following stochastic HJB equation [Kap05]

$$-\partial_t J(x^i, t) = \min_{u^i} \left(\frac{1}{2} u_i R u^i + V + (b_i + u_i) \partial_{x^i} J(x^i, t) + \frac{1}{2} \nu_{ij} \partial_{x^i x^j} J(x^i, t) \right)$$
$$= -\frac{1}{2} R^{-1} \partial_{x^i} J(x^i, t) \partial_{x^i} J + V + b_i \partial_{x^i} J(x^i, t) + \frac{1}{2} \nu_{ij} \partial_{x^i x^j} J(x^i, t), \quad (4.189)$$

where $b_i = (b^i)^T$, and $u_i = (u^i)^T$, and

$$u^i = -R^{-1} \partial_{x^i} J(x^i, t) \quad (4.190)$$

is the optimal control at the point (x^i, t). The HJB equation is nonlinear in J and must be solved with end boundary condition $J(x^i, t_f) = \phi(x^i)$.

Let us define $\psi(x^i, t)$ through the Log Transform

$$J(x^i, t) = -\lambda \log \psi(x^i, t), \quad (4.191)$$

and assume that there exists a scalar λ such that

$$\lambda \delta_{ij} = (R\nu)_{ij}, \quad (4.192)$$

with δ_{ij} the Kronecker delta. In the one dimensional case, such a λ can always be found. In the higher dimensional case, this restricts the matrices $R \propto (\nu_{ij})^{-1}$. Equation (4.192) reduces the dependence of optimal control on the nD noise matrix to a scalar value λ that will play the role of temperature, while (4.189) reduces to the linear equation (4.186) with

$$H = -\frac{V}{\lambda} + b_i \partial_{x^i} + \frac{1}{2} \nu_{ij} \partial_{x^i x^j} J(x^i, t).$$

Let $\rho(y^i, \tau | x^i, t)$ with $\rho(y^i, t | x^i, t) = \delta(y^i - x^i)$ describe a *diffusion process* for $\tau > t$ defined by the Fokker–Planck equation

$$\partial_\tau \rho = H^\dagger \rho = -\frac{V}{\lambda} \rho - \partial_{x^i}(b_i \rho) + \frac{1}{2} \nu_{ij} \partial_{x^i x^j} J(x^i, t) \rho \quad (4.193)$$

with $H^\dagger$ the Hermitian–conjugate of H. Then $A(\tau) = \int dy^i \rho(y^i, \tau | x^i, t) \psi(y^i, \tau)$ is independent of τ and in particular $A(t) = A(t_f)$. It immediately follows that

$$\psi(x^i, t) = \int dy^i \rho(y^i, t_f | x^i, t) \exp(-\phi(y^i)/\lambda) \quad (4.194)$$

4.5 Non–Physical Applications of Path Integrals

We arrive at the important conclusion that $\psi(x^i, t)$ can be computed either by backward integration using (4.186) or by forward integration of a diffusion process given by (4.193).

We can write the integral in (4.194) as a path integral. Following [Kap05] we can divide the time interval $t \to t_f$ in n_1 intervals and write $\rho(y^i, t_f | x^i, t) = \prod_{i=1}^{n_1} \rho(x_i^i, t_i | x_{i-1}^i, t_{i-1})$ and let $n_1 \to \infty$. The result is

$$\psi(x^i, t) = \int [dx^i]_{x^i} \exp\left(-\frac{1}{\lambda} S(x^i(t \to t_f))\right) \tag{4.195}$$

with $\int [dx^i]_{x^i}$ an integral over all paths $x^i(t \to t_f)$ that start at x^i and with

$$S(x^i(t \to t_f)) = \phi(x^i(t_f)) + \int_t^{t_f} d\tau \left(\frac{1}{2}(\dot{x}^i - b_i(x^i, \tau))R(\dot{x}^i - b^i(x^i, \tau)) + V(x^i, \tau)\right) \tag{4.196}$$

the Action associated with a path. From (4.191) and (4.195), the cost–to–go $J(x, t)$ becomes a log partition sum (i.e., a free energy) with temperature λ.

Monte Carlo Sampling

The path integral (4.195) can be estimated by stochastic integration from t to t_f of the diffusion process (4.193) in which particles get annihilated at a rate $V(x^i, t)/\lambda$ [Kap05]:

$$\begin{aligned} x^i &= x^i + b^i(x^i, t)dt + d\xi^i, & \text{with probability} \quad 1 - Vdt/\lambda \\ x^i &= \dagger, & \text{with probability} \quad Vdt/\lambda \end{aligned} \tag{4.197}$$

where $\dagger$ denotes that the particle is taken out of the simulation. Denote the trajectories by $x_\alpha^i(t \to t_f)$, $(\alpha = 1, \ldots, N)$. Then, $\psi(x^i, t)$ and u^i are estimated as

$$\hat{\psi}(x^i, t) = \sum_{\alpha \in \text{alive}} w_\alpha, \qquad u^i dt = \frac{1}{\hat{\psi}(x^i, t)} \sum_{\alpha \in \text{alive}}^N w_\alpha d\xi_\alpha^i(t), \tag{4.198}$$

with $\quad w_\alpha = \frac{1}{N} \exp(-\phi(x_\alpha^i(t_f))/\lambda),$

where 'alive' denotes the subset of trajectories that do not get killed along the way by the $\dagger$ operation. The normalization $1/N$ ensures that the annihilation process is properly taken into account. Equation (4.198) states that optimal control at time t is obtained by averaging the initial directions of the noise component of the trajectories $d\xi_\alpha^i(t)$, weighted by their success at t_f.

The above sampling procedure can be quite inefficient, when many trajectories get annihilated. One of the simplest procedures to improve it is by importance sampling. We replace the diffusion process that yields $\rho(y^i, t_f | x^i, t)$ by another diffusion process, that will yield $\rho'(y^i, t_f | x^i, t) = \exp(-S'/\lambda)$. Then (4.195) becomes, with $\Delta t = t_i - t_{i-1}$.

A time–evolution model for the asset price is strictly necessary in a theory of option pricing because the fair price at time $t = 0$ of an option $\mathcal{O}$, without possibility of anticipated exercise before the expiration date or maturity T (a so–called *European option*), is given by the scaled expectation value [Hul00]

$$\mathcal{O}(0) = e^{-rT} E[\mathcal{O}(T)], \tag{4.204}$$

where r is the risk–free interest and $E[\cdot]$ indicates the mean value, which can be computed only if a model for the asset underlying the option is understood. For example, the value $\mathcal{O}$ of an European call option at the maturity T will be $\max\{S_T - X, 0\}$, where X is the strike price, while for an European put option the value $\mathcal{O}$ at the maturity will be $\max\{X - S_T, 0\}$. It is worth emphasizing, for what follows, that the case of an European option is particularly simple, since in such a situation the price of the option can be evaluated by means of analytical formulae, which are get by solving the BSM partial differential equation with the appropriate boundary conditions [Hul00, PB99]. On the other hand, many further kinds of options are present in the financial markets, such as American options (options which can be exercised at any time up to the expiration date) and exotic options [Hul00], i.e., derivatives with complicated payoffs or whose value depend on the whole time evolution of the underlying asset and not just on its value at the end. For such options with path-dependent and early exercise features no exact solutions are available and pricing them correctly is a great challenge.

In the case of options with possibility of anticipated exercise before the expiration date, the above discussion needs to be generalized, by introducing a slicing of the time interval T. Let us consider, for definiteness, the case of an option which can be exercised within the maturity but only at the times $t_1 = \Delta t, t_2 = 2\Delta t, \ldots, t_n = n\Delta t = T$. At each time slice t_{i-1} the value $\mathcal{O}_{i-1}$ of the option will be the maximum between its expectation value at the time t_i scaled with $e^{-r\Delta t}$ and its value in the case of anticipated exercise $\mathcal{O}^Y_{i-1}$. If S_{i-1} denotes the price of the underlying asset at the time t_{i-1}, we can thus write for each $i = 1, \ldots, n$

$$\mathcal{O}_{i-1}(S_{i-1}) = \max\left\{\mathcal{O}^Y_{i-1}(S_{i-1}), e^{-r\Delta t} E[\mathcal{O}_i | S_{i-1}]\right\}, \tag{4.205}$$

where $E[\mathcal{O}_i | S_{i-1}]$ is the conditional expectation value of $\mathcal{O}_i$, i.e., its expectation value under the hypothesis of having the price S_{i-1} at the time t_{i-1}. In this way, to get the actual price $\mathcal{O}_0$, it is necessary to proceed backward in time and calculate $\mathcal{O}_{n-1}, \ldots, \mathcal{O}_1$, where the value $\mathcal{O}_n$ of the option at maturity is nothing but $\mathcal{O}^Y_n(S_n)$. It is therefore clear that evaluating the price of an option with early exercise features means to simulate the evolution of the underlying asset price (to get the $\mathcal{O}^Y_i$) and to calculate a (usually large) number of expectation conditional probabilities.

Standard Numerical Procedures

To value derivatives when analytical formulae are not available, appropriate numerical techniques have to be advocated. They involve the use of Monte Carlo (MC) simulation, binomial trees (and their improvements) and finite–difference methods [Hul00, WDH93].

A natural way to simulate price paths is to discretize (4.201) as

$$\ln S(t + \Delta t) - \ln S(t) = A\Delta t + \sigma\epsilon\sqrt{\Delta t},$$

or, equivalently,

$$S(t + \Delta t) = S(t) \exp\left[A\Delta t + \sigma\epsilon\sqrt{\Delta t}\right], \qquad (4.206)$$

which is correct for any $\Delta t > 0$, even if finite. Given the spot price S_0, i.e., the price of the asset at time $t = 0$, one can extract from a standardized normal distribution a value ϵ_k, $(k = 1, \ldots, n)$ for the random variable ϵ to simulate one possible path followed by the price by means of (4.206):

$$S(k\Delta t) = S((k - 1)\Delta t) \exp\left[A\Delta t + \sigma\epsilon_k\sqrt{\Delta t}\right].$$

Iterating the procedure m times, one can simulate m price paths $\{(S_0, S_1^{(j)}, S_2^{(j)}, \ldots, S_n^{(j)} \equiv S_T^{(j)}) : j = 1, \ldots, m\}$ and evaluate the price of the option. In such a MC simulation of the stochastic dynamics of asset price (Monte Carlo random walk) the mean values $E[\mathcal{O}_i|S_{i-1}], i = 1, \ldots, n$ are given by

$$E[\mathcal{O}_i|S_{i-1}] = \frac{\mathcal{O}_i^{(1)} + \mathcal{O}_i^{(2)} + \cdots + \mathcal{O}_i^{(m)}}{m},$$

with no need to calculate transition probabilities because, through the extraction of the possible ϵ values, the paths are automatically weighted according to the probability distribution function of (4.203). Unfortunately, this method leads to an estimated value whose numerical error is proportional to $m^{-1/2}$. Thus, even if it is powerful because of the possibility to control the paths and to impose additional constrains (as it is usually required by exotic and path-dependent options), the MC random walk is extremely time consuming when precise predictions are required and appropriate variance reduction procedures have to be used to save CPU time [Hul00]. This difficulty can be overcome by means of the method of the binomial trees and its extensions (see [Hul00] and references therein), whose main idea stands in a deterministic choice of the possible paths to limit the number of intermediate points. At each time step the price S_i is assumed to have only two choices: increase to the value $uS_i, u > 1$ or decrease to $dS_i, 0 < d < 1$, where the parameters u and d are given in terms of σ and Δt in such a way to give the correct values for the mean and variance of stock price changes over the time interval Δt. Also finite difference methods are known in the literature [Hul00] as an alternative to time-consuming MC simulations. They give the value of the derivative by solving the differential equation satisfied by the derivative, by converting it into a difference equation. Although tree approaches and finite difference methods are known to be faster than the MC random walk, they are difficult to apply when a detailed control of the history of the derivative is required and are also computationally time consuming when a number of stochastic variables is involved [Hul00]. It follows that the development of efficient and fast computational algorithms to price financial derivatives is still a key issue in financial analysis.

Option Pricing via Path Integrals

Recall that the path integral method is an integral formulation of the dynamics of a stochastic process. It is a suitable framework for the calculation of the transition probabilities associated to a given stochastic process, which is seen as the convolution of an infinite sequence of infinitesimal short-time steps [BRT99]. For the problem of option pricing, the path–integral method can be employed for the explicit calculation of the expectation values of the quantities of financial interest, given by integrals of the form [BRT99]

$$E[\mathcal{O}_i|S_{i-1}] = \int dz_i p(z_i|z_{i-1})\mathcal{O}_i(e^{z_i}), \qquad (4.207)$$

where $z = \ln S$ and $p(z_i|z_{i-1})$ is the transition probability. $E[\mathcal{O}_i|S_{i-1}]$ is the conditional expectation value of some functional $\mathcal{O}_i$ of the stochastic process. For example, for an European call option at the maturity T the quantity of interest will be $\max\{S_T - X, 0\}$, X being the strike price. As already emphasized, and discussed in the literature [Hul00, WDH93, PBS01, RT02, Mat02], the computational complexity associated to this calculation is generally great: in the case of exotic options, with path-dependent and early exercise features, integrals of the type (4.207) cannot be analytically solved. As a consequence, we demand two things from a path integral framework: a very quick way to estimate the transition probability associated to a stochastic process (4.201) and a clever choice of the integration points with which evaluate the integrals (4.207). In particular, our aim is to develop an efficient calculation of the probability distribution without losing information on the path followed by the asset price during its time evolution.

Transition Probability

The probability distribution function related to a SDE verifies the *Chapman–Kolmogorov equation* [PB99]

$$p(z''|z') = \int dz\, p(z''|z) p(z|z'), \qquad (4.208)$$

which states that the probability (density) of a transition from the value z' (at time t') to the value z'' (at time t'') is the 'summation' over all the possible intermediate values z of the probability of separate and consequent transitions $z' \to z$, $z \to z''$. As a consequence, if we consider a finite time interval $[t', t'']$ and we apply a time slicing, by considering $n + 1$ subintervals of length $\Delta t = (t'' - t')/n + 1$, we can write, by iteration of (4.208)

$$p(z''|z') = \int_{-\infty}^{+\infty} \cdots \int_{-\infty}^{+\infty} dz_1 \cdots dz_n p(z''|z_n) p(z_n|z_{n-1}) \cdots p(z_1|z'),$$

which, thanks to (4.202), can be written as [MNM02]

$$\int_{-\infty}^{+\infty} \cdots \qquad (4.209)$$

$$\cdots \int_{-\infty}^{+\infty} dz_1 \cdots dz_n \frac{1}{\sqrt{(2\pi\sigma^2 \Delta t)^{n+1}}} \exp\left\{-\frac{1}{2\sigma^2 \Delta t} \sum_{k=1}^{n+1} [z_k - (z_{k-1} + A\Delta t)]^2\right\}.$$

In the limit $n \to \infty$, $\Delta t \to 0$ such that $(n+1)\Delta t = (t'' - t')$ (infinite sequence of infinitesimal time steps), the expression (4.209), as explicitly shown in [BRT99], exhibits a Lagrangian structure and it is possible to express the transition probability in the path integral formalism as a convolution of the form [BRT99]

$$p(z'', t''|z', t') = \int_{\mathcal{C}} \mathcal{D}[\sigma^{-1} \tilde{z}] \exp\left\{-\int_{t'}^{t''} L(\tilde{z}(\tau), \dot{\tilde{z}}(\tau); \tau) d\tau\right\},$$

where L is the Lagrangian, given by

$$L(\tilde{z}(\tau), \dot{\tilde{z}}(\tau); \tau) = \frac{1}{2\sigma^2} \left[\dot{\tilde{z}}(\tau) - A\right]^2,$$

and the integral is performed (with functional measure $\mathcal{D}[\cdot]$) over the paths $\tilde{z}(\cdot)$ belonging to $\mathcal{C}$, i.e., all the continuous functions with constrains $\tilde{z}(t') \equiv z'$, $\tilde{z}(t'') \equiv z''$. As carefully discussed in [BRT99], a path integral is well defined only if both a continuous formal expression and a discretization rule are given. As done in many applications, the Itô prescription is adopted here (see subsection 4.1.3 above).

A first, naive evaluation of the transition probability (4.209) can be performed via Monte Carlo simulation, by writing (4.209) as

$$p(z'', t''|z', t') =$$

$$\int_{-\infty}^{+\infty} \cdots \int_{-\infty}^{+\infty} \prod_i^n dg_i \frac{1}{\sqrt{2\pi\sigma^2 \Delta t}} \exp\left\{-\frac{1}{2\sigma^2 \Delta t} [z'' - (z_n + A\Delta t)]^2\right\}, \quad (4.210)$$

in terms of the variables g_i defined by the relation

$$dg_k = \frac{dz_k}{\sqrt{2\pi\sigma^2 \Delta t}} \exp\left\{-\frac{1}{2\sigma^2 \Delta t} [z_k - (z_{k-1} + A\Delta t)]^2\right\}, \qquad (4.211)$$

and extracting each g_i from a gaussian distribution of mean $z_{k-1} + A\Delta t$ and variance $\sigma^2 \Delta t$. However, as we will see, this method requires a large number of calls to get a good precision. This is due to the fact that each g_i is related to the previous g_{i-1}, so that this implementation of the path integral approach can be seen to be equivalent to a naive MC simulation of random walks, with no variance reduction.

By means of appropriate manipulations [Sch81] of the integrand entering (4.209), it is possible, as shown in the following, to get a path integral expression which will contain a factorized integral with a constant kernel and a consequent variance reduction. If we define $z'' = z_{n+1}$ and $y_k = z_k - kA\Delta t$, $k = 1, \ldots, n$, we can express the *transition probability distribution* as

$$\int_{-\infty}^{+\infty} \cdots \int_{-\infty}^{+\infty} dy_1 \cdots dy_n \frac{1}{\sqrt{(2\pi\sigma^2 \Delta t)^{n+1}}} \cdot \exp\left\{-\frac{1}{2\sigma^2 \Delta t} \sum_{k=1}^{n+1} [y_k - y_{k-1}]^2\right\}, \quad (4.212)$$

in order to get rid of the contribution of the drift parameter. Now let us extract from the argument of the exponential function a quadratic form

$$\sum_{k=1}^{n+1} [y_k - y_{k-1}]^2 = y_0^2 - 2y_1 y_0 + y_1^2 + y_1^2 - 2y_1 y_2 + \ldots + y_{n+1}^2$$

$$= y^t M y + [y_0^2 - 2y_1 y_0 + y_{n+1}^2 - 2y_n y_{n+1}], \quad (4.213)$$

by introducing the nD array y and the nxn matrix M defined as [MNM02]

$$y = \begin{pmatrix} y_1 \\ y_2 \\ \vdots \\ \vdots \\ y_n \end{pmatrix}, \quad M = \begin{pmatrix} 2 & -1 & 0 & \cdots & \cdots & 0 \\ -1 & 2 & -1 & 0 & \cdots & 0 \\ 0 & -1 & 2 & -1 & \cdots & 0 \\ 0 & \cdots & -1 & 2 & -1 & 0 \\ 0 & \cdots & \cdots & -1 & 2 & -1 \\ 0 & \cdots & \cdots & \cdots & -1 & 2 \end{pmatrix}, \quad (4.214)$$

where M is a real, symmetric, non singular and tridiagonal matrix. In terms of the eigenvalues m_i of the matrix M, the contribution in (4.213) can be written as

$$y^t M y = w^t O^t M O w = w^t M_d w = \sum_{i=1}^{n} m_i w_i^2, \quad (4.215)$$

by introducing the orthogonal matrix O which diagonalizes M, with $w_i = O_{ij} y_j$. Because of the orthogonality of O, the Jacobian

$$J = \det \left| \frac{dw_i}{dy_k} \right| = \det |O_{ki}|,$$

of the transformation $y_k \to w_k$ equals 1, so that $\prod_{i=1}^{n} dw_i = \prod_{i=1}^{n} dy_i$. After some algebra, (4.213) can be written as

$$\sum_{k=1}^{n+1}[y_k - y_{k-1}]^2 = \sum_{i=1}^{n} m_i w_i^2 + y_0^2 - 2y_1 y_0 + y_{n+1}^2 - 2y_n y_{n+1} =$$

$$\sum_{i=1}^{n} m_i \left[w_i - \frac{(y_0 O_{1i} + y_{n+1} O_{ni})}{m_i} \right]^2 + y_0^2 + y_{n+1}^2 - \sum_{i=1}^{n} \frac{(y_0 O_{1i} + y_{n+1} O_{ni})^2}{m_i}. \quad (4.216)$$

Now, if we introduce new variables h_i obeying the relation

$$dh_i = \sqrt{\frac{m_i}{2\pi\sigma^2 \Delta t}} \exp\left\{ -\frac{m_i}{2\sigma^2 \Delta t} \left[w_i - \frac{(y_0 O_{1i} + y_{n+1} O_{ni})}{m_i} \right]^2 \right\} dw_i, \quad (4.217)$$

it is possible to express the *finite–time probability distribution* $p(z''|z')$ as [MNM02]

$$\int_{-\infty}^{+\infty} \cdots \int_{-\infty}^{+\infty} \prod_{i=1}^{n} dy_i \frac{1}{\sqrt{(2\pi\sigma^2 \Delta t)^{n+1}}} \exp\left\{ -\frac{1}{2\sigma^2 \Delta t} \sum_{k=1}^{n+1} [y_k - y_{k-1}]^2 \right\}$$

$$= \int_{-\infty}^{+\infty} \cdots \int_{-\infty}^{+\infty} \prod_{i=1}^{n} dw_i \frac{1}{\sqrt{(2\pi\sigma^2 \Delta t)^{n+1}}} e^{-(y_0^2 + y_{n+1}^2)/2\sigma^2 \Delta t}$$

$$\times \exp\left\{ -\frac{1}{2\sigma^2 \Delta t} \sum_{i=1}^{n} \left[m_i \left(w_i - \frac{(y_0 O_{1i} + y_{n+1} O_{ni})}{m_i} \right)^2 - \frac{(y_0 O_{1i} + y_{n+1} O_{ni})^2}{m_i} \right] \right\}$$

$$= \int_{-\infty}^{+\infty} \cdots \int_{-\infty}^{+\infty} \prod_{i=1}^{n} dh_i \frac{1}{\sqrt{2\pi\sigma^2 \Delta t \det(M)}} \quad (4.218)$$

$$\times \exp\left\{ -\frac{1}{2\sigma^2 \Delta t} \left[y_0^2 + y_{n+1}^2 + \sum_{i=1}^{n} \frac{(y_0 O_{1i} + y_{n+1} O_{ni})^2}{m_i} \right] \right\}.$$

The probability distribution function, as given by (4.218), is an integral whose kernel is a constant function (with respect to the integration variables) and which can be factorized into the n integrals

$$\int_{-\infty}^{+\infty} dh_i \exp\left\{ -\frac{1}{2\sigma^2 \Delta t} \frac{(y_0 O_{1i} + y_{n+1} O_{ni})^2}{m_i} \right\}, \quad (4.219)$$

given in terms of the h_i, which are gaussian variables that can be extracted from a normal distribution with mean $(y_0 O_{1i} + y_{n+1} O_{ni})^2/m_i$ and variance $\sigma^2 \Delta t / m_i$. Differently to the first, naive implementation of the path integral, now each h_i is no longer dependent on the previous h_{i-1}, and importance sampling over the paths is automatically accounted for.

It is worth noticing that, by means of the extraction of the random variables h_i, we are creating price paths, since at each intermediate time t_i the asset price is given by

$$S_i = \exp\left\{ \sum_{k=1}^{n} O_{ik} h_k + iA\Delta t \right\}. \quad (4.220)$$

Therefore, this path integral algorithm can be easily adapted to the cases in which the derivative to be valued has, in the time interval $[0, T]$, additional constraints, as in the case of interesting path–dependent options, such as Asian and barrier options [Hul00].

Integration Points

The above illustrated method represents a powerful and fast tool to calculate the transition probability in the path integral framework and it can be employed if we need to value a generic option with maturity T and with possibility of anticipated exercise at times $t_i = i\Delta t$ ($n\Delta t = T$) [MNM02]. As a consequence of this time slicing, one must numerically evaluate $n-1$ mean values of the type (9), in order to check at any time t_i, and for any value of the stock price, whether early exercise is more convenient with respect to holding the option for a future time. To keep under control the computational complexity and the time of execution, it is mandatory to limit as far as possible the number of points for the integral evaluation. This means that we would like to have a linear growth of the number of integration points with the time. Let us suppose to evaluate each mean value

$$E[\mathcal{O}_i|S_{i-1}] = \int dz_i\, p(z_i|z_{i-1}) \mathcal{O}_i(e^{z_i}),$$

with p integration points, i.e., considering only p fixed values for z_i. To this end, we can create a grid of possible prices, according to the dynamics of the stochastic process as given by (4.201)

$$z(t+\Delta t) - z(t) = \ln S(t+\Delta t) - \ln S(t) = A\Delta t + \epsilon \sigma \sqrt{\Delta t}. \qquad (4.221)$$

Starting from z_0, we thus evaluate the expectation value $E[\mathcal{O}_1|S_0]$ with $p = 2m+1$, $m \in \mathbb{N}$ values of z_1 centered on the mean value $E[z_1] = z_0 + A\Delta t$ and which differ from each other of a quantity of the order of $\sigma\sqrt{\Delta t}$

$$z_1^j = z_0 + A\Delta t + j\sigma\sqrt{\Delta t}, \qquad (j = -m, \ldots, +m).$$

Going on like this, we can evaluate each expectation value $E[\mathcal{O}_2|z_1^j]$ get from each one of the z_1's created above with p values for z_2 centered around the mean value

$$E[z_2|z_1^j] = z_1^j + A\Delta t = z_0 + 2A\Delta t + j\sigma\sqrt{\Delta t}.$$

Iterating the procedure until the maturity, we create a deterministic grid of points such that, at a given time t_i, there are $(p-1)i + 1$ values of z_i, in agreement with the request of linear growth. This procedure of selection of integration points, together with the calculation of the transition probability previously described, is the basis of the path integral simulation of the price of a generic option.

4.5 Non–Physical Applications of Path Integrals

By applying the results derived above, we have at disposal an efficient path integral algorithm both for the calculation of transition probabilities and the evaluation of option prices. In [MNM02] the application of the above path–integral method to European and American options in the BSM model was illustrated and comparisons with the results were get with the standard procedures known in the literature were shown. First, the path integral simulation of the probability distribution of the logarithm of the stock prices, $p(lnS)$, as a function of the logarithm of the stock price, for a BSM–like stochastic model, was given by (4.200). Once the transition probability has been computed, the price of an option could be computed in a path integral approach as the conditional expectation value of a given functional of the stochastic process. For example, the price of an European call option was given by

$$C = e^{-r(T-t)} \int_{-\infty}^{+\infty} dz_f\, p(z_f, T|z_i, t) \max[e^{z_f} - X, 0], \qquad (4.222)$$

while for an European put it will be

$$P = e^{-r(T-t)} \int_{-\infty}^{+\infty} dz_f\, p(z_f, T|z_i, t) \max[X - e^{z_f}, 0], \qquad (4.223)$$

where r is the risk–free interest rate. Therefore just 1D integrals need to be evaluated and they can be precisely computed with standard quadrature rules.

Continuum Limit and American Options

In the specific case of an *American option*, the possibility of exercise at any time up to the expiration date allows to develop, within the path integral formalism, a specific algorithm, which, as shown in the following, is precise and very quick [MNM02].

Given the time slicing considered above, the case of American options requires the limit $\Delta t \to 0$ which, putting $\sigma \to 0$, leads to a delta–like transition probability

$$p(z, t + \Delta t|z_t, t) \approx \delta(z - z_t - A\Delta t).$$

This means that, apart from volatility effects, the price z_i at time t_i will have a value remarkably close to the expected value $\bar{z} = z_{i-1} + A\Delta t$, given by the drift growth. In order to take care of the volatility effects, a possible solution is to estimate the integral of interest, i.e.,

$$E[\mathcal{O}_i|S_{i-1}] = \int_{-\infty}^{+\infty} dz\, p(z|z_{i-1})\mathcal{O}_i(e^z), \qquad (4.224)$$

by inserting in (4.224) the analytical expression for the $p(z|z_{i-1})$ transition probability

$$p(z|z_{i-1}) = \frac{1}{\sqrt{2\pi \Delta t \sigma^2}} \exp\left\{-\frac{(z-z_{i-1}-A\Delta t)^2}{2\sigma^2 \Delta t}\right\}$$
$$= \frac{1}{\sqrt{2\pi \Delta t \sigma^2}} \exp\left\{-\frac{(z-\bar{z})^2}{2\sigma^2 \Delta t}\right\},$$

together with a Taylor expansion of the kernel function $\mathcal{O}_i(e^z) = f(z)$ around the expected value $\bar{z}$. Hence, up to the second–order in $z - \bar{z}$, the kernel function becomes

$$f(z) = f(\bar{z}) + (z-\bar{z})f'(\bar{z}) + \frac{1}{2}f''(\bar{z})(z-\bar{z})^2 + O((z-\bar{z})^3),$$

which induces

$$E[\mathcal{O}_i|S_{i-1}] = f(\bar{z}) + \frac{\sigma^2}{2}f''(\bar{z}), + \ldots,$$

since the first derivative does not give contribution to (4.224), being the integral of an odd function over the whole z range. The second derivative can be numerically estimated as

$$f''(\bar{z}) = \frac{1}{\delta_\sigma^2}[f(\bar{z}+\delta_\sigma) - 2f(\bar{z}) + f(\bar{z}-\delta_\sigma)],$$

with $\delta_\sigma = O(\sigma\sqrt{\Delta t})$, as dictated by the dynamics of the stochastic process.

4.5.3 Dynamics of Complex Networks

Recall that many systems in nature, such as neural nets, food webs, metabolic systems, co–authorship of papers, the worldwide web, etc. can be represented as *complex networks*, or *small–world networks* (see, e.g., [WS98, DM03]). In particular, it has been recognized that many networks have scale–free topology; the distribution of the degree obeys the power law, $P(k) \sim k^{-\gamma}$. The study of the scale–free network now attracts the interests of many researchers in mathematics, physics, engineering and biology [Ich04].

Another important aspect of complex networks is their dynamics, describing e.g., the spreading of viruses in the Internet, change of populations in a food web, and synchronization of neurons in a brain. In particular, [Ich04] studied the synchronization of the random network of oscillators. His work follows the previous studies (see [Str00]) that showed that mean–field type synchronization, that Kuramoto observed in globally–coupled oscillators [Kur84], appeared also in the small–world networks.

Continuum Limit of the Kuramoto Net

Ichinomiya started with the standard network with N nodes, described by a variant of the *Kuramoto model*. Namely, at each node, there exists an oscillator and the phase of each oscillator θ_i is evolving according to

4.5 Non–Physical Applications of Path Integrals

$$\dot{\theta}_i = \omega_i + K \sum_j a_{ij} \sin(\theta_j - \theta_i), \qquad (4.225)$$

where K is the coupling constant, a_{ij} is 1 if the nodes i and j are connected, and 0 otherwise; ω_i is a random number, whose distribution is given by the function $N(\omega)$.

For the analytic study, it is convenient to use the *continuum limit equation*. We define $P(k)$ as the distribution of nodes with degree k, and $\rho(k,\omega;t,\theta)$ the density of oscillators with phase θ at time t, for given ω and k. We assume that $\rho(k,\omega;t,\theta)$ is normalized as

$$\int_0^{2\pi} \rho(k,\omega;t,\theta)d\theta = 1.$$

For simplicity, we also assume $N(\omega) = N(-\omega)$. Thus, we suppose that the collective oscillation corresponds to the stable solution, $\dot{\rho} = 0$.

Now we construct the continuum limit equation for the network of oscillators. The evolution of ρ is determined by the *continuity equation* $\partial_t \rho = -\partial_\theta(\rho v)$, where v is defined by the continuum limit of the r.h.s of (4.225). Because one randomly selected edge connects to the node of degree k, frequency ω, phase θ with the probability $kP(k)N(\omega)\rho(k,\omega;t,\theta)/\int dk k P(k)$, $\rho(k,\omega;t,\theta)$ obeys the equation

$$\partial_t \rho(k,\omega;t,\theta) = -\partial_\theta [\rho(k,\omega;t,\theta)(\omega +$$
$$+ \frac{Kk \int d\omega' \int dk' \int d\theta' N(\omega')P(k')k'\rho(k',\omega';t,\theta')\sin(\theta-\theta')}{\int dk' P(k')k'})].$$

The mean–field solution of this equation was studied by [Ich04].

Path–Integral Approach to Complex Networks

Recently, [Ich05] introduced the *path–integral* (see subsection 4.4.6 above) *approach* in studying the dynamics of complex networks. He considered the stochastic generalization of the Kuramoto network (4.225), given by

$$\dot{x}_i = f_i(x_i) + \sum_{j=1}^N a_{ij}g(x_i,x_j) + \xi_i(t), \qquad (4.226)$$

where $f_i = f_i(x_i)$ and $g_{ij} = g(x_i,x_j)$ are functions of network activations x_i, $\xi_i(t)$ is a *random force* that satisfies $\langle \xi_i(t) \rangle = 0$, $\langle \xi_i(t)\xi_j(t') \rangle = \delta_{ij}\delta(t-t')\sigma^2$. He assumed $x_i = x_{i,0}$ at $t=0$. In order to discuss the dynamics of this system, he introduced the so–called *Matrin–Siggia–Rose* (MSR) *generating functional* Z given by [Dom78]

$$Z[\{l_{ik}\},\{\bar{l}_{ik}\}] = \left(\frac{1}{\pi}\right)^{NN_t} \left\langle \int \prod_{i=1}^N \prod_{k=0}^{N_t} dx_{ik}d\bar{x}_{ik} e^{-S} \exp(l_{ik}x_{ik} + \bar{l}_{ik}\bar{x}_{ik})J \right\rangle,$$

where the *action* S is given by

$$S = \sum_{ik}[\frac{\sigma^2 \Delta t}{2}\bar{x}_{ik}^2 + i\bar{x}_{ik}\{x_{ik} - x_{i,k-1} - \Delta t(f_i(x_{i,k-1}) + \sum_j a_{ij}g(x_{i,k-1},x_{j,k-1}))\}],$$

and $\langle \cdots \rangle$ represents the average over the ensemble of networks. J is the functional Jacobian term,

$$J = \exp\left(-\frac{\Delta t}{2}\sum_{ijk}\frac{\partial(f_i(x_{ik}) + a_{ij}g(x_{ik},x_{jk}))}{\partial x_{ik}}\right).$$

Ichinomiya considered such a form of the network model in which

$$a_{ij} = \begin{cases} 1 & \text{with probability } p_{ij}, \\ 0 & \text{with probability } 1 - p_{ij}. \end{cases}$$

Note that p_{ij} can be a function of variables such as i or j. For example, in the 1D chain model, p_{ij} is 1 if $|i - j| = 1$, else it is 0. The average over all networks can be expressed as

$$\left\langle \exp\left[\sum_{ik} i\Delta t \bar{x}_{ik} \sum_j a_{ij}g(x_{i,k-1},x_{j,k-1})\right]\right\rangle =$$

$$\prod_{ij}\left[p_{ij}\exp\left\{\sum_k i\Delta t \bar{x}_{ik}g(x_{i,k-1},x_{j,k-1})\right\} + 1 - p_{ij}\right],$$

so we get

$$\langle e^{-S} \rangle = \exp(-S_0)\prod_{ij}\left[p_{ij}\exp\left\{\sum_k i\Delta t \bar{x}_{ik}g(x_{i,k-1},x_{j,k-1})\right\} + 1 - p_{ij}\right],$$

where $S_0 = \sum_{ik}\frac{\sigma^2 \Delta t}{2}\bar{x}_{ik}^2 + i\bar{x}_{ik}\{x_{ik} - x_{i,k-1} - \Delta t f_i(x_{i,k-1})\}.$

This expression can be applied to the dynamics of any complex network model. [Ich05] applied this model to analysis of the *Kuramoto transition in random sparse networks*.

4.5.4 Path–Integral Dynamics of Neural Networks

Let us return to the simplest setting in which to study the problem: single pattern recall in an attractor neural network with N binary neurons and $p = \alpha N$ stored patterns in the non–trivial regime, where $\alpha > 0$. We choose parallel dynamics, i.e., (2.205), with Hebbian synapses of the form (2.202) with $A_{\mu\nu} = \delta_{\mu\nu}$, i.e., $J_{ij} = N^{-1}\sum_\mu^p \xi_i^\mu \xi_j^\mu$, giving us the parallel dynamics

4.5 Non–Physical Applications of Path Integrals

version of the Hopfield model [Hop82]. Our interest is in the recall overlap $m(\boldsymbol{\sigma}) = N^{-1}\sum_i \sigma_i \xi_i^1$ between system state and pattern one. We saw above that for $N \to \infty$ the fluctuations in the values of the recall overlap m will vanish, and that for initial states where all $\sigma_i(0)$ are drawn independently the overlap m will obey:

$$m(t+1) = \int dz\, P_t(z)\tanh[\beta(m(t)+z)] : \qquad (4.227)$$

$$P_t(z) = \lim_{N\to\infty}\frac{1}{N}\sum_i \langle \delta[z - \frac{1}{N}\xi_i^1 \xi_i^\mu \xi_j^\mu \sigma_j(t)]\rangle,$$

and that all complications in a dynamical analysis of the $\alpha > 0$ regime are concentrated in the calculation of the distribution $P_t(z)$ of the (generally non–trivial) interference noise.

As a simple *Gaussian approximation* one could just assume [Ama77, Ama78] that the σ_i remain uncorrelated at all times, i.e., $\text{Prob}[\sigma_i(t) = \pm \xi_i^1] = \frac{1}{2}[1 \pm m(t)]$ for all $t \geq 0$, such that the argument given above for $t = 0$ (leading to a Gaussian $P(z)$) would hold generally, and where the map (4.227) would describe the overlap evolution at all times:

$$P_t(z) = [2\pi\alpha]^{-\frac{1}{2}}e^{-\frac{1}{2}z^2/\alpha} :$$
$$m(t+1) = \int Dz\, \tanh[\beta(m(t) + z\sqrt{\alpha})],$$

with the *Gaussian measure* $Dz = (2\pi)^{-\frac{1}{2}}e^{-\frac{1}{2}z^2}dz$. This equation, however, must be generally incorrect. Rather than taking all σ_i to be independent, a weaker assumption would be to just assume the interference noise distribution $P_t(z)$ to be a zero-average Gaussian one, at any time, with statistically independent noise variables z at different times. One can then derive (for $N \to \infty$ and fully connected networks) an evolution equation for the width $\Sigma(t)$, giving [AM88]:

$$m(t+1) = \int Dz\, \tanh[\beta(m(t) + z\Sigma(t))] :$$
$$P_t(z) = [2\pi \Sigma^2(t)]^{-\frac{1}{2}}e^{-\frac{1}{2}z^2/\Sigma^2(t)} :$$
$$\Sigma^2(t+1) = \alpha + 2\alpha m(t+1)m(t)h[m(t),\Sigma(t)]$$
$$\qquad\qquad + \Sigma^2(t)h^2[m(t),\Sigma(t)], \qquad \text{with}$$
$$h[m,\Sigma] = \beta\left[1 - \int Dz\,\tanh^2[\beta(m+z\Sigma)]\right].$$

These equations describe correctly the qualitative features of recall dynamics, and are found to work well when retrieval actually occurs. A final refinement of the Gaussian approach [Oka95] consisted in allowing for correlations between the noise variables z at different times (while still describing them by Gaussian distributions). This results in a hierarchy of macroscopic equations, which improve upon the previous Gaussian theories and even predict the correct stationary state and phase diagrams, but still fail to be correct at intermediate times.

In view of the non–Gaussian shape of the interference noise distribution, several attempts have been made at constructing non–Gaussian approximations. In all cases the aim is to arrive at a theory involving only macroscopic observables with a *single* time-argument. For a fully connected network with binary neurons and parallel dynamics a more accurate ansatz for $P_t(z)$ would be the sum of two Gaussian functions. In [HO90] the following choice was proposed:

$$P_t(z) = P_t^+(z) + P_t^-(z),$$

$$P_t^\pm(z) = \lim_{N\to\infty} \frac{1}{N} \sum_i \delta_{\sigma_i(t), \pm \xi_i^1} \langle \delta[z - \frac{1}{N} \xi_i^1 \xi_i^\mu \xi_j^\mu \sigma_j(t)] \rangle$$

$$P_t^\pm(z) = \frac{1 \pm m(t)}{2\Sigma(t)\sqrt{2\pi}} e^{-\frac{1}{2}[z \mp d(t)]^2/\Sigma^2(t)},$$

followed by a self–consistent calculation of $d(t)$ (representing an effective 're-tarded self–interaction', since it has an effect equivalent to adding $h_i(\boldsymbol{\sigma}(t)) \to h_i(\boldsymbol{\sigma}(t)) + d(t)\sigma_i(t)$), and of the width $\Sigma(t)$ of the two distributions $P_t^\pm(z)$, together with

$$m(t+1) = \frac{1}{2}[1+m(t)] \int Dz \; \tanh[\beta(m(t)+d(t)+z\Sigma(t))]$$
$$+ \frac{1}{2}[1-m(t)] \int Dz \; \tanh[\beta(m(t)-d(t)+z\Sigma(t))].$$

The resulting three–parameter theory, in the form of closed dynamic equations for $\{m, d, \Sigma\}$, is found to give a nice (but not perfect) agreement with numerical simulations.

A different philosophy was followed in [CS94] (for sequential dynamics). First (as yet exact) equations are derived for the evolution of the two macroscopic observables $m(\boldsymbol{\sigma}) = m_1(\boldsymbol{\sigma})$ and $r(\boldsymbol{\sigma}) = \alpha^{-1} \sum_{\mu>1} m_\mu^2(\boldsymbol{\sigma})$, with $m_\mu(\boldsymbol{\sigma}) = N^{-1} \sum_i \xi_i^1 \sigma_i$, which are both found to involve $P_t(z)$:

$$\dot{m} = \int dz \; P_t(z) \tanh[\beta(m+z)],$$

$$\dot{r} = \frac{1}{\alpha} \int dz \; P_t(z) z \tanh[\beta(m+z)] + 1 - r.$$

Next one closes these equations *by hand*, using a maximum–entropy (or, 'Occam's Razor') argument: instead of calculating $P_t(z)$ from (4.227) with the real (unknown) microscopic distribution $p_t(\boldsymbol{\sigma})$, it is calculated upon assigning equal probabilities to all states $\boldsymbol{\sigma}$ with $m(\boldsymbol{\sigma}) = m$ and $r(\boldsymbol{\sigma}) = r$, followed by averaging over all realisations of the stored patterns with $\mu > 1$. In order words: one assumes (i) that the microscopic states visited by the system

4.5 Non–Physical Applications of Path Integrals

are 'typical' within the appropriate (m,r) sub-shells of state space, and (ii) that one can average over the disorder. Assumption (ii) is harmless, the most important step is (i). This procedure results in an explicit (non-Gaussian) expression for the noise distribution in terms of (m,r) only, a closed 2–parameter theory which is exact for short times and in equilibrium, accurate predictions of the macroscopic flow in the (m,r)–plane, but (again) deviations in predicted time-dependencies at intermediate times. This theory, and its performance, was later improved by applying the same ideas to a derivation of a dynamic equation for the function $P_t(z)$ itself (rather than for m and r only).

If we now use the powerful *path–integral formalism* (see [II06b]), instead of working with the probability $p_t(\boldsymbol{\sigma})$ of finding a microscopic state $\boldsymbol{\sigma}$ at time t in order to calculate the statistics of a set of macroscopic observables $\boldsymbol{\Omega}(\boldsymbol{\sigma})$ at time t, we turn to the probability $\text{Prob}[\boldsymbol{\sigma}(0), \ldots, \boldsymbol{\sigma}(t_m)]$ of finding a microscopic *path* $\boldsymbol{\sigma}(0) \to \boldsymbol{\sigma}(1) \to \ldots \to \boldsymbol{\sigma}(t_m)$. W also add time–dependent external sources to the local fields, $h_i(\boldsymbol{\sigma}) \to h_i(\boldsymbol{\sigma}) + \theta_i(t)$, in order to probe the networks via perturbations and define a response function. The idea is to concentrate on the moment partition function $Z[\boldsymbol{\psi}]$, which, like $\text{Prob}[\boldsymbol{\sigma}(0), \ldots, \boldsymbol{\sigma}(t_m)]$, fully captures the statistics of paths:

$$Z[\boldsymbol{\psi}] = \langle e^{-i\sum_{t=0}^{t_m} \psi_i(t)\sigma_i(t)} \rangle.$$

It generates averages of the relevant observables, including those involving neuron states at different times, such as correlation functions $C_{ij}(t,t') = \langle \sigma_i(t)\sigma_j(t') \rangle$ and response functions $G_{ij}(t,t') = \partial \langle \sigma_i(t) \rangle / \partial \theta_j(t')$, upon differentiation with respect to the dummy variables $\{\psi_i(t)\}$:

$$\langle \sigma_i(t) \rangle = i \lim_{\boldsymbol{\psi}\to 0} \frac{\partial Z[\boldsymbol{\psi}]}{\partial \psi_i(t)}, \quad C_{ij}(t,t') = -\lim_{\boldsymbol{\psi}\to 0} \frac{\partial^2 Z[\boldsymbol{\psi}]}{\partial \psi_i(t)\partial \psi_j(t')},$$

$$G_{ij}(t,t') = i \lim_{\boldsymbol{\psi}\to 0} \frac{\partial^2 Z[\boldsymbol{\psi}]}{\partial \psi_i(t)\partial \theta_j(t')}.$$

Next one assumes (correctly) that for $N \to \infty$ only the statistical properties of the stored patterns will influence the macroscopic quantities, so that the partition function $Z[\boldsymbol{\psi}]$ can be averaged over all pattern realisations, i.e., $Z[\boldsymbol{\psi}] \to \overline{Z[\boldsymbol{\psi}]}$. As in replica theories (the canonical tool to deal with complexity in equilibrium) one carries out the disorder average *before* the average over the statistics of the neuron states, resulting for $N \to \infty$ in what can be interpreted as a theory describing a single 'effective' binary neuron $\sigma(t)$, with an effective local field $h(t)$ and the dynamics $\text{Prob}[\sigma(t+1) = \pm 1] = \frac{1}{2}[1 \pm \tanh[\beta h(t)]]$. However, this effective local field is found to generally depend on past states of the neuron, and on zero-average but temporally correlated Gaussian noise contributions $\xi(t)$:

$$h(t|\{\sigma\},\{\xi\}) = m(t) + \theta(t) + \alpha \sum_{t'<t} R(t,t')\sigma(t') + \sqrt{\alpha}\xi(t). \quad (4.228)$$

4 Nonlinear Dynamics of Path Integrals

The first comprehensive neural network studies along these lines, dealing with fully connected networks, were applied to asymmetrically and symmetrically extremely diluted networks [KZ91, WS91]. More recent applications include sequence processing networks [DCS98].[46] For $N \to \infty$ the differences between different models are found to show up only in the actual form taken by the effective local field (4.228), i.e., in the dependence of the 'retarded self-interaction' kernel $R(t,t')$ and the covariance matrix $\langle \xi(t)\xi(t')\rangle$ of the interference–induced Gaussian noise on the macroscopic objects $\mathbf{C} = \{C(s,s') = \lim_{N\to\infty} \frac{1}{N} C_{ii}(s,s')\}$ and $\mathbf{G} = \{G(s,s') = \lim_{N\to\infty} \frac{1}{N} G_{ii}(s,s')\}$. For instance [Coo01, SC00, SC01, CKS05]:

model	synapses J_{ij}	$R(t,t')$	$\langle \xi(t)\xi(t')\rangle$
fully connected, static patterns	$\frac{1}{N}\xi_i^\mu \xi_j^\mu$	$[(\mathbf{1}-\mathbf{G})^{-1}\mathbf{G}](t,t')$	$[(\mathbf{1}-\mathbf{G})^{-1}\mathbf{C}(\mathbf{1}-\mathbf{G}^\dagger)^{-1}](t,t')$
fully connected, pattern sequence	$\frac{1}{N}\xi_i^{\mu+1}\xi_j^\mu$	0	$\sum_{n\geq 0}[(\mathbf{G}^\dagger)^n \mathbf{C}\mathbf{G}^n](t,t')$
symm extr diluted, static patterns	$\frac{c_{ij}}{c}\xi_i^\mu \xi_j^\mu$	$G(t,t')$	$C(t,t')$
asymm extr diluted, static patterns	$\frac{c_{ij}}{c}\xi_i^\mu \xi_j^\mu$	0	$C(t,t')$

with the c_{ij} drawn at random according to $P(c_{ij}) = \frac{c}{N}\delta_{c_{ij},1} + (1-\frac{c}{N})\delta_{c_{ij},0}$ (either symmetrically, i.e., $c_{ij} = c_{ji}$, or independently) and where $c_{ii} = 0$, $\lim_{N\to\infty} c/N = 0$, and $c \to \infty$. In all cases the observables (overlaps and correlation– and response–functions) are to be solved from the following closed equations, involving the statistics of the single effective neuron experiencing the field (4.228):

$$m(t) = \langle \sigma(t)\rangle, \qquad C(t,t') = \langle \sigma(t)\sigma(t')\rangle, \qquad G(t,t') = \partial\langle \sigma(t)\rangle/\partial\theta(t').$$

It is now clear that Gaussian theories can at most produce exact results for asymmetric networks. Any degree of symmetry in the synapses is found to induce a non–zero retarded self–interaction, via the kernel $K(t,t')$, which constitutes a non–Gaussian contribution to the local fields. Exact closed macroscopic theories apparently require a number of macroscopic observables which grows as $\mathcal{O}(t^2)$ in order to predict the dynamics up to time t. In the case of sequential dynamics the picture is found to be very similar to the one above; instead of discrete time labels $t \in \{0, 1, \ldots, t_m\}$, path summations and matrices, there one has a real time variable $t \in [0, t_m]$, path–integrals, integral operators, and partition–functions.

Partition–Function Analysis for Binary Neurons

First we will define parallel dynamics, i.e., (2.205), driven as usual by local fields of the form $h_i(\boldsymbol{\sigma};t) = J_{ij}\sigma_j + \theta_i(t)$, but with a more general choice of

---

[46] In the case of sequence recall the overlap m is defined with respect to the 'moving' target, i.e., $m(t) = \frac{1}{N}\sigma_i(t)\xi_i^t$

Hebbian synapses, in which we allow for a possible random dilution (to reduce repetition in our subsequent derivations):

$$J_{ij} = \frac{c_{ij}}{c}\xi_i^\mu \xi_j^\mu, \quad \text{with} \quad p = \alpha c. \tag{4.229}$$

Architectural properties are reflected in the variables $c_{ij} \in \{0,1\}$, whereas information storage is to be effected by the remainder in (4.229), involving p randomly and independently drawn patterns $\boldsymbol{\xi}^\mu = (\xi_1^\mu, \ldots, \xi_N^\mu) \in \{-1, 1\}^N$. I will deal both with symmetric and with asymmetric architectures (always putting $c_{ii} = 0$), in which the variables c_{ij} are drawn randomly according to

$$c_{ij} = c_{ji}, \quad \text{(for all } i < j), \quad P(c_{ij}) = \frac{c}{N}\delta_{c_{ij},1} + (1 - \frac{c}{N})\delta_{c_{ij},0},$$

$$\text{(for all } i = j), \quad P(c_{ij}) = \frac{c}{N}\delta_{c_{ij},1} + (1 - \frac{c}{N})\delta_{c_{ij},0}.$$

Thus c_{kl} is statistically independent of c_{ij} as soon as $(k,l) \notin \{(i,j),(j,i)\}$. In leading order in N one has $\langle \sum_j c_{ij} \rangle = c$ for all i, so c gives the average number of neurons contributing to the field of any given neuron. In view of this, the number p of patterns to be stored can be expected to scale as $p = \alpha c$. The connectivity parameter c is chosen to diverge with N, i.e., $\lim_{N\to\infty} c^{-1} = 0$. If $c = N$ we get the fully connected (parallel dynamics) Hopfield model. Extremely diluted networks are got when $\lim_{N\to\infty} c/N = 0$.

For simplicity, we make the so-called 'condensed ansatz': we assume that the system state has an $\mathcal{O}(N^0)$ overlap only with a single pattern, say $\mu = 1$. This situation is induced by initial conditions: we take a randomly drawn $\boldsymbol{\sigma}(0)$, generated by

$$p(\boldsymbol{\sigma}(0)) = \prod_i \left\{ \frac{1}{2}[1 + m_0]\delta_{\sigma_i(0),\xi_i^1} + \frac{1}{2}[1 - m_0]\delta_{\sigma_i(0),-\xi_i^1} \right\},$$

so

$$\frac{1}{N}\xi_i^1 \langle \sigma_i(0) \rangle = m_0.$$

The patterns $\mu > 1$, as well as the architecture variables c_{ij}, are viewed as disorder. One assumes that for $N \to \infty$ the macroscopic behavior of the system is 'self-averaging', i.e., only dependent on the statistical properties of the disorder (rather than on its microscopic realisation). Averages over the disorder are written as $\overline{\cdots}$. We next define the *disorder-averaged partition function*:

$$\overline{Z[\boldsymbol{\psi}]} = \overline{\langle e^{-i\sum_t \psi_i(t)\sigma_i(t)} \rangle}, \tag{4.230}$$

in which the time t runs from $t = 0$ to some (finite) upper limit t_m. Note that $\overline{Z[\mathbf{0}]} = 1$. With a modest amount of foresight we define the macroscopic site-averaged and disorder-averaged objects $m(t) = N^{-1}\xi_i^1 \overline{\langle \sigma_i(t) \rangle}$, $C(t,t') = N^{-1}\overline{\langle \sigma_i(t)\sigma_i(t') \rangle}$ and $G(t,t') = N^{-1}\partial\overline{\langle \sigma_i(t) \rangle}/\partial\theta_i(t')$. They can be obtained from (4.230) as follows:

$$m(t) = \lim_{\psi \to 0} \frac{i}{N} \xi_j^1 \frac{\partial Z[\psi]}{\partial \psi_j(t)}, \qquad C(t,t') = -\lim_{\psi \to 0} \frac{1}{N} \frac{\partial^2 Z[\psi]}{\partial \psi_j(t) \partial \psi_j(t')},$$

$$G(t,t') = \lim_{\psi \to 0} \frac{i}{N} \frac{\partial^2 Z[\psi]}{\partial \psi_j(t) \partial \theta_j(t')}.$$

Now, as in equilibrium replica calculations, the hope here is that progress can be made by carrying out the disorder averages first. In equilibrium calculations we use the replica trick to convert our disorder averages into feasible ones; here the idea is to isolate the local fields at different times and different sites by inserting appropriate δ–distributions:

$$1 = \prod_{it} \int dh_i(t) \delta[h_i(t) - J_{ij}\sigma_j(t) - \theta_i(t)]$$

$$= \int \{d\mathbf{h} d\hat{\mathbf{h}}\} e^{i \sum_{it} \hat{h}_i(t)[h_i(t) - J_{ij}\sigma_j(t) - \theta_i(t)]},$$

with $\{d\mathbf{h} d\hat{\mathbf{h}}\} = \prod_{it}[d\hat{h}_i(t) dh_i(t)/2\pi]$, giving

$$\overline{Z[\psi]} = \int \{d\mathbf{h} d\hat{\mathbf{h}}\} e^{i \sum_{it} \hat{h}_i(t)[h_i(t) - \theta_i(t)]} \times$$

$$\times \langle e^{-i \sum_{it} \psi_i(t)\sigma_i(t)} \overline{\left[e^{-i \sum_{it} \hat{h}_i(t) J_{ij}\sigma_j(t)}\right]} \rangle_{\text{pf}},$$

in which $\langle \ldots \rangle_{\text{pf}}$ refers to averages over a constrained stochastic process of the type (2.205), but with prescribed fields $\{h_i(t)\}$ at all sites and at all times. Note that with such prescribed fields the probability of partition a path $\{\boldsymbol{\sigma}(0), \ldots, \boldsymbol{\sigma}(t_m)\}$ is given by

$$\text{Prob}[\boldsymbol{\sigma}(0), \ldots, \boldsymbol{\sigma}(t_m) | \{h_i(t)\}] =$$
$$p(\boldsymbol{\sigma}(0)) e^{\sum_{it} [\beta\sigma_i(t+1) h_i(t) - \log 2\cosh[\beta h_i(t)]]},$$

so

$$\overline{Z[\psi]} = \int \{d\mathbf{h} d\hat{\mathbf{h}}\} \sum_{\boldsymbol{\sigma}(0)} \cdots \sum_{\boldsymbol{\sigma}(t_m)} p(\boldsymbol{\sigma}(0)) e^{N\mathcal{F}[\{\boldsymbol{\sigma}\},\{\hat{\mathbf{h}}\}]} \times$$

$$\times \prod_{it} e^{i\hat{h}_i(t)[h_i(t) - \theta_i(t)] - i\psi_i(t)\sigma_i(t) + \beta\sigma_i(t+1) h_i(t) - \log 2\cosh[\beta h_i(t)]},$$

with $$\mathcal{F}[\{\boldsymbol{\sigma}\}, \{\hat{\mathbf{h}}\}] = \frac{1}{N} \log \overline{[e^{-i \sum_{it} \hat{h}_i(t) J_{ij}\sigma_j(t)}]}. \qquad (4.231)$$

We concentrate on the term $\mathcal{F}[\ldots]$ (with the disorder), of which we need only know the limit $N \to \infty$, since only terms inside $\overline{Z[\psi]}$ which are exponential in N will retain statistical relevance. In the disorder–average of (4.231) every site i plays an equivalent role, so the leading order in N of (4.231) should depend only on site–averaged functions of the $\{\sigma_i(t), \hat{h}_i(t)\}$, with no reference to any special direction except the one defined by pattern $\boldsymbol{\xi}^1$. The simplest such functions with a single time variable are

$$a(t;\{\boldsymbol{\sigma}\}) = \frac{1}{N}\xi_i^1\sigma_i(t), \qquad k(t;\{\hat{\mathbf{h}}\}) = \frac{1}{N}\xi_i^1\hat{h}_i(t),$$

whereas the simplest ones with two time variables would appear to be

$$q(t,t';\{\boldsymbol{\sigma}\}) = \frac{1}{N}\sigma_i(t)\sigma_i(t'), \qquad Q(t,t';\{\hat{\mathbf{h}}\}) = \frac{1}{N}\hat{h}_i(t)\hat{h}_i(t'),$$

$$K(t,t';\{\boldsymbol{\sigma},\hat{\mathbf{h}}\}) = \frac{1}{N}\hat{h}_i(t)\sigma_i(t').$$

It will turn out that all models of the type (4.229), have the crucial property that above are in fact the *only* functions to appear in the leading order of (4.231):

$$\mathcal{F}[\ldots] = \Phi[\{a(t;\ldots),k(t;\ldots),q(t,t';\ldots),Q(t,t';\ldots),K(t,t';\ldots)\}] + \ldots,$$

for $N \to \infty$ and some as yet unknown function $\Phi[\ldots]$. This allows us to proceed with the evaluation of $\overline{Z[\boldsymbol{\psi}]}$. We can introduce suitable δ-distributions (taking care that all exponents scale linearly with N, to secure statistical relevance). Thus we insert

$$1 = \prod_{t=0}^{t_m} \int da(t)\, \delta[a(t) - a(t;\{\boldsymbol{\sigma}\})]$$

$$= \left[\frac{N}{2\pi}\right]^{t_m+1} \int da d\hat{a}\, e^{iN\sum_t \hat{a}(t)[a(t) - \frac{1}{N}\xi_j^1\sigma_j(t)]},$$

$$1 = \prod_{t=0}^{t_m} \int dk(t)\, \delta[k(t) - k(t;\{\hat{\mathbf{h}}\})]$$

$$= \left[\frac{N}{2\pi}\right]^{t_m+1} \int d\mathbf{k} d\hat{\mathbf{k}}\, e^{iN\sum_t \hat{k}(t)[k(t) - \frac{1}{N}\xi_j^1\hat{h}_j(t)]},$$

$$1 = \prod_{t,t'=0}^{t_m} \int dq(t,t')\, \delta[q(t,t') - q(t,t';\{\boldsymbol{\sigma}\})]$$

$$= \left[\frac{N}{2\pi}\right]^{(t_m+1)^2} \int d\mathbf{q} d\hat{\mathbf{q}}\, e^{iN\sum_{t,t'} \hat{q}(t,t')[q(t,t') - \frac{1}{N}\sigma_j(t)\sigma_j(t')]},$$

$$1 = \prod_{t,t'=0}^{t_m} \int dQ(t,t')\, \delta[Q(t,t') - Q(t,t';\{\hat{\mathbf{h}}\})]$$

$$= \left[\frac{N}{2\pi}\right]^{(t_m+1)^2} \int d\mathbf{Q} d\hat{\mathbf{Q}}\, e^{iN\sum_{t,t'} \hat{Q}(t,t')[Q(t,t') - \frac{1}{N}\hat{h}_j(t)\hat{h}_j(t')]},$$

$$1 = \prod_{t,t'=0}^{t_m} \int dK(t,t') \, \delta[K(t,t') - K(t,t'; \{\boldsymbol{\sigma}, \hat{\mathbf{h}}\})]$$

$$= \left[\frac{N}{2\pi}\right]^{(t_m+1)^2} \int d\mathbf{K} d\hat{\mathbf{K}} \, e^{iN \sum_{t,t'} \hat{K}(t,t')[K(t,t') - \frac{1}{N}\hat{h}_j(t)\sigma_j(t')]}.$$

Using the short-hand

$$\Psi[\mathbf{a}, \hat{\mathbf{a}}, \mathbf{k}, \hat{\mathbf{k}}, \mathbf{q}, \hat{\mathbf{q}}, \mathbf{Q}, \hat{\mathbf{Q}}, \mathbf{K}, \hat{\mathbf{K}}] = i \sum_t [\hat{a}(t)a(t) + \hat{k}(t)k(t)]$$

$$+ i \sum_{t,t'} [\hat{q}(t,t')q(t,t') + \hat{Q}(t,t')Q(t,t') + \hat{K}(t,t')K(t,t')]$$

then leads us to

$$\overline{Z[\psi]} = \int d\mathbf{a} d\hat{\mathbf{a}} d\mathbf{k} d\hat{\mathbf{k}} d\mathbf{q} d\hat{\mathbf{q}} d\mathbf{Q} d\hat{\mathbf{Q}} d\mathbf{K} d\hat{\mathbf{K}} \times$$

$$e^{N\Psi[\mathbf{a},\hat{\mathbf{a}},\mathbf{k},\hat{\mathbf{k}},\mathbf{q},\hat{\mathbf{q}},\mathbf{Q},\hat{\mathbf{Q}},\mathbf{K},\hat{\mathbf{K}}]+N\Phi[\mathbf{a},\mathbf{k},\mathbf{q},\mathbf{Q},\mathbf{K}]+\mathcal{O}(\ldots)}$$

$$\times \int \{d\mathbf{h} d\hat{\mathbf{h}}\} \sum_{\boldsymbol{\sigma}(0)} \cdots \sum_{\boldsymbol{\sigma}(t_m)} p(\boldsymbol{\sigma}(0)) \times$$

$$\prod_{it} e^{i\hat{h}_i(t)[h_i(t)-\theta_i(t)] - i\psi_i(t)\sigma_i(t) + \beta\sigma_i(t+1)h_i(t) - \log 2\cosh[\beta h_i(t)]} \times$$

$$\prod_i e^{-i\xi_i^1 \sum_t [\hat{a}(t)\sigma_i(t) + \hat{k}(t)\hat{h}_i(t)] - i \sum_{t,t'} [\hat{q}(t,t')\sigma_i(t)\sigma_i(t') + \hat{Q}(t,t')\hat{h}_i(t)\hat{h}_i(t') + \hat{K}(t,t')\hat{h}_i(t)\sigma_i(t')]},$$

in which the term denoted as $\mathcal{O}(\ldots)$ covers both the non–dominant orders in (4.231) and the $\mathcal{O}(\log N)$ relics of the various pre–factors $[N/2\pi]$ in the above integral representations of the δ–distributions (note: t_m was assumed fixed). We now see explicitly that the summations and integrations over neuron states and local fields fully factorize over the N sites. A simple transformation

$$\{\sigma_i(t), h_i(t), \hat{h}_i(t)\} \to \{\xi_i^1 \sigma_i(t), \xi_i^1 h_i(t), \xi_i^1 \hat{h}_i(t)\}$$

brings the result into the form [Coo01, SC00, SC01, CKS05]

$$e^{N \Xi[\hat{\mathbf{a}},\hat{\mathbf{k}},\hat{\mathbf{q}},\hat{\mathbf{Q}},\hat{\mathbf{K}}]} = \int \{d\mathbf{h} d\hat{\mathbf{h}}\} \sum_{\boldsymbol{\sigma}(0)} \cdots \sum_{\boldsymbol{\sigma}(t_m)} p(\boldsymbol{\sigma}(0)) \times$$

$$\times \prod_{it} e^{i\hat{h}_i(t)[h_i(t) - \xi_i^1 \theta_i(t)] - i\xi_i^1 \psi_i(t)\sigma_i(t) + \beta\sigma_i(t+1)h_i(t) - \log 2\cosh[\beta h_i(t)]} \times$$

$$\prod_i e^{-i\xi_i^1 \sum_t [\hat{a}(t)\sigma_i(t) + \hat{k}(t)\hat{h}_i(t)] - i \sum_{t,t'} [\hat{q}(t,t')\sigma_i(t)\sigma_i(t') + \hat{Q}(t,t')\hat{h}_i(t)\hat{h}_i(t') + \hat{K}(t,t')\hat{h}_i(t)\sigma_i(t')]},$$

in which

$$\{dh d\hat{h}\} = \prod_t [dh(t) d\hat{h}(t)/2\pi], \qquad \pi_0(\sigma) = \frac{1}{2}[1+m_0]\delta_{\sigma,1} + \frac{1}{2}[1-m_0]\delta_{\sigma,-1}.$$

4.5 Non–Physical Applications of Path Integrals

At this stage $\overline{Z[\psi]}$ acquires the form of an integral to be evaluated via the saddle–point (or 'steepest descent') method,

$$\overline{Z[\{\psi(t)\}]} = \int d\mathbf{a}d\hat{\mathbf{a}}d\mathbf{k}d\hat{\mathbf{k}}d\mathbf{q}d\hat{\mathbf{q}}d\mathbf{Q}d\hat{\mathbf{Q}}d\mathbf{K}d\hat{\mathbf{K}}\ e^{N\{\Psi[\ldots]+\Phi[\ldots]+\Xi[\ldots]\}+\mathcal{O}(\ldots)}. \tag{4.232}$$

The disorder–averaged partition function (4.232) is for $N \to \infty$ dominated by the physical saddle–point of the macroscopic surface

$$\Psi[\mathbf{a},\hat{\mathbf{a}},\mathbf{k},\hat{\mathbf{k}},\mathbf{q},\hat{\mathbf{q}},\mathbf{Q},\hat{\mathbf{Q}},\mathbf{K},\hat{\mathbf{K}}] + \Phi[\mathbf{a},\mathbf{k},\mathbf{q},\mathbf{Q},\mathbf{K}] + \Xi[\hat{\mathbf{a}},\hat{\mathbf{k}},\hat{\mathbf{q}},\hat{\mathbf{Q}},\hat{\mathbf{K}}]. \tag{4.233}$$

It will be advantageous at this stage to define the following effective measure:

$$\langle f[\{\sigma\},\{h\},\{\hat{h}\}]\rangle_\star = \tag{4.234}$$

$$\frac{1}{N}\left\{\frac{\int\{dhd\hat{h}\}\sum_{\sigma(0)\cdots\sigma(t_m)} M_i[\{\sigma\},\{h\},\{\hat{h}\}]\ f[\{\sigma\},\{h\},\{\hat{h}\}]}{\int\{dhd\hat{h}\}\sum_{\sigma(0)\cdots\sigma(t_m)} M_i[\{\sigma\},\{h\},\{\hat{h}\}]}\right\},$$

with

$$M_i[\{\sigma\},\{h\},\{\hat{h}\}] =$$

$$\pi_0(\sigma(0))\ e^{\sum_t\{i\hat{h}(t)[h(t)-\xi_i^1\theta_i(t)]-i\xi_i^1\psi_i(t)\sigma(t)+\beta\sigma(t+1)h(t)-\log 2\cosh[\beta h(t)]\}}$$

$$\times e^{-i\sum_t[\hat{a}(t)\sigma(t)+\hat{k}(t)\hat{h}(t)]-i\sum_{t,t'}[\hat{q}(t,t')\sigma(t)\sigma(t')+\hat{Q}(t,t')\hat{h}(t)\hat{h}(t')+\hat{K}(t,t')\hat{h}(t)\sigma(t')]},$$

in which the values to be inserted for $\{\hat{m}(t),\hat{k}(t),\hat{q}(t,t'),\hat{Q}(t,t'),\hat{K}(t,t')\}$ are given by the saddle–point of (4.233). Variation of (4.233) with respect to all the original macroscopic objects occurring as arguments (those without the 'hats') gives the following set of saddle–point equations:

$$\hat{a}(t) = i\frac{\partial\Phi}{\partial a(t)}, \qquad \hat{k}(t) = i\frac{\partial\Phi}{\partial k(t)}, \qquad \hat{q}(t,t') = i\frac{\partial\Phi}{\partial q(t,t')},$$

$$\hat{Q}(t,t') = i\frac{\partial\Phi}{\partial Q(t,t')}, \qquad \hat{K}(t,t') = i\frac{\partial\Phi}{\partial K(t,t')}.$$

Variation of (4.233) with respect to the conjugate macroscopic objects (those with the 'hats'), in turn, and usage of our newly introduced short-hand notation $\langle\ldots\rangle_\star$, gives:

$$a(t) = \langle\sigma(t)\rangle_\star, \qquad k(t) = \langle\hat{h}(t)\rangle_\star, \qquad q(t,t') = \langle\sigma(t)\sigma(t')\rangle_\star,$$

$$Q(t,t') = \langle\hat{h}(t)\hat{h}(t')\rangle_\star, \qquad K(t,t') = \langle\hat{h}(t)\sigma(t')\rangle_\star$$

The above coupled equations have to be solved simultaneously, once we have calculated the term $\Phi[\ldots]$ that depends on the synapses. This appears to be a formidable task; it can, however, be simplified considerably upon first deriving the physical meaning of the above macroscopic quantities. We use identities such as

$$\frac{\partial \Xi[\ldots]}{\partial \psi_j(t)} = -\frac{i}{N}\xi_j^1 \left[\frac{\int\{dhd\hat{h}\}\sum_{\sigma(0)\cdots\sigma(t_m)} M_j[\{\sigma\},\{h\},\{\hat{h}\}]\sigma(t)}{\int\{dhd\hat{h}\}\sum_{\sigma(0)\cdots\sigma(t_m)} M_j[\{\sigma\},\{h\},\{\hat{h}\}]} \right],$$

$$\frac{\partial \Xi[\ldots]}{\partial \theta_j(t)} = -\frac{i}{N}\xi_j^1 \left[\frac{\int\{dhd\hat{h}\}\sum_{\sigma(0)\cdots\sigma(t_m)} M_j[\{\sigma\},\{h\},\{\hat{h}\}]\hat{h}(t)}{\int\{dhd\hat{h}\}\sum_{\sigma(0)\cdots\sigma(t_m)} M_j[\{\sigma\},\{h\},\{\hat{h}\}]} \right],$$

$$\frac{\partial^2 \Xi[\ldots]}{\partial \psi_j(t)\partial \psi_j(t')} = -\frac{1}{N}\left[\frac{\int\{dhd\hat{h}\}\sum_{\sigma(0)\cdots\sigma(t_m)} M_j[\{\sigma\},\{h\},\{\hat{h}\}]\sigma(t)\sigma(t')}{\int\{dhd\hat{h}\}\sum_{\sigma(0)\cdots\sigma(t_m)} M_j[\{\sigma\},\{h\},\{\hat{h}\}]} \right]$$
$$-N\left[\frac{\partial \Xi[\ldots]}{\partial \psi_j(t)}\right]\left[\frac{\partial \Xi[\ldots]}{\partial \psi_j(t')}\right],$$

$$\frac{\partial^2 \Xi[\ldots]}{\partial \theta_j(t)\partial \theta_j(t')} = -\frac{1}{N}\left[\frac{\int\{dhd\hat{h}\}\sum_{\sigma(0)\cdots\sigma(t_m)} M_j[\{\sigma\},\{h\},\{\hat{h}\}]\hat{h}(t)\hat{h}(t')}{\int\{dhd\hat{h}\}\sum_{\sigma(0)\cdots\sigma(t_m)} M_j[\{\sigma\},\{h\},\{\hat{h}\}]} \right]$$
$$-N\left[\frac{\partial \Xi[\ldots]}{\partial \theta_j(t)}\right]\left[\frac{\partial \Xi[\ldots]}{\partial \theta_j(t')}\right],$$

$$\frac{\partial^2 \Xi[\ldots]}{\partial \psi_j(t)\partial \theta_j(t')} = -\frac{i}{N}\left[\frac{\int\{dhd\hat{h}\}\sum_{\sigma(0)\cdots\sigma(t_m)} M_j[\{\sigma\},\{h\},\{\hat{h}\}]\sigma(t)\hat{h}(t')}{\int\{dhd\hat{h}\}\sum_{\sigma(0)\cdots\sigma(t_m)} M_j[\{\sigma\},\{h\},\{\hat{h}\}]} \right]$$
$$-N\left[\frac{\partial \Xi[\ldots]}{\partial \psi_j(t)}\right]\left[\frac{\partial \Xi[\ldots]}{\partial \theta_j(t')}\right],$$

and using the short–hand notation (4.234) wherever possible. Note that the external fields $\{\psi_i(t), \theta_i(t)\}$ occur only in the function $\Xi[\ldots]$, not in $\Psi[\ldots]$ or $\Phi[\ldots]$, and that overall constants in $\overline{Z[\psi]}$ can always be recovered *a posteriori*, using $\overline{Z[\mathbf{0}]} = 1$:

$$m(t) = \lim_{\psi \to 0} \frac{i}{N}\sum_i \xi_i^1 \frac{\int d\mathbf{a}\ldots d\hat{\mathbf{K}}\left[\frac{N\partial \Xi}{\partial \psi_i(t)}\right] e^{N[\Psi+\Phi+\Xi]+\mathcal{O}(\ldots)}}{\int d\mathbf{a}\ldots d\hat{\mathbf{K}}\, e^{N[\Psi+\Phi+\Xi]+\mathcal{O}(\ldots)}}$$
$$= \lim_{\psi \to 0} \langle \sigma(t) \rangle_\star,$$

$$C(t,t') =$$
$$-\lim_{\psi \to 0}\frac{1}{N}\sum_i \frac{\int d\mathbf{a}\ldots d\hat{\mathbf{K}}\left[\frac{N\partial^2 \Xi}{\partial \psi_i(t)\partial \psi_i(t')} + \frac{N\partial \Xi}{\partial \psi_i(t)}\frac{N\partial \Xi}{\partial \psi_i(t')}\right] e^{N[\Psi+\Phi+\Xi]+\mathcal{O}(\ldots)}}{\int d\mathbf{a}\ldots d\hat{\mathbf{K}}\, e^{N[\Psi+\Phi+\Xi]+\mathcal{O}(\ldots)}}$$
$$= \lim_{\psi \to 0} \langle \sigma(t)\sigma(t') \rangle_\star,$$

$$iG(t,t') =$$

$$-\lim_{\psi \to 0} \frac{1}{N} \sum_i \frac{\int d\mathbf{a} \ldots d\hat{\mathbf{K}} \left[\frac{N\partial^2 \Xi}{\partial \psi_i(t) \partial \theta_i(t')} + \frac{N\partial \Xi}{\partial \psi_i(t)} \frac{N\partial \Xi}{\partial \theta_i(t')} \right] e^{N[\Psi + \Phi + \Xi] + \mathcal{O}(\ldots)}}{\int d\mathbf{a} \ldots d\hat{\mathbf{K}} \ e^{N[\Psi + \Phi + \Xi] + \mathcal{O}(\ldots)}}$$

$$= \lim_{\psi \to 0} \langle \sigma(t) \hat{h}(t') \rangle_\star .$$

Finally we get useful identities from the seemingly trivial statements

$$N^{-1} \sum_i \xi_i^1 \partial \overline{Z[0]} / \partial \theta_i(t) = 0 \text{ and } N^{-1} \sum_i \partial^2 \overline{Z[0]} / \partial \theta_i(t) \partial \theta_i(t') = 0,$$

$$0 = \lim_{\psi \to 0} \frac{i}{N} \sum_i \xi_i^1 \frac{\int d\mathbf{a} \ldots d\hat{\mathbf{K}} \left[\frac{N\partial \Xi}{\partial \theta_i(t)} \right] e^{N[\Psi+\Phi+\Xi]+\mathcal{O}(\ldots)}}{\int d\mathbf{a} \ldots d\hat{\mathbf{K}} \ e^{N[\Psi+\Phi+\Xi]+\mathcal{O}(\ldots)}} = \lim_{\psi \to 0} \langle \hat{h}(t) \rangle_\star ,$$

$$0 = -\lim_{\psi \to 0} \frac{1}{N} \sum_i \frac{\int d\mathbf{a} \ldots d\hat{\mathbf{K}} \left[\frac{N\partial^2 \Xi}{\partial \theta_i(t) \partial \theta_i(t')} + \frac{N\partial \Xi}{\partial \theta_i(t)} \frac{N\partial \Xi}{\partial \theta_i(t')} \right] e^{N[\Psi+\Phi+\Xi]+\mathcal{O}(\ldots)}}{\int d\mathbf{a} \ldots d\hat{\mathbf{K}} \ e^{N[\Psi+\Phi+\Xi]+\mathcal{O}(\ldots)}}$$

$$= \lim_{\psi \to 0} \langle \hat{h}(t) \hat{h}(t') \rangle_\star .$$

The above identities simplify our problem considerably. The dummy fields $\psi_i(t)$ have served their purpose and will now be put to zero, as a result we can now identify our macroscopic observables *at the relevant saddle–point* as:

$$a(t) = m(t), \quad k(t) = 0, \quad q(t,t') = C(t,t'),$$
$$Q(t,t') = 0, \quad K(t,t') = iG(t',t).$$

Finally we make a convenient choice for the external fields, $\theta_i(t) = \xi_i^1 \theta(t)$, with which the effective measure $\langle \ldots \rangle_\star$ simplifies to

$$\langle f[\{\sigma\},\{h\},\{\hat{h}\}] \rangle_\star = \frac{\int \{dh d\hat{h}\} \sum_{\sigma(0) \cdots \sigma(t_m)} M[\{\sigma\},\{h\},\{\hat{h}\}] \ f[\{\sigma\},\{h\},\{\hat{h}\}]}{\int \{dh d\hat{h}\} \sum_{\sigma(0) \cdots \sigma(t_m)} M[\{\sigma\},\{h\},\{\hat{h}\}]}, \quad (4.235)$$

with $M[\{\sigma\},\{h\},\{\hat{h}\}] =$

$$\pi_0(\sigma(0)) \ e^{\sum_t \{i\hat{h}(t)[h(t) - \theta(t)] + \beta\sigma(t+1)h(t) - \log 2\cosh[\beta h(t)]\} - i \sum_t [\hat{a}(t)\sigma(t) + \hat{k}(t)\hat{h}(t)]}$$

$$\times e^{-i \sum_{t,t'} [\hat{q}(t,t')\sigma(t)\sigma(t') + \hat{Q}(t,t')\hat{h}(t)\hat{h}(t') + \hat{K}(t,t')\hat{h}(t)\sigma(t')]}.$$

Our final task is calculating the leading order of

$$\mathcal{F}[\{\boldsymbol{\sigma}\},\{\hat{\mathbf{h}}\}] == \frac{1}{N} \log \overline{\left[e^{-i \sum_{it} \hat{h}_i(t) J_{ij} \sigma_j(t)} \right]}. \quad (4.236)$$

Parallel Hopfield Network Near Saturation

The fully connected Hopfield [Hop82] network (here with parallel dynamics) is got upon choosing $c = N$ in the recipe (4.229), i.e., $c_{ij} = 1 - \delta_{ij}$ and $p = \alpha N$. The disorder average thus involves only the patterns with $\mu > 1$. Now (4.236) gives

$$\mathcal{F}[\ldots] = \frac{1}{N} \log \overline{\left[e^{-iN^{-1} \sum_t \xi_i^\mu \xi_j^\mu \hat{h}_i(t) \sigma_j(t)} \right]} \quad (4.237)$$

$$= i\alpha \sum_t K(t,t; \{\boldsymbol{\sigma}, \hat{\mathbf{h}}\}) - i \sum_t a(t) k(t) +$$

$$\alpha \log \overline{\left[e^{-i \sum_t [\xi_i \hat{h}_i(t)/\sqrt{N}][\xi_i \sigma_i(t)/\sqrt{N}]} \right]} + \mathcal{O}(N^{-1}).$$

We concentrate on the last term:

$$\overline{\left[e^{-i \sum_t [\xi_i \hat{h}_i(t)/\sqrt{N}][\xi_i \sigma_i(t)/\sqrt{N}]} \right]} = \int d\mathbf{x} d\mathbf{y} \, e^{-i\mathbf{x} \cdot \mathbf{y}} \times$$

$$\times \prod_t \left\{ \delta[x(t) - \frac{\xi_i \sigma_i(t)}{\sqrt{N}}] \, \delta[y(t) - \frac{\xi_i \hat{h}_i(t)}{\sqrt{N}}] \right\}$$

$$= \int \frac{d\mathbf{x} d\mathbf{y} d\hat{\mathbf{x}} d\hat{\mathbf{y}}}{(2\pi)^{2(t_m+1)}} \, e^{i[\hat{\mathbf{x}} \cdot \mathbf{x} + \hat{\mathbf{y}} \cdot \mathbf{y} - \mathbf{x} \cdot \mathbf{y}]} \overline{\left[e^{-\frac{i}{\sqrt{N}} \xi_i \sum_t [\hat{x}(t) \sigma_i(t) + \hat{y}(t) \hat{h}_i(t)]} \right]}$$

$$= \int \frac{d\mathbf{x} d\mathbf{y} d\hat{\mathbf{x}} d\hat{\mathbf{y}}}{(2\pi)^{2(t_m+1)}} \, e^{i[\hat{\mathbf{x}} \cdot \mathbf{x} + \hat{\mathbf{y}} \cdot \mathbf{y} - \mathbf{x} \cdot \mathbf{y}] + \sum_i \log \cos\left[\frac{1}{\sqrt{N}} \sum_t [\hat{x}(t) \sigma_i(t) + \hat{y}(t) \hat{h}_i(t)] \right]}$$

$$= \int \frac{d\mathbf{x} d\mathbf{y} d\hat{\mathbf{x}} d\hat{\mathbf{y}}}{(2\pi)^{2(t_m+1)}} \, e^{i[\hat{\mathbf{x}} \cdot \mathbf{x} + \hat{\mathbf{y}} \cdot \mathbf{y} - \mathbf{x} \cdot \mathbf{y}] - \frac{1}{2N} \sum_i \{ \sum_t [\hat{x}(t) \sigma_i(t) + \hat{y}(t) \hat{h}_i(t)] \}^2 + \mathcal{O}(N^{-1})} =$$

$$\int \frac{d\mathbf{x} d\mathbf{y} d\hat{\mathbf{x}} d\hat{\mathbf{y}}}{(2\pi)^{2(t_m+1)}} \, e^{i[\hat{\mathbf{x}} \cdot \mathbf{x} + \hat{\mathbf{y}} \cdot \mathbf{y} - \mathbf{x} \cdot \mathbf{y}] - \frac{1}{2} \sum_{t,t'} [\hat{x}(t) \hat{x}(t') q(t,t') + 2 \hat{x}(t) \hat{y}(t') K(t',t) + \hat{y}(t) \hat{y}(t') Q(t,t')]}.$$

Together with (4.237) we have now shown that the disorder average (4.236) is indeed, in leading order in N, with

$$\Phi[\mathbf{a}, \mathbf{k}, \mathbf{q}, \mathbf{Q}, \mathbf{K}] =$$

$$i\alpha \sum_t K(t,t) - i\mathbf{a} \cdot \mathbf{k} + \alpha \log \int \frac{d\mathbf{x} d\mathbf{y} d\hat{\mathbf{x}} d\hat{\mathbf{y}}}{(2\pi)^{2(t_m+1)}} \, e^{i[\hat{\mathbf{x}} \cdot \mathbf{x} + \hat{\mathbf{y}} \cdot \mathbf{y} - \mathbf{x} \cdot \mathbf{y}] - \frac{1}{2}[\hat{\mathbf{x}} \cdot \mathbf{q}\hat{\mathbf{x}} + 2\hat{\mathbf{y}} \cdot \mathbf{K}\hat{\mathbf{x}} + \hat{\mathbf{y}} \cdot \mathbf{Q}\hat{\mathbf{y}}]}$$

$$= i\alpha \sum_t K(t,t) - i\mathbf{a} \cdot \mathbf{k} + \alpha \log \int \frac{d\mathbf{u} d\mathbf{v}}{(2\pi)^{t_m+1}} \, e^{-\frac{1}{2}[\mathbf{u} \cdot \mathbf{q}\mathbf{u} + 2\mathbf{v} \cdot \mathbf{K}\mathbf{u} - 2i\mathbf{u} \cdot \mathbf{v} + \mathbf{v} \cdot \mathbf{Q}\mathbf{v}]}.$$

Now, for the single–time observables, this gives $\hat{a}(t) = k(t)$ and $\hat{k}(t) = a(t)$, and for the two–time ones:

4.5 Non–Physical Applications of Path Integrals

$$\hat{q}(t,t') = -\frac{1}{2}\alpha i \frac{\int d\mathbf{u}d\mathbf{v}\ u(t)u(t')e^{-\frac{1}{2}[\mathbf{u}\cdot\mathbf{q}\mathbf{u}+2\mathbf{v}\cdot\mathbf{K}\mathbf{u}-2i\mathbf{u}\cdot\mathbf{v}+\mathbf{v}\cdot\mathbf{Q}\mathbf{v}]}}{\int d\mathbf{u}d\mathbf{v}\ e^{-\frac{1}{2}[\mathbf{u}\cdot\mathbf{q}\mathbf{u}+2\mathbf{v}\cdot\mathbf{K}\mathbf{u}-2i\mathbf{u}\cdot\mathbf{v}+\mathbf{v}\cdot\mathbf{Q}\mathbf{v}]}},$$

$$\hat{Q}(t,t') = -\frac{1}{2}\alpha i \frac{\int d\mathbf{u}d\mathbf{v}\ v(t)v(t')e^{-\frac{1}{2}[\mathbf{u}\cdot\mathbf{q}\mathbf{u}+2\mathbf{v}\cdot\mathbf{K}\mathbf{u}-2i\mathbf{u}\cdot\mathbf{v}+\mathbf{v}\cdot\mathbf{Q}\mathbf{v}]}}{\int d\mathbf{u}d\mathbf{v}\ e^{-\frac{1}{2}[\mathbf{u}\cdot\mathbf{q}\mathbf{u}+2\mathbf{v}\cdot\mathbf{K}\mathbf{u}-2i\mathbf{u}\cdot\mathbf{v}+\mathbf{v}\cdot\mathbf{Q}\mathbf{v}]}},$$

$$\hat{K}(t,t') = -\alpha i \frac{\int d\mathbf{u}d\mathbf{v}\ v(t)u(t')e^{-\frac{1}{2}[\mathbf{u}\cdot\mathbf{q}\mathbf{u}+2\mathbf{v}\cdot\mathbf{K}\mathbf{u}-2i\mathbf{u}\cdot\mathbf{v}+\mathbf{v}\cdot\mathbf{Q}\mathbf{v}]}}{\int d\mathbf{u}d\mathbf{v}\ e^{-\frac{1}{2}[\mathbf{u}\cdot\mathbf{q}\mathbf{u}+2\mathbf{v}\cdot\mathbf{K}\mathbf{u}-2i\mathbf{u}\cdot\mathbf{v}+\mathbf{v}\cdot\mathbf{Q}\mathbf{v}]}} - \alpha\delta_{t,t'}.$$

At the physical saddle–point we can now express all non–zero objects in terms of the observables $m(t)$, $C(t,t')$ and $G(t,t')$, with a clear physical meaning. Thus we find $\hat{a}(t) = 0$, $\hat{k}(t) = m(t)$, and

$$\hat{q}(t,t') = -\frac{1}{2}\alpha i \frac{\int d\mathbf{u}d\mathbf{v}\ u(t)u(t')e^{-\frac{1}{2}[\mathbf{u}\cdot\mathbf{C}\mathbf{u}-2i\mathbf{u}\cdot[\mathbf{1}-\mathbf{G}]\mathbf{v}]}}{\int d\mathbf{u}d\mathbf{v}\ e^{-\frac{1}{2}[\mathbf{u}\cdot\mathbf{C}\mathbf{u}-2i\mathbf{u}\cdot[\mathbf{1}-\mathbf{G}]\mathbf{v}]}} = 0,$$

$$\hat{Q}(t,t') = -\frac{1}{2}\alpha i \frac{\int d\mathbf{u}d\mathbf{v}\ v(t)v(t')e^{-\frac{1}{2}[\mathbf{u}\cdot\mathbf{C}\mathbf{u}-2i\mathbf{u}\cdot[\mathbf{1}-\mathbf{G}]\mathbf{v}]}}{\int d\mathbf{u}d\mathbf{v}\ e^{-\frac{1}{2}[\mathbf{u}\cdot\mathbf{C}\mathbf{u}-2i\mathbf{u}\cdot[\mathbf{1}-\mathbf{G}]\mathbf{v}]}}$$

$$= -\frac{1}{2}\alpha i \left[(\mathbf{1}-\mathbf{G})^{-1}\mathbf{C}(\mathbf{1}-\mathbf{G}^\dagger)^{-1}\right](t,t'),$$

$$\hat{K}(t,t') + \alpha\delta_{t,t'} = -\alpha i \frac{\int d\mathbf{u}d\mathbf{v}\ v(t)u(t')e^{-\frac{1}{2}[\mathbf{u}\cdot\mathbf{C}\mathbf{u}-2i\mathbf{u}\cdot[\mathbf{1}-\mathbf{G}]\mathbf{v}]}}{\int d\mathbf{u}d\mathbf{v}\ e^{-\frac{1}{2}[\mathbf{u}\cdot\mathbf{C}\mathbf{u}-2i\mathbf{u}\cdot[\mathbf{1}-\mathbf{G}]\mathbf{v}]}}$$

$$= \alpha(\mathbf{1}-\mathbf{G})^{-1}(t,t'),$$

with $\mathbf{G}^\dagger(t,t') = G(t',t)$, and using standard manipulations of Gaussian integrals. Note that we can use the identity $(\mathbf{1}-\mathbf{G})^{-1} - \mathbf{1} = \sum_{\ell \geq 0}\mathbf{G}^\ell - \mathbf{1} = \sum_{\ell > 0}\mathbf{G}^\ell = \mathbf{G}(\mathbf{1}-\mathbf{G})^{-1}$ to compactify the last equation to

$$\hat{K}(t,t') = \alpha[\mathbf{G}(\mathbf{1}-\mathbf{G})^{-1}](t,t'). \quad (4.238)$$

We have now expressed all our objects in terms of the disorder–averaged recall Hopfield overlap $\mathbf{m} = \{m(t)\}$ and the disorder–averaged single–site correlation– and response–functions $\mathbf{C} = \{C(t,t')\}$ and $\mathbf{G} = \{G(t,t')\}$. We can next simplify the effective measure (4.235), which plays a crucial role in the remaining saddle–point equations. Inserting $\hat{a}(t) = \hat{q}(t,t') = 0$ and $\hat{k}(t) = m(t)$ into (4.235), first of all, gives us

$$M[\{\sigma\},\{h\},\{\hat{h}\}] = \pi_0(\sigma(0)) \times \quad (4.239)$$

$$e^{\sum_t \{i\hat{h}(t)[h(t)-m(t)-\theta(t)-\sum_{t'}\hat{K}(t,t')\sigma(t')]+\beta\sigma(t+1)h(t)-\log 2\cosh[\beta h(t)]\} - i\sum_{t,t'}\hat{Q}(t,t')\hat{h}(t)\hat{h}(t')}.$$

Secondly, causality ensures that $G(t,t') = 0$ for $t \leq t'$, from which, in combination with (4.238), it follows that the same must be true for the kernel $\hat{K}(t,t')$, since

$$\hat{K}(t,t') = \alpha[\mathbf{G}(\mathbf{1}-\mathbf{G})^{-1}](t,t') = \alpha\left\{\mathbf{G}+\mathbf{G}^2+\mathbf{G}^3+\ldots\right\}(t,t').$$

This, in turn, guarantees that the function $M[\ldots]$ in (4.239) is already normalised:
$$\int \{dh d\hat{h}\} \sum_{\sigma(0)\cdots\sigma(t_m)} M[\{\sigma\},\{h\},\{\hat{h}\}] = 1.$$

One can prove this iteratively. After summation over $\sigma(t_m)$ (which due to causality cannot occur in the term with the kernel $\hat{K}(t,t')$) one is left with just a single occurrence of the field $h(t_m)$ in the exponent, integration over which reduces to $\delta[\hat{h}(t_m)]$, which then eliminates the conjugate field $\hat{h}(t_m)$. This cycle of operations is next applied to the variables at time $t_m - 1$, etc. The effective measure (4.235) can now be written simply as

$$\langle f[\{\sigma\},\{h\},\{\hat{h}\}]\rangle_\star = \sum_{\sigma(0)\cdots\sigma(t_m)} \int \{dh d\hat{h}\}\, M[\{\sigma\},\{h\},\{\hat{h}\}]\, f[\{\sigma\},\{h\},\{\hat{h}\}],$$

with $M[\ldots]$ as given in (4.239). The remaining saddle–point equations to be solved, which can be slightly simplified by using the identity

$$\langle \sigma(t)\hat{h}(t')\rangle_\star = i\partial\langle\sigma(t)\rangle_\star/\partial\theta(t'), \quad \text{are}$$
$$m(t) = \langle\sigma(t)\rangle_\star, \qquad C(t,t') = \langle\sigma(t)\sigma(t')\rangle_\star, \qquad G(t,t') = \partial\langle\sigma(t)\rangle_\star/\partial\theta(t').$$

Here we observe that we only need to insert functions of spin states into the effective measure $\langle\ldots\rangle_\star$ (rather than fields or conjugate fields), so the effective measure can again be simplified. We get

$$\langle f[\{\sigma\}]\rangle_\star = \sum_{\sigma(0)\cdots\sigma(t_m)} \text{Prob}[\{\sigma\}]\, f[\{\sigma\}], \quad \text{with}$$

$$\text{Prob}[\{\sigma\}] = \pi_0(\sigma(0)) \int \{d\phi\}\, P[\{\phi\}] \times \qquad (4.240)$$
$$\times \prod_t \left[\frac{1}{2}[1 + \sigma(t+1)\tanh[\beta h(t|\{\sigma\},\{\phi\})]]\right],$$

in which $\pi_0(\sigma(0)) = \frac{1}{2}[1 + \sigma(0)m_0]$, and

$$h(t|\{\sigma\},\{\phi\}) = m(t) + \theta(t) + \alpha\sum_{t'<t}[\mathbf{G}(\mathbf{1}-\mathbf{G})^{-1}](t,t')\sigma(t') + \alpha^{\frac{1}{2}}\phi(t), (4.241)$$

$$P[\{\phi\}] = \frac{e^{-\frac{1}{2}\sum_{t,t'}\phi(t)[(\mathbf{1}-\mathbf{G}^\dagger)\mathbf{C}^{-1}(\mathbf{1}-\mathbf{G})](t,t')\phi(t')}}{(2\pi)^{(t_m+1)/2}\det^{-\frac{1}{2}}[(\mathbf{1}-\mathbf{G}^\dagger)\mathbf{C}^{-1}(\mathbf{1}-\mathbf{G})]}.$$

We recognize (4.240) as describing an effective single neuron, with the usual dynamics $\text{Prob}[\sigma(t+1) = \pm 1] = \frac{1}{2}[1 \pm \tanh[\beta h(t)]]$, but with the fields (4.241). This result is indeed of the form (4.228), with a retarded self–interaction kernel $R(t,t')$ and covariance matrix $\langle\phi(t)\phi(t')\rangle$ of the Gaussian $\phi(t)$ given by

$$R(t,t') = [\mathbf{G}(\mathbf{1}-\mathbf{G})^{-1}](t,t'),$$
$$\langle \phi(t)\phi(t')\rangle = [(\mathbf{1}-\mathbf{G})^{-1}\mathbf{C}(\mathbf{1}-\mathbf{G}^\dagger)^{-1}](t,t').$$

For $\alpha \to 0$ we loose all the complicated terms in the local fields, and recover the type of simple expression we found earlier for finite p: $m(t+1) = \tanh[\beta(m(t) + \theta(t))]$.

Note that always $C(t,t) = \langle \sigma^2(t)\rangle_\star = 1$ and $G(t,t') = R(t,t') = 0$ for $t \leq t'$. As a result the covariance matrix of the Gaussian fields can be written as

$$\langle \phi(t)\phi(t')\rangle = [(\mathbf{1}-\mathbf{G})^{-1}\mathbf{C}(\mathbf{1}-\mathbf{G}^\dagger)^{-1}](t,t')$$
$$= \sum_{s,s'\geq 0}[\delta_{t,s} + R(t,s)]C(s,s')[\delta_{s',t'} + R(t',s')]$$
$$= \sum_{s=0}^{t}\sum_{s'=0}^{t'}[\delta_{t,s} + R(t,s)]C(s,s')[\delta_{s',t'} + R(t',s')].$$

Considering arbitrary positive integer powers of the response function immediately shows that

$$(\mathbf{G}^\ell)(t,t') = 0, \quad \text{if} \quad t' > t - \ell, \quad \text{which gives}$$
$$R(t,t') = \sum_{\ell>0}(\mathbf{G}^\ell)(t,t') = \sum_{\ell=1}^{t-t'}(\mathbf{G}^\ell)(t,t').$$

Similarly we get from $(\mathbf{1}-\mathbf{G})^{-1} = \mathbf{1} + \mathbf{R}$ that for $t' \geq t$: $(\mathbf{1}-\mathbf{G})^{-1}(t,t') = \delta_{t,t'}$. To suppress notation we will simply put $h(t|..)$ instead of $h(t|\{\sigma\},\{\phi\})$; this need not cause any ambiguity. We notice that summation over neuron variables $\sigma(s)$ and integration over Gaussian variables $\phi(s)$ with time arguments s higher than than those occurring in the function to be averaged can always be carried out immediately, giving (for $t > 0$ and $t' < t$):

$$m(t) = \sum_{\sigma(0)...\sigma(t-1)} \pi_0(\sigma(0)) \int \{d\phi\}P[\{\phi\}] \tanh[\beta h(t-1|..)] \times$$
$$\times \prod_{s=0}^{t-2}\frac{1}{2}[1 + \sigma(s+1)\tanh[\beta h(s|..)]],$$

$$G(t,t') = \beta\{C(t,t'+1) - \sum_{\sigma(0)...\sigma(t-1)} \pi_0(\sigma(0)) \times$$
$$\times \int \{d\phi\}P[\{\phi\}] \tanh[\beta h(t-1|..)]\tanh[\beta h(t'|..)]$$
$$\times \prod_{s=0}^{t-2}\frac{1}{2}[1 + \sigma(s+1)\tanh[\beta h(s|..)]]\},$$

which we get directly for $t' = t - 1$, and which follows for times $t' < t - 1$ upon using the identity

$$\sigma[1 - \tanh^2(x)] = [1 + \sigma \tanh(x)][\sigma - \tanh(x)].$$

For the correlations we distinguish between $t' = t - 1$ and $t' < t - 1$,

$$C(t, t-1) = \sum_{\sigma(0)...\sigma(t-2)} \pi_0(\sigma(0)) \int \{d\phi\} P[\{\phi\}] \, \tanh[\beta h(t-1|..)] \times$$

$$\times \tanh[\beta h(t-2|..)] \prod_{s=0}^{t-3} \frac{1}{2} [1 + \sigma(s+1)\tanh[\beta h(s|..)]],$$

whereas for $t' < t - 1$ we have

$$C(t, t') = \sum_{\sigma(0)...\sigma(t-1)} \pi_0(\sigma(0)) \int \{d\phi\} P[\{\phi\}] \, \tanh[\beta h(t-1|..)] \times$$

$$\times \sigma(t') \prod_{s=0}^{t-2} \frac{1}{2} [1 + \sigma(s+1)\tanh[\beta h(s|..)]].$$

Now, the field at $t = 0$ is $h(0|..) = m_0 + \theta(0) + \alpha^{\frac{1}{2}}\phi(0)$, since the retarded self–interaction does not yet come into play. The distribution of $\phi(0)$ is fully characterized by its variance $\langle \phi^2(0) \rangle = C(0,0) = 1$. Therefore, with $Dz = (2\pi)^{-\frac{1}{2}} e^{-\frac{1}{2}z^2} dz$, we immediately find

$$m(1) = \int Dz \, \tanh[\beta(m_0 + \theta(0) + z\sqrt{\alpha})], \qquad C(1,0) = m_0 m(1),$$

$$G(1,0) = \beta \left\{ 1 - \int Dz \, \tanh^2[\beta(m_0 + \theta(0) + z\sqrt{\alpha})] \right\}.$$

For the self–interaction kernel this implies that $R(1,0) = G(1,0)$. We now move on to $t = 2$,

$$m(2) = \frac{1}{2} \sum_{\sigma(0)} \int d\phi(0) d\phi(1) P[\phi(0), \phi(1)] \, \tanh[\beta h(1|..)][1 + \sigma(0)m_0],$$

$$C(2,1) = \frac{1}{2} \sum_{\sigma(0)} \int d\phi(1) d\phi(0) P[\phi(0), \phi(1)] \times$$

$$\times \tanh[\beta h(1|..)] \tanh[\beta h(0|..)][1 + \sigma(0)m_0]$$

$$C(2,0) = \frac{1}{2} \sum_{\sigma(0)\sigma(1)} \int \{d\phi\} P[\{\phi\}] \, \tanh[\beta h(1|..)] \times$$

$$\sigma(0) \frac{1}{2} [1 + \sigma(1)\tanh[\beta h(0|..)]] [1 + \sigma(0)m_0],$$

4.5 Non–Physical Applications of Path Integrals

$$G(2,1) = \beta\{1 - \frac{1}{2}\sum_{\sigma(0)}\int d\phi(0)d\phi(1)P[\phi(0),\phi(1)] \times$$

$$\times \tanh^2[\beta h(1|..)][1+\sigma(0)m_0]\},$$

$$G(2,0) = \beta\{C(2,1) - \frac{1}{2}\sum_{\sigma(0)}\int d\phi(0)d\phi(1)P[\phi(0),\phi(1)] \times$$

$$\times \tanh[\beta h(1|..)]\tanh[\beta h(0|..)][1+\sigma(0)m_0]\} = 0.$$

We already know that $\langle\phi^2(0)\rangle = 1$; the remaining two moments we need in order to determine $P[\phi(0),\phi(1)]$ read [Coo01, SC00, SC01, CKS05]

$$\langle\phi(1)\phi(0)\rangle = \sum_{s=0}^{1}[\delta_{1,s}+\delta_{0,s}R(1,0)]C(s,0) = C(1,0)+G(1,0),$$

$$\langle\phi^2(1)\rangle = \sum_{s=0}^{1}\sum_{s'=1}^{1}[\delta_{1,s}+\delta_{0,s}R(1,0)]C(s,s')[\delta_{s',1}+\delta_{s',0}R(1,0)]$$

$$= G^2(1,0) + 2C(0,1)G(1,0) + 1.$$

We now know $P[\phi(0),\phi(1)]$ and can work out all macroscopic objects with $t = 2$ explicitly, if we wish. I will not do this here in full, but only point at the emerging pattern of all calculations at a given time t depending only on macroscopic quantities that have been calculated at times $t' < t$, which allows for iterative solution. Let us just work out $m(2)$ explicitly, in order to compare the first two recall overlaps $m(1)$ and $m(2)$ with the values found in simulations and in approximate theories. We note that calculating $m(2)$ only requires the field $\phi(1)$, for which we found $\langle\phi^2(1)\rangle = G^2(1,0) + 2C(0,1)G(1,0) + 1$:

$$m(2) = \frac{1}{2}\sum_{\sigma(0)}\int d\phi(1)P[\phi(1)]\ \tanh[\beta(m(1)+\theta(1)$$

$$+ \alpha G(1,0)\sigma(0) + \alpha^{\frac{1}{2}}\phi(1))][1+\sigma(0)m_0]$$

$$= \frac{1}{2}[1+m_0]\int Dz\ \tanh[\beta(m(1)+\theta(1)+\alpha G(1,0)$$

$$+ z\sqrt{\alpha[G^2(1,0)+2m_0\ m(1)\ G(1,0)+1]})]$$

$$+ \frac{1}{2}[1-m_0]\int Dz\ \tanh[\beta(m(1)+\theta(1)-\alpha G(1,0)$$

$$+ z\sqrt{\alpha[G^2(1,0)+2m_0\ m(1)\ G(1,0)+1]})]$$

Here we give a comparison of some of the approximate theories, the (exact) partition function (i.e., path–integral) formalism, and numerical simulations, for the case $\theta(t) = 0$ on the fully connected networks. The evolution of the recall overlap in the first two time–steps has been described as follows:

Naive Gaussian Approx: $m(1) = \int Dz\ \tanh[\beta(m(0) + z\sqrt{\alpha})]$
$m(2) = \int Dz\ \tanh[\beta(m(1) + z\sqrt{\alpha})]$

Amari-Maginu Theory: $m(1) = \int Dz\ \tanh[\beta(m(0) + z\sqrt{\alpha})]$
$m(2) = \int Dz\ \tanh[\beta(m(1) + z\Sigma\sqrt{\alpha})]$
$\Sigma^2 = 1 + 2m(0)m(1)G + G^2$
$G = \beta\left[1 - \int Dz\ \tanh^2[\beta(m(0) + z\sqrt{\alpha})]\right]$

Exact Solution: $m(1) = \int Dz\ \tanh[\beta(m(0) + z\sqrt{\alpha})]$
$m(2) = \tfrac{1}{2}[1 + m_0]\int Dz\ \tanh[\beta(m(1) + \alpha G + z\Sigma\sqrt{\alpha})]$
$+\quad \tfrac{1}{2}[1 - m_0]\int Dz\ \tanh[\beta(m(1) - \alpha G + z\Sigma\sqrt{\alpha})]$
$\Sigma^2 = 1 + 2m(0)m(1)G + G^2$
$G = \beta\left[1 - \int Dz\ \tanh^2[\beta(m(0) + z\sqrt{\alpha})]\right]$

We can now appreciate why the more advanced Gaussian approximation (Amari–Maginu theory, [AM88]) works well when the system state is close to the target attractor. This theory gets the moments of the Gaussian part of the interference noise distribution at $t = 1$ exactly right, but not the discrete part, whereas close to the attractor both the response function $G(1,0)$ and one of the two pre–factors $\tfrac{1}{2}[1 \pm m_0]$ in the exact expression for $m(2)$ will be very small, and the latter will therefore indeed approach a Gaussian shape. One can also see why the non–Gaussian approximation of [HO90] made sense: in the calculation of $m(2)$ the interference noise distribution can indeed be written as the sum of two Gaussian ones (although for $t > 2$ this will cease to be true).

Extremely Diluted Attractor Networks Near Saturation

The extremely diluted attractor networks were first studied in [DGZ87] (asymmetric dilution) and [WS91] (symmetric dilution). These models are got upon choosing $\lim_{N\to\infty} c/N = 0$ (while still $c \to \infty$) in definition (4.229) of the Hebbian synapses. The disorder average now involves both the patterns with $\mu > 1$ and the realisation of the 'wiring' variables $c_{ij} \in \{0,1\}$. Again, in working out the key function (4.236) we will show that for $N \to \infty$ the outcome can be written in terms of the above macroscopic quantities. We carry out the average over the spatial structure variables $\{c_{ij}\}$ first:

$$\mathcal{F}[\ldots] = \frac{1}{N}\log\overline{\left[e^{-\frac{i}{c}c_{ij}\xi_i^\mu\xi_j^\mu\sum_t \hat{h}_i(t)\sigma_j(t)}\right]}$$
$$= \frac{1}{N}\log\prod_{i<j}\overline{e^{-\frac{i}{c}\xi_i^\mu\xi_j^\mu[c_{ij}\sum_t \hat{h}_i(t)\sigma_j(t) + c_{ji}\sum_t \hat{h}_j(t)\sigma_i(t)]}}.$$

Now we have to distinguish between *symmetric* and *asymmetric dilution*. First we deal with the case of *symmetric dilution*: $c_{ij} = c_{ji}$ for all $i \neq j$. The average

4.5 Non–Physical Applications of Path Integrals

over the c_{ij} is trivial:

$$\overline{\prod_{i<j} e^{-\frac{i}{c}c_{ij}\sum_\mu \xi_i^\mu \xi_j^\mu \sum_t [\hat{h}_i(t)\sigma_j(t)+\hat{h}_j(t)\sigma_i(t)]}}$$

$$= \prod_{i<j}\left\{1 + \frac{c}{N}[e^{-\frac{i}{c}\xi_i^\mu \xi_j^\mu \sum_t [\hat{h}_i(t)\sigma_j(t)+\hat{h}_j(t)\sigma_i(t)]} - 1]\right\}$$

$$= \prod_{i<j} e^{-\frac{i}{N}\xi_i^\mu \xi_j^\mu \sum_t [\hat{h}_i(t)\sigma_j(t)+\hat{h}_j(t)\sigma_i(t)] - \frac{1}{2cN}[\xi_i^\mu \xi_j^\mu \sum_t [\hat{h}_i(t)\sigma_j(t)+\hat{h}_j(t)\sigma_i(t)]]^2 + \mathcal{O}(\frac{1}{N\sqrt{c}}) + \mathcal{O}(\frac{c}{N^2})}.$$

We separate in the exponent the terms where $\mu = \nu$ in the quadratic term (being of the form $\sum_{\mu\nu}\ldots$), and the terms with $\mu = 1$. We get (note: $p = \alpha c$):

$$\mathcal{F}[\ldots] = -i\sum_t a(t)k(t) - \frac{1}{2}\alpha \sum_{st}[q(s,t)Q(s,t) + K(s,t)K(t,s)] + \mathcal{O}(c^{-\frac{1}{2}}) + \mathcal{O}(c/N) +$$

$$\frac{1}{N}\log\{e^{-\frac{i}{N}\sum_t [\xi_i^\mu \hat{h}_i(t)][\xi_j^\mu \sigma_j(t)] - \frac{1}{4cN}\sum_{st}\xi_i^\mu \xi_j^\mu \xi_i^\nu \xi_j^\nu [\hat{h}_i(s)\sigma_j(s) + \hat{h}_j(s)\sigma_i(s)][\hat{h}_i(t)\sigma_j(t) + \hat{h}_j(t)\sigma_i(t)]}\}.$$

Our 'condensed ansatz' implies that for $\mu > 1$: $N^{-\frac{1}{2}}\xi_i^\mu \sigma_i(t) = \mathcal{O}(1)$ and $N^{-\frac{1}{2}}\xi_i^\mu \hat{h}_i(t) = \mathcal{O}(1)$. Thus the first term in the exponent containing the disorder is $\mathcal{O}(c)$, contributing $\mathcal{O}(c/N)$ to $\mathcal{F}[\ldots]$. We therefore retain only the second term in the exponent. However, the same argument applies to the second term. There all contributions can be seen as uncorrelated in leading order, so that $\sum_{i\neq j}\sum_{\mu\neq\nu}\ldots = \mathcal{O}(Np)$, giving a non-leading $\mathcal{O}(N^{-1})$ cumulative contribution to $\mathcal{F}[\ldots]$. Thus, provided $\lim_{N\to\infty} c^{-1} = \lim_{N\to\infty} c/N = 0$ (which we assumed), we have shown that the disorder average (4.236) is again, in leading order in N, with

Symmetric Case: $\quad \Phi[\mathbf{a},\mathbf{k},\mathbf{q},\mathbf{Q},\mathbf{K}] = -i\mathbf{a}\cdot\mathbf{k} - \frac{1}{2}\alpha\sum_{st}[q(s,t)Q(s,t) + K(s,t)K(t,s)].$

Next, we deal with the asymmetric case, where c_{ij} and c_{ji} are independent. Again, the average over the c_{ij} is trivial; here it gives

$$\overline{\prod_{i<j}\left\{e^{-\frac{i}{c}c_{ij}\xi_i^\mu \xi_j^\mu \sum_t \hat{h}_i(t)\sigma_j(t)} e^{-\frac{i}{c}c_{ji}\xi_i^\mu \xi_j^\mu \sum_t \hat{h}_j(t)\sigma_i(t)}\right\}}$$

$$= \prod_{i<j}\left\{1 + \frac{c}{N}[e^{-\frac{i}{c}\xi_i^\mu \xi_j^\mu \sum_t \hat{h}_i(t)\sigma_j(t)} - 1]\right\}\left\{1 + \frac{c}{N}[e^{-\frac{i}{c}\xi_i^\mu \xi_j^\mu \sum_t \hat{h}_j(t)\sigma_i(t)} - 1]\right\}$$

$$= \prod_{i<j}\left\{1 - \frac{c}{N}[\frac{i}{c}\xi_i^\mu \xi_j^\mu \sum_t \hat{h}_i(t)\sigma_j(t) + \frac{1}{2c^2}[\xi_i^\mu \xi_j^\mu \sum_t \hat{h}_i(t)\sigma_j(t)]^2 + \mathcal{O}(c^{-\frac{3}{2}})]\right\}$$

$$\times \left\{1 - \frac{c}{N}[\frac{i}{c}\xi_i^\mu \xi_j^\mu \sum_t \hat{h}_j(t)\sigma_i(t) + \frac{1}{2c^2}[\xi_i^\mu \xi_j^\mu \sum_t \hat{h}_j(t)\sigma_i(t)]^2 + \mathcal{O}(c^{-\frac{3}{2}})]\right\}$$

$$= \prod_{i<j} e^{-\frac{i}{N}\xi_i^\mu \xi_j^\mu \sum_t [\hat{h}_i(t)\sigma_j(t) + \hat{h}_j(t)\sigma_i(t)] - \frac{1}{2cN}[\xi_i^\mu \xi_j^\mu \sum_t \hat{h}_i(t)\sigma_j(t)]^2 - \frac{1}{2cN}[\xi_i^\mu \xi_j^\mu \sum_t \hat{h}_j(t)\sigma_i(t)]^2 + \mathcal{O}(\frac{1}{N\sqrt{c}}) + \mathcal{O}(\frac{c}{N^2})}.$$

Again, we separate in the exponent the terms where $\mu = \nu$ in the quadratic term and the terms with $\mu = 1$ and get

$$\mathcal{F}[\ldots] = -i\sum_t a(t)k(t) - \frac{1}{2}\alpha \sum_{st} q(s,t)Q(s,t) + \mathcal{O}(c^{-\frac{1}{2}}) + \mathcal{O}(c/n)$$

$$+ \frac{1}{N}\log\left\{e^{-\frac{i}{N}\sum_t[\xi_i^\mu \hat{h}_i(t)][\xi_j^\mu \sigma_j(t)] - \frac{1}{2cN}\xi_i^\mu \xi_j^\mu \xi_i^\nu \xi_j^\nu \sum_{st} \hat{h}_i(s)\sigma_j(s)\hat{h}_i(t)\sigma_j(t)}\right\}.$$

The scaling arguments given in the symmetric case, based on our 'condensed ansatz', apply again, and tell us that the remaining terms with the disorder are of vanishing order in N. We have again shown that the disorder average (4.236) is, in leading order in N with

Asymmetric Case: $\quad \Phi[\mathbf{a}, \mathbf{k}, \mathbf{q}, \mathbf{Q}, \mathbf{K}] = -i\mathbf{a}\cdot\mathbf{k} - \frac{1}{2}\alpha \sum_{st} q(s,t)Q(s,t).$

Now, asymmetric dilution corresponds to $\Delta = 0$, i.e., there is no retarded self–interaction, and the response function no longer plays a role. We now only retain $h(t|\ldots) = m(t) + \theta(t) + \alpha^{\frac{1}{2}}\phi(t)$, with $\langle \phi^2(t)\rangle = C(1,1) = 1$. We now get

$$m(t+1) = \sum_{\sigma(0)\ldots\sigma(t)} \pi_0(\sigma(0)) \int \{d\phi\}P[\{\phi\}] \tanh[\beta h(t|\ldots)] \times$$

$$\times \prod_{s=0}^{t-1} \frac{1}{2}[1 + \sigma(s+1)\tanh[\beta h(s|\ldots)]] = \int Dz \tanh[\beta(m(t) + \theta(t) + z\sqrt{\alpha})].$$

Similarly, for $t > t'$ equations for correlation and response functions reduce to

$$C(t,t') = \int \frac{d\phi_a d\phi_b}{2\pi\sqrt{1-C^2(t-1,t'-1)}} e^{-\frac{1}{2}\frac{\phi_a^2 + \phi_b^2 - 2C(t-1,t'-1)\phi_a\phi_b}{1-C^2(t-1,t'-1)}} \times$$
$$\times \tanh[\beta(m(t-1) + \theta(t-1) + \phi_a\sqrt{\alpha})] \times$$
$$\times \tanh[\beta(m(t'-1) + \theta(t'-1) + \phi_b\sqrt{\alpha})],$$

$$G(t,t') = \beta\delta_{t,t'+1}\left\{1 - \int Dz \tanh^2[\beta(m(t-1) + \theta(t-1) + z\sqrt{\alpha})]\right\}.$$

Let us also inspect the stationary state $m(t) = m$, for $\theta(t) = 0$. One easily proves that $m = 0$ as soon as $T > 1$, using

$$m^2 = \beta m \int_0^m dk[1 - \int Dz \tanh^2[\beta(k + z\sqrt{\alpha})]] \leq \beta m^2.$$

A continuous bifurcation occurs from the $m = 0$ state to an $m > 0$ state when $T = 1 - \int Dz \tanh^2[\beta z\sqrt{\alpha}]$. A parametrization of this transition line in the (α, T)–plane is given by

$$T(x) = 1 - \int Dz \, \tanh^2(zx), \qquad a(x) = x^2 T^2(x), \qquad \text{for } x \geq 0.$$

For $\alpha = 0$ we just jet $m = \tanh(\beta m)$ so $T_c = 1$. For $T = 0$ we get the equation $m = \text{erf}[m/\sqrt{2\alpha}]$, giving a continuous transition to $m > 0$ solutions at $\alpha_c = 2/\pi \approx 0.637$. The remaining question concerns the nature of the $m = 0$ state. Inserting $m(t) = \theta(t) = 0$ (for all t) into $C(t,t')$ tells us that $C(t,t') = f[C(t-1,t'-1)]$ for $t > t' > 0$, with 'initial conditions' $C(t,0) = m(t)m_0$, where

$$f[C] = \int \frac{d\phi_a d\phi_b}{2\pi\sqrt{1-C^2}} \, e^{-\frac{1}{2} \frac{\phi_a^2 + \phi_b^2 - 2C\phi_a\phi_b}{1-C^2}} \tanh[\beta\sqrt{\alpha}\phi_a] \tanh[\beta\sqrt{\alpha}\phi_b].$$

In the $m = 0$ regime we have $C(t,0) = 0$ for any $t > 0$, inducing $C(t,t') = 0$ for any $t > t'$, due to $f[0] = 0$. Thus we conclude that $C(t,t') = \delta_{t,t'}$ in the $m = 0$ phase, i.e., this phase is para–magnetic rather than of a spin–glass type.

On the other hand, physics of networks with symmetric dilution is more complicated situation. In spite of the extreme dilution, the interaction symmetry makes sure that the spins still have a sufficient number of common ancestors for complicated correlations to build up in finite time. We have

$$h(t|\{\sigma\},\{\phi\}) = m(t) + \theta(t) + \alpha \sum_{t'<t} G(t,t')\sigma(t') + \alpha^{\frac{1}{2}}\phi(t),$$

$$P[\{\phi\}] = \frac{e^{-\frac{1}{2}\sum_{t,t'} \phi(t)\mathbf{C}^{-1}(t,t')\phi(t')}}{(2\pi)^{(t_m+1)/2}\det^{\frac{1}{2}}\mathbf{C}}.$$

Thus the effective single neuron problem is found to be exactly of the form found also for the Gaussian model defined above (which, in turn, maps onto the parallel dynamics *Sherrington–Kirkpatrick model* (SK model) [SK75]) with the synapses $J_{ij} = J_0\xi_i\xi_j/N + Jz_{ij}/\sqrt{N}$ (in which the z_{ij} are symmetric zero–average and unit–variance Gaussian variables, and $J_{ii} = 0$ for all i), with the identification:

$$J \to \sqrt{\alpha}, \qquad \text{with} \qquad J_0 \to 1.$$

Since one can show that for $J_0 > 0$ the parallel dynamics SK model gives the same equilibrium state as the sequential one, we can now immediately write down the stationary solution of our dynamic equations which corresponds to the FDT regime, with $q = \lim_{\tau \to \infty} \lim_{t \to \infty} C(t, t+\tau)$:

$$q = \int Dz \tanh^2[\beta(m+z\sqrt{\alpha q})], \qquad m = \int Dz \tanh[\beta(m+z\sqrt{\alpha q})].$$

4.5.5 Cerebellum as a Neural Path–Integral

Recall that human motion is naturally driven by synergistic action of more than 600 skeletal muscles. While the muscles generate driving torques in the moving joints, subcortical neural system performs both local and global (loco)motion control: first reflexly controlling contractions of individual muscles, and then orchestrating all the muscles into synergetic actions in order to produce efficient movements. While the local reflex control of individual muscles is performed on the *spinal control level*, the global integration of all the muscles into coordinated movements is performed within the *cerebellum*.

All hierarchical subcortical neuro–muscular physiology, from the bottom level of a single muscle fiber, to the top level of cerebellar muscular synergy, acts as a *temporal $< out|in >$ reaction*, in such a way that the higher level acts as a command/control space for the lower level, itself representing an abstract image of the lower one:

A. At the *muscular level*, we have *excitation–contraction dynamics* [Hat77a, Hat78, Hat77b], in which $< out|in >$ is given by the following sequence of nonlinear diffusion processes: *neural-action-potential* $\leadsto$ *synaptic-potential* $\leadsto$ *muscular-action-potential* $\leadsto$ *excitation-contraction-coupling* $\leadsto$ *muscle-tension-generating* [Iva91, II06a]. Its purpose is the generation of muscular forces, to be transferred into driving torques within the joint anatomical geometry.

B. At the *spinal level*, $< out|in >$ is given by *autogenetic–reflex stimulus–response control* [Hou79]. Here we have a neural image of all individual muscles. The main purpose of the spinal control level is to give both positive and negative feedbacks to stabilize generated muscular forces within the 'homeostatic' (or, more appropriately, 'homeokinetic') limits. The individual muscular actions are combined into flexor–extensor (or agonist–antagonist) pairs, mutually controlling each other. This is the mechanism of *reciprocal innervation of agonists and inhibition of antagonists*. It has a purely mechanical purpose to form the so–called *equivalent muscular actuators* (EMAs), which would generate driving torques $T_i(t)$ for all movable joints.

C. At the *cerebellar level*, $< out|in >$ is given by *sensory–motor integration* [HBB96]. Here we have an abstracted image of all autogenetic reflexes. The main purpose of the cerebellar control level is integration and fine tuning of the action of all active EMAs into a synchronized movement, by *supervising* the individual autogenetic reflex circuits. At the same time, to be able to perform in new and unknown conditions, the cerebellum is continuously adapting its own neural circuitry by unsupervised (self–organizing) learning. Its action is subconscious and automatic, both in humans and in animals.

Naturally, we can ask the question: Can we assign a single $< out|in >$ measure to all these neuro–muscular stimulus–response reactions? We think that we can do it; so in this Letter, we propose the concept of *adaptive sensory–motor transition amplitude* as a unique measure for this temporal $< out|in >$ relation. Conceptually, this $< out|in > - amplitude$ can be formulated as the

4.5 Non–Physical Applications of Path Integrals

'neural path integral':

$$< out|in > \equiv \underset{amplitude}{\langle motor|sensory\rangle} = \int \mathcal{D}[w,x]\, e^{i\, S[x]}. \qquad (4.242)$$

Here, the integral is taken over all *activated* (or, 'fired') *neural pathways* $x^i = x^i(t)$ of the cerebellum, connecting its input *sensory*−state with its output *motor*−state, symbolically described by *adaptive neural measure* $\mathcal{D}[w,x]$, defined by the weighted product (of discrete time steps)

$$\mathcal{D}[w,x] = \lim_{n\to\infty} \prod_{t=1}^{n} w^i(t)\, dx^i(t),$$

in which the *synaptic weights* $w^i = w^i(t)$, included in all active neural pathways $x^i = x^i(t)$, are updated by the unsupervised *Hebbian–like learning* rule 4.290, namely

$$w^i(t+1) = w^i(t) + \frac{\sigma}{\eta}(w_d^i(t) - w_a^i(t)), \qquad (4.243)$$

where $\sigma = \sigma(t)$, $\eta = \eta(t)$ represent local neural *signal* and *noise* amplitudes, respectively, while superscripts d and a denote *desired* and *achieved* neural states, respectively. Theoretically, equations (4.242–4.243) define an ∞−*dimensional neural network*. Practically, in a computer simulation we can use $10^7 \le n \le 10^8$, roughly corresponding to the number of neurons in the cerebellum.

The exponent term $S[x]$ in equation (4.242) represents the *autogenetic–reflex action*, describing reflexly–induced motion of all active EMAs, from their initial *stimulus*−state to their final *response*−state, along the family of extremal (i.e., Euler–Lagrangian) paths $x_{\min}^i(t)$. ($S[x]$ is properly derived in (4.246–4.247) below.)

Spinal Autogenetic Reflex Control

Recall (from Introduction) that at the spinal control level we have the autogenetic reflex *motor servo* [Hou79], providing the local, reflex feedback loops for individual muscular contractions. A voluntary contraction force F of human skeletal muscle is reflexly excited (positive feedback $+F^{-1}$) by the responses of its *spindle receptors* to stretch and is reflexly inhibited (negative feedback $-F^{-1}$) by the responses of its *Golgi tendon organs* to contraction. Stretch and unloading reflexes are mediated by combined actions of several autogenetic neural pathways, forming the *motor servo*.

In other words, branches of the afferent fibers also synapse with with interneurons that inhibit motor neurons controlling the antagonistic muscles – *reciprocal inhibition*. Consequently, the stretch stimulus causes the antagonists to relax so that they cannot resists the shortening of the stretched muscle caused by the main reflex arc. Similarly, firing of the Golgi tendon receptors causes inhibition of the muscle contracting too strong and simultaneous *reciprocal activation* of its antagonist. Both mechanisms of reciprocal inhibition and activation performed by the autogenetic circuits $+F^{-1}$ and $-F^{-1}$, serve to generate the well–tuned EMA–driving torques T_i.

Now, once we have properly defined the symplectic musculo–skeletal dynamics [Iva04] on the biomechanical (momentum) phase–space manifold T^*M^N, we can proceed in formalizing its hierarchical subcortical neural control. By introducing the *coupling Hamiltonians* $H^m = H^m(q,p)$, selectively corresponding *only* to the $M \leq N$ *active joints*, we define the *affine Hamiltonian control function* $H_{aff} : T^*M^N \to \mathbb{R}$, in local canonical coordinates on T^*M^N given by (adapted from [NS90] for the biomechanical purpose)

$$H_{aff}(q,p) = H_0(q,p) - H^m(q,p)\,T_m, \qquad (m=1,\ldots,M \leq N), \qquad (4.244)$$

where $T_m = T_m(t,q,p)$ are affine feedback torque one–forms, different from the initial driving torques T_i acting in all the joints. Using the affine Hamiltonian function (5.63), we get the *affine Hamiltonian servo–system* [Iva04],

$$\dot{q}^i = \frac{\partial H_0(q,p)}{\partial p_i} - \frac{\partial H^m(q,p)}{\partial p_i} T_m, \qquad (4.245)$$

$$\dot{p}_i = -\frac{\partial H_0(q,p)}{\partial q^i} + \frac{\partial H^m(q,p)}{\partial q^i} T_m,$$

$$q^i(0) = q_0^i, \quad p_i(0) = p_i^0, \qquad (i=1,\ldots,N;\quad m=1,\ldots,M \leq N).$$

The affine Hamiltonian control system (4.245) gives our formal description for the autogenetic spinal motor–servo for all $M \leq N$ activated (i.e., working) EMAs.

Cerebellum – the Comparator

Having, thus, defined the spinal reflex control level, we proceed to model the top subcortical commander/controller, the *cerebellum*. It is a brain region anatomically located at the bottom rear of the head (the hindbrain), directly above the brainstem, which is important for a number of subconscious and automatic motor functions, including motor learning. It processes information received from the motor cortex, as well as from proprioceptors and visual and equilibrium pathways, and gives 'instructions' to the motor cortex and other subcortical motor centers (like the basal nuclei), which result in proper balance and posture, as well as smooth, coordinated skeletal movements, like walking, running, jumping, driving, typing, playing the piano, etc. Patients with cerebellar dysfunction have problems with precise movements, such as walking and balance, and hand and arm movements. The cerebellum looks *similar in all animals*, from fish to mice to humans. This has been taken as evidence that it performs a common function, such as regulating motor learning and the timing of movements, in all animals. Studies of simple forms of motor learning in the vestibulo–ocular reflex and eye–blink conditioning are demonstrating that timing and amplitude of learned movements are encoded by the cerebellum.

The cerebellum is responsible for coordinating precisely timed $<out|in>$ activity by integrating motor output with ongoing sensory feedback. It receives extensive projections from sensory–motor areas of the cortex and the periphery and directs it back to premotor and motor cortex [Ghe90, Ghe91]. This suggests a role in sensory–motor integration and the timing and execution of human movements. The cerebellum stores patterns of motor control for frequently performed movements, and therefore, its circuits are changed by experience and training. It was termed the *adjustable pattern generator* in the work of J. Houk and collaborators [HBB96]. Also, it has become the inspiring 'brain–model' in the recent robotic research [SA98, Sch98].

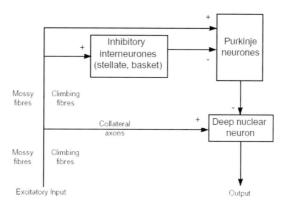

Fig. 4.27. Schematic $<out|in>$ organization of the primary cerebellar circuit. In essence, excitatory inputs, conveyed by collateral axons of Mossy and Climbing fibers activate directly neurones in the Deep cerebellar nuclei. The activity of these latter is also modulated by the inhibitory action of the cerebellar cortex, mediated by the Purkinje cells.

Comparing the number of its neurons ($10^7 - 10^8$), to the size of conventional neural networks, suggests that artificial neural nets *cannot* satisfactorily model the function of this sophisticated 'super–bio–computer', as its dimensionality is virtually infinite. Despite a lot of research dedicated to its structure and function (see [HBB96] and references there cited), the real nature of the cerebellum still remains a 'mystery'.

The main function of the cerebellum as a motor controller is depicted in Figure 4.28. A coordinated movement is easy to recognize, but we know little about how it is achieved. In search of the neural basis of coordination, a model of spinocerebellar interactions was recently presented in [AG05], in which the structural and functional organizing principle is a division of the cerebellum into discrete micro–complexes. Each micro–complex is the recipient of a spe-

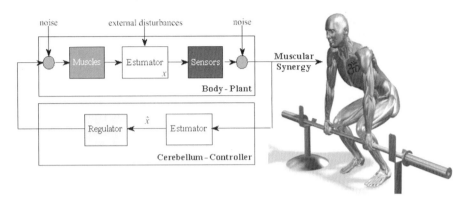

Fig. 4.28. The cerebellum as a motor controller.

cific motor error signal – that is, a signal that conveys information about an inappropriate movement. These signals are encoded by spinal reflex circuits and conveyed to the cerebellar cortex through climbing fibre afferents. This organization reveals salient features of cerebellar information processing, but also highlights the importance of systems level analysis for a fuller understanding of the neural mechanisms that underlie behavior.

Hamiltonian Action and Neural Path Integral

Here, we propose a *quantum–like adaptive control* approach to modelling the 'cerebellar mystery'. Corresponding to the affine Hamiltonian control function (5.63) we define the *affine Hamiltonian control action*,

$$S_{aff}[q,p] = \int_{t_{in}}^{t_{out}} d\tau \left[p_i \dot{q}^i - H_{aff}(q,p) \right]. \tag{4.246}$$

From the affine Hamiltonian action (4.246) we further derive the associated expression for the *neural phase–space path integral* (in normal units), representing the *cerebellar sensory–motor amplitude* $< out|in >$,

$$\langle q^i_{out}, p^{out}_i | q^i_{in}, p^{in}_i \rangle = \int \mathcal{D}[w,q,p] \, e^{i \, S_{aff}[q,p]} \tag{4.247}$$

$$= \int \mathcal{D}[w,q,p] \exp\left\{ i \int_{t_{in}}^{t_{out}} d\tau \left[p_i \dot{q}^i - H_{aff}(q,p) \right] \right\},$$

with

$$\int \mathcal{D}[w,q,p] = \int \prod_{\tau=1}^{n} \frac{w^i(\tau) dp_i(\tau) dq^i(\tau)}{2\pi},$$

where $w_i = w_i(t)$ denote the cerebellar synaptic weights positioned along its neural pathways, being continuously updated using the Hebbian–like self–organizing learning rule (4.243). Given the transition amplitude $< out|in >$

(4.247), the *cerebellar sensory–motor transition probability* is defined as its absolute square, $|<out|in>|^2$.

In (4.247), $q_{in}^i = q_{in}^i(t)$, $q_{out}^i = q_{out}^i(t)$; $p_i^{in} = p_i^{in}(t)$, $p_i^{out} = p_i^{out}(t)$; $t_{in} \leq t \leq t_{out}$, for all discrete time steps, $t = 1, ..., n \to \infty$, and we are allowing for the affine Hamiltonian $H_{aff}(q,p)$ to depend upon all the ($M \leq N$) EMA–angles and angular momenta collectively. Here, we actually systematically took a discretized differential time limit of the form $t_\sigma - t_{\sigma-1} \equiv d\tau$ (both σ and τ denote discrete time steps) and wrote $\frac{(q_\sigma^i - q_{\sigma-1}^i)}{(t_\sigma - t_{\sigma-1})} \equiv \dot{q}^i$. For technical details regarding the path integral calculations on Riemannian and symplectic manifolds (including the standard regularization procedures), see [Kla97, Kla00].

Now, motor learning occurring in the cerebellum can be observed using functional MR imaging, showing changes in the cerebellar action potential, related to the motor tasks (see, e.g., [MA02]). To account for these electro–physiological currents, we need to add the *source* term $J_i(t)q^i(t)$ to the affine Hamiltonian action (4.246), (the current $J_i = J_i(t)$ acts as a source $J_i A^i$ of the *cerebellar electrical potential* $A^i = A^i(t)$),

$$S_{aff}[q,p,J] = \int_{t_{in}}^{t_{out}} d\tau \left[p_i \dot{q}^i - H_{aff}(q,p) + J_i q^i \right],$$

which, subsequently gives the cerebellar path integral with the action potential source, coming either from the motor cortex or from other subcortical areas.

Note that the standard *Wick rotation*: $t \mapsto it$ (see [Kla97, Kla00]), makes all our path integrals real, i.e.,

$$\int \mathcal{D}[w,q,p] e^{i S_{aff}[q,p]} \quad \xrightarrow{Wick} \quad \int \mathcal{D}[w,q,p] e^{-S_{aff}[q,p]},$$

while their subsequent discretization gives the standard thermodynamic *partition functions*,

$$Z = \sum_j e^{-w_j E^j / T}, \tag{4.248}$$

where E^j is the energy eigenvalue corresponding to the affine Hamiltonian $H_{aff}(q,p)$, T is the temperature–like environmental control parameter, and the sum runs over all energy eigenstates (labelled by the index j). From (4.248), we can further calculate all statistical and thermodynamic system properties (see [Fey72]), as for example, *transition entropy* $S = k_B \ln Z$, etc.

4.5.6 Dissipative Quantum Brain Model

The *conservative brain* model was originally formulated within the framework of the quantum field theory (QFT) by [RU67] and subsequently developed in [STU78, STU79, JY95, JPY96]. The conservative brain model has been recently extended to the *dissipative quantum dynamics* in the work of G. Vitiello and collaborators [Vit95, AV00, PV99, Vit01, PV03, PV04].

The motivations at the basis of the formulation of the quantum brain model by Umezawa and Ricciardi trace back to the laboratory observations leading Lashley to remark (in 1940) that "masses of excitations... within general fields of activity, without regard to particular nerve cells are involved in the determination of behavior" [Las42, Pri91]. In 1960's, K. Pribram, also motivated by experimental observations, started to formulate his *holographic hypothesis*. According to W. Freeman [Fre90, Fre96, Fre00], "information appears indeed in such observations to be spatially uniform in much the way that the information density is uniform in a hologram". While the activity of the single neuron is experimentally observed in form of discrete and stochastic pulse trains and point processes, the 'macroscopic' activity of large assembly of neurons appears to be spatially coherent and highly structured in phase and amplitude.

Motivated by such an experimental situation, Umezawa and Ricciardi formulated in [RU67] the quantum brain model as a many–body physics problem, using the formalism of QFT with spontaneous breakdown of symmetry (which had been successfully tested in condensed matter experiments). Such a formalism provides the only available theoretical tool capable to describe long–range correlations such as the ones observed in the brain – presenting almost simultaneous responses in several regions to some external stimuli. The understanding of these long–range correlations in terms of modern biochemical and electrochemical processes is still lacking, which suggests that these responses could not be explained in terms of single neuron activity [Pri71, Pri91].

Lagrangian dynamics in QFT is, in general, invariant under some group G of continuous transformations, as proposed by the famous Noether theorem. Now, spontaneous symmetry breakdown, one of the corner–stones of Haken's synergetics [Hak83, Hak93], occurs when the minimum energy state (the ground, or vacuum, state) of the system is not invariant under the full group G, but under one of its subgroups. Then it can be shown [IZ80, Ume93] that collective modes, the so–called Nambu–Goldstone (NG) boson modes, are dynamically generated. Propagating over the whole system, these modes are the carriers of the *long–range correlation*, in which the order manifests itself as a global property dynamically generated. The long–range correlation modes are responsible for keeping the ordered pattern: they are coherently *condensed* in the ground state (similar to e.g., in the crystal case, where they keep the atoms trapped in their lattice sites). The long–range correlation thus forms a sort of net, extending over all the system volume, which traps the system components in the ordered pattern. This explains the "holistic" macroscopic collective behavior of the system components.

More precisely, according to the *Goldstone theorem* in QFT [IZ80, Ume93], the spontaneous breakdown of the symmetry implies the existence of long–range correlation NG–modes in the ground state of the system. These modes are massless modes in the infinite volume limit, but they may acquire a finite, non-zero mass due to boundary or impurity effects [ARV02]. In the quantum brain model these modes are called dipole–wave–quanta (DWQ). The density of their condensation in the ground states acts as a *code* classifying the state and the memory there recorded. States with different code values are unitarily inequivalent states, i.e., there is no unitary transformation relating states of different codes.[47]

Now, in formulating a proper mathematical model of brain, the conservative dynamics is not realistic: we cannot avoid to take into consideration the dissipative character of brain dynamics, since the brain is an intrinsically open system, continuously interacting with the environment. As Vitiello observed in [Vit01, PV03, PV04], the very same fact of "getting an information" introduces a partition in the time coordinate, so that one may distinguish between *before* "getting the information" (the past) and *after* "getting the information" (the future): the *arrow of time* is in this way introduced. ..."*Now* you know it!" is the familiar warning to mean that now, i.e. after having received a certain information, you are not the same person as before getting it. It has been shown that the psychological arrow of time (arising as an effect of memory recording) points in the same direction of the thermodynamical arrow of time (increasing entropy direction) and of the cosmological arrow of time (the expanding Universe direction) [AMV00].

The canonical quantization procedure of a dissipative system requires to include in the formalism also the system representing the environment (usually the heat bath) in which the system is embedded. One possible way to do that is to depict the environment as the time–reversal image of the system [CRV92]: the environment is thus described as the *double* of the system in the time–reversed dynamics (the system image in the mirror of time).

Within the framework of dissipative QFT, the brain system is described in terms of an *infinite collection of damped harmonic oscillators* A_κ (the simplest prototype of a dissipative system) representing the DWQ [Vit95]. Now, the collection of damped harmonic oscillators is ruled by the Hamiltonian [Vit95, CRV92]

$$H = H_0 + H_I, \quad \text{with}$$
$$H_0 = \hbar\Omega_\kappa(A_\kappa^\dagger A_\kappa - \tilde{A}_\kappa^\dagger \tilde{A}_\kappa), \qquad H_I = i\hbar\Gamma_\kappa(A_\kappa^\dagger \tilde{A}_\kappa^\dagger - A_\kappa \tilde{A}_\kappa),$$

where Ω_κ is the frequency and Γ_κ is the damping constant. The $\tilde{A}_\kappa$ modes are the 'time–reversed mirror image' (i.e., the 'mirror modes') of the A_κ modes. They are the doubled modes, representing the environment modes, in such a way that κ generically labels their degrees–of–freedom. In particular, we consider the damped harmonic oscillator (DHO)

---

[47] We remark that the spontaneous breakdown of symmetry is possible since in QFT there exist infinitely many ground states or vacua which are physically distinct (technically speaking, they are "unitarily inequivalent"). In quantum mechanics (QM), on the contrary, all the vacua are physically equivalent and thus there cannot be symmetry breakdown.

$$m\ddot{x} + \gamma \dot{x} + \kappa x = 0, \tag{4.249}$$

as a simple prototype for dissipative systems (with intention that thus get results also apply to more general systems). The damped oscillator (4.249) is a non–Hamiltonian system and therefore the customary canonical quantization procedure cannot be followed. However, one can face the problem by resorting to well known tools such as the *density matrix* ρ and the *Wigner function* $W = W(x, p, t)$.

Let us start with the special case of a *conservative particle* in the absence of friction γ, with the standard Hamiltonian,

$$H = -(\hbar \partial_x)^2/2m + V(x).$$

Recall (from the previous subsection) that the *density matrix equation of motion*, i.e., *quantum Liouville equation*, is given by

$$i\hbar \dot{\rho} = [H, \rho]. \tag{4.250}$$

The density matrix function ρ is defined by

$$\langle x + \tfrac{1}{2}y|\rho(t)|x - \tfrac{1}{2}y\rangle = \psi^*(x + \tfrac{1}{2}y, t)\psi(x - \tfrac{1}{2}y, t) \equiv W(x, y, t),$$

with the associated standard expression for the *Wigner function* (see, e.g., [II07b]),

$$W(p, x, t) = \frac{1}{2\pi\hbar} \int W(x, y, t) \, \mathrm{e}^{\left(-\mathrm{i}\frac{py}{\hbar}\right)} dy.$$

Now, in the coordinate x–representation, by introducing the notation

$$x_\pm = x \pm \tfrac{1}{2}y, \tag{4.251}$$

the Liouville equation (4.250) can be expanded as

$$i\hbar\, \partial_t \langle x_+|\rho(t)|x_-\rangle = \tag{4.252}$$

$$\left\{ -\frac{\hbar^2}{2m}\left[\partial_{x_+}^2 - \partial_{x_-}^2\right] + [V(x_+) - V(x_-)] \right\} \langle x_+|\rho(t)|x_-\rangle,$$

while the Wigner function $W(p, x, t)$ is now given by

$$i\hbar\, \partial_t W(x, y, t) = H_o W(x, y, t), \quad \text{with}$$

$$H_o = \frac{1}{m} p_x p_y + V(x + \tfrac{1}{2}y) - V(x - \tfrac{1}{2}y), \tag{4.253}$$

$$\text{and} \quad p_x = -i\hbar \partial_x, \quad p_y = -i\hbar \partial_y.$$

The new Hamiltonian H_o (4.253) may be get from the corresponding Lagrangian

4.5 Non–Physical Applications of Path Integrals

$$L_o = m\dot{x}\dot{y} - V(x + \frac{1}{2}y) + V(x - \frac{1}{2}y). \tag{4.254}$$

In this way, Vitiello concluded that the density matrix and the Wigner function formalism *required*, even in the conservative case (with zero mechanical resistance γ), the introduction of a 'doubled' set of coordinates, $x_\pm$, or, alternatively, x and y. One may understand this as related to the introduction of the 'couple' of indices *necessary* to label the density matrix elements (4.252).

Let us now consider the case of the *particle interacting* with a *thermal bath* at temperature T. Let f denote the *random force* on the particle at the position x due to the bath. The interaction Hamiltonian between the bath and the particle is written as

$$H_{int} = -fx. \tag{4.255}$$

Now, in the *Feynman–Vernon formalism* (see [Fey72]), the *effective action* $A[x,y]$ for the particle is given by

$$A[x,y] = \int_{t_i}^{t_f} L_o(\dot{x}, \dot{y}, x, y)\, dt + I[x,y],$$

with L_o defined by (4.254) and

$$e^{\frac{i}{\hbar}I[x,y]} = \langle (e^{-\frac{i}{\hbar}\int_{t_i}^{t_f} f(t)x_-(t)dt})_- (e^{\frac{i}{\hbar}\int_{t_i}^{t_f} f(t)x_+(t)dt})_+ \rangle, \tag{4.256}$$

where the symbol $\langle . \rangle$ denotes the average with respect to the thermal bath; '$(.)_+$' and '$(.)_-$' denote time ordering and anti–time ordering, respectively; the coordinates $x_\pm$ are defined as in (4.251). If the interaction between the bath and the coordinate x (4.255) were turned off, then the operator f of the bath would develop in time according to

$$f(t) = e^{iH_\gamma t/\hbar} f e^{-iH_\gamma t/\hbar},$$

where H_γ is the Hamiltonian of the isolated bath (decoupled from the coordinate x). $f(t)$ is then the force operator of the bath to be used in (4.256).

The interaction $I[x,y]$ between the bath and the particle has been evaluated in [SVW95] for a linear passive damping due to thermal bath by following Feynman–Vernon and Schwinger. The final result from [SVW95] is:

$$I[x,y] = \frac{1}{2}\int_{t_i}^{t_f} dt\, [x(t)F_y^{ret}(t) + y(t)F_x^{adv}(t)]$$
$$+ \frac{i}{2\hbar}\int_{t_i}^{t_f}\int_{t_i}^{t_f} dt ds\, N(t-s)y(t)y(s),$$

where the retarded force on y, F_y^{ret}, and the advanced force on x, F_x^{adv}, are given in terms of the retarded and advanced Green functions $G_{ret}(t-s)$ and $G_{adv}(t-s)$ by

$$F_y^{ret}(t) = \int_{t_i}^{t_f} ds\, G_{ret}(t-s) y(s), \qquad F_x^{adv}(t) = \int_{t_i}^{t_f} ds\, G_{adv}(t-s) x(s),$$

respectively. In (4.257), $N(t-s)$ is the *quantum noise* in the fluctuating random force given by

$$N(t-s) = \frac{1}{2} \langle f(t) f(s) + f(s) f(t) \rangle.$$

The real and the imaginary part of the action are given respectively by

$$\operatorname{Re}(A[x,y]) = \int_{t_i}^{t_f} L\, dt, \qquad (4.257)$$

$$L = m\dot{x}\dot{y} - \left[V\left(x + \frac{1}{2}y\right) - V\left(x - \frac{1}{2}y\right)\right] + \frac{1}{2}\left[x F_y^{ret} + y F_x^{adv}\right], \qquad (4.258)$$

and $$\operatorname{Im}(A[x,y]) = \frac{1}{2\hbar} \int_{t_i}^{t_f} \int_{t_i}^{t_f} N(t-s) y(t) y(s)\, dt ds. \qquad (4.259)$$

Equations (4.257–4.259), are *exact* results for linear passive damping due to the bath. They show that in the classical limit '$\hbar \to 0$' nonzero y yields an 'unlikely process' in view of the large imaginary part of the action implicit in (4.259). Nonzero y, indeed, may lead to a negative real exponent in the evolution operator, which in the limit $\hbar \to 0$ may produce a negligible contribution to the probability amplitude. On the contrary, at quantum level nonzero y accounts for quantum noise effects in the fluctuating random force in the system–environment coupling arising from the imaginary part of the action (see [SVW95]).

When in (4.258) we use

$$F_y^{ret} = \gamma \dot{y} \qquad \text{and} \qquad F_x^{adv} = -\gamma \dot{x} \qquad \text{we get,}$$

$$L(\dot{x}, \dot{y}, x, y) = m\dot{x}\dot{y} - V\left(x + \frac{1}{2}y\right) + V\left(x - \frac{1}{2}y\right) + \frac{\gamma}{2}(x\dot{y} - y\dot{x}). \qquad (4.260)$$

By using

$$V\left(x \pm \frac{1}{2}y\right) = \frac{1}{2}\kappa\left(x \pm \frac{1}{2}y\right)^2$$

in (4.260), the DHO equation (4.249) and its complementary equation for the y coordinate

$$m\ddot{y} - \gamma\dot{y} + \kappa y = 0. \qquad (4.261)$$

are derived. The y–oscillator is the time–reversed image of the x–oscillator (4.249). From the manifolds of solutions to equations (4.249) and (4.261), we could choose those for which the y coordinate is constrained to be zero, they simplify to

$$m\ddot{x} + \gamma\dot{x} + \kappa x = 0, \qquad y = 0.$$

4.5 Non–Physical Applications of Path Integrals 629

Thus we get the classical damped oscillator equation from a Lagrangian theory at the expense of introducing an 'extra' coordinate y, later constrained to vanish. Note that the constraint $y(t) = 0$ is *not* in violation of the equations of motion since it is a true solution to (4.249) and (4.261).

Therefore, the general scheme of the dissipative quantum brain model can be summarized as follows. The starting point is that the brain is permanently coupled to the environment. Of course, the specific details of such a coupling may be very intricate and changeable so that they are difficult to be measured and known. One possible strategy is to average the effects of the coupling and represent them, at some degree of accuracy, by means of some 'effective' interaction. Another possibility is to take into account the environmental influence on the brain by a suitable *choice* of the brain vacuum state. Such a choice is triggered by the external input (breakdown of the symmetry), and it actually is the end point of the internal (spontaneous) dynamical process of the brain (self–organization). The chosen vacuum thus carries the *signature* (memory) of the reciprocal brain–environment influence at a given time under given boundary conditions. A change in the brain–environment reciprocal influence then would correspond to a change in the choice of the brain vacuum: the brain state evolution or 'story' is thus the story of the trade of the brain with the surrounding world. The theory should then provide the equations describing the brain evolution 'through the vacua', each vacuum for each instant of time of its history.

The brain evolution is thus similar to a time–ordered sequence of photograms: each photogram represents the 'picture' of the brain at a given instant of time. Putting together these photograms in 'temporal order' one gets a movie, i.e. the story (the evolution) of open brain, which includes the brain–environment interaction effects.

The evolution of a memory specified by a given code value, say $\mathcal{N}$, can be then represented as a trajectory of given initial condition running over time–dependent vacuum states, denoted by $|0(t)>_\mathcal{N}$, each one minimizing the free energy functional. These trajectories are known to be *classical* trajectories in the infinite volume limit: transition from one representation to another inequivalent one would be strictly forbidden in a quantum dynamics.

Since we have now two–modes (i.e., non–tilde and tilde modes), the memory state $|0(t)>_\mathcal{N}$ turns out to be a two–mode coherent state. This is known to be an *entangled state*, i.e., it cannot be factorized into two single–mode states, the non–tilde and the tilde one. The physical meaning of such an entanglement between non-tilde and tilde modes is in the fact that the brain dynamics is permanently a dissipative dynamics. The entanglement, which is an unavoidable mathematical result of dissipation, represents the impossibility of cutting the links between the brain and the external world.[48]

---

[48] We remark that the entanglement is permanent in the large volume limit. Due to boundary effects, however, a unitary transformation could disentangle the tilde and non–tilde sectors: this may result in a pathological state for the brain. It is

In the dissipative brain model, noise and chaos turn out to be natural ingredients of the model. In particular, in the infinite volume limit the chaotic behavior of the trajectories in memory space may account for the high perceptive resolution in the recognition of the perceptual inputs. Indeed, small differences in the codes associated to external inputs may lead to diverging differences in the corresponding memory paths. On the other side, it also happens that codes differing only in a finite number of their components (in the momentum space) may easily be recognized as being the 'same' code, which makes possible that 'almost similar' inputs are recognized by the brain as 'equal' inputs (as in pattern recognition).

Therefore, the brain may be viewed as a complex system with (infinitely) many macroscopic configurations (the memory states). Dissipation is recognized to be the root of such a complexity.

QED Brain

In this subsection, mainly following [Sta95], we formulate a quantum electrodynamics brain model. Recall that quantum electrodynamics (extended to cover the magnetic properties of nuclei) is the theory that controls, as far as we know, the properties of the tissues and the aqueous (ionic) solutions that constitute our brains. This theory is our paradigm basic physical theory, and the one best understood by physicists. It describes accurately, as far as we know, the huge range of actual physical phenomena involving the materials encountered in daily life. It is also related to classical electrodynamics in a particularly beautiful and useful way.

In the low–energy regime of interest here it should be sufficient to consider just the classical part of the photon interaction defined in [Sta83]. Then the explicit expression for the unitary operator that describes the evolution from time t_1 to time t_2 of the quantum electromagnetic field in the presence of a set $L = \{L_i\}$ of specified classical charged–particle trajectories, with trajectory L_i specified by the function $x_i(t)$ and carrying charge e_i, is [Sta95]

$$U[L; t_2, t_1] = \exp <a^* \cdot J(L)> \exp <-J^*(L) \cdot a> \exp[-(J^*(L) \cdot J(L)/2)],$$

where, for any X and Y,

$$<X \cdot Y> \equiv \int d^4k (2\pi)^{-4} 2\pi \delta^+(k^2) X(k) \cdot Y(k),$$

---

known that forced isolation of a subject produces pathological states of various kinds. We also observe that the tilde mode is not just a mathematical fiction. It corresponds to a real excitation mode (quasi–particle) of the brain arising as an effect of its interaction with the environment: the couples of non–tilde/tilde dwq quanta represent the correlation modes dynamically created in the brain as a response to the brain–environment reciprocal influence. It is the interaction between tilde and non–tilde modes that controls the irreversible time evolution of the brain: these collective modes are confined to live *in* the brain. They vanish as soon as the links between the brain and the environment are cut.

$$(X \cdot Y) \equiv \int d^4k (2\pi)^{-4} i(k^2 + i\epsilon)^{-1} X(k) \cdot Y(k),$$

and $X \cdot Y = X_\mu Y^\mu = X^\mu Y_\mu$. Also,

$$J_\mu(L; k) \equiv \sum_i -ie_i \int_{L_i} dx_\mu \exp(ikx).$$

The integral along the trajectory L_i is

$$\int_{L_i} dx_\mu \exp(ikx) \equiv \int_{t_1}^{t_2} dt (dx_{i\mu}(t)/dt) \exp(ikx).$$

The $a^*(k)$ and $a(k)$ are the photon creation and annihilation operators:

$$[a(k), a^*(k')] = (2\pi)^3 \delta^3(k - k') 2k_0.$$

The operator $U[L; t_2, t_1]$ acting on the photon vacuum state creates the coherent photon state that is the quantum–theoretic analog of the classical electromagnetic field generated by classical point particles moving on the set of trajectories $L = \{L_i\}$ between times t_1 and t_2.

The $U[L; t_2, t_1]$ can be decomposed into commuting contributions from the various values of k. The general coherent state can be written [Sta95]

$$|q, p> \equiv \exp i(<q \cdot P> - <p \cdot Q>)|0>,$$

where $|0>$ is the photon vacuum state and

$$Q(k) = (a_k + a_k^*)/\sqrt{2} \quad \text{and} \quad P(k) = i(a_k - a_k^*)/\sqrt{2},$$

and $q(k)$ and $p(k)$ are two functions defined (and square integrable) on the mass shell $k^2 = 0$, $k_0 \geq 0$. The inner product of two coherent states is

$$<q, p|q', p'> = \exp -(<q - q' \cdot q - q'> + <p - p' \cdot p - p'> + 2i <p - p' \cdot q + q'>)/4.$$

There is a decomposition of unity

$$I = \prod d^4k (2\pi)^{-4} 2\pi \delta^+(k^2) \int dq_k dp_k/\pi$$

$$\times \exp(iq_k P_k - ip_k Q_k)|0_k><0_k| \exp -(iq_k P_k - ip_k Q_k).$$

Here meaning can be given by quantizing in a box, so that that the variable k is discretized. Equivalently,

$$I = \int d\mu(q, p)|q, p><q, p|,$$

where $\mu(q, p)$ is the appropriate measure on the functions $q(k)$ and $p(k)$. Then if the state $|\Psi><\Psi|$ were to jump to $|q, p><q, p|$ with probability density $<q, p|\Psi><\Psi|q, p>$, the resulting mixture would be [Sta95]

$$\int d\mu(q,p)|q,p\rangle\langle q,p|\Psi\rangle\langle\Psi|q,p\rangle\langle q,p|,$$

whose trace is

$$\int d\mu(q,p)\langle q,p|\Psi\rangle\langle\Psi|q,p\rangle=\langle\Psi|\Psi\rangle.$$

To represent the limited capacity of consciousness let us assume, in this model, that the states of consciousness associated with a brain can be expressed in terms of a relatively small subset of the modes of the electromagnetic field in the brain cavity. Let us assume that events occurring outside the brain are keeping the state of the universe outside the brain cavity in a single state, so that the state of the brain can also be represented by a single state. The brain is represented, in the path–integral method of Feynman, by a superposition of the trajectories of the particles in it, with each element of the superposition accompanied by the coherent–state electromagnetic field that this set of trajectories generates. Let the state of the electromagnetic field restricted to the modes that represent consciousness be called $|\Psi(t)\rangle$. Using the decomposition of unity one can write

$$|\Psi(t)\rangle = \int d\mu(q,p)|q,p\rangle\langle q,p|\Psi(t)\rangle.$$

Hence the state at time t can be represented by the function $\langle q,p|\Psi(t)\rangle$, which is a complex-valued function over the set of arguments $\{q_1, p_1, q_2, p_2, \ldots, q_n, p_n\}$, where n is the number of modes associated with $|\Psi\rangle$. Thus in this model the contents of the consciousness associated with a brain is represented in terms of this function defined over a $2n$D space: the ith conscious event is represented by the transition

$$|\Psi_i(t_{i+1})\rangle \longrightarrow |\Psi_{i+1}(t_{i+1})\rangle = P_i|\Psi_i(t_{i+1})\rangle,$$

where P_i is a projection operator.

For each allowed value of k the pair of numbers (q_k, p_k) represents the state of motion of the kth mode of the electromagnetic field. Each of these modes is defined by a particular wave pattern that extends over the whole brain cavity. This pattern is an oscillating structure something like a sine wave or a cosine wave. Each mode is fed by the motions of all of the charged particles in the brain. Thus each mode is a representation of a certain integrated aspect of the activity of the brain, and the collection of values $q_1, p_1, \ldots, p_n$ is a compact representation of certain aspects the over–all activity of the brain.

The state $|q,p\rangle$ represents the conjunction, or collection over the set of all allowed values of k, of the various states $|q_k, p_k\rangle$. The function

$$V(q,p,t) = \langle q,p|\Psi(t)\rangle\langle\Psi(t)|q,p\rangle$$

satisfies $0 \leq V(q,p,t) \leq 1$, and it represents, according to orthodox thinking, the 'probability' that a system that is represented by a general state $|\Psi(t)\rangle$

4.5 Non–Physical Applications of Path Integrals

just before the time t will be observed to be in the classically describable state $|q, p >$ if the observation occurs at time t. The coherent states $|q, p >$ can, for various mathematical and physical reasons, be regarded as the 'most classical' of the possible states of the electromagnetic quantum field.

To formulate a causal dynamics in which the state of consciousness itself controls the selection of the next state of consciousness one must specify a rule that determines, in terms of the evolving state $|\Psi_i(t) >$ up to time t_{i+1}, both the time t_{i+1} when the next selection event occurs, and the state $|\Psi_{i+1}(t_{i+1}) >$ that is selected and actualized by that event.

In the absence of interactions, and under certain ideal conditions of confinement, the deterministic normal law of evolution entails that in each mode k there is an independent rotation in the (q_k, p_k) plane with a characteristic angular velocity $\omega_k = k_0$. Due to the effects of the motions of the particles there will be, added to this, a flow of probability that will tend to concentrate the probability in the neighborhoods of a certain set of 'optimal' classical states $|q, p >$. The reason is that the function of brain dynamics is to produce some single template for action, and to be effective this template must be a 'classical' state, because, according to orthodox ideas, only these can be dynamically robust in the room temperature brain. According to the semi–classical description of the brain dynamics, only one of these classical–type states will be present, but according to quantum theory there must be a superposition of many such classical–type states, unless collapses occurs at lower (i.e., microscopic) levels. The assumption here is that no collapses occur at the lower brain levels: there is absolutely no empirical evidence, or theoretical reason, for the occurrence of such lower–level brain events.

So in this model the probability will begin to concentrate around various locally optimal coherent states, and hence around the various (generally) isolated points (q, p) in the $2n$D space at which the quantity [Sta95]

$$V(q, p, t) = < q, p | \Psi_i(t) >< \Psi_i(t) | q, p >$$

reaches a local maximum. Each of these points (q, p) represents a *locally–optimal solution* (at time t) to the search problem: as far as the myopic local mechanical process can see the state $|q, p >$ specifies an analog–computed 'best' template for action in the circumstances in which the organism finds itself. This action can be either intentional (it tends to create in the future a certain state of the body/brain/environment complex) or attentional (it tends to gather information), and the latter action is a special case of the former. As discussed in [Sta93], the intentional and attentional character of these actions is a consequence of the fact that the template for action actualized by the quantum brain event is represented as a projected body–world schema, i.e., as the brains projected representation of the body that it is controlling and the environment in which it is situated.

Let a certain time $t_{i+1} > t_i$ be defined by an (urgency) energy factor $E(t) = \hbar(t_{i+1} - t_i)^{-1}$. Let the value of (q, p) at the largest of the local–maxima of $V(q, p, t_{i+1})$ be called $(q(t_{i+1}), p(t_{i+1}))_{max}$. Then the simplest possible reasonable selection rule would be given by the formula

$$P_i = |(q(t_{i+1}), p(t_{i+1}))_{max} >< (q(t_{i+1}), p(t_{i+1}))_{max}|,$$

which entails that

$$\frac{|\Psi_{i+1} >< \Psi_{i+1}|}{< \Psi_{i+1}|\Psi_{i+1} >} = |(q(t_{i+1}), p(t_{i+1}))_{max} >< (q(t_{i+1}), p(t_{i+1}))_{max}|.$$

This rule could produce a tremendous speed up of the search process. Instead of waiting until all the probability gets concentrated in one state $|q, p>$, or into a set of isolated states $|q_i, p_i >$ [or choosing the state randomly, in accordance with the probability function $V(q, p, t_{i+1})$, which could often lead to a disastrous result], this simplest selection process would pick the state $|q, p>$ with the largest value of $V(q, p, t)$ at the time $t = t_{i+1}$. This process does not involve the complex notion of picking a random number, which is a physically impossible feat that is difficult even to define.

One important feature of this selection process is that it involves the state $\Psi(t)$ as a whole: the whole function $V(q, p, t_{i+1})$ must be known in order to determine where its maximum lies. This kind of selection process is not available in the semi–classical ontology, in which only one classically describable state exists at the macroscopic level. That is because this single classically describable macro–state state (e.g., some one actual state $|q, p, t_{i+1} >$) contains no information about what the probabilities associated either with itself or with the other alternative possibilities would have been if the collapse had not occurred earlier, at some micro-level, and reduced the earlier state to some single classically describable state, in which, for example, the action potential along each nerve is specified by a well defined classically describable electromagnetic field. There is no rational reason in quantum mechanics for such a micro–level event to occur. Indeed, the only reason to postulate the occurrence of such premature reductions is to assuage the classical intuition that the action–potential pulse along each nerve 'ought to be classically describable even when it is not observed', instead of being controlled, when unobserved, by the local deterministic equations of quantum field theory. But the validity of this classical intuition is questionable if it severely curtails the ability of the brain to function optimally.

A second important feature of this selection process is that the actualized state Ψ_{i+1} is the state of the entire aspect of the brain that is connected to consciousness. So the feel of the conscious event will involve that aspect of the brain, taken as a whole. The 'I' part of the state $\Psi(t)$ is its slowly changing part. This part is being continually re–actualized by the sequence of events, and hence specifies the slowly changing background part of the felt experience. It is this persisting stable background part of the sequence of templates for action that is providing the over–all guidance for the entire sequence of selection events that is controlling the on–going brain process itself [Sta95].

4.5 Non–Physical Applications of Path Integrals

A somewhat more sophisticated search procedure would be to find the state $|(q,p)_{max}>$, as before, but to identify it as merely a candidate that is to be examined for its concordance with the objectives imbedded in the current template. This is what a good search procedure ought to do: first pick out the top candidate by means of a mechanical process, but then evaluate this candidate by a more refined procedure that could block its acceptance if it does not meet specified criteria.

It may at first seem strange to imagine that nature could operate in such a sophisticated way. But it must be remembered that the generation of a truly random sequence is itself a very sophisticated (and indeed physically impossible) process, and that what the physical sciences have understood, so far, is only the mechanical part of nature's two–part process. Here it is the not–well–understood selection process that is under consideration. We have imposed on this attempt to understand the selection process the naturalistic requirement that the whole process be expressible in natural terms, i.e., that the universal process be a causal self–controlling evolution of the Hilbert–space state–vector in which all aspects of nature, including our conscious experiences, are efficacious.

It may be useful to describe the main features of this model in simple terms. If we imagine the brain to be, for example, a uniform rectangular box then each mode k would correspond to wave form that is periodic in all three directions: it would be formed as a combination of products of sine waves and cosine waves, and would cover the whole box–shaped brain. (More realistic conditions are needed, but this is a simple prototype.) Classically there would be an amplitude for this wave, and in the absence of interactions with the charged particles this amplitude would undergo a simple periodic motion in time. In analogy with the coordinate and momentum variables of an oscillating pendulum there are two variables, q_k and p_k, that describe the motion of the amplitude of the mode k. With a proper choice of scales for the variables q_k and p_k the motion of the amplitude of mode k if it were not coupled to the charges would be a circular motion in the (q_k, p_k)–plane. The classical theory would say that the physical system, mode k, would be represented by a point in q_k, p_k space. But quantum theory says that the physical system, mode k, must be represented by a wave (i.e., by a wave ψ–function) in (q_k, p_k) space. The reason is that interference effects between the values of this wave (function) at different points (q_k, p_k) can be exhibited, and therefore it is not possible to say the full reality is represented by any single value of (q_k, p_k): one must acknowledge the reality of the whole wave. It is possible to associate something like a 'probability density' with this wave, but the corresponding probability cannot be concentrated at a point: in units where Planck's constant is unity the bulk of the probability cannot be squeezed into a region of the (q_k, p_k) plane of area less that unity.

The mode k has certain natural states called 'coherent states', $|q_k, p_k>$. Each of these is represented in (q_k, p_k)–space by a wave function that has a 'probability density' that falls off exponentially as one moves in any direction away from the center–point (q_k, p_k) at which the probability density is maximum. These coherent states are in many ways the 'most classical' wave functions allowed by quantum theory [Gla63a, Gla63b], and a central idea of the present model is to specify that it is to one of these 'most classical' states that the mode-k component of the electromagnetic field will jump, or collapse, when an observation occurs. This specification represents a certain 'maximal' principle: the second process, which is supposed to pick out and actualize some classically describable reality, is required to pick out and actualize one of these 'most classical' of the quantum states. If this selection/actualization process really exists in nature then the classically describable states that are actualized by this process should be 'natural classical states' from some point of view. The coherent states satisfy this requirement. This strong, specific postulate should be easier to disprove, if it is incorrect, than a vague or loosely defined one.

If we consider a system consisting of a collection of modes k, then the generalization of the single coherent state $|q_k, p_k >$ is the product of these states, $|q, p >$. Classically this system would be described by specifying the values all of the classical variables q_k and p_k as functions of time. But the 'best' that can be done quantum mechanically is to specify that at certain times t_i the system is in one of the coherent states $|q, p >$. However, the equations of local quantum field theory (here quantum electrodynamics) entail that if the system starts in such a state then the system will, if no 'observation' occurs, soon evolve into a superposition (i.e., a linear combination) of many such states. But the next 'observation' will then reduce it again to some classically describable state. In the present model each a human observation is identified as a human conscious experience. Indeed, these are the same observations that the pragmatic Copenhagen interpretation of Bohr refers to, basically. The 'happening' in a human brain that corresponds to such an observation is, according to the present model, the selection and actualization of the corresponding coherent state $|q, p >$.

The quantity $V(q, p, t_{i+1})$ defined above is, according to orthodox quantum theory, the predicted probability that a system that is in the state $\Psi(t_{i+1})$ at time t_{i+1} will be observed to be in state $|q, p >$ if the observation occurs at time t_{i+1}. In the present model the function $V(q, p, t_{i+1})$ is used to specify not a fundamentally stochastic (i.e., random or chance–controlled) process but rather the causal process of the selection and actualization of some particular state $|q, p >$. And this causal process is controlled by features of the quantum brain that are specified by the Hilbert space representation of the conscious process itself. This process is a nonlocal process that rides on the local brain process, and it is the nonlocal selection process that, according to the principles of quantum theory, is required to enter whenever an observation occurs.

4.5.7 Action–Amplitude Psychodynamics

In this section, which is written in the fashion of the *quantum brain*, we present the top level of natural biodynamics, using geometrical generalization of the *Feynman path integral*. To formulate the basics of *force–field psychodynamics*, we use the *action–amplitude picture* of the $BODY \rightleftarrows MIND$ adjunction:

↓ **Deterministic (causal) world of *Human BODY*** ↓

$$Action : S[q^n] = \int_{t_{in}}^{t_{out}} (E_k - E_p + Wrk + Src^{\pm})\, dt$$

––––––––––––––––––––––

$$Amplitude : \langle out|in \rangle = \oint \mathcal{D}[w_n q^n]\, e^{iS[q^n]}$$

↑ **Probabilistic (fuzzy) world of *Human MIND*** ↑

In the action integral, E_k, E_p, Wrk and $Src^{\pm}$ denote the kinetic end potential energies, work done by dissipative/driving forces and other energy sources/sinks, respectively. In the amplitude integral, the peculiar sign $\oint$ denotes integration along smooth paths and summation along discrete Markov chains; i is the imaginary unit, w_n are synaptic–like weights, while $\mathcal{D}$ is the Feynman path differential (defined below) calculated along the configuration trajectories q^n. The action $S[q^n]$, through the *least action principle* $\delta S = 0$, leads to all biodynamic equations considered so far (in generalized Lagrangian and Hamiltonian form). At the same time, the action $S[q^n]$ figures in the exponent of the path integral $\oint$, defining the probability transition amplitude $\langle out|in \rangle$. In this way, the whole body dynamics is incorporated in the mind dynamics. This *adaptive path integral* represents an *infinite–dimensional neural network*, suggesting an infinite capacity of human brain/mind.

For a long time the cortical systems for *language and actions* were believed to be independent modules. However, according to the recent research of [Pul05], as these systems are reciprocally connected with each other, information about language and actions might interact in distributed neuronal assemblies. A critical case is that of action words that are semantically related to different parts of the body (e.g. 'pick', 'kick', 'lick',...). The author suggests that the comprehension of these words might specifically, rapidly and automatically activate the motor system in a somatotopic manner, and that their comprehension rely on activity in the action system.

Motivational Cognition in the Life Space Foam

Applications of nonlinear dynamical systems in psychology have been encouraging, if not universally effective [Met97]. Its historical antecedents can be traced back to Piaget's [PHE92] and Vygotsky's [Vyg82] interpretations of the dynamic relations between action and thought, Lewin's theory of social dynamics and cognitive–affective development [Lew97], and [Ber47] theory of self–adjusting, goal–driven motor action.

Now, both the original *Lewinian force–field theory* in psychology (see [Lew51, Gol99]) and modern decision–field dynamics (see [BT93, RBT01, BD02]) are based on the classical Lewinian concept of an individual's *life space*.[49] As a topological construct, Lewinian life space represents a person's psychological environment that contains *regions* separated by dynamical permeable *boundaries*. As a field construct, on the other hand, the life space is not empty: each of its regions is characterized by *valence* (ranging from positive or negative and resulting from an interaction between the person's *needs* and the dynamics of their *environment*). Need is an energy construct, according to Lewin. It creates *tension* in the person, which, in combination with other tensions, initiates and sustains behavior. Needs vary from the most primitive urges to the most idiosyncratic intentions and can be both internally generated (e.g., thirst or hunger) and stimulus–induced (e.g., an urge to buy something in response to a TV advertisement). Valences are, in essence, personal values dynamically derived from the person's needs and attached to various regions in their life space. As a field, the life space generates forces pulling the person towards positively–valenced regions and pushing them away from regions with negative valence. Lewin's term for these forces is *vectors*. Combinations of multiple vectors in the life space cause the person to move from one region towards another. This movement is termed *locomotion* and it may range from overt behavior to cognitive shifts (e.g., between alternatives in a decision–making process). Locomotion normally results in crossing the boundaries between regions. When their permeability is degraded, these boundaries become *barriers* that restrain locomotion. Life space model, thus, offers a meta–theoretical language to describe a wide range of behaviors, from goal–directed action to intrapersonal conflicts and multi–alternative decision–making.

In order to formalize the Lewinian life–space concept, a set of *action principles* need to be associated to Lewinian force–fields, (loco)motion paths (representing mental abstractions of biomechanical paths [II05]) and life space geometry. As an extension of the Lewinian concept, in this section we recall [IA07] a new concept of *life–space foam* (LSF, see Figure 4.29). According to this new concept, Lewin's life space can be represented as a *geometrical functor* with globally smooth macro–dynamics, which is at the same time underpinned by wildly fluctuating, non–smooth, local micro–dynamics, de-

---

[49] The work presented in this subsection has been developed in collaboration with Dr. Eugene Aidman, Senior Research Scientist, Human Systems Integration, Land Operations Division, Defence Science & Technology Organisation, Australia.

scribable by *Feynman's*: (i) *sum–over–histories* $\oint_{paths}$, (ii) *sum–over–fields* $\oint_{fields}$, and (iii) *sum–over–geometries* $\oint_{geom}$.

LSF is thus a two–level *geometrodynamical functor*, representing these two distinct types of dynamics within the Lewinian life space. At its *macroscopic spatio–temporal level*, LSF appears as a 'nice & smooth' geometrical functor with globally predictable dynamics – formally, a smooth $n-$dimensional manifold M with local Riemannian metrics $g_{ij}(x)$, smooth force–fields and smooth (loco)motion paths, as conceptualized in the Lewinian theory. To model the global and smooth macro–level LSF–paths, fields and geometry, we use the general physics–like *principle of the least action*.

Now, the apparent smoothness of the macro–level LSF is achieved by the existence of another level underneath it. This *micro–level* LSF is actually a collection of wildly fluctuating force–fields, (loco)motion paths, curved regional geometries and topologies with holes. The micro–level LSF is proposed as an extension of the Lewinian concept: it is characterized by uncertainties and fluctuations, enabled by microscopic time–level, microscopic transition paths, microscopic force–fields, local geometries and varying topologies with holes. To model these fluctuating microscopic LSF–structures, we use three instances of *adaptive path integral*, defining a multi–phase and multi–path (also multi–field and multi–geometry) *transition* process from *intention* to the goal–driven *action*.

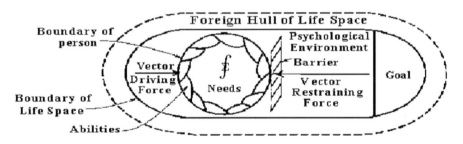

Fig. 4.29. Diagram of the *life space foam*: Lewinian life space with an adaptive path integral acting inside it and generating microscopic fluctuation dynamics.

We use the new LSF concept to develop modelling framework for motivational dynamics (MD) and induced cognitive dynamics (CD).

According to Heckhausen (see [Hec77]), *motivation* can be thought of as a process of *energizing* and *directing the action*. The process of energizing can be represented by Lewin's *force–field analysis* and Vygotsky's *motive formation* (see [Vyg82, AL91]), while the process of directing can be represented by *hierarchical action control* (see [Ber47, Ber96, Kuh85]).

Motivation processes both precede and coincide with every goal–directed action. Usually these motivation processes include the sequence of the following four feedforward *phases* [Vyg82, AL91]: (*)

A. *Intention Formation* $\mathcal{F}$, including: decision making, commitment building, etc.
B. *Action Initiation* $\mathcal{I}$, including: handling conflict of motives, resistance to alternatives, etc.
C. *Maintaining the Action* $\mathcal{M}$, including: resistance to fatigue, distractions, etc.
D. *Termination* $\mathcal{T}$, including parking and avoiding addiction, i.e., staying in control.

With each of the phases $\{\mathcal{F}, \mathcal{I}, \mathcal{M}, \mathcal{T}\}$ in (*), we can associate a *transition propagator* – an ensemble of (possibly crossing) feedforward paths propagating through the 'wood of obstacles' (including topological holes in the LSF, see Figure 4.30), so that the complete *transition functor* $\mathcal{T}A$ is a product of propagators (as well as sum over paths). All the phases–propagators are controlled by a unique *Monitor* feedback process.

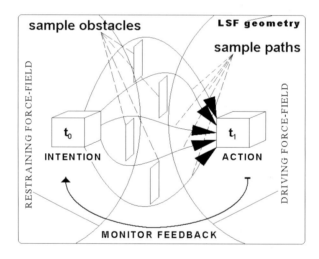

Fig. 4.30. *Transition–propagator* corresponding to each of the motivational phases $\{\mathcal{F}, \mathcal{I}, \mathcal{M}, \mathcal{T}\}$, consisting of an ensemble of feedforward paths propagating through the 'wood of obstacles'. The paths affected by driving and restraining force–fields, as well as by the local LSF–geometry. Transition goes from *Intention*, occurring at a sample time instant t_0, to *Action*, occurring at some later time t_1. Each propagator is controlled by its own *Monitor* feedback. All together they form the transition functor $\mathcal{T}A$.

In this subsection we propose an *adaptive path integral* formulation for the motivational–transition functor $\mathcal{T}A$. In essence, we sum/integrate over different paths and make a product (composition) of different phases–propagators. Recall that this is the most general description of the general *Markov stochastic process*.

We will also attempt to demonstrate the utility of the same LSF–formalisms in representing cognitive functions, such as memory, learning and decision making. For example, in the classical *Stimulus encoding* → *Search* → *Decision* → *Response* sequence [Ste69, Ash94], the environmental input–triggered *sensory memory* and *working memory* (WM) can be interpreted as operating at the micro–level force–field under the executive control of the *Monitor* feedback, whereas *search* can be formalized as a *control* mechanism guiding retrieval from the long–term memory (LTM, itself shaped by learning) and filtering material relevant to decision making into the WM. The essential measure of these mental processes, the *processing speed* (essentially determined by Sternberg's reaction–time) can be represented by our (loco)motion speed $\dot{x}$.

Six Faces of the Life Space Foam

The LSF has three forms of appearance: *paths + field + geometries*, acting on both macro–level and micro–level, which is six modes in total. In this section, we develop three least action principles for the macro–LSF–level and three adaptive path integrals for the micro–LSF–level. While developing our psycho–physical formalism, we will address the behavioral issues of motivational fatigue, learning, memory and decision making.

General Formalism

At both macro– and micro–levels, the total LSF represents a union of transition paths, force–fields and geometries, formally written as

$$LSF_{total} := LSF_{paths} \bigcup LSF_{fields} \bigcup LSF_{geom} \quad (4.262)$$
$$\equiv \oint_{paths} + \oint_{fields} + \oint_{geom}.$$

Corresponding to each of the three LSF–subspaces in (4.262) we formulate:

A. The *least action principle*, to model deterministic and predictive, macro–level MD & CD, giving a unique, global, causal and smooth path–field–geometry on the macroscopic spatio–temporal level; and
B. Associated *adaptive path integral* to model uncertain, fluctuating and probabilistic, micro–level MD & CD, as an ensemble of local paths–fields–geometries on the microscopic spatio–temporal level, to which the global macro–level MD & CD represents both time and ensemble *average* (which are equal according to the *ergodic hypothesis*).

In the proposed formalism, transition paths $x^i(t)$ are affected by the force–fields $\varphi^k(t)$, which are themselves affected by geometry with metric g_{ij}.

Global Macro–Level of LSF_{total}. In general, at the *macroscopic* LSF–level we first formulate the *total action* $S[\Phi]$, the central quantity in our formalism that has psycho–physical dimensions of $Energy \times Time = Effort$, with immediate cognitive and motivational applications: *the greater the action – the higher the speed of cognitive processes and the lower the macroscopic fatigue* (which includes all sources of physical, cognitive and emotional fatigue that influence motivational dynamics). The action $S[\Phi]$ depends on macroscopic paths, fields and geometries, commonly denoted by an abstract field symbol Φ^i. The action $S[\Phi]$ is formally defined as a temporal integral from the *initial* time instant t_{ini} to the *final* time instant t_{fin},

$$S[\Phi] = \int_{t_{ini}}^{t_{fin}} \mathfrak{L}[\Phi]\, dt, \qquad (4.263)$$

with *Lagrangian density* given by

$$\mathfrak{L}[\Phi] = \int d^n x\, \mathcal{L}(\Phi_i, \partial_{x^j}\Phi^i),$$

where the integral is taken over all n coordinates $x^j = x^j(t)$ of the LSF, and $\partial_{x^j}\Phi^i$ are time and space partial derivatives of the Φ^i–variables over coordinates.

Second, we formulate the *least action principle* as a minimal variation δ of the action $S[\Phi]$

$$\delta S[\Phi] = 0, \qquad (4.264)$$

which, using techniques from the calculus of variations gives, in the form of the so–called Euler–Lagrangian equations, a shortest (loco)motion path, an extreme force–field, and a life–space geometry of minimal curvature (and without holes). In this way, we effectively derive a *unique globally smooth transition functor*

$$\mathcal{TA} : INTENTION_{t_{ini}} \Rightarrow ACTION_{t_{fin}}, \qquad (4.265)$$

performed at a macroscopic (global) time–level from some initial time t_{ini} to the final time t_{fin}.

In this way, we get macro–objects in the global LSF: a single path described Newtonian–like equation of motion, a single force–field described by Maxwellian–like field equations, and a single obstacle–free Riemannian geometry (with global topology without holes).

For example, recall that in the period 1945–1949, John Wheeler and Richard Feynman developed their *action–at–a–distance electrodynamics* [WF49], in complete experimental agreement with the classical Maxwell's electromagnetic theory, but at the same time avoiding the complications of divergent self–interaction of the Maxwell's theory as well as eliminating its infinite number of field degrees–of–freedom. In Wheeler–Feynman view, "Matter consists of electrically charged particles," so they found a form for the action directly involving the motions of the charges only, which upon variation would give the Newtonian–like equations of motion of these charges. Here is the expression for this action in the flat space–time, which is in the core of quantum electrodynamics:

$$S[x; t_i, t_j] = \frac{1}{2} m_i \int (\dot{x}_\mu^i)^2 \, dt_i + \frac{1}{2} e_i e_j \int \int \delta(I_{ij}^2) \, \dot{x}_\mu^i(t_i) \dot{x}_\mu^j(t_j) \, dt_i dt_j$$

with (4.266)

$$I_{ij}^2 = \left[x_\mu^i(t_i) - x_\mu^j(t_j)\right] \left[x_\mu^i(t_i) - x_\mu^j(t_j)\right],$$

where $x_\mu^i = x_\mu^i(t_i)$ is the four–vector position of the ith particle as a function of the proper time t_i, while $\dot{x}_\mu^i(t_i) = dx_\mu^i/dt_i$ is the velocity four–vector. The first term in the action (4.266) is the ordinary mechanical action in Euclidean space, while the second term defines the electrical interaction of the charges, representing the Maxwell–like field (it is summed over each pair of charges; the factor $\frac{1}{2}$ is to count each pair once, while the term $i = j$ is omitted to avoid self–action; the interaction is a double integral over a delta function of the square of space–time interval I^2 between two points on the paths; thus, interaction occurs only when this interval vanishes, that is, along light cones [WF49]).

Now, from the point of view of Lewinian geometrical force–fields and (loco)motion paths, we can give the following life–space interpretation to the Wheeler–Feynman action (4.266). The mechanical–like locomotion term occurring at the single time t, needs a covariant generalization from the flat 4D Euclidean space to the nD smooth Riemannian manifold, so it becomes (see e.g., [II06b])

$$S[x] = \frac{1}{2} \int_{t_{ini}}^{t_{fin}} g_{ij} \, \dot{x}^i \dot{x}^j \, dt,$$

where g_{ij} is the Riemannian metric tensor that generates the total 'kinetic energy' of (loco)motions in the life space.

The second term in (4.266) gives the sophisticated definition of Lewinian force–fields that drive the psychological (loco)motions, if we interpret electrical charges e_i occurring at different times t_i as motivational charges – needs.

Local Micro–Level of LSF_{total}. After having properly defined macro–level MD & CD, with a unique transition map F (including a unique motion path, driving field and smooth geometry), we move down to the *microscopic* LSF–level of rapidly fluctuating MD & CD, where we cannot define a unique and smooth path–field–geometry. The most we can do at this level of *fluctuating uncertainty*, is to formulate an adaptive path integral and calculate overall probability amplitudes for ensembles of local transitions from one LSF–point to the neighboring one. This *probabilistic transition micro–dynamics* functor is defined by a multi–path (field and geometry, respectively) and multi–phase *transition amplitude* $\langle Action|Intention\rangle$ of corresponding to the globally–smooth transition map (4.265). This absolute square of this probability amplitude gives the *transition probability* of occurring the final state of *Action* given the initial state of *Intention*,

$$P(Action|Intention) = |\langle Action|Intention\rangle|^2.$$

The total transition amplitude from the state of *Intention* to the state of *Action* is defined on LSF_{total}

$$\mathcal{TA} \equiv \langle Action|Intention\rangle_{total} : INTENTION_{t_0} \Rrightarrow ACTION_{t_1}, \quad (4.267)$$

given by adaptive generalization of the Feynman's path integral [II07b]. The transition map (4.267) calculates the *overall probability amplitude* along a multitude of wildly fluctuating paths, fields and geometries, performing the *microscopic* transition from the micro–state $INTENTION_{t_0}$ occurring at initial micro–time instant t_0 to the micro–state $ACTION_{t_1}$ at some later micro–time instant t_1, such that all micro–time instants fit inside the global transition interval $t_0, t_1, ..., t_s \in [t_{ini}, t_{fin}]$. It is symbolically written as

$$\langle Action|Intention\rangle_{total} := \oint \mathcal{D}[w\Phi]\, e^{iS[\Phi]}, \quad (4.268)$$

where the Lebesgue integration is performed over all continuous $\Phi_{con}^i = paths + field + geometries$, while summation is performed over all discrete processes and regional topologies Φ_{dis}^j). The symbolic differential $\mathcal{D}[w\Phi]$ in the general path integral (4.288), represents an *adaptive path measure*, defined as a weighted product

$$\mathcal{D}[w\Phi] = \lim_{N\to\infty} \prod_{s=1}^{N} w_s d\Phi_s^i, \quad (i = 1, ..., n = con + dis), \quad (4.269)$$

which is in practice satisfied with a large N corresponding to infinitesimal temporal division of the four motivational phases (*). Technically, the path integral (4.288) calculates the *amplitude* for the transition functor $\mathcal{TA} : Intention \Rrightarrow Action$.

In the exponent of the path integral (4.288) we have the action $S[\Phi]$ and the imaginary unit $i = \sqrt{-1}$ (i can be converted into the real number -1 using the so–called *Wick rotation*, see next subsection).

In this way, we get a range of micro–objects in the local LSF at the short time–level: ensembles of rapidly fluctuating, noisy and crossing paths, force–fields, local geometries with obstacles and topologies with holes. However, by averaging process, both in time and along ensembles of paths, fields and geometries, we recover the corresponding global MD & CD variables.

Infinite–Dimensional Neural Network. The adaptive path integral (4.288) incorporates the *local learning process* according to the standard formula: $New\ Value = Old\ Value + Innovation$. The general *weights* $w_s = w_s(t)$

in (4.269) are updated by the $MONITOR$ feedback during the transition process, according to one of the two standard neural learning schemes, in which the micro–time level is traversed in discrete steps, i.e., if $t = t_0, t_1, ..., t_s$ then $t + 1 = t_1, t_2, ..., t_{s+1}$:

A. A *self–organized, unsupervised* (e.g., Hebbian–like [Heb49]) learning rule:

$$w_s(t+1) = w_s(t) + \frac{\sigma}{\eta}(w_s^d(t) - w_s^a(t)), \quad (4.270)$$

where $\sigma = \sigma(t)$, $\eta = \eta(t)$ denote *signal* and *noise*, respectively, while superscripts d and a denote *desired* and *achieved* micro–states, respectively; or

B. A certain form of a *supervised gradient descent learning*:

$$w_s(t+1) = w_s(t) - \eta \nabla J(t), \quad (4.271)$$

where η is a small constant, called the *step size*, or the *learning rate* and $\nabla J(n)$ denotes the gradient of the 'performance hyper–surface' at the t–th iteration.

Both Hebbian and supervised learning are used for the local decision making process (see below) occurring at the intention formation faze $\mathcal{F}$.

In this way, local micro–level of LSF_{total} represents an infinite–dimensional neural network. In the cognitive psychology framework, our adaptive path integral (4.288) can be interpreted as *semantic integration* (see [BF71, Ash94]).

Motion and Decision Making in LSF_{paths}

On the macro–level in the subspace LSF_{paths} we have the (loco)*motion action principle*

$$\delta S[x] = 0,$$

with the *Newtonian–like action* $S[x]$ given by

$$S[x] = \int_{t_{ini}}^{t_{fin}} dt \, [\frac{1}{2} g_{ij} \, \dot{x}^i \dot{x}^j + \varphi^i(x^i)], \quad (4.272)$$

where overdot denotes time derivative, so that $\dot{x}^i$ represents *processing speed*, or (loco)motion velocity vector. The first bracket term in (4.272) represents the kinetic energy T,

$$T = \frac{1}{2} g_{ij} \, \dot{x}^i \dot{x}^j,$$

generated by the *Riemannian metric tensor* g_{ij}, while the second bracket term, $\varphi^i(x^i)$, denotes the family of potential force–fields, driving the (loco)motions $x^i = x^i(t)$ (the *strengths* of the fields $\varphi^i(x^i)$ depend on their positions x^i in LSF, see LSF_{fields} below). The corresponding Euler–Lagrangian equation gives the Newtonian–like equation of motion

$$\frac{d}{dt}T_{\dot{x}^i} - T_{x^i} = -\varphi^i_{x^i}, \tag{4.273}$$

(subscripts denote the partial derivatives), which can be put into the standard Lagrangian form

$$\frac{d}{dt}L_{\dot{x}^i} = L_{x^i}, \quad \text{with} \quad L = T - \varphi^i(x^i).$$

In the next subsection we use the micro–level implications of the action S[x] as given by (4.272), for dynamical descriptions of the local decision–making process.

On the micro–level in the subspace LSF_{paths}, instead of a single path defined by the Newtonian–like equation of motion (4.273), we have an ensemble of fluctuating and crossing paths with weighted probabilities (of the unit total sum). This ensemble of micro–paths is defined by the simplest instance of our adaptive path integral (4.288), similar to the Feynman's original *sum over histories*,

$$\langle Action|Intention\rangle_{paths} = \oint \mathcal{D}[wx]\, \mathrm{e}^{\mathrm{i}S[x]}, \tag{4.274}$$

where $\mathcal{D}[wx]$ is a functional measure on the *space of all weighted paths*, and the exponential depends on the action $S[x]$ given by (4.272). This procedure can be redefined in a mathematically cleaner way if we Wick–rotate the time variable t to imaginary values $t \mapsto \tau = \mathrm{i}t$, thereby making all integrals real:

$$\oint \mathcal{D}[wx]\, \mathrm{e}^{\mathrm{i}S[x]} \Rightarrow^{Wick} \oint \mathcal{D}[wx]\, \mathrm{e}^{-S[x]}. \tag{4.275}$$

Discretization of (4.275) gives the *thermodynamic–like partition function*

$$Z = \sum_j \mathrm{e}^{-w_j E^j/T}, \tag{4.276}$$

where E^j is the motion energy eigenvalue (reflecting each possible motivational energetic state), T is the temperature–like environmental control parameter, and the sum runs over all motion energy eigenstates (labelled by the index j). From (4.276), we can further calculate all thermodynamic–like and statistical properties of MD & CD (see e.g., [Fey72]), as for example, *transition entropy* $S = k_B \ln Z$, etc.

From cognitive perspective, our adaptive path integral (4.274) calculates all (alternative) pathways of information flow during the transition *Intention* → *Action*.

In the language of transition–propagators, the integral over histories (4.274) can be decomposed into the product of propagators (i.e., Fredholm kernels or Green functions) corresponding to the cascade of the four motivational phases (*)

$$\langle Action|Intention\rangle_{paths} = \oint dx^{\mathcal{F}} dx^{\mathcal{I}} dx^{\mathcal{M}} dx^{\mathcal{T}}\, K(\mathcal{F},\mathcal{I})K(\mathcal{I},\mathcal{M})K(\mathcal{M},\mathcal{T}), \tag{4.277}$$

satisfying the Schrödinger–like equation

$$i\,\partial_t \langle Action|Intention\rangle_{paths} = H_{Action}\,\langle Action|Intention\rangle_{paths}, \qquad (4.278)$$

where H_{Action} represents the Hamiltonian (total energy) function available at the state of *Action*. Here our 'golden rule' is: the higher the H_{Action}, the lower the microscopic fatigue.

In the connectionist language, our propagator expressions (4.277–4.278) represent *activation dynamics*, to which our *Monitor* process gives a kind of *backpropagation* feedback, a version of the basic supervised learning (4.291).

Mechanisms of Decision–Making under Uncertainty. The basic question about our local decision making process, occurring under uncertainty at the intention formation faze $\mathcal{F}$, is: Which alternative to choose? (see [RBT01, Gro82, Gro99, Gro88, Ash94]). In our path–integral language this reads: Which path (alternative) should be given the highest probability weight w? Naturally, this problem is iteratively solved by the learning process (4.270–4.291), controlled by the $MONITOR$ feedback, which we term *algorithmic approach*.

In addition, here we analyze qualitative mechanics of the local decision making process under uncertainty, as a *heuristic approach*. This qualitative analysis is based on the micro–level interpretation of the Newtonian–like action $S[x]$, given by (4.272) and figuring both processing speed $\dot{x}$ and LTM (i.e., the force–field $\varphi(x)$, see next subsection). Here we consider three different cases:

A. If the potential $\varphi(x)$ is not very dependent upon position $x(t)$, then the more direct paths contribute the most, as longer paths, with higher mean square velocities $[\dot{x}(t)]^2$ make the exponent more negative (after Wick rotation (4.275)).

B. On the other hand, suppose that $\varphi(x)$ does indeed depend on position x. For simplicity, let the potential increase for the larger values of x. Then a direct path does not necessarily give the largest contribution to the overall transition probability, because the integrated value of the potential is higher than over another paths.

C. Finally, consider a path that deviates widely from the direct path. Then $\varphi(x)$ decreases over that path, but at the same time the velocity $\dot{x}$ increases. In this case, we expect that the increased velocity $\dot{x}$ would more than compensate for the decreased potential over the path.

Therefore, the most important path (i.e., the path with the highest weight w) would be one for which any smaller integrated value of the surrounding field potential $\varphi(x)$ is more than compensated for by an increase in kinetic–like energy $\frac{m}{2}\dot{x}^2$. In principle, this is neither the most direct path, nor the longest path, but rather a middle way between the two. Formally, it is the path along which the average Lagrangian is minimal,

$$< \frac{m}{2}\dot{x}^2 + \varphi(x) > \longrightarrow \min, \qquad (4.279)$$

i.e., the *path that requires minimal memory* (both LTM and WM, see LSF_{fields} below) and *processing speed*. This mechanical result is consistent with the 'filter theory' of *selective attention* [Bro58], proposed in an attempt to explain a range of the existing experimental results. This theory postulates a low level filter that allows only a limited number of percepts to reach the brain at any time. In this theory, the importance of conscious, directed attention is minimized. The type of attention involving low level filtering corresponds to the concept of *early selection* [Bro58].

Although we termed this 'heuristic approach' in the sense that we can instantly feel both the processing speed $\dot{x}$ and the LTM field $\varphi(x)$ involved, there is clearly a psycho–physical rule in the background, namely the averaging minimum relation (4.279).

From the decision making point of view, all possible paths (alternatives) represent the *consequences* of decision making. They are, by default, *short–term consequences*, as they are modelled in the micro–time–level. However, the path integral formalism allows calculation of the *long–term consequences*, just by extending the integration time, $t_{fin} \to \infty$. Besides, this *averaging decision mechanics* – choosing the optimal path – actually performs the 'averaging lift' in the LSF: from micro– to the macro–level.

Force–Fields and Memory in LSF_{fields}

At the macro–level in the subspace LSF_{fields} we formulate the *force–field action principle*

$$\delta S[\varphi] = 0, \qquad (4.280)$$

with the action $S[\varphi]$ dependent on Lewinian force–fields $\varphi^i = \varphi^i(x)$ ($i = 1, ..., N$), defined as a temporal integral

$$S[\varphi] = \int_{t_{ini}}^{t_{fin}} \mathfrak{L}[\varphi]\, dt, \qquad (4.281)$$

with Lagrangian density given by

$$\mathfrak{L}[\varphi] = \int d^n x\, \mathcal{L}(\varphi_i, \partial_{x^j}\varphi^i),$$

where the integral is taken over all n coordinates $x^j = x^j(t)$ of the LSF, and $\partial_{x^j}\varphi^i$ are partial derivatives of the field variables over coordinates.

On the micro–level in the subspace LSF_{fields} we have the Feynman–type sum over fields φ^i ($i = 1, ..., N$) given by the adaptive path integral

$$\langle Action|Intention\rangle_{fields} = \oint \mathcal{D}[w\varphi]\, e^{iS[\varphi]} \Rightarrow^{Wick} \oint \mathcal{D}[w\varphi]\, e^{-S[\varphi]}, \qquad (4.282)$$

with action $S[\varphi]$ given by temporal integral (4.281). (Choosing special forms of the force–field action $S[\varphi]$ in (4.282) defines micro–level MD & CD, in the LSF_{fields} space, that is similar to standard quantum–field equations, see e.g.,

[II06b].) The corresponding partition function has the form similar to (4.276), but with field energy levels.

Regarding topology of the force fields, we have in place *n−categorical Lagrangian–field* structure on the Riemannian LSF manifold M,

$$\Phi^i : [0,1] \to M, \quad \Phi^i : \Phi_0^i \mapsto \Phi_1^i,$$

generalized from the *recursive homotopy dynamics* [II06b], using

$$\frac{d}{dt} f_{\dot{x}^i} = f_{x^i} \longrightarrow \partial_\mu \left(\frac{\partial \mathcal{L}}{\partial_\mu \Phi^i} \right) = \frac{\partial \mathcal{L}}{\partial \Phi^i},$$

$$\text{with} \quad [x_0, x_1] \longrightarrow [\Phi_0^i, \Phi_1^i].$$

Relationship between Memory and Force–Fields. As already mentioned, the subspace LSF_{fields} is related to our *memory storage* [Ash94]. Its global macro–level represents the *long–term memory* (LTM), defined by the least action principle (4.280), related to *cognitive economy* in the model of *semantic memory* [Rat78, Col05]. Its local micro–level represents *working memory* (WM), a limited–capacity 'bottleneck' defined by the adaptive path integral (4.282). According to our formalism, each of Miller's 7 ± 2 units [Mil56] of the local WM are adaptively stored and averaged to give the global LTM capacity (similar to the physical notion of potential). This averaging memory lift, from WM to LTM represents *retroactive interference*, while the opposite direction, given by the path integral (4.282) itself, represents *proactive interference*. Both retroactive and proactive interferences are examples of the impact of cognitive contexts on memory. Motivational contexts can exert their influence, too. For example, a reduction in task–related recall following the completion of the task is one of the clearest examples of force–field influences on memory: the amount of details remembered of a task declines as the force–field tension to complete the task is reduced by actually completing it.

Once defined, the global LTM potential $\varphi = \varphi(x)$ is then affecting the locomotion transition paths through the path action principle (4.272), as well as general learning (4.270–4.291) and decision making process (4.279).

On the other hand, the two levels of LSF_{fields} fit nicely into the two levels of processing framework, as presented by [CL72], as an alternative to theories of separate stages for sensory, working and long–term memory. According to the *levels of processing framework*, stimulus information is processed at multiple levels simultaneously depending upon its characteristics. In this framework, our macro–level memory field, defined by the fields action principle (4.280), corresponds to the *shallow memory*, while our micro–level memory field, defined by the adaptive path integral (4.282), corresponds to the *deep memory*.

Geometries, Topologies and Noise in LSF_{geom}

On the macro–level in the subspace LSF_{geom} representing an n–dimensional smooth manifold M with the global Riemannian metric tensor g_{ij}, we formulate the *geometrical action principle*

$$\delta S[g_{ij}] = 0,$$

where $S = S[g_{ij}]$ is the n-dimensional *geodesic action* on M,

$$S[g_{ij}] = \int d^n x \sqrt{g_{ij}\, dx^i dx^j}. \tag{4.283}$$

The corresponding Euler–Lagrangian equation gives the *geodesic equation* of the *shortest path* in the manifold M,

$$\ddot{x}^i + \Gamma^i_{jk}\, \dot{x}^j\, \dot{x}^k = 0,$$

where the symbol Γ^i_{jk} denotes the so-called *affine connection* which is the source of *curvature*, which is geometrical description for *noise* (see [Ing97, Ing98]). The higher the local curvatures of the LSF–manifold M, the greater the noise in the life space. This noise is the source of our micro–level fluctuations. It can be internal or external; in both cases it curves our micro–LSF.

Otherwise, if instead we choose an n–dimensional Hilbert–like action (see [MTW73]),

$$S[g_{ij}] = \int d^n x \sqrt{\det|g_{ij}|}\, R, \tag{4.284}$$

where R is the scalar curvature (derived from Γ^i_{jk}), we get the n–dimensional Einstein–like equation:

$$G_{ij} = 8\pi T_{ij},$$

where G_{ij} is the Einstein–like tensor representing geometry of the LSF manifold M (G_{ij} is the trace–reversed Ricci tensor R_{ij}, which is itself the trace of the *Riemann curvature tensor* of the manifold M), while T_{ij} is the n–dimensional *stress–energy–momentum* tensor. This equation explicitly states that *psycho–physics of the LSF is proportional to its geometry*. T_{ij} is important quantity, representing motivational *energy*, geometry–imposed *stress* and *momentum* of (loco)motion. As before, we have our 'golden rule': *the greater the T_{ij}–components, the higher the speed of cognitive processes and the lower the macroscopic fatigue.*

The choice between the geodesic action (4.283) and the Hilbert action (4.284) depends on our interpretation of time. If time is not included in the LSF manifold M (non–relativistic approach) then we choose the geodesic action. If time is included in the LSF manifold M (making it a relativistic–like n–dimensional space–time) then the Hilbert action is preferred. The first approach is more related to the information processing and the working memory. The later, space–time approach can be related to the long–term memory: we usually recall events closely associated with the times of their happening.

On the micro–level in the subspace LSF_{geom} we have the adaptive *sum over geometries*, represented by the path integral over all local (regional) Riemannian metrics $g_{ij} = g_{ij}(x)$ varying from point to point on M (modulo diffeomorphisms),

4.5 Non–Physical Applications of Path Integrals

$$\langle Action|Intention\rangle_{geom} = \oint \mathcal{D}[wg_{ij}]\,e^{iS[g_{ij}]} \Rightarrow^{Wick} \oint \mathcal{D}[wg_{ij}]\,e^{-S[g_{ij}]}, \tag{4.285}$$

where $\mathcal{D}[g_{ij}]$ is diffeomorphism equivalence class of $g_{ij}(x) \in M$.

To include the topological structure (e.g., a number of holes) in M, we can extend (4.285) as

$$\langle Action|Intention\rangle_{geom/top} = \sum_{\text{topol.}} \oint \mathcal{D}[wg_{ij}]\,e^{iS[g_{ij}]}, \tag{4.286}$$

where the topological sum is taken over all connectedness–components of M determined by the *Euler characteristic* χ of M. This type of integral defines the *theory of fluctuating geometries*, a propagator between $(n-1)$–dimensional boundaries of the n–dimensional manifold M. One has to contribute a meaning to the integration over geometries. A key ingredient in doing so is to approximate (using simplicial approximation and Regge calculus [MTW73]) in a natural way the smooth structures of the manifold M by piecewise linear structures (mostly using topological simplices Δ). In this way, after the Wick–rotation (4.275), the integral (4.285–4.286) becomes a *simple statistical system*, given by partition function $Z = \sum_{\Delta} \frac{1}{C_{\Delta}} e^{-S_{\Delta}}$, where the summation is over all triangulations Δ of the manifold M, while C_T is the order of the automorphism group of the performed triangulation.

Micro–Level Geometry: the source of noise and stress in LSF. The subspace LSF_{geom} is the source of noise, fluctuations and obstacles, as well as psycho–physical stress. Its micro–level is adaptive, reflecting the human ability to efficiently act within the noisy environment and under the stress conditions. By averaging it produces smooth geometry of certain curvature, which is at the same time the smooth psycho–physics. This macro–level geometry directly affects the memory fields and indirectly affects the (loco)motion transition paths.

The Mental Force Law. As an effective summary of this section, we state that the psychodynamic transition functor $\mathcal{TA} : INTENTION_{t_{ini}} \Rightarrow ACTION_{t_{fin}}$, defined by the generic path integral (4.288), can be interpreted as a *mental force law*, analogous to our musculo–skeletal *covariant force law*, $F_i = mg_{ij}a^j$, and its associated *covariant force functor* $\mathcal{F}_* : TT^*M \to TTM$ [II05].

4.5.8 Joint Action Psychodynamics

Cognitive neuroscience investigations, including fMRI studies of human co-action, suggest that cognitive and neural processes supporting co-action include joint attention, action observation, task sharing, and action coordination [FFG05, KJ03, SBK06]. For example, when two actors are given a joint control task (e.g., tracking a moving target on screen) and potentially conflicting controls (e.g., one person in charge acceleration, the other – deceleration), their joint performance depends on how well they can anticipate each other's actions. In particular, better coordination is achieved when individuals receive real-time feedback about the timing of each other's actions [SBK06].

To model the dynamics of this joint action, we associate each of the actors with an n−dimensional (nD, for short) Riemannian Life-Space manifold, that is a set of their own time dependent trajectories, $M_\alpha = \{x^i(t_i)\}$ and $M_\beta = \{y^j(t_j)\}$, respectively. Following [IA07], we use the modelling machinery consisting of the joint psycho–physical action (4.294) and the corresponding adaptive path integral (4.288), visualized by the Feynman–like 1–loop diagram.[50]

Recently [IA07] we have suggested a generalized motivational/cognitive action, generating Lewinian force–fields [Lew51, Lew97] on smooth manifolds, extending and adapting classical Wheeler–Feynman *action–at–a–distance electrodynamics* [WF49]. Applying this approach to human co–action, we propose a two–term joint action:

$$A[x,y;t_i,t_j] = \frac{1}{2}\int_{t_i}\int_{t_j} \alpha_i \beta_j\, \delta(I_{ij}^2)\, \dot{x}^i(t_i)\dot{y}^j(t_j)\, dt_i dt_j + \frac{1}{2}\int_t g_{ij}\, \dot{x}^i(t)\dot{x}^j(t)\, dt$$

$$\text{with} \qquad I_{ij}^2 = \left[x^i(t_i) - y^j(t_j)\right]^2, \qquad (4.287)$$

where $IN \leq t_i, t_j, t \leq OUT$, while $\dot{x}^i(t_i) = dx^i/dt_i$ and $\dot{y}^j(t_j) = dy^j/dt_j$ are the corresponding nD (loco)motion velocities.

The first term in (4.294) represents *potential energy between the cognitive/motivational interaction* of the two agents α_i and β_j. It is a double integral over a delta function of the square of interval I^2 between two points on the paths in their Life–Spaces; thus, interaction occurs only when this interval, representing the motivational cognitive distance between the two agents, vanishes. Note that the cognitive (loco)motions of the two agents $\alpha_i[x^i(t_i)]$ and $\beta_j[y^j(t_j)]$, generally occur at different times t_i and t_j unless $t_i = t_j$, when *cognitive synchronization* occurs.

The second term in (4.294) represents *kinetic energy of the physical interaction*. Namely, when the cognitive synchronization in the first term actually takes place, the second term of physical kinetic energy is activated in the common manifold, which is one of the agents' Life Spaces, say $M_\alpha = \{x^i(t_i)\}$.

The adaptive path integral (see [IA07]) represents an infinite–dimensional neural network, corresponding to the psycho–physical action (4.294), reads

$$\langle OUT|IN \rangle := \oint \mathcal{D}[w,x,y]\, e^{iA[x,y;t_i,t_j]}, \qquad (4.288)$$

where the Lebesgue integration is performed over all continuous paths $x^i = x^i(t_i)$ and $y^j = y^j(t_j)$, while summation is performed over all associated

---

[50] The work presented in this subsection has been developed in collaboration with Dr Eugene Aidman and Mr Leong Yen, both Senior Research Scientists, Land Operations Division, Defence Science & Technology Organisation, Australia.

discrete Markov fluctuations and jumps. The symbolic differential in the path integral (4.288) represents an *adaptive path measure*, defined as a weighted product

$$\mathcal{D}[w,x,y] = \lim_{N\to\infty} \prod_{s=1}^{N} w_{ij}^s dx^i dy^j, \qquad (i,j=1,...,n). \qquad (4.289)$$

The adaptive path integral (4.288) incorporates the *local Bernstein's adaptation process* [Ber47, Ber96] according to Bernstein's discriminator concept

desired state $SW(t+1)$ = *current state* $IW(t)$ + *adjustment step* $\Delta W(t)$.

The robustness of biological motor control systems in handling excess degrees of freedom has been attributed to a combination of tight hierarchical central planning and multiple levels of sensory feedback-based self-regulation that are relatively autonomous in their operation [BLT]. These two processes are connected through a top-down process of action script delegation and a bottom-up emergency escalation mechanisms. There is a complex interplay between the continuous sensory feedback and motion/action planning to achieve effective operation in uncertain environments (in movement on uneven terrain cluttered with obstacles, for example).

Complementing Bernstein's motor control principles is Brooks' concept of computational *subsumption architectures* [Bro85, Bro90], which provides a method for structuring reactive systems from the bottom up using layered sets of behaviors. Each layer implements a particular goal of the agent, which subsumes that of the underlying layers.

For example, a robot's lowest layer could be "avoid an object", on top of it would be the layer "wander around", which in turn lies under "explore the world". The top layer in such a case could represent the ultimate goal of "creating a map". This way, the lowest layers can work as fast-responding mechanisms (i.e., reflexes), while the higher layers control the main direction to be taken in order to achieve a more abstract goal.

The substrate for this architecture comprises a network of finite state machines augmented with timing elements. The subsumption compiler compiles *augmented finite state machine* (AFSM) descriptions into a special-purpose scheduler to simulate parallelism and a set of finite state machine simulation routines. Their networked behavior can be described conceptually as:

final state $w(t+1)$ = *current state* $w(t)$ + *adjustment behavior* $f(\Delta w(t))$.

The Bernstein *weights*, or *Brooks nodes*, $w_{ij}^s = w_{ij}^s(t)$ in (4.298) are updated by the *Bernstein loop* during the joint transition process, according to one of the two standard neural learning schemes, in which the micro–time level is traversed in discrete steps, i.e., if $t = t_0, t_1, ..., t_s$ then $t+1 = t_1, t_2, ..., t_{s+1}$:

A. A *self-organized, unsupervised* (e.g., Hebbian–like [Heb49]) learning rule:

$$w_{ij}^s(t+1) = w_{ij}^s(t) + \frac{\sigma}{\eta}(w_{ij}^{s,d}(t) - w_{ij}^{s,a}(t)), \qquad (4.290)$$

where $\sigma = \sigma(t)$, $\eta = \eta(t)$ denote *signal* and *noise*, respectively, while new superscripts d and a denote *desired micro–states* and *achieved micro–states*, respectively; or

B. A certain form of a *supervised gradient descent learning*:

$$w_{ij}^s(t+1) = w_{ij}^s(t) - \eta \nabla J(t), \qquad (4.291)$$

where η is a small constant, called the *step size*, or the *learning rate*, and $\nabla J(n)$ denotes the gradient of the 'performance hyper–surface' at the t–th iteration.

Both Hebbian and supervised learning[51] are untilled in local decision making processes, e.g., at the intention formation phase (see [IA07]). Overall, the model presents a set of formalisms to represent time-critical aspects of collective performance in tactical teams. Its applications include hypotheses generation for real and simulation experiments on team performance, both in human teams (e.g., emergency crews) and hybrid human-machine teams (e.g., human-robotic crews). It is of particular value to the latter, as the increasing autonomy of robotic platforms poses non-trivial challenges, not only for the design of their operator interfaces, but also for the design of the teams themselves and their concept of operations.

4.5.9 General Adaptation Psychodynamics

Imagine three agents, $\alpha_i = \alpha_i(t_i), \beta_j = \beta_j(t_j)$ and $\gamma_k = \gamma_k(t_k)$, continually evolving and adapting in their own times t_i, t_j and t_k, performing a *goal–driven interaction*, that is a *joint action* of driving a car, so that α_i does the steering, β_j does the accelerating and γ_k does the braking.

To model this joint agents action, we associate to each of them an n–dimensional (nD, for short) configuration manifold, that is a set of their own–time dependent trajectories, $M_\alpha = \{x^i(t_i)\}$, $M_\beta = \{y^j(t_j)\}$ and $M_\gamma = \{z^k(t_k)\}$, respectively. Their associated tangent bundles contain their individual velocities, $TM_\alpha = \{\dot{x}^i(t_i) = dx^i/dt_i\}$, $TM_\beta = \{\dot{y}^j(t_j) = dy^j/dt_j\}$ and $TM_\gamma = \{\dot{z}^k(t_k) = dz^k/dt_k\}$.

The joint action happens in the common $3n$D *Finsler manifold* $M_J = M_\alpha \cup M_\beta \cup M_\gamma$, parameterized by the local joint coordinates dependent on the common time t. That is, $M_J = \{q^r(t), r = 1, ..., 3n\}$. Geometry of the

---

[51] Note that we could also use a reward–based, *reinforcement learning* rule [SB98], in which system learns its *optimal policy*:

$$innovation(t) = |reward(t) - penalty(t)|.$$

joint manifold M_J is defined by the *Finsler metric function* $ds = F(q^r, dq^r)$, defined by

$$F^2(q, \dot{q}) = g_{rs}(q, \dot{q})\dot{q}^r \dot{q}^s, \quad \text{(where } g_{rs} \text{ is the Riemann metric tensor)} \tag{4.292}$$

and the *Finsler tensor* $C_{rst}(q, \dot{q})$, defined by (see [Run59, II07b])

$$C_{rst}(q, \dot{q}) = \frac{1}{4} \frac{\partial^3 F^2(q, \dot{q})}{\partial \dot{q}^r \partial \dot{q}^s \partial \dot{q}^t} = \frac{1}{2} \frac{\partial g_{rs}}{\partial \dot{q}^r \partial \dot{q}^s}. \tag{4.293}$$

From the Finsler definitions (4.292)–(4.293), it follows that the partial interaction manifolds, $M_\alpha \cup M_\beta$, $M_\beta \cup M_\gamma$ and $M_\alpha \cup M_\gamma$ have Riemannian structures with the corresponding interaction kinetic energies, $T_{\alpha\beta} = \frac{1}{2}g_{ij}\dot{x}^i\dot{y}^j$, $T_{\alpha\gamma} = \frac{1}{2}g_{ik}\dot{x}^i\dot{z}^k$ and $T_{\beta\gamma} = \frac{1}{2}g_{jk}\dot{y}^j\dot{z}^k$

Now, following [IA07], we use the modelling machinery consisting of:

1. Adaptive joint action (4.294)–(4.296) at the top-master level, describing the externally-appearing deterministic, continuous and smooth dynamics, and

2. Corresponding adaptive path integral (4.288) at the bottom-slave level, describing a wildly fluctuating dynamics including both continuous trajectories and Markov chains.

At the *master level*, the adaptive joint action reads

$$A[t_i, t_j, t_k; t]$$
$$= \frac{1}{2}\int_{t_i}\int_{t_j}\int_{t_k} \alpha_i(t_i)\beta_j(t_j)\gamma_k(t_k)\delta(I^2_{ijk})\,\dot{x}^i(t_i)\,\dot{y}^j(t_j)\,\dot{z}^k(t_k)\,dt_i dt_j dt_k \tag{4.294}$$
$$+ \frac{1}{2}\int_t W^M_{rs}(t, q, \dot{q})\,\dot{q}^r \dot{q}^s\, dt \quad \text{(where } IN \leq t_i, t_j, t_k, t \leq OUT \text{)} \tag{4.295}$$

with $I^2_{ijk} = [x^i(t_i) - y^j(t_j)]^2 + [y^j(t_j) - z^k(t_k)]^2 + [z^k(t_k) - x^i(t_i)]^2$, (4.296)

The first term (4.294) in the joint action, contains the *cognitive intention Lagrangian* of the three agents coming into the joint action. It is a triple integral over their own timescales. The actual physical action (given by the second term (4.295)) would happen only if their timescales *synchronize*, that is in the case $t_i = t_j = t_k$. Otherwise, (4.294) is just the sum of their individual kinetic potentials. The sub-term $\delta(I^2_{ijk})$, given by (4.296) is the delta function of their mutual "cognitive distance" I^2_{ijk} that vanishes upon the synchronization.

The second action term (4.295) represents *adaptive kinetic energy of their physical interaction*. Namely, when the previous "cognitive synchronization" in the first term actually occurs, the second term of physical kinetic energy is activated in the common time t. Then we have the joint physical motions of the three agents, α_i, β_j and γ_k, physically moving in the joint Finsler coordinates $\{q^r(t), r = 1, ..., 3n\}$, along the common timescale t, in their joint $3nD$ manifold M_J. This joint kinetic energy is adaptive, represented by "master joint synaptic weights" $W^M_{rs}(t, q, \dot{q})$, which is the Riemannian metric tensor (4.292) allowed to evolve in time.

At the slave level, the adaptive path integral (see [IA07]), representing an infinite–dimensional neural network, corresponding to the adaptive joint action (4.294), reads

$$\langle OUT|IN\rangle := \int \mathcal{D}[w;x,y,z;q]\, e^{iA[t_i,t_j,t_k;t]}, \qquad (4.297)$$

where the Lebesgue integration is performed over all continuous paths $x^i = x^i(t_i)$, $y^j = y^j(t_j)$, $z^k = z^k(t_k)$ and $q^r = q^r(t)$ while the summation is performed over all associated discrete Markov fluctuations and jumps. The symbolic differential in the path integral (4.288) represents an *adaptive path measure*, defined as a weighted product

$$\mathcal{D}[w;x,y,z;q] = \lim_{N\to\infty}\prod_{S=1}^{N} w^S_{ijkr}\,dx^i dy^j dz^k dq^r, \quad (i,j,k=1,...,n;\ r=1,...,3n). \quad (4.298)$$

The "slave synaptic weights" $w^S_{ijkr} = w^S_{ijkr}(t)$ in (4.298) are updated according to a *self–organized, unsupervised* (e.g., Hebbian–like [Heb49]) learning rule (4.290), or a certain form of a *supervised gradient descent learning* (4.291), which are both naturally used for the local decision making process occurring at the intention formation faze (see [IA07]). In the cognitive psychology framework, our adaptive path integral (4.288) can be interpreted as *semantic integration* (see [BF71, Ash94]).

5

Complex Nonlinearity: Combining It All Together

This last Chapter puts all the previously developed techniques together and presents the *unified form of complex nonlinearity*. Here we have chaos, phase transitions, geometrical dynamics and topology change, all working together in the general path–integral form:

$$\langle \text{phase out} | \text{phase in} \rangle = \oint_{\text{topology change}} \mathcal{D}[x]\, e^{iS[x]}$$

The concluding section is devoted to discussion of hard vs. soft complexity, using the synergetic example of human bio-mechanics.

5.1 Geometrical Dynamics, Hamiltonian Chaos, and Phase Transitions

Recall that on the basis of the *ergodic hypothesis*, statistical mechanics describes the physics of many-degrees of freedom systems by replacing *time averages* of the relevant observables with *ensemble averages*. Therefore, instead of using statistical ensembles, we can investigate the Hamiltonian (microscopic) dynamics of a system undergoing a phase transition. The reason for tackling dynamics is twofold. First, there are observables, like Lyapunov exponents, that are intrinsically dynamical. Second, the geometrization of Hamiltonian dynamics in terms of Riemannian geometry provides new observables and, in general, an interesting framework to investigate the phenomenon of phase transitions [CCC97, Pet07]. The geometrical formulation of the *dynamics of conservative systems* [AM78] was first used by [Kry79] in his studies on the dynamical foundations of statistical mechanics and subsequently became a standard tool to study abstract systems in ergodic theory.

Consider classical many particle systems with N DOF (particles, classical spins, quasi-particles such as phonons, and so on), confined in a finite volume (therein free to move, or defined on a lattice), described by *standard Hamiltonians* [Pet07]

$$H(p,q) = \sum_{i=1}^{N} \frac{1}{2} p_i^2 + V(q_1, \ldots, q_N), \tag{5.1}$$

where the q's and the p's are, respectively, the coordinates and the conjugate momenta of the system. Our emphasis is on systems with a large number of degrees of freedom. The dynamics of the system (2.95) is defined in the $2ND$ phase space spanned by the q's and the p's.

Since the formulation of the kinetic theory of gases and then with the birth of statistical mechanics, Hamiltonian dynamics has had to cope with an *intrinsic dynamical instability*, which is usually called *Hamiltonian chaos*, a phenomenon that makes finite the predictability time scale of the dynamics. Cauchy's theorem of existence and uniqueness of the solutions of the differential equations of motion formalizes the *deterministic* nature of classical mechanics; however, *predictability* stems from the combination of determinism and *stability* of the solutions of the equations of motion. Roughly speaking, stability means that in phase space the trajectories group into bundles without any significant spread as time passes, or with an at most linearly growing spread with time. In other words, small variations of the initial conditions have limited consequences on the future evolution of the trajectories, which remain close to one another or at most separate in a nonexplosive fashion. Conversely, Hamiltonian chaos is synonymous with *unpredictability* of a *deterministic* but *unstable* Hamiltonian dynamics. A locally exponential magnification with time of the distance between initially close phase space trajectories is the hallmark of deterministic chaos [Pet07].

More specificaly, the geometrization of the dynamics of N DOF systems defined by a Lagrangian $L = T - V$, in which the kinetic energy is quadratic in the velocities,

$$T = \frac{1}{2} a_{ij} \dot{q}^i \dot{q}^j,$$

stems from the fact that the natural motions are the extrema of the *Hamiltonian action*

$$S_H = \int L\, dt,$$

or of the *Maupertuis' action*

$$S_M = 2 \int T\, dt.$$

In particular, from the Lagrangian

5.1 Geometrical Dynamics, Hamiltonian Chaos, and Phase Transitions

$$L = T - V = \sum_{i=1}^{N} \frac{1}{2}\dot{q}_i^2 - V(q_1, \ldots, q_N), \tag{5.2}$$

the equations of motion are derived in the Newtonian form

$$\ddot{q}_i = -\frac{\partial V}{\partial q^i}, \qquad i = 1, \ldots, N. \tag{5.3}$$

In fact also the geodesics (a line of stationary or minimum length joining the points A and B) of Riemannian and pseudo-Riemannian manifolds are the extrema of the *arc–length* functional

$$\ell = \int_A^B ds, \qquad \text{with} \qquad ds^2 = g_{ij} dq^i dq^j,$$

hence a suitable choice of the metric tensor allows for the identification of the arc-length with either S_H or S_M, and of the geodesics with the natural motions of the dynamical system. Starting from S_M the 'mechanical manifold' is the accessible configuration space endowed with the *Jacobi metric*

$$(g_J)_{ij} = [E - V(\{q\})]\, a_{ij},$$

where $V(q)$ is the potential energy and E is the total energy. Then Newton's equations (5.3) are retrieved from the *geodesic equations*

$$\frac{d^2 q^i}{ds^2} + \Gamma^i_{jk} \frac{dq^j}{ds}\frac{dq^k}{ds} = 0, \tag{5.4}$$

where Γ^i_{jk} are the so-called Christoffel symbols of the affine Levi-Civita connection of the Riemannian manifold in question (see, e.g., [II07b]).

A description of the extrema of Hamilton's action S_H as geodesics of a 'mechanical manifold' can be obtained using the *Eisenhart metric* [Eis29] on an enlarged configuration space-time ($\{q^0 \equiv t, q^1, \ldots, q^N\}$ plus one real coordinate q^{N+1}), whose arc-length is

$$ds^2 = -2V(q)(dq^0)^2 + a_{ij} dq^i dq^j + 2 dq^0 dq^{N+1}. \tag{5.5}$$

The manifold has a *Lorentzian structure* and the dynamical trajectories are those geodesics satisfying the condition $ds^2 = C dt^2$, where C is a positive constant. In the geometrical framework, the (in)stability of the trajectories is the (in)stability of the geodesics, and it is completely determined by the curvature properties of the underlying manifold according to the *Jacobi equation of geodesic deviation* [II07b]

$$\frac{D^2 J^i}{ds^2} + R^i{}_{jkm} \frac{dq^j}{ds} J^k \frac{dq^m}{ds} = 0, \tag{5.6}$$

whose solution J, usually called *Jacobi variation field*, locally measures the distance between nearby geodesics; D/ds stands for the *covariant derivative* along a geodesic and $R^i{}_{jkm}$ are the components of the *Riemann curvature tensor*.

No matter in which metric equation (5.6) is explicitly computed, it requires the simultaneous numerical integration of both the equations of motion and the (in)stability equation. Using the Eisenhart metric (5.5) the relevant part of the Jacobi equation (5.6) is [CCP96, CCC97]

$$\frac{d^2 J^i}{dt^2} + R^i{}_{0k0} J^k = 0, \quad i = 1, \ldots, N \tag{5.7}$$

where the only non-vanishing components of the curvature tensor are

$$R^i{}_{0k0} = \partial^2 V / \partial q_i \partial q_j.$$

Equation (5.7) is the standard *tangent dynamics equation* which is commonly used to measure *Lyapunov exponents* in standard Hamiltonian systems. Having recognized its geometric origin, a geometric reasoning was developed in [CCP96] to derive from (5.7) an *effective* scalar stability equation that *independently* of the knowledge of dynamical trajectories provides an average measure of their degree of instability. This is based on two main assumptions:

(i) The ambient manifold is *almost isotropic*, i.e., the components of the curvature tensor – that for an isotropic manifold (i.e., of constant curvature) are [CCC97]

$$R_{ijkm} = \kappa_0 (g_{ik} g_{jm} - g_{im} g_{jk}), \kappa_0 = const$$

can be approximated by

$$R_{ijkm} \approx \kappa(t)(g_{ik} g_{jm} - g_{im} g_{jk})$$

along a generic geodesic $\gamma(t)$; and

(ii) that in the large N limit the 'effective curvature' $\kappa(t)$ can be modeled by a Gaussian and δ−correlated stochastic process. The mean κ_0 and variance σ_κ of $\kappa(t)$ are given by the average and the r.m.s. fluctuation of the *Ricci curvature* $k_R = K_R/N$ along a geodesic: $\kappa_0 = \langle K_R \rangle / N$, and $\sigma_\kappa^2 = \langle (K_R - \langle K_R \rangle)^2 \rangle / N$, respectively. The Ricci curvature along a geodesic is defined as [CCC97]

$$K_R = R_{ij} \frac{dq^i}{dt} \frac{dq^j}{dt} / (\frac{dq^k}{dt} \frac{dq_k}{dt}),$$

where $R_{ij} = R^k{}_{ikj}$ is the *Ricci tensor*, which in the case of Eisenhart metric reduces to

$$K_R \equiv \Delta V = \sum_{i=1}^{N} \partial^2 V / \partial q_i^2.$$

The final result is the replacement of (5.7) with the effective stability equation which is independent of the dynamics and is in the form of a *stochastic oscillator equation* [CCP96]

5.1 Geometrical Dynamics, Hamiltonian Chaos, and Phase Transitions

$$\frac{d^2\psi}{dt^2} + \kappa(t)\,\psi = 0, \tag{5.8}$$

where $\psi^2 \propto |J|^2$. The exponential growth rate λ of the solutions of (5.8), which is therefore an estimate of the largest *Lyapunov exponent*, can be computed exactly:

$$\lambda = \frac{\Lambda}{2} - \frac{2\kappa_0}{3\Lambda}, \quad \Lambda = \left(2\sigma_\kappa^2 \tau + \sqrt{\frac{64\kappa_0^3}{27} + 4\sigma_\kappa^4 \tau^2}\right)^{\frac{1}{3}}, \tag{5.9}$$

where $\tau = \pi\sqrt{\kappa_0}/(2\sqrt{\kappa_0(\kappa_0 + \sigma_\kappa)} + \pi\sigma_\kappa)$; in the limit $\sigma_\kappa/\kappa_0 \ll 1$ one finds $\lambda \propto \sigma_\kappa^2$ [CCC97].

In this geometrical picture chaos is mainly originated by the *parametric instability* (which occurs when the parameters of a differential equation are suitably varied in time [Arn78]) activated by the fluctuating curvature 'felt' by the geodesics. On the other hand, the average curvature properties are statistical quantities like thermodynamic observables. This means that there exists a non-trivial relationship between *dynamical* properties (Lyapunov exponents) and suitable *static* observables. Generic thermodynamic observables have a non-analytic behavior as the system undergoes a phase transition. Hence the following question arises naturally: "Is there any peculiarity in the geometric properties associated with the dynamics, and thus in the chaotic dynamics itself, of systems which exhibit an equilibrium phase transition?" And in particular, do the curvature fluctuations and/or the Lyapunov exponent show any remarkable behavior in correspondence with the phase transition itself?

The remarkable properties of geodesic flows on hyperbolic manifolds (with negative curvature) have been known to mathematicians since the first decades of last century; it was Krylov who thought of using these results to account for the fast phase–space mixing of gases and thus for a dynamical justification of the *ergodic hypothesis in finite times* [Kry79]. Krylov's work has been very influential on the development of the so–called *abstract ergodic theory* [Sin89], where *Anosov flows* [Ano67] (e.g., geodesic flows on compact manifolds with negative curvature) play a prominent role. Ergodicity and mixing of these flows have been thoroughly investigated. To give an example, Sinai proved ergodicity and mixing for two hard spheres by just showing that such a system is equivalent to a geodesic flow on a negatively–curved compact manifold [Sin89].

A slightly more general version of the oscillator equation (5.8) is the *effective instability equation* [Pet07]

$$\frac{d^2\psi}{ds^2} + \langle k_R \rangle_\mu \psi + \langle \delta^2 k_R \rangle_\mu^{1/2} \eta(s)\,\psi = 0, \tag{5.10}$$

where ψ is such that $\|\psi^2(t)\| \sim \|J^2(t)\|$, k_R is the Ricci curvature of the mechanical manifold, $\langle \cdot \rangle_\mu$ stands for averaging on it, and $\eta(s)$ is a *Gaussian–distributed Markov process*. This equation is *independent of the dynamics*, it the strength of dynamical chaos, measured by the largest Lyapunov exponent λ_1, is affected by the existence of critical points of V. In particular, let us consider the possibility of a sudden variation, with the potential energy v, of the number of critical points (or of their indexes) in configuration space at some value v_c, it is then reasonable to expect that the pattern of $\lambda_1(v)$ – as well as that of $\lambda_1(E)$ since $v = v(E)$ – will be consequently affected, thus displaying jumps or cusps or other 'singular' patterns at v_c (this heuristic argument has been given evidence in the case of the XY–mean–field model, see [CPC00, CPC03]). On the other hand, *Morse theory* [Hir76] teaches us that the existence of critical points of V is associated with topology changes of the affects also the topology of the hypersurfaces $\{\Sigma_v\}_{v\in\mathbb{R}}$, provided that V is a good *Morse function* (that is: bounded below, with no vanishing eigenvalues of its Hessian matrix). Thus the existence of critical points of the potential V makes possible a conceptual link between dynamics and configuration space topology, which, on the basis of both direct and indirect evidence for a few particular models, has been formulated [CPC00] as a *topological hypothesis* about the relevance of topology for PTs phenomena.

More precisely, let $V_N(q_1,\ldots,q_N): R^N \to R$, be a smooth, bounded from below, finite-range and confining potential[6]. Denote by $\Sigma_v = V^{-1}(v)$, $v \in R$, its level sets, or equipotential hypersurfaces, in configuration space. Then let $\bar{v} = v/N$ be the potential energy per degree of freedom. If there exists N_0, and if for any pair of values $\bar{v}$ and $\bar{v}'$ belonging to a given interval $I_{\bar{v}} = [\bar{v}_0, \bar{v}_1]$ and for any $N > N_0$ then the sequence of the Helmoltz free energies $\{F_N(\beta)\}_{N\in\mathbb{N}}$ – where $\beta = 1/T$ (T is the temperature) and $\beta \in I_\beta = (\beta(\bar{v}_0), \beta(\bar{v}_1))$ – is uniformly convergent at least in $C^2(I_\beta)$ [the space of twice differentiable functions in the interval I_β], so that $\lim_{N\to\infty} F_N \in C^2(I_\beta)$ and neither first nor second order phase transitions can occur in the (inverse) temperature interval $(\beta(\bar{v}_0), \beta(\bar{v}_1))$, where the inverse temperature is defined as

$$\beta(\bar{v}) = \partial S_N^{(-)}(\bar{v})/\partial \bar{v}, \quad \text{while} \quad S_N^{(-)}(\bar{v}) = N^{-1}\log\int_{V(q)\leq \bar{v}N} d^N q$$

is one of the possible definitions of the microcanonical configurational entropy. The intensive variable $\bar{v}$ has been introduced to ease the comparison between quantities computed at different N-values [FP04].

This theorem means that a topology change of the $\{\Sigma_v\}_{v\in\mathbb{R}}$ at some v_c is a *necessary* condition for a phase transition to take place at the corresponding energy or temperature value. The topology changes implied here are those described within the framework of Morse theory through *attachment of handles* [Hir76, BM67].

Note that the topological condition of *diffeomorphicity* among all the hypersurfaces $\Sigma_{N\bar{v}}$ with $\bar{v} \in [\bar{v}_0, \bar{v}_1]$ has an analytical consequence: the absence of critical points of V in the interval $[\bar{v}_0, \bar{v}_1]$. For the proof, performed in the

---

[6] These requirements for V are fulfilled by standard interatomic and intermolecular interaction potentials, as well as by classical spin potentials.

spirit of the Yang-Lee theorem [YL52] and using Bott's 'critical neck theorem', see [FP04] and references therein.

5.4 Phase Transitions, Topology and the Spherical Model

Phase transitions (PTs) remain one of the most intriguing and interesting phenomena in physics. Recall that mathematically, a PT is signaled by the loss of analyticity of some thermodynamic function [YL52] in the thermodynamic limit.

Recently, a new characterization of PTs has been proposed, that conjectures that "at their deepest level PTs of a system are due to a change of the topology of suitable submanifolds in its configuration space" [CPC03]. PTs "would at a deeper level be related to a particular change in the topology of the configuration space of the system" [CPC00]. This is known as the Topological Hypothesis (TH) [CPC00]. In this new method one studies the topology of the configuration space Γ of the potential energy $V(\mathbf{x})$ of a system with N degrees of freedom, determining the changes that take place in the manifolds

$$M_v = \{\mathbf{x} \in \Gamma : V(\mathbf{x})/N < v\}$$

as the parameter v is increased. A topological transition (TT) is said to take place at c if $M_{c-\epsilon}$ and $M_{c+\epsilon}$ are not homeomorphic. The idea is that somehow TTs may be related to PTs.

The *necessity* of TTs at a phase transition point has been demonstrated for short ranged, confining models [FP04]. In the XY model [CPC03] TTs are found both in the mean field (MF) and unidimensional short range (SR) versions, whereas a PT is present only in the MF case. This led to a refinement of the TH: only sufficiently 'strong' TTs would be able to induce a PT. It was found that in the MF version a *macroscopic change* of the Euler characteristic happens at exactly the same point v_c where a PT appears. Several other models seem to be in agreement with this behavior [Gua04a]. But recently it was proved for a nonconfining potential that no topological criterion seems to be sufficient to induce a PT [Gua04b]. We show below that the same happens for the spherical model, which is a confining, short ranged potential.

In the spherical model it has been found that there is a direct correlation between the TT and the PT, in its mean field version [RS04]. Interestingly, in the case of non-vanishing external field there is no PT, but the configuration space displays a TT at energies that cannot be thermodynamically reached.

In this section, following [RGus05], we study the original *Berlin–Kac spherical model* [BK52] for spins placed on a $d-$dimensional lattice, interacting with their first neighbors. Using tools from topology theory we were able, for the case of vanishing field, to determine its topology exactly (up to homology). We show that the PT occurring for $d \geq 3$ cannot be related to any discontinuity in the homology of the manifolds at v_c. For non-vanishing field we cannot characterize the topology completely for all v, but show that a very abrupt change in the topology happens that does not have a corresponding PT. At variance with the MF version, the value of v at which this topological change occurs is thermodynamically accessible.

The spherical model is defined by a set of N spins ϵ_i lying on a d dimensional hyper-cubic lattice and interacting through the potential [RGus05]

$$V = -\frac{1}{2}\sum_{<ij>} J_{ij}\,\epsilon_i\epsilon_j - H\sum_i \epsilon_i,$$

where $J_{ij} = J$ gives the strength of the interaction between nearest-neighbor spins i and j, and H is an external field. The spin variables are real and constrained to lie on the sphere S^{N-1} (i.e., $\sum_i \epsilon_i^2 = N$). Periodic boundary conditions are imposed on the lattice.

In [BK52] it is shown that, at zero field, a continuous PT appears at a critical temperature $T_c(d)$ for $d \geq 3$, which is a strictly increasing function of d (see Table 1). On the other hand, no PT is possible in an external field.

As in previous works, the thermodynamic function we use to relate the statistical mechanical and topological approaches is the *average potential energy per particle* $\langle v \rangle$. Straightforwardly generalizing the results of [BK52] we obtain T_c and $\langle v_c \rangle$ for all dimensions (see Table 1). Although the specific details of $\langle v \rangle$ depend on d, some features are common to all hyper-cubic lattices: $\langle v \rangle \to 0$ for $T \to \infty$ and $\langle v \rangle \to -d$ (its lower bound) when $T \to 0$.

d	kT_c/J	$\langle v_c \rangle/J$
3	3.9573	-1.0216
4	6.4537	-0.7728
5	8.6468	-0.6759
6	10.7411	-0.6283

Table 5.1. Critical temperatures T_c and mean potential energies per particle $\langle v_c \rangle$ for hyper-cubic lattices in d dimensions. Values obtained from analytical expressions in [BK52].

In the topological approach one looks for changes in the topology of M_v as v is increased. A topological change happens at a certain value v_T if the manifolds $M_{v_T-\epsilon}$ and $M_{v_T+\epsilon}$ are not homeomorphic [Hat02] for arbitrarily small ϵ. To make a connection with statistical mechanics Casetti et al. [CPC00] proposed the nontrivial ansatz that, at the phase transition, v_T can be identified with $\langle v_c \rangle$, the thermodynamical average critical potential energy per particle. To study the topology of the configuration space of the spherical model it is most convenient to write the potential using the coordinates x_i that diagonalize the interaction matrix through an orthogonal transformation [RGus05]:

5.4 Phase Transitions, Topology and the Spherical Model

$$V = -\frac{1}{2}\sum_{i=1}^{N} \lambda_i x_i^2 - \sqrt{N} x_1 H \qquad (5.29)$$

where we set $J = 1$, and λ_i ($i = 1, \cdots, N$) are the eigenvalues of the interaction matrix, ordered from largest to smallest. We define the sets C_j, $j = 0, \cdots, \hat{N}$, where $\hat{N} + 1$ is the number of *distinct* eigenvalues. C_j is the set containing the indices of the eigenvalues that have the $(j+1)$th largest value. Therefore, $|C_j|$ gives the degeneracy associated to the $(j+1)$th largest eigenvalue. The *Frobenius–Perron theorem* ensures that the largest eigenvalue is not degenerated, i. e. $C_0 = \{1\}$.

The critical points of this potential on the sphere

$$\Gamma = S^{N-1} = \{\mathbf{x} \in \mathbb{R}^N : \sum_{j=1}^{N} x_j^2 = N\}$$

are found using Lagrange multipliers. Along with the spherical constraint, the critical point equations are [RGus05]:

$$x_1(2\mu + \lambda_1) + \sqrt{N} H = 0, \qquad (5.30)$$
$$x_i(2\mu + \lambda_i) = 0, \qquad (i = 2, \cdots, N),$$

where μ is the Lagrangian multiplier that results from enforcing the spherical constraint. From these equations and Eq. (5.29) $\hat{N} + 1$ critical values of v are obtained, denoted $v_k = -\lambda_l/2$, with $l \in C_k$, and $k = 0, \cdots, \hat{N}$ (ordered from smallest to largest). Notice that the degeneracy of the eigenvalues causes that the corresponding critical points be in fact *critical submanifolds*. This implies that in the directions tangent to the critical submanifolds the Hessian vanishes, which in turn implies that the potential is not a proper Morse function. Nevertheless, using Bott's extension of Morse theory the Euler characteristic can be found exactly. More precisely, the Hessian has $\sum_{i=0}^{j-1} |C_i|$ negative eigenvalues when restricted to the submanifold normal to the jth critical submanifold. However, profiting from the symmetries of the spherical model we took a more direct route to study its topology. As we show below, for vanishing external field it is possible to characterize *completely* the topology of the M_v, by explicitly giving the values of *all* the Betti numbers of the manifolds. Within Morse theory one can only obtain the alternate sum of the Betti numbers (i.e. the Euler characteristic), or bounds for them [Mil65].

For $H = 0$ the critical manifolds Σ_{v_j}, $j = 0, \cdots, \hat{N}$, are given by [RGus05]

$$\Sigma_{v_j} = \{\mathbf{x} \in \Gamma : \sum_{i \in C_j} x_i^2 = N\}$$

(see (5.30)). These are (hyper)spheres whose dimension is given by the degeneracy of the corresponding eigenvalues. To understand the nature of the topological change that happens at the critical values of v it is necessary to know the topology of the M_v for v between two critical values. We show below that in the interval (v_j, v_{j+1}) all the manifolds M_v are *homotopy equivalent* to S^{D-1}, where D is the number of eigenvalues larger than $-2v$. In fact we prove that S^{D-1} is a deformation retract of M_v, which in turn implies their homotopy equivalence [Hat02].

A submanifold $S \subset M$ is a deformation retract of a manifold M if there exists a series of maps $f^\nu : M \to M$ with $\nu \in [0,1]$, such that $f^0 = I$, $f^1(M) = S$ and $f^\nu|_S = I$ for all ν. The map when considered as $f : M \times [0,1] \to M$ must be continuous. Let us take $v \in (v_j, v_{j+1})$. The deformation retract that takes the manifold M_v onto its submanifold $S^{D-1} = \{\mathbf{x} \in M_v : \sum_{i=1}^D x_i^2 = N\}$ is given by $\mathbf{x}(\nu) = (f_1^\nu(\mathbf{x}), \cdots, f_N^\nu(\mathbf{x}))$ with [RGus05]

$$f_i^\nu(\mathbf{x}) = \begin{cases} x_i \sqrt{1 + \nu \sum_{k=D+1}^N x_k^2 / \sum_{k=1}^D x_k^2} & \text{for } i \leq D \\ x_i \sqrt{1 - \nu} & \text{for } i > D \end{cases} \quad (5.31)$$

This map can easily be shown to be continuous at all points $\mathbf{x} \in M_v$. The properties for $\nu = 0$ and $\nu = 1$ are evidently fulfilled. It is also easy to see that the retraction f^ν does not map any points outside M_v, since the image points always lie on the sphere S^{N-1}, and their potential energy does not exceed v. It can also be seen that no points are mapped outside M_v, as follows. It is easy to see that the image points always lie on the sphere S^{N-1}, but it must also be checked that their potential energy does not exceed v. For this, let us define a trajectory as the set of points resulting of applying all the maps f^ν to a single point in M_v. The potential energy of the points in the trajectory, $V^\nu(\mathbf{x}) = V(\mathbf{x}(\nu))$ is a linear function of ν. Thus, it must be bounded by the potential energy of the endpoints. The initial point, $\nu = 0$ has $V(\mathbf{x}) < v$ by definition. The final point is on the sphere S^{D-1}, where [RGus05]

$$\frac{V(\mathbf{x} \in S^{D-1})}{N} = \sum_{k=1}^D \frac{-\lambda_k}{2} \frac{x_k^2}{N} < \sum_{k=1}^D v \frac{x_k^2}{N} = v,$$

using the definition of v. We have used the fact that the trajectory is continuous, which depends on the continuity of the map, that can be readily checked. Indeed, (5.31) implies that the map can only be discontinuous in points $\mathbf{x}_d$ such that $x_i = 0$ for $i \leq D$, but these points satisfy

$$\frac{V(\mathbf{x}_d)}{N} = \sum_{k=D+1}^N \frac{-\lambda_k}{2} \frac{x_k^2}{N} \geq \sum_{k=D+1}^N v \frac{x_k^2}{N} = v$$

and thus they are outside M_v.

Homotopy equivalence implies that the Betti numbers of the M_v with $v \in (v_j, v_{j+1})$ are the same of S^{D-1}: $b_i(M_v) = 1$ for $i = 0$ and $i = D-1$ and $b_i(M_v) = 0$ otherwise. Thus at each v_j a *topological transition* occurs that changes the topology of the phase space from one homotopy equivalent

5.4 Phase Transitions, Topology and the Spherical Model

to $S^{D-|C_j|-1}$ to one homotopy equivalent to S^{D-1}. In terms of the Betti numbers, each transition changes two of them from 0 to 1 and from 1 to 0. Thus, at variance with other models, the *magnitude* of the Betti numbers is not a useful quantity in order to characterize the TT. It is better to look at changes in $D-1$, the highest *index* of the Betti number that changes at each transition, , i.e. the dimension of the deformation retract of the manifolds. Furthermore, as we have shown that the manifolds M_v are homotopy equivalent to (hyper)spheres, the information about their dimension $D-1$ completely characterizes their topology. Thus D is the relevant quantity to be studied. As shown above, the increase of D at each TT is given by the degeneracy of the corresponding eigenvalue.

In *Bott's extended Morse theory*, we define the *order* of the critical submanifolds as the number of negative eigenvalues of the Hessian of $V - \mu(\sum_{i=1}^{N} x_i^2 - N)$ when restricted to the submanifold normal to the jth critical submanifold. If the degeneracy $|C_j|$ is $o(N)$, given that $D = \sum_{i=0}^{j} |C_i|$, in the $N \to \infty$ limit D is equivalent to the *order* of the critical manifolds, which is defined as the number of negative eigenvalues of the Hessian ($\sum_{i=0}^{j-1} |C_i|$) when restricted to the submanifold normal to the jth critical submanifold [RGus05].

Furthermore, the Hessian of the potential has $\sum_{i=0}^{j-1} |C_i|$ negative eigenvalues when restricted to the submanifold normal to the jth critical submanifold. In particular, if the degeneracy $|C_j|$ is $o(N)$, given that $D = \sum_{i=0}^{j} |C_i|$, in the $N \to \infty$ limit D is equivalent to the *order* of the critical manifolds, defined as the number of negative eigenvalues of the Hessian of the normal submanifold. This generalizes to degenerate manifolds the definition of order of a saddle point.

For the spherical model it can be shown that the spectrum of eigenvalues is continuous in the infinite N limit. Thus, the set of $\hat{N}+1$ critical energies will be dense in $[-d, d]$, the interval of allowed potential energies. Consequently the model has a continuum of TTs. In this limit, and considering that D is $O(N)$, Because of this, for infinite size systems it is convenient to introduce a use the continuous and normalized version of D, $d(v) = D/N$, and also a *degeneracy density*, $c(v)$. They are related by $c(v) = \frac{\partial d(v)}{\partial v}$. In the following we search for singularities in these functions or their derivatives which could point to particularly strong TTs.

The spectrum of the adjacency matrix is given by [BK52]

$$\lambda_{\mathbf{p}} = 2 \sum_{i=1}^{d} \cos(2\pi p_i / N^{1/d}), \quad p_i = 0, \cdots, N^{1/d} - 1. \quad (5.32)$$

In the $N \to \infty$ limit, the degeneracy density is

$$c(v) = (2\pi)^d \int_0^{2\pi} (\Pi_{i=1}^d d\omega_i)\delta(v + \lambda(\boldsymbol{\omega})/2)$$
$$= \int_0^\infty \frac{dx}{\pi} \cos(x\,v)(J_0(x))^d, \qquad (5.33)$$

where
$$\lambda(\boldsymbol{\omega}) = 2\sum_{i=1}^d \cos(\omega_i).$$

It can be shown [RGus05] that the integral converges uniformly for all values of d and therefore $c(v)$ is a continuous function. The derivatives with respect to v can be obtained by performing the derivative inside the integral. But, as this is only valid if the resulting integral converges, this procedure allows us to obtain only the first $\lfloor (d-1)/2 \rfloor$ derivatives. All these derivatives are continuous except for the last, which is discontinuous *only* at the following points: at odd values of v if d is odd, at even values of $v/2$ if $d/2$ is odd and at odd values of $v/2$ if $d/2$ is even. But these values are clearly *different* from the ones at which a PT takes place, for all values of d. The manifolds M_v display TTs at the points $\langle v_c \rangle$ where PTs occur, since there is a continuum of TTs. These TTs, however, are not particularly abrupt. This non coincidence between the levels where a *special* TT (v_T) and a PT ($\langle v_c \rangle$) take place has also been observed in the ϕ^4 mean field model [AAR04].

The only possibility left to look for a relationship between TT's (in the sense of a discontinuity of some function of the topology) and PT's would be in the higher derivatives of $c(v)$, which cannot be studied by interchanging the integral and derivative operations. This possibility seems to us rather unreasonable, because it would imply not only that the derivative where discontinuities are to be looked for depends on the dimension of the lattice, but also that those discontinuities present in lower order derivatives should be disregarded.

We have thus shown that discontinuities in the derivatives of $c(v)$ are not sufficient to induce the PT present in the model. Nevertheless, it could be expected that if, for some model, discontinuities were present in $c(v)$ itself, this could be enough to induce a PT.

In the following, adding an external field H to the system, we give a counterexample to this possibility. Although this model presents no PT, we find discontinuities in its function $c(v)$. that this is not the case. Furthermore, we show in the following that, even though in the case of a nonzero external field there appear discontinuities in the function $c(v)$ itself, no connection between such TTs and PTs can exist, simply because the model does not display any PTs at all. Furthermore, we will show in the following that adding an external field H the model displays discontinuities in the function $c(v)$ itself. Nevertheless in this case the model does not have a phase transition and consequently no connection between this singular TT and a PT can exist.

5.4 Phase Transitions, Topology and the Spherical Model

With $H \neq 0$ it is not so easy to find the homotopy type of the submanifolds M_v, because of the breaking of the symmetry introduced by the field term in the Hamiltonian. Nevertheless, using Morse theory it is at least possible to establish the homotopy type of the submanifolds up to above the second smallest critical energy, where an abrupt topological change is shown to take place.

According to Morse theory, if there is one nondegenerate critical point at $c \in (a, b)$, the manifold $M_{c+\epsilon}$ is homeomorphic to $M_{c-\epsilon} \cup e_k$, where e_k is a k–cell. In other words, at the critical point, a k–cell (i.e. a kD open disk) is attached to the manifold, where k is the *index* of the critical point, defined as the number of negative eigenvalues of the Hessian at that point.

From the critical point equations (5.30) we obtain that the smallest critical energy is $v_+ = -(\lambda_1/2 + H)$, and the next is $v_- = -(\lambda_1/2 - H)$ [RGus05]. The Hessian of the potential on the sphere at the critical points $\mathbf{x}_\pm = (\pm\sqrt{N}, 0, \cdots, 0)$ is a diagonal matrix with $V_{ii}^\pm = \lambda_1 \pm H - \lambda_i$ $(i > 1)$. Therefore, at these two points the Hessian is not singular, which implies that $\mathbf{x}_\pm$ are nondegenerate critical points. But λ_1 is the largest eigenvalue, therefore for v_+ all the eigenvalues V_{ii} are positive. This was to be expected because this is the absolute minimum of the potential. Topologically this means that for $v < v_-$ M_v is homotopy equivalent to a disk. For $v = v_-$ the index of the critical point depends on the field: if $\lambda_2 < \lambda_1 - H$, v_- is a minimum. Thus, denoting the next critical value by v_2, if v_2 is the next critical point, M_v for $v \in (v_-, v_2)$ is homotopy equivalent to the union of two disjoint disks on the sphere. However, for large values of N the topological scenario is different. Since in this limit the spectrum of the adjacency matrix becomes dense, a certain number k of its eigenvalues will fall into the interval (v_+, v_-). This number becomes the *order* of the critical point at v_-, and gives the dimension of the k-cell that is attached to the disk. The manifold M_v for $v \in (v_-, v_2)$ is therefore homotopy equivalent to a sphere of k dimensions. For large N, k becomes proportional to N. In the interval (v_+, v_-) the manifolds M_v have the homotopy type of a point. At v_- an abrupt change in the topology takes place, and the M_v have now the homotopy type of a sphere with a macroscopic number of dimensions.

For higher values of V, the critical values are given by $v_j = -\lambda_j/2 + H^2/2(\lambda_1 - \lambda_j)$, but only for j such that $\lambda_1 - \lambda_j > H$. This is consistent with the absence of critical values between v_+ and v_-. The critical values v_j correspond to *critical submanifolds*, given by [RGus05]

$$\{\in \Gamma : x_1 = \sqrt{N}H/(\lambda_j - \lambda_1), \sum_{k \in C_j} x_k^2 = 1 - H^2/(\lambda_j - \lambda_1)^2\}.$$

Notice that, at variance with the case of vanishing H, there is a threshold energy below which the critical values have been suppressed. The critical submanifolds occurring at each v_j are again hyper-spheres whose dimension is given by the degeneracy of the corresponding eigenvalue λ_j. For all critical values we have calculated the order of the critical manifolds, $d(v)$, as well as the relative degeneracy, $c(v)$, for a few values of d (see Fig. 2). The main difference with the results for $H = 0$ is that now the connection between the order and the topology of the different manifolds is less obvious, and we have not been able to identify the homotopy types for all values of v. Nevertheless we have found exactly the topological change that takes place at v_-, and have shown that it is *macroscopic*. It may come as a surprise that this very abrupt change does not have a PT associated to it.

The use of tools from algebraic topology allowed us to characterize exactly the homology of the successive manifolds of the configuration space of the short range spherical model. We have shown that even though there is a continuum of topological transitions, a function of the topology can be defined whose derivatives display some discontinuities which could be associated to PTs. For vanishing field, however, they are not sufficient to induce a PT. We also explored the possibility that discontinuities in the function itself could be able to induce a PT. For the case of non–vanishing field we have shown that, even though such discontinuities are present, they do not have an associated PT. These results seem to be against the ubiquity of the topological hypothesis, at least in its present form.

We have shown that the manifolds of the configuration space of the short range spherical model display a continuum of topological transitions. Hence the necessity condition implied by the theorem in [FP04] is trivially met. Also, strong discontinuities have been found either in a function of the topology or in its derivatives. Although these discontinuities represent abrupt changes in the topology we have shown that they are not associated to PTs. Conversely, at the points where PTs take place no abrupt changes are observed in the topology. These are the first results on a short range confining potential to challenge the sufficiency of a topological mechanism in the origin of a phase transition, as proposed by the topological hypothesis in its present form.

5.5 Topology Change and Causal Continuity

In this section, following [DS98], we elaborate on Topology Change and Causal Continuity in the path–integral framework.

It is widely believed that any complete theory of quantum gravity must incorporate topology change. Indeed, within the particle picture of quantum gravity [Sor86a] the frozen topology framework for a generic spatial 3–manifold leads to the problem of spin–statistics violations and such wild varieties of quantum sectors that it seems that a frozen topology is unmaintainable. There is one result, however, that has been cited as counter–evidence for topology change: that of the singular propagation of a quantum field on a trousers space-time in $(1 + 1)$ dimensions. We will see how it may be possible to incorporate this result naturally in a framework which nevertheless allows topology change in general.

5.5 Topology Change and Causal Continuity

The most natural way of accommodating topology changing processes in quantum gravity is using the path–integral (path–integral) approach, although there have also been some efforts in this direction within the Hamiltonian picture [BK99]. We take a *history* in quantum gravity to be a pair (M,g), where M is a smooth nD manifold and g is a time–oriented Lorentzian metric on M.[7] The amplitude for the transition from an initial space (V_0, q_0) to a final space (V_1, q_1), where the V_i are closed $(n-1)$–manifolds and the q_i are Riemannian $(n-1)$–metrics, receives contributions from all compact interpolating histories (M, g), satisfying the boundary conditions [DS98]

$$\partial M = V_i \uplus V_f, \qquad g|_{V_{i,f}} = q_{i,f},$$

where $\uplus$ denotes disjoint union and V_0 and V_1 are initial and final space-like boundaries of (M, g). We call the manifold M such that $\partial M = V_i \uplus V_f$ a *topological cobordism* and (M, g) a *Lorentzian cobordism*. We will say that a topological cobordism or a history is *topology changing* if M is not a product $V_0 \times I$, where I is the unit interval. We will use the terminology *topology–changing transition* to refer to the transition from V_0 to V_1 when V_0 and V_1 are not diffeomorphic, without reference to any particular cobordism.

When V_0 and V_1 are not diffeomorphic, the existence of a topological cobordism, M, is equivalent to the equality of the *Stiefel-Whitney numbers* of V_0 and V_1 and is not guaranteed in arbitrary dimensions. If a topological cobordism does not exist we would certainly conclude that that transition is forbidden. In (3+1) and lower dimensions, however, a topological cobordism always exists. Then, given a topological cobordism, M, a Lorentzian cobordism based on M will exist iff [Rei63, Sor86b]: (1) n is even and $\chi(M) = 0$, or (2) n is odd and $\chi(V_0) = \chi(V_1)$. In (3+1) dimensions, a topological cobordism with $\chi(M) = 0$ always exists and thus any 3D V_0 and V_1 are Lorentz cobordant.

The theorem of Geroch [Ger67], extended to n–space-time dimensions, tells us that if a time oriented Lorentzian metric exists on a topology changing topological cobordism M then that metric must contain closed time-like curves. We consider these to be a worse pathology than the alternative which is to allow certain singularities in the metric i.e., to weaken the restriction that the metric be Lorentzian everywhere, and which, following the proposal of Sorkin [Sor97], is what we will choose to do in this section. The singularities which we need to admit in order to be able to consider all possible topological cobordisms are rather mild. Given any topological cobordism $(M; V_0, V_1)$, there exists an almost everywhere Lorentzian metric g on M which has singularities which take the form of degeneracies where the metric vanishes at (finitely many) isolated points. These degeneracies each take one of $(n+1)$ standard forms described by Morse theory as we shall relate. Allowing such singular metrics seems natural in light of the fact that within the path integral formulation, paths are not always required to be smooth; in fact they are

---

[7] Strictly speaking, a history is a geometry and only represented by (M,g).

known to be distributional. Moreover, such degeneracies are allowed within a vielbien formulation of gravity.

By allowing such mildly singular Lorentz cobordisms in the path–integral no topological cobordism is excluded and, in particular, every transition in 3+1 dimensions is viable at this level of the kinematics. We will refer to these cobordisms as *Morse cobordisms*. However it seems that dynamically some Morse cobordisms may be more equal than others. The $(1+1)$D case gives us an idea about a possible class of *physically desirable histories*. For a massless scalar quantum field on a fixed (flat) metric on the $(1+1)$–trousers topology there is an infinite burst of energy from the crotch singularity that propagates along the future light cone of the singularity. This tends to suggest that such a history would be suppressed in a full path–integral for $(1+1)$ quantum gravity. By contrast, the singular behavior of a quantum field on the background of a flat metric on the $(1+1)$ *yarmulke cobordism* (i.e., a hemisphere representing creation/destruction of an S^1 from/to nothing) is of a significantly different nature, in the sense that when integrated over the future null directions the stress–energy is finite. The singularity in the yarmulke case is therefore effectively 'squelched', while it propagates in the trousers. Indeed, there is a suppression of the trousers cobordism in the path–integral and an enhancement by an equal factor of the yarmulke cobordism (over the trivial cylinder) and separate from the suppression due to the backgrounds not being classical solutions [DS98].

What features of the trousers and yarmulke might account for the different behavior of quantum fields in these backgrounds? A closer look shows that in the Morse cobordism on the trousers manifold an observer encounters a discontinuity in the volume of her causal past as she traverses from the leg region into the body. Since such a discontinuity is absent in the yarmulke topology and cylinder topologies, Sorkin has conjectured that there may be an intimate connection between the discontinuity in the volume of the causal past/future of an observer (a *causal discontinuity*) and the physically undesirable infinite burst of energy for a scalar field propagating in such a background. And then further, that this could signal a suppression of the amplitude for a causally discontinuous spacetime in the full path–integral in quantum gravity [DS98].

5.5.1 Morse Theory and Surgery

Suppose M is an nD, compact, smooth, connected manifold such that ∂M has two disjoint $(n-1)$D components, V_0 and V_1, which are closed and correspond to the initial and final boundaries of the spacetime, respectively.

Any such M admits a *Morse function* $f : M \to [0, 1]$, with $f|_{V_i} = 0$, $f|_{V_f} = 1$ such that f possesses a set of critical points $\{p_k\}$ $(\partial_a f(p_k) = 0)$ which are nondegenerate (i.e., the Hessian $\partial_a \partial_f$ at these points is invertible). It follows that the critical points of f are isolated and that because M is compact, there are only a finite number of them.

5.5 Topology Change and Causal Continuity

Using this Morse function and any Riemannian metric h_{ab} on M, we may then construct an almost everywhere Lorentzian metric on M with a finite number of isolated degeneracies [DS98]

$$g_{ab} = h_{ab}(h^{cd}\partial_c f \partial_d f) - \zeta \partial_a f \partial_b f, \tag{5.34}$$

where the constant $\zeta > 1$. Clearly, g_{ab} is degenerate (zero) precisely at the critical points of f. We refer to these points as *Morse singularities*. Expressing a metric on M in terms of its Morse functions f relates the latter to the causal structure of the spacetime in an intimate manner, as we will see.

We now make the proposal that in the path–integral, for the amplitude for a topology changing process, for each topological cobordism, only metrics that can be expressed in the form (5.34) i.e., which can be constructed from some Morse function and some Riemannian metric) will be included. We call such metrics the *Morse metrics*. Note that since a Riemannian metric and Morse function always exist on a given topological cobordism, no cobordism is ruled out of the path–integral at this kinematical level.

A comment is in order here to relate this proposal to previous work on Lorentzian topology change and Morse theory. In work by [Yod72] the attitude was taken that the Morse singularities should not be considered as part of spacetime, in other words that the Morse points themselves were to be removed by sending them to infinity. In contrast, here we are suggesting that the Morse points should remain as part of the spacetime. Amongst other things, this entails extending the usual gravitational action to Morse metrics. Keeping the Morse points still allows a well-defined causal structure even at the Morse points and hence a well-defined causal ordering of all the spacetime points. This is something which ties in well with the idea that the fundamental underlying structure is a causal set.

Morse functions

Before proceeding any further, we briefly review some relevant properties of Morse functions which we will employ later. Morse Lemma reads: If $p \in M$ is a critical point of a Morse function $f : M \to [0,1]$, then there exists local coordinates $x_1, x_2 \cdots x_n$ in some neighborhood of p in terms of which f is given, in that neighborhood, by

$$f(x_1, ...x_n) = c - x_1^2 - x_2^2 \cdots - x_\lambda^2 + x_{\lambda+1}^2 \cdots + x_n^2$$

for $0 \leq \lambda \leq n$ and $c = const$.

The number of negative signs λ in the above expression is the number of maxima of f at the point p and is referred to as the *Morse index* of f at p. For example, the height function on the $(1+1)$–*yarmulke* topology has index 0 at the bottom point, while that on its time reversed counterpart has index 2. The height function on the trousers topology on the other hand has a Morse point of index 1 at the crotch as does its time reverse.

A vector field $\xi(p) : f \to \mathbb{R}$ on M is *gradient-like* for f if (a) $\xi(p)(f) > 0$ for every non-critical point p of f (b) For every index λ critical point p_c, the components of $\xi(p)$ in terms of the local coordinates $x_1 \cdots x_n$ about a neighborhood of p_c are $(-x_1, \cdots - x_\lambda, +x_{\lambda+1}, \cdots x_n)$. Note that (a) ensures that for p not critical $\xi(p)$ points in the direction of increasing f, while (b) ensures that $\xi(p)$ is well behaved at the critical point, i.e., $\xi(p_c) = 0$.

The *Morse number* of M on the other hand is defined to be the minimum over all Morse functions $f : M \to [0,1]$ of the number of critical points of f. Thus, for example, although the cylinder topology in $(1+1)$D *allows* Morse functions with any even number of critical points, its Morse number is nevertheless zero. We then refer to a topological cobordism of Morse number 0 as a *trivial* cobordism and that with Morse number 1 as an *elementary* cobordism. We have the following lemma: Any cobordism can be expressed as a composition of elementary cobordisms [MS74].

This decomposition is however not unique, as can be seen in the case of two dimensional closed universe S^2, shown in figure (5.1). Here we see that S^2 could be decomposed into (a) two elementary cobordisms, yarmulke and its time reverse, or (b) into four elementary cobordisms, namely, the yarmulke and an upside down trousers topology with two time reversed yarmulkes, one capping each leg. Clearly, the causal structure of the two resulting histories is very different.

Before introducing surgery we define D^k to be an open k ball and B^k to be the closed k ball (and $B^1 = I$).

A *surgery* of type λ on an $n-1$ dimensional manifold V is defined to be the following operation: Remove a thickened embedded $(\lambda - 1)$–sphere, $S^{\lambda-1} \times D^{n-\lambda}$ from V and replace it with a thickened $(n - \lambda - 1)$–sphere, $S^{n-\lambda-1} \times B^\lambda$ by identifying the boundaries using a diffeomorphism, $d : S^{\lambda-1} \times S^{n-\lambda-1} \to S^{n-\lambda-1} \times S^{\lambda-1}$.

In performing a surgery, effectively, a $(\lambda - 1)$–sphere is 'destroyed' and an $(n - \lambda - 1)$–sphere is 'created' in this process.[8] We then have the following theorem (which only depends on surgery type): If an $n-1$ dimensional manifold V_1 can be obtained from another $(n-1)$D manifold V_0 by a surgery of type λ, then there exists an elementary cobordism M, called the trace of a surgery, with boundary $V_0 \uplus V_1$ and a Morse function f on M, $f : M \to [0,1]$ which has exactly one Morse point of index λ [MS74].

As an example, consider $V_0 = S^2$ and $V_1 = S^1 \times S^1$ or a wormhole. Performing a type 1 surgery on S^2 can result in the manifold $S^1 \times S^1$, where an S^0 is 'destroyed' and an S^1 is 'created'. The above theorem then says that there exists an elementary cobordism M with boundary $S^2 \uplus S^1 \times S^1$ and a Morse function f on M with a single critical point of index $\lambda = 1$. The

---

[8] Note that this is not a unique operation — there may be many inequivalent ways to embed an $S^{\lambda-1}$ in V e.g., an S^1 may be knotted in three dimensions) as well as isotopically distinct gluing diffeomorphisms.

5.5 Topology Change and Causal Continuity

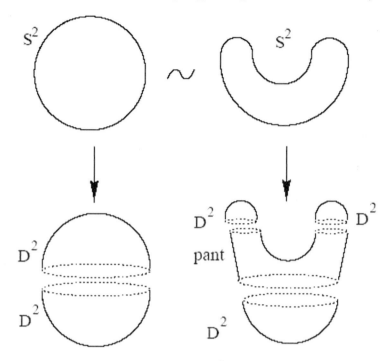

Fig. 5.1. Two ways of decomposing S^2 into elementary cobordisms.

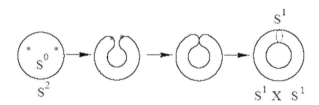

Fig. 5.2. 'Tracing out' a type 1 surgery on S^2, whereby an S^0 is destroyed and an S^1 is created to give the torus $S^1 \times S^1$.

manifold M may be visualized as shown in figure (5.2). We now explain how to construct the trace of a general surgery.

A λ surgery that turns V_0 into V_1 gives us an embedding $i : S^{\lambda-1} \to V_0$ and a neighborhood, N, of that embedded sphere whose closure, $\bar{N}$, is diffeomorphic to $S^{\lambda-1} \times B^{n-\lambda}$. Indeed, we have a diffeomorphism

$$d : \partial(\bar{N}) \to S^{\lambda-1} \times S^{n-\lambda-1},$$

the 'surgery diffeomorphism.' Now $S^{\lambda-1} \times S^{n-\lambda-1}$ is the boundary of $S^{\lambda-1} \times B^{n-\lambda}$ and we can extend d to a diffeomorphism, $\tilde{d} : \bar{N} \to S^{\lambda-1} \times B^{n-\lambda}$ such that $\tilde{d}$ restricts to d on the boundary. $\tilde{d}$ is unique up to isotopy since $B^{n-\lambda}$ is topologically trivial.

We construct the trace of the surgery by gluing together the two manifolds $M_1 = V_0 \times I$ and $M_2 = B^\lambda \times B^{n-\lambda}$ using a diffeomorphism from part of the boundary of one to part of the boundary of the other in the following way. $(\bar{N}, 1)$ is part of ∂M_1 and is diffeomorphic via $\tilde{d}$ to $S^{\lambda-1} \times B^{n-\lambda}$ which is part of the boundary of M_2. We identify all points $x \in (\bar{N}, 1)$ and $\tilde{d}(x)$. The resultant manifold clearly has one disjoint boundary component which is V_0. That the other boundary is diffeomorphic to V_1, i.e. the result of the surgery on V_0, takes a little more thought to see. Roughly speaking, in doing the gluing by $\tilde{d}$ we are eliminating $\bar{N}$ from V_0 and replacing it with the rest of the boundary of M_2 (the complement of $\text{Im}(\tilde{d})$ in ∂M_2) i.e. $B^\lambda \times S^{n-\lambda-1}$ exactly as in the original surgery.

Thus the trace of a surgery is a manifold with boundary with the property that one part of the boundary is the original manifold and the other part of the boundary is the surgically altered manifold (up to diffeomorphisms).

Finally, defining an $(n-1)$D submanifold, V_t, of M to be a *regular level* of a Morse function, f on M, if $V_t = f^{-1}(a)$ for some $0 \leq a \leq 1$ and V_t contains no Morse point, the converse to above theorem is: If the region between two regular levels, V_0 and V_1 of a Morse function on a manifold M contains only one critical point of index λ then V_0 and V_1 are related by a type λ surgery and M is diffeomorphic to the *trace* of that surgery and such that in this region of M an $S^{\lambda-1}$ is destroyed and an $S^{n-\lambda-1}$ is created.

We therefore see that there is a $1-1$ correspondence between the type λ of a surgery on an $n-1$D manifold V_0 and the index λ of an elementary cobordism from V_0 to V_1.

As an example, consider $V_0 = S^2$ and $V_1 = S^1 \times S^1$ or a wormhole. Performing a type 1 surgery on S^2 can result in the manifold $S^1 \times S^1$, where an S^0 is 'destroyed' and an S^1 is 'created'. The above theorem then says that there exists an elementary cobordism M diffeomorphic to the trace of this surgery, with boundary $S^2 \uplus S^1 \times S^1$ and a Morse function f on M with a single critical point of index $\lambda = 1$. The 'traced' surgery may be visualized as shown in figure (5.2).

Examples

The nD yarmulke cobordism and its time reverse hold a special place in our analysis since they are easy to characterize. If $f : M \to [0,1]$ has a single Morse point of index 0 then M is the trace of the surgery of type 0 in which an $S^{-1} = \Phi$ is destroyed and an S^{n-1} is created. If M is connected this implies that $M \cong B^n$. In other words, a cobordism can have a single index 0 point if and only if it is the yarmulke. This means that when a component of the universe is created from nothing (as opposed to being created by branching off from an already existing universe) its initial topology must be that of a sphere, no matter what the dimension: the big bang always results in an initially spherical universe. This might be thought of as a 'prediction' of this way of treating topology change. A similar argument for the time reversed case implies a connected cobordism can have a single Morse point of index n iff it is the time reversed yarmulke and the universe must be topologically spherical before it can finally disappear in a big crunch.

The trousers and its higher dimensional analogues are also important examples. There exists a Morse function on the $(1+1)$–trousers topology which possesses a single Morse point of index 1 and the trousers is therefore the trace of a surgery of type 1 in which an embedded $S^0 \times D^1$ is deleted from the initial $S^1 \uplus S^1$ and replaced with a $B^1 \times S^0$ to form a single S^1. In $(n-1)+1$ dimensions, the higher dimensional trousers (the manifold S^n with three open balls removed) for the process $S^{n-1} \uplus S^{n-1} \to S^{n-1}$ has an index 1 point and is the trace of a type 1 surgery in which an $S^0 \times D^{n-2}$, i.e., two balls, are removed and an $S^{n-2} \times B^1$, or wormhole, added. In these processes, parts of the universe which were spatially far apart suddenly become close (in these cases the parts of the universe are originally in disconnected components of the universe, but this isn't the defining characteristic of index 1 points). An index $n-1$ point is the time reverse of this and corresponds to a type $n-1$ surgery in which a wormhole is removed (or cut) and the ends 'capped off' with two balls, so that neighboring parts of the universe suddenly become distant.

It seems intuitively clear from these examples that there is something causally peculiar about the index 1 and $n-1$ points and in the next section we give a precise statement of a conjecture that encapsulates this.

5.5.2 Causal Discontinuity

Borde and Sorkin have conjectured that (M, g_{ab}) contains a *causal discontinuity* if and only if the Morse function f contains an index 1 or an index $n-1$ Morse point. What do we mean by causal discontinuity? There are many equivalent conditions for a Lorentzian spacetime to be causally discontinuous [HS74] and we define a Morse metric to be causally discontinuous iff the spacetime minus the Morse points (which is Lorentzian) is. Roughly speaking, a causal discontinuity results in the causal past or future of a point in spacetime jumping discontinuously as the point is continuously moved around. We see that behavior in the $(1+1)$–trousers. Clearly the same kind of thing will happen in the higher dimensional trousers, but not in the yarmulkes. Furthermore in the cases of index $\lambda \neq 1, n-1$, the spheres that are created and destroyed are all connected and so it seems that neighboring parts of the universe remain close and distant ones remain far part.

To lend further plausibility to the conjecture we will work out an example, the index 1 point in 1+1 dimensions, in detail. Choose a neighborhood of the Morse point p in which the Morse function has the standard form [DS98]:

$$f(x,y) = f(p) - x^2 + y^2 \tag{5.35}$$

in terms of some local coordinates (x, y). We take the flat Riemannian metric

5 Complex Nonlinearity: Combining It All Together

$$ds_R^2 = h_{\mu\nu} dx^\mu dx^\nu = dx^2 + dy^2 \tag{5.36}$$

Define the Morse metric $g_{\mu\nu}$ as in equation 5.34 with $\zeta = 2$ and $\partial_\mu f = (-2x, 2y)$ to obtain

$$ds_L^2 = -4(xdx - ydy)^2 + 4(xdy + ydx)^2 \tag{5.37}$$

This metric is actually flat since $2(xdx - ydy) = d(x^2 - y^2)$ and $2(xdy + ydx) = 2d(xy)$. The hyperbolae $xy = c$, c constant, are the integral curves of the vector field $\xi^\mu = h^{\mu\nu}\partial_\nu f$ and the spatial 'surfaces' of constant f are the hyperbolae $x^2 - y^2 = d$, d constant.

What are the null curves in the neighborhood of p? We have $ds_L^2 = 0$ which implies

$$d(x^2 - y^2) = \pm 2d(xy) \tag{5.38}$$
$$x^2 - y^2 = \pm 2xy + b \tag{5.39}$$

The null curves that pass through p are given by $b = 0$ so that there are four solutions: $y = (\pm 1 \pm \sqrt{2})x$. These are the straight lines through p at angles $\frac{\pi}{8}$, $\frac{3\pi}{8}$, $\frac{5\pi}{8}$, $\frac{7\pi}{8}$, to the x-axis. These are the past and future light 'cones' of p. The null curves which don't pass through p are given by the hyperbolae $x'y' = c'$ and $x'^2 - y'^2 = d'$ where (x', y') are rotated coordinates

$$x' = \cos\frac{\pi}{8} x + \sin\frac{\pi}{8} y \tag{5.40}$$
$$y' = -\sin\frac{\pi}{8} x + \cos\frac{\pi}{8} y. \tag{5.41}$$

The higher dimensional case can be similarly analyzed. Now we have [DS98]

$$f(\mathbf{x}, \mathbf{y}) = f(p) - x_1^2 - \cdots - x_\lambda^2 + y_1^2 + \cdots + y_{n-\lambda}^2 \tag{5.42}$$

Take the Cartesian metric in the local coordinates and let $r^2 = x_1^2 + \ldots x_\lambda^2$ and $\rho^2 = y_1^2 + \ldots y_{n-\lambda}^2$ so

$$ds_R^2 = dr^2 + r^2 d\Omega_{\lambda-1}^2 + d\rho^2 + \rho^2 d\Omega_{n-\lambda-1}^2 \tag{5.43}$$

The Morse metric we construct from these and $\zeta = 2$ is

$$ds_L^2 = 4(r^2 + \rho^2)[r^2 d\Omega_{\lambda-1}^2 + \rho^2 d\Omega_{n-\lambda-1}^2] \tag{5.44}$$
$$+ 4(\rho dr + r d\rho)^2 - 4(rdr - \rho d\rho)^2 \tag{5.45}$$

This is not flat for $n \geq 3$. We can now solve $ds_L^2 = 0$ for a fixed point on the $(\lambda - 1)$-sphere and $(n - \lambda - 1)$-sphere and find that the past and future light cones of p have base $S^{\lambda-1} \times S^{n-\lambda-1}$. Note that this base is disconnected for $\lambda = 1$ or $n - 1$. The light cones of other points are more complicated to calculate but a similar argument to that for the $1 + 1$ example shows that there is a causal discontinuity for $\lambda = 1$ or $n - 1$.

5.5 Topology Change and Causal Continuity

From now on we will assume that the Borde–Sorkin conjecture holds. Thus, we can search for causally continuous histories on M by asking if it admits any Morse function f which has no index 1 or $n-1$ critical points: a history corresponding to such an f would be causally continuous. If on the other hand, such an f does *not* exist, i.e., all Morse functions on M have critical points of index either 1 or $n-1$, then M *does not* support causally continuous histories.

We should remind ourselves that for a given Morse function f on M the number of index λ critical points m_λ, is not a topological invariant; in general different Morse functions will possess different sets of critical points. However there are lower bounds on the m_λ depending on the homology type of M. For the topological cobordism (M, V_0, V_1) we have the Morse relation [DS98],

$$\sum_\lambda (m_\lambda - \beta_\lambda(M, V_0))t^\lambda = (1+t)R(t), \qquad (5.46)$$

where $\beta_\lambda(M, V_0)$ are the Betti numbers of M relative to V_0 and $R(t)$ is a polynomial in the variable t which has positive coefficients [NS83, FDN92]. Letting $t = -1$, we immediately get the relative Euler characteristic of M in terms of the Morse numbers,

$$\chi(M, V_0) = \sum_\lambda (-1)^\lambda m_\lambda. \qquad (5.47)$$

Another consequence of (5.46) is,

$$m_\lambda \geq \beta_\lambda(M, V_0) \quad \forall \lambda, \qquad (5.48)$$

which places a lower bound on the m_λ.

5.5.3 General 4D Topology Change

As we have noted, in nD critical points of index 0 and n correspond to a big bang and big crunch, which allow causally continuous histories. It is only for $n \geq 4$ that other types of causally continuous histories can exist. For example, in 4 dimensions, elementary cobordisms with index 1 or 3 critical points correspond to causally discontinuous histories while those of index 2 are causally continuous.

For $n = 4$, we have already mentioned that any two 3 manifolds V_0 and V_1 are cobordant, i.e., there is a 4 dimensional M such that $\partial M = V_0 \uplus V_1$. However, we can ask whether, given a particular pair $\{V_0, V_1\}$, a cobordism M exists which admits a causally continuous metric. If not, then the Sorkin conjecture would imply that the transition $V_0 \to V_1$ would be suppressed. In other words, does a cobordism M exist which admits a Morse function with no index 1 or 3 points? The answer to this is supplied by a well known result in 3 manifold theory, the *Lickorish–Wallace theorem*, which states that any 3–manifold V_1 can be obtained from any other V_0 by performing a series of type 2 surgeries on V_0. Thus, by Theorem **1** there exists an interpolating cobordism M which is the trace of this sequence of surgeries and which therefore admits a Morse function with only index 2 points, so that M admits a causally continuous metric.

This result has the immediate consequence that even if the Sorkin and Borde–Sorkin conjectures hold and causally discontinuous histories are suppressed in the path–integral, no topological transition $V_0 \to V_1$ would be ruled out in 3+1 dimensions. Thus, in this sense, there is no 'causal' obstruction to any transition $V_0 \to V_1$ in 3+1 dimensions, just as there is no topological (nor Lorentzian) obstruction in 3+1 dimensions.

This is somewhat disappointing, however, since there are some transitions that we might hope would be suppressed. An important example is the process in which a single prime 3-manifold is produced. Quantized primes or topological geons occur as particles in canonical quantum gravity similar to the way skyrmions and other kinks appear in quantum field theory (see [Sor86a]). We would therefore not expect single geon production from the vacuum. However, the restriction of causal continuity will not be enough to rule this out and we'll have to wait for more dynamical arguments. This situation is in contrast to that for the *Kaluza–Klein monopole* where there's a purely topological obstruction to the existence of a cobordism for the creation of a single monopole [Sor86b] (though that case is strictly not within the regime of our discussion since the topology change involved is not local but changes the boundary conditions at infinity).

This result, however, says nothing about the status of any particular topological cobordism in the path–integral. In other words, it may not be true that a given topological cobordism, M, admits a causally continuous Morse metric [DS98].

5.5.4 A Black Hole Example

The pair creation of black holes has been investigated by studying Euclidean solutions of the equations of motion which satisfy the appropriate boundary conditions for the solution to be an instanton for false vacuum decay. One does not have to subscribe to the Euclidean path–integral approach to quantum gravity in order to believe that the instanton calculations are sensible. Indeed, we take the attitude that the instantons are not 'physical' but only machinery useful for approximately calculating amplitudes and that the functional integral is actually over Morse metrics. The issue of whether quantum fields can propagate in a non-singular way on these Morse geometries is therefore relevant and the question arises as to whether causally continuous Morse metrics can live on the instanton manifold.

The doubled instanton, or bounce, corresponding to the pair creation and annihilation of non-extremal black holes has the topology $S^2 \times S^2 - pt$ [GS91]. Let us compactify this to $S^2 \times S^2$. The fact that $S^2 \times S^2$ is closed implies that it will include at least one universe creation and one universe destruction, corresponding to Morse index 0 and 4 points, respectively. This can be seen from the Betti numbers, $\beta_0 = \beta_4 = 1$, $\beta_1 = \beta_3 = 0$ and $\beta_2 = 2$ so the Morse inequalities imply that $m_0 \geq 1$ and $m_4 \geq 1$. Although $\beta_1 = \beta_3 = 0$ we cannot conclude that there exists a Morse function that saturates the bounds of the inequalities (see the next section for an example). We will prove that such a Morse function exists (with $m_0 = m_4 = 1$, $m_1 = m_3 = 0$ and $m_2 = 2$) by explicit construction on the half-instanton, $S^2 \times B^2$.

Let (θ, ϕ) be standard polar coordinates on S^2 and (r, ψ) polar coordinates on B^2, where $\theta \in [0, \pi]$, $\phi \in [0, 2\pi]$, $0 \leq r \leq 1$ and $\psi \in [0, 2\pi]$. The boundary of $S^2 \times B^2$ is $S^2 \times S^1$ so that $S^2 \times B^2$ corresponds to the creation from nothing of an $S^2 \times S^1$ wormhole.

Define the function [DS98],

$$f(\theta, \phi, r, \psi) = \frac{1}{3}(1 + r^2 + \cos(1 - r^2)\theta). \tag{5.49}$$

Now, $f : S^2 \times B^2 \to [0, 1]$. The level surface $f^{-1}(1)$ satisfies the condition $r = 1$. This is easily seen to be the boundary $S^2 \times S^1$ of $S^2 \times B^2$. On the other hand, the level surface $f^{-1}(0)$ satisfies the condition $r = 0, \theta = \pi$ which is a point on $S^2 \times B^2$.

We find the critical points of f by noting that

$$\partial_r f = \frac{2}{3}r + \frac{2}{3}r\theta \sin(1 - r^2)\theta \quad \text{and} \quad \partial_\theta f = -\frac{1}{3}(1 - r^2)\sin(1 - r^2)\theta,$$

while $\partial_\phi f = \partial_\psi f = 0$ everywhere. Thus, there are only two (and therefore isolated) critical points of f, i.e., $p_1 = (r = 0, \theta = \pi)$ and $p_2 = (r = 0, \theta = 0)$ which are not on the boundary. In order to show the critical points are non-degenerate and to determine their indices we make use of the Morse Lemma and rewrite f in suitable local coordinate patches.

Near p_1: At p_1, $f = 0$. In the neighborhood of p_1, we may write $\theta = \pi - \epsilon$ where ϵ and r are both small and of the same order (note that the topology of this neighborhood is just $B^2 \times B^2$). Then,

$$\cos(1 - r^2)\theta \approx \cos(\pi - \epsilon) \tag{5.50}$$

$$\approx -1 + \frac{1}{2}\epsilon^2. \tag{5.51}$$

and putting $x_1 = \frac{r}{\sqrt{3}}\sin\psi$, $x_2 = \frac{r}{\sqrt{3}}\cos\psi$, $x_3 = \frac{\epsilon}{\sqrt{6}}\sin\phi$ and $x_4 = \frac{\epsilon}{\sqrt{6}}\cos\phi$, we see that

$$f \approx x_1^2 + x_2^2 + x_3^2 + x_4^2. \tag{5.52}$$

Thus, p_1 is an index 0 point.
Near p_2: At p_2, $f = \frac{2}{3}$. In the neighborhood of p_2, r and θ are small and of the same order. Then,

$$f \approx \frac{2}{3} + \frac{1}{3}r^2 - \frac{1}{6}\theta^2, \tag{5.53}$$

and using $y_1 = \frac{\theta}{\sqrt{6}} \sin\phi$ and $y_2 = \frac{\theta}{\sqrt{6}} \cos\phi$, $y_3 = \frac{r}{\sqrt{3}} \sin\psi$, $y_4 = \frac{r}{\sqrt{3}} \cos\psi$, we see that

$$f \approx \frac{2}{3} - y_1^2 - y_2^2 + y_3^2 + y_4^2. \tag{5.54}$$

So p_2 is an index 2 point.

The existence of such a Morse function with two critical points, one of index 0 and the other of index 2, shows that the black hole pair production topology can support histories that are causally continuous. The index 0 point is the creation of an S^3 from nothing and the index 2 point is the transition from S^3 to $S^2 \times S^1$. That this is means that a Morse function with the same Morse points exists on the original non-compact cobordism, half of $S^2 \times S^2$ − {point} was later shown in [DG98]. This result is evidence of consistency between the conclusion that the existence of an instanton implies that the process has a finite rate (approximated by $\tilde{e}^{-I}$ where I is the Euclidean action) and the idea that only causally continuous Morse histories contribute to the path–integral [DS98].

We note that a simple generalization of the above Morse function shows that the higher dimensional black hole pair creation-annihilation topological cobordism $S^{n-2} \times B^2$ admits a Morse function with one index 0 point and an index $(n-2)$ point and so supports histories that are causally continuous for any dimension $n > 4$ (though the actual instanton solution is unknown). It is also interesting that there is another simple cobordism for the transition from S^3 to $S^2 \times S^1$ which is $B^3 \times S^1$ with an embedded open four-ball deleted. This, however, by virtue of the Morse inequalities, admits no Morse function without an index 1 point and so is causally discontinuous. In some sense, this second causally discontinuous process is the way one might naturally imagine a wormhole forming: two distant regions of space coming 'close in hyperspace' and touching to form the wormhole. The index 2 cobordism for creation of a wormhole is harder to visualize.

5.5.5 Topology Change and Path Integrals

We have described a rather natural framework for considering topology change within the path–integral for quantum gravity based on Morse theory. Two key conjectures lead to the proposal that only causally continuous cobordisms be included in the Sum and that these are identified with Morse metrics with no index 1 or $n-1$ points. The *Lickorish–Wallace theorem* on surgery on 3-manifolds together with the Borde–Sorkin conjecture means that any topology changing transition in (3+1)D is achievable by a causally continuous cobordism. The higher dimensional statement is not known. We have shown that the black hole pair production instanton $S^2 \times S^2$ admits causally continuous Morse metrics whereas the 'U–tube' cobordism for pair production of topological geons of any sort is necessarily causally discontinuous [DS98].

A possible resolution that might save the geon spin–statistics result, is that there must be a weakness in the sequence of conjectures to which we have drawn attention and which form the framework in which causal continuity becomes so central. The *Borde–Sorkin conjecture*, that a Morse metric is causally continuous iff it contains no index 1 or $(n-1)$ points, seems to be the most solid. The Sorkin conjecture that infinite energy/particle production would occur in a Morse spacetime iff it contained a causal discontinuity seems plausible but would need to be verified by more examples than the 1+1 dimensional trousers and yarmulke studied so far. In particular, the first example of a causally continuous spacetime that is not the yarmulke occurs in 3+1 dimensions. Work on this second conjecture will be easier once the first is proved since then simple examples of causally continuous metrics can be written down using the Morse construction. Then finally, there is the idea that the singular behavior of quantum fields on a causally discontinuous background is a signal that it is infinitely suppressed in the path–integral. What one means by this is the following. Consider a scalar field minimally coupled to gravity. The path integral is [DS98]

$$\sum_{\text{all topologies}} \int [dg][d\phi] \exp^{i\int \sqrt{-g}R + i\int \sqrt{-g}(\partial\phi)^2}, \qquad (5.55)$$

where we have omitted the explicit statement about boundary conditions. We may integrate out the scalar field degrees of freedom, i.e.,

$$\int [d\phi] \exp^{i\int \sqrt{-g}(\partial\phi)^2} = F[g]. \qquad (5.56)$$

The functional $F[g]$ which is the path integral for a scalar field in a fixed background can now be regarded as an overall weight in the path integral over metrics,

$$\sum_{\text{all topologies}} \int [dg] F[g] \exp^{i\int \sqrt{-g}R}. \qquad (5.57)$$

The idea is that $F[g]$ is zero if g is causally discontinuous.

Perhaps, however, all the conjectures do hold at the continuum level and the simplest loophole of all is that the path–integral should be defined fundamentally as a sum over whatever discrete structure will prove to underly the differentiable manifold of general relativity. If it is a causal set then all quantities calculated will be regulated. The elimination altogether of the causally discontinuous cobordisms would then be too severe a truncation, and even if they are still suppressed, they might give a non-trivial contribution.

For more details, see [DS98].

5.6 'Hard' vs. 'Soft' Complexity: A Bio-Mechanical Example

In this concluding section we discuss the 'hard' vs. 'soft' approaches to complexity, using a paradigmatic example of a complex system: human/humanoid bio-mechanics (see subsection 2.2.2 above). Namely, recently, we have proposed the following *complexity conjecture* (see [IS08]): In a combined bio-mechanical system, where the action of Newtonian laws cannot be neglected, it is the mechanical part that determines the *lower limit of complexity* of the combined system, commonly defined as the *number of mechanical DOF*. The biological part of such system, as being 'more intelligent', naturally serves as a 'controller' for the 'non–intelligent' mechanical 'plant'. Although, in some special cases, the behavior of the combined system might have a 'simple' output, the realistic internal state–space analysis shows that the *total system complexity* represents either the *superposition*, or a kind of '*macroscopic entanglement*' of the two partial complexities. Neither 'mutual cancelling' nor 'averaging' of the mechanical degrees of freedom generally occurs in such bio-mechanical system. The combined system has both dynamical and control complexities. The 'realistic' computational model of such system also has its own *computational complexity*. We have demonstrated the validity of the above conjecture using the example of the physiologically realistic computer model. We have further argued that human motion is the simplest well–defined example of a general human behavior, and discussed issues of *simplicity* versus *predictability/controllability* in complex systems. Further, we have discussed *self–assembly* in relation to *conditioned training* in human/humanoid motion. It is argued that there is a significant difference in the *observational resolution* of human motion while watching 'subtle' movements of human hands playing a piano versus 'coarse' movements of human crowd at a football stadium from an orbital satellite. Techniques such as cellular automata can model the coarse crowd motion, but not the subtle hierarchical neural control of the dynamics of human hands playing a piano. Therefore, we have proposed the *observational resolution* as a new measure of bio-mechanical complexity. Finally, we have noted that there is a possible route to apparent simplicity in bio-mechanics, in the form of *oscillatory synchronization*, both external–kinematical, and internal–control.

5.6.1 Bio-Mechanical Complexity

Human (humanoid) bio-mechanics is a science of human (humanoid) motion. It is governed by both Newtonian dynamics and biological control laws [IB05, II05, II06a, II06b]. In its modern computational form, it also obeys computational rules. Thus, the human/humanoid bio-mechanics includes dynamical, control and computational complexities. This study shows that these three sources of complexity do not cancel each other. Instead, we have either their superposition or a kind of 'macro–entanglement' at work.

The mechanical part of a human bio-mechanical system determines the *lower limit of complexity*, which is simply defined by the *number of mechanical degrees of freedom*. The biological, in this case neuro–muscular, part of the combined system efficiently controls the complex dynamics of the mechanical skeleton.

5.6 'Hard' vs. 'Soft' Complexity: A Bio-Mechanical Example

The common complexity models are Cellular Automata (CA). It is common in nature to find systems whose overall behavior is extremely complex, yet whose fundamental component parts are each very simple. The complexity is generated by the cooperative effect of many simple identical components. Much has been discovered about the nature of the components in physical and biological systems; little is known about the mechanisms by which these components act together to give the overall complexity observed. According to S. Wolfram [Wol84, Wol02], what is needed is a general mathematical theory to describe the nature and generation of complexity.

CA are examples of mathematical systems constructed from many identical components, each simple, but together capable of complex behavior. From their analysis one may, on the one hand, develop specific models for particular systems, and, on the other hand, hope to abstract general principles applicable to a wide variety of complex systems.

However, our *bio-mechanical complexity* cannot be explained by CA, for the following reasons:

A. Human bones neither die nor grow during the simulation period, so there is an absence of any cancellation of the mechanical degrees of freedom like in CA.
B. Averaging of these degrees of freedom does not work in general either, as explained below.
C. Low–dimensional linear physical systems can be successfully modelled using CA (e.g, modelling a single linear 1D wave equation using a Margolus rule [B-Y97]). However, we are dealing with a system of 500 nonlinearly–coupled differential equations (see below), which has a completely different complexity level [9].
D. Human neural control (as well as humanoid–robotic control) has a natural hierarchical (multi–level) structure: spinal (reflex) level, cerebellar (synergetic) level, and cortical (planning) level. A system of this kind of complexity cannot be efficiently controlled using a single control level.

There are over 200 bones in the human skeleton driven by about 640 muscular actuators (see, e.g., [Mar98]). It is sufficient to have a glimpse at the structure and function of a single skeletal muscle to get an impression of the natural complexity at work in bio-mechanics. The efficient 'orchestration' of the whole musculo–skeletal dynamics is naturally performed by several levels of neural motor control: (i) spinal level of autogenetic reflexes, (ii) cerebellar level of muscular synergy, and (iii) cortical level of motion planing.

Here we need to emphasize that human joints are significantly more flexible than humanoid robot joints. Namely, each humanoid joint consists of a pair of coupled segments with only Eulerian rotational degrees of freedom. On the

---

[9] When solving partial differential equations using CA, in a way we emulate the classical finite element method (FEM). However, FEM, even in its most recent (and most expensive software) versions is simply an unsuitable tool for any kind of serious robotics.

other hand, in each human synovial joint, besides gross Eulerian rotational movements (roll, pitch and yaw), we also have some hidden and restricted translations along (X, Y, Z)–axes. For example, in the knee joint (see Figure 5.3), patella (knee cap) moves for about 7–10 cm from maximal extension to maximal flexion). It is well–known that even greater are translational amplitudes in the shoulder joint. In other words, within the realm of rigid body mechanics, a segment of a human arm or leg is not properly represented as a rigid body fixed at a certain point, but rather as a rigid body hanging on rope–like ligaments. More generally, the whole skeleton mechanically represents a system of flexibly coupled rigid bodies. This implies the more complex kinematics, dynamics and control then in the case of humanoid robots.

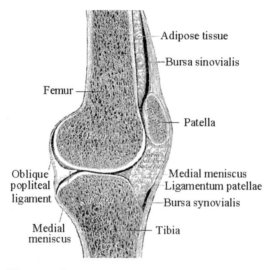

Fig. 5.3. Sagittal section through the knee joint.

We can immediately foresee here the increased problems of gait balance, stability and control [II05, II06b], but we still cannot neglect reality.

Modern unified geometrical basis for both human biomechanics and humanoid robotics represents the *constrained $SE(3)$–group*, i.e., the so–called *special Euclidean group of rigid–body motions in 3D space* (see [PC05, II05, II06a, II06b]). In other words, during human movement, in each movable human joint there is an action of a constrained $SE(3)$–group. In other words, constrained $SE(3)$–group represents *general kinematics* of human–like joints. The corresponding nonlinear dynamics problem (resolved mainly for aircraft and spacecraft dynamics) is called the *dynamics on $SE(3)$–group*, while the associated nonlinear control problem (resolved mainly for general helicopter control) is called the *control on $SE(3)$–group*.

The Euclidean $SE(3)$–group is defined as a semidirect (noncommutative) product of 3D rotations and 3D translations, $SE(3) := SO(3) \triangleright \mathbb{R}^3$. Its most important subgroups are the following:

Subgroup	Definition
$SO(3)$, group of rotations in 3D (a spherical joint)	Set of all proper orthogonal 3×3 − rotational matrices
$SE(2)$, special Euclidean group in 2D (all planar motions)	Set of all 3×3 − matrices: $\begin{bmatrix} \cos\theta & \sin\theta & r_x \\ -\sin\theta & \cos\theta & r_y \\ 0 & 0 & 1 \end{bmatrix}$
$SO(2)$, group of rotations in 2D subgroup of $SE(2)$ − group (a revolute joint)	Set of all proper orthogonal 2×2 − rotational matrices included in $SE(2)$ − group
$\mathbb{R}^3$, group of translations in 3D (all spatial displacements)	Euclidean 3D vector space

Using a 'realistic model' of human bio-mechanics comprising all above complexities (see [IB05, Iva04]), as a well–defined example of both a general bio–physical system and a general human behavior, we propose the following conjecture: In a combined bio–physical system, where the action of the physical laws (or engineering rules) cannot be neglected, it is the physical part that determines the lower limit of the total complexity. This complexity is commonly defined as the *number of mechanical degrees of freedom*. The biological part of the combined system, as being 'more intelligent', naturally serves as a 'controller' for the physical 'plant'. Although, in some special cases, the behavior of the combined system might appear 'simple' externally (i.e., have a low–dimensional output space), the realistic internal state–space analysis shows that the complexity of the total system equals the sum of the complexities of the two parts. Neither 'mutual cancelling' nor 'averaging' of the mechanical degrees of freedom generally occurs in such bio–physical system. We demonstrate the validity of the above conjecture using the example of the human bio-mechanical system and its realistic computer model. We further discuss simplicity versus predictability (and controllability) in a complex combined system. Then we identify self–assembly with training in human motion as a simple and well–defined example of general human behavior, and finally propose a new measure of complexity: the *observational resolution*.

Finally, we argue that there is a possible route to bio-mechanical simplicity in the form of oscillatory synchronization at the cost of long–term training.

5.6.2 Dynamical Complexity in Bio–Mechanics

This subsection briefly describes modern geometrical dynamics of human/ humanoid motion, to familiarize the reader with its complexity.

A physiologically realistic model of the human/humanoid bio-mechanics was developed in [Iva91, IB05, Iva05a, Iva04, II05] and implemented in a software package called Human Biodynamics Engine 'HBE' (for the preliminary,

Lagrangian version of the spinal only 'HBE–simulator', see [Iva05c]). The model was developed using generalized Hamiltonian mechanics and nonlinear control on Lie groups. It includes 264 active degrees–of–freedom, driven by 132 equivalent muscular actuators[10] (each with its own excitation and contraction dynamics), as well as two levels of neural–like control (stretch–reflex and cerebellum–like Lie–derivative stabilizer and target tracker). The cortical level of motion planning is currently under development, using adaptive fuzzy logic.

In this bio-mechanical $SE(3)$–based model (see Figure 5.4), rotational joint dynamics is considered 'active', driven by Newton–Euler type forces and torques, combined with neuro–muscular stretch–reflex and higher cerebellum control. Translational dynamics is considered 'passive', representing intervertebral discs, joint tendons and ligaments as a nonlinear spring–damper system. The model was initially applied for prediction of spinal injuries [IB03], representing the total motion of the human spine as a dynamical chain of 25 constrained $SE(3)$ groups (i.e., special Euclidean groups of rigid body motion).

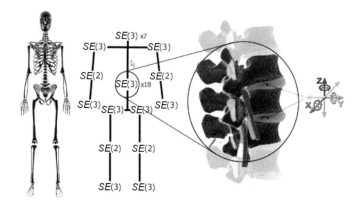

Fig. 5.4. Configuration manifold of human/humanoid skeleton, as modelled in the Human Biodynamics Engine.

Once the constrained $SE(3)$–based configuration manifold M^N is properly defined, we can define the full neuro–musculo–skeletal dynamics on its *momentum phase–space manifold* T^*M^N. The generalized Hamiltonian HBE–system is given, in a local canonical chart on T^*M^N, by (we skip here the

---

[10] An equivalent muscular actuator is a flexor–extensor pair of muscles, rotating a body segment (with all the masses attached to it) around a certain Euler axis. Each equivalent muscular actuator has its excitation dynamics, coming from the neural stimulus, as well as contraction dynamics, which generate the muscular torque in that joint. The muscular torque is the driving torque that counteracts inertial and gravity torques as well as joint elasticity and viscosity.

5.6 'Hard' vs. 'Soft' Complexity: A Bio-Mechanical Example

symplectic geometry derivations, see [Iva04, IB05, Iva05a, II05] for technical details)

$$\dot{q}^i = \frac{\partial H_0}{\partial p_i} + \frac{\partial R}{\partial p_i}, \tag{5.58}$$

$$\dot{p}_i = T_i - \frac{\partial H_0}{\partial q^i} + \frac{\partial R}{\partial q^i}, \tag{5.59}$$

$$q^i(0) = q_0^i, \qquad p_i(0) = p_i^0, \tag{5.60}$$

$$(i = 1, \ldots, N)$$

including the *contravariant velocity equation* (5.58) and the *covariant force equation* (5.59), with initial joint coordinates q_0^i and momenta p_i^0. Here the physical Hamiltonian function $H_0 : T^*M^N \to \mathbb{R}$ represents the total mechanical energy of the human motion

$$H_0(q, p) = \frac{1}{2} g^{ij} p_i p_j + V(q), \qquad (i, j = 1, \ldots, N),$$

where $g^{ij} = g^{ij}(q, m)$ denotes the contravariant *material metric tensor* (associated with Riemannian metrics $g : TM^N \to \mathbb{R}$ on M^N), relating internal joint coordinates q^i and external Cartesian coordinates x^r, and including n segmental masses m_μ

$$g^{ij}(q, m) = \sum_{\mu=1}^n m_\mu \delta_{rs} \frac{\partial q^i}{\partial x^r} \frac{\partial q^j}{\partial x^s},$$

$$(i, j = 1, \ldots, N), \qquad (r, s = 1, \ldots, 3n).$$

$R = R(q, p)$ denotes the Rayleigh nonlinear (usually biquadratic) dissipation function.

The driving covariant vector fields (i.e., one–forms), $T_i = T_i(t, q_{ang}^i, p_i^{ang}, u_i)$, are generalized muscular torques, depending on joint angles and angular momenta (not on translational coordinates and momenta), as well as on $u_i = u_i(t, q, p)$–corrections from the two neural control levels. Physiologically speaking, the torques T_i in the force equation (5.59) resemble neuro–muscular excitation dynamics, T_i^{EXC}, and contraction dynamics T_i^{CON}, of equivalent antagonistic muscular pairs in the i-th joint, i.e., $T_i = T_i^{MUS} = T_i^{EXC} \cdot T_i^{CON}$ (see [Iva04, IB05, Iva05a, II05] for technical details).

Now, to make the highly nonlinear and high–dimensional system (5.58–5.60) even closer to bio–physical reality, namely to account for ever–present external noise as well as imprecision of anthropometric and physiological measurements, we had to add to it [Iva91, IB05]:

A. *Stochastic forces*, in the form of diffusion fluctuations $B_{ij}[q^i(t), t]$ and discontinuous jumps as N–dimensional Wiener process $W^j(t)$; and

B. *Fuzzification* of the system parameters (segmental lengths, masses, inertia moments, joint dampings, tendon elasticities, etc.) and initial conditions (body configurations),

to get the *fuzzy–stochastic HBE system*:

$$dq^i = \left(\frac{\partial H_0(q,p,\sigma_\mu)}{\partial p_i} + \frac{\partial R}{\partial p_i}\right) dt, \quad (5.61)$$

$$dp_i = B_{ij}[q^i(t), t]\, dW^j(t) +$$
$$\left(\bar{T}_i - \frac{\partial H_0(q,p,\sigma_\mu)}{\partial q^i} + \frac{\partial R}{\partial q^i}\right) dt, \quad (5.62)$$

$$q^i(0) = \bar{q}_0^i, \qquad p_i(0) = \bar{p}_i^0,$$

where $\{\sigma\}_\mu$ (with $\mu \geq 1$) denote fuzzy sets of conservative parameters (segment lengths, masses and moments of inertia), dissipative joint dampings and actuator parameters (amplitudes and frequencies), while the bar $\bar{(.)}$ over a variable $(.)$ denotes the corresponding fuzzified variable.

It is clear that the fuzzy–stochastic HBE system (5.61–5.62) is even more complex and nonlinear and therefore harder to predict/control, compared to the crisp–deterministic system (5.58–5.59). However, it is much closer to the reality of human motion.

5.6.3 Control Complexity in Bio–Mechanics

As already stated, control of human motion is naturally and necessarily hierarchical, including three control levels: spinal, cerebellar and cortical. The first two levels have already been implemented in the software package HBE, while the cortical level is currently under the development. In this subsection, we briefly describe these three levels, so that the reader can get a 'feeling' for the control complexity involved.

Spinal–Like Reflex Force Control

The *force HBE servo–controller* is formulated as an affine control HBE–system. Introducing the coupling Hamiltonians $H^j = H^j(q,p)$, $j = 1, \ldots, M \leq N$, corresponding to the system's active joints, we define an *affine Hamiltonian function* $H_a : T^*M^N \to \mathbb{R}$, in local canonical coordinates on T^*M^N given as

$$H_a(q,p,u) = H_0(q,p) - H^j(q,p)\, u_j, \quad (5.63)$$

where $u_i = u_i(t,q,p)$ are feedback–controls. Using (5.63) we come to the affine Hamiltonian control HBE–system, in deterministic form

5.6 'Hard' vs. 'Soft' Complexity: A Bio-Mechanical Example

$$\dot{q}^i = \frac{\partial H_0(q,p)}{\partial p_i} - \frac{\partial H^j(q,p)}{\partial p_i} u_j + \frac{\partial R}{\partial p_i}, \quad (5.64)$$

$$\dot{p}_i = \bar{T}_i - \frac{\partial H_0(q,p)}{\partial q^i} + \frac{\partial H^j(q,p)}{\partial q^i} u_j + \frac{\partial R}{\partial q^i},$$

$$o^i = -\frac{\partial H_a(q,p,u)}{\partial u_i} = H^j(q,p),$$

$$q^i(0) = q_0^i, \qquad p_i(0) = p_i^0,$$

$$(i = 1, \ldots, N; \qquad j = 1, \ldots, M \leq N).$$

and in fuzzy–stochastic form

$$dq^i = \left(\frac{\partial H_0(q,p,\sigma_\mu)}{\partial p_i} - \frac{\partial H^j(q,p,\sigma_\mu)}{\partial p_i} u_j + \frac{\partial R(q,p)}{\partial p_i} \right) dt,$$

$$dp_i = B_{ij}[q^i(t),t]\, dW^j(t) \quad + \quad (5.65)$$

$$\left(\bar{T}_i - \frac{\partial H_0(q,p,\sigma_\mu)}{\partial q^i} + \frac{\partial H^j(q,p,\sigma_\mu)}{\partial q^i} u_j + \frac{\partial R(q,p)}{\partial q^i} \right) dt,$$

$$d\bar{o}^i = -\frac{\partial H_a(q,p,u,\sigma_\mu)}{\partial u_i} dt = H^j(q,p,\sigma_\mu)\, dt,$$

$$q^i(0) = \bar{q}_0^i, \qquad p_i(0) = \bar{p}_i^0.$$

Both affine control HBE–systems (5.64–5.65) resemble an *autogenetic motor servo* [Hou79], acting on the spinal–reflex level of the human locomotion control. A voluntary contraction force F of human skeletal muscle is reflexly excited (positive feedback $+F^{-1}$) by the responses of its *spindle receptors* to stretch and is reflexly inhibited (negative feedback $-F^{-1}$) by the responses of its *Golgi tendon organs* to contraction. Stretch and unloading reflexes are mediated by combined actions of several autogenetic neural pathways, forming the so–called 'motor servo.' The term 'autogenetic' means that the stimulus excites receptors located in the same muscle that is the target of the reflex response. The most important of these muscle receptors are the primary and secondary endings in the muscle–spindles, which are sensitive to length change – positive length feedback $+F^{-1}$, and the Golgi tendon organs, which are sensitive to contractile force – negative force feedback $-F^{-1}$.

The gain G of the length feedback $+F^{-1}$ can be expressed as the *positional stiffness* (the ratio $G \approx S = dF/dx$ of the force–F change to the length–x change) of the muscle system. The greater the stiffness S, the less the muscle will be disturbed by a change in load. The autogenetic circuits $+F^{-1}$ and $-F^{-1}$ appear to function as *servoregulatory loops* that convey continuously graded amounts of excitation and inhibition to the large *alpha skeletomotor neurons*. Small *gamma fusimotor neurons* innervate the contractile poles of muscle spindles and function to modulate spindle–receptor discharge.

Cerebellum–Like Velocity and Jerk Control

Nonlinear velocity and *jerk* (time derivative of acceleration) *servo–controllers*, developed in HBE using the Lie–derivative formalism, resemble self–stabilizing and adaptive tracking action of the cerebellum [HBB96]. By introducing the vector–fields f and g, given respectively by

$$f = \left(\frac{\partial H_0}{\partial p_i}, -\frac{\partial H_0}{\partial q^i}\right), \qquad g = \left(-\frac{\partial H^j}{\partial p_i}, \frac{\partial H^j}{\partial q^i}\right)$$

we obtain the affine controller in the standard nonlinear MIMO–system form (see [Isi89, NS90])

$$\dot{x}_i = f(x) + g(x)\, u_j. \tag{5.66}$$

Finally, using the *Lie derivative formalism* [Iva04][11] and applying the *constant relative degree* r to all HB joints, the *control law* for asymptotic tracking of the reference outputs $o_R^j = o_R^j(t)$ could be formulated as (generalized from [Isi89])

$$u_j = \frac{o_R^{(r)j} - L_f^{(r)} H^j + \sum_{s=1}^{r} c_{s-1}(o_R^{(s-1)j} - L_f^{(s-1)} H^j)}{L_g L_f^{(r-1)} H^j}, \tag{5.67}$$

where c_{s-1} are the coefficients of the linear differential equation of order r for the *error function* $e(t) = x^j(t) - o_R^j(t)$

$$e^{(r)} + c_{r-1} e^{(r-1)} + \cdots + c_1 e^{(1)} + c_0 e = 0.$$

---

[11] Let $F(M)$ denote the set of all smooth (i.e., C^∞) real valued functions $f : M \to \mathbb{R}$ on a smooth manifold M, $V(M)$ – the set of all smooth vector–fields on M, and $V^*(M)$ – the set of all differential one–forms on M. Also, let the vector–field $\zeta \in V(M)$ be given with its local flow $\phi_t : M \to M$ such that at a point $x \in M$, $\frac{d}{dt}|_{t=0}\, \phi_t x = \zeta(x)$, and ϕ_t^* representing the *pull–back* by ϕ_t. The *Lie derivative* differential operator $\mathcal{L}_\zeta$ is defined:
(i) on a function $f \in F(M)$ as

$$\mathcal{L}_\zeta : F(M) \to F(M), \qquad \mathcal{L}_\zeta f = \frac{d}{dt}(\phi_t^* f)|_{t=0},$$

(ii) on a vector–field $\eta \in V(M)$ as

$$\mathcal{L}_\zeta : V(M) \to V(M), \qquad \mathcal{L}_\zeta \eta = \frac{d}{dt}(\phi_t^* \eta)|_{t=0} \equiv [\zeta, \eta]$$

– the *Lie bracket*, and
(iii) on a one–form $\alpha \in V^*(M)$ as

$$\mathcal{L}_\zeta : V^*(M) \to V^*(M), \qquad \mathcal{L}_\zeta \alpha = \frac{d}{dt}(\phi_t^* \alpha)|_{t=0}.$$

In general, for any smooth tensor field $\mathbf{T}$ on M, the Lie derivative $\mathcal{L}_\zeta \mathbf{T}$ geometrically represents a directional derivative of $\mathbf{T}$ along the flow ϕ_t.

5.6 'Hard' vs. 'Soft' Complexity: A Bio-Mechanical Example

The affine nonlinear MIMO control system (5.66) with the Lie–derivative control law (5.67) resembles the self–stabilizing and synergistic output tracking action of the human cerebellum. To make it adaptive (and thus more realistic), instead of the 'rigid' controller (5.67), we can use the *adaptive Lie–derivative controller*, as explained in the seminal paper on geometrical nonlinear control [SI89, II06b].

Cortical–Like Fuzzy–Topological Control

For the purpose of our cortical control, the dominant, rotational part of the human configuration manifold M^N, could be first, reduced to an N–*torus*, and second, transformed to an N–*cube* ('hyper–joystick'), using the following topological techniques (see [II06b, II07b]).

Let S^1 denote the constrained unit circle in the complex plane, which is an Abelian Lie group. Firstly, we propose two reduction homeomorphisms, using the semidirect product $\ltimes$ of the constrained $SO(2)$–groups:

$$SO(3) \approx SO(2) \ltimes SO(2) \ltimes SO(2) \quad \text{and} \quad SO(2) \approx S^1.$$

Next, let I^N be the unit cube $[0,1]^N$ in $\mathbb{R}^N$ and '~' an equivalence relation on $\mathbb{R}^N$ obtained by 'gluing' together the opposite sides of I^N, preserving their orientation. Therefore, M^N can be represented as the quotient space of $\mathbb{R}^N$ by the space of the integral lattice points in $\mathbb{R}^N$, that is an oriented and constrained N–dimensional torus T^N:

$$\mathbb{R}^N/Z^N = I^N/\sim \approx \prod_{i=1}^{N} S_i^1 \equiv \{(q^i, i=1,\ldots,N) : \text{mod} 2\pi\}$$
$$= T^N. \tag{5.68}$$

Its *Euler–Poincaré characteristic* is (by the *De Rham theorem*) both for the configuration manifold T^N and its *momentum phase–space* T^*T^N given by (see [II07b])

$$\chi(T^N, T^*T^N) = \sum_{p=1}^{N} (-1)^p b_p,$$

where b_p are the *Betti numbers* defined as

$$b^0 = 1,$$
$$b^1 = N, \ldots b^p = \binom{N}{p}, \ldots b^{N-1} = N,$$
$$b^N = 1, \qquad (0 \le p \le N).$$

Conversely by 'ungluing' the configuration space we obtain the primary unit cube. Let '~*' denote an equivalent decomposition or 'ungluing' relation. According to Tychonoff's *product–topology theorem* [II07b], for every such quotient space there exists a 'selector' such that their quotient models are homeomorphic, that is, $T^N/\sim^* \approx A^N/\sim^*$. Therefore I_q^N represents a 'selector' for the configuration torus T^N and can be used as an N–directional '$\hat{q}$–command–space' for the *feedback control* (FC). Any subset of degrees of freedom on the configuration torus T^N representing the joints included in HB has its simple, rectangular image in the rectified $\hat{q}$–command space – selector I_q^N, and any joint angle q^i has its rectified image $\hat{q}^i$.

In the case of an end–effector, $\hat{q}^i$ reduces to the position vector in external–Cartesian coordinates z^r ($r = 1, \ldots, 3$). If orientation of the end–effector can be neglected, this gives a topological solution to the standard inverse kinematics problem.

Analogously, all momenta $\hat{p}_i$ have their images as rectified momenta $\hat{p}_i$ in the $\hat{p}$–command space – selector I_p^N. Therefore, the total momentum phase–space manifold T^*T^N obtains its 'cortical image' as the $\widehat{(q,p)}$–command space, a trivial $2N$–dimensional bundle $I_q^N \times I_p^N$.

Now, the simplest way to perform the feedback FC on the cortical $\widehat{(q,p)}$–command space $I_q^N \times I_p^N$, and also to mimic the cortical–like behavior, is to use the $2N$–dimensional fuzzy–logic controller, in much the same way as in the popular 'inverted pendulum' examples (see [Kos92]).

We propose the fuzzy feedback–control map Ξ that maps all the rectified joint angles and momenta into the feedback–control one–forms

$$\Xi : (\hat{q}^i(t), \hat{p}_i(t)) \mapsto u_i(t,q,p), \tag{5.69}$$

so that their corresponding universes of discourse, $\hat{Q}^i = (\hat{q}^i_{max} - \hat{q}^i_{min})$, $\hat{P}_i = (\hat{p}^{max}_i - \hat{p}^{min}_i)$ and $_i = (u_i^{max} - u_i^{min})$, respectively, are mapped as

$$\Xi : \prod_{i=1}^N \hat{Q}^i \times \prod_{i=1}^N \hat{P}_i \to \prod_{i=1}^N {_i}. \tag{5.70}$$

The $2N$–dimensional map Ξ (5.69,5.70) represents a *fuzzy inference system*, defined by (adapted from [IJB99b]):

A. *Fuzzification* of the crisp *rectified* and *discretized* angles, momenta and controls using Gaussian–bell membership functions

$$\mu_k(\chi) = exp[-\frac{(\chi - m_k)^2}{2\sigma_k}], \quad (k = 1, 2, \ldots, 9),$$

where $\chi \in D$ is the common symbol for $\hat{q}^i$, $\hat{p}_i$ and $u_i(q,p)$ and D is the common symbol for $\hat{Q}^i, \hat{P}_i$ and $_i$; the mean values m_k of the nine partitions of each universe of discourse D are defined as $m_k = \lambda_k D + \chi_{min}$, with partition coefficients λ_k uniformly spanning the range of D, corresponding to the set of nine linguistic variables $L = \{NL, NB, NM, NS, ZE, PS, PM, PB, PL\}$; standard deviations are kept constant $\sigma_k = D/9$. Using the linguistic vector L, the 9×9 FAM (fuzzy associative memory) matrix (a 'linguistic phase–plane'), is heuristically defined for each human joint, in a symmetrical weighted form

$$\mu_{kl} = \varpi_{kl}\, exp\{-50[\lambda_k + u(q,p)]^2\}, \qquad (k,l = 1, ..., 9)$$

with weights
$\varpi_{kl} \in \{0.6, 0.6, 0.7, 0.7, 0.8, 0.8, 0.9, 0.9, 1.0\}$.

B. *Mamdani inference* is used on each FAM–matrix μ_{kl} for all human joints:
(i) $\mu(\hat{q}^i)$ and $\mu(\hat{p}_i)$ are combined inside the fuzzy IF–THEN rules using AND (Intersection, or Minimum) operator,

$$\mu_k[\bar{u}_i(q,p)] = \min_l \{\mu_{kl}(\hat{q}^i),\, \mu_{kl}(\hat{p}_i)\}.$$

(ii) the output sets from different IF–THEN rules are then combined using OR (Union, or Maximum) operator, to get the final output, fuzzy–covariant torques,

$$\mu[u_i(q,p)] = \max_k \{\mu_k[\bar{u}_i(q,p)]\}.$$

C. *Defuzzification* of the fuzzy controls $\mu[u_i(q,p)]$ with the 'center of gravity' method

$$u_i(q,p) = \frac{\int \mu[u_i(q,p)]\, du_i}{\int du_i},$$

to update the crisp feedback–control one–forms $u_i = u_i(t,q,p)$.

Now, it is easy to make this top–level controller *adaptive*, simply by *weighting* both the above fuzzy–rules and membership functions, by the use of any standard competitive neural–network (see, e.g., [Kos92]). Operationally, the construction of the cortical $\widehat{(q,p)}$–command space $I_q^N \times I_p^N$ and the $2N$–dimensional feedback map Ξ (5.69,5.70), mimic the regulation of the *motor conditioned reflexes* by the motor cortex [HBB96].

5.6.4 Computational Complexity in Bio–Mechanics

A simplified version of the HBE system (5.61,5.62,5.65,5.66,5.67), with crisp parameters derived from the user anthropometry and physiology data, and simple random forces added to the crisp dynamics (5.58–5.60), has been developed at DSTO, Australia (together with a neural–like control described below), for the purpose of predicting the risk of musculo–skeletal injuries (see [IB03]). The system considered had 264 DOF (fingers and toes are not modelled), in the form of the set of 528 generalized Hamiltonian equations, with 132 Lie–derivative controllers. This huge set of nonlinearly–coupled nonlinear differential equations, were derived in Mathematica and then implemented in 'Delphi' compiler for MS Windows, using the specially developed *matrix–symplectic explicit integrator* of the 6th order.

It is practically *impossible to integrate* such a complex system of differential equations, even for 1 second, even with the best possible integrator, like Mathematica integrator NDSolve, the standard trick from modern mechanics and nonlinear control was adopted: *dynamical decoupling with simultaneous inertial* (static) *coupling* (see [II05]).[12] Once Hamiltonian equations are decoupled, they can be both numerically solved (using a matrix symplectic integrator) and efficiently controlled (using a linear or polynomial controller derived by Lie–derivative formalism described above).

A sample 'HBE' output is given in Figure 5.5, showing running with the speed of 6 m/s.

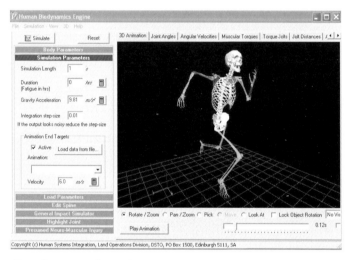

Fig. 5.5. Sample output from the Human Biodynamics Engine: running with the speed of 6 m/s – 3D animation view–port.

The 'HBE-simulator' has been kinematically validated against the standard biomechanical gait-analysis system 'Vicon'[Rob37] (see Figure 5.6).

The purpose of the simulator is prediction of the risk of soft spinal and other musculo-skeletal injuries.[13]

5.6.5 Simplicity, Predictability and 'Macro-Entanglement'

Here we argue that the simplification of the complex and realistic bio–mechanical model described above results in inaccurate prediction and control.

---

[12] The basic idea of geometrical decoupling is to 'free' the angular momentum (resp. angular velocity) and torque variables from the inertia matrix (i.e., metric tensor) g_{ij}, by putting it on the other side of Hamiltonian (resp. Lagrangian) equations [Isi89, NS90].

[13] Note that this subsection discusses the general bio-mechanical complexity issues, and explains the HBE complexity as an example only.

5.6 'Hard' vs. 'Soft' Complexity: A Bio-Mechanical Example

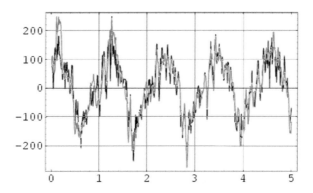

Fig. 5.6. Matching the 'Vicon' output with the 'HBE' output for the right-hip angular velocity around the dominant X-axis, while walking on a treadmill with a speed of 4 km/h.

Mutual Cancellation of the Model Components

Cancellation of the skeletal components is technically called 'amputation'. Clearly, this is not an option for solving the enormous complexity problem described in the previous sections. We cannot just cut–off human limbs to reduce the overall complexity of human motion.

5.6.6 Reduction of Mechanical DOF and Associated Controllers

It is possible to reduce the number of mechanical degrees of freedom, and therefore the bio-mechanical configuration manifold, by the total factor of six:

A. By replacing three–axial joints with uniaxial ones, which reduces the system's dimension by a factor of three; and
B. By neglecting all (restricted) joint translations, as is done in robotics, which reduces the system's dimension by a factor of two.

It is also possible to simplify the control subsystem:

A. by replacing nonlinear controllers with linear ones; and
B. by reducing a hierarchical, three–level control to the single level.

The overall result of these two simplifications is commonly known as 'dummy'. It can be very expensive and useful for crash–testing, but it cannot be used for any human–like performance.

Averaging of Mechanical DOF

Let us consider the possibility of averaging the degrees of freedom in bio-mechanics, using a technique similar to Maxwell's techniques in thermodynamics and statistical physics. In the past, it has been an old practice in bio-mechanics to use the body's 'center–of–mass' (CoM) motion as a simple representative of the full human musculo–skeletal dynamics (see, e.g., [McG99]), which is a systematic kinematical procedure of averaging the segmental trajectories. However, at present, it is used only for the low–resolution global positioning system (GPS) tracking of soldiers. It simply fails in simulating/predicting any realistic human movement, which is well–known to the researchers in robotics. For example, if we use the Cartesian vector trajectory of the CoM to simulate the motion of an athlete in a successful 'high–jump' event, we will see that the CoM trajectory passes under the bar while at the same time his whole body passes over the bar, which is represented by all segmental trajectories. This simple example shows that the averaging of mechanical degrees–of–freedom simply does not work if realistic representation of human motion is needed.

Superposition of Complexities or 'Macro–Entanglement'

From the standard engineering viewpoint, having two systems (biological and mechanical) combined as a single 'working machine', we can expect that the total 'machine' complexity equals the sum of the two partial ones. For example, electrical circuitry has been a standard modelling framework in neurophysiology.[14] Using the HH–approach for modelling human neuro–muscular circuitry as electrical circuitry, we get an electro–mechanical model for our bio-mechanical system, in which the superposition of complexities is clearly valid.

On the other hand, in a recent research on dissipative quantum brain modelling, one of the most popular issues has been *quantum entanglement*[15] between the *brain* and its *environment* (see [PV03, PV04]) where the brain–environment system has an entangled 'memory' state (identified with its ground state), that cannot be factorized into two single–mode states.[16] Similar to this microscopic brain–environment entanglement, we conjecture the

---

[14] Recall that A. Hodgkin and A. Huxley won a Nobel Prize for their circuit model of a single neuron, the celebrated HH–neuron [HH52]
[15] Entanglement is a term used in quantum theory to describe the way that particles of energy/matter can become correlated to predictably interact with each other regardless of how far apart they are; this is called a 'long–range correlation'.
[16] In the Vitiello–Pessa dissipative quantum brain model [PV03, PV04], the evolution of a memory system was represented as a trajectory of given initial condition running over time–dependent states, each one minimizing the free energy functional.

existence of a *macroscopic neuro–mechanical entanglement* between the operating modes of our neuro–muscular controller and purely mechanical skeleton (see [Eno01]).

In other words, we suggest that the *diffeomorphism* between the *brain motion manifold* (N−cube) and the *body motion manifold* M^N (which can be reduced to the constrained N−torus), described as the *cortical motion control*, can be considered a 'long–range correlation'.

Therefore, if the complexity of the two subsystems is not the 'expected' superposition of their partial complexities, then we have a macro–entanglement at work.

5.6.7 Self–Assembly, Synchronization and Resolution

Self–Assembly Versus Training

In the framework of human motion dynamics, *self–assembly* represents *adaptive motor control*, i.e., physiological motor training performed by *iteration of conditioned reflexes*. For this, a combination of *supervised* and *reinforcement training* is commonly used, in which a number of (nonlinear) *control parameters are iteratively adjusted* similar to the weights in neural networks, using either backpropagation–type or Hebbian–type learning, respectively (see, e.g., [Kos92]). Every human motor skill is mastered using this general method. Once it is mastered, it becomes *smooth and energy–efficient*, in accordance with *Bernstein's motor coordination and dexterity* (see [Ber67, Ber96]).

Therefore, bio-mechanical self–assembly clearly represents an 'evolution' in the parameter–space of human motion control. One might argue that such an evolution can be modelled using CA. However, this parameter–space, though being a dynamical and possibly even a contractible structure, is not an independent set of parameters – it is necessarily coupled to the mechanical skeleton configuration space, the plant to be controlled.

The system of 200 bones and 600 muscles can an produce infinite number of different movements. In other words, the *output–space dimension* of a skilled human motion dynamics equals *infinity* – there is no upper limit to the number of possible different human movements, starting with simple walk, run, jump, throw, play, etc. Even for the simplest motions, like walking, a child needs about 12 months to master it (and Honda robots took a decade to achieve this).

Furthermore, as human motion represents a simplest and yet well–defined example of a general human behavior, it is possible that other human behavioral and performance skills are mastered (i.e., self–assembled) in a similar way.

Observational Resolution

Similar to a GPS tracking of soldiers' motion being reduced to the CoM motion, the *observational resolution* represents a true criterion underlying the apparent external complexity. For instance, if we 'zoom–out' sufficiently enough to get to the 'satellite–level' observation, then the collective motion of a crowd of 100,000 people looks like a single 'soliton'. On the other hand, if we 'zoom–in' deep to get to the 'Earth–level', then the full bio-mechanical system complexity and possibly an infinite–dimensional output space of a single human member within the same crowd is seen. There is a significant difference in the resolution of human motion while watching 'subtle' hand movements playing a piano, or 'coarse' movements of the crowd (on a football stadium) from an orbital satellite. CA can be a good model for the crowd motion, but certainly not for hierarchical neural control of the dynamics of human hands playing a piano. Thus, the eventual criterion that determines apparent complexity is the observational resolution. In other words, the bio-mechanical complexity is a resolution–dependent variable: the higher the resolution, the higher the complexity. Note that, although apparently similar, this new concept is radically different from fractals. In case of fractals we have a similar pattern no matter how much we 'zoom-in' or 'zoom-out', that is we have roughly the constant complexity at all self-similar levels. On the other hand, in case of bio-mechanical observational resolution, each 'zoom-in' significantly increases the complexity. We conjecture that there is an exponential growth of complexity with increase of the observational resolution. For another approach to resolution/scale and complexity, defined in terms of the Shannon entropy from information theory, see [B-Y04].

Synchronization

Finally, there *is* also a possible route to simplicity in bio-mechanics. Namely, *synchronization* and *phase–locking* are ubiquitous in nature as well as in human brain (see [HI97, Izh99b, Izh04]). Synchronization can occur in *cyclic forms of human motion* (e.g., walking, running, cycling, swimming), both externally, in the form of *oscillatory dynamics*, and internally, in the form of *oscillatory cortical–control*. This oscillatory synchronization, e.g., in walking dynamics, has three possible forms: in–phase, anti–phase, and out–of–phase. The underlying phase–locking properties determined by type of oscillator (e.g., periodic/chaotic, relaxation, bursting[17], pulse-coupled, slowly connected, or

---

[17] Periodic bursting behavior in neurons is a recurrent transition between a quiescent state and a state of repetitive firing. Three main types of neural bursters are: (i) parabolic bursting ('circle/circle'), (ii) square–wave bursting ('fold/homoclinic'), and (iii) elliptic bursting ('subHopf/fold cycle'). Most burster models can be written in the singularly perturbed form:

$$\dot{x} = f(x, y), \qquad \dot{y} = \mu g(x, y),$$

where $x \in \mathbb{R}^m$ is a vector of fast variables responsible for repetitive firing (e.g., the membrane voltage and fast currents). The vector $y \in \mathbb{R}^k$ is a vector of slow variables that modulates the firing (e.g., slow (in)activation dynamics and changes in intracellular Ca^{2+} concentration). The small parameter $\mu \ll 1$ is a ratio of

5.6 'Hard' vs. 'Soft' Complexity: A Bio-Mechanical Example

connections with time delay) involved in the cortical control system (motion planner). According to Izhikevich–Hoppensteadt work (ibid), phase–locking is prominent in the brain: it frequently results in coherent activity of neurons and neuronal groups, as seen in recordings of local field potentials and EEG. In essence, the purpose of brain control of human motion is reduction of mechanical configuration space; brain achieves this through synchronization.

While cyclic movements indeed present a natural route to oscillatory bio-mechanical synchronization, both on the dynamical and cortical–control level, the various forms of synchronized group behavior in sport (such as synchronized swimming, diving, acrobatics) or in military performance represent the imperfect products of hard training. The synchronized team performance is achievable, but the cost is a difficult long–term training and sacrifice of one's natural characteristics.

For more details on bio–mechanical complexity, see [IS08].

---

fast/slow time scales. The synchronization dynamics between bursters depends crucially on their spiking frequencies, i.e., the interactions are most effective when the presynaptic interspike frequency matches the frequency of postsynaptic oscillations. The synchronization dynamics between bursters in the cortical motion planner induces synchronization dynamics between upper and lower limbs in oscillatory motions.

References

AA68. Arnold, V.I., Avez, A.: Ergodic Problems of Classical Mechanics. Benjamin, New York, (1968)

AA80. Aubry, S., André, G.: Colloquium on Computational Methods in Theoretical Physics. In Group Theoretical Methods in Physics, Horowitz, Ne'eman (ed), Ann. Israel Phys. Soc. **3**, 133–164, (1980)

AAC90. Artuso, R., Aurell, E., Cvitanovic, P.: Recycling of strange sets: I & II. Nonlinearity **3**, 325–359 and 361, (1990)

AAM76. Anderson, B.D., Arbib, M.A., Manes, E.G.: Foundations of System Theory: Finitary and Infinitary Conditions. Lecture Notes in Economics and Mathematical Systems Theory, Springer, New York, (1976)

AAR04. Andronico, A., Angelani, L., Ruocco, G., Zamponi, F.: Topological properties of the mean–field φ^4 model. Phys. Rev. E **70**, 041101, (2004)

AB84. Atiyah, M.F., Bott, R.: The moment map and equivariant cohomology. Topology, **23**, 1–28, (1984)

AB86. Aref, H., Balachandar, S.: Chaotic advection in a Stokes flow Phys. Fluids, **29**, 3515–3521, (1986)

ABC96. Aurell, E., Boffetta, G., Crisanti, A., Paladin, G., Vulpiani, A.: Predictability in Systems with Many Characteristic Times: The Case of Turulence. Phys. Rev. E **53**, 2337, (1996)

ABC97. Artale, V., Boffetta, G., Celani, A., Cencini, M., Vulpiani, A.: Dispersion of passive tracers in closed basins: Beyond the diffusion coefficient. Phys. Fluids A **9**, 3162, (1997)

ABG48. Alpher, R.A., Bethe, H., Gamow, G.: The Origin of Chemical Elements. Phys. Rev. **73**, 803, (1948)

ABS00. Acebrón, J.A., Bonilla, L.L., Spigler, R.: Synchronization in populations of globally coupled oscillators with inertial effects. Phys. Rev. E **62**, 3437–3454, (2000)

AC91. Ablowitz, M.J., Clarkson, P.A.: Solitons, nonlinear evolution equations and inverse scattering. London Math. Soc., **149**, CUP, Cambridge, UK, (1991)

AC94.	Connes, A.: Noncommutative Geometry. Academic Press, New York, (1994)
ACF98.	Arena, P., Caponetto, R., Fortuna, L., Porto, D.: Bifurcation and chaos in non-integer order cellular neural networks. Int. J. Bif. Chaos **8**, 1527–1539, (1998)
ACM98.	Ambjørn. J., Carfora, M., Marzuoli, A.: The Geometry of Dynamical Triangulations. Springer, Berlin, (1998)
ACP03.	Angelani, L., Casetti, L., Pettini, M., Ruocco, G., Zamponi, F.: Topological Signature of First Order Phase Transitions. Europhys. Lett. **62**, 775, (2003)
ACP05.	Angelani, L., Casetti, L. Pettini, M., Ruocco, G., Zamponi, F.: Topology and phase transitions: from an exactly solvable model to a relation between topology and thermodynamics. Phys. Rev. E **71**, 036152, (2005)
ACS85.	Arneodo, A., Coullet, P., Spiegel, E.A.: The dynamics of triple convection. Geophys. Astrophys. Fluid Dynamics **31**, 1–48, (1985)
AD86.	Anderson, A., Dewitt, B.: Does the topology of space fluctuate? Found. Phys. **16**, 91–105, (1986)
ADJ97.	Ambjørn. J., Durhuus, B., Jonsson, T.: Quantum geometry. Cambridge Monographs on Mathematical Physics, Cambridge Univ. Press, Cambridge, (1997)
AEH05.	Ahmed, E., Elgazzar, A.S., Hegazi, A.S.: An Overview of Complex Adaptive Systems. arXiv:nlin/0506059, (2005)
AF84.	Alfsen, K.H., Frøyland, J.: Systematics of the Lorenz model at $\sigma = 10$. Technical report, University of Oslo, Inst. of Physics, Norway, (1984)
AFH86.	Albeverio, S., Fenstat. J., Hoegh-Krohn, R., Lindstrom, T.: Nonstandard Methods in Stochastic Analysis and Mathematical Physics. Academic Press, New York, (1986)
AG05.	Apps, R., Garwicz, M.: Anatomical and physiological foundations of cerebellar information processing. Nature Rev. Neurosci., **6**, 297–311, (2005)
AG39.	Aurell, E., Gilbert, A.: Fast dynamos and determinants of singular integral operators. Geophys. Astrophys. Fluid Dyn. **73**, 5–32, (1993)
AGI98.	Amaral, L.A.N., Goldberger, A.L., Ivanov, P.Ch., Stanley, H.E.: Scale–independent measures and pathologic cardiac dynamics. Phys. Rev. Lett., **81**, 2388–2391, (1998)
AGM94.	Alekseevsky, D.V., Grabowski. J., Marmo, G., Michor, P.W.: Poisson structures on the cotangent bundle of a Lie group or a principle bundle and their reductions. J. Math. Phys., **35**, 4909–4928, (1994)
AGM97.	Alekseevsky, D., Grabowksi. J., Marmo, G., Michor, P.W.: Completely integrable systems: a generalization. Mod. Phys. Let. A, **12**(22), 1637–1648, (1997)
AGN94.	Antoniadis, I., Gava, E., Narain, K.S., Taylor, T.R.: Topological amplitudes in string theory. Nucl. Phys. B **413**, 162, (1994)

AGR81.	A. Aspect, P. Grangier, G. Roger: Experimental Test of Realistic Local Theories via Bell's Theorem, Phys. Rev. Lett. **47**, 460, (1981)
AGR82.	Aspect, A., Grangier, P., Roger, G.: Experimental realization of Einstein-Podolsky-Rosen-Bohm Gedankenexperiment: a new violation of Bell's inequalities. Phys. Rev. Lett., **48**, 91–94, (1982)
AGS85.	Amit, D.J., Gutfreund, H., Sompolinsky, H.: Spin-glass models of neural networks. Phys. Rev. A **32**, 1007–1018, (1985)
AH61.	Atiyah, M.F., Hirzebruch, F.: Vector bundles and homogeneous spaces. Proc. Symp. Pure Math. **3**, 7–38, (1961)
AH75.	Atherton, R.W., Homsy, G.M.: On the existence and formulation of variational principles for nonlinear differential equations, Stud. Appl. Math., **54**, 31–60, (1975)
AH88.	Atiyah, M.F., Hitchin, N.J.: The geometry and dynamics of magnetic monopoles, Princeton Univ. Press, Princeton, NJ, (1988)
AHC01.	Chamseddine, A.H.: Complexified gravity in noncommutative spaces. Comm. Math. Phys. **218**, 283, (2001)
AHS89.	Ashtekar, A., Husain, V., Samuel. J., Rovelli, C., Smolin, L.: 2+1 quantum gravity as a toy model for the 3+1 theory, Classical and Quantum Gravity **6**, L185, (1989)
AI92.	Ashtekar, A., Isham, C.J.: Representations of the holonomy algebras of gravity and non-Abelian gauge theories. Class. Quant. Grav. **9**, 1433–85, (1992)
AJ90.	Atiyah, M.F., Jeffrey, L.: Topological Lagrangians and Cohomology. J. Geom. Phys. **7**, 119, (1990)
AJ92.	Ambjørn. J., Jurkiewicz. J.: Four-dimensional simplicial quantum gravity. Phys. Lett. B **278**, 42–50, (1992)
AJ95.	Ambjørn. J., Jurkiewicz. J.: Scaling in 4D quantum gravity. Nucl. Phys. B **451**, 643–676, (1995)
AJK93.	Ambjørn. J., Jurkiewicz. J., Kristjansen, C.F.: Quantum gravity, dynamical triangulations and higher derivative regularization, Nucl. Phys. B **393**, 601–632, (1993)
AJL00a.	Ambjørn. J., Jurkiewicz. J., Loll, R.: Lorentzian and Euclidean quantum gravity – analytical and numerical results. In M-Theory and Quantum Geometry, eds. L. Thorlacius and T. Jonsson, NATO Science Series, Kluwer, 382–449, (2000)
AJL00b.	Ambjørn. J., Jurkiewicz. J., Loll, R.: A nonperturbative Lorentzian path integral for gravity. Phys. Rev. Lett. **85**, 924–927, (2000)
AJL01a.	Ambjørn. J., Jurkiewicz. J., Loll, R.: Dynamically triangulating Lorentzian quantum gravity. Nucl. Phys. B **610**, 347–382, (2001)
AJL01b.	Ambjørn. J., Jurkiewicz. J., Loll, R.: Nonperturbative 3d Lorentzian quantum gravity. Phys. Rev. D **64**, 044-011, (2001)
AJL01d.	Ambjørn. J., Jurkiewicz. J., Loll, R.: Computer simulations of 3d Lorentzian quantum gravity. Nucl. Phys. B **94**, 689–692, (2001)

AJL04.	Ambjørn. J., Jurkiewicz. J., Loll, R.: Emergence of a 4D world from causal quantum gravity. Phys. Rev. Lett. **93**, 131301, (2004)
AJL05C.	Ambjørn. J., Jurkiewicz. J., Loll, R.: Reconstructing the universe. Phys. Rev. D **72**, 064014, (2005)
AJL05a.	Ambjørn. J., Jurkiewicz. J., Loll, R.: Semiclassical universe from first principles. Phys. Lett. B **607**, 205–213, (2005)
AJL05b.	Ambjørn. J., Jurkiewicz. J., Loll, R.: Spectral dimension of the universe, Phys. Rev. Lett. **95**, 171301, (2005)
AJL06.	Ambjørn. J., Jurkiewicz. J., Loll, R.: The universe from scratch. Cont. Phys. **47**(2), 103–117, (2006)
AK02.	I.S. Aranson, L. Kramer, Rev. Mod. Phys. **74**, 99, (2002)
AK93.	Ambjørn. J., Kristjansen, C.F.: Nonperturbative 2D quantum gravity and Hamiltonians unbounded from below. Int. J. Mod. Phys. A **8**, 1259–1282, (1993)
AKO03.	Ansumali, S., Karlin, I.V., Öttinger, H.C.: Minimal Entropic Kinetic Models for Hydrodynamics, Europhys. Lett. **63**, 798–804, (2003)
AKS04.	Ansumali, S., Karlin, I.V., Succi, S.: Kinetic theory of turbulence modeling: smallness parameter, scaling and microscopic derivation of Smagorinsky model. Physica A **338**, 379, (2004)
AL04.	Ashtekar, A., Lewandowski. J.: Background Independent Gravity: A status report, Class. Quant. Grav. **21**, R53, (2004)
AL05.	Achimescu, S., Lipan, O.: Signal Propagation in Nonlinear Stochastic Gene Regulatory Networks. 3rd Int. Conf. Path. Netw., Sys. Rhodes, Greece 2005.
AL91.	Aidman, E.V., Leontiev, D.A.: From being motivated to motivating oneself: a Vygotskian perspective. Stud. Sov. Thought, **42**, 137–151, (1991)
AL95.	Ashtekar, A., Lewandowski. J.: Projective techniques and functional integration. J. Math. Phys. **36**, 2170, (1995)
AL98.	Ambjørn, J., Loll, R.: Non-perturbative Lorentzian quantum gravity, causality and topology change. Nucl. Phys. B **536**, 407–434, (1998)
ALM95.	Ashtekar, A., Lewandowski. J., Marolf, D., Mourao. J., Thiemann, T.: Quantization of diffeomorphism invariant theories of connections with local degrees of freedom. J. Math. Phys. **36**, 6456–6493, (1995)
AM78.	Abraham, R., Marsden. J.: Foundations of Mechanics. Benjamin, Reading, MA, (1978)
AM88.	Amari, S.I., Maginu, K.: Statistical neurodynamics of associative memory. Neu. Net. **1**, 63–73, (1988)
AM90.	Akbulut, S., McCarthy. J.D.: Casson's invariant for homological 3−spheres: An exposition, Mathematical Notes **36**, Princeton Univ. Press, Princeton, (1990)
AM91.	Aringazin, A., Mikhailov, A.: Matter fields in space–time with vector non–metricity. Clas. Quant. Grav. **8**, 1685, (1991)
AMR88.	Abraham, R., Marsden, J., Ratiu, T.: Manifolds, Tensor Analysis and Applications. Springer, New York, (1988)

AMV00.	Alfinito, E., Manka, R., Vitiello, G.: Vacuum structure for expanding geometry. Class. Quant. Grav. **17**, 93, (2000)
AN99.	Aoyagi, T., Nomura, M.: Oscillator Neural Network Retrieving Sparsely Coded Phase Patterns. Phys. Rev. Lett. **83**, 1062–1065, (1999)
AP96.	Aliev, R.R., Panfilov, A.V.: A simple two-variable model of cardiac excitation. Chaos, Solitons & Fractals **7**, 293–301, (1996)
AR95.	Antoni, M., Ruffo, S.: Clustering and relaxation in long-range Hamiltonian dynamics. Phys. Rev. E, **52**, 2361–2374, (1995)
ARV02.	Alfinito, E., Romei, O., Vitiello, G.: On topological defect formation in the process of symmetry breaking phase transitions. Mod. Phys. Lett. B **16**, 93, (2002)
AS04.	Albrecht, A., Sorbo, L.: Can the Universe Afford Inflation? Phys. Rev. D **70**, 063528, (2004)
AS63.	Atiyah, M.F., Singer, I.M.: The Index of Elliptic Operators on Compact Manifolds, Bull. Amer. Math. Soc. **69**, 322–433, (1963)
AS68.	Atiyah, M.F., Singer, I.M.: The Index of Elliptic Operators I, II, III. Ann. Math. **87**, 484–604, (1968)
AS71.	Atiyah, M.F., Segal, G.B.: Exponential isomorphisms for λ–rings. Quart. J. Math. Oxford Ser. **22**, 371–378, (1971)
AS72.	Abramowitz, M., Stegun, I.A.: Handbook of Mathematical Functions. Dover, New York, (1972)
AS82.	Albrecht, A., Steinhardt, P.J.: Cosmology For Grand Unified Theories With Radiatively Induced Symmetry Breaking. Phys. Rev. Lett. **48**, 1220, (1982)
AS92.	Abraham, R., Shaw, C.: Dynamics: the Geometry of Behavior. Addison–Wesley, Reading, (1992)
ASM03.	Alonso, S., Sagués, F., Mikhailov, A.S.: Winfree turbulence of scroll waves in excitable media. Science **299**, 1722, (2003)
ASM06.	Alonso, S., Sagués, F., Mikhailov, A.S.: Scroll Wave Instability Controlled by External Fluctuations. Chaos **16**, 023124, (2006)
ASS04.	Alonso, S., Sancho, J.M., Sagués, F.: Suppression of scroll wave turbulence by noise. Phys. Rev. E **70**, 067201, (2004)
AT01.	Anderson, C., Trayanova, N.A.: Success and failure of biphasic shocks: results of bidomain simulations. Math. Biosci. **174**, 91–109, (2001)
AT05.	Aguirre, A., Tegmark, M.: Multiple Universes, Cosmic Coincidences, and Other dark Matters, JCAP **0501**, 003, (2005)
ATT90.	Aihara, K., Takabe, T., Toyoda, M.: Chaotic Neural Networks, Phys.Lett. A **144**, 333–340, (1990)
AU82.	Alberti, P.M., Uhlmann, A.: Stochasticity and Partial Order: Doubly Stochastic Maps and Unitary Mixing. VEB Deutscher Verlag der Wissenschaften, Berlin, (1982)
AV00.	Alfinito, E., Vitiello, G.: Formation and life–time of memory domains in the dissipative quantum model of brain, Int. J. Mod. Phys. B, **14**, 853–868, (2000)

References

AW01. Anderson. J.D., Williams. J.G.: Long Range Tests of the Equivalence Principle, Class. Quant. Grav. **18**, 2447, (2001)

AW91. Axelrod, S., della Pietra, S., Witten, E.: Geometric Quantization of Chern-Simons Gauge Theory. J. Diff. Geom. **33**, 787, (1991)

Aba96. Abarbanel, H.D.I.: Analysis of Observed Chaotic Data. Springer–Verlag, Berlin, (1996)

Ach97. Acheson, D.: From Calculus to Chaos. Oxford Univ. Press, Oxford, (1997)

Ach97. Acharyya, M.: Nonequilibrium phase transition in the kinetic Ising model: Divergences of fluctuations and responses near the transition point. Phys. Rev. E **56**, 1234–1237, (1997)

Ada62. Adams. J.F.: Vector fields on spheres. Ann. Math. **75**, 603–632, (1962)

Ada78. Adams. J.F.: Infinite Loop Spaces. Princeton Univ. Press, Princeton, NJ, (1978)

Adl04. Adler, S.L.: Quantum Theory as an Emergent Phenomenon. Cambridge University Press, Cambridge, UK, (2004)

Aka82. Akama, K.: An Early Proposal of Brane World. Lect. Not. Phys. **176**, 267, (1982)

Ala99. Alaoui, A.M.: Differential Equations with Multispiral Attractors. Int. Journ. of Bifurcation and Chaos, **9**(6), 1009–1039, (1999)

Ale68. Alekseev, V.: Quasi-random dynamical systems I, II, III. Math. USSR Sbor. **5**, 73–128, (1968)

Ale94. Alexander, D.S.: A history of complex dynamics. From Schröder to Fatou and Julia. Aspects of Mathematics. Vieweg E24, (1994)

Alo03. Alon, U.: Biological Networks: The tinkerer as an engineer. Science **301**, 1866–1867, (2003)

Alt95. Alty, L.J.: Building blocks for topology change, J. Math. Phys. **36**, 3613–3618, (1995)

Ama72. Amari, S.I.: IEEE Trans. Characteristics of random nets of analog neuron-like elements. **SMC-2**, 643–657, (1972)

Ama77. Amari, S.I.: Neural theory of association and concept–formation. Biol Cybern. **26**(3), 175–185, (1977)

Ama78. Amari, S., Takeuchi, A.: Mathematical theory on formation of category detecting nerve cells. Biol. Cybern. **29**(3), 127–136, (1978)

Ama85. Amari, S.I.: Differential Geometrical Methods in Statistics. Springer, New York, (1985)

Ame93. Amemiya, Y.: On nonlinear factor analysis. Proc. Social Stat. Section. Ann. Meet. Ame. Stat. Assoc. 290–294, (1993)

And01. Andrecut, M.: Biomorphs, program for $Mathcad^{TM}$, Mathcad Application Files, Mathsoft, (2001)

And64. Anderson, P.W.: Lectures on the Many Body Problem. E.R. Caianiello (ed), Academic Press, New York, (1964)

Ano67. Anosov, D.V.: Geodesic flows on closed Riemannian manifolds with negative curvature. Proc. Steklov Math. Inst. **90**, 1 (1967).

Arb95.	Arbib, M.A.: The Handbook of Brain Theory and Neural Networks, MIT Press, (1995)
Arb98.	Arbib, M. (ed.): Handbook of Brain Theory and Neural Networks (2nd ed.) MIT Press, Cambridge, MA, (1998)
Are83.	Aref, H.: Integrable, chaotic and turbulent vortex motion in two-dimensional flows. Ann. Rev. Fluid Mech. **15** 345–89, (1983)
Are84.	Aref, H.: Stirring by chaotic advection. J. Fluid Mech. **143**, 1–21, (1984)
Ark01.	Arkin, A.P.: Synthetic cell biology. Curr. Opin. Biotech., **12**, 638–644, (2001)
Arn61.	Arnold, V.I.: On the stability of positions of equilibrium of a Hamiltonian system of ordinary differential equations in the general elliptic case. Sov. Math. Dokl. **2**, 247, (1961)
Arn65.	Arnold, V.I.: Sur une propriété topologique des applications globalement canoniques de la mécanique classique. C. R. Acad. Sci. Paris A, **261**, 17, (1965)
Arn78.	Arnold, V.I.: Ordinary Differential Equations. MIT Press, Cambridge, MA, (1978)
Arn78.	Arnold, V.I.: Mathematical Methods of Classical Mechanics. Springer, New York, (1978)
Arn88.	Arnold, V.I.: Geometrical Methods in the Theory of Ordinary differential equations. Springer, New York, (1988)
Arn92.	Arnold, V.I.: Catastrophe Theory. Springer, Berlin, (1992)
Arn93.	Arnold, V.I.: Dynamical systems. Encyclopaedia of Mathematical Sciences, Springer, Berlin, (1993)
Arn99.	Arnaud, M.C.: Création de connexions en topologie C^1. Preprint Université de Paris-Sud 1999. In Ergodic Theory and Dynamical Systems. C. R. Acad. Sci. Paris, Série I, **329**, 211–214, (1999)
Ash86.	Ashtekar, A.: New variables for classical and quantum gravity. Phys. Rev. Lett. **57**, 2244–2247, (1986)
Ash87.	Ashtekar, A.: New Hamiltonian formulation of general relativity. Phys. Rev. D **36**, 1587–1602, (1987)
Ash88.	Ashtekar, A.: New perspectives in canonical gravity. Bibliopolis, (1988)
Ash91.	Ashtekar, A.: Lecture notes on non-perturbative canonical gravity. Advaced Series in Astrophysics and Cosmology, Vol. 6, World Scientific, Singapore, (1991)
Ash94.	Ashcraft, M.H.: Human Memory and Cognition (2nd ed) HarperCollins, New York, (1994)
Ash97.	Ashby, N.: Relativistic effects in the global positioning system. Plenary lecture on quantum gravity at the GR15 conference, Puna, India, (1997)
Ati00.	Atiyah, M.F.: K–Theory Past and Present. arXiv:math/0012213, (2000)
Ati87.	Atiyah, M.F.: Magnetic Monopoles in hyperbolic spaces. In Proceedings of Bombay Colloquium 1984 on vector bundles in algebraic varieties. Oxford Univ. Press, 1–34, (1987)

Ati88a.   Atiyah, M.F.: Topological quantum field theory. Publ. Math. IHES **68**, 175–186, (1988)

Ati88b.   Atiyah, M.F.: New invariants of three and four dimensional manifols. In The Mathematical Heritage of Hermann Weyl, eds. R. Well *et al.*, Proc. Symp. Pure. Math. 48, Am. Math. Soc., Providence, (1988)

Ati89.   Atiyah, M.F.: The Geometry and Physics of Knots. Cambridge Univ. Press, Cambridge, (1989)

Att71.   Attneave, F.: Multistability in Perception. Scientific American, **225**, 62–71, (1971)

Aul00.   Auletta, G.: Foundations and Interpretations of Quantum Mechanics. World Scientific, Singapore, (2000)

Ave99.   Averin, D.V.: Solid–state qubits under control. Nature **398**, 748–749, (1999)

Axe97.   Axelrod, R.: The Dissemination of Culture: A Model with Local Convergence and Global. Polarization. J. of Conflict Resolut. **41**, 203, (1997)

B-Y97.   Bar-Yam, Y.: Dynamics of Complex Systems. Perseus Books, Reading, (1997)

B-Y04.   Bar-Yam, Y.: Multiscale Complexity / Entropy. Adv. Comp. Sys. **7**, 47–63, (2004)

BA90.   Braam, P.J. Austin, D.M.: Boundary values of hyperbolic monopoles, Nonlinearity **3**(3), 809–823, (1990)

BB03.   Brax, P., van de Bruck, C.: Cosmology and Brane Worlds: A Review. arXiv: hep-th/0303095, (2003)

BB04.   Ben-Bassat, O., Boyarchenko, M.: Submanifolds of generalized complex manifolds. J. Sympl. Geom. **2**(3), 309–355, (2004)

BB96.   Bialas, P., Burda, Z., Krzywicki, A., Petersson, B.: Focusing on the fixed point of 4d simplicial gravity, Nucl. Phys. B 472 (1996) 293-308.

BB98.   Bender, C.M., Boettcher, S.: Real Spectra in Non-Hermitian Hamiltonians Having PT Symmetry. Phys. Rev. Lett. **80**, 5243, (1998)

BBC93.   Bennett, C.H., Brassard, G., Crépeau, C., Jozsa, R., Peres, A., Wootters, W.K.: Teleporting an Unknown Quantum State via Dual Classical and Einstein–Podolsky–Rosen Channels, Phys. Rev. Lett. **70**, 1895, (1993)

BBD01.   Brax, P., van de Bruck, C., Davis, A.C.: Brane-world cosmology, bulk scalars and perturbations. JHEP **0110**, 026, (2001)

BBD98.   Boschi, D., Branca, S., DeMartini, F., Hardy, L. Popescu, S.: Experimental Realization of Teleporting an Unknown Pure Quantum State via Dual Classical and Einstein–Podolski–Rosen Channels. Phys. Rev. Lett. **80**, 1121, (1998)

BBJ02.   Bender, C.M., Brody, D.C., Jones, H.F.: Complex Extension of Quantum Mechanics. Phys. Rev. Lett. **89**, 270401, (2002)

BBK02.   Bena, I., van den Broeck, C., Kawai, R., Lindenberg, K.: Nonlinear response with dichotomous noise. Phys. Rev. E **66**, 045603, (2002)

BBM99.   Bender, C.M., Boettcher, S., Meisinger, P.N.: $\mathcal{PT}$–Symmetric Quantum Mechanics. J. Math. Phys. **40**, 2201, (1999)

BBP94.	Babiano, A., Boffetta, G., Provenzale, A., Vulpiani, A.: Chaotic advection in point vortex models and two-dimensional turbulence. Phys. Fluids A **6**(7), 2465–2474, (1994)
BBP96.	Bennett, C.H., Brassard, G., Popescu, B., Smolin, J.A., Wootters, W.K.: Purification of Noisy Entanglement and Faithful Teleportation via Noisy Channels. Phys. Rev. Lett. **76**, 722–725, (1996)
BBP96b.	Bennett, C.H., Bernstein, H.J., Popescu, S., Schumacher, B.: Concentrating Partial Entanglement by Local Operations. Phys. Rev. A **53**, 2046, (1996)
BBR91.	Birmingham, D., Blau, M., Rakowski, M., Thompson, G.: Topological field theory. Phys. Rep. **209**, 129, (1991)
BBV99.	Béguin, F., Bonatti, C., Vieitez, J.L.: Construction de flots de Smale en dimension 3. Ann. Fac. Sci. de Toulouse Math. **6**, 369–410, (1999)
BC85.	Benedicks, M., Carleson, L.: On iterations of $1 - ax^2$ on $(-1, 1)$, Ann. of Math. **122**, 1–25, (1985)
BC91.	Benedicks, M., Carleson, L.: Dynamics of the Hénon map, Ann. of Math. **133**, 73–169, (1991)
BC97.	Barret. J., Crane, L.: Relativistic spin networks and quantum gravity. arXiv:gr-qc/9709028, (1997)
BCB92.	Borelli, R.L., Coleman, C., Boyce, W.E.: Differential Equations Laboratory Workbook. Wiley, New York, (1992)
BCD06.	Bender, C.M, Chen J-H., Daniel W. Darg, D.W., Milton, K.A.: Classical Trajectories for Complex Hamiltonians. arXiv:math-ph/0602040, (2006)
BCE00.	Boffetta, G., Cencini, M., Espa, S., Querzoli, G.: Chaotic advection and relative dispersion in a convective flow. Phys. Fluids **12**, 3160–3167, (2000)
BCF02.	Boffetta, G., Cencini, M., Falcioni, M., Vulpiani, A.: Predictability: a way to characterize Complexity, Phys. Rep., **356**, 367–474, (2002)
BCG00.	Bowcock, P., Charmousis, C., Gregory, R.: General brane cosmologies and their global space–time structure, Class. Quant. Grav. **17**, 4745, (2000)
BCG91.	Bryant, R., Chern, S., Gardner, R., Goldscmidt, H., Griffiths, P.: Exterior Differential Systems. Springer, Berlin, (1991)
BCH99.	B.M. Boghosian, C.C. Chow, T. Hwa, Hydrodynamics of the Kuramoto–Sivashinsky Equation in Two Dimensions. Phys. Rev. Lett. **83**, 5262–5265, (1999)
BCI93.	Balmfort, N.J., Cvitanovic, P., Ierley, G.R., Spiegel, E.A., Vattay, G.: Advection of vector–fields by chaotic flows. Ann. New York Acad. Sci. **706**, 148, (1993)
BCO94.	Bershadsky, M., Cecotti, S., Ooguri, H., Vafa, C.: Kodaira–Spencer theory of gravity and exact results for quantum string amplitudes. Commun. Math. Phys. **165**, 311, (1994)
BCR82.	Borsellino, A., Carlini, F., Riani, M., Tuccio, M.T., Marco, A.D., Penengo, P., Trabucco, A.: Effects of visual angle on perspective reversal for ambiguous patterns. Perception, **11**, 263–273, (1982)

BCV95.	Biferale, L., Crisanti, A., Vergassola, M., Vulpiani, A.: Eddy diffusivities in scalar transport. Phys. Fluids **7**, 2725, (1995)
BD02.	Busemeyer. J.R., Diederich, A.: Survey of decision field theory. Math. Soc. Sci., **43**, 345–370, (2002)
BD82.	Birrel, N.D., Davies, P.C.W.: Quantum fields in curved space. Cambridge Univ. Press, Cambridge, UK, (1982)
BD95.	Baez. J., Dolan. J.: Higher dimensional algebra and topological quantum field theory. J. Math. Phys. **36**, 6073–6105, (1995)
BD99.	Bonatti, C., Díaz, L.J.: Connexions heterocliniques et genericité d'une infinité de puits ou de sources. Ann. Sci. École Norm. Sup. **32**, 135–150, (1999)
BDB00.	Van de Bruck, C., Dorca, M., Brandenberger, R.H., Lukas, A.: Cosmological perturbations in brane-world theories: Formalism. Phys. Rev. D **62**, 123515, (2000)
BDB07.	Baronchelli, A., Dall'Asta, L., Barrat, A., Loreto, V.: Nonequilibrium phase transition in negotiation dynamics. Phys. Rev. E **76**, 051102, (2007)
BDE96.	A. Barenco, D. Deutsch, A. Ekert, R. Jozsa: Conditional Quantum Dynamics and Logic Gates. Phys. Rev. Lett. **74**, 4083–4086, (1995)
BDG04.	Banks, T., Dine, M., Gorbatov, E.: Is There a String Theory Landscape? JHEP **0408**, 058, (2004)
BDG93.	Bielawski, S., Derozier, D., Glorieux, P.: Experimental characterization of unstable periodic orbits by controlling chaos. Phys. Rev. A, **47**, 2492, (1993)
BDG99.	Borde, A., Dowker, H.F., Garcia, R.S., Sorkin, R.D., Surya, S.: Causal continuity in degenerate space-times. Class. Quant. Grav. **16**, 3457–3481, 1999.
BDM00.	Van de Bruck, C., Dorca, M., Martins, C.J., Parry, M.: Cosmological consequences of the brane/bulk interaction. Phys. Lett. B **495**, 183, (2000)
BDP99.	Bonatti, C., Díaz, L.J., Pujals, E.: A C^1–dichotomy for diffeomorphisms: weak forms of hyperbolicity or infinitely many sinks or sources. Preprint, (1999)
BDR01.	Barré, J., Dauxois, T., Ruffo, S.: Clustering in a Hamiltonian with repulsive long range interactions. Physica **A 295**, 254 (2001)
BE93.	Bär, M., Eiswirth, M.: Turbulence due to spiral breakup in a continuous excitable medium. Phys. Rev. E **48**, R1635–R1637 (1993)
BEZ00.	Bouwmeester, D., Ekert, A., Zeilinger, A. (eds.): The Physics of Quantum Information. Springer, Heidelberg, (2000)
BF01.	Banks, T., Fischler, W.: M–Theory Observables for Cosmological space-times. arXiv:hep-th/0102077, (2001)
BF71.	Bransford. J.D., Franks. J.J.: The Abstraction of Linguistic Ideas. Cogn. Psych., **2**, 331–350, (1971)
BF79.	Boldrighini, Franceschini, V.: A Five-Dimensional Truncation of the Plane Incompressible Navier–Stokes Equations. Commun. Math. Phys., **64**, 159–170, (1979)

BFC06.	Baronchelli, A., Felici, M., Caglioti, E., Loreto, V., Steels, L.: Sharp Transition towards Shared Vocabularies in Multi–Agent Systems. J. Stat. Mech. P06014, (2006)
BFL02.	Barbara, P. Filatrella, G., Lobb, C.J., Pederson, N.F.: Studies of High Temperature Superconductors, NOVA Sci. Pub. **40**, Huntington, (2002)
BFM00.	Bertoldi, G., Faraggi, A., Matone, M.: Equivalence principle, higher dimensional Moebius group and the hidden antisymmetric tensor of quantum mechanics. Class. Quant. Grav. **17**, 3965, (2000)
BFR04.	Bagnoli, F., Franci, F., Rechtman, R.: Chaos in a simple cellular automata model of a uniform society. In Lec. Not. Comp. Sci., Vol. 3305, 513–522, Springer, London, (2004)
BG03.	Bassi, A., Ghirardi, G.C.: Dynamical Reduction Models. Phys. Rep. **379**, 257–426, (2003)
BG04.	Berselli L.C., Grisanti C.R.: On the consistency of the rational large eddy simulation model. Comput. Vis. Sci. **6**, N 2–3, 75–82, (2004)
BG79.	Barrow-Green, J.: Poincaré and the Three Body Problem. American Mathematical Society, Providence, RI, (1997)
BG88.	Baulieu, L., Grossman, B.: Monopoles and topological field theory. Phys. Lett. B **214**(2), 223, (1988)
BG90.	Bouchaud, J.P., Georges, A.: Anomalous diffusion in disordered media – statistical mechanisms, models and physical applications. Phys. Rep. **195**, 127, (1990)
BG96.	Baker, G.L., Gollub, J.P.: Chaotic Dynamics: An Introduction (2nd ed) Cambridge Univ. Press, Cambridge, (1996)
BGG03.	Bryant, R., Griffiths, P., Grossman, D.: Exterior Differential Systems and Euler–Lagrange partial differential equations. Univ. Chicago Press, Chicago, (2003)
BGG80.	Benettin, G., Giorgilli, A., Galgani, L., Strelcyn, J.M.: Lyapunov exponents for smooth dynamical systems and for Hamiltonian systems; a method for computing all of them. Part 1: theory, and Part 2: numerical applications. Meccanica, **15**, 9–20 and 21–30, (1980)
BGG89.	Batlle, C., Gomis, J., Gràcia, X., Pons, J.M.: Noether's theorem and gauge transformations: application to the bosonic string and CP_2^{n-1} model. J. Math. Phys. **30**, 1345, (1989)
BGL00.	Boccaletti, S., Grebogi, C., Lai, Y.-C., Mancini, H., Maza, D.: The Control of Chaos: Theory and Applications. Physics Reports **329**, 103–197, (2000)
BGM02.	Bruzzo, U., Gorini, V., Moschella, U. (eds.): Geometry and Physics of Branes. Institute of Physics, Bristol, (2002)
BGS76.	Benettin, G., Galgani, L., Strelcyn, J.M.: Kolmogorov Entropy and Numerical Experiments. Phys. Rev. A **14**, 2338, (1976)
BGT95.	Bucher, M., Goldhaber, A.S., Turok, N.: An Open Universe from Inflation. Phys. Rev. D **52**, 3314, (1995)

BH00. D. ben-Avraham and S. Havlin: Diffusion and reactions in fractals and disordered systems, Cambridge Univ. Press, Cambridge, UK (2000)
BH01. Brody, D., Hughston, L.: Geometric quantum mechanics. J. Geom. Phys. **38**, 19, (2001)
BH01. Benjamin, S.C., Hayden, P.M.: Multiplayer quantum games. Phys. Rev. A **64**, 030301(R) (2001)
BH05. Bosse, A.W., Hartle, J.B.: Representations of space-time Alternatives and Their Classical Limits. Phys. Rev. **A72**, 022105, (2005)
BH83. Bhatnagar, K.B., Hallan, P.P.: The effect of perturbation in coriolis and centrifugal forces on the non-linear stability of equilibrium points in the restricted problem of three bodies. Celest. Mech. & Dyn. Astr. **30**, 97, (1983)
BH93. Bohm, D., Hiley, B.J.: The Undivided Universe. Routledge, London, (1993)
BH95. Bestvina, M., Handel, M.: Train-tracks for surface homeomorphisms. Topology, **34**(1), 109–140, (1995)
BH96. Banaszuk, A., Hauser. J.: Approximate feedback linearization: a homotopy operator approach. SIAM J. Cont. & Optim., **34**(5), 1533–1554, (1996)
BH99. Brunel, N., Hakim, V.: Fast Global Oscillations in Networks of Integrate-and-Fire Neurons. Neural Computation **11**, 1621–1671, (1999)
BK52. Berlin, T.H., Kac, M.: The Spherical Model of a Ferromagnet. Phys. Rev. **86**, 821–835, (1952)
BK64. Bellman, R., Kalaba, R.: Selected papers on mathematical trends in control theory. Dover, New York, (1964)
BK92. Berleant, D., Kuipers, B.: Qualitative–Numeric Simulation with Q3, in Recent Advances in Qualitative Physics. eds. Boi Faltings and Peter Struss, MIT Press, Cambridge, (1992)
BK99. Balasubramanian, V., Kraus, P.: A Stress Tensor for Anti-de Sitter Gravity. Commun. Math. Phys. 208, 413, (1999)
BKR03. Ben-Naim, E., Krapivsky, P.L., Redner, S.: Bifurcations and Patterns in Compromise Processes. Physica D **183**, 190, (2003)
BKT90. Barkley, D., Kness, M., Tuckerman, L.S.: Spiral wave dynamics in a simple to compound rotation. Phys. Rev. A **42**, 2489–2492, (1990)
BL92. Blackmore, D.L., Leu, M.C.: Analysis of swept volumes via Lie group and differential equations. Int. J. Rob. Res., **11**(6), 516–537, (1992)
BLM87. Boiti, M., Leon, J.J., Manna, M., Penpinelli, F.: On a spectral transform of a KdV–like equation related to the Schrodinger operator in the plane. Inverse Problems **3**, 25, (1987)
BLM87. Bombelli, L. Lee, J.-H., Meyer, D. Sorkin, R.: Space-time as a causal set. Phys. Rev. Lett. **59**, 521, (1987)
BLM88. Boiti, M., Leon, J.J., Martina, L., Penpinelli, F.: Scattering of Localized Solitons in the Plane. Phys. Lett. A **132**, 432–439, (1988)

BLR01.	Boffetta, G., Lacorata, G., Redaelli, G., Vulpiani, A.: Barriers to transport: a review of different techniques. Physica D, **159**, 58–70, (2001)
Ber47.	Bernstein, N.A.: On the structure of motion (in Russian) Medgiz, Moscow, (1947)
Ber67.	Bernstein, N.A.: The Coordination and Regulation of Movements, Pergamon Press, Oxford, (1967)
Ber96.	Bernstein, N.A.: On Dexterity and Its Development. In M.L. Latash and M.T. Turvey, (ed.), Dexterity and Its Development. Lawrence Erlbaum Associates, Mahwah, NJ, (1996)
BLT.	Bernstein, N.A., Latash, M.L., Turvey, M.T. (Eds): Dexterity and its development. Hillsdale, NJ, (1996)
BLV01.	Boffetta, G., Lacorata, G., Vulpiani, A.: Introduction to chaos and diffusion. Chaos in geophysical flows, ISSAOS, (2001)
BM82.	Choquet–Bruhat, Y., DeWitt–Morete, C.: Analysis, Manifolds and Physics (2nd ed) North–Holland, Amsterdam, (1982)
BM00.	Choquet–Bruhat, Y., DeWitt–Morete, C.: Analysis, Manifolds and Physics, Part II: 92 Applications (rev. ed) North–Holland, Amsterdam, (2000)
BMA72.	Borsellino, A., Marco, A. D., Allazatta, A., Rinsei, S. Bartolini, B.: Reversal time distribution in the perception of visual ambiguous stimuli. Kybernetik, **10**, 139–144, (1972)
BMB82a.	Van den Broeck, C., Mansour, M., Baras, F.: Asymptotic Properties of Coupled Non-linear Langevin Equations in the Limit of Weak Noise, I: Cusp Bifurcation. J. Stat. Phys. **28**, 557–575, (1982)
BMB82b.	Van den Broeck, C., Mansour, M., Baras, F.: Asymptotic Properties of Coupled Non-linear Langevin Equations in the Limit of Weak Noise, II: Transition to a Limit Cycle. J. Stat. Phys. **28**, 577–587, (1982)
BMM95.	Biró, T.S., Matinyan, S.G., Müller, B.: Chaos and gauge field theory. World Scientific, Singapore, (1995)
BMP97.	Breckenridge. J.C., Myers, R.C., Peet, A.W., Vafa, C.: D–branes and spinning black holes. Phys. Lett. B **391**, 93, (1997)
BMW01.	Bridgman, H.A., Malik, K.A., Wands, D.: Cosmic vorticity on the brane. Phys. Rev. D **63**, 084012, (2001)
BMW02.	Bridgman, H.A., Malik, K.A., Wands, D.: Cosmological perturbations in the bulk and on the brane. Phys. Rev. D **65**, 043502, (2002)
BM67.	Bott, R., Mather, J.: Topics in Topology and Differential Geometry. In Battelle Rencontres, Eds. C.M. De Witt and J.A. Wheeler, (1967)
BO95.	Basar, T., Olsder, G.J.: Dynamic Noncooperative Game Theory (2nd ed.), Academic Press, New York, (1995)
BOA85.	Bondeson, A., Ott, E., Antonsen, T.M.: Quasiperiodically Forced Damped Pendula and Schrodinger Equations with Quasi-periodic Potentials: Implications of Their Equivalence. Phys. Rev. Lett. **55**, 2103, (1985)

BP00. Bousso, R., Polchinski. J.: Quantization of four-form fluxes and dynamical neutralization of the cosmological constant. JHEP **06**, 006, (2000)

BP02. Benvenuto, N., Piazza, F.: On the complex backpropagation algorithm. IEEE Trans. Sig. Proc., **40**(4), 967–969, (1992)

BP02. Barahona, M., Pecora, L.M.: Synchronization in Small–World Systems. Phys. Rev. Lett. **89**, 054101, (2002)

BP82. Barone, A., Paterno, G.: Physics and Applications of the Josephson Effect. Wiley, New York, (1982)

BP97. Badii, R. Politi, A: Complexity: Hierarchical Structures and Scaling in Physics, Cambridge Univ. Press, Cambridge, (1997)

BPM97. D. Bouwmeester, J.-W. Pan, K. Mattle, M. Eibl, H.Weinfurter, A. Zeilinger, Experimental Quantum Teleportation. Nature **390**, 575, (1997)

BPS75. Belavin, A.A., Polyakov, A.M., Swartz, A.S., Tyupkin, Yu.S.: SU(2) instantpons discovered. Phys. Lett. B **59**, 85, (1975)

BPS98. Blackmore, D.L., Prykarpatsky, Y.A., Samulyak, R.V.: The Integrability of Lie-invariant Geometric Objects Generated by Ideals in the Grassmann Algebra. J. Nonlin. Math. Phys., **5**(1), 54–67, (1998)

BPT94. Van den Broeck, C., Parrondo, J.M.R., Toral, R.: Noise–Induced Non–equilibrium Phase Transition. Phys. Rev. Lett. **73**, 3395–3398, (1994)

BPT97. Van den Broeck, C., Parrondo, J.M.R., Toral, R., Kawai, R.: Non–equilibrium phase transitions induced by multiplicative noise. Phys. Rev. E **55**, 4084–4094, (1997)

BPV03. Batista, A.M., de Pinto, S.E., Viana, R.L., Lopes, S.R.: Mode locking in small–world networks of coupled circle maps. Physica A **322**, 118, (2003)

BR75. Bowen, R., Ruelle, D.: The ergodic theory of Axiom A flows. Invent. Math. **29**, 181–202, (1975)

BRR95. Botina, J., Rabitz, H., Rahman, N.: Optimal control of chaotic Hamiltonian dynamics. Phys. Rev. A **51**, 923–933, (1995)

BRT89. Birmingham, D., Rakowski, M., Thompson, G. BRST quantization of topological field theories. Nucl. Phys. B **315**, 577, (1989)

BRT99. Bennati, E., Rosa-Clot, M., Taddei, S.: A Path Integral Approach to Derivative Security Pricing: I. Formalism and Analytical Results, Int. Journ. Theor. Appl. Finance **2**, 381, (1999)

BS00. Becskei, A., Serrano, L.: Engineering stability in gene networks by autoregulation. Nature, **405**, 590–593, (2000)

BS02. Bornholdt, S., Schuster, H.G. (eds): Handbook of Graphs and Networks. Wiley–VCH, Weinheim, (2002)

BS04. Banos, B., Swann, A.: Potentials for hyper–Kähler metrics with torsion, Class. Quant. Grav. **21**, 3127–3135, (2004)

BS04. Breuer, J., Sinha, S.: Controlling spatio-temporal chaos in excitable media by local biphasic stimulation. arXiv:nlin.CD/0406047, (2004)

BS04.	Bashkirov, D., Sardanashvily, G.: Covariant Hamiltonian Field Theory. Path Integral Quantization. Int. J. Theor. Phys. **43**, 1317-1333, (2004)
BS70.	Bhatia, N.P., Szego, G.P.: Stability theory of dynamical systems. Springer–Verlag, Heidelberg, (1979)
BS85.	Bamber, D., van Santen. J.P.H.: How many parameters can a model have and still be testable? J. Math. Psych., **29**, 443–473, (1985)
BS93.	Beck, C., Schlögl, F.: Thermodynamics of chaotic systems. Cambridge Univ. Press, Cambridge, (1993)
BS95.	Bakker, B.V., Smit. J.: Curvature and scaling in 4D dynamical triangulation. Nucl. Phys. **B439**, 239, (1995)
BSM97.	Breitenbach, G., Schiller, S., Mlynek. J.: Measurement of the quantum states of squeezed light. Nature, **387**, 471–475 (1997)
BT00.	Bellet, L.R., Thomas, L.E.: Asymptotic Behavior of Thermal Non-Equilibrium Steady States for a Driven Chain of Anharmonic Oscillators. Commun. Math. Phys. **215**, 1–24, (2000)
BT03.	Bauchau, O.A., Trainelli, L.: The Vectorial Parameterization of Rotation. Non. Dyn., **31**(1), 71–92, (2003)
BT86.	Barrow. J., Tipler, F.: The Anthropic Cosmological Principle. Oxford University Press, Oxford, (1986)
BT87.	Binney, J., Tremaine, S.: Galactic Dynamics. Princeton Univ. Press, Princeton, NJ, (1987)
BT88.	Brown. J.D., Teitelboim, C.: Neutralization of the Cosmological Constant by Membrane Creation. Nucl. Phys. **B297**, 787, (1988)
BT93.	Busemeyer. J.R., Townsend. J.T.: Decision field theory: A dynamic–cognitive approach to decision making in an uncertain environment. Psych. Rev., **100**, 432–459, (1993)
BW92.	C.H. Bennett, S.J. Wiesner: Communication via 1– and 2–Particle Operators on Einstein–Podolsky–Rosen States, Phys. Rev. Lett. **69**, 2881, (1992)
Bae96a.	Baez. J.C.: Spin Network States in Gauge Theory. Adv. Math. **117**, 53, (1996)
Bae96b.	Baez. J.C.: Spin Networks in Nonperturbative Quantum Gravity. In Kauffman, LH, (ed.) The Interface of Knots and Physics. Am. Math. Soc., Providence, Rhode Island, (1996)
Bae96c.	Baez. J.C.: 4-Dimensional BF Theory as a Topological Quantum Field Theory, Lett. Math. Phys. **38**. 128, (1996)
Bae98.	Von Baeyer, H.C.: All Shook Up. The Sciences, **38**(1), 12–14, (1998)
Bai89.	Bai-Iin, H.: Elentary Symbolic Dynamics and Chaos in Dissipative Systems. World Scientific, Singapore, (1989)
Bal06.	Balakrishnan, J.: A geometric framework for phase synchronization in coupled noisy nonlinear systems. Phys. Rev. E **73**, 036206, (2006)
Bal87.	Ballentine, L.E.: Foundations of Quantum Mechanics Since the Bell Inequalities. Amer. J. Phys. **55**, 785, (1987)
Ban07.	Banos, B.: Monge–Ampere equations and generalized complex geometry. J. Geom. Phys. **57**, 841, (2007)

728 References

Ban84. Bando, S.: On the three dimensional compact Kähler manifolds of nonnegative bisectional curvature. J. Diff. Geom. **19**, 283–297, (1984)

Ban05. Banks, T., Johnson, M: Regulating Eternal Inflation. arXiv:hep-th/0512141, (2005)

Bar79. Barker, J.A.: A quantum–statistical Monte Carlo method; path integrals with boundary conditions. J. Chem. Phys. **70**, 2914, (1979)

Bar94. Barbero, F.: Real-polynomial formulation of general relativity in terms of connections. Phys. Rev. **D49**, 6935–6938, (1994)

Bar95a. Barbero, F.: Real Ashtekar Variables for Lorentzian Signature Space–Times. Phys. Rev. **D 51**, 5507–5510, (1995)

Bar95b. Barbero, F.: Reality Conditions and Ashtekar Variables: a Different Perspective. Phys. Rev. **D 51**, 5498–5506, (1995)

Bar97. Barbieri, A.: Quantum tetrahedra and simplicial spin networks. arXiv:gr-qc/9707010, (1997)

Bau00. Baum, H.: Twistor and Killing spinors on Lorentzian manifolds and their relations to CR and Kaehler geometry. Int. Congr. Diff. Geom. In memory of Alfred Gray, Bilbao, Spain, (2000)

Bax82. Baxter, R.J.: Exactly solved models in statistical mechanics. Academic Press, (1982)

Bea95. Bear, M. F. et. al. (eds): Neuroscience: Exploring The Brain. Williams and Wilkins, Baltimore, (1995)

Bel64. Bell, J.S: On the Einstein Podolsky Rosen Paradox. Physics **1**, 195, (1964)

Bel66. Bell, J.S: On the problem of hidden variables in quantum mechanics. Rev. Mod. Phys. **38**, 447, (1966)

Bel81. Belbruno, E.A.: A new family of periodic orbits for the restricted problem, Celestial Mech., **25**, 397–415, (1981)

Bel87. Bell, J.S: Speakable and Unspeakable in Quantum Mechanics. Cambridge Univ. Press, Cambridge, (1987)

Ben67. Bénabou. J.: Introduction to bicategories. In: Lecture Notes in Mathematics. Springer, New York, (1967)

Ben73. Bennett, C.H.: Logical Reversibility of Computation, IBM J. Res. Develop. **17**, 525, (1973)

Ben80. P. Benioff: Quantum mechanical Hamiltonian models of discrete processes. J. Stat. Phys. **22**, 563–591, (1980)

Ben84. Benettin, G.: Power law behavior of Lyapunov exponents in some conservative dynamical systems. Physica D **13**, 211–213, (1984)

Ben96. Bennett, C. *et al.*: 4–Year COBE DMR Cosmic Microwave Background Observations: Maps and Basic Results, Ap. J. **464**, L1, (1996)

Ber00. De Bernardis, P. et. al.: A Flat Universe from High–Resolution Maps of the Cosmic Microwave Background Radiation, Nature **404**, 955, (2000)

Ber65. Berger, M.: Sur les *variétés* d'Einstein compactes. C. R. IIIe *Réunion* Math. Expression latine, Namur, 35–55, (1965)

Ber74.	Berezin, F.: Sov. Math. Izv. 38, 1116, (1974); Sov. Math. Izv. 39, 363, (1975); Comm. Math. Phys. 40, 153, (1975); Comm. Math. Phys. 63, 131, (1978)
Ber84.	Berry, M.V.: Quantal phase factor accompanying adiabatic change. Proc. R. Soc. Lond. A **392**, 45–57, (1984)
Ber85.	Berry, M.V.: Classical adiabatic angles and quantal adiabatic phase. J. Phys. A **18**, 15–27, (1985)
Ber91.	Bernier, O: Stochastic analysis of synchronous neural networks with asymmetric weights. Europhys. Lett. **16**, 531–536, (1991)
Bil65.	Billingsley, P.: Ergodic theory and information, Wiley, New York, (1965)
Bir15.	Birkhoff, G.D.: The restricted problem of three–bodies, Rend. Circolo Mat. Palermo, **39**, 255–334, (1915)
Bir17.	Birkhoff, G.D.: Dynamical Systems with Two Degrees of Freedom. Trans. Amer. Math. Soc. **18**, 199–300, (1917)
Bir27.	Birkhoff, G.D.: Dynamical Systems, Amer. Math. Soc., Providence, RI, (1927)
Bir82.	Birrell, N.D., Davies, P.C.W.: Quantum Fields in Curved Space. Cambridge Univ. Press, Cambridge, UK, (1982)
Bla83.	Blattner, R.: Nonlinear Partial Differential Operators and Quantization Procedure, in Proceedings, Clausthall 1981, Springer–Verlag, New York, 209-241, (1983)
Bla84.	Blanchard, P.: Complex analytic dynamics on the Riemann sphere. Bull. AMS **11**, 85–141, (1984)
Bob92.	Bobbert, P.A.: Simulation of vortex motion in underdamped twoo-dimensional arrays of Josephson junctions. Phys. Rev. B **45**, 7540–7543, (1992)
Boc04.	Boccara, N.: Modeling complex systems. Springer, Berlin, (2004)
Boh35.	Bohr, N.: Can Quantum Mechanical Description of Physical Reality be Considered Complete? Phys. Rev. **48**, 696, (1935)
Boh61.	Bohr, N.: Atomic Theory and the Description of Nature. Cambridge Univ. Press, Cambridge, (1961)
Boh92.	Bohm, D.: Thought as a System. Routledge, London, (1992)
Bol01.	Bollobás, B.: Random Graphs, (2nd ed) Cambridge Univ. Press, Cambridge, (2001)
Bon95.	Bontempi, G.: Modelling with uncertainty in continuous dynamical systems: the probability and possibility approach. IRIDIA – ULB Technical Report, 95–16, (1995)
Boo86.	Boothby, W.M.: An Introduction to Differentiable Manifolds and Riemannian Geometry, Academic Press, New York, (1986)
Bot59.	Bott, R.: The Stable Homotopy of the Classical Groups, Ann. Math. **70**, 313–337, (1959)
Bot82.	R. Bott, Bull. Amer. Math. Soc. **7**, 331, (1982)
Bou05.	Bousso, R.: Cosmology and the S-matrix. Phys. Rev. D **71**, 064024, (2005)
Bow70.	Bowen, R.: Markov partitions for Axiom A diffeomorphisms. Amer. J. Math. **92**, 725–747, (1970)

Bow73.	Bowen, R.: Symbolic dynamics for hyperbolic flows. Amer. J. Math. **95**, 429–460, (1973)
Bow75.	Bowen, R.: Equilibrium states and the ergodic theory of Anosov diffeomorphisms. Lecture Notes in Math. **470**, Springer–Verlag, Berlin, (1975)
Bp00.	Bousso, R., Polchinski. J.: Quantization of Four Form Fluxes and Dynamical Neutralization of the Cosmological Constant. JHEP **0006**, 006, (2000)
Bre04.	Brent, R.: A partnership between biology and engineering. Nature, **22**, 1211–1214, (2004)
Bro58.	Broadbent, D.E.: Perception and communications. Pergamon Press, London, (1958)
Bro73.	Bröcker, T.H., Jänich, K.: Introduction to differential topology. Cambridge University Press, Cambridge, (1973)
Bro85.	Brooks, R.A.: A Robust Layered Control System for a Mobile Robot. IEEE Trans. Rob. Aut., **2**(1), 14—23, (1986)
Bro90.	Brooks, R.A.: Elephants Don't Play Chess. Rob. Aut. Sys. **6**, 3—15, (1990)
Bru00.	Brunel, N.: Dynamics of sparsely connected networks of excitatory and inhibitory spiking neurons. J. Comp. Neurosci. **8**, 183–208, (2000)
Bry04.	Bryant, R.: Gradient Kähler Ricci Solitons. arXiv:math/0407453, (2004)
Buc02.	Bucher, M.: A braneworld universe from colliding bubbles. Phys. Lett. B **530**, 1, (2002)
Buc41.	Buchanan, D.: Trojan satellites – limiting case. Trans. Roy. Soc. Canada Sect. III, **35**(3), 9–25, (1941)
Byt72.	Bytev, V.O.: Group–theoretical properties of the Navier–Stokes equations, Numerical Methods of Continuum Mechanics, **3**(3), 13–17 (in Russian), (1972)
CA93.	Chialvo, D.R., Apkarian, V.: Modulated noisy biological dynamics: Three examples. Journal of Statistical Physics, **70**, 375–391, (1993)
CA97.	Chen, L., Aihara, K.: Chaos and asymptotical stability in discrete-time neural networks, Physica D **104**, 286–325, (1997)
CC99.	Cao, H.D., Chow, B.: Recent Developments on the Ricci Flow. Bull. Amer. Math. Soc. **36**, 59–74, (1999)
CAM05.	Cvitanovic, P., Artuso, R., Mainieri, R., Tanner, G., Vattay, G.: Chaos: Classical and Quantum. ChaosBook.org, Niels Bohr Institute, Copenhagen, (2005)
CB88.	Casson, A., Bleiler, S.: Automorphisms of surfaces, after Nielsen and Thurston, volume 9 of London Mathematical Society Student Texts. Cambridge Univ. Press, (1988)
CB89.	Cao, J.S., Berne, B.J.: On energy estimators in path integral Monte Carlo simulations: Dependence of accuracy on algorithm. J. Chem. Phys. **91**, 6359, (1989)
CB97.	Cleve, R., Buhrman, H.: Substituting Quantum Entanglement for Communication, Phys. Rev. A **56**, 1201, (1997)

CBP05.	Catanzaro, M., Boguñá, M., Pastor-Satorras, R.: Generation of uncorrelated random scale-free networks. Phys. Rev. E **71**, 027103, (2005)
CBS91.	Chan, C.K., Brumer, P., Shapiro, M.: Coherent radiative control of IBr photodissociation via simultaneous (w1,w3) excitation. J. Chem. Phys. **94**, 2688–2696, (1991)
CC70.	Chapman, S., Cowling, T.: Mathematical Theory of Non-Uniform Gases. Cambridge Univ. Press, Cambridge, (1970)
CC77.	Callan, C., Coleman, S.: Fate of the false vacuum. II. First quantum corrections. Phys. Rev. D **16**, 1762, (1977)
CC95.	Christini, D.J., Collins, J.J.: Controlling Nonchaotic Neuronal Noise Using Chaos Control Techniques. Phys. Rev. Lett. **75**, 2782–2785, (1995)
CC95.	Christini, D.J., Collins, J.J.: Using noise and chaos control to control nonchaotic systems. Phys. Rev. E **52**, 5806–5809 (1995)
CC96.	Cheeger. J., Colding, T.H.: Lower bounds on Ricci curvature and almost rigidity of wraped products. Ann. Math. **144**, 189–237, (1996)
CC96.	Casetti, L., Clementi, C., Pettini, M.: Riemannian theory of Hamiltonian chaos and Lyapunov exponents. Phys. Rev. E **54**, 5969–5984, (1996)
CC96.	Chamseddine, A.H., Connes, A.: The Spectral Action Principle. Phys. Rev. Lett. **24**, 4868–4871, (1996)
CCC97.	L. Caiani, L. Casetti, C. Clementi, M. Pettini: Geometry of Dynamics, Lyapunov Exponents, and Phase Transitions. Phys. Rev. Lett. **79**, 4361–4364 (1997)
CCC97.	Caiani, L., Casetti, L., Clementi, C., Pettini, M.: Geometry of dynamics, Lyapunov exponents and phase transitions. Phys. Rev. Lett. **79**, 4361, (1997)
CCC98.	Caiani, L., Casetti, L., Clementi, C., Pettini, G., Pettini, M., Gatto, R.: Geometry of dynamics and phase transitions in classical lattice φ^4 theories Phys. Rev. E **57**, 3886, (1998)
CCI91.	Cariñena. J., Crampin, M., Ibort, L.: On the multisymplectic formalism for first order field theories. Diff. Geom. Appl. **1**, 345, (1991)
CCN85.	Conway, J.H., Curtis, R.T., Norton, S.P., Parker, R.A., Wilson, R.A: Atlas of Finite Groups: Maximal Subgroups and Ordinary Characters for Simple Groups. Clarendon Press, Oxford, (1985)
CCP02a.	Casetti, L., Cohen, E.G.D., Pettini, M.: Phase transitions as topology changes in configuration space: an exact result. Phys. Rev. E **65**, 036112 (2002)
CCP02b.	Casetti, L., Cohen, E.G.D., Pettini, M.: Exact result on topology and phase transitions at any finite N. Phys. Rev. E **65**, 036112 (2002)
CCP96.	Christiansen, F., Cvitanovic, P., Putkaradze, V.: Hopf's last hope: spatio-temporal chaos in terms of unstable recurrent patterns. Nonlinearity, **10**, 1, (1997)

CCP96. Casetti, L., Clementi, C., Pettini, M.: Riemannian theory of Hamiltonian chaos and Lyapunov exponents. Phys. Rev. E **54**, 5969–5984, (1996)

CCP99. Casetti, L., Cohen, E.G.D., Pettini, M.: Topological origin of the phase transition in a mean-field model. Phys. Rev. Lett. **82**, 4160–4163 (1999)

CCT87. Chaiken, J., Chu, C.K., Tabor, M., Tan, Q.M.: Lagrangian Turbulence in Stokes Flow. Phys. Fluids, **30**, 687, (1987)

CCV98. Castellanos, A., Coolen, A.C., Viana, L.: Finite-size effects in separable recurrent neural networks. J. Phys. A: Math. Gen. **31**, 6615–6634, (1998)

CD98. Chen, G., Dong, X.: From Chaos to Order. Methodologies, Perspectives and Application. World Scientific, Singapore, (1998)

CDM01. Colless, M., Dalton, G., Maddox, S., Sutherland, W.: The 2dF Galaxy Redshift Survey: Spectra and Redshifts, MNRAS 328, 1039–1063, (2001)

CDS98. Connes, A., Douglas, M., Schwarz, A.: Noncommutative Geometry and Matrix Theory: Compactification on Tori. JHEP **9802**, 003, (1998)

CE75. Cheeger, J., Ebin, D.G.: Comparison theorems in Riemannian geometry. North Holland, Amsterdam, (1975)

CE91. Cvitanovic, P., Eckhardt, B.: Periodic orbit expansions for classical smooth flows. J. Phys. A **24**, L237, (1991)

CE97. Chandre, C., Eilbeck, J.C.: Does the existence of a Lax pair imply integrability. Preprint, (1997)

CEG00. Csaki, C., Erlich. J., Grojean, C., Hollowood, T.J.: General properties of the self-tuning domain wall approach to the cosmological constant problem. Nucl. Phys. B **584**, 359, (2000)

CEH98. Cirac, J.I., Ekert, A., Huelga, S.F., Macchiavello, C.: On the Improvement of Frequency Stardards with Quantum Entanglement. Phys. Rev. A **59**, 4249, (1999)

CF93. Cariñena, J., Fernández–Núñez, J.: Geometric theory of time-dependent singular Lagrangians, Fortschr. Phys. **41**, 517, (1993)

CF94. Crane, L., Frenkel, I.: Four dimensional topological quantum field theory, Hopf categories, and the canonical bases. J. Math. Phys. **35**, 5136–5154, (1994)

CFD86. Coalson, R.D., Freeman, D.L., Doll, J.D.: Partial Averaging Approach to Fourier. Coefficient Path Integration. J. Chem. Phys. **85**, 4567–4583, (1986)

CFL97. Crisanti, A. Falcioni, M., Lacorata, G., Purini, R., Vulpiani, A.: Characterization of a periodically driven chaotic dynamical system. J. Phys. A, Math. Gen. **30**, 371–383, (1997)

CFP90. Crisanti, A. Falcioni, M., Provenzale, A., Vulpiani, A.: Passive advection of particles denser than the surrounding fluid. Phys. Lett. A **150**, 79, (1990)

CFP91. Crisanti, A., Falcioni, M., Paladin, G., Vulpiani, A.: Lagrangian Chaos: Transport, Mixing and Diffusion in Fluids. Riv. Nuovo Cim. **14**, 1, (1991)

CFP94.	Crisanti, A., Falcioni, M., Paladin, G., Vulpiani, A.: Stochastic Resonance in Deterministic Chaotic Systems. J. Phys. A **27**, L597, (1994)
CFR79.	Clark, R.A., Ferziger, J.H., Reynolds, W.C.: Evaluation of Subgrid–Scale Turbulence Models Using an Accurately Simulated Turbulent Flow. J. Fluid. Mech. **91**, 1–16, (1979)
CG83.	Cohen, M.A., Grossberg, S.: Absolute stability of global pattern formation and parallel memory storage by competitive neural networks. IEEE Trans. Syst., Man, Cybern., **13**(5), 815–826, (1983)
CG94.	Crisanti, A., Grassberger, P.: Critical behaviour of non-equilibrium q-state systems. J. Phys. A: Math. Gen. **27**, 6955–6962, (1994)
CGC98.	Chakravarty, C., Gordillo, M.C., Ceperley, D.M.: Comparing Fourier– and Bisection–Path Integral Monte Carlo. J. Chem. Phys. **109**, 2123, (1998)
CGC99.	Chakravarty, C., Gordillo, M.C., Ceperley, D.M.: A Comparison of the Efficiency of Fourier– and Discrete Time–Path Integral Monte Carlo. J. Chem. Phys. **111**, 7687, (1999)
CGI95.	Cariñena. J., Gomis. J., Ibort, L. and Román, N.: Canonical transformation theory for presymplectic systems. J. Math. Phys. **26**, 1961, (1985)
CGK99.	Csaki, C., Graesser, M., Kolda, C.F., Terning. J.: Cosmology of one extra dimension with localized gravity. Phys. Lett. B **462**, 34, (1999)
CGP88.	Cvitanovic, P., Gunaratne, G., Procaccia, I.: Topological and metric properties of Hénon-type strange attractors. Phys. Rev. A **38**, 1503–1520, (1988)
CGP99.	Casetti, L., Gatto, R., Pettini, M.: Geometric approach to chaos in the classical dynamics of Abelian lattice gauge theory. J. Phys. A **32**, 3055, (1999)
CGR00.	Csaki, C., Graesser, M., Randall, L., Terning. J.: Cosmology of brane models with radion stabilization. Phys. Rev. D **62**, 045015, (2000)
CGR00.	Charmousis, C., Gregory, R., Rubakov, V.A., Wave function of the radion in a brane world.. Phys. Rev. D **62**, 067505, (2000)
CGS99.	Cline. J.M., Grojean, C., Servant, G.: Cosmological expansion in the presence of extra dimensions. Phys. Rev. Lett. **83**, 4245, (1999)
CH01.	De Carvalho, A., Hall, T.: Pruning theory and thurston's classification of surface homeomorphisms. J. Eur. Math. Soc., **3**(4), 287–333, (2001)
CH02.	De Carvalho, A., Hall, T.: The forcing relation for horseshoe braind types. Experimental Math., **11**(2), 271–288, (2002)
CH03.	De Carvalho, A., Hall, T.: Conjugacies between horseshoe braids. Nonlinearity, **16**, 1329–1338, (2003)
CH04.	Craig, D., Hartle, J.B.: Generalized Quantum Theories of Recollapsing, Homogeneous Cosmologies. Phys. Rev. D **69**, 123525–123547, (2004)

CH64. Conway, E.D., Hopf, E.: Hamilton's Theory and Generalized Solutions of the Hamilton–Jacobi Equation. J. Math. Mech. **13**, 939–986, (1964)

CH93. Cross, M.C., Hohenberg, P.C.: Pattern formation outside of equilibrium. Rev. Mod. Phys. **65**, 851– 1112, (1993)

CH95. Chow, C.C., Hwa, T.: Defect-mediated Stability: An Effective Hydrodynamic Theory of Spatiotemporal Chaos. Physica D **84**, 494, (1995)

CHR00. Chamblin, A., Hawking, S.W., Reall, H.S.: Brane-world black holes. Phys. Rev. D **61**, 065007, (2000)

CHS85. Candelas, P., Horowitz, G.T., Strominger, A., Witten, E.: Vacuum configurations for superstrings, Nucl. Phys. B **258**, 46, (1985)

CJP93. Crisanti, A, Jensen, M.H., Paladin, G., Vulpiani, A.: Intermittency and Predictability in Turbulence. Phys. Rev. Lett. **70**, 166, (1993)

CK05. Cannone, M., Karch, G.: About the regularized Navier–Stokes equations, J. Math. Fluid Mech., **7**, 1–28, (2005)

CK97. Cutler, C.D., Kaplan, D.T.: (eds): Nonlinear Dynamics and Time Series. Fields Inst. Comm. **11**, American Mathematical Society, (1997)

CKN82. Caffarelli, L., Kohn, R., Nirenberg, L.: Partial regularity of suitable weak solutions of the Navier-Stokes equations, Comm. Pure Appl. Math. **35**, 771–831, (1982)

CKS05. Coolen, A.C.C., Kuehn, R., Sollich, P.: Theory of Neural Information Processing Systems. Oxford Univ. Press, Oxford, (2005)

CL72. Craik, F., Lockhart, R.: Levels of processing: A framework for memory research. J. Verb. Learn. & Verb. Behav., **11**, 671–684, (1972)

CL80. Coleman, S., De Luccia, F.: Gravitational effects on and of vacuum decay. Phys. Rev. D **21**, 3314, (1980)

CL81. Caldeira, A.O., Leggett, A.J.: Influence of Dissipation on Quantum Tunneling in Macroscopic Systems. Phys. Rev. Lett. **46**, 211, (1981)

CL84. Cheng, T.-P., Li, L.-F.: Gauge Theory of Elementary Particle Physics. Clarendon Press, Oxford, (1984)

CL85. Caldeira, A.O., Leggett, A.J.: Influence of damping on quantum interference: An exactly soluble model. Phys. Rev. A **31**, 1059–1066, (1985)

CL90. Connes, A., Lott. J.: Particle models and non-commutative geometry. Nucl. Phys. **B18**, 29–47, (1990)

CLL01. Copeland, E.J., Liddle, A.R., Lidsey. J.E.: Steep inflation: Ending braneworld inflation by gravitational particle production. Phys. Rev. D **64**, 023509, (2001)

CLM94. Chinea, D., de León, M., Marrero. J.: The constraint algorithm for time-dependent Lagrangians. J. Math. Phys. **35**, 3410, (1994)

CLP95.	Casetti, L., Livi, R., Pettini, M.: Gaussian Model for Chaotic Instability of Hamiltonian Flows. Phys. Rev. Lett. **74**, 375–378, (1995)
CLS96.	Colombo, G. Lautman, D., Shapiru I.I.: The Earth's dust belt: Fact or fiction? Gravitational focusing and Jacobi capture. J. Geophys. Res. **71**, 5705–5717, (1996)
CM97.	Carpinteri, A., Mainardi, F.: Fractals and Fractional Calculus in Continuum Mechanics. Springer Verlag, Wien & New York, (1997)
CMP00.	Carballo, C.M., Morales, C.A., Pacifico, M.J.: Homoclinic classes for generic C^1 vector fields. Preprint MAT. **07**, PUC-Rio, (2000)
CMP98a.	Claussen, J.C., Mausbach, T., Piel, A. Schuster, H.G.: Improved difference control of unknown unstable fixed–points: Drifting parameter conditions and delayed measurement. Phys. Rev. E, **58**(6), 7256–7260, (1998)
CMP98b.	Claussen, J.C., Mausbach, T., Piel, A. Schuster, H.G.: Memory difference control of unknown unstable fixed–points: Drifting parameter conditions and delayed measurement. Phys. Rev. E **58**(6), 7260–7273, (1998)
CMV00.	Castellano, C., Marsili, M., Vespignani, A.: Nonequilibrium Phase Transition in a Model for Social Influence. Phys. Rev. Lett. **85**, 3536–3539, (2000)
CN01.	Cottrill-Shepherd, K., Naber M.: Fractional Differential Forms. J. Math. Phys. **42**, 2203–2212, (2001)
CN95.	Carlip, S., Nelson. J.E.: Comparative Quantizations of (2+1)–Dimensional Gravity. Phys. Rev. D **51**, 5643, (1995)
CNH03.	Chiorescu, I., Nakamura, Y., Harmans, C., Mooij. J.: Coherent Quantum Dynamics of a Superconducting Flux Orbit, Science, **299**, 1869, (2003)
CP02.	Clementi, C. Pettini, M.: A geometric interpretation of integrable motions. Celest. Mech. & Dyn. Astr., **84**, 263–281, (2002)
CP93.	Casetti, L., Pettini, M.: Analytic computation of the strong stochasticity threshold in Hamiltonian dynamics using Riemannian geometry. Phys. Rev. E **48**, 4320–4332, (1993)
CP95.	Cerruti-Sola, M., Pettini, M.: Geometric description of chaos in self-gravitating systems. Phys. Rev. E **51**, 53–64, (1995)
CP96.	Cerruti-Sola, M., Pettini, M.: Geometric description of chaos in two-degrees-of-freedom Hamiltonian systems. Phys. Rev. E **53**, 179–188, (1996)
CPC00.	Casetti, L., Pettini, M., Cohen, E.G.D.: Geometric Approach to Hamiltonian Dynamics and Statistical Mechanics. Phys. Rep. **337**, 237—341, (2000)
CPC03.	Casetti, L., Pettini, M., Cohen, E.G.D.: Phase transitions and topology changes in configuration space. J. Stat. Phys., **111**, 1091, (2003)
CPP02.	García–Compeán, H., Plebański. J., Przanowski, M., Turrubiates, F.: Deformation Quantization of Geometric Quantum Mechanics. J. Phys. **A35**, 4301, (2002)

CPS93.	Coolen, A.C., Penney, R.W., Sherrington, D.: Coupled dynamics of fast spins and slow interactions: An alternative perspective on replicas. Phys. Rev. B **48**, 16116–16118, (1993)
CR00.	Cao, H.D., Hamilton, R.S.: Gradient Kähler-Ricci solitons and periodic orbits, Comm. Anal. Geom. **8**, 517–529, (2000)
CR88.	Cohen, E.G.D., Rondoni, L.: Note on phase–space contraction and entropy production in thermostatted Hamiltonian systems. Chaos **8**, 357–365, (1998)
CR89.	Cariñena. J., Rañada, M.: Poisson maps and canonical transformations for time-dependent Hamiltonian systems. J. Math. Phys. **30**, 2258, (1989)
CR93.	Cariñena. J., Rañada, M.: Lagrangian systems with constraints. A geometric approach to the method of Lagrange multipliers. J. Phys. A. **26**, 1335, (1993)
CR99.	Chamblin, H.A., Reall, H.S.: Dynamic dilatonic domain walls, Nucl. Phys. B **562**, 133, (1999)
CRV92.	Celeghini, E., Rasetti, M., Vitiello, G.: Quantum Dissipation. Annals Phys. **215**, 156, (1992)
CS83.	Coppersmith, S.N., Fisher, D.S.: Pinning transition of the discrete sine–Gordon equation. Phys. Rev. B **28**, 2566–2581, (1983)
CS90.	Clauser, J.F., Shimony, A.: Bell's Theorem: Experimental Tests and Implications, Rep. Prog. Phys. **41**, 1131, (1990)
CS94.	Coolen, A.C.C., Sherrington, D.: Order–parameter flow in the fully connected Hopfield model near saturation. Phys. Rev. E **49** 1921–934; and Erratum: Order-parameter flow in the fully connected Hopfield model near saturation, 5906, (1994)
CS95.	Chernikov, A.A., Schmidt, G.: Conditions for synchronization in Josephson-junction arrays. Phys. Rev. E **52**, 3415–3419, (1995)
CS96.	Constantine, G.M., Savits, T.H.: A multivariate Faà di Bruno formula with applications. Trans. Amer. Math. Soc. **348**(2), 503–520, (1996)
CS98.	Claussen, J.C., Schuster, H.G.: Stability borders of delayed measurement from time–discrete systems. arXiv nlin. CD/0204031, (1998)
CSO99.	Chen, H., Succi, S., Orszag, S.: Analysis of Subgrid Scale Turbulence Using the Boltzmann Bhatnagar–Gross–Krook kinetic equation. Phys. Rev. E **59**, R2527–R2530, (1999)
CST86.	Cremmer, E., Schwimmer, A., Thorn, C.: The Vertex Function in Witten's Formulation of String Field Theory. Phys. Lett. **B179**, 57, (1986)
CT02.	Chen, X., Tian, G.: Ricci flow on Kahler-Einstein surfaces. Invent. Math. **147**, 487–544, (2002)
CT88.	Cardoso, O., Tabeling, P.: Anomalous diffusion in a linear array of vortices. Europhys. Lett. **7**(3), 225–230, (1988)
CV91.	Cecotti, S., Vafa, C.: Topological – antitopological fusion. Nucl. Phys. B **367**, 359, (1991)

CY89.	Crutchfield. J.P., Young, K.: Computation at the onset of chaos. In Complexity, Entropy and the Physics of Information, p. 223, SFI Studies in the Sciences Complexity, Vol. VIII, W.H. Zurek (Ed.), Addison–Wesley, Reading, MA, (1989)
CY93.	Crane, L., Yetter, D.: A categorical construction of 4D topological quantum field theories. In Quantum Topology, R. Baadhio and L.H. Kauffman ed. World Scientific, Singapore, (1993)
Cal57.	Calabi, E.: On Kähler manifolds with vanishing canonical class. In Algebraic geometry and topology: a symposium in honor of S. Lefschetz, eds. R.H. Fox, D.C. Spencer and A.W. Tucker, Princeton Univ. Press, (1957)
Can78.	Cantwell B.J.: Similarity transformations for the two-dimensional, unsteady, stream-function equation. J. Fluid Mech. **85**, 257–271, (1978)
Can86.	Canarutto, D.: Bundle splittings, connections and locally principle fibred manifolds. Bull. U.M.I. Algebra e Geometria, Seria VI **V-D**, 18, (1986)
Cao85.	Cao, H.D.: Deformation of Kähler metrics to Kähler-Einstein metrics on compact Kähler manifolds. Invent. Math., 81, 359–372, (1985)
Cao94.	Cao, H.D.: Existence of gradient Kähler-Ricci solitons, Elliptic and parabolic methods in geometry. Minneapolis, MN, (1994)
Cao97.	Cao, H.D.: Limits of solutions to the Kähler-Ricci flow. J. Diff. Geom. **45**, 257–272, (1997)
Cap74.	Capper, D.M., Duff, M.J.: Trace Anomalies in Dimensional Regularization, Nuovo Cimento **23A**, 173, (1974)
Car92.	Do Carmo, M.P.: Riemannian geometry. Birkhäuser, Boston, (1992)
Cas05.	Castellano, C.: Effect of network topology on the ordering dynamics of voter models. AIP Conf. Proc. **779**, 114, (2005)
Cas92.	Cassidy, D.: Uncertainty: The Life and Science of Werner Heisenberg. Freeman, New York, (1992)
Cas95.	Casetti, L.: Efficient symplectic algorithms for numerical simulations of Hamiltonian flows. Phys. Scr., **51**, 29, (1995)
Cav05.	Cavalcanti, G.R.: New aspects of the dd^c–lemma. arXiv:math.DG/0501406, (2005)
Cav86.	Caves, C.: Quantum Mechanics and Measurements Distributed in Time I: A Path Integral Approach. Phys. Rev. D **33**, 1643, (1986)
Cep95.	Ceperley, D.M.: Path integrals in the theory of condensed helium· Rev. Mod. Phys. **67**, 279–355, (1995)
Ch53.	Courant, R., Hilbert, D.: Methods of Mathematical Physics, Vol 1, Interscience Pub. (1953)
Ch99.	Chu, C.-S., Ho, P.-M.: Noncommutative Open String and D-brane, Nucl. Phys. **B550**, 151, (1999)
Cha02.	Charmousis, C.: Dilaton space–times with a Liouville potential. Class. Quant. Grav. **19**, 83, (2002)
Cha03.	Chamley, C.: Rational Herds: Economic Models of Social Learning, Cambridge Univ. Press, Cambridge, (2003)

References

Cha05. Chamseddine, A.H.: Hermitian Geometry and Complex Space-Time. arXiv:hep-th/0503048, (2005)

Cha43. Chandrasekhar, S.: Stochastic Problems in Physics and Astronomy. Rev. Mod. Phys. **15**, 1–89, (1943)

Cha48. Chandra, H.: Relativistic Equations for Elementary Particles. Proc. Roy. Soc. London A, **192**(1029), 195–218, (1948)

Cha97. Chalmers, D.: The Conscious Mind. Oxford Univ. Press, Oxford, (1997)

Che02. Chen, X.: Recent progress in Kähler geometry, in Proceedings of the International Congress of Mathematicians, Vol. II (Beijing, 2002), Higher Ed. Press, Beijing, 273–282, (2002)

Che46. Chern, S.S.: Characteristic classes of Hermitian manifolds, Ann. of Math. **47**, 85-121, (1946)

Che55. Chevalley, C.: Theorie differential equations groupes de Lie. Vol. 1–3. Hermann D.C., Paris, (1955)

Che70. Chernikov, Yu. A.: The photogravitational restricted three body problem. Sov. Astr. AJ. **14**(1), 176-181, (1970)

Che96. Chern, S.S.: Riemannian Geometry as a Special Case of Finsler Geometry, Cont. Math., 51–58, Vol. 196, Amer. Math. Soc. Providence, RI, (1996)

Chi79. Chirikov, B.V.: A universal instability of many-dimensional oscillator systems. Phys. Rep. **52**, 264-379, (1979)

Cho91. Chow, B.: The Ricci flow on the 2-sphere. J. Diff. Geom. **33**, 325–334, (1991)

Chu94. Chua, L.O: Chua's Circuit: an overview ten years later. J. Circ. Sys. Comp. **4**(2), 117-159, (1994)

Cla02a. Claussen, J.C.: Generalized Winner Relaxing Kohonen Feature Maps. arXiv cond–mat/0208414, (2002)

Cla02b. Claussen, J.C.: Floquet Stability Analysis of Ott-Grebogi-Yorke and Difference Control. arXiv:nlin.CD/0204060, (2002)

Cle95. Clementi, C.: Laurea Thesis in Physics. Università di Firenze, Italy, (1995)

Col02. Collins, P.: Symbolic dynamics from homoclinic tangles. Intern. J. Bifur. Chaos, **12**(3), 605–617, (2002)

Col05. Collins, P.: Forcing relations for homoclinic orbits of the Smale horseshoe map, Experimental. Math. **14**(1), 75-86, (2005)

Col69. Coleman, S.: Acausality in Theory and Phenomenology in Particle Physics, ed. A. Zichichi, New York, (1969)

Col77. Coleman, S.: Fate of the false vacuum: Semiclassical theory. Phys. Rev. D15, 2929–36, (1977)

Col80. Coleman, S., De Luccia, F.: Gravitational Effects on and of Vacuum Decay. Phys. Rev. D **21**, 3305, (1980)

Con63. Conley, C.C.: Some new long period solutions of the plane restricted body problem of three–bodies, Comm. Pure Appl. Math., **16**, 449–467, (1963)

Con94. Connes, A.: Noncommutative Geometry. Academic Press, New York, (1994)

Coo01. Coolen, A.C.C.: Statistical Mechanics of Recurrent Neural Networks. In F. Moss, S. Gielen (eds.) Handbook of Biological Physics, **4**, Elsevier, (2001)

Coo89.	Cook, J.: The mean–field theory of a Q-state neural network model. J. Phys. A **22**, 2057, (1989)
Cop35.	Copson, E.T.: Theory of Functions of a Complex Variable. Oxford Univ. Press, London, (1935)
Cox43.	Coxeter, H.S.M.: A geometrical background for de Sitter's world. Am. Math. Mont. **50**, 217–228, (1943)
Cox92.	Cox, E.: Fuzzy Fundamentals, IEEE Spectrum, 58–61, (1992)
Cox94.	Cox, E.: The Fuzzy Systems Handbook. AP Professional, (1994)
Cra04.	Crainic, M.: Generalized complex structures and Lie brackets. math.DG/0412097, (2004)
Cra91.	Crawford. J.: Clifford algebra: Notes on the spinor metric and Lorentz, Poincaré and conformal groups. J. Math. Phys. **32**, 576, (1991)
Cro80.	Croke, C.: Some Isoperimetric Inequalities and Consequences. Ann. Sci. E. N. S., Paris, **13**, 419–435, (1980)
Cve01.	Cveticanin, L.: Analytic approach for the solution of the complex valued strong nonlinear differential equation. Physica A **297**, 348-360, (2001)
Cve05.	Cveticanin, L.: Approximate solution of a strongly nonlinear complex differential equation. J. Sound Vibr. **284**, 503-512, (2005)
Cve92.	Cveticanin, L.: Approximate analytical solutions to a class of nonlinear equations with complex functions. J. Sound Vibr. **157**, 289-302, (1992)
Cve92.	Cveticanin, L.: An approximate solution for a system of two coupled differential equations. J. Sound Vibr. **152**, 375-380, (1992)
Cve93.	Cveticanin, L.: An asymptotic solution for weak nonlinear vibrations of the rotor. Mech. Mach. The. **28**, 495-505, (1993)
Cve98.	Cveticanin, L.: Analytical methods for solving strongly nonlinear differential equations. J. Sound Vibr. **214**, 325-328, (1998)
Cvi00.	Cvitanovic, P.: Chaotic field theory: a sketch. Physica A **288**, 61–80, (2000)
Cvi91.	Cvitanovic, P.: Periodic orbits as the skeleton of classical and quantum chaos. Physica, D **51**, 138, (1991)
DBO01.	DeShazer, D.J., Breban, R., Ott, E., Roy, R.: Detecting Phase Synchronization in a Chaotic Laser Array. Phys. Rev. Lett. **87**, 044101, (2001)
DC76.	Dowker. J.S., Critchley, R.: Effective Lagrangian and Energy–Momentum Tensor in de Sitter Space. Phys. Rev. D **13**, 3224, (1976)
DCS98.	Düring, A, Coolen, A.C., Sherrington, D.: Phase diagram and storage capacity of sequence processing neural networks. J. Phys. A: Math. Gen. **31**, 8607, (1998)
DD04.	Denef, F., Douglas, M.R.: Distributions of Flux Vacua. arXiv:hep-th/0404116, (2005)
DD06.	Denef, F., Douglas, M.R.: Computational complexity of the landscape. arXiv:hep-th/0602072, (2006)

740 References

DD67. Deprit, A., Deprit-Bartholomé, A.: Stability of the triangular Lagrangian points. Astron. J. **72**, 173-179, (1967)

DDK01. Deruelle, N., Dolezel, T., Katz. J.: Perturbations of brane worlds. Phys. Rev. D **63**, 083513, (2001)

DDT01. Dorey, P., Dunning, C. Tateo, R.: Supersymmetry and the spontaneous breakdown of PT-symmetry. J. Phys. A: Math. Gen. **34**, L391, (2001)

DDT03. Daniels, B.C., Dissanayake, S.T.M., Trees, B.R.: Synchronization of coupled rotators: Josephson junction ladders and the locally coupled Kuramoto model. Phys. Rev. E **67**, 026216 (2003)

DEF99. Deligne, P., Etingof, P., Freed, D.S., Jeffrey, L.C., Kazhdan, D., Morgan. J.W., Morrison, D.R., Witten, E.: Quantum Fields and Strings: A Course for Mathematicians, Am. Math. Soc., (1999)

DEH93. Derrida, B., Evans, M.R., Hakim, V., Pasquier, V.: Exact solution of a 1d asymmetric exclusion model using a matrix formulation. J. Phys. A: Math. Gen. A **26**, 1493–1517, (1993)

DK79. Duru, I.H., Kleinert, H.: Solution of the path integral for the H–atom. Phys. Lett. B **84**, 185–188, (1979)

DF84. Doll, J.D., Freeman, D.L.: A Monte-Carlo Method for Quantum-Boltzmann Statistical Mechanics. J. Chem. Phys. **80**, 2239–2240, (1984)

Deu85. Deutsch, D.: Quantum theory, the Church-Turing principle and the universal quantum computer. Proc. Roy. Soc. (London), A **400**, 97–117, (1985); also, Feynman R.P., Quantum mechanical Computers, Found. of Phys., **16**(6), 507-31, (1985)

DF99. Doll, J.D., Freeman, D.L.: A Monte–Carlo Method for Quantum–Boltzmann Statistical Mechanics. J. Chem. Phys. **111**, 7685, (1999)

DFM94. Dotsenko, V., Franz, S., Mézard, M.: Memorizing polymers' shapes and permutations. J. Phys. A: Math. Gen. **27**, 2351, (1994)

DFN92. Dubrovin, B.A., Fomenko, A.N., Novikov, S.P.: Modern Geometry – Methods and Applications. Part I. Springer-Verlag, New York, (1992)

DFN95. B.A.Dubrovin, A.T.Fomenko, and S.P.Novikov, Modern Geometry-Methods and applications. Part II. Springer-Verlag, New York, 1995)

DFR94. Doplicher, S., Fredenhagen, K., Roberts. J.E.: Space-time quantization inducedby classical gravity. Phys. Lett. **B331**, 39-44, (1994)

DG03. Dragovic, V., Gajic, B.: The Wagner Curvature Tensor in Nonholonomic Mechanics. Reg. Chaot. Dyn., **8**(1), 105–124, (2003)

DG98. Dowker, H.F., Garcia, R.S.: A handle-body calculus for topology change. Class. Quant. Grav. **15**, 1859–1879, 1998.

DGG94. Dowker, H.F., Gauntlett, J.P., Giddings, S.B., Horowitz, G.T.: On pair creation of extremal black holes and kaluza-klein monopoles. Phys. Rev. **D50**, 2662–2679, (1994)

DGG95.	Dowker, H.F., Gauntlett, J.P., Giddings, S.B., Horowitz, G.T.: The decay of magnetic fields in kaluza-klein theory. Phys. Rev. **D52**, 6929–6940, 1995.
DGG96.	Dowker, H.F., Gauntlett, J.P., Giddings, S.B., Horowitz, G.T.: Nucleation of p-branes and fundamental strings. Phys. Rev. **D53**, 7115–7128, 1996.
DGO89a.	Ding, M., Grebogi, C., Ott, E.: Evolution of attractors in quasi–periodically forced systems: From quasiperiodic to strange nonchaotic to chaotic. Phys. Rev. A **39**, 2593–2598, (1989)
DGO89b.	Ding, M., Grebogi, C., Ott, E.: The Dimension of Strange Nonchaotic Attractors. Phys. Lett. A **137**, 167–172, (1989)
DGS00a.	H. F. Dowker, R. S. Garcia and S. Surya: Morse index and causal continuity: A criterion for topology change in quantum gravity, Class. Quant. Grav. 17 (2000) 697-712.
DGS00b.	Dowker, H.F., Garcia, R.S., Surya, S.: K–causality and degenerate space-times. Class. Quant. Grav. **17**, 4377–4396, (2000)
DGS99.	Dowker, H.F., Garcia, R.S., Surya, S.: Morse index and causal continuity: A criterion for topology change in quantum gravity. Class. Quant. Grav. **17**, 697–712, (2000)
DGV95.	Diekmann, O., van Gils, S.A., Verduyn-Lunel, S.M., Walther, H.-O.: Delay Equations. Appl. Math. Sci. **110**, Springer–Verlag, New York, (1995)
DGY97.	Ding, M., Grebogi, C., Yorke, J.A.: Chaotic dynamics. In The Impact of Chaos on Science and Society, Eds. C. Grebogi and J.A. Yorke, 1–17, United Nations Univ. Press, Tokyo, (1997)
DGZ87.	Derrida, B., Gardner, E., Zippelius, A.: An exactly solvable asymmetric neural network model. Europhys. Lett. **4**, 167, (1987)
DH04.	Dowker, F., Henson. J.: A Spontaneous Collapse Model on a Lattice. J. Stat. Phys. **115**, 1349, (2004)
DH68.	Deprit, A., Henrard, J.: A manifold of periodic solutions, Advan. Astron. Astr. **6**, 6–12, (1968)
DH85.	Douady, A., Hubbard. J.: On the dynamics of polynomial-like mappings. Ann. scient. Éc. Norm. Sup., 4e ser. **18**, 287–343, (1985)
DH90.	Ding, M.-Z., Hao, B.-L.: Systematics of the periodic windows in the Lorenz model and its relation with the antisymmetric cubic map. Commun. Theor. Phys. **9**, 375, (1988)
DHR00.	Dauxois, T., Holdsworth, P., Ruffo, S.: Violation of ensemble equivalence in the antiferromagnetic mean-field XY model. Eur. Phys. J. **B 16**, 659, (2000)
DHS91.	Domany, E., van Hemmen. J.L., Schulten, K. (eds.): Models of Neural Networks. Springer, Berlin, (1991)
DKS02.	Dyson, L., Kleban, M., Susskind, L.: Disturbing Implications of a Cosmological Constant. JHEP **0210**, 011, (2002)
DL01.	Dasgupta, A., Loll, R.: A proper-time cure for the conformal sickness in quantum gravity. Nucl. Phys. B **606**, 357–379, (2001)

742   References

DL05. Dittrich, B., Loll, R.: Counting a black hole in Lorentzian product triangulations, preprint Utrecht, June (2005)
DLL01. Duff, M.J., Liu. J.T., Lu. J. (eds.): Strings: Proceedings of the 2000 International Superstrings Conference, World Scientific, Singapore, (2001)
DLS02. Dyson, L., Lindesay. J., Susskind, L.: Is There Really a de Sitter/CFT Duality? arXiv:hep-th/0202163, (2002)
DM03. Dorogovtsev, S.N., Mendes. J.F.F.: Evolution of Networks. Oxford Univ. Press, (2003)
DM06. Dolotin, V., Morozov, A.: Universal Mandelbrot Set. Beginning of the Story. World Scientific, Singapore, (2006)
DM07a. Dolotin, V., Morozov, A.: On the shapes of Elementary Domains, or Why Mandelbrot Set is Made from Almost Ideal Circles, arXiv:hep-th/0701234, (2007)
DM07b. Dolotin, V., Morozov, A.: Introduction to Non-Linear Algebra. World Scientific, Singapore, 2007.
DN79. Devaney, R., Nitecki, Z.: Shift automorphisms in the Hénon mapping. Comm. math. Phys. **67**, 137–48, (1979)
DN93. Dixit, A.K., Nalebuff, B.: Thinking Strategically: The Competitive Edge in Business, Politics, and Everyday Life. W.W. Norton & Company, (1993)
DNA00. G. Deffuant, D. Neau, F. Amblard, G. Weisbuch: Mixing beliefs among interacting agents. Adv. Compl. Syst. **3**, 87–98, (2000)
DP80. Dubois, D., Prade, H.: Fuzzy Sets and Systems. Academic Press, New York, (1980)
DP97. Dodson, C.T.J., Parker, P.E.: A User's Guide to Algebraic Topology. Kluwer, Dordrecht, (1997)
DR94. Dittrich, W., Reuter, M.: Classical and Quantum Dynamics. Springer Verlag, Berlin, (1994)
DR96. DePietri, R., Rovelli, C.: Geometry Eigenvalues and Scalar Product from Recoupling Theory in Loop Quantum Gravity. Phys. Rev. **D54**, 2664–2690, (1996)
DS74. Davey, A., Stewartson, K.: On Three-Dimensional Packets of Surface Waves. Proc. R. Soc. A **338**, 101–110, (1974)
DS98. Dowker, H.F., Surya, S.: Topology change and causal continuity. Phys. Rev. D **58**, 124019, (1998)
DS98b. Dowker, H.F., Sorkin, R.D.: A spin-statistics theorem for certain topological geons. Class. Quant. Grav. **15**, 1153–1167, (1998)
DSR05. Dullin, H.R., Schmidt, S., Richter, P.H., Grossmann, S.K.: Extended Phase Diagram of the Lorenz Model. Chaos (to appear) (2005)
DSS90. Ditto, W.L., Spano, M.L., Savage, H.T., Rauseo, S.N., Heagy, J., Ott, E.: Experimental observation of a strange nonchaotic attractor. Phys. Rev. Lett. **65**, 533—536, (1990)
DT89. Ditzinger, T., Haken, H.: Oscillations in the perception of ambiguous patterns: A model based on synergetics. Biol. Cybern. **61**, 279–287, (1989)

References

DT90. Ditzinger, T., Haken, H.: The impact of fluctuations on the recognition of ambiguous patterns. Biol. Cybern. **63**, 453–456, (1990)

DT92. Ding, W., Tian, G.: Kähler–Eistein metric and the generalized Futaki invariant. Inv. Math. **110**, 315–335, (1992)

DT95. Denniston, C., Tang, C.: Phases of Josephson Junction Ladders. Phys. Rev. Lett. **75**, 3930, (1995)

DTP02. Dauxois, T., Theodorakopoulos, N., Peyrard, M.: Thermodynamic instabilities in one dimension: correlations, scaling and solitons. J. Stat. Phys. **107**, 869, (2002)

DVV91. Dijkgraaf, R., Verlinde, H., Verlinde, E.: Notes On Topological String Theory And 2D Quantum Gravity, in String Theory and Quantum Gravity, Proceedings of the Trieste Spring School 1990, (eds.) M. Green *et al.*, World Scientific, 91–156, (1991)

DW95. Dormayer, P., Lani–Wayda, B.: Floquet multipliers and secondary bifurcations in functional differential equations: Numerical and analytical results, Z. Angew. Math. Phys. **46**, 823–858, (1995)

Dan67. Danilov, Yu.A.: Group properties of the Maxwell and Navier–Stokes equations, Khurchatov Inst. Nucl. Energy, Acad. Sci. USSR, (in Russian) (1967)

Das02. Dasgupta, A.: The real Wick rotations in quantum gravity. JHEP, **0207**, (2002)

Dav02a. Davis, S.C.: Cosmological brane world solutions with bulk scalar fields. JHEP **0203**, 054, (2002a)

Dav02b. Davis, S.C.: Brane cosmology solutions with bulk scalar fields. JHEP **0203**, 058, (2002b)

Dav81. Davydov, A.S.: Biology and Quantum Mechanics. Pergamon Press, New York, (1981)

Dav89. Davies, E.B.: Heat Kernels and Spectral Theory. Cambridge Univ. Press, (1989)

Dav91. Davydov, A.S.: Solitons in Molecular Systems. (2nd ed), Kluwer, Dordrecht, Ger, (1991)

DeP97. DePietri, R.: On the relation between the connection and the loop representation of quantum gravity, Class. and Quantum Grav. **14**, 53–69, (1997)

Dea90. Dearnaley, R.: The Zero-Slope Limit of Witten's String Field Theory with Chan-Paton Factors. Nucl. Phys. **B334**, 217, (1990)

Deb84. Debreu, G.: Economic theory in the mathematical mode. In Les Prix Nobel 1983. Reprinted in Am. Ec. Rev. **74**, 267–278, (1984)

Des91. Descartes, R.: Discourse on Method and Meditations on First Philosophy (tr. by D.A. Cress) Cambridge, (1991)

Deu85. Deutsch, D.: Quantum theory, the Church-Turing principle and the universal quantum computer. Proc. Roy. Soc. London A **400**, 97, (1985)

Deu89. Deutsch, D.: Quantum Computational Networks Proc. Roy. Soc. London A **425**(1868), 73–90, (1989)

Deu92. Deutsch, D., Jozsa, R.: Rapid solution of problems by quantum computation. Proc. Roy. Soc. (London), A **439**, 553–558, (1992)

Dev89. Devaney, R.: An Introduction to Chaotic Dynamical Systems. Addison Wesley Publ. Co. Reading MA, (1989)

Die69. Dieudonne, J.A.: Foundations of Modern Analysis (in four volumes) Academic Press, New York, (1969)

Die88. Dieudonne, J.A.: A History of Algebraic and Differential Topology 1900–1960. Birkháuser, Basel, (1988)

Dim59. Dimentberg, F.M.: Izgibnije Kolabanija Vrashchajushihsja Valov. Izd. Akad. Nauk SSSR, Moscow, (1959)

Dir25. Dirac, P.A.M.: The Fundamental Equations of Quantum Mechanics. Proc. Roy. Soc. London A, **109**(752), 642–653, (1925)

Dir26a. Dirac, P.A.M.: Quantum Mechanics, a Preliminary Investigation of the Hydrogen Atom. Proc. Roy. Soc. London A, **110**(755), 561–579, (1926)

Dir26b. Dirac, P.A.M.: The Elimination of the Nodes in Quantum Mechanics. Proc. Roy. Soc. London A, **111**(757), 281–305, (1926)

Dir26c. Dirac, P.A.M.: Relativity Quantum Mechanics with an Application to Compton Scattering. Proc. Roy. Soc. London A, **111**(758), 281–305, (1926)

Dir26d. Dirac, P.A.M.: On the Theory of Quantum Mechanics. Proc. Roy. Soc. London A, **112**(762), 661–677, (1926)

Dir26e. Dirac, P.A.M.: The Physical Interpretation of the Quantum Dynamics. Proc. Roy. Soc. London A, **113**(765), 1–40, (1927)

Dir28a. Dirac, P.A.M.: The Quantum Theory of the Electron. Proc. Roy. Soc. London A, **117**(778), 610–624, (1928)

Dir28b. Dirac, P.A.M.: The Quantum Theory of the Electron. Part II. Proc. Roy. Soc. London A, **118**(779), 351–361, (1928)

Dir29. Dirac, P.A.M.: Quantum Mechanics of Many-Electron Systems. Proc. Roy. Soc. London A, **123**(792), 714–733, (1929)

Dir32. Dirac, P.A.M.: Relativistic Quantum Mechanics. Proc. Roy. Soc. London A, **136**(829), 453–464, (1932)

Dir36. Dirac, P.A.M.: Relativistic Wave Equations. Proc. Roy. Soc. London A, **155**(886), 447–459, (1936)

Dir49. Dirac, P.A.M.: The Principles of Quantum Mechanics. Oxford Univ Press, Oxford, (1949)

Dir58. Dirac, P.A.M.: Generalized Hamiltonian Dynamics. Proc. Roy. Soc. London A, **246**(1246), 326–332, (1958)

Dom78. de Dominicis, C.: Dynamics as a substitute for replicas in systems with quenched random impurities. Phys. Rev. B **18**, 4913–4919, (1978)

Don84. Donaldson, S.K.: The Nahm equations and the classification of monopoles, Commun. Math. Phys. 96, 387–407, (1984)

Don87. Donaldson, S.: The Orientation of Yang–Mills Moduli Spaces and Four–Manifold Topology. J. Diff. Geom. **26**, 397, (1987)

Don96. Donoghue. J.F.: The Quantum Theory of General Relativity at Low Energies. Helv. Phys. Acta. **69**, 269–275, (1996)

Dor94.	Dorcak, L.: Numerical Models for Simulation the Fractional–Order Control Systems, UEF SAV, The Academy of Sciences, Inst. of Exper. Phys., Kosice, Slovak Republic, (1994)
Dou03.	Douglas, M.: The statistics of string / M theory vacua. JHEP **0305**, 46, (2003)
Dow02.	Dowker, F.: Topology change in quantum gravity. In Proceedings of Stephen Hawking's 60th birthday conference. Cambridge, UK, (2002)
Dow05.	Dowker, F.: Causal Sets and the Deep Structure of spacetime. In 100 Years of Relativity, ed. by A. Ashtekar, World Scientific, Singapore, (2005)
Dra06.	Drake, G.W.F (Ed.): Springer Handbook of Atomic, Molecular, and Optical Physics. Springer, New York, (2006)
Dri77.	Driver, R.D.: Ordinary and delay differential equations. Applied Mathematical Sciences 20, Springer Verlag, New York, (1977)
Dri86.	Drinfeld, V.G.: Quantum groups. Proc. ICM, Berkeley, 798, (1985)
Du02.	Du, J. et al.: Experimental Realization of Quantum Games on a Quantum Computer. Phys. Rev. Lett. **88**, 137902 (2002)
Duf18.	Duffing, G.: Erzwungene Schwingungen bei veränderlicher Eigenfrequenz. Vieweg Braunschweig, (1918)
Duf94.	Duff, M.: Twenty years of the Weyl anomaly. Class. Quant. Grav. **11**, 1387, (1994)
Dum01.	Dummett, M.: Origini della Filosofia Analitica. Einaudi. ISBN 88-06-15286-6, (2001)
Dun99.	Dunne, G.V.: Aspects of Chern–Simons Theory. arXiv:hep-th/9902115, (1999)
Dus84.	Dustin, P.: Microtubules. Springer, Berlin, (1984)
EB01.	Endy, D., Brent, R.: Modelling cellular behaviuor. Nature **409**, 391–395, (2001)
EHM00.	Emparan, R., Horowitz, G.T., Myers, R.C.: Exact Description of Black Holes on Branes II: Comparison with BTZ Black Holes and Black Strings. JHEP 0001, 007, (2000)
EIS67.	Eccles, J.C., Ito M., Szentagothai J.: The Cerebellum as a Neuronal Machine. Springer, Berlin, (1967)
EJ96.	Ekert, A., Jozsa, R.: Quantum Computation and Shor's Factoring Algorithm. Rev. Mod. Phys. **68**, 733, (1996)
EJM99.	Emparan, R., Johnson, C.V., Myers, R.C.: Surface terms as counterterms in the AdS–CFT correspondence. Phys. Rev. D **60**, 104001, (1999)
EJP00.	Eisert, J., Jacobs, K., Papadopoulos, P., Plenio, M.B.: Nonlocal content of quantum operations. Phys. Rev. A **62**, 052317 (7 pages), (2000)
EKL02.	Evans, M.R., Kafri, Y., Levine, E., Mukamel, D.: Phase transition in a non-conserving driven system. J. Phys. A: Math. Gen. A **35**, L433, (2002)
EL00.	Elowitz, M.B., Leibler, S.: A synthetic oscillatory network of transcriptional regulators. Nature, **403**, 335–338, (2000)

EL78.	Eells. J., Lemaire, L.: A report on harmonic maps. Bull. London Math. Soc. **10**, 1–68, (1978)
EM90.	Evans, D.J., Morriss, G.P.: Statistical mechanics of nonequilibrium fluids. Academic Press, New York, (1990)
EMN92.	Ellis. J., Mavromatos, N., Nanopoulos, D.V.: String theory modifies quantum mechanics. CERN-TH/6595, (1992)
EMN99.	Ellis. J., Mavromatos, N., Nanopoulos, D.V.: A microscopic Liouville arrow of time. Chaos, Solit. Fract., **10**(2–3), 345–363, (1999)
EMR91.	Echeverría-Enríquez, A., Muñoz-Lecanda, M., Román-Roy, N.: Geometrical setting of time–dependent regular systems. Alternative models, Rev. Math. Phys. **3**, 301, (1991)
EMR95.	Echeverría-Enríquez, A., Muñoz-Lecanda, M., Román-Roy, N.: Non-standard connections in classical mechanics. J. Phys. A **28**, 5553, (1995)
EMR98.	Echeverría-Enríquez, A., Muñoz-Lecanda, M., Román-Roy, N.: Multivector fields and connections. Setting Lagrangian equations in field theories. J. Math. Phys. **39**, 4578, (1998)
EP98.	Ershov, S.V., Potapov, A.B.: On the concept of Stationary Lyapunov basis. Physica D, **118**, 167, (1998)
EPG95.	Ernst, U., Pawelzik, K., Geisel, T.: Synchronization induced by temporal delays in pulse-coupled oscillators. Phys. Rev. Lett. **74**, 1570, (1995)
EPR35a.	A. Einstein, B. Podolsky, N. Rosen: Can Quantum–Mechanical Description of Physical Reality Be Considered Complete? Phys. Rev. **47**, 777, (1935)
EPR35b.	Einstein, A., Podolsky, B., Rosen, N.: Quantum theory and Measurement. Zurek, W.H., Wheeler, J.A. (ed.), (1935)
EPR99.	Eckmann, J.P., Pillet, C.A., Rey-Bellet, L.: Non-equilibrium statistical mechanics of anharmonic chains coupled to two heat baths at different temperatures. Commun. Math. Phys., **201**, 657–697, (1999)
ER59.	Erdös, P., Rényi, A.: On random graphs. Pub. Math. **6**, 290–297, (1959)
ER85.	Eckmann, J.P., Ruelle, D.: Ergodic theory of chaos and strange attractors, Rev. Mod. Phys., **57**, 617–630, (1985)
ES46.	Einstein, A., Strauss, E.: A generalization of the relativistic theory of gravitation II. Ann. Math. **47**, 731–741, (1946)
ESA05.	European Space Agency. Payload and Advanced Concepts: Superconducting Tunnel Junction (STJ) February 17, (2005)
ESH98.	Elson, R.C., Selverston, A.I., Huerta, R. et al.: Synchronous Behavior of Two Coupled Biological Neurons. Phys. Rev. Lett. **81**, 5692–5695, (1998)
EW00.	Eisert, J., Wilkens, M.: Quantum Games. J. Mod. Opt. **47**, 2543–2556, (2000)
EW06.	Eisert, J., Wolf, M.M.: Quantum computing. A chapter in Handbook of Nature-Inspired and Innovative Computing, Springer, Berlin, (2006)
EWL99.	Eisert, J., Wilkens, M., Lewenstein, M.: Quantum Games and Quantum Strategies. Phys. Rev. Lett. **83**, 3077–3080, (1999)

Ecc64.	Eccles, J.C.: The Physiology of Synapses. Springer, Berlin, (1964)
Ein05.	Einstein, A.: Die von der Molekularkinetischen Theorie der Warme Gefordete Bewegung von in ruhenden Flussigkeiten Suspendierten Teilchen. Ann. Phys. **17**, 549, (1905)
Ein48.	Einstein, A.: Quantum Mechanics and Reality (Quanten–Mechanik und Wirklichkeit) Dialectica **2**, 320–324, (1948)
Eis01.	Eisert, J.: Entanglement in quantum information theory. PhD thesis, Univ. Potsdam, (2001)
Eis29.	Eisenhart, L.P.: Dynamical trajectories and geodesics. Math. Ann. **30**, 591–606, (1929)
Eke91.	Ekert, A.K.: Quantum cryptography based on Bell's theorem. Phys. Rev. Lett. **67**, 661–663, (1991)
Elk99.	Elkin, V.I.: Reduction of Nonlinear Control Systems. A Differential Geometric Approach, Kluwer, Dordrecht, (1999)
Elw82.	Elworthy, K.D.: Stochastic Differential Equations on Manifolds. Cambridge Univ. Press, Cambridge, (1982)
Eno01.	Enoka, R.M.: Neuromechanics of Human Movement. Human Kinetics, Champaign, IL (3rd ed), (2001)
Ens05.	Enss, C. (ed): Cryogenic Particle Detection. Topics in Applied Physics **99**, Springer, New York, (2005)
Erm96.	Ermentrout, G.B.: Type I membranes, phase resetting curves, and synchrony. Neural Computation, **8**(5), 979–1001, (1996)
FA02.	Flitney, A.P., Abbott, D.: Quantum version of the Monty Hall problem. Phys. Rev. A **65**, 062318 (4 pages) (2002)
FA04.	Field, T., Anandan. J.: Geometric phases and coherent states. J. Geom. Phys. **50**, 56–78, (2004)
FA04.	Flitney, A.P., Abbott, D.: Quantum two and three person duels. J. Optics B **6**, S860–S866, (2004)
FCH02.	Fenton, F.H., Cherry, E.M., Hastings, H.M., Evans, S.J.: Multiple mechanisms of spiral wave breakup in a model of cardiac electrical activity. Chaos **12**, 852–892, (2002)
FCS99.	Franzosi, R., Casetti, L., Spinelli, L., Pettini, M.: Topological aspects of geometrical signatures of phase transitions. Phys. Rev. E, **60**, 5009–5012, (1999)
FD84.	Freeman, D.L., Doll, J.D.: Statistical Mechanics Using Fourier Representations of Path Integrals. J. Chem. Phys. **80**, 5709, (1984)
FD94.	Feldman, D.E., Dotsenko, V.S.: Partially annealed neural networks. J. Phys. A: Math. Gen. **27**, 4401-4411, (1994)
FDN92.	Fomenko, Dubrovin and Novikov. Modern Geometry– Methods and Applications (two volumes). Springer-Verlag, New York, (1992)
FFF94.	Fatibene, L., Ferraris, M., Francaviglia, M.: Nöther formalism for conserved quantities in classical gauge field theories II. J. Math. Phys., **35**, 1644, (1994)
FFG05.	Fogassi, L., Ferrari, P.F., Gesierich, B., Rozzi, S., Chersi, F., Rizzolatti, G.: Parietal lobe: From action organization to intention understanding. Science, **29**, 662–667, (2005)

FG94. Fröhlich. J., Gawedzki, K.: Conformal field theory and the geometry of strings. CRM Proceedings and Lecture Notes **7**, 57–97, (1994)

FGK94. Dowker, F., Gauntlett, J.P., Kastor, D.A., Traschen, J.: Pair creation of dilaton black holes. Phys. Rev. D **49**, 2909–2917 (1994)

FGR98. Fajstrup, L., Goubault, E., Raussen, M.: Detecting Deadlocks in Concurrent Systems, In D. Sangiorgi R. de Simone, editor, CONCUR '98; Concurrency Theory, number 1466 in Lecture Notes in Computer Science, 332–347. Springer, Berlin, (1998)

FH65. Feynman, R.P., Hibbs, A.: Quantum Mechanics and Path Integrals. McGraw-Hill, New York, (1965)

FH90. Friedman, J.L., Higuchi, A.: State vectors in higher dimensional gravity with kinematic quantum numbers of quarks and leptons. Nuclear Physics, B **339**, 491–515, (1990)

FH96. Fang, H.-P., Hao, B.-L.: Symbolic dynamics of the Lorenz equations. Chaos, Solitons, and Fractals, **7**, 217–246, (1996)

FHH79. Fischetti, M.V., Hartle, J.B., Hu, B.L.: Quantum Effects in the Early Universe I. Influence of Trace Anomalies on Homogeneous, Isotropic, Classical Geometries. Phys. Rev. D **20**, 1757, (1979)

FHI00. Fujii, K., Hayashi, D., Inomoto, O., Kai, S: Noise-Induced Entrainment between Two Coupled Chemical Oscillators in Belouzov-Zhabotinsky Reactions. Forma **15**, 219, (2000)

FHT03. Freed, D., Hopkins, M. Teleman, C.: Twisted K–theory and loop group representations. arXiv:math.AT/0312155, (2003)

FJ66. Fosdick, L.D., Jordan, H.F.: Path-Integral Calculation of the Two-Particle Slater Sum for He^4. Phys. Rev. **143**, 58–66, (1966)

FK39. Frenkel, J., Kontorova, T.: On the theory of plastic deformation and twinning. Phys. Z. Sowiet Union, **1**, 137–49, (1939)

FK82. Ferraris, M., Kijowski. J.: On the equivalence of the relativistic theories of gravitation. GRG **14**, 165, (1982)

FK83. Frölich, H., Kremer, F.: Coherent Excitations in Biological Systems. Springer, New York, (1983)

FK88. Fontanari, J.F., Köberle, R.: Information Processing in Synchronous Neural Networks. J. Physique **49**, 13, (1988)

FKL05. Freivogel, B., Kleban, M., Susskind, L.: Observational Consequences of a Landscape. arXiv:hep-th/0505232, (2005)

FKN95. Frittelli, S., Kozameh, C., Newman, E.T.: General Relativity via Characteristic Surfaces. J. Math. Phys. **36**, 4984, (1995)

FKN97. Frittelli, S., Kozameh, C., Newman, E.T., Rovelli, C., Tate, R.S.: Fuzzy space–time points from the null–surface formulation of general relativity, Class. Quantum Gravity, **14**, A143, (1997)

FKP95. Feudel, U., Kurths, J., Pikovsky, A.S.: Strange Nonchaotic Attractor in a Quasiperiodically Forced Circle Map. Physica D **88**, 176-186, (1995)

FLL00.	Forste, S., Lalak, Z., Lavignac, S., Nilles, H.P.: A comment on self-tuning and vanishing cosmological constant in the brane world. Phys. Lett. B **481**, 360, (2000)
FM03.	Fujii, Y., Maeda, K-I.: The Scalar–Tensor Theory of Gravitation. Cambridge Univ. Press, (2003)
FM93.	Franks, J., Misiurewicz, M.: Cycles for disk homeomorphisms and thick trees. In Nielsen Theory and Dynamical Systems, **152** in Contemporary Mathematics, 69–139, (1993)
FMV01.	Festa, R., Mazzino, A., Vincenzi, D.: An analytical approach to chaos in Lorenz–like systems. A class of dynamical equations. Europhys. Lett. **56**, 47–53, (2001)
FMV02a.	Festa, R., Mazzino, A., Vincenzi, D.: Lorenz deterministic diffusion. Europhys. Lett., **60**(6), 820–826, (2002)
FMV02b.	Festa, R., Mazzino, A., Vincenzi, D.: Lorenz–like systems and classical dynamical equations with memory forcing: An alternate point of view for singling out the origin of chaos. Phys. Rev. E **65**, 046205, (2002)
FNS77.	Forster, D., Nelson, D.R., Stephen, M.J.: Large-distance and long-time properties of a randomly stirred fluid. Phys. Rev. A **16**, 732–749, (1977)
FP04.	Franzosi, R., Pettini, M.: Theorem on the origin of Phase Transitions. Phys. Rev. Lett. **92**(6), 060601, (2004)
FP94.	Fushchych, W.I., Popowych, R.O.: Symmetry reduction and exact solutions of the Navier–Stokes equations. J. Nonl. Math. Phys. **1**, 75–113, 156–188, (1994)
FPR04.	Forger, M., Paufler, C., Römer, H.: Hamiltonian Multivector Fields and Poisson Forms in Multisymplectic Field Theory. arXiv:math-ph/0407057.
FPS00.	Franzosi, R., Pettini, M., Spinelli, L.: Topology and phase transitions: a paradigmatic evidence. Phys. Rev. Lett. **84**, 2774–2777, (2000)
FPS92.	Friedman. J.L., Papastamatiou, N.J., Simon. J.Z.: Unitarity of Interacting Fields in Curved space-time. Phys. Rev. D **46**, 4441, (1992)
FPU55.	Fermi, E., Pasta, J., Ulam, S.: Studies of nonlinear problems. Los Alamos Report LA-1940 (1955). in E. Segré (ed.), Collected Papers of Enrico Fermi, 2, 978, Univ. Chicago, (1965)
FPV88.	Falcioni, M., Paladin, G., Vulpiani, A.: Regular and chaotic motion of fluid particles in a two-dimensional fluid. J. Phys. A: Math. Gen., **21**, 3451–3462, (1988)
Fri90.	Friedkin, N.E.: A Guttman Scale for the Strength of an Interpersonal Tie. Soc. Net. 12, 239–252, (1990)
FR04.	Forger, M., Römer, H.: Currents and the energy–momentum tensor in classical field theory: a fresh look at an old problem. Ann. Phys. (N.Y.) **309**, 306–389, (2004)
FR84.	Fuks, D.B., Rokhlin, V.A.: Beginner's Course in Topology. Springer-Verlag, Berlin, (1984)
FS80.	Friedman, J.L., Sorkin, R.D.: Spin-1/2 from gravity. Phys. Rev. Lett. **44**, 1100–1103, 1980.

FS92. Freeman. J.A., Skapura, D.M.: Neural Networks: Algorithms, Applications, and Programming Techniques. Addison-Wesley, Reading, MA, (1992)

FSY75. Fuji, Y., Shirane, G., Yamada, Y.: Study of the 123-K. phase transition of magnetite by critical neutron scattering. Phys. Rev. B **11**, 2036–2041, (1975)

FT82. Fradkin, E.S., Tseytlin, A.A.: Higher Derivative Quantum Gravity: One Loop Counterterms and Asymptotic Freedom, Nucl. Phys. **B201**, 469, (1982)

FT84. Fradkin, E.S., Tseytlin, A.A.: Conformal Anomaly in Weyl Theory and Anomaly Free Superconformal Theories. Phys. Lett. **134B**, 187, (1984)

FT87. Faddeev, L.D., Takhtajan, L.A.: Hamiltonian Methods in the Theory of Solitons. Springer-Verlag, Berling, (1987)

FTR01. Fujisaka, H., Tutu, H., Rikvold, P.A.: Dynamic phase transition in a time-dependent Ginzburg–Landau model in an oscillating field. Phys. Rev. E **63**, 036109, (2001)

FTW00. Flanagan, E.E., Tye, S.H., Wasserman, I.: Cosmological expansion in the RS brane world scenario. Phys. Rev. D **62**, 044039, (2000)

FU84. Freed, D., Uhlenbeck, K.K.: Instantons and four manifolds, Springer, New York, (1984)

FW80. Franks, J., Williams, R.F.: Anomalous Anosov flows. 158-174 in Lecture Notes in Math. **819**, Springer, Berlin, (1980)

FW95. Filatrella, G., Wiesenfeld, K.: Magnetic–field effect in a two-dimensional array of short Josephson junctions. J. Appl. Phys. **78**, 1878–1883, (1995)

FY83. Fujisaka, H., Yamada, T.: Amplitude Equation of Higher-Dimensional Nikolaevskii Turbulence. Prog. Theor. Phys. **69**, 32 (1983)

Fat19. Fatou, P.: Sur les équations fonctionnelles. Bull. Soc. math. France **47**, 161–271, (1919)

Fat22. Fatou, P.: Sur les fonctions méromorphes de deux variables and Sur certaines fonctions uniformes de deux variables. C.R. Acad. Sc. Paris **175**, 862–65 and 1030–33, (1922)

Fed69. Federer, H.: Geometric Measure Theory. Springer, New York, (1969)

Fei78. Feigenbaum, M.J.: Quantitative universality for a class of nonlinear transformations. J. Stat. Phys. **19**, 25–52, (1978)

Fei79. Feigenbaum, M.J.: The universal metric properties of nonlinear transformations. J. Stat. Phys. **21**, 669–706, (1979)

Fei80. Feigenbaum, M.: Universal Behavior in Nonlinear Systems. Los Alamos Science. **1**, 4, (1980)

Fer23. Fermi, E.: Beweis dass ein mechanisches Normalsysteme im Allgemeinen quasi–ergodisch ist, Phys, Zeit. **24**, 261–265, (1923)

Fer98. Fersht, A.: Structure and Mechanism in Protein Science. W.H.Freeman, New York, (1998)

Fer99.	Ferber. J.: Multi-Agent Systems. An Introduction to Distributed Artificial Intelligence. Addison-Wesley, Reading, MA, (1999)
Fey48.	Feynman, R.P.: Space-time Approach to Non-Relativistic Quantum Mechanics. Rev. Mod. Phys. **20**, 267, (1948)
Fey51.	Feynman, R.P.: An Operator Calculus Having Applications in Quantum Electrodynamics. Phys. Rev. **84**, 108–128, (1951)
Fey72.	Feynman, R.P.: Statistical Mechanics, A Set of Lectures. WA Benjamin, Inc., Reading, Massachusetts, (1972)
Fey96.	Feynman, R.P.: Feynman lectures on computation. Addison-Wesley, Reading, MA, (1996)
Fey98.	Feynman, R.P.: Quantum Electrodynamics. Advanced Book Classics, Perseus Publishing, (1998)
Fit61.	FitzHugh, R.A.: Impulses and physiological states in theoretical models of nerve membrane. Biophys J., **1**, 445–466, (1961)
Fla63.	Flanders, H.: Differential Forms with Applications to the Physical Sciences. Acad. Press, (1963)
Fla76.	Flaherty, E.J.: Hermitian and Kählerian Geometry in General Relativity, Lecture Notes in Physics, Vol. 46, Springer, Heidelberg, (1976)
Fla80.	Flaherty, E.J.: Complex Variables in Relativity. In General Relativity and Gravitation, One Hundred Years after the Birth of Albert Einstein, ed A. Held Plenum, New York, (1980)
Flo87.	Floer, A.: Morse theory for fixed points of symplectic diffeomorphisms. Bull. AMS. **16**, 279, (1987)
Flo88.	Floer, A.: Morse theory for Lagrangian intersections. J. Diff. Geom., **28**(9), 513–517, (1988)
Fly04.	Flynn, T.: J.P. Sartre. In The Stanford Encyclopedia of Philosophy, Stanford, (2004)
Fok29.	Fokker, A.D.: Ein invarianter Variationssatz fuer die Bewegung. Z. Phys., **58**, 386–393, (1929)
Fol75.	Follesdal, D.: Meaning, Experience. In 'Mind and Language', ed. S. Guttenplan. Oxford, Clarendon, 25–44, (1975)
For90.	Fordy, A.P. (ed.): Soliton Theory: A Survey of Results. MUP, Manchester, UK, (1990)
For92.	Ford, J.: The Fermi-Pasta-Ulam problem: paradox turns discovery. Phys. Rep. **213**, 271–310, (1992)
Fos62.	Fosdick, L.D.: Numerical estimation of the partition function in quantum statistics. J. Math. Phys. **3**, 1251–1264, (1962)
Fos68.	Fosdick, L.D.: The Monte Carlo method in quantum statistics. SIAM Rev. **10**, 315–328, (1968)
Fre00.	Freeman, W.J.: Neurodynamics: An exploration of mesoscopic brain dynamics. Springer, Berlin, (2000)
Fre01.	Freed, D.S.: The Verlinde Algebra is Twisted Equivariant K–Theory. Turk. J. Math **25**, 159–167, (2001)
Fre90.	Freeman, W.J.: On the the problem of anomalous dispersion in chaotic phase transitions of neural masses, its significance for the management of perceptual information in brains. In

	H.Haken, M.Stadler (ed.) Synergetics of cognition 45, 126–143. Springer Verlag, Berlin, (1990)
Fre91.	Freeman, W.J.: The physiology of perception. Sci. Am., **264**(2), 78–85, (1991)
Fre92.	Freeman, W.J.: Tutorial on neurobiology: from single neurons to brain chaos. Int. J. Bif. Chaos. **2**(3), 451–82, (1992)
Fre96.	Freeman, W.J.: Random activity at the microscopic neural level in cortex sustains is regulated by low dimensional dynamics of macroscopic cortical activity. Int. J. of Neural Systems 7, 473, (1996)
Fri03.	Friedman. J. et. al.: Quantum Superposition of Distinct Macroscopic States, Nature, **406**, 43–46, (2000)
Fri88.	Friberg, O.: A Set of Parameters for Finite Rotations and Translations. Comp. Met. App. Mech. Eng., **66**, 163–171, (1988)
Fri99.	Frieden R.B.: Physics from Fisher Information: A Unification, Cambridge Univ. Press, (1999)
Ful89.	Fulling, S.: Aspects of quantum field theory in curved space–time. Cambridge Univ. Press, (1989)
Fur61.	Furstenberg, H.: Strict ergodicity and transformation of the torus. Am. J. Math. **83**, 573–601, (1961)
Fut83.	Futaki, A.: An obstruction to the existence of Einstein Kähler metrics. Inv. Math. **73**(3), 437–443, (1983)
GBL03.	Gardner, T.S., di Bernardo, D., Lorenz, D., Collins. J.J.: Inferring genetic networks and identifying compound mode of action via expression profiling. Science, **301**, 102–105, (2003)
GC00.	Gollub, J.P., Cross, M.C.: Many systems in nature are chaotic in both space and in time. Nature, **404**, 710-711, (2000)
GC95a.	Gallavotti, G., Cohen, E.G.D.: Dynamical ensembles in nonequilibrium statistical mechanics. Phys. Rev. Letters **74**, 2694-2697, (1995)
GC95b.	Gallavotti, G., Cohen, E.G.D.: Dynamical ensembles in stationary states. J. Stat. Phys. **80**, 931-970, (1995)
GE01.	Golomb, D., Ermentrout, G.B.: Bistability in Pulse Propagation in Networks of Excitatory and Inhibitory Populations. Phys. Rev. Lett. **86**, 4179–4182, (2001)
GEH02.	Guet, C., Elowitz, M.B., Hsing, W., Leibler, S.: Combinatorial synthesis of genetic networks. Science, **296**(5572), 1466-1470, (2002)
GG03.	Gaucher, P., Goubault, E.: Topological Deformation of Higher Dimensional Automata. Homology, Homotopy and Applications, **5**(2), 39–82, (2003)
GGK67.	Gardner, C.S., Greene, J.M., Kruskal, M.D., Miura, R.M.: Method for solving the Korteweg-de Vries equation. Phys. Rev. Let., **19**, 1095–97, (1967)
GGS81.	Guevara, M.R., Glass, L., Shrier, A.: Phase locking, period-doubling bifurcations, and irregular dynamics in periodically stimulated cardiac cells. Science **214**, 1350-53, (1981)

GH00.	Gade, P.M., Hu, C.K.: Synchronous chaos in coupled map lattices with small-world interactions. Phys. Rev. E **62**, 6409–6413, (2000)
GH02.	Gibbons, G., Hartnoll, S.A.: A gravitational instability in higher dimensions, Phys. Rev. D **66**, 064024, (2002)
GH77.	Gibbons, G.W., Hawking, S.W.: Action integrals and partition functions in quantum gravity. Phys. Rev. D **15**, 2752, (1977)
GH81.	Greenberg, M.J., Harper, J.R.: Algebraic Topology, a first course. Benjamins Cummings, Reading, MA, (1981)
GH83.	Guckenheimer, J., Holmes, P.: Nonlinear Oscillations, Dynamical Systems, and Bifurcations of Vector Fields. Springer-Verlag, Berlin, (1983)
GH90.	Gibbons, G.W., Hartle, J.B.: Real Tunneling Geometries and the Large-Scale Topology of the Universe. Phys. Rev. D **42**, 2458, (1990)
GH92a.	Gibbons, G.W., Hawking, S.W.: Selection rules for topology change. Commun. Math. Phys. **148**, 345–352, 1992.
GH92b.	Gibbons, G.W., Hawking, S.W.: Kinks and topology change. Phys. Rev. Lett. **69**, 1719–1721, 1992.
GH93.	Gibbons, G.W., Hawking, S.W. (eds.): Euclidean quantum gravity. World Scientific, Singapore, (1993)
GH94.	Griffiths, P., Harris. J.: Principles of Algebraic Geometry, Wiley Interscience, New York, (1994)
GH96.	Gérard, R., Tahara, H.: Singular nonlinear partial differential equations, Aspects of Mathematics, Friedr. Vieweg & Sohn, Braunschweig, (1996)
GHO02.	Gauthier, D.J., Hall, G.M., Oliver, R.A. *et al.*: Issue on Mapping and Control of Complex Cardiac Arrhythmias. Chaos **12**, 952, (2002)
GHP78.	Gibbons, G.W., Hawking, S.W., Perry, M.J.: Path Integrals and the Indefiniteness of the Gravitational Action, Nucl. Phys. **B138**, 141, (1978)
GHT00.	Gratton, S., Hertog, T., Turok, N.: An Observational Test of Quantum Cosmology. Phys. Rev. D **62**, 063501 (2000)
GIM04.	Gotay, M.J., Isenberg. J., Marsden. J.E., Montgomery, R.: Momentum Maps and Classical Relativistic Fields II: Canonical Analysis of Field Theories. arXiv:math-ph/0411032.
GIM98.	Gotay, M.J., Isenberg. J., Marsden. J.E.: Momentum Maps and the Hamiltonian Structure of Classical Relativistic Fields. arXiv:hep/9801019.
GIT02.	Gen, U., Ishibashi, A., Tanaka, T.: Brane big-bang brought by bulk bubble. Phys. Rev. D **66**, 023519, (2002)
GJ87a.	Gross, D.J., Jevicki, A.: Operator Formulation of Interacting String Field Theory (I) Nucl. Phys. **B283**, 1, (1987)
GJ87b.	Gross, D.J., Jevicki, A.: Operator Formulation of Interacting String Field Theory (II) Nucl. Phys. **B287**, 225, (1987)
GJ94.	van Groesen E., De Jager E.M. (ed): Mathematical Structures in Continuous Dynamical Systems. Studies in Mathematical Physics, vol. 6 North-Holland, Amsterdam, (1994)

GK92.	Georgiou, G.M., Koutsougeras, C.: Complex domain backpropagation, IEEE Trans. Circ. Sys., **39**(5), 330–334, (1992)
GKE89.	Gray, C.M., König, P., Engel, A.K., Singer, W.: Oscillatory responses in cat visual cortex exhibit intercolumnar synchronization which reflects global stimulus properties. Nature **338**, 334, (1989)
GKP98.	Gubser, S.S., Klebanov, I.R., Polyakov, A.M.: Gauge Theory Correlators from Noncritical String Theory. Phys. Lett. **B428**, 105, (1998)
GKS02.	Goheer, N., Kleban, M., Susskind, L.: The Trouble with de Sitter space. arXiv:hep-th/0212209, (2002)
GL93.	Gregory, R., Laflamme, R.: Black Strings and P–Branes Are Unstable. Phys. Rev. Lett. **70**, 2837, (1993)
GLW93.	Geigenmuller, U., Lobb, C.J., Whan, C.B.: Friction and inertia of a vortex in an underdamped Josephson array. Phys. Rev. B **47**, 348–358, (1993)
GM01.	Gordon, C., Maartens, R.: Density perturbations in the brane world. Phys. Rev. D **63**, 044022, (2001)
GM04.	Grinza, P., Mossa, A.: Topological origin of the phase transition in a model of DNA denaturation. Phys. Rev. Lett. **92**(15), 158102, (2004)
GM79.	Guyer, R.A., Miller, M.D.: Commensurability in One Dimension at $T \neq 0$. Phys. Rev. Lett. **42**, 718–722, (1979)
GM84.	De Groot, S.R., Mazur, P.: Nonequilibrium thermodynamics. Dover, New York, (1984)
GM87.	Gross, D., Mende, P.: The High-Energy Behavior Of String Scattering Amplitudes. Phys. Lett **B197**, 129, (1987)
GM88.	Garland, H., Murray, M.K.: Kac-Moody monopoles and periodic instantons, Commun. Math. Phys. 120, 335–351, (1988)
GM90.	Giachetta, G., Mangiarotti, L.: Gauge-invariant and covariant operators in gauge theories. Int. J. Theor. Phys., **29**, 789, (1990)
GM92.	Gotay, M.J., Marsden. J.E.: Stress-energy–momentum tensors and the Belinfante-Rosenfeld formula. Contemp. Math. **132**, AMS, Providence, 367–392, (1992)
GM94.	Gell-Mann, M.: The Quark and the Jaguar. W. Freeman, San Francisco, (1994)
GMH05.	Giddings, S.B., Marolf, D., Hartle, J.B.: Observables in Effective Gravity. arXiv:hep-th/0512200, (2005)
GMH06.	Giddings, S., Marolf, D., Hartle. J.: Observables in Effective Gravity. arXiv:hep-th/0512200, (2005)
GMH90.	Gibbons, G.W., Hartle, J.B.: Real Tunneling Geometries and the Large–scale Topology of the Universe. Phys. Rev. D **42**, 2458, (1990)
GMH93.	Gell-Mann, M., Hartle, J.B.: Classical Equations for Quantum Systems. Phys. Rev. D **47**, 3345, (1993)
GMP05.	Grunwald, P., Myung, I.J., Pitt, M.A.(eds.): Advances in Minimum Description Length: Theory and Applications. MIT Press, Cambridge, MA (2005)

GMS02a.	Giachetta, G., Mangiarotti, L., Sardanashvily, G.: Covariant geometric quantization of non-relativistic Hamiltonian mechanics. J. Math. Phys. **43**, 56, (2002)
GMS02b.	Giachetta, G., Mangiarotti, L., Sardanashvily, G.: Geometric quantization of mechanical systems with time-dependent parameters. J. Math. Phys., **43**, 2882, (2002)
GMS05.	Giachetta, G., Mangiarotti, L., Sardanashvily, G.: Lagrangian supersymmetries depending on derivatives. Global analysis and cohomology. Commun. Math. Phys. **259**, 103–128, (2005)
GMS95.	Brian R. Greene, David R. Morrison, and Andrew Strominger. Black hole condensation and the unification of string vacua. Nucl. Phys. **B451**, 109–120, 1995.
GMS95.	Golubitsky, M., Marsden, J., Stewart, I., Dellnitz, M.: The constrained Lyapunov–Schmidt procedure and periodic orbits. Fields Institute Communications, **4**, 81–127, (1995)
GMS97.	Giachetta, G., Mangiarotti, L., Sardanashvily, G.: New Lagrangian and Hamiltonian Methods in Field Theory, World Scientific, Singapore, (1997)
GMS99.	Giachetta, G., Mangiarotti, L., Sardanashvily, G.: Covariant Hamilton equations for field theory. J. Phys. A **32**, 6629, (1999)
GN80.	Gotay, M.J., Nester. J.M.: Generalized constraint algorithm and special presymplectic manifolds. In G. E. Kaiser. J. E. Marsden, Geometric methods in mathematical physics, Proc. NSF-CBMS Conf., Lowell/Mass. 1979, Berlin: Springer-Verlag, Lect. Notes Math. **775**, 78–80, (1980)
GN90.	Gaspard, P., Nicolis, G.: Transport properties, Lyapunov exponents, and entropy per unit time. Phys. Rev. Lett. **65**, 1693-1696, (1990)
GNH78.	Gotay, M., Nester. J., Hinds, G.: Presymplectic manifolds and the Dirac-Bergmann theory of constraints. J. Math. Phys. **19**, 2388, (1978)
GNR91.	Gross, P., Neuhauser, D., Rabitz, H.: Optimal Control of Unimolecular Reactions in the Collisional Regime. J. Chem. Phys. **94**, 1158, (1991)
GNR92.	Gross, P., Neuhauser, D., Rabitz, H.: Optimal control of curve-crossing systems. J. Chem. Phys. **96**, 2834–2845, (1992)
GNR93.	Gross, P., Neuhauser, D., Rabitz, H.: Teaching lasers to control molecules in the presence of laboratory field uncertainty and measurement imprecision. J. Chem. Phys. **98**, 4557–4566, (1993)
GOP84.	Grebogi, C., Ott, E., Pelikan, S., Yorke, J.A.: Strange attractors that are not chaotic. Physica D **3**, 261–268, (1984)
GOY87.	Grebogi, C., Ott, E., Yorke, J.A.: Chaos, strange attractors, and fractal basin boundaries in nonlinear dynamics. Science, **238**, 632–637, (1987)
GP00.	Grantcharov, G., Poon, Y.S.: Geometry of hyper-Kähler connections with torsion Comm. Math. Phys. **213**(1), 19–37, (2000)

GP74. V. Guillemin, A. Pollack, Differential Topology. Prentice Hall, Englewood Cliffs NJ, (1974)

GP78. Gibbons, G.W., Perry, M.J.: Quantizing Gravitational Instantons, Nucl. Phys. **B146**, 90, (1978)

GP83a. Grassberger, P., Procaccia, I.: Measuring the Strangeness of Strange Attractors. Phys. D **9**, 189–208, (1983)

GP83b. Grassberger, P., Procaccia, I.: Characterization of Strange Attractors. Phys. Rev. Lett. **50**, 346–349, (1983)

GPJ98. Gray, R.A., Pertsov, A.M., Jalife, J.: Spatial and temporal organization during cardiac fibrillation. Nature **392**, 75, (1998)

GPS95. Gordon, R., Power, A.J., Street, R.: Coherence for tricategories. Memoirs Amer. Math. Soc. **117**(558), (1995)

GR93. Golomb, D., Rinzel, J.: Phys. Dynamics of globally coupled inhibitory neurons with heterogeneity. Rev. E **48**, 4810–4814, (1993)

GRS01. Gorbunov, D.S., Rubakov, V.A., Sibiryakov, S.M.: Gravity waves from inflating brane or mirrors moving in adS(5). JHEP **0110**, 015, (2001)

GRS96. Gilks, W.R., Richardson, S., Spiegelhalter, D.J.: Markov Chain Monte Carlo in Practice. Chapman & Hall, (1996)

GRT02. Gisin N, Ribordy G, Tittel W, Zbinden H: Quantum cryptography Rev. Mod. Phys. **74**, 145–195, (2002)

GRW86. Ghirardi, G.C., Rimini, A., Weber, T.: Unified dynamics for microscopic and macroscopic systems. Phys. Rev. D, **34**(2), 470–491, (1986)

GS02. Gen, U., Sasaki, M.: Quantum radion on de Sitter branes. arXiv:gr-qc/0201031, (2002)

GS88. Giansanti, A., Simic, P.D.: Onset of dynamical chaos in topologically massive gauge theories. Phys. Rev. D **38**, 1352–1355, (1988)

GS91. Garfinkle, D., Strominger, A.: Semiclassical Wheeler wormhole production, Phys. Lett. B **256**, 146–149, (1991)

GS96. Giachetta, G., Sardanashvily, G.: Stress–Energy–Momentum of Affine–Metric Gravity. Class.Quant.Grav. **13** L, 67–72, (1996)

GS97. Giachetta, G., Sardanashvily, G.: Dirac Equation in Gauge and Affine-Metric Gravitation Theories. Int. J. Theor. Phys. **36**, 125–142, (1997)

GS98. Grosche, C., Steiner, F.: Handbook of Feynman path integrals. Springer Tracts in Modern Physics 145, Springer, Berlin, (1998)

GSD92. Garfinkel, A., Spano, M.L., Ditto, W.L., Weiss, J.N.: Controlling cardiac chaos. Science **257**, 1230–1235, (1992)

GSV05. Garriga. J., Schwartz-Perlov, D., Vilenkin, A., Winitzki, S.: Probabilities in the Inflationary Multiverse. arXiv:hep-th/0509184, (2005)

GSW87. Green, M.B., Schwarz. J.H., Witten, E.: Superstring Theory, 2 Vols., Cambridge Univ. Press, Cambridge, (1987)

GT00. Garriga. J., Tanaka, T.: Gravity in the brane-world. Phys. Rev. Lett. **84**, 2778, (2000)

GT01.	Gratton, S., Turok, N.: Homogeneous Modes of Cosmological Instantons. Phys. Rev. D **63**, 123514, (2001)
GT77.	Grossmann, S., Thomae, S.: Invariant distributions and stationary correlation functions of one-dimensional discrete processes. Z. Naturforsch. A **32**, 1353–1363, (1977)
GT87.	Gozzi, E., Thacker, W.D.: Classical adiabatic holonomy and its canonical structure. Phys. Rev. D **35**, 2398, (1987)
GT99.	Gratton, S., Turok, N.: Cosmological Perturbations from the No Boundary Path Integral. Phys. Rev. D **60**, 123507, (1999)
GV93.	Gasperini, M., Veneziano, G.: Pre-Big–Bang in String Cosmology, Astropart. Phys. **1**, 317, (1993)
GW82.	Gibbons, G.W., Wiltshire, D.L.: space-time as a membrane in higher dimensions. Nucl. Phys. B **717**, (1987)
GW99.	Goldberger, W.D., Wise, M.B.: Modulus stabilization with bulk fields. Phys. Rev. Lett. **83**, 4922, (1999)
GZ83.	Gates, S.J., Zwiebach, B.: Gauged $N = 4$ Supergravity Theory with a New Scalar Potential. Phys. Lett. **B123**, 200, (1983)
GZ94.	Glass, L., Zeng, W.: Bifurcations in flat-topped maps and the control of cardiac chaos. Int. J. Bif. Chaos **4**, 1061–1067, (1994)
Gal83.	Gallavotti, G.: The Elements of Mechanics. Springer-Verlag, Berlin, (1983)
Gar72.	García, P.: Connections and 1-jet fibre bundles. Rend. Sem. Univ. Padova, **47**, 227, (1972)
Gar77.	García, P.: Gauge algebras, curvature and symplectic structure. J. Diff. Geom., **12**, 209, (1977)
Gar85.	Gardiner, C.W.: Handbook of Stochastic Methods for Physics, Chemistry and Natural Sciences, (2nd ed) Springer-Verlag, New York, (1985)
Gas96.	Gaspard, P.: Hydrodynamic modes as singular eigenstates of the Liouvillian dynamics: Deterministic diffusion. Phys. Rev. E **53**, 4379–4401, (1996)
Gau84.	Gauduchon, P.: La 1-forme de torsion d'une variété hermitienne compacte. Math. Ann. **267**, 495, (1984)
Gef06.	Gefter, A.: Stephen Hawking's Strange New Universe. New Sci. 22 April, (2006)
Geo88.	Georgii, H.O.: Gibbs Measures and Phase Transitions. Walter de Gruyter, Berlin, (1988)
Ger67.	Geroch, R.P.: Topology in general relativity. J. Math. Phys. **8**(4), 782, (1967)
Ghe90.	Ghez, C.: Introduction to motor system. In: Kandel, E.K. and Schwarz. J.H. (eds.) Principles of neural science. 2nd ed. Elsevier, Amsterdam, 429–442, (1990)
Ghe91.	Ghez, C.: Muscles: Effectors of the Motor Systems. In: Principles of Neural Science. 3rd Ed. (Eds. E.R. Kandel. J.H. Schwartz, T.M. Jessell), Appleton and Lange, Elsevier, 548–563, (1991)
Gia92.	Giachetta G.: Jet manifolds in nonholonomic mechanics. J. Math. Phys. **33**, 1652, (1992)

Gib85. Gibbons, G.W.: Aspects of Supergravity Theories. In Supersymmetry, Supergravity, and Related Topics, eds. F. del Aguila. J.A. de Azcarraga and L.E. Ibanez. World Scientific, Singapore, 346-351, (1985)

Gil81. Gilmor, R.: Catastrophe Theory for Scientists and Engineers. Wiley, New York, (1981)

Gla63. Glauber, R.J. Time-dependent statistics of the Ising model. J. Math. Phys. **4**, 294, (1963)

Gla63a. Glauber, R.J.: The Quantum Theory of Optical Coherence. Phys. Rev. **130**, 2529–2539, (1963)

Gla63b. Glauber, R.J.: Coherent and Incoherent States of the Radiation Field. Phys. Rev. **131**, 2766–2788, (1963)

Gle87. Gleick. J.: Chaos: Making a New Science. Penguin–Viking, New York, (1987)

Gol56. Goldberg, S.I.: Construction of universal bundles. Ann. Math. **63**, 64, (1956)

Gol65. Goldberg, S.I.: Curvature and Homology. Dover, New York, (1965)

Gol99. Goldberger, A.L.: Nonlinear Dynamics, Fractals, and Chaos Theory: Implications for Neuroautonomic Heart Rate Control in Health and Disease. In: Bolis CL, Licinio J, eds. The Autonomic Nervous System. World Health Organization, Geneva, (1999)

Gol99. Gold, M.: A Kurt Lewin Reader, the Complete Social Scientist. Am. Psych. Assoc., Washington, (1999)

Gom94. Gómez. J.C.: Using symbolic computation for the computer aided design of nonlinear (adaptive) control systems. Tech. Rep. EE9454, Dept. Electr. and Comput. Eng., Univ. Newcastle, Callaghan, NSW, AUS, (1994)

Goo98. Goodwine. J.W.: Control of Stratified Systems with Robotic Applications. PhD thesis, California Institute of Technology, Pasadena, Cal, (1998)

Gor26. Gordon, W.: Der Comptoneffekt nach der Schrödingerschen Theorie. Z. Phys., **40**, 117–133, (1926)

Got71. Goto, T.: Relativistic quantum mechanics of one-dimensional mechanical continuum and subsidary condition of dual resonance model. Prog. Theor. Phys, **46**, 1560, (1971)

Got82. Gotay, M.J.: On coisotropic imbeddings of presymplectic manifolds. Proc. Amer. Math. Soc. **84**, 111, (1982)

Got91a. Gotay, M.J.: A multisymplectic framework for classical field theory and the calculus of variations. I: Covariant Hamiltonian formalism. In M. Francaviglia (ed.), Mechanics, analysis and geometry: 200 years after Lagrangian. North-Holland, Amsterdam, 203–235, (1991)

Got91b. Gotay, M.J.: A multisymplectic framework for classical field theory and the calculus of variations. II: Space + time decomposition. Differ. Geom. Appl. **1**(4), 375–390, (1991)

Got91c. Gotay, M.: A multisymplectic framework for classical field theory and the calculus of variations. I. Covariant Hamiltonian

	formalism. In: M.Francaviglia (ed.) Mechanics, Analysis and Geometry: 200 Years after Lagrange. 203-235. North-Holland, Amsterdam, (1991)
Got96.	Gottlieb, H.P.W.: Question #38. What is the simplest jerk function that gives chaos? Am. J. Phys., **64**(5), 525, (1996)
Got97.	Gottesman, D.: Stabilizer codes and quantum error correction. PhD thesis, CalTech, Pasadena, (1997)
Gou95.	Goubault, E.: The Geometry of Concurrency. PhD thesis, Ecole Normale Supérieure, (1995)
Goz83.	Gozzi, E.: Functional-integral approach to Parisi–Wu stochastic quantization: Scalar theory. Phys. Rev. D **28**, 1922, (1983)
Gra90.	Granato, E.: Phase transitions in Josephson–junction ladders in a magnetic field. Phys. Rev. B **42**, 4797–4799, (1990)
Gre00.	Greene, B.R.: The Elegant Universe: Superstrings, Hidden Dimensions, and the Quest for the Ultimate Theory. Random House, (2000)
Gre51.	Green, M.S.: Brownian motion in a gas of non–interacting molecules. J. Chem. Phys. **19**, 1036–1046, (1951)
Gre79.	Greene, J.M.: A method for determining a stochastic transition. J. Math. Phys. **20**, 1183–1201, (1979)
Gre96.	Greene, B.R.: String Theory on Calabi–Yau Manifolds. Lectures given at the TASI-96 summer school on Strings, Fields and Duality, (1996)
Gri02.	Griffiths, R.B.: Consistent Quantum Theory. Cambridge University Press, Cambridge, (2002)
Gri83a.	Griffiths, P.A.: Exterior Differential Systems and the Calculus of Variations, Birkhauser, Boston, (1983)
Gri83b.	Griffiths, P.A.: Infinitesimal variations of Hodge structure. III. Determinantal varieties and the infinitesimal invariant of normal functions. Compositio Math., **50**(2-3), 267–324, (1983)
Gri84.	Griffiths, R.B.: Consistent Histories and the Interpretation of Quantum Mechanics. J. Stat. Phys. **36**, 219, (1984)
Gro69.	Grossberg, S.: Embedding fields: A theory of learning with physiological implications. J. Math. Psych. **6**, 209–239, (1969)
Gro82.	Grossberg, S.: Studies of Mind and Brain. Dordrecht, Holland, (1982)
Gro83.	Grothendieck, A.: Pursuing stacks (Unpublished manuscript, distributed from UCNW) Bangor, UK, (1983)
Gro88.	Grossberg, S.: Neural Networks and Natural Intelligence. MIT Press, Cambridge, MA, (1988)
Gro99.	Grossberg, S.: How does the cerebral cortex work? Learning, attention and grouping by the laminar circuits of visual cortex. Spatial Vision **12**, 163–186, (1999)
Gru00.	Grunwald, P.: The minimum description length principle. J. Math. Psych., **44**, 133–152, (2000)
Gru99.	Grunwald, P.: Viewing all models as 'probabilistic'. Proceedings of the Twelfth Annual Conference on Computational Learning Theory (COLT' 99), Santa Cruz, CA, (1999)
Gua04a.	Gualtieri, M.: Generalized complex geometry. arXiv:math.DG/0401221, (2004a)

Gua04b.	Gualtieri, M.: Generalized geometry and the Hodge decomposition. arXiv:math.DG/04090903, (2004b)
Gun03.	Gunion. J.F.: Class Notes on Path–Integral Methods. U.C. Davis, 230B, (2003)
Gut77.	Gutzwiller, M.C.: Chaos in Classical and Quantum Mechanics. J. Math. Phys. **18**, 806, (1977)
Gut81.	Guth, A.H.: The Inflationary Universe: A Possible Solution to the Horizon and Flatness Problems. Phys. Rev. D **23**, 347, (1981)
Gut90.	Gutzwiller, M.C.: Chaos in Classical and Quantum Mechanics. Springer, New York, (1990)
Gut98.	Gutkin, B.S., Ermentrout, B.: Dynamics of membrane excitability determine interspike interval variability: A link between spike generation mechanisms and cortical spike train statistics. Neural Comput., **10**(5), 1047–1065, (1998)
HB87.	Havlin, S., Ben-Avraham, D.: Diffusion in disordered media. Adv. Phys. **36**, 695-798, (1987)
HBB96.	Houk, J.C., Buckingham. J.T., Barto, A.G.: Models of the cerebellum and motor learning. Behavioral and Brain Sciences, **19**(3), 368–383, (1996)
HC99.	Hilbert, D., Cohn-Vossen, S.: Geometry and the Imagination. (Reprint ed.), Amer. Math. Soc, (1999)
HCK02a.	Hong, H., Choi, M.Y., Kim, B.J.: Synchronization on small-world networks. Phys. Rev. E **65**, 26139, (2002)
HCK02b.	Hong, H., Choi, M.Y., Kim, B.J.: Phase ordering on small-world networks with nearest-neighbor edges. Phys. Rev. E **65**, 047104, (2002)
HCT97.	Hall, K., Christini, D.J., Tremblay, M., Collins, J.J., Glass, L., Billette, J.: Dynamic Control of Cardiac Alternans. Phys. Rev. Lett. **78**, 4518–4521, (1997)
HDD98.	Arkani-Hamed, N., Dimopoulos, S., Dvali, G.: Phenomenology, Astrophysics and Cosmology of theories with sub-millimeter dimensions and TeV scale quantum gravity. Phys. Lett. **B429**, 263, (1998)
HDK00.	Arkani-Hamed, N., Dimopoulos, S., Kaloper, N., Sundrum, R.: A small cosmological constant from a large extra dimension.. Phys. Lett. B **480**, 193, (2000)
HDR02.	Hasty. J., Dolnik, M., Rottschafer, V., Collins. J.J.: A synthetic gene network for entraining and amplifying cellular oscillations. Phys. Rev. Lett. **88**(14), 1–4, (2002)
HE73.	Hawking, S.W., Ellis, G.F.R.: The large scale structure of space-time. Cambridge Univ. Press, (1973)
HE79.	Hawking, S.W., Israel, W. (ed.): General relativity: an Einstein centenary survey. Cambridge Univ. Press, Cambridge, (1979)
HF67.	Hohl, F., Feix, M.R.: Numerical Experiments with a One-Dimensional Model for a Self–Gravitating Star System. Astrophys. J. **147**, 1164–1180, (1967)

HGM94.	Hartle, J.B., Gell-Mann, M.: Time Symmetry and Asymmetry in Quantum Mechanics and Quantum Cosmology in Physical Origins of Time Asymmetry ed. by J. Halliwell. J. Perez-Mercader, and W. Zurek. Cambridge Univ. Press, Cambridge, (1994)
HH01.	Horodecki, P., Horodecki, R.: Distillation and bound entanglement. Quant. Inf. Comp. **1**(1), 45, (2001)
HH02.	Hawking, S.W., Hertog, T.: Why Does Inflation Start at the Top of the Hill? Phys. Rev. D **66**, 123509, (2002)
HH04.	Hertog, T., Horowitz, G.T.: Towards a Big–Crunch Dual. JHEP **0407**, 073, (2004)
HH06.	Hawking, S.W., Hertog, T.: Populating the Landscape: A Top Down Approach. Phys. Rev. D **73**, 123527, (2006)
HH52.	Hodgkin, A.L., Huxley, A.F.: A quantitative description of membrane current and application to conduction and excitation in nerve. J. Physiol., **117**, 500–544, (1952)
HH81.	Hartle, J.B., Horowitz, G.T.: Ground–State Expectation value of the Metric in the $1/N$ or Semiclassical Approximation to Quantum Gravity. Phys. Rev. D **24**, 257, (1981)
HH83.	Hartle, J.B., Hawking, S.W.: The Wave Function of the Universe. Phys. Rev. D **28**, 2960, (1983)
HH90.	Halliwell. J.J., Hartle, J.B.: Integration Contours for the No-Boundary Wave Function of the Universe. Phys. Rev. D **41**, 1815, (1990)
HH94.	Heagy, J.F., Hammel, S.M.: The birth of strange nonchaotic attractors. Physica D **70**, 140-153, (1994)
HHL97.	Heinz, U., Hu, C., Leupold, S., Matinyan, S. Müller, B.: Thermalization and Lyapunov exponents in Yang-Mills-Higgs theory. Phys. Rev. D **55**, 2464, (1997)
HHR00.	Hawking, S.W., Hertog, T., Reall, H.S.: Brane New World. Phys. Rev. D **62**, 043501, (2000)
HHR01.	Hawking, S.W., Hertog, T., Reall, H.S.: Trace Anomaly Driven Inflation. Phys. Rev. D **63**, 083504, (2001)
HHT00.	Hawking, S.W., Hertog, T., Turok, N.: Gravitational Waves in Open de Sitter Space. arXiv:hep-th/0003016, (2000)
HI97.	Hoppensteadt, F.C., Izhikevich, E.M.: Weakly Connected Neural Networks. Springer, New York, (1997)
HKK03.	Hori, K., Katz, S., Klemm, A., Pandharipande, R., Thomas, R., Vafa, C., Vakil, R., Zaslow, E.: Mirror symmetry. Clay Mathematics Monographs 1, American Mathematical Society, Providence, Clay Mathematics Institute, Cambridge, MA, (2003)
HKM76.	Hawking, S.W., King, A.R., McCarthy, P.J.: A new topology for curved space-time which incorporates the causal, differential, and conformal structures. J. Math. Phys. **17**, 174–181, 1976.
HKS01.	Hellerman, S., Kaloper, N., Susskind, L.: String Theory and Quintessence. arXiv:hep-th/0104180, (2001)
HL84.	Horsthemke, W., Lefever, R.: Noise–Induced Transitions. Springer, Berlin, (1984)

HL84. Hamoui, A., Lichnerowicz, A.: Geometry of dynamical systems with time–dependent constraints and time-dependent Hamiltonians: An approach towards quantization. J. Math. Phys. **25**, 923, (1984)

HL86a. Hale, J.K., Lin, X.B.: Symbolic dynamics and nonlinear semiflows, Ann. Mat. Pur. Appl. **144**(4), 229–259, (1986)

HL86b. Hale, J.K., Lin, X.B.: Examples of transverse homoclinic orbits in delay equations, Nonlinear Analysis **10**, 693–709, (1986)

HL90. Halliwell, J.J., Louko, J.: Steepest-Descent Contours in the Path Integral Approach to Quantum Cosmology, III. A General Method with Applications to Anisotropic Minisuperspace Models. Phys. Rev. D **42**, 3997, (1990)

HL93. Hale, J.K., Lunel, S.M.V.: Introduction to Functional Differential Equations. Springer, New York. (1993)

HLQ95. Hartley, T.T., Lorenzo, C.F., Qammer, H.K.: Chaos on a Fractional Chua's System, IEEE Trans. Circ. Sys. Th. App. **42**, 485–490, (1995)

HLR86. Horowitz, G.T., Lykken. J., Rohm, R., Strominger, A.: Purely Cubic Action for String Field Theory. Phys. Rev. Lett. **57**, 283, (1986)

HM82. Hawking, S.W., Moss, I.G.: Supercooled phase transitions in the very early universe. Phys. Lett. **B110**, 35–8, (1982)

HM88. Hitchin, N.J., Murray, M.K.: Spectral curves and the ADHMN method, Commun. Math. Phys. 114, 463–474, (1988)

HM89. Hurtubise. J., Murray, M.K.: On the construction of monopoles for the classical groups, Commun. Math. Phys. 122, 35–89, (1989)

HM90. Hurtubise. J., Murray, M.K.: Monopoles and their spectral data, Commun. Math. Phys. 133, 487–508, (1990)

HM97. Hartle, J.B., Marolf, D.: Comparing Formulations of Generalized Quantum Mechanics for Reparametrization Invariant Systems. Phys. Rev. D **56**, 6247–6257, (1997)

HMM95. Hitchin, N.J., Manton, N.S., Murray, M.K.: Symmetric Monopoles, Nonlinearity 8, 661–692, (1995)

HMM95. Hehl, F., McCrea. J., Mielke, E., Ne'eman, Y.: Metric–affine gauge theory of gravity. Phys. Rep. **258**, 1, (1995)

HMP97. S.F. Huelga, C. Macchiavello, T. Pellizzari, A.K. Ekert, M.B. Plenio, J.I. Cirac: Improvement of Frequency Standards with Quantum Entanglement. Phys. Rev. Lett. **79**, 3865–3868, (1997)

HMS82. Hawking, S.W., Moss, I.G., Stewart. J.M.: Bubble Collisions in the Very Early Universe. Phys. Rev. D **26**, 2681, (1982)

HN54. Huxley, A.F., Niedergerke, R.: Changes in the cross–striations of muscle during contraction and stretch and their structural interpretation. Nature, **173**, 973–976, (1954)

HNT01. Haskell, E., Nykamp, D.Q., Tranchina, D.: Network: Computation in Neural Systems **12**, 14, (2001)

HO01. Hunt, B.R., Ott, E.: Fractal Properties of Robust Strange Nonchaotic Attractors. Phys. Rev. Lett. **87**, 254101, (2001)

HO90.	Henkel, R.D., Opper, M.: Distribution of internal fields and dynamics of neural networks. Europhys. Lett. **11**, 403-408, (1990)
HP02.	Horowitz, G.T., Polchinski. J.: Instability of spacelike and null orbifold singularities. Phys. Rev. D **66**, 103512, (2002)
HP93.	Hameroff, S.R., Penrose, R.: Conscious events as orchestrated space-time selections. Journal of Consciousness Studies, **3**(1), 36-53, (1996)
HP96.	Hameroff, S.R., Penrose, R.: Orchestrated reduction of quantum coherence in brain microtubules: A model for consciousness. In: Hameroff, S. R., Kaszniak, A.W. and Scott, A.C. Eds: Toward a Science of Consciousness: the First Tucson Discussion and Debates, 507–539. MIT Press, Cambridge, MA, (1996)
HP96.	Hawking, S.W., Penrose, R.: The Nature of Space and Time. Princetone Univ. Press, Princetone, NJ, (1996)
HPS70.	Hirsch, M., Pugh, C., Shub, M.: Stable manifolds and hyperbolic sets. Proc. of Symposia in Pure Mathematics-Global Analysis, **14**, 133–163, (1970)
HR83.	Hut, P., Rees, M.J.: How stable is our vacuum? Nature, 508–509, (1983)
HR83.	Hänggi, P., Riseborough, P.: Activation rates in bistable systems in the presence of correlated noise. Phys. Rev. A **27**, 3379–3382, (1983)
HS01.	Himemoto, Y., Sasaki, M.: Brane–world inflation without inflaton on the brane. Phys. Rev. D **63**, 044015, (2001)
HS74.	Hawking, S.W., Sachs, R.K.: Causally Continuous Space-times. Comm. Math. Phys. **35**, 287, (1973)
HS74.	Hirsch, M.W., Smale, S.: Differential Equations, Dynamical Systems and Linear Algebra. Academic Press, New York, (1974)
HS88.	Hale, J.K., Sternberg, N.: Onset of Chaos in Differential Delay Equations, J. Comp. Phys. **77**(1), 221–239, (1988)
HS88.	Harsanyi, J.C., Selten, R.: A general theory of equilibrium selection in games. MIT Press, Cambridge, (1988)
HS98.	Hofbauer. J., Sigmund, K.: Evolutionary games and population dynamics, Cambridge Univ. Press, U.K., (1998)
HS98.	Henningson, M., Skenderis, K.: The Holographic Weyl Anomaly. JHEP **9807**, 023, (1998)
HSK92.	Hauser. J., Sastry, S., Kokotovic, P.: Nonlinear control via approximate input–output linearization: The ball and beam example, IEEE Trans. Aut. Con., AC–37, 392–398, (1992)
HT00.	Hertog, T., Turok, N.: Gravity Waves from Instantons. Phys. Rev. D **62**, 083514, (2000)
HT01.	Hawking, S.W., Hertog, T.: Living with ghosts. arXiv:hep-th/0107088, (2001)
HT85.	Hopfield. J.J., Tank, D.W.: Neural computation of decisions in optimisation problems. Biol. Cybern., **52**, 114–152, (1985)
HT92.	Henneaux, M., Teitelboim, C.: Quantization of Gauge systems. Princeton Univ. Press, (1992)

HT93. Hunt, L, Turi. J.: A new algorithm for constructing approximate transformations for nonlinear systems. IEEE Trans. Aut. Con., AC–38,1553–1556, (1993)

HT98. Hawking, S.W., Turok, N.: Open Inflation without False Vacua. Phys. Lett. **B425**, 25, (1998)

HTS02. Himemoto, Y., Tanaka, T., Sasaki, M.: A bulk scalar in the braneworld can mimic the 4d inflaton dynamics. Phys. Rev. D **65**, 104020 (2002)

HV93. Hale, J.K., Verduyn-Lunel, S.M.: Introduction to Functional Differential Equations, Applied Math. Sci. **99**, Springer Verlag, New York, (1993)

HW04. Hamber, H., Williams, R.M.: Non-perturbative Gravity and the Spin of the Lattice Gravition. Phys. Rev. D **70**, 124007, (2004)

HW82. Hameroff, S.R., Watt, R.C.: Information processing in microtubules. J. Theo. Bio. **98**, 549-561, (1982)

HW83. Van der Heiden, U., Walther, H.-O.: Existence of chaos in control systems with delayed feedback, J. Diff. Equs. **47**, 273–295, (1983)

HW83. Hameroff, S.R., Watt, R.C.: Do anesthetics act by altering electron mobility? Anesth. Analg., **62**, 936-40, (1983)

HW96. Horava, P., Witten, E.: Heterotic and type I string dynamics from eleven dimensions, Nucl. Phys. B **460**, 506, (1996)

HWM97. Hartle, J.B., Williams, R.M., Miller, W.A., Williams, R.: Signature of the Simplicial Supermetric, Class. Quant. Grav. **14**, 2137-2155, (1997)

HYW03. Hao, N., Yildirim, N., Wang, Y., Elston, T.C., Dohlman, H.G.: Regulators of G protein signaling and transient activation of signaling: experimental and computational analysis reveals negative and positive feedback controls on G protein activity. J. Biol. Chem., 278, 46506–46515, (2003)

Haa92. Haag, R.: Local Quantum Physics. Springer, Berlin, (1992)

Hac03. Hackermüller, L. *et al.*: The Wave Nature of Biomolecules and Flurofullerenes. Phys. Rev. Lett. **91**, 090408, (2003)

Had75. Hadjidemetriou, J.D.: The continuation of periodic orbits from the restricted to the general three–body problem, Celestial Mech., **12**, 155–174, (1975)

Hak00. Haken, H.: Information and Self-Organization: A Macroscopic Approach to Complex Systems. Springer, Berlin, (2000)

Hak02. Haken, H.: Brain Dynamics, Synchronization and Activity Patterns in Pulse-Codupled Neural Nets with Delays and Noise, Springer, New York, (2002)

Hak83. Haken, H.: Synergetics: An Introduction (3rd ed) Springer, Berlin, (1983)

Hak91. Haken, H.: Synergetic Computers and Cognition. Springer-Verlag, Berlin, (1991)

Hak93. Haken, H.: Advanced Synergetics: Instability Hierarchies of Self-Organizing Systems and Devices (3nd ed.) Springer, Berlin. (1993)

Hak96.	Haken, H.: Principles of Brain Functioning: A Synergetic Approach to Brain Activity, Behavior and Cognition, Springer, Berlin, (1996)
Hal94.	Hall, T.: The creation of horseshoes. Nonlinearity, **7**(3), 861–924, (1994)
Ham00.	Hamber, H.W.: On the gravitational scaling dimensions, Phys. Rev. D 61 (2000) 124008.
Ham82.	Hamilton, R.: Three-manifolds with positive Ricci curvature. J. Diff. Geom. **17**, 255–306, (1982)
Ham86.	Hamilton, R.: Four-manifolds with positive curvature operator. J. Diff. Geom. **24**, 153–179, (1986)
Ham87.	Hameroff, S.R.: Ultimate Computing: Biomolecular Consciousness and Nanotechnology. North-Holland, Amsterdam, (1987)
Ham88.	Hamilton, R.: The Ricci flow on surfaces. Contem. Math. **71**, 237–261, (1988)
Ham93.	Hamilton, R.: The formation of singularities in the Ricci flow, volume II. Internat. Press, (1993)
Ham98.	Hameroff, S.: Quantum computation in brain microtubules? The Penrose-Hameroff Orch OR model of consciousness. Philos. Trans. R. Soc. London Ser. A **356**, 1869-1896, (1998)
Ham82.	Hamilton, R.S.: Three-manifolds with positive Ricci curvature. J. Diff. Geom. **17**, 255–306, (1982)
Han99.	Handel, M.: A fixed-point theorem for planar homeomorphisms. Topology, **38**(2), 235–264, (1999)
Har03.	Hartle, J.B.: The State of the Universe, in The Future of Theoretical Physics and Cosmology: Stephen Hawking 60th Birthday Symposium, ed. by G.W. Gibbons, E.P.S. Shellard, and S.J. Ranken. Cambridge Univ. Press, UK, (2003)
Har03.	Hartle, J.B.: Theories of Everything and Hawking's Wave Function of the Universe. In The Future of Theoretical Physics and Cosmology, ed. by G.W. Gibons, E.P.S. Shellard and S.J. Rankin. Cambridge Univ. Press, Cambridge, (2003)
Har03.	Hartnoll, S.A.: Compactification, topology change and surgery theory. arXiv:hep-th/0302072v2, (2003)
Har04.	Hartle, J.B.: Bohmian Histories and Decoherent Histories. Phys. Rev. **A69**, 042111, (2004)
Har04a.	Hartle, J.B.: Linear Positivity and Virtual Probability. Phys. Rev. **A70**, 022104, (2004)
Har04b.	Hartle, J.B.: Anthropic Reasoning and Quantum Cosmology. The New Cosmology, ed. R. Allen *et al.*, AIP, (2004)
Har05a.	Hartle, J.B.: Anthropic Reasoning and Quantum Cosmology, Proceedings of Strings and Cosmology Conference, Texas A&M, AIP Conf. Proc. 743, 298, (2005)
Har05b.	Hartle, J.B.: Generalizing Quantum Mechanics for Quantum space-time. Proc. of the 23rd Solvay Conference, The Quantum Structure of Space and Time, Brussels, (2005)
Har06.	Hartle, J.B.: Generalizing Quantum Mechanics for Quantum Gravity. Int. J. Theor. Phys. **45**, 1390-1396, (2006)

Har85a. Hartle, J.B.: Simplicial Minisuperspace I. General Discussion. J. Math. Phys. **26**, 804, (1985)

Har85b. Hartle, J.B.: Unruly Topologies in Two Dimensional Quantum Gravity, Class. Quant. Grav. **2**, 707, (1985)

Har90. Hartle, J.B.: Excess Baggage. In Elementary Particles and the Universe: Essays in Honor of Murray Gell-Mann ed. by J. Schwarz. Cambridge Univ. Press, Cambridge, (1990)

Har91b. Hartle, J.B.: space-time Coarse Grainings in Non-Relativistic Quantum Mechanics. Phys. Rev. D **44**, 3173-3196, (1991)

Har92. Harris-Warrick, R.M. (ed): The Stomatogastric Nervous System. MIT Press, Cambridge, MA, (1992)

Har94a. Hartle, J.B.: Unitarity and Causality in Generalized Quantum Mechanics for Non-Chronal space-times. Phys. Rev. D **49**, 6543, (1994)

Har94c. Hartle, J.B.: Unitarity and Causality in Generalized Quantum Mechanics for Non–Chronal space-times, Phys Rev. D **49**, 6543, (1994)

Har96. Hart, W.D.: Dualism. In A Companion to the Philosophy of Mind, Blackwell, Oxford, (1996)

Hat02. Hatcher, A.: Algebraic Topology. Cambridge Univ. Press, Cambridge, (2002)

Hat77a. Hatze, H.: A myocybernetic control model of skeletal muscle. Biol. Cyber. **25**, 103–119, (1977)

Hat77b. Hatze, H.: A complete set of control equations for the human musculoskeletal system. J. Biomech. **10**, 799–805, (1977)

Hat78. Hatze, H.: A general myocybernetic control model of skeletal muscle. Biol. Cyber., **28**, 143–157, (1978)

Haw75. Hawking, S.W.: Particle creation by black holes. Commun. Math. Phys, **43**, 199-220, (1975)

Haw78. Hawking, S.W.: Quantum Gravity And Path Integrals. Phys. Rev. D **18**, 1747–1753, (1978)

Haw78b. S. W. Hawking. Space-time foam. Nucl. Phys. **B144**, 349–362, 1978.

Haw84a. Hawking, S.W.: The Cosmological Constant in Probably Zero. Phys. Lett. **B134**, 403, (1984)

Haw84b. Hawking, S.W.: The Quantum State of the Universe, Nucl. Phys. **B239**, 257, (1984)

Haw88. Hawking, S.W.: Wormholes in space-time. Phys. Rev. **D37**, 904–910, 1988.

Haw92. Hawking, S.W.: The chronology protection conjecture. Phys. Rev. **D46**, 603–611, 1992.

Hay00. Hayashi, S.: A C^1 make or break lemma. Bull. Braz. Math. Soc., **31**, 337–350, (2000)

Hay94. Haykin, S.: Neural Networks: A Comprehensive Foundation. Macmillan, (1994)

Hay97. Hayashi, S.: Connecting invariant manifolds and the solution of the C^1 stability and Ω-stability conjectures for flows. Annals of Math., **145**, 81–137, (1997)

Hay98.	Hayashi, S.: Hyperbolicity, stability, and the creation of homoclinic points. In Documenta Mathematica, Extra Volume ICM, Vol. II, (1998)
Heb49.	Hebb, D.O.: The Organization of Behavior, Wiley, New York, (1949)
Hec77.	Heckhausen, H.: Achievement motivation and its constructs: a cognitive model. Motiv. Emot, **1**, 283–329, (1977)
Hec87.	Hecht-Nielsen, R.: Counterpropagation networks. Applied Optics, **26**(23), 4979–4984, (1987)
Hee90.	Heermann, D.W.: Computer Simulation Methods in Theoretical Physics. (2nd ed), Springer, Berlin, (1990)
Hel01.	Helgason, S.: Differential Geometry, Lie Groups and Symmetric Spaces. Am. Math. Soc., Providence, (2001)
Hel986.	Helmholtz, H.: Uber der physikalische Bedeutung des Princips der kleinsten. J. Reine Angew. Math. **100**, 137-166, (1886)
Hen06.	Henson. J.: The Causal Set Approach to Quantum Gravity. arXiv:gr-qc/0601121, (1980)
Hen66.	Hénon, M.: Sur la topologie des lignes de courant dans un cas particulier. C. R. Acad. Sci. Paris A, **262**, 312-314, (1966)
Hen69.	Hénon, M.: Numerical study of quadratic area preserving mappings. Q. Appl. Math. **27**, (1969)
Hen76.	Hénon, M.: A two-dimensional mapping with a strange attractor. Com. Math. Phys. **50**, 69–77, (1976)
Hig87.	Higuchi, A.: Symmetric tensor spherical harmonics on the N-sphere and their application to the de Sitter group $SO(N,1)$. J. Math. Phys. **28** (7), 1553, (1987)
Hil00.	Hilfer, R. (ed): Applications of Fractional Calculus in Physics. World Scientific, Singapore, (2000)
Hil38.	Hill, A.V.: The heat of shortening and the dynamic constants of muscle, Proc. R. Soc. B, **76**, 136–195, (1938)
Hil94.	Hilborn, R.C.: Chaos and Nonlinear Dynamics: An Introduction for Scientists and Engineers. Oxford Univ. Press, Oxford, (1994)
Hir66.	Hirzebruch, F.: Topological Methods in Algebraic Geometry, Springer-Verlag, Berlin, (1966)
Hir71.	Hirota, R.: Exact Solution of the Korteweg–de Vries Equation for Multiple Collisions of Solitons. Phys. Rev. Lett. **27**, 1192–1194, (1971)
Hir76.	Hirsch, M.W.: Differential Topology. Springer, New York, (1976)
Hit03.	Hitchin, N.J.: Generalized Calabi–Yau manifolds. Quat. J. Math. **54**, 281-308, (2003)
Hit05.	Hitchin, N.J.: Instantons, Poisson structures and generalized Kähler geometry. arXiv:math.DG/0503432, (2005)
Hit82.	Hitchin, N.J.: Monopoles and Geodesics, Commun. Math. Phys. 83, (1982), 579–602.
Hit83.	Hitchin, N.J.: On the construction of monopoles, Commun. Math. Phys. 89, (1983), 145–190.
Hod64.	Hodgkin, A.L.: The Conduction of the Nervous Impulse. Liverpool Univ. Press, Liverpool, (1964)

Hoo86. Hoover, W.G.: Molecular dynamics. Lecture Notes in Physics **258**, Springer, Heidelberg, (1986)

Hoo95. Hoover, W.G.: Remark on Some Simple Chaotic Flows. Phys. Rev. E, **51**(1), 759–760, (1995)

Hop82. Hopfield. J.J.: Neural networks and physical systems with emergent collective computational activity. Proc. Natl. Acad. Sci. USA., **79**, 2554–2558, (1982)

Hop82. Hopfield, J.J.: Neural networks and physical systems with emergent collective computational abilities. Proc. Natl. Acad. Sci. USA **79**, 2554, (1982)

Hop84. Hopfield. J.J.: Neurons with graded response have collective computational properties like those of two–state neurons. Proc. Natl. Acad. Sci. USA, **81**, 3088–3092, (1984)

Hor89. Horowitz, G.T.: Exactly soluble diffeomorphism invariant theories. Comm. Math. Phys. **125**(3), 417, (1989)

Hor91. Horowitz, G.T.: Topology change in classical and quantum gravity. Class. Quant. Grav. **8**, 587–602, 1991.

Hou79. Houk. J.C.: Regulation of stiffness by skeletomotor reflexes. Ann. Rev. Physiol., **41**, 99–114, (1979)

Hul00. Hull, J.C.: Options, Futures, and Other Derivatives. (4th ed.), Prentice Hall, New Jersey, (2000)

Hul84. Hull, C.M.: New Gauging of $N=8$ Supergravity. Phys. Rev. D **30**, 760, (1984)

Hur93. Hurmuzlu, Y.: Dynamics of bipedal gait. J. Appl. Mech., **60**, 331–343, (1993)

Hux57. Huxley, A.F.: Muscle structure and theories of contraction. Progr. Biophys. Chem., **7**, 255–328, (1957)

Hux898. Huxley, T.H.: On the Hypothesis that Animals are Automata, its History. Reprinted in Method, Results: Essays by Thomas H. Huxley. D. Appleton, Company, New York (1898)

IA07. Ivancevic, V., Aidman, E.: Life–space foam: A medium for motivational and cognitive dynamics. Physica, A **382**, 616–630, (2007)

IAG99. Ivanov, P. Ch., Amaral, L. A. N., Goldberger, A. L., Havlin, S., Rosenblum, M. B., Struzik, Z. & Stanley, H.E.: Multifractality in healthy heartbeat dynamics. Nature, **399**, 461–465, (1999)

IB03. Ivancevic, V., Beagley, N.: BMathematical twist reveals the agony of back pain. New Scientist, 9 Aug. (2003)

IB05. Ivancevic, V., Beagley, N.: Brain–like functor control machine for general humanoid biodynamics. Int. J. Math. Math. Sci. **11**, 1759–1779, (2005)

IGF99. Ioffe, L.B., Geshkenbein, V.B., Feigel'man, M.V., Fauchère, A.L., Blatter, G.: Environmentally decoupled sds-wave Josephson junctions for quantum computing. Nature **398**, 679-681, (1999)

II05. Ivancevic, V., Ivancevic, T.: Human–Like Biomechanics. Springer, Mechanical Engineering Ser., (2005)

II06a. Ivancevic, V., Ivancevic, T.: Natural Biodynamics. World Scientific, Series: Mathematical Biology, (2006)

II06b.	Ivancevic, V., Ivancevic, T.: Geometrical Dynamics of Complex Systems. Springer, Series: Microprocessor-Based and Intelligent Systems Engineering, Vol. 31, (2006)
II07a.	Ivancevic, V., Ivancevic, T.: High-Dimensional Chaotic and Attractor Systems. Springer, Series: Springer, Intelligent Systems, Control and Automation: Science and Engineering, Vol. 32, (2007)
II07b.	Ivancevic, V., Ivancevic, T.: Applied Differential Geometry: A Modern Introduction. World Scientific, Series: Mathematics, (2007)
IJB99a.	Ivancevic, T., Jain, L.C., Bottema, M.: New Two–feature GBAM–Neurodynamical Classifier for Breast Cancer Diagnosis. Proc. KES'99, IEEE Press, (1999)
IJB99b.	Ivancevic, T., Jain, L.C., Bottema, M.: A New Two–Feature FAM–Matrix Classifier for Breast Cancer Diagnosis. Proc. KES'99, IEEE Press, (1999)
IJK02.	Ichiki, K., Yahiro, M., Kajino, T., Orito, M., Mathews, G.J.: Observational constraints on dark radiation in brane cosmology. Phys. Rev. D **66**, 043521, (2002)
IJL02.	Iliescu T., John V., Layton W.: Convergence of finite element approximations of large eddy motion. Numer. Methods Partial Differential Equations, **18**, 689–710, (2002)
IJL03.	Iliescu T., John V., Layton W.J., Matthies G., Tobiska L.: A numerical study of a class of LES models, Int. J. Comput. Fluid Dyn. **17**, 75–85, (2003)
IK04.	Ibragimov, N.H., Kolsrud, T.: Lagrangian approach to evolution equations: symmetries and conservation laws, Nonlinear Dyn. **36**, 29–40, (2004)
IK80.	Iyanaga, S., Kawada, Y. (eds.): Pontryagin's Maximum Principle. In Encyclopedic Dictionary of Mathematics. MIT Press, Cambridge, MA, 295-296, (1980)
IKH87.	Ichikawa, Y.H., Kamimura, T., Hatori, T.: Physica D **29**, 247, (1987)
IKK97.	Ishibashi, N., Kawai, H., Kitazawa, Y., Tsuchiya, A.: A Large-N Reduced Model as Superstring. Nucl. Phys. **B498**, 467, (1997)
ILI95.	Ivancevic, V., Lukman, L., Ivancevic, T.: Selected Chapters in Human Biomechanics. Textbook (in Serbian) Univ. Novi Sad Press, Novi Sad, (1995)
ILW92.	Ivanov, A., Lani–Wayda, B., Walther, H.-O.: Unstable hyperbolic periodic solutions of differential delay equations, Recent Trends in Differential Equations, ed. R.P. Agarwal, 301–316, World Scientific, Singapore, (1992)
IN92.	Igarashi, E., Nogai, T.: Study of lower level adaptive walking in the saggital plane by a biped locomotion robot. Advanced Robotics, **6**, 441–459, (1992)
IN96.	Inoue, M., Nishi, Y.: Dynamical Behavior of Chaos Neural Network of an Associative Schema Model. Prog. Theoret. Phys. **95**, 837–850, (1996)

IP01a.	Ivancevic, V., Pearce, C.E.M.: Poisson manifolds in generalised Hamiltonian biomechanics. Bull. Austral. Math. Soc. **64**, 515–526, (2001)
IP01b.	Ivancevic, V., Pearce, C.E.M.: Topological duality in humanoid robot dynamics. ANZIAM J. **43**, 183–194, (2001)
IP05b.	Ivancevic, V., Pearce, C.E.M.: Hamiltonian dynamics and Morse topology of humanoid robots. Gl. J. Mat. Math. Sci. (to appear)
IS01.	Ivancevic, V., Snoswell, M.: Fuzzy–stochastic functor machine for general humanoid–robot dynamics. IEEE Trans. on Sys, Man, Cyber. B, **31**(3), 319–330, (2001)
IS08.	Ivancevic, V., Sharma, S.: Complexity in Human and Humanoid Bio-Mechanics. Int. J. Hum. Rob. (to appear), (2008)
IU94.	Ibragimov, N.H., Ünal G.: Equivalence transformations of Navier–Stokes equations, Istanbul Tek. Üniv. Bül., **47**, 203–207, (1994)
IZ80.	Itzykson, C., Zuber, J.: Quantum field theory. McGraw-Hill, New York, (1980)
Ich04.	Ichinomiya, T.: Frequency synchronization in random oscillator network. Phys. Rev. E **70**, 026116, (2004)
Ich05.	Ichinomiya, T.: Path-integral approach to the dynamics in sparse random network. Phys. Rev. E **72**, 016109, (2005)
Ida00.	Ida, D.: Brane–world cosmology. JHEP **0009**, 014, (2000)
Ike90.	Ikeda, S.: Some remarks on the Lagrangian Theory of Electromagnetism. Tensor, N.S. **49**, (1990)
Ila01.	Ilachinski, A.: Cellular automata. World Scientific, Singapore, (2001)
Imm97.	Immirzi, G.: Quantum Gravity and Regge Calculus. Nucl. Phys. Proc. Suppl. **57**, 65–72, (1997)
Ing97.	Ingber, L.: Statistical mechanics of neocortical interactions: Applications of canonical momenta indicators to electroencephalography, Phys. Rev. E, **55**(4), 4578–4593, (1997)
Ing98.	Ingber, L.: Statistical mechanics of neocortical interactions: Training and testing canonical momenta indicators of EEG, Mathl. Computer Modelling **27**(3), 33–64, (1998)
Ion97.	Ionicioiu, R.: Topology change from Kaluza-Klein dimensions, gr-qc/9709057, (1997)
Ion97.	Ionicioiu, R.: Building blocks for topology change in 3D, arXiv:gr-qc/9711069, (1997)
Ish84.	Isham, C.J.: In B. DeWitt, R.Stora, (ed.), Relativity, Groups and Topology, Les Houches Session XL. North Holland, Amsterdam, (1984)
Ish94.	Isham, C.J.: Quantum Logic and the Histories Approach to Quantum Theory. J. Math. Phys. **35**, 2157, (1994)
Ish97.	Isham, C.J.: Structural issues in quantum gravity, in General Relativity and Gravitation: World Scientific, Singapore, (1997)
Ish97.	Ishwar, B.: Nonlinear stability in the generalised restricted three body problem. Celest. Mech. Dyn. Astr. **65**, 253–289, (1997)

Isi03.	Isidro, J.M.: Duality, Quantum Mechanics and (Almost) Complex Manifolds. Mod. Phys. Lett. **A18**(28), 1975, (2003)
Isi04a.	Isidro, J.M.: Quantum Mechanics in Infinite Symplectic Volume. Mod. Phys. Lett. **A19**(5), 349, (2004)
Isi04b.	Isidro, J.M.: Quantum-Mechanical Dualities on the Torus. Mod. Phys. Lett. **A19**, 1733, (2004)
Isi89.	Isidori, A.: Nonlinear Control Systems. An Introduction, (2nd ed) Springer, Berlin, (1989)
Isi92.	Isichenko, M.B.: Percolation, statistical topography, and transport in random media. Rev. Mod. Phys. **64**, 961–1043, (1992)
Isr66.	Israel, W.: Singular Hypersurfaces and Thin Shells in General Relativity, Nuovo Cim. B **44S10**, 1, (1966)
Ito60.	Ito, K.: Wiener Integral and Feynman Integral. Proc. Fourth Berkeley Symp. Math., Stat., Prob., **2**, 227–238, (1960)
Iva02.	Ivancevic, V.: Generalized Hamiltonian biodynamics and topology invariants of humanoid robots. Int. J. Mat. Mat. Sci., **31**(9), 555–565, (2002)
Iva04.	Ivancevic, V.: Symplectic Rotational Geometry in Human Biomechanics. SIAM Rev., **46**(3), 455–474, (2004)
Iva05a.	Ivancevic, V.: Dynamics of Humanoid Robots: Geometrical and Topological Duality. In Biomathematics: Modelling and simulation (ed. J.C. Misra), World Scientific, Singapore (to appear)
Iva05b.	Ivancevic, V.: A Lagrangian model in human biomechanics and its Hodge–de Rham cohomology. Int. J. Appl. Math. Mech., **3**, 35–51, (2005)
Iva05c.	Ivancevic, V.: Lie–Lagrangian model for realistic human biodynamics. Int. J. Humanoid Robotics 3(2), 205–218, (2006)
Iva91.	Ivancevic, V.: Introduction to Biomechanical Systems: Modelling, Control and Learning (in Serbian) Scientific Book, Belgrade, (1991)
Iva95.	Ivancevic, T.: Some Possibilities of Multilayered Neural Networks' Application in Biomechanics of Muscular Contractions, Human Motion and Sport Training. Master Thesis (in Serbian), University of Novi Sad, YU, (1995)
Iye81.	Iyengar, B.K.S.: Light on Yoga. Unwin Publishers, London, (1981)
Izh99a.	Izhikevich, E.M.: Class 1 neural excitability, conventional synapses, weakly connected networks, and mathematical foundations of pulse-coupled models. IEEE Trans. Neu. Net., **10**, 499–507, (1999)
Izh99b.	Izhikevich, E.M.: Weakly Connected Quasiperiodic Oscillators, FM Interactions and Multiplexing in the Brain. SIAM J. Appl. Math., **59**(6), 2193–2223, (1999)
Izh01a.	Izhikevich, E.M.: Resonate-and-fire neurons. Neu. Net., **14**, 883–894, (2001)
	1. Izhikevich, E.M.: Synchronization of Elliptic Bursters, SIAM Rev. **43**(2), 315–344, (2001)

Izh04. Izhikevich, E.M.: Which model to use for cortical spiking neurons? IEEE Trans. Neu. Net., **15**, 1063–1070, (2004)

JAD00. Jozsa, R., Abrams, D.S., Dowling, J.P., Williams, C.P.: Quantum Clock Synchronization Based on Shared Prior Entanglement. Phys. Rev. Lett. 85, 2010, (2000)

JAD99. Jalife, J., Anumonwo, J.M., Delmar, M., Davidenko, J.M.: Basic Cardiac Electrophysiology for the Clinician. Futura Pub. Armonk, (1999)

JBC98. Jongen, G., Bollé, D., Coolen, A.C.: The XY spin–glass with slow dynamic couplings. J. Phys. A **31**, L737-L742, (1998)

JBO97. Just, W., Bernard, T., Ostheimer, M., Reibold, E., Benner, H.: Mechanism of time–delayed feedback control. Phys. Rev. Lett., **78**, 203–206, (1997)

JF68. Jordan, H.F., Fosdick, L.D.: Three–Particle Effects in the Pair Distribution Function for He^4 Gas. Phys. Rev. **171**, 129–149, (1968)

JG07. Jafarpour, F.H., Ghavami, B.: Phase transition in a three-states reaction–diffusion system. archiv:cond-mat/0703198, (2007)

JHP93. Jayaprakash, C., Hayot, F., Pandit, R.: Universal properties of the two–dimensional Kuramoto–Sivashinsky equation. Phys. Rev. Lett. **71**, 12, (1993)

JK79. W. Janke, H. Kleinert, Lett. Nuovo Cimento **25**, 297, (1979)

JL02. Joseph, B., Legras, B.: Relation between kinematic boundaries, stirring and barriers for the Antarctic polar vortex. J. Atmos. Sci. **59**, 1198-1212, (2002)

JLR90. Judson, R., Lehmann, K., Rabitz, H., Warren, W.S.: Optimal Design of External Fields for Controlling Molecular Motion – Application to Rotation J. Mol. Struct. **223**, 425-446, (1990)

JP98. Jakšić, V., Pillet, C.-A.: Ergodic properties of classical dissipative systems I. Acta mathematica **181**, 245–282, (1998)

JPY96. Jibu, M., Pribram, K.H., Yasue, K.: From conscious experience to memory storage and retrieval: the role of quantum brain dynamics and boson condensation of evanescent photons, Int. J. Mod. Phys. B, **10**, 1735, (1996)

JS92. Janke, W., Sauer, T.: Path Integrals in Quantum Mechanics, Statistics and Polymer Physics. Phys. Lett. A **165**, 199, (1992)

JS93. Janke, W., Sauer, T.: Path integral Monte Carlo using multigrid techniques. Chem. Phys. Lett. **201**, 499–505, (1993)

JS94. Janke, W., Sauer, T.: Multicanonical multigrid Monte Carlo method. Phys. Rev. E **49**, 3475–3479, (1994)

JS96. Janke, W., Sauer, T.: Multigrid Method versus Staging Algorithm for PIMC Simulations. Chem. Phys. Lett. **263**, 488, (1996)

JS97. Janke, W., Sauer, T.: Optimal energy estimation in path–integral Monte Carlo simulations. J. Chem. Phys. **107**, 5821–5839 , (1997)

JT80. Jaffe, A., Taubes, C.: Vortices and monopoles. Birkhäuser, Boston, MA, (1980)

JT00.	Jones, J.L., Tovar, O.H.: Electrophysiology of ventricular fibrillation and defibrillation. Crit. Care Med. **28** (Suppl.), N219–N221, (2000)
JY95.	Jibu, M., Yasue, K.: Quantum brain dynamics and consciousness. John Benjamins, Amsterdam, (1995)
JY95.	Jiang, Q., Yang, H.-N., Wang, G.-C.: Scaling and dynamics of low-frequency hysteresis loops in ultrathin Co films on a Cu(001) surface. Phys. Rev. B **52**, 14911, (1995)
JZ85.	Joos, E., Zeh, H.D.: Emergence of Classical Properties through Interaction with the Environment, Zeit. Phys. B **59**, 223, (1985)
Jac82.	Jackson, F.: Epiphenomenal Qualia. Reprinted in Chalmers and David (ed., 2002) Philosophy of Mind: Classical, Contemporary Readings. Oxford Univ. Press, Oxford, (1982)
Jac91.	Jackson, E.A.: Perspectives of Nonlinear Dynamics **2**. Cambridge Univ. Press, Cambridge, (1991)
Jar98a.	Jarvis, S.: Euclidean monopoles and rational maps, Proc. London Math. Soc. 3(77), 170–192, (1998a)
Jar98b.	Jarvis, S.: Construction of Euclidean monopoles, Proc. London Math. Soc. 3(77), 193–214, (1998b)
Jon85.	Jones, V.F.R.: A new polynomial invariant of knots and links. Bull. Amer. Math. Soc. **12**, 103, (1985)
Jos74.	Josephson, B.D.: The discovery of tunnelling supercurrents. Rev. Mod. Phys. **46**(2), 251-254, (1974)
Jul18.	Julia, G.: Mémoires sur l'itération des fonctions rationelles. J. Math. **8**, 47–245, (1918)
Jun93.	Jung, P.: Periodically driven stochastic systems. Phys. Reports **234**, 175, (1993)
KA83.	Krinsky, V., Agladze, K.I.: Interaction of rotating waves in an active chemical medium. Physica D **8**, 50–56, (1983)
KA85.	Kawamoto, A.H., Anderson, J.A.: A Neural Network Model of Multistable Perception. Acta Psychol. **59**, 35–65, (1985)
Kar93.	Karma, A.: Spiral breakup in model equations of action potential propagation in cardiac tissue. Phys. Rev. Lett. **71**, 1103–1106, (1993)
KD99.	Klages, R., Dorfman, J.R.: Simple deterministic dynamical systems with fractal diffusion coefficients. Phys. Rev. E **59**, 5361–5383, (1999)
KET03.	Klemm, K., Eguíluz, V.M., Toral, R., San Miguel, M.: Global culture: A noise–induced transition in finite systems. Phys. Rev. E **67**, 045101 (2003)
KF86.	Klein, M.V., Furtak, T.E.: Optics (2nd ed), Wiley, New York, (1986)
KF99.	Karlin, I.V., Ferrante, A., Öttinger, H.C.: Perfect Entropy Functions of the Lattice Boltzmann Method. Europhys. Lett. **47**, 182-188, (1999)
KFA69.	Kalman, R.E., Falb, P., Arbib, M.A.: Topics in Mathematical System Theory. McGraw Hill, New York, (1969)

KFP91.	Kaplan, D.T., Furman, M.I., Pincus, S.M., Ryan, S.M., Lipsitz, L.A., Goldberger, A.L.: Aging and the complexity of cardiovascular dynamics. Biophys. J., **59**, 945–949, (1991)
KGI97.	Kim, Y.H., Garfinkel, A., Ikeda, T. *et al.*: Spatiotemporal complexity of ventricular fibrillation revealed by tissue mass reduction in isolated swine right ventricle. Further evidence for the quasiperiodic route to chaos hypothesis. J. Clin. Invest. **100**, 2486–2500, (1997)
KH95.	Katok, A., Hasselblatt, B.: Introduction to the Modern Theory of Dynamical Systems. Cambridge Univ. Press., Cambridge, (1995)
KHL79.	Kitahara, K., Horsthemke, W., Lefever, R.: Coloured–noise–induced transitions: exact results for external dichotomous Markovian noise. Phys. Lett. **A70**, 377, (1979)
KHS93.	Koruga, D.L., Hameroff, S.I., Sundareshan, M.K., Withers. J., Loutfy, R.: Fullerence C60: History, Physics, Nanobiology and Nanotechnology. Elsevier Science Pub, (1993)
KI04.	Kushvah, B.S., Ishwar, B.: Triangular equilibrium points in the generalised photogravitational restricted three body problem with Poynting–Robertson drag. Rev. Bull. Cal. Math. Soc. **12**, 109–114, (2004)
KJ03.	Knoblich, G., Jordan, S. Action coordination in individuals and groups: Learning anticipatory control. J. Exp. Psych.: Learning, Memory & Cognition, **29**, 1006–1016, 2003.
KK00.	Kye, W.-H., Kim, C.-M.: Characteristic relations of type-I intermittency in the presence of noise. Phys. Rev. E **62**, 6304–6307, (2000)
KK01.	Kawahara, G., Kida, S.: Periodic motion embedded in plane Couette turbulence: regeneration cycle and burst. J. Fluid Mech. **449**, 291–300, (2001)
KK01.	Koyama, H., Konishi, T.: Emergence of power–law correlation in 1–dimensional self–gravitating system. Phys. Lett. **A 279**, 226–230, (2001)
KK02a.	Koyama, H., Konishi, T.: Hierarchical clustering and formation of power–law correlation in 1–dimensional self–gravitating system. Euro. Phys. Lett. **58**, 356, (2002)
KK02b.	Koyama, H., Konishi, T.: Long–time behavior and relaxation of power–law correlation in one–dimensional self–gravitating system. Phys. Lett. **A 295**, 109, (2002)
KK06.	Koyama, H., Konishi, T.: Formation of fractal structure in many–body systems with attractive power–law potentials. Phys. Rev. E **73**, 016213, (2006)
KKH03.	Kim, J.-W., Kim, S.-Y. Hunt, B., Ott, E.: Fractal properties of robust strange nonchaotic attractors in maps of two or more dimensions. Phys. Rev. E **67**, 036211, (2003)
KKL01.	Kallosh, R., Kofman, L., Linde, A.D.: Pyrotechnic universe. Phys. Rev. D **64**, 123523, (2001)
KKL03.	Kachru, S., Kallosh, R., Linde, A., Trivedi, S.P.: De Sitter Vacua in String Theory. arXiv:hep-th/0301240, (2003)

KKV99.	Katz, S., Klemm, A., Vafa, C.: M–theory, topological strings and spinning black holes. Adv. Theor. Math. Phys. **3**, 1445, (1999)
KL99.	Keener, J.P., Lewis, T.J.: The biphasic mystery: Why a biphasic shock is more effective than a monophasic shock for defibrillation. J. Theoret. Biol. **200**, 1–17, (1999)
KLO03.	Kim, S.-Y., Lim, W., Ott, E.: Mechanism for the intermittent route to strange nonchaotic attractors. Phys. Rev. E **67**, 056203, (2003)
KLR03.	Kushner, A., Lychagin, V., Roubtsov, V.: Contact geometry and Non-linear Differential Equations. Cambridge Univ. Press, Cambridge, (2003)
KLR03.	Kye, W.-H., Lee, D.-S., Rim, S., Kim, C.-M., Park, Y.-J.: Periodic Phase Synchronization in coupled chaotic oscillators. Phys. Rev. E **68**, 025201(R), (2003)
KLS94.	Kofman, L., Linde, A., Starobinsky, A.: Reheating after Inflation. Phys. Rev. Lett. **73**, 3195–3198, (1994)
KLS99.	Kraus, P., Larsen, F., Siebelink, R.: The gravitational action in asymptotically AdS and flat space–times. Nucl. Phys. B **563**, 259, (1999)
KLV85.	Krasil'shchik, I., Lychagin, V., Vinogradov, A.: Geometry of Jet Spaces and Nonlinear Partial Differential Equations, Gordon and Breach, Glasgow, (1985)
KM63.	Kervaire, M.A., Milnor, J.W.: Groups of homotopy spheres – I. Ann. of Math. **77**, 504, (1963)
KM86.	Konstantinov, M.Y., Melnikov, V.N.: Topological Transitions in the Theory of Space-Time. Class. Quant. Grav. **3**, 401, (1986)
KM89.	Kivshar, Y.S., Malomend, B.A.: Dynamics of solitons in nearly integrable systems. Rev. Mod. Phys. **61**, 763–915, (1989)
KN00.	Kotz, S., Nadarajah, S.: Extreme Value Distributions. Imperial College Press, London, (2000)
KO04.	Kim, P., Olver, P.J.: Geometric integration via multi-space, Regul. Chaotic Dyn. **9**(3), 213–226, (2004)
KO89.	Kamron, N., Olver, P.J.: Le probléme d'equivalence à une divergence prés dans le calcul des variations des intégrales multiples. C. R. Acad. Sci. Paris,t. **308**, Sèrie I, 249–252, (1989)
KOS01.	Khoury. J., Ovrut, B.A., Seiberg, N., Steinhardt, P.J., Turok, N.: From Big Crunch to Big–Bang. arXiv:hep-th/0108187, (2001)
KOS01b.	Khoury. J., Ovrut, B.A., Steinhardt, P.J., Turok, N.: The ekpyrotic universe: Colliding branes and the origin of the hot Big–Bang. Phys. Rev. D **64**, 123522, (2001)
KP02.	Katz, N., Pavlovic, N.: A cheap Caffarelli–Kohn–Nirenberg inequality for the Navier-Stokes equation with hyper-dissipation. Geom. Funct. Anal. **12**(2), 355-379, (2002)
KP05.	Kobes, R., Peles, S.: A Relationship Between Parametric Resonance and Chaos. arXiv:nlin/0005005, (2005)

KP95. Kocarev, L., Parlitz, U.: General Approach for Chaotic Synchronization with Applications to Communication. Phys. Rev. Lett. **74**, 5028–5031, (1995)

KP96. Kocarev, L., Parlitz, U.: Generalized Synchronization, Predictability, and Equivalence of Unidirectionally Coupled Dynamical Systems. Phys. Rev. Lett. **76**, 1816–1819, (1996)

KPV95. Krinsky, V., Plaza, F., Voignier, V.: Quenching a rotating vortex in an excitable medium. Phys. Rev. E **52**, 2458–2462, (1995)

KR03. Kock, A., Reyes, G.E.: Some calculus with extensive quantities: wave equation. Theory and Applications of Categories, **11**(14), 321–336, (2003)

KR95. Kauffman, L.H., Radford, D.E.: Invariants of 3-manifolds derived from finite–dimensional Hopf algebras. J. Knot Theory Ramifications **4**, 131–162, (1995)

KR99. Kauffman, L.H., Radford, D.E.: Quantum algebra structures on $n \times n$ matrices. J. Algebra **213**, 405–436, (1999)

KRG89. Kosloff, R., Rice, S.A., Gaspard, P., Tersigni, S., Tannor, D.J.: Wavepacket Dancing: Achieving Chemical Selectivity by Shaping Light Pulses. Chem. Phys. **139**, 201–220, (1989)

KRL03. Kachru, S., Kallosh, R., Linde, A., Trivedi, S.P.: De Sitter Vacua in String Theory. Phys. Rev. D **68**, 046005, (2003)

KS00. Koyama, K., Soda. J.: Evolution of cosmological perturbations in the brane world. Phys. Rev. D **62**, 123502, (2000)

KS02. Koyama, K., Soda. J.: Bulk gravitational field and cosmological perturbations on the brane. Phys. Rev. D **65**, 023514, (2002)

KS03. Kobayashi, H., Shimomura, Y.: Inapplicability of the Dynamic Clark Model to the Large Eddy Simulation of Incompressible Turbulent Channel Flows. Phys. Fluids **15**, L29–L32, (2003)

KS03. Kanamaru, T., Sekine, M.: Analysis of globally connected active rotators with excitatory and inhibitory connections using the Fokker-Planck equation. Phys. Rev. E **67**, 031916, (2003)

KS04. Karczmarek. J., Strominger, A.: Matrix Cosmology. JHEP **0404**, 055, (2004)

KS75. Kogut, J., Susskind, L.: Hamiltonian formulation of Wilson's lattice gauge theories. Phys. Rev. D **11**, 395–408, (1975)

KS76. Kijowski. J., Szczyrba, W.: A Canonical Structure for Classical Field Theories. Commun. Math. Phys. **46**, 183, (1976)

KS77. R.C.Kirby and L.C.Siebemann, Foundational essays on topological manifolds, smoothings and triangulations (Princeton Univ. Press, Princeton, New Jersey, 1977)

KS90. Kirchgraber, U., Stoffer, D.: Chaotic behavior in simple dynamical systems, SIAM Review **32**(3), 424–452, (1990)

KS97. Ketoja, J.A., Satija, I.I.: Harper equation, the dissipative standard map and strange non-chaotic attractors: Relationship between an eigenvalue problem and iterated maps. Physica D **109**, 70–80, (1997)

KS97. K. Krebs, S. Sandow, J. Phys. A: Math. Gen. A **30**, 3165, (1997)

KS98.	Keener, J., Sneyd, J.: Mathematical Physiology. Springer-Verlag, New York, (1998)
KSI06.	Kushvah, B.S., Sharma, J.P., Ishwar, B.: Second Order Normalization in the Generalized Photogravitational Restricted Three Body Problem with Poynting–Robertson Drag. arXiv:math.DS/0602528, (2006)
KT01.	Kye, W.-H., Topaj, D.: Attractor bifurcation and on–off intermittency. Phys. Rev. **E 63**, 045202(R), (2001)
KT75.	Kamber, F., Tondeur, P.: Foliated Bundles and Characteristic Classes. Lecture Notes in Math. 493, Springer, Berlin, (1975)
KT75.	Kuramoto, Y., Tsuzuki, T.: On the formation of dissipative structures in reaction–diffusion systems. Prog. Theor. Phys. **54**, 687, (1975)
KT79.	Kijowski. J., Tulczyjew, W.: A Symplectic Framework for Field Theories. Springer-Verlag, Berlin, (1979)
KT85.	Kubo, R., Toda, M., Hashitsume, N.: Statistical Physics. Springer-Verlag, Berlin, (1985)
KTL88.	Kono, H., Takasaka, A., Lin, S.H.: Monte Carlo calculation of the quantum partition function via path integral formulations. J. Chem. Phys. **88**, 6390, (1988)
KTW00.	Kirklin, K., Turok, N., Wiseman, T.: Singular Instantons Made Regular. Phys. Rev. D **63**, 083509, (2000)
KV03a.	Katic, D., Vukobratovic, M.: Advances in Intelligent Control of Robotic Systems, Book series: Microprocessor-Based and Intelligent Systems Engineering, Kluwer Acad. Pub., Dordrecht, (2003)
KV03b.	Katic, D., Vukobratovic, M.: Survey of intelligent control techniques for humanoid robots. J. Int. Rob. Sys, **37**, 117–141, (2003)
KV98.	Katic, D., Vukobratovic, M.: A neural network-based classification of environment dynamics models for compliant control of manipulation robots. IEEE Trans. Sys., Man, Cyb., B, **28**(1), 58–69, (1998)
KW83.	Knobloch, E., Weisenfeld, K.A.: Bifurcations in fluctuating systems: The. center manifold approach. J.Stat.Phys. 33, 611, (1983)
KWR01.	Korniss, G., White, C.J., Rikvold, P.A., Novotny, M.A.: Dynamic phase transition, universality, and finite–size scaling in the two–dimensional kinetic Ising model in an oscillating field . Phys. Rev. E **63**, 016120, (2001)
KY75.	Kaplan, J.L., Yorke, J.A.: On the stability of a periodic solution of a differential delay equation, SIAM J. Math. Ana. **6**, 268–282, (1975)
KY79.	Kaplan, J.L., Yorke, J.A.: Preturbulence: a regime observed in a fluid flow of Lorenz. Commun. Math. Phys. **67**, 93-108, (1979)
KYR98.	Kim, C.M., Yim, G.S., Ryu, J.W., Park, Y.J.: Characteristic Relations of Type-III Intermittency in an Electronic Circuit. Phys. Rev. Lett. **80**, 5317–5320, (1998)

KZ91. Kree, R., Zippelius, A.: In Models of Neural Networks, I, Domany, R., van Hemmen, J.L., Schulten, K. (Eds), 193,, Springer, Berlin, (1991)

KZH02. Kiss, I.Z., Zhai, Y., Hudson, J.L.: Emerging coherence in a population of chemical oscillators. Science **296**, 1676–1678, (2002)

Kac51. Kac, M.: On Some Connection between Probability Theory and Differential and Integral Equations. Proc. 2nd Berkeley Sympos. Math. Stat. Prob. 189–215, (1951)

Kai97. Kaiser, D.: Preheating in an expanding universe: Analytic results for the massless case. Phys. Rev. D **56**, 706–716, (1997)

Kai99. Kaiser, D.: Larger domains from resonant decay of disoriented chiral condensates. Phys. Rev. D **59**, 117901, (1999)

Kal21. Kaluza, T.: Sitzungsber. Preuss. Akad. Wiss. Berlin (Math. Phys.) K1, 966, (1921)

Kal60. Kalman, R.E.: A new approach to linear filtering and prediction problems. Transactions of the ASME, Ser. D. J. Bas. Eng., **82**, 34–45, (1960)

Kal99. N. Kaloper. Bent domain walls as braneworlds. Phys. Rev. D **60**, 123506, (1999)

Kam92. Van Kampen, N.G.: Stochastic Processes in Physics and Chemistry, (2nd ed.) North-Holland, Amsterdam, (1992)

Kan58. Kan, D.M.: Adjoint Functors. Trans. Am. Math. Soc. **89**, 294–329, (1958)

Kan98. Kanatchikov, I.: Canonical structure of classical field theory in the polymomentum phase space. Rep. Math. Physics, **41**(1), 49—90, (1998)

Kap00. Kaplan, D.T.: Applying Blind Chaos Control to Find Periodic Orbits. arXiv:nlin.CD/0001058, (2000)

Kap05. Kappen, H.J.: A linear theory for control of nonlinear stochastic systems. Phys. Rev. Let. (to appear)

Kar00. Karplus, M.: Aspects of Protein Reaction Dynamics: Deviations from Simple Behavior. J. Phys. Chem. B **104**, 11–27, (2000)

Kar84. Kardar, M.: Free energies for the discrete chain in a periodic potential and the dual Coulomb gas. Phys. Rev. B **30**, 6368–6378, (1984)

Kat75. Kato, T.: Quasi–linear equations of evolution, with applications to partial differential equations, Lecture Notes in Math. **448**, Springer-Verlag, Berlin, 25–70, (1975)

Kat83. Kato, T.: On the Cauchy Problem for the (generalized) Korteweg–de Vries Equation, Studies in Applied Math. Adv. in Math. Stud. **8**, 93–128, (1983)

Kau94. Kauffman, L.H.: Knots and Physics, (2nd ed.) World Scientific, Singapore (1994)

KdV95. Korteweg, D.J., de Vries, H.: On the change of form of long waves advancing in a rectangular canal, and on a new type of long stationary waves. Philosophical Magazine, **39**, 422–43, (1895)

Kha92.	Khalil, H.K.: Nonlinear Systems. New York: MacMillan, (1992)
Khi57.	Khinchin, A.I.: Mathematical foundations of Information theory. Dover, (1957)
Kie04.	Kiefer, C.: Quantum gravity. Oxford Univ. Press, Oxford, UK (2004)
Kim95a.	Kim, J.: Problems in the Philosophy of Mind. Oxford Companion to Philosophy. Ted Honderich (ed.) Oxford Univ. Press, Oxford, (1995)
Kim95b.	Kim, J.: Mind–Body Problem. Oxford Companion to Philosophy. Ted Honderich (ed.) Oxford Univ. Press, Oxford, (1995)
Kir89.	Kirby, R.C.: The topology of 4–manifolds. Springer-Verlag, Berlin, (1989)
Kit97.	Kitaev, A.Y.: Quantum computations: Algorithms and error correction. Rus. Math. Surv. **52**, 1191–1249, (1997)
Kla00.	Klauder. J.R.: Beyond Conventional Quantization, Cambridge Univ. Press, Cambridge, (2000)
Kla97.	Klauder, J.R.: Understanding Quantization. Found. Phys. **27**, 1467–1483, (1997)
Kle94.	Kleinert, H.: Path integrals in quantum mechanics, statistics and polymer physics, (2nd ed.) World Scientific, Singapore (1994)
Kna87.	Knauf, A.: Ergodic and topological properties of Coulombic periodic potentials. Comm. Math. Phys. **110**, 89–112, (1987)
Kob07.	Kobayashi, N.: Equivalence between quantum simultaneous games and quantum sequential games. arXiv:quant-ph 0711.0630, (2007)
Koc81.	Kock, A.: Synthetic Differential Geometry. London Math.Soc. Lecture Notes Series No. 51, Cambridge Univ. Press, Cambridge, (1981)
Kof03.	Kofman, L.: Probing String Theory with Modulated Cosmological Fluctuations. astro-ph/0303641, (2003)
Kog79.	Kogut, J.: An introduction to lattice gauge theory and spin systems. Rev. Mod. Phys. **51**, 659–713, (1979)
Kol02.	Kol, B.: Topology change in general relativity and the black-hole black-string transition, arXiv:hep-th/0206220.
Kos04.	Kosmann-Schwarzbach, Y.: Derived brackets. Lett. Math. Phys. **69**, 61–87, (2004)
Kos92.	Kosko, B.: Neural Networks and Fuzzy Systems, A Dynamical Systems Approach to Machine Intelligence. Prentice–Hall, New York, (1992)
Kos96.	Kosko, B.: Fuzzy Engineering. Prentice Hall, New York, (1996)
Kra76.	Kraichnan, R.H.: Eddy Viscosity in Two and Three dimensions J. Atmos. Sci. **33**, 1521-1536, (1976)
Kra97.	Krasnov, K.: Geometrical entropy from loop quantum gravity. Phys. Rev. D **55**, 3505, (1997)
Kre84.	Krener, A.: Approximate linearization by state feedback and coordinate change, Systems Control Lett., 5, 181–185, (1984)

Kru97. Krupkova, O.: The Geometry of Ordinary Variational Equations. Springer, Berlin, (1997)

Kry79. Krylov, N.S.: Works on the foundations of statistical mechanics. Princeton Univ. Press, Princeton, (1979)

Kub57. Kubo, R.: Statistical-mechanical theory of irreversible processes. I. J. Phys. Soc. (Japan) **12**, 570–586, (1957)

Kuc92. Kuchař, K.: Time and Interpretations of Quantum Gravity. In Proceedings of the 4th Canadian Conference on General Relativity and Relativistic Astrophysics, ed. by G. Kunstatter, D. Vincent, and J. Williams, World Scientific, Singapore, (1992)

Kuh85. Kuhl. J.: Volitional Mediator of cognition-Behaviour consistency: Self-regulatory Processes and action versus state orientation (pp. 101-122) In: J. Kuhl & S. Beckman (Eds.) Action control: From Cognition to Behaviour. Springer, Berlin, (1985)

Kur68. Kuratowski, K.: Topology II. Academic Press PWN-Polish Sci. Publishers Warszawa, (1968)

Kur76. Kuramoto, Y., Tsuzuki, T.: Persistent propagation of concentration waves in dissipative media far from thermal equilibrium. Progr. Theor. Physics **55**, 365, (1976)

Kur84. Kuramoto, Y.: Chemical Oscillations. Waves and Turbulence. Springer, New York, (1984)

Kur91. Kuramoto, Y.: Collective synchronization of pulse-coupled oscillators and excitable units. Physica D, **50**, 15, (1991)

LAT83. Lobb, C.J., Abraham, D.W., Tinkham, M.: Theoretical. interpretation of resistive transitions in arrays of superconducting. weak links. Phys. Rev. **B27**, 150, (1983)

LAV01. Lacorata, G., Aurell, E., Vulpiani, A.: Drifter dispersion in the Adriatic Sea: Lagrangian data and chaotic model. Ann. Geophys. **19**, 1-9, (2001)

LB87. Li, X.-P., Broughton, J.Q.: High–order correction to the Trotter expansion for use in computer simulation. J. Chem. Phys. **86**, 5094, (1987)

LC97. Lo, H.K., Chau, H.F.: Is Quantum Bit Commitment Really Possible? Phys. Rev. Lett. **78**, 3410–3413, (1997)

LCD87. Leggett, A.J., Chakravarty, S., Dorsey, A.T., Fisher, M.P.A., Chang, A., Zwerger, W.: Dynamics of the dissipative two–state system. Rev. Mod. Phys. **59**, 1, (1987)

LD05. Levine, M., Davidson, E.H.: Gene regulatory networks for development. Proc Natl. Acad. Sci. USA, **102**(14), 4936–4942, (2005)

LD98. Loss, D., DiVincenzo, D.P.: Quantum computation with quantum dots. Phys. Rev. A **57**(1), 120–126, (1998)

LDC02. Leong, B., Dunsby, P., Challinor, A., Lasenby, A.: 1+3 covariant dynamics of scalar perturbations in braneworlds. Phys. Rev. D **65**, 104012, (2002)

LGC01. Lee, K.J., Goldstein, R.E., Cox, E.C.: Resetting Wave Forms in Dictyostelium Territories. Phys. Rev. Lett. **87**, 068101, (2001)

LGS98. Lygeros. J., Godbole, D.N., Sastry, S.: Verified hybrid controllers for automated vehicles. IIEEE Trans. Aut. Con., **43**, 522–539, (1998)

LH94. Lou, S., Hu, X.: Symmetries and algebras of the integrable dispersive long wave equations in $(2+1)$-dimensional spaces. J. Phys. A: Math. Gen. **27**, L207, (1994)

LJ03. Lee, C.F., Johnson, N.F.: Efficiency and formalism of quantum games. Phys. Rev. A **67**, 022311 (5 pages) (2003)

LJS69a. Lawande, S.V., Jensen, C.A., Sahlin, H.L.: Monte Carlo integration of the Feynman propagator in imaginary time. J. Comp. Phys. **3**, 416, (1969)

LJS69b. Lawande, S.V., Jensen, C.A., Sahlin, H.L.: Monte Carlo evaluation of Feynman path integrals in imaginary time and spherical polar coordinates. J. Comp. Phys. **4**, 451, (1969)

LL00. Liddle, A.R., Lyth. D.: Cosmological Inflation and Large Scale Structure. Cambridge Univ. Press, (2000)

LL03. Landau, L.D., Lifshitz, E.M.: Course of Theoretical Physics, v.6, Fluid Mechanics, (2003)

LL59. Landau, L.D., Lifshitz, E.M.: Fluid Mechanics. Pergamon Press, (1959)

LL78. Landau, L.D., Lifshitz, E.M.: Statitsical Physics. Pergamon Press, Oxford, (1978)

LL92. Lichtenberg, A.J., Lieberman, M.A.: Regular and Chaotic Dynamics (2nd ed). Springer-Verlag, Berlin, (1992)

LL98. Labastida. J.M.F., Lozano, C.: Lectures on topological quantum field theory, in Proceedings of the CERN–Santiago de Compostela–La Plata Meeting on 'Trends in Theoretical Physics', eds. H. Falomir, R. Gamboa, F. Schaposnik, Amer. Inst. Physics, New York, (1998)

LM03. Lopez, M.C., Marsden. J.E.: Some remarks on Lagrangian and Poisson reduction for field theories. J. Geom. Phys., **48**, 52—83, (2003)

LM87. Libermann, P., Marle, C.M.: Symplectic Geometry and Analytical Mechanics, Reidel, Dordrecht, (1987)

LM93. de León, M., Marrero. J.: Constrained time-dependent Lagrangian systems and Lagrangian submanifolds. J. Math. Phys. **34**, 622, (1993)

LM96. de León, M., Martín de Diego, D.: On the geometry of nonholonomic Lagrangian systems. J. Math. Phys. **37**, 3389, (1996)

LM97. Lewis, A.D., Murray, R.M.: Controllability of simple mechanical control systems, SIAM J. Con. Opt., **35**(3), 766–790, (1997)

LM99. Lewis, A.D., Murray, R.M.: Configuration controllability of simple mechanical control systems, SIAM Review, **41**(3), 555–574, (1999)

LMD04. de León, M., Martin, D., de Diego, A.: Santamaria-Merino: Symmetries in Classical Field Theory. arXiv:math-ph/0404013.

LMM97. de León, M., Marrero. J., Martín de Diego D.: Nonholonomic Lagrangian systems in jet manifolds. J. Phys. A. **30**, 1167, (1997)

LMS01. Langlois, D., Maartens, R., Sasaki, M., Wands, D.: Large-scale cosmological perturbations on the brane. Phys. Rev. D **63**, 084009, (2001)

LMS02. Liu, H., Moore, G.W., Seiberg, N.: Strings in a Time-Dependent Orbifold. JHEP **0206**, 045, (2002)

LMW00. Langlois, D., Maartens, Wands, D.: Gravitational waves from inflation on the brane. Phys. Lett. B **489**, 259, (2000)

LMW02. Langlois, D., Maeda, K.I., Wands, D.: Conservation laws for collisions of branes (or shells) in general relativity. Phys. Rev. Lett. **88**, 181301, (2002)

LO97. Lenz, H., Obradovic, D.: Robust control of the chaotic Lorenz system. Int. J. Bif. Chaos **7**, 2847–2854, (1997)

LOJ04. Lindgren B., Österlund J., Johansson A.: Evaluation of scaling laws derived from Lie group symmetry methods in zero-pressure-gradient turbulent boundary layers. J. Fluid Mech. **502**, 127–152, (2004)

LOS00. Lukas, A., Ovrut, B.A., Stelle, K.S., Waldram, D.: Boundary inflation. Phys. Rev. D **61**, 023506, (2000)

LOS88. Larkin, A.I., Ovchinnikov, Yu.N., Schmid, A.: Physica B **152**, 266, (1988)

LOS99. Lukas, A., Ovrut, B.A., Stelle, K.S., Waldram, D.: The universe as a domain wall. Phys. Rev. D **59**, 086001, (1999); ibid Heterotic M–theory in five dimensions, Nucl. Phys. B **552**, 246, (1999)

LP90. Lo, W.S., Pelcovits, R.A.: Ising Model in a Time–Dependent Magnetic Field. Phys. Rev. A **42**, 7471, (1990)

LP92. V.S. L'vov, I. Procaccia, Phys. Rev. Lett. **69**, 3543, (1992)

LP94. Langer. J., Perline, R.: Local geometric invariants of integrable evolution equations. J. Math. Phys., **35**(4), 1732–1737, (1994)

LP94. L'vov, V., Procaccia, I.: Comment on "Universal properties of the two–dimensional Kuramoto–Sivashinsky equation" Phys. Rev. Lett. **72**, 307, (1994)

LPV67. Lebowitz, J.L., Percus, J.K., Verlet, L.: Ensemble Dependence of Fluctuations with Application to Machine Computations. Phys. Rev. **153**, 250–254, (1967)

LR01. Langlois, D., Rodriguez-Martinez, M.: Brane cosmology with a bulk scalar field. Phys. Rev. D **64**, 123507, (2001)

LR05. Lauscher, O., Reuter, M.: Fractal space-time structure in asymptotically safe gravity. JHEP 0510 (2005) 050.

LR91. Luo, C.H., Rudy, Y.: A model of the ventricular cardiac action potential. Depolarization, repolarization, and their interaction. Circ. Res. **68**, 1501–1526, (1991)

LR93. Lychagin, V.V., Roubtsov, V.N., Chekalov, I.V.: A classification of Monge–Ampère equations. Ann. Scient. Ec. Norm. Sup. 4 ème série, 26, 281–308, (1993)

LRR02.	Lee, T.I., Rinaldi, N.J., Robert, F. et al.: Transcriptional regulatory networks in Saccharomyces cerevisiae. Science, **298**, 799–804, (2002)
LS02.	Langlois, D., Sorbo, L.: Effective action for the homogeneous radion in brane cosmology.. Phys. Lett. B **543**, 155, (2002)
LS82.	Ledrappier, F., Strelcyn, J.-M.: A proof of the estimation from below in Pesin's entropy formula. Ergod. Th. and Dynam. Syst. **2**, 203-219, (1982)
LS97.	Louko, J., Sorkin, R.D.: Complex actions in two-dimensional topology change. Class. and Quant.Grav. **14**, 179–204, (1997)
LSA03.	Louzoun, Y., Solomon, S., Atlan, H., Cohen, I.R.: Proliferation and competition in discrete biological systems. Bull. Math. Biol. **65**, 375, (2003)
LSR84.	Luwel, M., Severne, G., Rousseeuw, P.J.: Numerical Study of the Relaxation of One-Dimensional Gravitational Systems. Astrophys. Space Sci. **100**, 261–277, (1984)
LTZ01.	Lou, S., Tang, X., Zhang, Y.: Chaos in soliton systems and special Lax pairs for chaos systems. arXiv:nlin/0107029, (2001)
LV97.	Li, M., Vitanyi, P.: An Introduction to Kolmogorov Complexity and its Applications. Springer, New York, (1997)
LVW89.	Lerche, W., Vafa, C., Warner, N.P.: Chiral Rings In N=2 Superconformal Theories, Nucl. Phys. B **324**, 427, (1989)
LW03.	Loll, R., Westra, W.: Space–time foam in 2D and the sum over topologies. Workshop on Random Geometry, Krakow, May 15–17, 2003.
LW05.	Lipan, O., Wong, W.H.: The use of oscillatory signals in the study of genetic networks. Proc. Natl. Acad. Sci. USA, **102**, 7063-7068, (2005)
LW76.	Lasota, A., Wazewska–Czyzewska, M.: Matematyczne problemy dynamiki ukladu krwinek czerwonych, Mat. Stosowana **6**, 23–40, (1976)
LW95.	Lani–Wayda, B, Walther, H.-O.: Chaotic motion generated by delayed negative feedback, Part I: A transversality criterion. Diff. Int. Equs. **8**(6), 1407–52, (1995)
LW96.	Lani–Wayda, B, Walther, H.-O.: Chaotic motion generated by delayed negative feedback, Part II: Construction of nonlinearities. Math. Nachr. **180**, 141–211, (1996)
LWX97.	Liu, Z.J., Weinstein, A., Xu, P.: Manin triples for Lie bialgebroids. J. Diff. Geom. **45**(3), 547–574, (1997)
LY77.	Li, T.Y., Yorke, J.A.: Period three implies chaos. Am. Math. Monthly **82**, 985–992, (1977)
LY79.	Li, P., Yau, S.T. Estimates of eigenvalues of a compact riemannian manifold. In Proceedings of Symposia in Pure Mathematics **36**, 205–239, (1979)
LY85.	Ledrappier, F., Young, L.S.: The metric entropy of diffeomorphisms: I. Characterization of measures satisfying Pesin's formula, II. Relations between entropy, exponents and dimension. Ann. of Math. **122**, 509-539, 540–574, (1985)

LYA04.	Leonov, I., Yaresko, A.N., Antonov, V.N. et al.: Charge and Orbital Order in Fe_3O_4. Phys. Rev. Lett. **93**, 146404, (2004)
LZJ95.	Liou, J.C., Zook, H.A., Jackson, A.A.: Radiation pressure, Poynting-Robertson drag and solar wind drag in the restricted three body problem. Icarus **116**, 186–201, (1995)
Lai94.	Lai, Y.-C.: Controlling chaos. Comput. Phys. **8**, 62–67, (1994)
Lak03.	Lakshmanan, M., Rajasekar, S: Nonlinear Dynamics: Integrability, Chaos and Patterns, Springer-Verlag, New York, (2003)
Lak97.	Lakshmanan, M.: Bifurcations, Chaos, Controlling and Synchronization of Certain Nonlinear Oscillators. In Lecture Notes in Physics, **495**, 206, Y. Kosmann-Schwarzbach, B. Grammaticos, K.M. Tamizhmani (ed), Springer-Verlag, Berlin, (1997)
Lam45.	Lamb, H.: Hydrodynamics. Dover, New York, (1945)
Lam76.	Lamb, G.L. Jr.: Bäcklund transforms at the turn of the century. In Bäcklund Transforms, (ed) by R.M. Miura, Springer, Berlin, (1976)
Lan00.	Landsberg, A.S.: Disorder–induced desynchronization in a 2×2 circular Josephson junction array. Phys. Rev. B **61**, 3641–3648, (2000)
Lan00.	Langlois, D.: Brane cosmological perturbations. Phys. Rev. D **62**, 126012, (2000)
Lan01.	Langlois, D.: Evolution of cosmological perturbations in a brane–universe. Phys. Rev. Lett. **86**, 2212, (2001)
Lan02.	Langlois, D.: Brane cosmology: an introduction. In the proceedings of YITP Workshop: Braneworld: Dynamics of Spacetime Boundary. (2002)
Lan08.	Langevin, P.: Comptes. Rendue **146**, 530, (1908)
Lan75.	Lanford, O.E.: Time evolution of large classical systems. 1-111 in Lecture Notes in Physics **38**, Spinger-Verlag, Berlin, (1975)
Lan86.	Lanczos, C.: The variational principles of mechanics. Dover Publ. New York, (1986)
Lan91.	Landauer, R.: Information is Physical, Phys. Today **5**, 23, (1991)
Lan95.	Landsman, N.P.: Against the Wheeler–DeWitt equation. Class. Quan. Grav. L **12**, 119–124, (1995)
Lan95a.	Lani–Wayda, B: Persistence of Poincaré maps in functional differential equations (with application to structural stability of complicated behavior) J. Dyn. Diff. Equs. **7**(1), 1–71, (1995)
Lan95b.	Lani–Wayda, B: Hyperbolic Sets, Shadowing and Persistence for Noninvertible Mappings in Banach spaces, Research Notes in Mathematics **334**, Longman Group Ltd., Harlow, Essex, (1995)
Lan98.	Landi, G.: An introduction to Noncommutative Spaces and Their Geometries, Springer, Berlin, (1998)
Lan99.	Lani–Wayda, B: Erratic solutions of simple delay equations. Trans. Amer. Math. Soc. **351**, 901–945, (1999)
Lap51.	Laplace, P.S.: A Philosophical Essay on Probabilities, translated from the 6th French edition, Dover publications, New York, (1951)

Las42.	Lashley, K.S.: The problem of cerebral organization in vision. In Biological Symposia, VII, Visual mechanisms, 301–322. Jaques Cattell Press, Lancaster, (1942)
Las77.	Lasota, A.: Ergodic problems in biology. Asterisque **50**, 239–250, (1977)
Laz94.	Lazutkin, V.A.: Positive Entropy for the Standard Map I, 94-47, Université de Paris-Sud, Mathématiques, Bâtiment **425**, 91405 Orsay, France, (1994)
Leb93.	Lebowitz, J.L.: Boltzmann's entropy and time's arrow. Physics Today **46**(9), 32–38, (1993)
Led81.	Ledrappier, F.: Some Relations Between Dimension and Lyapunov Exponents. Commun. Math. Phys. **81**, 229–238, (1981)
Lee97.	Lee, K.J.: Wave Pattern Selection in an Excitable System. Phys. Rev. Lett. **79**, 2907, (1997)
Leg02.	Leggett, A.J.: Testing the Limits of Quantum Mechanics: Motivation, State-of-Play, Prospects. J. Phys. Cond. Matter, **14**, R415, (2002)
Leg86.	Leggett, A.J.: In The Lesson of Quantum Theory. Niels Bohr Centenary Symposium 1985; J. de Boer, E. Dal, O. Ulfbeck (ed) North Holland, Amsterdam, (1986)
Lei02.	Leinster, T.: A survey of definitions of $n-$category. Theor. Appl. Categ. **10**, 1–70, (2002)
Lei03.	Leinster, T.: Higher Operads, Higher Categories. London Mathematical Society Lecture Notes Series, Cambridge Univ. Press, (2003)
Lei04.	Leinster, T.: Operads in higher–dimensional category theory. Theor. Appl. Categ. **12**, 73–194, (2004)
Lei714.	Leibniz, G.W.: Monadology, (1714)
Lei75.	Leith, C.A.: Climate response and fluctuation dissipation. J. Atmos. Sci. **32**, 2022, (1975)
Lei78.	Leith, C.A.: Predictability of climate. Nature **276**, 352, (1978)
Leo62.	Leontovic, An.M.: On the stability of the Lagrange periodic solutions for the reduced problem of three body. Sov. Math. Dokl. **3**, 425-429, (1962)
Leo74.	Leonard, A.: Energy Cascade in Large-Eddy Simulations of Turbulent Fluid Flows. Adv. Geophys. **18**, 237-248, (1974)
Ler34.	Leray, J.: Sur le mouvement d'un liquide visquex emplissant l'espace, Acta Math. **63**, 193-248, (1934)
Lev87.	Levichev, A.V.: Prescribing the conformal geometry of a lorentz manifold by means of its causal structure. Soviet Math. Dokl. **35**, 452–455, 1987.
Lev88.	Levinthal, C.F.: Messengers of Paradice, Opiates and the Brain. Anchor Press, Freeman, New York, (1988)
Lew00a.	Lewis, A.D.: Simple mechanical control systems with constraints, IEEE Trans. Aut. Con., **45**(8), 1420–1436, (2000)
Lew00b.	Lewis, A.D.: Affine connection control systems. Proceedings of the IFAC Workshop on Lagrangian and Hamiltonian Methods for Nonlinear Control 128–133, Princeton, (2000)
Lew51.	Lewin, K.: Field Theory in Social Science. Univ. Chicago Press, Chicago, (1951)

Lew95. Lewis, A.D.: Aspects of Geometric Mechanics and Control of Mechanical Systems. Technical Report CIT–CDS 95–017 for the Control and Dynamical Systems Option, California Institute of Technology, Pasadena, CA, (1995)

Lew97. Lewin, K.: Resolving Social Conflicts: Field Theory in Social Science, American Psych. Assoc., New York, (1997)

Lew98. Lewis, A.D.: Affine connections and distributions with applications to nonholonomic mechanics, Reports on Mathematical Physics, **42**(1/2), 135–164, (1998)

Lew99. Lewis, A.D.: When is a mechanical control system kinematic?, in Proceedings of the 38th IEEE Conf. Decis. Con., 1162–1167, IEEE, Phoenix, AZ, (1999)

Li04. Li, Y.: Chaos in Partial Differential Equations. Int. Press, Sommerville, MA, (2004)

Li04a. Li, Y.: Persistent homoclinic orbits for nonlinear Schrödinger equation under singular perturbation. Dyn. PDE, **1**(1), 87–123, (2004)

Li04b. Li, Y.: Existence of chaos for nonlinear Schrödinger equation under singular perturbation. Dyn. PDE. **1**(2), 225–237, (2004)

Li04c. Li, Y.: Homoclinic tubes and chaos in perturbed sine-Gordon equation. Cha. Sol. Fra., **20**(4), 791–798, 2004)

Li04d. Li, Y.: Chaos in Miles' equations. Cha. Sol. Fra., **22**(4), 965–974, (2004)

Li05. Li, Y.: Invariant manifolds and their zero-viscosity limits for Navier-Stokes equations. Dynamics of PDE. **2, no.2**, 159–186, (2005)

Li06. Li, Y.: On the True Nature of Turbulence. arXiv:math.AP/0507254, (2005)

Li80. Li, P.: On the sobolev constant and the $p-$ spectrum of a compact Riemannian manifold. Ann. Sci. E.N.S., Paris, **13**, 451–468, (1980)

Lia83. Liao, S.T.: On hyperbolicity properties of nonwandering sets of certain 3-dimensional differential systems. Acta Math. Sc., **3**, 361–368, (1983)

Lig58. Lighthill, M.J.: An Introduction to Fourier Analysis and Generalised Functions. Campridge Univ. Press, (1958)

Lig85. Liggett, T.M., Interacting Particle Systems. Springer, New-York, (1985)

Lil92. Lilly D.: A proposed modification of the Germano subgrid-scale closure method, Phys. Fluids **4**, 633–635, (1992)

Lin82. Linde, A.D.: A New Inflationary Universe Scenario: A Possible Solution Of The Horizon, Flatness, Homogeneity, Isotropy And Primordial Monopole Problems. Phys. Lett. **B108**, 389, (1982)

Lin83. Linde, A.D.: Chaotic Inflation. Phys. Lett. **B129**, 177, (1983)

Lin86a. Linde, A.D.: Eternal Chaotic Inflation. Mod. Phys. Lett. **A1**, 81, (1986)

Lin86b. Linde, A.D.: Eternally Existing Selfreproducing Chaotic Inflationary Universe. Phys. Lett. **175B**, 395, (1986)

Lin88.	Linde, A.D.: Inflation and Axion Cosmology. Phys. Lett. B201, 437, (1988)
Lin97.	Linz, S.J.: Nonlinear Dynamical Models and Jerky Motion. Am. J. Phys., **65**(6), 523–526, (1997)
Lo97.	Lo, H.K.: Insecurity of quantum secure computations. Phys. Rev. A **56**, 1154–1162, (1997)
Lol01.	Loll, R.: Discrete Lorentzian quantum gravity. Nucl. Phys. B **94**, 96–107, (2001)
Lol03.	Loll, R.: A discrete history of the Lorentzian path integral, Lecture Notes in Physics 631 (2003) 137-171.
Lol98.	Loll, R.: Discrete approaches to quantum gravity in four dimensions, Living Rev. Rel. **1**, 13, (1998)
Loo00.	Loo, K.: A Rigorous Real Time Feynman Path Integral and Propagator. J. Phys. A: Math. Gen, **33**, 9215–9239, (2000)
Loo99.	Loo, K.: A Rigorous Real Time Feynman Path Integral. J. Math. Phys., **40**(1), 64–70, (1999)
Lor63.	Lorenz, E.N.: Deterministic Nonperiodic Flow. J. Atmos. Sci., **20**, 130–141, (1963)
Lu90.	Lu, J.-H.: Multiplicative and affine Poisson structures on Lie groups. PhD Thesis, Berkeley Univ., Berkeley, (1990)
Lu91.	Lu, J.-H.: Momentum mappings and reduction of Poisson actions. in Symplectic Geometry, Groupoids, and Integrable Systems, eds.: P. Dazord and A. Weinstein, 209–225, Springer, New York, (1991)
Lya47.	Lyapunov, A.: Problème générale de la stabilité du mouvement. Ann. Math. Studies 17, Princeton Univ. Press, Princeton, NJ, (1947)
Lya56.	Lyapunov, A.M.: A general problem of stability of motion. Acad. Sc. USSR, (1956)
Lyc79.	Lychagin, V.V.: Contact geometry and nonlinear second order differential equations. Uspèkhi Mat. Nauk, **34**, 137–165 (in Russian), (1979)
Lyo92.	Lyons, G.W.: Complex Solutions for the Scalar Field Model of the Universe. Phys. Rev. D **46**, 1546, (1992)
MA02.	Mascalchi, M. *et al.*: Proton MR Spectroscopy of the Cerebellum and Pons in Patients with Degenerative Ataxia, Radiology, **223**, 371, (2002)
MA94.	Miron, R., Anastasiei, M.: The Geometry of Lagrangian Spaces: Theory and Applications, Kluwer Academic Publishers, (1994)
MB01.	Mennim, A., Battye, R.A.: Cosmological expansion on a dilatonic brane-world. Class. Quant. Grav. **18**, 2171, (2001)
MBP00.	Myung, I.J., Balasubramanian, V., Pitt, M.A.: Counting probability distributions: differential geometry and model selection. Proceedings of the National Academy of Science, USA, **97**, 11170–11175, (2000)
MC06.	Mazzoni, L.N., Casetti, L.: Curvature of the energy landscape and folding of model proteins. Phys. Rev. Lett. **97**, 218104, (2006)

MCD88. Manogue, C.A., Copeland, E., Dray, T.: The trousers problem revisited. Pramana, **30**, 279, 1988.

MCR00. Martin, X., O'Connor, D., Rideout, D.P., Sorkin, R.D.: On the 'renormalization' transformations induced by cycles of expansion and contraction in causal set cosmology. Phys. Rev. **D63**, 084026, 2001.

MDC85. Martinis, J.M., Devoret, M.H., Clarke, J.: Energy–Level Quantization in the Zero–Voltage State of a Current–Biased Josephson Junction. Phys. Rev. Lett. **55**, 1543–1546, (1985)

MDT00. Mangioni, S.E., Deza, R.R., Toral, R., Wio, H.S.: Non–equilibrium phase ransitions induced by multiplicative noise: effects of self–correlation. Phys. Rev. E, **61**, 223–231, (2000)

MF53. Morse, M., Feshbach, H.: Methods of Theoretical Physics. Part I, McGraw–Hill, New York, (1953)

MFB00. Myung, I.J., Forster, M., Browne, M.W.: Special issue on model selection. J. Math. Psych., **44**, 1–2, (2000)

MFV90. Morandi, G., Ferrario, C., Lo Vecchio, G., Marmo, G., Rubano, C.: The inverse problem in the calculus of variations and the geometry of the tangent bundle, Phys. Rep. **188**, 147, (1990)

MG77. Mackey, M.C., Glass, L.: Oscillation and chaos in physiological control systems. Science **197**, 287–295, (1977)

MGB96. Mira, C., Gardini, L., Barugola, A., Cathala, J.-C.: Chaotic Dynamics in Two-Dimensional Noninvertible Maps. World Scientific, Singapore, (1996)

MH38. Morse, M., Hedlund, G.: Symbolic Dynamics. Am. J. Math. **60**, 815–866, (1938)

MH92. Meyer, K.R., Hall, G.R.: Introduction to Hamiltonian Dynamical Systems and the N–body Problem. Springer, New York, (1992)

MI95. Mishra, P., Ishwar, B: Second-Order Normalization in the Generalized Restricted Problem of Three Bodies, Smaller Primary Being an Oblate Spheroid. Astr. J. **110**(4), 1901–1904, (1995)

MK05. Moon, S.J., Kevrekidis, I.G.: An equation-free approach to coupled oscillator dynamics: the Kuramoto model example. Submitted to Int. J. Bifur. Chaos, (2005)

MKA88. Miron, R., Kirkovits, M.S., Anastasiei, M.: A Geometrical Model for Variational Problems of Multiple Integrals. Proc. Conf. Diff. Geom Appl. Dubrovnik, Yugoslavia, (1988)

ML81. Morris, C., Lecar, H.: Voltage oscillations in the barnacle giant muscle fiber. Biophys. J., **35**, 193–213, (1981)

ML92. Méais, O., Lesieur, M.: Spectral large-eddy simulation of isotropic and stably stratified turbulence. J. Fluid Mech. **256**, 157–194, (1992)

MLS94. Murray, R.M., Li, X., Sastry, S.: Robotic Manipulation, CRC Press, Boco Raton, Fl, (1994)

MM75. Marsden, J.E., McCracken, M.: The Hopf bifurcation and its applications. MIT Press, Cambridge, MA, (1975)

MM92.	Marathe, K., Martucci, G.: The Mathematical Foundations of Gauge Theories. North-Holland, Amsterdam, (1992)
MM95.	Matsui, N., Mori, T.: The efficiency of the chaotic visual behavior in modeling the human perception-alternation by artificial neural network. In Proc. IEEE ICNN'95. **4**, 1991–1994, (1995)
MMG05.	Motter, A.E., de Moura, A.P.S., Grebogi, C., Kantz, H.: Effective dynamics in Hamiltonian systems with mixed phase space. Phys. Rev. E **71**, 036215, (2005)
MMZ99.	Manasevich, R., Mawhin. J., Zanolin, F.: Periodic solutions of some complex–valued Lienard and Rayleigh equations. Nonl. Anal. **36**, 997–1014, (1999)
MN01.	Maldacena. J., Nunez, C.: Supergravity Description of Field Theories on Curved Manifolds and No Go Theorem, Int. J. Mod. Phys. **A16**, 822, (2001)
MN95a.	Mavromatos, N.E., Nanopoulos, D.V.: A Non-critical String (Liouville) Approach to Brain Microtubules: State Vector reduction, Memory coding and Capacity. ACT-19/95, CTP-TAMU-55/95, OUTP-95-52P, (1995)
MN95b.	Mavromatos, N.E., Nanopoulos, D.V.: Non-Critical String Theory Formulation of Microtubule Dynamics and Quantum Aspects of Brain Function. ENSLAPP-A-524/95, (1995)
MNM02.	Montagna, G., Nicrosini, O., Moreni, N.: A path integral way to option pricing. Physica A **310**, 450 – 466, (2002)
MNT00.	Morosawa, S., Nishsimura, Y., Taniguchi, M., Ueda, T.: Holomorphic dynamics. Cambridge Univ. Press, (2000)
MOS99.	Mangiarotti, L., Obukhov, Yu., Sardanashvily, G.: Connections in Classical and Quantum Field Theory. World Scientific, Singapore, (1999)
MP01.	Morales, C.A., Pacifico, M.J.: Mixing attractors for 3-flows. Nonlinearity, **14**, 359–378, (2001)
MP70.	Meyer, K.R.: Palmore, J.I.: A new class of periodic solutions in the restricted three–body problem, J. Diff. Eqs. **44**, 263–272, (1970)
MP92.	De Melo, W., Palis, J.: Geometric Theory of Dynamical Systems-An Introduction. Springer Verlag, Berlin, (1982)
MP94.	Massa, E., Pagani, E.: Jet bundle geometry, dynamical connections and the inverse problem of Lagrangian mechanics. Ann. Inst. Henri Poincaré **61**, 17, (1994)
MP97.	Muñuzuri, A.P., Pérez-Villar, V., Markus, M.: Splitting of Autowaves in an Active Medium. Phys. Rev. Lett. **79**, 1941, (1997)
MPS98.	Marsden. J.E., Patrick, G.W., Shkoller, S.: Multisymplectic Geometry, Variational Integrators, and Nonlinear PDEs. Comm. Math. Phys., **199**, 351–395, (1998)
MPT98.	Milanović, Lj., Posch, H.A., Thirring, W.: Statistical mechanics and computer simulation of systems with attractive positive power-law potentials. Phys. Rev. E **57**, 2763–2775, (1998)
MPV87.	Mézard, M., Parisi, G., Virasoro, M.A.: Spin Glass Theory and Beyond. World Scientific, Singapore, (1987)

MQ86. Mathai V., Quillen, D. Super-connections, Thom classes and equivariant differential forms. Topology **25**, 85-110, (1986)
MR90. Miller, B.N., Reidl, C.J. Jr.: Gravity in one dimension - Persistence of correlation. Astrophys. J. **348**, 203-211, (1990)
MR91. Mathur, V.S., Rajeev, S.G.: What Are the Anti-Particles of $K(L,S)$? Mod. Phys. Lett. A **6**, 2741, (1991)
MR92. Muñoz-Lecanda, M. and Román-Roy, N.: Lagrangian theory for presymplectic systems. Ann. Inst. Henrí Poincaré **57**, 27, (1992)
MR99. Marsden. J.E., Ratiu, T.S.: Introduction to Mechanics and Symmetry: A Basic Exposition of Classical Mechanical Systems. (2nd ed), Springer, New York, (1999)
MRR53. Metropolis, N., Rosenbluth, A.W., Rosenbluth, M.N., Teller, A.H., Teller, E.: Equations of State calculations by fast computing machines. J. Chem. Phys. **21**, 1087–1092, (1953)
MRS04. Munteanu, F., Rey, A.M., Salgado, M.: The Günther's formalism in classical field theory: momentum map and reduction. J. Math. Phys. **45**(5), 1730–1750, (2004)
MS00. Meyer, K.R., Schmidt, D.S.: From the restricted to the full three-body problem. Trans. Amer. Math. Soc., **352**, 2283–2299, (2000)
MS00. Murray, M.K., Singer, M.A.: On the complete integrability of the discrete Nahm equations Communications in Mathematical Physics, 210(2) 497–519, (2000)
MS00a. Mangiarotti, L., Sardanashvily, G.: Connections in Classical and Quantum Field Theory, World Scientific, Singapore, (2000)
MS00b. Mangiarotti, L., Sardanashvily, G.: Constraints in Hamiltonian time-dependent mechanics. J. Math. Phys. **41**, 2858, (2000)
MS03. Moroianu, A., Semmelmann, U.: Twistor forms on Kähler manifolds. Ann. Scuola Norm. Sup. Pisa Cl. Sci. **2**(4), 823–845, (2003)
MS71. Meyer, K.R., Schmidt, D.S.: Periodic orbits near L_4 for mass ratios near the critical mass ratio of Routh, Celest. Mech. **4**, 1971, 99–109, (1971)
MS74). Milnor, J.W., Stasheff, J.D.: Characteristic Classes. Princeton Univ. Press, Princeton, NJ, (1974)
MS78. Modugno, M., Stefani, G.: Some results on second tangent and cotangent spaces. Quadernidell' Instituto di Matematica dell' Universit a di Lecce Q., **16**, (1978)
MS90. Mallet–Paret, J., Smith, H.L.: The Poincaré–Bendixson theorem for monotone cyclic feedback systems, J. Dyn. Diff. Eqns. **2**, 367–421, (1990)
MS90. Mirollo, R.E., Strogatz, S.H.: Synchronization of pulse-coupled biological oscillators. SIAM J. Appl. Math. **50**, 1645, (1990)
MS95. Marmo, G., Simoni, A., Stern, A.: Poisson Lie group symmetries for the isotropic rotator. Int. J. Mod. Phys. A **10**, 99–114, (1995)

MS95.	Müllers, J., Schmid, A.: Resonances in the current-voltage characteristics of a dissipative Josephson junction. cond-mat/9508035, (1995)
MS96.	Mallet–Paret, J., Sell,G.: The Poincaré–Bendixson theorem for monotone cyclic feedback systems with delay, J. Diff. Eqns. **125**, 441–489, (1996)
MS96.	Murray, M.K., Singer, M.A.: Spectral curves of non-integral hyperbolic monopoles, Nonlinearity 9, 973–997, (1996)
MS97.	Madore, J., Saeger, L.A.: Topology at the Planck Length. gr-qc/9708053, (1997)
MS98.	Mangiarotti, L., Sardanashvily, G.: Gauge Mechanics. World Scientific, Singapore, (1998)
MS99.	Mangiarotti, L., Sardanashvily, G.: On the geodesic form of non-relativistic dynamic equations. arXiv: math-ph/9906001.
MSM00.	Mukohyama, S., Shiromizu, T., Maeda, K.I.: Global structure of exact cosmological solutions in the brane world. Phys. Rev. D **62**, 024028, (2000)
MT03.	Minic, D., Tze, C.: Background independent quantum mechanics and gravity. Phys. Rev. D **68**, 061501, (2003)
MT80.	Marsden. J.E., Tipler, F.: Maximal Hypersurfaces and Foliations of Constant Mean Curvature in General Relativity, Physics Reports **66**, 109, (1980)
MT92.	Müller, B., Trayanov, A.: Deterministic chaos in non-Abelian lattice gauge theory. Phys. Rev. Lett. **68**, 3387 (1992)
MTW73.	Misner, C.W., Thorne, K.S., Wheeler. J.A.: Gravitation. Freeman, San Francisco, (1973)
MTY88.	Morris, M., Thorne, K.S., Yurtsver, U.: Wormholes, Time Machines, and the Weak Energy Condition. Phys. Rev. Lett. **61**, 1446, (1988)
MU49.	Metropolis, N., Ulam, S.: The Monte Carlo Method. J. Am. Stat. Ass. **44**, 335–341, (1949)
MUW85.	Mazenko, G.F., Unruh, W.G., Wald, R.M.: Does a Phase Transition in the Early Universe Produce the Conditions Needed for Inflation? Phys. Rev. D **31**, 273, (1985)
MW00.	Marinatto, L., Weber, T.: A Quantum Approach to Static Games of Complete Information. Phys. Lett. A **272**, 291 (2000)
MW00.	Maeda, K.I., Wands, D.: Dilaton–gravity on the brane. Phys. Rev. D **62**, 124009, (2000)
MW74.	Marsden, J., Weinstein, A.: Reduction of symplectic manifolds with symmetries. Rep. Math. Phy. **5**, 121–130, (1974)
MW89.	Marcus, C.M., Westervelt, R.M.: Dynamics of iterated-map neural networks. Phys. Rev. A **40**(1), 501–504, (1989)
MW94.	Mallet–Paret, J., Walther, H.-O.: Rapid oscillations are rare in scalar systems governed by monotone negative feedback with a time lag. Math. Inst. Univ. Giessen, (1994)
MWB00.	Maartens, R., Wands, D., Bassett, B.A., Heard, I.: Chaotic inflation on the brane. Phys. Rev. D **62**, 041301, (2000)
MWH01.	Michel, A.N., Wang, K., Hu, B.: Qualitative Theory of Dynamical Systems (2nd ed) Dekker, New York, (2001)

MZA03. Mangan, S., Zaslaver, A., Alon, U.: The coherent feedforward loop serves as a sign-sensitive delay element in transcription networks. JMB, **334**(2), 197–204, (2003)

Maa00. Maartens, R.: Cosmological dynamics on the brane. Phys. Rev. D **62**, 084023, (2000)

Maa01. Maartens, R.: Geometry and dynamics of the brane-world. arXiv:gr-qc/0101059, (2001)

Mac59. Mach, E.: The Analysis of Sensations and the Relation of Physical to the Psychical. (5th ed.) Dover, New York, (1959)

Mac00. MacKenzie, R.: Path integral methods and applications. arXiv:quant-ph/0004090, (2000)

Mag54. Magnus, W.: On the exponential solution of differential equations for a linear operator. Commun. Pure Appl. Math. **7**, 649-673, (1954)

Mah98. Mahmoud, G.M.: Approximate solutions of a class of complex nonlinear dynamical systems. Physica A **253**, 211-222, (1998)

Maj84. Majda, A.: Compressible Fluid Flow and System of Conservation Laws in Several Space Variables. Springer-Verlog, Berlin, (1984)

Mal00. Malasoma, J.M.: What is the Simplest Dissipative Chaotic Jerk Equation which is Parity Invariant? Phys. Lett. A, **264**(5), 383–389, (2000)

Mal01. Maldacena. J.M.: Eternal Black Holes in AdS. arXiv:hep-th/0106112, (2001)

Mal77. Malament, D.B.: The class of continuous time-like curves determines the topology of space-time. J. Math. Phys. **18**, 1399–1404, (1977)

Mal798. Malthus, T.R.: An essay on the Principle of Population. Originally published in 1798. Penguin, (1970)

Mal88. Mallet–Paret, J.: Morse decompositions for delay differential equations, J. Diff. Equs. **72**, 270–315, (1988)

Mal98. Maldacena. J.: The Large N Limit of Superconformal Field Theories and Gravity, Adv. Theor. Math. Phys. **2**, 231, (1998)

Man77. Manton, N.S.: The force between 't Hooft-Polyakov monopoles, Nucl. Phys. B126, 525–541, (1977)

Man80a. Mandelbrot, B.: Fractal aspects of the iteration of $z \mapsto \lambda z(1-z)$ for complex λ, z, Annals NY Acad. Sci. **357**, 249–259, (1980)

Man80b. Mandelbrot, B.: The Fractal Geometry of Nature. WH Freeman and Co., New York, (1980)

Man82. Manton, N.S.: A remark on the scattering of BPS monopoles, Phys. Lett. B110, 54–56, (1982)

Man82. Mañé, R.: An ergodic closing lemma. Annals of Math., **116**, 503–540, (1982)

Man98. Manikonda, V.: Control and Stabilization of a Class of Nonlinear Systems with Symmetry. PhD Thesis, Center for Dynamics and Control of Smart Structures, Harvard Univ., Cambridge, MA, (1998)

Mar03.	Marshall, W., Simon, C., Penrose, R., Bouwmeester, D.: Quantum Superposition of a Mirror. Phys. Rev. Lett. **91**, 13, (2003)
Mar86.	Margein, C: Positive Pinched manifolds are space forms. Proceedings of Symp. Pure Math. **44**, 307–328, (1986)
Mar98.	Marieb, E.N.: Human Anatomy and Physiology. (4th ed.), Benjamin/Cummings, Menlo Park, CA, (1998)
Mar99.	Marsden, J.E.: Elementary Theory of Dynamical Systems. Lecture notes. CDS, Caltech, (1999)
Mat02.	Matacz, A.: Path Dependent Option Pricing: the Path Integral Partial Averaging Method. J. Comp. Finance, **6**(2), (2002)
Mat82.	Matsumoto, M.: Foundations of Finsler Geometry and Special Finsler Spaces. Kaisheisha Press, Kyoto, (1982)
Mau70.	Maunder, C.R.F.: Algebraic topology. Dover, New York, (1970)
Mau.	Maunder, C.R.F.: Introduction to Algebraic Topology. Van Nostrand Reinhold, London, (1970)
May73.	May, R.M. (ed.): Stability and Complexity in Model Ecosystems. Princeton Univ. Press, Princeton, NJ, (1973)
May76.	May, R.M. (ed.): Theoretical Ecology: Principles and Applications. Blackwell Sci. Publ. (1976)
May76.	May, R.: Simple Mathematical Models with Very Complicated Dynamics. Nature, **261**(5560), 459-467, (1976)
May81.	Mayer, P.A.: A Differential Geometric Formalism for the Ito Calculus. Lecture Notes in Mathematics, Vol. 851, Springer, New York, (1981)
May97.	Mayers, D.: Unconditionally Secure Quantum Bit Commitment is Impossible. Phys. Rev. Lett. **78**, 3414–3417, (1997)
McCul87.	McCullagh, P.: Tensor methods in statistics. Monographs in statistics and applied probability. Chapman & Hall, Cambridge, UK, (1987)
McG99.	McGinnis, P.M.: Biomechanics of Sport and Exercise. Human Kinetics, Champaign, IL, (1999)
Men98.	Mendes, R.V.: Conditional exponents, entropies and a measure of dynamical self-organization. Phys. Let. A **248**, 167–1973, (1998)
Mer73.	Merton, R.: Theory of Rational Option Pricing. Bell J. Econom. Managem. Sci. **4**, 141–183, (1973)
Mer93.	Mermin, D.: Hidden Variables and the Two Theorems of John Bell. Rev. Mod. Phys. **65**, 803, (1993)
Mes00.	Messiah, A.: Quantum Mechanics (two volumes bound as one) Dover Pubs, (2000)
Met97.	Metzger, M.A.: Applications of nonlinear dynamical systems theory in developmental psychology: Motor and cognitive development. Nonlinear Dynamics, Psychology, and Life Sciences, **1**, 55–68, (1997)
Mey70.	Meyer, K.R.: Generic bifurcation of periodic points. Trans. Amer. Math. Soc. **149**, 95–107, (1970)
Mey71.	Meyer, K.R.: Generic stability properties of periodic points, Trans. Amer. Math. Soc. **154**, 273–277, (1971)

Mey73.	Meyer, K.R.: Symmetries and integrals in mechanics. In Dynamical Systems (M. Peixoto, Ed.), 259–272, Academic Press, New York, (1973)
Mey81a.	Meyer, K.R.: Periodic solutions of the N-body problem. J. Dif. Eqs. **39**(1), 2–38, (1981)
Mey81b.	Meyer, K.R.: Periodic orbits near infinity in the restricted N-body problem, Celest. Mech. **23**, 69–81, (1981)
Mey94.	Meyer, K.R.: Comet like periodic orbits in the N–body problem, J. Comp. and Appl. Math. **52**, 337–351, (1994)
Mic01.	Michor, P.W.: Topics in Differential Geometry. Lecture notes of a course in Vienna, (2001)
Mil56.	Miller, G.A.: The magical number seven, plus or minus two: Some limits on our capacity for processing information, Psych. Rev., **63**, 81–97, (1956)
Mil61.	Milnor. J.: A procedure for killing homotopy groups of differentiable manifolds, Proc. Sympos. Pure Math. Vol. III 39-55, (1961)
Mil62.	Milnor. J.: A survey of cobordism theory, Enseign. Math. **8**, 16, (1962)
Mil63.	Milnor. J.: Morse Theory. Princeton Univ. Press, Princeton, (1963)
Mil65.	Milnor. J.: Lectures on the h–cobordism theorem, Princeton Univ. Press, New Jersey, (1965)
Mil65a.	Milnor. J.: Topology from the differentiable viewpoint, The University Press of Virginia, Charlottesville, (1965)
Mil99.	Milinković, D.: Morse homology for generating functions of Lagrangian submanifolds. Trans. Amer. Math. Soc. **351**(10), 3953–3974, (1999)
Mla91.	Mladenova, C.: Mathematical Modelling and Control of Manipulator Systems. Int. J. Rob. Comp. Int. Man., **8**(4), 233–242, (1991)
Mla99.	Mladenova, C.: Applications of Lie Group Theory to the Modelling and Control of Multibody Systems. Mult. Sys. Dyn., **3**(4), 367–380, (1999)
Mof83.	Moffat, H.K.: Transport effects associated with turbulence, with particular attention to the influence of helicity. Rep. Prog. Phys. **46**, 621-664, (1983)
Moh91.	Mohler, R.R.: Nonlinear systems, Vol. 2. Applications to bilinear control. Prentice Hall, Inc. (1991)
Mok88.	Mok, N.: The uniformization Theorem for compact Kähler manifolds of non-negative holomorphic bisectional curvature. J. Diff. Geom. **27**, 179–214, (1988)
Mor07.	Morozov, A.: Universal Mandelbrot Set as a Model of Phase Transition Theory. arXiv:0710.2315, (2007)
Mor34.	Morse, M.: The Calculus of Variations in the Large. Amer. Math. Soc. Coll. Publ. No. 18, Providence, RI, (1934)
Mor73.	Morita, T.: Solution of the Bloch Equation for Many–Particle Systems in Terms of the Path Integral. J. Phys. Soc. Jpn. **35**, 980–984, (1973)

Mor79.	Mori, S.: Projective manifolds with ample tangent bundles. Ann. Math., **76**(2), 213–234, (1979)
Mos62.	Moser, J.: On invariant curves of area - preserving mappings of an annulus., Nach. Akad. Wiss., Gottingen, Math. Phys. Kl. **2**, 1, (1962)
Mos73.	Moser. J.: Stable and Random Motions in Dynamical Systems. Princeton Univ. Press, Princeton, (1973)
Mos96.	Mosekilde, E.: Topics in Nonlinear Dynamics: Application to Physics, Biology and Economics. World Scientific, Singapore, (1996)
Mou06.	Moulton, F.R.: A class of periodic solutions of the problem of three–bodies with applications to lunar theory. Trans. Amer. Math. Soc. **7**, 537–577, (1906)
Mou12.	Moulton, F.R.: A class of periodic orbits of the superior planets, Trans. Amer. Math. Soc. **13**, 96–108, (1912)
Mou95.	Mould, R.: The inside observer in quantum mechanics. Found. Phys. **25**(11), 1621–1629, (1995)
Mou98.	Mould, R.: Consciousness and Quantum Mechanics. Found. Phys. **28**(11), 1703–1718, (1998)
Mou99.	Mould, R.: Quantum Consciousness. Found. Phys. **29**(12), 1951–1961, (1999)
Muk00.	Mukohyama, S.: Gauge-invariant gravitational perturbations of maximally symmetric space–times. Phys. Rev. D **62**, 084015, (2000)
Mun75.	Munkres, J.R.: Topology a First Course. Prentice-Hall, New Jersey, (1975)
Mur01.	Murray, M.K.: Monopoles. arXiv:math-ph/0101035, (2001)
Mur02.	Murray, J.D.: Mathematical Biology, Vol. I: An Introduction (3rd ed.), Springer, New York, (2002)
Mur84.	Murray, M.K.: Non-Abelian Magnetic Monopoles, Commun. Math. Phys. 96, 539–565, (1984)
Mur94.	Murray, C.D.: Dynamical effect of drag in the circular restricted three body problem 1. Location and stability of the Lagrangian equilibrium points. Icarus **112**, 465-484, (1994)
Mus99.	Mustafa, M.T.: Restrictions on harmonic morphisms. Confomal Geometry and Dynamics (AMS), **3**, 102–115, (1999)
Mye91.	Myerson, R.B.: Game Theory: An Analysis of Conflict. MIT Press, Cambridge, MA, (1991)
NAY60.	Nagumo, J., Arimoto, S., Yoshizawa, S.: An active pulse transmission line simulating 1214-nerve axons, Proc. IRL **50**, 2061–2070, (1960)
NB99.	Newman, M.E.J., Barkema, G.T.: Monte Carlo methods in statistical physics, Oxford Univ. Press, Oxford, UK (1999)
NC00.	Nielsen, M., Chuang, I.L.: Quantum computation and information. Springer, Berlin, (2000)
NE99.	Nimmo, S., Evans, A.K.: The Effects of Continuously Varying the Fractional Differential Order of Chaotic Nonlinear Systems. Chaos, Solitons & Fractals **10**, 1111–1118, (1999)
NHP89.	Noakes, L., Heinzinger, G., Paden, B.: Cubic splines on curved spaces, IMA J. Math. Con. Inf. , **6**(4), 465–473, (1989)

NK96. Nishikawa, T., Kaneko, K.: Fractalization of a torus as a strange nonchaotic attractor. Phys. Rev. E **54**, 6114–6124, (1996)

NKF97. Nishimura, H., Katada, N., Fujita, Y.: Dynamic Learning and Retrieving Scheme Based on Chaotic Neuron Model, In: R. Nakamura *et al.* (eds): Complexity and Diversity, Springer-Verlag, New York, 64–66, (1997)

NLG00. Newrock, R.S., Lobb, C.J., Geigenmüller, U., Octavio, M.: Solid State Physics. Academic Press, San Diego, Vol. 54, (2000)

NM79. Nayfeh, A.H., Mook, D.T.: Nonlinear Oscillations. Wiley, New York, (1979)

NML03. Nishikawa, T., Motter, A.E., Lai, Y.C., Hoppensteadt, F.C.: Heterogeneity in Oscillator Networks: Are Smaller Worlds Easier to Synchronize? Phys. Rev. Lett. **91**, 014101, (2003)

NMP84. Novikov, S.P., Manakov, S.V., Pitaevskii, L.P., Zakharov, V.E.: Theory of Solitons, Plenum/Kluwer, Dordrecht, (1984)

NNM00. Nagao, N., Nishimura, H., Matsui, N.: A Neural Chaos Model of Multistable Perception. Neu. Proc. Lett. **12**(3): 267–276, (2000)

NNM97. Nishimura, H., Nagao, N., Matsui, N.: A Perception Model of Ambiguous Figures based on the Neural Chaos, In: N. Kasabov *et al.* (eds): Progress in Connectionist-Based Information Systems, **1**, 89–92, Springer-Verlag, New York, (1997)

NP62. Newman, E.T., Penrose, R.: An Approach to Gravitational Radiation by a Method of Spin Coefficients. J. Math. Phys. **3**(3), 566–768, (1962)

NPT99. Nakamura, Y., Pashkin, Yu.A., Tsai, J.S.: Coherent control of macroscopic quantum states in a single-Cooper-pair box, Nature, **398**, 786–788, (1999)

NS90. Nijmeijer, H., van der Schaft, A.J.: Nonlinear Dynamical Control Systems. Springer, New York, (1990)

NS98. Nelson, D.R., Shnerb, N.M.: Non-Hermitian localization and population biology. Phys. Rev. E **58**, 1383, (1998)

NS83. Charles Nash and Siddhartha Sen. Topology and Geometry for Physicists. Academic Press, New York, (1983)

NT92. T. Nattermann, L.-H. Tang, Phys. Rev. A **45**, 7156, (1992)

NU00a. Neagu, M. Udriște, C.: Multi–Time Dependent Sprays and Harmonic Maps on $J^1(T,M)$, Third Conference of Balkan Society of Geometers, Politehnica University of Bucarest, Romania, (2000)

NU00b. Neagu, M. Udriște, C.: Torsions and Curvatures on Jet Fibre Bundle $J^1(T,M)$. arXiv:math.DG/0009069.

Nag74. Nagel, T.: What is it like to be a bat? Philos. Rev. **83**, 435–456, (1974)

Nah82. Nahm, W.: The construction of all self–dual monopoles by the ADHM method, in Monopoles in Quantum Field Theory, Proceedings of the monopole meeting in Trieste 1981, World Scientific, Singapore, (1982)

Nan04.	Nanayakkara, A.: Classical trajectories of 1D complex non-Hermitian Hamiltonian systems. J. Phys. A: Math. Gen. **37**, 4321, (2004)
Nan95.	Nanopoulos, D.V.: Theory of Brain Function, Quantum Mechanics and Superstrings. CERN-TH/95128, (1995)
Nas50c.	Nash, J.F.: The bargaining problem. Econometrica **18**, 155–162, (1950c)
Nas51.	Nash, J.F.: Non–cooperative games. Annals of Mathematics **54**, 286–295, (1951)
Nas53.	Nash, J.F.: Two–person cooperative games. Econometrica **21**, 128–140, 1953.
Nas83.	Nash, C., Sen, S.: Topology and Geometry for Physicists. Academic Press, London, (1983)
Nas97.	Nash, J.F.: Essays on Game Theory. Publ. Edward Elgar, (1997)
Nay73.	Nayfeh, A.H.: Perturbation Methods. Wiley, New York, (1973)
Nea00.	Neagu, M.: Upon h−normal Γ−linear connection on $J^1(T, M)$. arXiv:math.DG/0009070, (2000)
Nea02.	Neagu, M.: The Geometry of Autonomous Metrical Multi-Time Lagrange Space of Electrodynamics. Int. J. Math. Math. Sci. **29**, 7-16, (2002)
New00.	Newman, M.E.J.: Models of the small world. J. Stat. Phys. **101**, 819, (2000)
New61.	Newman, E.: J. Math. Phys. **2**, 324, (1961)
New74.	Newell, A.C. (ed): Nonlinear Wave Motion. Am. Mat. Soc., Providence, R.I., (1974)
New80.	Newhouse, S.: Lectures on dynamical systems. In Dynamical Systems, C.I.M.E. Lectures, 1–114, Birkhauser, Boston, MA, (1980)
Nie94.	Niedzielska, Z.: Nonlinear stability of the libration points in the photo-gravitational restricted three body problem. Celest. Mech. Dyn. Astron., **58**(3), 203-213, (1994)
Nih99.	Nihei, T.: Dynamics of Scalar field in a Brane World. Phys. Lett. B **465**, 81, (1999)
Nik95.	Nikitin, I.N.: Quantum string theory in the space of states in an indefinite metric. Theor. Math. Phys. **107**(2), 589–601, (1995)
Nis86.	Nishikawa, S.: Deformation of Riemannian metrics and manifolds with bounded curvature ratios. Proceedings of Symp. Pure Math. **44**, 345–352, (1986)
OC99.	Osipov, G.V., Collins, J.J.: Using weak impulses to suppress traveling waves in excitable media. Phys. Rev. E **60**, 54–57, (1999)
OGY90.	Ott, E., Grebogi, C., Yorke, J.A.: Controlling chaos. Phys. Rev. Lett., **64**, 1196–1199, (1990)
OHF06.	Ouchi, K., Horita, T., Fujisaka, H.: Critical dynamics of phase transition driven by dichotomous Markov noise. Phys. Rev. E **74**, 031106, (2006)
OR94.	Osborne, M., Rubinstein, A.: A Course in Game Theory. MIT Press, Cambridge, MA, (1994)

OS74. Oldham, K.B, Spanier. J.: The Fractional Calculus. Acad. Press, New York, (1974)

OS94. Ovchinnikov, Yu.N., Schmid, A.: Resonance phenomena in the current-voltage characteristic of a Josephson junction. Phys. Rev. B **50**, 6332–6339, (1994)

OSV04. Ooguri, H., Strominger, A., Vafa, C.: Black hole attractors and the topological string. Phys. Rev. D **70**, 106007, (2004)

OTK02. Ozbudak, E.M., Thattai, M., Kurtser, I., Grossman, A.D., van Oudenaarden, A.: Regulation of noise in the expression of a single gene. Nature Gen., **31**,69-73, (2002)

OTL04. Ozbudak, E.M., Thattai, M., Lim, H.N., Shraiman, B.I., van Oudenaarden, A.: Multistability in the lactose utilization network of Escherichia coli. Nature, **427**(6976), 737–40, (2004)

Obe00. Oberste-Vorth, R.: Horseshoes among Hénon mappings. In Mastorakis, N. (ed.) Recent Advances in Applied and Theoretical Mathematics, WSES Press, 116–121, (2000)

Obe01. Oberlack, M.: A unified approach for symmetries in plane parallel turbulent shear flows. J. Fluid Mech. **427**, 299–328, (2001)

Obe87. Oberste-Vorth, R.: Complex Horseshoes and the Dynamics of Mappings of Two Complex Variables. PhD thesis, Cornell Univ., (1987)

Obe99. Oberlack M.: Symmetries, invariance and scaling-laws in inhomogeneous turbulent shear flows. Flow, Turbulence and Combustion **62**, 111–135, (1999)

Odo04. Odor, G. : Universality classes in nonequilibrium lattice systems. Rev. Mod. Phys. **76**, 663–724, (2004)

Oht98. Ohta, Y.: Topological Field Theories associated with Three Dimensional Seiberg-Witten monopoles. Int.J.Theor.Phys. **37**, 925-956, (1998)

Oka95. Okada, M.: A hierarchy of macrodynamical equations for associative memory. Neu. Net. **8**, 833-838, (1995)

Oku81. Okubo, S.: Canonical quantization of some dissipative systems and nonuniqueness of Lagrangians. Phys. Rev. A **23**, 2776, (1981)

Olv01. Olver, P.J.: Geometric foundations of numerical algorithms and symmetry. Appl. Algebra Engrg. Comm. Comput. **11**, 417–436, (2001)

Olv86. Olver, P.J.: Applications of Lie Groups to Differential Equations (2nd ed.) Graduate Texts in Mathematics, Vol. 107, Springer, New York, (1986)

Omn94. Omnès, R.: Interpretation of Quantum Mechanics, Princeton Univ. Press, Princeton, (1994)

Omo86. Omohundro, S.M.: Geometric Perturbation Theory in Physics. World Scientific, Singapore, (1986)

Ons35. Onsager, L.: Reciprocal relations in irreversible processes. II. Phys. Rev. **38**, 2265-2279, (1931)

Oog92a. Ooguri, H.: Topological Lattice Models in Four Dimensions. Mod. Phys. Lett. **A7**, 2799-2810, (1992)

Oog92b.	Ooguri, H.: Partition Functions and Topology-Changing Amplitudes in the 3D Lattice Gravity of Ponzano and Regge. Nucl. Phys. **B382**, 276, (1992)
Ord86.	P. Ordeshook, Game Theory and Political Theory: An Introduction. Cambridge Univ. Press, Cambridge, (1986)
Ori01.	Oriti, D.: space-time geometry from algebra: Spin foam models for non-perturbative quantum gravity. Rept. Prog. Phys. **64**, 1489–1544, (2001)
Ose68.	Oseledets, V.I.: A Multiplicative Ergodic Theorem: Characteristic Lyapunov Exponents of Dynamical Systems. Trans. Moscow Math. Soc., **19**, 197–231, (1968)
Ott89.	Ottino, J.M.: The kinematics of mixing: stretching, chaos and transport. Cambridge Univ. Press, Cambridge, (1989)
Ott93.	Ott, E.: Chaos in dynamical systems. Cambridge University Press, Cambridge, (1993)
PB89.	Peyrard, M., Bishop, A.R.: Statistical mechanics of a nonlinear model for DNA denaturation. Phys. Rev. Lett. **62**, 2755, (1989)
PB94.	Plischke, M. Bergersen, B.: Equilibrium Statistical Mechanics. World Scientific, Singapore, (1994)
PB99.	Paul, W., Baschnagel. J.: Stochastic Processes: from Physics to Finance. Springer, Berlin, (1999)
PBK62.	Pontryagin, L., Boltyanskii, V., Gamkrelidze, R., Mishchenko, E.: The mathematical theory of optimal processes. Interscience, (1962)
PBR97.	Prasad, A., Mebra, V., Ramaswamy, R.: Intermittency Route to Strange Nonchaotic Attractors. Phys. Rev. Lett. **79**, 4127–4130, (1997)
PBS01.	Potters, M., Bouchaud. J.P., Sestovic, D.: Hedged Monte-Carlo: low variance derivative pricing with objective probabilities. Physica A **289**, 517—25, (2001)
PC05.	Park. J., Chung, W.-K.: Geometric Integration on Euclidean Group With Application to Articulated Multibody Systems. IEEE Trans. Rob. **21**(5), 850–863, (2005)
PC84.	Pollock, E.L., Ceperley, D.M.: Simulation of quantum many-body systems by path–integral methods. Phys. Rev. B **30**, 2555–2568, (1984)
PC89.	Parker, T.S., Chua, L.O.: Practical Numerical Algorithm for Chaotic Systems, Springer-Verlag, New York, (1989)
PC90.	Pecora, L.M., Carroll, T.L.: Synchronization in chaotic systems. Phys. Rev. Lett. **64**, 821–824, (1990)
PC91.	Pecora, L.M., Carroll, T.L.: Driving systems with chaotic signals. Phys. Rev. A, **44**, 2374—2383, (1991)
PC91.	Pettini, M., Cerruti-Sola, M.: Strong stochasticity threshold in nonlinear large Hamiltonian systems: Effect on mixing times. Phys. Rev. A **44**, 975–987, (1991)
PC98.	Pecora, L.M., Carroll, T.L.: Master stability functions for synchronized coupled systems. Phys. Rev. Lett. **80**, 2109—2112, (1998)

PCC05. Pettini, M., Casetti, L., Cerruti-Sola, M., Franzosi, R., Cohen, E.G.D.: Weak and strong chaos in FPU models and beyond. Chaos **15**, 015106 (2005)

PCS93. Penney, R.W., Coolen, A.C., Sherrington D.: Coupled dynamics of fast and slow interactions in neural networks and spin systems. J. Phys. A: Math. Gen. **26**, 3681, (1993)

PE02. Popper, K., Eccles, J.: The Self and Its Brain. Springer, (2002)

PE34. Pauli, W., Weisskopf, V.: ´Uber die Quantisierung der skalaren relativistischen. Helv. Phys. Acta, **7**, 708-731, (1934)

PG81. Page, D.N., Geilker, C.D.: Indirect Evidence for Quantum Gravity. Phys. Rev. Lett. **47**, 979–982, (1981)

PGY06. Politi, A., Ginelli, F., Yanchuk, S., Maistrenko, Y.: From synchronization to Lyapunov exponents and back. arXiv:nlin.CD/0605012, (2006)

PH93. Panfilov, A.V., Hogeweg, P.: Spiral break–up in a modified FitzHugh–Nagumo model. Phys. Lett. A **176**, 295–299, (1993)

PHE92. Piaget. J., Henriques, G., Ascher, E.: Morphisms and categories. Erlbaum Associates, Hillsdale, NJ, (1992)

PHV86. Posch, H.A., Hoover, W.G., Vesely, F.J.: Canonical Dynamics of the Nosé Oscillator: Stability, Order, and Chaos. Phys. Rev. A, **33**(6), 4253–4265, (1986)

PI03. Pearce, C.E.M., Ivancevic, V.: A generalised Hamiltonian model for the dynamics of human motion. In Differential Equations and Applications, Vol. 2, Eds. Y.J. Cho. J.K. Kim and K.S. Ha, Nova Science, New York, (2003)

PI04. Pearce, C.E.M., Ivancevic, V.: A qualitative Hamiltonian model for the dynamics of human motion. In Differential Equations and Applications, Vol. 3, Eds. Y.J. Cho. J.K. Kim and K.S. Ha, Nova Science, New York, (2004)

PJL92. Procaccia, I., Jensen, M.H., L'vov *et al.*: Surface roughening and the long-wavelength properties of the Kuramoto–Sivashinsky equation. Phys. Rev. A **46**, 3220–3224, (1992)

PK93. Panfilov, A.V., Keener, J.P.: Effects of high frequency stimulation on cardiac tissue with an inexcitable obstacle. J. Theor. Biol. **163**, 439, (1993)

PK97. Pikovsky, A., Kurth, J.: Coherence Resonance in a Noise-Driven Excitable Systems. Phys. Rev. Lett. **78**, 775–778, (1997)

PL90. Pettini, M., Landolfi, M.: Relaxation properties and ergodicity breaking in nonlinear Hamiltonian dynamics. Phys. Rev. A **41**, 768–783, (1990)

PLK97. Parlinski, K., Li, Z.Q., Kawazoe, Y.: First–Principles Determination of the Soft Mode in Cubic ZrO_2. Phys. Rev. Lett. **78**, 4063–4066, (1997)

PLS00. Pappas, G.J., Lafferriere, G., Sastry, S.: Hierarchically consistent control systems. IEEE Trans. Aut. Con., **45**(6), 1144–1160, (2000)

PMZ02. Pitt, M.A., Myung, I.J., Zhang, S.: Toward a method of selecting among computational models of cognition Psychological Review, **109**(3), 472–491, (2002)

PO82. Ponomarev, V., Obukhov, Yu.: Generalized Einstein–Maxwell theory. GRG **14**, 309, (1982)

POR97. Pikovsky, A., Osipov, G., Rosenblum, M., Zaks, M., Kurths, J.: Attractor–repeller collision and eyelet intermittency at the transition to phase synchronization. Phys. Rev. Lett. **79**, 47–50, (1997)

PPO06. Piekarz, P., Parlinski, K., Oleś, A.M.: Mechanism of the Verwey Transition in Magnetite. Phys. Rev. Lett. **97**, 156402, (2006)

PPO07. Piekarz, P., Parlinski, K., Oleś, A.M.: Origin of the Verwey transition in magnetite: Group theory, electronic structure, and lattice dynamics study. Phys. Rev. B **76**, 165124, (2007)

PPO07. Piekarz, P., Parlinski, K., Oleś, A.M.: Order parameters in the Verwey phase transition. J. Phys. Conf. Ser. 92, 012164, (2007)

PPS02. Pandit, R., Pande, A., Sinha, S., Sen, A.: Spiral turbulence and spatiotemporal chaos: characterization and control in two excitable media. Physica A **306**, 211–219, (2002)

PR02. Paufler, C., Römer, H.: The geometry of Hamiltonian n–vector–fields in multisymplectic field theory. J. Geom. Phys. **44**(1), 52–69, (2002)

PR68. Ponzano, G., Regge, T.: Spectroscopy and Group Theoretical Methods. In F. Block (ed.), North-Holland, Amsterdam, (1968)

PR84. Parrinello, M., Rahman, A.: Study of an F center in molten KCl. J. Chem. Phys. **80**, 860, (1984)

PR86. Penrose, R., Rindler, W.: Spinors and Space–Time. Cambridge University Press, Cambridge, (1986)

PRK01. Pikovsky, A., Rosenblum, M., Kurths, J.: Synchronization: A Universal Concept in Nonlinear Sciences. Cambridge Univ. Press, Cambridge, UK, (2001)

PS02. Pappas, G.J., Simic, S.: Consistent hierarchies of affine nonlinear systems. IEEE Trans. Aut. Con., **47**(5), 745–756, (2002)

PS62. Perring, J.K., Skyrme, T.R.H.: A model unified field equation. Nuc. Phys., **31**, 550–55, (1962)

PS75. Prasad, M.K., Sommerfield, C.M.: Exact classical solutions for the 't Hooft monopole and Julia-Zee dyon. Phys. Rev. Lett. **35**, 760–762, (1975)

PS89. Pugh, C.C., Shub, M.: Ergodic attractors. Trans. Amer. Math. Soc. **312**, 1-54, (1989)

PS94. Penney, R.W., Sherrington, D.: Slow interaction dynamics in spin-glass models. J. Phys. A: Math. Gen. **27**, 4027–4041, (1994)

PSS96. Pons. J.M., Salisbury, D.C., Shepley, L.C.: Gauge transformations in the Lagrangian and Hamiltonian formalisms of generally covariant theories. arXiv gr-qc/9612037, (1996)

PST86. Park. J.K., Steiglitz, K. and Thurston, W.P.: Soliton–like Behavior in Automata. Physica D, **19**, 423–432, (1986)

PSV87. Paladin, G., Serva, M., Vulpiani, A.: Complexity in Dynamical Systems with Noise. Phys. Rev. Lett. **74**, 66–69, (1995)

PT93.	Palis, J., Takens, F.: Hyperbolicity and sensitive–chaotic dynamics at homoclinic bifurcations. Cambridge University Press, (1993)
PTB86.	Pokrovsky, V.L., Talapov, A.L., Bak, P.: In Solitons, 71–127, ed. Trullinger, Zakharov, Pokrovsky. Elsevier Science, (1986)
PV03.	Pessa, E., Vitiello, G.: Quantum noise, entanglement and chaos in the quantum field theory of mind/brain states. Mind and Matter, **1**, 59–79, (2003)
PV04.	Pessa, E., Vitiello, G.: Quantum noise induced entanglement and chaos in the dissipative quantum model of brain. Int. J. Mod. Phys. B, **18** 841–858, (2004)
PV95.	Pettini, M., Valdettaro, R.: On the Riemannian description of chaotic instability in Hamiltonian dynamics. Chaos **5**, 646, (1995)
PV98.	Plenio, M.B., Vedral, V.: Entanglement in Quantum Information Theory. Contemp. Phys. **39**, 431, (1998)
PV98.	Plenio, M.B., Vedral, V.: Teleportation, entanglement and thermodynamics in the quantum world. Contemp. Phys. **39**, 431–446, (1998)
PV99.	Pessa, E., Vitiello, G.: Quantum dissipation and neural net dynamics. Biolectrochem. Bioener. **48**, 339–342, (1999)
PY97.	Pan, S., Yin, F.: Optimal control of chaos with synchronization. Int. J. Bif. Chaos **7**, 2855–2860, (1997)
PZR97.	Pikovsky, A., Zaks, M., Rosenblum, M., Osipov, G., Kurths, J.: Phase synchronization of chaotic oscillations in terms of periodic orbits. Chaos **7**, 680, (1997)
Pal67.	Palmore, J.I.: Bridges and Natural Centers in the Restricted Three–Body Problem, University of Minnesota Report, Minneapolis, MN, (1969)
Pal88.	Palmer, K.J.: Exponential dichotomies, the shadowing lemma and transversal homoclinic points, U. Kirchgraber and H.–O. Walther (eds), 265–306, Dynamics Reported, vol. I, Teubner-Wiley, Stuttgart/Chichester, (1988)
Pal97.	Palais, R.S.: The symmetries of solitons. Bull. Amer. Math. Soc. **34**, 339–403, (1997)
Pal59.	Palais, R.S.: Natural operations on differential forms. Trans.Amer.Math.Soc. **92**, 125–141, (1959)
Pan98.	Panfilov, A.V.: Spiral breakup as a model of ventricular fibrillation. Chaos **8**, 57–64, (1998)
Pan99.	Panfilov, A.V.: Three–dimensional organization of electrical turbulence in the heart. Phys. Rev. E **59**, 6251, (1999)
Par84.	Parker, L., Toms, D.J.: Renormalization-group Analysis of Grand Unified Theories in Curved space-time. Phys. Rev. D **29**, 1584, (1984)
Pee80.	Peebles, P.J.E.: The Large Scale Structure of the Universe. Princeton Univ. Press, (1980)
Pen89.	Penrose, R.: The Emperor's New Mind. Oxford Univ. Press, Oxford, (1989)

Pen00.	Penrose, R.: Wavefunction Collapse as a Real Gravitational Effect, in Mathematical Physics 2000, ed. by A. Fokas, T.W.B. Kibble, A. Grigourion, and B. Zegarlinski, Imperial College Press, London, 266–282, (2000)
Pen04.	Penrose, R.: The Road to Reality. Jonathan Cape, London, (2004)
Pen67.	Penrose, R.: Twistor algebra. J. Math. Phys., **8**, 345–366, (1967)
Pen71a.	Penrose, R.: Angular momentum: an approach to combinatorial space–time, in Bastin, T. (ed.), Quantum Theory and Beyond, 151–180. Cambridge Univ. Press, Cambridge, UK, (1971)
Pen71b.	Penrose, R.: Applications of negative dimensional tensors, in Welsh, D. (ed.) Combinatorial Mathematics and its Application, 221–243, (1971)
Pen94.	Penrose, R.: Shadows of the Mind. Oxford Univ. Press, Oxford, (1994)
Pen97.	Penrose, R.: The Large, the Small and the Human Mind. Cambridge Univ. Press, (1997)
Per37.	Perron, O.: Neue periodische Lösungen des ebenen Drei und Mehrkörperproblem, Math. Z., **42**, 593–624, (1937)
Per84.	Peretto, P.: Collective properties of neural networks: a statistical physics approach. Biol. Cybern. **50**, 51, (1984)
Per86.	Perelomov, A.: Generalized Coherent States and their Applications. Springer, Berlin, (1986)
Per92.	Peretto, P.: An Introduction to the Theory of Neural Computation. Cambridge Univ. Press, Cambridge, (1992)
Per97.	Pert, C.B.: Molecules of Emotion. Scribner, New York, (1997)
Per98.	Percival, I.: Quantum State Diffusion. Cambridge Univ. Press, Cambridge, UK, (1998)
Pes75.	Peskin, C.: Mathematical Aspects of Heart Physiology. Courant Inst. Math. Sci., New York Univ, (1975)
Pes76.	Pesin, Ya.B.: Invariant manifold families which correspond to non-vanishing characteristic exponents. Izv. Akad. Nauk SSSR Ser. Mat. **40**(6), 1332–1379, (1976)
Pes77.	Pesin, Ya.B.: Lyapunov Characteristic Exponents and Smooth Ergodic Theory. Russ. Math. Surveys, **32**(4), 55–114, (1977)
Pet02.	Petras, I.: Control of Fractional-Order Chua's System. J. El. Eng. **53**(7–8), 219–222, (2002)
Pet07.	Pettini, M.: Geometry and Topology in Hamiltonian Dynamics and Statistical Mechanics. Springer, New York, (2007)
Pet93.	Peterson, I.: Newton's Clock: Chaos in the Solar System. W.H. Freeman, San Francisco, (1993)
Pet93.	Pettini, M.: Geometrical hints for a nonperturbative approach to Hamiltonian dynamics. Phys. Rev. E **47**, 828–850, (1993)
Pet96.	Petrov, V., Showalter, K.: Nonlinear Control from Time-Series. Phys. Rev. Lett. **76**, 3312, (1996)
Pet98.	Petersen, P.: Riemannian Geometry. Springer, New York, (1998)

Pet99. Petersen, P.: Aspects of Global Riemannian Geometry. Bull. Amer. Math. Soc., **36**(3), 297–344, (1999)
Pet99. Petras, I.: The Fractional–order controllers: Methods for their synthesis and application. J. El. Eng. **9-10**, 284–288, (1999)
Pic86. Pickover, C.A.: Computer Displays of Biological Forms Generated From Mathematical Feedback Loops. Computer Graphics Forum, **5**, 313, (1986)
Pic87. Pickover, C.A.: Mathematics and Beauty: Time–Discrete Phase Planes Associated with the Cyclic System. Computer Graphics Forum, **11**, 217, (1987)
Pine97. Pinel, J.P.: Psychobiology. Prentice Hall, New York, (1997)
Pink97. Pinker, S. How the Mind Works. Norton, New York, (1997)
Pod99. Podlubny, I.: Fractional Differential Equations. Academic Press, San Diego, (1999)
Poi899. Poincaré, H.: Les méthodes nouvelles de la mécanique céleste. Gauther–Villars, Paris, (1899)
Pol95. Polchinski. J.: Dirichlet–branes and Ramond-Ramond charges. Phys. Rev. Lett. **75**, 4724, (1995)
Pol98. Polchinski. J.: String theory. 2 Vols. Cambridge Univ. Press, (1998)
Pol99. Polchinski J.: String Theory, Two Volumes. Cambridge Univ. Press, (1999)
Pom78. Pommaret. J.: Systems of Partial Differential Equations and Lie Pseudogroups, Gordon and Breach, Glasgow, (1978)
Pop00. Pope, S.B.: Turbulent Flows. Cambridge Univ. Press, Cambridge, (2000)
Pop59. Popper, K.R: The Logic of Scientific Discovery. New York, NY: Basic Books, (1959)
Pos86. Postnikov, M.M.: Lectures in Geometry V, Lie Groups and Lie Algebras, Mir Publ., Moscow, (1986)
Poy03. Poynting, J.H.: Radiation in the solar system: its effect on temperature and its pressure on small bodies. Mon. Not. Roy. Ast. Soc. **64**, 1–5, (1903)
Pra91. Pratt, V.: Modelling concurrency with geometry. in: Proc. of the 18th ACM Symposium on Principles of Programming Languages, (1991)
Pre98. Preskill, J.: Quantum information and computation. Lecture notes for Physics 229, CalTech, Pasadena, (1998)
Pri62. Prigogine, I.: Introduction to thermodynamics of irreversible processes. John Wiley, New York, (1962)
Pri71. Pribram, K.H.: Languages of the brain. Prentice-Hall, Englewood Cliffs, N.J., (1971)
Pri91. Pribram, K.H.: Brain and perception. Lawrence Erlbaum, Hillsdale, N.J., (1991)
Pry96. Prykarpatsky, A.K.: Geometric models of the Blackmore's swept volume dynamical systems and their integrability, In: Proc. of the IMACS-95, Hamburg 1995, ZAMP, **247**(5), 720–724, (1996)
Pul05. Pulvermüller, F.: Brain mechanicsms kinking language and action. Nature Rev. Neurosci. **6**, 576–582, (2005)

Put93.	Puta, M.: Hamiltonian Mechanical Systems and Geometric Quantization, Kluwer, Dordrecht, (1993)
Pyr92.	Pyragas, K.: Continuous control of chaos, by self-controlling feedback. Phys. Lett. A, **170**, 421–428, (1992)
Pyr95.	Pyragas, K.: Control of chaos via extended delay feedback. Phys. Lett. A, **206**, 323–330, (1995)
RAP01.	Wright, J.P., Attfield, J.P., Radaelli, P.G.: Long Range Charge Ordering in Magnetite Below the Verwey Transition. Phys. Rev. Lett. **87**, 266401, (2001)
RBO87.	Romeiras, J., Bodeson, A., Ott, E., Antonsen, T.M. Jr., Grebogi, C.: Quasiperiodically forced dynamical systems with strange nonchaotic attractors. Physica D **26**, 277–294, (1987)
RBT01.	Roe, R.M., Busemeyer. J.R., Townsend. J.T.: Multi-alternative decision field theory: A dynamic connectionist model of decision making. Psych. Rev., **108**, 370–392, (2001)
REB03.	Rani, R., Edgar, S.B., Barnes, A.: Killing Tensors and Conformal Killing Tensors from Conformal Killing Vectors. Class.Quant.Grav., **20**, 1929–1942, (2003)
RFK99.	Rappel, W.J., Fenton, F., Karma, A.: Spatiotemporal Control of Wave Instabilities in Cardiac Tissue. Phys. Rev. Lett. **83**, 456–459, (1999)
RG98.	Rao, A.S., Georgeff, M.P.: Decision Procedures for BDI Logics. Journal of Logic and Computation, **8**(3), 292—343, (1998)
RGL99.	Rodriguez, E., George, N., Lachaux, J., Martinerie, J., Renault, B., Varela, F.: Long-distance synchronization of human brain activity. Nature, **397**, 430, (1999)
RGO92.	Romeiras, F.J., Grebogi, C., Ott, E., Dayawansa, W.P.: Using small perturbations to control chaos. Physica D **58**, 165-192, (1992)
RGus05.	S. Risau-Gusman, A.C. Ribeiro-Teixeira, D.A. Stariolo: Topology, Phase Transitions, and the Spherical Model. Phys. Rev. Lett. **95**, 145702 (2005)
RH06.	Razafindralandy, D., Hamdouni, A.: Consequences of symmetries on the analysis and construction of turbulence models. Sym. Int. Geo. Met. Appl. **2**, 052, (2006)
RH89.	Rose, R.M., Hindmarsh. J.L.: The assembly of ionic currents in a thalamic neuron. I The three-dimensional model. Proc. R. Soc. Lond. B, **237**, 267–288, (1989)
RKH88.	Riedel, U., Kühn, R., van Hemmen, J.L.: Temporal sequences and chaos in neural nets. Phys. Rev. A **38**, 1105, (1988)
RM86.	Rumelhart, D. E., McClelland, J. L. and the PDP Research Group: Parallel Distributed Processing. vol. 1, MIT Press, (1986)
RMR04.	Bou-Rabee, N.M., Marsden, J.E., Romero, L.A.: Tippe Top Inversion as a Dissipation-Induced Instability. SIAM J. Appl. Dyn. Sys. **3**, 352, (2004)
RMS90.	Riani, M., Masulli, F., Simonotto, E.: Stochastic dynamics and input dimensionality in a two-layer neural network for modeling multistable perception. In: Proc. IJCNN. 1019–1022, (1990)

RN03.	Russell, Norvig: Artificial Intelligence. A modern Approach. Prentice Hall, (2003)
RO87.	Romeiras, F.J., Ott, E.: Strange nonchaotic attractors of the damped pendulum with quasiperiodic forcing. Phys. Rev. A **35**, 4404–4413, (1987)
ROH98.	Rosa, E., Ott, E., Hess, M.H.: Transition to Phase Synchronization of Chaos. Phys. Rev. Lett. **80**, 1642–1645, (1998)
RPK96.	Rosenblum, M., Pikovsky, A., Kurths, J.: Phase synchronization of chaotic oscillators. Phys. Rev. Lett. **76**, 1804, (1996)
RPK97.	Rosenblum, M., Pikovsky, A., Kurths, J.: From Phase to Lag Synchronization in Coupled Chaotic Oscillators. Phys. Rev. Lett. **78**, 4193–4196, (1997)
RR83.	De Raedt, H., De Raedt, B.: Applications of the generalized Trotter formula Phys. Rev. A **28**, 3575–3580, (1983)
RR97.	Reisenberger, M., Rovelli, C.: 'Sum over surfaces' form of loop quantum gravity. Phys. Rev. D **56**, 3490–3508, (1997)
RRR02.	Robinson, P.A., Rennie, C.J., Rowe, D.L.: Dynamics of large-scale brain activity in normal arousal states and epileptic seizures. Phys. Rev. E **65**, 041924, (2002)
RS00.	Rideout, D.P., Sorkin, R.D.: A classical sequential growth dynamics for causal sets. Phys. Rev. **D61**, 024002, 2000.
RS04.	Ribeiro-Teixeira, A.C., Stariolo, D.A.: Topological hypothesis on phase transitions: The simplest case. Phys. Rev. E **70**, 16113, (2004)
RS72.	Rourke, C.P., Sanderson, B.J.: Introduction to piecewise–linear topology, Springer-Verlag, Berlin, (1972)
RS75.	Reed, M., Simon, B.: Methods of modern mathematical physics, vol.2: Fourier analysis, self–adjointness, Academic Press, San Diego (1975)
RS87.	Rovelli, C., Smolin, L.: A new approach to quantum gravity based on loop variables. Int. Conf. Gravitation and Cosmology, Goa, Dec 14-19 India, (1987)
RS88.	Rovelli, C., Smolin, L.: Knot theory and quantum gravity. Phys. Rev. Lett. **61**, 1155, (1988)
RS90.	Rovelli, C., Smolin, L.: Loop representation of quantum general relativity, Nucl. Phys., **B331**:(1), 80–152, (1990)
RS94.	Rovelli, C., Smolin, L.: The physical Hamiltonian in non–perturbative quantum gravity. Phys. Rev. Lett. **72**, 446, (1994)
RS95.	Rovelli, C., Smolin, L.: Spin Networks and Quantum Gravity. Phys. Rev. **D52**, 5743–5759, (1995)
RS99a.	Randall, L., Sundrum, R.: A large mass hierarchy from a small extra dimension. Phys. Rev. Lett. **83**, 3370, (1999)
RS99b.	Randall, L., Sundrum, R.: An alternative to compactification. Phys. Rev. Lett. **83**, 4690, (1999)
RSS00.	Rajewsky, N., Sasamoto, T., Speer, E.R.: Spatial particle condensation for an exclusion process an a ring. Physica A **279**, 123, (2000)
RSW90.	Richter, P.H., Scholz, H.-J., Wittek, A.: A breathing chaos. Nonlinearity **3**, 45-67, (1990)

RT02.	Rosa-Clot, M., Taddei, S.: A Path Integral Approach to Derivative Security Pricing. Int. J. Theor. Appl. Finance, **5**(2), 123–146, (2002)
RT71.	Ruelle, D., Takens, F.: On the nature of turbulence. Comm. Math. Phys., **20**, 167–192, (1971)
RT90.	Reshetikhin, N., Turaev, V.: Ribbon graphs and their invariants derived from quantum groups. Comm. Math. Phys. **127**, 1–26, (1990)
RT91.	Reshetikhin, N.Yu, Turaev, V.G.: Invariants of four-manifolds via link polynomials and quantum groups. Invent. Math. **10**, 547, (1991)
RT99.	Reshetikhin, N., Takhtajan, L.: Deformation Quantization of Kahler Manifolds. arXiv:math.QA/9907171, (1999)
RU67.	Ricciardi, L.M., Umezawa, H.: Brain physics and many-body problems, Kibernetik, **4**, 44, (1967)
RVS02.	Riazuelo, A., Vernizzi, F., Steer, D., Durrer, R.: Gauge invariant cosmological perturbation theory for braneworlds. arXiv:hep-th/0205220, (2002)
RYD96.	Ryu, S., Yu, W., Stroud, D.: Dynamics of an underdamped Josephson-junction ladder. Phys. Rev. E **53**, 2190–2195, (1996)
Raj07.	Rajeev, S.G.: Dissipative Mechanics Using Complex-Valued Hamiltonians. arXiv:quant-ph/0701141, (2007)
Raj82.	Rajaraman, R.: Solitons, Instantons. North-Holland, Amsterdam, (1982)
Ram90.	Ramond, P.: Field Theory: a Modern Primer. Addison–Wesley, Reading, MA, (1990)
Ras79.	Rasetti, M.: Topological Concepts in Phase Transition Theory. In: Differential Geometric Methods in Mathematical Physics, H.D. Döbner, (ed.) Springer, New York, (1979)
Rat69.	Ratner, M.: Markov partitions for Anosov flows on 3–dimensional manifolds. Mat. Zam. **6**, 693-704, (1969)
Rat78.	Ratcliff, R.: A theory of memory retrieval. Psych. Rev., **85**, 59–108, (1978)
Ree01.	Rees, M.J.: The State of Modern Cosmology. In N.G. Turok, (ed.), Critical Dialogues in Cosmology, World Scientific, Singapore, (1997)
Reg61.	Regge, T.: General relativity without coordinates. Nuovo Cim. A **19**, 558–571, (1961)
Rei48.	Reidemeister, K.: Knotentheorie, Chelsea Pub., New York, (1948)
Rei63.	Reinhart, B.L.: Cobordism and the Euler number. Topology **2**, 173, (1963)
Reu98.	Reuter, M.: Nonperturbative evolution equation for quantum gravity, Phys. Rev. D **57**, 971–985, (1998)
Rha84.	de Rham, G.: Differentiable Manifolds. Springer, Berlin, (1984)
Ric01.	Richter, P.H.: Chaos in Cosmos. Rev. Mod. Ast. **14**, 53–92, (2001)

Ric77. Ricciardi, L.M.: Diffusion Processes and Related Topics on Biology. Springer-Verlag, Berlin, (1977)

Ric93. Ricca, R.L.: Torus knots and polynomial invariants for a class of soliton equations, Chaos, **3**(1), 83–91, (1993)

Rie84. Riegert, R.J.: Non-Local Action for the Trace Anomaly. Phys. Lett. **134B**, 56, (1984)

Ris84. Risken, H.: The Fokker–Planck Equation. Springer, Berlin, (1984)

Rob37. Robertson, H.P.: Dynamical effects of radiation in the solar system, Mon. Not. Roy. Ast. Soc. **97**, 423-437, (1937)

Rob79. Robbins, K.: Periodic solutions and bifurcation structure at high r in the Lorenz system. SIAM J. Appl. Math. **36**, 457–472, (1979)

Rob83. Robinson, H.: Aristotelian dualism. Oxford Studies in Ancient Philosophy 1, 123-44, (1983)

Ros30. Rosenfeld, L.: Über die Gravitationswirkungen des Lichtes, Z. Phys. **65**, 589–599, (1930)

Ros63. Rosenfeld, L.: On Quantization of Fields. Nucl. Phys. **40**, 353–356, (1963)

Ros66. Rosenfeld, L.: Quantentheorie und Gravitation in Entstehung, Entwicklung, und Perspektiven der Einsteinschen Gravitationstheorie, Akademie-Verlag, Berlin, 185–197, (1966) [English translation in Selected Papers of Léon Rosenfeld, ed. by R.S. Cohen and J. Stachel, D. Reidel, Dordrecht, (1979]

Ros76. Rössler, O.E.: An Equation for Continuous Chaos. Phys. Lett. A, **57**(5), 397–398, (1976)

Ros98. Rosenberg, S.: Testing Causality Violation on space-times with Closed Timelike Curves. Phys. Rev. D **57**, 3365, (1998)

Rot01. Roth, G.: The brain and its reality. Cognitive Neurobiology and its philosophical consequences. Frankfurt a.M.: Aufl. Suhrkamp, (2001)

Rot88. Rotman, J.J.: An introduction to algebraic topology. Springer-Verlag, New York, (1988)

Rov93. Rovelli, C.: The basis of the Ponzano-Regge-Turaev-Viro-Ooguri model is the loop representation basis. Phys. Rev. D **48**, 2702–2707, (1993)

Rov96a. Rovelli, C.: Black Hole Entropy from Loop Quantum Gravity. Phys. Rev. Lett. **14**, 3288–3291, (1996)

Rov96b. Rovelli, C.: Loop Quantum Gravity and Black Hole Physics, Helv. Phys. Acta. **69**, 582–611, (1996)

Rov97. Rovelli, C. Half way through the woods, in Earman, J, and Norton, J, (eds.) The Cosmos of Science, 180–223, Univ. Pittsburgh Press, Konstanz, (1997)

Rov97. Rovelli, C.: Strings, loops and others: a critical survey of the present approaches to quantum gravity. Plenary lecture on quantum gravity at the GR15 conference, Puna, India, (1997)

Rov98. Rovelli, C.: Loop quantum gravity. Living Rev. Rel. **1**, (1998)

Rub01. Rubakov, V.A.: Large and infinite extra dimensions: An introduction. Phys. Usp. **44**, 871, (2001)

Rue76.	Ruelle, D.: A measure associated with Axiom A attractors. Am. J. Math. **98**, 619-654, (1976)
Rue78.	Ruelle, D.: An inequality for the entropy of differentiable maps. Bol. Soc. Bras. Mat. **9**, 83-87, (1978)
Rue78.	Ruelle, D.: Thermodynamic formalism. Addison-Wesley, Reading, MA, (1978)
Rue78.	Ruelle, D.: Thermodynamic formalism. Encyclopaedia of Mathematics and its Applications. Addison–Wesley, Reading, MA, (1978)
Rue79.	Ruelle, D.: Ergodic theory of differentiable dynamical systems. Publ. Math. IHES **50**, 27–58, (1979)
Rue89.	Ruelle, D.: The thermodynamical formalism for expanding maps. Commun. Math. Phys. **125**, 239–262, (1989)
Rue98.	Ruelle, D.: Nonequilibrium statistical mechanics near equilibrium: computing higher order terms. Nonlinearity **11**, 5-18, (1998)
Rue99.	Ruelle, D.: Smooth Dynamics and New Theoretical Ideas in Non–equilibrium Statistical Mechanics. J. Stat. Phys. **95**, 393–468, (1999)
Rul01.	Rulkov, N.F.: Regularization of Synchronized Chaotic Bursts. Phys. Rev. Lett. **86**, 183–186, (2001)
Run59.	Rund, H.: The Differential Geometry of Finsler Spaces, Springer, Berlin, (1959)
Rus18.	Russell, B.: Mysticism, Logic and Other Essays, London: Longmans, Green, (1918)
Rus844.	Russell, J.S.: Report on Waves, 14th meeting of the British Association for the Advancement of Science, BAAS, London, (1844)
Rus885.	Russell, J.S.: The Wave of Translation in the Oceans of Water, Air and Ether. Trübner, London, (1885)
Ryb71.	Rybicki, G.B.: Exact statistical mechanics of a one-dimensional self-gravitating system. Astrophys. Space Sci. **14**, 56–72, (1971)
Ryd96.	Ryder, L.: Quantum Field Theory. Cambridge Univ. Press, (1996)
Ryl49.	Ryle, G.: The Concept of Mind. Chicago: Chicago Univ. Press, (1949)
SA98.	Schaal, S., Atkeson, C.G.: Constructive incremental learning from only local information. Neural Comput., **10**, 2047–2084, (1998)
SAB87.	Schultz, C.L., Ablowitz, M.J., BarYaacov, D.: Davey-Stewartson I System: A Quantum (2+1)-Dimensional Integrable System. Phys. Rev. Lett. **59**, 2825–2828, (1987)
SB89.	Shapiro, M., Brumer, P.: Coherent Radiative Control of Unimolecular Reactions: Selective Bond Breaking with Picosecond Pulses. J. Chem. Phys. **90**, 6179, (1989)
SB98.	Sutton, R.S., Barto, A.G.: Reinforcement Learning: An Introduction. MIT Press, Cambridge, MA, (1998)
SBK06.	Sebanz, N., Bekkering,H., Knoblich, G.: Joint action: bodies and minds moving together. Tr. Cog. Sci. **10**(2), 70–76, 2006.

SC00. Skantzos, N.S., Coolen, A.C.C.: $(1+\infty)$-Dimensional Attractor Neural Networks. J. Phys. A **33**, 5785-5807, (2000)

SC01. Segel, L.A., Cohen, I.R.(eds.): Design principles for the immune system and other distributed autonomous systems, Oxford Univ. Press, (2001)

SC01. Skantzos, N.S., Coolen, A.C.C.: Attractor Modulation and Proliferation in $(1+\infty)$−Dimensional Neural Networks. J. Phys. A **34**, 929–942, (2001)

SDG92. Shinbrot, T., Ditto, W., Grebogi, C., Ott, E., Spano, M., Yorke, J.A.: Using the sensitive dependence of chaos (the "butterfly effect") to direct trajectories in an experimental chaotic system. Phys. Rev. Lett. **68**, 2863–2866, (1992)

SDK53. Seeger, A., Donth, H., Kochendörfer, A.: Theorie der Versetzungen in eindimensionalen Atomreihen. Zeitschrift für Physik, **134**, 173–93 (1953)

SF03. Sakaguchi, H., Fujimoto, T.: Elimination of spiral chaos by periodic force for the Aliev–Panfilov model. Phys. Rev. E **67**, 067202, (2003)

SFC02. Succi, S., Filippova, O., Chen, H., Orszag, S.: Towards a Renormalized Lattice Boltzmann Equation for Fluid Turbulence. J. Stat. Phys. **107**, 261–278, (2002)

SFM98. Skiffington, S., Fernandez, E., McFarland, K.: Towards a validation of multiple features in the assessment of emotions. Eur. J. Psych. Asses. **14**(3), (1998)

SG05. Saveliev V., Gorokhovski M., Group-theoretical model of developed turbulence and renormalization of the Navier–Stokes equation. Phys. Rev. E **72**, 016302, (2005)

SG88. Solomon, T.H., Gollub, J.P.: Passive transport in steady Rayleigh-Benard convection. Phys. Fluids **31**, 1372, (1988)

SGL93. Shih, C.L., Gruver, W. Lee, T.: Inverse kinematics and inverse dynamics for control of a biped walking machine. J. Robot. Syst., **10**, 531–555, (1993)

SH02. Stokes, H.T., Hasch, D.M.: ISOTROPY software, stokes.byu.edu/isotropy.html, (2002)

SHT02. Shimizu-Sato, S., Huq, E., Tepperman. J.M., Quail, P.H.: A light-switchable gene promoter system. Nature, Biotech., **20**, 1041–1044, (2002)

SHW94. Stephens, C.R., 't Hooft, G., Whiting, B.F.: Black Hole Evaporation without Information Loss. Class. Quant. Grav. **11**, 621-648, (1994)

SI89. Sastry, S.S., Isidori, A.: Adaptive control of linearizable systems. IEEE Trans. Aut. Con., **34**(11), 1123–1131, (1989)

SIT95. Shea, H.R., Itzler, M.A., Tinkham, M.: Inductance effects and dimensionality crossover in hybrid superconducting arrays Phys. Rev. B **51**, 12690–12697, (1995)

SJD94. Schiff, S.J., Jerger, K., Duong, D.H., Chang, T., Spano, M.L., Ditto, W.L.: Controlling chaos in the brain. Nature, **370**, 615–620, (1994)

SK05.	Sakaguchi, H., Kido, Y.: Elimination of spiral chaos by pulse entrainment in the Aliev-Panfilov model. Phys. Rev. E **70**, 052901, (2005)
SK75.	Sherrington, D. and Kirkpatrick S.: Solvable Model of a Spin-Glass. Phys. Rev. Lett. **35**, 1792–1796, (1975)
SK86.	Sompolinsky, H., Kanter, I.: Temporal Association in Asymmetric Neural Networks. Phys. Rev. Lett. **57**, 2861, (1986)
SK86.	Shinomoto, S., Kuramoto, Y.: Phase transitions in active rotator systems. Prog. Theor. Phys. **75**, 1105, (1986)
SK93.	Shih, C.L., Klein, C.A.: An adaptive gait for legged walking machines over rough terrain. IEEE Trans. Syst. Man, Cyber. A, **23**, 1150–1154, (1993)
SK98a.	Shabanov, S.V., Klauder. J.R.: Path Integral Quantization and Riemannian–Symplectic Manifolds. Phys. Lett. B4 **35**, 343–349, (1998)
SK98b.	Shabanov, S.V., Klauder. J.R.: Path Integral Quantization and Riemannian–Symplectic Manifolds. Phys. Lett. B **435**, 343-349, (1998)
SKC00.	Succi, S., Karlin, I.V., Chen, H., Orszag, S.: Resummation Techniques in the Kinetic–Theoretical Approach to Subgrid Turbulence Modeling. Physica A **280**, 92–98, (2000)
SKC85.	Sprik, M., Klein, M.L., Chandler, D.: Staging: A sampling technique for the Monte Carlo evaluation of path integrals. Phys. Rev. B **31**, 4234–4244, (1985)
SKJ92.	Sneppen, K., Krug, J., Jensen, M.H., Jayaprakash, C., Bohr, T.: Dynamic scaling and crossover analysis for the Kuramoto–Sivashinsky equation. Phys. Rev. A **46**, R7351–R7354, (1992)
SKM93.	Samko, S.G., Kilbas, A.A., Marichev, O.I.: Fractional Integrals and Derivatives Theory and Applications. Gordon & Breach, New York, (1993)
SKW95.	Sakai, K., Katayama, T., Wada, S., Oiwa, K.: Chaos causes perspective reversals for ambiguous patterns. In: Advances in Intelligent Computing IPMU'94, 463–472, (1995)
SL00.	Sprott, J.C., Linz, S.J.: Algebraically Simple Chaotic Flows. Int. J. Chaos Theory and Appl., **5**(2), 3–22, (2000)
SM71.	Siegel, C.L., Moser, J.K.: Lectures on Celestial Mechanics. Springer–Verlag, New York, (1971)
SM80.	Sivashinsky, G.I., Michelson, D.M.: Flow of a Liquid Film Down a Vertical Plane. Prog. Theor. Phys. **63**, 2112, (1980)
SMS00.	Shiromizu, T., Maeda, K.I., Sasaki, M.: The Einstein equations on the 3–brane world. Phys. Rev. D **62**, 024012, (2000)
SN79.	Shimada, I., Nagashima, T.: A Numerical Approach to Ergodic Problem of Dissipative Dynamical Systems. Prog. The. Phys. **61**(6), 1605–1616, (1979)
SNM05.	Seikh, M.M., Narayana, C., Metcalf, P.A. *et al.*: Brillouin scattering studies in Fe_3O_4 across the Verwey transition. Phys. Rev. B **71**, 174106, (2005)
SOC02.	Stamp, A.T., Osipov, G.V., Collins, J.J.: Suppressing arrhythmias in cardiac models using overdrive pacing and calcium channel blockers. Chaos **12**, 931–940, (2002)

SOG90. Shinbrot, T., Ott, E., Grebogi, C., Yorke, J.A.: Using chaos to direct trajectories to targets. Phys. Rev. Lett. **65**, 3215–3218, (1990)

SPP01. Sinha, S., Pande, A., Pandit, R.: Defibrillation via the Elimination of Spiral Turbulence in a Model for Ventricular Fibrillation. Phys. Rev. Lett. **86**, 3678–3681, (2001)

SR90. Shi, S., Rabitz, H.: Quantum mechanical optimal control of physical observables in microsystems. J. Chem. Phys. **92**(1), 364-376, (1990)

SR91. Schwieters, C.D., Rabitz, H.: Optimal control of nonlinear classical systems with application to unimolecular dissociation reactions and chaotic potentials. Phys. Rev. A **44**, 5224–5238, (1991)

SR91. Shi, S., Rabitz, H.: Optimal Control of Bond Selectivity in Unimolecular Reactions. Comp. Phys. Com. **63**, 71, (1991)

SRK98. Schäfer, C., Rosenblum, M.G., Kurths, J., Abel, H.-H: Heartbeat Synchronized with Ventilation. Nature, **392**, 239–240 (1998)

SRM98. Sides, S.W., Rikvold, P.A., Novotny, M.A.: Kinetic Ising Model in an Oscillating Field: Finite-Size Scaling at the Dynamic Phase Transition. Phys. Rev. Lett. **81**, 834–837, (1998)

SS00. Shraiman, B.I., Siggia, E.D.: Scalar Turbulence. Nature **405**, 639-46, (2000)

SS01. Shapiro, I.L., Sola. J.: Massive Fields Temper Anomaly–Induced Inflation: The Clue to Graceful Exit? arXiv:hep-th/0104182, (2001)

SS07. Sinha, S., Sridhar, S.: Controlling spatiotemporal chaos and spiral turbulence in excitable media: A review. In Handbook of Chaos Control, 2nd ed. (Eds.) E Scholl and H G Schuster, Wiley-VCH, Berlin, (2007)

SSK87. Sakaguchi, H., Shinomoto, S., Kuramoto, Y.: Local and global self-entrainments in oscillator-lattices. Prog. Theor. Phys. **77**, 1005–1010, (1987)

SSP07. Shajahan, T.K., Sinha, S., Pandit, R.: Spiral–wave dynamics depend sensitively on inhomogeneities in mathematical models of ventricular tissue. Phys. Rev. E **75**, 011929, (2007)

SSR93. Shen, L., Shi, S., Rabitz, H., Lin, C., Littman, M., Heritage, J.P., Weiner, A.M.: Optimal control of the electric susceptibility of a molecular gas by designed nonresonant laser pulses of limited amplitude. J. Chem. Phys. **98**, 7792–7803, (1993)

STP95. Schaub, H., Tsiotras, P., Junkins. J.: Principal Rotation Representations of Proper N x N Orthogonal Matrices. Int. J. Eng. Sci., **33**,(15), 2277–2295, (1995)

STU78. Stuart, C.I.J., Takahashi, Y., Umezawa, H.: On the stability and non-local properties of memory. J. Theor. Biol. **71**, 605–618, (1978)

STU79. Stuart, C.I.J., Takahashi, Y., Umezawa, H.: Mixed system brain dynamics: neural memory as a macroscopic ordered state, Found. Phys. **9**, 301, (1979)

STU93.	Susskind, L., Thorlacius, L., Uglum. J.: The Stretched Horizon and Black Hole Complementarity. Phys. Rev. D **48**, 3743-3761, (1993)
STZ93.	Satarić, M.V., Tuszynski. J.A., Zakula, R.B.: Kinklike excitations as an energy-transfer mechanism in microtubules. Phys. Rev. E, **48**, 589–597, (1993)
SV96a.	Strominger, A., Vafa, C.: On the microscopic origin of the Bekenstein–Hawking entropy. Phys. Lett. B3 **79**, 99, (1996)
SV96b.	Strominger, A., Vafa, C.: Microscopic Origin of the Bekenstein-Hawking Entropy. Phys. Lett. **B379**, 99–104, (1996)
SVW95.	Srivastava, Y.N., Vitiello, G., Widom, A.: Quantum dissipation and quantum noise, Annals Phys., **238**, 200, (1995)
SW72.	Sulanke R., Wintgen P.: Differential geometry und faserbundel; bound 75, Veb. Deutscher Verlag der Wissenschaften, Berlin, (1972)
SW87.	Shapere, A. Wilczek, F.: Two Applications of Axion Electrodynamics. Phys. Rev. Lett. **58**, 2051, (1987)
SW89a.	Shapere, A. Wilczek, F.: Geometry of self-propulsion at low Reynold's number. J. Fluid. Mech. **198**, 557, (1989)
SW89b.	Shapere, A. Wilczek, F.: Geometric Phases in Physics. World Scientific, Singapore, (1989)
SW90.	Sassetti, M., Weiss, U.: Universality in the dissipative two-state system. Phys. Rev. Lett. **65**, 2262–2265, (1990)
SW93.	Schleich, K., Witt, D.: Generalized Sums over Histories for Quantum Gravity: I. Smooth Conifolds, Nucl. Phys. **402**, 411, (1993); II. Simplicial Conifolds, ibid. **402**, 469, (1993)
SW94a.	Seiberg, N., Witten, E.: Monopoles, duality and chiral symmetry breaking in $N = 2$ supersymmetric QCD. Nucl. Phys. **B426**, 19-52, (1994)
SW94b.	Seiberg, N., Witten, E.: Electric-magnetic duality, monopole condensation, and confinement in $N = 2$ supersymmetric Yang–Mills theory. Nucl. Phys. **431**, 484-550, (1994)
SW99.	Seiberg, N., Witten, E.: String Theory and Noncommutative Geometry. JHEP **9909**, 032, (1999)
SWV99.	Srivastava, Y.N., Widom, A., Vitiello, G.: Quantum measurements, information and entropy production. Int. J. Mod. Phys. B13, 3369–3382, (1999)
SY94.	Schoen, R., S.T. Yau, Lectures on differential geometry, in Conference Proceedings and Lecture Notes in Geometry and Topology, 1, International Press, Cambridge, MA, (1994)
SZ97.	Scully, M.O., Zubairy, M.S.: Quantum Optics. Cambridge Univ. Press, (1997)
SZT98.	Satarić, M.V., Zeković, S., Tuszynski. J.A., Pokorny. J.: Mössbauer effect as a possible tool in detecting nonlinear excitations in microtubules. Phys. Rev. E **58**, 6333–6339, (1998)
Sag04.	Sagaut, P.: Large eddy simulation for incompressible flows. An introduction. Scientific Computation, Springer, (2004)
Sah04.	Sahni, V.: Dark matter and dark energy. Lecture Notes in Physics **653**, 141–180, (2004)

Sak02. Sakaguchi, H.: Stochastic synchronization in globally coupled phase oscillators. Phys. Rev. E **66**, 056129, (2002)

Sak04. Sakaguchi, H.: Oscillatory phase transition and pulse propagation in noisy integrate–and–fire neurons. Phys. Rev. E **70**, 022901, (2004)

Sam01. Samal, M.K.: Speculations on a Unified Theory of Matter and Mind. Proc. Int. Conf. Science, Metaphysics: A Discussion on Consciousness, Genetics. NIAS, Bangalore, India, (2001)

Sam89. Davies, P.: The New Physics. Cambridge Univ. Press, Cambridge (1989)

Sam99. Samal, M.K.: Can Science 'explain' Consciousness? Proc. Nat. Conf. Scientific, Philosophical Studies on Consciousness, ed. Sreekantan, B.V. *et al.*, NIAS, Bangalore, India, (1999)

Sar03. Sardanashvily, G.: Geometric quantization of relativistic Hamiltonian mechanics. Int. J. Theor. Phys. **42**, 697–704, (2003)

SZ92. Sardanashvily, G., Zakharov, O.: Gauge Gravitation Theory. World Scientific, Singapore, (1992)

Sar92. Sardanashvily, G.: On the geometry of spontaneous symmetry breaking. J. Math. Phys. **33**, 1546, (1992)

Sar93. Sardanashvily, G.: Gauge Theory in Jet Manifolds. Hadronic Press, Palm Harbor, FL, (1993)

Sar94. Sardanashvily, G.: Constraint field systems in multimomentum canonical variables. J. Math. Phys. **35**, 6584, (1994)

Sar98. Sardanashvily, G.: Hamiltonian time-dependent mechanics. J. Math. Phys. **39**, 2714, (1998)

Sau01. Sauer, T.: The Feynman Path Goes Monte Carlo. arXiv:physics/0107010, (2001)

Sau89. Saunders, D.J.: The Geometry of Jet Bundles. Lond. Math. Soc. Lect. Notes Ser. **142**, Cambr. Univ. Pr., (1989)

Sch01. Schlichenmaier, M.: Berezin–Toeplitz Quantization and Berezin's Symbols for Arbitrary Compact Kähler Manifolds, in Coherent States, Quantization and Gravity, M. Schlichenmaier *et al.* (eds.) Polish Scientific Publishers PWN, Warsaw, (2001)

Sch01. Schleich, W.: Quantum Optics in Phase Space. Wiley, New York, (2001)

Sch01. G.M. Schütz Phase Transitions and Critical Phenomena, vol 19, C. Domb, J. Lebowitz eds. Academic, London, (2001)

Sch02. Schmaltz, T.: Nicolas Malebranche, The Stanford Encyclopedia of Philosophy, Stanford, (2002)

Sch60. Schelling, T.C.: The Strategy of Conflict. Harvard Univ. Press, Cambridge, MA, (1960)

Sch74. Schmidt, D.S.: Periodic solutions near a resonant equilibrium of a Hamiltonian system, Celest. Mech. **9**, 81–103, (1974)

Sch78). Schwarz, A.S.: New topological invariants in the theory of quantised fields. Lett. Math. Phys. **2**, 247, (1978)

Sch78. Schot, S.H.: Jerk: The Time Rate of Change of Acceleration. Am. J. Phys. **46**(11), 1090–1094, (1978)

Sch80.	Schuerman, D.W.: The restricted three body problem including radiation pressure. Astrophys. J. **238**(1), 337-342, (1980)
Sch81.	Schulman, L.S.: Techniques and Applications of Path Integration, John Wiley & Sons, New York, (1981)
Sch88.	Schuster, H.G.(ed): Handbook of Chaos Control. Wiley-VCH, (1999)
Sch90.	Schmidt, D.S.: Transformation to versal normal form. In Computer Aided Proofs in Analysis, 235-240, (ed. K.R. Meyer, D.S. Schmidt), IMA Series **28**, Springer–Verlag, (1990)
Sch91.	Schroeder, M.: Fractals, Chaos, Power Laws. Freeman, New York, (1991)
Sch93.	Schwarz, M.: Morse Homology, Birkhäuser, Basel, (1993)
Sch94.	Schmidt, D.S.: Versal normal form of the Hamiltonian function of the restricted problem of three–bodies near L_4, J. Com. Appl. Math. **52**, 155–176, (1994)
Sch95.	Schumacher, B.: Quantum coding. Phys. Rev. A **51**, 2738–2747, (1995)
Sch96.	Schafer, R.D.: An Introduction to Nonassociative Algebras. Dover, New York, (1996)
Sch96.	Schellekens, A.N.: Introduction to conformal field theory. Fortsch. Phys., **44**, 605, (1996)
Sch98.	Schaal, S.: Robot learning. In M. Arbib (ed) Handbook of Brain Theory and Neural Networks (2nd ed.), MIT Press, Cambridge, (1998)
Sch99.	Schaal, S.: Is imitation learning the route to humanoid robots?. Trends Cogn. Sci., **3**, 233–242, (1999)
Sco04.	Scott, A. (ed): Encyclopedia of Nonlinear Science. Routledge, New York, (2004)
Sco99.	Scott, A.C.: Nonlinear Science: Emergence and Dynamics of Coherent Structures, Oxford Univ. Press, Oxford, (1999)
Sei06.	Seiberg, N.: Emergent space-time. arXiv:hep-th/0601234, (2006)
Sei95.	Seiler, W.M.: Involution and constrained dynamics II: The Faddeev-Jackiw approach. J. Phys. A, **28**, 7315–7331, (1995)
Sem02.	Semmelmann, U.: Conformal Killing forms on Riemannian manifolds. arXiv math.DG/0206117, (2002)
Sha06.	Sharma, S.: An Exploratory Study of Chaos in Human-Machine System Dynamics. IEEE Trans. SMC B. 36(2), 319-326, (2006)
She05.	Shen, Z.: Riemann–Finsler geometry with applications to information geometry, unpublished material, (2005)
Shi65.	Shilnikov, L.P.: On the generation of a periodic motion from a trajectory which leaves and re-enters a saddle-saddle state of equilibrium. Soc. Math. Dokl. **6**, 163–166, (in Russian), (1965)
Shi87.	Shinomoto, S.: Memory maintenance in neural networks. J. Phys. A: Math. Gen. **20**, L1305-L1309, (1987)
Shi98.	Shishikura, M.: The Hausdorff dimension of the boundary of the Mandelbrot set and Julia sets. Ann. of Math. **147**, 225–267, (1998)

Shi99.	Shimomura, Y.: A Family of Dynamic Subgrid-Scale Models Consistent with Asymptotic Material Frame Indifference, J. Phys. Soc. Jap. **68**, 2483–2486, (1999)
Sho97.	Shor, P.W.: Polynomial-Time Algorithms for Prime Factorization and Discrete Logarithms on a Quantum Computer. SIAM J. Comput. **26**, 1484–1509, (1997)
Shu93.	Shuster, M.D.: A Survey of Attitude Representations. J. Astr. Sci., **41**(4), 316–321, (1993)
Sie50.	Siegel, C.L.: Über eine periodische Lösung im Dreikörperproblem, Math. Nachr. **4**, 28–64, (1950)
Sin68a.	Sinai, Ya.G.: Markov partitions and C-diffeomorphisms. Funkts. Analiz i Ego Pril. **2**(1), 64–89, (1968)
Sin68b.	Sinai, Ya.G.: Constuction of Markov partitions. Funkts. Analiz i Ego Pril. **2**(3), 70–80, (1968)
Sin72.	Sinai, Ya.G.: Gibbsian measures in ergodic theory. Uspehi Mat. Nauk **27**(4), 21–64, (1972)
Sin89.	Sinai, Ya.G.: Dynamical Systems II, Encyclopaedia of Mathematical Sciences, **2**, Springer-Verlag, Berlin, (1989)
Siu80.	Siu, Y.T., Yau, S.T.: Compact Kähler manifolds of positive bisectional curvature. Invent. Math. **59**, 189–204, (1980)
Siv77.	Sivashinsky, G.I.: Nonlinear analysis of hydrodynamical instability in laminar flames – I. Derivation of basic equations. Acta Astr. **4**, 1177, (1977)
Ski72.	Skinner, B.F.: Beyond Freedom, Dignity. New York: Bantam/Vintage Books, (1972)
Sm88.	Strogatz, S.H., Mirollo, R.E.: Phase-locking and critical phenomena in lattices of coupled nonlinear oscillators with random intrinsic frequencies. Physica D **31**, 143, (1988)
Sma60.	Smale, S.: The generalized Poincaré conjecture in higher dimensions, Bull. Amer. Math. Soc., **66**, 373–375, (1960)
Sma63.	Smagorinsky, J.: General Circulation Experiments with the Primitive Equations: I. The Basic Equations. Mon. Weath. Rev. **91**, 99–164, (1963)
Sma67.	Smale, S.: Differentiable dynamical systems, Bull. Amer. Math. Soc., **73**, 747–817, (1967)
Sma99.	van der Smagt, P.: (ed.) Self–Learning Robots. Workshop: Brainstyle Robotics, IEE, London, (1999)
Smi82.	Smith, J.: Evolution and the Theory of Games. Cambridge Univ. Press, Cambridge, (1982)
Smo00.	Smolin, L.: Three roads to quantum gravity, Weidenfeld and Nicolson, London, UK (2000)
Smo92.	Smolin, L.: Did the universe evolve? Class. Quant. Grav. **9**, 173–192, 1992.
Smo97a.	Smolin, L.: Loops and Strings. Living Reviews, September, (1997)
Smo97b.	Smolin, L.: The Life of the Cosmos. Oxford Univ. Press, Oxford, (1997)
Sni80.	Sniatycki. J.: Geometric Quantization and Quantum Mechanics, Springer-Verlag, Berlin, (1980)

Sni80.	Sniatycki. J.: Geometric Quantization and Quantum Mechanics. Springer-Verlag, Berlin, (1980)
Sny86.	Snyder, S.H.: Drugs and the Brain. Scientific American Library, W.H. Freeman, Co., New York, (1986)
Soc91.	Socolovsky, M.: Gauge transformations in fibre bundle theory. J. Math. Phys. **32**, 2522, (1991)
Sok78.	Sokol'skii, A.: On the stability of an autonomous Hamiltonian system. J. Appl. Math. Mech. **38**, 741–49, (1978)
Sor86a.	Sorkin, R.D.: Introduction to topological geons. In P.G. Bergmann and 1986 V. de Sabbata, editors, Topological Properties and Global Structure of Space-Time, pages 249–270, Erice, Italy, May 1985. Plenum Press, (1986)
Sor86b.	Sorkin, R.D.: Topology change and monopole creation. Phys. Rev. D **33**, 978 (1986)
Sor86c.	Sorkin, R.D.: Non–Time–Orientable Lorentzian Cobordism Allows for Pair Creation. Int. J. Theor. Phys. **25**, 877–881, (1986)
Sor89.	Sorkin, R.D.: Consequences of space-time topology. In A. Coley, F. Cooperstock, and B. Tupper, editors, Proceedings of the third Canadian Conference on General Relativity and Relativistic Astrophysics, Victoria, Canada, May 1989, 137–163. World Scientific, Singapore, 1990.
Sor89.	Sorkin, R.D.: Classical topology and quantum phases: Geons. In G. Marmo S. de Filippo, M. Marinaro, editor, Geometrical and Algebraic Aspects of Nonlinear Field Theories, 201–218, Amalfi, Italy, May 1988. Elsevier, Amsterdam, 1989.
Sor90.	Sorkin, R.D.: First steps with causal sets. In R. Cianci, R. de Ritis, M. Francaviglia, G. Marmo, C. Rubano, and P. Scudellaro, editors, Proceedings of the ninth Italian Conference on General Relativity and Gravitational Physics, Capri, Italy, September 1990, pages 68–90. World Scientific, Singapore, 1991.
Sor91.	Sorkin, R.D.: Space-time and causal sets. In J. C. D'Olivo, E. Nahmad-Achar, M. Rosenbaum, M. P. Ryan, L. F. Urrutia, and F. Zertuche, editors, Relativity and Gravitation: Classical and Quantum, Proceedings of the SILARG VII Conference, Cocoyoc, Mexico, December 1990, pages 150–173. World Scientific, Singapore, 1991.
Sor97.	Sorkin, R.D.: Forks in the road, on the way to quantum gravity. Int. J. Theor. Phys. **36**, 2759–2781, 1997.
Sor98.	Sorkin, R.D.: Indications of causal set cosmology. Int. J. Theor. Phys. **39**, 1731–1736, 2000.
Spa82.	Sparrow, C.: The Lorenz Equations: Bifurcations, Chaos, and Strange Attractors. Springer, New York, (1982)
Spr93a.	Sprott, J.C.: Automatic Generation of Strange Attractors. Comput. & Graphics, **17**(3), 325–332, (1993)
Spr93b.	Sprott, J.C.: Strange Attractors: Creating Patterns in Chaos. M&T Books, New York, (1993)
Spr94.	Sprott, J.C.: Some Simple Chaotic Flows. Phys. Rev. E, **50**(2), R647–R650, (1994)

Spr97.	Sprott, J.C.: Some Simple Chaotic Jerk Functions. Am. J. Phys., **65**(6), 537–543, (1997)
Sta00.	Stanislavsky, A.A.: Memory effects and macroscopic manifestation of randomness. Phys. Rev. E **61**, 4752, (2000)
Sta63.	Stasheff. J.D.: Homotopy associativity of H−spaces I & II. Trans. Amer. Math. Soc., **108**, 275–292, 293–312, (1963)
Sta80.	Starobinsky, A.A.: A New Type of Isotropic Cosmological Models without Singularity. Phys. Lett. **B91**, 99, (1980)
Sta83.	Starobinsky, A.A.: The Perturbation Spectrum Evolving from a Nonsingular, Initially de Sitter cosmology, and the Microwave Background Anisotropy. Sov. Astron. Lett. **9**, 302, (1983)
Sta83.	Stapp, H.P.: Exact solution of the infrared problem. Phys. Phys. Rev. D **28**, 1386–1418, (1983)
Sta93.	Stapp, H.P.: Mind, Matter and Quantum Mechanics. Spinger-Verlag, Heidelberg, (1993)
Sta95.	Stapp, H.P.: Chance, Choice and Consciousness: The Role of Mind in the Quantum Brain. arXiv:quant-ph/9511029, (1995)
Sta97.	Stark, J.: Invariant Graphs for Forced Systems. Physica D **109**, 163-179, (1997)
Ste36.	Steuerwald, R.: Über Enneper'sche Flächen und Bäcklund'sche Transformation. Abhandlungen der Bayerischen Akademie der Wissenschaften München, 1–105, (1936)
Ste69.	Sternberg, S.: Memory-scanning: Mental processes revealed by reaction-time experiments. Am. Sci., **57**(4), 421–457, (1969)
Ste72.	Steenrod, N.: The Topology of Fibre Bundles, Princeton Univ. Press, Princeton, (1972)
Ste77.	Stelle, K.S.: Renormalization of Higher Derivative Quantum Gravity. Phys. Rev. D **16**, 953, (1977)
Ste90.	Stewart, J.M.: Perturbations of Friedmann-Robertson-Walker cosmological models. Class. Quant. Grav. **7**, 1169, (1990)
Ste93.	Stengel, R.: Optimal control and estimation. Dover, New York, (1993)
Ste98.	Steane, A.: Quantum Computing. Rep. Prog. Phys. **61**, 117, (1998)
Sto05.	Stoljar, D.: Physicalism. The Stanford Encyclopedia of Philosophy, Stanford, (2005)
Sto68.	Stong, R.E.: Notes on Cobordism Theory. Princeton Univ. Press, Princeton, (1968)
Sto90.	Stonier, T.: Information and the internal structure of the Universe. Springer, New York, (1990)
Str00.	Strogatz, S.: From Kuramoto to Crawford: exploring the onset of synchronization in populations of coupled oscillators. Physica D, **143**, 1–20, (2000)
Str01.	Strogatz, S.H.: Exploring complex networks. Nature, **410**, 268, (2001)
Str68.	Strang, G.: On the Construction and Comparison of Difference Schemes. SIAM J. Num. Anal. **5**, 506–517, (1968)

Str90.	Strominger, A.: Special Geometry. Commun. Math. Phys. **133**, 163, (1990)
Str94.	Strogatz, S.: Nonlinear Dynamics and Chaos. Addison-Wesley, Reading, MA, (1994)
Stu99.	Stuart. J.: Calculus (4th ed.) Brooks/Cole Publ., Pacific Grove, CA, (1999)
Sug00.	Sugino, F.: Witten's open string field theory in constant B-field background. J. High Energy Phys. JHEP03, 017, (2000)
Sus83.	Sussmann, H.J.: Lie brackets and local controllability: a sufficient condition for scalar–input systems, SIAM J. Con. Opt., **21**(5), 686–713, (1983)
Sus87.	Sussmann, H.J.: A general theorem on local controllability, SIAM J. Con. Opt., **25**(1), 158–194, (1987)
Sus95.	Susskind, L.: The World as a Hologram. J. Math. Phys. **36**, 6377-6396, (1995)
Sut97.	Sutcliffe, P.: BPS monopoles, Int. J. Mod. Phys. A12, 4663–4706, (1997)
Suz90.	Suzuki, M.: Fractal decomposition of exponential operators with applications to many-body theories and Monte Carlo simulation. Phys. Lett. A **146**, 319–323, (1990)
Suz91.	Suzuki, M.: General theory of fractal path integrals with applications to many-body theories and statistical physics. J. Math. Phys. **32**, 400–407, (1991)
Swi75.	Switzer, R.K.: Algebraic Topology – Homology and Homotopy. (in Classics in Mathematics), Springer, New York, (1975)
Syn61.	Synge, J.L.: On a certain nonlinear differential equation. Proc. R. Irish. Acad. **62**, 1, (1961)
tHo93.	't Hooft, G.: Dimensional Reduction in Quantum Gravity. arXiv:gr-qc/9310026, (1993)
tHo06.	't Hooft, G.: Determinism Beneath Quantum Mechanics. In Quo Vadis Quantum Mechanics, ed. by A. Elitzur, S. Dolen, and N. Kolenda, Springer Verlag, Heidelburg, (2005)
TBG98.	Tittel, W., Brendel, J., Gisin, B. *et al.*: Experimental Demonstration of Quantum Correlations Over More Than 10 km. Phys. Rev. A **57**, 3229, (1998)
TEM89.	Tonomura, A., Endo. J., Matsuda, T., Kawasaki, T.: Demonstration of Single-electron Build-up of an Interference Pattern, Am. J. Phys. **57**, 117–120, (1989)
TF91.	Tsue, Y., Fujiwara, Y.: Time–Dependent Variational Approach to (1+1)-Dimensional Scalar-Field Solitons, Progr. Theor. Phys. **86**(2), 469–489, (1991)
TH98.	Turok, N., Hawking, S.W.: Open Inflation, the Four Form and the Cosmological Constant. Phys. Lett. **B432**, 271, (1998)
TLO97.	Tanaka, H.A., Lichtenberg, A.J., Oishi, S.: Self-synchronization of coupled oscillators with hysteretic response. Physica D **100**, 279, (1997)
TLO97.	Tanaka, H.A., Lichtenberg, A.J., Oishi, S.: First Order Phase Transition Resulting from Finite Inertia in Coupled Oscillator Systems. Phys. Rev. Lett. **78**, 2104–2107, (1997)

TLZ02.	Tang, X., Lou, S., Zhang, Y.: Chaos in soliton systems and special Lax pairs for chaos systems. Localized excitations in (2+ 1)–dimensional systems. Phys. Rev. E **66**, 046601, (2002)
TM01.	Trees, B.R., Murgescu, R.A.: Phase locking in Josephson ladders and the discrete sine-Gordon equation: The effects of boundary conditions, current-induced magnetic fields. Phys. Rev. E **64**, 046205, (2001)
TMO00.	Trías, E., Mazo, J.J., Orlando, T.P.: Discrete Breathers in Nonlinear Lattices: Experimental Detection in a Josephson Array. Phys. Rev. Lett. **84**, 741, (2000)
TMS93.	Tsodyks, M., Mitkov, I., Sompolinsky, H.: Pattern of synchrony in inhomogeneous networks of oscillators with pulse interactions. Phys. Rev. Lett. **71**, 1280–1283, (1993)
TO90.	Tomé, T., de Oliveira, M.J.: Dynamic phase transition in the kinetic Ising model under a time-dependent oscillating field. Phys. Rev. A **41**, 4251–4254, (1990)
TP01.	Tabuada, P., Pappas, G.J.: Abstractions of Hamiltonian Control Systems. Proceedings of the 40th IEEE Conf. Decis. Con., Orlando, FL, (2001)
TPS98.	Tomlin, C., Pappas, G.J., Sastry, S.: Conflict resolution for air traffic management: A case study in multi-agent hybrid systems, IEEE Trans. Aut. Con., **43**, 509-521, (1998)
TR85.	Tannor, D.J., Rice, S.A.: Control of selectivity of chemical reaction via control of wave packet evolution. J. Chem. Phys. **83**, 5013–5018, (1985)
TRW98.	Tass, P., Rosenblum, M.G., Weule, J., Kurths, J. et al.: Detection of $n:m$ Phase Locking from Noisy Data: Application to Magnetoencephalography. Phys. Rev. Lett. **81**, 3291–3294, (1998)
TS01.	Thompson. J.M.T., Stewart, H.B.: Nonlinear Dynamics and Chaos: Geometrical Methods for Engineers and Scientists. Wiley, New York, (2001)
TS99.	Trees, B.R., Stroud, D.: Two-dimensional arrays of Josephson junctions in a magnetic field: A stability analysis of synchronized states. Phys. Rev. B **59**, 7108–7115, (1999)
TSS05.	Trees, B.R., Saranathan, V., Stroud, D.: Synchronization in disordered Josephson junction arrays: Small-world connections and the Kuramoto model. Phys. Rev. E **71**, 016215, (2005)
TT04.	Teramae, J.N., Tanaka, D.: Robustness of the noise-induced phase synchronization in a general class of limit cycle oscillators. Phys.Rev.Lett. **93**, 204103, (2004)
TV92.	Turaev, V., Viro, O.: State Sum Invariants of 3- Manifolds and Quantum 6-J Symbols. Topology **31**, 865-902, (1992)
TVP99.	Tabony. J., Vuillard, L., Papaseit, C.: Biological self-organisation and pattern formation by way of microtubule reaction-diffusion processes. Adv. Complex Syst. **2**(3), 221–276, (1999)
TW82.	Turner, M.S., Wilczek, F.: Is our vacuum metastable? Nature, **298**, 633–634, (1982)

TW95.	Tucker, R., Wang, C.: Black holes with Weyl charge and non-Riemannian waves. Class. Quant. Grav. **12**, 2587, (1995)
TWG02.	Timme, M., Wolf, F., Geisel, T.: Prevalence of unstable attractors in networks of pulse-coupled oscillators. Phys. Rev. Lett. **89**, 154105, (2002)
TWG03.	Timme, M., Wolf, F., Geisel, T.: Unstable attractors induce perpetual synchronization and desynchronization. Chaos, **13**, 377, (2003)
TZ00.	Tian, G., Zhu, X.: Uniqueness of Kähler-Ricci solitons, Acta Math. **184**, 271–305, (2000)
TZ05.	Tarasov, V.E., Zaslavsky, G.M.: Fractional Ginzburg-Landau equation for fractal media. Physica A **354**, 249–261, (2005)
Tap74.	Tappert, F.: Numerical Solutions of the Korteweg-de Vries Equations and its Generalizations by the Split-Step Fourier Method. In Nonlinear Wave Motion, 215–216, Lectures in Applied Math. **15**, Amer. Math. Soc., (1974)
Tar04a.	Tarasov, V.E.: Fractional Generalization of Liouville Equation. Chaos **14**, 123–127, (2004)
Tar04b.	Tarasov, V.E.: Path integral for quantum operations. J. Phys. A **37**, 3241–3257, (2004)
Tar05a.	Tarasov, V.E.: Fractional Systems and Fractional Bogoliubov Hierarchy Equations. Phys. Rev. E **71**, 011102, (2005)
Tar05b.	Tarasov, V.E.: Fractional Liouville and BBGKI equations. J. Phys.: Conf. Ser. **7**, 17–33, (2005)
Tar05c.	Tarasov, V.E.: Continuous Medium Model for Fractal Media. Phys. Lett. A **336**, 167–174, (2005)
Tar05d.	Tarasov, V.E.: Possible Experimental Test of Continuous Medium Model for Fractal Media. Phys. Let. A **341**, 467-472, (2005)
Tar05e.	Tarasov, V.E.: Fractional Hydrodynamic Equations for Fractal Media. Ann. Phys. **318**, 286–307, (2005)
Tar05f.	Tarasov, V.E.: Fractional Fokker-Planck Equation for Fractal Media. Chaos **15**, 023102, (2005)
Tar05g.	Tarasov, V.E.: Fractional Generalization of Gradient and Hamiltonian Systems. J. Phys. A **38**(26), 5929–5943, (2005)
Tau90.	Taubes, C.H.: Casson's invariant and gauge theory. J. Diff. Geom. **31**, 547–599, (1990)
Tau94.	Taubes, C.H.: The Seiberg–Witten invariants and symplectic forms, Math. Res. Lett. **1**, 809–822, (1994)
Tau95a.	Taubes, C.H.: More constraints on symplectic manifolds from Seiberg-Witten invariants, Math. Research Letters, **2**, 9-14, (1995)
Tau95b.	Taubes, C.H.: The Seiberg–Witten invariants and The Gromov invariants, Math. Research Letters **2**, 221–238, (1995)
Tay21.	Taylor, G.I.: Diffusion by continuous movements. Proc. London Math. Soc. **20**, 196, (1921)
Tay38.	Taylor, G.I.: Frozen–in Turbulence Hypothesis. Proc. R. Soc. London A **164**, 476, (1938)
Tay79.	Taylor, P.: Evolutionarily stable strategies with two types of players. Journal of Applied Probability **16**, 76–83, (1979)

References

Tei83. Teitelboim, C.: Causality versus gauge invariance in quantum gravity and supergravity. Phys. Rev. Lett. **50**, 705–708, (1983)

Tel90. Tél, T.: Transient Chaos. In Directions in Chaos, H. Bai-Lin (ed.), World Scientific, Singapore, (1990)

Teu73. Teukolsky, S.: Perturbations of a rotating black hole. Astrophys. J. **185**, 635–647, (1973)

Thi03. Thiemann, T.: Lectures on loop quantum gravity, Lect. Notes Phys. **631**, 41–135, (2003)

Thi79. Thirring, W.: A Course in Mathematical Physics (in four volumes) Springer, New York, (1979)

Thi96. Thiemann, T.: Anomaly-Free Formulation of Nonperturbative Four–Dimensional Lorentzian Quantum Gravity. Phys. Lett. **B380**, 257–264, (1996)

Tho75. Thom, R.: Structural Stability and Morphogenesis. Addison–Wesley, Reading, (1975)

Tho79. Thorpe. J.A.: Elementary Topics in Differential Geometry. Springer, New York, (1979)

Tho89. Thom, R.: Structural Stability and Morphogenesis: An Outline of a General Theory of Models. Addison-Wesley, Reading, MA, (1989)

Tho93. Thompson, G.: Non-uniqueness of metrics compatible with a symmetric connection. Class. Quant. Grav. **10**, 2035, (1993)

Tia97. Tian, G.: Kähler-Einstein metrics with positive scalar curvature. Invent. Math. **130**, 1–39, (1997)

Tia98. Tian, G.: Some aspects of Kähler Geometry. Lecture note taken by M. Akeveld, (1997)

Tin75. Tinkham, M.: Introduction to Superconductivity. McGraw-Hill, New York, (1975)

Tip70. Tipler, F.J.: Singularities and causality violation. Ann. Phys. **108**, 1 (1970)

Tip85. Tipler, F.J.: Topology Change In Kaluza–Klein And Superstring Theories, Phys. Lett. B **165**, 67, (1985)

Tod67. Toda, M.: Vibration of a chain with nonlinear interactions. J. Phys. Soc. Jap. **22**, 431–36; also, Wave propagation in anharmonic lattices. J. Phys. Soc. Jap. **23**, 501–06, (1967)

Tom77. Tomboulis, E.: 1/N Expansion and Renormalization in Quantum Gravity. Phys. Lett. **B70**, 361, (1977)

Tri88. Tritton, D.J.: Physical fluid dynamics. Oxford Science Publ., Oxford, (1988)

Tur00. Turok, N.: Before Inflation. Lecture at CAPP2000. arXiv:astro-ph/0011195, (2000)

UC02. Uezu, T., Coolen, A.C.C.: Hierarchical Self–Programming in Recurrent Neural Networks. J. Phys. A **35**, 2761-2809, (2002)

UJ98. Unruh, W.G., Jheeta, M.: Complex Paths and the Hartle-Hawking Wave Function for Slow Roll Cosmologies. arXiv:gr-qc/9812017, (1998)

UMM93. Ustinov, A.V., Cirillo, M., Malomed, B.A.: Fluxon dynamics in one-dimensional Josephson-junction arrays. Phys. Rev. B **47**, 8357–8360, (1993)

UN99.	Udriste, C., Neagu, M.: Extrema of p-energy functional on a Finsler manifold. Diff. Geom. Dyn. Sys. 1(1), 10–19, (1999)
Udr00.	Udriste, C.: Geometric dynamics. Kluwer Academic Publishers, (2000)
Ula91.	Ulam, S.M.: Adventures of a Mathematician. Univ. Calif. Press, (1991)
Ume93.	Umezawa, H.: Advanced field theory: micro, macro and thermal concepts. American Institute of Physics, New York, (1993)
Una94.	Ünal, G.: Application of equivalence transformations to inertial subrange of turbulence, Lie Groups Appl. **1**, 232–240, (1994)
Una97.	Ünal, G.: Constitutive equation of turbulence and the Lie symmetries of Navier–Stokes equations. In Modern Group Analysis VII, Editors N.H. Ibragimov, K. Razi Naqvi and E. Straume, Trondheim, Mars Publishers, 317–323, (1997)
Ush99.	Ushio, T.: Synthesis of Synchronized Chaotic Systems Based on Observers. Int. J. Bif. Chaos **9**, 541–546, (1999)
VAE94.	C. van Vreeswijk, L.F. Abbott, G.B. Ermentrout, J.Comp.Neuro. **1**, 313, (1994)
VW94.	Vafa, C., Witten, E.: A strong coupling test of S duality. arXiv:hep-th/9408074, (1994)
Vaa95.	Van der Vaart, N.C. *et al.*: Resonant Tunneling Through Two Discrete Energy States. Phys. Rev. Lett. **74**, 4702–4705, (1995)
Vaf97.	Vafa, C.: Lectures on Strings and Dualities. arXiv:hep-th/9702201, (1997)
Vai94.	Vaisman, I.: Lectures on the Geometry of Poisson Manifolds. Birkhäuser Verlag, Basel, (1994)
Van88.	Vance, J.M.: Rotor Dynamics of Turbomachinery. Wiley, New York, (1988)
Ved96.	V. Vedral, A. Barenco, A. Ekert: Quantum networks for elementary arithmetic operations. Phys. Rev. A **54**, 147–153, (1996)
Ven91.	Veneziano, G.: Scale factor duality for classical and quantum strings. Phys. Lett. **B265**, 287, (1991)
Ver39.	Verwey, E.J.W.: Nature, **144**, 327, (1939)
Ver838.	Verhulst, P. F.: Notice sur la loi que la population pursuit dans son accroissement. Corresp. Math. Phys. **10**, 113-121, (1838)
Ver845.	Verhulst, P.F.: Recherches Mathematiques sur La Loi D'Accroissement de la Population (Mathematical Researches into the Law of Population Growth Increase) Nouveaux Memoires de l'Academie Royale des Sciences et Belles-Lettres de Bruxelles, **18**(1), 1-45, (1845)
Ver88.	Verlinde, E.: Fusion rules and modular transformations in 2D conformal field theory. Nucl. Phys. **B300**, 360–376, (1988)
Ver98.	Vergassola, M.: In Analysis and Modelling of Discrete Dynamical Systems, eds. D. Benest & C. Froeschlé, 229, Gordon & Breach, (1998)

Via97.	Viana, M.: Multidimensional non–hyperbolic attractors. Publ. Math. IHES **85**, 63–96, (1997)
Vil03.	Vilenkin, A.: Quantum cosmology and eternal inflation, in The future of theoretical physics and cosmology, eds. G.W. Gibbons, E.P.S. Shellard and S.J. Rankin, Cambridge Univ. Press (2003)
Vil83.	Vilenkin, A.: Birth of Inflationary Universes. Phys. Rev. D **27**, 2848, (1983)
Vil85.	Vilenkin, A.: Classical and Quantum Cosmology of the Starobinsky Inflationary Model. Phys. Rev. D **32**, 2511, (1985)
Vil94.	Vilenkin, A.: Approaches To Quantum Cosmology. Phys. Rev. D **50**, 2581, (1994)
Vio83.	Dubois–Violette, M.: Structures Complexes au-dessus des Variétés, Applications, in Mathématique et Physique, Séminaire de l'Ecole Normale Supérieure 1979–1982, L. Boutet de Monvel *et al.* (eds.), Birkhäuser, Boston, (1983)
Vit01.	Vitiello, G.: My Double Unveiled. John Benjamins, Amsterdam, (2001)
Vit95.	Vitiello, G.: Dissipation and memory capacity in the quantum brain model. Int. J. Mod. Phys. B, **9**, 973–989, (1995)
Voi02.	Voisin, C.: Hodge Theory and Complex Algebraic Geometry I. Cambridge Univ. Press, Cambridge, (2002)
Von05.	Vonk, M.: A mini-course on topological strings. arXiv: hep-th/0504147.
Vre96.	Van Vreeswijk, C.: Partial synchronization in populations of pulse-coupled oscillators. Phys. Rev. E **54**, 5522, (1996)
Vyg82.	Vygotsky, L.S. Historical meaning of the Psychological crisis. Collected works. Vol. 1. Pedag. Publ., Moscow, (1982)
WA00.	Wall, M.M., Amemiya, Y.: Estimation for polynomial structural equation models. Journal of American Statistical Association, **95**, 929–940, (2000)
WA98.	Wall, M.M., Amemiya, Y.: Fitting nonlinear structural equation models. Proc. Social Stat. Section. Ann. Meet. Ame. Stat. Assoc. 180–185, (1998)
WBB94.	Wiesenfeld, K., Benz, S.P., Booi, P.A.: Phase-locked oscillator optimization for arrays of Josephson junctions. J. Appl. Phys. **76**, 3835–3846, (1994)
WBI92.	D.J. Wineland, J.J. Bollinger, W.M. Itano, F.L. Moore: Spin squeezing and reduced quantum noise in spectroscopy. Phys. Rev. A **46**, R6797–R6800, (1992)
WC02.	Wang, X.F., Chen, G.: Synchronization in small-world dynamical networks. Int. J. Bifur. Chaos **12**, 187, (2002)
WC72.	Wilson, H.R., Cowan, J.D.: Excitatory and Inhibitory Interactions in Localized Populations of Model Neurons. Biophys. J. **12**, 1–24, (1972)
WC72.	Wilson, H.R., Cowan, J.D.: A Mathematical Theory of the Functional Dynamics of Cortical and Thalamic Nervous Tissue. Kybernetik **13**, 55–80, (1973)

WCL96.	Whan, C.B., Cawthorne, A.B., Lobb, C.J.: Synchronization and phase locking in two-dimensional arrays of Josephson junctions. Phys. Rev. B **53**, 12340, (1996)
WCS96.	Wiesenfeld, K., Colet, P., Strogatz, S.H.: Synchronization Transitions in a Disordered Josephson Series Array. Phys. Rev. Lett. **76**, 404–407, (1996)
WCS98.	Wiesenfeld, K., Colet, P., Strogatz, S.H.: Frequency locking in Josephson arrays: Connection with the Kuramoto model. Phys. Rev. E **57**, 1563–1569, (1998)
WDH93.	Wilmott, P., Dewynne. J., Howinson, S.: Option Pricing: Mathematical Models and Computation. Oxford Financial Press, (1993)
WF49.	Wheeler. J.A., Feynman, R.P.: Classical Electrodynamics in Terms of Direct Interparticle Action. Rev. Mod. Phys. **21**, 425–433, (1949)
WFP97.	Witt, A., Feudel, U., Pikovsky, A.S.: Birth of Strange Nonchaotic Attractors due to Interior Crisis. Physica D **109**, 180-190, (1997)
WMW01.	Weibert, K., Main, J., Wunner, G.: Periodic orbit quantization of chaotic maps by harmonic inversion. Phys. Lett. A **292**, 120, (2001)
WS91.	Watkin, T.L.H., Sherrington, D.: The parallel dynamics of a dilute symmetric Hebb-rule network. J. Phys. A: Math. Gen. **24**, 5427-5433, (1991)
WS97.	Watanabe, S., Swift, J.W.: Stability of periodic solutions in series arrays of Josephson junctions with internal capacitance. J. Nonlinear Sci. **7**, 503, (1997)
WS98.	Watts, D.J., Strogatz, S.H.: Collective dynamics of 'small-world' networks. Nature, **393**, 440–442, (1998)
WSZ95.	Watanabe, S., Strogatz, S.H., van der Zant, H.S.J., Orlando, T.P.: Whirling Modes, Parametric Instabilities in the Discrete Sine-Gordon Equation: Experimental Tests in Josephson Rings. Phys. Rev. Lett. **74**, 379–382, (1995)
WT92.	Williams, R.M., Tuckey, P.: Regge Calculus: A bibliography and brief review. Class. Quant. Grav. **9**, 1409, (1992)
WW83a.	Wehner, M.F., Wolfer, W.G.: Numerical evaluation of path–integral solutions to Fokker–Planck equations. I., Phys. Rev. A **27**, 2663–2670, (1983)
WW83b.	Wehner, M.F., Wolfer, W.G.: Numerical evaluation of path–integral solutions to Fokker–Planck equations. II. Restricted stochastic processes, Phys. Rev. A, **28**, 3003–3011, (1983)
WWV01.	Winckelmans, G.S., Wray, A., Vasilyev, O.V., Jeanmart, H.: Explicit filtering large-eddy simulation using the tensor-diffusivity model supplemented by a dynamic Smagorinsky term. Phys. Fluids, **13**, 1385–1403, (2001)
WZ98.	Waelbroeck, H., Zertuche, F.: Discrete Chaos. J. Phys. A **32**, 175, (1998)
Wal00.	Van der Wal, C. *et al.*: Quantum Superposition of Macroscopic Persistent-current States, Science, **290**, 773–777, (2000)

Wal60.	Wallace, A.H.: Modifications and cobounding manifolds. Canad. J. Math. **12**, 503, (1960)
Wal68.	Wallace, A.H.: Differential Topology: first steps. Benjamin, New York, (1968)
Wal84.	Wald, R.: General Relativity. University of Chicago Press, (1984)
Wal89.	Walther, H.-O.: Hyperbolic periodic solutions, heteroclinic connections and transversal homoclinic points in autonomous differential delay equations. Memoirs AMS **402**, (1989)
Wal94.	Wald, R.: Quantum field theory in curved space–time and black hole thermodynamics, Chicago Univ. Press, (1994)
Wat99.	Watts, D.J.: Small Worlds. Princeton Univ. Press, Princeton, (1999)
Wei00.	Weidlich, W.: Sociodynamics; A Systematic Approach to Mathematical Modelling in Social Sciences, Harwood Academic Publishers (2000)
Wei05.	Weisstein, E.W.: MathWorld–A Wolfram Research Web Resource. http://mathworld.wolfram.com, (2005)
Wei05.	Weinberg, S.: The Cosmological Constant Problems. astro-ph/0005265, (2005)
Wei36.	Weiss, P.: Proc. Roy. Soc., A **156**, 192–220, (1936)
Wei64.	Weinberg, S.: Photons and Gravitons in S-Matrix Theory: Derivation of Charge Conservation and Equality of Gravitational and Inertial Mass. Phys. Rev. **B135**, 1049, (1964)
Wei79.	Weinberg, E.: Parameter counting for multimonopole solutions. Phys. Rev. D **20**, 936–944, (1979)
Wei80.	Weinberg, S.: Conceptual foundations of the unified theory of weak and electromagnetic interactions. Rev. Mod. Phys. **52**, 515–523, (1980)
Wei87.	Weinberg, S.: Anthropic Bound on the Cosmological Constant. Phys. Rev. Lett. **59**, 2607, (1987)
Wei90.	Weinstein, A.: Affine Poisson structures. Internat. J. Math., **1**, 343–360, (1990)
Wei92.	Weinberg, S.: Dreams of a Final Theory, Pantheon Books, New York, (1992)
Wei94.	Weibull, J.: Evolutionary Game Theory. MIT Press, Cambridge, (1994)
Wei99.	Weiss, U.: Quantum Dissipative Systems. World Scientific, Singapore, (1999)
Wer98.	Werner, R.F.: Optimal cloning of pure states. Phys. Rev. A **58**, 1827–1832, (1998)
Whe61.	Wheeler. J.A.: Geometrodynamics and the Problem of Motion. Rev. Mod. Phys. **33**, 63–78, (1961)
Whe62.	Wheeler. J.A.: Geometrodynamics. Academic Press, New York, (1962)
Whe86.	Wheeler. J.A.: How Come the Quantum? In New Techniques and Ideas in Quantum Measurement Theory, ed. by D. Greenberger, Ann. N.Y. Acad. Sci 480, 304–316, (1986)

Whi27.	Whittaker, E.T.: A Treatise on the Analytical Dynamics of Particles and Rigid Bodies, Cambridge Univ. Press, Cambridge, (1927)
Whi84.	Whitt, B.: Fourth-Order Gravity as General Relativity Plus Matter. Phys. Lett. **B145**, 176, (1984)
Whi87.	Whitney, D.E.: Historical perspective and state of the art in robot force control. Int. J. Robot. Res., **6**(1), 3–14, (1987)
Wie61.	Wiener, N.: Cybernetics. Wiley, New York, (1961)
Wig90.	Wiggins, S.: Introduction to Applied Dynamical Systems and Chaos. Springer, New York, (1990)
Wik07.	Wikipedia, the free encyclopedia. (2007) http://wikipedia.org.
Wil00.	Wilson, D.: Nonlinear Control, Advanced Control Course (Student Version), Karlstad Univ., (2000)
Wil93.	Willmore, T.J.: Riemannian Geometry. Oxford Univ. Press, Oxford, (1993)
Wil97.	Williams, R.M.: Recent progress in Regge calculus. Nucl. Phys. B **57**, 73–81, (1997)
Wil99.	Wilczek, F.: Getting its from bits, Nature **397**, 303–036, (1999)
Win01.	Winfree, A.T.: The Geometry of Biological Time. Springer, New York, (2001)
Win67.	Winfree, A.T.: Biological rhythms and the behavior of populations of coupled oscillators. J. Theor. Biol. **16**, 15, (1967)
Win80.	Winfree, A.T.: The Geometry of Biological Time. Springer, New York, (1980)
Win87.	Winfree, A.T.: When Time Breaks Down. Princeton Univ. Press, Princeton, (1987)
Wit02.	Witten, E.: The Universe on a String. Astronomy magazine, June, (2002)
Wit82.	Witten, E.: Supersymmetry and Morse theory. J. Diff. Geom., **17**, 661–692, (1982)
Wit82.	Witten, E.: Instability of the kaluza-klein vacuum. Nucl. Phys. **B195**, 481, 1982.
Wit86a.	Witten, E.: Interacting Field Theory of Open Superstrings. Nucl. Phys. **B276**, 291, (1986)
Wit88a.	Witten, E.: 2+1 Gravity as an Exactly Soluble Model. Nucl. Phys. **B311**, 46–78, (1988)
Wit88b.	Witten, E: Topological quantum field theory, Commun. Math. Phys., **117**, 353, (1988)
Wit88c.	Witten, E.: Space–Time and Topological Orbifolds. Phys. Rev. Lett. **61**, 670–673, (1988)
Wit88d.	Witten, E.: Topological Sigma Models. Commun. Math. Phys. **118**, 411, (1988)
Wit88e.	Witten, E.: Topological Gravity. Phys. Lett. B **206**, 601, (1988)
Wit89.	Witten, E.: Quantum field theory and the Jones polynomial. Commun. Math. Phys. **121**, 351, (1989)
Wit89.	Witten, E.: Topology changing amplitudes in (2+1)-dimensional gravity. Nucl. Phys. **B323**, 113, 1989.

Wit90.	Witten, E.: On the structure of the topological phase of two–dimensional gravity. Nucl. Phys. B **340**, 281, (1990)
Wit91.	Witten, E.: Introduction To Cohomological Field Theories. Int. J. Mod. Phys. A **6**, 2775, (1991)
Wit92.	Witten, E.: Mirror manifolds and topological field theory. In Essays on mirror manifolds, ed. S.-T. Yau, International Press, 120–158, (1992)
Wit94.	Witten, E.: Monopoles and four manifolds. Math. Res. Lett. **1**, 769–796, (1994)
Wit95.	Witten, E.: Chern–Simons gauge theory as a string theory. Prog. Math. **133**, 637, (1995)
Wit95.	Witten, E: String Theory Dynamics in Various Dimensions. Nucl. Phys. B **443**, 85, (1995)
Wit96.	Witten, E.: Bound States of Strings and p-Branes. Nucl. Phys. **B460**, 335, (1996)
Wit98a.	Witten, E.: Magic, mystery, and matrix. Notices AMS, **45**(9), 1124–1129, (1998)
Wit98b.	Witten, E: Anti-de Sitter space and holography. Adv. Theor. Math. Phys., **2**, 253, (1998)
WM02.	Woltering, M., Markus, M.: Control of Spatiotemporal Disorder in an Excitable Medium. Science Asia 28, 43–48, (2002)
Wol98.	Wolf, F.A.: The Timing of Conscious Experience: A Causality–Violating, Two–Valued, Transactional Interpretation of Subjective Antedating and Spatial-Temporal Projection. J. Sci. Expl. **12**(4), 511–542, (1998)
Wol84.	Wolfram, S.: Cellular Automata as Models of Complexity. Nature, **311**, 419–424, (1984)
Wol02.	Wolfram, S.: A New Kind of Science. Wolfram Media, (2002)
Woo82.	Wootters, W.K., Zurek, W.H.: A single quantum cannot be cloned. Nature **299**, 802,(1982)
Woo92.	Woodhouse, N.: Geometric Quantization. Clarendon Press, Oxford, (1992)
XH94.	Xu, Z., Hauser. J.: Higher order approximate feedback linearization about a manifold. J. Math. Sys. Est. Con., **4**, 451–465, (1994)
XH95.	Xu, Z., Hauser. J.: Higher order approximate feedback linearization about a manifold for multi–input systems. IEEE Trans. Aut. Con, AC–**40**, 833–840, (1995)
XQW99.	Xie, F., Qu, Z., Weiss, J.N., Garfinkel, A.: Interactions between stable spiral waves with different frequencies in cardiac tissue. Phys. Rev. E **59**, 2203, (1999)
YA01.	Yalcin, I., Amemiya, Y.: Nonlinear factor analysis as a statistical method. Statistical Science, **16**, 275–294, (2001)
YAS96.	Yorke, J.A., Alligood, K., Sauer, T.: Chaos: An Introduction to Dynamical Systems. Springer, New York, (1996)
YC98.	Yang, T., Chua, L.O.: Control of chaos using sampled-data feedback control. Int. J. Bif. Chaos **8**, 2433–2438, (1998)
YL52.	Yang, C.N., Lee, T.D.: Statistical theory of equation of state and phase transitions I: Theory of condensation. Phys. Rev. **87**, 404–409, (1952)

YL52.	Yang, C.N., Lee, T.D.: Statistical Theory of Equations of State and Phase Transitions. I. Theory of Condensation. Phys. Rev. **87**, 404–409, (1952)
YL96.	Yalcinkaya, T., Lai, Y.-C.: Blowout bifurcation route to strange nonchaotic attractors. Phys. Rev. Lett. **77**, 5039–5042, (1996)
YL97.	Yalcinkaya, T., Lai, Y.-C.: Phase Characterization of Chaos. Phys. Rev. Lett. **79**, 3885–3888, (1997)
YML00.	Yanchuk, S., Maistrenko, Yu., Lading, B., Mosekilde, E.: Effects of a parameter mismatch on the synchronizatoin of two coupled chaotic oscillators. Int. J. Bifurcation and Chaos **10**, 2629-2648, (2000)
YST96.	Yamamoto, K., Sasaki, M., Tanaka, T.: Quantum fluctuations and CMB anisotropies in one-bubble open inflation models. Phys. Rev. D54, 5031–5048, (1996)
YTY02.	Yasui, T., Tutu, H., Yamamoto, M., Fujisaka, H.: Dynamic phase transitions in the anisotropic XY spin system in an oscillating magnetic field. Phys. Rev. E **66**, 036123, (2002)
YZ99.	Yong. J., Zhou, X.: Stochastic controls. Hamiltonian Systems and HJB Equations. Springer, New York, (1999)
Yag87.	Yager, R.R.: Fuzzy Sets and Applications: Selected Papers by L.A. Zadeh, Wiley, New York, (1987)
Yak81.	Yakhot, V.: Large–scale properties of unstable systems governed by the Kuramoto–Sivashinksi equation. Phys. Rev. A **24**, 642, (1981)
Yan52.	Yano, K.: Some remarks on tensor fields and curvature. Ann. of Math. **55**(2), 328–347, (1952)
Yan65.	Yano, K.: Differential Geometry on Complex and Almost Complex Manifolds, Pergamon Press, New York, (1965)
Yau78.	Yau, S.T.: On the Ricci curvature of a compact Kähler manifold and the complex Monge–Ampere equation, I^*. Comm. Pure Appl. Math. **31**, 339–441, (1978)
Yeo92.	Yeomans, J.M.: Statistical Mechanics of Phase Transitions. Oxford Univ. Press, Oxford, (1992)
Yoc93.	Yoccoz, J.C. : Introduction to hyperbolic dynamics, Proceeding of the NATO Advanced Study Intitute in Real and Complex Dynamical Systems, Hillerod, Denmark: Kluwer, (1993)
Yod72.	Yodzis, P.: Lorentz cobordism. Comm. Math. Phys. **26**, 39–52, (1972)
Yod73.	Yodzis, P.: Lorentz cobordisms. Gen. Rel. Grav. **4**, 299, 1973.
You00.	Youm, D.: Bulk fields in dilatonic and self-tuning flat domain walls, Nucl. Phys. B **589**, 315, (2000)
ZCW05.	Zhang, H., Cao, Z., Wu, N.J., Ying, H.P., Hu, G.: Suppress Winfree Turbulence by Local Forcing Excitable Systems. Phys. Rev. Lett. **94**, 188301, (2005)
ZF71.	Zakharov, V.E., Faddeev, L.D.: Korteweg-de Vries equation is a fully integrable Hamiltonian system, Funkts. Anal. Pril. **5**, 18–27, (1971)

ZGD91. Zhang, R.B., Gould, M.D., Bracken, A.J.: Quantum Group Invariants and Link Polynomials. Commun. Math. Phys, **137**, 13, (1991)

ZHH03. Zhang, H., Hu, B., Hu, G.: Suppression of spiral waves and spatiotemporal chaos by generating target waves in excitable media. Phys. Rev. E **68**, 026134, (2003)

ZK65. Zabusky, N.J., Kruskal, M.D.: Interactions of solitons in a collisionless plasma and the recurrence of initial states. Phys. Rev. Let., **15**, 240–43, (1965)

ZOC95. Zhang, R.B., Wang, B.L., Carey, A.L., McCarthy. J.: Topological Quantum Field Theory and Seiberg–Witten Monopoles. arXiv:hep-th/9504005, (1995)

ZS72. Zakharov, V.E., Shabat, A.B.: Exact theory of two–dimensional self–focusing and one–dimensional self-modulation of waves in nonlinear media. Sowiet Physics, JETP **34**, 62–69, (1972)

ZTG04. Zumdieck, A., Timme, M., Geisel, T.: Long Chaotic Transients in Complex Networks. Phys. Rev. Lett. **93**, 244103, (2004)

Zak92. Zakharov, O.: Hamiltonian formalism for nonregular Lagrangian theories in fibred manifolds. J. Math. Phys. **33**, 607, (1992)

Zal89. Zaleski, S.: A stochastic model for the large scale dynamics of some fluctuating interfaces. Physica D **34**, 427–438, (1989)

Zas05. Zaslavsky, G.M.: Hamiltonian Chaos and Fractional Dynamics. Oxford Univ. Press, Oxford, (2005)

Zin05. Zinn-Justin, J.: Path integrals in quantum mechanics. Oxford Univ. Press, Oxford, UK (2005)

Zin93. Zinn-Justin, J.: Quantum Field Theory and Critical Phenomena. Oxford Univ. Press, Oxford, (1993)

Zwi04. Zwiebach, B.: A first course in string theory. Cambridge Univ. Press, Cambridge, UK (2004)

Index

absolute covariant derivative, 330, 337
abstract ergodic theory, 661
abstraction of Euclidean geometry, 368
achieved micro–states, 654
action, 316, 466, 511
action of a new theory, 425
action potential, 155
action principle, 316
action–amplitude formalism, 425
action–amplitude picture, 425, 637
action–angle coordinates, 344
action–at–a–distance electrodynamics, 652
active polarized state, 226
adaptive, 705
adaptive Lie–derivative controller, 703
adaptive motor control, 709
adaptive path integral, 437, 637, 639, 640
adaptive path measure, 653, 656
adiabatic elimination, 203
affine connection, 312
affine Hamiltonian function , 700
affine transformation, 39
agreement, 225
algebraic topology, 380
Aliev–Panfilov model, 159
almost Lorentzian metric, 404
alpha skeletomotor neurons, 701
Ambrose–Singer theorem, 117
American option, 593
amplitude, 432
amplitude of the driving force, 3

Andronov, 33
Andronov–Hopf bifurcation, 33
Anosov diffeomorphism, 33
Anosov flow, 34, 661
Anosov map, 33
anti–control of chaos, 7
approach to equilibrium, 77
arc–element, 331
arc–length, 382, 659
area–preserving map, 53
Arnold cat map, 33
arrow of time, 625
asymmetric dilution, 614
atlas, 306, 318
atmospheric convection, 45
attaching a handle, 396
attachment of handles, 672
attractor, 4, 44, 79, 102
attractor neural networks, 245
autocatalator, 64
autogenetic motor servo, 701
autonomous system, 3
average, 348, 352
average degree of chaoticity, 348
average energy, 193
Axelrod model, 224
Axiom A, 79
Axiom–A systems, 35

Bär–Eiswirth model, 159
background–independent, 487
Baker map, 55
Banach manifold, 320

Banach space, 320, 471
Barkley model, 159
Basic consequence: a different way for calculating probabilities., 431
basic sets, 79
basin of attraction, 44, 80, 145
basins of attraction, 4, 53
Bekenstein–Hawking entropy, 574
Belouzov–Zhabotinski reaction, 127
Belusov–Zhabotinsky reaction, 155
Berlin–Kac spherical model, 673
Bernoulli map, 33
Bernoulli shift dynamics, 109, 113
Bernoulli systems, 32
Bernstein's motor coordination and dexterity, 709
Bessel functions, 268, 275
Betti number, 417
Betti numbers, 309, 375, 377, 667, 703
Bianchi identity, 314
Bianchi symmetry condition, 336
bifurcation, 6
bifurcation diagram, 51
bifurcation point, 17
Big–Bang, 397, 398
bijection, 306
billiard, 32
binary systems, 101
bio–diversity, 50
biomechanical force–velocity relation, 233
biomorph, 66
biomorphic systems, 66
biphasic pacing, 165
Birkhoff, 26, 30, 32
Birkhoff curve shortening flow, 32
black hole dynamics, 572
Black–Scholes–Merton formula, 584
block entropy, 21
body motion manifold, 709
Boltzmann, 25
Boltzmann constant, 192
Boltzmann entropy, 72
Bolzmann constant, 71
Borde–Sorkin conjecture, 693
Bose–Einstein condensate, 433
bosonic string theory, 535
Bott's extended Morse theory, 677
boundaries, 377

boundary condition, 468
boundary conditions, 463
boundary of a boundary, 377
boundary operator, 377
bra–covectors, 433
bra–ket, 433
brain, 708
brain dynamics, 7
brain motion manifold, 709
brane, 532
Brouwer degree, 341
Brownian dynamics, 428
Brownian motion, 243, 585
BRST–operator, 517
brute–force, 143
building blocks, 394
Burgers dynamical system, 116, 579
butterfly effect, 46, 110

calculus of variations, 316
Campbell–Baker–Hausdorff formula, 335
Cantor set, 37, 103
Cantori, 98
capacity dimension, 65, 130
carrying capacity, 49
Cartesian product, 307
Cartwright, 33
Catalan numbers, 501
catastrophes, 6
category, 43
Cauchy Theorem, 558
causal continuity, 404
causal discontinuity, 408, 682, 687
Central–limit–theorem, 352
chain, 376
chain complex, 377
chain coupling, 234
chaos, 27
chaos control, 128, 158
chaos theory, 29, 108, 200
chaotic, 3, 102
chaotic attractor, 4, 102, 145
chaotic behavior, 109
chaoticity, 343
Chapman–Kolmogorov equation, 429, 588
Chapman–Kolmogorov integro–differential equation, 430

Chapman–Kolmogorov law, 328
characteristic equation, 195
characteristic Lyapunov exponents, 129
chart, 306
chemical kinetics, 64
Chern class, 576
Chern–Simons gauge theory, 512, 571
Chern–Simons theory, 515
Christoffel symbols, 86, 313, 331
Chua's circuit, 62
Chua–Matsumoto circuit, 127
circle, 306
circle map, 54
Clifford, 309
closed string theories, 529
closed timelike curves, 395
co–area formula, 665
coarse graining, 72
coarse system, 33
cobordism, 393
code, 625
cognitive intention Lagrangian, 655
collapse of wave packets, 77
colored multiplicative noise, 202
combinatorial route, 434
compact manifold, 307
compact manifolds, 375
complex number, 433
complexity conjecture, 694
computational complexity, 694
conatural projection, 322
condensed, 624
conditioned training, 694
configuration manifold, 317, 321
configuration–space path integral, 442
configurational partition function, 362
conformal, 350
conformal z–map, 65
conformal field theory, 534
congruence of geodesics, 348
connection homotopy, 334
consensus, 224
conservation law, 48
constant relative degree, 702
continuity, 368
continuous deformation, 368
continuous phase transition, 231
continuous phase transitions, 176
continuous–time regularization, 452

contravariant velocity equation, 699
control law, 702
control parameter, 180, 231
control parameters, 229, 230
control parameters are iteratively
    adjusted, 709
convective Bénard fluid flow, 45
conventional chaos theory, 344
convolution, 463
coordinate ball, 319
coordinate chart, 319
coordinate domain, 319
coordinate map, 319
correlation function, 243, 464
correspondence principle, 431
cortical motion control, 709
cotangent bundle, 322
cotangent space, 322
countably additive measure, 451
coupling, 706
covariant derivative, 659
covariant differentiation, 330
covariant force equation, 699
covariant force functor, 86, 651
covariant force law, 233, 651
covariant formalism on smooth
    manifolds, 84
cover, 306
critical phenomena, 178
critical point, 176, 381, 382, 404, 671
critical points, 370
critical slowing down, 231, 232, 472
critical submanifolds, 675, 679
critical value, 381
cross–section, 321
cube, 703
cumulative distribution function, 427
Curie–Weiss law, 260
curvature, 309, 348
curvature singularities, 395
curved space, 490
CW complexes, 375
cycle, 22
cycles, 377
cyclic forms of human motion, 710

damped pendulum, 10, 56
damping parameter, 3
damping rate, 145

834   Index

De Rham theorem, 703
Defuzzification, 705
degeneracy density, 677
degenerate critical point, 371
degree of order, 230
desired micro–states, 654
determinism, 343
deterministic, 658
deterministic chaos, 1, 45
deterministic chaotic system, 26
diffeomorphicity, 672
diffeomorphism, 37, 79, 320, 665, 709
diffeomorphism invariant, 666
difference equation, 50
different, 666
diffusion equation, 450, 463
diffusion fluctuations, 429
dilaton field, 535
dimensionality, 428
direct strategy, 225
directed attachment, 225
Dirichlet branes, 537
discontinuous phase transition, 232
discontinuous phase transitions, 176
discrete–time models, 50
discrete–time steps, 30
discretized, 704
disjoint union, 104, 369
disorder $\Rightarrow$ order, 232
disorder–averaged partition function, 601
dissipation, 428, 630
dissipative structures, 70, 425
distribution function, 426, 450
Donaldson theory, 516
double-scaling limit, 503
drift, 429
driven nonlinear pendulum, 3
driven pendulum, 57
Duffing, 32
Duffing map, 54
Duffing oscillator, 60
Duffing–Van der Pol equation, 60
dynamical, 661
dynamical edge of chaos, 173
dynamical intuition, 317
dynamical invariant, 32
dynamical phase transition, 206
dynamical similarity, 109

dynamical system, 16
dynamical systems, 26
dynamics, 8
dynamics of conservative systems, 657
Dyson–Wyld diagrammatic analysis, 119

eddy, 113
edges, 309, 368
effective gaussian process, 367
effective instability equation, 661
Ehrenfest classification scheme, 175
eigenvalue relation, 221
eigenvalues, 195
eigenvectors, 195
Einstein equation, 475
Einstein tensor, 314
Einstein–Hilbert action, 475
Eisenhart metric, 659
elastic pendulum, 63
electron–phonon interaction, 181
elementary cobordism, 396
elliptic geometry, 309
energy functionals, 382
energy surface, 98
ensemble averages, 657
entangled state, 629
entropic function, 259
entropy, 32, 70, 174, 229
environment, 708
equilibrium point, 4
equilibrium statistical mechanics, 245, 257
equipartition, 428
ergodic hypothesis, 25, 32, 71, 78, 657, 661
ergodic hypothesis of Boltzmann, 25
ergodic system, 248
ergodicity breaking, 245
error function, 702
escape rate, 25
Euclidean, 495
Euclidean 3D space, 308
Euclidean action, 462
Euclidean chart, 318
Euclidean geometry, 305
Euclidean image, 318
Euclidean metric, 320
Euclidean nD space, 309

Euclidean path integral, 462
Euclidean path–integral, 489
Euclidean quantum gravity, 496, 503
Euclidean spaces, 305
Euclidean triangulations, 484
Euler, 25, 309
Euler character, 528
Euler characteristic, 27, 309, 328, 341, 368, 371, 651, 666
Euler class, 527
Euler's formula, 369
Euler–Lagrange equations, 315
Euler–Lagrangian equations, 334
Euler–Poincaré characteristic, 662, 703
European option, 586
excitable media, 166
excited state, 155
existence & uniqueness theorems for ODEs, 16
expanding Jacobian, 80
expansion principle, 434
exponential law, 227
extended Hamilton oscillator, 237
external configuration manifold, 84
extract order from chaos, 229
extrinsic view, 308

faces, 309, 368
Faddeev–Popov procedure, 519
Farey construction, 99
feedback control, 704
Feigenbaum cascade, 47
Feigenbaum constant, 47
Feigenbaum number, 51
ferromagnet, 231
Feynman diagram, 531, 578
Feynman kernel, 436
Feynman path integral, 194, 425, 430, 434, 637
Feynman–Vernon formalism, 627
fibre, 321
field, 8
field action functional, 194
finite–time probability distribution, 591
finitely additive, 451
Finsler manifold, 654
Finsler metric function, 655
Finsler tensor, 655
first–order phase transitions, 176

Fitzhugh–Nagumo model, 156
fixed–point, 36, 41, 44
flame front, 116
Floquet multiplicator, 136
Floquet stability analysis, 136
flow, 16, 38, 79, 114, 327
flow line, 324
flow pattern, 11
flow property, 327
fluctuating force, 243
fluctuation theorem, 77
fluctuation–dissipation theorem, 244, 295
fluctuation–dissipation theorems, 296
foamlike structure, 394
Fock space, 537
Fock state, 432
Fokker–Planck equation, 188, 200, 429, 430, 582
folding, 4, 35
force equation, 233
force HBE servo–controller, 700
force–field psychodynamics, 637
forced nonlinear oscillators, 11
forced Van der Pol oscillator, 58
formal exponential, 327
formalism of jet bundles, 11
forward–Euler integration scheme, 167
Fourier decomposition, 469
Fourier transform, 108
fractal attractor, 4, 46
fractal dimension, 44
fractal microstructure, 511
fractal pattern, 51
fractal set, 37
fractals, 65
fractional dimension, 65
fragmentation, 224
Fredholm integral equation, 464
Fredholm operator, 525
Fredholm theory, 464
free Abelian group, 376
free energy, 193, 257
free energy potential, 174, 231
free particle, 442
free string, 529
frequency, 3
friction, 428
Frobenius–Perron theorem, 675

from a classical point of view, 506
full coupling, 234
functional integral, 455
functional manifold, 116
functor, 380
Fuzzification, 700, 704
fuzzy inference system, 704

gamma fusimotor neurons, 701
gauge condition, 458
Gauss, 308
Gauss map, 341
Gauss' Theorema Egregium, 349
Gauss–Bonnet formula, 328, 341
Gauss–Bonnet theorem, 369, 662
Gauss–Bonnet–Hopf theorem, 667
Gauss–Kronecker curvature, 667
Gaussian approximation, 597
Gaussian curvature, 328, 480, 663
Gaussian integrals, 441
Gaussian measure, 597
Gaussian multiplicative noise, 199
Gaussian saddlepoint approximation, 578
Gaussian–distributed Markov process, 661
general sense, 127
generalized Hénon map, 53
generalized solution, 127
generalized SRB measure, 81
generating functional, 194, 466
genus, 369
geodesic, 310, 325, 334
geodesic equation, 313, 334
geodesic equations, 659
geodesic spray, 325
geodetic separation field, 348
geometrical intuition, 318
geometrodynamical functor, 639
ghost number, 537
Gibbs entropy, 71
Ginzburg–Landau equation, 114, 180, 579
Ginzburg–Landau model, 180
global chaos control, 158
globular cluster, 52
goal–driven interaction, 654
golden numbe, 98
Goldstone theorem, 624

Golgi tendon organs, 701
gradient-like, 684
gradient-like vector–field, 419
graphs, 369
Green's function, 241, 463
grows exponentially, 5
grows linearly, 5
growth of the entropy, 493
growth rate, 49

H–cobordism theorem, 34
Hénon map, 30, 52, 142
Hénon strange attractor, 52
Hadamard, 32
Haken's synergetics, 111
halo orbit, 31
Hamilton, 25
Hamilton oscillator, 236
Hamilton's principle, 316
Hamiltonian, 317
Hamiltonian action, 658
Hamiltonian chaos, 95, 144, 658
Hamiltonian dynamics, 323
Hamiltonian system, 53
Hamming distance, 101, 102, 106
handle, 369, 405
handle decomposition, 403
handle-body decomposition, 415
harmonic oscillator, 443
Hausdorff dimension, 504
Hausdorff space, 319
Hayashi, 32
heat bath, 193, 428
Hebbian synapses, 601
Heisenberg commutation relations, 452
Heisenberg picture, 453
Hermitian inner product, 433
Hessian, 333, 404
high–dimensional chaos theory, 44
Hilbert manifold, 320
Hilbert space, 26, 320
hippocampus, 187
history, 431, 681
Hodgkin–Huxley, 160
holographic hypothesis, 624
homeomorphic, 368
homoclinic point, 29, 38, 42
homoclinic tangle, 27, 41
homology group, 376, 377

homology theory, 368
homomorphism, 377
homotopy, 379
homotopy equivalent, 371, 676
Hopf, 32
Hopf bifurcation, 145
Hopf–like bifurcation, 62
Hopf–Rinow Theorem, 310
Hopfield discrete–time network, 239
Hopfield overlap, 280
Hopfield overlaps, 262
human–like locomotor system, 233
hurricane, 112
hybrid dynamical system of variable structure, 126
hybrid dynamics, 428
hybrid systems, 127
hyperbolic fixed–point, 43
hyperbolic geometry, 309
hyperbolic system, 17
hysteresis effect, 231, 232

ideal thermostat, 75
imaginary time, 465
Implantable Cardioverter–Defibrillator, 167
impossible to integrate, 706
index, 370, 381, 396, 404, 677, 679
index of the critical point, 671
infinite–dimensional neural network, 637
infinity, 709
information, 129
information theory, 33
inner product, 309
instability, 11, 343
instability sequence, 231
instanton vacua, 578
integral curve, 324
integrate–and–fire neuron, 188
integration, 16
interaction, 225
internal configuration manifold, 84
intrinsic curvature, 490
intrinsic definition for differentiable manifolds, 309
intrinsic dynamical instability, 658
intrinsic view, 308
intuition, 26

invariant distribution, 427
invariant set, 37, 39
invariant torus, 97
inverted driven pendulum, 63
irregular and unpredictable, 1
irreversibility, 70
Ising–spin Hamiltonian, 257
isolated closed trajectory, 5
isometric, 310
isotropic, 349
Itô lemma, 585
iterated map, 30
iteration of conditioned reflexes, 709
iterative maps, 17
Ito stochastic differential equation, 429
Ito stochastic integral, 429
Ito stochastic integral equation, 429

Jacobi equation, 659
Jacobi equation of geodesic deviation, 337
Jacobi fields, 337, 383
Jacobi metric, 659
Jacobi variation field, 659
Jacobi–Levi-Civita equation, 347
jerk, 702
jerk function, 144
jet space, 332
joint action, 654

Kähler gauge, 572
Kähler potential, 574
Kaluza–Klein monopole, 690
Kaluza–Klein theories, 395
KAM–torus, 99
Kaplan–Yorke dimension, 131, 157
Kaplan–Yorke map, 55
Karder–Parisi–Zhang equation, 119
Karma model, 159, 171
Kepler, 27
ket–vectors, 433
kick equation, 241
Killing vector–field, 423
kinetic energy, 194
kinetic–energy estimator, 473
Klein–Gordon Lagrangian, 458
Kolmogorov, 32, 33
Kolmogorov–Arnold–Moser (KAM) Theorem, 97

Kolmogorov–Sinai, 131
Kolmogorov–Sinai entropy, 20, 129, 131
Kosterlitz–Thouless transition, 662
Kramers–Moyal expansion, 277
Krylov, 32
Kunneth formula, 417
Kuramoto–Sivashinsky equation, 579

lack of memory, 427
Lagrange, 25
Lagrange's equations, 315
Lagrange's points, 31
Lagrangian chaos, 87
Lagrangian density, 458, 642
Lagrangian mechanics, 315
laminar flow, 111
Landau, 229
Landau free energy, 181
Landau's theory of phase transitions, 179
Landau–Ginzburg equation, 158
Langevin rate equation, 243
Langevin rate equations, 428
Laplace–Beltrami operator, 333
large system, 74
largest Lyapunov exponent, 129, 343
lattice regularization, 449
laws of probability, 433
learning, 225
learning rate,, 654
Legendre transformation, 317
Levi–Civita connection, 86, 312, 330
Lewinian force–field theory, 638
Lickorish–Wallace theorem, 689, 692
Lie, 309
Lie bracket, 312, 330, 334, 335, 702
Lie derivative, 125, 330, 348, 667, 702
Lie derivative formalism, 702
Lie group, 307
Lie structure equations, 117
Lie super–algebra, 517
limit cycle, 12, 44, 57
limit–cycle attractor, 5
linear homotopy ODE, 126
linear homotopy segment, 127
linear operator equation, 463
linearization, 3
Liouville equation, 429
Liouville measure, 230

Liouville theorem, 98, 230
Liouville tori, 98
Lipschitz condition, 326
Lissajous curves, 64
Littlewood, 33
local Bernstein's adaptation process, 653
local chaos control, 158
locally stable, 80
locally unstable, 80
locally–optimal solution, 633
logistic equation, 49, 200
logistic growth, 49
logistic map, 50, 52, 104, 133
long–range correlation, 624
Lorentz–invariant theories, 530
Lorentzian cobordism, 395, 412, 681
Lorentzian dynamical triangulations, 477
Lorentzian space-time, 492, 495
Lorentzian structure, 659
Lorenz attractor, 5, 125
Lorenz equations, 64, 143
Lorenz flow, 110
Lorenz mask, 46, 110
Lorenz system, 18, 46, 52
lower limit of complexity, 694
Luo–Rudy I model, 160
Lyapunov, 33
Lyapunov dimension, 130
Lyapunov exponent, 2, 19, 68, 660, 661, 671
Lyapunov exponents, 44, 81
Lyapunov function, 279
Lyapunov spectrum, 157
Lyapunov stability, 17
Lyapunov time, 2

Möbius strip, 378
Möbius band, 397
macroscopic change, 673
macroscopic entanglement, 694
macroscopic neuro–mechanical entanglement, 709
magnitude, 677
major topology change, 664
Malthus model, 49
Malthusian parameter, 49, 51
Mamdani inference, 705

Mandelbrot and Julia sets, 65
manifold, 305, 317
manifold structure, 319
manifoldness, 309
map, 50
map sink, 44
Markov assumption, 430
Markov chain, 21, 247, 427, 470
Markov partitions, 34, 79
Markov process, 21
Markov stochastic process, 427, 430, 641
Markovian networks, 227
Master equation, 430
material metric tensor, 699
Mathai–Quillen formula, 525
mathematical induction, 309
Matrix Product Formalism, 219
matrix–symplectic explicit integrator, 705
Maupertius action principle, 333
Maupertuis' action, 658
Maupertuis' principle, 346
Maurer–Cartan connection, 527
maximal geodesic, 325
maximal integral curve, 324
Maxwell, 25
Maxwell–Haken laser equations, 64, 111
mean first–passage time, 207
mean kinetic energy, 244
mean square limit, 429
mean–field, 236
mean–field approximation, 199
mean–field theory, 101, 176
measure theory, 666
Mel'nikov integrals, 344
membrane potential, 156
memory, 225
memory term, 146
mental force law, 651
Mermin–Wagner theorem, 668
metal–insulator transition, 181
metastable state, 77
metric space, 310
metric tensor, 313
Metropolis algorithm, 470
Mexican–hat coupling, 190
micro–canonical ensemble, 71, 78
Minkowski action, 462

Mittag–Leffler, 28
mixing, 81
mixing Anosov diffeomorphism, 79
model space, 320
momentum phase–space, 703
Monte Carlo method, 52, 468
Monte Carlo simulation, 206
Monte–Carlo simulations, 492
Morse, 32
Morse cobordisms, 682
Morse function, 381, 382, 404, 413, 672, 682
Morse index, 413, 683
Morse inequalities, 404
Morse Lemma, 671
Morse lemma, 381, 663
Morse metric, 413
Morse metrics, 407, 683
Morse number, 381, 684
Morse singularities, 683
Morse structure, 414
Morse theory, 32, 35, 380, 381, 404, 406, 412, 413, 511, 672
Morse trousers, 407
motor conditioned reflexes, 705
multi–kick equation, 241
multi–spiral strange attractor, 62
multicanonical sampling, 473
multigrid method, 472
multiplicative zero–mean Gaussian white noise, 200

Nambu–Goto action, 534
natural projection, 322
Navier–Stokes equations, 45, 91, 108, 115
nearest neighbor coupling, 234
Necker cube, 238
Neumann boundary condition, 538
neural information processing, 187
neural path integral, 619
neutral strategy, 225
Newton, 25
Newton's Second Law, 315
Newtonian mechanics, 16, 315
no–flux Neumann boundary conditions, 167
noble numbers, 98
Noether Theorem, 317

noise, 654
noise–free limit, 253
noise–induced phase transitions, 210
non–autonomous 2D continuous
    systems, 11
non–autonomuous system, 9
non–conserving dynamics, 218
non–equilibrium phase transition, 218
non–equilibrium statistical mechanics,
    245
non–equilibrium steady state, 77
Non–Euclidean geometry, 309
non–wandering set, 37, 79
non–degenerate, 381
nonholonomic coordinates, 332
nonlinear control theory, 531
nonlinear non–equilibrium systems, 199
nonlinear oscillators, 25
nonlinear Schrödinger equation, 180
nonlinear sigma model, 534
Nonlinear velocity, 702
normal vector–field, 323
number of mechanical degrees of
    freedom, 694, 697
number of mechanical DOF, 694

observational resolution, 694, 697, 709
ODEs, 17
one–parameter group of diffeomor-
    phisms, 327
open string theories, 529
opinion dynamics, 225
opinion–dynamics, 227
optimal policy, 654
orbit, 16, 22, 38
order, 677, 679
order parameter, 177, 203, 205, 229, 231
order parameter equation, 205, 232, 264
order parameters, 181, 245
ordered symmetry–breaking state, 199
ordering chaos, 7
oriented cobordism, 402
oriented strings, 530
Ornstein–Uhlenbeck noise, 203
Ornstein–Uhlenbeck process, 289
oscillatory cortical–control, 710
oscillatory dynamics, 710
oscillatory synchronization, 694
Oseledec theorem, 81

Ott–Grebogi–Yorke map, 55
output–space dimension, 709
overdrive pacing, 165

pacing response diagrams, 165
Panfilov model, 159, 170
parabolic Einstein equation, 341
parallel transport, 312
parametric instability, 350, 661, 662
partition function, 34, 192, 220, 257,
    460
path integral, 425, 580
path–integral approach, 405
path–integral expression, 456
path–integral formalism, 454, 599
path–integral formulation, 317, 453
path–integral quantization, 452
pendulum angle, 3
period doubling bifurcations, 51
period–doubling bifurcation, 47, 133
periodic orbit, 29, 41
periodic orbit theory, 24, 116
periodic solutions, 11
perturbative path integral, 474
perturbative string theory, 535
Pesin formula, 82
phase, 4, 432
phase point, 11
phase portrait, 4, 11
phase space, 308
phase transition, 6, 173, 174, 231
phase transition of first order, 232
phase transition of second order, 231
phase transitions, 472
phase–flow, 10, 16
phase–locking, 710
phase–space, 4
phase–space flow, 30
phase–space path integral, 441, 449
phase–transition effects, 231
phase–transition theory, 229
physical Hamiltonian function, 699
physically desirable histories, 682
Pickover's biomorphs, 67
pinball game, 1
pinned Wiener measure, 451
Planck length, 529
playground swing, 3
Poincaré, 26, 309

Poincaré conjecture, 34, 309, 378
Poincaré duality, 369
Poincaré section, 5, 27, 30, 60, 135
Poincaré map, 53
Poincaré section, 24, 52
Poincaré–Bendixson theorem, 6, 11, 27
Poincaré–Birkhoff Theorem, 96
Poincaré–Birkhoff theorem, 344
Poincaré–Fermi theorem, 353
Poincaré–Hopf index theorem, 27
Poincaré section, 147
Poisson detection statistics, 432
Poisson process, 243
polarization, 224
Polyakov action, 534
Pontryagin, 33
Pontryagin Maximum Principle, 580
population models, 48
positional stiffness, 701
positive leading Lyapunov exponent, 130
potential energy, 194, 198
Prandtl number, 46
predictability, 343, 658
predictability time, 130
predictability/controllability, 694
Principle of stationary action, 317
probabilistic description, 426
probability amplitude, 453
probability density, 200
probability density function, 430
probability of acknowledged influence, 225
product topology, 114
product–topology theorem, 703
propagator, 436, 438, 449, 460, 488
protozoan morphology, 66
pruning, 22
pull–back, 702
pulse–coupled oscillators, 188
Pyragas control, 136

quantum behavior, 509
quantum brain, 637
quantum coherent state, 431
quantum entanglement, 708
quantum field theory, 194
quantum gravity, 474
quantum modifications, 510

quantum probability, 430
quantum space-time, 487
quantum statistical mechanics, 192
quantum theory, 507
quantum universe, 503
quantum–mechanics propagator, 454
quasi–isotropic manifold, 351
quasi-isotropy, 662
quotient space, 377

Rössler, 143
random thermostat, 75
random variable, 426
random walk, 427
rate of error growth, 129
Rayleigh–Bénard convection, 18, 127
reaction–diffusion systems, 218
recovery period, 155
rectified, 704
recurrent neural networks, 244
recursive homotopy dynamics, 649
reduced curvature 1–form, 118
reentrant excitations, 156
Regge calculus, 479, 490
Regge geometries, 476
Regge geometry, 490
Regge simplicial action, 481
regular level, 686
reinforcement learning, 654
reinforcement training, 709
relative degree, 125
relaxation oscillator, 57
reliable predictor, 49
repeller, 4, 24
representative point, 318
resting state, 155
return map, 24
reverse strategy, 225
Reynolds number, 109, 111
ribbon, 397
Ricci curvature, 660
Ricci flow, 341
Ricci scalar curvature, 314
Ricci tensor, 314, 329, 336, 660
Riemann, 309
Riemann curvature tensor, 313, 328, 334, 650, 660
Riemann sphere, 311
Riemann surface, 310

Riemannian geometry, 311
Riemannian manifold, 309
Riemannian manifolds, 308
Riemannian metric tensor, 328
Rossler, 61
Rossler system, 61
route to chaos, 6, 34, 145
route to turbulence, 113
Rudolphine Tables, 27
Ruelle, 34

saddle point, 4
saddle–point integration, 259
Sard's Theorem, 383
scalar curvature, 329, 336
scalar Gaussian curvature, 337
scale factor, 509
scatterers, 663
Schrödinger equation, 431, 448
Schrödinger picture, 453
scroll waves, 156
Second Law of thermodynamics, 70
second variation formula, 338
second–countable space, 319
second–order phase transition, 199
second–order phase transitions, 176
sectional curvature, 329
self–assembly, 694, 709
self–consistency relation, 204
self–limiting process, 49
self–organized, 653, 656
semantic integration, 656
sensitive dependence on initial conditions, 102
sensitivity to initial conditions, 5
sensitivity to parameters, 5
sequence of period doublings, 145
servo–controllers, 702
servoregulatory loops, 701
set, 43
Shannon, 33
shape operator, 667
Sherrington–Kirkpatrick model, 617
short–term predictability, 5
short–time evolution, 205
signal, 654
signature, 492
simplicial complexes, 375
simplicial homology, 376

simplicity, 694
Sinai, 33
sine–Gordon equation, 113
singular homology, 376
singularities, 395
Smale, 32
Smale horseshoe, 99
Smale horseshoe map, 35
Smale–Zehnder Theorem, 99
small system, 74
smooth manifold triad, 412
smooth manifolds, 308
solution, 127
source, 466
sources and sinks, 425
space entropy, 80
space–time, 308
spaces, 492
spatio-temporal chaos, 156
specific heat capacity, 174
spectral decomposition, 79
spectral dimension, 510
sphere, 307
spherical modification, 402
spin glass, 77
spin networks, 475
spindle receptors, 701
spiral turbulence, 157
spiral waves, 156
spontaneous rotational symmetry breaking, 120
squeezing, 4, 35
stability, 11, 343, 658
stable and unstable manifold, 38
stable eigen–direction, 99
stable manifold, 53, 99
staging algorithm, 472
standard Hamiltonian systems, 671
standard Hamiltonians, 658
standard map, 53
state, 8
state vector, 453
stationary probability density, 200
statistical error, 470
Steifel–Whitney and Pontrjagin numbers, 412
step size, 654
Stiefel–Whitney numbers, 681
stochastic force, 428

Stochastic forces, 699
stochastic influence, 428
stochastic integral, 428
stochastic oscillator equation, 660
stochastic oscillator problem, 354
stochastic system, 10
strange attractor, 4, 7, 27, 34, 44–46, 60, 110
Stratonovitch interpretation, 200
stream function, 88
stretch–and–fold, 60
stretching, 4, 35
string corrections, 529
string tension, 534
string–field, 571
string–field–theory action, 537
stroboscopic section, 41
structural instability, 230
structural stability, 17
structurally stable, 33
Sturm–Liouville problem, 463
subsumption architectures, 653
sum–over–histories, 431
sum–over–topologies, 493
super–field, 563
super–space, 564
supercell thunderstorms, 112
superposition, 488, 694
superstring theory, 535
supersymmetry, 535
supervised, 709
supervised gradient descent learning, 654, 656
support of a vector–field, 326
surface of Earth, 305
surgery, 402, 684
survival probability, 25
symbolic dynamics, 22, 34, 38, 40, 79, 101
symmetric affine connection, 330
symmetric dilution, 614
symmetry, 463
symmetry breaking instability, 231
symmetry–breaking, 177
symmetry–breaking oscillation, 206
symmetry–breaking transition, 668
symmetry–restoring oscillation, 206
symplectic manifolds, 308
synchronization, 710

synergetics, 229, 238
system parameters, 233
systematic error, 469

tachyon field, 536
Takens, 34
tangent bundle, 310, 322
tangent dynamics equation, 660
tangent dynamics equation , 671
tangent map, 322
tangent space, 309, 321
tangent vector–field, 323
tensor–field, 16
theoretical ecology, 50
theory of quantum gravity, 486
theory of turbulence, 34
thermal equilibrium, 229
thermodynamics, 70
three–body problem, 25, 27
threshold, 155
time averages, 657
time entropy, 80
time–dependent Schrödinger equation, 430
time–dependent vector–field, 324, 327
time–flow, 395
time–ordered product, 463
time–phase plot, 11
time–reversal invariance, 72
time–reversal symmetry, 177
topological cobordism, 395, 412, 681
topological entropy, 2
topological hypothesis, 665, 669, 672
topological invariant, 309, 368, 669
topological manifold, 308
topological property, 309
topological quantum field theory, 511
topological theorem, 671
topological transition, 676
topologically transitive, 102
topology, 26, 368, 493
topology change, 394, 412, 496, 671
topology changing, 681
topology changing processes, 416
topology–changing transition, 681
tornado, 112
torsion, 312
torus, 307, 703
total system complexity, 694

trace, 686
trajectory, 11, 16, 22
transient chaos, 113
transition amplitude, 405, 425, 431, 433, 447
transition energy, 671
transition functions, 318
transition map, 306
transition probability, 431, 433
transition probability amplitude, 438
transition probability distribution, 590
transition temperature, 671
transitive Anosov flow, 79
trapping region, 102
Trotter decomposition formula, 471
trousers, 394, 397, 398, 400, 407
turbulence, 17, 108, 159
turbulent flow, 111
twist map, 96
two–point function, 464

umbrella sampling, 473
uncorrelated configuration model, 227
undamped pendulum, 55
universality class, 179
unpredictability, 27, 658
unstable eigen–direction, 99
unstable manifold, 53, 99
unstable periodic orbits, 25
unsupervised, 653, 656

vacuum state, 433, 629
Van der Pol, 32
Van der Pol oscillator, 36, 125
vector–field, 16
velocity equation, 233
velocity phase–space manifold, 321
velocity vector–field, 321
ventricular fibrillation, 166

Verhulst model, 200
vertices, 309, 368
virial estimator, 473
virtual particles, 428
virtual paths, 425, 488
visual cortex, 187
volatility, 585
Volume(T), 493
von Neumann, 26
vortex, 111
vortices, 156
vorticity dynamics, 112

W–cycle, 473
wandering point, 79
water vapor, 230
wave–particle duality, 432
Weierstrass, 28
weights, 653
Weyl, 309
Whitney, 309
Wick rotation, 483, 540
Wiener, 33
Wiener measure, 450
Wiener process, 429
Wigner function, 626
winding number, 95
Witten's TQFT, 511
world–sheet, 529
wormhole creation, 400

Yang–Baxter equation, 515
Yang–Mills gauge theory, 512
Yang–Mills relation, 578
yarmulke, 397, 398, 407, 683
yarmulke cobordism, 682

Zamolodchikov metric, 568
Zorn's lemma, 311

Printing: Krips bv, Meppel, The Netherlands
Binding: Stürtz, Würzburg, Germany